6th International Conference on Research, Implementation, and Education Mathematics and Science (ICRIEMS 6)

Integrating Science, Technology, Engineering, & Mathematics (STEM) and Education for Disaster Risk Reduction and Mitigation

Journal of Physics: Conference Series Volume 1397

Yogyakarta, Indonesia
12 - 13 July 2019

Editors:

Restu Widiatmono
Antuni Wiyarsi
Kuwat Triyana

ISBN: 978-1-7138-1166-4
ISSN: 1742-6588

Printed from e-media with permission by:

Curran Associates, Inc.
57 Morehouse Lane
Red Hook, NY 12571

Some format issues inherent in the e-media version may also appear in this print version.

Institute of Physics
Dirac House, Temple Back
Bristol BS1 6BE UK

Phone: 44 1 17 929 7481
Fax: 44 1 17 920 0979

techtracking@iop.org

Additional copies of this publication are available from:

Curran Associates, Inc.
57 Morehouse Lane
Red Hook, NY 12571 USA
Phone: 845-758-0400
Fax: 845-758-2633
Email: curran@proceedings.com
Web: www.proceedings.com

TABLE OF CONTENTS

Journal of Physics: Conference Series **1397** (2019) 011001 doi:10.1088/1742-6596/1397/1/011001

The 6th International Conference on Research, Implementation, and Education of Mathematics and Science (ICRIEMS 6)

Universitas Negeri Yogyakarta, Indonesia

12-13 July 2019

Editors:

Dr. Restu Widiatmono

Universitas Negeri Yogyakarta, Indonesia

restu@uny.ac.id

Dr. Antuni Wiyarsi

Universitas Negeri Yogyakarta, Indonesia

antuni_w@uny.ac.id

Dr. Eng. Kuwat Triyana, M. Si.

Universitas Gadjah Mada, Indonesia

triyana@ugm.ac.id

1397 (2019) 011001 doi:10.1088/1742-6596/1397/1/011001

Peer Review Statement

All papers published in this volume of *IOP Conference Series: Journal of Physics Conference Series* have been peer reviewed through processes administered by the proceedings Editors. Reviews were conducted by expert referees to the professional and scientific standards expected of a proceedings journal published by IOP Publishing.

Journal of Physics: Conference Series **1397** (2019) 011001 doi:10.1088/1742-6596/1397/1/011001

Preface: Proceedings of 6th International Conference on Research, Implementation, and Education of Mathematics and Science (ICRIEMS 6)

R Widiatmono[1], A Wiyarsi[2], and K Triyana[3]

[1]Physics Education Department, Universitas Negeri Yogyakarta, Indonesia
[2]Chemistry Education Department, Universitas Negeri Yogyakarta, Indonesia
[3]Physics Department, Universitas Gadjah Mada, Indonesia

We are pleased to present you the 6th International Conference on Research, Implementation, and Education of Mathematics and Science (ICRIEMS 6) Conference Proceeding. The 6th ICRIEMS was held on 12-13th July 2019 and organized by the Faculty of Mathematics and Science, Universitas Negeri Yogyakarta, Indonesia. We are raising a theme **"Integrating Science, Technology, Engineering, & Mathematics (STEM) and Education for Disaster Risk Reduction and Mitigation"**. We hope that the theme will promote STEM and its education to enhance society knowledge and awareness on disasters. The knowledge could help people and government agencies to reduce and prevent the emergence of a larger disaster impact.

The conference also was an event where prominent practitioners, researchers, students and educators shared their research findings and exchange ideas. Five keynote speakers were given their talk: Prof. Dr. Gultekin Cakmakci, from Hacettepe University, Turkey; Prof. Dr. Wing Mui Winnie So from The Education University of Hong Kong; Martianus Frederic Ezerman, Ph.D, from Nanyang Technological University, Singapore; Prof. Dwikorita Karnawati, Ph.D from Meteorological, Climatological, and Geophysical Agency of Indonesia; and Dr. Insih Wilujeng from Universitas Negeri Yogyakarta. In addition, two invited speakers also contributed: Dr. Hanik Humaida from the Center for Volcanology and Geological Hazard Mitigation of Indonesia; and Dr. Azlan Kamari from Universiti Pendidikan Sultan Idris, Malaysia.

The conference has accepted 210 papers from six countries, i.e. Turkey, Indonesia, Hongkong, Singapore, Malaysia, and Thailand. After an intensive review process, finally 95 papers have been selected for publication. We believe the proceedings will provide the scientists of the world with an excellent reference and be an impetus to stimulate further studies in all related areas. We address big appreciation and many thanks to all authors and participants who have been actively involved in the conference. We also wish to thanks to our scientific committee for their invaluable comments and suggestions.

24 October 2019

Editors
Restu WIDIATMONO
Antuni WIYARSI
Kuwat TRIYANA

List of Committees

STEERING COMMITTE

Prof. Dr. Wim T van HORSSEN (Applied Mathematics, TU Delft, Netherland)
Prof. Dr. Muammer CALIK (Chemistry Education, Trabzon University, Turkey)
Prof. Fang Ying YANG Ed.D (Science Education, National Taiwan Normal University, Taiwan)
Allen PRICE, Ph.D (Physics Education, Emmanuel College Boston, USA)
Guillermo DÁVILA-RASCÓN, Univesidad de Sonora, Mexico
Osval Antonio Montesinos LOPEZ, Ph.D. (Statistic, Universidad de Colima, Mexico)
Dr. Azmi MOHAMMED (Chemistry, Universiti Pendidikan Sultan Idris, Malaysia)
Dr. Eng. Kuwat TRIYANA (Physcis, Uninersitas Gadjah Mada, Indonesia)
Dr. HARTONO (Mathematics, Universitas Negeri Yogyakarta, Indonesia)
Dr. Slamet SUYANTA (Biology, Universitas Negeri Yogyakarta, Indonesia)

ORGANIZING COMMITTEE

Dr. Restu WIDIATMONO
Dr. Antuni WIYARSI
Dr. Kun Sri BUDIASIH
Dr. KARYATI
Dr. PUJIANTO
Dr. Retno ARIANINGRUM
Ilham RIZKIANTO, M.Sc
Kuswari HERNAWATI, M.Kom

Journal of Physics: Conference Series **1397** (2019) 011002 doi:10.1088/1742-6596/1397/1/011002

Peer review statement

All papers published in this volume of *Journal of Physics: Conference Series* have been peer reviewed through processes administered by the proceedings Editors. Reviews were conducted by expert referees to the professional and scientific standards expected of a proceedings journal published by IOP Publishing.

The effects of annealing temperature and angular velocity variation on microstructure and optical properties of barium titanate ($BaTiO_3$) using chemical solution deposition method

R. P. Rini[1], F. Nurosyid[1,2], Y. Iriani[1,2*]

[1]Graduate Physics Department, Universitas Sebelas Maret, Jl. Ir. Sutami 36A Kentingan, Surakarta 57126
[2]Physics Department, Faculty of Mathematics and Natural Sciences, Universitas Sebelas Maret, JL. Ir. Sutami 36A Kentingan, Surakarta 57126

*Email: yofent_iriani@staff.uns.ac.id

Abstract. This study is aimed to find out the energy gap value of $BaTiO_3$ thin film since this study has a purpose to reduce the energy gap so that it can be used for photovoltaic applications. The making of Barium Titanate ($BaTiO_3$) thin film was carried out on the quartz substrate using the Chemical Solution Deposition (CSD) method with the technique of depositing spin coating. The making of $BaTiO_3$ thin film was carried out using variation of angular velocity at 4000 rpm and 5000 rpm and variation of annealing temperature at 750°C, 800°C, 850°C. The samples in this study were annealed at 750°C, 800°C, 850°C with a holding time of 1 hour and an increase in temperature of 3°C/minute. The microstructure characterization of the samples was carried out using X-Ray Diffraction (XRD) equipment, and the thickness of the thin film characterization and grain size of samples using Scanning Electron Microscopy (SEM) equipment. The annealing temperature increase, the intensity becomes higher. As the annealing temperature increase, then the crystallinity, crystallite size and grain size of $BaTiO_3$ thin film will get bigger. The annealing temperature increase is resulted the thickness and energy gap are smaller. The intensity value of the XRD remains decrease along with the angular velocity increase. The higher angular velocity causes the crystallinity and crystallite size of $BaTiO_3$ thin film becomes higher. However, the angular velocity increase then the grain size will become smaller. The angular velocity higher, then the $BaTiO_3$ thin film gets thinner. The angular velocity gets higher, then the smaller energy gap produced. The energy gap value was produced between 3.00 – 3.50 eV.

1. Introduction

This very rapid technological development can occur due to the existence of certain material. The forms of development in the material are advanced materials, one of which is ferroelectric material. Ferroelectric is a material that has good potential as a solar cell, lighting where the optical sensor is affected by anomalous photovoltaic effects [1]. The photovoltaic effect is the result of the conversion of light energy into electrical energy that occurs directly [2]. Ferroelectric has a higher photovoltaic voltage of two to four times (10^3-10^5 V/ cm) than a bandgap (Eg), and the photovoltaic characteristics can be regulated and controlled by an electric field [3].

Several types of ferroelectric materials which are often used including Barium Titanate [4], Barium Strontium Titanate (BST) [5], Barium Zirconium Titanate (BZT) [6], Lead Zirconat Titanate (PZT) [7], and Strontium Titanate ($SrTiO_3$) [8]. These ferroelectric materials can be made with various methods such as Pulsed Laser Deposition (PLD) [9], Chemical Vapour Deposition (CVD) [10], Chemical Solution Deposition (CSD) [11] and Co precipitation [12].

The well-known ferroelectric materials have ABO_3 perovskite structures are Barium Titanate (BT), Barium Strontium Titanate (BST), and Barium Zirconium Titanate (BZT). ABO_3 perovskite structure is a structure characterized by the diameter of atom A which is bigger than atom B. Atom A is located at the end of the rib cube, whereas B atom is located in the diagonal space, and O atom is located in the diagonal plane [13]. Barium Titanate (BT) has a perovskite structure of ABO_3, which has five types of crystal structures, they are: hexagonal, cubic, tetragonal, orthorhombic, and rhombohedral, each of which depends on the temperature of the phase [14]. Barium titanate has no response to the sun since it has a wide bandgap ($E_g > 3$ eV) [1]. In order that barium titanate can be used as one of the ferroelectric solar cells, it must be doped to reduce the semiconductor bandgap. The doping material given can affect the crystal structure, microstructure, and optical structure of the barium titanate. Barium titanate can be doped by adding La^{3+}, Pb^{2+}, Na^{+}, Ga^{2+}, Nd^{3+}, Th Cr^{4+}, for replacing the position of A or Ba^{2+}, meanwhile, Zr^{4+}, Fe^{3+}, Al^{3+}, In^{3+}, Cr^{3+}, Nb^{5+}, Ta^{5+}, Sb^{5+}, W^{6+} are used for replacing the position of B or Ti^{4+} [13].

In this study, $BaTiO_3$ was made using the Chemical Solution Deposition (CSD) method using the spin coating process. The samples annealing process was carried out with a temperature of 750°C-850°C, which was held for 1 hour with a heating rate of 3°C/minute. Before conducting the annealing process on the samples, the solutions were deposited on a quartz substrate with an angular velocity of 4000 and 5000 rpm. The variations carried out in this study were annealing temperature and angular velocity variation. The characterization of the $BaTiO_3$ thin film was carried out using XRD equipment characterization to determine the crystal structure of the $BaTiO_3$ thin film. Meanwhile, the characterization of $BaTiO_3$ thickness and grain size was carried out using SEM equipment.

2. Experiment

The thin film of $BaTiO_3$ was grown using Chemical Solution Deposition (CSD) method. The raw materials were used in this study were Barium Acetate [$Ba(CH_3COO)_2$] 99.00%, Acetic Acid [(CH_3COOH)] 100%, Titanium Isopropoxide [$Ti(OC_3H_7)_4$] 97%, Ethylene Glycol [($HOCH_2CH_2OH$)] 100%. The solution of BaTiO3 was made with several stages. The solutions were deposited on the quartz substrate with a spin coating technique at 4000 rpm and 5000 rpm for 30 seconds. This process was repeated with the specified angular velocity to obtain the desired number of films. Then, the substrate that has been deposited was annealed with varying temperatures of 750°C, 800°C, 850°C with a holding time of 1 hour and a heating rate of 3°C/ minute. The samples were characterized using XRD equipment to identify the structural phase and crystallite size of the $BaTiO_3$ thin film. The samples testing was carried out using SEM equipment to determine the surface morphology and grain size. The samples were characterized using a UV-VIS spectrophotometer to determine the amount of absorbance and transmittance of the thin film.

3. Result and Discussion

The thin film of $BaTiO_3$ that has been created then was characterized using XRD equipment to find out the crystal structure with λ Cu 1.5406Å. The data of XRD test results were obtained by displaying a graph of the relationship between 2θ with intensity. The diffraction peaks were compared to the database ICDD #8318880, and by using this procedure, it can be indicated that the peaks are the diffraction peaks of the samples. Based on the results of XRD characterization of samples with variations of annealing temperature and variations of angular velocity produce seven diffraction peaks as can be seen in Figure 1. All the angle 2θ from the graphic of XRD has a different value of intensity.

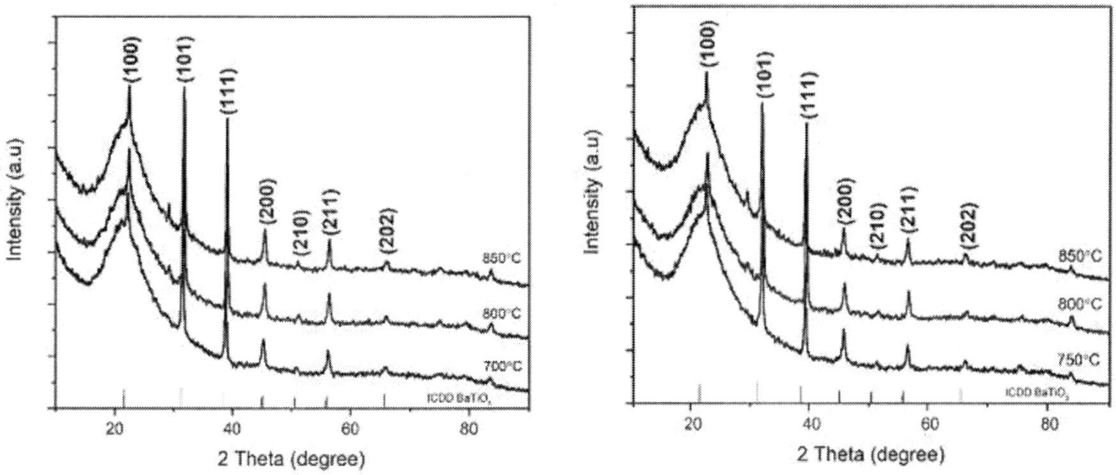

(a) (b)

Figure 1. XRD patterns of $BaTiO_3$ using annealing temperature and angular velocity variations
(a) 4000rpm, (b) 5000rpm

Based on the results of the characterization carried out using the XRD equipment, it can be seen that as the annealing temperature increase, the intensity value will get higher. Annealing temperature does not affect the angle of diffraction, but it affects the peak intensity. It can be seen in Figure 1 (a) and (b) that when the angular velocity gets higher, then the intensity will reduce. The result of this study is consistent with the results observed by Naat et al., when the angular velocity increase then the intensity of crystal orientation area will reduce [15].

Table 1. Crystallinity and crystallite size of $BaTiO_3$ using annealing temperature and angular velocity variations

Sample	Crystallinity (%)	Crystallite Size (nm)
4000rpm, 750°C	9.8	35
4000rpm, 800°C	9.9	49
4000rpm, 850°C	10.3	56
5000rpm, 750°C	9.8	44
5000rpm, 800°C	10.4	52
5000rpm, 850°C	10.5	58

The crystallinity was calculated from the intensity of the peak diffraction formed. The crystallinity formation of samples with annealing temperature variations is an overview of the regularity crystals formed in each film. The calculation of crystallinity formation from the samples can be seen in Table 1. As the annealing temperature increase, the crystallinity of the films will get higher. This condition influenced by the annealing temperature increase then there is a vibration of atomic, so the atomic become more organized, which cause the crystallinity gets higher. At an angular velocity of 5000 rpm, the crystallinity of samples gets higher too. This result is the same as the result observed by Darmasetiawan et al., when the angular velocity increase, then the crystallinity film increase [16]. The crystallinity was produced between 9.8% - 10.5%.

The crystallite size was obtained by measuring the half-width of the highest peak or Full-Width Half Maximum (FWHM). The crystallite size of the $BaTiO_3$ thin film can be calculated using the equation of Scherrer as follow:

$$D = \frac{k\,\lambda}{\beta \cos\theta} \tag{1}$$

Where, the value of β is the value of FWHM, θ which is the diffraction angle, k is the Scherrer constant where 0.9, and λ is the wavelength of the X-ray [1]. The measurement of the sample items can be seen in Table 1.

Based on the variations of annealing temperature when made the samples, it can be seen that when the annealing temperature increase, the crystallite size of the $BaTiO_3$ thin film will get bigger. In accordance with the increase of annealing temperature, when the angular velocity increase, the crystallite size will get bigger. The crystallite size was produced between 35 - 58 nm.

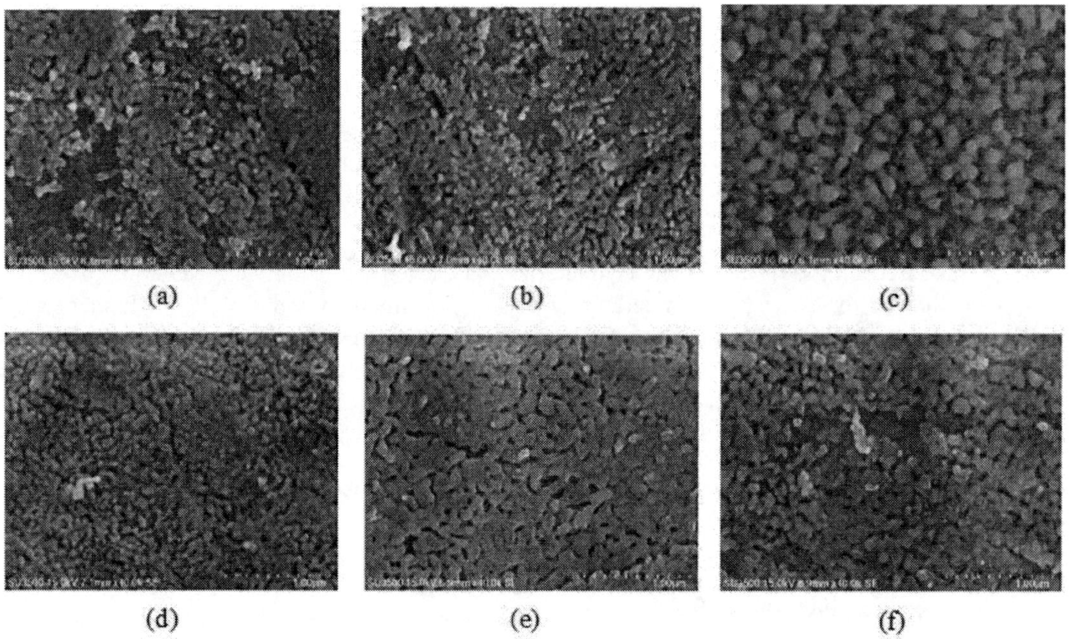

Figure 2. SEM image of $BaTiO_3$ (a) 750°C, 4000 rpm, (b) 750°C, 5000 rpm, (c) 800°C, 4000 rpm, (d) 800°C, 5000 rpm, (e) 850°C, 4000 rpm, (f) 850°C, 5000 rpm

The morphology of the $BaTiO_3$ thin film it can be seen in Figure 2. The thin film of $BaTiO_3$ with annealing temperature variation higher then the porosity of thin film decrease however, the grain boundaries are visible but not clear. This is happened because between the grain in the thin film has experienced diffusion. As the angular velocity increase, then the resulting thin film is thinner so the porosity decrease. However, when the angular velocity gets higher, the grain boundaries are visible and clear.

Based on the result of the characterization of SEM, obtained the grain size and the thickness of the samples of $BaTiO_3$ as can be seen in the following Table 2.

Table 2. Grain Size and thickness with annealing temperature and angular velocity variations

Sample	Grain Size (nm)	Thickness (nm)
4000rpm, 750°C	123	595
4000rpm, 800°C	198	478
4000rpm, 850°C	227	417
5000rpm, 750°C	104	443
5000rpm, 800°C	126	426
5000rpm, 850°C	197	395

The annealing temperature increase, then the grain size increase. The angular velocity increase, the grain size will be smaller and more homogeneous. This result is matched with the research by Naat et al., [15].

The results of cross-section measurements on samples with variations of annealing temperature and angular velocity obtained thickness, as seen in Table 2. The thickness of the $BaTiO_3$ thin film gets thinner when the annealing temperature gets higher. This condition occurs since the higher annealing temperature causes more water to be evaporated. Thus, at an annealing temperature of 850°C the thickness of the resulting film is smaller. The higher of angular velocity, the thickness of the film becomes thinner. The result is matched with the result observed by Naat et al., when the angular velocity increase, then the thickness of thin film decrease [15].

The calculation of band gap values using the method of tauch-plot by using the equation:

$$(\alpha h\upsilon)^2 = A(h\upsilon - E_g) \quad (2)$$

Where $h\upsilon$ is the photon energy, A is constant, α is the absorption coefficient dan E_g is the optical band-gap [1].

The transmittance spectrum of the quartz substrate was grown with $BaTiO_3$ is shown in Figure 3. In this figure, it can be seen that the annealing temperature and the angular velocity higher, the transmittance value get higher. Based on Figure 3, it can be seen that at an annealing temperature of 850°C, the transmittance value was highest. This result is matched with the research Bao et al., when the annealing temperature increase, the absorbance value is lower, and the peak of absorbance has shifted towards the right (towards the visible area) [17]. The transmittance value increases at a wavelength of 400 – 800 nm, where the wavelength is a visible area. Besides thickness, the absorption coefficient value also affects the amount of the energy gap value. The energy gap value was produced between 3.00 – 3.50 eV.

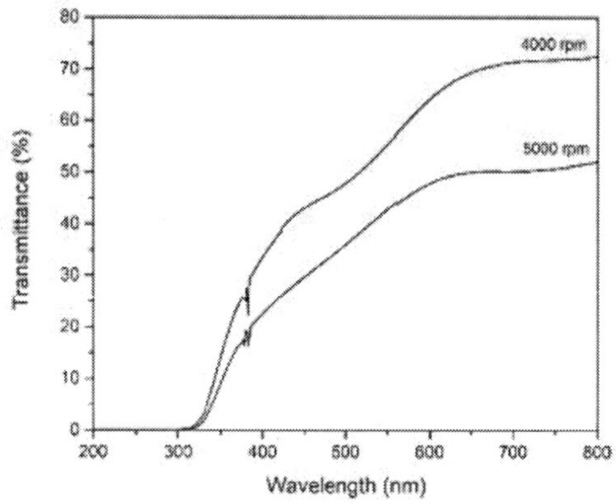

(a)

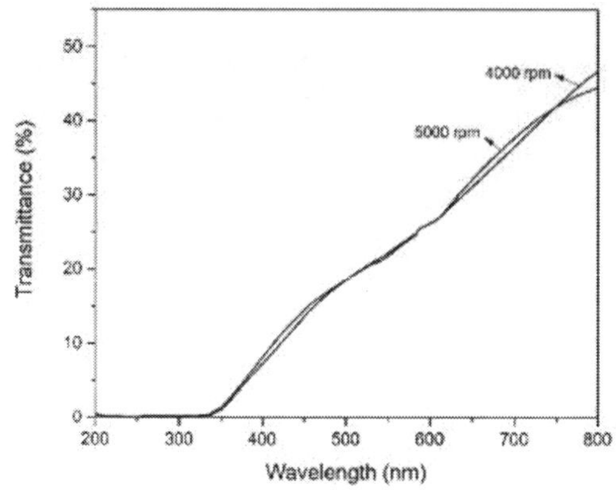

(b)

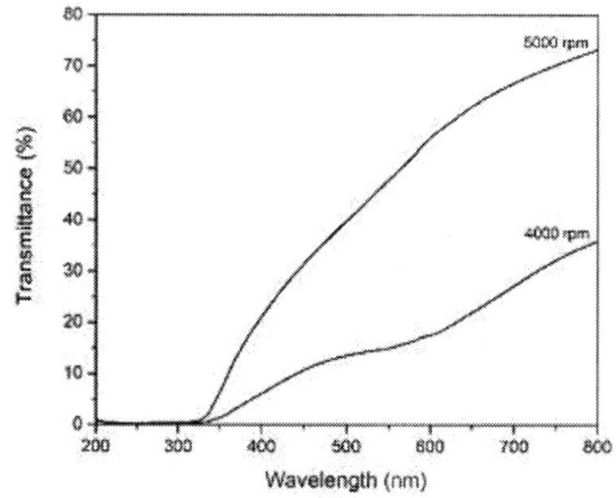

(c)

Figure 3. Transmittance of $BaTiO_3$ (a) 750°C, (b) 800°C and (c) 850°C

Table 3. Energy Gap with annealing temperature and angular velocity variations

Sample	Energy Gap (eV)
4000rpm, 750°C	3.50
4000rpm, 800°C	3.25
4000rpm, 850°C	3.15
5000rpm, 750°C	3.25
5000rpm, 800°C	3.15
5000rpm, 850°C	3.00

As the annealing temperature increase, then the energy gap of the $BaTiO_3$ thin film will get smaller. This is because as the annealing temperature increase the crystallinity and crystallite size were increase. These results are matched with the result observed by Xu & Shen, where the energy gap is affected by the microstructure of the film [18]. The angular velocity gets higher than the smaller energy gap produced.

4. Conclusions

The annealing temperature increase, the intensity becomes higher. As the annealing temperature increase, then the crystallinity, crystallite size and grain size of $BaTiO_3$ thin film will get bigger. The annealing temperature increase results in the thickness and energy gap are smaller.

The intensity value of the XRD remains decrease along with the angular velocity increase. The higher angular velocity causes the crystallinity and crystallite size of $BaTiO_3$ thin film becomes higher. However, the angular velocity increase then the grain size will become smaller. The angular velocity higher, then the $BaTiO_3$ thin film gets thinner. The angular velocity gets higher, then the smaller energy gap produced. The energy gap value was produced between 3.00 – 3.50 eV.

5. Acknowledgments

The authors wish to thank Kementrian Riset, Teknologi & Pendidikan Tinggi for financial support through Penelitian Dasar 2019 No 719/UN27.21/PP/2019.

6. References

[1] Jiang W, Cai W, Lin Z, Fu C 2013 *Journal Materials Research Bulletin* **48** 3092-3097
[2] Miles R W 2006 *Science Direct* **80** 1090-1097
[3] Yang X, Su X, Shen M, Zheng F, Y Xin, L Zhang, Hua M, Chen Y, Harris V G 2012 *Advanced Materials* **24** 1202–1208
[4] Ramakanth S, Raju K C J 2014 *Journal Solid State Communications* **187** 59-63
[5] Retnasamy V, Sauli Z, Vairavan R 2014 *Advanced Materials Research* **896** 211-214
[6] Iriani Y, Saputri D F, Hadiati S, Ramelan A H, Variani V I 2014 *Advanced Materials Research* **911** 205-209
[7] Goh W C, Yao K, Ong C K 2005 *Applied Physics A Materials Science & Processing* **81** 1089-1093
[8] Cowley R A 1964 *Physical Review* **134** A981-A997
[9] Yang Y, Wang Z, Fang Li J, Vichland D 2010 *Journal of Nanomaterials* **2010** 756319
[10] Besserguenev V G, Pereira R J F, Mateus M C, Khmelinskii, I V, Nicula R C, Burkel E 2003 *International Journal of Photoenergy* **05** 99-105
[11] Yang C, Lin Y, Tang T 2004 *Proc of SPIE* **5774** 224-227
[12] Subarwanti Y, Safitri R D, Supriyanto A, Iriani Y, Jamaludin A 2017 *Material Science and Engineering* **176** 012044
[13] Iriani Y, Jamaludin A, Nurhadi N 2016 *Journal of Physics: Conference Series* **776** 012064
[14] Li W, Xu Z, Chu R, Fu P, Hao J 2009 *Journal of Alloys and Compounds* **482** 137-140
[15] Naat J N, Tjahjanto R T, Masruroh 2014 *Chimica et Natura Acta* **2** 115-119
[16] Darmasetiawan H, Irzaman, Indro M N, Sukaryo S G, Hikam M, Bo N P 2002 *Physica Status Solidi* **193** 53-60
[17] Bao D, Yang H, Zhang L, Yao X 1998 *Phys. Stat. Sol* **169** 227
[18] Xu Y & Shen M 2008 *Journal Materials Letter* **62** 3600-3602

Effect of pre-annealing and annealing temperature on microstructural and optical properties of multiferroic $BiFeO_3$ thin films prepared by chemical solution deposition (CSD)

E B Agustina[1], Y Iriani[2*] and R Suryana[2]

[1]Department of Physics, Postgraduade, Universitas Sebelas Maret Jl. Ir. Sutami 36A Kentingan, Surakarta 57126
[2]Department of Physics, Faculty of Mathematics and Sciences, Universitas Sebelas Maret Jl. Ir. Sutami 36A Kentingan, Surakarta 57126

*E-mail: yofent_iriani@staff.uns.ac.id

Abstract. $BiFeO_3$ thin films has been successfully grown on a quartz substrate using the chemical solution deposition (CSD) method. X-ray diffraction (XRD) used to determine the microstructure of BFO. The morphology and transmittance of BFO thin film were studied using SEM and UV-VIS. Deposition of the $BiFeO_3$ thin film with pre-annealing in air and variation of annealing temperature showed that the increase in annealing temperature caused crystallinity to be higher and crystallite size to increase. High annealing temperature results in a larger grain size and lower thickness. Deposition of the $BiFeO_3$ thin film with pre-annealing variation showed that $BiFeO_3$ with pre-annealing in air had a higher crystallinity compared with in O_2. The crystallite size of $BiFeO_3$ with pre-annealing in O_2 is greater than pre-annealing in air. $BiFeO_3$ with pre-annealing in air has a smaller grain size and greater thickness compared to pre-annealing in O_2. $BiFeO_3$ band gap with pre-annealing in air and variation of annealing temperature were obtained 1.4-3.5 eV. While the $BiFeO_3$ band gap with pre-annealing variation was obtained from 2.3-2.7 eV.

1. Introduction

Perovskite ferroelectric is a large group of ferroelectric material, that has a significance and wide application in electronic industry [1]. Bismuth ferrite oxide ($BiFeO_3$) with perovskite structure is one of the most studied multiferroic materials and a potentially important lead-free ferroelectric material because its transition temperatures are well above room temperature, which has a high Neel temperature (TN) of ~ 370°C and a Curie temperature (TC) of 830°C [2-4].

Perovskite-oxide has a structural formula of ABO_3, in which A is a large cation such as Bi^{3+}, Ba^{2+}, or Pb^{2+}, and B is a medium-sized cation such as Fe^{3+}, Ti^{4+}, or Zr^{4+}. These cations are located in cages formed by a network of oxygen anions[5]. Perovskite-type unit cell with a rhombohedral structure has a lattice parameter of $a = b = c = 0.3965$ nm and $\alpha = 89.3°$ at RT [6].

Some various of techniques for the synthesis of $BiFeO_3$ are plasma and ion-beam sputter deposition [7], electron beam or oven-induced evaporation for molecular beam epitaxy (MBE) [8], pulsed laser ablation deposition [9], (metal organic) chemical vapor deposition (CVD) [10], chemical solution methods or chemical solution deposition (CSD) [11]. Among above methods, CSD has the advantage of easy compositional control in films with high quality and microchemical homogeneity, which is favorable for obtaining high optical transparent film [12].

A significant visible-light photovoltaic effect in BFO diode structures was recently investigated by researchers, which has further confirmed the photovoltaic application of BFO thin films [13–15]. Recently, Choi et al. and Yang et al. demonstrated photovoltaic effects in bulk and epitaxial films of the BFO and obtained low band gaps of about 2.7 eV [16]. The effect of annealing atmosphere and annealing temperature on microstructural of multiferroic $BiFeO_3$ thin films prepared by chemical solution deposition was recently investigated by researchers [17]. Only $Bi_2Fe_4O_9$ were observed as the secondary phase during deposition and the fraction of secondary phase depended upon the annealing temperature as well as the annealing atmosphere.

In this paper, we report on the effect of pre-annealing atmosphere and temperature annealing on microstructural and optical properties of multiferroic $BiFeO_3$ thin films deposited on quartz substrates prepared by chemical solution deposition (CSD) technique.

2. Experimental

$BiFeO_3$ thin films were fabricated by chemical solution deposition (CSD) on quartz substrate. Bismuth nitrate pentahydrate $Bi(NO_3)_3.5H_2O$ and iron nitrate nanohydrate $Fe(NO_3)_3.9H_2O$ were used as starting materials. 2-methoxyethanol and acetic acid with a ratio of 1:1 were used as solvents. 0.2 ml acetyleacetone was added as stabilizing agent. The films were deposited on quartz substrate by spin coated at 3000 rpm for 30 second, dried at 150°C for 1 minute. This process was repeated several times to obtain films of desired thickness and then the films were pre-annealed at 400°C for 10 minutes in air atmosphere and heated by rapid thermal annealing at 550, 600 and 650°C for 1 hour in O_2 atmosphere. This process was repeated at 600°C with pre-annealing and annealing process in O_2 atmosphere.

X-ray diffraction (XRD) equipment (Philips PW 3710/40 kV with Cu-Kα radiation, λ = 1.5406 Å) was used for structural phase identification and the crystallite size of BFO thin films.

Crystallite size is determined using the equation:

$$D = \frac{k\lambda}{\beta cos\theta} \quad (1)$$

Where k is the Schere constant, λ is the wavelength, θ is the diffraction angle, and β is the full width at half maximum (FWHM) and D is the crystallite size.

The surface morphology and grain size of BFO films were observed by Scanning Electron Microscopy (SEM) model SU3500. The transmittance of the films was measured using an UV–VIS light spectrophotometer.

3. Result and Discussion

The characterization results using XRD equipment are shown in Figure 1. The peaks that appear are matched to the ICDD database # 200169.

XRD patterns of $BiFeO_3$ thin films with pre-annealing in air and variations of annealing temperature is shown in Figure 1(a). $BiFeO_3$ thin films have rhombohedral perovskite structures with peaks (101), (110), (111), (020), (120) and (121) that can be observed. The $BiFeO_3$ thin film was annealed at 550°C and 600°C showed a single phase $BiFeO_3$. The thin film annealed at 650°C presence of other phases, such as $Bi_2Fe_4O_9$ and $Bi_2O_{2.33}$. This results are match with the results observed by Xu et al. [2]. The crystallinity and crystallite size of $BiFeO_3$ with variations of annealing temperature can be seen in Table 1. Increasing the annealing temperature causes crystallinity to increase. High temperatures in annealing process produce a larger crystallite size. These results indicate that the crystallite size and crystallinity of $BiFeO_3$ thin film is very dependent on the temperature of annealing.

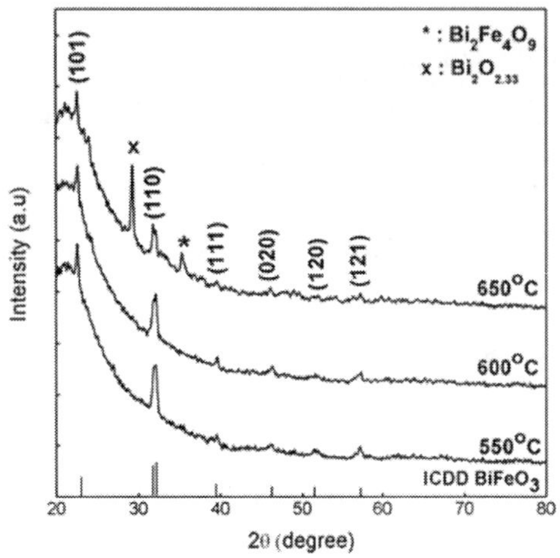

(a)

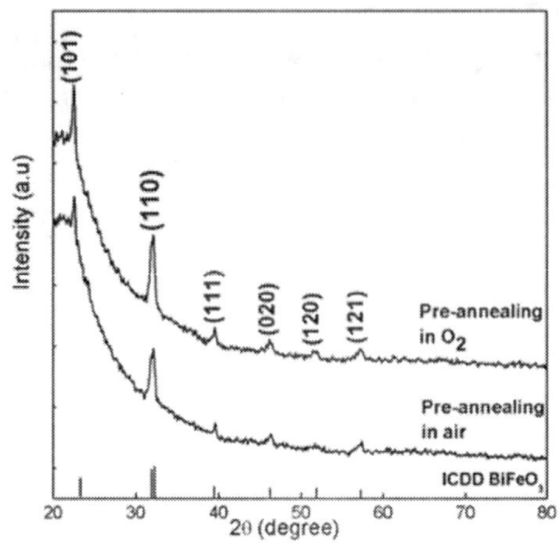

(b)

Figure 1. XRD patterns of $BiFeO_3$ thin films (a) with pre-annealing in air and variations of annealing temperature. (b) with variation pre-annealing (annealing at 600°C).

Table 1. Crystallite size and crystallinity of $BiFeO_3$

Pre-annealing at 400°C	Annealing (°C)	Crystallite size (nm)	Crystallinity (%)
In air	550	20.2	9.6
In air	600	21.6	10.0
In air	650	21.3	9.9
In O_2	600	23.5	9.8

Figure 1(b) shows the XRD pattern of $BiFeO_3$ thin film with variation pre-annealing in air and O_2 (annealing at 600°C). XRD results indicate the presence of a single phase $BiFeO_3$ without presence of non-perovskite phases. The difference treatment in the pre-annealing process causes differences in the width of the diffraction peak. BFO with pre-annealing in O_2 shows a higher diffraction peak compared to pre-annealing in the air [12]. The difference treatment in pre-annealing process causes a difference crystallinity. Table 1 shows that $BiFeO_3$ with pre-annealing in air has a higher crystallinity compared to that in O_2. The crystallite size of $BiFeO_3$ with pre-annealing in O_2 is greater than pre-annealing in air. Crystallite size obtained from broad diffraction peaks, this result is in accordance with the Scherer method. These results indicate that crystallite size and crystallinity of the $BiFeO_3$ thin film is very dependent on the pre-annealing treatment [9].

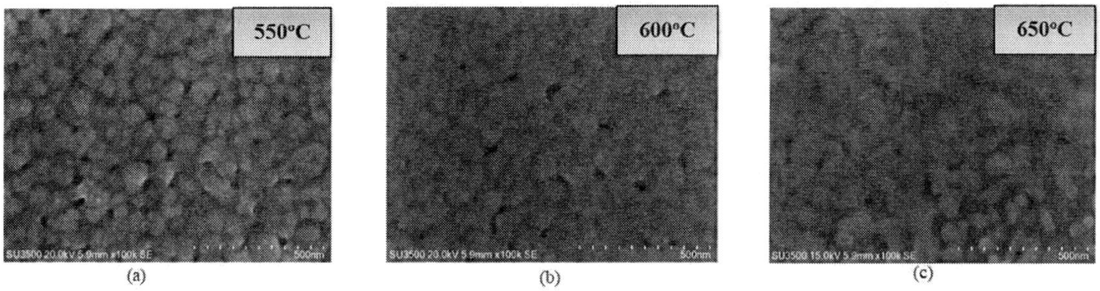

Figure 2. SEM images of $BiFeO_3$ with pre-annealing in air and variation of annealing temperature

Surface morphology, grain size and thickness were studied using Scanning Electron Microscopy (SEM). Figure 2 shows the morphological surface of $BiFeO_3$ thin film on quartz substrate with variations of annealing temperature. In plain view, annealing temperature increases resulting in low homogeneity of grain. Low homogeneity causes the distance between grains to be more tenuous and the boundary between grains cannot be clearly seen. This results in a higher porosity. The grain size and thickness of $BiFeO_3$ with variations of annealing temperature can be seen in Table 2. These results indicate that high annealing temperatures produce a larger grain size [2].

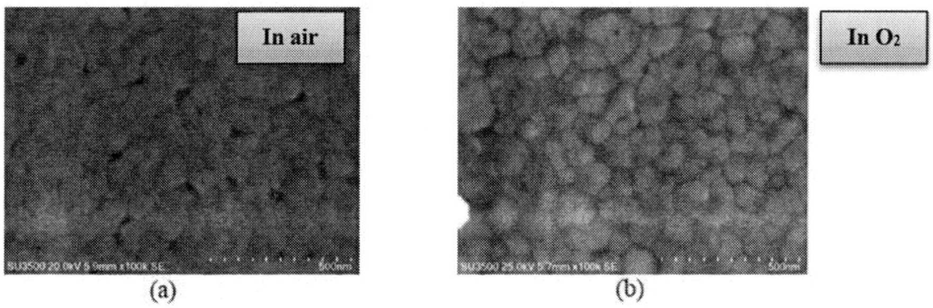

Figure 3. SEM images of $BiFeO_3$ prepared with variation pre-annealing (annealing at 600°C).

Table 2. Grain size and thickness of $BiFeO_3$

Pre-annealing at 400°C	Annealing (°C)	Grain size (nm)	Thickness (nm)
In air	550	66	325
In air	600	82	170
In air	650	93	120
In O_2	600	103	162

Figure 3 shows the surface morphology of the BFO thin film on a quartz substrate with variation pre-annealing treatments. The difference treatment in pre-annealing process affects the morphological structure of film. The $BiFeO_3$ thin film with pre-annealing treatment in O_2 looks more homogeneous with a greater thickness. The homogeneity can be seen from the distance between more dense grains, the boundaries between grains are more visible, and the grain size is larger. This is caused by a good atomic vibration process. The $BiFeO_3$ thin film with pre-annealing in air has a smaller grain size compared to pre-annealing in O_2. This results in a low homogeneity of film and resulting in high porosity. The low homogeneity results in changes in the morphological structure of the layer.

The transmittance spectra of films are shown in Figure 4. It can be seen that the highest peak transmittance at 650°C. The optical band gap E_g can be determined from absorption coefficient α, which is calculated according to Tauc's relation using the following equation: [18]

$$(\alpha hv)^2 = A(hv - E_g) \quad (2)$$

Where A is a constant, α is the absorption coefficient, hv is the photon energy and E_g is the optical band gap. E_g can be obtained by plotting $(\alpha hv)^2$ versus hv and extrapolating the linear portion of the plot to $(\alpha hv)^2 = 0$.

The band gaps of the $BiFeO_3$ thin film with variation annealing at 550°C, 600°C and 650°C were 1.4 eV, 2.3 eV, and 3.5 eV, respectively. While the band gaps of the $BiFeO_3$ thin films with variation pre-annealed in air and O_2 were 2.3 - 2.7 eV. This results are match with the other research, that band gap of BFO is 2.2–2.8 eV [2,16,19].

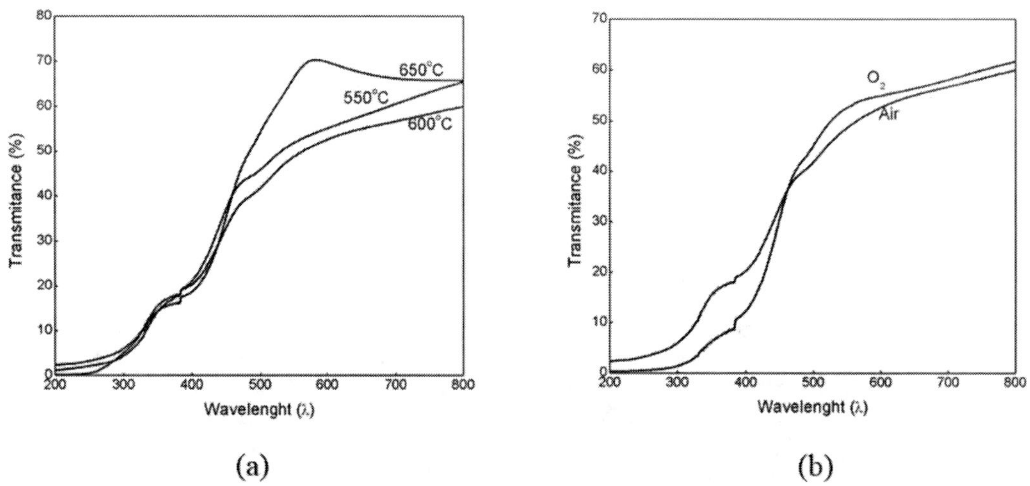

Figure 4. Transmittance of $BiFeO_3$ prepared (a) with pre-annealing in air and variation of annealing temperature **(b)** with variation pre-annealing (annealing at 600°C).

4. Conclusions

$BiFeO_3$ thin films has been successfully grown on a quartz substrate using the chemical solution deposition (CSD) method. Deposition of the $BiFeO_3$ thin film with pre-annealing in air and variation of annealing temperature showed that the increase in annealing temperature caused crystallinity to be higher and crystallite size to increase. High annealing temperature results in a larger grain size and lower thickness. Deposition of the $BiFeO_3$ thin film with pre-annealing variation showed that $BiFeO_3$ with pre-annealing in air had a higher crystallinity compared with in O_2. The crystallite size of $BiFeO_3$ with pre-annealing in O_2 is greater than pre-annealing in air. $BiFeO_3$ with pre-annealing in air has a smaller grain size and greater thickness compared to pre-annealing in O_2. $BiFeO_3$ band gap with pre-annealing in air and variation of annealing temperature were obtained 1.4 - 3.5 eV. While the $BiFeO_3$ band gap with pre-annealing variation was obtained from 2.3 - 2.7 eV.

Acknowledgments

The authors wish to thank Penelitian Unggulan Terapan 2019 PNBP UNS No.516/UN27.21/PP/2019 for financial support.

References

[1] Iriani Y, Jamaludin J, Nurhadi N 2014 *Journal of Physics: Conference Series* **776** 012064
[2] Yu X and Mingrong S 2008 *Materials Letters* **62** 3600-3602

[3] Chang H W, Yuan F T, Yu Y C, Cheng P C, Wang C R, Tu C S, Jen S U 2013 *Journal of alloys and compounds* **574** 402-406
[4] Catalan G, Scott J F 2009 *Journal Advanced Materials* **21** 2463–2485
[5] Shima H, Naganuma H, Okamura S 2013 *Material Science-Advanced Topics* **33**-61
[6] Kubel K and Schmid H 1990 *Acta Crystallogr. B* Vol. **46** 698-702
[7] Zhu G, Yang Z, Xu H 2012 *Journal homepage: Vacuum* **86** 1883e1885
[8] Ramesh R, Anggarwal S, Auciello O 2001 *Materials Science and Engineering* **32** 191–236
[9] Prashanthi K, Gupta M, Tsui Y Y, Thundat T 2013 *Appl phys A* **110** 903-907
[10] Kartavtseva M S, Gorbenko O Y, Kaul A R, Murzina T V, Savinov S A, Barthelemy A 2007 *Thin Solid Films* **515** 6416–6421
[11] Yun K Y, Ricinschi D, Kanashima T, Noda M, Okuyama M 2004 *Jpn. J. Appl. Phys* **43** L647–L648
[12] Wordenweber R, Julich F, Julich 2011 *Elsevier B.V. All rights reserved* **406** 177-203
[13] Choi T, Lee S, Choi Y J, Kiryukhin V, Cheong S W 2009 *Science* **324** 63–66
[14] Ji W, Yao K, Liang Y C 2010 *Adv. Material* **22** 1763
[15] Yi H T, Choi T, Choi S G, Oh Y S, Cheong S W 2011 *Adv. Material* **23** 3403–3407
[16] Yang S Y, Martin L W, Byrnes S J, Conry T E, Basu S R, Paran D, Reichertz L, Ihlefeld J, Adamo C, Melville A, Chu Y H, Yang C H, Musfeldt J L, Schlom D G, Ager J W, Ramesh R 2009 *Appl. Phys. Lett* **95** 062909
[17] Singh S K, Funakuba H, Uchida H and Ishiwara1 H 2005 *Integrated Ferroelectrics* **76** 139–146
[18] Cai W, Fu C, Gao J C, Guo Q, Deng X L and Zhang Y C 2011 *Physica B* **406** 3583–3587
[19] Santika M, Iriani Y and Suryana R 2019 *Materials Today: Proceedings* **13** 82–86

Journal of Physics: Conference Series **1397** (2019) 012004 doi:10.1088/1742-6596/1397/1/012004

The effect of immersion temperature using chlorophyll sensitizer (*Amaranthus hybridus L.*) on the performance of dye-sensitized solar cells

F. Nurosyid[1,2], D. D. Pratiwi[1] and K. Kusumandari[1,2]

[1]Physics Department of Graduate Program, Universitas Sebelas Maret, Indonesia

[2]Department of Physics, Faculty of Mathematics and Natural Sciences, Universitas Sebelas Maret, Indonesia

Corresponding author e-mail: fahrunurosyid@staff.uns.ac.id

Abstract. This article investigated the effect of immersion temperature on the TiO_2 film using chlorophyll extract for dye-sensitized solar cells (DSSC). The chlorophyll was extracted from spinach leaves with ethanol solvent. The solution of spinach extract was used for the immersion process on the TiO_2 film. The immersion temperatures were varied of 283 K, 303 K and 323 K. The optical properties of TiO_2/dye films were analyzed by UV-Vis and Fourier Transform Infrared (FTIR) spectrophotometers. The DSSCs performance were characterized by I-V meter. The highest absorbance was revealed from the immersion temperature of 303 K. Characteristic of FTIR spectra resulted a maximum transmittance of 67.2% at the frequency peak of 3341 cm^{-1}. Additionally, these results directly proportional with the maximum performance of DSSC at immersion temperature of 303 K. The best performance of DSSCs was obtained the V_{oc}, I_{sc}, and conversion efficiency values of 0.43 V, 0.16 mA and 0.04%, respectively.

1. Introduction

A dye-sensitized solar cell (DSSC) is an energy device based on the sensitization that converts light into electrical energy [1]. One of the advantages of DSSC is simple and low cost fabrication, so it does not require the high and complicated technology [2,3].

The DSSC composed of a transparent conducting oxide (TCO) glass substrate, a semiconductor photoanode (usually TiO_2 film which is coated on the TCO glass substrate), a dye sensitizer, a counter electrode and an electrolyte solution containing iodide/tri-iodide (I^-/I_3^-) redox couple [4,5]. The dye sensitizer is one of the most important component that affects the performance of DSSC. The dye sensitizer has a major role in capturing the photons and converting it into electrical energy [6,7]. The number of dye molecules that can be adsorbed on the surface of TiO_2 film affects the DSSC efficiency. The more dye molecules absorbed by the TiO_2 film, the more effective the absorption of photons to excite the electrons.

The synthetic dyes are the most common dyes used in DSSC. These dyes provide great electrical output and high efficiency. However, these dyes are heavy metals and environmental threat, complex manufacturing procedure, hight cost, uneconomical for large scale production, etc [8,9]. These problems can be overcome by applying natural dyes. The advantages of natural dyes are low production cost, abudantly available and enviromental friendliness [10-12]. The natural dye utilized from plant sources, such as leaves, flowers, fruits, rind, etc [3]. The natural dye that is often applied as DSSC sensitizer is chlorophyll [13]. The chlorophyll absorbs the visible light spectrum from red, blue, and violet with maximum absorption at the wavelength of 670 nm. Chlorophyll is an attractive pigment as a sensitizer for DSSC application [14].

Many studies have been carried out on the chlorophyll as sensitizer of DSSC. Anita et al. (2013) reported the conversion efficiency of 0.002% using chlorophyll extract from long bean leaves [15]. The utilization of green cabbage and papaya leaves extract as DSSC sensitizers were resulted the conversion efficiency of 0.1% and 0.094%, respectively [13,16]. However, research on the adsorption properties of natural dye for DSSCs have been very limited. The temperature is one of the major parameter of adsorption properties for immersion process on the TiO_2 surface. In addition, the study of immersion temperature based on the stability aspects of chlorophyll which are easily degraded by condition such as light, heat, temperature, pH and oxygen [17]. This article presented the effect of immersion temperature on the performance of DSSC using chlorophyll extract from *Amaranthus hybridus L.* as a sensitizer. This work is our efforts to obtain the better TiO_2/chlorophyll dye film in the fabrication of DSSC.

2. Experiments

Fluorine-doped Tin Oxide (FTO) conductive glass from Dyesol were used as the substrates with size of 2 cm x 2 cm. The TiO_2 paste was deposited on the FTO glass surface with effective cell area 0.75 cm^2 by spin coating technique. The counter electrodes were coated using Pt catalyst by brush painting technique. The electrolyte solution was obtained from the mixture of the polyethylene glycol (PEG) 400, potassium iodide (KI) and iodine (I_2). KI and I_2 were prepared by dissolving into the PEG 400 and stirred for 30 minutes.

The chlorophyll dye was extracted from spinach leaves. The leaves were washed using distilled water then crushed using mortar and pestle. The crushed mass were dissolved into the ethanol solvent. The solution was stirred at room temperature. Then extract was filtrated and used as dye solution. The TiO_2 electrode was immersed into dye solution for 24 hours with immersion temperature of 283 K, 303 K and 323 K. Fabrication of DSSCs were arranged into the sandwich structure. The working and counter electrodes were assembled using scotch tape which has thickness of 76 μm [18]. The electrolyte solution was injected in the space between the two electrodes.

Optical phenomena of dye adsorption on the TiO_2 electrodes were analyzed using UV-Vis Spectrophotometer (Lambda 25 UV-Vis Spectrophotometer, PE Electronics). While, functional groups for bonding between TiO_2 and dye molecules were characterized using fourier transform infrared spectroscopy (FTIR, Shimadzu Prestige 21). Efficiency of DSSCs were evaluated from current-voltage (*I-V*) curves measured using Keithley *I-V* meter 2602A.

3. Results and Discussion

3.1. Optical absorption spectrum of chlorophyll dye after adsorbed on the TiO_2 film

Fig.1 revealed the UV-Vis absorption spectra of TiO_2/dye photoelectrodes. The absorbance peaks indicated the chlorophyll pigment that playing a role in the process of photon absorption. The highest absorbance can be exhibited at immersion temperature of 303 K. It showed that TiO_2/dye photoelectode at room temperature able to absorb the photon energy maximally, so that the more electrons are excited and produced the better electricity. A higher temperature involved the instability of chlorophyll dye. This condition inhibited the immersion process on the TiO_2 surface. Reported that the treatment of immersion temperature at high temperature will increase activity of chlorophyllase enzyme and formed the pheophytin [19]. It will replace magnesium ions that contained in the chloropyll with hydrogen ions. As consequence, the number of chloropyll will decrease.

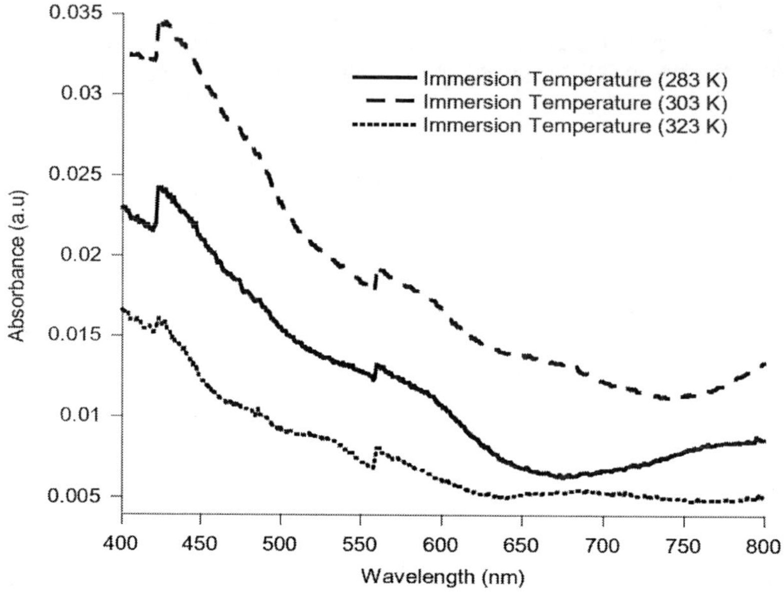

Figure 1. Absorption spectra of TiO_2/dye photoelectrodes with different immersion temperatures

3.2. FTIR spectral analysis

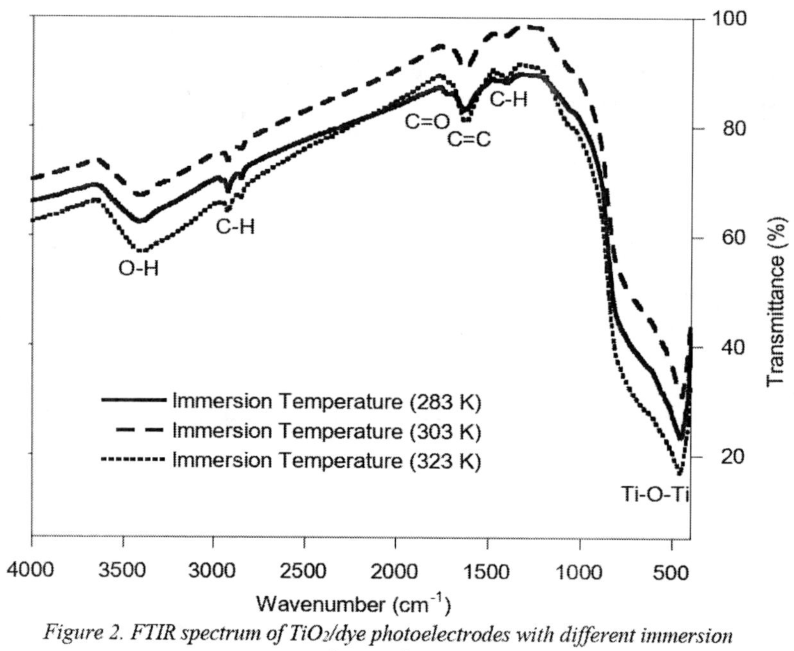

Figure 2. FTIR spectrum of TiO_2/dye photoelectrodes with different immersion temperatures

FTIR spectrum of TiO_2/dye photoelectrodes were recorded at wavenumber range of 4000 - 400 cm^{-1} as shown in Fig.2. The curve from Fig.2 can be interpreted in Table 1. It revealed the characteristic of functional groups and absorption frequency.

Table 1. FTIR peak values and functional groups of TiO_2/dye photoelectrodes with different immersion temperatures

283 K		303 K		323 K		Functional group
Peak	T (%)	Peak	T (%)	Peak	T (%)	
421	27.5			421	23	Ti-O-Ti
458	23.3	448	30.9	440	18.6	
				461	16.8	
1402	87.9	1403	96.6	1403	88.9	C-H
1624	83.2	1622	90.8	1621	81	C=C
1640	82.9					C=O
		1724	93.7			
2321	79.6					
2853	70.3	2854	75.7	2854	66.9	C-H
2923	67.9	2924	73.61	2959	65.5	
3395	62.5	3411	67.2	3389	56.9	O-H

The FTIR spectrum presented the O-H stretching vibration of ethanol (C_2H_5OH) at wavenumber range of 3200-3400 cm^{-1}. The spectra peak in the range of 2800-3000 cm^{-1} corresponds to the C-H stretching vibration of chlorophyll [20]. The carbonyl bond (C=O) appeared at 1630-1725 cm^{-1}. The functional group of C=O carbonyl and O-H hydroxyl at immersion temperature of 303 K were higher than other temperatures. It indicated that the carboxyl group (-COOH) at immersion temperature of 303 K has a bond with hydroxyl of TiO_2. It enhanced the effect of electrons coupling from dye molecules to the conduction band of TiO_2 that allows fast electron transfer. A higher immersion temperature showed the degradation of C-H dan C=O groups. The degradation of these groups affected the weak contact between dye molecule with the TiO_2 surface, thus blocking electron transfer.

3.3. *Performance of DSSCs*

The current-voltage (*I-V*) response of DSSCs were measured by Keithley *I-V* meter under light intensity of 100 mW/cm^2 from a xenon lamp. The *I-V* plot of DSSCs with different immersion temperatures are shown in Fig.3. A current-voltage characteristic was used to determine short-circuit current (I_{sc}), open-circuit voltage (V_{oc}), *fill factor* (*FF*), and efficiency (η). The photovoltaic parameters of the DSSCs are listed in Table 2. The fill factor *FF* and efficiency η were calculated using Eq. (1) and (2).

$$FF = \frac{V_{max} \times I_{max}}{V_{oc} \times I_{sc}} \tag{1}$$

$$\eta = \frac{V_{oc} \times I_{sc} \times FF}{P_{in}} \times 100\% \tag{2}$$

where P_{in} is the power of incident light. It calculated from multiplication between the light intensity and effective cell area. I_{max} and V_{max} are photocurrent and photovoltage for maximum power output P_{out}.

The highest performance of DSSC was showed at immersion temperature of 303 K. This result corresponded to the optical properties of TiO_2/dye photoelectrodes that revealing the highest absorbance and FTIR transmittance peaks at 303 K. When dye absorb more photon it will excite more electrons and resulting more currents. Also enhancement of C=O and O-H groups showed by FTIR transmittance will effect to electrons coupling. This is suggesting the more efficient of electron transfer, as reported by Chang and Lo [21]. Consequently, the resulting current and efficiency are higher with maximum FTIR transmittance of 67.2% as compared to other immersion temperatures.

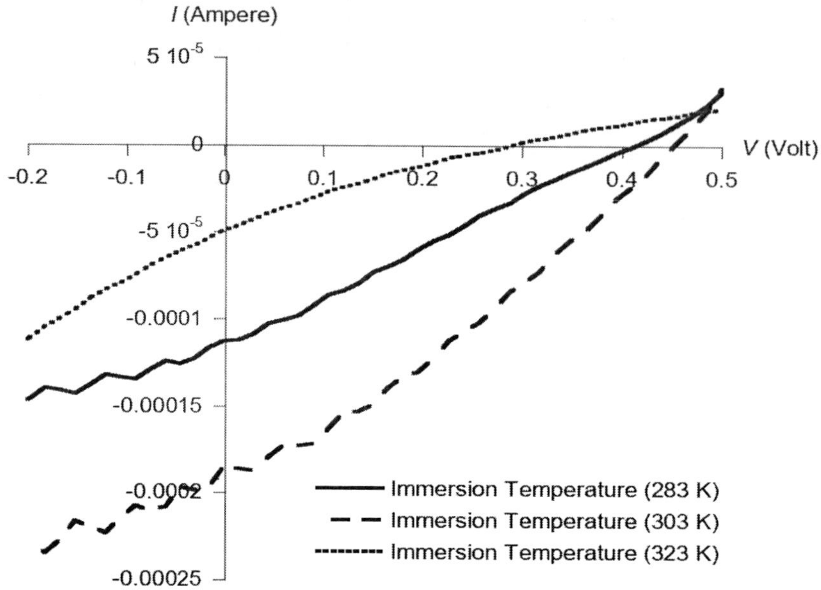

Figure 3. I-V plot of DSSCs with different immersion temperatures

Table 2. Photovoltaic parameters of DSSCs with different immersion temperatures

Temperature	V_{oc} (V)	I_{sc} (mA)	*FF*	η (%)
283 K	0.42	0.12	0.28	0.018
303 K	0.43	0.16	0.43	0.04
323 K	0.28	0.04	0.26	0.004

The immersion temperature of 323 K resulted in degradation of dye that showing the low performance of DSSC. This is caused by the reduced number of molecules dye on the TiO_2 surface, thus the number of electrons is also low. This analysis also can be absorved from the FTIR results that degradation of the -COOH group occured at the immersion temperature of 323 K. It prevented the electron injection and finally the output current decreased.

4. Conclusion

The immersion temperature of chlorophyll sensitizer affected the optical and electrical properties of DSSC. The characteristic of absorption spectra showed the maximum photon, while the FTIR spectrum exhibited the highest peak percent transmittance at the immersion temperature of 303 K (ambient temperature), respectively. Hence, the chlorophyll sensitizer succesfully maintain its stability, which the rate of electron injection in the TiO_2 conduction band was faster and more efficient. This condition was also supported by the highest efficiency of DSSC at the immersion temperature of 303 K, with the efficiency value was 0.04%. Eventually, the ambient temperature as the best condition to optimize the DSSC efficiency.

Acknowledgments
This research was supported by Higher Education and Institute for Research and Community services. Universitas Sebelas Maret by PU UNS Grant Program (Contract Number: 516/UN27.21/PP/2019).

References

[1] Grätzel M 2005 *J. Inor. Chems.* **44**, 6841-6851
[2] Nogueira A F, Longo C and De Paoli M-A 2004. *J. Coord. Chems. Rev.* **248**, 1455–1468
[3] Bhogaita M, Shukla A D and Nalini R P 2016 *J. Sol. Energy.* **137** 212-224
[4] Shalini S, Balasundaraprabhu R, Prasanna S, Mallick T K and Senthilarasu S 2015 *J. Renew. Sustain. Energy. Rev.* **51** 1306–1325

[5] Sulaeman U and Abdullah A Z 2017 *J. Renew. Sustain. Energy. Rev.* **74** 438-452
[6] Calogero G, Citro I, Crupi C and Marco G D 2014 *J. Spectrochim Acta Part A: Mol. Biomol. Spectrosc.* **132** 477-484
[7] Blaszczyk A 2018 *J. Dyes. Pigments.* **149** 707-718
[8] Wongcharee K, Meeyoo V and Chavadej S 2007 *J. Sol. Energy. Mat. Solar Cells* **91** 566-571
[9] Chang H, Wu H M, Chen T L, Huang K D, Jwo C S and Lo Y J 2010 *J. Alloys Compd.* **495** 606-610
[10] Shahid M, Shahid-ul-Islam and Mohammad F 2013 *J. Cleaner. Produc.* **53** 310-331
[11] Ludin N A, Mahmoud A M A, Mohamad A B, Kadhum A A H, Sopian K and Karim N S A 2014 *J. Renew. Sustain. Energy. Rev.* **31** 386-396
[12] Tripathi M and Chawla P 2015 *Ionics* **21**, 541-546
[13] Arifin Z, Soeparman S, Widhiyanuriyawan D and Suyitno 2017 *Int. J. Photoenergy.* **2017** 2704864 5 pages
[14] Amao Y, Yamada Y and Aoki K 2004 *J. Photochems. Photobio. A: Chems.* **164** 47-51
[15] Anita, Boisansi, Nurussaniah, Cari, Supriyanto A and Suryana R 2013 *Karakteristik Klorofil pada Daun Kacang Panjang (Vigna sinensis) sebagai Dye-Sensitized Solar Cells* (Semarang: PGRI Semarang University), pp LPF1353, ISBN: 978-602-8047-80-7
[16] Amadi L, Jenny S S, Ahmed A, Brown N, Yadav S, Brown D, Ghann W, Gayrama A, Jiru M and Uddin J 2015 *Nanosci. Nanoeng.* **3** 25-32
[17] Tonucci L H and Von E J H 1992 *J. Agric. Food Chem.* **40** 2341-2344
[18] Alfidharisti S R, Nurosyid F, Supriyanto A, Suryana R and Iriani Y 2017 *J. Phys: Conf. Ser.* **909,** 012026 (2017)
[19] Hortensteiner S 2006 *Annu. Rev. Plant. Biol.* **57** 55-57
[20] Das R K, Gogoi N, Babu P J, Sharma P, Mahanta C and Bora U 2012 *J. Adv. Mater. Phys. Chem.* **2** 275–281
[21] Chang H and Lo Y. 2010 *J. Sol. Energy.* **84** 1833-1837

Numerical and Experimental Study of Thermal Response of an Electrified Nickel Wire

N. F. Lubis, P. M. Widartiningsih, S. Viridi

Department of Physics, Faculty of Mathematics and Natural Science
Bandung Institute of Technology, Jl. Ganesha no. 10
Bandung, Indonesia, 40132.

Corresponding author e-mail: nurainafikalubis@gmail.com

Abstract. An electrified nickel wire can produce heat. The heat is transferred to the environment around the wire radiation. However, the temperature produced by the wire will differ depending on the number of voltages supplied by the power source. Here, we do an experiment to measure the temperature of stand-up wire at each voltage of 4.5 V, 6.0 V, 7.5 V, 9.0 V, and 12 V. As a comparison, numerical measurements are done by JavaScript program. We show that the highest temperature of the wire approaches 314 K at 12 V. To equalize the temperature in both methods, the numerical scale factor of time and current in numerical measurements are changed according to the voltage. As a result, temperatures produced in both methods have minimal gap difference with error percentage below 0.5%. In conclusion, the experiment results were used to determine the scaling parameters to approach the ideal thermal response curve in actual values.

1. Introduction

Heat is a physical phenomenon that commonly occurs every day. Heat is also a form of energy because it can move from one medium to another. Also, the heat effect is equivalent to some works of a system [1]. The simple method of heat transfer is a medium that has higher temperature transfers heat to a medium that has lower temperature [2,3]. One method of heat transfer discussed in this study is radiation.

Radiation does not require a medium to receive heat. It caused by scattering of electromagnetic waves that depend on size, shape and medium properties [4]. The amount of heat produced by a medium depends on the thermal emissivity of the medium. Thermal emissivity is the ability of a medium to deliver heat by radiation. Thermal emissivity values of some materials have been obtained from previous experiments so that they become a general reference. One of the experiments is the determination of thermal emissivity of coated copper and aluminum placed on liquid-nitrogen insulated dewar, a container that has same working principles as vacuum flask [5]. We should identify all thermal emissivity values, especially for metals, because they are essential for material science concept [6]. However, the accuracy of thermal emissivity depends on radiation heat fluxes from the heat source that works as a temperature distributor [7].

A simple experiment for proving the effect of thermal emissivity is by hotwire. A hotwire which is connected to a voltage source (i.e. variable adaptor) can discharge electric current. Yu et al. in Semyonov et al. used hotwire method to measure heat flux between a metallic filament (such as nickel) and a rarefied gas [8]. Hotwire also used in anemometer that based on the measurement of heat transfer [9]. In this research, we use nickel wire as hotwire to know the radiation effect on its temperature at a time. A voltage source which is connected to nickel wire flows electric current, then it converts to heat on this wire. Then, we measure nickel wire temperatures and analyze the correlation between the amount of voltages and temperatures.

However, obtaining the ideal curve of the temperature gradient is difficult due to sensitivity of measurement devices. So, we do numerical simulation as a reference to nickel wire experiment. In this simulation, we use time scale in order to minimize the gap between experimental and simulation data. Time scale is a form of time which is scaled in simulation, so the results fit with observation time of the experiment. Also, a dynamic comparison of different components is part of time scale characteristic [10]. A reference of an experiment involving time scale is detection of evolutionary changes in complex system configuration by generating intervals accordingly [11]. Then, ranges in complex system configuration adjust to the relation between time and space dependencies [12]. Regardless of that, heat transfer phenomena are included in complex system because they have time scale factor which plays very important role in small scale devices [13,14]. We can also add current scale factor, besides time scale factor, to support heat transfer in our nickel wire simulation. Then, we determine both time scale and current scale values towards experimental temperatures.

This study is our original research about comparison of numerical and experimental methods to determine the exact temperature of electrified nickel wire. So far, we cannot discover the previously similar simple research that states the precise temperature of electrified nickel wire at a certain voltage. When its temperature can change because environmental condition, numerical simulation becomes a comparison whether the temperature is accurate. We hope that our research can be developed to other electrified materials, i.e. metals for device manufacturing.

2. Theories

The current wire which is flowed by electric current can deliver heat at a specific temperature. Radially delivered heat will go to the environment. The heat emitted through radiation is examined in sections 2.1 to 2.5 below.

2.1. Dissipation Power

Heat is one type of energy. Thus, the energy quantity can be calculated per unit time. The amount of energy per unit of time is power. The electric wire current can generate power, which is called dissipative power. The dissipative power is lost due to an electric current that flows through electrical resistance on the wire. Dissipative power P_{dis} that occurs at the voltage source V and current wire as electrical resistance R are as follows.

$$P_{dis} = \frac{V^2}{R} \quad (1)$$

2.2. Heat Source

The heat source of the current wire comes from an electric voltage source. The heat spread is influenced by the mass of the wire m, the specific heat of the wire c and the temperature difference between the two ends of the wire ΔT. The heat generated from the wire is equal to Q and is formulated in the following Equation (2).

$$Q = mc\Delta T \quad (2)$$

Derivation of Equation (2) regarding time t will produce Equation (3).

$$\frac{dQ}{dt} = mc\frac{dT}{dt} \quad (3)$$

The derivation of heat energy Q regarding time t can be defined as power P. Therefore, the left side of Equation (3) is changed to wire power P_{wire} in Equation (4) below.

$$P_{wire} = \frac{dQ}{dt} \quad (4)$$

2.3. Stefan-Boltzmann Law

The radiation process that occurs in the universe is stated in Stefan-Boltzmann's Law. In everyday case, this process happens when a substance or material emits temperature T to the environment temperature T_{env}. This legal formula can be expressed by the radiation power of a material P_{rad} in the following equation (5).

$$P_{rad} = A\sigma\varepsilon\left(T^4 - {T_{env}}^4\right) \quad (5)$$

Surface area A is found in Equation (5) because radiation occurs per certain area unit. Another factor that affects the heat amount is Stefan-Boltzmann constant σ of 5.67×10^{-8} W/m²·K⁴ and the emissivity of material ε.

2.4. Calculation of Wire Temperature

To calculate wire temperature, a wire detail variable is needed such as wire diameter D and the wire sheath area A. The wire diameter is formulated in the equation

$$D = 2\left(\frac{A}{\pi}\right)^{\frac{1}{2}} \quad (6)$$

and the wire sheath area A_L is formulated in Equation (7).

$$A_L = \pi DL = 2L(\pi A_D)^{\frac{1}{2}} \quad (7)$$

Index L means lateral (along the long direction L) and index D means the cross-sectional area of the wire with diameter D. Substitution of Equation (7) in Equation (5) produces Equation (8) below.

$$\frac{P_{rad}}{2L(\pi A_D)^{\frac{1}{2}}} = \sigma\varepsilon\left(T^4 - T_{env}{}^4\right) \quad (8)$$

Through Equation (8), the power entering the wire P_{in} is

$$P_{in} = \frac{V^2}{\rho L/A_D} = \frac{V^2 A_D}{\rho L} \quad (9)$$

2.5. Energy Conservation
In general, the amount of heat energy transferred from a medium to another medium is equal. This case is included in energy conservation. One important aspect of energy conservation is heat storage from any heat transfer types (including radiation) [15]. When radiation occurs, the power found on the wire P_{wire} will be equal as the difference in the input power P_{in} with the radiation power P_{rad}. This statement formulated in the following Equation (10).

$$P_{wire} = P_{in} - P_{rad} \quad (10)$$

Next, substitution of Equations (4), (9) and (8) in Equation (10) becomes the following equation.

$$mc\frac{dT}{dt} = \frac{V^2 A_D}{\rho L} - 2L(\pi A_D)^{\frac{1}{2}}\sigma\varepsilon\left(T^4 - T_{env}{}^4\right) \quad (11)$$

3. Research Method

3.1. Experiment
The first step in this research is identifying all properties of nickel wire. Those properties can affect the final experiment results, which are given in Table 1.

Table 1. Properties of nickel wire.

Quantities	Value
Radius R_W (m)	2.5×10^{-4}
Wire cross-sectional area A_d (m^2)	1.96×10^{-7}
Wire length that used as experiment indicator L (m)	0.6
Wire resistivity ρ_W (Ω.m)	6.99×10^{-8}
Wire thermal conductivity κ_W (W/m.K)	90.7
Wire thermal emissivity ε_W	0.12
Wire heat specific c_W (J/kg.K)	440
Wire mass m_W (kg)	2.5×10^{-3}
Wire density m_W / V_W (kg/m^3)	8,908

The experimental setup is shown in Figure 1. The nickel wire with a length of 60 cm was enclosed by two layers of isolator, a tissue paper and a plastic tube. The enclosed wire was placed on a stand as illustrated in Figure 1a. In order to retain the wire shape, a metal load was attached to the lower end of the wire. The electrical potential difference was given at both ends of the wire. A digital thermometer' sensor was put on nickel wire, as well as another sensor was put down around the experiment area. Both sensors are used to measure nickel wire temperatures and environment temperatures, respectively. This system was examined in five values of voltage 4.5 V, 6.0 V, 7.5 V, 9.0 V and 12 V. The temperature data were extracted at every minute for 35 minutes.

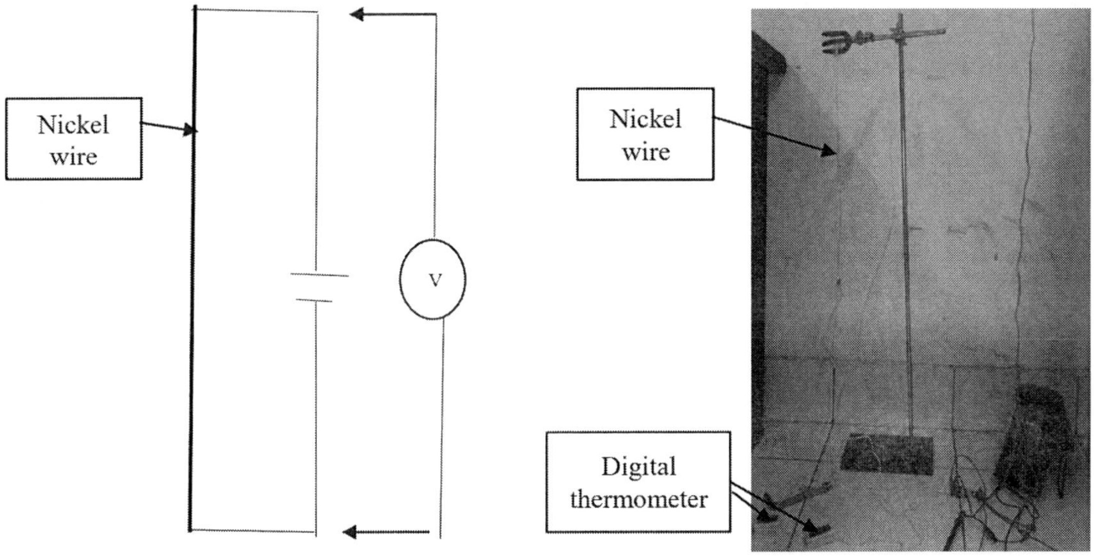

(a) (b)

Figure 1. (a) Scheme and (b) picture of the experiment to measure nickel wire and environment temperatures.

3.2. Simulation

We also do research by simulation to obtain numerical temperatures of nickel wire. The simulation is useful as a comparison on experimental temperatures of nickel wire. The simulation is made by JavaScript program shown by HTML program in Google Chrome browser as shown in Figure 2. This program involves forward finite difference method to solve differential equation, i.e. Equation (11) numerically. In order to minimize the difference between experimental and numerical temperatures, we must change SCALE1 and SCALE 2 variable with the appropriate value. SCALE1 and SCALE2 represent time scale and current scale respectively.

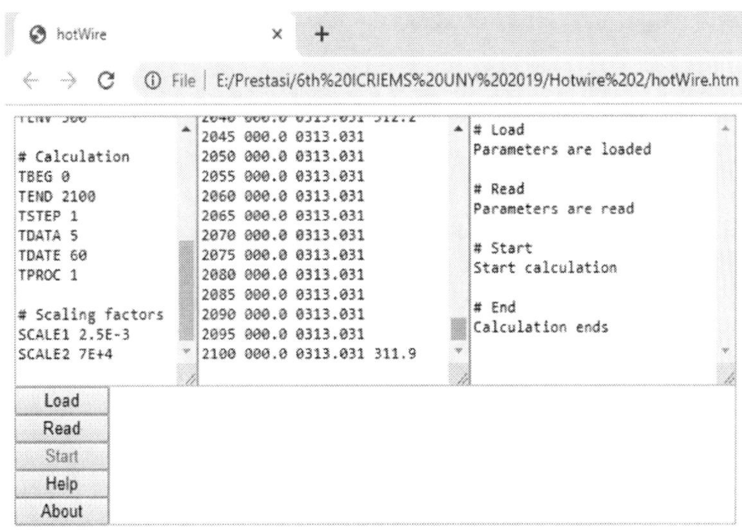

Figure 2. Display of JavaScript program to determine numerical temperatures of nickel wire.

4. Result and Discussion

Research results by experiment and numeric both are shown by two different curves in Figure 3 until Figure 7. Both methods take time 35 minutes. T_{exp} curve represents experimental temperatures in every minute. Otherwise, T_{num} curve represents numerical temperatures in every second for 2,100 seconds, equal as 35 minutes. Variations of voltage in each experiment produce different temperature data.

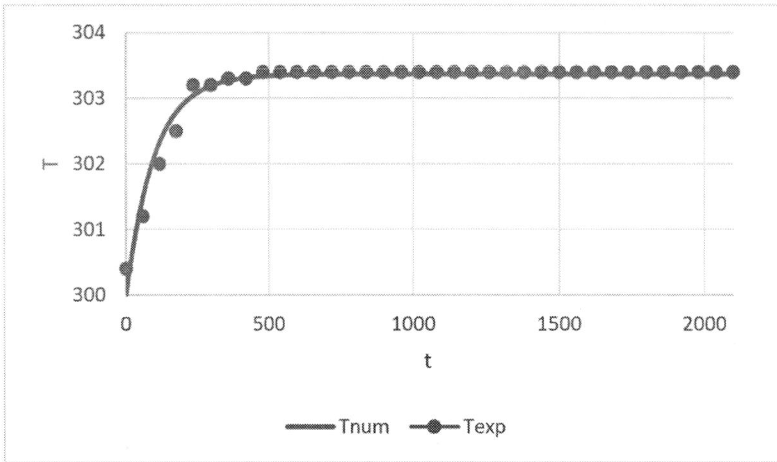

Figure 3. Comparison of numerical and experimental temperatures of nickel wire for voltage 4.5 V.

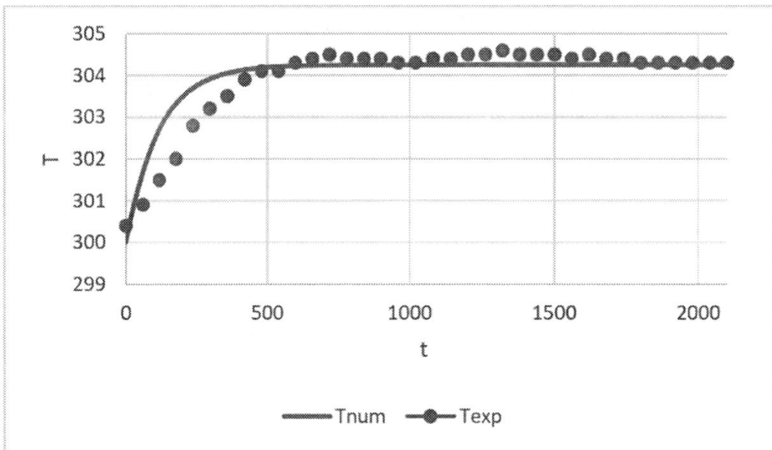

Figure 4. Comparison of numerical and experimental temperatures of nickel wire for voltage 6.0 V.

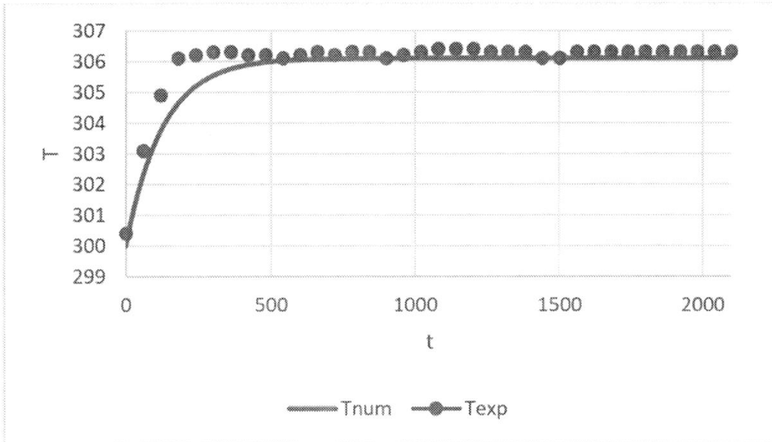

Figure 5. Comparison of numerical and experimental temperatures of nickel wire for voltage 7.5 V.

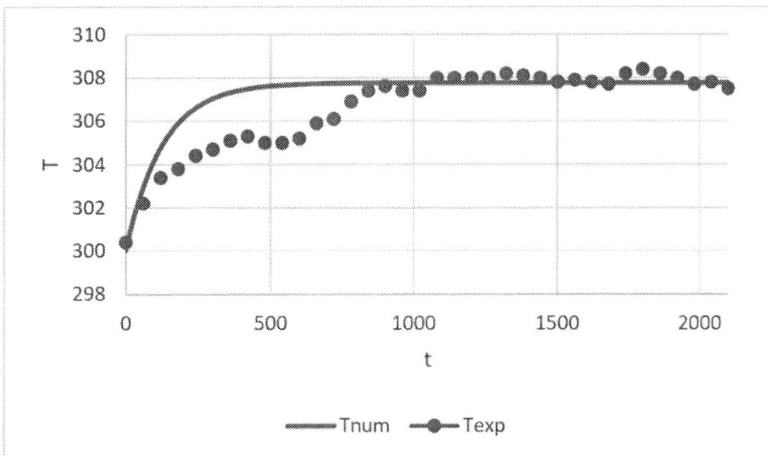

Figure 6. Comparison of numerical and experimental temperatures of nickel wire for voltage 9.0 V.

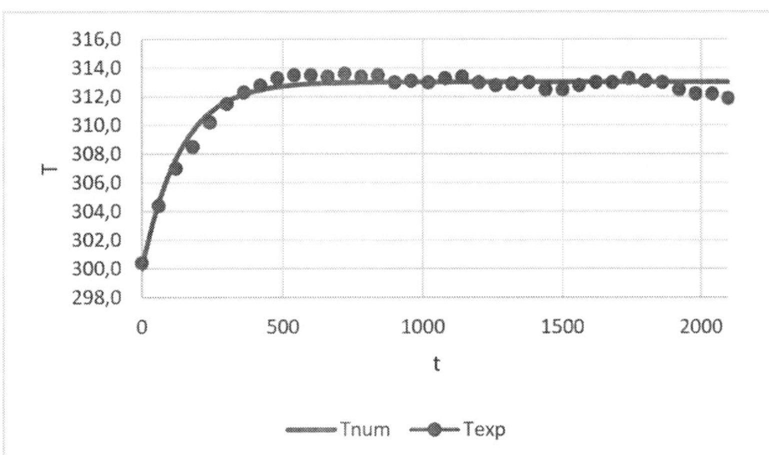

Figure 7. Comparison of numerical and experimental temperatures of nickel wire for voltage 12 V.

In Figures 3 - 7, we can analyze when the voltage is greater, then the temperature of nickel wire will increase. The highest temperature rising is at the largest voltage, which is 12 V with a maximum temperature reaching 314 K in both methods. The reason is the greater the voltage set on the variable adapter (12 V), the more heat energy is generated. Heat energy per unit time or power is also directly proportional to the voltage stated in Equation (4). Power connected to electric current is also called dissipation. This power causes the actual voltage (voltage on the variable adapter) to fall because it converts almost half of the voltage to heat.
To minimalize the difference between experimental temperatures and numerical temperatures, we set SCALE1 (time scale) and SCALE2 (current scale) factors. The purpose of setting both factors is minimizing the gap difference between T_{exp} and T_{num} curve. Thus, all error percentages between experimental temperatures and numerical temperatures are below 0.5%. In case to keep this error value, experiment and simulation results in Figure 6 is slightly away between zero to 1,000 seconds. Then, they return to approach each other after 1,000 seconds. From Figure 8 below, the higher voltage requires a higher value of SCALE 2. In contrary, the value of SCALE1 decreases with increasing voltage.

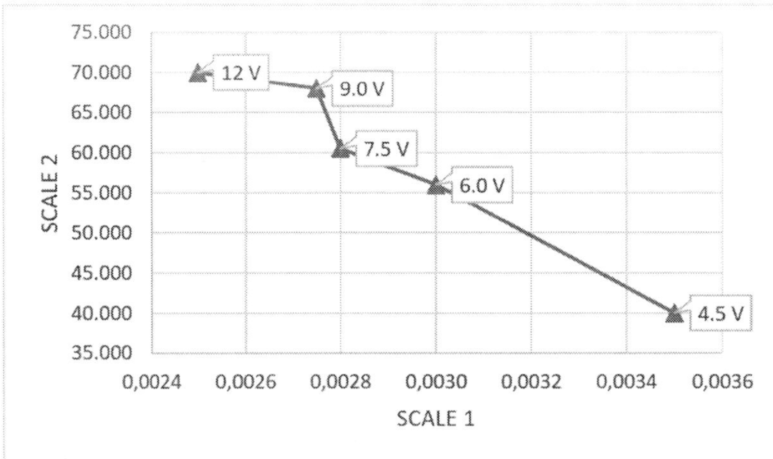

Figure 8. Comparison between SCALE 2 vs SCALE 1 at each voltage.

As a reference to both experimental and numerical methods, all equations in Section 2 are not used. For example, Equation (11) which is the formula towards nickel wire, is not used in experimental method because T_{env} changes during experiment. Otherwise, Equation (11) is applied for numerical method because some formulas in Section 2 are used in JavaScript simulation. As a consequence, SCALE1 and SCALE2 factors are enough for proving the accuracy of experimental temperatures at certain voltages and times.

5. Conclusion
Time scale and current scale factors in the nickel wire simulation can prove the accuracy for experimental method. Both factors can scale accurately on nickel wire temperatures in this simulation. They also minimize the gap difference between experimental and numerical temperatures with below 0.5% error percentage. Through the statements above, we can say that our nickel wire experiment is accurate.

6. Acknowledgment
The authors would like to thank Physics Department – Bandung Institute of Technology for providing Students Activity Grants (Bantuan Kegiatan Mahasiswa) 2019.

7. References
[1] Adkins C J 1983 *Equilibrium Thermodynamics: Third Edition* (Cambridge: Cambridge University Press)
[2] Utomo D K, Kurniawati A, Hendro and Viridi S 2019 *J. Phys. Conf. Series.* **1127** 1
[3] Hecht E 2008 *Phys. Teach.* **46**, 77 80
[4] Golyk V A, Kruger M and Kardar M 2012 *Phys. Rev.* E. **85** 1
[5] Kim S H, Conway Z A, Ostroumov P N and Shepard K W 2014 *AIP Conf. Proc.* **1573** 500-6
[6] Filep E, Kutasi D N and Kenez L 2018 *Acta Mater. Transylv.* **1/1** 31

[7] Fu T, Tan P and Duan M 2015 *Mea. Sci. Technol.* **26** 8
[8] Chalabi H, Buchina O, Saraceno L, Lorenzini M, Valaugeorgis D and Morini G R 2012 *J. Phys. Conf. Series.* **362** 2
[9] Kielbasa J 2007 *Rev. Sci. Instrum.* **78** 1
[10] Pan Z, Guo Q and Sun H 2016 *Appl. Energy* **167** 230
[11] Darst R K, Granell C, Arenas A, Gomez S, Saramaki J and Fortunato S 2016 *Sci. Rep.* **6** 1
[12] Li J, Zhang J, Ge W and Liu X 2004 *Chem. Eng. Sci.* **59** 1697
[13] Marin E 2010 *Lat. Am. J. Phys. Educ.* Vol. 4, no. 1 56
[14] Bergamasco L, Alberghini M, Fasano M, Cardellini A, Chiavazzo E and Asinari P 2018 *Entropy* **20**, 126 1
[15] Demirbas A 2002 *Energ. Explor. Exploit.* Vol. 20, no. 5 393

Optical Activity Effect on Planar Chiral Metamaterials

Juliasih Partini[1], Restu Widiatmono[2]
[1]*Department of Physics, Universitas Gadjah Mada, Yogyakarta, Indonesia*
[2]*Department of Physics, Universitas Negeri Yogyakarta, Yogyakarta, Indonesia*

juliasih@ugm.ac.id

Abstract. Metamaterial is an artificial material which has a negative refractive index. Both of permitivitty and permeability have a negative value to possess negative refraction. The breakthrough of metamaterial development was present as the invention of planar chiral metamaterials. This research was focussed on the development of a new design of planar chiral metamaterial with different depth. The developed metamaterial design was analyzed by chirality analysis in relation to the optical activity characteristics.
The planar chiral metamaterial development was following a design, a structure examination, and a fabrication process using a focused ion beam system. Chirality characterization were measured using a femtosecond Ti:Sapphire pumped terahertz emission spectroscopy system. THz signal was occurred due to the optical rectification process as a result of an intense femtosecond laser pulses radiation on planar chiral metamaterial sample. The samples were capable to generate a THz emission over a frequency range of 0 – 2.24 THz, while the THz signal was measured as ~5 ps. The effect of optical activity was able to rotate the polarization plane on a frequency range of 0 – 1.4 THz for both of the clockwise and the counter-clockwise samples. The maximum polarization plane rotation were +13° and −18° at a frequency 1.35 THz. It can be concluded that the new design planar chiral metamaterial with different depth has been successfully obtained with its elliptic polarized THz emission characteristic.

Key words: Optical Activity, Planar Chiral Metamaterials, Polarization Plane

1. Introduction

Metamaterials can be describe as the arrangement of artificial dielectric elements or magnetic elements in order to get unusual electromagnetic properties [1]. Unlike natural materials, metamaterials must be builded to gain negative permittivity and negative permeability simultaneously. Negative refraction can be constructed from the dielectric metamaterials (as a negative permittivity) and the magnetic metamaterial (as a negative permeability). Chiral metamaterials, in the other hand, has a different way to get the negative refraction [2]. The chiral metamaterials possess a difference response for a left circularly polarized wave and a right circularly polarized wave [3]. Chiral structure will do significant effect when they interact with the electromagnetic field, such as polarized the field. For example, the optical activity, which is caused by a difference in indices of refraction between left/right-handed circularly polarized wave [4].

To satisfy the increasing demand of chiral materials with strong optical properties avalaibility, nanostructured metamaterials are used to construct the chiral materials. We can obtain negative refraction in chiral metamaterials with large chirality without the need to have negative permittivity and negative permeability simultanously. The polarization plane of a linearly polarized wave will rotate when it pass through the planar chiral metamaterials. The phase between two circularly polarized wave will increase as their superpositioned. The polarization change from linear to circular are characterized with its ellipticity [5]. When the ellipticity is large, the linearly polarized wave will strongly distorted then become elliptically polarized, as shown is fig.1.

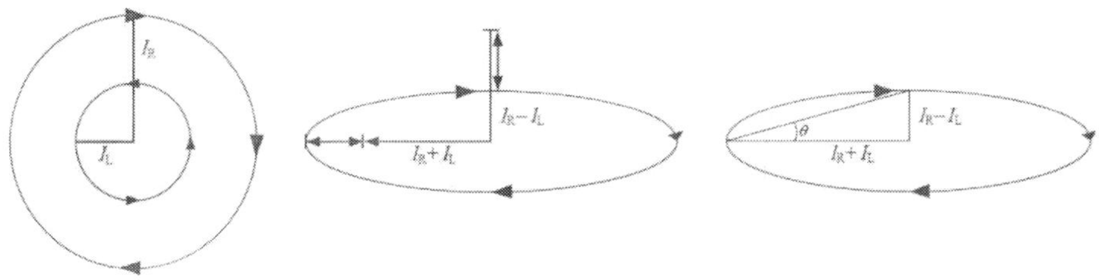

Figure 1. Schematic of ellipticity

Chiral metamaterials will rotate the polarization plane of a linearly polarized light which passes through it. These metamaterials optical properties are called optical activity and the polarization mechanism is shown in fig. 2. Optical activity is shown by the polarization angle of elliptically polarized wave. The polarization angle can be determined from the angle between the polarization ellipsoid major axis and the horizontal axis [6].

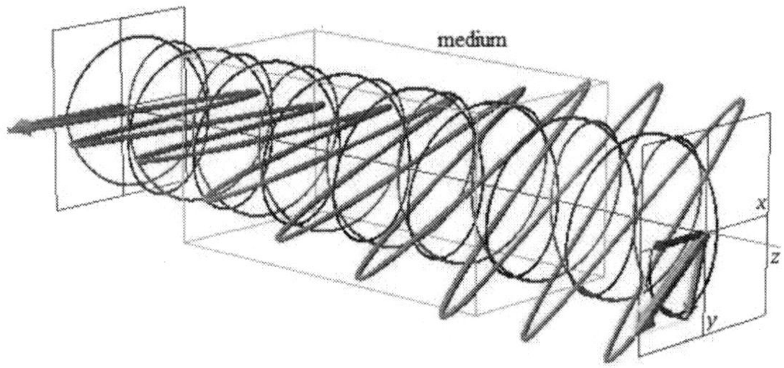

Figure 2. Mechanism of optical activity

Research on planar chiral metamaterials also emerge at the terahertz region and the developed metamaterials are consider as a new class of metamaterials [7]. Todays advancement of a high-field terahertz source lead to the chance to explore many nonlinear effect including optical rectification. Rectification will generate terahertz frequencies from optical pulses [8]. Nevertheless, the experiments in planar chiral metamaterials at the terahertz region still face major challenges due to the the geometry complexity [9]. Nowaday, advances in nanoscale fabrication have made the possibility to create artificially nanostructured materials with extraordinary optical properties [10]. In this paper, we report the analysis of planar chiral metamaterial with different depth. The structure was experimented using terahertz emission and its polarization angle spectra analysis in relation to the optical activity effect.

2. Sample preparation & experimental setup

The planar chiral metamaterial development was conducted following the steps as follows: first, we begin with a design mechanism, followed by a structure examination, then a fabrication process using an FB2200 Focused Ion Beam system. THz emission and chiral metamaterials optical activity effect were determined using THz emission spectroscopy. The planar chiral metamaterial samples which were used have dimensions 1.2 μm × 0.3 μm × 0.3 μm. We have developed two type of a 1200 periodic square pattern samples of a planar chiral metamaterials. Samples was builded with a period formed by a square pattern of chiral with different depth on a silver thin film with a clock-wise (CW) and a counter clock-wise pattern . The single pattern of a chiral unit and its dimension were shown in Figure 3 [11].

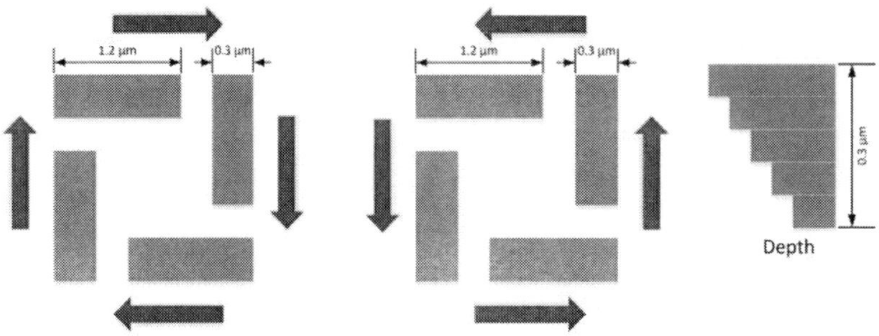

Figure 3. Pattern of (a) sample 1 (CW) (b) sample 2 (CCW) and (c) Sample depth

A standard THz generation setup as shown in fig. 4. were using to the experiments. The technical spesification of the Laser source were as follows: Ti:Sapphire oscillator (Scientific XL, Femtolasers), generates pulses with 50 fs duration, wavelength 800 nm, repetition rate 11 MHz, and average power output 800 mW. Laser beam was directed to the beam splitter which transmit 80% intensity to use as the pumping beam and 20% use as the sampling beam. The pumping beam will focused to an approximately 2 mm spot size on the sample surface, at 45° angle of incidence. The generated THz radiation were gained using an off-axis paraboloidal mirror. The reflection were focused to a ZnTe detection crystal. At the same time, the copropagating sampling pulse was synchronized to the detection crystal. The THz electric field will elliptically polarizes the probe beam. The ellipticity was determined using a differential detection setup that consist of a quarter wave plate, a Wollaston prism, and a differential detector.

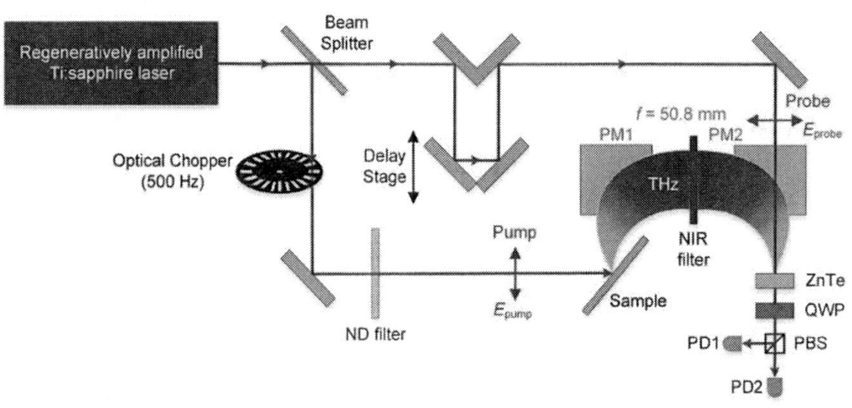

Figure 4. Schematic of the THz setup used in the experiment

As described before, the polarization was transform from linear polarization to circularly polarization. The intensity difference between probe pulse s and p wave was determined after the probe pulse transmit through the quarter wave plate. The polarization states then were analysed using polarization state experiment, as shown in figure 5.

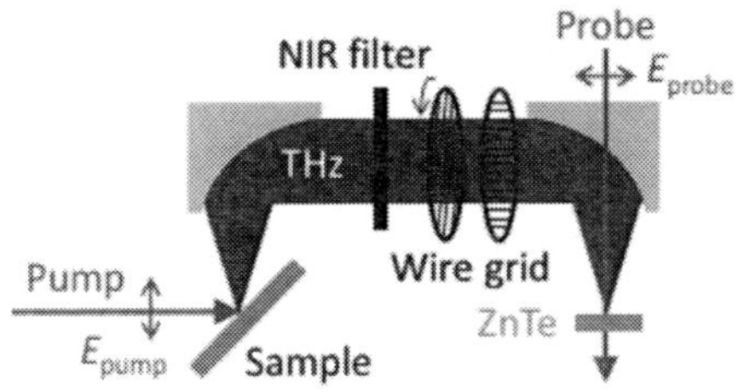

Figure 5. Schematic of the Polarization states experiment

We could describe left-hand circular polarization as an electric field vector that do an anti-clockwise rotation at a fixed point, as observed along the beam. The ellipticity was yielded as [12]

$$\eta = \frac{|E_L|-|E_R|}{|E_L|+|E_R|} \tag{1}$$

E_R shows the amplitude of right-hand circularly polarization and E_L shows the left-hand. The gained polarization angle was the angle between the polarization ellipsoid the major axis and the horizontal axis.

3. Discussion

Femtosecond laser pulses which were radiated to a planar chiral metamaterials sample will generate THz signal. THz signal occurred due to the non linear optical rectification process. The developed chiral metamaterial samples capable to generate THz emission signal at ~5 ps. A Fourier transform process to the signal's time-domain traces has sucessfully determined that the signal was emitted with a frequency of 2 THz. The emitted THz electric field from the planar chiral metamaterial have a single-cycle shape followed by an oscillating tail as shown in Fig. 6. The weaked oscillations were suspected originate from absorption and re-emission of THz light by water vapor molecules in the atmosphere.

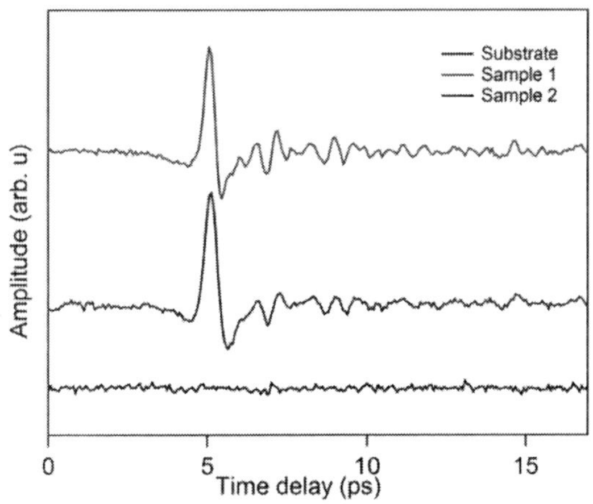

Figure 6. Waveform of the THz pulse

Second order dependence in these experiments are interesting to explore. The THz amplitude were convincingly non linearly increase and the second-order non linear process was consider involve with the THz emission. Figure 7 shows the rotation angle of THz emission spectra from every sample. It can be stated that the chirality of the sample chiral structure will determine the polarization rotation and its rotation direction. The strong oscillating appearance for the right-hand and the left-hand sample was raised to define the angle for the almost elliptic polarized wave.

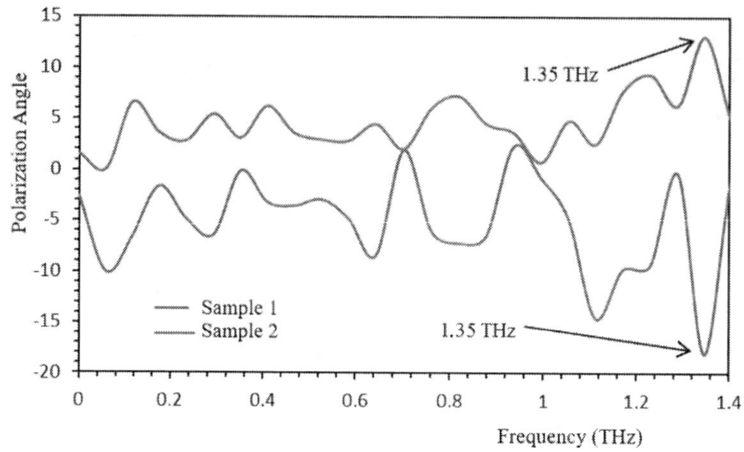

Figure 7. Polarization angle spectra

The sample optical activity was able to rotate the polarization plane on a frequency range of 0 – 1.4 THz for both of the clockwise and the counter-clockwise samples. The maximum polarization plane rotation were +13° and −18° at a frequency 1.35 THz. More importantly, we are succesfully observed polarization reversal of ellipticity in each sample. The opposite polarity stated that the absorbance value of a clock-wise and a counter clock-wise sample was different. The planar chiral metamaterial has been constructed and posses THz emission with elliptically polarization.

4. Conclusion

This new structure design of a chiral metamaterial has shown a promising result as it simple and easy to fabricate. The developed sample were successfully performed that the optical activity was depend on their chirality. The planar chiral metamaterial samples were capable to generate THz signal under an intense femtosecond laser pulses radiation. The generated THz emission were ranging from a frequency of 0 to 2.24 THz, and its THz signal was ~5 ps. The optical activity effect was able to rotate the polarization plane on a frequency range of 0 to 1.4 THz for both of the clockwise and the counter-clockwise samples. The maximum polarization plane rotation were +13° and −18° at a frequency 1.35 THz. We have the conclusion that the planar chiral metamaterial has been builded and posses THz emission with elliptically polarization.

Acknowledgments :

We are gratefully express our appreciation to Dr. Satoshi Tomita, Matsui Takahiro from Quantum Materials Science Laboratory, Nara Institute of Science Technology and the team of Laser Terahertz Division at Osaka University for their supporting work.

References

[1] Pendry, J.B., 2000, Negative Refraction Makes a Perfect Lens, *Physical Review Letters*, 85, 3966 – 3969.

[2] Pendry, J.B., 2004, A Chiral Route to Negative Refraction, *Science,* 306, 1353 – 1355.

[3] Dincer, F., Karaaslan, M., Akgol, O., Unal, E., and Demirel, E., 2015, New generation planar chiral metamaterials with small and constant chirality over a certain frequency band, *Modern Phyical Letters B*, 29, 1450257, 1 - 12

[4] Hua, L., Yan, G., and Lin, Y., 2014, Experimental and simulated study of dual-band chiral metamaterials with strong optical activity, *Microw. Opt. Technol. Lett.*, Vol. 56, No. 10, 2381-

2385

[5] Ma, X., Huang, C., Pu, M., Pan, W., Wang, Y., and Luo, X., 2013, Circular Dichroism and Optical Rotation in Twisted Y-Shaped Chiral Metamaterial, *Applied Physics Express*, 6, 022001, 1 – 3

[6] Ganzel, J.K., Thiel, M., Rill, S., Decker, M., Bade, K., Saile, V., Freyman, G., Linden, S., and Wegener, M., 2009, Gold helix photonic metamaterial as broadband circular polarizer, *Science*, 325, 1513, 1 – 5

[7] Yoo, S. and Park, Q., 2019, Metamaterials and chiral sensing: a review of fundamentals and applications, *Nanophotonics,* 8(2), 249 - 261

[8] Kadlec, F., Kuzel, P., and Countaz, J.L., 2005, Study of terahertz radiation generated by optical rectification on thin gold films, *Optics Letters*, Vol. 30, No. 11, 1402 – 1404

[9] Konstantin, L. 2006, Optical generation of narrow-band terahertz packets in periodically-inverted electro-optic crystals: conversion efficiency and optimal laser pulse format. *Optics Express*, Vol. 14, No. 6, 2263 - 2276

[10] Kanda, N., Konishi, and Kuwata-Gonokami, M., 2007, Terahertz wave polarization rotation with double layered metal grating of complimentary chiral patterns, *Optics Express*, Vol. 15, No 18, 11117 – 111125.

[11] Juliasih, P., Kamsul, A., and Arief, H., 2017, Chirality Analysis on A Square Chiral Metamaterial, *Material Science Forum*, vol. 901, 65 – 68

[12] Suo, H., Takano, K., Ohno, S., Kurasawa, H., Nakayama, K., Ishihara, T., and Hangyo, M., 2013, Polarization property of terahertz wave emission from gammadion-type photoconductive antennas, *Applied Physics Letters*, 103, 111106, 1 – 4

Identification of Subsurface Lithology in Sendang Mulyo, Purwoharjo Village, Samigaluh Subdistrict, Kulon Progo Regency

L. Katriani*, D. Darmawan, B. Ruwanto, H. Lutfiana, and H H. Prameswari

Department of Physics Education, Universitas Negeri Yogyakarta, Jl. Colombo No. 1, Karang Malang 55281, Indonesia

*Corresponding author e-mail: laila_katriani@uny.ac.id

Abstract. Research aimed to determining the lithological conditions subsurface of landslide areas in Sendang Mulyo, Purwoharjo Village, Samigaluh Subdistrict, Kulon Progo Regency by geoelectric and microseismic methods. The research had been conducted after knowing the subsurface lithology conditions, it could anticipate landslides as disaster mitigation. The location of research is on a hillside in Purwoharjo Village. The research located on that area because the history landslides. The area is the longest landslide and had a major impact on the villagers that located below the hill. The research used a set of resistivity geoelectric devices with Syscal brands and microseismic devices Digital Portable Seismograph brands TDL-3035 type Taide brands. In geoelectric method used 5 lines, 2 lines 150 m from north to south, 2 lines 120 m and 1 line 100 m from east to west, while the microseismic method used 30 points with spaces of 10-100 meters. The results of the geoelectric method are obtained lithology of clay material with resistivity of 0.16 - 160 Ω.m, sand material with resistivity 160 - 1093 Ω.m, sandstone material with resistivity 1093 - 8000 Ω.m, and a andesite material with resistivity of 8000 - 122165 Ω.m, while the results of the microseismic method obtained clay rock material having a value of V_s between 406.33 m/s and 742.52 m/s represented by the dark blue spectrum. In the middle landslide has a depth of clay material up to 33 meters. The results showed that subsurface lithology in the area is prone to landslides, because it is suspected that there is a sliding field at a depth of 33 meters. This is supported by the presence of andesite rocks which spread at a depth of 20 meters and above the andesite material, there is clay material that quite thick.

1. Introduction

In the past few years there had been a landslide in the hills of Menoreh Kulon Progo. In 2004, at least 15 people died and material damage occurred such as houses, agricultural land and transportation routes. Inclined of landslides in the area is supported by the incidence of Cempaka Tropical Cyclone at 28 - 30 November 2017 on the southern island of Java. This incidence causes the potential for landslides in Kulon Progo Regency to be higher. The Geospatial Information Agency (BIG) in collaboration with the Regional Disaster Prevention Agency (BPBD) of Kulon Progo Regency made rapid mapping in parts of Samigaluh, Kalibawang and Girimulyo Districts. According to historical landslides there are longest landslide in Samigaluh by the storm, exactly in Sendang Mulyo, Purwoharjo village [1].

One of natural disaster is landslide. This disaster impact in damage to residential land, disruption of traffic lanes, damage to agricultural land, damage to bridges, irrigation and other physical infrastructure. Landslides in general are a geological phenomenon which land movements as rock falls or land lumps [5]. Landslides happen when the incentive force on the slope is greater than the barrier force. Barrier forces are generally influenced by rock strength and soil density, while the incentive force is influenced by the magnitude of slope, water, load and rock type specific gravity.

Knowledge of subsurface lithology is very necessary to anticipate of the inclined area to natural disasters, especially for landslides and as a landslide mitigation. The method that can be used to determine subsurface lithology is geoelectric resistivity and microtremor. Geoelectric method is one of the geophysical methods that knows the nature of electricity in the earth by injecting electric current, then measuring the potential difference and current that occurs used electrode configuration, so that the rock layer beneath the resistivity value is obtained [10]. Subsurface lithology is obtained from the distribution of layers resistivity, which has been compared with resistivity tables [9]. Microtremor is a very small and continuous ground vibration that originates from a variety of vibrations such as, traffic, wind, human activities and others [3]. Microtremor is based on recording microseismic signals to determine the parameters of dynamic characteristics (damping ratio and natural frequency) and the transformation function (frequency and amplification) of buildings. Microtremor observation can be used to determine the characteristics of the soil layer based on the parameters of the dominant period and the amplification factor. Nakamura [7] states that the value of the reinforcement factor (amplification) of the soil is related to the contrast ratio of the surface layer impedance to the layer below it. If the contrast impedance ratio of the two layers is high, the reinforcement factor value is also high, and vice versa. Microtremors can also be used to determine the type of soil based on the level of violence [8], if the smaller the dominant period of land, the greater the level of violence or the land that has a dominant period the greater the soft or soft nature. Marjiyono [4] states that, amplification is directly proportional to the value of the horizontal and vertical spectral ratio (H/V). HVSR techniques (Horizontal to Vertical Fourier Amplitude Spectral Ratio) in microtremor data analysis have been widely used for studies of local effects and microzonation. HVSR is one of the easiest and most inexpensive ways to understand the structural properties of subsurface layers without causing disturbance to these structures [6].

2. Method

This research used dipole-dipole geoelectric resistivity method with dipole-dipole configuration. The research used resistivity data on five lines. To confirm the result, this research used microseismic method with microtremor 30 data of measurement points . The research location is in the Sendang Mulyo, Purwo Harjo village, Samigaluh subdistrict, Kulon Progo regency of Yogyakarta. The site was located at 7°41'47.26" S - 7°41'44.63" S and 110°10'29.37" E - 110°10'31.19" E. Based on the geological structure, Purwoharjo Village has the Kebo Butak Formation. This formation generally consists of conglomerates, sandstones and clays [2].

The measurement data from the geoelectric method are potential values (ΔV), the current strength (I), and the space (a). These data used to knowing the geometry factor (K) and resistance (R) values so that the value of apparent resistivity (ρ_a) is obtained by the equation $\rho_a = k\frac{\Delta V}{I}$, k is geometric factor of dipole-dipole configuration. The equation to get the value of geometric factor is $k=n(n+1)(n+2)\pi a$, with $n = 1,2,3,4,5,$ and 6. To obtained the true resistivity, the next step is processing the data using *Res2DinV* software. Microtremor data is obtained by measuring microtremor signals and then analyzed using HVSR (Horizontal to Vertical Spectra Ratio) to get the *H/V* curve . The *H/V* curve is used to determine subsurface lithology with predominant frequency values and amplification values. V_s (shear wave velocity) values obtained in V_s ground profiles using the Dinver program to classify the types of subsurface layers of the study area.

3. Result and Discussion

a. Geoelectric Method

1) First Line

The first line is 150 m long with the northeast to the southwest with located at 7°41'44.78" S, 110°10'33.11" E to 7°41'46.58" S, 110°10'28.84"E. The space between electrode is 10 meters, 76 datum points are obtained. The depth of the subsurface structure is up to 21.8 meters. The results of the interpretation are obtained by changing the iteration several times until an interpretation is approaching the actual field conditions. On the first line, it is carried out until the 5th iteration with a value of 10.1 rms.

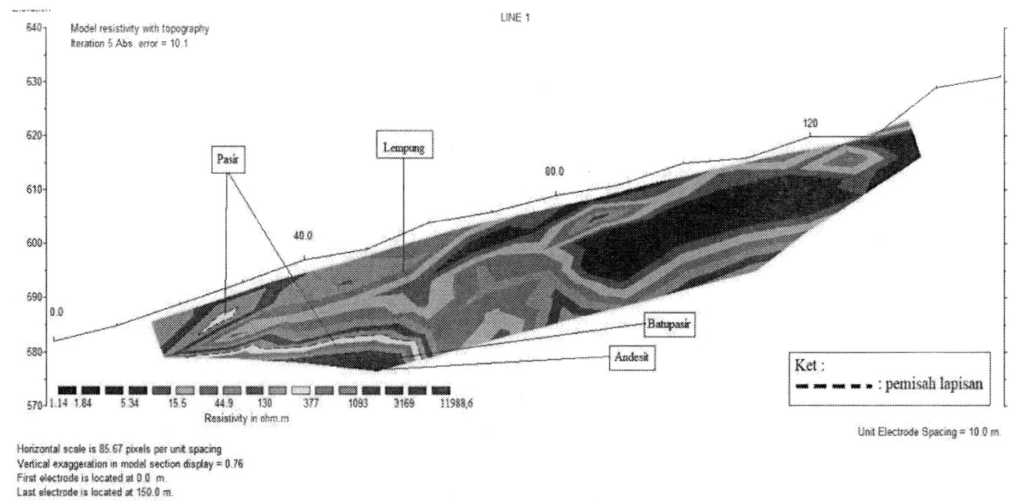

Figure 1. Cross Section Resistivity of the First Line

Based on Figure 1 there are four layers of material, clay material with resistivity of 1.14 - 130 Ω.m. There are a few layers of sand at a depth of 2 meters and 7 meters at a distance of 45 meters from the starting point with resistivity of 160 - 1093 Ω.m. At a depth of 17 to 18 meters, sandstone layers were found with resistivity from 1093-1923.5 Ω.m. At a depth of 20 to 21 meters, bedrocks were found with andesite material which had a resistivity of 11988.6 Ω.m. On this line a sample such as clay is taken at a distance of 40 meters from the starting point in 1 meter depth.

2) Second Line

The second line is at 7°41'43.79" S 110°10'31.55" E to 7°41'47.16" E 110°10'28.09" E. The length of this line is 150 m with the northeast to the southwest while the spacing between the electrodes is 10 meters. This line obtained 76 datum points data. The target of the depth is 21.8 meters below.

Journal of Physics: Conference Series **1397** (2019) 012008 doi:10.1088/1742-6596/1397/1/012008

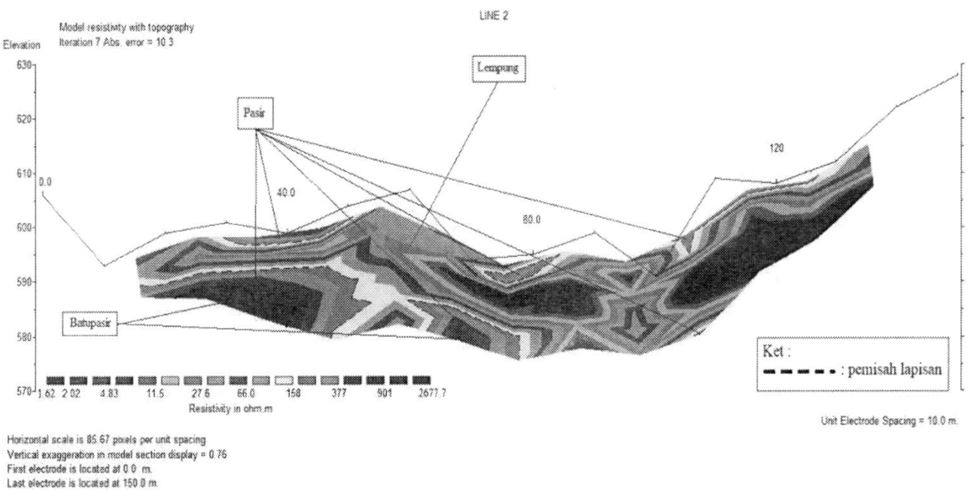

Figure 2. Cross Section Resistivity of the Second Line

On the second line (Figure 2) there are 3 layers of materials. Clay layer with resistivity 1.62 - 158 Ω.m. Sand material is dominant on the surface with resistivity 160 - 901 Ω.m. This sand material is also found in depths of 10 to 20 meters. At a certain depth, around 15 meters there are sandstones that have a resistance of 1093 to 2677.7 Ω.m. Sample such as clay rock found at a distance of 40 meters from the starting point and 30 cm depth.

3) Third Line

The length of the third line is 120 m from the north to the south. The distance between the electrodes is 10 meter. The depth target is 21.8 meters below. In this line there are fewer data from the previous lines, which are 52 datum points. Model iterations are needed to get the interpretation results that are close to field conditions and in this research, iterated 6 times with value of rms is 7.7.

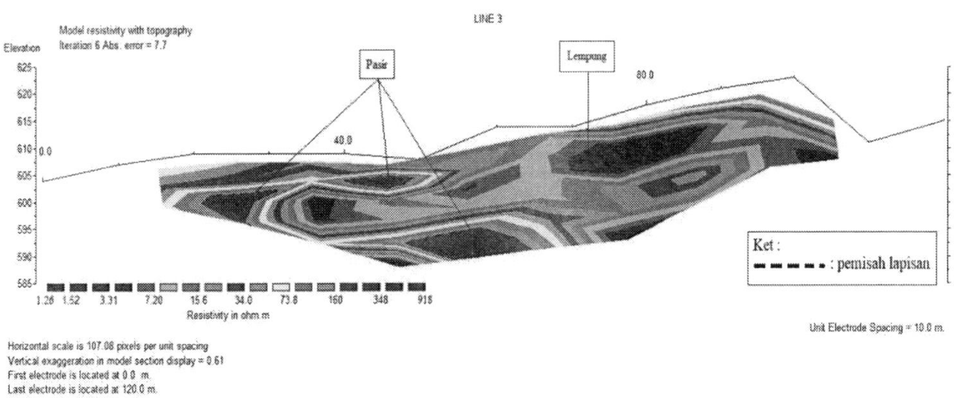

Figure 3. Cross Section Resistivity of the Third Line

On the third path (Figure 3) there are only 2 layers. There is a clay layer with resistivity 1.26 - 160 Ω.m and a few sand materials found. Located at a depth of 5 meters with a distance of 25 meters from the starting point of the research, at a depth of 10 meters at a distance of 25 meters from the first electrode, and there is also a meter to 55 meters in depth. There is also a layer of sand with a resistivity of 160 to 918 Ω.m. In this line, samples are taken at 90 meters from starting point and 1 meter depth such as clay rocks.

4) Fort Line

The length of the fourth line is 120 m from the north to the south. Located at 7°41'44.65" S 110°10'28.93" E to 7°41'48.17" S 110°10'29.93" E. Space between electrode is 10 meters. The amount of data obtained is the same as the third line, which is 52 datum points. The iterations used on this line are 4 with the value of rms is 8.6.

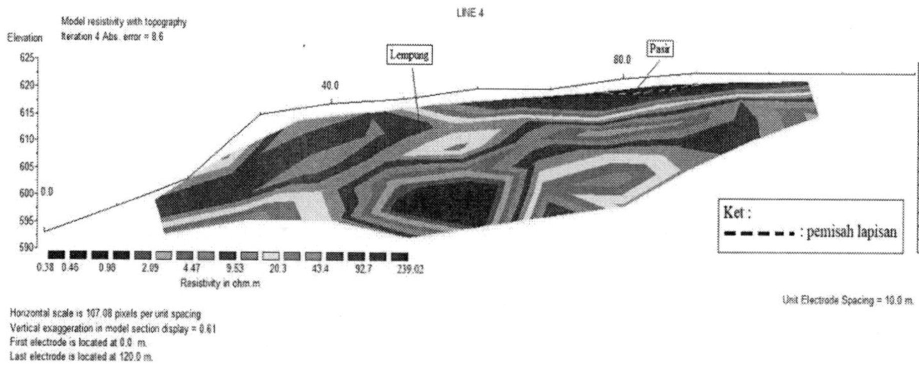

Figure 4. Cross Section Resistivity of the Forth Line

On this fourth line there are only 2 types of layers. The resistivite of clay layer is 0.38 – 160 Ω.m. There found sand material on the surface with resistivity value of 160 - 239.02 Ω.m. On this line, samples taken such as darker clay at a distance of 80 meters from the starting point and a depth is 20 cm.

5) Fifth Line

Line along 100m stretch from northeast to southwest with the position of 7°41'46.55" S 110°10'28.24" E to 7°41'46.55" S 110°10'31.40" E. The spacing between electrodes is 10 meters. The data obtained are fewer than the previous lines, which are as many as 36 datum points. The depth target obtained on this line is the same as the previous lines, which are 21.8 meters. In this fifth line, it is iterated 5 times and produces a large value of rms which is 23.4. More iterations can be done, so that smaller rms are obtained, but this does not produce an interpretation that is close to the field conditions.

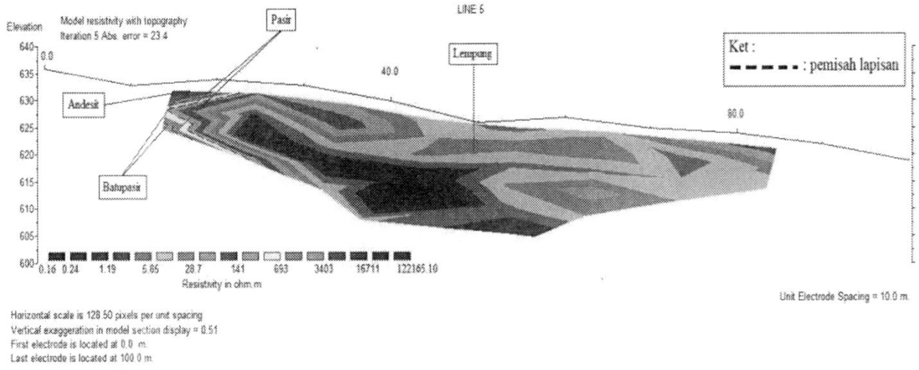

Figure 5. Cross Section Resistivity of the Fifth Line

On the fifth track (Figure 5) there are 4 layers. The resistivity of clay is 0.16 – 141 Ω.m and there are sandstone layers at a depth of 5 - 10 meters with resistivity 160 - 404 Ω.m. At a depth of 2 meters sandstones can be found with a resistivity of 3532 Ω.m. There is a bed rock at the end of the first meter. The bed rock form of andesite material with the depth 1 meter that has high resistivity as 122165.10 Ω.m. On this line a sample such as clay is taken at the 25th meter, one meter of the depth.

The correlation showed that till the depth of 21.8 meters, there is not found the slip plane. Based on the combined results of the line can be identified that the material is clay with a bond between weak particles and has a resistivity <160 Ω.m, then the area is called weak zone. This zone can be used as a reference for vulnerability to landslides in an area. However, if you look at the field conditions, it can be predicted that there is a skid field because some sandstone is found in the first line. The result of the correlation with the field for the first line, it is possible that there is a slip plane because some sandstone is found along the end of the first meter on the first track. However, it cannot be proven by the parallel correlation of the line in this research because the second line is difficult to reach.

b. Microseismic Method

The microceismic method was carried out by measuring microtremor signals at 30 research points. The result of the measurement is the value of V_s 3D modeling from the ellipticity curve method (Figure 6) . The location has lithology composed of soil layers having a value of V_s between 51 m/s to 309.86 m/s which is represented by the purple spectrum in almost all surfaces of the research location with a thickness of about 10 meters . Clay materials has a value of V_s between 406.33 m/s and 742.52 m/s which is represented by the dark blue spectrum. In the middle part which is the location of the landslide in 2017 has a depth of clay material up to 33 meters. In addition to these two materials (soil and clay), breccia and pumice materials were found to a depth of 70 meters with a value of V_s with a range of values between 814.17 m/s to 1478.64 m/s represented by a green spectrum, with a fairly thick material distribution dominated in the western region of the research point. Bed rock material as dacite and andesite has a value V_s between 1628.34 m/s to 2283.84 m/s are represented by yellow to red color spectrum and only found on the southwest side of the hill.

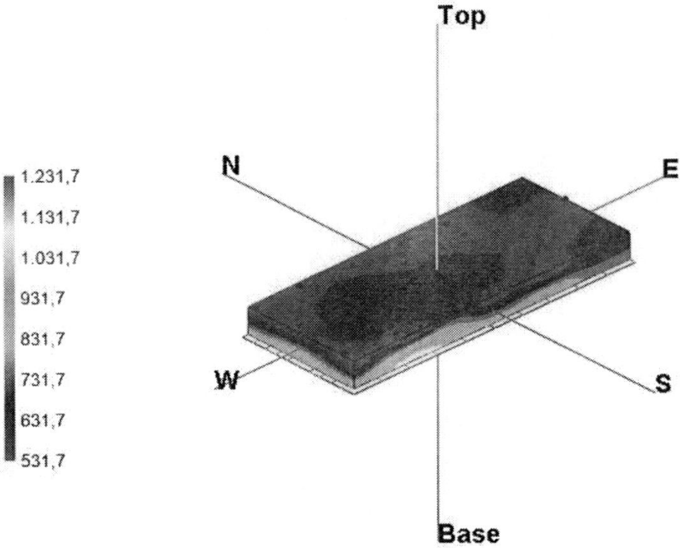

Figure 6. 3D Model of Subsurface Lithology

Viewed from the results of the incision (Figure 7) on the location of the landslide there is still a layer of clay to a depth of about 33 meters. This can be one of the reasons that the location is still prone to landslides, in addition with other landslide factors such as rainfall, earthquakes and others that caused the vulnerability of the area to landslides to increase.

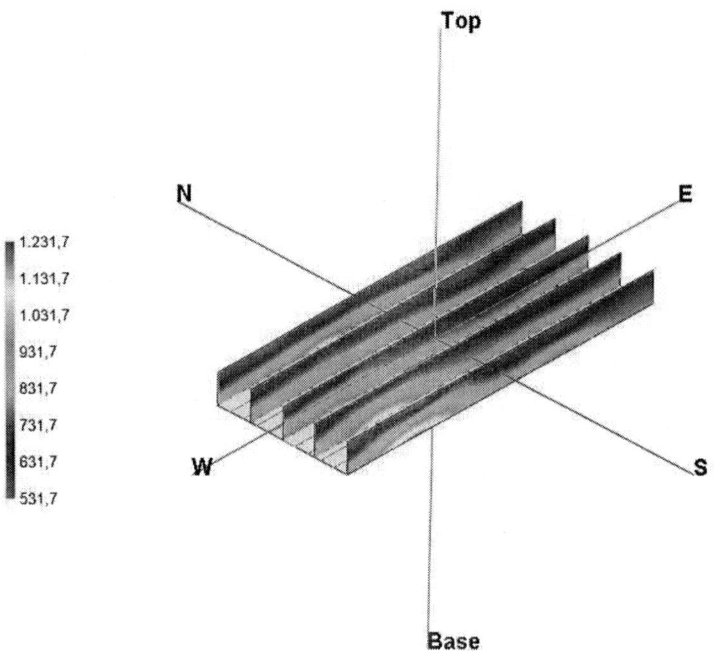

Figure 7. Distribution Incision of V_s

Based on the results of the geoelectric method, it was found that the area had lithology which was dominated by clay material, besides that in some points there were materials such as sand, sandstone, and andesite rocks exposed at the bottom of the hill. The lithology resulted were supported by the resulted of the microtremor method. In lithological mikrotremor research resulted of the research area is dominated by a layer of clay soil and rock, in addition to the material contained brecci, pumice, desit and andesite on the hill the other with varying depths.

4. Conclusion

The conclusions of this research were Sendang Mulyo, Purwoharjo Village, Samigaluh Subdistrict, Kulon Progo Regency, which had subsurface lithology dominated by clay material with resistivity (0.16 - 160) Ω.m and te value of V_s 51 m/s up to 309.86 m/s . In addition there is sand on the surface material with resistivity (160 - 1093) Ω.m and value V_s 406.33 m/s to 742.52 m/s. At several points of the research there were also sandstones with resistivity (1093 - 8000) Ω.m and value of V_s 814.17 m/s to 1478.64 m/s. The most basic layers predicted is the slip plane such as andesite material with resistivity (8000 - 122165) Ω.m and value of V_s 1628.34 m/s to 2283.84 m/s. The research area has the potential for landslides. The results of the research stated that the area was dominated by clay which in the phenomenon of landslide would have a major impact on the population. This was worsened through historical facts of landslides in the previous year. In addition, the area has a dominant underground structure with clay, with a sediment thickness estimated at 20 m and supported by geographical factors that showed the slope is > 45° as well as fracture of land on the hill during the dry season.

5. References

[1] BIG. (2017). *Pemetaan Cepat Kejadian Longsor Kulonprogo*. Diunduh pada tanggal 27 2018 dari http://www.big.go.id/pemetaan-cepat-kejadian-longsor-kulonprogo-2/.
[2] BPS Kabupaten Kulon Progo. (2009). *Kecamatan Samigaluh dalam Angka: 2009*. Kulon Progo: BPS Kabupaten Kulon Progo.
[3] Kanai, Kiyoshi. 1983. *Engineering seismology*. University of Tokyo Press.
[4] Marjiono, 2010, Estimasi karakteristik dinamika tanah dari data mikrotremor (Studi kasus Kota Bandung), Tesis, Program Studi Geofisika Terapan, ITB, Bandung.
[5] Muntohar, A.S. (2015). *Tanah Longsor Analisis-Prediksi-Mitigasi*. Yogyakarta: Teknik Sipil UMY.
[6] Nakamura, Yutaka. 1989. A Method for Dynamic Characteristics Estimation of Subsurface using Microtremor on Ground Surface. *Quarterly Report of Railwat Technical Research Institute (RTRI) Vol. 30 No. 1.*
[7] Nakamura, Yutaka. 2000. Clear Identification of Fundamental Idea of Nakamura's Technique and Its Applications. *System and Data Research, Japan.*
[8] Subardjo., 2008, *Parameter Gempabumi*, Pusat Pendidikan dan Latihan BMKG, Jakarta.T.M.
[9] Sutasoma, M., Susilo, A., Suryo, E.A., Fitriah, F., & Bahtiar. (2018). Preliminary Study Of Landslide In Sri Mulyo, Malang, Indonesia Using Resistivity Method And Drilling Core Data. *International Journal of GEOMATE, Aug., 2018 Vol.15, Issue 48, pp.161-168.*
[10] Wahid, U & Prayogo, S. (2003). Survei Resistivitas 3 Dimensi untuk Menentukan Distribusi Tahanan Jenis Tanah Bawah Permukaan Daerah Rawan Longsor di Desa Lumbang Rejo, Prigen, Jawa Timur. Jakarta: The Proc Joint Convention HAGI-IAGI.

Conceptual Framework of Reflective-Inquiry Learning Model to Promote Critical Thinking Ability of Preservice Physics Teachers

N N S P Verawati[1*], Hikmawati[1] and S Prayogi[2]

[1] Universitas Mataram, Jln Majapahit No 62 Mataram 83115, Indonesia.
[2] Institut Keguruan dan Ilmu Pendidikan (IKIP) Mataram, Jln Pemuda No 59A Mataram 83126, Indonesia.

*Corresponding author e-mail: veyra@unram.ac.id

Abstract. Critical thinking has been a crucial competence in 21st century learning and encouraging critical thinking ability at the university level is assumed as an important achievement at higher education. This study was aimed at developing Reflective-Inquiry Learning (RIL) model to promote critical thinking ability for preservice teachers in particular. This study was a pre-developmental stage in which the developed RIL model was constructed in line with the supporting theories and empirical findings. With the result that, it produced a hypothetical framework from RIL model itself. Then, the product of the RIL model was validated with involving 7 experts as validators through a focus group discussion (FGD) process. The assessed aspect of the product consisted of the content validity and construct validity. The results of the product validity were analyzed through a descriptive analysis viewed from the average score of the validity. The result of the validity showed the validity level (Va) of the RIL model was 4.28 and it was said very valid. The conceptual framework of the developed model and the validity result would be further elaborated in this article.

1. Introduction

Critical thinking includes a higher order thinking domain and becomes one of the essential skills that should be achieved by learners in the 21st learning century [1]. A critical thinking learning ought to be internalized by preservice teachers since they come in a university in order to be able to treat their students when they become a real teacher in the future [2]. In addition, encouraging learners' critical thinking development is an important achievement at higher education [3]. The same tune is echoed by Innabi & Elsheikh [4] in which educational institutions should provide and facilitate preservice teachers to develop their critical thinking ability. For instance, in Indonesia, the critical thinking ability become a demand of the learning need for learners. It is stated in Regulation of Ministry of Education and Culture, number 73 in the year of 2013 in accordance with Indonesia National Qualification Framework and Regulation of Ministry of Research, Technology, and Higher Education, number 44 in the year of 2015 associating with Higher Educational National Standard. However, the study by Prayogi et al [1] revealed that the critical thinking ability of teachers who teach science are still relatively low, and this is one of the challenges of education in Indonesia.

Universities and higher educations have made serious efforts to put critical thinking dimensions into the instructional curriculum. Unfortunately, based on a number of researchers, learners still are not able to think critically because most of the lecturers do not integrate between learning processes of the critical thinking acquisition and learning practices requiring a various reflection [5, 6, 7, 8]. Besides, various instructions such as inquiry learning activities have been implemented in higher educations to promote learners' critical thinking abilities focused on science learning activities [9]. However, the teaching practice in the class often avoids the core of a critical thinking process, the reflective process [10].

Based on the research findings, the data show some indicators of the critical thinking ability (e.g. analysis, inference, and evaluation) are low by applying inquiry learning model [11, 12]. The reflective thinking concept as a precursor to train learners to think critically is not thoroughly investigated. Most of the studies do not give a solution for effective learning of how lecturers combine both reflective and inquiry learning. In fact, the great success of training learners to think critically is not found [13]. The way the lecturers apply the inquiry learning model tends to emphasize the aspect of "learn to find" and "testing concept or fact simply", whereas the core of learning processes is how learners are able to organize their thinking theoretically and practically in a further complicated context. Therefore, the reflective attributes of every teaching step of the inquiry learning model are important to be done in order that the improvement of critical thinking ability would be optimal.

Integrating the reflective concept and inquiry model as a set of a learning model is called hereafter Reflective-Inquiry Learning (RIL) Model. It is serial learning processes that utilize the inquiry learning model attributed to reflective activities such as providing anomaly phenomena, monitoring, evaluating performance, and continued reflection. These activities are aimed at improving learners' critical thinking ability. In this study, the RIL model is developed to enhance preservice teachers' critical thinking abilities.

1.1. Theoretical Framing

The conceptual framework of the RIL model is constructed from the supporting theories and empirical findings relating to the reflective, inquiry, and critical thinking concepts. The researchers arranged the three concepts started from; the concept of the critical thinking, inquiry learning as the foundation of the critical thinking, and the reflective learning framework.

1.1.1. Critical Thinking Concept. Thinking is a cognitive process or mental activity to attain knowledge. According to Solso [14] in his book of Cognitive Psychology, thinking is a common process considering problems of the mind and generating a form of new mental representation. Critical thinking cannot be amended naturally. Therefore, it should be underpinned by an environmental stimulus and the various atmospheres. Some experts have been exploring the inclination of the individual thinking way and relating to achievement gained. Gallagher in Webb [15] mentioned learners in solving problems have two manners of thinking, convergent and divergent. The convergent thinking is identical with the critical thinking patterns. Meanwhile, divergent thinking is associated with the creative thinking. John Dewey, a philosopher and psychologist, is well-known as a person introducing the critical thinking concepts. Dewey [16] announces critical thinking as reflective thinking. According to Facione [17], critical thinking is basically a detail description from types of characters consisting of the interpretation, analysis, evaluation, inference, explanation, and self-regulation. One of the outstanding contributors to the critical thinking tradition is Robert Ennis. Ennis [18] and Hassard [19] have a similar point of view. Both define the critical thinking concept as a type of logical and reflective thinking focusing on deciding what to believe and what to do.

Almost all people working on the critical thinking field have generated lists of critical thinking skills as a foundation to think critically. Lipman in Jeevanantham [20] argues the critical thinking as a type of critical skills with having a responsibility that facilitates to take a right decision because (a) it depends on criteria, (b) it regulates self-correction, and (c) it has a sense of contexts. Furthermore, Rudinow & Barry [21] state the critical thinking is similar to a set of dainty tools with connecting an intellectual and strategic ability to make reasonable decisions about what to believe and what to do. According to Lai [22] after doing a number of literature reviews and analyzing some experts' judgments of the critical thinking, he argues the critical thinking can be seen from three main approaches based on its definition. Those are a philosophical approach, cognitive psychological approach, and educational approach. In the philosophical approach, critical thinking more emphasizes the quality and character of being a critical thinker. In cognitive psychology approach, the critical thinking more emphasizes the real action and behavior that can be performed by a critical thinker so that in the critical thinking definition it contains lists of critical thinking skills [23]. Last but not least, in educational approach, critical thinking emphasizes a process to make learners think at a higher level. It is called in the term 'Higher Order Thinking Skills' (HOTs).

1.1.2. Inquiry as a Teaching Foundation of the Critical Thinking. Learning is an impact resulted from a thinking process. The retention, understanding, and using the active knowledge can be created through learning experiences in which learners think. A number of experts conclude that human being does not have a natural inclination to think critically. People who have high motivation often are similar to those who have low motivation where both people do not think critically [24]. Critical thinking is a higher order thinking domain that should be taught [25]. According to Thompson [26], in the teaching and learning critical thinking, holistic approaches are necessary to be applied and it should involve a set of an appropriate learning model and be oriented on a learning goal that enables learners to manipulate cognitive skills. The learning goals indicating the critical thinking in the curriculum of the elementary level, junior and senior high school level, and higher education level seem inconsistent with how the learning goals are interpreted in practice.

Developing the critical thinking ability in recent decades has been directed through inquiry activities. According to Bailin [27], the learning objectives of inquiry activities are dominantly focused on critical thinking ability such as identifying assumption, using logical thinking, analyzing direct experiences and phenomena, analyzing secondary sources, analyzing arguments by reviewing current scientific knowledge, considering pieces of evidence, and examining logical aspects. National Science Education Standards state that the critical thinking is the most important dimension of the science education in which its main learning activities are through inquiry learning. A scientific inquiry is an activity that refers to diverse ways in which scientist study natural aspects and make explanations of those based on pieces of evidence obtained from the results of an investigation. Besides, inquiry refers to learning activities where they develop knowledge and understanding of the scientific ideas, as well as an understanding of how they learn nature. The inquiry as an activity involves observation, asking questions, checking information sources to confirm what they already know, planning investigations, conducting experiments, using tools to collect, analyze, and interpret the data, proposing answers, explanations, and predictions, and communicating the results [28].

Experts and researchers have developed and modified the inquiry instruction with different terms including the traditional inquiry, guided inquiry, structured inquiry, open inquiry, directed inquiry, inquiry learning, inquiry teaching, authentic inquiry, scientific inquiry, partial inquiry, and full inquiry [29]. National Research Council (NCR) depicts that the inquiry has three level namely: (1) structured inquiry, in which a teacher in the inquiry activities prepares or provides problems and processes for learners to solve those; (2) guided inquiry, in which a teacher raises problems and learners should determine the process and solution of those; and (3) open inquiry, in which a teacher only provides a context of solving problems and learners go to identify and solve the provided problems [30].

Inquiry activities in the learning process have long been introduced by experts ever since the beginning of the 20th century, namely John Dewey who introduced the teaching steps that imply the inquiry activities of it [31]. However, many researchers argue that the scientific inquiry was based on the Atkin-Karplus learning cycles that were popularized in 1962 [32]. In the learning phase, it was introduced the investigation steps that become the forerunners of the inquiry process. Arend [33] explains the inquiry is as a teaching model that aims to teach learners how to think. It means that the inquiry is a teaching foundation to train the higher order thinking skills for learners. The learning task in the inquiry learning plan is oriented to the purposes of content and process. The purpose of a goal means that a teacher plans for learners to attain new knowledge related to the focus of an investigation.

Meanwhile, the purpose of a process means that a teacher also wants the students to study the investigation process especially a process related to scientific investigation and to develop a positive attitude towards the investigation and process applied to investigate. The terms of an inquiry instruction as inquiry-based lesson that has 6 teaching steps that consist of gaining attention and explaining the inquiry process, presenting the inquiry problem or discrepant event, having students formulate hypotheses to explain the problem or event, encouraging students to collect data to test the hypothesis, formulating explanations and/or conclusions, and reflecting the problem situation and the thinking processes used to inquire into it [33].

1.1.3. Reflective Concept in Learning. Reflection as a process of thinking used by Socrates more than 2,000 years ago, but the current approach used to apply the reflection as a regulation in the learning derived from the work of John Dewey [34]. Reflection is defined as the cognitive processes that are conducted to learn from experiences [16]. Reflection is based on the concept of reflective thinking. The most common understanding of the cognitive process of reflection is to analyze and find a way that will lead to the production of new knowledge and experience based on the context of prior knowledge, and the development of alternative ways [35]. John Dewey proposes the reflective approach in learning because it is the most important needs of a learner. Dewey defines reflective thinking as an active, persistent, and careful consideration in the structure of knowledge that supports the belief, knowledge, and results to be achieved. A reflection is a form of checking the process that has been done. Reflection affects the way the teacher plans lessons, the types of a decision made, and general learning practices [36].

Reflection leads students to deeper learning [37] and the achievement of knowledge which is more complex, integrated, and useful [38]. Some studies show that reflection is important for successful learning processes [39, 40]. For instance, Davis [39] presents that the reflection helps learners create new relationships between the initial and acquired knowledge and it makes the learning process more effective. Reflection is relevant to the learning process, but it is also a challenging activity because what learners think and feel about their experiences may be different from the actual event [41]. In addition, some studies show that instead of evaluating their own experiences, learners tend to wait for the teacher to present the results of an evaluation [42]. That is why there is a need to guide learners to reflect their learning. The reflection can be guided in many ways, for example providing the guided questions to show specific elements of an activity [43, 44], using portfolios to record important events during activities of at the end of activities [45, 46], recording actions to be further evaluated [47, 48], and requesting a feedback from other friends who can provide alternative views of the activities carried out [49].

The process of the reflection is based on the type of the reasoning where the analysis phase is associated with the awareness, acceptance, action processes. This awareness process is important because pedagogical practices are based on trust at the awareness level of the context being learnt. The acceptance in the learning context doesn't convince the learners about the concept of the truth or untruth, but it creates conditions for how learners convince themselves of the truth of the concepts being studied [36].

Based on the literature study there are several important points that connect the concepts of critical thinking, inquiry, and reflection in learning which are the basic frameworks for developing RIL models which then become the framework of teacher mindset, there are the reflective thinking as a precursor to training critical thinking, reflection influences the types of decisions that are made, helps create relationships between acquired knowledge, and makes the learning process more effective. Furthermore, reflective thinking acts as a driver of critical thinking during the process of problem solving in inquiry. Moreover, learner who are active in reflection can encourage the development of a better understanding of the inquiry process and support critical thinking.

2. Methods

This study was a part of the development research. Based on the theoretical review and empirical findings, the conceptual framework of the RIL model was developed and subsequently validated. The research procedure was adapted from the previous study [1, 9, 50]. The results of the validation process were used to measure the quality of the developed model. The RIL model was categorized as a product of instructional models. According to Nieeven [51], a product can be said to be qualified if it meets a valid, practical, and effective category. This study was aimed at formulating the validity of the conceptual framework of the RIL model to promote the critical thinking ability of pre-service physics teacher.

The validation method was employed to know the validity of the RIL model. The validation was done with considering two aspects of validities which are content validity and construct validity. Content validity refers to all components of the model that should be based on the state-of-the-art of acknowledgment. Meanwhile, the construct validity is all components that should be consistently and logically linked to each other [52]. The validation process of the RIL model was carried out through the processes of Focus Group Discussion (FGD) involving 7 validators that consist of some experts and the practitioners as users of the model. The assessment of the validity used validation sheets with a Likert scale. Then, the assessment determines the validity level of the RIL model. The validity criteria of the model can be seen in Table 1.

Table 1. The validity criteria of the RIL model

Interval (Va = Validity Level)	Criteria
Va > 4,21	Very valid
3,40 < Va ≤ 4,21	Valid
2,60 < Va ≤ 3,40	Quite valid
1,79 < Va ≤ 2,60	Less valid
Va ≤ 1,79	Invalid

3. Results and Discussion

Reflective-inquiry learning (RIL) model is the developed learning model in this study by integrating the reflective attributions into the specific inquiry learning model phases to train learners' critical thinking ability. The framework of the development and RIL model hypothetic provided in Figure 1 and Table 2 as follows.

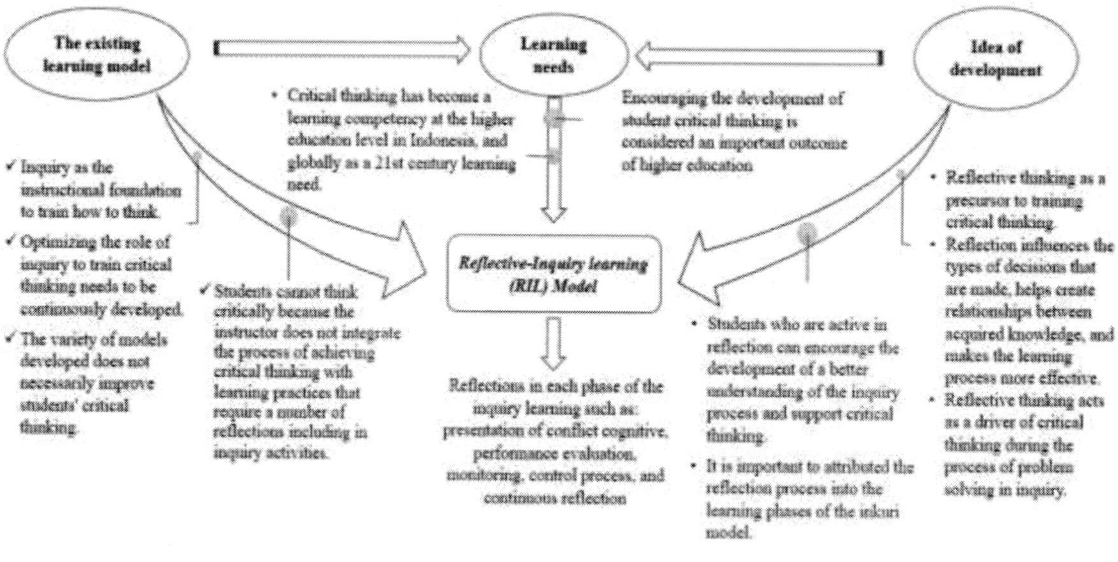

Journal of Physics: Conference Series **1397** (2019) 012009 doi:10.1088/1742-6596/1397/1/012009

Figure 1. The Framework of RIL Model Development

Table 2. Hypothetic Framework of Reflective-Inquiry Learning (RIL) Model

Learning Phase	Learning Process with Integrating Reflective Process
Phase 1. Orientation	• Preparing pre-service teacher to learn and describe the process and learning objectives
Phase 2. Providing Problems	• Presenting cognitive conflict with authentic phenomena and requesting pre-service teachers' responses • Monitoring pre-service teachers' responses toward the provided phenomena (*monitoring process*). • Conducting correction if the pre-service teachers' responses are inappropriate with the context being studied (*control process*)
Phase 3. Formulating hypothesis	• Encouraging pre-service teachers to hypothesize problem situation stated initially • Examining the correlation between hypothesis and problem condition stated to be confirmed with each proposed hypothesis (*performance evaluation*)
Phase 4. Examining hypothesis	• Asking pre-service teacher to examine hypothesis through an experiment • Asking pre-service teacher to explain how they collect data to examine the hypothesis through experimental activities (*control process, performance evaluation*)
Phase 5. Formulating explanation	• Asking pre-service teacher to formulate explanations and making a generalization (*control process*) • Confirming each explanation arranged (*performance evaluation*)
Phase 6. Reflection	• Involving pre-service teacher to check the processes that they have done and identifying mistakes to be continuously corrected (*continuous reflection*).

The framework of RIL model hypothetic arranged was validated by experts and users. The validation was done through FGD activities involving 7 validators. The validity assessment was focused on two aspects, the content validity and construct validity. The result of validation can be seen in Table 3 as follows.

Table 3. Expert validation results on the ICP learning model

Nu	Aspects of validation		Average score	Category
1	Content validity	The need for developing the model	4,42	Very valid
		The model designed in line with the state-of-the-art of knowledge.	4,10	Valid
2	Construct validity	Consistency and logic of all arrangement components of the model.	4,32	Very valid
		Validity Level (Va)	*4.28*	*Very valid*

The validity level of the developed RIL model, in general, is very valid with 4.28 of the validity level (Va). It is said very valid because Va > 4,21. The content validity seen from the aspects of the need for model development was very valid with 4,42 of the validity level. Some descriptions that show the validity of aspects of the need of model development include that the development of the RIL model aims to enhance the critical thinking ability as a need of the 21st century learning skills, global learning needs, and learning needs in accordance with the Indonesian National Qualification Framework and Higher Education National Standard. The development of the RIL model is in line with science learning needs that emphasize the process of knowledge acquisition through explorations and experiments in general. In addition, developing the RIL model also bridges the gap between the expected competencies of the 21st century learning outcomes that have critical thinking ability. In fact, the pre-service teachers' ability to think critically is low and teachers are not fully successful to train them.

The content validity seen from the designed model based on state of the art of knowledge was valid with 4.10 of the validity level. Descriptions of the RIL model showing the great aspects include that the development of the RIL model is a recommendation to improve the inquiry learning model which aims to train the critical thinking ability with integrating reflection aspects in the teaching process. Intervening the reflection process in the inquiry activities is a new paradigm of the knowledge transferability process. Through the reflection process, the structuring cognitive process happens and it helps learners to develop their better understanding of instructional materials and the inquiry process itself. In addition, the development of the RIL model includes a new model that can be used to treat learners' critical thinking ability. Seen from the phase of the process analysis and decision making, the inquiry learning activities can be conducted through reflective thinking processes.

Assessing the construct validity of the RIL model is viewed from the consistency and logical aspects of the model. The validation result shows that those aspects are very valid with 4,32 of the validity level. The criteria showing the consistency and logicality of the RIL model include that the model arrangement is based on the theoretical review and empirical support related to the model arrangement purposes, the teaching phases show the logic learning activity orders and are consistent with the scientific inquiry learning activities. In addition, the teaching phases integrate the reflective processes in the learning activities.

In the FGD activities, the validators agreed that the hypothetical framework of the RIL model is consistent in arranging the teaching steps to treat learners to think critically. The RIL model construction shows an integrative process between inquiry activities and reflective activities to promote learners' critical thinking ability. Critical thinking is a reflective process to analyze and evaluate information to make decisions on what to believe and what to do [10]. From here, critical thinking ability can be learnt through learning models that have activities that require learners to learn autonomously [53]. For example, the inquiry learning focuses on two aspects which are the content and process. It aims to train how to learn to think [33]. Reflection can be linked to various learning methods including inquiry learning [40]. The inquiry learning is a process of finding new activities in which learners formulate the hypothesis and examine it through experimental way or observation [54].

Teaching critical thinking through inquiry learning is not only limited to finding facts but how to interpret the facts [50]. Reflection in learning requires learners to be curious, open-minded, and responsible for the knowledge they have or they are exploring. Furthermore, the reflective learning requires learners to carry out a systematic cycle of self-evaluation through an open discussion or written analysis during the learning process guided by the teacher [55]. In the context of the inquiry teaching and learning, the reflective inquiry is the thinking individual process to measure learners' own experiences to gain an understanding of the assumptions and implications from an event in everyday life. The general phenomenon that people understand about inquiry learning is limited to learning models that facilitate students to learn to find and examine concepts or facts in a simple way. Due to the absence of inquiry learning models that integrate the reflection attribution of inquiry learning steps, the statement of the inquiry learning is able to increase learners' critical thinking ability is just a theory, but is practically weak.

The inclination of the cognitive, psychomotor, and psychological development works in conditions of insufficient information sources to process information related to the environment and one's self [56]. The relevant information on the specific phenomena also tends to have irrelevant information.

This is in accordance with the logic that the relevant information in certain circumstances can be irrelevant in different conditions. Therefore, the attribution-reflection is very important to make learners perform the learning activities in line with the learning objectives to be achieved. It is critical thinking as the demand for 21st century skills. An efficient process requires activities that allow learners to keep reflecting the learning and thinking process so that they remain in the desired learning activities.

The strong relationship between the reflection and the inquiry process is seen when learners actively do reflection process that can encourage developing their better understanding about the instructional materials content and the inquiry process [57]. Reflection can be done in the scientific inquiry activities by providing the conflict phenomena (cognitive conflict) [58]. The cognitive conflict helps the assimilation process become more effective and meaningful to form the learners' intelligence. Providing cognitive conflict helps learners reflect the concepts and explanations of the phenomena being learnt. This process is provided to develop their learning activities and encourage their critical thinking to understand the whole concepts [59]. Reflection can also be done by a monitoring process and process control [13]. Monitoring is important because it is in line with the decision making [60, 61, 62, 63]. Reflection is carried out consistently and continuously in every learning process. This activity leads learners to think about what they have done as a manifestation of the critical reflection [64].

4. Conclusion

RIL model is a learning model based on scientific inquiry activities that intervene with reflective processes. The development of the model in specific is aimed at training the pre-service physics teacher to think critically. The validation results showed that RIL model was valid theoretically. Besides, the content and construct validity were valid. It means that the RIL model enables to promote pre-service physics teachers' critical thinking ability. From the theoretical basis that has been described, experts also suggest that the aspects of critical thinking that need to be developed are analysis, inference, evaluation, and decision making. However, the continuous study in the model implementation is important to be conducted to evaluate the practicality and effectiveness of the RIL model.

References

[1] Prayogi S, Yuanita L and Wasis 2017 *J. Phys: Conf. Ser.* **947**, 1-6.
[2] Warburton E C 2008 *J. Edu. Hum. Dev.* **2(1)**, 1-16.
[3] Tiruneh DT, DeCock M, Weldeslassie A G, Elen J and Janssen R 2017 *Int. J. Sci. Math. Edu.* **15**, 663-682.
[4] Innabi H and Elsheikh O 2007 *Edu. Stud. in Math.* **64(1)**, 45-68.
[5] Choy S C and Cheah 2009 *Int. J. Teac. Learn. in High. Edu.* **20(2)**, 196-204.
[6] Rudd R D 2007 *Techniques*, **82(7)**, 46-49.
[7] Black S 2005 *The Edu. Dig.* **70(6)**, 42-47.
[8] Vaske 2001 *Critical thinking in adult education: An elusive quest for a definition of the field.* Unpublished doctoral thesis, Drake University, Des Moines, Iowa.
[9] Prayogi S, Yuanita L and Wasis 2018 *J. of Turk. Sci. Edu.* **15(1)**, 43-56.
[10] Ennis R H 2011 *The nature of critical thinking: An outline of critical thinking dispositions and abilities.* Presentation at the Sixth International Conference on Thinking at MIT, Cambridge, MA.
[11] Qing Z, Jing G and Yan W 2010 *Proc. Soc. Behav. Sci.* **2**, 4597-4603.
[12] Miri B, Ben-Chaim D and Zoller U 2007 *Res. in Sci. Edu.* **37(4)**, 353-369.
[13] Choy S C and Oo P S 2012 *Int. J. of Instruct.* **5(1)**, 167-182.
[14] Solso R, Maclin O and Maclin K 2008 *Cognitive psychology* (8th ed). Boston, MA: Pearson Inc.
[15] Webb M E, Little D R, Cropper S J and Roze K 2017 *Thin. & Reas.* **2017**, 1-24.
[16] Dewey J 1933 *How we think.* Buffalo, NY: Prometheus Books.

[17] Facione P 2011 *Critical thinking. What it is and why its counts*. Millbrae, CA: The California Academic Press.
[18] Ennis R H 1996 *Critical thinking*. New York: Prentice-Hall.
[19] Hassard J 2005 *The art teaching science*. New York: Oxford University Press.
[20] Jeevanantham L S 2005 *Afr. Edu. Rev.* **2(1)**, 118-129.
[21] Rudinow J and Barry V E 2008 *Invitation to critical thinking*. New York: Thomson Higher Education.
[22] Lai E 2011 *Critical thinking: A literatur review*. Pearson Research Reports. Retrieved from http://images.pearsonassessments.com/CriticalThinking ReviewFINAL.pdf
[23] Lewis A and Smith D 1993 *Theo. In. Prac.* **32(3)**, 131-137.
[24] Macpherson R and Stanovich K E 2007 *Lear. & Indi. Diff.* **17**, 115-127.
[25] Woolfolk A 2009 *Educational Psychology*. New York: Pearson.
[26] Thompson C 2011 *Int. J. of Hum. & Soc. Sci.* **1(9)**, 1-7.
[27] Bailin S 2002 *Sci. & Edu.* **11(4)**, 361-375.
[28] Hussain A, Azeem M, and Shakoor A 2011 *Int. J. of Hum. & Soc. Sci.* **1(19)**, 269-276.
[29] Buck L B, Bretz L and Towns M H 2008 *J. of Coll. Sci. Teac.* **38(1)**, 52-58.
[30] Zion M and Sadeh I 2007 *J. of Bio. Edu.* **41(4)**, 162-169.
[31] Rodger W B, Joseph A T, April G, Pamela V S, Janet C P, Anne W and Nancy L 2006 *The BSCS 5E instructional model: Origins and effectiveness*. Report by Science Education National Institutes of Health.
[32] Bailin S 2002 *Sci. & Edu.* **11(4)**, 361-375.
[33] Arends R 2012 *Learning to teach*. (9th Edition). New York: McGraw-Hill.
[34] Leijen Ä, Valtna K, Leijen D A J and Pedaste M 2012 *Stud. in High. Edu.* **37(2)**, 203 – 217.
[35] Sünbül A M and Kurnaz A 2016 *Reflective thinking and teaching reflective thinking* (Book Chapter). Ankara: Çözüm Egitim Yayincilik.
[36] Tugui C 2011 *Proc. – Soc. & Beha. Sci.* **29**, 533-538.
[37] Moon J A 2004 *A handbook of reflective and experiential learning: Theory and practice*. London: Routledge Falmer.
[38] Billing D 2007 *Hig. Edu.* **53**, 483-516.
[39] Davis E A 2003 *J. of Lear. Sci.* **12**, 91-142.
[40] Kori K, Mäeots M and Pedaste M 2014 *Proc. – Soc. & Beha. Sci.* **112**, 242-251.
[41] Argyris C and Schön D 1974 *Theory in practice: Increasing professional effectiveness*. San Francisco: Jossey-Bass.
[42] Leijen Ä, Lam I, Wildschut L, Simons P R J and Admiraal W 2009 *Comp. & Edu.* **52**, 169-176.
[43] Hsieh S W, Jang Y R, Hwang G J, and Chen N S 2011 *Comp. & Edu.* **57**, 1194-1201.
[44] Winchester T M and Winchester M 2011 *Int. J. Acad. Dev.* **16(2)**, 119-131.
[45] Roberts A 2009 *Refl. Prac.* **10(5)**, 633-644.
[46] Paulus T and Spence M 2010 *Tech. Tren.* **54(5)**, 62-68.
[47] Bannik A and VanDam J 2007 *Teac. & Teac.: theo. & prac.* **13(6)**, 565-586.
[48] Calandra B, Brantley-Dias L, Lee J K and Fox D L 2009 *J. of Rese. on Tech. in Edu.* **42(1)**, 73-94.
[49] Chen N S, Wei C W, Wua K T and Uden L 2009 *Comp. & Edu.* **52**, 283-291.
[50] Prayogi S, Muhali, Verawati N N S P and Asy'ari M 2016 *J. Peng. MIPA.* **21(2)**, 148-153.
[51] Nieveen N 1999 *Prototyping to reach product quality*. Netherlands: Kluwer Academic Publisher.
[52] Nieveen N 2007 *Formative evaluation in educational design research*. Proceedings of the seminar conducted at the East China Normal University: Shanghai (PR China).
[53] Mitrevski B and Zajkov O 2011 *Bulg. J. of Phys.* **38**, 318-324.
[54] Mäeots M, Pedaste M, and Sarapuu T 2011 *Interactions between inquiry processes in a web-based learning environment*. In Proceedings of the 2011 11th IEEE International Conference

on Advanced Learning Technologies: 11th IEEE International Conference on Advanced Learning Technologies (pp. 331-335). Athens, Georgia, USA.
[55] Choy S C, Yim J S C and Tan P L 2017 *Iss. in Edu. Rese.* **27(2)**, 234-251.
[56] Demetriou E and Holtzer R 2017 *J. of the Int. Neur. Soc.* **23(1)**, 44-55.
[57] White B and Frederiksen J 2005 *Edu. Psyc.* **40(4)**, 211-223.
[58] Kahan D M 2013 *Judg. & Dec. Making* **8(4)**, 407-424.
[59] Akmam A, Anshari R, Amir H, Jalinus N and Amran A 2018 *J. Phys: Conf. Ser.* **335**, 1-7.
[60] Cokely E T Y and Kelly C M 2009 *Judg. & Dec. Making* **4**, 20-33.
[61] Koehler D J and James G 2010 Probability matching and strategy availability. Memory & Cognition, 38(6), 667-676.
[62] Toplak M E, West R F and Stanovich K E 2011 *Memo. & Cog.* **39(7)**, 1275-289.
[63] Toplak M E, West R F and Stanovich K E 2014 *Thin. & Reas.* **20(2)**,147-168.
[64] Boody R M 2008 *Education* **128(3)**, 498-506.

Acknowledgments
This study is a part of qualified applied research of higher education financially sponsored by Ministry of research, technology, and higher education, Indonesian Republic. We would like to thank those people who contribute to support this study including the research partners. We hope this study can be useful for educational development in Indonesia in particular and for anyone in general.

Development of The Android-Based Interactive Physics Mobile Learning Media (IPMLM) to Improve Higher Order Thinking Skills (HOTS) of Senior High School Students

Beatrix Elvi Dasilva* **and Suparno**

Graduate School in Physics Education, Yogyakarta State University, Yogyakarta, Indonesia

*Corresponding author e-mail: beatrix.elvi@gmail.com

Abstract. This study aims to find out: 1) generating useful android aplication IPMLM in termodynamic material on physics, 2) examining the impact of IPMLM toward students' higher order thinking skill (HOTS). In order to find out the answer, the researcher used research and development method in 4-D model (defining, designing, developing and disseminating). The IPMLM application was assessed by media experts and physics expert. The limited test stage was conducted in SMAN 1 Kota Kupang with 32 students of 11th grade science program. Subjects were chosen by purposive sampling technique. The extensive test stage was conducted in SMAN 2 Kota Kupang, SMAN 3 Kota Kupang and SMAN 5 Kota Kupang with 210 students of 11th grade science program. Subjects were chosen by purposive sampling technique. The research design of extensive test was pretest-posttest control group design. Data collection was done by questionnaire and test. Instrument test to measure students' HOTS instrument was reasoning multiple choice questions. The result shows that 1) the IPMLM application was eligible to be used in physics learning about thermodynamics, 2) using IPMLM in physics learning increased effectively the capability on HOTS aspect.

1. Introduction

Competencies based on the 21st century of learning paradigm are include the skills in learning and innovation, skills in using media and information, and life and career skills [1]. Kurikulum 2013 has tried to facilitate learning in accordance by demands of the 21st century of learning paradigm. One of them is the thinking ability must be achieved by students in learning especially learning physics. Peraturan Menteri Pendidikan Nasional No 21 Tahun 2016 about content standards for primary and secondary education has suggested in learning physics students must be able to analyze concepts, design or modify projects, create simple products, design experiments, carry out experiments, present experimental results, to conclude, and report an experimental results [2]. These competencies lead to high-level thinking abilities of students. In other words, the 2013 curriculum is requires students to achieve high order thinking skills. The results of research have shown that students with high level of thinking ability have a level of creativity in advanced level [3].

Physics is one branch of science so that it has in accordance with the nature of science that physics is learned through by scientific processes. Physics must be learned in such a way that the physics knowledge can be used to solve problems [4]. Physics learning must facilitate students to build their own knowledge and thinking skills [5]. However, physics learning process does not facilitate students to train their thinking skills [6]. The learning process has tend a teacher centered [7], [8]. Teachers are less creative and less able to develop learning media based on ICT (specific instructional goals) are varied and fun for students [9]. Most teachers is only inform about how to solve physics problems using existing equations [10], [11]. Informative learning models will be difficult to train thinking skills, especially HOTS students [4]. This is evidenced by the research data of the Trends of the International Mathematics and Science Study (TIMSS) which shows that reasoning (in physics) ability of Indonesian students is ranked 40 out of 42 countries has been researched [12]. Other research conducted by Faizaha, Suparmi and Aminah showed students' HOTS in 11th grade MIA in Sragen Regency was still low with 19.01% in the very low of HOTS category [13]. The data that has been described shows that HOTS students still need to be improved by using models, methods, approaches, strategies and learning media that are in accordance with the nature of physics.

Theoretical and empirical studies show that the 21st century learning paradigm has not been fulfilled. One of the ways to answer the demands of the 21st century learning paradigm is to integrate technology in the learning process. Several studies have proven is the use of technology in learning can improve students' abilities in cognitive, affective, and psychomotor aspects. By integrating with certain learning methods, for example by Group Investigation, this media can improve critical thinking skills [14]; by method of inqury can improve the performance of students [15]. In addition, by utilizing the android application, the teacher can use application-based comics to visualize abstract physics concepts [16]. An application for Android physics mobile learning media (IPMLM) was developed on the topic of thermodynamics. IPMLM is expected to be able to improve the ability of students, especially Higher Order Thinking Skills (HOTS).

2. Methods

Research had done included the type of Research and Development (R & D). The product developed is an android application on the subject of thermodynamics. The development procedure was adapted from the 4D development model (Defining, Designing, Developing, and Disseminating) [17]. The first stage of defining are consisted of preliminary studies in the form of field observations and literature reviews. The second stage of designing was in designing IPMLM applications. The third is developing is an application development consisting of several stages. Draft 1 in the form of the initial product is assessed for its eligibility by media experts and material experts.The results of the assessment are used to improve the application. The results of the repairs are in the form of Draft III. Draft III is then used in learning at the broad test stage.The final stage is disseminating namely the dissemination of applications and research results.

The subjects of limited trial were 11th grade science program students of SMAN 2 Kota Kupang has amount 32 students. The subject of empiric test for HOTS measurement instrument was 323 students of 12th grade science program in Kota Kupang. The subjects were chosen by stratified sampling technique. The subjects of the extensive trial were 210 11th grade science program students from SMAN 2 Kota Kupang, SMAN 3 Kota Kupang and SMAN 5 Kota Kupang. The subject for extensive trial were chosen by cluster sampling technique. The research was conducted in an even semester on January 2019.

Limited trials are concluded readability tests of IPMLM applications and empirical tests of HOTS measurement instrument. The content validity of each item can be found through Aiken's validity coefficient and empiric validity and realiability through the QUEST output by looking at the compatibility of item with the PCM model or MNSQ infit. The question item is said to be compatible with the PCM model, if the MNSQ infit value is ≥ 0.77 and ≤ 1.30 [18]. Instrument reliability can be determined through the estimated reliance of items on QUEST output.

The research design has used in the extensive trial is the Pretest-Posttest Control Group Design as shown in Table 1 [19].

Table 1. Research design[a].

O_1[b]	X_1[c]	O_2[d]
O_3[e]	X_2[f]	O_4[g]

[a] pretest-posttest control group design
[b] pretest score experiment group
[c] learning assisted by IPMLM
[d] posttest score experiment group
[e] pretest score control group
[f] Using conventional products commonly used by teachers in research schools (physics package book)
[g] posttest score control group

Data collection technique are test and non-tests. The test technique is used to measure HOTS with a reasoning multiple choice questions. Non-test techniques are used to validate HOTS questions, assess media eligibility and respond to questionnaires by students. Validation of HOTS questions, assessment of media eligibility, and student responses were obtained by questionnaires with rating scales 1-4 (Likert Scale).

The eligibility of media has assessed by several aspects are namely software engineering, easiness and flexibility in accessing, presentation, interactivity and social communication. The eligibility of the material contained in IPMLM has assessed by two components are namely the learning component and the material component. In addition, students provide responses to the media. The aspects that are responded to are aspects of content, presentation, graphical and language.

Data on the results of the eligibility assessment were analyzed used by descriptive analysis. The average score for each aspect of the eligibility assessment of the learning device is converted to scale 5 [20].The average score for each aspect of the assessment is converted to a value with the criteria as shown in Table 2. The eligibility of media is concluded according those criteria.

Tabel 2. Score quality classification

Score of Respondent	Criteria	Categories
$X \geq X_i + 1{,}8\ SB_i$	$X \geq 85$	Very high
$X_i + 0{,}6\ SB_i < X \leq X_i + 1{,}8\ SBi$	$70 < X \leq 85$	High
$X_i - 0{,}6\ SB_i < X \leq X_i + 0{,}6\ SB_i$	$55 < X \leq 70$	Fair
$X_i - 1{,}8\ SB_i < X \leq X_i - 0{,}6\ SB_i$	$40 < X \leq 55$	Low
$X \leq X_i - 1{,}8\ SB_i$	$X \leq 40$	Very low

X_i = ½ (maximum ideal score + minimum ideals score); = 1/6 (maximum ideal score - minimum ideals score)

The HOTS measurement instrument was validated by 7 experts. The results of the assessment are then analyzed using Aikens' validity coefficient (V). The validity coefficient V of single item by n raters calculated by the Equation 1 [21].

$$V = \frac{\sum s}{[n\,(c-1)]} \tag{1}$$

S = r-lo; *lo* = lowest validity rating; *r* = the number given by a rater; *c* = highest validity rating.

The validity coefficient is compared to the Aiken' table. The item is concluded to be valid if the validity coefficient (V) is greater or equal to the minimum value stated in the Aiken table. The empirical validation of the questions was analyzed using the QUEST program to determine the validity and reliability of items according to item response theory (IRT).

Students' HOTS achievements are grouped according to the 4 categories, which are very good, good, fair good, less good. Increased HOTS is known by calculating the standard gain (g) obtained through Equation 2.

$$Std\ gain\ (g) = \frac{\bar{X}_{posttest} - \bar{X}_{pretest}}{X - \bar{X}_{pretest}} \quad (2)$$

$\bar{X}_{posttest}$: average score after learning; $\bar{X}_{pretest}$: average score before learning; X: maximum score

The standard gain criteria are shown in Table 3 [22].

Tabel 3. Standard gain category

Standard gain (g) score	Categories
(g) ≥ 0.7	High
0,7 > (g) ≥ 0.3	Medium
(g) < 0.3	Low

3. Results and disscusion

3.1. Defining Stage

The results of the initial analysis have shown that SMAN 2 Kota Kupang, SMAN 3 Kota Kupang and SMAN 5 Kota Kupang have not integrated technology in learning physics yet. However, students are permitted to use mobile phones at school. SMAN 2 Kota Kupang and SMAN 3 Kota Kupang provide wifi facilities for students and teachers, while SMAN 5 Kota Kupang does not provide wifi facilities. The teaching material analyzed is thermodynamic material. Core competencies (Kompetensi Inti/KI) and basic competencies (Kompetensi Dasar/KD) from the Kurikulum 2013 are reviewed and analyzed. Kompetensi Dasar is described again as an Indicator of Learning Achievement (IPK) and learning objectives. The description refers to HOTS abilities. HOTS capabilities that are honed and measured are the ability to analyze (C4) and the ability to evaluate (C5). Table 4 shows the results of curriculum analysis.

Table 4. Results of curriculum analysis

Basic Competencies (KD)	Indicators of Competence Achievement (IPK)
3.7 Analyzing changes of the ideal gas state using the laws of Thermodynamics	**First meeting** 3.7.1 Explain the Zeroth Law of Thermodynamics 3.7.2 Differentiates thermodynamic processes: isothermic, isobaric, isochhoric, and adiabatic 3.7.3 Explaining the First Law of Thermodynamics **Second meeting** 3.7.4 Formulate the First Law of Thermodynamics in gases by various thermodynamic processes. 3.7.5 Solve problems related to the the First Law of Thermodynamics **Third Meeting** 3.7.6 Analyze the working principle of a carnot engine, heat engine and cooling machine 3.7.7 Assessing the Second Law of Thermodynamics 3.7.8 Solve problems related to the working principle of the Carnot engine, heat engine and cooling engine

4.7 Creating a work or model about apply of the First Law of Thermodynamics and the Second Law of Thermodynamics	**Second meeting** 4.7.1 Designing tools in experiments to prove the First Law of Thermodynamics Conduct an experiment 'burning balloons' to prove the First Law of Thermodynamics 4.7.2 Doing experiments to prove the First Law of Thermodynamics Conduct an experiment 'burning balloons' to prove the First Law of Thermodynamics

2.1. Designing stage

Product design is designed flowcharts and story boards which contained an overview of the applications developed. Flowchart and story board are used to determine what will be displayed on the learning media.

The IPMLM application is contained several menus has namely Petunjuk Penggunaan, Kompetensi, Materi, LKPD, Evaluasi and Profil Pengembang. HOTS evaluation menu is equipped with a timer and when students finish working on the questions, the score will be immediately displayed. In addition, the application is also equipped with a chat room feature that allows teachers and students to give feedback to each other regarding learning materials. Figure 1, Figure 2, Figure 3, and Figure 4 show the example of the IPMLM appearance.

Figure 1. Appearance of the IPMLM initial display

Figure 2. Appearance of the IPMLM menus.

Figure 3. Appearance of the IPMLM materials menu (content)

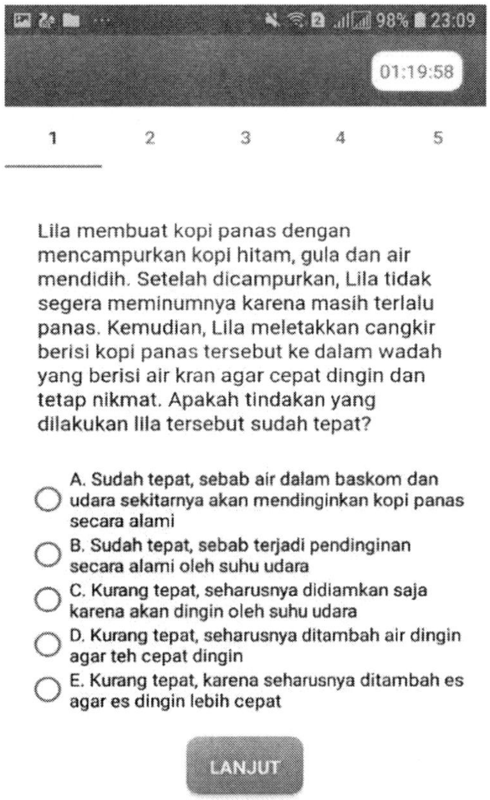

Figure 4. Appearance of the IPMLM HOTS evaluation menu

2.2. Developing stage

IPMLM was developed through several stages. The development begins with the creation of content about thermodynamics as outlined in the application. The content developed is material consisting of experimental videos, animated videos, illustrations, images and text. In addition, 24 items of HOTS were also developed. The results of the development are described as follows.

2.2.1. *Product validation*
The eligibility of IPMLM is assessed by material expert lecturers and media experts. Tabel 5 shows the results of the assessment by media expert and material expert.

Tabel 5. The results of the IPMLM eligibility assessment by physics material experts.

Aspects	Score	Categories
	Media expert	
Software engineering	75	High
Easiness and flexibility in accessing	72	High
Presentation	71	High
Interactivity	81	High
Soacial communication	75	High
Average	**74.80**	**High**
	Material expert	
Learning	92	Very high
Material	91	Very high
Average	**91.5**	**Very high**

Table 5 shows that every aspect of the media reaches "high" category and every aspects of material in IPMLM reaches "very high" category. So, IPMLM is eligible to be used in physics learning about thermodynamics in accordance with the suggested revision given by the assessor.

Similar development research has been done by Mardiani and Kuswanto [23]. The product developed is an interactive android application. The aspects of the android based media that assessed by experts were learning aspect, material aspect, audio-visual appearance aspect and software technology aspect. The results of the development show that every aspect of the application achieves the "very good" category. The application is used in learning and proven to reduce students' misconceptions in temperature and heat material [23]. Application development based on local wisdom developed by Lestari and Kuswanto has shown that the combination of physics material, local wisdom (rowing boat), and technology (android) is valid and easy to use in physics learning [24].

HOTS measurement instruments are validated through two stages, namely validation by experts and empirical tests. Referring to the Aiken's validity coefficient (V) table, the minimum value received with a 1% error rate is 0.86 (Aiken, 1985). The results of content validation have shown that the validity coefficient (V) of each item HOTS was more than a minimum value of 0.86 with the lowest value V = 0.90 and the highest value V = 1.00. Thus all HOTS items are valid in terms of content.

The result of empiric test shows that each item of HOTS instruments are fit to partial credit model (PCM). So can be cocluded that the instrument is valid. Meanwhile, the reliability of instrument is known by reliability of estimate in QUEST program output. The results show that person reliability of person estimate is 0,00 dan reliability of item estimate is 0,72. The coefficient reliability show that the HOTS items are quite reliable however, the consistency of students' answers is weak. In spite of that HOTS instruments is still acceptable because the items' reliability is good enough.

3.3.2 IPMLM revision

Results assessment of material experts and media experts is used to improve the application. Table 6 shows the comments/suggestions from media experts and material experts as well as improvements made to the application.

Table 6. Revision of IPMLM applications

No	Comments/Suggestions	Revised Results
1	The initial login should only use autenfication once during registration / initial use, then the user can simply log in with a username and password.	It is easier for users to register and login, using only a username and password.
2	The teacher has not been able to give feedback (there are no facilities for feedback or teacher comments).	There is an additional "chat room" feature so students can ask questions and the teacher can answer the question. 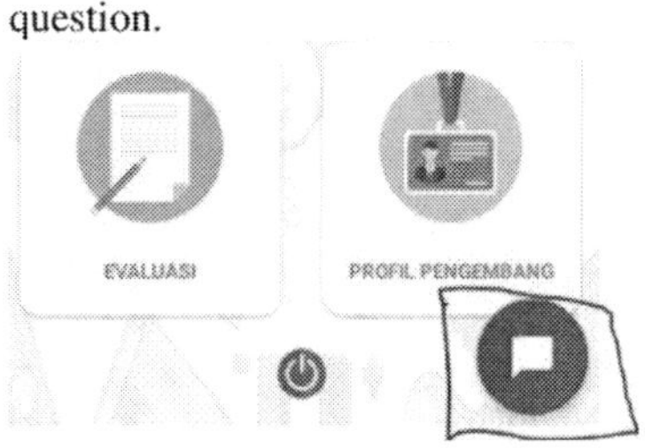
3	Material layout does not showed hierarchy	The composition of the material sub-menu is arranged that shows the level of difficulty of the material from the easiest to the most difficult material. 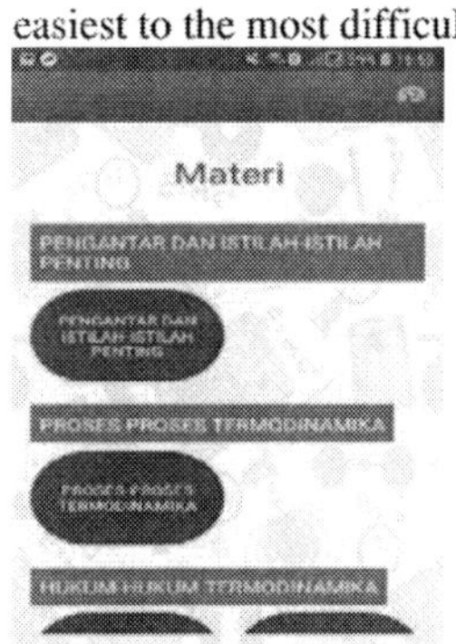
4	New media has the role of replacing books, not increasing physical understanding. Add narration to the animation to clarify the physics process.	There are additional narratives on videos and animations to facilitate physical understanding by pupils.

		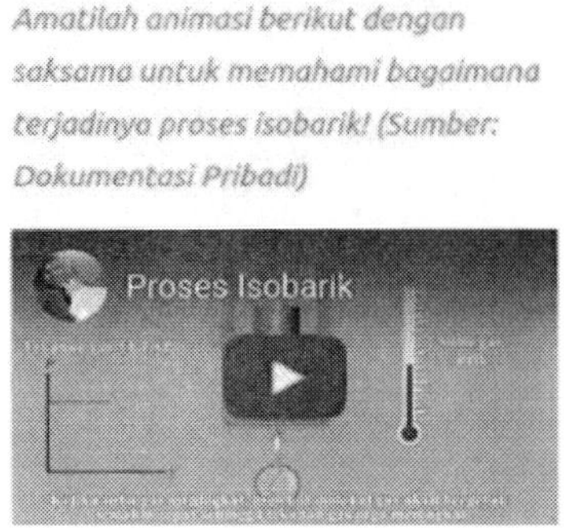
5	Some equations are italicized and vector quantities are bold.	Physics equations are italicized. In addition, scalar and vector quantities have not been differentiated.

3.3.3. Limited trial

The limited trial is about empirical validation of the HOTS measurement instrument and readability of IPMLM by 32 students. The QUEST analysis shows that all items are within the acceptance limit area ≥ 0.77 and ≤ 1.30. So it can be concluded that all the items tested are fit to the PCM model **(valid)** based on four categories of politomus data. In addition, the QUEST output has shown the coefficient reliability is 0.72. This value states that the items in the instrument are **quite reliable**.

The research on HOTS problem development was also done by Istiyono, Mardapi, and Suparno are namely PhysTHOTS instruments which were developed in the form of reasoning multiple choice on the ability to analyze, evaluate, and create for physics material [12]. PhysTHOTS instruments have fulfilled content validity by expert judgment and have obtained empirical evidence of fit construct validity in the Partial Credit Model (PCM) based on four categories of politomus data. In addition, PhysTHOTS reliability is high, by a reliability coefficient of more than 0.90 [12].

The IPMLM application is used by students in a limited trials to test readability. Readability test results are shown in Figure 5.

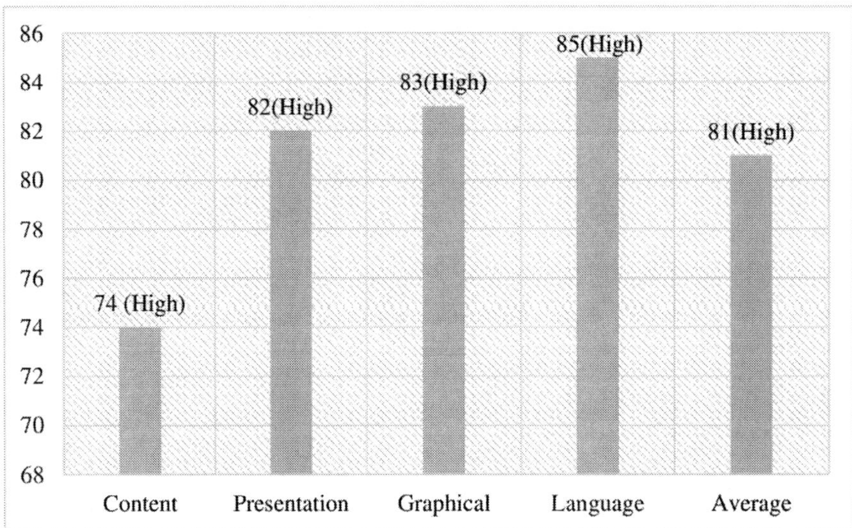

Figure 5. Results of readability of IPMLM applications on limited trial

Figure 5 showed that the readability of IPMLM media reaches the category of "high" with an average score is 81. The results of student responses indicate that the application is eligible to use in physics learning. Some students stated that IPMLM applications are very interesting and applications should also be developed for other materials. However, sometimes the LKPD sheet still has an error. Figure 6 and Figure 7 show examples of limited trial subjects responses to the application.

Bagian yang PerluDiperbaiki	Komentar/Saran
Kolom jawaban sering kali error.	penambahan materi lain.
	memperbanyak latihan soal & fitur menarik.

Figure 6. Response of student "A" in limited test

Bagian yang PerluDiperbaiki	Komentar/Saran
Warna	Sebenarnya warnanya sudah bagus, tapi kalau bisa lebih menggunakan warna cerah agar lebih semangat.

Figure 7. Response of student "B" in limited test

Another suggestion from students is said that the media should be made offline and also made for other Physics material. However, fully offline mode cannot be realized because the evaluation and upload process of student answers to the LKPD must be uploaded (requires internet connection) so the results can be recapitulated by the teacher on the application dashboard.

3.3.4. Extensive Trials

Extensive trials were carried out at SMAN 2 Kota Kupang, SMAN 3 Kota Kupang, SMAN 5 Kota Kupang. Each school consists of one experimental class and one control class. The learning process in the experimental class uses IPMLM, while in the control class uses the physics package books that are commonly used by teachers. Table 7 showed the results of the assessment of the application by students of extensive.

Table 7. Results of the assessment of IPMLM applications by students of extensive trial

Aspect	Score	Categories
Content	82	Good
Presentation	82	Good
Graphical	83	Good
Linguistik	83	Good
Average	**82,50**	**Good**

Table 7 showed that IPMLM applications are considered good by students. So, IPMLM media is good to be used as a media of learning in the classroom.

The effectiveness of media use is known through HOTS achievement of students. Figure 8 shows HOTS achievement of control class pupils.

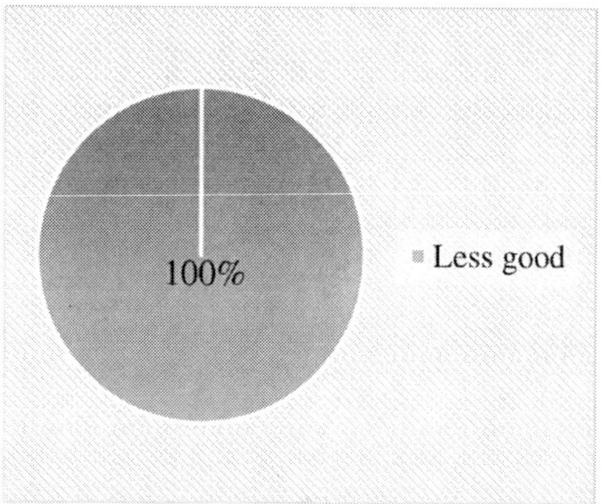

(a)

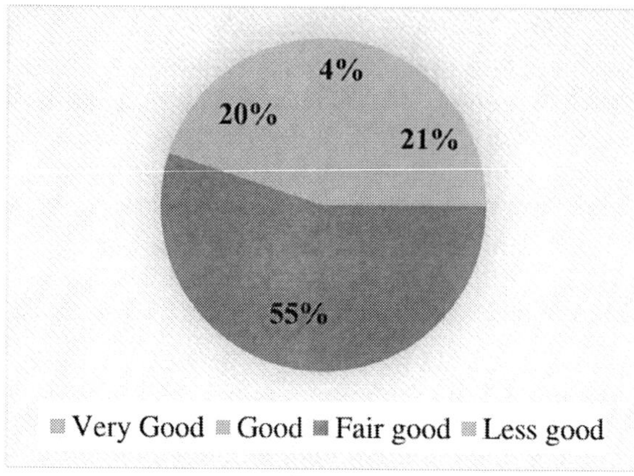

(b)

Figure 8. Results of HOTS achievement in control class students before learning (a) and after learning (b)

Figure 8 showed that after participating in learning only 4% of students achieved a very good category (A). There are even 20% of students who are still in the less good category (D). Figure 9 shows the achievement of HOTS experimental class students.

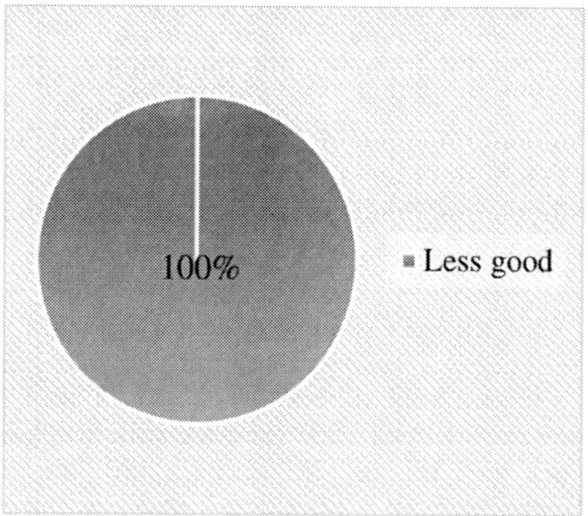

(a)

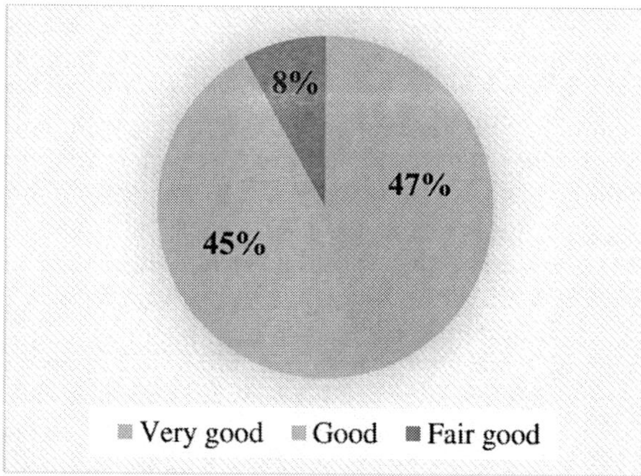

(b)

Figure 9. Results of HOTS achievement in experiment class students before learning (a) and after learning (b)

Figure 9 showed that students who used IPMLM are mostly reach very good category (A) which is as much as 47% and only 8% that reaches the fair good category (C). HOTS students experienced an increase in analyse aspect and evaluate aspect. The increase of those aspects is shown by Figure 10.

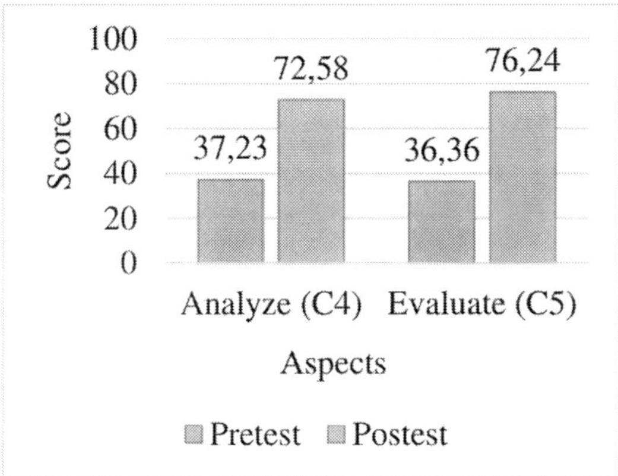

(a)

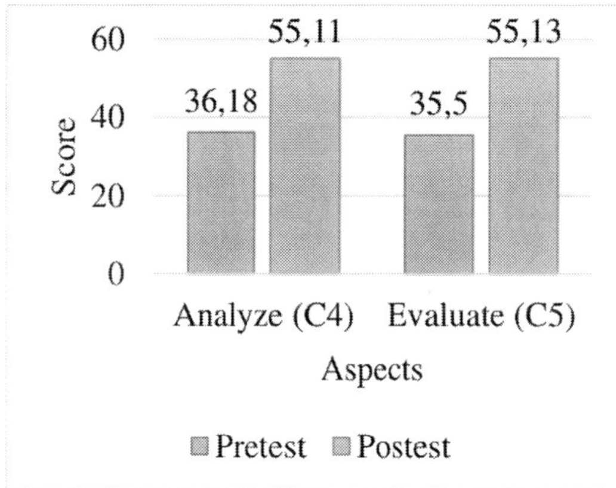

(b)

Figure 10. Results of HOTS achievement in experiment class students (a) and control class students (b)

The increase in each aspect is also indicated by the achievement of the gain standard as show in Table 8.

Table 8. Comparison of Gain Standards (g) Experimental Classes and Control Classes

HOTS Aspects	Control group		Experiment group	
	Gain Std (g)	Categories	Gain Std (g)	Categories
Analyze (C4)	0,30	Medium	0,56	Medium
Evaluate (C5)	0,30	Medium	0,63	Medium

Figure 10 and Table 8 showed that the increase in HOTS in the experimental class is better than the control class. The standard gain value in the control class only reached 0.30 for each aspect of HOTS. So it can be concluded that increasing HOTS of students using IPMLM is better than not using IPMLM. These results support the research findings that there are differences in the increase of students' creativity and cognitive learning outcomes between using the android-based chemistry instructional media integrated to learning together and learning together without the android-based instructional media [25].

Similar applications have also been used in learning namely ThinkLearn applications on the subject of energy transfer [15]. The results have shown that secondary school students using the ThinknLearn application experienced a significant increase in knowledge and have shown more critical thinking than the control group that did not use the ThinknLearn application. The results of follow-up tests show that students using the ThinknLearn application maintain information/material better than students who did not [15]. Learning using mobile devices provides benefits for students and also helps teachers in learning. The developed IPMLM is equipped with images. This facility helps understand the concept of physics using images that are familiar with students. The cognitive process that occurs is that students connect thermodynamic concepts to familiar environments or situations. The process can develop students' scientific understanding. Developing students' scientific understanding is one way to improve high-level thinking skills (HOTS) [26]. Besides images, IPMLM applications are equipped with videos. Video facilities in the application help students to develop students' scientific skills and understanding of a process [27]. Both of these are aspects related to high-level thinking skills [28].

The learning process has used technology enables the creation of a fun physics learning process. Some students stated that the learning process with IPMLM was very interesting and not boring. Such conditions, help students to stay focused and also increase motivation to learn material. Figure 11, Figure 12 and Figure 13 show examples of student responses to the application.

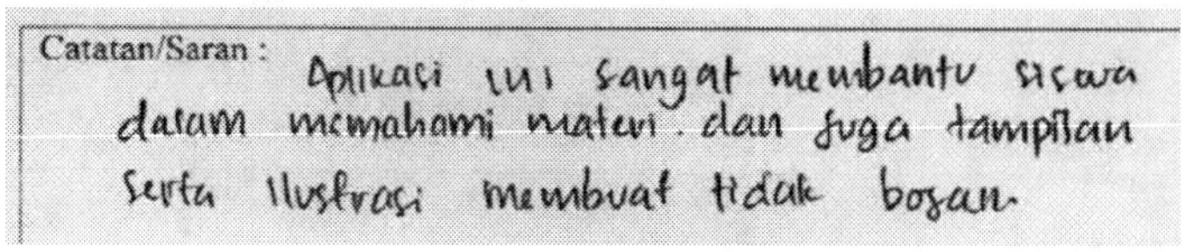
Catatan/Saran : Aplikasi ini sangat membantu siswa dalam memahami materi. dan juga tampilan serta ilustrasi membuat tidak bosan.

Figure 11. Response of first student

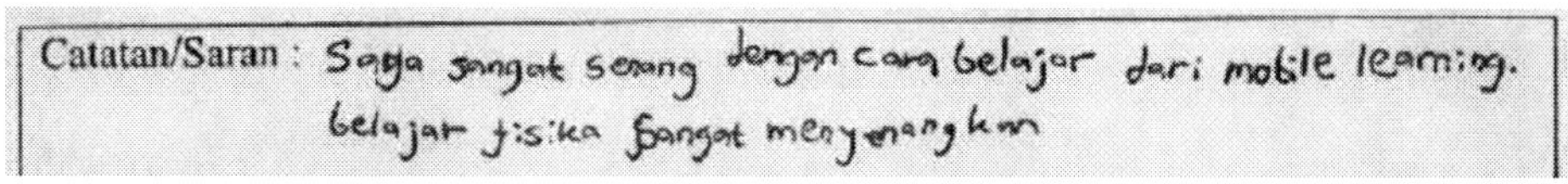
Catatan/Saran : Saya sangat senang dengan cara belajar dari mobile learning. belajar fisika sangat menyenangkan

Figure 12. Response of second Student

Catatan/Saran : Aplikasinya sangat cocok untuk kami

Figure 13. Response of Third Student

3.4. Disseminating stage

The last stage is disseminating, namely the dissemination of application download links to students at SMAN 2 Kota Kupang, SMAN 3 Kota Kupang dan SMAN 5 Kota Kupang and presenting the results of research in scientific publication.

4. Conclusion

It has been successfully developed an android-based Interactive Physics Mobile Learning Media (IPMLM) that is suitable for use in learning in thermodynamic material. The results of media eligibility by media experts reached "high" category with an average score of 74.80. The media eligibility by material experts reached "very high" category with an average score of 91.5. The results of student responses in limited tests indicate that IPMLM is "eligible" with an average score of 81.

The extensive test results have shown that the IPMLM application is effective in increasing HOTS. The result shows that the more dominant aspect of HOTS is evaluating aspect. Standard gain in the experimental class for the analyzing aspect reach the "medium" category ($g = 0.56$) and for the evaluating aspect reach the "medium" category ($g = 0.63$). Standard gain (g) in the control class reaches the "medium" category, with g of 0.30 for each aspect. So, it could be concluded that learning uses IPMLM be more effective than using Physics package books in improving HOTS students.

References

[1] Saputri A A and Wilujeng I 2017 Developing physics e-scaffolding teaching media to increase the 11th-grade students' problem solving ability and scientific attitude *International Journal of Environmental & Science Education* **12** (4) pp 729-745. Available from http://www.ijese.net/makale/1841

[2] Kementerian Pendidikan dan Kebudayaan 2016 *Peraturan Menteri Pendidikan Nasional No 21 Tahun 2016* tentang Standar Isi Pendidikan Dasar dan Menengah (Jakarta: Kemendikbud)

[3] Ramadhani I and Motlan 2015 Efek model pembelajaran berbasis proyek dengan strategi think talk write dan kreativitas ilmiah terhadap hasil belajar kognitif tingkat tinggi siswa SMA pada pelajaran fisika *Jurnal Pendidikan Fisika* **4** (1) pp 17-24 ISSN 2252-732X. Available from http://jurnal.unimed.ac.id/2012/index.php/jpf/article/view/Irham%20Ramadhani%20dan%20Motlan
[4] Mundilarto 2002 *Kapita Selekta Pendidikan Fisika*. Yogyakarta: Universitas Negeri Yogyakarta.
[5] Gedgrave I 2009 *Modern Teaching of Physics* (Delhi: Global Media)
[6] Rofiah E, Aminah N S and Ekawati E Y 2013 Penyusunan tes kemampuan berpikir tingkat tinggi fisika pada siswa SMA *Jurnal Pendidikan Fisika* **1** (2) pp 17-22. ISSN: 2338-0691. Available from http://www.jurnal.fkip.uns.ac.id/index.php/pFisika/article/view/2797
[7] Marlina L, Liliasari, Tjasyono B and Hendayana 2018 Improving the critical thinking skills of junior high school students on earth and space science (ESS) materials *Journal of Physics: Conf. Series* **1013**. https://doi.org/10.1088/1742-6596/1013/1/01206
[8] Rusnayati H and Prima EC 2011 Penerapan model pembelajaran problem based learning dengan pendekatan inkuiri untuk meningkatkan keterampilan proses sains dan penguasaan konsep elastisitas pada siswa SMA *Pros. Seminar Nasional Penelitian, Pendidikan dan Penerapan MIPA* (Yogyakarta: Universitas Negeri Yogyakarta) pp 331-338
[9] Mardiana N and Kuswanto H 2017 Android-assisted physics mobile learning to improve senior high schools students' divergent thinking skills and physics HOTS *In AIP Conf. Proc.* **1868** (AIP Publishing) pp. 070005(1)- 070005 (12). https://doi.org/10.1063/1.4995181
[10] Suryani W, Harahap M B and Sinulingga K 2017 Efek model pembelajaran scientific inquiry menggunakan mind mapping dan keterampilan berpikir kritis terhadap keterampilan proses sains siswa sekolah menengah atas *Jurnal Pendidikan Fisika* **6** (2) pp 86-90
[11] Subekti Y and Ariswan A 2016 Pembelajaran fisika dengan metode eksperimen untuk meningkatkan hasil belajar kognitif dan keterampilan proses sains *Jurnal Inovasi Pendidikan IPA* **2** (2) pp 252-261. http://dx.doi.org/10.21831/jipi.v2i2.6278
[12] Istiyono E, Mardapi D and Suparno 2014 Pengembangan tes kemampuan berpikir tingkat tinggi fisika (PhysTHOTS) peserta didik SMA *Jurnal Penelitian dan Evaluasi Pendidikan* **1** pp 1-12.
[13] Faizaha AW A A, Suparmi and Aminah N S 2018 Students' conceptions on momentum and impulse toward higher order thinking *skills Proc. Int. Conf. on Research, Implementation and Education of Mathematics and Science (5th ICRIEMS)* ISBN: 978-602-74529-3-0 pp 50-55
[14] Kartikawati S and Pratama H 2017 Pengaruh penggunaan whatsapp messenger sebagai mobile learning terintegrasi metode group investigation terhadap kemampuan berpikir kritis *Jurnal Pendidikan Teknik Elektro* **2** (2), ISSN: 2477-8354. Available from http://ejournal.unipma.ac.id/index.php/JUPITER/article/view/1797/1391
[15] Ahmed S and Parsons D 2013. Abductive science inquiry using mobile devices in the classroom. *Computers and Education* **63**, pp 62-72. http://dx.doi.org/10.1016/j.compedu.2012.11.017
[16] Hadi W S and Dwijananti P 2015 Pengembangan komik fisika berbasis android sebagai suplemen pokok bahasan radioaktivitas untuk sekolah menengah atas *Unnes Physics Education Journal* **4**(2) ISSN: 2252-6935. Available from http://lib.unnes.ac.id/23253/1/4201410097.pdf
[17] Thiagarajan S, Semmel, D S and Semmel M I 1974 *Instructional development for training teachers of expectional children* Minneapolis, Minnesota: leadership training institute/special education, University of Minnesota
[18] Adams R J & Khoo S T 1996 *QUEST: the interactive test analysis system version 2.1* (Victoria: The Australian Council for Educational Research)
[19] Kothari C R 2004 *Researh Methodology: Methods and Techniques (*New Delhi : New Age International Publisher)

[20] Sukardjo 2012 *Buku pegangan kuliah evaluasi pembelajaran IPA* (Program Pascasarjana UNY: Program Studi Pendidikan Sains)

[21] Aiken L R 1985 *Three coefficients for analysing the reliability and validity of ratings Educational and Psychological Measurement* **45** pp 131-142

[22] Hake R R, Wakeland R, Bhattacharyya A and Sirochman R 1994 Assessment of individual student performance in an introductory mechanics course *AAPT Announcer* **24** (4) p 76

[23] Mardiani N and Kuswanto H 2018 Developing the android-assisted physics interactive learning media to reduce senior high school students' misconception about physics an improve their attitude towards it. *Proc. Int. Conf. on Research, Implementation and Education of Mathematics and Science (5th ICRIEMS)* ISBN: 978-602-74529-3-0 pp 82-89

[24] Lestari M L and Kuswanto H 2018 Development of physics instructional media by using the cultural theme (rowboat) based on android mobile phone for junior high school students *Proc. Int. Conf. on Research, Implementation and Education of Mathematics and Science (5th ICRIEMS)* ISBN: 978-602-74529-3-0pp 30-35

[25] Ulfa A M, Sugiyarto K H and Ikhsan J 2016 The effect of the use of android-based application in learning together to improve students' academic performance *In AIP Conf. Proc.* **1847** (AIP Publishing) pp. 050008 (1) - 050008 (6) http://dx.doi.org/10.1063/1.4983910

[26] Madhuri G V, Kantamreddi V S S N and Goteti L N S P 2012 Promoting higher order thinking skills using inquiry-based learning *European Journal of Engineering Education* **37** (2) pp 117-123 http://dx.doi.org/10.1080/03043797.2012.661701

[27] Ekanayake S Y and Wishart J 2014 Mobile phones images and video in science teaching and learning. *Learning, Media and Technology* **39** pp 229-249, http://dx.doi.org/10.1080/17439884.2013.825628

[28] Brookhart S M and Nitko A J 2011 *Assessment and grading in classroom* (New Jersey: Pearson Education Inc)

Correlation between increasing mastery concepts of wave and optics and habits of mind prospective physics teacher students

E Susilowati[1*], Suyidno[1], T Mayasari[2], N Winarno[3], D Rusdiana[4], I Kaniawati[5], P H Santoso[6]

[1]Department of Physics Education, FKIP, Universitas Lambung Mangkurat, Banjarmasin, Indonesia
[2]Department of Physics Education, PMIPA, Universitas PGRI Madiun, Madiun, Indonesia
[3]Department of Science Education, Universitas Pendidikan Indonesia, Bandung, Indonesia
[4,5]Department of Physics Education, Universitas Pendidikan Indonesia, Bandung, Indonesia
[6] Department of Physics Education, Universitas Sulawesi Barat, Majene, Indonesia

*Corresponding author e-mail: titis_pfis@ulm.ac.id

Abstract. The purpose of this study was to determine the relationship between increasing mastery wave concepts and optics with habits of mind of physics prospective teacher students. This study involved 46 students who programed the Wave and Optics courses in one of the LPTK Physics Education Study Programs in Banjarmasin. The method used is mixed method with an embedded experimental model. Data analysis techniques used N Gain to determine the increase in mastery of concepts and habits of mind, then carried out normalization tests, and correlation tests to determine the relationship between the two. Analysis of the results of the study of increasing the mastery of wave concepts and optics with habits of mind of prospective physics teacher students showed a positive, strong, and very significant relationship with $r(46) = 0.64$; $p = 0.01$. This explains that the better the mastery of the concepts held by prospective teacher students, the better habits of mind of prospective teacher students, or vice versa.

1. Introduction

Mastery of concepts is cognitive knowledge that supports science and critical analysis is a standard element that prospective teacher students must have that is recommended by the National Science Teachers Association [1]. This is stated in the content knowledge standard to form an effective teacher. Teacher profession is one of the professions that participate in competing in free trade between ASEAN countries that are members of the ASEAN Economic Community [2].

Mastery of concepts is the ability of students to understand concepts after learning activities. Mastery of concepts is defined as students' ability to understand scientifically meaning both theory and its application in daily life [3]. Mastery of concepts becomes the main means to improve students' thinking habits. Mastery of concepts to solve problems is the main prerequisite in developing thinking habits [4].

The most important goal of education is to develop mental habits that allow individuals to learn about everything they want or need to understand everything related to their lives. Every individual in his life is inseparable from problems, both problems are easy or difficult to solve that are related to personal problems or social problems. In situations where individuals do not know how to overcome or respond to these problems, intelligent behavior is needed as a solution, in the sense that it not only knows information but also how to act and behave. The ability to behave intelligently is called habits of mind [5].

Dimension of learning consists of: (a) first dimension, attitude and perception of learning, (b) second dimension, obtaining and integrating knowledge, (c) third dimension, expanding and smoothing knowledge, (d) the fourth dimension, using meaningful knowledge, (e) the fifth dimension, utilizing habits of mind. The first dimension and the fifth dimension are the main factors that need to be used in the learning process because the two dimensions that determine the success of other dimensions [6].

Some cognitive psychology researchers suggest that humans have the ability to think, but also have the ability to control their behavior through habits of mind. These thinking are placed through three categories, namely self regulation, critical thinking, and creative thinking [6]. Whereas other researchers [8,9,10] categorize habits from indicators that are similar to habits of mind developed by Marzano[6].

When examined, aspects that exist in mind of mind can be seen that these aspects provide a provision for someone to develop mental habits which are the main purpose of education [6,7,9]. In addition, behavior and behavior are behaving intelligently at the highest level of thinking and solving problems and are a determinant of success in academics, work, and social interaction [8,10].

Thinking skills that support a person who can be better explored, trained and developed than before. In online learning compared to other aspects [11]. Carter [9] in the book he compiled with the title Keys to Effective Learning Developing Powerful Habits of Mind Put forward various types of strategies to explore, train, and develop individual habits of mind.

After observing, aspects of habits of mind as explained earlier, it is considered important to train habits of mind in students in order to produce students who can behave and act intelligently so that students succeed in academic, work, and social interactions as students' supplies. in facing their lives as personal beings and social beings. In line with the opinion expressed by Rustaman [12] who said that habits of mind as intelligent behavior are far more important than providing high-level thinking skills to students through science education. Therefore, students need to be given provisions in developing mastery of concepts and habits of mind. Both of these are competencies needed by students in LPTK that can be done in all subjects. But to achieve this requires good guidance. By learning appropriate concepts, students practice thinking habits [13]. Therefore, to practice the habit of thinking and mastering the concepts, the subjects with suitable characteristics are chosen.

In the physics education curriculum at the LPTK, students are given supplies with study program expertise courses namely Wave and Optics in the fifth semester. Students who take the Wave and Optics program have passed Basic Physics courses as a prerequisite. But in reality, based on preliminary studies, students who have passed Basic Physics cannot be used as a benchmark for mastering the mastery competencies of the Wave and Optics concepts so that they need to be improved again. One way to improve mastery of concepts is to provide learning resources. The ability to master concepts and habits of mind is a competency that must be mastered by students [14]. Students 'competencies are expected to be able to increase students' self-potential. Therefore both of these competencies need to be applied to this course.

2. Method

2.1. Research design

The design of this study uses mixed method with an embedded experimental model as shown in Figure 1.

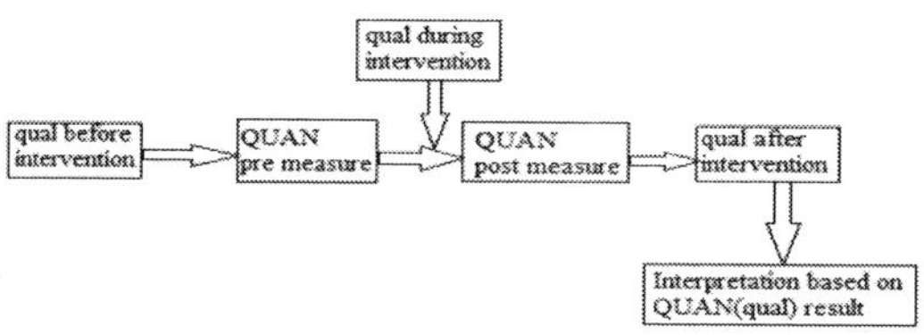

Figure 1. Mixed method design with embedded experimental model.

2.2. *Data Analysis Technique*

Increasing the average mastery score of concepts and critical thinking habits is calculated using N Gain as follows:

$$N\ Gain = \frac{score\ posttest - score\ pretest}{score\ maximum - score\ pretest}$$

Interpretation of N Gain data obtained from calculations is categorized in Table 1.

Tabel 1. N Gain Category

N Gain	Category
0,7 < *N Gain*	High
0,3 ≤ *N Gain* ≤ 0,7	Medium
0,3 < *N Gain*	Low

[15]

The purpose of this data analysis is to find out the correlation between the increase in mastery of concepts and an increase in habits of mind. The data used is the average data of increase in N Gain mastery of concepts and mean improvement data N Gain habits of mind. Before using the correlation test, first use the normality test. To find out the strength of the relationship of mastery of concepts and habits of mind used the Pearson Correlation correlation test with the assumption:

(1) There is no significant correlation if the significance value of pk> 0.01

(2) There is a significant correlation if the significance value is <0.01

3. Result

Concept mastery data were obtained from the results of the pre-test and post-test about Wave and Optics questions consisting of 30 items. Students' pre-test and post-test scores were analyzed to find out how much the mastery of concepts increased by calculating N gain as shown in Table 2.

Table 2. Data on increasing average N Gain mastery of student concepts

N	*Mean*		Percentage (%)	Ideal score	N gain	Category
	Pre test	*Post test*				
46	18.63	24.98	21.17	30	0.57	Medium

The average mastery value of the wave and optic concepts of students before the lecture in the experimental class amounted to 18.63. This means that before being given treatment, students have mastered the concept of 62.1%. After being given treatment, the average student mastery concept value becomes 24.98 from the ideal score of 30 or students have mastered the optical wave concept of 83.27%. From the difference in the average value of the post test and pre test which is 6.35, the students experienced a 21.17% increase in mastering the wave and optical concepts.

The value of increasing data N mastery of wave and optic concepts of 0.57 is in the medium category.Students' habits of mind are measured through habits of mind questionnaires given at the beginning and end of lectures. To determine the increase, N Gain calculations are used as shown in Table 3.

Table 3. Average increase in N Gain habits of mind

N	*Pretest*	*posttest*	Percentage (%)	Ideal Score	Mean N Gain	Category
46	2.24	3.16	23	4	0.52	Medium

The increase in the average N Gain of mind of students in the Wave and Optics lectures was 0.52 in the moderate category which came from the initial score (mean = 2.24; 56%) and the final score (mean = 3.16, 79%) with an ideal score of 4. The percentage of the increase is 23% overall.

The relationship between mastery of concepts and habits of mind was tested using a correlation test, where the correlation test aims to see the strength or magnitude of the relationship between two variables, namely the mastery of concepts and variables of habits of mind. The data used in this correlation test is the average data N Gain mastery of the concept and the average N Gain habits of mind students from ordinal data that has been synchronized into interval data. Before the correlation test is conducted, to find out whether the data is normally distributed, a normality test is carried out. The results of the N normality test for mastery of concepts and students' habits of mind are shown in Table 4.

Table 4. N Gain normality test results mastery of concepts and habits of mind

Score	*N*	*Mean*	*SD*	Distribution (α=0,05)	
				p_n	Note
PK	46	0.57	0.15	0.85	Normal
HoM	46	0.50	0.12	0.73	Normal

From Table 4, the mastery of concept data has a significance value (p_n) of 0.85> 0.05. That is, the data is normally distributed. For habits of mind data, the significance value (p_n) is 0.73. This means that habits of mind data are also normally distributed. Because both data are normally distributed, the requirements for conducting Pearson correlation can be carried out. The results of the correlation test of the two variables are presented in Table 5.

Table 5. Average correlation test N gain mastery of concepts and habits of mind

Variable	*N*	*mean*	*SD*	*r*	*P*	Note
PK	46	0.57	0.15	0.64	0.01	There is correlation
HoM	46	0.50	0.12			

Based on the results of Pearson Correlation, the variable mastery of mind and students showed a positive, strong, and very significant relationship, r (46) = 0.64; p <0.01 between masters of concepts (mean = 0.57; SD = 0.15) and habits of mind (mean = 0.50; SD = 0.12). This means that stronger students have, the stronger the habits of mind of students. The increase in the mean of gain from mastery of concepts is higher than that of students. The class Increased Mean N Gain mastery concept is higher than students' habits of mind is shown in Figure 2, and the increase in mean N Gain mastery is the same as the increase in mean N is shown in Figure 3, the class Increased mean N Gain mastery concept is the same as the increase in mean N Gain habits of mind of students is shown in Figure 4.

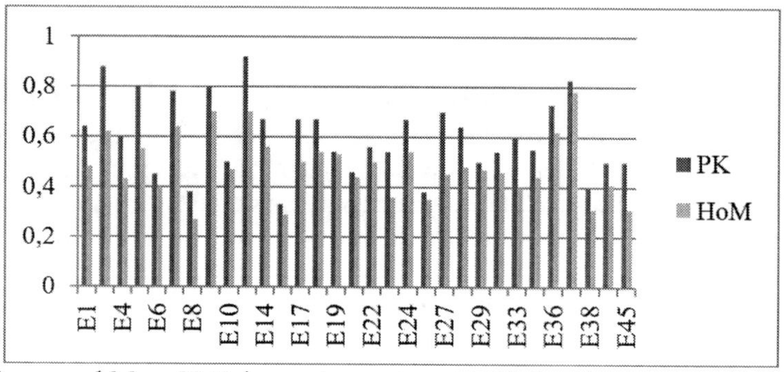

Figure 2. Increased Mean N Gain mastery concept is higher than students' habits of mind

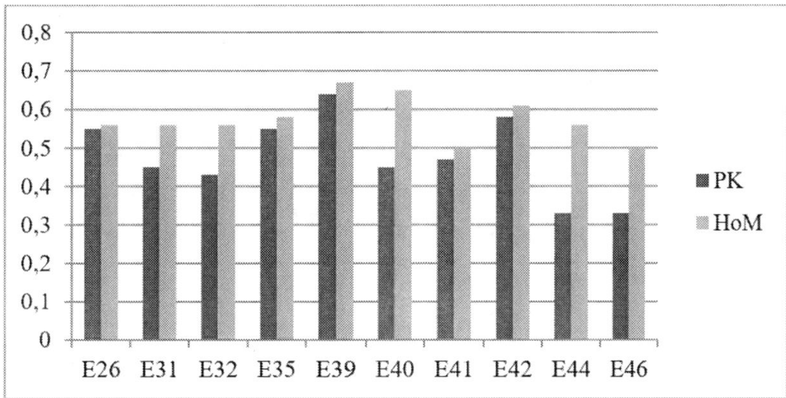

Figure 3. Increased mean N Gain mastery concept is lower than students' habits of mind

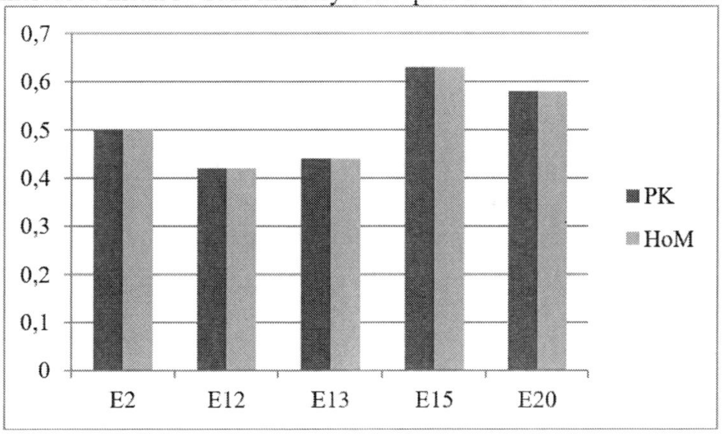

Figure 4. Increased mean N Gain mastery concept is the same as the increase in mean N Gain habits of mind of students

The increase in Mean N Gain mastery concept is higher than the habits of mind of students in the experimental class as many as 31 students (67%), while the increase in the average N Gain mastery of concepts is lower than the students' habits of mind in the experimental class as many as 10 students (22%), and the increase in the mean N Gain mastery concept was the same as the increase in the mean N Gain of the habits of mind of the students in the experimental class as many as 5 students (11%).

4. Discussion

Analysis of the results of the study of mastery of concepts and habits of mind of students showed a positive, strong, and very significant relationship, $r(46) = 0.64$; $p < 0.01$ between variable mastery (mean = 0.57; SD = 0.15) and habits of mind (mean = 0.51; SD = 0.11). This means that the stronger the mastery of the concepts students have, the stronger the habits of mind of students. This finding corroborates the results study which explains that students who have high mastery of concepts will have a habit of thinking better [16]. Furthermore, the other research also states that students' thinking habits help improve mastery of concepts, discuss and work together actively about scientific concepts in learning [17,18].

The results of the study regarding the reciprocal relationship between habitual thinking and mastery of concepts that occur showed that students' thinking habits were influenced by mastery of concepts. Students who have good thinking habits, the mastery of the concept is also good [19]. The results [11] study agree with the study that mastery of students' concepts of learning content influences thinking habits. Mastery of concepts and habits of thought are related to one another [11,20]. By learning appropriate scientific concepts, students can practice the thinking habits [21]. This means, mastery of concepts affects someone's thinking habits [21,23].

5. Conclusion

Based on the results, findings, and discussion it can be concluded that there is a positive, strong and very significant relationship between increasing the mastery of concepts wave and optics with the habits of mind (self regulation, critical thinking, and creative thinking) of prospective teacher students. This explains that the better the mastery of the concepts held by prospective teacher students, the better habits of mind of prospective teacher students. Or conversely, the better habits of mind possessed by student teachers, the better the mastery of the concepts held by prospective teacher students.

6. References

[1] NSTA. (2012). *2012 NSTA Preservice science standards*. www.nsta.org.

[2] BBC. (2014). *What you should know about the Asean Economic Community*.

[3] Dahar, R. W. (2003). *Learning theories*. Jakarta: Penerbit Erlangga.

[4] Elyousif, Y. A. K., & Abdelhamied, N. E. (2013). Assessing secondary school teachers' performance in developing habits of mind for the students. *International Interdisciplinary Journal of Education, 2(2),* 168-180.

[5] Costa, A.L. & Kallick, B. (2000). *Describing 16 habits of mind: Habits of mind developmental series. Alexandria, VA*. http://www.ccsnh.edu/documents/CCSNH MLC.

[6] Marzano, R.J., Pickering, & McTighe. (1993). *Assesing student outcomes: Performance assesment using the dimension of learning model*. Alexandria, Virginia:ASCD

[7] Costa, A.L. & Kallick, B. (2000). *Describing 16 habits of mind: Habits of mind developmental series. Alexandria, VA*. http://www.ccsnh.edu/documents/CCSNH MLC.

[8]Costa, A.L. & Kallick, B (2000). *Assesing and reporting on habits of mind. Alexandria*: Association for Supervision and Curriculum Development.

[9] Carter, C., Bishop, J., Kravits, S., L. (2005). *Keys to effectve learning developing powerful habits of mind. Australia: Pearson Prentice Hall.*

[10] Campbell, J. (2006). *Theorising habits of mind as a framework for learning*. Diakses dari: http://www.aare.edu.au/06pap/cam06102.pdf.

[11]Cheung, W.S. & Hew, K.F. (2008). Examining facilitators' habits of mind and learners' participation. *In Hello! Where are you in the landscape of educational technology?Proceedings ascilite Melbourne 2008.* Access http://www.ascilite.org.au/conferences/melbourne08/procs/cheung.pdf

[12] Rustaman, N., Y. (2011). Science education and research in developing high-level thinking skills for character development. Proceedings of the National Seminar on Biology Biology Education Study Program FKIP University Eleven March. Surakarta, UNS, 15-34.

[13] Owusu, A. M. (2014). Reading habits among students and its effect on academic performance: A study of students of koforidua polytechnic. *Library Phylosophy and Practice (e-journal).* Paper 1130.

[14] AACTE. (2010). *21st Century knowledge and skill in educator preparation.* Acces http://www.p21.org/storage/documents/aacte_p21_whitepaper2010.pdf

[15] Hake, R., R. (1998). Interactive-engagement versus traditional methods: A six-thousand students survey of mechanincs test data for introductory physics course. *American Journal Physics 66(1),* 64-74.

[16] Alkhateeb, O. (2015). *The impact of using model of Marzano gain students the ability to configure and integrated conceptual structure in islamic concepts. Journal of Education and Practice, 6(5), 146-160.*

[17] Claxton, G. (2006). Expanding the capacity to learn: A new end for education. *In British Educational Research Association annual conference.*

[18] Wiyarsi, A., Sutrisno, H., & Rohaeti, E. (2018, September). The effect of multiple representation approach on students' creative thinking skills: A case of 'Rate of

Reaction'topic. In *Journal of Physics: Conference Series* (Vol. 1097, No. 1, p. 012054). IOP Publishing.
[19]Smith, J. I., & Tanner, K. (2010). "The problem of revealing how students think: Concept inventories and beyond ". *Journal of Life Science Education 9,* 1-5.
[20] Elyousif, Y. A. K., & Abdelhamied, N. E. (2013). Assessing secondary school teachers' performance in developing habits of mind for the students. *International Interdisciplinary Journal of Education, 2(2),* 168-180.
[21] Santiago, H. C. (2011). Visual mapping to enhance learning and critical thinking skills. *Optometric Education Journal 36*(3), 125-139.
[22] Paterson, C., & Chapman, J. (2013). Enhancing skills of critical reflection to evidence learning in professional practice. *Physical Therapy in Sport Journal 14*(3), 133-138.
[23] Susilowati, E., Hartini, S., Mayasari, T., & Winarno, N. (2018, November). Profile Habits of Mind Students in Physics Learning. In *Journal of Physics: Conference Series* (Vol. 1120, No. 1, p. 012055). IOP Publishing.

Student's Problem Solving Skills in Collaborative Inquiry Learning Supplemented by Formative E-Assessment : Case of Static Fluids

Ernila Siringo Ringo[1], Sentot Kusairi[2], Eny Latifah[2], dan Awal M R Tumanggor[3]

[1]Graduate Student, Physics Department, Universitas Negeri Malang, Indonesia,
[2]Lecturer, Physics Department, Universitas Negeri Malang, Indonesia,
[3]raduate Student, Physics Department, Universitas Negeri Yogyakarta, Indonesia

*Corresponding author e-mail: ernilasr94@gmail.com

Abstract. This research aims to analyze the development of the static fluid problem-solving skills of high school students in collaborative inquiry supplemented by formative e-assessment. This research used a mixed-method approach with the research design of an embedded experimental design model involving 34 students of class XI MIPA in SMA Negeri 4 Malang. The Data on the problem-solving skills were obtained through pretest and posttest using three problem-solving questions in the form of descriptions and supplemented by interviews. The calculation results of N-gain score showed the improvement of student's problem solving skills for a moderate category and the calculation result of Cohen's d-effect size in for strong effect category, as many as 79% of the students are categorized as experts. The expert students have good skills in the useful description indicator followed sequentially on the indicator of physics approach, specific application of physics, mathematical procedures, and logical proggression. Meanwhile, the novice students are only capable on the indicator of useful description and physics approach. Students' skill to solve problems still need to be trained intensively using contectual problems in daily life.

1. Introduction

Problem-solving is the individuals or groups skills to finding solutions for a problem in physics learning that is conceptual and related in daily life, which involves the knowledge and actions possessed [1–4]. This skill has been a core part of education for decades in developing knowledge and understanding in various learning contexts [5], as well as the most important learning outputs in the field of science especially physics [6–7]. One physics concept that is often found in the application of daily life is static fluid [8]. Some researches still find student difficulties in solving problems in Hydrostatic Pressure, Pascals' Law, and Archimedes' Law

Students have difficulty in solving static fluid problems such as the use of buoyant force formulas, the resultant determination of forces and forces on the plane in the fluid, and the determination of the depth of field in fluid [9]. The difficulty is also found in the research by Datur (2016) which states, students have not been able to describe the problem correctly and still use solutions with a wrong physics approach, thus influencing students' mathematical procedures [10]. These difficulties cause the problem-solving skills possessed by students who are still in the low category. This is shown in the research by Purwanto & Yuliati (2017) who found that the average problem solving of students was only

48.88 on a scale of 0-100 which was categorized as still low [11]. The same results were also obtained in the research by Purnamasari (2018) who found that student's problem solving skills in the good category were less than 50% of the number of students with the rest of the students with not good problem-solving categories. Therefore, it takes a series of learning processes that can improve problem-solving skills, namely through collaborative inquiry learning [12–15].

Collaborative inquiry learning is collaborative learning process for students to construct their knowledge through investigations to learn strategies in finding solutions of a problem [16–18]. This is supported by socio-constructivist learning theory which says that knowledge and cognitive conflict arise by finding collaborative solutions to problems, based on information distributed among group members on ongoing learning [13, 15]. Also, a system is needed to monitor student learning both before, during, and after learning through formative assessment [19–20]. However, in the formative assessment, giving feedback quickly to students during learning is still not effective, because it takes a long time to be implemented for many students in the class [21]. Therefore, alternatives are used to support their implementation, namely through formative e-assessment [22]. Formative e-assessment is applied by teachers outside of school hours and provides broad opportunities for students to help students relearn and has a repetitive exercise on the material that has been taught [23–26]. Based on this, so that students can investigate and solve problems in daily life with the right concepts and solutions, and receive feedback quickly, it is necessary to combine the two variables. The lack of research results that examine the combination of collaborative inquiry learning supplemented by formative e-assessment on the analysis of students' problem-solving skills in the static fluid, so that this research is important.

2. Method

This study used a mixed-methods approach with an embedded experimental model design by Creswell (2011) as presented in Figure 1.

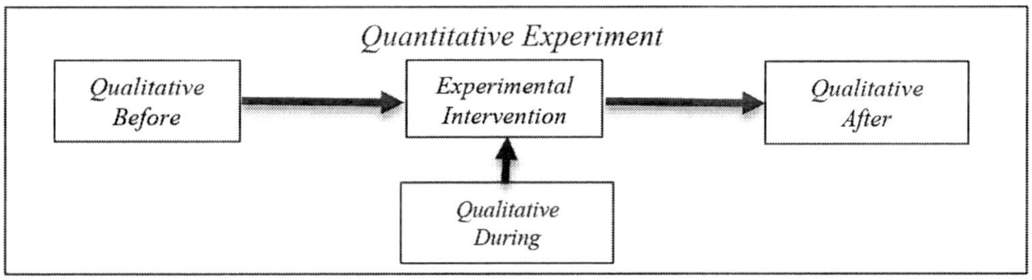

Figure 1. Embedded Experimental Model (Creswell, 2011)

The design of this study consisted of quantitative research methods as main and qualitative data as supporting data carried out simultaneously [27]. The research phase begins with giving a pretest to find out the students' initial problem-solving skills. The results of the pretest were analyzed quantitatively to provide information on the student scores and qualitatively for grouping categories of students based on students' answers according to the criteria on the five indicators of problem-solving, and strengthened from the results of interviews to representatives of student categories. The second stage is implementing collaborative inquiry learning supplemented by formative e-assessment. Some learning activities are documented in the form of photos and videos. After that, the posttest is given to find out the student's final skills. The results were analyzed quantitatively and qualitatively to determine the increase in the problem-solving categories of students based on student answers reinforced from the results of interviews. All data are interpreted to make conclusions in answering research questions.

The subjects in this study consisted of 34 students of class XI MIPA in SMA Negeri 4 Malang, East Java, Indonesia, who studied static fluid and were selected by sampling technique using purposive sampling. The research instrument used was a problem-solving skills test in the form of three description questions with Cronbach's Alfa Reliability of 0.783. The student answer assessment rubric refers to Doctor (2016) rubric with the range of scores set in this study is a score of 0 to 4 for each indicator on each question with a maximum score of 4. Data analysis was carried out with quantitative analysis in the form on non-parametric analysis using further tests Wilcoxon, N-gain test, and Cohen's d-Effect Size and qualitative analysis by categorizing the novice and expert student's problem solving skills as presented in Table 1 [28–31].

Table 1. Categorization of Criteria for Novice and Expert Students

No	Indicator	Criteria	
		Novice	Expert
1.	*Useful Description*	Describe the problem by writing an influential variable that is incomplete, partly missing or contains an error	Describe the problem by summarizing relevant infromation in a symbolic form of variables that are influential, images, and verbally appropriate and complete
2.	*Physics Approach*	Some of the physics approaches that are written are not right, are still wrong, even past this step	Writing down a physics approach that is useful as a solution to the problem correctly and completely
3.	*Specific Application of Physics*	Only write general equations without applying to the problem, not complete, containing errors, and not writing it down	Choosing the relevant equation as a solution to be applied to the problem correctly and completely
4.	*Mathematical Procedures*	Processing and obtaining data is still inaccurate, incomplete, not even doing calculations at all	Perform calculations according to the procedures to obtain the right and complete results
5.	*Logical Progression*	The resolution process used is unclear, unfocused, only rewrites the results obtained, and does not link the results to the process used as a solution	The settlement process that is used is clear, focused, and precise so that it is able to prove the suitability of the results obtained with the solutions used

This study applied collaborative inquiry learning supplemented by formative e-assessment in the learning process with the stages as presented in Table 2.

Table 2. Syntax of Collaborative Inquiry Learning with Formative E-Assessment

Syntax	Activities
Formative e-assessment is carried out through *http://e-learning.fmipa.um.ac.id/*	Students follow *quizzes* and discussion forums, receive *feedback*, and learn learning videos.
Phase 1 Attention and explain the learning process	Students respond to questions related to learning objectives that are displayed through interactive powerpoint (*Learning goals*) and pay attention to problems through learning videos.
Phase 2 *Orienting and asking question*	Students listen and understand questions related to the problem given.
Phase 3 *Hypothesis generation*	Students discuss making hypothesis.
Phase 4 *Planning*	Students design experiments with their groups.
Phase 5 *Investigation*	Students investigate and collect experimental data.
Phase 6 *Analysis*	Students independently analyze experimental data and interpret it in graphical form to train students's problem-solving skills. Every student has a record during learning (*Active involvement of students in their own learning*).
Phase 7 *Conclusion*	Students make conclusions to prove the previous hypothesis
Phase 8 *Communication*	Students pair outside the group members to present their results (*rich conversations*). One group member presents the results of the discussion in front of the class and the teacher gives applause (*learning needs and strenght*). Students respond to learning through teacher questions and reinforcing examples of application in the daily life (*timely feedback*).

Formative e-assessment is carried out through *http://e-learning.fmipa.um.ac.id/*	Students take *formative tests* and discussion forums and receive *feedback*.

3. Results and Discussion

Students' problem solving skills experience better changes after being given collaborative inquiry learning supplemented by formative e-assessment, namely in the acquisition of the average posttest value of students of 2.87 has increased from the average pretest value of 0.96. Obtaining the score for the problem-solving skills oh the students has approached the expected score with a maximum score of 4 [29]. This can be seen in the highest posttest score achieved by students at 3.67 with the criteria for the expert category, but there are still the lowest posttest scores of some students at 1.67 with the criteria for novice category answers to static fluid problems [25, 28, 29, 30]. The percentage of students in the novice and expert categories based on data on the distribution of scores for problem-solving skills after learning is shown in the form of a diagram as shown in Figure 2.

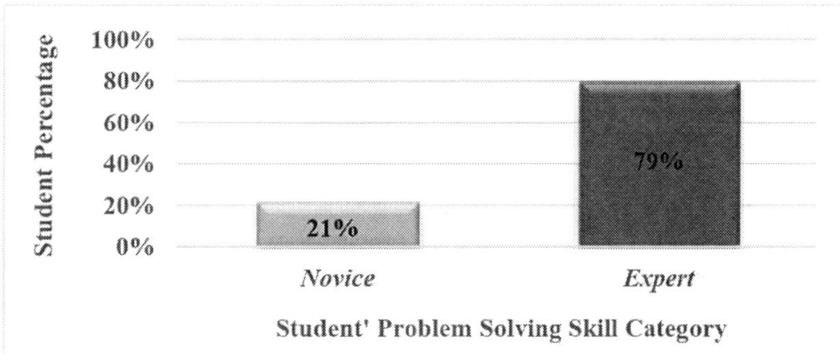

Figure 2. Percentage of Categories of Problem-Solving Skills for Novice dan Expert Students

Figure 2 shows, as many as 79% of students have been categorized as experts before all students were in the novice category. This shows, more and more students have better problem-solving skills, but there are also skills of students who are not appropriate in solving problems after being given learning. This is supported by the N-gain data of 0.63 which is in the medium category [32] and the effect size of 3.79 which is included in the strong effect [33], so collaborative inquiry learning supplemented by formative e-assessment significantly influences students' problem-solving skills. The results of this study are in line with the findings made by Yuliati (2018) which states that through inquiry learning skill to solve problems increases in the experimental class by 64.99 which is higher than the ability of students in the control class at 54.67 [34].

This is following the theory which said that when collaborative inquiry learning is implemented, the cognitive conflict of the students is built to assist students in carrying out investigations of the problems presented [13]. The process of the investigation carried out can train students' skill to solve problems based on relevant concepts as a solution to the problem [35]. The learning helped train students' skills through a collaborative inquiry process in experiments with Hydrostatic Pressure, Pascal's Law, and Archimedes' Law, to prove the hypothesis of a problem based on the solution found [12]. Students are also trained to solve problems through formative e-assessment based on quiz and formative tests related to static fluid problems to determine the appropriate solution. If the student's solution is incorrect, students may immediately improve it based on the feedback that students receive immediately during the online test [9]. Also, students are trained to work on problem-solving questions to understand the process of the five problem-solving indicators [22].

Students' problem-solving skills after learning can also be determined based on the percentage of students in the novice and expert categories in each problem-solving indicator as shown in the form of a diagram as shown in Figure 3.

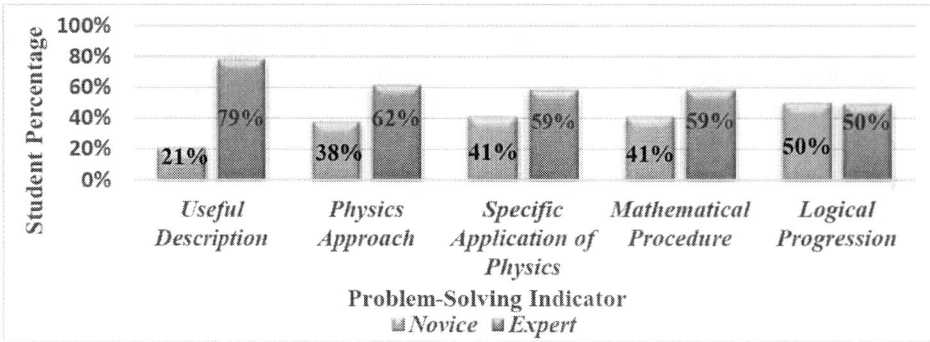

Figure 3. Percentage of Novice and Expert Student Categories on Each Problem Solving Indicator

Based on Figure 3 it can be seen that, after collaborative inquiry learning supplemented by formative e-assessment, the percentage of expert category students on each indicator of problem-solving has reached more than 50% of the number of students perviously only on indicators useful description. This shows, most students have been able to solve static fluid problems according to expert criteria for each problem-solving indicator [25, 28, 29, 30]. The highest problem-solving skills of students is in the first indicator, namely usefull description, which is that most students can describe the problem with the missing variables. The skill to describe good problems can help students determine problem-solving strategies [36–37]. The highest percentage of expert students is also on the physics approach indicator, which shows that most students can determine the appropriate physics concepts on the given problem so that students will help apply the principles of physics as a solution to problems [29, 38].

The percentage of expert students on indicators of the specific application of physics and mathematical procedures obtained is the same. This shows that most students can determine the right and complete equation. This shows that expert students identify variables that influence the problem given, connect with mathematical procedures and consider theories, concept, laws, principles that underlie problems, analyze, and solve with the right concept [3, 30, 39, 40]. Also, the logical progression indicator shows that the percentage of students in the expert and novice categories is the same. The expert student has carried out the process of solving the problem correctly, completely, and giving appropriate conclusions from the results obtained from the solution used. Meanwhile, the novice student does not understanding the meaning of the results obtained from the solutions that are applied. This shows that novice students still experience difficulties in the previous problem-solving process because they have not mastered the application of static fluid concepts to the problems of daily life. Therefore, the skill to solve students' static fluid problems after being given treatment is far better than before, because there are already expert students in each indicator that reaches a percentage above 50%.

Students' problem-solving skills after learning can be known more deeply based on the analysis of criteria for the answers of novice students and experts on the five indicators of problem-solving and strengthened from the results of interviews with two students from both categories. As for one example of the problem solving Hydrostatic Pressure as presented in Figure 4 and example of student answers categorized as novice and expert as presented in Table 3.

1. Carles, Doni, and Eko are students of SMA Negeri 4 Malang who are currently following the Environmental Recognition Study to Wisata Bahari Lamongan. They want to prove the law of hydrostatic principle when swimming. Carles and Doni decide to swim in the sea, while Eko prefers to swim in a swimming pool. After they finished swimming they gathered to discuss the depth they were swimming and the hydrostatic pressure they felt. (Remarks: Their depth swim is 3 meters, $\rho_{seawater} = 1{,}025\ kg/m^3$, and $\rho_{water} = 1{,}000\ kg/m^3$). Based on the data collected, analyze the relationship between density and fluid depth to hydrostatic pressure on the three students while swimming!

Figure 4. Problem Solving Hydrostatic Pressure Question

Table 3. Example of Novice and Expert Students's Answer of Hydrostatic Pressure Problem

No	Indicator	Kategori Kemampuan Pemecahan Masalah	
		Novice	*Expert*
1.	***Useful Description*** *N* = 12% *E* = 88%	Known: $\rho_{seawater} = 1.025\ kg/m^3$ $\rho_{water} = 1.000\ kg/m^3$ Their depth swim 3 meters	Known : $\rho_{seawater} = 1.025\ kg/m^3$ $\rho_{water} = 1.000\ kg/m^3$ Their depth is Carles, Doni, Eko, which is 3 meters Carles and Doni swim in the sea Eko swim in the swimming pool Asked : The relationship between density (ρ) and depth (h) of fluid to the phydrostatic pressure (ρ_h) of the three students?
2.	***Physics Approach*** *N* = 24% *E* = 76%	The deeper and wider the place, the greater the hydrostatic pressure	The concept of hydrostatic pressure is the greater the density of the fluid and the depth in the fluid, the greater the hydrostatic pressure. Because, the relationship is directly proportional.
3.	***Specific Application of Physics*** *N* = 26% *E* = 74%	Formula : $P = \rho\ .g\ .h$	The relationship between hydrostatic pressure is : $P \approx h\ ;\ P \approx \rho$ Then the equation used : $P = \rho\ .g\ .h$ At sea : $P_{\text{Carles}} = \rho_{\text{seawater}} \cdot g\ .h_{Carles}$ $P_{\text{Doni}} = \rho_{\text{seawater}} \cdot g\ .h_{Doni}$ At swimming pool : $P_{Eko} = \rho_{\text{air}} \cdot g\ .h_{Eko}$
4.	***Mathematical Procedures*** *N* = 21% *E* = 79%	$P_{water} = \rho_{\text{water}} \cdot g\ .h$ $= 1.000 \times 10 \times 3$ $= 30.000\ Pa$ $P_{\text{seawater}} = \rho_{seawater} \cdot g\ .h$ $= 1.025 \times 10 \times 3$ $= 30.750\ Pa$	Swim in the sea $P_{\text{Carles}} = \rho_{\text{seawater}} \cdot g\ .h_{Carles}$ $P_{\text{Carles}} = (1.025\ kg/m^3)(10\ m/s^2)(3\ m)$ $= 30.750\ Pa$ P_{Doni} is the same as P_{Carles} is $30.750\ Pa$, because h and its density are the same Swim in the swimming pool $P_{Eko} = \rho_{\text{water}} \cdot g\ .h_{Eko}$ $P_h = (1.000\ kg/m^3)(10\ m/s^2)(3\ m)$ $= 30.000\ Pa$
5.	***Logical Progression*** *N* = 41% *E* = 59%	$P_{\text{seawater}} = 30.750\ Pa$ dan $P_{seawater} = 30.000\ Pa$	The greater pressure experienced by Carles and Doni than Eko is because the density of seawater is greater than the density of water in the pool so that the relationship of hydrostatic pressure is directly proportional to the density.

Based on the data in Table 3, it can be seen that in the problem-solving indicators useful description students can describe the problem correctly and completely, only four students are still incomplete in describing it. Student difficulties are no longer found in this indicator, except that there are still a few shortcomings in some novice student answers. The lack of novice students found in the variable in questions in not written down, the units in the physics scale are also not included, and students have not applied the quantities known to the problem given. This result was also found in De Cock (2012) & Doctor (2015) 's study that, a small number of students in the novice category did not fully describe the problem because they did not write down some quantities and satan of the variables known and asked [41–42]. The second highest problem-solving skill of students is found in the fourth indicator, namely mathematical procedures. Inaccurate processes are still found in the novice category students, which only do calculations from known values without based on concepts or similarities. This can be seen in indicators of specific application of physics, namely novice students do not write down the equations that will be used as solutions and still write incorrect equations. The ability of some students is still based on mathematical knowledge, because novice students have not mastered physcis approaches and concepts well [43–45].

The problem-solving process of the students on the physics approach indicator has been done correctly namely, students use the concept of hydrostatic pressure related to the relationship of fluid density to the amount of hydrostatic pressure experienced by someone swimming in the sea and the pool at the same depth. This is confirmed by the opinions of expert students during interviews, namely the deeper the object is from the surface of the water, the greater the hydrostatic pressure it experiences. The same is true for the events of the three people who are swimming at the same depth, except that the difference in pressure is large because there is a different mass factor in the type of fluid. However, there are still students who are in the novice category who have not mastered the concepts correctly and have difficulties in determining students' solutions to problem-solving. Students still consider the volume of water to also affect the amount of hydrostatic pressure. This was also expressed by students at the interview who assumed that the hydrostatic pressure experienced by people swimming in the sea was of greater value than in the swimming pool because the volume of seawater was more. This is in line with previous research which found students' difficulties in explaining the magnitudes that affect hydrostatic pressure, so students consider the vessel area and fluid volume to be influential [43, 46]. However, student statements regarding the size of the vessel that affected hydrostatic pressure were not found in this study. Also, in the indicator of logical progression, the except student can explain the results obtained from the physics concepts used. However, the novice student only writes the hydrostatic pressure in the sea gretaer than the number in the pool based on the value obtained. The students did not associate the results obtained with the physics concept, so they were unable to prove the truth of the approach and application of physics used as a solution to the problem.

The results of these studies indicate, broadly speaking students already have better problem-solving skills, so they can solve complex problems in daily life [6, 39]. The students' skills are supported by collaboration between students when organizing important information in the form of symbols and physical quantities on the problems given [16]. Students who study collaboratively in small groups compared to individuals will have cognitive benefits. Students' knowledge grows through reciprocity of sometimes conflicting opinions and ideas or through learning from more experienced ones [15]. The feedback element through rapid formative e-assessment will help correct student conceptual errors that students use to solve physics problems [47, 48]. Therefore, collaborative inquiry learning supplemented by formative e-assessment can help train and improve students' skills in solving physics problems related to application of daily life.

4. Conclusion

The students' problem-solving skill on the subject of static fluid develops better after collaborative inquiry supplemented by formative e-assessment. About 79% of students have been categorized by expert problem solving skills. The data shows that, when students are given complex problems, more students have been able to analyze the factors that influence, determine, and prove the truth of the appropriate physics approach as a solution to solving problems. This research is recommended for further research that wants to analyze deeper problem-solving skills on the object of statis fluid especially the difficulties experienced by novice and students when solving problems.

References

[1] Argaw A S, "The Effect of Problem Based Learning (PBL) Instruction on Students' Motivation and Problem Solving Skills of Physics," *EURASIA Journal of Mathematics, Science and Technology Education*, vol. 13, no. 3, Mar 2017.

[2] Ceberio M, Almudí J M, and Franco A, "Design and Application of Interactive Simulations in Problem-Solving in University-Level Physics Education," *Journal of Science Education and Technology*, vol. 25, no. 4, hlm. 590–609, Agu 2016.

[3] Docktor J L and M. Mestre J P, "Synthesis of Discipline-Based Education Research in Physics," *Physical Review Special Topics - Physics Education Research*, vol. 10, no. 2, Sep 2014.

[4] Mansyur J, "Teachers' and Students' Preliminary Stages in Physics Problem Solving," *International Education Studies*, vol. 8, no. 9, Agu 2015.
[5] Bahar and Maker C J, "Cognitive Backgrounds of Problem Solving: A Comparison of Open-ended vs. Closed Mathematics Problems," *EURASIA Journal of Mathematics, Science & Technology Education*, Okt 2015.
[6] Wismath S L, "Shifting The Teacher-Learner Paradigm: Teaching for The 21 st Century," *College Teaching*, vol. 61, no. 3, hlm. 88–89, Jul 2013.
[7] Zuza K, Almudí J M, Leniz A, and J. Guisasola, "Addressing Students' Difficulties with Faraday's Law: A Guided Problem Solving Approach," *Physical Review Special Topics - Physics Education Research*, vol. 10, no. 1, Jun 2014.
[8] Chen Y, Irving P W, and Sayre E C, "Epistemic Game for Answer Making in Learning About Hydrostatics," *Physical Review Special Topics - Physics Education Research*, vol. 9, no. 1, 2013.
[9] Koes-H S, Muhardjito, and Wijaya C P, "Scaffolding for Solving Problem in Static Fluid: A Case Study," International Conference for Science Educators and Teachers (ISET), 2018, hlm. 030028.
[10] Datur I S, L. Yuliati, and N. Mufti, "Exploration of Physics Students' Problem Solving Skills in The Static Fluid," *National Conference Proceeeding of Science Pascasarjana UM*, 2016.
[11] Purwanto R and Yuliati L, "Exploration of High School Students' Problem Solving Skills in Static Fluid," *National Conference Proceeeding of Science Pascasarjana UM*, vol. 2, hlm. 7, 2017.
[12] Adams R, Aleong R, Goldstein M, and Solis F, "Rendering a Multi-Dimensional Problem Space as an Unfolding Collaborative Inquiry Process," *Design Studies*, Apr 2018.
[13] Bell T, Urhahne D, Schanze S, and Ploetzner R, "Collaborative Inquiry Learning: Models, Tools, and Challenges," *International Journal of Science Education*, vol. 32, no. 3, hlm. 349–377, 2010.
[14] Ontario Ministry of Education, "Achieving excellence: A renewed vision for education in Ontario," *ON: Queens Printer for Ontario*, 2014.
[15] Urhahne D, Schanze S, Bell T, Mansfield A, and Holmes J, "Role of the Teacher in Computer-supported Collaborative Inquiry Learning," *International Journal of Science Education*, 2009.
[16] Laal M and Khattami, "Teaching and Education; Collaborative Style," *Educational Research Review*, vol. 116, hlm. 4057–4061, Feb 2014.
[17] Nichols K, Gillies R, and Hedberg J, "Argumentation-Based Collaborative Inquiry in Science Through Representational Work: Impact on Primary Students' Representational Fluency," *Research in Science Education*, vol. 46, no. 3, hlm. 343–364, Jun 2016.
[18] Sipayung H D, Sani, and Bunawan, "Collaborative Inquiry For 4C Skills," *Proceedings of The 3rd Annual International Seminar on Transformative Education and Educational Leadership*, 2018.
[19] Gogri M H, Shaikh S A, and Iyengar V, "Evaluation of Students Performance Based on Formative Assessment Using Data Mining," *International Journal of Computer Applications*, 2013.
[20] Heitink M C, Van der Kleij V M, Veldkamp B P, Schildkamp K, and Kippers W B, "A Systematic Review of Prerequisites for Implementing Assessment for Learning in Classroom Practice," *Educational Research Review*, vol. 17, hlm. 50–62, Feb 2016.
[21] Kusairi S, Alfad H, and Zulaikah S, "Development of Web-Based Intelligent Tutoring (iTutor) to Help Students Learn Fluid Statics.," *Journal of Turkish Science Education (TUSED)*, 2017.
[22] Bhagat K K and Spector J M, "Formative Assessment in Complex Problem-Solving Domains: The Emerging Role of Assessment Technologies," *Journal of Educational Technology & Society*, vol. 20, no. 4, hlm. 312–317, 2017.
[23] Ediyanto, "Development of Web Based Formative Assessment Model to Improve Understanding of Student Physics Concepts, " *Journal of Science Education*: State University of Malang, 2014.
[24] Kusairi S, "Supporting Physcis Student Learning with Web-Based Assessment for Learning.," *National Conference and Reunion:Education in Nation Character Development*, 2014.
[25] Lin J W and Lai Y C, "Online Formative Assessments with Social Network Awareness," *Computers & Education*, vol. 66, hlm. 40–53, Agu 2013.
[26] Zou X and Zhang X, "Effect of Different Score Reports of Web-Based Formative Test on Students' Self-Regulated Learning," *Computers & Education*, vol. 66, hlm. 54–63, Agu 2013.

[27] Creswell J W and Clark V L V, *Designing And Conducting Mixed Methods Research*, 2ND Edition. United States of America: SAGE Publications, Inc, 2011.

[28] J. Docktor and K. Heller, "Robust assessment instrument for student problem solving," dalam *Proceedings of the NARST 2009 Annual Meeting, Garden Grove, CA*, 2009, hlm. 1–19.

[29] Docktor J L, Dornfeld J, Frodermann E, and Heller K, "Assessing Student Written Problem Solutions: A problem-solving rubric with application to introductory physics," *Physical Review Physics Education Research*, vol. 12, no. 1, Mei 2016.

[30] Hull M M, Kuo E, Gupta A, and Elby A, "Problem-Solving Rubrics Revisited: Attending to the blending of informal conceptual and formal mathematical reasoning," *Physical Review Special Topics - Physics Education Research*, vol. 9, no. 1, Feb 2013.

[31] Lin S Y and Singh C, "Effect of Scaffolding on Helping Introductory Physics Students Solve Quantitative Problems Involving Strong Alternative Conceptions," *Physical Review Special Topics - Physics Education Research*, vol. 11, no. 2, Agu 2015.

[32] Hake R R, "Interactive-Engagement Versus Traditional Methods: A sixthousand- student survey of mechanics test data for introductory physics courses," *American Journal of Physics*,1998.

[33] Tomczak M and Tomczak E, "The Need to Report Effect Size Estimates Revisited: An overview of some recommended measures of effect size," *Trends in Sport Sciences*, hlm. 21(1), 2014.

[34] Yuliati L, Parno, A. A. Hapsari, F. Nurhidayah, and L. Halim, "Building Scientific Literacy and Physics Problem Solving Skills through Inquiry-Based Learning for STEM Education," *Journal of Physics: Conference Series*, vol. 1108, hlm. 012026, Nov 2018.

[35] Saab N, van Joolingen, and van Hout-Wolters, "Support of The Collaborative Inquiry Learning Process: Influence of support on task and team regulation," *Metacognition and Learning*, vol2012.

[36] Ganina S and Soord Ly, "Opportunities for increasing learning efficiency through problem solving," *Physics Education*, vol. 4, hlm. 46, 2011.

[37] M. A. Murray and R. M. . Byrne, "Cognitive change in insight problem solving: Initial model errors and counterexamples," *Journal of Cognitive Psychology*, vol. 25, no. 2, hlm. 210–219, 2013.

[38] Yu K C, Fan S C, and Lin K Y, "Enhancing students' problem-solving skills through context-based learning," *International Journal of Science and Mathematics Education*, 2015.

[39] Adams W K and Wieman C E, "Analyzing The Many Skills Involved in Solving Complex Physics Problems," *American Journal of Physics*, vol. 83, no. 5, hlm. 459–467, Mei 2015.

[40] Leak A E, Rothwell S L, and Olivera J, "Examining Problem Solving in Physics-Intensive Ph.D. research," *Physical Review Physics Education Research*, vol. 13, no. 2, Jul 2017.

[41] De Cock M, "Representation Use and Strategy Choice in Physics Problem Solving," *Physical Review Special Topics - Physics Education Research*, vol. 8, no. 2, Nov 2012.

[42] Docktor J L, Strand N E, Mestre J P, and Ross B H, "Conceptual Problem Solving in High School Physics," *Physical Review Special Topics - Physics Education Research*, vol. 11, no. 2, Sep 2015.

[43] Kuczmann I, "The Structure of Knowledge and Students' Misconceptions in Physics," American Institute of Physics, 2017, hlm. 050001.

[44] Kurniawan B R, Handayanto S K, and Parno, "Profil of Graduate Student' Problem Solving Skills in The State University of Malang." National Conference Proceeding of Science Education Pascasarjana UM, 2016.

[45] Radovanović J and Sliško J, "Applying A Predict–Observe–Explain Sequence in Teaching of Buoyant Force," *Physics Education*, vol. 48, no. 1, hlm. 28, 2013.

[46] Yuliati L and Parno, "Exploration of physics problem-solving skills within phenomenon-based learning in senior high school students," International Conference on Education, 2018.

[47] Barana A and Marchisio M, "Ten Good Reasons to Adopt an Automated Formative Assessment Model for Learning and Teaching Mathematics and Scientific Disciplines," *Procedia - Social and Behavioral Sciences*, vol. 228, hlm. 608–613, Jul 2016.

[48] de-Miguel M, San-Fabián, and Belver, "Assessment of The Satisfaction of Participants in Training for Employment," *Electronic Journal of Educational Research, Assessment & Evaluation*, 2011.

Journal of Physics: Conference Series **1397** (2019) 012013 doi:10.1088/1742-6596/1397/1/012013

Developing Physics Comic Media a Local Wisdom: *Sulamanda* (*Engklek*) Traditional Game Chapter of Impulse and Momentum

Fita Permata Sari*, Syafridatun Nikmah, Heru Kuswanto, and Ratna Wardani

Graduate Program, Universitas Negeri Yogyakarta, Indonesia
Colombo Street No 1 Yogyakarta 55281, Indonesia

*Corresponding author e-mail: fitapermata.2017@student.uny.ac.id

Abstract. Comics have output values and great influence on the entertainment market in accordance with the times. Comics can also be used in the field of education and are not only used as an entertainment media. This research aims to produce the local wisdom physics comics media: *sulamanda* or *engklek* traditional game impulse and momentum chapter and to know the feasibility of the product. Research and Development (R & D) research uses the Borg & Gall model. The research began with preliminary research, research design, developing initial products, limited trials, limited revised trials, initial field trials, revisions to the initial field trials. The research instruments used included product assessment sheets, student response questionnaires, and test questions. The results of the study show that the local wisdom physics comics media: *sulamanda* or *engklek* traditional game is assisted by an android impulse and momentum chapter containing impulse and momentum material, learning videos, examples and problem exercises and can be operated using a smartphone or computer. Products included in the feasible category are used in the learning process of physics for students based on good categories from the expert judegement and this media is can be implemented in the learning process inside and outside the classroom.

1. Introduction

Comics, games, and animations have a big influence and the value of output in the media market is in accordance with the development of today's life [1]. Comics can be used in the field of education and are not only used as an entertainment tool [2]. Conversations in comics use language that can turn rigid science into simple science. Comics are very popular among teenagers because they feature humor, narrative, and visual imagery [3]. The benefits of comics today are not widely known including the effects on skills, and communication skills [4].

Students can experiment and make a comic to encourage the development of creativity and critical thinking [5]. Educators use comics to display students' creativity and intelligence. The story in comics needs a good plot, a strong character of development, and writing techniques to convey the message clearly and accurately [6].

Scientific literacy products to study science are worthy of being used as learning media. Educators are advised to conduct extensive experiments to examine their effectiveness in improving scientific literacy and increasing students' awareness of the problem of crisis in the environment [7]. The development of comics is needed to invite other students who have diverse abilities, because textbooks are less attractive

to middle-class students [8]. Readers prefer to read physics comics rather than reading physics textbooks [3].
Comics can be used as effective learning media at various levels of Education [9], [10]. Comics are introduced to elementary school students, high schools, and students to assess the ability of comics to increase their interest and understanding. Comics are expected to install curiosity and encourage students to learn more about their own subjects effectively [10]. Physics learning can balance the value of local wisdom to build the nation's character with a scientific attitude in knowledge of physics [11], [12]. Educators feel physics looks difficult and abstract for students. Students must be able to understand that physics plays a role in real life, so they will be literate in physics [13].
The values of local wisdom are maintained so that they continue to exist following the times and remain a hallmark of every region in Indonesia [14]. Local culture used in learning physics can change the rigid view of physics into "science for the future", "science for daily living", dan "science for all" [15]–[17]. *Sulamanda* or *engklek* traditional game contain problem solving values consisting of understanding problems, compiling problem solving strategies, and exploring solutions. These results can be followed up by compiling modules and applying them to students to improve children's problem solving skills through *Sulamanda* traditional games [18]. Physics includes learning that is important in various professions and fields regardless of gender [19], [20].
Local wisdom is important for connecting scientific theories and local conditions, so science can be accessed by the public. Local wisdom is used as a science learning model to build contextual learning, so the learning process becomes meaningful and easily accessible to students [15]. *Engklek* or also called sunda manda, sulamanda, sondah, ingkling, jlong jling, lempeng, or dampu are games commonly found in Sumatra, Java, Bali, Kalimantan and Sulawesi. *Sulamanda* traditional game is a traditional game that emerged during the Dutch era and is now rarely played. This game is played by 2 to 5 people and is played individually not a team. *Sulamanda* or *engklek* traditional game has the content of problem solving values [18]. *Sulamanda* traditional game contain several physics concepts, including: Motion Dynamics (parabolic motion and displacement), Sound Waves, Equilibrium of Tegar Objects (center of gravity), and Impulses and Momentum. Educators can use the media in the learning process in the classroom [21].

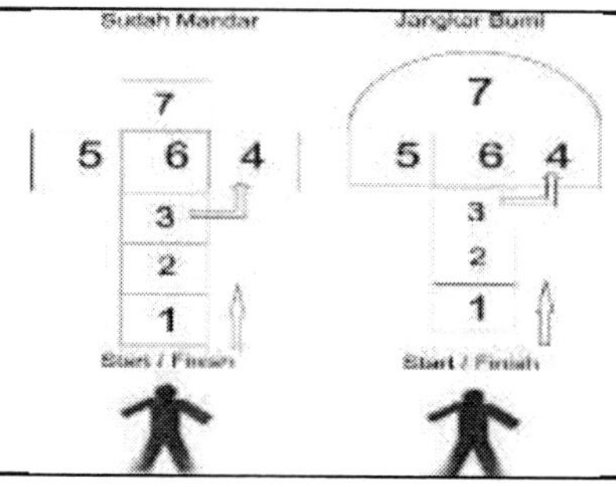

Figure 1. *Sulamanda* Traditional Game Plots

Based on the background that has been explained, there is a need for research on developing physics comics media of the local wisdom: *sulamanda (engklek)* traditional game in the chapter of impulse and momentum. These products can later be used in the learning process in the classroom or outside the classroom by the teacher to students.

2. Methodology
The type of research and development (R & D) research uses the development method of Borg & Gall. Research on the development of local wisdom physics comics media: *Sulamanda* traditional game of impulse and momentum chapter using the Borg & Gall model. Development of physics comics media of local wisdom: Sulamanda traditional game of impulse and momentum chapter starts from (1)

preliminary research, (2) research design, (3) developing initial products, (4) limited trials, (5) limited revised trials, (6) initial field trials, (7) revision of the initial field trials.
The first stage of product development is to analyze the physical aspects of local wisdom: *Sulamanda* traditional game. The first stage is finished, then draw characters, make a plot, storyboard, and combine stories with pictures for comics. Images are made manually using paper and pencil, the results of the description are scanned, the editing process uses Adobe Photoshop CS5 software. The finishing stage uses Sigil software in the form of *.epub files and the file can be accessed on Android using the Himawari Reader/Reasily-EPUB Reader application.

Table 1. Physics Concept of Local Wisdom: *Sulamanda* Traditional Game

Activities	Theory
When the gaco is motionless or moving	**1) Momentum** When the gaco has mass m in stationary condition ($v = 0$) or moves ($v \neq 0$). Gaco has momentum that indicates the state of motion of the engine. Momentum is a measure of difficulty in stopping the motion of an object. Momentum is influenced by mass and speed, these equations can be written like Equation 1. $p = mv$ (i)
When the gaco comes in contact with another gaco or sulamanda game plot within a certain interval of time	**2) Impulses** Force when the gaco touch each other with another gaco or Sulamanda traditional game plot in a short period of time. Players give an average impulsive force (F) to another gaco or game plot in a short interval of time (Δt). So that the impulse can be written like Equation 2. $I = F\Delta t$ (ii) Impulses are proportional to the impulsive force - average and time interval, the greater the impulsive force - the greater the impulse. Impulse is also a change of momentum, can be seen from changes in the velocity of the motion. The formula can be written like Equation 3. $I = \Delta p$ $I = m(v_2 - v_1)$ (iii)
When there is a gaco pounding another gaco or about another player.	**3) Collision** When the gaco hits another gaco, there will be a collision process between the gaco. In any collision process, momentum is always eternal as long as there is no external force acting (total zero external force). So as written in Equation 4. $m_1v_1 + m_2v_2 = m_1v_1' + m_2v_2'$ (iv) Here we consider the gaco to experience collisions each does not change. In the collision process, in addition to speed, the mass of each object before and after the impact can change. For example, after the collision, the two gaco join, or after the collision there is a broken gaco The collision process is divided into 3 types, namely: a) perfectly resilient collision ($e = 1$), b) partial resilient collisions ($0 < e < 1$), and

c) collisions are not resilient at all $(e = 0)$.

Where, e is the value of the restitution coefficient obtained from Equation 5.

$$e = \frac{-(v_1' - v_2')}{v_1 - v_2} \qquad \text{(v)}$$

[22]

Figure 2. The Process of Making Products with Paper and Pencil

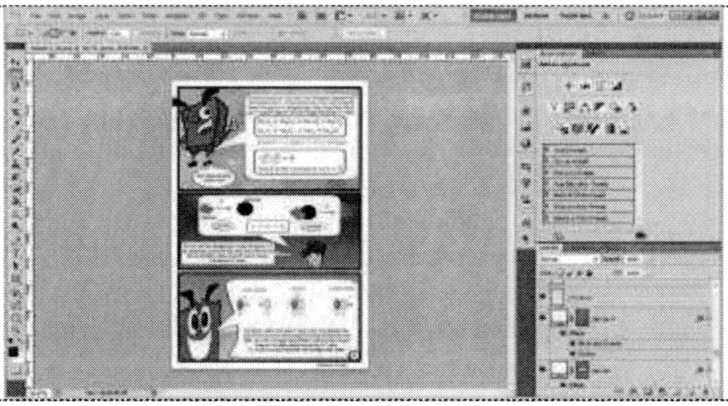

Figure 3. Product Making Process with Adobe Photoshop CS5 Software

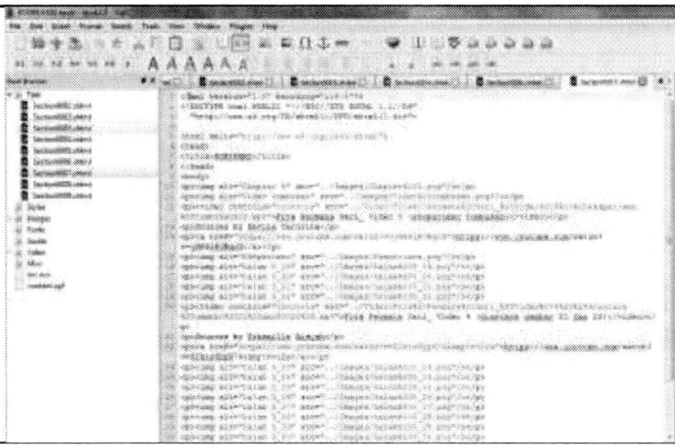

Figure 4. Display of Sigil on Product Creation

The product to be validated is analyzed using Aiken's V and calculated using the Aiken's V formula with product criteria such as Table 2.

$$V = \sum \frac{s}{[n(c-1)]}$$

s = r - lo
r = number given by the assessor
lo = lowest validity rating
n = number of assessors
c = highest validity rating

Table 2. Criteria for Assessment of Aiken's V Products

Score Range	Quality Category
$0{,}75 < V \leq 1{,}00$	Very Good
$0{,}50 < V \leq 0{,}75$	Good
$0{,}25 < V \leq 0{,}50$	Pretty Good
$0{,}00 < V \leq 0{,}25$	Not Good

Limited testing is the product validation stage to experts. The finished products are then validated by expert lecturers, namely media experts and theory experts, physics teachers, and peer reviewers. The task of the expert lecturer, physics teacher, and peer reviewer is to assess the product from the theory feasibility and presentation of the media. Products that have been validated receive an assessment and input to be used as a comic repair before the initial field trial is carried out. The initial field trial uses assessment tests and non-tests. Validation is done by using a validation sheet that has been validated by an expert lecturer.

3. Result and Disscusion

3.1. Result

The product trial in this study consisted of 2 stages, namely: limited trials (expert validation and practitioners) and initial field trials. This stage aims to produce a viable product. The data obtained consisted of data from the results of the validator's assessment and data from the results of the trial responses to students. Limited trials were carried out by material expert lecturers, media experts, physics teachers, and peer reviewers.

Table 3. Results of Product Assessment from Media Aspects

No	Rated Aspect	*Aiken V*	Information
1	Comic Design	1.00	Very Good
2	Visual Quality	0.88	Very Good
3	Content	0.83	Very Good
4	Organization, Language, and Readability	0.92	Very Good

Table 4. Results of Product Assessment from Material Aspects

No	Rated Aspect	*Aiken V*	Information
1	Presentation	0.88	Very Good

2	Coverage of Material	0.96	Very Good

Table 5. Product Assessment Results from Practical Aspects

No	Rated Aspect	*Aiken V*	Information
1	Perception of Ease of Use	0.90	Very Good
2	Usability felt	0.95	Very Good
3	Attitude	0.90	Very Good
4	Use of the actual product	0.90	Very Good

The results of data analysis are based on evaluations by media experts, material experts, physics teachers, and peer reviewers. Table 3, Table 4, Table 5 shows the results of the assessment of physics comics media of local wisdom: *Sulamanda* traditional game based on the aspects of media, learning materials, and practicality so that they are expressed in very good categories.
The initial field test was carried out to see and measure the implementation of local wisdom physics comics media: *Sulamanda* traditional game developed in the learning process, in the form of clarity, readability, effectiveness and usability in the impulse and momentum chapters.

Table 6. Results of Product Appraisal at Initial Field Trials

Theory	Media	Mean
3.75	4.71	4.07
Good	Very Good	Good

Evaluation of the feasibility of the local wisdom physics comics media: *Sulamanda* traditional games assisted by Android, obtained an average rating of 4.07 in the good category. Media of local wisdom physics comics: *Sulamanda* traditional games are concluded to be feasible to be used and continued in the main field trials. Media on local wisdom physics comics: *Sulamanda* traditional games produced can be seen in Figure 5.

Figure 5. Final Products

3.2. Disscusion

The product manufacturing phase starts from analyzing the physical aspects of local wisdom: *Sulamanda* or *Engklek* traditional game. Next draw characters, make grooves, storyboards, and combine stories with pictures for comics. Stories in comics need good plot, strong development character, and writing techniques to convey messages clearly and accurately [6]. Images are made manually using paper and pencil, the results of picture on paper are scanned, the editing process uses Adobe Photoshop CS5 software. The finishing stage uses Sigil software and produces the output of *.epub file. The file can be accessed on a smartphone using the Himawari Reader or Reasily-EPUB Reader or other EPUB Reader application or using a computer with an internet browser.

The physics comics media of local wisdom: *Sulamanda* or *Engklek* traditional game of impulse and momentum chapter has been said to be valid after going through a limited trial process with validation processes from expert lecturers, physics teachers, and peer reviewers. The Aiken V value from the media aspect includes comic design 1.00; visual quality 0.88; content 0.83; organization, language, and readability of 0.92. The results of the Aiken V assessment of material aspects include presentation and material coverage of 0.88 and 0.96. Whereas the results of the product assessment from the practical aspects include perceptions of ease, perceived usefulness, attitude, and product usage of 0.90; 0.95; 0.90; 0.90. So, the assessment category is very good. Assessments produced from the validation process by expert lecturers, physics teachers, and peer reviewers were generated to proceed to the field trial stage in high school students.

The physics comics media of local wisdom: *Sulamanda* or *Engklek* traditional game is said to be feasible after being applied to the initial field testing process for high school students. The results of product assessment in the initial field trial amounted to 4.07 from standard scale 1-5 with categories both in terms of material and media. Products included in the feasible category are used in the learning process of physics for students. This is in accordance with the results of previous studies from [10] that comic science education is an educational tool that is feasible to use in the field of education. The physics comics media of local wisdom: *Sulamanda* or *Engklek* traditional game of impulse and momentum

chapter can be used on smartphones with the Reasily-EPUB Reader application or using a computer with internet browser application.
Comics can be used as effective learning media at various levels of Education [9], [10]. Scientific literacy products to study science are worthy of being used as learning media [7]. The development of Android-based physics comic media is appropriate to be used in physics learning in class [23]. Conversations in comics use language that can turn rigid science into simple science so students can understand it [3]. Physics comics media is can be implemented in the learning process inside and outside the classroom.

4. Conclussion

The results showed that the physics comics media of local wisdom: *Sulamanda* (*Engklek*) traditional game of impulse and momentum chapter is a media that is operated using a smartphone with the application EPUB Reader or by using a computer with an internet browser. Products included in the feasible category are used in the learning process of physics for students based on good categories and this media is can be implemented in the learning process inside and outside the classroom.

5. References

[1] J. J. Sheu and K. T. Chu, "Mining association rules between positive word-of-mouth on social network sites and consumer acceptance: A study for derivative product of animations, comics, and games," *Telemat. Informatics*, vol. 34, no. 4, pp. 22–33, 2017.
[2] P. D. Widyastuti, M. Mardiyana, and D. R. S. Saputro, "An instructional media using comics on the systems of linear equation an instructional media using comics on the systems of linear equation," *J. Phys. Conf. Ser.*, 2017.
[3] S.-F. Lin, H. Lin, L. Lee, and L. D. Yore, "Are science comics a good medium for science communication? the case for public learning of nanotechnology," *Int. J. Sci. Educ. Part B*, vol. 5, no. 3, pp. 1–19, Jul. 2015.
[4] C. S. Babaian and A. A. Chalian, "'The thyroidectomy story': Comic books, graphic novels, and the novel approach to teaching head and neck surgery through the genre of the comic book," *J. Surg. Educ.*, vol. 71, no. 3, pp. 413–418, 2014.
[5] J. L. P. Ribeiro, "A plane mirror experiment inspired by a comic strip," *Phys. Educ.*, vol. 51, no. 1, 2016.
[6] M. Bitz and O. Emejulu, "Creating comic books in Nigeria: International reflections on literacy, creativity, and student engagement," *J. Adolesc. Adult Lit.*, vol. 59, no. 4, pp. 431–441, 2016.
[7] R. Aisyah, I. A. Zakiyah, I. Farida, and M. A. Ramdhani, "Learning Crude Oil by Using Scientific Literacy Comics," *J. Phys. Conf. Ser.*, pp. 1–7, 2017.
[8] S. F. Lin and H. S. Lin, "Learning nanotechnology with texts and comics: the impacts on students of different achievement levels," *Int. J. Sci. Educ.*, vol. 38, no. 8, pp. 1373–1391, 2016.
[9] D. di Fuccia, T. Witteck, S. Markic, and I. Eilks, "Trends in practical work in German science education," *Eurasia J. Math. Sci. Technol. Educ.*, vol. 8, no. 1, pp. 59–72, 2012.
[10] J. Kim, M. S. Chung, H. G. Jang, and B. S. Chung, "The use of educational comics in learning anatomy among multiple student groups," *Anat. Sci. Educ.*, vol. 10, no. 1, pp. 1–8, 2017.
[11] M. N. Sholakhudin, Sutarto, and Subiki, "Paket Sumber Belajar (PSB) dengan Analisis Foto Kejadian Fisika (AFKF) berbasis kearifan lokal pada pembelajaran fisika di SMK (Kajian pengembangan pada pokok bahasan fluida untuk SMK jurusan perikanan dan kelautan)," *J. Pembelajaran Fis.*, vol. 5, no. 3, pp. 253–260, 2016.
[12] I. W. Suastra and L. P. B. Yasmini, "Model pembelajaran fisika untuk mengembangkan kreativitas berpikir dan karakter bangsa berbasis kearifan lokal Bali," *J. Pendidik. Indones.*, vol. 2, no. 2, pp. 221–235, 2013.
[13] P. T. Oon and R. Subramaniam, "On the declining interest in physics among students-from the perspective of teachers," *Int. J. Sci. Educ.*, vol. 33, no. 5, pp. 727–746, 2011.

[14] U. Fajarini, "Peranan kearifan lokal dalam pendidikan karakter," *Sosio Didakt. Soc. Sci. Educ. J.*, vol. 1, no. 2, pp. 123–130, 2014.
[15] A. Anwari, M. S. Nahdi, and E. Sulistyowati, "Biological science learning model based on Turgo's local wisdom on managing biodiversity," *AIP Conf. Proc.*, vol. 1708, pp. 1–6, 2016.
[16] I. Meliono, "Understanding the Nusantara thought and local wisdom as an aspect of the Indonesian education," *TAWARIKH Int. J. Hist. Stud.*, vol. 2, no. 2, pp. 221–234, 2011.
[17] I. P. W. Y. Pramadi, I. W. Suastra, and I. M. Candiasa, "Pengaruh penggunaan komik berorientasi kearifan lokal bali terhadap motivasi bePramadi, I. P. W. Y., Suastra, I. W., & Candiasa, I. M. (2013). Pengaruh penggunaan komik berorientasi kearifan lokal bali terhadap motivasi belajar dan pemahaman konsep fisika," *e-Journal Progr. Pascasarj. Univ. Pendidik. Ganesha*, vol. 3, 2013.
[18] Iswinarti, "Nilai-nilai problem solving permainan tradisional engklek," *Semin. Nas. dan Gelar Prod.*, no. October 2016, pp. 1–9, 2017.
[19] M. Baran, "An analysis on high school students' perceptions of physics courses in terms of gender (A sample from Turkey)," *J. Educ. Train. Stud.*, vol. 4, no. 3, pp. 150–160, 2016.
[20] S. Bates, R. Donnelly, C. Macphee, D. Sands, M. Birch, and N. R. Walet, "Gender differences in conceptual understanding of Newtonian mechanics: A UK cross-institution comparison," *Eur. J. Phys.*, vol. 34, no. 2, pp. 421–434, 2013.
[21] Darmawan, D. (2016). *Mobile learning* sebuah aplikasi teknologi pembelajaran. Jakarta: Rajawali Press
[22] Abdullah, M. (2016). Fisika Dasar 1. Bandung: Institut Teknologi Bandung.
[23] F. P. Sari, L. Ratnaningtyas, I. Wilujeng, Jumadi, and H. Kuswanto, "Development of android comics media on thermodynamic experiment to map the science process skill for senior high school," *J. Phys. Conf. Ser.*, no. 012052, pp. 0–9, 2019.

Journal of Physics: Conference Series **1397** (2019) 012014 doi:10.1088/1742-6596/1397/1/012014

Synthesis of the Cognitive Aspects' Science Literacy and Higher Order Thinking Skills (HOTS) in Chapter Momentum and Impulse

Himawan Putranta[*] and Supahar

Department of Physics Education, Graduate School of Yogyakarta State University, Colombo Street No 1, Yogyakarta 55281, Indonesia

[*]Corresponding author e-mail: himawanputranta.2017@student.uny.ac.id

Abstract. This paper is based on the background of the problem in the form of low order thinking skills and science literacy in students. The science literacy that students have has an important role in developing higher-order thinking skills (HOTS) in solving problems in everyday life. The method used in this research is descriptive qualitative research method with literature studies in the form of primary sources (books) and secondary sources (scientific articles). The results of this study obtained a synthesis of cognitive aspects of science literacy and HOTS in momentum and impulses used as a guide in developing overall learning devices combined with physical problems that occur in real life experienced by students in a realistic manner. In addition, the synthesis of the cognitive aspects of science literacy is in the form of several important aspects, such as scientific knowledge and scientific competence. The main aspects in HOTS are discussed in the form of analysis, evaluate, and create. These aspects must be actualized by students in their lives in the present and in the future that are useful for mutual benefit.

1. Introduction

One of the goals achieved in learning activities in schools is to create intelligent and skilled people in various fields. Therefore, students at the secondary school order are directed to achieve maximum educational goals in terms of developing higher-order thinking skills (HOTS) by studying every problem that occurs in everyday life using appropriate physics concepts/science literacy [1]. Meanwhile, the learning objectives can be achieved well, if in carrying out learning activities combined with problems that occur in life that are packaged thoroughly in a learning device [2,3]. Learning tools that combine problems that occur in everyday life must be developed with careful planning so that the implementation of learning runs smoothly and in particular so that aspects of science literacy and HOTS of students can be measured properly [4]. Therefore, with the aim of learning physics and learning devices combined with problems that occur in everyday life, students are expected to be able to optimize science literacy and HOTS. Regarding science literacy and HOTS are two of the most basic things students must have in learning physics. This is because science literacy is the ability of students to solve problems that occur in their lives using appropriate physical concepts [5,6].

If students have understood the concept of physics appropriately, it is possible for students to solve various problems in life from the easiest to the most difficult. After students are able to solve various problems in life using physics concepts from the easiest to the most difficult, they can develop shorter and easier ways to solve problems using HOTS [7,8]. HOTS are basically skills that arise because they often solve problems from the easiest to the most difficult [9]. Therefore, one effort that needs to be done to improve science literacy and HOTS is often practice solving physics problems in everyday life from simple problems to complex problems. Therefore, in general, the achievement of the results of physics learning for Indonesian secondary school students in the international arena in terms of science literacy and HOTS is still low [10]. In general, the achievement of the results of studying physics in Indonesian high school students in the international arena in terms of science literacy and HOTS is still low [10]. Furthermore, based on the achievement of 2012 PISA science literacy from 65 countries, Indonesia is ranked second from the bottom with a science literacy score on average students 382 below the PISA average score, i.e 501 [11]. [12] states that the low reading and writing skills of Indonesian students are influenced by many things, including the education curriculum, the selection of learning methods and models, learning resources, and teaching materials.

Furthermore, the achievement of low science literacy can be caused by physics learning activities or inappropriate learning tools [13]. Thus, it needs to be evaluated in learning that prioritizes solving problems in everyday life so that science literacy of students increases. In addition, [14] also states that most middle school students have not been able to find the right solution or find it difficult to solve physics problems correctly, so that their HOTS are still quite low. Therefore, changes in student-centered physics learning activities are needed, so as to improve science literacy and HOTS of students [15]. In addition, the application of learner-centered learning can be done with a variety of innovations, both innovation in learning media, learning devices, or assessment instruments. Meanwhile, the physics chapter used in the synthesis analysis of cognitive aspects of science literacy and HOTS is physics in the scope of mechanics, i.e momentum and impulse. This chapter is adapted to the results of research which states that the momentum and impulse is one of the material in physics subjects that still often occurs in student errors and is one of the physics chapters that are not easy [16]. In this study only limited to the cognitive domain because it is in accordance with [17,18] which states that someone who has good science literacy skills, especially in the aspect of knowledge, will have good HOTS. Therefore, a new author's synthesis or view of cognitive aspects in science literacy and aspects of HOTS in the momentum and impulse can be used as a basis for the development of learning tools or assessment instruments.

2. Methods

The method used in this study is descriptive qualitative method, so that in this paper the data discussed is a synthesis of several primary and secondary sources of scientific books and articles which each of the major publications after 2010 from qualitative data about cognitive aspects of science literacy and HOTS. Qualitative descriptive method itself is a research method that analyzes data in the form of qualitative information which is then connected with other data to obtain clarity about the truth, so that new views or views are obtained by combining existing views [19,20]. In this study, the new points to be discussed are new views on the cognitive aspects of science literacy and HOTS in momentum and impulse. After the data from primary and secondary sources are analyzed, the next step is to conclude the data in general so that new views or new syntheses can be obtained about the cognitive aspects of science literacy and HOTS which can be shown in Figure 1.

Journal of Physics: Conference Series **1397** (2019) 012014 doi:10.1088/1742-6596/1397/1/012014

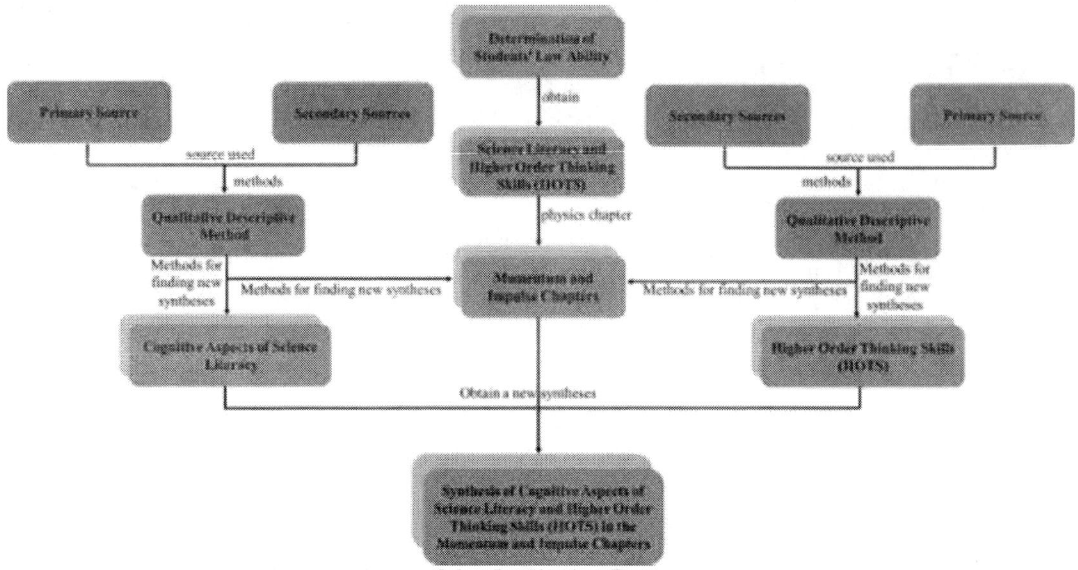

Figure 1. Steps of the Qualitative Descriptive Methods

3. Results and disscusion

3.1. The Concept of Cognitive Aspect' Science Literacy

In accordance with what has been discussed in the background of the problem, taking into account some weaknesses in the abilities possessed by senior high school students in Indonesia, science literacy and HOTS are selected, which are also related to each other. Furthermore, science literacy is one of the important abilities students must have in learning. According to OECD states that the Program for International Student Assessment or PISA defines that science literacy is the ability of students to use scientific knowledge, identify problems and explain scientific facts in order to understand regarding natural phenomena [21]. According to Greenhow, Gibbins, and Menzer that science literacy is an understanding of the concepts and scientific processes needed for making decision [22]. In addition, literacy skills in science are also the ability to understand the nature of science and scientific development; understand and use science intelligently; use scientific processes to solve problems; and understand the relationship between science and technology [23]. The most important thing in developing science literacy of students includes knowledge of science, science processes, development of scientific attitudes, and understanding of students towards science [24].

Someone who has science literacy is someone who uses science concepts and has science process skills to be able to judge in making everyday decisions [25]. Furthermore, students who have science literacy is characterized by the ability to solve problems by using scientific concepts [26]. Science literacy can also be divided into four major dimensions, i.e the dimensions of the science context, science content, science processes, and the dimensions of scientific attitude [27]. The science context is recognizing everyday life situations involving science and technology, and science content is understanding natural phenomena and changes [28]. Meanwhile, the science process is the ability to use understanding and scientific knowledge to solve various kinds of problems in everyday life, the attitude of science is a marker of interest in science and the motivation to act responsibly [29]. In addition, PISA also suggests that there are three competencies for measuring science literacy skills, i.e identifying scientific issues, explaining scientific phenomena, and using scientific evidence [30]. Identifying scientific issues requires students to recognize scientific problems

[31]. Using scientific evidence requires students to interpret evidence in drawing conclusions and communicating [32]. Meanwhile, in 2006 the aspects measured in PISA included scientific knowledge, scientific competencies, context, and attitudes [33]. In PISA 2015, science literacy is also divided into four aspects of assessment, covering aspects of the context, aspects of scientific knowledge, scientific competencies, and aspects of attitudes [34]. The context aspect focuses on personal, local and global related situations [35].

The scientific aspect of knowledge refers to an understanding of facts, concepts, and theories that form the basis of scientific knowledge [34]. The scientific aspects of knowledge include content knowledge is the knowledge needed to understand natural phenomena; procedural knowledge is the knowledge needed to form various kinds of ideas about science; and epistemic knowledge is knowledge needed to rationally understand the various things that underlie the occurrence of [28]. Furthermore, the scientific aspects of competencies are the ability to apply knowledge in accordance with scientific phenomena [36]. Meanwhile, aspects of attitudes include interest in science and technology [37]. Based on the description of several experts regarding science literacy, it can be concluded that science literacy is the ability to use the knowledge and understanding possessed as a basis for solving problems in daily life scientifically. Science literacy can also be said as the ability of students to understand science as a whole in everyday life. Therefore, students are expected to be able to apply the knowledge gained during classroom learning to be applied in daily life so that students can have sensitivity and concern for their surrounding environment.

3.2 The Concept of Higher Order Thinking Skills (HOTS)

HOTS is the ability to use the mind more broadly to find new challenges [38]. HOTS is also a thought process that involves mental activities that are carried out to gain knowledge which includes thinking of analyzing, evaluating, and creating [39]. HOTS require someone to apply new information or prior knowledge to reach possible answers in new situations [40]. Furthermore, HOTS is a concept of change in the world of education based on Bloom's Taxonomy [41]. Bloom's Taxonomy is a classification of educational goals in behaviors that emphasize intellectual aspects such as knowledge, understanding, and is the order of one's thinking ability [42]. Meanwhile, Anderson and Krathwohl revised Bloom's cognitive taxonomy, from what was originally composed of verbs (knowledge, understanding, application, analysis, synthesis, and evaluation) replaced into verbs (remembering, understanding, applying, analyze, evaluate, and create) [43].

The cognitive domain in the revised Bloom taxonomy is also the basis HOTS with thoughts based on several types of learning that require more cognitive processes than others [44]. Furthermore, in Bloom's taxonomy the ability to analyze, evaluate, and create is considered part of HOTS [45]. Based on Bloom's revised taxonomy, cognitive processes are divided into Lower Order Thinking and Higher Order Thinking [43]. Meanwhile, analyzing is a problem-solving ability by looking for interrelationships from each part [43]. The ability to analyze is a type of ability that is much demanded by students in learning [46]. Demands for students to have the ability to analyze often tend to be more important than other dimensions of cognitive processes [47]. Evaluating includes checking and criticizing activities [43]. Checking leads to testing activities that are inconsistent or failures of an operation or product [48]. When associated with the thought process of planning and implementing, checking will lead to the determination of the extent to which a plan runs well [49]. Meanwhile, criticizing leads to the assessment of a product based on predetermined criteria [50]. Creating includes activities to generalize and produce things [43]. Generalizing is an activity representing problems and finding alternative hypotheses that are needed [47].

Meanwhile, production leads to planning to solve problems [49]. The difference in creating with other dimensions of cognitive thinking is in other dimensions students working with known information, while creating students working and producing something new [51]. Furthermore, details of indicators to measure HOTS such as analyzing, in the form of analyzing information that enters and divides information in smaller parts [52]. Evaluate, in the form of evaluating solutions and methodologies using appropriate criteria [53]. Meanwhile, creating is in the form of generalizing the perspective on things and organizing things into new structures that have never [54]. Thus, based on some expert opinions on HOTS, it can be concluded that HOTS is thinking at a higher order. However, high-order thinking ability is the ability to connect, manipulate, and transform the knowledge and experience that is already possessed to think critically and creatively in an effort to solve problems in new situations. This is so that students are able to solve various kinds of scientific problems that require a higher cognitive process than scientific problems that they usually solve especially the problems of physics in everyday life.

3.3 Student Problems in the Momentum and Impulse

Momentum and impulse is one of the chapter in physics subjects that still often occurs in students' mistakes and is one of the physics chapter that are not easy [16]. In addition, the concept of momentum and impulse is included in concepts whose phenomena tend to be abstract because this concept cannot be observed with the naked eye. This is in line with the opinion of Sekercioglu and Kocakula which states that the concept of momentum and impulse is considered simple even though it is actually a complex concept [55]. Meanwhile, a study was conducted by Ivowi who examined misconceptions in physics from 128 students from two secondary schools in Nigeria [56]. He asked students to explain clearly about conservation of momentum and about half of the sample gave the wrong answer to the question of momentum conservation [56]. In the study, Ivowi revealed that (although the concept of momentum relates to mass and speed) students associate conservation of momentum erroneously only with the concept of speed.

In another study, Graham and Berry examined students' understanding of momentum, the result is that students' ability to gain momentum can be grouped into four categories, i.e the first they are confused with the concept of momentum. The second is those who do not know the relationship between momentum and impulse and momentum conservation law. The third is those who can understand momentum as a vector quantity. The last group is students who fully understand the concept of momentum [57]. Furthermore, Dalaklioğlu, Demirci, and Şekercioğlu stated in their research that high school students tend to have many misunderstandings related to momentum, impulses, and energy concepts. For example, students experience difficulties in connecting between momentum and energy conservation in real life [58]. The research conducted by İngeç, Ünlü and Taşar to investigate students 'conceptual understanding of momentum, states that the order of understanding students' concepts of momentum is much higher without the need for understanding the concepts of mass and speed [59]. Singh and Rosengrant investigated students' understanding of the concepts of energy and momentum and obtained results that most students had difficulty interpreting conceptually the basic principles relating to energy and momentum [60].

Furthermore, Güneş conducted a study of the order of students' ability to explain the concepts of impulses and momentum that they knew firsthand, the result was that only one-third of students were able to correctly explain the concept of momentum [60]. In addition, some students feel very confused about the concept of momentum and impulse. Students' misconceptions in the impulse and momentum that often occur are that students think that impulse is equal to momentum. Thus, based on some results of research conducted by experts on understanding students' concepts of momentum and impulse, it can be concluded that momentum and impulse is one part of mechanics which is abstract and difficult for students to understand. In addition, students also often experience errors in understanding the concepts of momentum, changes in momentum, and conservation of momentum.

3.4 Results of the Synthesis of Cognitive Aspects' Science Literacy and HOTS in Chapter Momentum and Impulse

Based on the opinions of several experts on the concept of science literacy, especially the cognitive and HOTS aspects, students need to have the ability to use and understand science, especially in momentum and impulse as a basis for solving problems in daily life scientifically with HOTS. In addition, the cognitive aspects of science literacy and indicators in HOTS used in this study are science knowledege and science competencies; analysis, evaluate and create. Meanwhile, the opinions of some experts on the concept of science literacy are mainly cognitive aspects and Higher Order Thinking Skills (HOTS), hence the overall synthesis of this research can be shown in table 1.

Tabel 1. Synthesis of cognitive aspects' science literacy and higher order thinking skills (HOTS) in chapter momentum and impulse

Aspect	Reference	Expert	Expert' Opinion or Research	
The Concept of Cognitive Aspect' Science Literacy	Book	OECD	The ability of students to use scientific knowledge, identify problems, and explain scientific facts.	**Synthesis of Cognitive Aspects Science Literacy** Science literacy is the ability to use and understand the science that is owned as a basis for solving problems in daily life scientifically. In addition, there are four cognitive aspects in science literacy, i.e scientific knowledge, scientific competence, context, and attitude.
		Spellman and Bayer	The most important things in science literacy are knowledge of science, the process of science, the development of scientific attitudes, and students' understanding of science.	
		Douglas, Klentschy, Worth, and Binder	Someone who has science literacy is someone who uses science concepts and has science process skills in making everyday decisions.	
		OECD	Divided into four main dimensions, i.e the dimensions of the science context, science content, science processes, and scientific attitude dimensions.	
		OECD	Three scientific competencies in science literacy, i.e identifying scientific problems, explaining scientific phenomena, and using scientific evidence.	
		Lau	PISA 2006, aspects of science literacy; scientific knowledge, scientific competence, context, and attitude.	
		OECD	PISA 2015, science literacy is divided into four aspects of assessment, i.e, aspects of context, scientific knowledge, scientific competence, and attitude.	
	Scientific articles	Greenhow, Gibbins, and Menzer	Understanding of scientific concepts and processes needed for personal decision making.	
		Zen and Secken	The ability to understand the nature of science and scientific development, understand and use science intelligently, and also use scientific processes to solve problems.	

Aspect	Reference	Expert	Expert' Opinion or Research	
		Olander	Students who have science literacy are able to solve problems using scientific concepts, are able to use and maintain technological products, and be creative in making technological innovations.	
The Concept of Higher Order Thinking Skills (HOTS)	Book	Arwood	The process of thinking is carried out to acquire knowledge that includes thinking of analyzing, evaluating, and creating.	**Synthesis of Higher Order Thinking Skills (HOTS)** HOTS are thinking at a higher order than just memorizing and repeating information that is already known. In addition, important indicators in HOTS are analysis, evaluate, and create.
		Conklin and Williams	The ability to analyze, evaluate, and create is part of HOTS.	
		Anderson and Krathwohl	Capabilities that include HOTS are the ability to analyze, evaluate, and create.	
		Anderson and Krathwohl	Bloom's cognitive taxonomy is revised to remember, understand, apply, analyze, evaluate, and create.	
	Scientific articles	Budsankom, Sawangboon, Damrongpanit, and Chuensirimongkol	HOTS are carried out by applying new information or prior knowledge by manipulating information to achieve possible answers in new situations.	
		Jensen, McDaniel, Woodard, and Kummer	The concept of change in the world of education is based on Bloom's taxonomy.	
		Bagarukayo	Bloom's Taxonomy is a classification of educational goals that emphasizes intellectual aspects and is a sequence of one's thinking abilities.	
		Bolotin and Nashon	The ability to use the mind more broadly to find new challenges.	
		Vijayaratnam,	Cognitive domains in the revised Bloom taxonomy are the basis for HOTS that require more cognitive processes.	
Student Problems in the Momentum and Impulse Chapters	Scientific articles	McNeill, Lizotte, and Krajcik	Momentum and impulse is a matter of physics that still often misconception in student.	**Synthesis Student Problems in the Momentum and Impulse Chapters** Momentum and impulse is one part of mechanics that is abstract and difficult for students to understand. In addition, students also often experience errors in
		Sekercioglu and Kocakula	The concept of momentum and impulse is considered simple even though it is actually a complex concept.	
		Ivowi	Students associate the conservation of momentum erroneously, that is, only with the concept of speed.	
		Dalaklioğlu, Demirci, and Şekercioğlu	High school students tend to have many misconceptions related to	

Aspect	Reference	Expert	Expert' Opinion or Research	
			momentum, impulses, and concepts of energy.	understanding the concepts of momentum, changes in momentum, and conservation of momentum.
		Singh and Rosengrant	Most students have difficulty interpreting conceptually the basic principles relating to energy and momentum.	
		Güneş	Students think that impulses are equal to momentum and scalar quantities.	
Synthesis of Cognitive Aspects' Science Literacy and Higher Order Thinking Skills (HOTS) in Chapter Momentum and Impulse				
Students need to have the ability to use and understand science especially in momentum and impulse as a basis for solving problems in everyday life scientifically with higher order thinking. In addition, cognitive aspects in science literacy and indicators in HOTS used in this study are science and science competencies, respectively; analysis, evaluation and creation.				

Furthermore, as for the details of each cognitive aspect of science literacy combined with momentum and impulse chapter, it can be shown in Table 2.

Tabel 2. Operational indicators for the cognitive aspects of science literacy

Cognitive Aspects of Science Literacy		Operational Indicator
Aspects	**Sub Aspects**	
Science Knowledge	Content Knowledge	Identify physics events in everyday life related to momentum and impulse.
		Describe the formulation of the problem in accordance with physical events in everyday life about momentum and impulses.
		Arrange hypotheses in accordance with physics events in everyday life related to momentum and impulse.
	Procedural Knowledge	Determine the simulation or experiment title related to momentum and impulse.
		Determine variables related to simulation or experiment of momentum and impulse.
		Arrange the right work steps to do simulation or experiment of momentum and impulse.
	Empystemist Knowledge	Analyzing physical events that occur in simulations or experiment related to momentum and impulse.
Science Competence	Explain phenomena scientifically	Analyze other events in daily life that apply the concepts of momentum and impulse.
		Matching several solutions to problems in physics in everyday life that apply the concepts of momentum and impulse.
	Interpreting scientific data and evidence	Arrange the scheme or mechanism of other events in everyday life that apply the concepts of momentum and impulses.
	Evaluating scientific investigations	Arrange universal conclusions about momentum and impulse based on simulation or experiment results and the results of scientific investigations.

Furthermore, for details of each indicator of HOTS combined with momentum and impulses, it can be shown in Table 3.

Tabel 3. Operational indicators for HOTS

HOTS Aspect	HOTS Sub Aspect	Operational Indicator
Analysis	Distinguish	Analyze the amount of momentum carried out by an object. Analyze the amount of impulses on an object.
	Sort	Analyzing momentum is done by an object based on the law of conservation of momentum. Analyze amount of impulses on objects associated with changes in momentum.
	Give Special Characteristics	Analyze the momentum of a colliding object. Analyze the amount of impulses on objects that collide. Analyze the coefficient of restitution on a colliding object.
Evaluate	Check	Projecting the relationship between force and time changes in objects moving on the graph.
	Criticize	Predict an event based on the law of conservation of momentum. Predict events based on perfect resilience collisions. Predicting an event based on a collision is not resilient.
Create	Bring up the idea	Form images of phenomena in everyday life that apply the concept of momentum. Form images of phenomena in everyday life that apply the concept of impulses. Form images of phenomena in everyday life that apply the concept of collision.
	Planning	Planning the scheme of a phenomenon in life that applies momentum and impulse. Planning events in life that apply momentum and impulses by determining their physical quantities.
	producing	Make imitations of new events that apply momentum and impulse.

3. Conclusions

Based on the expert's view of the basic concepts of science literacy especially in the cognitive and HOTS aspects of momentum and impulses, it can be concluded that students need to have the ability to use and understand science, especially in momentum and impulses as a basis for solving problems everyday life scientifically with HOTS. There are four cognitive aspects in science literacy, i.e the context, scientific competencies, science knowledge, and attitudes, but in this study only scientific and scientific competencies are used. Meanwhile, there are three aspects in HOTS which are all used in this study, i.e analysis, evaluate, and create.

4. Acknowledgments

The authors thanks to the guidance lecturers, Postgraduate Program of Yogyakarta State University, and research institutes and community service, UNY due to the funding of DIPA UNY 2019 for this research.

References

[1] Minister of Education and Culture of the Republic of Indonesia. (2013). Minister of Education and Culture Regulation Number 65, 2013, concerning Basic and Secondary Education Process Standards.

[2] Newcombe, N. S., & Shipley, T. F. (2014). Studying visual and spatial reasoning for design creativity: Thinking about spatial thinking. Dordrecht: Springer.

[3] Loewenthal, K., & Lewis, C. A. (2018). *An introduction to psychological tests and scales*. London: Psychology Press.

[4] Aizikovitsh-Udi, E., & Cheng, D. (2015). Developing critical thinking skills from dispositions to abilities: Mathematics education from early childhood to high school. *Creative Education, 6*(1), 455-462.

[5] Kurniawati, E., Hartanto, & Zamzaili. (2017). The effect of assurance, relevance, interest, assessment, satisfaction (ARIAS) integrative and early ability model in improving the understanding of mathematical concept and troubleshooting the first middle school students in Kepahiang. *Jurnal*

Pendidikan Matematika Raflesia, 2(2), 174-187.
[6] Mądrala, M., Wąsik, M., & Małoszewski, P. (2017). Interpretation of environmental tracer data for conceptual understanding of groundwater flow: An application for fractured aquifer systems in the Kłodzko Basin, Sudetes, Poland. *Isotopes in Environmental and Health Studies, 2*(1), 1-19.
[7] Furberg, A. (2016). Teacher support in computer-supported lab work: Bridging the gap between lab experiments and students' conceptual understanding. *International Journal of Computer-Supported Collaborative Learning, 2*(1), 1-25.
[8] Sadiqin, I. K., Santoso, U. T., & Sholahuddin, A. (2017). Understanding the natural science concept of middle school students through learning problem solving on topics of changing objects around us. *Jurnal Inovasi Pendidikan IPA, 3*(1), 52-62.
[9] Kiryak, Z., & Çalik, M. (2017). Improving grade 7 students' conceptual understanding of water pollution via common knowledge construction model. *International Journal of Science and Mathematics Education, 1*(1), 1-22.
[10] Wibowo, F. C., & Suhandi, A. (2013). Application of project-based science creative learning (SCL) model for improving cognitive learning outcomes and creative thinking skills. *Jurnal Pendidikan IPA Indonesia, 2*(1), 67-75.
[11] OECD. (2014). PISA 2012 result in focus why 15-year-old know and what they can do with what they know. Paris: OECD Publishing.
[12] Hume, A., & Coll, R. (2014). Student experiences of carrying out a practical science investigation under direction. *International Journal of Science Education, 30*(9), 1201-1228.
[13] Rahmatan, H., Liliasari, & Redjeki, S. (2012). Development of computer-based biochemical learning models to collect creative thinking skills of biological teacher students. *Jurnal Pendidikan IPA Indonesia, 1*(2), 178-182.
[14] Splitter, L. J. (2014). Critical thinking: What, why, when, and how. *Educational Philosophy And Teory, 23*(1), 89-109.
[15] Permanasari, A. (November 2011). Science learning: A potential vehicle for learning soft skills and characters. Paper presented at the National Science Education Seminar at the University of Lampung.
[16] McNeill, K., Lizotte, D. J., & Krajcik, J. (2016). Supporting student"s construction of scientific explanation by fading scaffolds in instructional materials. *The Journal of The Learning Science, 15* (2), 153 – 191.
[17] Rahayuni, G. (2016). Hubungan keterampilan berpikir kritis dan literasi sains pada pembelajaran IPA terpadu dengan model PBM dan STM. *Jurnal Penelitian dan Pembelajaran IPA, 2*(2), 131-146.
[18] Kardemir, E., & Ulucinar, U. (2017). Examining the relationship between middle school students' critical thinking skills, science literacy skills, and attitudes: A structural equation modelling. *Journal of Education in Science, Environment and Health (JESEH), 3*(1), 29-39.
[19] Neuman, W. L. (2011). Social research methods: Qualitative and quantitative approaches. Boston: Allyn & Bacon.
[20] Grbich, C. (2012). *Qualitative data analysis: An introduction*. London, England: Sage.
[21] OECD. (2016). PISA 2015 Assessment and Analytical Framework: Science, Reading, Mathematic and Financial Literacy. Paris: OECD Publishing.
[22] Greenhow, C., Gibbins, T., & Menzer, M. M. (2015). Re-thinking science literacy out-of-school: Arguing science issues in a niche Facebook application. *Computers in Human Behavior, 53*, 593-604.
[23] Zen, N., & Secken, N. (2015). Monthly chemistry newspaper in vocational high schools. *International Journal of Social Sciences and Education, 5*(4), 746-766.
[24] Spellman, F. R., & Bayer, J. P. (2011). *In defense of science: Why science literacy matters?* Maryland: The scarecrow press, Inc.
[25] Douglas, R., Klentschy, M. P., Worth, K., & Binder, W. (2006). *Linking science: In the K–8*

classroom. Arlington: National Science Teachers Association Press.

[26] Olander, C., Hajer, M., Johansson, S., Jakobsson, A., Karlsson, A., Kouns, M., Lyngfelt, A., & Larsson, P. N. (2017). Development of science literacy: A multidisciplinary study in multilingual secondary schools. *European Science Education Research Association Conference, Dublin City University, Dublin, Ireland*, 1-3.

[27] OECD. (2010). PISA 2009 result: What students know and can do, student performance in reading, mathematics, and science. Paris: OECD Publishing.

[28] Dori, Y. J., Mevarech, Z. R., & Baker, D. R. (2018). *Cognition, metacognition, and culture in STEM education: Learning, teaching and assessment*. Switzerland: Springer International Publishing AG.

[29] Romine, W. L., Sadler, T. D., & Kinslow, A. T. (2016). Assessment of science literacy: Development and validation of the quantitative assessment of socio-scientific reasoning (QuASSR). *Journal of Research in Science Teaching, 54*(1), 274-295.

[30] OECD. (2007). Executive summary PISA 2006: Science competencies for tomorrow's world. Paris: OECD Publishing.

[31] Benjamin, T. E., Marks, B., Demetrikopoulos, M. K., Rose, J., Pollard, E., Thomas, A., & Muldrow, L. L. (2017). Development and validation of science literacy scale for college preparedness in STEM with freshmen from diverse institutions. *International Journal of Science and Mathematics Education, 15*(4), 607-623.

[32] Bradshaw, R. E. A. (2015). Teaching for science literacy? An examination of instructional practices in secondary schools in Barbados. *Research in Science Education, 47*(1), 67-93.

[33] Lau, K. C. (2009). A critical examination of PISA's assessment on science literacy. *International Journal of Science and Mathematics Education, 7*(1), 1061-1088.

[34] Aisyah, R., Zakiyah, I. A., Farida, I., & Ramdhani, M. A. (2017). Learning crude oil by using science literacy comics. *Journal of Physics: Conference Series*, 895, 1-8.

[35] Steinbrink, C., Zimmer, K., Lachmann, T., Dirichs, M., & Kammer, T. (2014). Development of rapid temporal processing and its impact on literacy skills in primary school children. *Child Development, 85*(4), 1711–1726.

[36] Lindner, M., & Kubat, C. (2014). Science camps in Europe-collaboration with companies and school, implications and results on science literacy. *Science Education International, 25*(1), 79-85.

[37] Bolotin, M. M., & Nashon, S. M. (2011). The essence of student visual–spatial literacy and higher order thinking skills in undergraduate biology. *Protoplasma, 241*, 25-30.

[38] Arwood, E. L. (2011). Language function: An introduction to pragmatic assessment and intervention for higher order thinking and better literacy. London: Jessica Kingsley Publishers.

[39] Budsankom, P., Sawangboon, T., Damrongpanit, S., & Chuensirimongkol, J. (2015). Factors affecting higher order thinking skills of students: A meta-analytic structural equation modeling study. *Educational Research and Reviews, 10*(19), 2639-2652.

[40] Jensen, J. L., McDaniel, M. A., Woodard, S. M., & Kummer, T. A. (2014). Teaching to the test...or testing to teach: Exams requiring higher order thinking skills encourage greater conceptual understanding. *Educational Psychology Review, 26*(2), 307-329.

[41] Bagarukayo, E. (2012). An approach to learning by construction. International Journal of Education and Development using Information and Communication Technology (IJEDICT), 8(3), 43-61.

[42] Anderson, L. W., & Krathwohl, D. R. (2014). *Framework for learning, teaching and assessment: Revision of Bloom's taxonomy*. Yogyakarta: Pustaka Pelajar.

[43] Vijayaratnam, P. (2012). Developing higher order thinking skills and team commitment via group problem solving: A bridge to the real world. *Procedia-Social and Behavioral Sciences, 66*, 53-63.

[44] Conklin, W., & Williams, R. B. (2012). H*igher order thinking skills to develop 21st century learners*. California: Shell Educational Publishing, Inc.

[45] Yen, T. S., & Halili, S. H. (2015). Effective teaching of higher-order thinking (HOT) in education. *The Online Journal of Distance Education and e-Learning, 3*(2), 41-47.
[46] Ahraria, S., Samah, B. A., Hassan, M. S. B., Wahat, N. W. A., & Zaremohzzabieh, Z. (2016). Deepening critical thinking skills through civic engagementin Malaysian higher education. *Thinking Skills and Creativity, 22*, 121-128.
[47] Lee, S. M. (2014).The relationships between higher order thinking skills, cognitive density, and social presence in online learning. *Internet and Higher Education, 21*, 41-52.
[48] Chiu, T. K. F. (2016). Effects of prior knowledge on mathematics different order thinking skills in mobile multimedia environments. *Mobile Learning Design*, 373-386.
[49] Garcia, L. C. (2015). Environmental science issues for higher-order thinking Skills (HOTS) development: A case study in the Philippines. *Biology Education and Research in a Changing Planet.* 45-54.
[50] Simon, N. (2015). Improving higher-order learning and critical thinking skills using virtual and simulated science laboratory experiments. *Systems Sciences, and Engineering, 312*. 187-192.
[51] Brookhart, S. M. (2010). How to assess higher-order thinking skills in your classroom. Virginia: ASCD.
[52] Heong, Y. M., Yunos, J. M., Othman, W., Hassan, R., Kiong, T. T., & Mohamad, M. M. (2012). The needs analysis of learning higher order thinking skills for generating ideas. *Procedia-Social and Behavioral Sciences, 59*, 197-203.
[53] Madhuri, G. V., Kantamreddi, V. S. S. N., & Prakash, G. (2012). Promoting higher order thinking skills using inquiry-based learning. *European Journal of Engineering Education, 37*(2), 117-123.
[54] Sekercioglu, A. G., & Kocakula, M. S. (2008). Grade 10 students' misconceptions about impulse and momentum. *Journal of Turkish Science Education, 5*(2), 47-59.
[55] Ivowi, U., (1984). Misconceptions in physics amongst Nigerian secondary school students. *Physics Education, 19*, 279-285.
[56] Graham, T., & Berry, J. (1996). A hierarchical model of the development of student understanding of momentum. *International Journal of Science Education, 18*(1), 75-91.
[57] Dalaklioğlu, S., Demirci, N., & Şekercioğlu, A. (2015). Eleventh grade students' difficulties and misconceptions about energy and momentum concepts. *International Journal on New Trends in Education and Their Implications, 6*(1), 13-21.
[58] İngeç, Ş., Ünlü, P., & Taşar, M. F. (2003).Students identify a problem-pulse and lame applications. V. National Science and Mathematics Education Congress, Ankara.
[59] Singh, C., & Rosengrant, D. (2003). Multiple-choice test of energy and momentum concepts. *American Journal of Physics, 71*(6), 607-617.
[60] Güneş, P., İngeç, Ş., & Taşar, M. (2001). Öğretmen adaylarının açık uçlu sorularla momentum ve impulsu nasıl tanımladıklarının belirlenmesi. *Gazi Üniversitesi Eğitim Fakültesi Dergisi, 22*(3), 121-138.

Journal of Physics: Conference Series **1397** (2019) 012015 doi:10.1088/1742-6596/1397/1/012015

Developing physics test instrument in the context of ocean literacy

Purwoko Haryadi Santoso*, Mutmainna

Department of Physics Education, Universitas Sulawesi Barat, Majene, 91412, Sulbar, Indonesia

*E-mail: purwokoharyadisantoso@unsulbar.ac.id

Abstract. Ocean literacy is an understanding that should be integrated into physics learning for the coastal area. Ocean issues will help students understanding physics concepts. Indonesian government through Kemendikbud and Kemenko Maritim have been developing ocean-relate learning. To support government effort improving ocean literate-education system, it needs a sustainable process. Author choose to develop physics test instrument as one step which addresses the ocean literacy to the Indonesian curriculum. This article presents task analysis of kinematics and dynamics concepts to the 7 essential principles and 24 fundamental concepts of ocean literacy. The results will be described with the matrices relating those. It will be a framework aligning 2013 curriculum to the principles of ocean literacy. Some developed item tests with ocean issues will be described in the end of section.

1. Introduction

Physics learning in the context of ocean literacy is relatively new. Even though West Sulawesi (Sulbar) geographically is one of maritime province in Indonesia. It was found that teacher and students still have poor understanding of ocean sciences [1]. Most of public services are located near the sea. Yet, this maritime potential of Sulbar have not been explored by teacher while they are discussing physics content in their class. Arief Rachman on LIPI said that ocean literacy has a strong relation especially with physics [2]. Therefore, student should study linear motion by means of ocean issues like tides, wind, and surface current.

LIPI argues that ocean-oriented learning should be focused for school in coastal area [2]. This idea is contradicted with the circumstances that maritime education hasn't been applied to the learning process in there nowadays [3] On the other hand, physics teacher in Sulbar have satisfied the requirements to execute this idea obviously. They should have integrated ocean literacy in their class.

The future goal of nowadays government is establishing Indonesia as Global Maritime Axis [4]. Responsible institutions must take cooperative strategy to realize this dream. Initially, it may start with enhancing the education quality by policy makers like ministry, university, and schools. They must promote ocean literacy to all of educational faculty members. This vision will be achieved if each of them support each other improving the effective ocean-literate education.

Creating physics learning in the context of ocean literacy is the teacher duty. They should infuse this context into physics problems. It doesn't mean that teacher must make a curriculum document separately. They just engage the essential concept of ocean literacy in their physics instruction. Ocean issues can be a scaffolding for student to understand physics concepts [2].

Since 2004, Center of Ocean Sciences Education Excellence (COSSEE) with another education communities have carried out workshop to identify essential principles of ocean literacy. The event produces a framework as teachers guide applying ocean literacy to be implemented in every school. They are motivated to do this grassroots effort to create ocean-literate society. They have been presented their discussion to align physics concepts to the seven essential principles of ocean literacy [5] The essential principles have also been fit to the *Next Generation Science Standards* (NGSS). It is the existing science standard that guide science teacher to achieve their learning expectation [6].

Furthermore, ocean-literate learning has also been taught in other countries, such as Chile, Japan, Portugal, and Spain [5]. Since 2017, Indonesia has also been establishing policy to implement the ocean-literate curriculum. The Ministry of Education and Culture of Republic of Indonesia (Kemendikbud) and The Coordinating Ministry of Maritime Affairs (Kemenko Maritim) has been administering this project. It is not an effort to make ocean literacy as a single subject taught by a single teacher like religious education or citizenship education. Yet, they want the science concepts can be taught through the ocean issues [2]. In 2018, a year trial has been held in 48 schools spread across 12 provinces in Indonesia. They began to implement in early childhood until high school education. The forty-eight schools are distributed including 12 early childhood schools, 12 elementary schools, 12 secondary schools, and 12 high schools. Learning material is developed and evaluated simultaneously. According to the Monitoring Team of Kemenko Maritim, this implementation has not had a good quality equally [7] It needs further assessment to benchmark the implementation quality of ocean learning.

Aligning physics learning in the context of ocean literacy must be matched with the 2013 curriculum. We know that it doesn't take a short time to gain the best quality implementation. Strang [8] affirm that this implementation must align science concepts with curriculum, analyze science concepts to the ocean literacy, redesign teachers guide to make ocean-literate learning, construct standardized assessment in each grade, and develop curricular material. Until now, there is no standard document that can be used as a guide to implement ocean literacy in Indonesia. It causes there is no valid and reliable instrument to measure ocean understanding for Indonesian citizen. This research is considered to develop physics test instrument in the context of ocean literacy. The whole of article will illustrate our initial stage doing some matrix aligning the essential principles to the 2013 curriculum. This matrix will guide us to construct physics item test for further research steps.

2. Physics Instruction in the Context of Ocean Literacy

According to COSEE [5], ocean literacy is a knowledge about ocean influence towards mankind and how they influence ocean. This definition will be described into 7 essential principles in the table below. It was developed through community-wide consensus involving 100 members of those communities in USA. This most important ideas were derived and explained by 45 fundamental concepts of ocean literacy. They are described in Table 1 below.

Table 1. Essential Principle and Fundamental Concept of Ocean Literacy

Essential Principle (P)	Fundamental Concepts (K)
1. The earth has one big ocean with many features	A. Physical feature of earth B. Lithosphere C. Ocean circulation system D. Sea level E. Physical feature of seawater F. Water cycle G. Ocean as the estuary H. Limitation of ocean resources
2. The ocean and life in the ocean shape the features of earth	A. Earth material B. Tectonic plate C. Erosion D. Carbon cycle

Essential Principle (P)	Fundamental Concepts (K)
	E. Impact physical force to the landforms
3. The ocean is a major influence on weather and climate	A. Weather and climate B. Solar radiation C. Ocean thermodynamics D. Hurricanes and cyclones E. Ocean absorption to carbon cycles F. Climate change G. Climate change impact
4. The ocean made earth habitable	A. Oxygen cycle B. The earliest evidence of life C. Ocean for life
5. The ocean supports a great diversity of life and ecosystems	A. Ocean life range B. Microbes C. Exclusive major groups D. Cycles, adaptations, and relationships among organisms E. Ocean living space F. Environmental factors G. Ocean energy H. Zonation patterns I. Estuaries
6. The ocean and humans are inextricably interconnected	A. Human life B. Food, medicines, mineral, and energy C. Source of inspiration, recreation, rejuvenation, and discovery D. Pollution E. Extinction F. Natural hazards G. Ocean sustainability
7. The ocean is largely unexplored	A. Discovery, innovation, and investigation opportunity B. Human curiosity of sea C. Ocean resources D. Technology in ocean life E. Mathematical models F. Interdisciplinary research

2.1. *Alignment National Science Education Standards (NSES) to The Ocean Literacy*

Concepts in the physical science must be in line with the essential principles. Bridging students with ocean problems in class is the first step to understanding physics with problem solving approach. However, each concept has a particular definition and example in the observed physical phenomena. It can't be inferred that all of ocean literacy principles has been appropriate to it. One physical concept has been taught with some essential problems, it doesn't mean that the concept can be aligned with another principles. Teacher must select the appropriate ocean problems which will be engaged into physics problems. Consensus results of some maritime institutions in United States have compared all of the physics concept with the possible principles of ocean literacy. The results shown in Table 2 below. It will describe how the physical concepts can be worked with the essential principles according to science education standards applied in there.

Table 2. Distribution of Physical Science Concepts in The Essential Principles [8]

Physical Science	Essential Principle Ocean Literacy

	P1	P2	P3	P4	P5	P6	P7
Structure of Atoms							
Structure and Properties of Matter	E, F		E				
Chemical Reaction					G	B	
Motion and Forces	C, E	E					
Conservation in Energy and Increase in Disorder	D, F		B, C, F, G				
Interaction of Energy and Matter	B, C, D, E	C, D, E					

2.2. Alignment 2013 Curriculum to The Ocean Literacy

Content, teaching, dan assessment in ocean-literate physics learning should be aligned with the national curriculum [5,9]. According to Table 2, Motion and Forces concepts can be taught through essential principles code 1C, 1E, and 2E. This next analysis will make a match between basic competence in 2013 curriculum and those. Each of basic competence will derive indicators that should be achieved among the student after joining the class.

The description of the selected principles (P1.C, P1.E, P2.E) is shown below.

P1.C Throughout the ocean there is one *interconnected circulation system* powered by **wind, tides, the force of Earth's rotation** (Coriolis effect), the Sun and water density differences. The *shape of ocean basins and adjacent land masses influence the path of circulation*. This "global ocean conveyor belt" **moves water** throughout all of the ocean basins, transporting energy (heat), matter, and organisms around the ocean. **Changes in ocean circulation** have a large impact on the climate and cause changes in ecosystems.

P1.E Most of Earth's water (97%) is in the ocean. Seawater has unique properties. It is salty, its freezing point is slightly lower than fresh water, its density is slightly higher, its **electrical conductivity** is much higher, and it is slightly basic. Balance of pH is vital for the health of marine ecosystems, and important in controlling the rate at which the ocean will absorb and buffer changes in atmospheric carbon dioxide.

P2.E Tectonic activity, **sea level changes**, and the **force of waves** influence the physical structure and landforms of the coast.

Table 3. Alignment Basic Competence (KD) with The Principle P1.C, P.1E, P2.E

Basic Competence (KD)	Indicator	Essential Principles		
		P1.C	P1.E	P2.E
3.4 Analyzing physical quantities in linear motion with constant velocity and constant acceleration as well as its application in daily life	1. Understanding the concept of linear motion.	√		
	2. Understanding the concept of linear motion with constant acceleration.			√
	3. Describing the difference between distance and displacement as well as between speed and velocity.	√		
	4. Using the equation of linear motion with constant velocity and with constant acceleration correctly.			√
	5. Analyzing physical quantities through linear motion example in daily life.			√
3.5 Analyzing projectile motion using vectors, understanding the physical quantities,	1. Explaining velocity and position vector of two-dimensional motion	√		
	2. Analyzing position, velocity, and acceleration vector of an object.	√		

Basic Competence (KD)	Indicator	Essential Principles		
		P1.C	P1.E	P2.E
and applying those in daily life	3. Analyzing position and velocity vector of projectile motion at time *t*	√		
	4. Analyzing maximum height and maximum distance of projectile motion.	√		
3.6 Analyzing physical quantities of angular motion with constant velocity and its application in daily life	1. Explaining quantities on circular motion	√		
	2. Explaining the equation of circular motion with constant velocity.	√		
	3. Applying the equations to solve problems related circular motion.	√		
	4. Explaining the system of wheels.	√		
3.7 Analyzing force interaction and the relationship between force, mass, and linear motion, as well as the application into daily life	1. Explaining Newton's First Law.			√
	2. Explaining Newton's Second Law about the relationship between force, mass, and acceleration.			√
	3. Explaining Newton's Third Law about action and reaction.			√
	4. Analyzing forces acting on an object.			√
	5. Analyzing frictional force.	√		
	6. Analyzing dynamics problems through applying Newton's Laws to determine certain quantity.	√		

3. Physics Test Instrument in the Context of Ocean Literacy

Table 3 becomes a framework for teachers to design the learning outcome and to construct test items. Test items in Table 4 must obey this guideline if it is developed to benchmark the ocean-relate physics learning. Several examples of them has been developed. It can be represented below in the Table 4. This research remains some items that must be developed in the further agenda. It is distributed into 30 kinematics (KD 3.4 and 3.5) dan 30 dynamics problems (KD 3.6 and 3.7).

Table 4. The Example of Physics Item Test in the Context of Ocean Literacy

Indicator	Essential Principles	Problem
3.4.1	P1.C	1. Wind will move surface waters when it blows over the sea. The surface current speed is typically about 3 % of the wind speed. A fisherman will measure the wind speed blowing there. He found that his boat moves 18 meters in a minute. The measured wind speed is ... A. 0,3 m/s D. 10 m/s B. 1 m/s E. 18 m/s C. 3 m/s *Solution*: Surface current speed $v_{surface} = \frac{s}{t} = \frac{18}{60} = 0{,}3$ m/s Wind speed $v_{surface} = 3\% v_{wind}$

<table>
<tr><th>Indicator</th><th>Essential Principles</th><th>Problem</th></tr>
<tr><td></td><td></td><td>

$$0{,}3 = \frac{3}{100} v_{wind}$$
$$v_{wind} = 10 \text{ m/s}$$

Answer: D
</td></tr>
<tr><td>3.4.2</td><td>P2.E</td><td>

2. Wind is able to move the bouyant body in the sea. It will affect surface waters and cause some areas have different current speed in the sea. Keramba uses this phenomena to cultivate fishes. This method must be placed in proper conditions. The surface waters may not greater than 100 cm/s. A floating object is being observed. The displacement at time *t* is summarized below.

<table>
<tr><th>time (s)</th><th>displacement (cm)</th></tr>
<tr><td>1</td><td>5</td></tr>
<tr><td>2</td><td>20</td></tr>
<tr><td>3</td><td>45</td></tr>
</table>

According to this data, the kind of motion and the possibility to put karamba in there is....
A. Linear motion with constant velocity, possible
B. Linear motion with constant velocity, impossible
C. Linear motion with constant acceleration, possible
D. Linear motion with constant acceleration, impossible
E. It can't be determined the motion and the opportunity to cultivate fish

Solution:

The motion type can be determined by calculating the speed. We know, speed can be calculated with this equation :

$$v = \frac{s}{t}$$

and the result listed in the table below :

<table>
<tr><th>time (s)</th><th>displacement (cm)</th><th>speed (cm/s)</th></tr>
<tr><td>1</td><td>5</td><td>5</td></tr>
<tr><td>2</td><td>20</td><td>10</td></tr>
<tr><td>3</td><td>45</td><td>15</td></tr>
</table>

We see that speed changes uniformly and the magnitude is less than 100 cm/s. Therefore, the surface waters move with constant velocity and it is likely to make keramba in this area.

Answer: C
</td></tr>
</table>

4. Conclusion

Ocean literacy is an appropriate context for physics learning in the coastal area. Student is engaged to recognize ocean life during studying physics. This article has explained how 2013 curriculum in Indonesia aligned with essential principles of ocean literacy. Some ocean problems will be discussed in development process of item test. They will be constructed that satisfied with the analysis before. Furthermore, empirical study must be executed to check the validity and reliability of the test.

Acknowledgments
This study is funded by PDP research grant from Directorate General of Research and Development (DRPM) under Research, Technology, and Higher Education Ministry of Republic of Indonesia (Kemenristekdikti). Therefore, we would thank to policy makers in Ministry and especially for Institute of Research, Community Service, and Quality Assurance (LPPM & PM) of Universitas Sulawesi Barat so author can propose and get this financial support.

References
[1] Guest H, Lotze H, and Wallace D. (2015). Youth and the sea: Ocean literacy in Nova Scotia, Canada. *Marine Policy,* 58, 98-107.
[2] LIPI. (2016). *Kurikulum Kemaritiman Masuk dalam Pelajaran*. (http://lipi.go.id/lipimedia/lipi-:-kurikulum-kemaritiman-masuk-dalam-pelajaran/16687. 2019 [accessed02.05.19]
[3] Pratama, W., & Pardjono, P. (2016). Model Pembelajaran Karakter Pelaut. *Jurnal Pendidikan Vokasi*, *6*(3), 318-334.
[4] Rustam, I. (2016). Tantangan ALKI dalam Mewujudkan Cita- cita Indonesia sebagai Poros Maritim Dunia. *Indonesian Perspective*, *1*(1), 1-21.
[5] COSEE (Centers for Ocean Sciences Education Excellence). Ocean literacy: the essential principles and fundamental concepts for learners of all ages. Pamphlet resulting from the 2-week on-line workshop on ocean literacy through science standards; published by the National Geographic Society, National Oceanic and Atmospheric Administration, National Marine Educators Association, College of Exploration. ⟨http://oceanliteracy.wp2.coexplo ration. org⟩; 2013[accessed06.04.15]
[6] National Marine Education Association. (2015). Alignment of Ocean Literacy Framework to the NGSS for Grades 9 – 12.
[7] Kemenko Maritim. (2018). Kemenko Maritim Mengawal Implementasi Kurikulum Kemaritiman. (http://maritim.go.id/kemenko-maritim-mengawal-implementasi-kurikulum-kemaritiman/; [accessed 03.05.19]
[8] Strang C, deCharon A, and Schoedinger S. (2007). Can You be Science Literate without being Ocean Literate? . *The Journal od Marine Education*, 23(1), 7–9.
[9] Cava F, Schoedinger S, Strang C, and Tuddenham P. (2005). Science Content and Standards for Ocean Literacy: A Report on Ocean Literacy. National Geographic Society, National Oceanic and Atmospheric Administration, National Marine Educators Association, College of Exploration.
[10] Schoedinger S, Cava F, and Jewell B. (2006). The Need for Ocean Literacy in the Classroom. *The Science Teacher*. September 2006

Journal of Physics: Conference Series **1397** (2019) 012016 doi:10.1088/1742-6596/1397/1/012016

Analysis of Physics Aspects of Local Wisdom: *Long Bumbung* (Bamboo Cannon) in Media Development for Android-Based Physics Comics in Sound Wave Chapter

Rahmad Hudan Ramadhan*, Latifah Ratnaningtyas, Heru Kuswanto, Ratna Wardani

Physics Education, Graduate School, Yogyakarta State University
Colombo Street 1, DI Yogyakarta 55281, Indonesia

*Corresponding author e-mail: hudanrr@gmail.com

Abstract. Local wisdom in Indonesia is a property that should be maintained. One way to instill local wisdom in students is to link physics learning. This study aims to develop a learning media for android physics comics based on local wisdom, android-assisted *long bumbung*. This article presents how researchers develop physics comics by exploring physical aspects that exist in local wisdom *long bumbung*. Android is used as a basis for using comics so that it can be used to learn physics whenever and wherever. This physics comic is also sought as enrichment material to strengthen the concepts and material students learn in class. The method in the development of physics comics is R & D (Educational Research and Development) with 4-D models. The process of developing comic physics media is through stages (1) Define consisting of Introduction, Analysis of students, Concept Analysis, and Analysis of learning objectives. (2) Design consists of Material Determination Study and Making Comic Media Component Specifications. Comic Media Physics in the form of electronic digital Book (EPUB) that can be read on the Android system with the help of Reader Reasily, Lithium, Himawari Reader, and others

1. Introduction

The failure of students in solving problems is more due to failure to activate the relevant knowledge that they already have, not because they do not have the right knowledge related to the problem [1] [2]. Therefore there is a need for direction for students to activate knowledge quickly and precisely when solving physics problems. Learning physics has a good meaning if the physical concept is studied in depth. Physics learning more often applies mathematical mindset than deep concept, this brings the atmosphere of learning physics which leads to the memorization of formulas only, so that conceptual and high-level thinking abilities of students are poorly trained [3] [4]. So it is important for teachers to provide high-level thinking stimuli in learning

Local wisdom in each region in Indonesia is an asset or hidden asset for the Indonesian nation that must be maintained. One way to instill local wisdom in students is by linking integrated physics learning. In addition, physics learning can easily be done by using the surrounding items as teaching materials [5] [6]. Utilization of local culture is still rarely noticed by teachers in exploring related physics. Local wisdom as a model for learning science enables students to build contextual learning, and can be easily accessed by local communities [7] [10]. With all the many local cultures in

Indonesia, teachers should be able to apply local culture in learning. thus providing good contextual learning among students

Comics or manga are so close to teenagers, especially students. Manga is a Japanese comic that is spread throughout the world. Comics become an opium for most students in various countries, especially Asia. The instructor is also not looking at comics for studying science [10]. In addition to print media, comics in this era can easily be read through mobile devices in web content [11]. Comic strips and the internet are one of the most accessible materials that can be used in science classes as learning tools. Developing physics comics to learn gives a good contribution to the concept of sound in lower middle school students.

2. Literature Review

The comic contains a unique narrative code consisting of a mixture of images and text in telling a story. comics have the potential to be good teaching materials by paying attention to aspects that are relevant to the topic of learning. In addition, comics can also be used for learning media that are proven to be able to convey incoming information in a clear, coherent, and fun way [12] [13]. In addition to presenting information, comic books have a strong appeal to provoke reading interest through features of humor, narrative, and visual representation [14] [15]. In addition, comic books have advantages over textbooks in general, which can lead to many perspectives from visual and verbal readers' thoughts, predict the beginning and end of the storyline, find climax moments, and feel the complexity of complex stories [17]. Comics in the form of mobile applications also allow paper savings, especially comics in the form of applications can have more than one feature, which is available online tests, animations, and material [18]. So the development of physics comics that are tailored to the context of the material in learning can provide good learning and increase student motivation and attractiveness

Local wisdom is important to stay close to school and interact with learners in the local context in the development of knowledge that can be adapted to school potential [5] [19]. Local wisdom as a model for learning science enables students to build contextual learning, which makes learning more meaningful and becomes more accessible for students in local communities [7] [20]. Product development and media based on local wisdom must pay attention to the core part which consists of: media explanation intro; making local wisdom (products); concept maps; chapter title; keywords; material description; product character; worksheets, and summaries; and the closing section in the form of practice questions, competency tests, bibliography, key answers [21] [8]. Through contextual specifications with physical aspects, learning media can be a good source of information for learning physics concepts.

Long Bumbung (bamboo cannon) is one of the traditional Malay traditional games that was popular in its time and is almost known throughout the archipelago. Various regions recognize the game of bamboo cannons with different names. In Java, especially around Yogyakarta and Central Java, bamboo cannons are commonly known as "Long Bumbung". In addition, in other places this game is known as bamboo rifle, mercon bumbung, and others. This *long bumbung* game is usually played by boys when commemorating religious / customary holidays and enlivening the month of Ramadhan (fasting) before the feast day.

Long bumbung requires fuel that can help produce a boom that is in the form of kerosene or carbide mixed with water with a certain dose. Bamboo cannons are played by pouring kerosene or carbide liquid into the hole where the ignition is made on the bamboo ends. A piece of wood wrapped in a cloth dampened using kerosene is used as an ignition cannon (long bumbung). Until finally the ignition is directed towards the hollow hole of a bamboo cannon which will produce a thumping sound.

Figure 1. Local Wisdom: *Long Bumbung*

Students often avoid difficult questions posed by the teacher. Though deep questions can facilitate good learning, even leading to high level of knowledge [3] [22]. The use of conceptual questions can help students learn to verify the concepts they understand [23]. Questions and dialogues that have been arranged in order to make it easier in the practice of making comics have good advantages in delivering physics material. Memdalam learning can be done by introducing general concepts and bringing on more specific concepts, and qualitative learning [3]. As an effort to stimulate the thinking of students to train HOTS in deep learning, it can be done by means of free and structured interviews. The form of free interviews can be adapted into writing in the form of structured questions. So that structured interviews can be expressed in a deep question in the physics comics media.

3. Methodology

This research is included in development research. The final product of this research is learning media in the form of physics wisdom based on local wisdom: *long bumbung* on contextual sound material. The method of this research is the development of R & D education (Educational Research and Development) with the Thiagajaran 4-D (Four-D Models) model (in 1974).

The product development procedure with a 4-D model in this study consists of 4 stages, namely: Define (Definition), Design (Develop), Develop (Development) and Disseminate (Distribution).

This article presents the design of physics comics through the Define and Design stage. This study presents how the physics comics media are developed based on local wisdom *long bumbung* with the subject matter of sound wave chapter.

3.1 D*efine*

The definition process is carried out to determine the terms and components of the physics comics media so that they can be used in the learning process to the fullest. The definition phase is carried out through the following stages: Preliminary Study, Learner Analysis, Analysis of Learning Objectives, Analysis of Physics Concepts, and Task Analysis.

3.2 *Design*

The design phase is the stage of designing product media, namely the physics-based comics based on formulating indicators of student achievement in accordance with the physics material in *Long Bumbung*'s local wisdom. The indicators are then pinned into the basis of making a prototype design or initial product design consisting of the following stages: Determination of Material and Making Specifications of Media Components Comics. The next step is developing is product realization, preparation of instruments for validation, product validation, and product testing. Products are developed using the Photoshop application. Comic images that have been finished are then arranged into several chapters using the sigil application and will produce comic files in .EPUB format that can be opened using the upub reader on an android smartphone.

4. Result

The development of comic physics is done through the following steps:

1.1. Preliminary Study

The preliminary study was conducted to detect problems in learning in an effort to improve the quality of learning through android-based comic physics media, namely the low level of physical representation ability and high-level thinking ability (HOTS): analysis among high school students. A preliminary study was also conducted to see the curriculum being implemented at school, as well as the selection of *Long Bumbung* as Yogyakarta's local wisdom integrated in the teaching media..

1.2. Analysis of students

This stage is done to find out the background of students related to the learning process, which consists of: academic abilities, cognitive, psychological, and social development of students in classroom learning. The target of media products is 11th grade high school students in the Physics class.

1.3. Physics Concept Analysis

The physics concept analysis is carried out by balancing the reality and workings of *Long Bumbung*. The appropriate concept of physics is then explored based on the existing context. Sound wave concept analysis is carried out by formulating facts/ physical phenomena, physical concepts, physical principles, physical laws, and physical theories based on local cultural games *long bumbung*.

1.4. Task Analysis

Analysis of learning objectives is the result of task analysis and analysis of concepts which are summarized as indicators of student learning outcomes that refer to KI and KD in related physics material. In addition, this study focuses on honing HOTS skills and the physical representation of students on sound material. Table 1 presents the learning objectives with the physics comics media based on the 2013 curriculum applicable in Indonesia.

1.5. Material Determination

The process of identifying physical concepts is carried out on the phenomenon of *Long Bumbung*'s local wisdom. The process of determining the material is also carried out by analyzing physics concepts that are in accordance with the problem of research variables. So that there will be a suitability of the case or plot of the story in the comic physics media relating to the achievement of learning objectives. The material of physics that is set in comics is sound waves which consist of sub chapters of sound characteristics, hearing and seeing sounds; sound propagation; sound intensity; and sound intensity level. In this study, the focus of variable achievement focuses on sound material.

1.6. Making Media Component Specifications for Comics

The making of comic physics media specifications is done through the following stages: 1) the process of making characters and characters; 2) determine the storyline; 3) design illustration images and conversation dialogue; 4) designing problems and solving physical phenomena; 5) designing sound material questions.

1.7. Product Making: Media Physics Comics

Media Physics Comics consists of 4 parts, where each activity conveys sub topics. In addition, 4 parts of the story in the comics of physics are arranged systematically through the most basic phenomena, events, dialogues and sub-questions. It aims to make the learning process through 5M (observing, asking, trying, associating, and communicating) and carried out well, and directs learning that is coherent so that the material can be easily digested by students. In addition, dialogue in physical phenomena is arranged on the basic (factual and procedural) questions that are proposed, which include the nature of teaching and emerge "astonishment" (understanding, prediction, detecting anomalies, applications, and planning) that can lead to deep learning [22] An in-depth question can be

arranged by introducing the concept of "key" which is accompanied by an explanation that will make learning facilities good in the general learning environment [3]
The developer integrates deep questions in the comic physics media to help students gain the depth of learning. The following is a sequence of chapters in comic physics:Introduction: local wisdom-*Long Bumbung* 1) Chapter I: Hearing & Seeing Sound; 2) Chapter II: Sound Fast propagate; 3) Chapter III: Sound phenomenon; 4) Chapter IV: Intensity and Sound Intensity Level.

Table 1. Analysis of Learning Objectives

No	Analysis		Result
1	Core Competence	3	Understand, apply, analyze factual, conceptual, procedural knowledge based on their curiosity about science, technology, art, culture, and humanities with humanity, nationality, state and civilization insights regarding the causes of phenomena and events, and apply procedural knowledge to the field of study specific ones according to their talents and interests to solve problems.
2	Basic Competencies	3.10.	Applying the concepts and principles of sound and light waves in technology
3	Indicator	3.10.1.	Comparing sound strength (TI sound intensity level) that is heard by the human ear because of the difference in the distance of the listener from the sound source that arises in the explosion of the *long bumbung*
		3.10.2.	Large grouping of sound propagation based on propagation medium
		3.10.3.	Make a parable about the nature of sound and the process of sound propagation that occurs in the *long bumbung* to the sense of hearing
		3.10.4.	Concludes the classification of sound waves based on pitch and sound strength
		3.10.5.	Sort out the variables that affect the speed of sound in the event of a bamboo cannon sound
		3.10.6.	Choose the right way to manipulate large sound velocity
		3.10.7.	Find methods for dealing with physical problems related to the weakness of sound in *long bumbung*
		3.10.8.	Make an outline about the intensity of sound waves (I) in *long bumbung* which is viewed from the size of bamboo material
		3.10.9.	Describes the role of variable area (A) and power (P) on the sound wave intensity (I)
		3.10.10.	Deconstructing the arrangement of bamboo materials (in terms of length and magnitude) to show the truth of physical phenomena about the intensity of sound waves

The process of making physics comics with local wisdom *long bumbung* consists of several stages with the following steps:

1. Draw self-contained comics with hand sketches
   Drawing sketches made with hand painting with half-clear strokes. The researcher must adjust the right scene for each page of the comic. For maximum results, the story plot must be adjusted to the material analysis, learning objectives, and physical analysis that was made at the beginning. Figure 2 (left-side) is an example of a comic sketch that has been made.

2. Physics Comic Editing Process
   The results of the comic drawing sketch are then scanned in the form of an image (.png; .jpeg). The scanned comic results are then imported into Adobe Photoshop software. In this stage, the results of comic sketches are processed through the process of coloring and tidying, as well as giving balloons to comic dialogue. Figure 2 (right-side) is an example of a comic sketch that has gone through the final editing process using Adobe Photoshop.

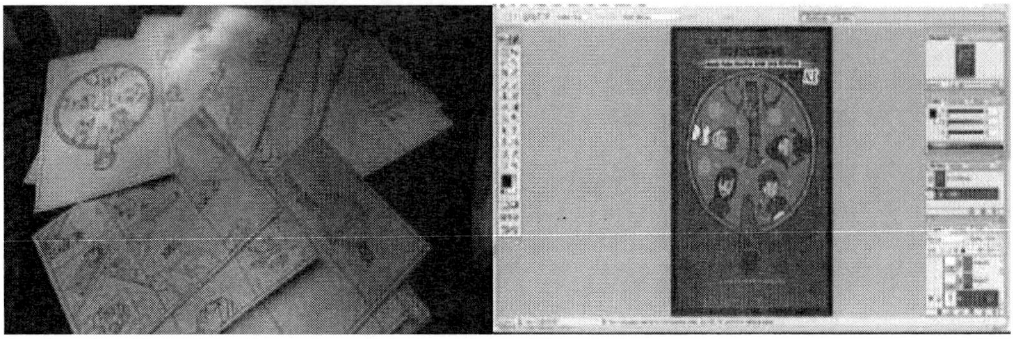

Figure 2. Sketch And Comic Aditing Procces

3. Compilation of comics in the form of EPUB using SIGIL software
   Comics that have gone through the final editing phase will be ready to be packaged using SIGIL software. In this process, each comic part is arranged in a sequence that has been determined. The advantage, researchers can include video features that can support students' understanding of the sound concepts associated. So that the integrity of comic physics can be poured with the final product in the form of EPUB (digital book).

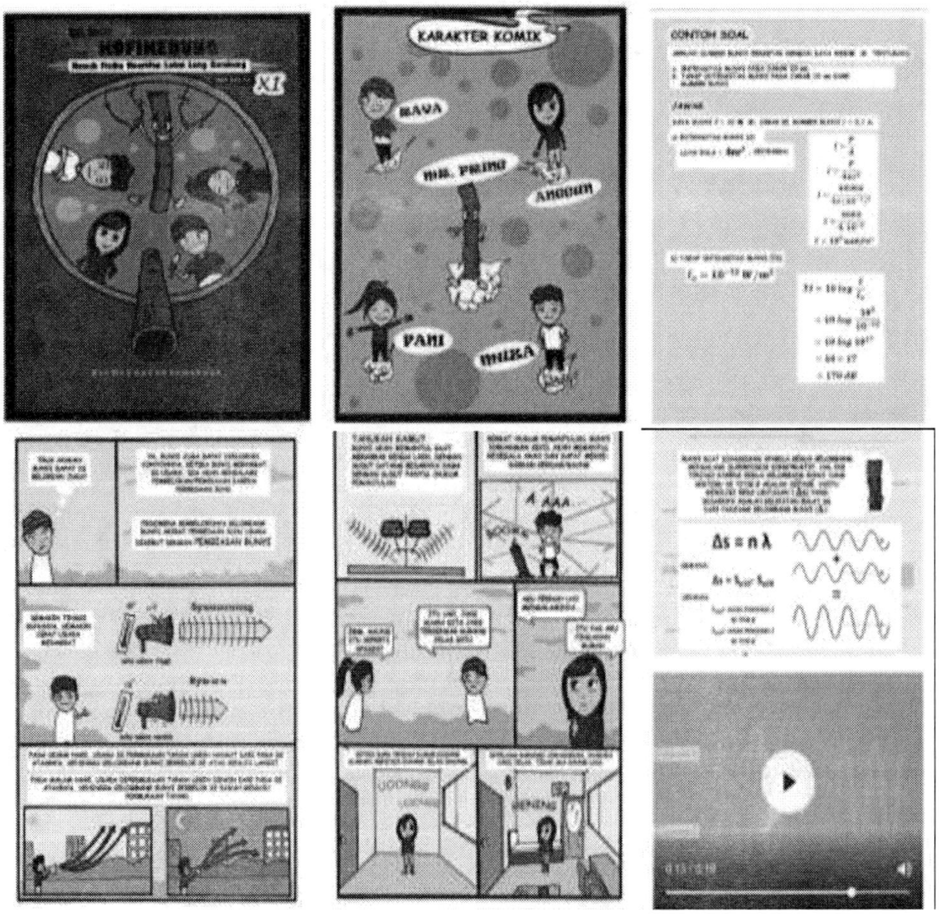

Figure 3. Display of Physics Comics

4. Product of *Long Bumbung* Local Wisdom Physics Comics

Final format physics comics are packaged in EPUB format (.epub), which is a digital book format. Comics can be read on mobile Android devices using the EPUB reader application such as: Reasily, Himawari Reader; lithium: Epub Reader, and others. Figure 3 shows the local wisdom physics comics *long bumbung* through the application Reasily.

5. Discussion

The product that will be developed by the author is an android-based physics comic with the utilization of local wisdom *long bumbung*. Comics are designed to attract students' interest in learning physics. In addition, the use of android as a learning medium can be used by students to study physics whenever and wherever. This physics comic is also sought as enrichment material to strengthen the concepts and material students learn in class. Here are the Android-based comic physics specifications designed by the author: 1) Physics comics are designed with the theme of local wisdom *long bumbung* in the form of an android application. 2) Physics comics are designed that focus on sound physics material. 3) Physics comics are designed as material for student learning enrichment. 4) Physics comics are designed with the theme of science fiction with interesting character characters that display humor and culture. 5) Each dialogue on the comic story plot is based on an approach to depth-questions about related material. 6) Physics comics are equipped with several physical support videos, sample questions, answer discussions, and practice questions in each event plot.

6. Conclusion

Given the importance of local cultural values in Indonesia, educators must have a stake in preserving them. Physics learning by seeking local wisdom is very good to do. The development of android physics comics learning media based on local wisdom of *long bumbung* is one example of efforts to preserve local cultural values. Researchers developed physics comics by exploring physical aspects that exist in local wisdom *long bumbung*. The method in the development of physics comics is R & D (Educational Research and Development) with 4-D models. The manufacturing process through stages (1) Define consists of Introduction, Analysis of students, Concept Analysis, and Analysis of learning objectives. (2) Design consists of Material Determination Study and Making Comic Media Component Specifications. The content of comic physics contains a treatment that can facilitate students to hone the ability of physics representation and HOTS, comics present the phenomenon pisycs sound waves, problems and examples of problems, supporting videos.

7. Reference

[1] L. Buteler and E. Coleoni, "Solving problems to learn concepts , how does it happen ? A case for buoyancy," vol. 020144, pp. 1–12, 2016.

[2] J. L. Docktor, N. E. Strand, J. P. Mestre, and B. H. Ross, "Conceptual problem solving in high school physics," vol. 020106, pp. 1–13, 2015.

[3] S. D. Craig, B. Gholson, J. K. Brittingham, J. L. Williams, and K. T. Shubeck, "Promoting vicarious learning of physics using deep questions with explanations," *Comput. Educ.*, vol. 58, no. 4, pp. 1042–1048, 2012.

[4] G. V. Madhuri, V. S. S. N. Kantamreddi, and L. N. S. Prakash Goteti, "Promoting higher order thinking skills using inquiry-based learning," *Eur. J. Eng. Educ.*, vol. 37, no. 2, pp. 117–123, 2012.

[5] C. Pornpimon, A. Wallapha, and C. Prayuth, "Strategy Challenges the Local Wisdom Applications Sustainability in Schools," *Procedia - Soc. Behav. Sci.*, vol. 112, pp. 626–634, 2014.

[6] A. Rusilowati, Supriyadi, and A. Widiyatmoko, "NATURAL DISASTER vision LEARNING SETS INTEGRATED IN SUBJECT OF PHYSICS-BASED LOCAL WISDOM," *J. Pendidik. Fis. Indones. J. Phys. Educ.*, vol. 11, no. 1, pp. 42–48, 2015.

[7] Anwari, M. S. Nandi, and E. Sulistyowati, "Biological Science Learning Model Based on Turgo's Local Wisdom on Managing Biodiversity," in *PROCEEDINGS OF INTERNATIONAL SEMINAR ON MATHEMATICS, SCIENCE, AND COMPUTER SCIENCE EDUCATION*

(MSCEIS 2015), 2016, vol. 1708.

[8] P. Z. Kun, "Pembelajaran Sains Berbasis Kearifan Lokal," *Pros. Semin. Nas. Fis. dan Pendidik. Fis.*, vol. 2, no. 1, pp. 246–256, 2013.

[9] B. Setiawan, D. K. Innatesari, W. B. Sabtiawan, and S. Sudarmin, "The development of local wisdom-based natural science module to improve science literation of students," *J. Pendidik. IPA Indones.*, vol. 6, no. 1, pp. 49–54, 2017.

[10] R. Egusa *et al.*, "Improving the usability of manga-on-a-tablet for collaborative learning," in *CSEDU 2014 - Proceedings of the 6th International Conference on Computer Supported Education*, 2014, vol. 1, pp. 446–451.

[11] H. Tolle and K. Arai, "Manga content extraction method for automatic mobile comic content creation," in *2013 International Conference on Advanced Computer Science and Information Systems, ICACSIS 2013*, 2013, pp. 321–328.

[12] D. Yulianti, S. Khanafiyah, and S. Sulistyorini, "Inquiry-based science comic physics series integrated with character education," *J. Pendidik. IPA Indones.*, vol. 5, no. 1, pp. 38–44, 2016.

[13] Widyastuti, Mardiyana, and Saputro, "An Instructional Media using Comics on the Systems of Linear Equation," *J. Phys. Conf. Ser.*, vol. 895, p. 012039, 2017.

[14] C. Guérin, C. Rigaud, K. Bertet, and A. Revel, "An ontology-based framework for the automated analysis and interpretation of comic books' images," *Inf. Sci. (Ny).*, vol. 378, pp. 109–130, 2017.

[15] J. Kim, M. S. Chung, H. G. Jang, and B. S. Chung, "The use of educational comics in learning anatomy among multiple student groups," *Anat. Sci. Educ.*, vol. 10, no. 1, pp. 79–86, 2017.

[16] J. Hosler and K. B. Boomer, "Are comic books an effectiveway to engage nonmajors in learning and appreciating science?," *CBE Life Sci. Educ.*, vol. 10, no. 3, pp. 309–317, 2011.

[17] C. S. Babaian and A. A. Chalian, "'The thyroidectomy story': Comic books, graphic novels, and the novel approach to teaching head and neck surgery through the genre of the comic book," *J. Surg. Educ.*, vol. 71, no. 3, pp. 413–418, 2014.

[18] W. S. Hadi and P. Dwijananti, "PENGEMBANGAN KOMIK FISIKA BERBASIS ANDROID SEBAGAI SUPLEMEN POKOK BAHASAN RADIOAKTIVITAS UNTUK SEKOLAH MENENGAH ATAS," *Unnes Phys. Educ. J.*, vol. 2, no. 1, 2015.

[19] A. Rusilowati and A. Widiyatmoko, "PEMBELAJARAN KEBENCANAAN ALAM BERVISI SETS TERINTEGRASI DALAM MATA PELAJARAN FISIKA BERBASIS KEARIFAN LOKAL NATURAL DISASTER VISION LEARNING SETS INTEGRATED IN SUBJECT OF PHYSICS-BASED LOCAL WISDOM," *Januari J. Pendidik. Fis. Indones.*, vol. 11, no. 1, pp. 1693–1246, 2015.

[20] B. Subali, A. Sopyan, and E. Ellianawati, "Developing local wisdom based science learning design to establish positive character in elementary school," *J. Pendidik. Fis. Indones.*, vol. 11, no. 1, pp. 1–7, 2015.

[21] J. Widodo, "Urban Environment and Human Behaviour: Learning from History and Local Wisdom," *Procedia - Soc. Behav. Sci.*, vol. 42, pp. 6–11, 2012.

[22] C. Chin, D. E. Brown, and B. C. Bruce, "Student-generated questions: A meaningful aspect of learning in science," *Int. J. Sci. Educ.*, vol. 24, no. 5, pp. 521–549, 2002.

[23] I. Labutov, S. Basu, and L. Vanderwende, "Deep Questions without Deep Understanding," *Proc. 53rd Annu. Meet. Assoc. Comput. Linguist. 7th Int. Jt. Conf. Nat. Lang. Process. (Volume 1 Long Pap.*, pp. 889–898, 2015.

Journal of Physics: Conference Series **1397** (2019) 012017 doi:10.1088/1742-6596/1397/1/012017

Developing of Bloomian HOTS Physics Test: Content and Construct Validation of The PhysTeBloHOTS

Edi Istiyono[1*], Wipsar Sunu Brams Dwandaru[2], Muthmainah[3]

[1,2] Department of Mathematics Education, Faculty of Mathematics and Natural Sciences, Universitas Negeri Yogyakarta

[3] Graduate School, Universitas Negeri Yogyakarta

*Corresponding author e-mail: edi_istiyono@uny.ac.id

Abstract. Higher order thinking skills (HOTS) of the students in physics learning need to be assessed. There are several kinds of HOTS, one of which is Bloomian HOTS. The Bloomian HOTS Physics test for Senior High School Students called PhysTeBloHOTS has been developed to assess students' HOTS. The purpose of this study is to prove validation of the PhysTeBloHOTS. There were two steps of validation: proving the content validity and construct validity. The content validity analysis was conducted using Aiken's V and the analysis of construct validation was conducted using exploratory factor analysis (EFA). The PhysTeBloHOTS consisted of four test sets: A, B, C, and D. The each of test set consists of 40 items with 8 anchor items. For the test trial, PhysTeBloHOTS were administrated to 301 Senior High School Students in Yogyakarta. The result of the study shows that PhysTeBloHOTS is valid according to the content and construct validity. The PhysTeBloHOTS items are valid according to the expert judgments with Aiken's V index ranged from 0.83 to 0.97, respectively. The result of EFA and CFA show that PhysTeBloHOTS consist of three aspects: analysis, evaluate, and create.

1. Introduction

Assessment is a very important process in learning, especially in physics. Jensen, et.al state that assessment informs what the students expect from the learning process [1]. This statement is corroborated by Kusairi who underlines the importance of assessment done by teachers in learning process, assessment activity can help teachers to understand the strengths and weaknesses possessed by the students during studying time [2]. The more quality is shown in the assessment process, the better is the understanding ofstudents' strengths and weaknesses. As it is still rare that teachers do such assessment to measure students' high order thinking skill, it is important to design a test for that level.

The assessment instrument used by teachers to score students' learning result is a cognitive assessment instrument. Assessment category used is in the cognitive area as revised in Bloom's Taxonomy (Bloomian) which is divided into Lower Order Thinking Skills (LOTS) and Higher Order Thinking Skills (HOTS). So far the most frequently used is Bloomian Lower Order Thinking Skills (LOTS) that includes test assessment using questions in knowledge level C1 (remember), C2 (understand), and C3 (apply), which means the students are not used to answering questions on levelC4(analyze),C5 (evaluate), and C6 (create) [3]. The questions in level C4, C5, and C6 are still rarely encountered. Thus, it is important to develop a test assessment up to the level of Bloomian HOTS.

Teachers' incompetence to measure the students' higher order thinking becomes the reason why it is very important to develop a test to measure those particular skills. The investigation done by Jensen et al, argues that there are many teachers who fail to give a statement on their students' knowledge and thinking skills, instead, teachers only give statement on the students' memorizing aspect that is included in the Lower Order Thinking Skills [1]. Thus it is important to make questions that really measure the higher order thinking skills. Madhuri, Kantamreddi & Goteti argues that active learning that promotes HOTS plays an important role in education system [4].

Instruments used in assessment are usually test and non-test instruments. In this research, test instrument is used to measure the students' skills. Test instruments include multiple choice and essay questions, each has strengths and weaknesses. Multiple choice questions are the most frequent form of test in the assessment process in high school level, while in general most of multiple choice questions are seldom able to measure the students' higher order tinking skills. This becomes a problem as the semester tests and national tests are served in the form of multiple choice [5]. To measure the students' higher order thinking skills, it is better if the question used is reasoning multiple choice, as the development of the regular multiple choice questions. Thus, reasoning multiple choice has to be developed. Cullinane argues that attaching reasons in the second tier of two-tier multiple-choice question can be used to increase the higher order of thinking skill and it can be used to see the students' ability in reasoning [6]. Winarti et al Thus, in choosing the answer, the students have to think about the reasoning of their answers, and it is a straight forward thinking process to decide the proper reasoning so that it is a proper exercise to higher order thinking skills [7].

The frequently used test execution and scoring procedure are step-by-step scoring and item-based scoring, whose result is then summarized. This scoring procedure is used for regular multiple choice questions. This scoring model is not reliable because each question's unique difficulty index is not considered. Alternative approach that can be used is the approach of item response theory for polytomous scoring. This scoring is adapted to the reasoning multiple choice questions. One of the models frequently used in polytomous item scoring is Partial Credit Model (PCM) [8]. PCM is developed to analyze test items that need several steps to do, in which each item follows a partial credit pattern, thus if an individual has a higher skill it is expected that he or she would get a higher score than those who have lower skill. PCM application in modified multiple choice as an alternative model in physics learning assessment is considered effective and fair [9].

The objective to introduce HOTS as the form of the revision of assessment system started in the 21st Century. Even, many developed countries have applied HOTS-based assessment in their education system. Besides, the development of HOTS assessment instrument, according to Hassan, Rosli & Zakaria also can bring benefits, such as ; (1) to increase awareness that there is a benchmark to achieve by students, (2) HOTS guides the students to seek solution and invention (analyze, evaluate, and create), (3) to introduce the students to scientific research, (4) to increase understanding of the subject, especially physics, that is proven to be more effectively taught with HOTS, and (5) to increase the student's competence in investigating and exploring ideas in learning, especially in physics [10].

Not only assessment development, technology development also requires a more effective and efficient assessment process. In the 21st century, various computer-based media are invented to complete classic paper-based test process and helped the advancing of assessment process [11]. Computer-based test is designed using adaptive system in which the next questions given are adapted to the ability level of the students shown by the performance when answering the previous questions. Right now the popular media using adaptive system in test scoring is Computerized Adaptive Test (CAT). Kantrowitz, Dawson, &Fetzer argues that CAT is a part of technology development that helps the teachers as well as the students to improve the learning process, especially a more organized assessment method that adapts to the students' learning ability [12].

The research done by Veeravag, Muthusamy, & Marimuthu shows that the level of thinking as stated by Bloom taxonomy influences the work of the students [13]. HOTS influences the students' ability to answer questions of the higher level. Midyarto, et.al stated that the assessment in students' cognitive competence can encourage and enable the teacher to monitor the students' competence achievement [14]. Thus it is easier to indicate what to teach and what to prepare for the students' further learning.

Cognitive area in popular Revised Bloom's Taxonomy is referred to as Higher Order Thinking Skills, or what is usually abbreviated as HOTS. The students' higher order thinking skills include the ability to analyze, to evaluate, and to create, in accordance to the levels of C4, C5, and C6 revised Bloom Taxonomy. In this study, the discussed HOTS is HOTS in physics, in which the students are expected to be able to answer Physics questions by analyzing, evaluating, and creating. Physics is usually regarded as a difficult subject by students. Learning the subject needs logical thinking competence, critical thinking, creativity, and objectivity. Drilling the students with the above stated thinking habits will familiarize them to think in higher order thinking skills. Several reasons stated above support the objective of the study, which is to develop the Bloomian HOTS instrument assessment in the form of reasoning multiple choice with the scoring based on PCM to measure students' higher order thinking skills using CAT.

2. Methods

This research is research and development (R & D) which aims to develop of Bloomian HOTS Physics Test (PhyTeBloHOTS). The research procedure is carried out through three stages, namely: (a) planning, (b) try out, and (c) measurement and interpretation. The purpose of this study is to prove validation of the PhyTeBloHOTS. The draft of the instruments have been validated to get an expert judgment. Then the instruments tested to 300 first year students in three senior high schools in Yogyakarta. The product trial design in this study included content validation steps and construct validation steps, which were carried out to measure the validity of each item developed.

The content validation is carried out by expert judgement by filling in the validation sheet. The sheet was composed with interval scale of 1 to 4. The data analysis of the validation questionnaire was done in one by the following steps:

1) The first step was to find Aiken's V index using formula [15]:

$$V = \frac{\sum S}{[n(c-1)]} \tag{1}$$

Note :
s = r - lo, lo = lowest validity score,
c = highest validity score,
r = score given by rater

2) The second step is to convert Aiken's V index of every questionnaire item into qualitative data with V index range between 1 to 0. Validation result is declared valid if Aiken's V index results > 0.37.

Construct validity states the extent to which the performance of the test is consistent with the construct in certain theoretical considerations. This study also investigates the construct validity of the instrumen to verify the consistency instrument with the construct of higher order thinking skills on physics subjects. The construct validity can use factor analysis procedures. The exploratory factor analysis (EFA) is a method of factor analysis to identify the relationship between manifest variables or indicator variables in constructing a construct. Besides EFA, confirmatory factor analysis (CFA) is a method of factor analysis to test the indicators that have been grouped according to their latent variables (constructs). The latent variables are consistent in their constructs. The standardized loading factor (SLF) resulting from confirmatory factor analysis (CFA) is expressed as a result of construct validity.

3. Result and Discussion

This development research results a product of 62 test items, each is a question fit for Bloomian HOTS with the material of elasticity and Hooke law, static fluid, temperature and calorie, and optical instruments with 10 anchor items. The test (PhysTeBloHOTS) are divided into two packages, which are 36 questions in Package A and 36 questions in Package B.

3.1. Content Validation

In this step, the expert validates the test instrument that was developed. The test developed was PhysTeBloHOTS. The validation score of the PhysTeBloHOTS based on material aspect, construction aspect, language aspect, and appearance aspect. Each test item is scored based on the four scoring aspects. Based on the response and scoring of two experts of physic education scoring and physics expert. Each test item is scored based on the four scoring aspect. Based on the response and scoring of two physics education expert and physics expert, then it resulted Aiken's V index of 62 items from 0.71 to 0.78 shown completely by Tabel 1, so that the entire question items developed is considered valid according to expert and can be used for the further step of trial, which is an empirical trial. This is in accordance to the interpretation steps done by Kowsalya et al, 2012), all of the items are valid if it is in the range of 0.37-1.00.

Table 1. Aiken's V of Set A and Set B tems

Set A				***Set B***			
Item	Aiken's V	*Item*	Aiken's V	*item*	Aiken's V	*Item*	Aiken's V
Item 1	0,78	Item 19	0,72	item 1*)	0,72	Item 19	0,72
Item 2	0,73	Item 20	0,72	Item 2*)	0,72	Item 20	0,72
Item 3	0,73	Item 21	0,74	Item 3*)	0,72	Item 21	0,74
Item 4	0,73	Item 22	0,72	Item 4*)	0,72	Item 22	0,73
Item 5	0,75	Item 23	0,72	Item 5*)	0,74	Item 23	0,73
Item 6	0,72	Item 24	0,72	Item 6*)	0,72	Item 24	0,72
Item 7	0,72	Item 25	0,72	Item 7*)	0,72	Item 25	0,72
Item 8	0,72	Item 26	0,72	Item 8*)	0,73	Item 26	0,73
Item 9	0,72	Item 27*)	0,72	Item 9*)	0,73	Item 27	0,71
Item 10	0,72	Item 28*)	0,72	Item 10*)	0,73	Item 28	0,73
Item 11	0,73	Item 29*)	0,72	Item 11	0,74	Item 29	0,73
Item 12	0,72	Item 30*)	0,72	Item 12	0,74	Item 30	0,73
Item 13	0,72	Item 31*)	0,74	Item 13	0,74	Item 31	0,73
Item 14	0,72	Item 32*)	0,72	Item 14	0,75	Item 32	0,74
Item 15	0,73	Item 33*)	0,72	Item 15	0,73	Item 33	0,73
Item 16	0,72	Item 34*)	0,73	Item 16	0,73	Item 34	0,73
Item 17	0,72	Item 35*)	0,73	Item 17	0,73	Item 35	0,73
Item 18	0,72	Item 36*)	0,73	Item 18	0,72	Item 36	0,77

3.2. Construct Validation

After going through the expert validation process, the PhysTeBloHOTS was packed into two question packages, i.e. package A and package B, each containing 36 questions with 10 anchor items that would be used in the construct validation stage.

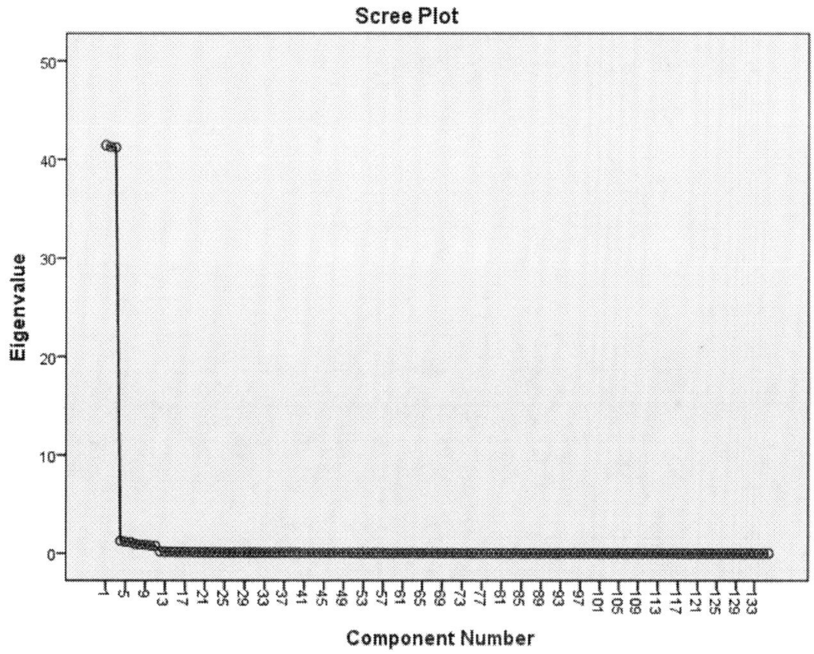

Figure 1. Scree Plot of Component Number

Table 2. Total Variance Explained

Total Variance Explained

Component	Initial Eigenvalues			Extraction Sums of Squared Loadings			Rotation Sums of Squared Loadings		
	Total	% of Variance	Cumulative %	Total	% of Variance	Cumulative %	Total	% of Variance	Cumulative %
1	41.425	30.460	30.460	41.425	30.460	30.460	41.411	30.450	30.450
2	41.252	30.332	60.792	41.252	30.332	60.792	41.252	30.332	60.782
3	41.200	30.294	91.086	41.200	30.294	91.086	41.214	30.305	91.086
4	1.293	.951	92.037						
5	1.186	.872	92.909						

Base on Tabel 2, the results of factor analysis using EFA, PhysTeBloHOTS is formed 3 factors. The percentage of variance explained is 88,057%. The first factor explains the variance of 43,234%, this indicates that the instrument meets the dimensional assumptions. This is in accordance with Reckase (Smits, Cujpers & van Straten, 2011) that if the output of factor analysis produced by the first factor is able to explain the variance of more than 20%, then the dimensional assumption has been fulfilled. Grouping items into factors formed based on the value of loading factors from each item. The critical value of loading factors with a sample size of 300 is 0.35. Based on the guideline for determining the factor loading significance presented by SOLO Power Analysis, BMDP Statistical Software, Inc. 1993 [16].

Table 3. Guidelines for Identifying Factor Loading Values Based on Sample Size

The factor loading value is considered significant	Required of Sample size
0,3	350
0,35	250
0,4	200
0,45	150
0,5	120
0,55	100
0,6	85
0,65	70
0,7	60
0,75	50

Table 4. Factor Grouping

	Factors		
	1	2	3
Number of Items	9A, 10A, 11A, 12A, 13A, 14A, 15A, 16A, 17A, 18A, 19A, 20A, 21A, 22A, 18B, 19B, 20B, 21B, 22B, 23B, 24B, 25B, 26B, 27B, 28B.	1B, 2B, 3B, 4B, 5B, 6B, 7B, 8B, 9B, 10B, 11B, 12B, 13B, 14B, 23A, 24A, 25A, 26A, 27A, 28A, 29A, 30A.	1A, 2A, 4A, 6A, 8A, 15B, 31A, 32A, 33A, 34A, 35A, 36A, 37A, 38A, 39A, 40A, 16B, 17B, 29B, 30B, 31B, 32B.
Name of Factor	Analyze	Evaluate	Create

CFA is similar to EFA, but it is quitely different. At CFA, first the researcher must specify the number of factors and which factor of each variable before analyze it. Researchers can get a better comprehension of the measurement quality if they combined construct validity with CFA (Hair et al., 2006). The result of confirmatory factor analysis of Bloomian HOTS Physics Test shows in Figure 2.

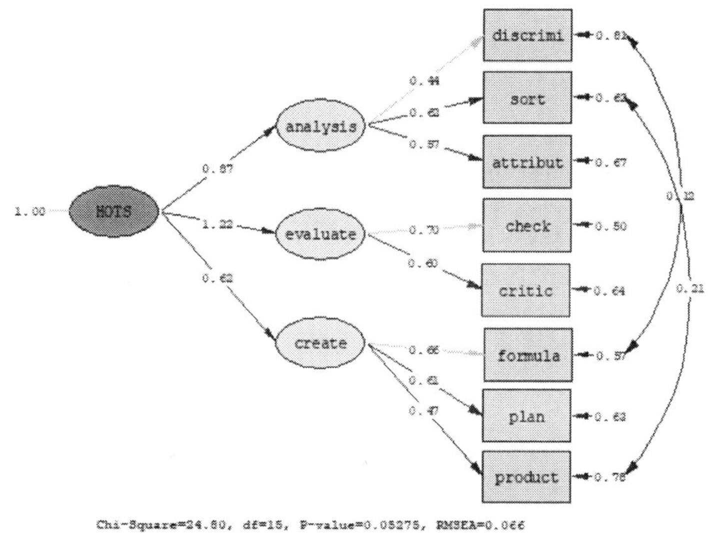

Figure 2. Model of the PhysTeBloHOTS

The first step to analyzing CFA was directed to assessing the goodness of fit (GOF) between the data and the model. In Figure 1, the Chi Square obtained = 24.80, df = 15, p. value = 0.05275 and RMSEA = 0,066. It is indicates that the model has a fit. The results and criteria of the model-fit can be seen in the Table 5.

Table 5. The Results of Model Fit

GOF Measure	Degree Target of Fit	Estimation Results	Degree of Fit
χ^2	$\chi^2 \geq 24.99579$	24.80	Marginal fit
p value	$p \geq 0.05$	0.05275	Good fit
RMSEA	RMSEA ≤ 0.08	0.066	Good fit
NFI	NFI ≥ 0.90	0.94	Good fit
CFI	CFI ≥ 0.92	0.97	Good fit
GFI	GFI ≥ 0.90	0.96	Good fit

Tabel 5 shows that the fit between the data and the model was in general good, the next step was looking the construct validity of the model [17]. Construct validity refers to a theoretical view to explain constructs in a psychological theory [18]. The results of the second order CFA for the value of t-value and Standarized Loading Factor (SLF) to evaluating the construct are presented in Table 6

Table 6. The Results of Second Order CFA

Variabel	Aspect	Sub aspect	Second Order CFA	
			t-value	SLF
HOTS	Analysis	Discriminate	-	0.44
		Sort	4.37	0.62
		Attribution	4.21	0.57
	Evaluate	Check	-	0.70
		Criticize	6.32	0.60
	Create	Formulate	-	0.66
		Plan	4.77	0.61
		Product	4.14	0.47

Table 6 shows that t-value of all sub aspect was bigger than 1.96. this indicates that all sub aspect are significant in supporting the construct of HOTS. The Standarized Loading Factor (SLF) value also shows that all sub aspect meet the criteria of good construct validity.

4. Conclusion

PhysTeBloHOTS is valid according to the content and construct validity. The PhysTeBloHOTS items are valid according to the expert judgments with Aiken's V index ranged from 0.83 to 0.97, respectively. The result of EFA shows that PhysTeBloHOTS consist of three aspects: analysis, evaluate, and create. The result of CFA shows that three aspect of PhysTeBloHOTS are significant in supporting the construct.

References

[1] Jensen, J. L.; McDaniel, M. A.; Woodard, S. M., &Kummer, T. A. 2014. Teaching to the Test or Testing to Teach: Exams Requiring Higher Order Thinking Skills Encourage Greater Conceptual Understanding. *Journal EducPsychol Rev,* 1-24.

[2] Kusairi, S. 2012. Analisis Asesmen Formatif Fisika SMA Berbantuan komputer. Jurnal Penelitian dan Evaluasi Pendidikan, 16/Edisi Dies Natalis ke-48 UNY, 68-82.

[3] Anderson, L.W., &Krathwohl, D.R. 2001. *A Taxonomy of Learning, Teaching, and Assessing: A Revision of Bloom's Taxonomy of Educational Objectives*. New York: Longman.

[4] Madhuri, G.V., Kantamreddi, V.S.S.N., &Goteti, L.N.S.P. 2013. Promoting higher order thinking skills using inquiry-based learning. *European Journal of Engineering Education*, 37/2, 117-123.

[5] Istiyono, E. 2013. *Pengembangan Instrument Untuk Mengukur Kemampuan Berpikir Tingkat Tinggi Dalam Mata Pelajaran Fisika Di SMA*.Disertasi Doktor, tidak diterbitkan, Universitas Negeri Yogyakarta, Yogyakarta.

[6] Cullinane, Alison, & Liston, M. 2011. *Two-tier Multiple Choice Question: An Alternative Methodof Formatif Assessment for FirstYear Undergraduate BiologyStudents*. Limerick: National Center for Excellence In Mathematics and Education Science Teaching and Learning (NCE-MSTL).Faster, Smarter, And More Secure Approach To Pre-Employment Testing [Versielektronik]. *Journal of Business Psychology*, 26, 227–232.

[7] Winarti.; Cari.; Suparmi.; Sunarno, W. &Istiyono, E. 2017. Development of two tier test to assess conceptual understanding in heat and temperature. *Journal of Physics: Conf. Series,* 1-5.

[8] Widhiarso, W. 2010. *AplikasiTeoriRespon Item untuk Pemodelan Respons Menipu pada Skala Kepribadian*. Laporan Hasil Penelitian UGM.

[9] Istiyono, E., Mardapi, D., & Suparno. 2014. Pengembangan Tes Kemampuan Berpikir Tingkat Tinggi Fisika (PhysTHOTS) Peserta Didik SMA. Jurnal Penelitian dan Evaluasi Pendidikan, 18/1, 1-12.

[10] Hassan, S.R., Rosli, R., &Zakaria, E. 2016. The Use of i-Think Map and Questioning to Promote Higher-Order Thinking Skills in Mathematics. *Creative Education,* 7, 1069-1078.

[11] Hambleton, R.K., Swaminathan, H., & Roger H.J. 1991. *Fundamentals of Item Response Theory*. California: SAGE Publications, Inc.

[12] Kantrowitz, T.M., Dawson, C.R., &Fetzer M.S. 2011. Computer adaptive testing (CAT): A Faster, Smarter, And More Secure Approach To Pre-Employment Testing [Versielektronik]. *Journal of Business Psychology*, 26, 227–232.

[13] Veeravag, J., Muthusamy, C &Marimuthu, R. 2010. Using Bloom's Taxonomy To Gauge Students' Reading Comprehension Performance. *Canadian Social Science,* 6/3, 2010, 205-212.

[14] Mindyarto, B.N.; Rusilowati, A.; Kartono&Sugiyanto. 2011. PengembanganSistem Diagnosis Kognitif Fisika Online Untuk SMP. *JurnalPenelitianPendidikan*, 29/2, 86-95.

[15] Kowsalya, D.N., Lakshmi, V., & Suresh, K.P. 2012. Development and Validation of a Scale to Asses SelfConcept in Mild IntellecctualyDisabled Children. *International Journal Soc. Sci and Education,* 2/4, 669-709.
[16] Hair, J.F, Black, W.C., Babin, B.J., & Anderson, R.E. 2014. Multivariate data analysis (7thed). Edinburg Gate: Pearson Education Limited.
[17] Hair, J. F., Black, W. C., Babin, B. J., & Anderson, R. E. 2006. Multivariate data analysis (7th ed.). Prentice Hall
[18] Wiersma, W. (2000). Research methods in education: an introduction (7th ed.). Boston: Allyn & Bacon

Analysis of Students' Critical Thinking Skills at Junior High School in Science Learning

A. Febri*, Sajidan, and Sarwanto
Science Education, Postgraduate Program, Sebelas Maret University, Ir. Sutami St. No. 36A, Kentingan Jebres, Surakarta, Indonesia

*Corresponding author e-mail: adefebri1@student.uns.ac.id

Abstract. Critical thinking skills is one of students' skills needed to face globalization in 21st century challenge. Critical thinking skills are active processes to thinking deeply due to topic or materials through meaningful way. The indicators of critical thinking skills are thinking about certain topic in depth, asking questions, and finding relevant information. This study aimed to analyze students' critical thinking skills at junior high school in science learning. Instruments used in this study were created according to Facione indicators of critical thinking skills namely to interpret, to analyze, to evaluate, to infer, to explain, and to do self-control. Subjects in this study were 256 students of seventh grade at junior high school in Karanganyar. Purposive sampling were used as sampling technique. Results showed that two indicators have gained good category namely to analyze that was 62% and to evaluate that was 70%; while four other indicators have gained sufficient category namely to interpret that was 35%, to infer that was 49%, to explain that was 40%, and to do self-control that was 46%. The conclusion of this study was there were two indicators of students' critical thinking skills gained good category and four indicators of students' critical thinking skills gained sufficient category. Thus proper follow-up of this study is needed to explore ways to improve the achievements of each indicators in students' critical thinking skills.

1. Introduction

The development of the 21st century has brought an advanced technology by encouraging various fields including education [1]. In 21st century, human resources began to be replaced by technology. The era of the 21st century has become easier because of technology [2]. Technological developments encourage humans to change their habits so that it can be said that humans cannot live without the existence of technology. Technology can help people in communication, transportation, and education [3]. The 21st century learning has been accepted and has become a focus on several skills that must be mastered by students. Students in the 21st century must have new learning goals through different and new teaching [4]. The intended learning goal is to improve the skills needed in the 21st century. Throughout history, critical thinking and problem solving skills are the main components of human development [5]. The 21st century skills represent characteristics that students should have to overcome difficulties and achieve success [6].

The 21st century learning is firstly characterized by encouragement to students in finding knowledge out of information sources. Second, learning is directed at formulating problems. Third, learning trains students in making decisions. Fourth, the 21st century learning emphasizes collaboration in problem solving [7]. The 21st century learning focuses on four skills (4C) that must be mastered by students namely communication, critical thinking, collaboration, and creativity. One of the most important things in facing the 21st century era is critical thinking skills. Critical thinking skills are active processes to thinking deeply due to topic or materials through meaningful way. The indicators of critical thinking skills are thinking about certain topic in depth, asking questions, and finding relevant information [8]. Critical thinking skills include: 1) actively looking for all sides of the argument, 2) testing the strength of claims made, and 3) testing the strength of the evidence used to support the claims. Critical thinking helps in creating strong arguments [9]. Students who are active in learning can train their scientific reasoning in developing critical thinking skills [10]. Based on this background, this study aims to analyze students' critical thinking skills at junior high school in science learning.

2. Research Method

Instruments used in this study were created according to Facione indicators of critical thinking skills namely to interpret, to analyze, to evaluate, to infer, to explain, and to do self-control. Subjects in this study were 256 students of seventh grade at junior high school in Karanganyar. Purposive sampling were used as sampling technique. Data were collected from students' answer toward questions using evaluation rubric that has been validated.

3. Results and Discussion

Critical thinking processes can improve activities in identifying problems, investigating data, discussing findings, evaluating findings, making solutions, presenting solutions, and presenting learning outcomes [11]. Results showed that two indicators have gained good category namely to analyze that was 62% and to evaluate that was 70%; while four other indicators have gained enough category namely to interpret that was 35%, to infer that was 49%, to explain that was 40%, and to do self-control that was 46%. Results of students' critical thinking skills per indicators in percentage were shown in Figure 1.

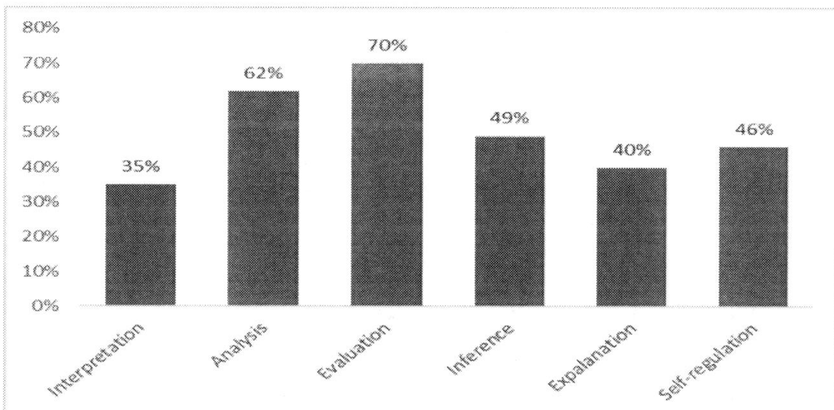

Figure 1. Results of Indicators in Students' Critical Thinking Skills

The best percentage result of indicator in critical thinking skills is to evaluate with a percentage of 70%. To evaluate is an indicator including commenting on successes and failures of something, or evaluating something [12]. Evaluation activities that have been mastered by students can be presentation skills, performance evaluation, and applying knowledge about product making, synthesizing information, and giving reasons based on former knowledge and information [13]. Evaluation can also be in the form of interpreting statements about the description / experience of analytical activities [14]. Evaluation activities can be used to make judgments about information [15].

Another indicator that is classified as good is to analyze with a percentage of 62%. This indicates that students can analyze evidence and facts that differentiate from presumptions [16]. Analysis, synthesis, and evaluation are the results associated with critical thinking activities, seeking knowledge, understanding, and application [17]. The next indicator which is classified as sufficient is to infer with a percentage of 49%. This indicates that students have not been maximal in inferring their findings. To infer is the ability to use two or more information from a text to arrive at the third information implied

[18]. Conclusion is one aspect of understanding in student literacy and reading skills toward learning materials [19]. The activity of drawing conclusions requires consideration of the analysis that has been made [20].

The fourth indicator that is classified as sufficient is the indicator to do self-control with a percentage of 46%. This indicates that students are not maximal in indicator of self-control. Self-control activities are monitoring the activities of individuals' knowledge through questions and correction of results [14]. The fifth aspect which is quite sufficient is to explain with a percentage of 40%. This indicates that students are not maximal in the indicator of explanation. Explanatory indicator require students to think critically to present their thoughts and justify arguments regarding consideration of evidence [21]. Students' critical thinking skills are accustomed to assessing any information, explaining the reasons, and being able to solve unknown problems [22]. The last indicator that is quite sufficient is to interpret with a percentage of 35%. This indicates that students are not maximal in aspects of interpretation. This indicator is an activity that understands the meaning of various experiences / events [14]. Interpretation can be an explanation of the meaning of something [9]. Interpretation includes sharing information into smaller parts to achieve greater understanding [15]. Indicators of critical thinking that are still relatively sufficient can be increased and trained by using inquiry. Inquiry learning is designed to help students gain an in-depth understanding of the scientific method while developing critical thinking [23]. Science learning is carried out by inquiry-based to empower critical thinking [24].

4. Conclusion

Results of each indicators were different and unique from one to another. Results showed that two indicators have gained good category namely to analyze that was 62% and to evaluate that was 70%; while four other indicators have gained sufficient category namely to interpret that was 35%, to infer that was 49%, to explain that was 40%, and to do self-control that was 46%. The conclusion of this study was there were two indicators of students' critical thinking skills gained good category and four indicators of students' critical thinking skills gained sufficient category. Thus proper follow-up of this study is needed to explore ways to improve the achievements of each indicators in students' critical thinking skills.

References

[1] Amornkitpinyo T and Wannapiroon P 2015 *Proc. Soc. and Behav. Scie.* **174** 2091

[2] Hidayah R, Salimi M, and Susiani T S 2017 *J. Tam. Cend.* **1** 127

[3] Fatimah A S and Santiana 2017 *J. Ling. and Eng. Teach.* **2** 126

[4] Mishra P and Mehta R 2017 *J. Dig. Learn. Teach. Educ.* **33** 7

[5] Boyaci S and Atalay N 2016 *Inter. J. Instruc.* **9** 134

[6] Ball A, Joyce H D and Butcher D A 2016 *Inter. J. Sch. Soc. Work* **1** 2

[7] Wijaya E Y, Sudjimat D A, Nyoto A 2016 Pros. *Sem. Nas. Pend. Mat.* **1** 266

[8] Fisher A 2014 *Berpikir Kritis: Sebuah Pengantar* (Jakarta: Erlangga)

[9] The Open University 2008 *Skills for OU Study Thinking Critically* (London: Thanet Press)

[10] Nichols M, Cator K and Torres M 2016 *Challenge Based Learner User Guide* (California: Digital Promise)

[11] Changwong K, Sukkamart A, and Sisan B 2018 *J. Int Studies* **11** 45

[12] University of Leeds. (n/d). *Understanding critical thinking*

[13] Sumarni W, Supardi K I, Widiarti N 2018 *Conf. Ser.: Mater. Sci. Eng.* p 2

[14] Facione P A 2015 *Critical Thinking; A Statement of Expert Concensus for Purpose of Educational Assessment and Instruction* (Newark: American Philosophical Association)

[15] Bloom B S 1956 *Taxonomy of Educational Objectives. Handbook 1, Cognitive Domain* (London: Longman)

[16] Belecina R R and Ocampo J M 2018 *Int. J. Educ. Stud.* **10** 112

[17] Styron R A 2014 *Systemics, Cybernetics, and Informatics.* **12** 26

[18] Kispal A 2008 *Effective teaching of inference skills for reading: Literature review* (National Foundation for Educational Research)

[19] Rumainah 2018 *Ta'dib: J. of Islam. Educ.* **23** 61

[20] Ennis R H 1985 *Critical Thinking* (Urbana: University of Illinois)

[21] Piaw C Y 2004 *Creative and Critical Thinking Styles* (Malaysia: Universiti Putra Malaysia)

[22] Thomas T 2011 *Asian Soc. Scie.* **7** 30

[23] Eggen P and Kauchak D 2012 *Strategi dan Model Pembelajaran Mengajarkan Konten dan Keterampilan Berpikir Edisi Keenam* (translated by Satrio Wahono) (Jakarta: Indeks)

[24] Damanik D P and Bukit N 2013 *J. Pend. Fis.* **2** 24

Analysis of Students' Misconception Based on the Use of Learning Objectives in Classification of Materials and Their Properties

M. A. S. Cahyanto*, Ashadi, and S. Saputro

Science Education, Postgraduate Program, Sebelas Maret University, Ir. Sutami St. No. 36A, Kentingan Jebres, Surakarta, Indonesia

*Corresponding author e-mail: muh_agung2407@student.uns.ac.id

Abstract. The objectives of this study are to: (1) know the percentage of students' misconception on materials and their properties in terms of the use of learning objectives at junior high school, (2) know the flaws of learning objectives that lead to students' misconception and (3) be the supportive material from learning objectives. This study was carried out using descriptive-quantitative research method. The technique of collecting data was done by interview. The conclusions of this study were: (1) percentage of student' level of understanding of classification of material and their properties was 38.43% for distinguishing elements, compounds, and mixtures indicator; 52.31% for distinguishing homogeneous and heterogeneous mixtures indicator and 48.71% for distinguishing physical or chemical changes indicator, (2) the components in the book were not complete because there was no indicator of competency achievement and (3) there is a conceptual error in writing hydrogen elements as a semilogical element, some confusing sentences lead to different interpretations and less depth in material presentation of learning objectives lead to students' misconception as students' concepts different from experts from material and language content.

1. Introduction

Education in Indonesia at its time requires a discussion in the field of learning in order to improve the quality of education and anticipate the weaknesses of conventional learning, so the combination of learning methods, learning resources, and learning media should be better. Learning is something that are external and intentionally designed to support the occurrence of internal learning processes in individuals. This is in line with the explanation who stated that learning is a process that is intentionally designed to create learning activities in individuals in certain condition and environment [1]. Teaching materials as learning objectives are an important requirement in the teaching and learning process because the presence of proper teaching materials can lead to good learning so that proper and correct can be held. Teaching materials can be interpreted as all forms of material arranged systematically that allow students to learn by designing according to the applicable curriculum [2]. Technically, learning objectives can be designed as a representation of the explanation of the teacher, lecturer, or instructor in front of the class in addition to acting as a guideline for learning activities including targets and means to be achieved. Information, descriptions, and messages that should be delivered and information that should be presented can be collected in learning objectives. Teaching and learning objectives have various types, both printed and non-printed. Printed teaching materials that are often encountered include handouts, books, modules, brochures, and student worksheets. Teaching and learning objectives must contain proven experience as well as students' interests and needs. The latter gives the implication that schools need to create emergency curriculum to meet students' interests and needs. Learning objectives for students cannot be taken solely from textbooks classified in the form of strict disciplines and provide stimulation to children to experiment. One of the subjects in the teaching and learning objectives is science subjects. Science learning aims to develop curiosity and a positive attitude towards science, technology and society. In science learning, there are still many students who do not understand the contents of the material that already exists in the book, even many science books are still incomplete. It is not clarified with images so that students experience problems in understanding the material and there is still material that is not in accordance with the concepts described, giving rise to misconceptions in the material. These students are said to experience misconceptions, if the conceptions students have are not in accordance with the conception of scientists [3]. Misconception is the source of students' difficulties in learning science. Learning that does not consider students' initial knowledge lead to more complex and more stable misconceptions. Misconceptions are seen as important inhibiting factors for students as equal as teachers in learning and teaching science [4]. The importance of the misconception problem is the solution by analyzing teaching materials so that students further enhance their broader insights. If we observe today, there are so many teaching books that are widely circulated, whether they are books published by the government or the private sector. But currently there has never been a deeper study of the content in the teaching materials.

2. Experimental Method

This study was carried out using descriptive-quantitative research method to know the percentage of students' misconception on materials and their properties in terms of the use of learning objectives at junior high school, know the flaws of learning objectives that lead to students' misconception and be the supportive material from learning objectives. This type of research was a case study. This study used descriptive-quantitative analysis techniques. Descriptive-quantitative analysis was done by analyzing and presenting facts systematically so that it was easier to understand and conclude. By giving a review or interpretation of the data obtained, the results become clear and meaningful compared to just the numbers. In this study, researchers only use primary data. Primary data is data that is obtained or collected directly from the object of research by the person conducting the research. Primary data is also called original data or new data. The primary data to be obtained were data on the analysis of teaching materials as well as data on the percentage level of students in understanding the concept. The research subjects in this study were students' learning objectives of class VII SMP / MTs in Natural Science subjects published by the Ministry of Education and Culture 2016 within 3rd chapter namely classification of material and their properties.

2.1 Data collection technique

2.1.1 Interview

The interview is a method of collecting answers or information that is done by giving respondents direct questions related to the problems [5]. Interviews were conducted on classroom teachers who taught Natural Sciences subjects to obtain information on the quality of Natural Sciences teaching materials as learning objectives. Interviews were noted through the documents during the process as the results.

2.1.2 Data analysis technique

Interpretation of data in this study was started from the beginning until the end of data collection. Interpretation of data was needed to help researchers to describe the events or situations that take place in the class overcome by researchers when conducting research. Data from the results of research in the field were processed and analyzed descriptively and qualitatively using statistics and numerical interpretation.

3. Result and Discussion

3.1Results

Table 1. Students' Learning Achievement

No.	Indicator	Percentage (%)
1	Distinguish elements, compounds and mixtures	38.43
2	Distinguish homogeneous and heterogeneous mixtures	52.31
3	Distinguish physical / chemical changes from changes in substances	48.71

Table 1 showed that students' learning achievement on materials and their classification were defined through three sub-chapters, including: distinguish elements, compounds and mixtures; distinguish homogeneous and heterogeneous mixtures; and distinguish physical / chemical changes from changes in substances. Based on the results, the highest percentage of students' learning achievement was on sub-chapter distinguish homogeneous and heterogeneous mixtures.

3.2 Discussion

This study was carried out to determine students' misconception. Misconceptions were caused by several things: 1) students, 2) teachers, 3) textbooks, 4) contexts, and 5) teaching methods [6]. It can be seen from the results of the learning outcomes based on students' learning achievement percentage. In this study, the causes of misconceptions in terms of teaching materials used in learning. Teaching materials used in schools that are observing by researchers are students' book published by the Ministry of Education and Culture in 2016.

In the students' book published by the Ministry of Education and Culture in 2016, it does not contain indicators developed so that it cannot be compared between the ones in the book with the curriculum 2013 curriculum due to the main competence and fundamental competence contained in the syllabus. Regarding the material contained in the book, it is in accordance with the syllabus, only the material contained in it is not very deep in the explanation so that sometimes the concepts listed in the book do not reach the reader who are students. This is also reinforced by interviews with teachers who explained that the book used as a learning resource was less profound in discussing the concept of classification of materials and their properties so that students often experience misconceptions in certain sub-sections which is to distinguish elements, compounds and mixtures .

Concept is a series of statements, ideas that are interrelated about various events and become the basis / guidance in conducting research. Based on Table 1, it is known that there are several indicators on material classification and changes that students still do not understand the concept or can experience misconceptions in it [7]. To be able to find out about misconceptions or not concept understanding, researchers interviewed a number of students to ensure that these students really experienced misconceptions. These indicators include the following:

3.2.2 Distinguish elements, compounds and mixtures

In this section, it is known that only 38.43% of students have completed the indicator while the rest still do not understand or experience misconceptions. 52% of students have still experienced misconceptions due to the concept of distinguishing elements, compounds, and mixtures [8]. In this book, there is a sentence that makes students experience misconceptions in distinguishing between elements and compounds, namely in terms of compounds, meaning that it is a single / pure substance that can be broken down into two or more simpler substances. This understanding indicates that compounds with elements are the same, namely the same and single substance together, this is what often makes students experience misconceptions in distinguishing the two. To make students not confused in distinguishing it, it can be written that elements are substances that cannot be broken down into simpler substances, whereas compounds are substances that consist of two or more elements [9]. Eliminating a single word from the previous understanding can make students understand more. The concepts in chemical science materials are presented in three levels, namely macroscopic, submicroscopic, and symbolic levels [10]. According to Johnstone, a proper understanding of the concept on chemistry can be achieved if students are able to link the three levels of understanding [11]. If not, students will experience confusion in understanding a concept and do not rule out the possibility that students will construct alternative concepts to make it easier to understand the real concept. The formation of alternative concepts shows that there are real differences in the individual in understanding reality and experience [12]. There are also fewer examples in the book. In this part of the book, especially in the periodic part of the element, there are conceptual errors, namely the element hydrogen is a semi-metallic element. Semilogam elements are elements that have a suffix between metals and non-metals. In this case the element hydrogen is a nonmetallic element because hydrogen is gas and does not have metal properties in it [9]. This concepts error or alternative concepts are very dangerous because it includes a basic concept. If students experience misconceptions in one of the basic concepts, then the possibility of the emergence of misconceptions in more complex concepts will be even greater as long as they learn science both in contextual way or theoretical [13].

3.2.3 Distinguish homogeneous and heterogeneous mixtures

Indicator to distinguish homogeneous and heterogeneous mixtures has been observed too. Based on the results of the daily test, it is known that the percentage of indicator completeness is 52.31% of students. For this indicator, the percentage is greater than the indicator distinguishing elements, compounds, and mixtures. Approximately 34% of students still experience misconceptions in this indicator [8]. In this section, the contents of the book are still not very detailed, what is a homogeneous mixture and what is a heterogeneous mixture. It can be further deepened by providing differences from each one so that students here as readers can understand the whole. The location of misconceptions that occur here based on interviews with students is that students are still confused in distinguishing between homogeneous and heterogeneous mixtures, as one example of the wrong concept here is that the mixture of water and salt is a suspension. Suspense itself belongs to a heterogeneous mixture. This is the reason why students understand why a mixture of water and salt includes suspense because both are composed of solid and liquid substances. This is certainly wrong because even though it is composed of solid and liquid substances, a mixture of water and salt is a homogeneous mixture because the substances are not able to be seen again or differentiated. For example, if the amount of sugar is dissolved in water and stirred well, the mixture will basically be uniform in all parts [14].

3.2.4 Distinguish physical / chemical changes from changes in substances

In this indicator, the percentage of completeness is 48.71%, which means that only about 48.71% of students have completed the indicator. The rest are still do not understand the concept and happen to experience misconceptions. Misconceptions occur because students assume that all changes in the form of substances produce new substances even though this is not true because the changes in the form of substances, the constituent elements remain the same as the original substances, but only change their form. After seeing the location of the misconceptions in this indicator section, the researcher then looked at the contents of the book in the sub section of the physics and chemical changes. It is known that the contents of the material in it are good because they are presented with experiments that require students to conduct the experiment, so the learning process has gone well because here students experience their own learning experiences. It's just that here it is not explained as to why the process of changing matter includes physical changes. Students are sometimes still confused about the concept, they assume that the evaporating process is a change in the form of a substance from water to steam. Students argue that water and steam here are different substances, whereas between water and steam here are the same substances, namely water, it's just a different form, one in the form of a liquid while when it has undergone the process of evaporation into gas.

4. Conclusion

The conclusions of this study were: (1) percentage of student' level of understanding of classification of material and their properties was 38.43% for distinguishing elements, compounds, and mixtures indicator; 52.31% for distinguishing homogeneous and heterogeneous mixtures indicator and 48.71% for distinguishing physical or chemical changes indicator, (2) the components in the book were not complete because there was no indicator of competency achievement and (3) there is a conceptual error in writing hydrogen elements as a semilogical element, some confusing sentences lead to different interpretations and less depth in material presentation of learning objectives lead to students' misconception as students' concepts different from experts from material and language content.

References

[1] Pribadi, B.A 2010 *Model Desain Sistem Pembelajaran* (Jakarta: Dian Aksara)
[2] Lestari, I 2013 *Pengembangan Bahan Ajar Berbasis Kompetensi : Sesuai dengan Kurikulum Tingkat Satuan Pendidikan* (Padang : Akademia)
[3] Tayubi, Y R 2005 *Jurnal Mimbar Pendidikan* **3** 5
[4] Suratno, T 2007 *Peranan Konstruktivisme dalam Pembelajaran dan Pengajaran Sains* Seminar Internasional Pendidikan IPA FITK UIN Jakarta
[5] Norman K D & Yvonna S L Eds 2000 *Handbook of Qualitative Research* Terj oleh Dariyatno 2009 (Yogyakarta: Pustaka Pelajar)
[6] Suparno, P 2013 *Miskonsepsi dan Perubahan Konsep dalam Pendidikan Fisika* (Jakarta: Gramedia Widiasarana)
[7] Handhika J, Cari C, Suparmi S and Sunarno W 2015 *Int. Conf. on Math. Sci. and Ed.* **4** 34-37
[8] Tuysuz, C 2009 *Journal Scientific Research and Essay* **4** 626 - 31
[9] Brady, J 2005 *Kimia Universitas* (Tangerang: Binarupa Aksara)
[10] Nahum. T.L., Hofstein, A., Naaman, R.M., dan Bardov, Z 2004 *Can Final Examination Amplify Students' Misconception in Chemistry?* Chemistry Education: Research and Practice **5** (3): 301- 25.
[11] Rahayu, S & Kita, M 2010 *International Journal of Science and Mathematics Education* **(8)** 667 - 88.
[12] Marton, F. 1981. *Phenomenography – Describing Conceptions of the World Around Us. Instructional Science* **10** : 177-200.
[13] A'yun, Q., Harjito, Nuswowati, M 2018 *Jurnal Inovasi Pendidikan Kimia.* **12** : 2108 - 17.
[14] Setyarini, L. W 2012 *Jurnal Teknik Pomits* **1**: 1-5.

Profile of Students' Creative Thinking Skills using Open-ended Multiple Choice Test in Science Learning

R. D. Pratiwi*, Ashadi, and Sukarmin
Science Education, Postgraduate Program, Sebelas Maret University, Ir. Sutami St. No. 36A, Kentingan Jebres, Surakarta, Indonesia

*Corresponding author e-mail: rintadian01@student.uns.ac.id

Abstract. This study aimed to know the profile of students' creative thinking skills in grade VII of junior high school during the process of science learning. This study was carried out using quantitative-descriptive method. Data collection were done by using open-ended multiple-choice tests based on students' creative thinking skills indicators. The test was previously validated by learning evaluation experts. Validity of test items were determined using Karl-Pearson formula related to product moment correlation coefficient. Reliability of tests were determined using Alpha-Cronbach formula. Difficulty level and discriminating power were also used as item parameters development of test instruments in this study. The sample taken in this study was 256 students which consisted of all seventh grade students in SMP Negeri 1 Jaten selected through purposive sampling technique due to identify students' creative thinking skills. The results of this study showed that students' creative thinking skills on each indicator were 49% of fluency which was in the low category, 72% of flexibility which was in the medium category, 34% of originality which is in the low category, and 35% of elaboration which is in the low category. The conclusion from this research is that creative thinking skill of the students still was in low category, so that needed a way to enhance it on some indicators.

1. Introduction

Science has experienced rapid development and progress. It was in line with the development of increasingly sophisticated human thinking. Indonesia is one of the developing countries as this country will not be able to progress rapidly as long as it has not improved the quality of human resources. The quality of life of the nation can increase if it is supported by an established education system. In the 21st century, high-level thinking skills have strategic values to be applied. Skills in the 21st century skills rainbow knowledge that learning and innovation skills [1]. Learning and innovation skills include critical thinking, collaborative skill, communication, and creative thinking. All countries from poor, developing, and developed categories need the ability to develop creative thinking [2]. The ability to think creatively has a strategic role towards the advancement of science and technology. This is based on the development of times that require individuals to be able to face global competition, so that an individual has the ability to create new ideas or ideas obtained from the ability to think creatively. Within the 21st century competency framework, it shows that students must have life and career skills, learning and innovation skills, the ability to use information and communication [3]. Creative thinking skills are defined as skills needed in most or almost all subjects [4]. The importance of each individual to enhance creative thinking skills. Thus important creative thinking skills are possessed by students from learning activities to complete the learning objectives.

Creative thinking skills are defined as a thought process that allows students to apply their imagination to generate ideas, questions and hypotheses, experiment with alternative conceptual and to evaluate ideas, processes and products implications [5]. Creative thinking skills is related to think and find something, about things that produce something new by using something that already exists [6]. The results of these thoughts can be given in the form of real and abstract ideas. Creative thinking skills are always associated with a unique or different from the others because creative thinking skills is purely the result of a persons' thinking, therefore, creative thinking skills are synonymous with originality [7]. Sub dimensions to the creative thinking skills in learning process are introduced by Guilford, which includes fluency (ability to generate lots of ideas, answers, problem solving, or questions), flexiblity (ability that produces ideas vary from the information that has been obtained), originality (generating ideas or ideas that are new and different from before) and elaboration (the ability to develop and add ideas in detail that seems more interesting) [8]. The creative thinking skills can be formed through divergent thinking stages. The process stage in creative thinking includes preparation, incubation, illumination, and verification or evaluation [9]. Creativity can be maintained by teachers and learning environments that encourage questions, open to new ideas, and learn from mistakes and failures [1].

The study of measuring creative thinking skills has not been done much. Studies related to creative thinking skills in the countries of India, Israel, Hong Kong, Germany, and Romania were still relatively low, so it is important to conduct further studies as the contribution to educators, students, and education [10]. Although it is difficult to assess creativity, there are several instruments and assessments that have been designed to measure creativity in a particular field [11]. Studies to find out the initial profile of students can be done in various ways depending on the category or dimension to learn. The study to find out students' creative thinking skills as a initial profile is very important as it has strong connection with the existence of strategic contributions. An initial assessment of student competence in overcoming learning difficulties can be done by analyzing the initial profile of students' creative thinking skills. The data obtained can be used as a consideration in evaluating the learning process in class, especially in science learning for seventh grade students in SMP Negeri 1 Jaten in Karanganyar, Central Java, Indonesia. The low level of students' creative thinking skills can be caused by one of indicators because the learning process is not even optimal. Science is a subject that is difficult to learn for students because science are consisted of branches namely biology, physics, and chemistry. Science learning difficulties are also caused by difficulties in finding relationships between concepts and misconceptions that occur regularly. The aim of this study was to know the initial profile of students' creative thinking skills in science learning.

2. Research Method

This research is a quantitative-descriptive method. The test method was chosen to describe the percentage to the ability to think creatively in seventh grade students of SMP Negeri 1 Jaten in Karanganyar, Central Java, Indonesia. The sample taken in this study was 256 students which consisted of all seventh grade students in SMP Negeri 1 Jaten selected through purposive sampling technique due to identify students' creative thinking skills. Students are given tests to analyze their ability to think creatively, especially in science learning. The test form is open-ended multiple-choice test and the questions have been developed by indicators of creative thinking skills that previously performed the validation of evaluation learning experts. Validity of test items were determined using using Karl-Pearson formula related to product moment correlation coefficient. Reliability of tests were determined using Alpha-Cronbach formula. Difficulty level and discriminating power were also used as item parameters or the test used. Science topics in the problem include everyday life phenomena regarding temperature and heat. Scoring from open-ended multiple-choice test and rubric that has been validated by experts from each indicator. Furthermore, the test data were analyzed with percentage to find the indicator creative thinking skills which the student has not passed by using the formula:

$$Percentage\ (\%) = \frac{score\ of\ each\ indicator}{total\ score\ of\ each\ indicator} \times 100\%$$

The data obtained in the form of the results of creative thinking skills tests are processed by calculating the percentage of scores obtained by students on each indicator of creative thinking skills and then the percentage is interpreted according to the criteria of the level of students' creative thinking skills as it can be seen on Table 2.

3. Results and Discussion

3.1 Results

This research has been conducted to analyze the initial profile to students' creative thinking skills. The initial profile of students' creative thinking skills is represented by four indicators contained in the questions, including fluency, flexibility, originality, elaboration. The test questions used in the form of open-ended multiple-choice test based on creative thinking skills that have been validated by the expert team of evaluation. Validity of test items were detemined using using Karl-Pearson formula related to product moment correlation coefficient. Reliability of tests were determined using Alpha-Cronbach formula. Difficulty level and discriminating power were also used as item parameters. The results of the students' initial profile test are presented in Table 1 while analysis of the test results refers to the creative thinking level criteria presented in Table 2 [12].

Table 1. Profile of Students' Creative Thinking Skills

Indicator	Percentage (%)	Criteria
Fluency	49	Low
Flexibility	72	Medium
Originality	34	Low
Elaboration	35	Low
Mean	47.5	

Table 2. Criteria of Students' Creative Thinking Skills Level

Percentage (%)	Criteria
< 55	Low
≤ 55 - < 75	Medium
≥ 75	High

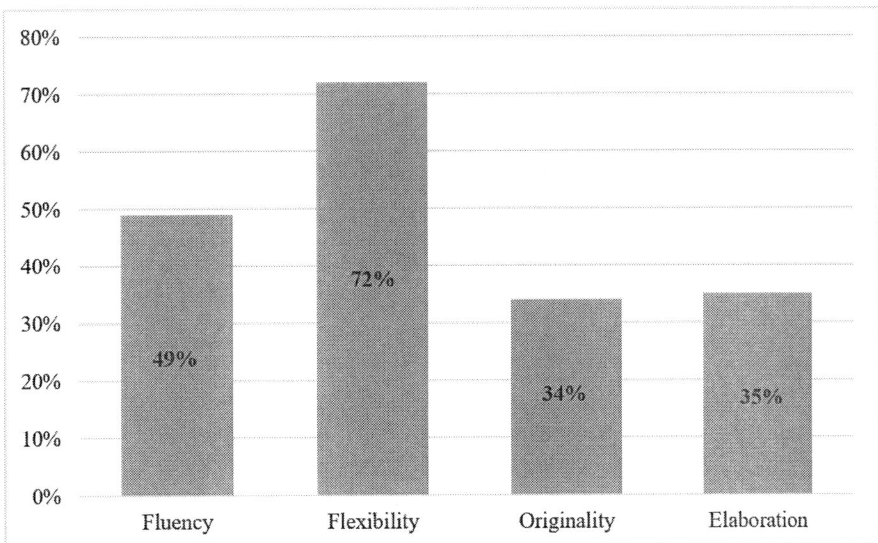

Figure 1. Percentage of the Students' Creative Thinking Skills

This is a sample of answer.

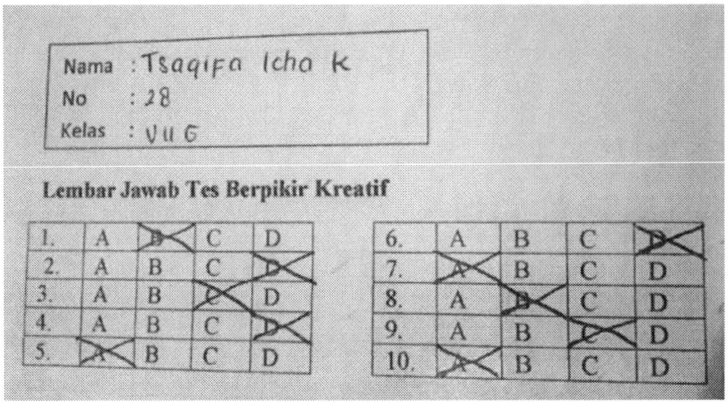

Figure 2. Sample of students answer on creative thinking skill test

3.2 Discussion

Table 1 showed the level of achievement of students' creative thinking skills in each indicator. Results showed that students have medium and low creative thinking skills due to the creative thinking skills of Table 2. Fluency indicator has been owned by 49 % of students who have been able to answer the test by getting the maximum score on this indicator so that it is interpreted low. Flexibility indicator is still being interpreted with a percentage of 72%. The interpretation of the originality indicator is still low because students who are able to answer with a maximum score of only 34 %. Elaboration indicator has gained 35% of the percentage so it is interpreted low. The achievement of indicators of students' creative thinking skills in SMP Negeri 1 Jaten is still low, especially in the indicators of fluency, originality, and elaboration. Based on the results of the initial profile test, the ability to think creatively has not reached an ideal score. This finding is in line with some of the previous studies' results that showed the low category of students' creative thinking skills in Indonesia. Meanwhile, low category of students' creative thinking skills has apparently not only appeared on a national scale. The other study has also found a decrease of students' creative thinking skills in France [13]. The assessment of the ability to think creatively using the Guilford indicator shows that most students almost reach the target in flexibility indicator as it can be seen in the results of the initial profile test of students within percentage of 72%. In contrary, the percentage of achievement of students' creative thinking skills that is very far from the target is originality indicator. According to previous research, limited or unsatisfactory achievement on the other three criteria includes flexibility, elaboration, and originality were passed by originality [14]. In addition, the most important indicator of creative thinking skills is originality [15]. Originality is the ability to produce new and different ideas or ideas. Students can achieve originality indicators if they are accustomed to practicing their creative thinking skills with different points of view because these sub-dimensions depend on several factors [14]. Based on these results indicated, there is a result that the way students think have not been optimally trained,especially when they are asked to think creatively. Therefore, it is important to empower creative thinking skills in learning, especially in natural science subjects as it has been given the strategic value.

4. Conclusion

The conclusion of this study is the initial profile of the creative thinking ability of class VII SMP N 1 Jaten in science learning which is still relatively low. The results of this study showed that students' creative thinking skills on each indicator were 49% of fluency which was in the low category, 72% of flexibility which was in the medium category, 34% of originality which is in the low category, and 35% of elaboration which is in the low category. These results indicate that students' creative thinking skills need to be empowered. Efforts to empower creative thinking skills are important to be considered given the low percentage of each indicator.

References

[1] Trilling B and Fadel C 2009 *21st Century Skills : Learning for Life in our Times* (San Francisco: Jossey-Bass)
[2] Freudenberg B, Brimble M and Cameron C 2011 Work-integrated learning and generic skill development: The development of business students' generic skills through work-integrated learning *Asia-Pacific Journal of Cooperative Education* **12** no.2 pp 79-93
[3] Partnership for 21st Century 2009 *Framework for 21st century learning* Retrieved 28th March 2019 from http://www. p21.org/*about*-us/p21-framework
[4] Heilman G and Korte W B 2010 *The Role of Creativity and Innovation in School Curricula in the eu2* (Luxembourg: Publications Office of the European Union)
[5] Kampylis P and Berki E 2014 *Nurturing Creative Thinking* (Belle France: UNESCO)
[6] Dariyanto 2009 *Panduan Proses Belajar Kreatif dan Inovatif* (Jakarta: Buku yang Cerdas dan Mencerdaskan)
[7] Massoudil M 2003 Can scientific writing be creative? *Journal of Science Education and Technology* **12** pp 115–28
[8] Guilford J P 1975 Varieties of creative giftedness, their measurement and development *Gifted Child Quaterly* **19** pp 107-21
[9] Sitorus J and Masrayati 2016 *Thinking skills and Creativity* **22** pp 111-20
[10] Hürsen C, Kaplan A and Özdal H 2014 *Procedia-Social and Behavioral Sciences* **143** pp 1177-85
[11] Pacific Policey Research Center 2010 *21st Century Skills for Students and Teachers* (Honolulu: Kamehameha Schools, Reasearch and Evaluation Divition) p7
[12] Shriki A 2013 A model for assessing the development of students' creativity in the context of problem posing *Creative Education* **4** no.7 pp 430–39
[13] Georgsdottir A S and Lubart T I 2003 *Psychologies Francaise* **48** pp 29-40
[14] Tran T B L, Ho T N, Mackenzie S V and Le L K 2017 *Thinking Skills and Creativity* **25** pp 20-6
[15] Siswono T Y E 2011 Level of student's creative thinking in classroom mathematics *Educational Research and Review* **6** no.7 pp 548-53

Misconception Analysis Based on Feedback of Computational Thinking Result of College Students

S Masfuah* and F Fakhriyah

Primary Educational Teacher Department, Universitas Muria Kudus, Indonesia

*Corresponding author e-mail: siti.masfuah@umk.ac.id

Abstract. This research has purposes to analyse misconceptions of college students based on feedback of their computational thinking test. This descriptive quantitative research involves fourth semester college students of 4C class of primary school education teacher program by using random sampling. The instruments are test instrument consisting of computational thinking indicators and questionnaires semi structured to identify and analyse of factor causing misconception of the students. The results were then analysed descriptively by using cluster analysis and crosschecked to questionnaire and interview results. The findings show computational thinking of the students was within algorithm stage meanwhile misconception is found in movement system material. The causal reasons of the misconception are grouped into 3 clusters. Cluster 1, the students have already understood the concept but their misconception is 33.3%. Cluster 2, misconception of the students is 38.1%. Cluster 3, the students have not understood the concept with percentage 28.6%. The contributing factors to misconception are students' arguments that animals/plants have similar natures to human, handbook, lecturer's explanation, scientific name, mathematics calculation, formula, graphic understanding, influences of peer in answering, and students' experiences in daily life.

1. Introduction

Human Resources and technology are great problems in 4.0 industrial revolution era. Its mastery by human becomes competitive power of Indonesian people to face the era. In this era, people are demanded to think comprehensively by using technology to prevent us becoming slaves of technology. Therefore, there is a need to improve skills to keep irreplaceable human resource position from technology advancement. Skill improvement may be done through educational field.

Primary school education teacher of UMK is an institution to prepare primary school teacher to have further contribution because they have roles to prepare Science, Technology, and Culture to future generation. Thinking ability, technology mastery, and life skill are abilities to have in this era. The future generations will have those skills when teachers are capable and skillful. Therefore, improvement is always done to have better and qualified skill graduate outputs.

The required skills of 21st century are scientific literacy and computational thinking [1]. Scientific literacy is an ability to understand and implement science in an individual's life [2]. It must be owned by students since it contributes to technological mastery. Technology advancement is caused by scientific innovation found by researchers. Scientific literacy is part of science, technology, and social developments whose high thinking level components, scientific attitudes, social attitudes, and integrated studies [3]. The implementation of scientific literacy will be maximal when an individual has computational thinking ability. It is used to solve complex problem by involving ideas, models, simulations as if it were computer [4]. It is needed since not every problem can be solved only by thinking but it needs to integrate various studies by using technology, model, and simulation [5],[6]. Computational thinking is also needed to innovate in science, technology, enginery, and mathematics fields (STEM) [7]. Computational thinking as scientific domain and scientific literacy as its implementation are needed by experts and practitioners to innovate. However, the students have not had those skills maximally.

Scientific literacy of the students are still on nominal level [8]. This level allows to identify a conceptual formation but not to understand the concept clearly [3]. Students will have better literacy skill when they have better CT skills because the skill is a meant to implement scientific literacy. After measuring the scientific level of the students, they were given learning material based on scientific literacy completed by CT competences to improve both scientific literacy and computational thinking. Based on the posttest, the students were in level of finding problems and had not had capability to promote algorithm [9]. The students had found solution but could not solve problems orderly and systematically. In fact, based on score of given book's readability, the students argued the book was readable and understandable [9]. The students felt the book was readable but the posttest showed low ability. The previous study conducted on applied science course had not shown any maximum ability [10]. On the course, the students were demanded to create a project by creating tool or mini research based on the already given science concept. Basically, the students had owned good ideas dealing with mini research and manipulative demonstrative tool to create. However, the students still had difficulties to apply the concept and to relate the concept o other concepts, and to analyze the project. Those proved that students had understood science concept but they could not think comprehensively in solving given problem. It identified students had misconception or missing the concepts.

Initially, students had owned pre-concept. This concept functions as basic initial thinking of the students. When they join learning process, this concept experienced changes. It might be concept addition, concept modification, and event concept changes contributing to latest resulted concept. This change is as process of bringing students' thought to proper scientific concept [11]. In this new concept formation, mistakes, misses, or unsynchronized initial concepts to what is learnt occur, called as misconception.

Misconception occurs due to inappropriateness of the understood concept to learnt theory. Misconception is permanent within students' thoughts and difficult to change [12]. Misconception occurs in students' thoughts in reviewing concept into conception and becoming facts [13]. If the formed intuition by students is not correct, then it is difficult to fix because consistently a wrong concept has become the guidance [14]. Therefore, it needs to seek misconception causes of the students because initial understanding of them influences further learning.

Misconception of the students is difficult to evade but it can be prevented when its causes are figured out [15]. Misconception measurement is important for a department or a program because it is used to evaluate and improve the program [16]. Many previous studies analyzing misconception of science teacher candidates but they have not reached until level of finding out misconception experienced by students. Therefore, it is important to investigate.

2. Method

It is a descriptive quantitative research with purposes to analyze misconception of students based on feedback of computational thinking result by the students. It is also a preliminary research to identify misconception of the students. The population is fourth semester students in academic year 2018/2019. The study uses C class as a research sample taken by random sampling. The researcher gave computational thinking test. Then, it was evaluated. The investigated computational thinking indicators are to analyze and find data, elaborate problems, create abstraction, find patterns of problems, algorithm, automation, simulation, and conclusion. The task was given to the students then followed by semi structured questionnaire and interview to identify and analyze the misconception The results were then descriptively analyze by using cluster analysis technique with SPSS program. SPPS analysis uses the SPSS 25 application. Cluster analysis is organizing set of patterns into cluster or group based on characteristic similarities. Cluster analysis uses non-hierarchical grouping method. The analysis is started by determining numbers of demanded clusters. In this analysis, the researcher uses three clusters. After the clusters are determined, then the subsequent process was done without adhering hierarchical process. Conclusion drawing is subjective and depends on purposes of the research [17].

3. Results and Discussion

This preliminary research about test development is to measure misconception of the students. At the beginning, the students were given task consisting of computational thinking indicators. The test was given to measure the students' CT skills. They were given essay tasks based on computational thinking indicators covering from 16 numbers. This research is specialized on movement system of low organism movement, high organism movement, influences of movement to matter and Newton law. Based on questionnaire results and learning achievement of the students, movement system material had the lowest average score. It should be evaluated by lecturer to find out the cause and to improve learning. Basically, both lecturers and teachers have been aware of the causes. Therefore, evaluation of misconception should be done periodically. Measured evaluation activity done by teachers in accordance to misconception experienced by his students done periodically [18]. The findings show basically most of teacher had been aware of misconception but they could not find out the causes. In fact, the causes are used to improve learning.

Based on CT test, the students were on finding solution level but were not in algorithm level where they should have known the answers but could not solve problems systematically and orderly. The results are indicated the students experiencing misconception because they had known solutions but they could not solve the problems well. This misconception is most frequently experienced but they could not explain in detail about it since misconception is closely related to initial concept instilled in their thoughts [19].

To analyze misconception existence, the students were given feedback questionnaire of the already done CT test. It consisted of reasons and possibilities causing mistake in answering the task. There are 21 options becoming the reasons. They are confusion to answer, pre-conception of the students, confusion to understand the tasks, confusion to mathematically formulate, confusions about the formula, confusion in conversing the measurement, confusion in equality principle, assumption of students about human-like natures of animals and plants, difficulties to calculate, meet the phenomenon daily, guess the answers, difficulties to understand graphics/diagrams, carelessness of the students, problems of memorizing scientific/Latin names, influenced answers by peers, and other reasons to write by the students.

The questionnaire is analyzed by using cluster analysis plus and SPSS program. At the beginning, 16 tasks and 21 options were input. Then, 3 clusters were determined. The data is then analyzed and the results as follow on Table 1.

Table 1. Final Cluster Centers of Students' Questionnaire

Item Z Skor	Cluster		
	1	2	3
Zscore: A. I am confused to understand the task	-.55004	.30558	.12223
Zscore: B. I do not understand the concept	-.86605	1.24770	-.19082
Zscore: C. I am confused in mathematics formulation	-.47108	.87487	-.20189
Zscore: D. I do not memorize the formulas	-.33597	.95415	-.30909
Zscore: E. I am confused in coversing the measurements or symbols	-.43301	-.43301	.43301
Zscore: F. I do not understand equality principle	-.44379	-.08876	.26628
Zscore: G. That is my pre-conception	-.24738	-.44528	.34633
Zscore: H. I assume animals/plants are like human	1.21731	-.52171	-.34780
Zscore: I. I assume things are like human	.36596	-.36596	.00000

Item Z Skor	Cluster		
	1	2	3
Zscore: J. That is what I learn from book	1.20073	-.80777	-.19648
Zscore: K. That is what my lecturer says	.86674	-.86674	.00000
Zscore: L. I believe and argue so	.41692	-1.14318	.36313
Zscore: M. I guess the answer	-.43391	-.43391	.43391
Zscore: N. I cannot calculate	.36596	.36596	-.36596
Zscore: O. I meet the phenomenon daily	-.32545	-.91718	.62131
Zscore: P. I do not hear my lecturer well	-.50833	-.69318	.60075
Zscore: Q. I am careless	-.14002	.18003	-.02000
Zscore: R. My answers are influenced	-.25000	.95000	-.35000
Zscore: S. I do not understand the graphs	-.30121	.70282	-.20080
Zscore: T. I do not memorize scientific or latin names well	.84740	-.14123	-.35308
Zscore: U. Other reasons	.23146	-.11573	-.05786

Based on Z score which is positive on the table, it can be categorized that cluster 1 consisting of reason of assuming animals/plants are like human, assuming things are like human, that is what I learn from book, that is what my lecturer says, that is my opinion, I do not memorize the Latin or scientific names, and other reasons. Cluster 2 is categorized by options consisting of my confusion to understand the concept, I do not understand the concept, I am confused to mathematically formulate, I do not memorize the formula, I cannot calculate, I am careless, my answers are influenced by my friends, and I do not understand the graphs or diagrams. Cluster 3 consists of my confusion to converse the measurement, I do not understand equality principle, it is my initial concept, I just guess the answer, I see the phenomenon, I do not listen my lecturer well. The Z scores are categorized into 3 clusters on the Figure 1.

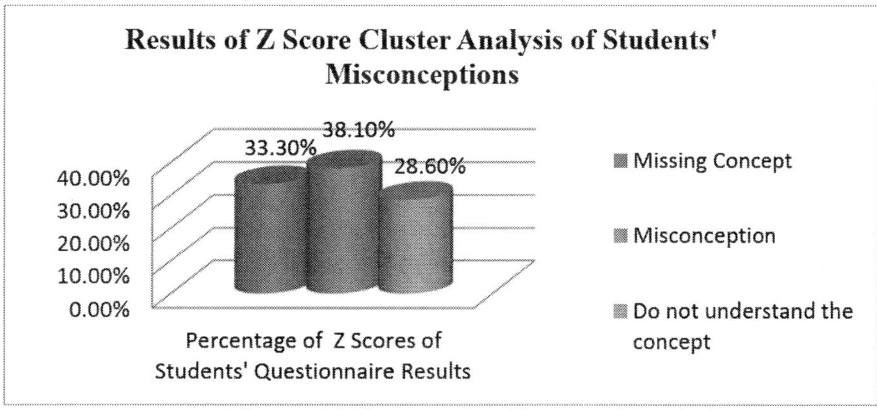

Figure 1. Misconception Clusters of the Students

Based on the analysis and procedures by using non-hierarchical method, it can be concluded that cluster 1 consisting of students whom have understood the concept but having missing concept, 33%. Cluster 2 consists of students whom understand the concept but having several weaknesses or misconceptions, 38.1%. Cluster 3 consists of students whom have not understood the concept and tend to guess the answers, 28.6%.

In cluster 1, the students actually have understood the concept but some of the concepts are missing. They were able to answer but limited by technical terms such as assuming animals, plants, or non-living things were like human. They assumed both animals and plants had similar natures to human. It means they assumed based on what they saw without having strong initial concept. They might have understood the natures of plants and animals but not in detail. They assumed that plants and animals were living creature, thus they had similar natures to human. They also assumed that non-living things owned humans' attributes because human was also a thing, without considering other parameters. Those are the misconceptions where students miss several concepts of certain material. Aufschnaiter and Christian [20] explains that missing concepts occurs when students cannot explain a specific concept explicitly. Their answers could be gained from any phenomenon they see. They do not adhere to instruction well and do not control possible affective variable. Besides that, theories of plant, seed, and movement system also cause misconception of teacher candidates [21].

Besides that, the students experienced difficulties in naming. The scientific language style tends to be more difficult to understand. Therefore, it contributes to misconception of the students. This misconception influences learning process and even learning achievement. It is strengthened by [22], finding that misconception occurs when students have problems of mastering certain language phrase related to scientific language [22]. The difference of daily language to scientific language causes confusion and misconception [23]. Language as communication meant has important role in scientific literacy [22].

Cluster 2 consists of group with misconception. They were sure of their answers but their assumptions were contradicted to current theories. Besides that they were not able to answer properly due to their limited mathematics calculation, formulas, and poor understanding about graphs and diagrams. Meanwhile, cluster 3 consists of students with very poor understanding of the concept. They tended to guess the answers without considering the background or reasons. This misconception was caused by measurement conversion, equality, pre-conception, guess, and less attention. This misconception experienced by students was caused by many factors such as environment, handbook, personal argument, and the given learning [24].

Based on Z score results, the contributive factors to misconceptions are incorrect assumption of animals/plants to own similar natures as human, handbook, lecturer's explanation, scientific names, mathematics calculation, formulas, graphic understandings, peer influences in answering, and students' experiences in daily life. It is strengthened by [18] and [25] that causes of misconceptions are students' daily life experiences, teacher's handbook, environment, friends, learning strategy, and teacher. Similar research telling that teacher is main factor to determine success of learning who sometimes triggers any misconception [26],[27]. Therefore, teachers and lecturers must be capable of predicting any misconception which may appear during learning.

After the questionnaire was given, students were interviewed about the causes of misconception and what materials or concepts are assumed to be most difficult and have most misconceptions. Based on descriptive data analysis, the most difficult materials and contributes to misconception are Newton law, effect of force to moving matter, and movement system of low level creatures

4. Conclusion and Suggestion

It can be concluded that students have misconception from the test result. Based on the data, 33.3% of them have missing concept, 38.1% have misconception, and 28.6% do not understand the concept. The most contributive factors are students' assumptions of animals/plants to have similar natures as human, the learnt textbooks, lecturer's explanation, scientific name, mathematics calculation, formula, graphic understanding, peer influences in answering, students' experiences in daily life. Based on the questionnaire and interview, the most contributive materials to misconception are Newton law, effect of force to moving matter, and movement system of low level creatures.

It is suggested to create more detail questionnaire for the students related to misconception causes experienced by them so that clustering step is more accurate.

Acknowledgments

Thanks to research team of Universitas Muria Kudus, Dr. Insih Wilujeng, M.Pd and Prof. Djemari Mardapi, M.Pd., Ph.D from Universitas Negeri Yogyakarta, to cooperate in this research. Thanks to the chief of Primary School Education Teacher Department of Universitas Muria Kudus to support and allow this research. Thanks to Direktorat Riset dan Pengabdian Masyarakat Kemenristek Dikti to financially support this research. Thanks to all parties to have helped this research.

References

[1] Wing J M 2006 *Communication of the ACM* **49(3)** 33-35
[2] R Bybee, B McCrae, R Laurie 2009 *Journal of Research in Scince Teaching* **46(8)** 865-883
[3] Holbrook and Rannikmae 2009 *International Journal of Environmental & Science Education* **4(3)**: 275-288
[4] Qualls J A and Linda B S 2010 *Journal of Computing Sciences in Colleges* **25(5)** 66-71
[5] Liu J and Wang L 2010 *IEEE 2nd International Workshop on Education Technology and Computer Science* 413-416
[6] Voskoglou M and Sherly B 2012 *Egyptian Computer Science Journal ECS* **36(4)** 45-46
[7] Swaid S I 2015 *Procedia Manufacturing* **3** 3657-3662
[8] Fakhriyah F, S Masfuah, M Roysa, A Rusilowati, E S Rahayu 2017 *Indonesian Journal of Science Education* **6(1)** 81-87
[9] Fakhriyah F, S Masfuah, M Roysa 2018 *Proceeding: Advances in Social Science, Education and Humanities Research* **262** 165-169
[10] Masfuah S and F Fakhriyah 2017 *UNNES Science Eduction Journal* **6(3)** 1708-1716
[11] Agiande D U, Williams J W, Dunnamah A Y, Tumba D P 2015 *European Scientific Journal* **11 (35)** 395-408
[12] Morgil I, Seyhan H G, Secken N 2009 *Chemistry Education* **18(3)** 53-60
[13] Wiyono F, M Sugiyanto, E Yulianti 2016 *Jurnal Penelitian Fisika dan Aplikasinya* **6(2)** 61-69
[14] Tayubi Y R 2005 *Mimbar Pendidikan* **24(3)** 4-9
[15] Sadler P M and Sonnert G 2016 *American Educator* **1** 26-31
[16] B A Couch, W B Wood, J K Knight 2015 *CBE-Life Science Education* **14** 1-11.
[17] Ghozali I 2018 *Aplikasi Analisis Multivariate dengan IBM SPSS 25 edisi 9.* (Badan Penerbit Universitas Diponegoro: Semarang)
[18] Anam I M S 2018 *International Journal for Cross-Disciplinary Subjects in Education (IJCDSE)* **9(1)** 3323-3328
[19] Schmidt A L 2011 *Creative Education* **2(5)** 435
[20] Aufschnaiter C V and Christian R 2010 *Eurasia Journal Mathematics, Science and Technology Education* **6(1)** 3-18
[21] Yangin S, Sabri S, Yasin G 2014 *Journal of Baltic Science Education* **13(3)** 105-117
[22] Chrzanowski M M, Grajkowski W, Zuchowski S, Spalik K, Ostrowska E B 2018 *Journal of Turkish Science Education* **15(4)** 29-54
[23] Page K 2012 *Science Scope* **35(8)** 12-15
[24] Thompson F 2006 *International Education Journal* **7(4)** 553-559
[25] Oberoi M 2017 *IJSRE* **5(3)** 6274-6280
[26] Lin J W, Yen M H, Liang J C, Chiu M H, Guo C J 2016 *Eurasia Journal of Mathematics, Science & Technology Education* **12(9)** 2617-2646
[27] Gomez-Zweip S 2008 *Journal of Science Teacher Education* **19(5)** 437-454

Journal of Physics: Conference Series **1397** (2019) 012022 doi:10.1088/1742-6596/1397/1/012022

Treatment of electronics industry effluent using low-cost and commercial adsorbents: A comparative study

A Kamari, R Shamsudin, A A A Eljiedi, S N M Yusoff, S T S Wong, S Ishak, J Jumadi, M M Abdulrasool and S Kumaran

Department of Chemistry, Faculty of Science and Mathematics, Universiti Pendidikan Sultan Idris, 35900, Tanjung Malim, Perak, Malaysia.

E-mail: azlan.kamari@fsmt.upsi.edu.my

Abstract. In the present study, the potential of two low-cost adsorbents namely kenaf (*Hibiscus cannabinus* L.) bast fiber and golden apple snail (*Pomacea canaliculata*) shell to treat heavy metals concentration in electronics industry effluent has been investigated. The physicochemical properties and adsorption mechanisms were determined using several analytical instruments such as Fourier Transform Infrared (FTIR) Spectrometer, Scanning Electron Microscope (SEM) and Energy Dispersive X-ray (EDX) Spectrometer. With an initial metal ion concentration of 50 mg/L, the adsorption of Cd(II), Cr(VI), Cu(II), Ni(II), Pb(II) and Zn(II) was optimized at acidic media (solution pH of 4.0-6.0), 60 minutes of contact time and 0.5 g of adsorbent dosage. The equilibrium data of metal ions adsorption onto kenaf bast fiber and golden apple snail shell followed the Langmuir isotherm model. Based on Langmuir isotherm model, the adsorption capacity of adsorbents for metal ions was in the order of Pb(II) > Cu(II) > Zn(II) > Ni(II) > Cd(II) > Cr(VI). Golden apple snail shell was able to remove Cr(VI), Ni(II) and Zn(II) ions from the electronics industry effluent at a higher capacity as compared to kenaf bast fiber and commercial mangrove stem derived activated carbon. The presence of carbonate functional group in green apple snail shell favored the adsorption of metal ions. Overall, results from this study suggest that kenaf bast fiber and golden apple snail shell have great potential to be used as alternative adsorbents to treat industrial effluents.

1. Introduction

Industrial activities such as mining, smelting, electroplating, electronics, and battery manufacturing have caused significant water pollution, worldwide. It is known that effluents from aforementioned industries are normally contained high level of heavy metals such as Cd, Cr, Cu, Ni, Pb and Zn [1]. The amount of heavy metals discharged into water environment is continue to increase, year by year [2]. In fact, in many cases, the concentration of toxic metals exceeds the permissible limits set by World Health Organization (WHO) [2,3]. A high concentration of heavy metals in aquatic environment is of great concern. Heavy metals can easily enter the food web and therefore may create serious problems and as well as adverse effects to human [4]. For instance, excessive intake of Pb may cause toxic effects to human kidney and nervous system. According to WHO, the safe blood level of Pb is 0.1 g/m^3 [5]. In addition, anemia begins to occur when Pb level in blood greater than 0.3 g/m^3 (children) and 0.6 g/m^3 (adults) [5], respectively. At blood Pb concentration as low as 5 µg/dL, the human intelligent quotient (IQ) will be affected [6]. In this context, children are at high risk.

The remediation of water polluted by heavy metals remains a great challenge to environmental scientists. Although modern treatment techniques such as ion exchange, flocculation, electrodialysis, and electrodeposition are reported to be effective to treat contaminated water, their operational costs are expensive [7]. Adsorption has been regarded as an economical method for lowering heavy metals concentration in contaminated water [8]. Activated carbon is the most widely used adsorbent for this purpose. In spite of its high porosity and capacity to remove pollutants, activated carbon is known costly [7,9]. Great efforts have been directed towards the development of cost-effective adsorbent for adsorption technique particularly for application in developing countries such as Malaysia, Indonesia, and Thailand. The ultimate aim of this study was to evaluate the feasibility of kenaf bast fiber and golden apple snail shell as low-cost adsorbents for water treatment.

2. Materials and methods

2.1. Preparation of adsorbents
The kenaf bast fiber was kindly supplied by Malaysian National Kenaf and Tobacco Board, Kubang Kerian, Kelantan. The golden apple snail shell was obtained from a paddy farmer in Pendang, Kedah. Both materials were soaked in tap water overnight and rinsed several times with deionized water. The materials were then dried in an oven at 70 °C for 24 hours and ground to a fine powder (150-250 μm) using a laboratory mill. To avoid moisture effect, the adsorbents were kept in self-sealing bags until ready for characterization and adsorption experiments.

2.2. Characterization studies
In this study, Fourier Transform Infrared (FTIR) spectra, Scanning Electron Microscope (SEM) images, and Energy Dispersive X-ray (EDX) spectra were obtained before and after the interaction of adsorbents with heavy metals. A Thermo Nicolet 6700 ATR-FTIR Spectrometer and a Hitachi SU 8020 UHR SEM were used in characterization studies. For FTIR analysis, the scanning range was from 4000 to 400 cm^{-1} with 35 cumulative scans. Meanwhile, for SEM and EDX analyses the samples were coated with a thin layer of platinum. This procedure is important in order to avoid electron charging which might reduce the quality of SEM image and EDX spectra. The adsorbents were observed at several magnifications.

2.3. Adsorption studies
The 250 mL stock solutions of 1000 mg/L were prepared by dissolving the exact amount of metal salts ($Cd(NO_3)_2.4H_2O$, $K_2Cr_2O_7$, $Cu(NO_3)_2.3H_2O$, $Ni(NO_3)_2.6H_2O$, $Pb(NO_3)_2$ and $Zn(NO_3)_2.6H_2O$) in deionized water, separately. The stock solutions were then diluted to the required concentrations. Metal salts used in this study were of analytical grade and used as received.

Prior to application on real electronics effluent, the ability of both kenaf bast fiber and golden apple snail shell to remove heavy metals was tested in batch adsorption studies. The effects of several experimental parameters such as solution pH, contact time and adsorbent dosage on adsorption of heavy metals by both adsorbents were studied in details. The effect of solution pH was conducted in the pH range of 1.0 to 6.0. The initial solution pH of 50 mL of 50 mg/L metal ion solution was adjusted using 0.1 mol/L HCl and 0.1 mol/L NaOH solutions. The pH was measured using a Thermo Scientific Orion 2-Star pH meter. The effect of contact time was performed at 15, 30, 45, 60, 75 and 90 minutes, while the effect of dosage was carried out using 0.125, 0.250, 0.500 and 0.750 g of adsorbent.

Adsorption experiments were conducted by shaking an amount of adsorbent with 50 mL of metal ion solution in 250 mL conical flasks by using a Protech Orbital shaker (model 720) set at 100 rpm. Following equilibration, the samples were then separated through filter paper (Advantec 231, 0.18 mm). The concentration of heavy metals in the supernatant was measured using a PerkinElmer AAnalyst 400 Atomic Absorption Spectrometer (AAS). Calibration curve was plotted (R^2 = 0.9989) prior to AAS measurement. All experiments were carried out in triplicates. The equilibrium adsorption capacity and the removal percentage were calculated using Equations (1) and (2), respectively [4,5]:

$$q_e = \left(\frac{C_i - C_f}{M}\right) \times V \tag{1}$$

$$\text{Removal (\%)} = \left(\frac{C_i - C_f}{C_i}\right) \times 100 \tag{2}$$

where q_e is the amount of metal ion adsorbed by adsorbent at equilibrium (mg/g), C_i and C_f are the initial and final concentrations of metal ions (mg/L), M is the mass of adsorbent (g) and V is the volume of metal ion solution (L).

2.4. Effectiveness and comparative studies

The effectiveness of kenaf bast fiber and golden apple snail shell to treat metal laden contaminated water was examined using an electronics industrial effluent. The effluent sample was obtained from an electronics factory in Shah Alam, Selangor. The industrial effluent was brought to the research lab, filtered through a filter paper for dirt and suspended solid removal. The concentrations of heavy metals in the effluent sample were then measured using AAS.

In this study, the applicability of kenaf bast fiber and golden apple snail shell to adsorb heavy metals from industrial effluent was compared with commercial mangrove stem derived activated carbon. The industrial effluent was treated using three adsorbents at optimum experimental conditions determined earlier in Section 2.3.

3. Results and discussion

3.1. Characterization studies

The Brunauer–Emmett–Teller (BET) surface area of kenaf bast fiber and golden apple snail shell were determined as 4.73 and 2.64 m^2/g, while the average pore diameter of kenaf bast fiber and golden apple snail shell were measured as 1.53 and 2.38 Å, respectively. Based on porosity classification set by International Union of Pure and Applied Chemistry (IUPAC), both adsorbents are micropores.

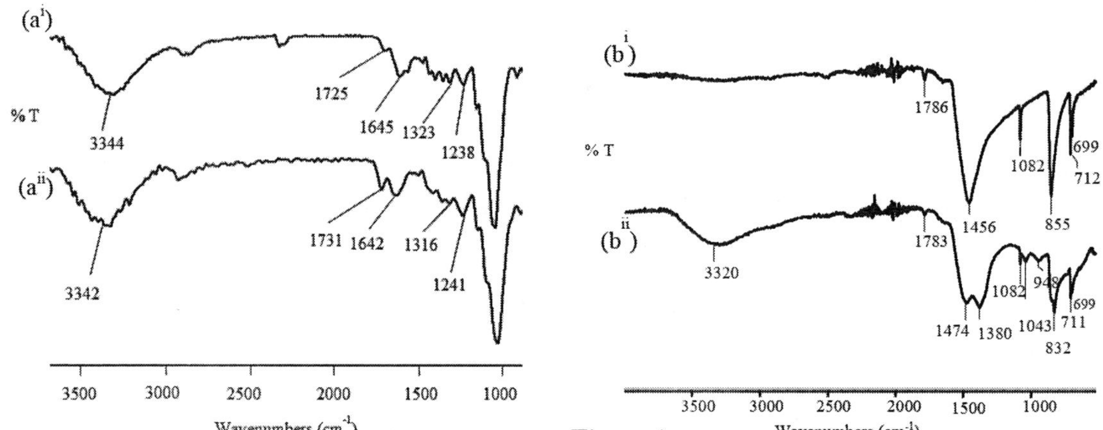

Figure 1. FTIR spectra of kenaf bast fiber before (a^i) and after Ni(II) adsorption (a^{ii}), and FTIR spectra of golden apple snail shell before (b^i) and after Ni(II) adsorption (b^{ii}).

The presence of functional groups in both adsorbents was confirmed by FTIR analysis. Figure 1 presents the FTIR spectra of kenaf bast fiber and golden apple snail shell, before and after Ni(II) adsorption, as an example. Figure 1(a^i) shows the main characteristics of kenaf bast fiber at 3344, 1725 and 1645 cm^{-1} which correspond to hydroxyl (OH) group, C-O stretching vibration and N-H bending vibration of amine group, respectively. Following interaction with Ni(II), the wavenumber of these bands shifted to 3342, 1731 and 1642 cm^{-1} (Figure 1(a^{ii})). This phenomenon suggests the involvement of O-H, C-O and N-H in binding metal ions which might be through the formation of dative bond via sharing of valence electron [10,11].

From Figure 1(b^{i}), discernible absorption bands at 1786, 1456, 1082, 855 and 712 cm^{-1} represent the properties of the carbonate (CO_3^{2-}) group [12]. After adsorption of Ni(II), a new absorption peak appeared at 3320 cm^{-1} was due to OH group. Meanwhile, the prominent band at 1456 cm^{-1} split into two bands at wavenumbers of 1474 and 1380 cm^{-1} (Figure 1(b^{ii})). Another set of new bands were observed at 1043 and 948 cm^{-1}. These findings indicate that carbonate groups play an important role as binding site for metal ions mainly through electrostatic attraction [13].

The change in surface morphology following interaction with metal ions was studied using SEM analysis. Figure 2 shows the SEM images of kenaf bast fiber and golden apple snail shell, before and after interaction with Pb(II), as an example. From Figures 2(a^{i}) and 2(b^{i}), it is apparent that both adsorbents have dense and uneven surface morphology with no porous texture. Adsorption of Pb(II) onto both adsorbents changed their surface morphology significantly. After Pb(II) adsorption, it was observed that there were deposits on the surface of fiber (Figure 2(a^{ii})), meanwhile rod-like crystals with sharp and oval shape were clearly seen on the surface of snail shell (Figure 2(b^{ii})). These deposits and crystals could represent metal ion precipitates.

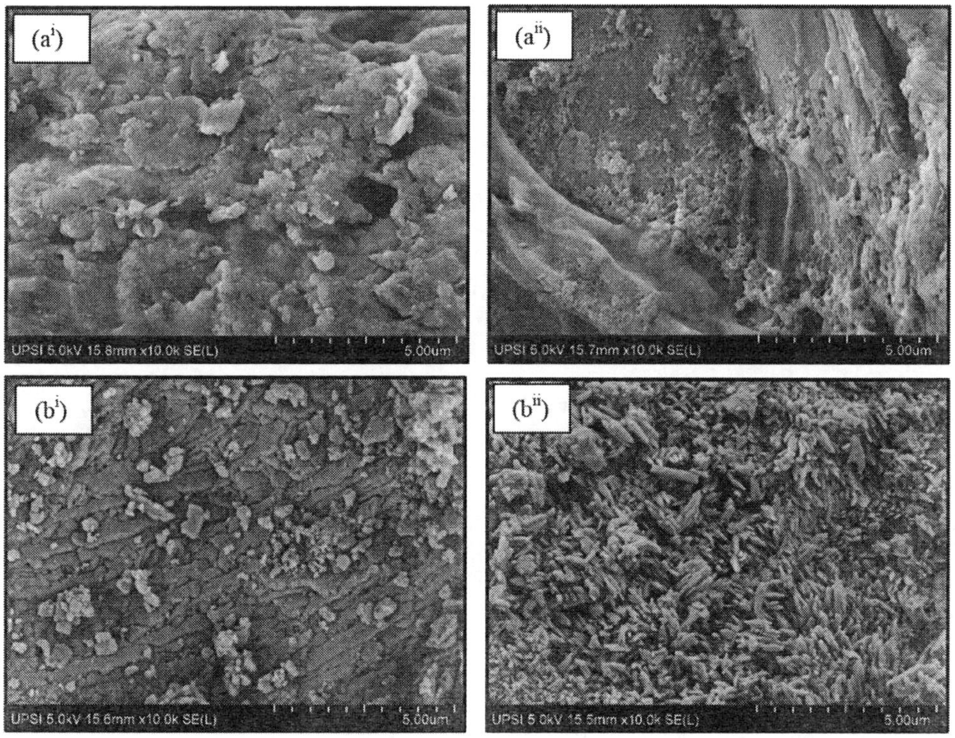

Figure 2. SEM images of kenaf bast fiber before (a^{i}) and after Pb(II) adsorption (a^{ii}), and SEM images of golden apple snail shell before (b^{i}) and after Pb(II) adsorption (b^{ii}) at 10,000x magnification.

In order to confirm the successful of heavy metals adsorption onto the surface of adsorbents, EDX analysis was performed. EDX analysis enables the determination of the elemental composition of adsorbents, particularly after the adsorption process. From Figure 3, C, O and K are the main elements in kenaf bast fiber, meanwhile Ca and O are the dominant elements in snail shell. The features of Cd

$L\alpha_1$ and Cd $L_{\beta1}$ were observed at 3.13 and 3.32 keV of EDX spectrum of fiber after interaction with Cd (Figure 3(a)). In the case of snail shell, the features of Cu $K\alpha_1$, Cu $K_{\beta1}$, Cu $L\alpha_1$ and Cu $L_{\beta1}$ were appeared at 8.05, 8.90, 0.93 and 0.95 keV following adsorption of Cu (Figure 3(b)). These findings support speculation made earlier in SEM analysis, of which deposits and crystals observed on the surface of adsorbents belong to metal ion precipitates.

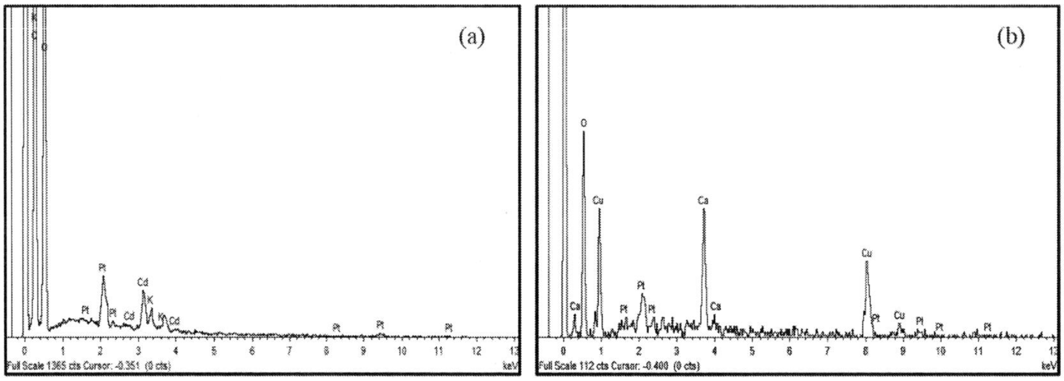

Figure 3. EDX spectrum of kenaf bast fiber after Cd(II) adsorption (a) and EDX spectrum of golden apple snail shell after Cu(II) adsorption (b).

3.2. Adsorption studies

The solution pH greatly affects the surface charge of the adsorbent and the degree of ionization and speciation of adsorbate in aqueous solutions [13,14]. Therefore, it is imperative to study such effect on the adsorption capacity of both kenaf bast fiber and golden apple snail shell. As stated in Section 2.3, the effect of solution pH was performed in the pH range of 1.0 to 6.0. Based on our preliminary studies, it was found that several metal ions started to form hydroxide precipitates at pH above than 6.0. To avoid any possible interference from hydroxide precipitates, we have set pH 6.0 as the maximum solution pH.

Figure 4(a) depicts the amount of Cd(II), Cu(II) and Pb(II) adsorbed onto snail shell as a function of solution pH, as an example. From Figure 4(a), it is clear that the adsorption of metal ions increased with an increase in solution pH and optimum pH was obtained at pH 6.0. A similar trend was obtained for kenaf bast fiber. In the case of snail shell, a lower amount of metal ions adsorbed at acidic media particularly at pH 1.0 and 2.0 can be related to a high amount of H^+ or H_3O^+ ions available, which has caused a competitive scenario with metal ions for adsorption sites (CO_3^{2-}). In contrast, at a higher pH values less H^+ or H_3O^+ ions were available. Therefore, metal ions were able to interact well with adsorption sites causing adsorption capacity to increase.

Of metal ions studied, Cr(VI) ions showed a unique adsorption behavior of which the adsorption capacity was initially increased from solution pH of 1.0 to 3.0 and decreased from pH 4.0 to 6.0 particularly onto kenaf bast fiber. Depending on the concentration of chromium and the pH of the solution, Cr(VI) normally exists in anionic forms in aqueous solution as $Cr_2O_7^{2-}$, $HCrO_4^-$, CrO_4^{2-} and $HCr_2O_7^-$ [15,16]. These anions can interact effectively with protonated amine functional groups available in kenaf bast fiber.

From the environmental engineering point of view, an ideal adsorption process should not take much time to treat contaminated water. In this study, the effect of contact time was studied up to 90 minutes. The adsorption of metal ions onto both adsorbents increased with contact time and attained equilibrium at about 60 minutes for an initial concentration of 50 mg/L of metal ions.

Figure 4(b) shows the effect of the adsorbent dosage on the amount of Pb(II) ions adsorbed by snail shell, as an example. As can be seen from Figure 4(b), the adsorption of Pb(II) decreased significantly from 39.25 mg/g to 7.11 mg/g when the amount of dosage was increased from 0.125 to 0.750 g. This scenario can be explained by the fact that at the lower dosage, the ratio of adsorption sites of adsorbent to adsorbate was low. Therefore, all adsorption sites were occupied by adsorbate. In contrast, at higher dosage the aforementioned ratio was high. In other words, more adsorptions sites were available than adsorbate, causing many of these sites were unoccupied. Therefore, the amount of metal ions adsorbed was decreased. Sessarego et al. [17] and Sharma et al. [18] have also obtained and discussed similar explanation.

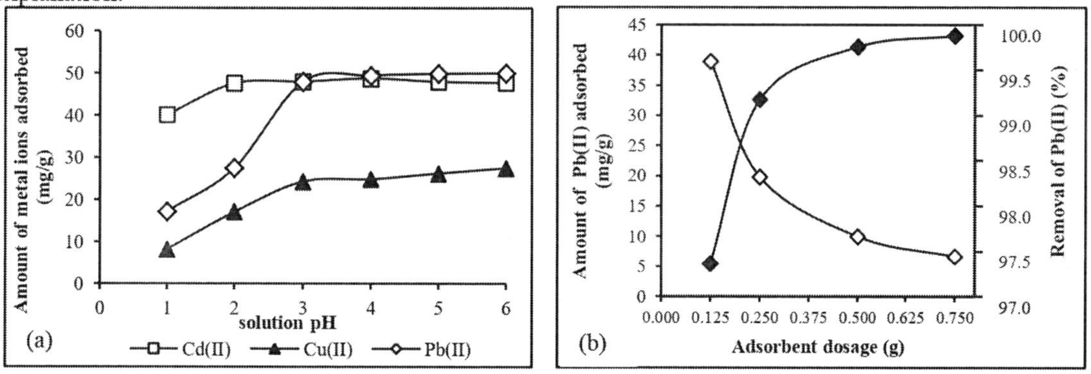

Figure 4. Effect of solution pH (a) and adsorbent dosage (b) on the adsorption of metal ions onto golden apple snail shell.

3.3. Adsorption isotherms

In such adsorption study, the interaction and distribution of metal ions on the surface of adsorbents can be described using adsorption isotherm models [18,19]. In this study, the adsorption equilibrium data were fitted to the Freundlich and Langmuir isotherm models. Freundlich isotherm is an empirical model and was developed to describe adsorption process that occurs at multilayer heterogeneous surface which involve interaction between molecules of adsorbates and active sites of adsorbents [20]. The linear form of the Freundlich isotherm can be represented by Equation (3) [21]:

$$\log q_e = \log K_f + \frac{1}{n} log\, C_e \tag{3}$$

where q_e is the amount of metal ion adsorbed per unit weight of adsorbent (mg/g), K_f (mg/g) and n (g/L) are the Freundlich constants. C_e is the equilibrium concentration of metal ions (mg/L). Meanwhile, Langmuir isotherm is a theoretical model for monolayer adsorption with the homogenous surface without interaction between adsorbed molecules [22]. The linear form of the Langmuir isotherm can be expressed in Equation (4) [23]:

$$\frac{C_e}{q_e} = \frac{C_e}{Q} + \frac{1}{Qb} \tag{4}$$

where Q is the maximum adsorption capacity at monolayer surface (mg/g) and b is the Langmuir constant (mL/mg). The validity of both isotherm models was checked and depicted in Figures 5(a) and 5(b). Table 1 lists the Freundlich and Langmuir constants for metal ions adsorption by the golden apple snail shell. Based on the correlation coefficient (R^2) values, the adsorption equilibrium data were well fitted to the Langmuir isotherm model. Based on Q value determined from Langmuir isotherm model, the affinity of snail shell towards metal ions was in order of Pb(II) (120.47 mg/g) > Cu(II) (94.18 mg/g) > Zn(II) (85.09 mg/g) > Ni(II) (78.34 mg/g) > Cd(II) (52.99 mg/g) > Cr(VI) (31.62 mg/g).

Table 1. Freundlich and Langmuir constants for metal ions adsorption by golden apple snail shell.

Metal ion	Freundlich			Langmuir		
	K_f (mg/g)	*1/n* (g/L)	R^2	*Q* (mg/g)	*b* (L/mg)	R^2
Cd(II)	6.17	0.63	0.8015	52.99	0.06	0.9978
Cr(VI)	34.94	0.28	0.7399	31.62	0.24	0.9880
Cu(II)	10.93	0.49	0.9028	94.18	0.47	0.9762
Ni(II)	22.95	0.82	0.7430	78.34	0.15	0.9598
Pb(II)	40.80	0.60	0.8127	120.47	0.29	0.9761
Zn(II)	25.36	0.17	0.8305	85.09	0.08	0.9905

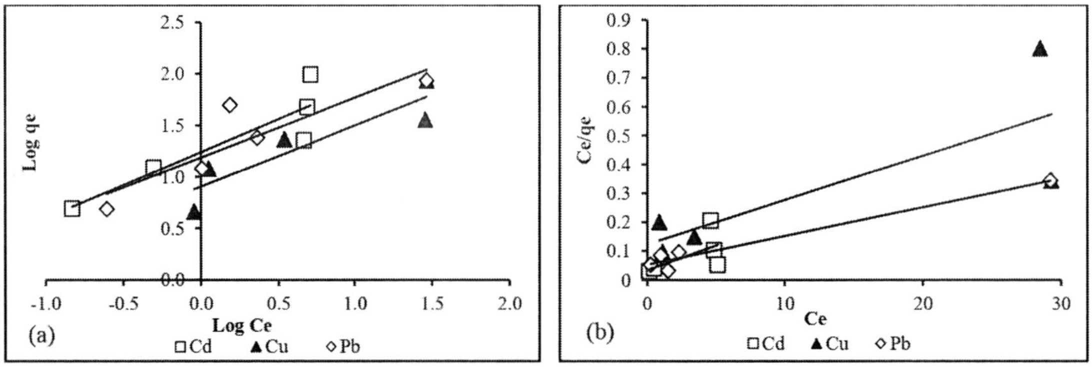

Figure 5. Test of linearized Freundlich equation (a) and linearized Langmuir equation (b) for adsorption of Cd(II), Cu(II) and Pb(II) onto golden apple snail shell.

3.4 Effectiveness and comparative studies

The feasibility of kenaf bast fiber and golden apple snail shell to treat metal contaminated water was further evaluated using an electronics industry effluent. The adsorption performance of both low-cost adsorbents was compared with mangrove stem derived activated carbon (MSDAC). MSDAC is a commercial adsorbent and has been widely used for water treatment in Malaysia. The removal of metal ions from electronics industry effluent is given in Table 2.

From Table 2, it is clear that the adsorption capacity of kenaf bast fiber and golden apple snail shell was comparable to that of MSDAC. In fact, the snail shell was able to adsorb Cr(VI), Ni(II) and Zn(II) at a higher capacity than MSDAC and kenaf bast fiber under experimental conditions studied. It is also important to note that the three adsorbents studied were successfully reduced Pb(II) and Cu(II) concentration from 8.72 and 5.25 mg/L to less than 0.10 and 0.20 mg/L, respectively, and thereby meeting the Malaysian Environmental Quality Act 1974 maximum limit of 0.10 mg/L and 0.20 mg/L for Pb and Cu in industrial effluent to be discharged into inland surface water [24].

Table 2. Removal of metal ions from electronics industry effluent.

Adsorbent	Metal ion	Concentration of metal ion (mg/L)		Removal (%)
		Before adsorption	After adsorption	
Kenaf bast fiber	Cd(II)	0.41	0.32	21.95
	Cr(VI)	0.64	0.18	71.88
	Cu(II)	5.25	0.18	88.38
	Ni(II)	1.86	0.67	63.98
	Pb(II)	8.72	0.09	96.22
	Zn(II)	3.25	0.88	72.92
Golden apple snail shell	Cd(II)	0.41	0.29	29.27
	Cr(VI)	0.64	0.09	85.94

	Cu(II)	5.25	0.16	96.95
	Ni(II)	1.86	0.38	79.57
	Pb(II)	8.72	0.04	99.54
	Zn(II)	3.25	0.39	88.00
Mangrove stem derived activated carbon	Cd(II)	0.41	0.20	51.22
	Cr(VI)	0.64	0.17	73.44
	Cu(II)	5.25	0.10	98.10
	Ni(II)	1.86	0.47	74.73
	Pb(II)	8.72	0.06	99.31
	Zn(II)	3.25	0.44	86.46

4. Conclusion

In the present work, the efficacy of kenaf bast fiber and golden apple snail shell to reduce heavy metals concentration in electronics industry effluent has been successfully evaluated. The physicochemical properties and adsorption mechanism were studied using several analytical instruments. Results obtained from this study highlight the applicability of both materials as alternative low-cost and environmentally friendly adsorbents for water treatment, particularly to be used in developing countries such as Malaysia, Indonesia, and Thailand, where similar resources are available at a huge amount. In order to evaluate the versatility of kenaf bast fiber and golden apple snail shell to be used as universal adsorbents for water treatment, it is therefore worth to investigate their performance using different type of industrial effluents. There is ready supply of kenaf bast fiber and golden apple snail shell. Their utilisation in water treatment as adsorbents will not only solve water pollution issue, but also reduce the risk of solid waste disposal from agricultural activity.

References

[1] Bardestani R, Roy C and Kaliaguine S 2019 *J. Environ. Manage.* **240** 404-420
[2] Basu M, Arun K G and Ray L 2019 *Bioresour. Technol.* **283** 86-95
[3] Dong J, Du Y, Duyu R, Shang Y, Zhang S and Han R 2019 *Bioresour. Technol. Reports* **6** 96-102
[4] Ebisike K, Okoronkwo A E and Alaneme KK 2019 *Environ. Technol. Innov.* **14** 100337
[5] El Boundati Y, Ziat K, Naji A and Saidi M 2019 *J. Mol. Liq.* **276** 15-26
[6] Gorman-Lewis D, Martens-Habbena W and Stahl D A 2019 **255** 127-143
[7] Jiang C, Wang X, Wang G, Hao C, Li X and Li T 2019 *Composites* **169** 45-54
[8] Dąbrowski A 2010 *Adv. Colloid Interface Sci.* **93** 135-224
[9] Kameda T, Honda R, Kumagai S, Saito Y and Yoshioka T 2019 *J. Solid State Chem.* **277** 143-148
[10] Kong W, Yue Q, Li Q and Gao B 2019 *Sci. Total. Environ.* **668** 1165-1174
[11] Liu R and Lian B 2019 *Sci. Total Environ.* **659** 122-130
[12] Liu J, Hu C and Huang Q 2019 *Bioresour. Technol.* **271** 487-491
[13] Mahar F K, He L, Wei K, Mehdi M, Zhu M, Gu J, Zhang K, Khatri Z and Kim I 2019 *Chemosphere* **225** 360-367
[14] Nie T, Hao P, Zhao Z, Zhou W and Zhu L 2019 *Sci. Total. Environ.* **673** 522-532
[15] Pawlaczyk M and Schroeder G 2019 *J. Mol. Liq.* **281** 176-185
[16] Sato T, Abe S, Ito S and Abe T 2019 *J. Environ. Chem. Eng.* **7(2)** 102958
[17] Sessarego S, Rodrigues S C G, Xiao Y, Lu Q and Hill J M 2019 *Carbohydr. Polym.* **211** 249-256
[18] Sharma M, Singh J, Hazra S and Basu S 2019 *Microchem. J.* **145** 105-112
[19] Xie X, Gao H, Luo X, Su T, Zhang Y and Qin Z 2019 *J. Environ. Chem. Eng.* **7(3)** 103183
[20] Guo X and Wang J 2019 *J. Mol. Liq.* **288** 111100
[21] Freundlich H M F 1906 *Z. Phys. Chem.* **57** 385-470
[22] Yu J, Zhu S, Chen P, Zhu G-T, Jiang X and Di S 2019 *Appl. Surf. Sci.* **484** 124-134
[23] Langmuir I 1918 *J. Am. Chem. Soc.* **40(9)** 1361-1403
[24] Department of Environment Malaysia (DOE) 1974 Environmental Quality Act 1974 (Act 127)

Acknowledgements
The authors express their gratitude to a paddy farmer in Pendang, Kedah for providing golden apple snail shell for this project. Cooperation and support from Malaysian National Kenaf and Tobacco Board, Kubang Kerian, Kelantan especially for supplying raw kenaf bast fiber are gratefully acknowledged. The authors also thank laboratory assistants from Department of Chemistry, Faculty Science and Mathematics, Universiti Pendidikan Sultan Idris, Malaysia for their technical assistance.

Adsorption pattern on the optimization of interaction time for the reduction of phenol with charcoal activated $ZnCl_2$

S Kristianingrum, A Fillaeli*, ED Siswani, Sulistyani, Parwanti

Departmentof Chemistry Education, Faculty of Mathematics and Natural Sciences, Universitas Negeri Yogyakarta, Indonesia, Jl. Colombo 1, Yogyakarta55281, Indonesia

*Email: annisa_fillaeli@uny.ac.id

Abstract. Interaction time significantly influenced the pattern of adsorption.This study aims to determine the characteristic of activated charcoal adsorbent, the decrease in phenol levels and its adsorption capacity, and test the adsorbent applicability to remove phenol from batik wastewater. The concentration of phenol was measured by UV-Vis spectrophotometer. The results showed that the characters of activated charcoal adsorbent in accordance with Indonesian National Standard (INS) 06-3730-1995. Its characters were water content, ash content and iodine adsorption power. Characterization using Surface Area Analyser (SAA) showed that the surface area of activated charcoal was 68.54 m^2/g. The percentage decrease in phenol concentration will be greater along with the increase in concentration where 92.90%; 95.60%; and 97.15% of phenol decrease were influenced by the increasing level concentration of 0.6 mg/L; 0.8 mg/L; and 1.0 respectively with 24 hours optimal contact time. The adsorption capacity of activated charcoal was 0.048575 mg/g while the isotherm adsorption pattern of adsorbent fulfilled the pattern of Freundlich isotherm. Therefore, the availability of high surface area makes the activated charcoal of marine pandan leaves has significant impact to the adsorption of phenol where it was contacted optimum at 24 hours interaction with Freundlich adsorption pattern.

1. Introduction

The development of the batik industry in Indonesia has increased along with the increase in demand for clothing products using the batik process. This has caused the waste generated from the batik process to also increase. Wastewater from the batik industry is one of the main sources of water pollution because batik wastewater contains very high organic compounds and also a variety of dangerous chemical compounds, phenol [1]. Phenol is produced from an organic waste originating from the wax dissolving process or coloring and is included in the category of B3 waste [2].The content of phenol in batik waste will be dangerous and cause an adverse effect on the environment if it is thrown into the environment without prior processing. It will enter the food chain and does not rule out the possibility of reaching the human body, skin and respiratory tract. It can also affect the central nervous system, liver and kidneys, causing coma, gastric damage and respiratory failure. Phenol acts as carcinogenic properties for humans if the concentration is 5-25 mg/L [2]. Phenol content in textile industry waste according to the Living Environment Ministry ordinance No. 5 of 2014 is at a maximum of 0.5 mg/L [3]. Therefore it is necessary to treat phenol in wastewater. One of them is through the adsorption method using activated charcoal adsorbents. The method of adsorption with activated charcoal is a physical-chemical treatment process that is effective for removing contaminants in wastewater [4]. Activated charcoal preparation is easier, cost effective and has a high adsorption capacity. The use of other methods has the disadvantage of operating conditions that are insensitive, less efficient and for advanced disposal require high costs [5]. The selection of activator types that can be used for activation will greatly affect the quality of activated charcoal obtained, including KOH, H_2SO_4 [6], HNO_3 [7], NaOH [8] and $ZnCl_2$ [9,10,11]. The use of $ZnCl_2$ activators can produce activated charcoal which is quite effective for the adsorption process. Efforts to reduce phenol levels in batik waste were carried out with a batch system and the phenol content in the waste was analyzed using UV-Vis spectrophotometry.

The proof was done in research conducted by Kristianingrum and Siswani [12] about making adsorbents from marine pandan leaves which were used for heavy metal adsorption processes such as Cr and Pb. The results showed that the synthesized adsorbent was in accordance with INS and was effectively used as an adsorbent. Activated charcoal that has been activated using $ZnCl_2$ will adsorb phenol compounds in the waste so that the levels drop and safe if the waste is disposed into the environment. The use of $ZnCl_2$ activator aims to improve the quality of the adsorbent so that the adsorbent with optimal adsorptivity will be obtained [13]. This study will analyze the characterization of the adsorbent from the activated charcoal of marine pandan leaves (*Pandanus tectorius*).

Marine pandan leaves are rich in various kinds of chemical compounds, including cellulose which has major content of carbon as sources to synthesize activated charcoal. The optimal adsorbent will be applied to batik industry liquid waste containing phenol. Some studies showed that the most ideal time for the chemical activation process is 24 hours [14]. In this research, activated charcoal was chemically activated (using 5% $ZnCl_2$ solution) and activation of physics (heating at high temperatures). The treatment of physical activation greatly influenced the formation of micro-pore structures on activated charcoal so as to produce activated charcoal with large surface areas [15]. In addition, the adsorption efficiency of phenol was calculated (percentage reduction in phenol) to the addition of activated adsorbents and their adsorption capacity. The method used in this study is the characterization of the adsorbent including: moisture content, ash content, volatile substance content, absorption rate of I_2 and charcoal content. To find out the surface area of the adsorbent and the pore size distribution, characterization used SAA (Surface Area Analyzer).

2. Experimental Section

2.1. Materials and Instruments

The materials used marine pandan leaves, batik industrial wastewater taken from Godean area, Yogyakarta, 5% $ZnCl_2$ solution, 0.1M $Na_2S_2O_3$ solution, starch indicator, 0.1M iodine solution, phenol mother liquor, distilled water, pH paper and filter paper.

The tools used in this study:UV-Vis spectrophotometry (Shimadzu), porosimeter (Quantachrome, NOVAe Series), furnace, magnetic stirrer, drying oven, buchner filter, vacuum device, desiccator, analytic balance, 80 mesh sieve, porcelain exchange rate, and glass-wares.

2.2. Synthesis of activated charcoal

A total of 8.5 kg of marine pandan leaves were dried and then carbonated at 500°C for 6 hours, then the yielded charcoal was smoothed and sifted using an 80 mesh sieve. After that, chemical activation is carried out by taking 200 gr charcoal size 80 mesh with a 5% $ZnCl_2$ solution for 24 hours at ambient temperature. The next step is washed to neutral pH and continued with physics activation at temperature of 700 ° C with NO_2 gas flowed for 2 hours. This activation process is carried out to expand the surface of activated charcoal in order to increase the adsorption of activated charcoal against the adsorbate. Activated charcoal will experience changes in physical and chemical properties if it has been activated and will increase in adsorption and surface area. When the surface area of activated charcoal is increasing, the adsorption efficiency also increases [16].

2.3. Characterization of Charcoal (INS 06-3730-1995)[16]
The size of porous were identified by porosimeter for both raw and activated charcoals. According to INS 06-3730-1995, the activated charcoal was tested in finding the content of water, ash level, volatile degree, and the capacity adsorption of activated charcoal to I_2.The properties of adsorption were investigated by interacting activated charcoal with solutions of artificial waste of phenol (0.2; 0.3; 0.4; 0.6; 0.8 and 1.0 mg/L) in a ratio of 2:100 (g/mL) for 24 hours in medium pace respectively to find the capacity, the efficiency and the isotherm models.

2.4. Adsorption
Adsorption of activated charcoal activated by 5% $ZnCl_2$ in batik simulation waste solution was done by making a hundred mililitre of artificial waste of phenol with 6 variations of concentration (0.2; 0.3; 0.4; 0.6; 0.8 and 1.0 mg / L) contacted with activated charcoal as much as 2 grams during 24 hour time variation , 6 hours and 3 hours with constant stirring. The mixture was filtered and the filtrate was taken to analyze the phenol content using a UV-Vis spectrophotometer.

The phenol levels (INS 06-6989.21-2004)[17] were investigated by testing the sample of 100 mL added by 2.5 mL NH_4OH 0.5N reagent and the pH was 7.9 with the addition of a phosphate buffer solution. Added 1 mL 4-aminoanthipyrin solution while stirring, then added 1 mL of potassium ferisianida solution and allowed to stand for 15 minutes. Measurements were carried out by spectrophotometer at a wavelength of 500 nm [18].

Determination of the surface area of activated charcoal used the SAA (Surface Area Analyzer) tool. This is to determine the surface area of the adsorbent, the size of the pore distribution of charcoal before the activation process is carried out and after going through the activation process. All is done by supporting installed program of SAA device.

Decrease in phenol content and adsorption capacity of activated charcoal in marine pandan leaves was calculated by the equation:

$$Ea\ (\%) = (C - Ce) / C \times 100\%$$

$$qm\ (mg / g) = (C - Ce) / m \times V$$

where Ea is a decrease in phenol level (%) while qm is adsorption capacity (mg/g), C is the initial phenol concentration (mg/L), Ce is the concentration of phenol after adsorption (mg/L), m is the mass of the adsorbent for adsorption (g) and V is the volume of adsorbed (L) wastewater [19].

3. Result and Discussion

3.1. Characterization of Activated Charcoal
The characterization of activated charcoal in accordance with INS 06-3730-1995 includes moisture content, ash content, volatile substances, charcoal content and iodine absorption. Characterization of activated charcoal data before and after activation with $ZnCl_2$ is shown in Table 1.

Table 1. Data on charcoal characterization before and after activation with $ZnCl_2$

Marine pandan leaves charcoal	Water content (%)	Ash content (%)	Volatile content (%)	Carbon content (%)	Adsorption to Iodium (mg/g)
Before activation process	5.42	19.62	70.50	4.46	2157.55
After activation process	4.24	8.19	45.49	42.15	2223.26
SNI No. 06-3730-1995	Max 15	Max 10	Max. 25	Min. 65	Min. 750

From the table 1, it can be seen that the water content, ash content and charcoal adsorption of I_2 after activation has been in accordance with INS 06-3730-1995 because of charcoal activation process. From the results, the levels of volatile substances and charcoal content not yet in accordance with INS standards are likely to be caused during the carbonization process. Air leaks still occur even though they have been minimized. Ash is also produced due to oxidized samples from outside air. The incomplete carbonization process results in the majority of non-charcoal elements being eliminated as volatile gas species [20].

The decrease in water content from 5.42% to 4.24% in table 1 is likely caused by the surface of the activated charcoal containing fewer polar groups, so that the interaction between water vapor which is polar is also small. In addition, it is assumed that only water is a volatile compound, because it is possible there is still water trapped in the cavity and covering the pores of activated charcoal. The lower water content shows that less water is left behind and covers the active charcoal pore. If the water content is low then many places in the pore of activated charcoal can be occupied by I_2 molecules [21]

The results of characterization using a Surface Area Analyzer (SAA) showed that the surface area of activated charcoal in marine pandan leaves activated by 5% $ZnCl_2$ was 68.543 m^2/g. The activation process for activated charcoal causes the breakdown of hydrocarbon bonds on charcoal so that the porous surface area increases [22], whereas the original surface area was only 3.7719 m^2/g. Activated charcoal adsorbent of marine pandan leaves that have been activated by $ZnCl_2$ have pore size in the meso-porous category because they have a surface area between 200-100 m^2/g [23]. The surface area is related to the crystal structure, number and size of pores present in the adsorbent. A large number of pores and widths will widen the adsorption capacity.

However, when compared to the results of both studies, the charcoal after activation using $ZnCl_2$ has better quality than before the activation process, so that activated charcoal can be applied as an adsorbent to adsorb phenol.

3.2. Decreasing Phenol (Ea) and Adsorption Capacity (qm) in Batik Simulation Waste

The data for decreasing phenol and its adsorption capacity are shown in Table 2.

Table2.Data on decreasing phenol content and its adsorption capacity of activated charcoal

The initial phenol content (mg/L)	The decrease of phenol content (%)			Adsorption capacity of activated charcoal (mg/g)		
	3 hours	6 hours	24 hours	3 hours	6 hours	24 hours
0.2	41.7	49.10	90.35	0.00417	0.00491	0.00904
0.3	50.03	72.23	91.40	0.00751	0.01084	0.01371

0.4	74.33	83.80	92.45	0.01487	0.01676	0.01849
0.6	81.5	92.50	92.90	0.02445	0.02775	0.02787
0.8	87.05	94.94	95.60	0.03482	0.03796	0.03874
1.0	89.82	95.39	97.15	0.04491	004770	0.04858

It can be clearly observed that the difference between each relationship of the initial concentration of phenol to the decrease in phenol levels as in Figure 1. Both percentage of phenol decrease and the adsorption capacity of activated charcoal significantly increase. The maximum phenol percentage of decrease for 24 hours interaction with the adsorption capacity is 0.04858 mg/g. It shows the more contact time, the interaction of particles more intense. The intensity of time interaction increases the bond strength of phenol and the adsorbent.

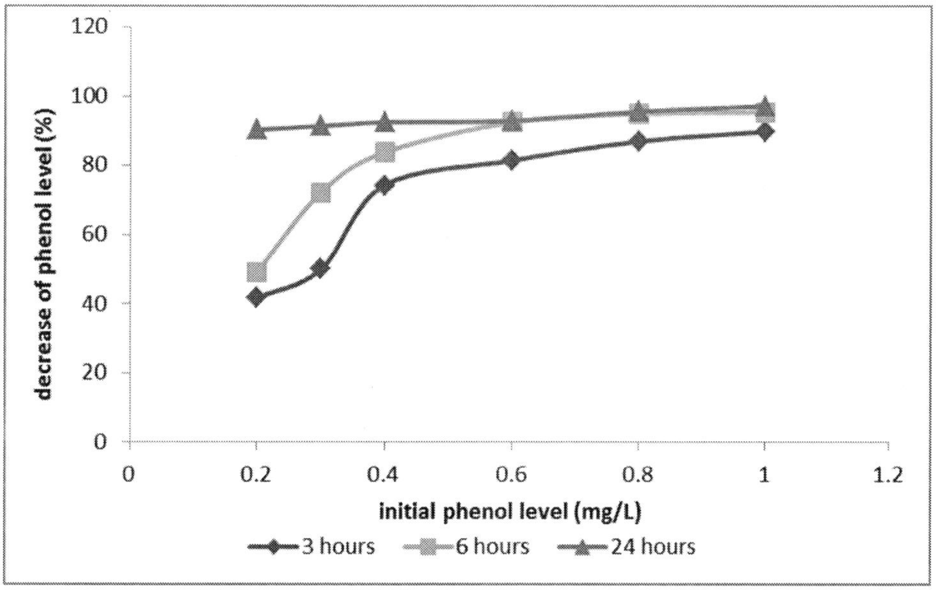

Figure 1. The graph of initial phenol content relation toward its decrease

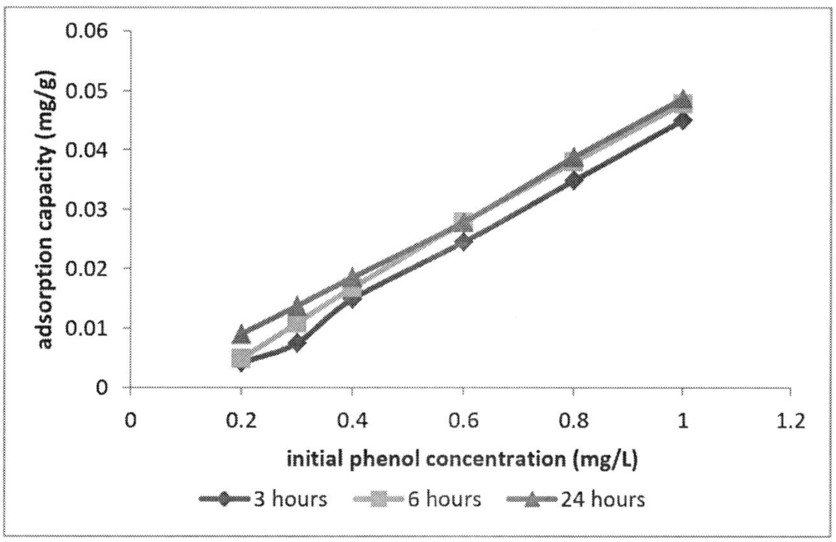

Figure2. The graph initial phenol concentration toward adsorption capacity

The differences of relation in every initial concentration of phenol to the adsorption capacity value are shown in Figure 2.

3.3. Decreasing Phenol (Ea) and Adsorption Capacity (qm) in Real Batik Waste

Data for decreasing phenol levels and adsorption capacity in real batik industry waste are shown in Table 3.

Table3. Data for decreasing phenol levels and its adsorption capacity to batik wastewater

Initial phenol level (mg/L)	The decrease of phenol level (%)			Adsorption capacity of activated charcoal (mg/g)		
	3 hours	6 hours	24 hours	3 hours	6 hours	24 hours
0.3515	9.87	19.00	69.02	0.00174	0.00334	0.01213

The maximum decrease in phenol content was 69.02% with adsorption capacity of 0.01213 mg/g in batik waste water. The value is slightly different (lower) with results in simulated batik wastewater. This may be due to the condition of the batik industry's real wastewater is viscous so that the adsorption process is less optimal and less maximal due to the presence of disturbing compounds contained in batik waste including the presence of heavy metals, chrome and cobalt [24]. This may be also caused by using difference reagent for analysis phenol by spectrophotometry UV-Vis. In this study, 4-aminoanthipyrin reagents were used as in the INS 06-6989.21-2004 procedure, but in another study the researchers used powder pillow reagents [24] and fenton reagents [26].The measurement of phenol level by colorimetric method is all phenol in water will react with 4-aminoanthipyrin at pH 7.9 ± 0.1 with the addition of potassium solution hexacyanate forms quinonymin brownish red [27].

3.4. Isotherm Type of Phenolic Adsorption

There are two types of adsorption isotherms used in the adsorption process of activated charcoal in the solution, namely Freundlich and Langmuir adsorption isotherms [28]. By making the relation of log Xm/m and log Ce, it is resulted graph of Freundlich and Langmuir adsorption patterns.

The graph of Freundlich model of 3, 6, and 24 hours of contact time are shown in figure 3. Based on the figure 3 combined with the linearity factors on table 3, it is clearly seen that there were various linearity numbers. From the plotting of linearity formula, the R^2 of determination coefficient were 0.7115, 0.9248, and 0.4053 for 24, 6, and 3 hours interaction respectively so that are one formula exceed the R^2 as the linearity requirement.

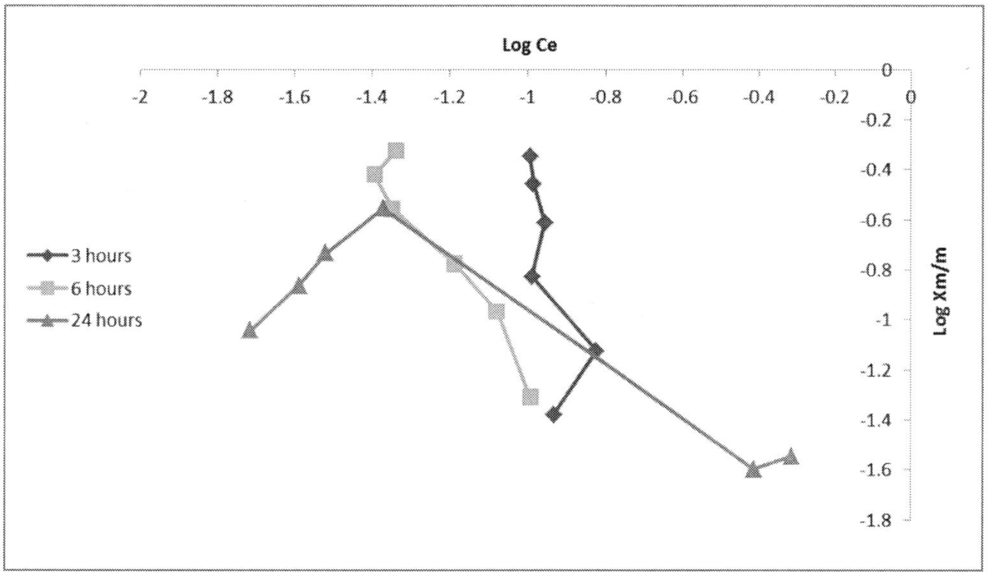

Figure 3. Freundlich adsorption isotherm of phenol in 3, 6, and 24 hours interaction time on 5% $ZnCl_2$ activated charcoal.

The linearity value are shown in table 4.

Table 4. Data of R^2 of phenol adsorption

Contact time	Isotherm data	Phenol adsorption tipe of isotherm	
		Freundlich	Langmuir
24 hours	Linear equation	y = - 0.5828x – 1.7285	y = - 1.1144x + 0.1722
	R^2	0.7115	0.0183
6 hours	Linear equation	y = - 2.191x – 3.4035	y = 28.9x – 1.2385
	R^2	0.9248	0.8641
3 hours	Linear equation	y = - 3.9736x – 4.5517	y = 36.843x – 3.1329
	R^2	0.4053	0.4043

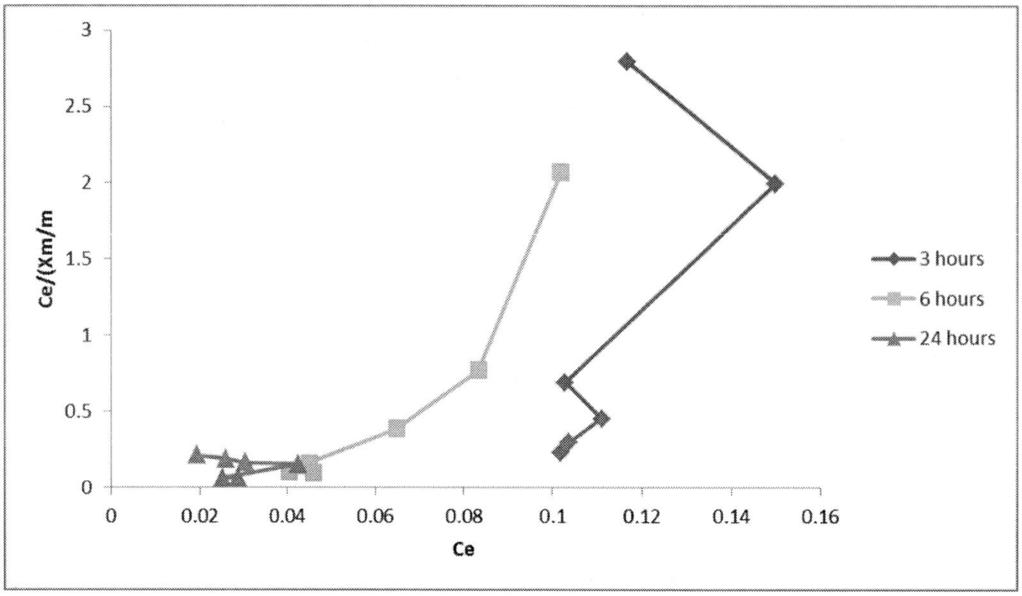

Figure 4. Langmuir adsorption isotherm of phenol in 3, 6, and 24 hours interaction time on 5% $ZnCl_2$ activated charcoal.

Based on figure 4 and table 4, it can be inferred that all the data for investigating Langmuir pattern cannot support the conclusion of its model. Therefore, it can be concluded that type of activated charcoal adsorption isotherm of marine pandan leaves tends to meet the pattern of Freundlich model because Freundlich and Langmuir equations can be applied if the value of R^2 approaches 1 or> 0.9 [24, 28].

Freundlich isotherms can be applied to the adsorption of phenolic organic substance by various adsorbents. This model assumes heterogeneous adsorption on the surface and occurs with more than one layer (multilayer) adsorbent so that each molecule has different adhesion intensity [29, 30].

4. Conclusion

Based on the results and discussions above, it can be concluded that the characters of charcoal that had been activated by $ZnCl_2$ explained in details: for water content, ash content and iodine adsorption power levels have met the required standards, while for levels of volatile substances and charcoal content have not met INS 06-3730-1995 standards. Characterization using SAA showed that the surface area of the adsorbent of activated charcoal was 68.543 m2 / g. Moreover, the decrease of phenol concentration was greater along with the increase in concentration. The values were 92.90%; 95.60%; and 97.15% for the level concentration of 0.6 mg/L; 0.8 mg/L; and 1.0 respectively. The highest decrease of phenol reached at 24 hours of contact time. Also, the largest capacity of adsorbent adsorptions was 0.048575 mg / g occurred at 24 hours contact time. And so, type of phenol adsorption isotherm from the activated charcoal adsorbent of marine pandan leaves tends to meet the pattern of Freundlich isotherms.

References

[1] El-Naas M H, Al-Muhtaseb S A, & Makhlouf S 2009 *J. Hazard. Mater.164*(2–3)720–725
[2] Government Regulation No. 85 Year 1999 about: The changes on Government Regulation No. 18 year 1999 about *Poisonous and Hazardous Material Waste Management*
[3] Ministry of Natural Environment Regulation No. 5 year 2014 about *Quality standard of*

wastewater. Lampiran XLII Jakarta
[4] Ahluwalia S S, & Goyal D 2005 *Engineering in Life Sciences5*(2) 158–162
[5] Sembiring M T, & Sinaga T S 2003*Arang Aktif (Pengenalan dan Proses Pembuatannya) USU Digital Library* 1–9
[6] EsterlitaMO dan Herlina N 2015 *Jurnal Teknik Kimia USU* 4(1) 1-6
[7] A Fillaeli, S Kristianingrum, ED Siswani, Sulistyani and SD Fatimah 2019 *J. Phys.: Conf. Ser.*1156 012001
[8] Sulistyani, S Kristianingrum, ED Siswani, and A Fillaeli 2019 *J. Phys.: Conf. Ser.*1156 012005
[9] H Kristianto 2017 *Jurnal Integrasi Proses* 6(3) 104 – 111
[10] Chen C, Zhao P, Li Z, & Tong Z 2016 *Desalination and Water Treatment* 57(27)12572–12584
[11] Bouguettoucha A, Reffas A, Chebli D, Mekhalif T dan Amrane A. 2016 *Desalination and Water Treatment*57(50) 24056–24069
[12] Kristianingrum S, & Siswani E D2014 *Jurnal Sains Dasar* 3(1) 48–55
[13] AL-Othman Z A, Ali R, & NaushadM 2012 *Chem. Eng. J.*184 238– 247
[14] Anirudhan S 2009 *J. Environ. Sci.* 21(2)1459–1466
[15] Laeliocattleya R A, Muchlisyiyah J, &Putri W D 2017 *Kimia Fisik Pangan*, Malang: Brawijaya University Press
[16] Indonesian National Standard No.06-3730-1995 *Arang Aktif Teknis* Jakarta
[17] Indonesian National Standard No.06-6989.21-2004 *Prosedur Pemeriksaan Fenol Spektro* Jakarta
[18] Aloma I, MartinLara M A 2012 *Journal of The Taiwan Institute of Chemical Engineers* 43 (2) 275 – 281
[19] Bansal R C, & Meenakshi G2005*Activated Charcoal Adsorption*. United States of America: Taylor & Francis Group
[20] Meisrilestari Y, KhomainiR, &Wijayanti H 2013 *Konversi* 2(1) 46-51
[21] Ghafarunnisa, D, Rauf, A, and Rukmana, BTS 2017 *Prosiding Seminar Nasional XII"Rekayasa Teknologi Industri & Informasi*". STTNAS Yogyakarta
[22] Sasongko, Dwi, Wildan Panji Tresna 2010 *Jurnal Ilmu Pengetahuan dan Teknologi TELAAH* Volume 27
[23] Gribble, C., Matthews, G.P., Laudone, G.,Turner A, Ridgway, C 2011 *Journal Chemical Engineering Science.*1-9
[24] Handayani, M., &Sulistiyono, E. 2009 *Posiding Seminar Nasional Sains dan Teknologi Nuklir PTNBR–Batan* 130-136 Bandung: Pusat Penelitian Metalurgi – LIPI
[25] Yulvicar C 2011 *Jurnal Ilmiah Pendidikan Bio Edukasi* 3(2) 9-15
[26] SetyaningsihT, Riyani K, Dwiasi DW., dan Rahayu, EB 2018 *Jurnal Kimia VALENSI: Jurnal Penelitian & Pengembangan Ilmu Kimia.* 4(1)26-33
[27] Kolliopoulos AV, Kampouris DK, Banks CE 2015 *Analyst* 140(9) 3244-3250
[28] Ngapa YD 2017 *Jurnal Kimia dan Pendidikan Kimia* 2(2) 90-96
[29] Triyono 2017 *Kesetimbangan Kimia* Yogyakarta: UGM Press
[30] Ismaya 2018 *Bioteknologi Inseminasi Buatan pada Sapi dan Kerbau* Yogyakarta:UGM Press

ACKNOWLEDGEMENT

Thanks to the Dean of Mathematics and Natural Science Faculty who fully funded this research in the scheme of faculty Research Group 2018.

Identification and antioxidant activity of phenolic compound from leaves of *Scurrula parasitica L*

Sri Atun[*], **Asri Nur Innayati, Herlambang HW, Sri Handayani**

Departement of Chemistry education, Faculty Mathematic and Natural Science, Universitas Negeri Yogyakarta, Indonesia, Jl. Colombo No. 1 Depok, Sleman, Yogyakarta, Indonesia, 55281

[*]Corresponding author: sriatun@uny.ac.id

Abstract: *Scurrula parasitica L* is one of the Loranthus plant species which is an herbal plant recorded in ancient documents and widely used as traditional medicine. The purpose of this study was to identify phenolic compounds and to test the antioxidant activity of compounds from *Scurrula parasitica* L. leaf extract. The research method was carried out by extracting dried powder of *S. parasitica* leaves with ethanol. Ethanol extract was partitioned using *n*-hexane, chloroform, and ethyl acetate solvents respectively. The ethyl acetate fraction obtained is then concentrated to form deposits. The precipitate obtained was separated and washed using *n*-hexane and a mixture of ethyl acetate: acetone (1: 1) in sequence. The precipitate obtained is then purified by recrystallization with ethanol as solvent. The crystals obtained were tested for their purity by thin layer chromatography (TLC) and identified using UV-Vis, IR and NMR spectrophotometers. The method for testing antioxidant activity using 2,2- diphenyl-1-picrylhydrazyl (DPPH). The results of spectroscopic identification showed that the isolation compound obtained was quercitrin (quercetin-3-O-rhamnoside). Quercitrin shows high antioxidant activity with IC50 5.00 µg / mL. Thus, *S. parasitica* plant is very potential as a source of natural antioxidants

1. Introduction

Indonesia is a tropical country that has a high level of plant diversity. Some of these plants are widely used as medicinal plants. One of the plants that can be used as medicine is Loranthus or its local name is "Benalu". Loranthus is a parasitic plant that attaches to the stem of host plants, such as *Mangifera indica* (Anacardiaceae), *Melia azedarach* (Meliaceae), and *Psidium guajava* (Myrtaceae) [1; 2]. Several studies have shown that Loranthus plants contain compounds that show cytotoxic properties of cancer cells [3], can inhibit polio virus [4], antimicrobial and anti-inflammatory [5], neuroprotective [6], antioxidant [7], and antidiabetic [8]. One type of Loranthus plant found in Indonesia is *Scurrula parasitica* L. The plant is attached to the stem of the plant *Lagerstroemia speciosa* L. Press or flower plant whose local name is "bungur". Polysaccharide compounds from the leaves of *Scurrula parasitica* are reported to inhibit cancer cells [9]. Several studies have shown the presence of several secondary metabolites from Loranthus plants including quercetin [10] and quersitrine (quercetin-3-O-rhamnosida) [11].

Lagerstroemia speciosa is one of the plants as a host for Loranthus. The leaf part of *L. speciosa* contains favonol groups namely 5.3 ', 4'-trihydroxy flavonol, saponins, glycosides, tannins and terpenoids [12]. *L. speciosa* are usually used as antidiabetic and anti-obesity [13; 14], anticariogenic [15]. The secondary metabolite content of the plant Loranthus is usually influenced by the type of host plant, so there is a possibility that the secondary metabolites found in plants attached to the stem are similar to those found in *L. speciosa*. In this article we will report the identification and testing of antioxidant activity of compounds obtained from the leaves of the *Scurrula parasitica* plant which is hemiparasit on *L. speciose.*

2. Experimental Section

2.1 Apparatus and reagent

The equipment used in this study included evaporator Buchi Rotavapor R-114, spectrophotometer, vortex mixers, and glassware. The materials used include organic solvents such as ethanol, methanol, *n*-hexane, ethyl acetate, and acetone, 2,2-diphenyl-1-picrilhidrazil (DPPH, Aldrich), ascorbic acid (Aldrich), leaves of *Scurrula parasitica* plant that hemiparasite on *L. speciosa*, and distilled water.

2.2 Extraction, isolation, and identification of phenolic compounds

Extraction was carried out by maceration method of leaf dry powder *Scurrula parasitica* for 24 hours and repeated twice. The maceration results were concentrated with the Buchi evaporator, then partitioned with *n*-hexane, chloroform, ethyl acetate solvents in succession. The results of the partition from ethyl acetate solvents were then concentrated with the Buchi evaporator, to obtain concentrated ethyl acetate fractions of 84.42 g. Solids formed from concentrated fractions of ethyl acetate are separated and purified. Purification was carried out by washing solids from concentrated ethyl acetate fraction using *n*-hexane and a mixture of ethyl acetate: acetone (1: 1) in sequence. The precipitate obtained is recrystallized by dissolving it in ethanol and left for several days to form crystals. The obtained crystals are filtered and left to dry at room temperature. The crystal obtained is shiny yellow as much as 0.27 g. Furthermore, the purity test was carried out with TLC using ethyl acetate eluent: methanol = 4: 1 and 4.5: 0.5. The TLC chromatogram shows a single compound which means that the compound isolated is a pure compound. Then pure compounds from the isolation were identified using UV-Vis, IR, and NMR spectroscopy.

2.3 Determination antioxidant activity test by DPPH method

The antioxidant activity of the isolated compounds was tested using the DPPH (2,2-diphenyl-1-picrylhydrazyl) method [16]. Samples were dissolved in ethanol at various concentrations (50; 25; 12.5; 6.25; 3,125 ug / mL). 5 mL of sample solution in various concentrations were mixed with 5 mL DPPH solution in methanol (0.12 mM) and incubated at room temperature for 30 minutes. The absorbance of each sample was determined using a spectrophotometer at a wavelength of 516 nm. As a positive control in this experiment ascorbic acid was used. Each sample was tested with three replications. Inhibition activity was calculated as a percentage decrease in sample solution compared to DPPH solution without sample. The inhibitory activity obtained from each sample was used to calculate the antioxidant activity expressed as IC_{50} (inhibitory activity at a concentration of 50%).

3. Results and discussion

L. speciose or Bungur is a purple flowering plant every September-October. Many of these plants are planted on the roadside or garden. One parasitic plant attached to the plant is *S. parasitica*. The results of isolation and purification of compounds in the ethyl acetate fraction of *S. parasitica* leaves were obtained one compound in the form of yellow crystalline solids. The UV-VIS spectrum shows the maximum wavelength (λmax) at 351.80; 258.00; 217.60 nm as in Fig.1. The UV-VIS spectrum data shows that the isolated compounds have sinamoil, benzoyl, and conjugated phenol chromophore. The IR spectrum shows absorption at (ν max) 3286.70; 2924.09; 1658.78; 1604.77 and 1505.77; 1357.02 cm^{-1} (Fig. 2.), which indicates the presence of hydroxyl groups, carbonyl groups, and aromatic C = C bonds.

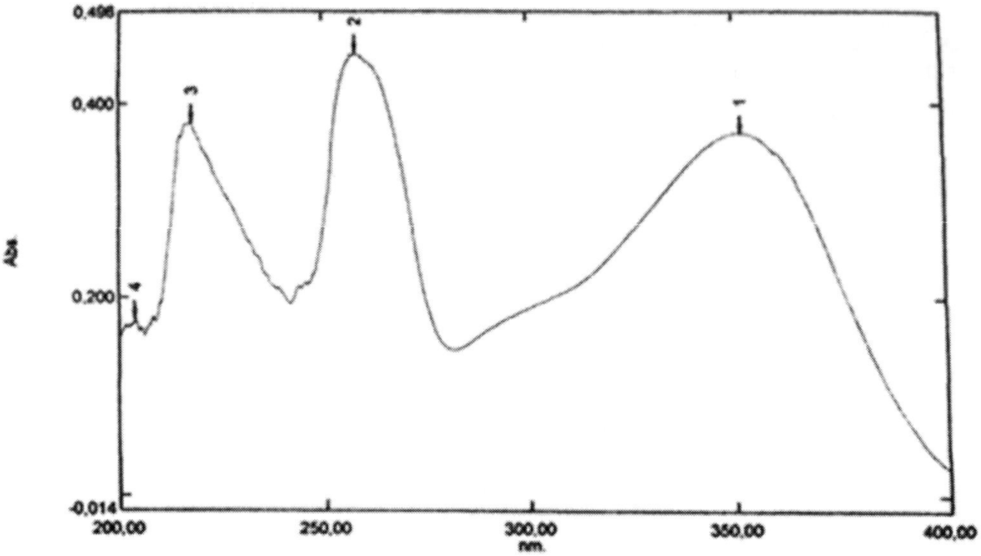

Fig. 1. Spectrum UV-VIS (in MeOH) of isolated compound in the ethyl acetate fraction of *S. parasitica* leaves

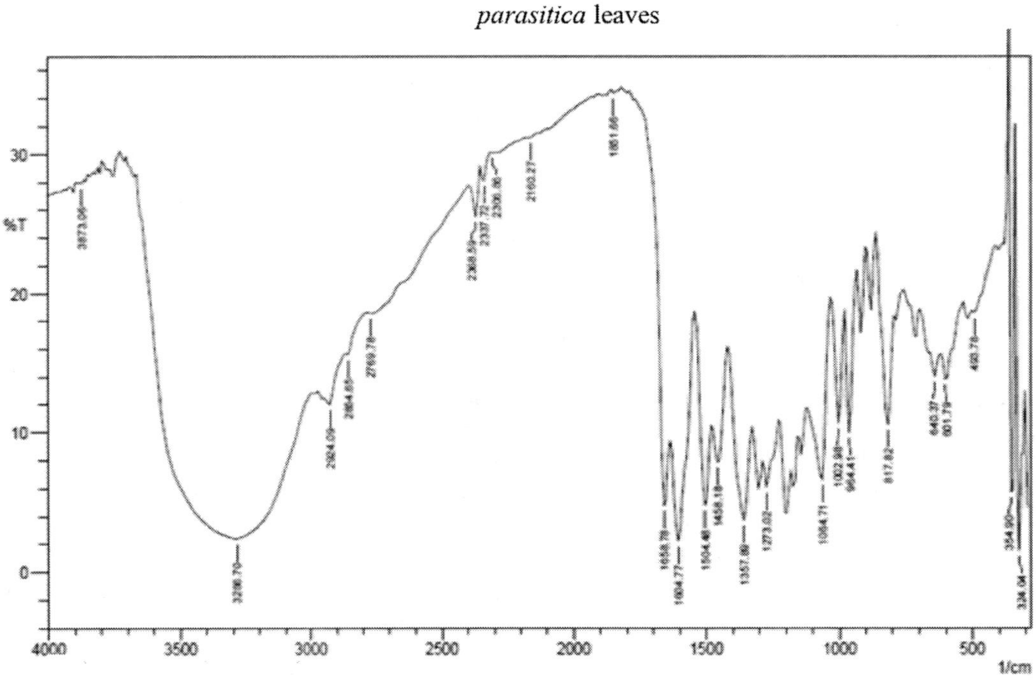

Fig. 2. Spectrum IR (KBr pellet) of isolated compound in the ethyl acetate fraction of *S. parasitica* leaves

The ^{1}H NMR spectrum data of the isolated compound shows 14 protons. Aromatic proton signals at δ 6.36 (1H, *d*, *J*=2.3 Hz) and δ 6.19 (1H, *d*, *J*=2.3 Hz) ppm indicate the presence of benzene rings substituted at positions 1,2,3 and 5. Aromatic proton signals at δ 7.30 (1H, *dd*, *J*= 8.3 and 2.1 Hz); δ 6.90 (1H, *d*, *J*=8.3 Hz); and δ 7.32 (1H, *d*, *J*= 2.1 Hz) ppm indicating the presence of substituted benzene rings at position 1 '; 3 '; 4 '. Aliphatic proton signal at δ 5.34 (1H, *d*, *J*=1.7 Hz); δ 4.22 (1H, *d*, *J*=3.3); δ 3.75 (1H, *dd*, *J*= 3.3 and 3.4 Hz); δ 3.41 (1H, *m*); δ 3.34 (1H, *d*, *J*= 6,1 Hz); and δ 0.94 (3H, *d*, *J*=6.1 Hz) ppm showing protons from the rhamnosil group. The presence of the ramnosil group is strengthened by the presence of a methyl group proton at δ 0.94 (3H, *d*, *J*=6.1 Hz) The hydroxyl proton signal from the rhamnosil group is indicated by a signal that is high and wide in the area of δ 4.90 ppm. The ^{1}H NMR spectrum data indicates that the isolated compounds are flavonoids with the rhamnosil group substitued. Spectrum ^{1}H NMR of this isolated compound can be seen at Fig. 3.

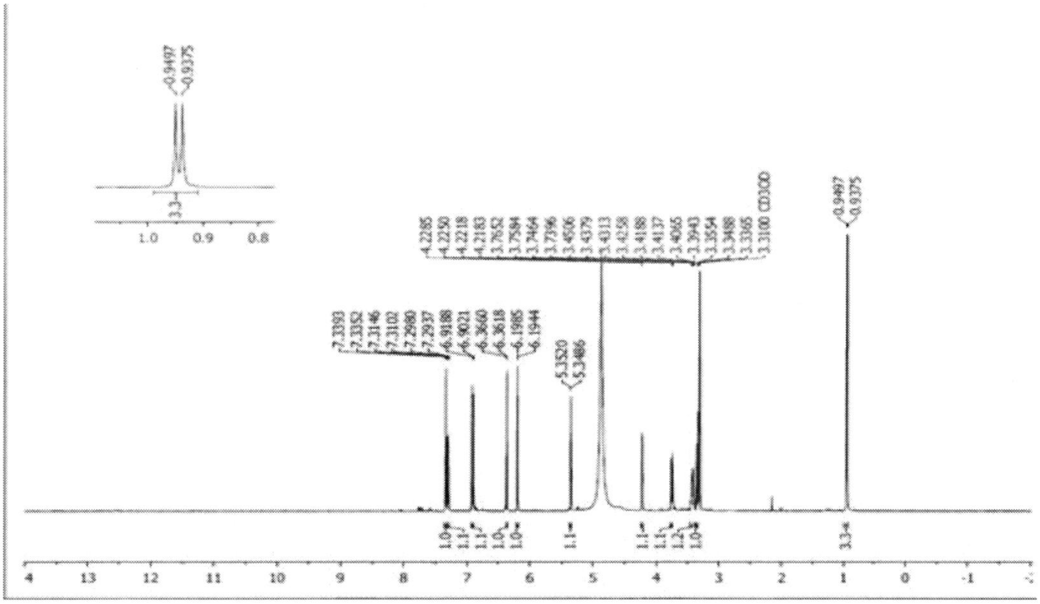

Fig. 3. Spectrum ^{1}H NMR (CD_3OD) of isolated compound in the ethyl acetate fraction of *S. parasitica* leaves

The ^{13}C-NMR spectrum data can be seen in Fig. 4. From the spectrum data, the isolated compound has two benzene rings. Carbon signals at δ 163.1 (C1), δ 103.6 (C2), δ 158.4 (C3), δ 99, 8 (C4), δ 165.8 (C5), and δ 94.7 (C6) ppm which shows the presence of a substituted benzene ring with two hydroxyl groups at positions 3 and 5. Then the second aromatic ring is shown from the carbon signal at δ 122.9 (C1 '), δ 116.9 (C2 '), δ 146, 3 (C3'), δ 149.7 (C4 '), δ 116.9 (C5'), and δ 122.8 (C6) ppm which are substituted benzene rings at position 1 ', 3 ', and 4 '. The carbon signal at δ 179.6 is a carbonyl group. The carbon signal at δ 159.2 and δ 136.2 ppm shows the alkene group. Other carbon signals are at δ 103.6, δ 71.8, δ 72.0, δ 73.2, δ 72.1, and δ 17.6 ppm which indicates signals carbon from the rhamnosil group. Furthermore, to ascertain the structure of the isolated compound used two-dimensional NMR spectrum data including HMQC (Heteronuclear Multiple Quantum Coherence) and HMBC (Heteronuclear Multiple Bond Coherence) and by comparing spectroscopic data with compounds found in the *Dendrophthoe falcata* plant [11]. From these data indicate that the compound isolated is quercitrine which has the structure as in Fig. 5. Quercitrin is the main compound found from *D. falcata* which is an antioxidant [11].

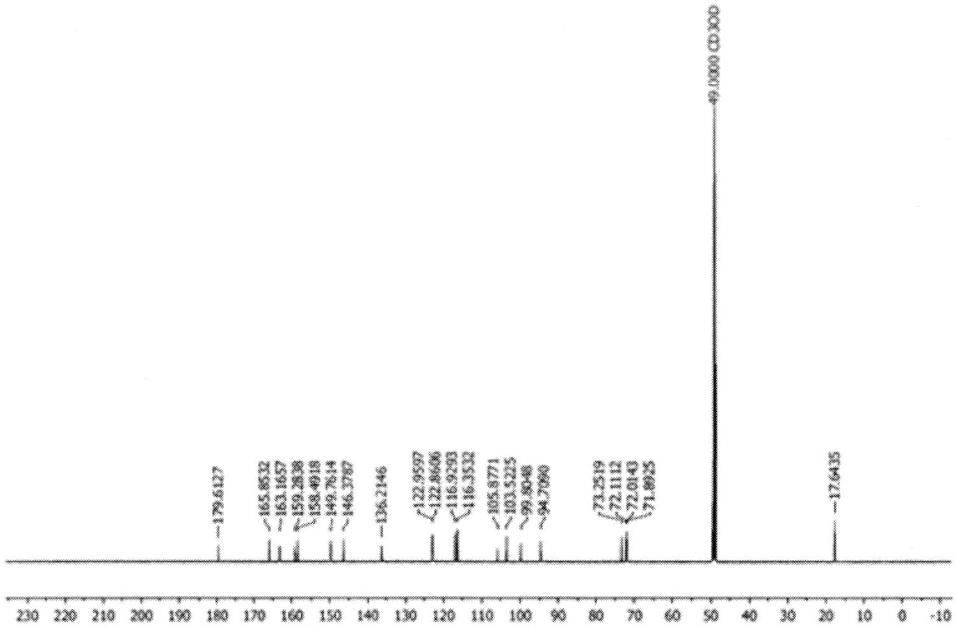

Fig. 4. Spectrum ^{13}C NMR (CD_3OD) of isolated compound in the ethyl acetate fraction of *S. parasitica* leaves

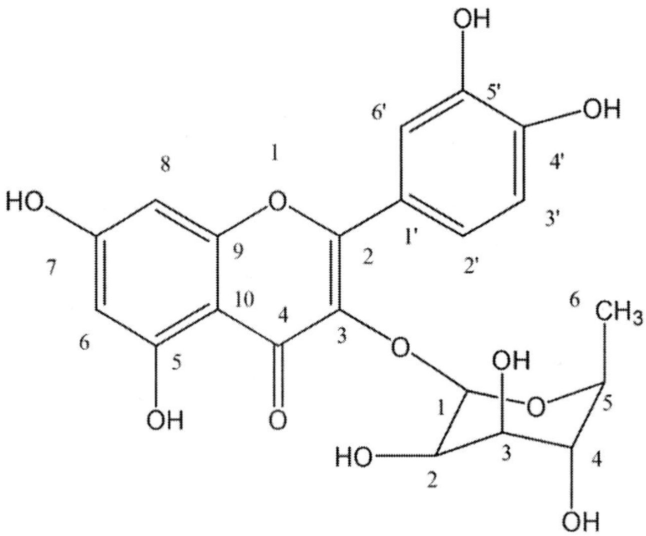

Fig. 5. Structure of isolated compound in the ethyl acetate fraction of *S. parasitica* leaves

Test of antioxidant activity from isolated compounds and plant extracts can be done by various methods. One of the simplest and easiest methods to do is to use DPPH. The antioxidant activity test using the DPPH method is based on radical capture reactions by antioxidant compounds with the mechanism of transfer of oxygen atoms, which will produce a stable DPPH-H molecule [17]. Reduction of DPPH concentration is indicated by a decrease in the intensity of the sample solution compared to a blank solution (DPPH solution without test sample). The absorbance of each solution was measured using spectrophotometry at a wavelength of 516 nm. Antioxidant capacity is expressed as IC_{50} (µg/ml) from DPPH capture activity at various concentrations. The absorbance data of each sample is then made into a linear regression equation. From the regression equation used to calculate IC50. The inhibitory activity of each sample and calculation of antioxidant activity data expressed as IC_{50} are shown in table 1.

Table 1. Antioxidant activity (IC_{50}) test leaves of isolated compound

Sample	Data of antioxidant test			
	Concentration (µg/ml)	Inhibition activity (%)	The regression equation	Antioxidant activity (IC_{50} µg/ml)
Isolated compounds (Quercitrin)	10	87.87	y = 8,2225x + 8,8838 R^2 = 0,981	5.00
	5	55.87		
	2.5	32.82		
	1.25	16.49		
	0.625	10.68		
Ascorbic acid (positive control)	5.00	95.59	y = 13.76x + 32.51 R^2 = 0.883	1.27
	2.50	79.82		
	1.25	51.45		
	0.625	32.28		

Quercitrin is a compound that has been found in the flower of *Allamanda cathartica* [18], *Zanthoxylum bungeanum* [19], *Anarcardium occidentale* L [21]. Quercitrin is a major compound in several species of loranthus plants [11]. Quercitrin showed strong antioxidant and moderate antibacterial activities [18]. Ascorbic acid is a widely known antioxidant compound, exhibiting very high antioxidant activity and is commonly used as a positive control [17]. The results showed that IC_{50} qurcitrin from compounds isolated from ethyl acetate fraction from *S. parasitica* leaves by 5.00 µg/ml was higher than positive control (ascorbic acid), but still in very active criteria.

4. Conclusion

From this study it can be concluded that the leaves of the plant *S. parasitica* contain quercitrin (quercetin-3-O-rhamnoside). Quercitrin showed high antioxidant activity with IC_{50} of 5.00 µg/mL. Therefore, *S. parasitica* leaves can be developed as natural antioxidants.

References

[1] Pattanayak SP, Sunita P, 2008 *J. of Ethnopharmacol* **120**, 241-247.
[2] Dashora N, Sodde, V, Bhagat J, Phabhu KS, Lobo R. 2017 *Pharmaceutical Crops* **2**, 1-7.
[3] Patil S, Anarthe R, Jadhav R, Surana S, 2011 *Iran J Pharm Res.* **10** (2), 253-9.
[4] De've'hat R, Tomasi S, Fontanel D, Boustie J. 2002 *Z. Naturforsch.* **57**(c), 1092-95.
[5] Wong DZH, Kadir HA, Lee CL, Goh BH. 2012 *J. Ethnopharmacol* **139**, 256-264.
[6] Govindappa M, Channabasava R, Sowmya DV, Meenakshi J, Shreevidya MR, Lavanya A, Gustavo S, Sadananda TS. 2011 *Pharmacognosy Journal* **3** (25): 82-90.
[7] Wong DZH, Kadir HA, Lee CL, Goh BH. 2012 *J. Nat. Med.* **66**, 544–551.
[8] Chandrakasan L, Neelamegam R. 2012 *Plant Archives* **12** (1), 31-40.

[9] Xiao Y, Chen B, Zeng H, Fan Y, Zhang Q. 2010 *China J. of Chinese Materia Med.* **35** (3), 381-384.
[10] Manthri S, Kota CS, Talluri M. 2011 *Journal of Phytology,* **3** (3), 10-25.
[11] Hasan MS, Ahmed MI, Mondal S, Uddin S J, Masud M M, Sadhu SK, Ishibashi M. 2006 *Oriental Phar. and Exp. Med.* **6**(4), 355-360.
[12] Roni A, Suganda, AG, Hartati R. 2018 *Jurnal Farmasi Galenika,* **5** (2), 5-11.
[13] Klein G, Kim J, Himmeldirk K., Cao, Y., Chen, X. 2007. *eCAM,* **4**(4), 401-407.
[14] Nurcahyanti AD, Arieselia Z, Kurniawan SV, Sofyan F, Wink M. 2018. *Pharmacognosy Review,* **12**, 40-45.
[15] Vivek M, Pramod N, Prashith K, Mukunda S, Mallikarjun N. 2012 *Science, Technology and Arts Research Journal,* **1**(1), 53-56.
[16] Liu CT, Wu CY, Weng YM, Tseng CY 2005 J. Ethnopharmacol. **99**, 293–300.
[17] Molyneux P, Songklanakarin P 2004 *J.Sci. Technol*, **26**(2), 211-219.
[18] Hema K, Sukumar D. 2013 *Int J Pharm Bio Sci.* **4**(4): 519-524.
[19] Zhang YJ, Wang DM, Yang LN, Zhou D., Zhang JF. 2014 *PLoS ONE.* **9**: e105725 doi: 10.1371/journal.pone.0105725.
[20] Ajileye OO., Obuotor EM., Akinkunmi EO., Aderogba M.A., **2015** *J. of King Saud University Science.* 27: 244-252.
[21] Sinoriya P., Sharma V., Sinoriya A.,2011 *Asian J. of Pharma. and Clin. Res.***4** (2):1-5.

Acknowledgements
This research was funded by the Minister of Research and Technology of the Directorate of Higher Education, Indonesia to fundamental research grand 2019. The author would like to express his gratitude for the funding.

Brief review on materials used as carrier agents for larvicide formulations

S T S Wong, A Kamari, S N M Yusoff, J Jumadi, M M Abdulrasool, S Kumaran and S Ishak

Department of Chemistry, Faculty of Science and Mathematics, Universiti Pendidikan Sultan Idris, 35900 Tanjong Malim, Perak, Malaysia.

Email: susanawongsiewtin@hotmail.com

Abstract. Mosquito-borne diseases are great concerns of people in the worldwide, especially dengue. There are 390 million people infected with dengue in a year. According to Dengue Situation Update Report for Western Pacific Region, there is an increment in the number of the infected areas as compared to the same period of previous year. World of Health Organization (WHO) and Ministry of Health (MOH) from every country pay the highest attention towards this issue. Other than inventing new formulation of adulticides, application of larvicide is newly emerging as an effective method to control mosquitoes from spreading virus. In this brief review, several recent studies about ways of the virus transmitted through mosquito biting behaviour, larvicide formulation and types of carrier agents being used to carry larvicide will be discussed. In order to sustain the aquatic ecosystem, materials for carrier agents should be environmental-friendly. This review paper will give insight into the recent development in carrier system for mosquitoes larvicide formulations.

1. Introduction

Weather in Southeast Asia region which located in the Equator is hot and humid throughout the whole year. This makes the countries in this region desired habitats and breeding sites of female mosquitoes and it leads to a proliferation of mosquito-borne diseases in these countries [1]. Dengue and malaria transmitted by *Aedes aegypti* bring high mortality rate around the world. There are approximately 390 million infection cases around the world in a year. Mosquito-borne diseases take away lives more than a war, terrorism, gun violence or other human activities [2]. Table 1 shows the statistics of dengue cases in Southeast Asia countries from 1 January to 20 June 2019.

The proliferation of mosquito-borne diseases after war rises alarm of society towards the reduction of diseases transmittance in the shortest time. Therefore, insecticides and mosquito repellent start to emerge in markets at the late few decades of the 20th century. The spread of the diseases was well controlled after the precaution mentioned were taken. However, rapid urbanisation, dense population growth and modernisation of transportation system prior to economic growth after World War II have driven the diseases to another high new level of uncontrollable infection [3].

Mosquito control is the main way to prevent the transmission of mosquito-borne diseases. After years of application of pyrethrin-based products on adult mosquitoes, there are findings stated that killing larvae is a more effective method to reduce the number of mosquitoes because larvae are living in water and they are not as mobile as adults [4, 5]. The breeding sites are easier to be banished by removal of stagnant water. Mosquito control is carried out mostly by using insecticides on their breeding sites such as poor drainage system, tyre or rubbish in urban areas. However, this leads to environmental pollution and harms other non-targeted organisms [6]. In addition, extensive usage of synthetic insecticide could lead to insecticide resistance and undesirable effects on non-target organism and human as well as the environment.

There is a variety of formulations in larvicides available on the commercial market. Essential oil (EO) from plants emerge as an important role in larvicide formulation in these few decades. Despite there are many active ingredients which also known as secondary metabolites in EO, insects react slower towards insecticides that synthesised from plants and those secondary metabolites are biodegradable. EO is much cheaper than those synthetic larvicides. EO has low water solubility, thus emulsification by oil in water (o/w) has to be done in terms of stabilising the structure [7].

In this brief review, the transmittance of diseases virus via mosquito biting behaviour, larval source management and properties of nanoparticles material for carrying larvicides has been compiled.

Table 1. Statistics of dengue cases and number of death cases in Southeast Asia countries [8].

Country	Number of dengue cases	Number of death cases
Malaysia	56,819	88
Cambodia	2,490	5
China	142	(N/A)[a]
Lao PDR	4,216	14
Philippines	83,570	354
Singapore	4,720	(N/A)[a]
Vietnam	70,491	4

[a] Not available.

2. Virus transmitted through mosquito biting behaviour

Male mosquitoes only feed on flower nectar due to the absence of proboscis. But female mosquitoes feed on the blood of animals including human to obtain protein for oviposition purpose. They take blood from bloodstream after it pierces skin using its proboscis and this is shown in figure 1. A female mosquito could live up to a month. They are active during dawn and dusk and they feed every two or three days. Blends of volatile compounds and respective ratio and spatiotemporal of the hosts lead female mosquitoes to them to bite. They will do close-range navigation by using odorant receptor to detect odour of hosts surrounding them. After that, they will select the host that fulfill most of requirements. After selection, landing on the skin will be done. These actions might be because of odour alone or olfactory receptors of mosquitoes [9].

Besides that, the concentration of carbon dioxide during exhalation of the host could attract mosquitoes to bite because carbon dioxide could rise in plumes. Mosquitoes can sense the plumes from hundreds of metres away. Body temperature and skin emanation are preferences of mosquitoes in biting [10]. For example, *Aedes* species like to bite on the face as carbon dioxide emits from nose and mouth. However, it is not by carbon dioxide alone to attract mosquitoes for biting but it is a synergist in biting matter [11]. The mechanism in sweat glands of hosts will give stimuli towards antennae of mosquitoes in detecting hear emanating from warm-blooded bodies. After capillaries closest to the surface are found, proboscis will be pierce through skin and saliva will be released to numb the area and prevents the blood from clotting. Virus of the diseases will be transmitted during piercing of proboscis into bloodstream of host if the mosquito carries the virus. The host does not realise about the bite only if the itchiness caused by the minor allergic reaction is detected [12].

There is a study shows that those who have larger body size and males have higher tendency to get bitten because they have larger skin emanation area and higher carbon dioxide concentration in their exhaled air. When there are presence of additional odorants such as lactic acid, ammonia and carboxylic acid mixtures, this would increase attraction from mosquitoes [13].

Biting behaviour of mosquitoes is also affected by external factors like humidity, temperature, visual stimuli or movement. Mosquitoes could detect their hosts within range of 100 m due to the movements and colours. If the host is moving, the waves of light would be changing and this concept is also applicable to those who wear dark-coloured clothes [10]. Some species of mosquitoes such as *Anopheles* or *Culex* which are night bitters would use different light orientation to locate the host [14].

By doing these prevention steps, the frequency of getting mosquitoes bites will be less and the percentage of infected with mosquito-borne diseases will be lower.

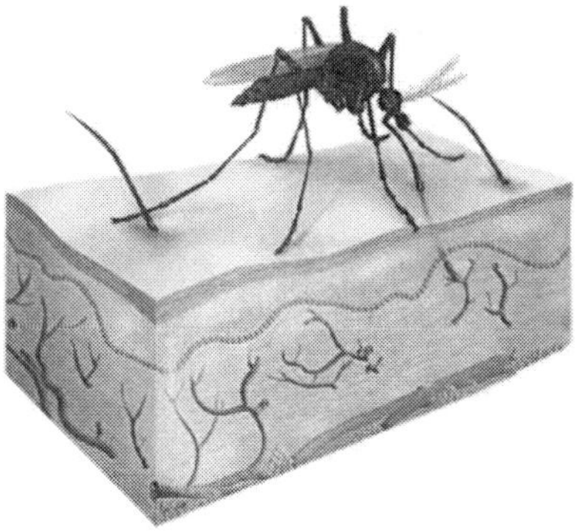

Figure 1. A mosquito is taking blood from bloodstream after it pierces skin using its proboscis. Saliva carries viruses into blood streams of host once it starts to take blood [15].

3. Larval Source Management (LSM)

Mosquito surveillance has been a highlighted issue in order to detect abnormal changes of the mosquito population from time to time for the early warning of mosquitoes-borne diseases. Breeding rate of mosquitoes is highly influenced by meteorological factors like precipitation, temperature variation and relative humidity based on mosquitoes' preference. These could lead to a change in life cycles of mosquitoes and susceptibility of some pathogen [16]. High occurrence of mosquito-borne diseases will be observed when the population density of mosquitoes in the area is dense. Thus, larval source management (LSM) is also one of the surveillance method chosen to monitor the change of mosquito population density [5]. *Aedes aegypti* and *Aedes albopictus* are the main vectors of dengue because they prefer to lay eggs in household and artificial containers which are available vastly in urbanised area [17]. As mentioned above, mosquito larvae live underneath the water. Mobility of larvae is only in water and this makes attacking the breeding sites of larvae much easier than habitats of adults. Adults could fly away from any encountered intervention measures like application of insecticide in the compound.

In year 2004, World Health Organization (WHO) introduced a programme named Integrated Vector Management (IVM) for controlling vector-borne diseases. More mosquitoes could be killed by targeting the larval stage before they are dispersed into the human environment. The type and scale of LSM application are defined by the ecology of the area and available financial support [18]. LSM requires high human source in surveillance and this could create job opportunities to locals [4]. Besides that, there are many hidden places to be mosquito breeding sites.

In Malaysia, the commonly used larvicides controls are application of temephos (Abate), house inspection and enforcement of Destruction of Disease-Bearing Insect Act 1975. Abate is low toxicity to organisms even human. Larval surveys are successful and this can be seen when Singapore government implemented this together with public education and enforced the law. The results show that the house index reduced from 16% to 2% [19]. In short, larval control should not be implemented alone and should implement together with others commonly used approaches such as bed nets or indoor-spraying [4].

3.1 Synthetic larvicides

There are four categories of insecticides used for indoor vector control which are organochlorides, pytheroids, organophosphates and carbamates. Organophosphates is one of commonly used larvicide formulation in worldwide. Temephos and fenthion are examples of the organophosphates. Other than that, insect growth regulators are also used to control larvae. However, abuse of drugs and extensive usage of the synthetic larvicides brings disruption of ecosystem in that area. Commonly used larvicide types have their own benefits of application and limitations and this is shown in table 2.

Table 2. Type of larvicide, benefits of application and limitation [20].

Type of larvicide	Benefits of application	Limitation
Temephos	• Long-term efficacy • Can be applied in occasional water flow	• Highly toxic to non-targeted organisms • Incompatible with alkali • Not effective against mosquito pupae
Bacillus thuringienesis israelensis (B.t.i)	• Highly toxic towards larvae of mosquito and black fly • Not harmful to non-target organisms	• Only effective for early stages of larval • Re-treatment at regular intervals is required
Bacillus sphaericus (B.s)	• More effective towards Culex larvae	• Not effective towards Aedini larvae
Larvicidal oils	• Convenient • Applicable to water with limited surface area	• Not effective if the oil film did not cover water surface fully • Not suitable in occasional water flow • Reapplication is reqluired after heavy rain • Harm aquatic flora and fauna
Insect growth regulators (IGR)	• Little or no toxic towards non-targeted organisms	• Could not estimate the appropriate dosage • Difficulty in monitoring until emergence of adults
Monomolecular films	• Very effective towards pupae • Environmental friendly	• Might fail kill larvae if not fully cover water surface

3.2 Essential oil based larvicides

In ancient, people could use plants in their daily life for generating smoke, decorating their habitats and applying plant extract on their bodies. In recent years, application of EO has been widely discovered in larvicide formulation. EO can be extracted from entire plant or a specific part of the plant which has abundant active ingredients needed. EO contain variety of active ingredients. Secondary metabolites of respective plants act as synergists in larvicides formulation. According to Bhatia [21], secondary metabolites of plants are extremely useful phytochemicals and they possess high chemical or structural complexity and this makes synthesis not available by using recent technology. Ingredients in EO will interfere feeding behaviour of insects by acting as an insect growth regulator or toxify their neuro system. This interference may causes protein denaturation, enzymatic inhibition or membrane disintegration and this is not adapted easily by insects. Thus, those EO that have been used as mosquito repellent also have potential to be used as larvicides (table 3) [17].

However, most EO are not soluble in water. Hence, nanoparticles acts as carrier agents to carry essential oil into water. Other than that, another limitation of EO as larvicides is application has to be focused on the most productive breeding site in an area.

Table 3. Essential oil that extracted from plants with respective active ingredients been used as larvicide [18].

Essential oil	Plant	Active ingredient (s)
Basil	*Ocimum spp*	Linalool
Catnip	*Nepeia cataria(*L.*)*	Nepetalactone
Cinnamon	*Cinnamomum zeylanicum*	Cinnamaldehdye
Citronella	*Cympobogon sp*	Citronellal
Citronella	*Cympobogon winterianus*	Citronellal, geraniol, citronellol
Clove	*Syzygium aromaticum*	Eugenol
Eucalyptus	*Eucalyptus globules*	1,8 cineole
Garlic	*Alium sativum*	Ajoene
Lemon eucalyptus	*Corymbia citriodora*	Cintronellal, p-methane-3,4-diol, citronellol, limonene
Lemon eucalyptus	*Eucalyptus citriodora*	p-methane-3,8-diol (PMD)
Lemongrass	*Cymbopogan citratus*	Citral
Neem	*Azadirachta indica*	Azadirachitin
Orange	*Citrus sinensis*	d-Limonene
Peppermint	*Mentha piperita*	Menthol
Wild tomato plant	*Lycopersicon hisatum*	2-undecanone

4. Nanoparticles as carrier agents for drugs

Nanotechnology is a new branch of technology since early of the 21st century. Most of technology can be invented in nanosize to achieve similar or higher impacts towards the target. Nanotechnology is most useful in nanomedicine, nanocarrier and so forth [21]. Advantages of nanoparticles are chosen as carrier agents are high stability, small size, high surface area and high tuneable surface [22]. In addition, Hussien et al. [23] reported that nanomaterial as drug delivery is the most potential drug delivering agents due to their high loading capacity, efficient release and targeted delivery.

Drug loading in nanoparticles can be done by entrapment, encapsulation, nanoprecipitation or others. Application of nanoparticles for larvicide delivery becomes important because less amount of larvicide needed to have similar effects with the usual amount of larvicides use. Besides that, some larvicide are not compatible in water, hence water-soluble nanoparticles could carry oil-soluble larvicide into water which is the site of larvae grow.

Nanoparticles are formulated based on release mechanism. Nanoparticles will encapsulate EO and EO will be released by trigger of surroundings such as light, heat or pH. In recent years, natural polymer-based carrier agents such as chitosan, sodium alginate or gelatin become highlights in drug loading and release because of their low cost and toxicity towards non-targeted organisms and environment. Properties of nanoparticles as carrier agents will be discussed below and advantages of the carrier agents are tabulated in table 4.

4.1 Polymer microcapsules

Polymer microcapsules can be made from several techniques such as multiple emulsion-solvent evaporation, nanoprecipitation, dialysis or supercritical fluid technology. Encapsulation, entrapment, dissolvation or adsorption of active ingredients will be done in polymer microcapsule. The most commonly used biodegradable polymer are chitosan, gelatin, polylactic acid and so forth. Besides that, polyethylene glycol and poloxamers are also used as copolymers to provide steric hindrance in order to give higher half-life of active compounds. The surface of polymer microcapsules can be manipulated by certain ligands to deliver the active ingredients to target sites more effectively or improve their circulation half-life [24].

4.2 Solid lipid micro and nanoparticles

Triglycerides, steroids, fatty acids and waxes are commonly used materials to make lipid of micro and nanoparticles and they will be in the solid state. Dispersions of solid lipid micro and nanoparticles can be stabilised by several types of emulsifier and the combinations. Examples of techniques that have been used to develop solid lipid micro and nanoparticles are high shear homogenisation, ultrasonication, solvent emulsification or evaporation, spray drying, microemulsion-based method and so forth. The extent of solubility of the drug in lipid melt, preparation method, structure and polymeric state of lipid matrix lead to loading efficiency of solid lipid micro and nanoparticles. Poor capacity to load hydrophilic molecules due to partitioning effects is usually a limit of solid lipid micro and nanoparticles but this could be solved by preparation of liquid-drug conjugate bulk via salt formation or covalent linking that later processed with an aqueous phase containing a surfactant such as Tween [24].

4.3 Nanoemulsion

Nanoemulsion is a system that consists of two immiscible liquids and disperses in three ways: oil in water (o/w), water in oil (w/o) or bi-continuous. The dispersions are stabilised by an amphiphilic surfactant or co-surfactant. They are thermodynamically stable after addition of a droplet of nanoemulsion which has a size range of between 20 to 200 nm. Size and shape of particles dispersed in the continuous phase differ emulsion from nanoemulsion. The carriers are solid spheres which it is amorphous and lipophilic that carries with a negative charge. It provides better encapsulation effect for those oil-soluble active ingredients [25].

4.4 Liposomes

Liposomes is also known as phospho-lipid vesicles. They are self-assembled phospholipid bilayers formed in aqueous media. They have a hydrophilic core and hydrophobic outer layer, hence they could encapsulate both hydrophilic and lipophilic active constituents in a single platform. Performance of the loaded active ingredients will be enhanced by increasing protection, solubility and control release while undesirable interaction with other molecules is avoided. However, they do have drawbacks: poor loading efficiency of hydrophilic materials and in-vivo stability [26].

4.5 Polymeric micelles

Micelles are formed from amphiphilic polymers which have hydrophobic and hydrophilic properties in their molecular structures and unique molecules arrangement form when they are in contact with an aqueous solution. The hydrophobic portion can be encapsulated in an interior structure and a hydrophilic portion remains facing outwards the aqueous medium. Hydrophobic drugs can be encapsulated into micelles core by hydrophobic interaction or covalent bonding [24].

4.6 Nanostructured hydrogel

A nanostructured hydrogel is a three-dimensional network of hydrophilic cross-linked polymer that does not dissolve but can swell in water. They can react to the fluctuations of the environmental stimuli because they have good compatibility with bio-or other hydrophilic molecules. They are highly absorbent and can hold up to 600 times their weight in water. It will shrink when it is dried. It is highly stimuli towards the environment. Nanostructured hydrogels are swollen polymeric networks that have drugs inside it. Swelling properties are influenced by type and composition of monomers, environmental factors like temperature, pH or ionic strength and cross-linking between molecules [27].

Table 4. Types of nanoparticles, its advantages and example [22, 24].

Type of nanoparticle	Advantages	Example
Polymer microcapsule	• Protect active ingredients from environment that might causes degradation. • Increases security of active ingredients.	Sodium alginate with calcium chloride
Solid lipid micro and nanoparticles	• Slow release of active ingredients. • Simple production method.	Lipid-core nanocapsules (LNC)
Nanoemulsion/ microemulsion	• Provide higher stability compared with classical emulsion. • Thermodynamically stable	Oil in water emulsion (o/w), water in oil emulsion (w/o)
Liposomes	• Biocompatible • Prolonged release • Reduction of skin permeation and toxicity	Phospholipids
Polymeric micelles	• Provide long-lasting reliability	N,N-diethyl-2-phenylacetamie (DEPA)
Nanostructured hydrogel	• Increase duration of reaction • Safer formulation	Poly(ethylene glycol)

5. Conclusion

Materials for carrier agents should be environmental-friendly for not harming the ecosystem in an aquatic environment. Besides that, the dosage of larvicides should be controlled so that larvicide resistance is not developed in mosquito larvae in coming few decades. Type of nanocarriers used to carry larvicide should be taken into consideration when new water-insoluble larvicide is developed. This brief review is beneficial to scientists who are working on formulation of larvicide for controlling mosquito larvae.

References

[1] Hamdan H, Sofian-Azirun M, Nazni W A and Lee H L 2005 *Trop Biomed.* **22** 45-52
[2] Leal W S 2014 *Curr. Opin. Insect Sci.* **6** 93-98
[3] Gubler D J 2011 *Trop. Med. Health* **39** 3-11
[4] Fillinger U and Lindsay S W 2011 *Malar. J.* **10** 353
[5] Killeen G F, Fillinger U and Knols B G 2002 *Malar J.* **1** 1-8
[6] Bradberry S M, Cage S A, Proudfoot A T and Vale J A 2005 *Toxicol. Rev.* **24** 93-106
[7] Ahn Y J , Kim N J, Byun S G, Cho J E and Chung K 2008 *Pest Manag Sci.* **64** 857-62
[8] World Health Organization. Regional Office for the Western Pacific 2019 *Dengue Situation Updates Number 571* (Manila: WHO Regional Office for the Western Pacific)
[9] Suh E, Bohbot J D and Zwiebel L J 2014 *Curr. Opin. Insect Sci.* **6** 86-92
[10] Fradin M S 1998 *Ann. Intern. Med.* **128** 931
[11] Webster B and Cardé R T 2017 *Biol. Rev. Camb. Philos. Soc.* **92** 1241-49
[12] Anuar A A and Yusof N 2016 *Fash. Text.* **3** 12
[13] Kelly D W 2001 *Trends Parasitol.* **17** 578-81
[14] Islam J, Zaman K, Duarah S, Raju P S and Chattopadhyay P. 2017 *Acta Trop.* **167** 216-30
[15] Spector T 2016 *Why Mosquitoes Bite Some People and Not Others* (Available from https://www.nzherald.co.nz/lifestyle/news/article.cfm?c_id=6&objectid=11590839)
[16] Guo S, Ling F, Hou J, Wang J, Fu G and Gong Z. 2014 *PLoS One* **9** e112975
[17] Baldacchino F, Caputo B, Chandre F, Drago A, della Torre A, Montarsi F and Rizzoli A. 2015 *Pest Manag Sci.* **71** 1471-85

[18] Tavares M, da Silva M R, de Siqueira L B, Rodrigues R A, Bodjolle-d'Almeida L, dos Santos E P and Ricci-Junior E 2018 *Int J Pharm.* **539** 190-209
[19] Ong S Q 2016 *Sains Malays.* **45** 777-85
[20] World Health Organization. Regional Office for South-East Asia 2011 *Comprehensive Guidelines for Prevention and Control of Dengue and Dengue Haemorrhagic Fever Revised and Expanded Edition* (New Delhi: Regional Office for South-East Asia)
[21] Bhatia S 2016 Nanoparticles Types, Classification, Characterization, Fabrication Methods and Drug Delivery Applications *Natural Polymer Drug Delivery* Systems (Switzerland: Springer) pp 33-93
[22] Ding C and Li Z 2017 *Mater Sci Eng C Mater Biol Appl.* **76** 1440-53
[23] Hussien N A, Isiklan N and Turk M 2018 *Artif Cells Nanomed Biotechnol.* **46** 264-73
[24] Mukherjee B, Satapathy B S, Bhattacharya S, Chakraborty R and Mishra V P 2017 Pharmacokinetic and Pharmacodynamic Modulations of Therapeutically Active Constituents From Orally Administered Nanocarriers Along With a Glimpse of Their Advantages and Limitations *Nano- and Microscale Drug Delivery Systems* ed A M Grumezescu (Amsterdam: United Kingdom/United State Elsevier) chapter 19 pp 357-75
[25] Singh Y, Meher J G, Raval K, Khan F A, Chaurasia M, Jain N K and Chourasia M K. 2017 *J Control Release* **252** 28-49
[26] Kumari P, Ghosh B and Biswas S 2016 *J Drug Target.* **24** 179-91
[27] Vashist A, Kaushik A, Vashist A, Jayant R D, Tomitaka A, Ahmad S, Gupta Y K and Nair M 2016 *Biomater. Sci.* **4** 1535-53

Acknowledgement
This research is supported by Fundamental Grant Research Scheme (FRGS, 2019-0003-102-02) from Ministry of Education Malaysia.

Synthesis and characterization of thymol-loaded lauryl glycol chitosan for pesticide formulation

S N M Yusoff, A Kamari, S Ishak, J Jumadi, M M Abdulrasool, S Kumaran and S T S Wong

Department of Chemistry, Faculty of Science and Mathematics, Universiti Pendidikan Sultan Idris, 35900, Tanjung Malim, Perak, Malaysia.

E-mail: najiahyusoff@gmail.com

Abstract. Thymol is a hydrophobic active ingredient and widely used as an alternative pesticide for pest control. However, due its low solubility in water, a water-soluble carrier is required in its formulation. Therefore in this study, the potential of an amphiphilic chitosan, namely lauryl glycol chitosan (LGC) as a carrier to encapsulate and release thymol was evaluated. The physical and chemical properties of LGC were characterized using Fourier Transform Infrared (FTIR) Spectrometer, Proton Nuclear Magnetic Resonance (^{1}H NMR) Spectrometer, UV-Visible Spectrophotometer, Thermogravimetric Analyser (TGA) and Transmission Electron Microscope (TEM). A Fluorescence Spectrometer was used to determine the critical micelle concentration (CMC) value of LGC in aqueous solution. The encapsulation efficiency of LGC for thymol was determined by using a High Performance Liquid Chromatography (HPLC). The FTIR and ^{1}H NMR analyses confirmed the structure of the synthesized LGC. It was noted that the addition of lauryl and glycol groups to the chitosan backbone has enhanced the solubility properties of chitosan in neutral and basic media. TEM observation confirmed that LGC could form self-aggregates in the solution with a spherical shape. The CMC value determined for LGC was 0.008 mg/mL. LGC exhibited good affinity towards thymol with an encapsulation efficiency of 58%. The findings from in vitro release study showed that the LGC could prolong the release of thymol from its micelles. Results from this study highlight that LGC possesses great characteristics to be further developed as a promising carrier in the pesticide formulation.

1. Introduction

Thymol, a natural monoterpene phenol isolated from certain *Lamiaceae* families such as thyme and oregano has good acaricidal, antimicrobial, bactericidal and insecticidal properties [1]. Due to its low level of residue and risks in the environment, it has been employed as an alternative pesticide to fights insects, mites, ticks and bacteria [1,2]. However, the application of thymol as a pesticide has a drawback as it exhibits low water solubility (0.9 mg/mL) [3,4]. Therefore, in order to dissolve thymol, a large amount of organic solvents is frequently utilized in its current formulation [3]. Nevertheless, there are various negative side effects on the usage of organic solvents in pesticide production [5]. Organic solvents in the pesticide formulations can leach and volatile to the environment thus can end up in the atmosphere, surface water and adsorb into soil or sediment [5,6]. Consequently, it can contaminated the environment and caused health hazard to human and animals [5,6].

New trends of pesticides formulation have lean towards reducing the application of organic solvent, control the release of pesticide and improve the activity of pesticide active ingredients [7]. In recent years, the application of polymeric micelles particularly amphiphilic chitosan derivatives as a carrier in controlled release formulation for hydrophobic pesticides has received a lot of attention from scientist [5,8]. By conjugating hydrophilic and hydrophobic segments on the polymer, the polymer will exhibits amphiphilic properties which can self-assemble in aqueous solution. The hydrophobic segment of the amphiphilic chitosan derivatives can encapsulate the hydrophobic pesticide in its core, while the hydrophilic segment can interact with water [8,9]. Therefore, its solubility in aqueous solution can be enhanced and the excessive use of organic can be reduced. In addition, the encapsulation of thymol in amphiphilic chitosan derivative could protect the pesticide from degradation and premature leakage before arrival at the target site [8,9]. The aims of the present study were to assess the ability of an amphiphilic chitosan derivative, lauryl glycol chitosan (LGC) to encapsulate thymol and its potential as a carrier for pesticide formulation.

2. Materials and methods

2.1. Preparation of LGC

Glycol chitosan (GC) was prepared following a procedure from our previous work [10]. Meanwhile, lauryl glycol chitosan (LGC) was synthesized using similar methods to that proposed by Mobarak and Abdullah [11] and Zhang *et al.* [12], with some modifications. First, 2 g of GC was dissolved in the 100 mL mixture of deionized water and methanol (1:1 v/v) to form GC solution. Then, about 2 g of lauric aldehyde was slowly added to the GC solution. The mixture was reacted under horizontal agitation (100 rpm) for 24 hours. After 24 hours, 0.5 g of sodium borohydride that was dissolved in 5 mL of water was slowly dropped to the solution. The solution was reacted for further 24 hours, before neutralized with HCl solution. The resulting solution was dialyzed against ethanol solution (25% v/v) for 3 days before lyophilized to produce solid LGC.

2.2. Preparation of thymol-loaded LGC

Thymol-loaded LGC was prepared by a modified version of the procedure suggested by Lao *et al.*, using reverse micelles technique [13]. For this experiment, thymol solution (10 mg/mL) was prepared by dissolving an appropriate amount of thymol crystal in methanol. Briefly, the synthesized LGC was dissolved in the thymol solution at the weight ratio of 100 to 1 (w/w). Then, deionized water at the volume of five-folds of the initial volume of thymol solution was added slowly to the solution. The solution was gently agitated for 2 hours before centrifuged for 10 minutes at speed of 1,000 rpm. The resulting solution was then filtered with 0.45 μm pore-size membrane.

An Agilent 1200 Infinity High Performance Liquid Chromatography (HPLC) was employed to determine the amount of thymol that successfully encapsulated by LGC. The mobile phase used for the analysis was a mixture of acetonitrile (HPLC grade) and deionized water at a ratio of 70 to 30 (v/v). The column used for the analysis was an Agilent Poroshell 120 C18 (50 mm x 4.6 mm, 2.7 μm) and its temperature was set at 40 °C. The injection volume for the sample was set at 10.0 μL, while the flow rate of the system was set at 1.0 mL/min. The analysis was carried out at a wavelength of 204 nm. The percentage of encapsulation efficiency (%EE) of thymol by LGC was calculated following Equation 1 [14]:

$$\%EE = \text{Amount of thymol in LGC/Total amount of thymol used} \times 100 \quad (1)$$

2.3. Characterization study

Fourier Transform Infrared (FTIR) spectra of chitosan and LGC were acquired from a Thermo Nicolet 6700 ATR-FTIR Spectrometer. The scanning range was from 4000 to 400 cm^{-1}, at 32 cumulative scans. A Jeol JNM-ECX-500 Nuclear Magnetic Resonance Spectrometer was utilized to record the Proton Nuclear Magnetic Resonance (^1H NMR) spectrum of LGC. For the ^1H NMR analysis, acetic acid-d was used as a solvent to dissolve LGC. Thermogravimetry analysis (TGA) was carried out by using a TGA/DSC 1 Mettler Toledo Analyser to identify the thermal decomposition pattern of chitosan and LGC. The analysis was performed from 25 to 900 °C, at a heating rate of 10 °C/min under argon gas (flow rate of 20 mL/min). The solubility properties of chitosan and LGC in different pH was determined as a function of the percentage of transmittance according to a method proposed by Kubota et al. [15], with some modification. For this analysis, the sample solution (2 mg/mL) was prepared by dissolving LGC in 2% (v/v) HCl. The analysis was conducted using an Agilent Cary 60 UV-Visible Spectrophotometer at a wavelength of 600 nm. The critical micelle concentration (CMC) of LGC was determined by using a spectrofluorometer (Agilent Cary Eclipse Fluorescence Spectrometer), with pyrene was utilized as a fluorescence probe [14]. For the CMC study, pyrene was first dissolved in the methanol to prepare a pyrene solution with a concentration of 1.25 x 10^{-3} mg/mL. Then, pyrene solution (1 mL) was transferred into a vial, before methanol was completely eliminated by evaporation. After that, 6 mL of LGC solution at concentration varied from 1.0 x 10^{-4} to 1.0 mg/mL was added individually to the vial. The mixture was then sonicated for 10 minutes. Lastly, the pyrene emission spectra of the samples were scanned from a wavelength of 344 to 700 nm, with excitation wavelength was set at 334 nm. For this experiment, both opening slits of the emission and excitation was set at 5 nm. Transmission electron microscopy (TEM) was used to examine internal morphology of LGC. The study was carried out using a Hitachi SU 8020 UHR High Resolution Field Emission Scanning Electron Microscope, operated at 20 kV.

2.4. In vitro release kinetics studies for thymol-loaded LGC micelles

The in vitro release of thymol from LGC micelles was studied through dialysis method in 0.1 M phosphate buffered saline (PBS) solution at pH 7.4. For this study, 5 mg of thymol-loaded LGC was dissolved in 3 mL PBS solution and placed in a dialysis bag that has molecular weight cut off 14,000 D. Then, it was introduced into a beaker containing 250 mL of PBS solution. The system was maintained at room temperature under continuous stirring (100 rpm). To determine the amount of thymol released from LGC micelles, 3 mL of PBS solution was periodically withdrawn from the beaker. After each withdrawal, 3 mL of fresh PBS was added into the beaker to maintain the volume of PBS solution at 250 mL. The thymol released was assayed by using an Agilent Cary 60 UV-Visible Spectrophotometer at 204 nm. The release study of thymol was carried out in triplicates.

The release data of thymol can be mathematically modeled to obtain information concerning the release mechanism. Four kinetic models, namely zero-order, first-order, Higuchi and Korsemeyer-Peppas kinetic models were applied to the thymol's release data to deduce the mechanism involved in the release of thymol from LGC micelles [16,17].

3. Results and discussion

3.1. FTIR analysis

The FTIR analysis was used to confirm the chemical structure of chitosan and LGC. Figure 1 presents the FTIR spectra of chitosan and LGC. Figure 1(a) shows the basic characteristics of chitosan at 3464 cm^{-1} (OH and N-H stretches), 2951 and 2887 cm^{-1} (–C-H stretches), 1664 cm^{-1} (C=O stretch), 1596 cm^{-1} (N-H bend), 1438 and 1383 cm^{-1} (C-N stretches) and 1153 and 1081 cm^{-1} (C-O stretches) [14,18]. Meanwhile for LGC, when comparing to the FTIR spectrum of chitosan (figure 1(a)), the intensity of peaks related to –C-H stretching vibration (2923 and 2853 cm^{-1}) in the FTIR spectrum of LGC (figure 1(b)) were evidently increased, revealing the presence of the long alkyl chain from the hydrophobic lauryl group and methylene groups of glycol in the chitosan derivatives [11,19]. It was observed that the absorption bands related to hydroxyl groups (3308 and 1054 cm^{-1}) shifted and became more intense due to the conjugation of glycol groups [14]. It also can be noted that the intensity of absorption bands related to the amino group (3308, 1456 and 1416 cm^{-1}) became relatively weaker. It may be due to the formation of amide linkage between GC and lauryl group, which proved the substitution of lauryl to amino group of chitosan [13,19].

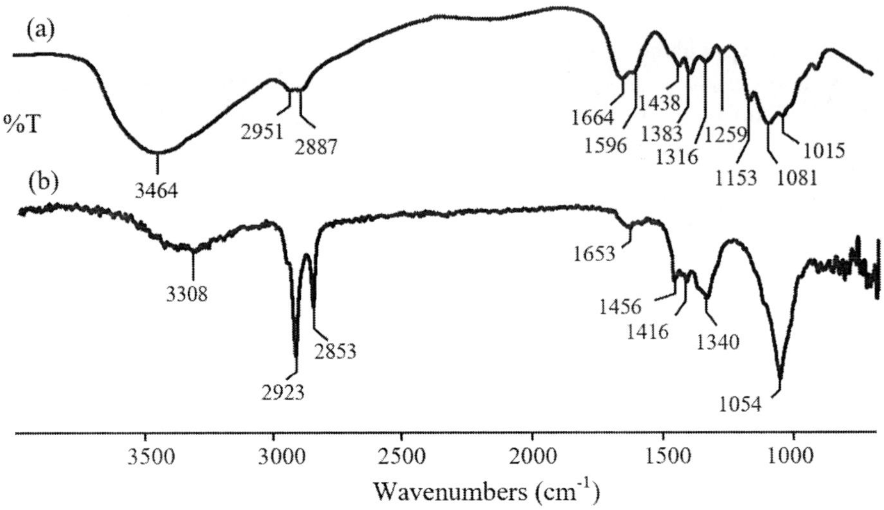

Figure 1. FTIR spectra of (a) Chitosan and (b) LGC.

3.2. ¹H NMR analysis

The formation of LGC was further confirmed by characterization with ^{1}H NMR analysis. Figure 2 shows the ^{1}H NMR spectrum of LGC. To interpret the spectrum, the solvent peak that appeared at 2.02 ppm was used as a reference. As presented in figure 2, the spectrum of LGC exhibited a signal at the chemical shift of 0.87 ppm, which attributed to the methyl hydrogen of lauryl (-NH$(CH_2)_{11}$**CH_3**) [11,20]. The signals that detected at the chemical shifts of 1.52 to 1.57 ppm could be associated with methane hydrogen of lauryl, (-NH(**CH_2**)$_{11}CH_3$) [11,20]. Besides that, the peaks observed at the chemical shifts of 3.63 to 3.99 ppm could be to overlap signals from proton related to glycol group and chitosan [14].

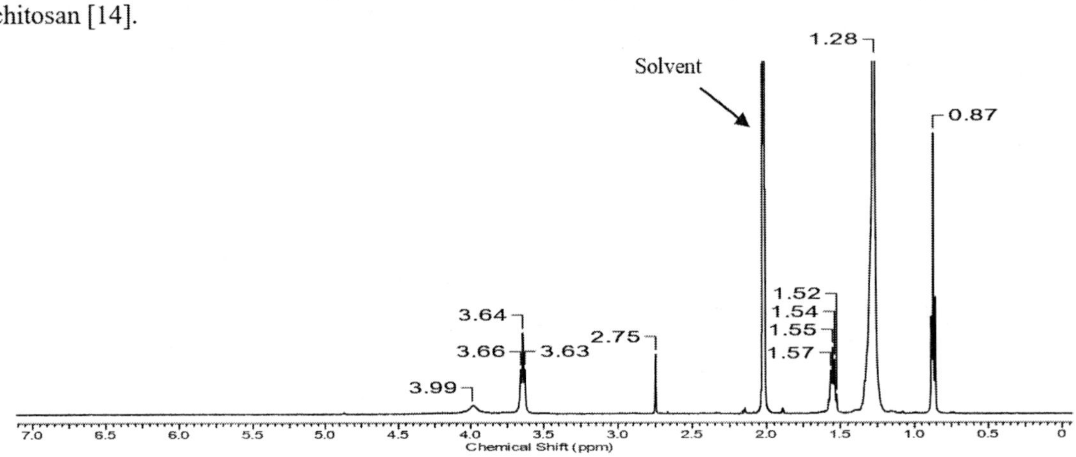

Figure 2. ^{1}H NMR spectrum of LGC.

3.3. TGA analysis

TGA analysis was conducted to study the thermal decomposition pattern of chitosan and LGC. The TGA thermograms of chitosan and LGC are displayed in figure 3. The thermal decomposition of chitosan progressed through two decomposition stages. The initial decomposition event was observed from 27 to 141 °C, due to vaporization of water bound to chitosan [21]. The second decomposition event occurred at the temperature between 235 to 449 °C, with 46% of the weight was a loss at this stage. This might due to the dehydration of the polysaccharide ring and decomposition of chitosan polymer [21]. Similar to chitosan, the thermal decomposition of LGC also takes place at two stages. At the initial decomposition stage that occurred at the temperature between 30 and 81 °C, LGC showed a weight loss of 4% due loss of residue water in LGC [21,22]. Meanwhile, the LGC exhibits rapid weight loss (76%) at the second stage (165 to 560 °C) was due to the degradation of the polymer and glycosidic bond [21,22]. The results from TGA analysis demonstrate that the introduction of lauryl and glycol groups onto chitosan have enhanced its thermal stability.

3.4. Solubility study

The solubility of chitosan and LGC in acidic, neutral and basic media were assessed based on their light transmittance (%) at different pH. Figure 4 shows the solubility trends on chitosan and LGC at pH 1.0 to 13.0. Based on figure 4, the solubility of chitosan was notably depended on the pH of the dissolution media. For example, chitosan has good solubility in the acidic medium as the values of percentage transmittance (%T) at pH 1.0 to 6.0 are nearly 100%. However, chitosan showed poor solubility in both neutral and basic media, as %T at pH 7.0 to 13.0 varies from 29 to 48%. The poor solubility of chitosan at aforementioned media maybe due to the rigidity of the chitosan structure, which owed to the presence of strong intra- and/or intermolecular hydrogen bonds [23]. In contrast to chitosan, LGC exhibits excellent solubility in all media with %T ranged from 70 to 82%. The conjugation of the hydrophilic segment (glycol group) to chitosan has advocated the formation of hydrogen bonds between the polymer and hydrogen atom in water, thus subsequently leading to the enhancement of solubility for chitosan [23]. Nonetheless, slight solubility observed on LGC at pH 4.0. A probable reason for the observation is the formation of aggregation caused by highly acetylated chain segments in LGC solution [24]. This also could occur due to the formation of the amide bond between alkyl groups [24]. Overall, these results highlighted that the LGC endowed important characteristic as an ideal carrier for hydrophobic pesticide active ingredient in aqueous solution.

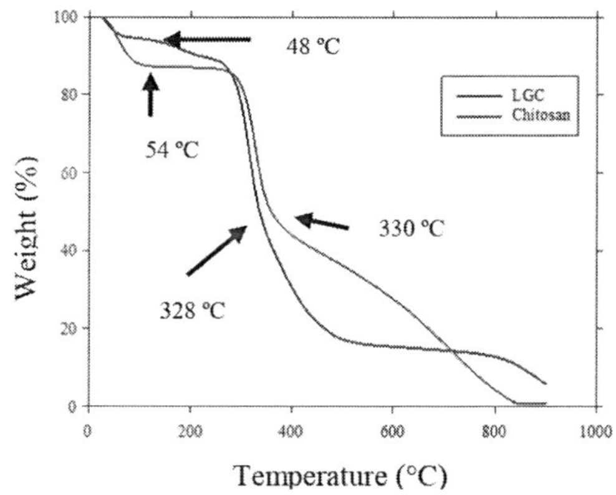

Figure 3. TGA thermograms of chitosan and LGC.

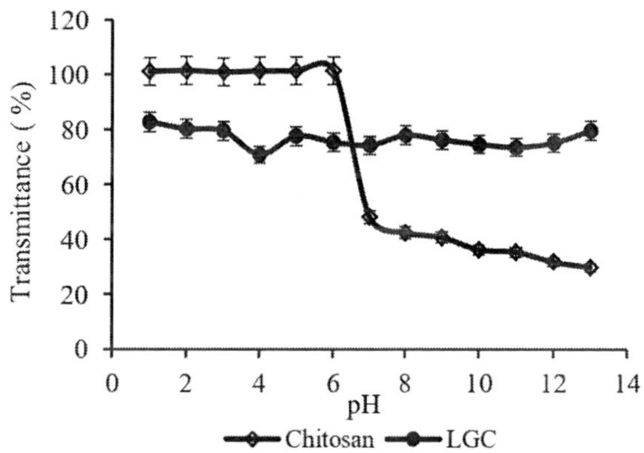

Figure 4. Solubility of chitosan and LGC at pH 1.0 to 13.0.

3.5. Critical micelles concentration

The ability of LGC to form micelles in aqueous solution was monitored by using fluorescence probe studies. Pyrene has a strong hydrophobic characteristic, thus resulted in poor solubility and self-quenching in an aqueous solution [25]. However, pyrene will emits strong fluorescence radiation if the hydrophobic micro-domains are a presence in aqueous solution as it has tendency to stay close to or lies inside these micro-domains [25]. In general, CMC is a threshold concentration when an amphiphilic molecule possibly undergoes self-aggregation by intra- and/or intermolecular association [25]. The intensity of the first and third peaks of pyrene emission spectra at 373 and 392 nm, respectively are highly sensitive to the change in polarity of its microenvironment [13,25]. Hence, the ratio of the I_{373}/I_{392} versus logarithm concentrations of LGC in aqueous solution can be used to determine CMC value [13,25]. The CMC value of LGC that was determined by interception of two line from the aforementioned graph was 0.008 mg/mL. This value was lower than the CMC value of low molecular-weight surfactant such as sodium dodecyl sulfate (2.3 mg/mL) [26]. In addition, the CMC value of LGC is comparable or lower than some others polymeric micelles likes *N*-lauryl-*N,O*-

carboxymethyl chitosan (0.063 mg/mL) [27] and *N*-mercapto acetyl-*N'*-octyl-*O,N'*-glycol chitosan (0.03278 ± 0.00589 mg/mL) [28]. The result suggests that LGC micelles are stable in aqueous solution, even after dilution and might be able to retain thymol in its micelles for a longer period.

3.6. TEM analysis
TEM analysis was adapted in order to examine the internal morphology and directly measure the size of the LGC micelles. Figure 5 displays the TEM micrographs of LGC at different magnifications, namely 10,000x and 100,000x. From figure 5, the TEM images showed that LGC micelles have a core-shell configuration, with the shape of the micelles are almost spherical. The size of the LGC micelles (dry state) are ranged from 39.7 to 53.6 nm.

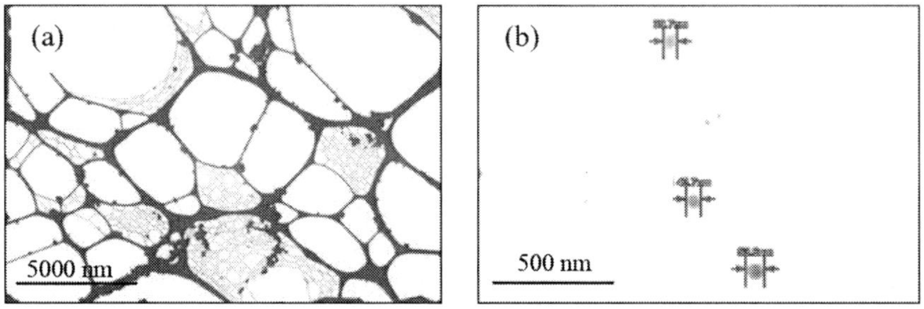

Figure 5. TEM images of LGC at (a) 10,000x and (b) 100,000x magnifications.

3.7. Encapsulation efficiency of thymol into LGC micelles
The efficacy of LGC to encapsulate thymol in its micelles was evaluated in terms of percentage of encapsulation efficiency (%EE). In this study, thymol was successfully encapsulated in the LGC micelles using reverse micelle technique. The LGC shows good %EE as it can successfully encapsulated thymol in its micelles for about 58%. The good %EE value suggests that thymol has a good affinity towards hydrophobic segments (lauryl group) in LGC [14]. The encapsulation efficiency of LGC to encapsulate thymol was comparable with other polymers. For example, thymol encapsulated by polylactide microcapsules has an encapsulation efficiency of 55.02% [29]. Meanwhile, sodium caseinate has encapsulation efficiency of 72.1 to 92.3% [3].

3.8. In vitro release and kinetics studies of thymol-loaded LGC micelles
To evaluate the ability of LGC as a carrier for controlled release system, the release study of thymol was carried out. The in vitro release profile of free thymol and thymol-loaded LGC in PBS solution is presented in figure 6. Notably, the encapsulation of thymol by LGC can effectively improve the release profile of the hydrophobic pesticide. The LGC was able to prolong the release of thymol in PBS solution up to 90 hours. In contrast, free thymol solution showed complete release from dialysis bag in 38 hours. The controlled release properties of thymol from LGC micelle suggest that LGC might be able to protect and improve the pesticide availability on target for a longer period.

The mechanism for the release of a solute from polymeric matrices usually influenced by several factors such as the desorption of the solute from the surface, the diffusion of the solute through the pore or polymeric wall and dissolution of the polymeric wall [30]. The mathematical models can be applied to provide information on the release mechanism of pesticide from the polymer matrices [16,30]. In this study, the release data of thymol from LGC micelles were individually fitted to zero-order, first-order, Higuchi and Korsmeyer-Peppas kinetic models. The kinetic model with correlation coefficients closer to 1 was chosen as the best model to explain the mechanism of thymol from LGC. Table 1 displayed the correlation coefficients values obtained from the fitted kinetics models. Based on table 1, the thymol release data was correlated well to the Korsmeyer-Peppas kinetic model, as it gave the best correlation coefficient value ($R^2 = 0.9940$) than three other kinetic models.

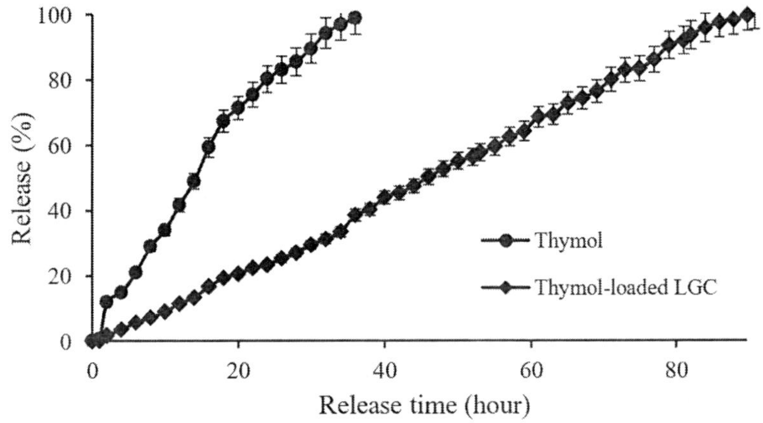

Figure 6. In vitro release of thymol and thymol-loaded LGC in PBS solution at room temperature.

Table 1. The correlation coefficient (R^2) of the fitted kinetic models for the release of thymol.

Zero-order model	First-order model	Higuchi model	Korsmeyer-Peppas models
0.9160	0.7271	0.8255	0.9940

The Korsmeyer-Peppas kinetic model can be described as Equation 2 [17]:

$$M_t / M_\infty = kt^n \quad (2)$$

where M_t represents the amount thymol release at time t and M_∞ is the maximum amount of thymol release. Meanwhile, k is the kinetic constant that integrates the characteristic of LGC and thymol system, and n is the release exponent. The application of the Korsmeyer-Peppas model to the thymol release data gave values of release constant (k) and release exponent (n) of 0.005 hr^{-1} and 1.266, respectively. It was noted that the value of the release exponent (n) of thymol from LGC, which corresponds to the type of the release mechanism involved, was higher than 0.85. Thus, this finding suggests that the release of thymol from LGC micelles govern by the Case II type transport, which associates with stress and state transition of the hydrophilic group in the polymer (relaxation of polymer chain) in water and biological fluids [17,30].

4. Conclusion

In the present work, an amphiphilic chitosan derivative (LGC) was successfully synthesized by conjugating glycol and lauryl groups onto chitosan backbone. The introduction of the glycol group as a hydrophilic segment and the lauryl group as hydrophobic segment have facilitated the self-assemble properties of LGC, thus was able to form micelles or self-aggregates in aqueous solution. The fluorescence and TEM analyses confirmed that LGC was able form stable core-shell self-aggregate in aqueous solution as LGC has low CMC value. TGA analysis showed that LGC is thermally stable, hence could protect thymol from thermal degradation. The LGC micelle showed good affinity with thymol and was able to enhance the solubility of thymol for about 6.48 times higher than its inherent water solubility (0.9 mg/mL). The release of thymol from LGC micelles was controlled by relaxation of the hydrophilic group in polymer chains in aqueous solution. From this research, it is concluded that LGC can offer promising applicability to be utilized as a carrier in pesticide formulation.

5. References

[1] Chauhan K R, Le T C, Chintakunta P K and Lakshman D K 2017 *J. Agric. Chem. Environ.* **6** 175-185
[2] Pandey S K, Upadhyay S and Tripathi A K 2009 *Parasitol. Res.* **105** 507-512
[3] Pan K, Chen H, Davidson P M and Zhong Q 2014 *J. Agric. Food Chem.* **62** 1649-1657

[4] Yalkowsky S H, He Y and Jain P 2010 *Handbook of aqueous solubilty data* 2nd ed. (Boca Raton: CRC Press, Taylor & Francis Group) p 710
[5] Zhao X, Cui H, Wang Y, Sun C, Cui B and Zeng Z 2018 *J. Agric. Food Chem.* **66** 6504-6512
[6] Roy W R 2001 Environmental impact of solvents: The environmental fate and movement of organic solvents in water, soil, and air *Handbook of Solvent* ed G Wypych (Toronto: ChemTec Publishing) pp 1150-1162
[7] Hazra D K 2015 *Int. J. Res. Rev.* **2** 35-40
[8] Adak T, Kumar J, Shakil N A and Walia S 2012 *J. Environ. Sci. Health B* **47** 217-225
[9] Kuskov A N, Kulikov P P, Goryachaya A V, Tzatzarakis M N, Tsatsakis A M, Velonia K and Shtilman M I 2017 *J. Appl. Polym. Sci.* **135** 45637
[10] Yusoff S N M and Kamari A 2018 *J. Appl. Polym. Sci.* **135** 46855
[11] Mobarak N N and Abdullah M P 2010 *Malaysian J. Anal. Sci.* **14** 82-99
[12] Zhang C, Qineng P and Zhang H 2004 *Colloids Surf. B Biointerfaces* **39** 69-75
[13] Lao S-B, Zhang Z-X, Xu H-H and Jiang G-B 2010 *Carbohydr. Polym.* **82** 1136-1142
[14] Huo M, Zhang Y, Zhou J, Zou A, Yu D, Wu Y, Li J and Li H 2010 *Int. J. Pharm.* **394** 162-173
[15] Kubota N, Tatsumoto N, Sano T and Toya K 2000 *Carbohydr. Res.* **324** 268-274
[16] Costa P and Lobo J M S 2001 *Eur. J. Pharm. Sci.* **13** 123–133
[17] Korsmeyer R W, Gurny R, Doelker E, Buri P and Peppas N A 1983 *Int. J. Pharm.* **15** 25-35
[18] He L, Wang H, Xia G, Sun J and Song R 2014 *Appl. Surf. Sci.* **314** 510-515
[19] Rekha M R and Sharma C P 2009 *J. Control. Release* **135** 144-151
[20] Rabea E I, Badawy M E I, Rogge T M, Stevens C V, Höfte M, Steurbaut W and Smagghe G 2005 *Pest Manag. Sci.* **61** 951-960
[21] Pereira F S, da Silva Agostini D L, Job A E and González E R P 2013 *J. Therm. Anal. Calorim.* **114** 321–327
[22] Qu D, Lin H, Zhang N, Xue J and Zhang C 2013 *Carbohydr. Polym.* **92** 545-554
[23] Pillai C K S, Paul W and Sharma C P 2009 *Prog. Polym. Sci.* **34** 641–678
[24] Mourya V K, Inamdar N N and Tiwari A 2010 *Adv. Mater. Lett.* **1** 11-33
[25] Sukamporn P, Baek S J, Gritsanapan W, Chirachanchai S, Nualsanit T and Rojanapanthu P 2017 *Mater. Sci. Eng. C* **77** 1068-1077.
[26] Rahman A and Brown C W 2003 *J. Appl. Polym. Sci.* **28** 1331–1334
[27] Fei X, Yu M, Zhang B, Cao L, Yu L, Jia G and Zhou J 2016 *Spectrochim. Acta A Mol. Biomol. Spectrosc.* **152** 343-351
[28] Huo M, Fu Y, Liu Y, Chen Q, Mu Y, Zhou J, Li L, Xu W and Yin T 2018 *Carbohydr. Polym.* **181** 477-488
[29] Martins I M, Rodrigues S N, Barreiro M F and Rodrigues A E 2012 *Ind. Eng. Chem. Res.* **51** 11565-11571
[30] Silva M d S, Cocenza D S, Grillo R, de Melo N F S, Tonello P S, de Oliveira L C, Cassimiro D L, Rosa A H and Fraceto L F 2011 *J. Hazard. Mater.* **190** 366-374

Acknowledgements
The authors acknowledge the financial support from the Ministry of Education Malaysia (FRGS 2014-0105-101-02), Universiti Pendidikan Sultan Idris (GPU 2017- 0187-102-01) and Islamic Educational, Scientific and Cultural Organization (ISESCO). Siti Najiah Mohd Yusoff would like to thank the Ministry of Education Malaysia (Higher Education) for granting a scholarship under MyBrain15 (MyPhD) program. The authors also would like to express their gratitude to laboratory assistants from Department of Chemistry, Faculty Science and Mathematics, Universiti Pendidikan Sultan Idris, Malaysia for their technical assistances.

Removal of methylene blue and congo red by magnetic chitosan nanocomposite: Characterization and adsorption studies

J Jumadi, A Kamari, N A Rahim, S T S Wong, S N M Yusoff, S Ishak, M M Abdulrasool and S Kumaran

Department of Chemistry, Faculty of Science and Mathematics, Universiti Pendidikan Sultan Idris, 35900, Tanjung Malim, Perak, Malaysia.

E-mail: julianajumadi@gmail.com

Abstract. Dyes are stable organic pollutants that have been used in many industries and become one of the toxic contaminants for the environment. This research aimed at preparing a magnetic chitosan nanocomposite (MCN) as an effective adsorbent for cationic and anionic dye removal from aqueous solution. The MCN was synthesized using a coprecipitation method. The physical properties of MCN were characterized using elemental analysis (CHN-O), vibrating sample magnetometry (VSM), scanning electron microscopy (SEM), thermogravimetric analysis (TGA) and X-ray diffractometry (XRD). The results confirm that chitosan has been embedded on the surface of magnetic material and shows superparamagnetic properties behavior. A series of adsorption were conducted to investigate dosage of adsorbent, contact time and initial dye concentration to obtain the optimum condition of adsorption for both methylene blue (MB) and congo red (CR) dyes. The experimental equilibrium adsorption were then analyzed by the Langmuir and Freundlich isotherm models. Based on the correlation coefficient (R^2) values, the equilibrium adsorption data were fitted to the Langmuir isotherm model. The maximum adsorption capacity (q_m) obtained from the Langmuir isotherm model for MB and CR onto MCN was 0.1308 and 1.8257 mg/g, respectively. Results highlight the potential of MCN application as an adsorbent for water contaminated by dyes.

1. Introduction

In this era of globalization, the rapidly growing of industries is in tandem with increasing of water demand that making water contamination become a serious problem in recent years. Textile industry is one of the industries that contribute to significant water pollution problems. It is known, textile production involves the use of many chemicals and dyes that are resistant to aerobic digestion due to their stability to light and oxidation [1]. In most situation, the wastewater from textile industry contain large amounts of suspended solids, oil and grease, high concentration of chemical oxygen demand (COD) and biochemical oxygen demand (BOD), organic and inorganic dyes and heavy metals that caused negative effects to the environment when discharged to the river without proper treatment [2]. Therefore, among a number of water treatment methods that are available to be applied for the removal of color from the wastewater, adsorption has been found to be the most effective in the elimination of color with cheap and simple method [3].

Chitosan is obtained from deacetylation of chitin and has reactive hydroxyl and amino group that makes chitosan received much attention as an excellent adsorbent for removal of dye as well as heavy metal [4,5]. To enhance the properties of the chitosan adsorption in term separation of the adsorbent from the solution after the treatment, the chitosan-magnetic Fe_3O_4 nanocomposite in the form core shell microsphere structure was studies. The adsorbent can be easily removed from the solution/effluent in a short time when the external magnetic fields are introduced [6]. The most important feature of magnetic chitosan nanocomposite adsorbent was it can be recyclable, high removal efficiency and low cost of production which economical for industry [7]. Although many researchers have been tested MCN to adsorb a number of dyes, there are no reports on preparing Fe_3O_4/chitosan hybrid nanocomposite with minimum ingredient using simple coprecipitation method to removal of dyes and characterization before and after the removal were also been studied.

Therefore, the target of this study was to prepare and characterize the magnetic chitosan nanoparticle for the removal of anionic congo red and cationic methylene blue from aqueous solution. Experiment was done on a series of adsorption test to figure out the optimum adsorbent dosage, contact time and initial dye concentration to achieve a maximum reduction of cationic and anionic dyes removal. Hybridization between chitosan and magnetic material is applicable and ease to be applying in dye removal by showing features such as good compatibility, low cost and easy operation.

2. Materials and methods

2.1. Preparation of magnetic chitosan nanocomposite (MCN)

Chitosan-coated magnetic Fe_3O_4 nanocomposite was synthesized using a technique suggested by Pham *et al.* [8] with some modifications. Basically, 2.0 g of sodium dodecyl sulfate (SDS) was dissolved in 400 mL of ionized water which then 0.25 g of iron oxide nanoparticles was added into the SDS solution with intensive stirring at room temperature. Subsequently, 100 mL chitosan solution (0.5 g) was added to obtain the MCN nanoparticle. The mixture was continuously stirred at a constant speed for one hour at room temperature. Then, the magnetic nanoparticle coated with chitosan obtained was magnetically separated from solution by using a magnet bar and thoroughly washed several times with deionized water and ethanol to remove any impurities. Finally, the synthesized MCN was then dried overnight at 60 °C. The ratio of expected MCN is 1:5 (w/w) of chitosan to Fe_3O_4 nanoparticle.

2.2. Characterization study

Elemental analyze was carried out by using Thermo-Finnigan Flash EA 1112 Elemental Analyzer. The magnetic property was measured on a LakeShore-7404 vibrating sample magnetometer at room temperature for 5 to 25 min. Surface morphology of nanocomposite was examined on a Hitachi SU 8020 UHR FESEM and elemental composition by using Horiba EDX Spectrometer. Thermal behavior of nanocomposite was performed with a TGA/DSC 1 Mettler Toledo that carried out under argon atmosphere in the range of heating temperature 25-1000°C with a 20 mL/min of flow rate. The structure phases of the sample before and after the adsorption were investigated by Rigaku Miniflex II in the range of 3 - 90° with 2° per min of rate.

2.3. Adsorption study

A series of adsorption experiment was carried out in order to determine the adsorption efficiency of magnetic chitosan nanocomposite. The adsorption experiments were conducted by adding 50 mg of MCN into 50 mL of required dye solution in Erlenmeyer flask followed by mechanically agitating by using a Protech Orbital shaker model 720 at 100 rpm for 1 hour. The test was performed at ambient temperature. The external magnet was used to separate adsorbent from the supernatant and then small amount of sample were collected. Finally, the concentration of dyes in the supernatant after magnetic separation was determined using an Agilent Cary 60 UV-Visible Spectrophotometer at the maximum wavelength (λ_{max}) of 498 nm and 665 nm for CR and MB, respectively. The adsorption experiments were conducted as a function of three operational parameters namely, dose of adsorbent, contact time and initial dye concentration. The percentage of dyed uptake onto MCN was calculated using equation (1) [18]:

Journal of Physics: Conference Series **1397** (2019) 012027 doi:10.1088/1742-6596/1397/1/012027

$$removal(\%) = \frac{(C_i - C_e)}{C_i} \times 100 \quad (1)$$

where C_i is an initial concentration of dyes and C_e are the concentrations of dye solution at time t (mg/L).

3. Results and discussion

3.1. Characterization of MCN

Elemental analysis: The percentage of elemental composition analysis of carbon (C), hydrogen (H), nitrogen (N) and oxygen (O) for chitosan and MCN are presented in Table 1.

Table 1. Elemental analysis result of chitosan and MCN.

Sample	Weight percent (%)							
	Experimental value				Theoretical value			
	C	H	N	O	C	H	N	O
Chitosan	45.22 ± 0.8	7.81 ± 0.2	7.37 ± 0.8	38.85 ± 0.3	45.65	7.66	7.60	39.09
MCN	29.16 ± 0.3	4.95 ± 0.6	5.68 ± 0.5	36.82 ± 0.2	29.49	4.81	5.75	37.11

*values represent mean ± standard deviation from three replicates.

The experimental value for elemental composition was in the coherence to the theoretical values with difference within ± 0.4%.for both chitosan and MCN. For that reason, it may be concluded that with small difference elemental analysis results proved that the MCN was successfully synthesized.

Magnetic properties analysis: Vibrating sample magnetometry was used to identify the magnetic properties of the magnetite and MCN, as shown in Figure 1(a). From the results, both samples exhibited superparamagnetic properties with zero remanence and coercivity at room temperature was observed in the hysteresis loop. The saturation magnetization values (Ms) of the Fe_3O_4 and MCN were 46.901 and 9.2315 emu/g, respectively. The decreased of saturation magnetization of MCN due to the presence of the chitosan coated onto the surface of Fe_3O_4 nanoparticles or due to decrease in particle size of nanocomposite [8,9].

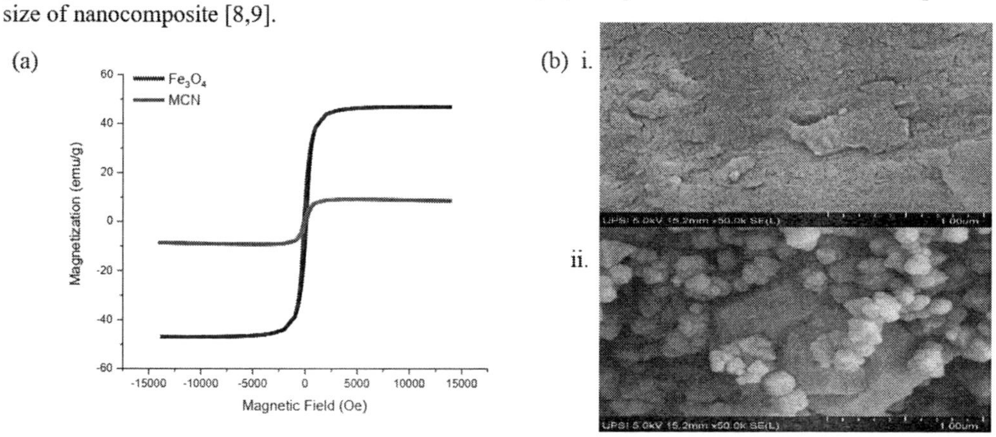

Figure 1. (a) Magnetization hysteresis loop of Fe_3O_4 and MCN at room temperature, and (b) SEM image of (i) chitosan and (ii) MCN at 50,000x magnifications.

SEM analysis: The morphology of pristine chitosan and synthesized magnetic chitosan nanoparticle are presented in Figures 1(b) i and 1(b) ii, respectively. SEM image of chitosan shows a nonporous and flat surface while MCN image shows the spherical shape represents the Fe_3O_4 nanoparticle which illustrates a high degree of agglomeration caused by the dipole-dipole interaction [10]. It was observed that after coating of chitosan at the surface of magnetic particle, the small cluster was formed within the spherical shape due to the presence of chitosan layer on the surface of magnetic material. On the other hand, the EDX spectra has confirmed the formation MCN (Figure 2(a)), where displays three sharp peaks at the energy values of 0.3, 0.5 and 6.4 KeV, which assigned as carbon, oxygen and iron element, respectively.

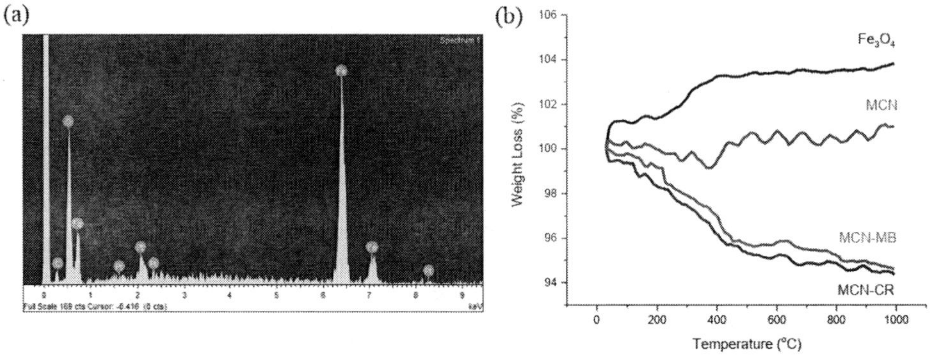

Figure 2. (a) EDX spectra of MCN and (b) TGA thermograms of uncoated Fe_3O_4, MCN, MCN-CR and MCN-MB.

TGA analysis: Figure 2(b) presents the TGA thermograms pattern of Fe_3O_4, MCN and MCNs after interaction with MB and CR dyes denoted as MCN-MB and MCN-CR that were used to identify the stability of the four particles at high temperature. It was observed that the uncoated Fe_3O_4 shows refractory material properties that can resist heat at high temperature were no significant change in weight loss or stable up to 1000 °C. On the other hand, MCN shown that about 4% of weight loss from uncoated Fe_3O_4 may be due to the loss of physically adsorbed moisture or OH group on chitosan was adsorbed onto the Fe_3O_4 surface [8]. In addition, the weight loss found around 380-425 °C of 4.5% for samples MCN-MB and MCN-CR related to the presence of organic compound of dye. These result of electrostatic interaction or hydrogen bonding interaction and chemical reaction of protonated dyes molecule attach on the surface of MCN [15].

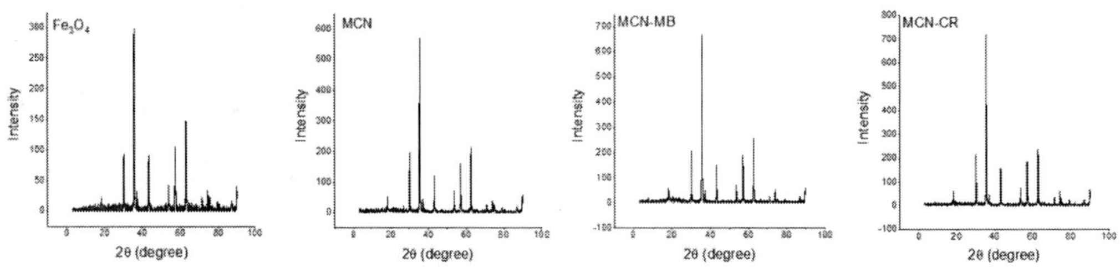

Figure 3. XRD patterns of uncoated Fe_3O_4 and MCN before and after dyes adsorption.

XRD analysis: The crystal structures of Fe_3O_4, MCN and MCNs before and after adsorption of dyes were examined, and are shown in Figure 2(a). The X-ray diffraction patterns for the uncoated Fe_3O_4 shows six main sharp peaks at 30.02, 35.58, 42.95, 53.37, 57.00 and 62.82° which matched with the standard Fe_3O_4 (ICSD, file n° 01-086-1340) [16]. The narrow and sharp peaks indicated that the Fe_3O_4 particles were associated with cubic crystalline magnetite structure and patterns peaks observed in all MCN, MCN-CR and MCN-MB nanoparticle showed the crystalline structure of magnetite. The average crystallite size Fe_3O_4, MCN, MCN-MB and MCN-CR were calculated using the Scherrer's equation was 31.12, 28.38, 28.07 and 27.73 nm, respectively.

3.2. Adsorption Studies

Effect of dosage on adsorption: The adsorbent dose is the most essential parameter in adsorption studies to evaluate the effect of the amount of adsorbent at the minimum value on the adsorption efficiency for removal of dyes. Figures 4(a) and (b) show the adsorption removal of MB and CR, respectively, onto chitosan, Fe_3O_4 and MCN at the various dosage of adsorbent from 20-100 mg at dye initial concentration 10 mg/L. According to the result, it was clearly seen that with increasing amount of three types of adsorbent, removal of both dyes percentage was also increase. It may be due to available surface area of the adsorbent that has more vacant adsorption sites on its surface to penetrate the dyes. Fe_3O_4 has the lowest removal efficiency on MB and CR with maximum removal is less than 3% and 30% at 80 and 100 mg dose of adsorbent, respectively. The pristine chitosan shows the higher adsorption in all different dosage for both MB and CR removal compared to MCN adsorbent. The sorption of MCN is lower may be because no hydrogen/electrostatic bonding was formed between negatively charged iron oxide nanoparticle with the protonated MB and CR dyes molecules that do not contain a hydroxyl group in the structure, which lead to less adsorption efficiency [14].

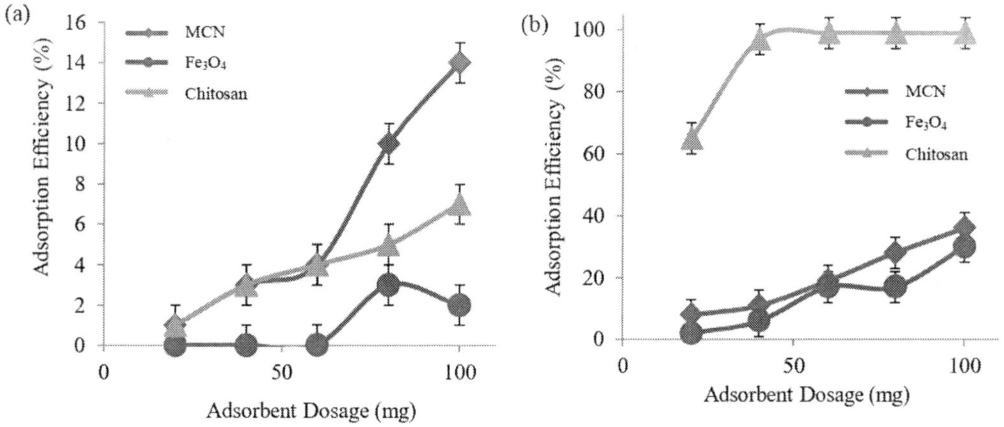

Figure 4. Effect of adsorbent dosage for the adsorption of (a) MB and (b) CR onto MCN (initial concentration: 10 mg/L in 50 mL of dyes solution, at 26 ± 2 °C).

Effect of contact time on adsorption: Effect of contact time is also a vital parameter for determining the maximum time consumed for the outstanding dyes adsorption onto adsorbent especially in industry perceptive. For the evaluation of adsorption as a function of time, graph adsorption efficiency for the removal of MB and CR against contact time was plotted as presented in Figures 5(a) and 5(b). The dye removal percentage increased with the increase in contact time. As time progress the surface coverage of the adsorbent is high and furthers no adsorption take place. The decreased adsorption rate indicates the possible monolayer formation of dyes on the adsorbent surface caused by lacking active site for further adsorption [12]. As for MCN, by increasing the contact time the adsorption of both dyes was keep increase which shows the availability of the MCN surface to interact with the dyes molecule.

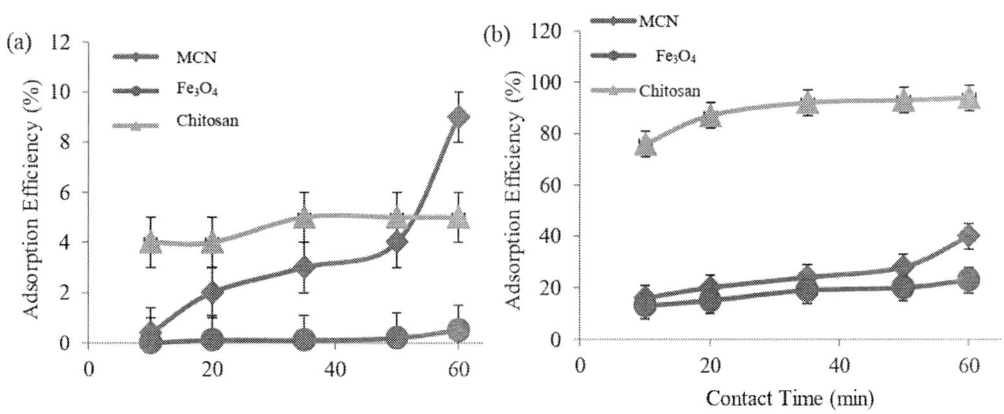

Figure 5. Effect of contact time on the adsorption of (a) MB and (b) CR onto MCN (adsorbent dosage: 50 mg, at 26 ± 2 °C).

Effect of initial concentration on adsorption: The effect of the initial dye concentration factor depends on the immediate relation between the dye concentration and the available sites on the adsorbent surface. The effect of the initial MB and CR concentration, on the adsorption efficiency is shown in Figure 6(a) and (b). This may be attributed to the fact that at low concentration of dyes solution the number of vacant active sites adsorbent is higher and hence all dye molecules may interact with the adsorbent. It might be explained that during the preliminary stage, an abundant number of vacant surface sites of MCN was available for dyes adsorption. However, with longer the adsorption time, the remaining vacant surface sites were difficult to be occupied due to steric barrier between MB adsorbed on the surface of MCN and solution phase [11]. The adsorption efficiency increased with increase in the initial dye concentration for MB due to the increasing in the driving force for mass transfer.

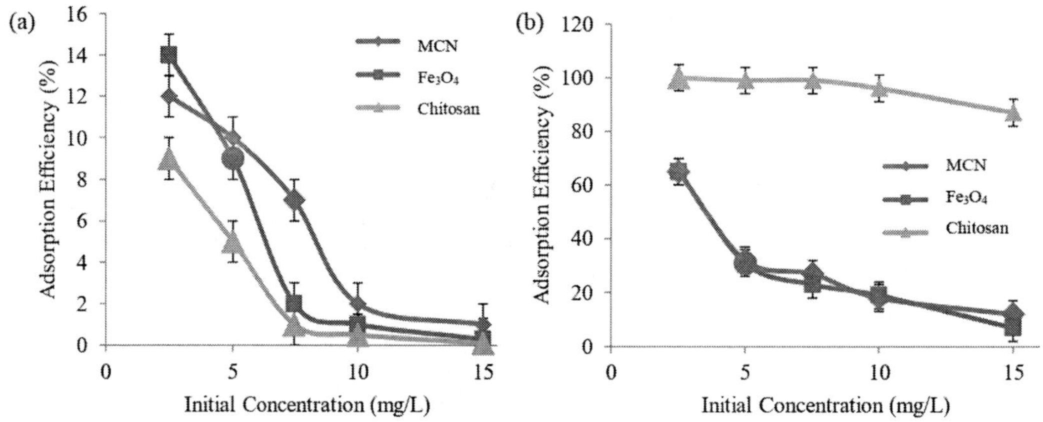

Figure 6. Effect of initial concentration on the adsorption of (a) MB and (b) CR onto MCN (adsorbent dosage: 50 mg, at 26 ± 2 °C).

3.3 Adsorption Isotherm

Langmuir isotherm: The Langmuir model assumes that the possible monolayer adsorptions of the dyes occur at homogeneous sites on the chitosan, Fe_3O_4 and MCN surface until equilibrium adsorption takes place [17]. The linear form of saturated monolayer isotherm represented by the equation (2) and can be expressed in terms of dimensionless equilibrium parameter called equilibrium parameter (R_L) as equation (3):

$$\frac{C_e}{q_e} = \frac{1}{q_m} C_e + \frac{1}{K_L q_m} \quad (2)$$

$$R_L = \frac{1}{1 + K_L C_e} \quad (3)$$

where C_e is the equilibrium concentration of dyes (mg/L), q_e is the amount of dye adsorbed per unit mass of MCN (mg/g), q_m is the maximum adsorption efficiency (mg/g) and K_L is Langmuir constant (L/mg).

Value of Langmuir parameter and correlation coefficients are summarized in Table 2. The Langmuir plot of C_e/q_e against C_e give a R^2 value of CR adsorption for three adsorbents are very close to 1 compared to adsorption on MB, considering that the adsorption isotherm of CR is better described in Langmuir model than MB. Thus, these results show that both dyes ware adsorbed onto MCN by a monolayer adsorption process. On the other hand, the R_L value of all adsorbent in MB and CR system was found between 0 and 1 which indicate the adsorption of the dyes on to adsorbent is favorable. Based on another researcher at low concentration of adsorbate, the surface of adsorbent becomes homogeneous that make the system difficult to obey the Langmuir model due to the energetic heterogeneity of adsorption sites [13].

Table 2. Langmuir and Freundlich Isotherm Models for the Adsorption of MB and CR onto MCN, Fe_3O_4 and Chitosan.

Dye	Adsorbent	Langmuir				Freundlich		
		q_m (mg/g)	K_L (L/mg)	R_L	R^2	K_F (mg/g)	n	R^2
MB	MCN	0.1308	0.3316	0.2868	0.8805	0.6354	2.4310	0.0699
	Fe_3O_4	0.0351	0.1976	0.4029	0.7986	1.3409	0.8610	0.7180
	Chitosan	0.0088	0.2154	0.3823	0.6803	1.4796	0.621	0.7257
CR	MCN	1.8257	13.906	0.0095	0.9919	1.6429	20.756	0.0569
	Fe_3O_4	1.1091	0.7420	0.1523	0.9032	1.6989	16.765	0.2140
	Chitosan	12.466	18.859	0.0072	0.9968	12.012	4.0178	0.9678

Freundlich isotherm: The Freundlich model equation indicates the empirical relationship between the loadings of adsorbent on the heterogeneous surface to the concentration of dye that involved in not uniform adsorption energies of each site [17]. The logarithmic form Freundlich equation is expressed as equation (4):

$$\ln q_e = \ln K_F + \frac{1}{n} \ln C_e \tag{4}$$

where q_e and C_e same meaning as in Langmuir equation, K_F is a Freundlich adsorption capacity constant (mg/g) and n is an intensity of adsorption. Table 2 shows the Freundlich adsorption of the MB and CR dye onto the three adsorbents by linear analysis. The experimental data of n value for dyes adsorption onto MCN less than 1 indicating chemical process was occur in adsorption while n value more than 1 point out physical process which represent good adsorption due to a distribution of surface sites or any factor that causes a decrease in adsorbent-adsorbate interaction with increasing surface density [16]. However, the R^2 for all adsorbent in MB and CR in Freundlich is lower than Langmuir. The extremely lower R^2 is at MB and CR adsorption onto MCN at 0.0699 and 0.0569, respectively which worst fitting agreement in the isotherm model.

4. Conclusion

In this research, the hybrid Fe_3O_4/chitosan nanocomposite was prepared with a simple coprecipitation method. The characterization using several analysis instruments revealed that the chitosan successfully coated to the Fe_3O_4 surface and exhibited superparamagnetic behavior. The removal efficiency of the MB and CR dyes was increased as the adsorbent dosage and contact time increased at 14 and 36 %, respectively, at an initial adsorbent dosage of 100 mg. The adsorption process was best described by the Langmuir isotherm model which indicated the dyes occur at monolayer surface of MCN. CR had greater affinity towards adsorbents as compared to MB. The maximum adsorption capacities (q_m) of MCN, Fe_3O_4 and chitosan for CR were determined as 1.8257, 1.1091 and 12.466 mg/g, respectively. The MCN has great potential to be used as an environmentally friendly and cost-effective adsorbent. The influence of chitosan to magnetite ratio in the production of MCN on adsorption capacity of MCN for cationic and anionic dyes is being studied.

References

[1] Mohan D, Singh K P, Singh G and Kumar K 2001 *Ind. Eng. Chem. Res.* **41** 3688-3695
[2] Liang C-Z, Sun S-P, Zhao B-W and Chung T-S 2015 *Ind. Eng. Chem. Res.* **54** 11159-11166
[3] Cheng M, Zheng G, Huang D, Lai C, Liu Y, Zhang C, Wang R, Qin L, Xue W, Song B, Ye S and Yi H 2018 *J. Colloid Interface Sci.* **515** 232-239
[4] Zhang Lei, Zeng Y and Cheng Z 2016 *J. Mol. Liq.* **214** 175-191
[5] Wan Ngah W S, Teong L C and Hanafiah M A K M 2011 *Carbohyd. Polym.* **83** 1446-1456
[6] Reddy D H K and Lee S-M 2013 *Adv. Colloid. Interfac.* **201-202** 68-93
[7] Liu X, Hu Q, Fang Z, Zhang X and Zhang B 2009 *Langmuir* **25** 3-8
[8] Pham X N, Nguyen T P, Pham T N, Tran T T N T and Tran T V T 2016 *Adv. Nat. Sci.: Nanosci. Nanotechnol.* **7** 045010
[9] Malwal D and Gopinath P 2017 *Colloid Interfac. Sci. Commun.* **19** 14-19
[10] Shete P B, Patil R M, Thorat N D, Prasad A, Ningthoujam R S, Ghosh S J and Pawar S H 2014 *Appl. Surf. Sci.* **288** 149-157
[11] Shi H, Li W, Zhong L and Xu C 2014 J *Ind. Eng. Chem.* **53** 1108-1118
[12] Pathania D, Sharma S and Singh P 2017 *Arab. J. Chem.* **10** S1445-S1451
[13] Bhatt A, Sakaria P L, Vasudevan M, Pawar R R, Sudheesh N, Bajaj H C and Mody H M 2012 *RSC Adv.* **2** 8663.
[14] Saha B, Das S, Saikia J and Das G 2011 *J. Phys. Chem. C* **115** 8024-8033
[15] Wang P, Yan T and Wang L 2013 *Bioresources* **8** 6026-6043
[16] Freire T M, Dutra L M U, Quiroz D C, Ricordo N M P S, Barreto K, Denardin J C, Wurm F R, Sousa C P, Correia A N, Lima-Neto P D and Fechine P B A 2016 *Carbohydr. Polym.* **151** 760-769
[17] Robita D, Mirza B, Rajabi M, Moradi O, Tyagi I, Agarwal S and Gupta V K 2016 *Chem. Eng. J.* **284** 687-697
[18] Aljeboree A M, Alshirifi A N and Alkaim A F 2017 *Arab. J. Chem.* **10** S3381-S3393

Acknowledgement
The authors would like to express their sincere gratitude to Department of Chemistry, Faculty of Science and Mathematics, Universiti Pendidikan Sultan Idris for providing the instrumentation services for this research.

Journal of Physics: Conference Series **1397** (2019) 012028 doi:10.1088/1742-6596/1397/1/012028

Synthesis of birnessite-type manganese oxide from two different reducing agents via solvent-free method and the catalytic activity in degradation of POME

A Awaluddin, Muhdarina, W Chaerani, M A Pratiwi and S S Siregar
Department of Chemistry, Universitas Riau, Kampus Bina Widya, Km 12.5 Simpang Baru, Pekanbaru, Riau 28293 Indonesia

Corresponding author: amirawaluddin01@gmail.com

Abstract. Birnessite-type manganese oxide is multifunction material that has been used in various application and the characteristic depends heavily upon its preparation method. In this study, birnessite-type manganese oxide was synthesized using the reaction between $KMnO_4$ (oxidant) and two different reducing agents i.e., citric acid and oxalic acid via a solvent-free method. The characterization of the birnessite was carried out using X-Ray Diffraction (XRD), Scanning Electron Microscopy-Energy Dispersive X-ray Spectroscopy (SEM-EDX), and Surface Area Analyzer (SAA). Birnessite from this variation of reducing agents showed different levels of crystallinity, diverse morphology, crystal size, and surface area. The birnessite-type was then applied as a catalyst for the degradation of Palm Oil Mill Effluent (POME) with hydrogen peroxide (H_2O_2) as an oxidant to reduce the concentrations of Chemical Oxygen Demand (COD). The maximum reduction in COD concentration was obtained using the catalyst concentration of 400 and 800 mg/L for reducing agent of as-synthesized catalyst citric acid and oxalic acid. Birnessite-type manganese oxide shows potential as a catalyst in the degradation of POME with citric acid as reducing agents showed better degradation in reducing COD values.

1. Introduction

Palm oil is one of the world's most produced and consumed oils that can be used in a wide variety of food, cosmetic, hygiene products, and a source for bio-fuel or biodiesel. According to Indonesian Statistic Agency 2019, the production of palm oil in Indonesia has increased significantly from 26 million tonnes in 2012 to 41 million tonnes in 2018. In spite of the benefits obtained from high production and consumption of palm oil, environmental problems such as high amounts of water consumption and high toxicity of resulting effluent generated from palm oil processing mills should be concerned and treated before being released.

Palm oil mill effluent (POME) has been classified as highly pollutant-containing wastewater with thick brownish viscous liquid waste and has an unpleasant odour. The characteristics of POME typically contains large amounts of total solids (11,500–79,000 mg/L); suspended solids contents (5,000–54,000 mg/L); total volatile solid (9,000-72,000 mg/L); oil and grease (130–18,000 mg/L); total nitrogen (180–1,400 mg/L); ammoniacal nitrogen (4-80 mg/L); and high BOD and COD contents which are in the range 10,250–43,750 mg/L and 15,000–100,000 mg/L, respectively and also amounts of N, P, K, Mg and Ca which are the vital nutrient elements for plant growth [1,2]. These characteristics depend on the quality of the raw material and production processes.

Therefore, POME requires highly efficient processing technology to ensure that waste meets the disposal requirements set by the Indonesian Ministry of Environment and Forestry. Generally, oily wastewater is discharged in open dumping ponds and it takes a very long time to reach the allowable level of COD and BOD, so these common conventional treatments (anaerobic and aerobic) cannot meet the regulations set. The main reason is due to slow microbial action, requires a large area of land even though the operating costs are low [2]. The chemical reaction is generally faster and suitable to speed up the treatment of POME. POME with a large amount of organic matter is enabled and easy to oxidized compared to wastewater containing inorganic compounds and heavy metals from heavy industrial wastewater. Hence, oxidative processes as a potential alternative technology should be developed in recovering POME.

Several chemical processes have been developed such as advanced oxidation processes (AOPs) that generate highly reactive radicals such as the hydroxyl radical (HO•) in sufficient quantities to degrade organic matter and nutrients present in the wastewater effluents [3]. Superoxide radical anions (O_2^{-}•), hydroperoxyl radicals (HO_2^{-}•), sulfate radicals (SO_4^{-}•), and organic peroxyl radicals (ROO•) are other radicals and active oxygen species generated in AOPs. Hydroxyl radicals are an extremely reactive species, with the redox potential of 2.80 V that attack most of the organic molecules to destroy and even mineralize them to a certain extent [4,5]. Among the AOPs treatment, Fenton and Fenton-like process are extensively studied and used in industry because of their low costs, ease of operation, rapid and highly efficient.

The application of manganese oxide with porous and layered structures are being developed in recent years as a catalyst to replace Fe metal ions in Fenton reagents in the AOPs method [6]. Manganese oxide can react with H_2O_2 to produce various types of active radicals as an intermediary to degrade organic compounds, one of which is dyes [7,8]. Formation of OH•, O_2•$^-$, and HO_2• to degrade methylene blue above 90% by layered and porous manganese oxide structures reported by Siregar and Awaluddin, Cheng *et al.*, and Cui *et al.* [9,10,11], which synthesized manganese oxide by solvent-free, hydrothermal, and reflux methods. Manganese oxide prepared by the sol-gel [12,13,14] and precipitation method [15,16] also shows good catalytic activity for degradation of methylene blue more than 90% in less than 30 minutes using H_2O_2 as an oxidant. In addition, manganese oxide is also used for the degradation of organic compounds, one of which is POME. Manganese oxides synthesized with different reducing agents show different properties. Therefore, in this study, manganese oxide was synthesized using two different reducing agents to see the properties effect for POME degradation. The advantages of free-solvent methods are very simple, does not require solvents, does not take a long time, and uses organic acids as a precursor that do not have toxic effects so that it is environmentally friendly.

2. Experimental

2.1. *Materials*
All the chemicals such as hydrogen peroxide, potassium permanganate, citric acid, oxalic acid, and reagents for COD determination were purchased from Merck and used without further purification.

2.2. *Synthesis*
The manganese oxide was prepared via a ceramic or solvent-free method as described by Siregar and Awaluddin but using different reducing agent [9]. The appropriate amount of $KMnO_4$ and organic acid as a reducing agent (citric or oxalic acid) with mole ratio 3:1 were mixed and ground in one mortar. The mixture was calcined at the 700°C in 5 h for citric acid and 7 h for oxalic acid (The different temperature of calcination according to our previous experiment that showed relatively good in crystallinity with small impurities). The final product was cooled up to room temperature, washed using HCl 0.1 M and distilled water. The black product was dried at 110°C.

2.3. Characterization
X-ray powder diffraction (XRD) was performed on a Rigaku X-ray diffractometer operating at 40 kV and a current of 30 mA with Cu Kα radiation (λ = 1.54060 Å). The data were collected with scan speed 2° per minute in 2θ, from 2θ = 10° to 80°. SEM-EDX were taken on a JSM-6510LV (*low vacuum*) instrument. The calcined samples were degassed at 423 K for 2 h prior to measurement. The Brunauer-Emmett-Teller (BET) surface area measurements and Barret-Joyner-Halenda (BJH) adsorption were carried out using a BELSORP-max instrument (BEL Japan) at 77 K.

2.4. Sampling of POME
The POME samples were obtained in Mei 2018 from the pond of the Palm Oil Mill in Kampar, Riau. Sampling is done by grab sampling in the palm oil mill effluent pool inlet area of 10 L. The initial stage is to analyze the pH and temperature of the waste. The handling sample method refers to Indonesian National Standard, SNI 6989.59:2008. The samples were then transported to the laboratory and kept in a refrigerator.

2.5. COD analysis
The COD analysis was performed according to the Indonesian National Standard, SNI 6989.73:2009.

2.6. Degradation of POME with H_2O_2 using manganese oxide as catalyst
The catalytic reaction was carried out in the reactor, contained 10 mL POME and an amount of catalyst. The suspension was continuously stirred magnetically for 30 minutes to achieve adsorption/desorption equilibrium. Following adding H_2O_2 (30 wt%) solution, the mixture was allowed to react at room temperature. At given time intervals, the mixture was pipetted and ready to analysis.

3. Result and discussion

3.1. Characterization of manganese oxide

The diffractogram presented in figure 1 showed the XRD patterns of as-synthesized material using potassium permanganate (oxidant) with two different reducing agents (citric and oxalic acid). The diffraction peaks of these manganese oxides match well with the standard pattern of a monoclinic phase of birnessite (JCPDS card no. 80-1098). The diffraction peaks at 2θ values of 12.52 (7.07 Å); 25.21 (3.53 Å); 37.20 (2.41 Å); 39.70 (2.26 Å); and 52.44 (2.12 Å). Diffractogram in figure 1 shows sharp peaks with high intensity. This explains that birnessite synthesized by the solvent-free method of $KMnO_4$ and citric acid (3: 1) at a temperature of 700°C has good crystallinity. While the use of the oxalic acid as reducing agent produced diffraction peak at 2θ = 12.46° (79.97 Å); 25.17° (35.37 Å); 37.11° (24.12 Å); 40.06° (22.65 Å); and 42.30° (21.24 Å) with the reflection fields (001), (002), (1-1-1), (201) and (1-1-2).

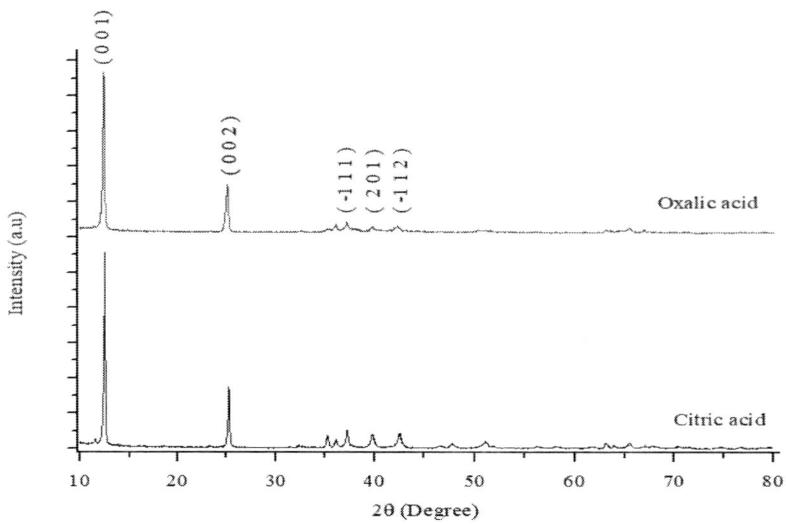

Figure 1. Diffractogram of as-synthesized birnessite-type manganese oxide

The average crystal size was determined using the Scherrer equation. The average size of the as-synthesized birnessite was 37.89 nm and 21.22 nm for citric and oxalic acid as a reducing agent, respectively. This crystal size is greater than birnessite conducted by Ye *et al.* with precursor $KMnO_4$ and glucose using the sol-gel method (crystal size of 20.9 nm) [17] and Cheney *et al.* with $KMnO_4$ and KCl as a precursor by hydrothermal method (12.38 nm) [18]. The different of crystal size may be due to differences in the precursor, mole ratio, and synthesis method.

The morphology of the as-synthesized birnessite was characterized using an SEM instrument. SEM photos of as-synthesized birnessite-type manganese oxides are shown in figure 2 with 1000x, 5000x and 10000x magnifications. The birnessite-type produced in the form of irregularly shaped granules of different sizes. These morphological results have little in common with the birnessite that synthesized by Ching *et al.*[19]; Siregar and Awaluddin [9].

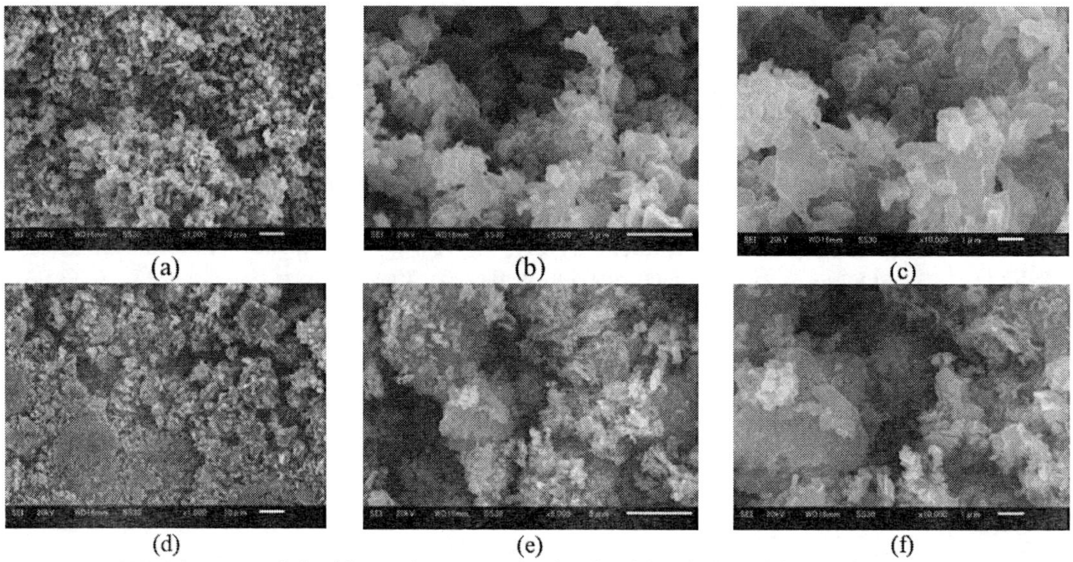

Figure 2. SEM images of the birnessite as-synthesized using citric acid as reducing agent, Mag.x (a) 1000 (b) 5000 dan (c) 10000 and using oxalic acid as a reducing agent, Mag.x (d) 1000 (e) 5000 dan (f) 10000

The chemical composition of the as-synthesized material was analyzed using EDX analysis. EDX results show the presence of elements K, Mn, and O in the as-synthesized manganese oxide. Accordingly, the composition of the materials was shown in table 1. Atoms K and Mn are constituent atoms of the structure of manganese oxide, Mn is in the octahedral framework and K is in the interlayer which binds to H_2O.

Table 1. Result of analysis using Energy Dispersive X-Ray (EDX)

No.	Elements	Citric Acid		Oxalic acid	
		% Atom	% Mass	% Atom	% Mass
1.	Mn	16.39	41.37	10.32	28.52
2.	O	38.10	27.99	45.63	36.71
3.	C	36.96	20.39	38.06	22.99
4.	K	4.12	7.40	5.99	11.78

Surface area, pore-volume, average pore diameter, and pore size distribution produced using the BET and BJH methods are shown in table 2. These surface area are smaller when compared to the surface area of other studies of birnessite which also uses the BET method, as research conducted by Ye *et al.* with precursor of $KMnO_4$ and glucose using a sol-gel method which produced a surface area of 42.61 m^2/g and a small pore size of 8.71 nm [17]. However, these surface area almost the same as the birnessite conducted by Siregar and Awaluddin with precursor $KMnO_4$ and glucose which obtained a smaller surface area of 3.56 m^2/g and pore size of 13.96 nm [9]. This is because the high calcination temperature in the solvent-free method causes a high level of crystallinity so that the resulting surface area getting smaller. Another factor that also causes the surface area of birnessite is relatively small, namely the presence of K^+ ions which cover the active site on the surface of the birnessite so that N_2 cannot be entered when analyzed. The surface area of birnessite generally ranges from 10-45 m^2/g, but there is also a surface area of more than 100 m^2/g [20].

Table 2. Textural property of the synthesized manganese oxide

Parameter	Citric acid	Oxalic acid
Surface area (m^2/g)	5.725	10.544
Total pore volume (cc/g)	0.028	0.054
Average pore radius (nm)	19.658	20.369
Pore distribution (nm)	78.010	167.750

The small surface area of birnessite in this study is one of the factors that influence the performance of the catalyst in the process of degradation of palm oil mill effluent with the efficiency of decreasing the COD concentration almost 70%. It is assumed that the larger the surface area will increase catalytic activity. However, after comparing the results in other studies, the surface area is not the main reason in catalytic activity. Liu *et al.* reported that crystalline β-MnO_2 catalytic activity with a small surface area was better than amorphous δ-MnO_2 (birnessite) [21].

3.2. Catalytic degradation of POME

The variation of catalyst concentration was carried out to determine the maximum conditions for adding birnessite-type manganese oxide as a catalyst for the degradation process of palm oil mill effluent (POME). The variation of catalyst concentration added in this study was 400; 800; 1,600 mg/L; and without the addition of a catalyst. Whereas other conditions such as the concentration of

H_2O_2 maintained in 17,000 mg/L. The initial conditions of POME for the degradation process, as listed in table 3.

Table 3. Initial conditions of POME

Parameter	Value
Temperature	47°C
pH	5.0
COD	39,571-56,288 mg/L
BOD_5	14,496-14,500 mg/L
Suspended solid content	28,102 mg/L
Oil and grease	15,355 mg/L
Total nitrogen	350 mg/L

Effect of variations in catalyst concentration on decreasing COD values can be seen in figure 3 and 4. Maximum degradation obtained in addition 400 mg/L catalyst, with 61% efficiency of decreasing COD concentration in 60-minute reaction using citric acid as a reducing agent, while for oxalic acid reach out 45%. In the addition of 800 mg catalyst/L, COD value achieved 56% if using a catalyst synthesized using oxalic acid of 800 mg/L compared to citric acid which only reached 59% in the 45-minute reaction. From this result, it shows that citric acid as reducing agent shows better results. Although seen from the extent of the surface, the results show little, but the catalytic activity is much better. Surface area is not the main factor affecting degradation. The small percentage in COD degradation of POME, which is <70%, maybe influenced by the characteristics of the catalyst which has a high level of crystallinity and a small surface area, because crystallinity can reduce catalytic activity.

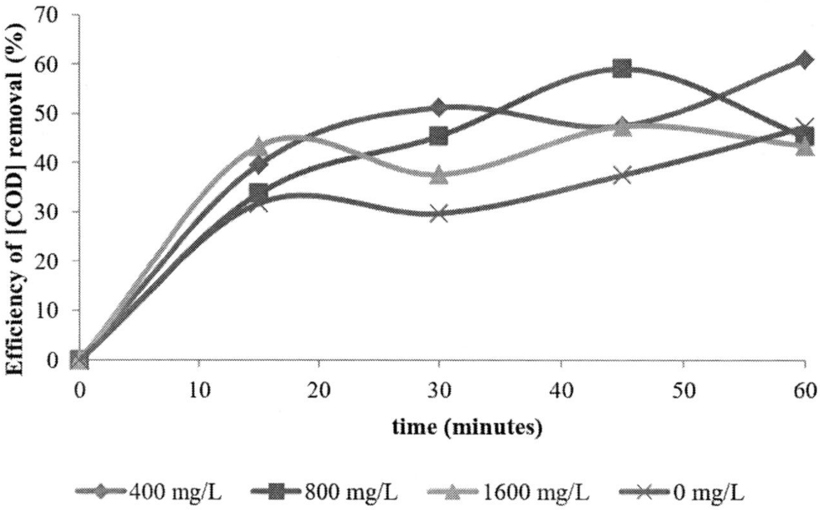

Figure 3. Effect of catalyst concentration (citric acid as a reducing agent) in COD removal of POME degradation

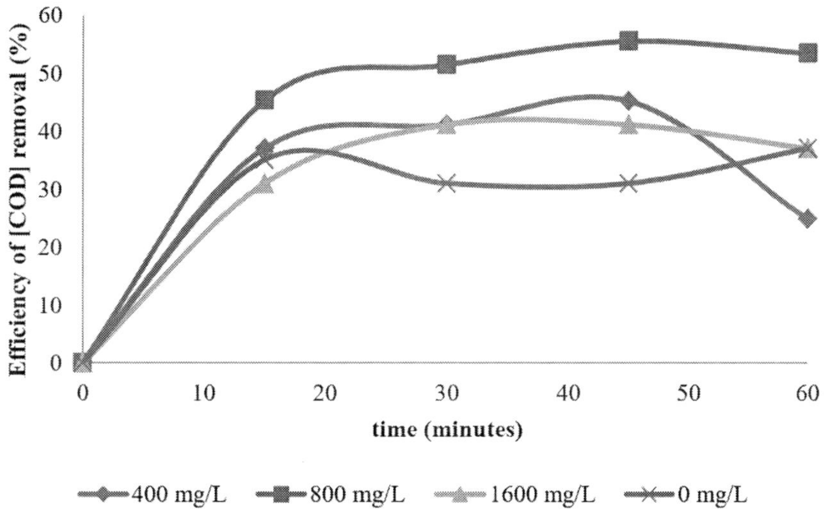

Figure 4. Effect of catalyst concentration (oxalic acid as a reducing agent) in COD removal of POME degradation.

The efficiency of decreasing COD concentration occurs at a catalyst concentration of 400 mg/L. This is quite efficient because in a small amount of catalyst can reduce the COD concentration by up to 70%. The results obtained are the same as the results of several studies that show the condition of the addition of large H_2O_2 and the small amount of catalyst give better results in degradation. Increasing the catalyst concentration causes in decreasing the percentage of degradation because it passes the critical concentration of the catalyst. So that, hydroxyl radicals are not able to oxidize further because the presence of Mn (II) ions in the solution is already excessive. This condition causes competition between excess Mn (II) ions and organic components to react with hydroxyl radicals. This condition is called recombination of hydroxyl radical. A significant decrease in COD concentration occurred in 15 minutes of reaction. A typical general reduction in COD waste can be divided into two stages. First, a rapid phase of decline, after that the second stage occurs a turning point where the reaction speed decreases due to the formation of organic carbon as an intermediate result. Therefore, in 30, 45, and 60 minutes of reaction, there was no significant decrease in COD concentration.

The mechanism of waste degradation, according to Yu *et al.* through the adsorption-oxidation-desorption stage[22]. The initial stage of the degradation process is the adsorption of H_2O_2 on the surface of manganese oxide (1.1) followed by the decomposition of H_2O_2 into free radicals such as $HO\bullet$, $O_2\bullet^-$ or $HOO\bullet$ (1.2) - (1.4). The high oxidative ability of free radicals ($HO\bullet$, $O_2^-\bullet$) to decompose adsorbed organic waste molecules to produce CO_2, H_2O, or other small molecules (1.5). The small molecules produced from this degradation process are absorbed from the surface of the catalyst in the final stages of degradation [23].

$$\text{Mn (IV)} + H_2O_2 \leftrightarrow \text{Mn (IV)}\cdot H_2O_2 \quad (1.1)$$

$$\text{Mn (IV)}\cdot H_2O_2 \rightarrow \text{Mn (II)} + HO_2\bullet + H \quad (1.2)$$

$$\text{Mn (II)} + H_2O_2 \rightarrow \text{Mn (IV)} + HO\bullet + OH^- \quad (1.3)$$

$$HO_2\bullet \leftrightarrow H^+ + O_2\bullet^- \quad (1.4)$$

$$HO\bullet\ (O_2\bullet^-) + \text{POME (adsorbed)} \rightarrow CO_2 + H_2O + \ldots \quad (1.5)$$

$$2HO\bullet + 2HO\bullet \rightarrow O_2 + 2H_2O \quad (1.6)$$

Conclusion

Birnessite-type manganese oxide has been successfully synthesized through the solvent-free method with variations in reducing agents (citric acid and oxalic acid). Variations of reducing agents produce the same type of manganese oxide but show different levels of crystallinity, diverse morphology, crystal size, and surface area. The results of the degradation POME using manganese oxide-type birnessite shows the potential based on the decrease in the value of COD. The efficiency of decreasing the value of COD was obtained in reaction conditions of H_2O_2 17,000 mg/L, catalyst concentration: 400 mg/L and 800 mg/L, respectively for reducing agent of as-synthesized catalyst citric acid and oxalic acid. Citric acid as reducing agents showed better degradation in reducing COD values, although further research is still needed to get the best conditions for degradation of POME using H_2O_2 and birnessite-type manganese oxide as a catalyst.

References

[1] Kamyab H, Chelliapan S, Rezania S, Din M F M, Khademi T and Kumar A 2018 *Palm Oil Mill Effluent as an Environmental Pollutant* (*Palm Oil)* ed V Waisundara (London: Intechopen) chapter 2 pp 13–28
[2] Iskandar M J, Baharum A, Anuar F H and Othaman R 2017 *Environ. Technol. Innov.* **9** 169–185
[3] Covinich L G, Bengoechea D I, Fenoglio R J and Area M C 2014 *Am. J. Environ. Eng.* **4** 56–70
[4] Asghar A, Aziz A, Raman A, Mohd W and Wan A *J.* 2014 *Clean. Prod.* **87** 826-838
[5] Nidheesh P V, Gandhimathi R and Ramesh S T 2013 *Environ. Sci. Pollut. Res.* **20** 2099–2132
[6] Bokare A D and Choi W 2014 *J. Hazard. Mater.* **275** 121–135
[7] Ma Z, Wei X, Xing S and Li J 2015 *Catal. Commun. J.* **67** 68–71
[8] Yu T T, Li K L, Guo X L, Li F, Huang J M and Zhang Y X 2015 *J. Phys. Chem. Solids* **87** 196–202
[9] Siregar S S and Awaluddin A 2018 *IOP Conf. Ser., Mater. Sci. Eng.* **345** 012005
[10] Cheng G, Yu L, Yang R, Sun M, Lan B, Yang L and Deng F 2014 *J. Solid State Chem.* **217** 57–63
[11] Cui H J, Huang H Z, Fu M L, Yuan B L and Pearl W 2011 *Catal. Commun.* **12** 1339–1343
[12] Awaluddin A, Agustina M, Aulia R R and Muhdarina 2017 *AIP Conf. Proc.* **1823** 020108
[13] Awaluddin A, Anggraini R, Siregar S S, Muhdarina and Prasetya 2019 *MATEC Web Conf.* **276** 06005
[14] Awaluddin A, Astuti L, Linggawati A, Siregar S S and Saputra L 2018 *AIP Conf. Proc.* **2026** 020075
[15] Awaluddin A, Zulfa R, Absus S, Nurhayati, Linggawati A and Siregar S S 2019 *IOP Conf. Ser., Mater. Sci. Eng.* **509** 012011
[16] Absus S, Zulfa R, Awaluddin A, Anita S and Siregar S S 2018 *AIP Conf. Proc.* **2049** 020009
[17] Ye Q, Lu H, Zhao J, Cheng S, Kang T, Wang D and Dai H 2014 *Appl. Surf. Sci.* **317** 892–901
[18] Cheney M A, Bhowmik P K, Moriuchi S, Villalobos M, Qian A and Joo S W 2008 *J. Nanomater.* **2008** 1-9
[19] Ching S, Landrigan J A, Jorgensen M L, Duan N, Suib S L and O'Young C L 1995 *Chem. Mater* **7** 1604–1606
[20] Händel M, Händel M, Rennert T and Totsche K U 2013 *Geoderma* **193–194** 117–121
[21] Liu Y, Chen Z, Shek C, Wu C M L and Lai J K L 2014 *Appl. Mater. Interfaces* **6** 9776–9784
[22] Yu C, Li G, Wei L, Fan Q, Shu Q and Yu J C 2014 *Catal. Today* **224** 154–162
[23] Zhang W, Yang Z, Wang X, Zhang Y, Wen X and Yang S 2006 *Catal. Commun.* **7** 408–412

Acknowledgment

This study was funded by PDUPT Grant Research from The Ministry of Research, Technology and Higher Education of the Republic of Indonesia on behalf of Prof. Dr. Amir Awaluddin, M.Sc.

The Potential of Chitosan-TiO_2 Nanocomposite for Methyl Orange and Rhodamine B Removal

M M Abdulrasool, A Kamari, S Kumaran, S Ishak, S N M Yusoff, J Jumadi and S T S Wong

Department of Chemistry, Faculty of Science and Mathematics, Universiti Pendidikan Sultan Idris, 35900, Tanjung Malim, Perak, Malaysia.

E-mail: mustafabioch.1991@gmail.com

Abstract. The overall objective of present work was to prepare and use chitosan-titanium dioxide (CS-TiO_2) nanocomposite as an adsorbent to remove two toxic dyes, namely Methyl Orange (MO) and Rhodamine B (RhB) from the aqueous solution. This work consists of two main parts, namely adsorption and characterization studies. For adsorption studies, several experimental parameters were studied such as the effect of initial concentration, solution pH and adsorbent dosage. Meanwile, Scanning Electron Microscope (SEM), Thermogravimetric Analyser (TGA), and X-ray Diffractometer (XRD) were used for characterization studies. The outcomes have shown that the equilibrium data correlated well to Freundlich isotherm model with adsorption capacity at equilibrium (K_F) was determined as 1.608 mg/g for MO and 7.646 mg/g for RhB, respectively. Based on SEM analysis, the rod- and needle-like deposits on the surface of CS-TiO_2 nanocomposite were disappeared following interaction with MO and RhB. There were cube-like deposits observed on the surface of CS-TiO_2 nanocomposite after adsorption of MO, while a dense and uneven surface texture were observed after adsorption of RhB. From TGA analysis, CS had two degradation stages, while CS-TiO_2 nanocomposite had three steps of decomposition. Furthermore, XRD results observed diffraction peaks for CS-TiO_2 at $2\theta = 34.075°, 29.235°, 25.47°$ and $20.26°$.

1. Introduction

Excessive release of dyes and heavy metals into the environment due to rapid industrialization and urbanization has caused significant water contamination, worldwide [1]. Textile, leather, plastic, printing, paper, cosmetics, pharmaceuticals and food industries used methyl orange (MO) and rhodamine B (RhB) dyes in large quantities to colour their products [2]. Approximately 700,000 tonnes of the world dyes was produced per year and from the numbers about 56% was widely utilized by the textile industry [3,4]. The presence of toxic dyes in water has been a major concern to many life forms due to their adverse effects. There are various techniques and processes that had been developed to eradicate the toxic contaminants from aqueous solutions. These processes include adsorption, membrane-based filtration, ion exchange, photodegradation, solvent extraction, chemical precipitation, reverse osmosis and evaporation [5]. Among all, adsorption has been claimed to be the most commonly used method to treat the contaminated water [6]. This is mainly due to various factors such as low cost, can be attained easily, simple designs, easy to be operated and able to remove the contaminants in many forms of concentrations [7]. In addition, the effectiveness of adsorption relies on the capacity of adsorbent to adsorb the contaminants.

Chitosan (CS), poly β-(1-4)-2-amino-2-deoxy-D-glucopyranose is a polysaccharide that produced from deacetylation of chitin, which can be isolated from exoskeletons of marine organisms such as crabs, clams and shrimps or cell walls of fungi [8,9]. The polysaccharide has great characteristics which are non-toxic, biodegradable, biocompatible and have good antimicrobial properties [10]. Chitosan has two important functional groups namely hydroxyl group (-OH) and animo group ($-NH_2$) in its chemical structure. The presence of these functional groups makes chitosan soluble in acidic medium and therefore limits its application for water treatment. To improve its physical property, several chemical and physical modifications including cross-linking, grafting and blending with other materials can be performed [10,11].

In this study, CS-TiO_2 nanocomposites was synthesized, characterized and applied to adsorb Methyl Orange (MO) and Rhodamine B (RhB) in aqueous solution. The performance of CS-TiO_2 nanocomposite to remove MO and RhB was evaluated using several experimental parameters. Meanwhile, the physical and chemical characterisatics of CS-TiO_2 nanocomposite were studied using several analytical instruments.

2. Materials and Methods

2.1. Materials

Chitosan (CS) with a molecular weight of 600,000-800,000 Daltons was purchased from Acros Organics. Titanium(IV) oxide (TiO_2) nanopowder that has a particle size of 21 nm was supplied by Sigma Aldrich. Methyl orange (MO) and Rhodamine B (RhB) were obtained from Bendosen and Merck, respectively. Figure 1 displays the chemical structures of CS, MO and RhB, respectively.

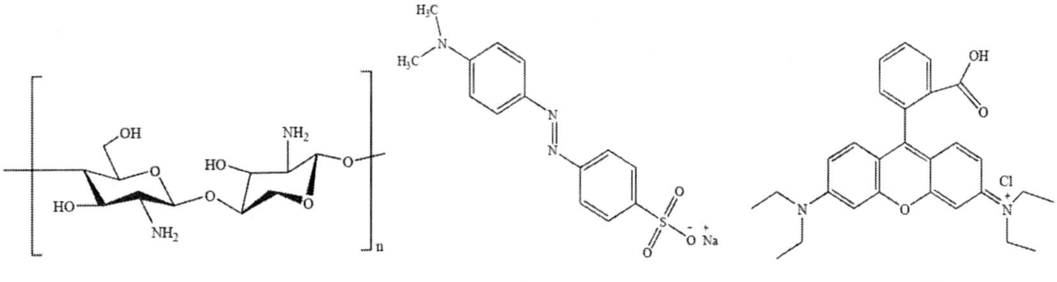

(a) (b) (c)

Figure 1. The chemical structure of (a) CS, (b) MO and (c) RhB.

2.2. Preparation of CS-TiO_2

To prepare the CS-TiO_2 nanocomposite, 0.15 g of TiO_2 nanopowder was first dissolved in 100 mL of 1% (v/v) acetic acid solution. Next, about 2 g of CS was added to the aforementioned solution. The mixture was left to homogenise using a magnetic stirrer for 2 hours. Then, Cs-TiO_2 nanocomposite beads were prepared through neutralization process by adding the resultant mixture dropwise in a bath that contained 500 mL of 0.5 mol/L NaOH. The NaOH solution was stirred with a magentic stirrer to prevents the CS-TiO_2 nanocomposite beads from sticking to the glassware surfaces. The wet CS-TiO_2 nanocomposite beads were then filtered and rinsed thoroughly with deionized water to remove the remaining NaOH residue. Finally, CS-TiO_2 nanocomposite beads were left air-dried. Prior to the last step, a mortar and pestle were used to grind the dried beads. The fine powder was kept in an air-tight container until ready for analysis [12].

2.3. Adsorption study

The concentration of each stock solution of MO and RhB was 100 mg/L. The preparation of stock solution was done separately by dissolving 0.25 g of each dye in 250 mL of deionized water. Adsorption studies of MO and RhB involved three different parameters such as initial concentration

(2.5-20 mg/L), pH of solution (2.0-8.0) and dosage of adsorbent (0.025, 0.050, 0.075, 0.1 and 0.2 g). The experiments were conducted in a 250 mL conical flasks. In order to equilibrate the solution, a Protech Orbital Shaker "model 720" was used to shake the samples at 100 rpm for 1 hour. Following adsorption experiment, CS-TiO_2 nanocomposite were carefully filtered using a filter paper (FiltreFioroni 601, 110 mm, slow speed). The concentration of MO and RhB in aqueous solution was measured by using an Agilent Cary 60 UV-Visible Spectrophotometer at 462 and 554 nm, respectively. All experiments were carried out in triplicates. Equation (1) shows the calculation of the adsorption capacity at equilibrium (q_e) [13]:

$$q_e = \frac{C_i - C_e}{W} V \tag{1}$$

where q_e is the maximum of adsorption capacity at equilibrium (mg/g), C_i represents the initial concentration MO and RhB (mg/L), C_e signifies the final concentration of MO and RhB at equilibrium (mg/L), W refers to the mass of the CS-TiO_2 (g) and V represents the volume of MO and RhB (L). By using equation (2), the percentage of removal of each MO and RhB from aqueous solution was calculated [13]:

$$\text{Removal (\%)} = \frac{C_i - C_e}{C_i} \times 100 \tag{2}$$

2.4. *Adsorption isotherms*

The adsorption isotherm data at equilibrium is designed based on two commonly used models which are Langmuir and Freundlich isotherm models. These models are used to explain the connection between the adsorbent (stationary phase) and adsorbate (mobile phase). Freundlich isotherm described the adsorption process between the adsorbents and contaminants is likely occur at heterogeneous or multilayer surface [9]. The following equation (3) presents the linearized equation of Freundlich isotherm [14]:

$$\log q_e = \log K_F + \frac{1}{n} \log C_e$$

(3)

q_e refers to the amount of dyes adsorbed per unit weight of adsorbent (mg/g) (CS-TiO_2 nanocomposite), K_F (mg/g) represents the adsorption capacity, and n (g/L) refers to the strength of adsorption process, and C_e refers to equilibrium concentration (mg/L) (dyes).

Meanwhile, Langmuir isotherm model is described as an adsorption process for monolayer at homogenous surface [15] without presenting the interaction between adsorbent and adsorbate particles. Equation (4) expresses the linear model of Langmuir [15]:

$$\frac{C_e}{q_e} = \frac{C_e}{Q} + \frac{1}{Qb}$$

(4)

C_e refers to the concentration at equilibrium (mg/L), q_e signifies the amount of adsorbed (dyes) per unit weight of adsorbent (mg/g), q_e refers to the amount of adsorbed (dyes) per unit weight of adsorbent (mg/g), and Q and b are the Langmuir constants.

2.5. *Characterization study*

In present study, several characteristics related to the physical and chemical properties of CS and CS-TiO_2 nanocomposites were studied. The changes in physical and chemical properties of CS-TiO_2 nanocomposites after adsorption with dyes were investigated using a number of analytical techniques. A Hitachi SU 8020 UHR FESEM was used to conduct scanning electron microscope (SEM) analysis to observe the morphology of CS-TiO_2 nanocomposites before and after adsorption with dyes. Prior to analysis, the surface of CS- TiO_2 nanocomposites was coated with a thin layer of platinum to avoid the charging of electron. The thermal properties of CS and CS-TiO_2 nanocomposites were examined using a TGA/DSC 1 Mettler Toledo Analyser. For the thermal analysis, the samples was heated at temperature range between 30 to 900 °C under argon gas with a flow rate of 20 mL/min. Lastly, the crystallinity properties of the nanocomposite was performed using the X-ray diffraction (XRD) analysis.

3. Results and discussion

3.1. Effect of initial concentration

One of the most important factors which effects dyes adsorption process is the initial concentration of dyes. Figure 2(a) displays a function of initial concentration for percentages of removal of each MO and RhB. Based on Figure 2(a), it is clearly observed that when the initial concentration of MO increased, the percentage removal of MO had decreased from 90% to 75%. This is because the adsorbent has limited number of active sites and could be saturated by few dye concentrations. In this matter, the CS-TiO_2 nanocomposite exhibits lesser number than the dye molecules. Meanwhile, as for RhB, when the initial concentration increased, the percentage of CS-TiO_2 nanocomposite adsorbent will also increase from 6% to 28%. This can be explained by the fact that at high initial dye concentration, it caused high driving force of dye mass transfer from aqueous and solid phase caused all the available active sites of nanocomposits to be occupied with dye molecule [16-18].

3.2. Effect of solution pH

In the adsorption process, the pH of dye solution is crucial as it may have effects onto the degree and surface charges of adsorbent [19]. Figure 2(b) represents the effect of solution pH (2.0-8.0) for each MO and RhB adsorbed onto CS-TiO_2 nanocomposite. The adsorption capacity (q) of CS-TiO_2 nanocomposite for RhB was found to be decreased as as pH of solution increases. This may be due to the fact that RhB exists as small cations at the pH below 4.0, while at a higher pH it exists as large zwitterions [19]. Therefore, adsorption capacity of CS-TiO_2 nanocomposites for RhB at pH higher than 4.0 decreased as the large zwitterionic RhB was unable to enter the pore of CS-TiO_2 nanocomposites. A similar trend was observed by Fang *et al.* [20] for the adsorption of RhB onto montmorillonite. On the other hand, the adsorption capacity of CS-TiO_2 nanocomposite for MO increased gradually as pH of the solution increased from 2.0 to 5.0. Meanwhile, as the pH increased from 6.0 to 8.0, the adsorption capacity was decreased. This scenario might be due to more H^+ and H_3O^+ ions were available in system at the pH of 2.0 to 5.0, thus causing a protonation effect to CS-TiO_2 nanocomposite. Therefore, it increased electrostatic attraction between anionic dyes and the positive charge site on the surface of CS-TiO_2 nanocomposite. As the solution pH increased from 6.0 to 8.0, the adsorption system has more negative charged sites which led to electrostatic repulsion between negative site of the adsorbent and anionic MO. Gao *et al.* [21] have reported a similar observation for removal MO by NiFe-layered double hydroxides (LDHs). From figure 2(b), it was noted that the maximum adsorption capacity (q) of CS-TiO2 nanocomposite for MO was 1.203 mg/g.

3.3. Effect of absorbent dosage

Figure 2(c) shows the effect dosages of adsorbent on removal of MO and RhB in aqueous solution. For this study, the pH of MO and RhB solution was adjusted to the optimum pH. As displayed in Figure 2(c), a decline in adsorption capacity was observed for both MO and RhB as the mass of CS-TiO_2 nanocomposite was increased. The adsorption capacity for MO and RhB was decreased from 1.638 to 0.117 mg/g and 0.536 to 0.202 mg/g, respectively. This might be due to the increase in amount of CS-TiO_2 has caused the ratio of actives sites of CS-TiO_2 to MO and RhB molecules increased. Thus, low adsorption capacity onto nanocomposites. Rattanapan *et al.* [22] have studied MO removal by using activated carbon derived from coffee grounds and reported a similar trend of MO adsorption. In addition, Mukhtar *et al.* [23] have mentioned the same trend for the removal of RhB using vanadium pentoxide/titanium butyl oxide hybrid xerogels.

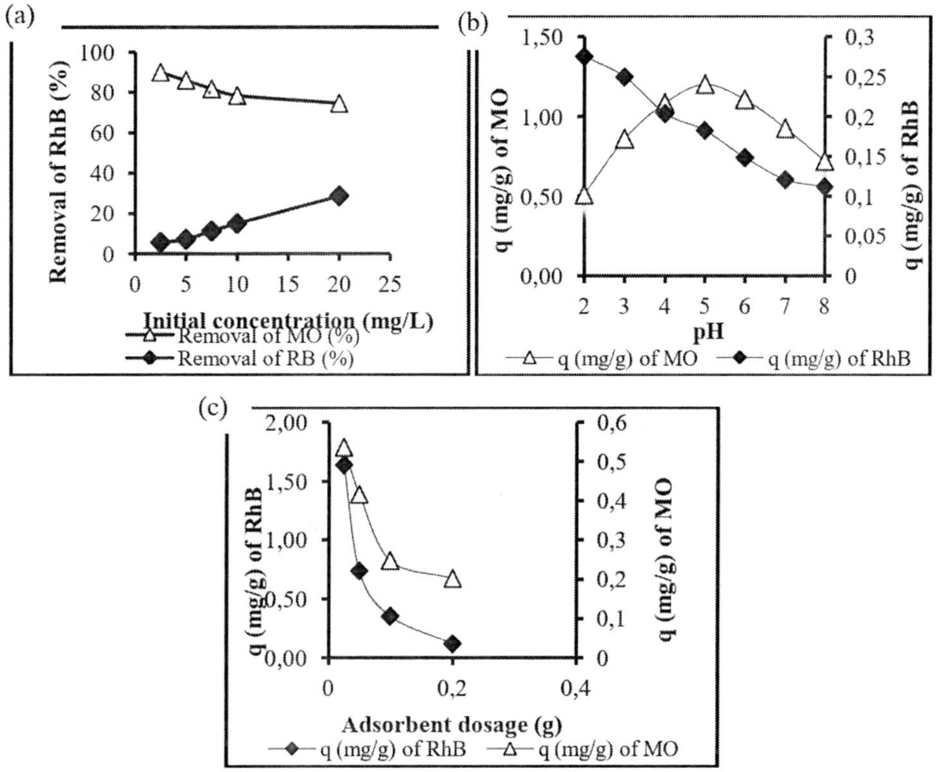

Figure 2. The effect of initial concentration (a), effect of solution pH (b) and effect of adsorbent dosage (c) on adsorption capacity if MO and RhB onto CS-TiO_2 nanocomposite.

3.4. Adsorption isotherms

In this study, the Langmuir and Freundlich isotherm models were used to fit the adsorption equilibrium data. This present work was performed under following conditions, namely the initial concentrations for each dyes range between 2.5 to 20 mg/L, the mass of the nanocomposites was 0.1 g and the solution pH was set at optimum pH (2.0 for RhB and 5.0 for MO). Table 1 displays the isotherm constant and correlation coefficient values determined from Freundlich and Langmuir isotherm models. From the table, the adsorption data for dyes correlated well with to the Freundlich isotherm model. Based on Freunlich isotherm model, the maximum adsorption capacity values was determined as 1.608 and 7.646 mg/g for MO and RhB, respectively. The n value from Freundlich isotherm model indicate adsorption affinity of the system. For instance, $n > 1$ represents the favourable adsorption, whereas $n = 1$ is attributed to linear adsorption and $n < 1$ corresponds to the unfavorable adsorption [24]. Thus, MO adsorption is favourable and RhB is unfavourable for adsorption.

Table 1. The Freundlich (K_F) and Langmuir (Q) constants and correlation coefficient (R^2).

Dyes	Freundlich			Langmuir		
	K_F (mg/g)	n	R^2	Q (mg/g)	b (L/mg)	R^2
Methyl orange (MO)	1.608	1.642	0.996	13.386	0.308	0.875
Rhodamine B (RhB)	7.646	0.481	0.978	0.410	0.067	0.886

The adsorption capacity (q_{max}) obtained from this study were compared with other adsorption studies using several adsorbents (table 2).

Table 2. Comparison of q_{max} from other adsorption studies.

Adsorbent	Freundlich isotherm model				References
	MO		RhB		
	K_F (mg/g)	n	K_F (mg/g)	n	
Spent Tea Leaves	0.088	1.370			[25]
Egussi peeling	3.979×10^{-2}	0.955			[26]
Carbon Nanospheres (CNSs)			1.946	0.36	[27]
Water chestnut peel			0.4– 1.7		[28]
Razor Clam (*Ensis directus*)	0.010	0.660			[29]

3.5. SEM analysis

The surface morphology of CS-TiO_2 nanocomposites before and after interaction with MO and RhB dyes was observed using SEM analysis at 5.0 k magnification. Figure 3(a) displays SEM image of pure CS that have nonporous surface and smooth texture. Figure 3(b) exhibits CS-TiO_2 nanocomposite image, where rod- and needle-like aggregations were observed on the surface of TiO_2 nanocomposite. The deposits for MO and RhB can be observed on the outer layer of CS-TiO_2 nanocomposite, as shown in figure 3(c and d). As presented in figure 3(c), CS-TiO_2 showed the appearance of cubic-shape deposit after adsorption with MO. Meanwhile, a dense and uneven surface texture were observed on CS-TiO_2 following RhB adsorption. The changes in surface morphology of CS-TiO_2 nanocomposites after interaction with dyes might due to the reorganization of surface's functional group that bind with dyes.

3.6. TGA analysis

The thermal stability and decomposition temperature of pure chitosan (CS) and CS-TiO_2 nanocomposite were assessed by TGA analysis. Figure 4 shows TGA thermograms of CS and CS-TiO_2 nanocomposite. Based on thermogravimetric curve, CS exhibited two decomposition steps, where the first step was observed at temperature of 30 to 152 °C. At this step, a weight loss around 9.67% was observed due to removal of moisture content in the nanocomposite [30]. Second decomposition step began at 190 to 330 °C with weight loss of 65.44% due to dehydration and decomposition of chitosan polymer [30]. Meanwhile, CS-TiO_2 nanocomposite exhibited three decomposition steps, which started with the first stage at 87 °C to 133 °C and weight loss of 11.74% was observed due to evaporation of water [31]. The second stage observed at 221 °C and went on to 288 °C with a weight loss of 30.20%, which attributed to thermal decomposition of CS main chain [31]. Third stage occurred at 630 °C till 993 °C with 28.70 % weight loss associated with further decomposition of polysaccharide ring and glycolsidic bonds [31].

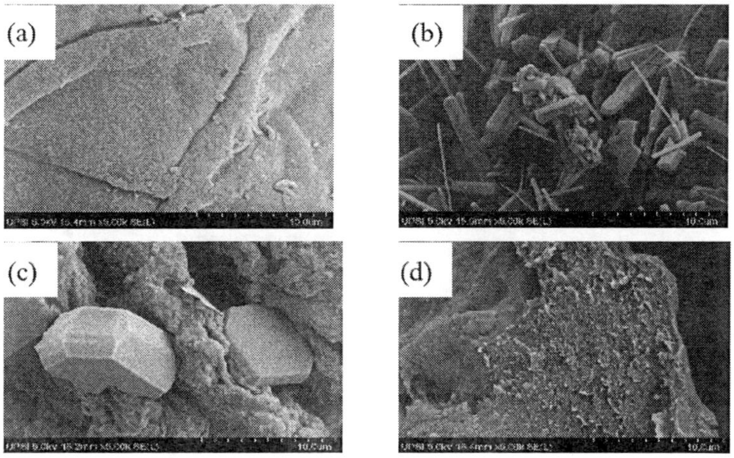

Figure 3. (a) SEM images of CS, (b) CS-TiO_2 (c) CS-TiO_2-MO (d) CS-TiO_2-RhB at 5.00k magnifications

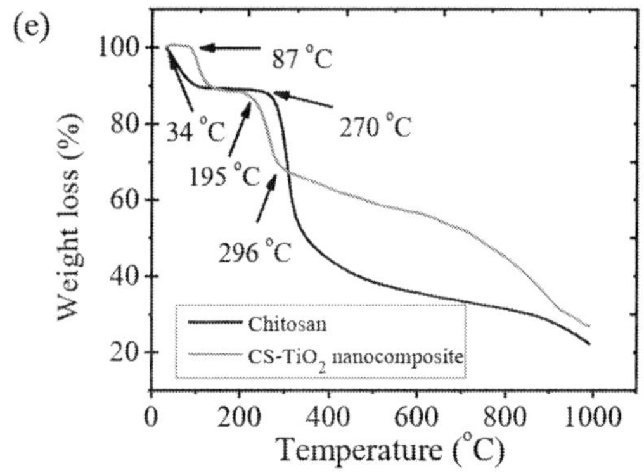

Figure 4. TGA thermograms of CS and CS-TiO_2.

3.7. XRD analysis

Figure 5 illustrates XRD diffractograms for CS, TiO_2 nanopowder, CS-TiO_2 nanocomposite and CS-TiO_2 nanocomposites after adsorption with MO and RhB, respectively. XRD analysis confirmed that CS presented in amorphous state and there was no diffraction peaks appeared except the band that exhibited at 2θ = 23.33°, as displayed in figure 5(a) [32]. TiO_2 nanopowder diffractogram in figure 5(b) exhibits a sharp diffraction peak, clearly observed at 2θ = 25.37°[33]. Meanwhile, figure 5(c) presents XRD diffractogram of CS-TiO_2 nanocomposite, where diffraction peaks that can attributed for each CS and TiO_2 was observed at 2θ = 34.075°, 29.235° 25.47° and 20.26°. It was noted that the intensity of diffraction peak of CS was decreased after composite with TiO_2, and the introduction of TiO_2 has improved the properties of chitosan. Following adsorption of MO and RhB, there was a marginal effect to crystallinity of TiO_2 nanocomposite. The intensity of diffraction peaks were slightly decreased after the interaction of nanocomposite with dye molecules (figure 5 (d and e). This suggests, the penetration of dye molecules into the internal structure of nanocomposite did not significantly change the crystallinity property of the nanocomposite.

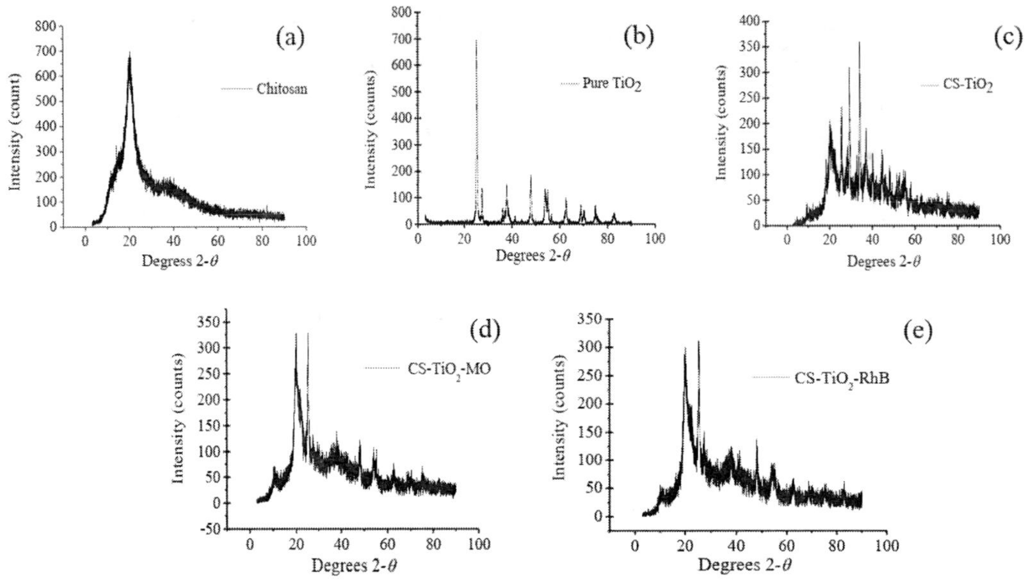

Figure 5. XRD diffractograms of (a) CS, (b) TiO_2 nanopowder (c) CS-TiO_2 nanocomposite (d) CS-TiO_2-MO and (e) CS-TiO_2-RhB.

4. Conclusion

The potential of CS-TiO_2 nanocomposite for adsorption of MO and RhB from aqueous solution has been investigated. The Freundlich isotherm was the best model to explain the adsorption equilibrium data. The adsorption capacity of the nanocomposites was greatly influenced by the initial concentration, solution pH and adsorbent dosage. The adsorption of MO and RhB onto CS-TiO_2 nanocomposite has caused a significant change in surface morphology of the nanocomposites. In conclusion, the CS-TiO_2 feasible to be used as adsorbent for water treatment.

5. References

[1] Anitha A, Kohilavani K and Murugalakshmi R 2018 *Int. J. Appl. Eng. Res.* **13** 14669-14674
[2] Huang J H, Huang K L, Liu S Q, Wang A T and Yan C 2008 *Colloids Surf. A Physicochem. Eng. Asp.* **330** 55-61
[3] Neethu N and Choudhury T 2018 *Recent Pat. Nanotechnol.* **12** 200-207
[4] Razzaz A, Ghorban S, Hosayni L, Irani M and Aliabadi M 2016 *J. Taiwan Inst. Chem. Eng.* **58** 333-343
[5] Krishnan S, Rawindran H, Sinnathambi C M and Lim J W 2017 *IOP Conf. Ser. Mater. Sci. Eng.* **206** 012089
[6] Nicomel N, Leus K, Folens K, Van Der Voort P and Du Laing G 2016 *Int. J. Environ. Res. Public Health* **13** 62
[7] Ong C B, Ng L Y and Mohammad A W 2018 *Renew. Sustain. Energy Rev.* **81** 536-551
[8] Al-Taweel S S, Saud H R, Kadhum A A H and Takriff M S 2019 *Results Phys.* **13** 102296
[9] Elieh-Ali-Komi D and Hamblin M R 2016 *Int. J. Adv. Res.* **4** 411-427
[10] Sutirman Z A, Sanagi M M, Karim K J A, Naim A A and Ibrahim W A W 2018 *Malays. J. Anal. Sci.* **22** 839-850
[11] Tchuifon D R, Anagho S G, Njanja E, Ghogomu J N, Ndifor-Angwafor N G and Kamgaing T 2014 *Int. J. Chem. Sci.* **12** 741-761
[12] Kamari A 2011 *Chitosan as Soil Amendment for the Remediation of Metal Contaminated Soil* Doctoral Dissertation (Glasgow: University of Glasgow)

[13] Kusrini E, Wicaksono B, Yulizar Y, Prasetyanto E A and Gunawan C 2018 *IOP Conf. Ser. Mater. Sci. Eng.* **316** 012029
[14] Freundlich H M F 1906 *Z. Phys. Chem.* **57** 385-470
[15] Langmuir I 1918 *J. Am. Chem. Soc.* **40** 1361-1403
[16] Yagub M T, Sen T K, Afroze S and Ang H M 2014 *Adv. Colloid Interface Sci.* **209** 172-184
[17] Jethave G, Fegade U, Rathod R and Pawar J 2018 *J. Dispers. Sci. Technol.* **40** 563-573
[18] Oluwasogo Dada A, Inyinbor A A, Adekola F A, Odebunmi E O, Bello O S and Ayo-Akere S 2018 *Int. J. Civ. Eng. Technol.* **9** 1591–1605
[19] Banerjee S and Chattopadhyaya M C 2017 *Arab. J. Chem.* **10** S1629-S1638
[20] Fang Y, Zhou A, Yang W, Araya T, Huang Y, Zhao P, Wang J, Johnson D and Ren Z J 2018 *Sci. Rep.* **8** 229
[21] Gao J, Lu Y, Fang L, Wu F, Ling F, Guan T, Shi L, Hu B and Meng F 2018 *Environ. Eng. Sci.* **35** 373-381
[22] Rattanapan S, Srikram J and Kongsune P 2017 *Energy* Procedia **138** 949-954
[23] Mukhtar S, Liu M, Han J and Gao W 2017 *Chin. Phys. B* **26** 058202
[24] Deka J R, Lin Y H and Kao H M 2014 *RSC Adv.* **4** 49061-49069
[25] Pei L Y, Suhaidi A N, Zulkifli S M, Hassim S H, Kanakaraju D and Chin L Y 2017 *Pertanika J. Sci. Technol.* **25** 73-84
[26] Tchuifon D R, Anagho S G, Njanja E, Ghogomu J N, Ndifor-Angwafor N G and Kamgaing T 2014 *Int. J. Chem. Sci.* **12** 741-761
[27] Qu J, Zhang Q, Xia Y, Cong Q and Luo C 2015 *Environ. Sci. Pollut. Res.* **22** 1408-1419
[28] Khan T A, Nazir M and Khan E A 2013 *Toxicol. Environ. Chem.* **95** 919-931
[29] Areibat L E M and Kamari A 2017 *AIP Conf. Proc.* **1847** 040004
[30] Gedam A H and Dongre R S 2015 *RSC Adv.* **5** 54188-54201
[31] John S, Salam A, Baby A M and Joseph A 2019 *Prog. Org. Coat.* **129** 254-259
[32] Dongre R S 2016 *Bull. Mater. Sci.* **39** 865-874
[33] Jayakumar R, Ramachandran R, Divyarani V V, Chennazhi K P, Tamura H and Nair S V 2011 *Int. J. Biol. Macromol.* **48** 336-344

Acknowledgement
The authors are thankful to the Ministry of Education Malaysia for providing a research grant (2019-0003-102-02) and to the Department of Chemistry, Faculty of Science and Mathematics, Universiti Pendidikan Sultan Idris (UPSI) for their valuable technical assistance.

Synthesis, Characterisation and Mechanism of Novel Ca-ZSM-5 Zeolite Nanocomposite from Eggshell using Simple Co-Precipitation Method

S Kumaran, A Kamari, M M Abdulrasool, S T S Wong, J Jumadi, S N M Yusoff and S Ishak

Department of Chemistry, Faculty of Science and Mathematics, Universiti Pendidikan Sultan Idris, 35900, Tanjung Malim, Perak, Malaysia.

E-mail: k.sathiakumaran@gmail.com

Abstract. In this study, Ca-ZSM-5 zeolite nanocomposite was synthesised as controlled-released fertiliser due to the excess of nutrients leaching in the soil which has caused serious environmental problem such as eutrophication. The zeolite nanocomposite was first synthesised using a simple co-precipitation template method. The zeolite nanocomposite was then impregnated with calcium, originated from chicken eggshell powder, at a molar ratio of 1:1. The physical and chemical properties of Ca-ZSM-5 zeolite nanocomposite were characterised using several analytical instruments such as Fourier Transform Infrared (FTIR) Spectrometer, Scanning Electron Microscope (SEM), Energy Dispersive X-ray (EDX) Spectrometer and Transmission Electron Microscopy (TEM). Based on SEM analysis, the porosity of mesoporous H-ZSM-5 nanozeolite was in the range of 31.7 to 190 nm. Meanwhile, the porosity of macroporous eggshell was 71.4 to 230 nm and mesoporous H-ZSM-5 nanozeolite material was impregnated into the macroporous eggshell due to porosity of eggshell was larger than porosity of H-ZSM-5 nanozeolite. Ca-ZSM-5 zeolite nanocomposite with porosity of 71.4-198 nm was successfully synthesised by simple co-precipitation method. While the results obtained from FTIR analysis confirmed the presence of tetrahedral (TOx, x = Si or Al) stretching. Overall, the synthesised Ca-ZSM-5 zeolite nanocomposite possess key characteristic as a precursor for various applications.

1. Introduction

Organic waste like eggshell can be used as natural biomaterials in various applications. For an example, it is used as catalysts in industrial field while dental, bone implantation and wound dressings in biomedical field. Meanwhile, in agricultural sector the residue of eggshell is suitable for microbial application because it can supply nutrients required by plants for growth [1]. Furthermore, eggshell waste is used in various applications because of high content of calcium carbonate ($CaCO_3$) in the form of calcite which can be utilised as nutrients [1]. Moreover, organic waste from eggshell contains many tiny holes which called as pore canals on the surface layer which can allow air to follow through easily [2]. An interesting template synthetic routes show high level interest of scientists in cutting edge technology to prepare and impregnate synthetic alumina-silicate materials in nanoscale [3]. Zeolite is well defined as crystalline aluminosilicate with different pore sizes which classified as microporous (pore size smaller than 2 nm), mesoporous (pore size between 2 to 50 nm) and macroporous (pore size larger than 50 nm) [3]. Ethylene glycol is commonly used as a 'soft' material to create the template of H-ZSM-5 due its ability to form inter and intramolecular forces of hydrogen bonding (-OH) with HCl [3]. Many researchers have used several materials such as NPK and Ag for impregnation of zeolite [4]. Chicken eggshell was used as an impregnation agent in this study because it contains ceramic materials such as cuticle, calcareous and mammillary which are considered as environmentally friendly [1]. In this study, we present the synthesis of nanozeolite using simple co-precipitation method using ethylene glycol as organic soft template as well as impregnation of calcium derived from white chicken eggshell powder with H-ZSM-5 nanozeolite.

2. Materials and methods

2.1 Materials

Sodium silicate (Fisher Scientific), ethylene glycol (Merck), aluminium sulfate 98 % (Sigma-Aldrich), commercial zeolite (Merck), commercial ZSM-5 zeolite (Merck), sodium hydroxide (Bendosen) and acid hydrochloric 37 % (R&M Chemicals) were used to synthesise H-ZSM-5 nanozeolite.

2.2 Preparation of H-ZSM-5 and Ca-ZSM-5 Zeolite Nanocomposite

H-ZSM-5 nanozeolite was prepared by simple co-precipitation method using ethylene glycol as organic soft template [3]. White precipitates of sodium silicate solution (55 g/100 mL deionised water) and 25 mL of ethylene glycol were mixed together and stirred for 2 h while the temperature maintained at 50-60 °C to obtain a homogenous mixture. Aluminium sulfate (39.35 g/100 mL) and sodium hydroxide (15 g/100 mL) were added dropwise until the final pH of the mixture reached to 12 and adjusted to neutral using 1 M HCl. The mixture was then filtered using a filter paper (Smith 102, 110 mm) and white colour precipitate was obtained. The precipitate underwent drying process in an oven set at 105 °C for 1 h and purification using 25 mL of ethanol solution. Grey colour H-ZSM-5 nanozeolite was produced after calcination at 650 °C for 2 h. H-ZSM-5 nanozeolite was impregnated with eggshell powder at a molar ratio of 1:1 using simple suspension method in 20 mL deionised water. The mixture was stirred for 3 h to attain maximum impregnation of calcium. The resulting suspension underwent vacuum filtration using a filter paper (Double Rings 203, 90 mm) followed by drying in an oven set at 50 °C for 24 h. The light grey coloured Ca-ZSM-5 zeolite nanocomposite was obtained after the drying process was completed.

2.3 Characterisation study

In this study, materials were characterised using several analytical instruments. The FTIR spectra were recorded using a Thermo Nicolet 6700 FTIR spectrometer at wavenumbers range between 4000 and 400 cm^{-1} with over 32 cumulative scans. A Hitachi SU8020 UHR was used to obtain SEM images, TEM images and EDX spectra of the materials. The SEM was operated at an acceleration of 5.0 kV.

3. Results and discussion

3.1. FTIR analysis

The FTIR spectra of commercial zeolite, H-ZSM-5 nanozeolite and commercial ZSM-5 zeolite which was purchased from Merck are presented in figure 1. From figure 1, absorption bands observed at 788-618 cm^{-1} can be assigned to internal and external of asymmetrical tetrahedral lingkage [3]. On the other hand, strong absorption bands at wavenumber of 1040 cm^{-1} (figure 1(b)) and 1062 cm^{-1} (figure 1(c)) can be attributed to tetrahedral $_{v}(TO_x$, x = Si or Al) stretches [4]. The discernible peaks at 788 cm^1 (figure 1(c)) and 785 cm^{-1} (figure 1(c)) are correspond to external symmetry (-OTO-) [4]. The appearance of these peaks is imperative in order to proof that the occurrence of nucleation process in simple co-precipitation method, which was conducted at low temperature. Meanwhile, the weak bands at 1652 cm^{-1} (figure 1(a)) and 1634 cm^{-1} (figures 1(b) and 1(c)) showed deformation mode of water. This observation suggests that Si was not bonded with OH in H-ZSM-5 nanozeolite [3]. Overall, the absorption bands observed in FTIR spectra of commercial zeolite and ZSM-5 zeolite were well matched with FTIR spectrum of synthesised H-ZSM-5 nanozeolite. Figure 2 shows FTIR spectra of H-ZSM-5 nanozeolite, eggshell and Ca-ZSM-5 zeolite nanocomposite. A shown in figure 2(b), the prominent peaks observed at 1393, 871 and 711 cm^{-1} was due to the presence of carbonate minerals in the eggshell [5]. Following impregnation with eggshell, there was a significant change in FTIR spectrum of H-ZSM-5 nanozeolite (figure 2(c)), of which the features of carbonate minerals were appeared at absorption bands at 1404, 872 and 711 cm^{-1} [6]. This strongly supports that calcium was successfully impregnated to H-ZSM-5 nanozeolite. Additionally significant peaks at 3200-3600, 1652 and 1393 cm^{-1} which expected the presences of positively charged functional group of amine and amide [5]. Overall the synthesised Ca-ZSM-5 zeolite nanocomposite possesses key characteristics as a precursor for various applications.

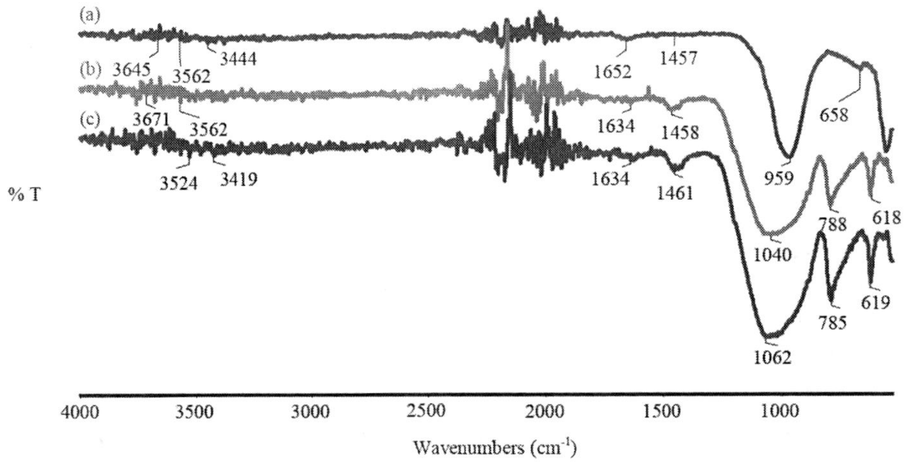

Figure 1. FTIR spectra of (a) commercial zeolite, (b) H-ZSM-5 nanozeolite and (c) commercial ZSM-5 zeolite.

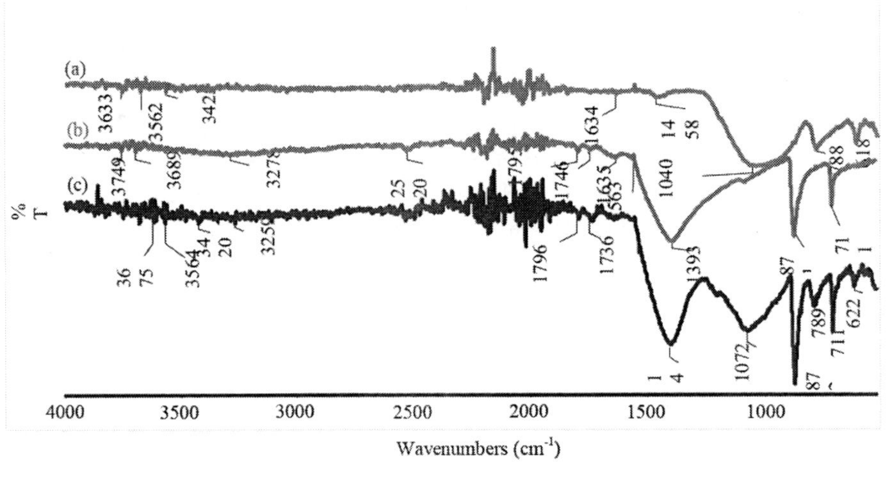

Figure 2. FTIR spectra of (a) H-ZSM-5 nanozeolite (b) eggshell and (c) Ca-ZSM-5 zeolite nanocomposite.

3.2. SEM analysis

The surface morphology of commercial ZSM-zeolite, H-ZSM-5 nanozeolite, eggshell and Ca-ZSM-5 zeolite nanocomposite was observed using SEM analysis at 25,000x magnification as presented in figure 3. It was found that commercial ZSM-5 shows a well-defined structure (figure 3(a)). After conducting simple co-precipitation procedure of which ethylene glycol (25 mL, 0.72 M) was added as a soft template material to the synthesised H-ZSM-5 nanozeolite , the size of porosity of this material was developed into mesoporous structure [7][8]. As shown in figure 3(b), the size of porosity of H-ZSM-5 nanozeolite was estimated between 31.7 and 190 nm. In the context of eggshell (figure 3(c)), shell unit between calcite and membrane was well-organised [3][9]. Furthermore, SEM analysis showed that the porosity of eggshell can be classified as macroporous with the size ranging from 71.4 to 230 nm. The surface morphology and porous structure of Ca-ZSM-5 zeolite are shown in figure 3(d) [10]. It is apparent that the mesoporous H-ZSM-5 nanozeolite material was impregnated into the macroporous eggshell due to porosity of eggshell was larger than porosity of H-ZSM-5 nanozeolite [8]. Moreover, smooth calcite membrane was formed when ethylene glycol was used as a soft template material in H-ZSM-5 nanozeolite [11][12].

Figure 3. SEM images of (a) commercial ZSM-5 zeolite (b) H-ZSM-5 nanozeolite (c) eggshell and (d) Ca-ZSM-5 zeolite nanocomposite at 25,000x magnification.

3.3. EDX analysis

Energy Dispersive X-ray (EDX) analysis was performed in order to identify the element present in H-ZSM-5 nanozeolite and Ca-ZSM-5 zeolite nanocomposites. As shown in figure 4(a), the EDX spectrum of H-ZSM-5 nanozeolite showed strong peaks represent Si, Na and Al. The presence of these three elements suggests the similarity in composition of commercial ZSM-5 zeolite [13]. Meanwhile additional peaks of 56.40 % O and 2.59 % Cl correspond to ethylene glycol and HCl, which were used to synthesise H-ZSM-5 nanozeolite. Furthermore, figure 4(b) presents the elemental composition of H-ZSM-5 nanozeolite following impregnation with eggshell. From figure 4(b), it is clear that 86.84 % Ca mineral from eggshell already impregnated into mesoporous structure of H-ZSM-5 nanozeolite [5]. In addition, 29.69 % of Si composition (figure 4(a)) reduced to 11.59 % suggesting that 18.10 % of Si was used to substitute Ca in order to produce Ca-ZSM-5 zeolite nanocomposite. Results from EDX analysis corroborate the results obtain from SEM analysis conforming that Ca has successfully impregnated into H-ZSM-5 nanozeolite (figure 3(d)) [14].

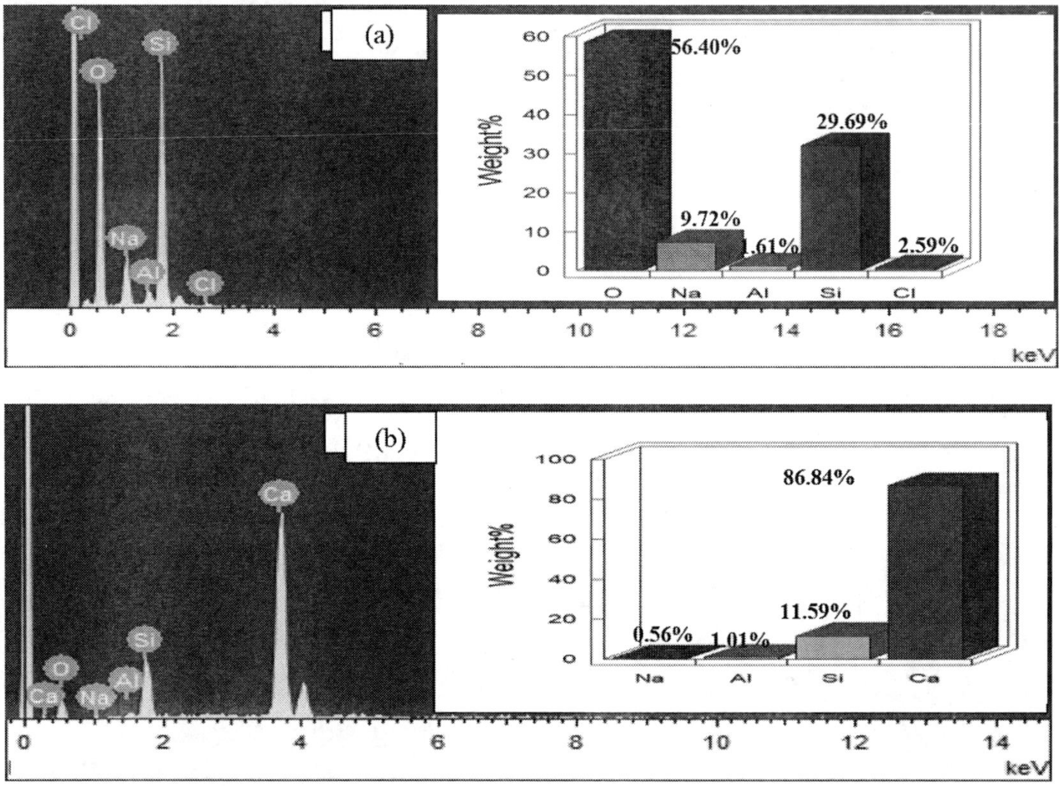

Figure 4. EDX spectra of (a) H-ZSM-5 nanozeolite, (b) Ca-ZSM-5 zeolite nanocomposite at 1 micrometer magnification.

3.4. TEM analysis

Figure 5 shows the TEM images of H-ZSM-5 at 200,000x and 500,000x magnifications. It is obvious that the particles of H-ZSM-5 zeolite nanocomposite were distributed so uniformly. This was expected because the H-ZSM-5 nanozeolite was surrounded by soft template which was prepared during simple co-precipitation step [15]. From figures 5(a) and 5(b), the size of these particles was in the range between 2.38 nm and 26.8 nm. In addition, from figure 5(b) it can be observed that disaggregate crystallites were formed in H-ZSM-5 nanozeolite [16].

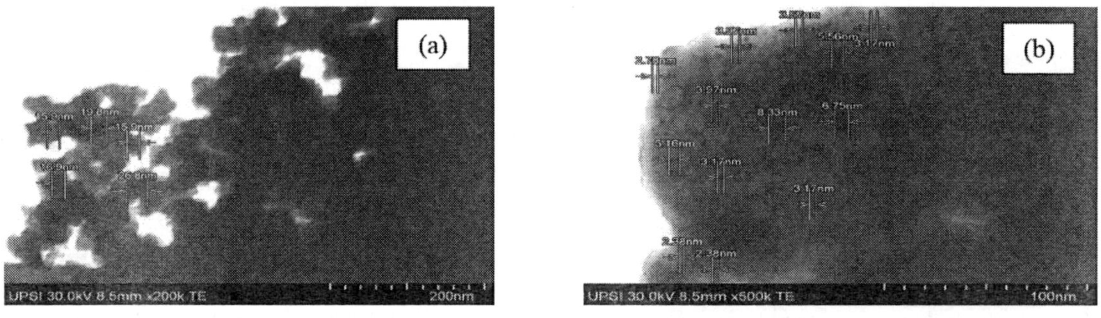

Figure 5. TEM images of H-ZSM-5 nanozeolite at (a) 200,000x, and (b) 500,000x magnifications.

3.5. XRD analysis

Figure 6 shows phase study of synthesised (a) H-ZSM-5 nanozeolite and (b) Ca-ZSM-5 zeolite nanocomposite diffractograms. After undergo calcination at 600 °C, 2 major peaks in H-ZSM-5 nanozeolite can be observed at 22.0°, 25.0° and 28.5° which characteristic primary nanocrystals structure of Mordenite Framework Inverted (MFI) [17]. Furthermore synthesised H-ZSM-5 nanozeolite show a high intensity peak at 25.0°. Characteristic peaks of XRD patterns Ca-ZSM-5 zeolite nanocomposite were identify existence of calcium from eggshell as crystalline phase from $CaCO_3$ which present in the form of calcite [18]. Moreover, major peaks in Ca-ZSM-5 zeolite nanocomposite observed at 29.0° well-fitted with characteristic of $CaCO_3$ in a stable calcite phase [19]. This result confirmed that solid phase between H-ZSM-5 nanozeolite and $CaCO_3$ in calcite phase formed at the surface layer of eggshell with saturated Ca^{2+} [20]. Furthermore, figures 3(b) and (d) and XRD patterns in figure 6 showed that the framework of H-ZSM-5 nanozeolite remains unchanged after undergoing Ca^{2+} ion-exchange process.

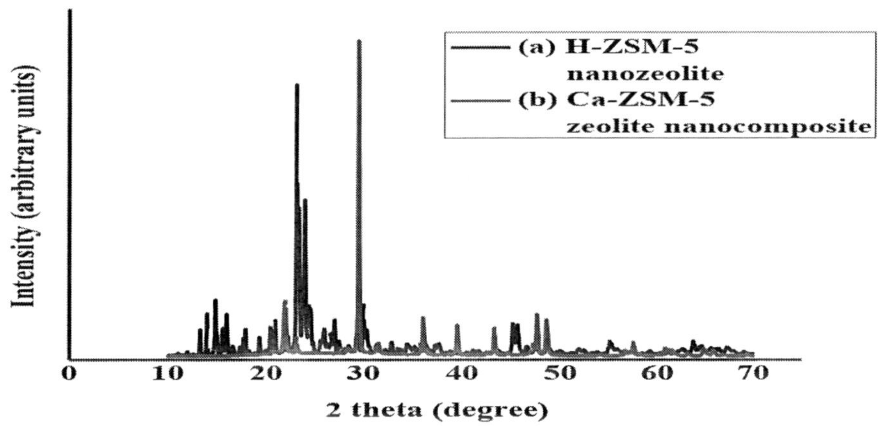

Figure 6. XRD patterns of (a) H-ZSM-5 nanozeolite and (b) Ca-ZSM-5 zeolite nanocomposite.

3.6. Reaction mechanism

Scheme 1 shows that in ZSM-5 framework where two primary unit of SiO_4 and AlO_4^- will gathered in particular structures named Secondary Building Unit (SBU) based on Lowenstein rule to form aluminosilicate polyhedral structure [21]. The polyhedral aluminosilicate framework showed in tetrahedral which substituted by Si^{4+} and Al^{3+} occupied as central atoms with negative charge while in the vertices contains four oxygen atoms which can be balanced by monovalent or divalent cations and can be used as cation exchange site [21].

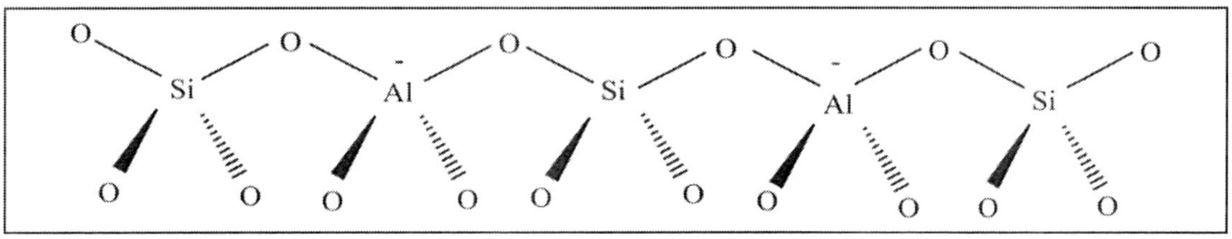

Scheme 1. Schematic of Possible Structure of Polyhedral Aluminosilicate with Cation Exchange Site

Possible mechanism for the formation of H-ZSM-5 nanozeolite is shown in scheme 2. Figures 1(b) and (c) show hydrogen-bonded interaction of ZSM-5 in acidic sites (1040 and 1062 cm^{-1}) [21]. Due to the porous structure in the ZSM-5 framework refer to figure 3(a), and scheme 1, H^+-ZSM-5 nanozeolite was prepared which can act as ion-exchange sites for mono or divalent cations and the chemical equation is given below [22]:

$$\text{ZSM-5 zeolite (s)} + H^+ \text{(aq)} \longleftrightarrow H^+\text{-ZSM-5 nanozeolite (s)} \quad (1)$$

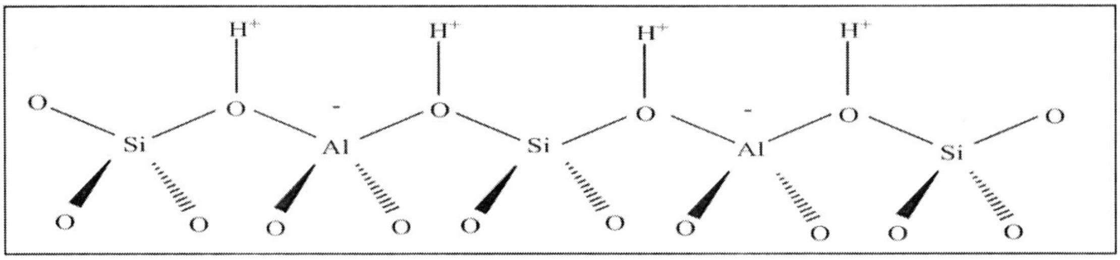

Scheme 2. Schematic of Possible Mechanism for the Formation of H^+-ZSM-5 Nanozeolite (NZ)

Scheme 3 illustrates the possible mechanism for the formation of Ca^{2+}-ZSM-5 Zeolite Nanocomposite (ZNC). Ion-exchange process will occur between the H^+ from HCl and calcium ion from eggshell in calcite form and act as cation which going to exchange [22][23]. Chemical equation calcium ion-exchange is given in Equation 2 below:

$$2H^+\text{-ZSM-5 nanozeolite (s)} + Ca^{2+} \text{(aq)} \longleftrightarrow \text{Ca-ZSM-5 zeolite nanocomposite (s)} + 2H^+ \text{(aq)} \quad (2)$$

When the ion-exchange process took place, the Ca^{2+} ions are substituted with H^+ ions. EDX spectrum displayed in figure 4(b) reveals that 86.84 % of Ca^{2+} ions were successfully exchanged into H-ZSM-5 nanozeolite. Furthermore, this calcium ion-exchange level was supported by the data from figure 6 in XRD patterns showed that high intensity peak at 29.0° which corresponding calcium ions from $CaCO_3$ in calcite crystalline phase from eggshell.

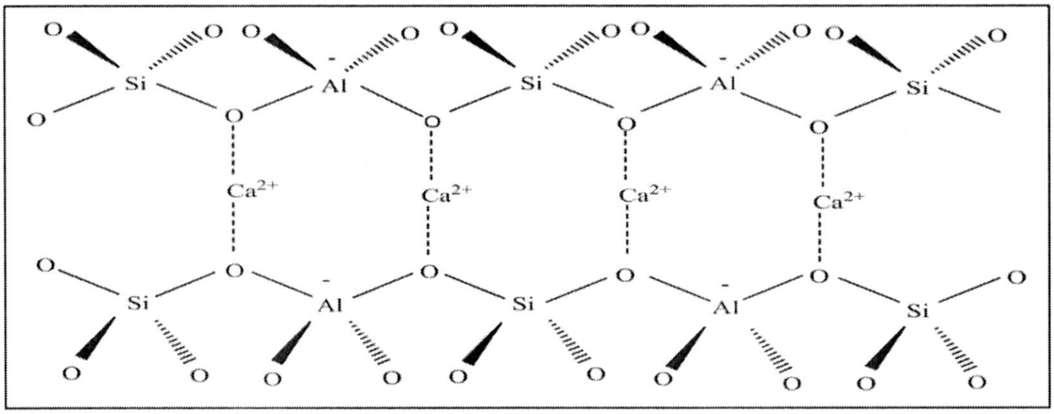

Scheme 3. Schematic of Possible Mechanism for the Formation of Ca^{2+}-ZSM-5 Zeolite Nanocomposite (ZNC)

4. Conclusion

Ca-ZSM-5 zeolite nanocomposite with porosity of 71.4-198 nm has successfully prepared by simple co-precipitation method using ethylene glycol as organic soft template. The FTIR, SEM, EDX, TEM and XRD results proved that calcium from chicken eggshell was impregnated into H-ZSM-5 nanozeolite with particle size of 2-26 nm. Zeolites are commonly used as catalysts for hydrocarbon application. Due to the ion-exchange capability, Ca-ZSM-5 zeolite nanocomposite has been developed as controlled-release fertiliser for agricultural usage particularly in plant growth. This study also emphasises the utilisation of waste-based materials that can be converted into innovative high technology materials. This is in line with national policy on waste-to-wealth.

5. References

[1] Zaman T, Mostari M, Mahmood M A A and Rahman M S 2018 *Cerâmica* **64** 236-41
[2] Krishna D S R, Siddharthan A, Seshadri S K and Kumar T S 2007 *J. Mater. Sci.: Mater. Med.* **18** 1735-43
[3] Lateef A, Nazir R, Jamil N, Alam S, Shah R, Khan M N and Saleem M 2016 *Microporous Mesoporous Mater.* **232** 174-183
[4] Aznan N A N, Malek N A N N, Noor N S M and Javed M A 2018 *Asian J. Agric. Bio.* **6** 55-65
[5] Hosseini B, Mirhadi S M, Mehrazin M, Yazdanian M and Motanedi M R K 2017 *Trauma Mon.* **22** e36139
[6] Gackowski M, Podobiński J and Hunger M 2019 *Microporous Mesoporous Mater.* **273** 67-72
[7] Garnjanagoonchorn W and Changpuak A 2007 *Int. J. Food Prop.* **10** 497-503
[8] Shankar V and Jambulingam R 2017 *Sustain. Environ. Res.* **27** 273-78
[9] Chung K H, Jeong S, Kim H, Kim S J, Park Y K and Jung S C 2017 *Catalysts* **7** 321
[10] Zhang K and Ostraat M L 2015 *Catal. Today* **264** 3-15
[11] Yang X Y, Chen L H, Li Y, Rooke J C, Sanchez C and Su B L 2017 *Chem. Soc. Rev.* **46** 481-558
[12] Tsintskaladzea G P, Eprikashcilia L G, Urushadzeb T F, Kordzakhiaa T N, Sharashenidzea T V, Zautashvilia M U and Burjanadzea M N 2016 *Ann. Agrar. Sci.* **14** 163-168
[13] Li X, Zhang H, Fang Y, Lin S, Song L, Li H, Xin H and Li X 2014 *Energy Environ. Focus* **3** 257–265
[14] Tangboriboon N, Kunanuruksapong R and Sirivat A 2012 *Mater. Sci. Poland* **30** 313-22
[15] Koohsaryan E and Anbia M 2016 *Chinese J Catal.* **37** 447-467
[16] Nada M H and Larsen S C 2017 *Microporous Mesoporous Mater.* **239** 444-52
[17] Feliczak-Guzik A 2018 *Microporous Mesoporous Mater.* **259** 33-45
[18] Alueshima B M, Eterigho E J and Friday O D 2018 *IOP Conf. Ser.: Earth Environ. Sci.* **173** 012041
[19] Hamidi A A, Salimi M N and Yusoff A H M 2017 *AIP Conf. Proc.* **1835** 020045
[20] Shaikh I R, Shaikh R A, Shaikh A A, War J A, Hangirgekar S P, Shaikh A L and Shaikh R R 2015 *J. Catal.* **2015** 1-14
[21] Wu W and Weitz E 2014 *Appl. Surf. Sci.* **316** 405-15
[22] Aziz A, Park H, Kim S and Kim K S 2016 *Int. J. Environ.* **13** 2805-16
[23] Kianfar E, Salimi M, Pirouzfar V and Koohestani B 2018 *Int. J. Chem. React. Eng.* **16** 20170229

Acknowledgement

This work was supported by Universiti Pendidikan Sultan Idris (UPSI) under research grant (GPU 2017-0187-102-01). The authors thank technical staffs of Department of Chemistry UPSI for their technical assistance. Sathia Kumaran thanks Ministry of Education Malaysia for providing a study leave for his MSc programme.

Science generic skills of 'chemistry'? prospective teachers: A study on collaborative learning using Exe-media

N Harefa[1,a] **and R D Suyanti**[2]
[1]Department of Chemistry Education, Faculty of Education and Teacher Training, Universitas Kristen Indonesia, Jakarta, Indonesia
[2]Department of Chemistry Education, Faculty of Mathematic and Natural Science, Universitas Negeri Medan, Medan, Indonesia

[a]E-mail: harefa.nelius1991@gmail.com

Abstract. Collaborative learning is a learning model that guides students to gather ideas and data to solve common problems. In this study, collaborative learning was integrated with Exe-media on reaction rate and salt hydrolysis to improve prospective teacher generic science skills. Specifically, this study aims to investigate Exe-media effectiveness to improve prospective teacher generic science skill. This study employed methods of a quasi experiment with a nonequivalent control group design. The subject of this research are prospective teachers of class Kim'B as experiment class and Kim'C as control class of Department of Chemistry Education, Faculty of Mathematic and Natural Science, Universitas Negeri Medan. Experiment class received treatment in form of learning used collaborative learning using Exe-media. Control class received treatment in form of learning used conventional model using power point as media. Data was collected through pretest and posttest results from prospective teacher in both classes. A t-test was used and it is found that level of significance is 0.002 (sig. < 0.05) proving that collaborative learning using Exe-media is effective to improve prospective teachers' generic science skill.

1. Introduction

The Indonesian National Qualifications Framework (KKNI) based learning process is expected to be a solution in order to develop a better learning process in Indonesia. Because of this goal, the improvement of human quality resources must be a top priority, especially the teachers or prospective teacher who will be the main foundation in the learning process so that the goal of improving can be achieved as expected. In order to improve the quality of prospective teacher, many studies have been conducted which aim to hone the ability of teachers candidate so that later they become teachers who can help their students to better understand the essence of what they are learning so that learning becomes more beneficial for students. One of the capabilities that must be owned by students in the process of learning science is generic skills.

Learning science substantially has several aspects: equality between processes and products, actively taking part in investigations, the ability to think inductively and deductively, and enhancing behavior [18], [19]. Science is a human discipline for understanding nature through proper observation of goals, together with using procedures and explaining logically to get conclusions [25]. Science is a part of knowledge about nature that produces bright visions, ideas, intellectuals, and insights, but nowadays, science is not only part of knowledge, but also a way of discovery or scientific method [10].

Collaborative learning is a study model that guides students to gather ideas and data to solve common problems [4]. Learning with collaborative emphasized the thought thinking process skills [13], [14]. Collaborative learning enhances students' critical thinking [7] and creative thinking [4]. Besides, the aid of using collaborative learning consist of: (1) expansion of higher-level thinking, verbal communication, self-management, and leadership skill; (2) encouragement of students interplay; (3) escalate in students conservation, self-esteem, and commitment; (4) exposure to and an upgrading in understanding of disparate perspectives; and (5) arrangement for real life and employment circumstances [4], [14], [18].

Audio visual is a media that involves the sense of hearing and vision in one process. Nature reservations that can be sent through the media are in the form of verbal and non-verbal messages. Reservations that are heard and seen can be presented through audio-visual programs such as films, videos, and television [17]. Video technology can help teachers to identify propositions faced in class and as an effective solution to present learning in a structured way that provides a real opportunity to pay attention and review endless learning [29].

Based on its characteristics, there are benefits in applying learning media in video displays such as: (1) overcoming the limitations of distance and time; (2) the video can be repeated if needed to add an explanation; (3) instructions are delivered immediately and are comfortable to remember; (4) broadening students' thoughts and opinions; (5) broadening students' imagination; (6) clarifying things that are noetic and giving illustrations that are even more realistic; (7) has a high influence on students' emotions; (8) explains the process and skills, able to show stimuli that reflect the expected goals and responses; (9) all students can learn from video, both students who are smart or who are not smart; and (10) develop their interest and drive for learning [17].

Exe-media is one of the audio visual media. Learning using Exe-media can make consisted of communicating in both writing and oral, planning the experiment, predicting hypothesis, interpreting the observation, observing, measuring, and using tools and material [12]. Implemented of this media can improve students' thinking process like critical thinking and creative thinking skills [26]. And also, can improve the students' problem solving skills [22], [30].

Generic science skills are cognitive aspects related to affective and psychomotor aspects that can be explored further by prospective teachers. Generic science skills can be used to learn many concepts and solve any science problem [23]. Generic science skills are skills that can be used to learn various concepts and solve various scientific problems [27]. Generic science is a skill produced by several structures of the learning process using a scientific approach [20]. There is not one standard definition of generic science skill indicators [15]. Generic science skills are part of critical thinking skills, problem solving skills, interpersonal skills, curiosity, creativity skills, and integrity [2]. Generic science skills can be categorized into 10 indicators including: (1) direct observation; (2) indirect observation; (3) scale awareness; (4) symbolic language; (5) logical frame; (6) logical consistency; (7) the law of cause and effect; (8) modeling; (9) logical inference; and (10) abstraction [24].

Based on some expert elucidation, it can be concluded that science generic skills are the ability to think based on knowledge related to cognitive aspects, affective aspects, and psychomotor aspects possessed by prospective teachers through the learning process. In science, generic skills can be used to learn many concepts and solve problems.

The purpose of this study is to established effectiveness of the use of collaborative models using Exe-media to improve prospective teacher science generic skills.

2. Research Methodology

2.1. Location and Time

This study was conducted in Department of Chemistry Education, Faculty of Mathematic and Natural Science, Universitas Negeri Medan, Medan. The study was conducted in the even semester of the academic year in 2016/2017.

2.2. Type of Research

This study used quasi experimental design with nonequivalent control group design. This study consist two groups both are experiment and control class with purposive sampling technique. In experiment class, prospective teacher was taught used collaborative learning using Exe-media and control class prospective teacher learned conventionally using power point media.

2.3. Population and Sample

Population of this study are prospective teachers of class Kim'B as experiment class and Kim'C as control class. Each class consisted of 38 prospective teachers. The experiment class of this research was given learning Exe-media about reaction rate and salt hydrolysis. Sample was purposed to improve prospective teachers' science generic skills.

2.4. Data Collection Technique

Data is collected in some steps, (1) giving pretest to all class groups, (2) giving Exe-media to experiment class group and power point to control class group, (3) giving posttest to all class groups. Data collection format is served in Table 1.

Table 1. Experiment Design

Group	Pretest	Treatment	Posttest
EC	X_1	P	X_2
CC	Y_1	Q	Y_2

Note:

EC : Experimental class
CC : Control class
X_1 : The initial skills of the experiment class
X_2 : Final experiment class skills
Y_1 : The initial skills of the control class
Y_2 : Final control class skills
P : The use of collaborative learning using Exe-media
Q : The use of conventional learning using power point

The treatment process was held in four sessions for 12 hours of meeting. The first meeting discussed the reaction rate using Exe-media in 3 hours. The second meeting discussed a number of subjects that could be changed by the process in 3 hours. The third meeting discussed about salt hydrolysis using Exe-media in 3 hours. The fourth meeting discussed a number of subjects that could be changed by the process in 3 hours. The objects discussed are reviewed physically and chemically reviewed. Through this process, students from the experimental class receive worksheets to observe and understand all information collected using Exe-media but on the other hand, prospective teachers of control class receive learning using power points as media.

Data collection based on generic science skill indicators instrument [3], [24], [26]. This study consists of 10 facets consisting of 26 indicators as shown in Table 2. According to generic science skills, experimental data can be collected through science generic skills tests. Science generic skills tests consist of 30 multiple choice questions and 20 essays. Data on the results of science generic skills were obtained from the pretest as the initial skill and the posttest as the final skill.

Table 2. Science Generic Skill Indicators

Science Generic Skill Facet	Indicator
Direct Observation	a. Using as many senses as possible in observing natural experiment/phenomena b. Gathering the facts of the experiment or the natural phenomenon c. Searching for differences and similarities
Indirect Observation	a. Using a measuring instrument as a sensing device in observing experiment or natural phenomenon b. Collecting facts from experimental result of physics or natural phenomena c. Looking for difference and similarities
Awareness of the Scale	Recognizing natural objects and high sensitivity to numerical scales as microscopic or macroscopic scale
Symbolic Language	a. Understanding symbol and terms b. Understanding the quantitative meaning of units and magnitudes of equations c. Using mathematical rules to solve problems or phenomena of natural phenomena d. Reading a diagram, table, and mathematical mark
Logical Frame	Look for a logical relationship between two concepts
Logical Consistency	a. Understanding the rules b. Rule-based arguments c. Described the problem based on rules d. Drawing the conclusion of a phenomenon based on rules presiding laws
Cause and Effect Law	a. Expressing the relations between two or more variables in a specific natural phenomena b. Estimating the causes of natural phenomenon c. Revealing phenomena or problems in the form of sketches of image or graphs d. Expressing formula in formulas e. Applying alternative problem solving
Modeling	a. Make a table of available data b. Suggest alternative solutions to problems
Logical Inference	Added a new concepts
Abstraction	a. Describing or abstracting concepts or events in to the real-life every-day life b. Creating visual animations of abstract microscopic events.

2.5. Data Analysis Technique

The research data were analyzed based on the gain score, normality test, homogeneity test, and t-test using SPSS 21. The gain score was used to see the difference between the posttest and pretest scores. Scores are used to understand the increase in students' science generic skills after undergoing treatment. Science generic skills acquisition scores can be calculated using this formula:

$$gainscore = \frac{score_{postest} - score_{pretest}}{score_{maksimal} - score_{pretest}} \tag{1}$$

The gain score values can be categorized as shown in Table 3.

Normality test is used to see whether science generic skills data is normally distributed or not. The data used in the normality test was obtained from the pretest. The normality test results can be concluded as distributed by "Kolmogorov-Smirnov" if sig. > 0.05.

Table 3. Conversion of N-gain Into Category

Scale	Category
0.01 – 0.29	Low
0.30 – 0.69	Medium
0.70 – 1.00	High

Homogeneity test is used to see the similarity of variance in the experimental and control classes. The data used in the homogeneity test was obtained from the pretest. Homogeneity test results are based on "Levene Statistics" if sig. <0.05, the variance can be concluded is not the same and if sig. > 0.05, the variance can be concluded to be the same.

Hypothesis testing in this study was carried out using an independent t-test. Independent t-test is used to determine the difference in the average of generic skills between the experimental class and the control class. The basis for decision making in any independent effort that survives this test can be based on the sig. (2-tailed). If the value of sig. (2-tailed)> 0.05 then H_0 will be accepted and H_a will be rejected but if the value of sig. (2-tailed) <0.05 then H_0 is rejected and H_a is accepted.

H_0 : Collaborative learning used Exe-media not effective to improve the science generic skill of prospective teachers

H_a : Collaborative learning used Exe-media is effective to improve the generic science skill of prospective teacher

3. Result and Discussion

This research was conducted at the Department of Chemistry Education, Faculty of Mathematics and Natural Sciences, Medan State University, Medan, Indonesia. This study consisted of two classes, both the experimental class and the control class. In the experimental class, prospective teachers are taught to use collaborative learning using Exe-media to improve generic science skills. In the control class, teacher candidates learn conventionally by using power points.

The study began with a pretest and continued with learning sessions in 4 meeting sessions both in the control class and experiment, and ended with a posttest. Pretest aims to see the prospective teacher's generic science skills begin. The learning process is held in four sessions for 12 hours in total. The first meeting in 3 hours discussed the rate of reaction using Exe-media. The second meeting in 3 hours discussed a number of subjects that could be changed by the process. The third meeting in 3 hours discussed about salt hydrolysis using Exe-media. Four meetings in 3 hours discuss several subjects that can be changed by the process. All sessions are intended to provide treatment to the control and experiment classes. Posttest aims to determine the final generic science skills of prospective teachers after receiving treatment. The average posttest and pretest results are shown in Table 4.

Table 4. Average Score of Posttest and Pretest

No	Class	Pretest	Posttest
1	Control	40.43	56.34
2	Experiment	41.14	70.28

According to the table, after getting treatment, control and experimental class showed an increase in generic science skills. However, the average increase (pretest and posttest scores) of the experimental class was higher than the control class. So it can be concluded that the increase in the generic science skills of prospective teachers in the experimental class is higher than the control class. This shows that learning using collaborative learning using Exe-media is more effective than conventional learning using power points. Learning using Exe-media can make consisted of communicating in both writing and oral, planning the experiment, predicting hypothesis, interpreting the observation, observing, measuring, and using tools and material [12].

Analysis of improvement in science generic skills can be seen and categorized based on acquisition scores as shown in Table 5.

Table 5. Average Gains Score

No	Class	Gain	Category
1	Control	0.3654	Medium
2	Experiment	0.7115	High

The gain score analysis in this study shows that the experimental class gain score is higher than the control class gain score. According to table 5 it can be categorized that the increase in generic science skills of teacher candidates in the control class is low but on the other hand it is high in the experimental class.

Testing the hypothesis in this study using an independent t-test. Parametric statistical testing conditions on independent t-tests have gone through normality and homogeneity. The results of the normality test are shown in Table 6 and the homogeneity test is shown in Table 7.

Table 6. Test of Normality

	Class	Kolmogorov-Smirnov[a]		
		Statistic	Df	Sig.
Pretest	Control	.093	38	.256*
Score	Experiment	.144	38	.256*

*. "This is a lower bound of the true significance"
"a. Lilliefors significance correction"

Normality tests are held to determine whether generic science skills data in control and experimental classes are normally distributed or not. The results of the normality test in table 6 indicate that the significance (sig.) Of the control and experimental class is higher than 0.05 (0.256> 0.05). It can be concluded that control and experimental class data are normally distributed.

Table 7. Test of Homogeneity of Variances

	Levene Statistic	df1	df2	sig.
Pretest Score	3.678	1	74	.076

Homogeneity test is done to see whether generic science skills about control and variant of experimental class are based on homogeneous Levene Statistics. According to the homogeneity test in table 7 shows that the significance (sig.) Is higher than 0.05 (0.076> 0.05). This shows that the control and distribution of the experimental class are homogeneous.

According to the results of normality and homogeneity tests, it shows that the data are normally distributed and homogeneous. Data that is normally distributed and homogeneous can be tested using independent t-test. Independent t-tests are presented in Table 8.

Table 8. Independent Samples t-test

	T	Df	Mean Difference	sig. (2-tailed)
Equal variances Assumed	12.635	74	31.67345	.002
Equal variances not assumed	12.635	74	31.67345	.002

According to the table, independent t-test results can be used to decide whether H_0 will be accepted or rejected. Sig value (2-tailed) of the independent t-test is 0.002. This shows that the value of sig. (2-tailed) 0.002 <0.05, so H_0 is rejected and H_a is accepted.

Based on experimental results and control class analysis, it can be concluded that collaborative learning using Exe-media about reaction rates and salt hydrolysis is effective for improving the science generic skills of prospective teachers. This consists of the theory that science generic skill facets of the skills of prospective teachers developed in this study can be achieved through science media developed. The use of technology has a positive effect especially for millennial prospective teachers [9]. Exe-media as an information and communication technology tool can consist of communication both in writing and orally, planning experiments, predicting hypotheses, interpreting observations, observing, measuring, and using tools and materials [12]. Science Exe-media are developed according to the strengths expressed [14]. In addition, the science learning using Exe-media developed has its own advantages in fostering learning motivation of prospective teachers [8]. Learning through direct experience, deliberate learning, simulation, and represent reality in daily life that can be effectively accepted by prospective teachers [6]. In addition, Exe-media used in learning can clarify and simplify the delivery of messages so that they do not become verbal and learning varies so that prospective teachers are not bored in accepting lessons [5]. It is also appropriate that science learning integrated with Exe-media can influence prospective science generic skills [16].

Generic science skills can help prospective teachers to understand concepts and solve problems in learning [28]. Generic science skills are part of the ability to communicate in problem solving [11], [18]. Generic science skills have a very important role in supporting learning, especially in science learning because it can emphasize aspects of the process [1].

4. Conclusion

Based on the results of data analysis and discussion it can be concluded that the learning used collaborative learning using Exe-media about reaction rates and salt hydrolysis is effective for improving the science generic skills of prospective teachers. This result is based on the acquisition of the average score of the experimental class which is better than the average score of the control class with a high category. Learn used Exe-media can make complete communication both in writing and orally, plan experiments, predict hypotheses, interpret observations, observe, measure, and use tools and materials. In addition, learning by using Exe-media is easier to remember and can increase the attention of prospective teachers about the reaction rate and salt hydrolysis.

References

[1] Agustina S, Muslim M, and Taufik 2016 *JIPF* **3** (1) 1-7
[2] Badcock P B, Pattison P E, and Harris, K L 2010 *High Euc* **60** 451-458
[3] Brotosiswoyo B S 2000 Hakekat Pembelajaran Fisika di Perguruan Tinggi (Dalam Hakekat Pembelajaran MIPA & Kiat Pembelajaran Fisika di Perguruan Tinggi) (Jakarta: DEPDIKNAS)
[4] Chandra R 2015 *IJRME* **5** (2) 1-4
[5] Riyana C 2007 *Pedoman Pengembangan Media Video* (Bandung: Program P3AI UPI)
[6] Davis B and Summers M 2015 *Qscience Proceedings* (USA: Purdue University) 1-7
[7] Gokhale A A 2015 *JTE* **13** (1) 22-30
[8] Hanik A N, Sudarmin, and Supardi K I 2018 *JISE* **7** (2) 320-329
[9] Harefa N, Gultom S, and Purba L S L 2019 *Jurnal Dinamika Pendidikan* **12** (1) 17-28
[10] Hewit, Paul G, Suzanne L, and John Suchoki 2006 *Conceptual Integrated Science* (San Fransisco: Pearcon Education)
[11] Hockey A, Bescos C J, and Maclean, J. 2010 *TPR* **81** (5) 523-540
[12] Juniar A, Silalahi A, and Suyanti R D 2018 *3rd Annual International Seminar on Transformative Education and Educational Leadership (AISTEEL 2018)* 200 500-503
[13] Kennedy-Clark S, Kearney S, and Galstaun V 2007 *Education Sciences* **7** (80) 1-14

[14] Laal M and Laal M 2012 *Procedia - Social Behavioral Sciences* **31** 491-495
[15] Maknun J 2015 *American Journal of Education Research* **21** (6) 742-748
[16] Mulyani S, Liliasari, Hana W M N, and Nursa'adah M 2016 *JPII* **5** (1) 150-156
[17] Munadi Y 2013 *Media Pembelajaran Sebuah Pendekatan Baru* (Jakarta: GP Press Group)
[18] Murti D H, Prasetyo Z K, and Wilujeng I 2018 *Journal of Physics: Conference Series* (Vol. 1097, No. 1, p. 012005)
[19] Muslichah Asy'ari 2006 *Penerapan Pendekatan Sains-Teknologi-Masyarakat* (Jakarta: Departemen Pendidikan Nasional)
[20] Ratna I S, Yamtinah S, Masykuri M, and Shidiq, A S 2017 *Advance in Social Science, Education and Humanities Research (ASSEHR)* **158** 958-963
[21] Retno Dwi Suyanti 2006 Pembekalan Kemampuan Generik Bagi Calon Guru Melalui Pembelajaran Kimia Anorganik Berbasis Multimedia Komputer (UPI Bandung: Disertasi-Unpublished)
[22] Saprudin, Sutarno, and Liliasari 2010 *4th International Conference on Teacher Education* 669-675
[23] Selvianti, Ramdani, and Jusniar 2013 *Jurnal Chemica* **14** (1) 55-65
[24] Sudarmin and Haryani S 2014 *IJSR* **4** (5) 468-475
[25] Susanto A 2013 *Teori Belajar dan Pembelajaran di Sekolah Dasar* (Jakarta: Kencana Prenada Media Grup)
[26] Suyanti R D and Sianturi K M 2017 *Prosiding Seminar Nasional Kimia UNY 2017* 115-122
[27] Tawil M and Liliasari 2013 *Berpikir kompleks dan implementasinya dalam pembelajaran IPA* (Makassar: Badan Penerbit Universitas Negeri Makassar)
[28] Tricot A and Sweller J 2014 *Educ Physchol* **26** 265-283
[29] Wang J and Hartley K 2013 *Journal of Technology and Teacher Education* **11** (1) 105-138
[30] Zirconia A, Supriyanti F M T, and Supriatna A 2008 *12th Joint Conference on Chemistry* **349** 1-6

Self-Efficacy of Students Senior High School in Problem Based Learning Model of Chemical Equilibrium Topic

M Handayani[1,a], **ISY Louise**[2,b]

[1]*Postgraduate of Chemistry Education, Yogyakarta State University, Indonesia*
[2]*Chemistry Education Department, Faculty of Mathematics and Natural Science, Yogyakarta State University, Indonesia*

[a)]meliahandayani7@gmail.czom

Abstract. The contextual learning model used in chemical equilibrium topic is problem based learning. Problem based learning facilitates students to analyze problem in daily life, using appropriate learning resources, and organize information into a conceptual framework. Self-efficacy is a feeling of confidence or student's belief to carry out the task and achieved good result. This study aims to analyses students' self-efficacy skills profile in chemical equilibrium topic using problem based learning model. This study used quasi experiment with post-test only design and used cluster random sampling. Self-efficacy of students was measured by 30 item statements questionnaire sheets. The result of this study indicated that student's self-efficacy in problem based learning model achieved good category in some indicator. The higher score of students' self-efficacy contained in indicator overcome the difficulties, complete the assignment, solve the problem, getting good result, apply efficacy in daily life, cooperate and communicate in team.

Keyword: problem based learning, self-efficacy, scientific approach, chemical equilibrium topic

1. Introduction

The 21st century is a transformation of technology and industry in all aspect. Challenges in 21st century require students to develop competitive skills that focus on creativity, solve the problems, communicate and cooperate [1]. Soft skills are the center of attention in education and industry. Neff explain the determination of success is not on technical skill/ hard skill, but in the quality of soft skill [2]. Students learn how to cooperate and communicate in self-efficacy.

Self-efficacy is a person's belief in his or her own ability to do a task with a certain level of proficiency [3]. Self-efficacy becomes the main determinant in measuring self-belief of student [4]. Minister of education number 23 in 2016 about the standard assessment, result of the study include aspect of attitude, knowledge, and skill [5]. The result study in 2014 is one of senior high school in Yogyakarta showed 42.69% or below 50% of students who had high level of self-efficacy [6]. From survey data, students' self-efficacy still has a chance to be increase. Lack of self-efficacy make students have a low struggling ability, in other side the importance of self-efficacy was not felt by some students.

Students' self-efficacy should better in some condition, so self-belief in person ability still can be improved. The development of self-efficacy on students is important, because when students have high self-efficacy they will have better performance and have the ability to solve problems in learning [7]. High self-efficacy has the opportunity to carry out an activity and develop in positive attitude. Students who have a high self-efficacy are gotten the opportunity to choose the type of question that more challenging [8]. Student with low efficacy tend to like the types of question that are not challenging, and usually will take the opportunity to work on questions with the same level of difficulty. Students with problem based learning model are expected to feel challenge in analyze case and provide solution to solve problem. In an educational context, self-efficacy is important variable that have positive effect on the level of students' motivation [9]. Students who have high motivation tend to have high belief about what students can do, anticipate possibilities will occur through action and plan programs for the future.

The dimension of self-efficacy in each individual is consists of three components that have important implication performance. The first is magnitude level, this dimension relates with difficulty level of the task. Individual will try behavior that they feel able to do and avoid beyond the limits of abilities. Second is the strength level, this dimension related to the level of strength an individual or expectation regarding of his abilities. Weak hope is easily broken by experience that not support, strong expectation encourage individual to survive in effort. The third is generality level, this dimension relates with area behavior where individual can survive in effort. Generality can be assessed from the same level of activities, abilities, affective, type of situation that faced and individual characteristic in according life goal [3].

Problem solving skills are needed in order to prepare a generation that is ready to work and career in the era of global economic [10]. In the global era, the state continues to strive and improve the quality of education, especially for developing countries. For supporting problem solving abilities, the teacher can vary with the learning model. One of the interactive model and has student center character in learning process is problem based learning model [11].

Problem based learning is a learning model which has facilitates students to analyze problems in daily life, use appropriate learning resource, and organize information into a conceptual framework [12]. Problem based learning has the same characteristic as self-efficacy which is able to guide students to feel confident and more desire to try assignment and solve problem as well as possible [13]. Akinoglu explained the activities with problem based learning are carry out by dividing students into small class and the teacher act as a facilitator in the discussion activities [14]. The existence of a problem based learning model is expected to have a positive impact on self-efficacy.

Problem based learning model have five stage of learning process, are (1) given orientation about the problem, (2)organize students to study, (3)conduct individual and group investigation, (4)develop and present result, (5)analyze the problem solving process [15]. The advantage of problem based learning are: being able to increase learning motivation, able to encourage high level of thinking, encourage students to optimize metacognition abilities, making learning more meaningful [16] so, expected students can enthusiast in learning process especially focus on investigation, giving solution, and collaborative in a group.

Contextualization is one part of creating character education that is useful, culture, and beneficial for the formation of a civilized community [17]. Contextual learning is a learning concept that helps teachers associates the content of subject matter with real world condition. The subject matter of this study is chemical equilibrium. Chemical equilibrium has characteristic with the material which real and requires understanding concept, need to high logical reasoning in solving the problems [18]. For example correlation the principle of chemical equilibrium on making ammonia, formation stalagmite and stalactite, how to make the ammonia and sulfuric acid can produce a maximum product, and what factor must be analyze.

One of learning models which compatible with the characteristics of chemical equilibrium material is problem based learning. Learning model is a technique in designing and applying to obtain information, skill, concept, and ideas [19]. Problem based learning model is expected to provide positive result for self-efficacy of students. The objective of this study is to analyze profile of students' self-efficacy abilities in chemical equilibrium topic using problem based learning model

2. Research methods

2.1. Research design and samples

Quasi-experiment with posttest only design was used in this study. In this study two class were used, one class an experimental class and one class as a control class. Student in experimental class was given treatment with problem based learning, while the control class was given treatment by learning approach that used in the school, scientific approach.

Determination of the sample was done by using two stages cluster random sampling technique. The first stage is to determine the school that used to study. The populations in this study are all students of eleven grades from science department in senior high school of special region Yogyakarta with cluster category in moderate abilities. There are six senior high school which has characteristic: (1) have A grade for accreditation, (2) use the 2013 curriculum, (3) an average national examination score of 61-63, (4) each class consist of 28-30 students, and (5) the school have completely laboratory facility. Among six senior high schools in Yogyakarta, one school was chosen randomly. From these criteria so selected 11th Senior High School.

The second cluster random sampling stage is to determine the class used as a sample study. Some of science class in 11th senior high schools were conducted an equality analysis to measure differences based on students' initial knowledge. ANOVA analysis result is sig.1.27 > 0.05, it is mean no difference in the initial knowledge of students' science class in 11th grade. Furthermore, because it does not have difference from the seven classes could be randomly, and then it obtained XI science 6 and XI science 7 as a sample study. Number of students in xi science 6 consists of 31 students and xi science 7 consists of 30 students.

2.2. Techniques and data collection instruments

Self-efficacy was measured by using scale. Respondents were ask to give a check mark on answer category. Developing self-efficacy questionnaire based on likert scale modified. The self-efficacy scale contains 30 statement that consist 15 positive statement and 15 negative statement. Each statement on the questionnaire has used 5 scales for positive statement and negative statement was presented on Table 1.

Table 1. Scoring of questionnaire of self-efficacy

Positive statement	Negative statement	Score
Never	Always	1
Rarely	Often	2
Sometimes	Sometimes	3
Often	Rarely	4
Always	Never	5

The distribution of questionnaire aims to find out information about a problem that can provide answer according to statement given. The questionnaire is given at the end of meeting after students take part in learning process. The dimension in self-efficacy in each individual lies in three components, are magnitude, strength, and generality [3]. Three components made into six indicators, from indicator made into five statements in both positive and negative forms that shown in Table 2.

The quality of learning can be achieved through measurement which is measured by the level of validity and reliability. Theoretical validity will be done after the instrument is developed, then consult with expert judgment. The lecturer will examine the instrument based on aspect of self-efficacy according the criteria in scoring. The instrument then had been revised based on suggestion from validator. Empirical validity carry out by testing instrument on 280 students. The reliability estimation analysis at 0.69 which mean the reliability were sufficient.

Table 2. Indicator of self-efficacy

Dimension	Indicator	Item Number	
		Positive	Negative
Magnitude	Feeling and being able to overcome the difficulties of working on assignment	1, 2, 3	4, 5
	Feeling of belief to solve problems related to equilibrium material	6, 7, 8	9, 10
Strength	Feeling sure to be able to do the task in accordance with order	11, 12, 13	14, 15
	Feeling of belief to get good result	16, 17	18, 19, 20
Generality	Feeling of beliefs able to cooperate and communicate well during learning process	21, 22	23, 24, 25
	Feeling of belief able to apply self-efficacy in daily life	26, 17	28, 29, 30

2.3. Data analysis techniques

Total students which follow this study are 61, but there are 3 students not allowing in data analysis because the student not follow completely during learning process. Data analysis of this study took 29 students from experimental class and 29 students from control class. The data analysis use to determine profile self-efficacy based on the score of questionnaire.

Categories of self-efficacy are done by classifying the score obtained from questionnaire converted into a category based on ideal assessment. The steps in quantitative descriptive analysis are as follows: (1)recap the questionnaire data score, (2)make ideal score ranges and rating categories, (3) categories the score according to the ideal score. Self-efficacy category made into five categories, are very good, good, sufficient, less good, and bad. Determination of ideal rating categories refers to the suggestion of Widoyoko [20] with the highest score is 150 and the lowest score is 30, ideal of rating categories presented in Table 3.

Table 3. Ideal rating categories

Category	Score Range
Very good	$126 < X$
Good	$102 < X \leq 126$
Sufficient	$78 < X \leq 102$
Less good	$54 < X \leq 78$
Bad	$X \leq 54$

3. Result and discussion

Learning activities with problem based learning model was implemented based on syntax of Arends [15], started by giving problems to students, the purpose of giving problem is students are able to analyze problems in daily life related with material. Students are divided into six groups for each class to make good condition in discussion and teamwork session. During the learning process students allowed to look for references from various sources to facilitate in solving problem. The result discussion in small group will be discussed together in large group then drawn a conclusion. At the other meeting, students began to explore the concept by carry out practical activities, then students were given time to associate and communicate observation result with theories. At the end of meeting, students were given time to fill out self-efficacy questionnaire.

3.1. Category of self-efficacy

Self-efficacy category was viewed according ideal rating category. The result of self-efficacy was presented in Table 4.

Table 4. Category of self-efficacy

Category	Experimental Class		Control Class	
	f	%	f	%
Very good	3	10.34	-	-
Good	18	62.06	19	65.51
Sufficient	8	27.58	9	31.03
Less good	-	-	1	3.44
Bad	-	-	-	-

Based on Table 4 it can be seen that students in experiment and control class have good self-efficacy categories with the value of each class respectively 62.06 and 65.51. Although in experimental class on good category was lower score than control class, but in experimental class also has very good category.

3.2. Students' self-efficacy in several indicator

Self-efficacy score in each class has been divided into each indicator of self-efficacy. The score of each indicator obtained from measurement in percentage form that presented in Figure 1.

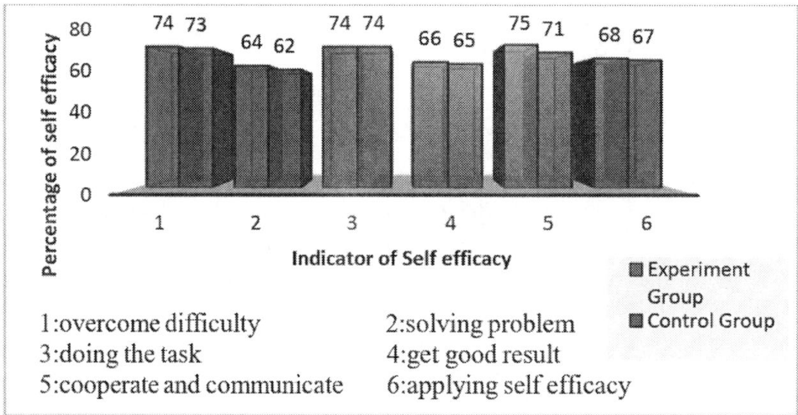

Figure 1. The percentage self-efficacy of students in each indicator

Figure 1 six indicator will be describe one by one. The first is overcome difficulty. Student in experimental class and control class have ability in to do assignment till hardest part and able to make lab report completely without asking friend each other. Both of them almost have the same score in 74 and 73. This show the ability in two class is not much different. The second is solving the problem. In this part student have ability to make good plan in compiling lab tool and material carefully. Student score in this indicator not much different with score 64 and 62.

The third indicator is doing the task. In this part measure ability to do practicum correctly, students try to lab work in accordance with procedure. Students in experiment and control class have same abilities with score 74. The fourth is get good result, students' optimist can do chemical equilibrium test correctly. Students in both of class have the same belief with score 66 and 65. The five cooperate and communicate with the score 75 and 71. The experiment class has a high score in ability to cooperate and communicate, this occurred before the implementation of practical activities, students in experimental class conduct discussion in preparing practical work. The last is applying self-efficacy in daily life. Students have belief able to solve problem in daily life with score 68 and 67.

The main score for each indicator in experiment class more supreme in several indicator including the ability to overcome the difficulties, ability to solve the problem, ability for getting good result, and ability to cooperate and communicate. In this study shown that problem based learning model has a positive impact on students' self-efficacy, even though some students are still shy in asking or answering question. This study accordance with result study from Mataka & Kowalske [21], students which participate with problem based learning class increase in self-efficacy beliefs through mastery experience. Study of self-efficacy is supported by Bandura theory which explain if individual are faced on tasks that arrange according to the level of difficulty, the possibility of students to choose on easy level, medium, even difficult depend on ability[3]. Measuring of self-efficacy must use specific domain such as how well students can work on assignment to completion [22].

Bodner, Savery & Duffy explain problem based learning using constructivist principle can encourage in terms of: (1) conducting authentic assignment, (2) supporting learning environment to bring up ideas, (3) encouraging students to develop themselves in completing assignment [20]. In other study, there is a positive correlation between self-efficacy and performance of students after students follow problem based learning [23].

4. Conclusion

The conclusion for this study is profile of students' self-efficacy in experimental class have a good category. Students' in experimental class more supreme in several aspect, are in overcoming difficulties, solving problems, working the task, getting good result, cooperation and communicating.

References

[1] Griffin P, Mc Graw B, and Care E 2012 *Assessment and teaching of 21st century skills* (Dordrecht: Springer)
[2] Sailah I 2008 *Pengembangan softskill di perguruan tinggi* (Jakarta: LPPM-IPB)
[3] Bandura 1997 *Self efficacy*: the exercise of control (New York: W H Freeman and Company)
[4] Kurbanoglu N I 2010 The relationships between university students' chemistry laboratory anxiety, attitudes, and self efficacy beliefs *AJTE 35*(8) 48-59
[5] Kementrian Pendidikan dan Kebudayaan Republik Indonesia 2016 Peraturan Mentri Pendidikan dan Kebudayaan republik Indonesia Nomor 23 tentang Standar Penilaian Pendidikan
[6] Wulandari S and Rachmawati M A 2014 Efikasi diri dan stress akademik pada siswa sekolah menengah atas program akselerasi *Psikologika 19*(2) 246-252
[7] Ferrel B and Barbera J 2015 Analysis of students self-efficacy, interest, and effort beliefs in general chemistry *CERP 16*(2), 318-337
[8] Villafane S M, Garcia C A, and Lewis J E 2014 Exploring diverse students trends in chemistry self efficacy through out a semester of college level preparatory chemistry *CERP 15*(2) 114-127
[9] Hassankhani H, Aghdam, A M, Rahmani A, and Mohammadpoorfard Z 2015 The relationship between learning motivation and self efficacy among nursing students *Res.and Dev.Med.Educ, 4* 97-101
[10] Sani A R 2014 *Pembelajaran saintifik untuk implementasi kurikulum 2013* (Jakarta: Bumi Aksara)
[11] Aryanti F, Surtikanti H, and Riandi 2017 Penerapan model problem based learning (PBL) berbantuan teknologi informasi dan komunikasi untuk meningkakan kemampuan berpikir kritis siswa pada konsep pencemaran lingkungan. *Biosfer: Jurnal Biologi dan Pendidikan Biologi 2*(1) 14-20
[12] Gunter T and Alpat S K 2017 The effects of problem based learning (PBL) on the academic achievement of students studying "electrochemistry" *CERP 18*(1) 78-98
[13] Ramrain U and Ramaila S 2017 The relationship between chemistry self efficacy of south african first year university students and their academic performance *CERP 19*(1) 60-67
[14] Krisnawati E, Susilo H, and Gufur A 2018 ICT based-problem based learning on students cognitive learning outcomes *Jurnal Pendidikan Sains 6*(2) 38-42
[15] Arends R I 2012 *Learning to teach* 9th edition (New York: The McGraw-Hill Companies)

[16] Abidin Y 2014 *Desain sistem pembelajaran dalam konteks kurikulum 2013* (Bandung: PT Refika Aditama)
[17] Sahlan and Prasetyo 2012 *Desain pembelajaran berbasis pendidikan karakter* (Yogyakarta: Ar-Ruzz Media)
[18] Sari and Muchlis 2014 Implementasi model pembelajaran tipe kooperatif tipe TPS dengan pendekatan kontekstual untuk melatih sikap jujur dan tanggung jawab siswa kelas XI IPA pada materi kesetimbangan *Unesa Journal of Chemical Education 3*(1) 10-20
[19] Joyce B, Weil M, and Calhoun E 2009 *Model of Teaching* (New York: Pearson Education Inc)
[20] Widoyoko E P 2009 *Evaluasi program pembelajaran* (Yogyakarta: Pustaka Pelajar)
[21] Mataka L M and Kowalske M G 2015 The influence of pbl on students' self-efficacy beliefs in chemistry *CERP 16*(4) 929-938
[22] Villafane S M, Xu X, and Raker J R 2016 Self efficacy and academic performance in first semester organic chemistry: testing a model of reciprocal causation *CERP 17*(4), 973-984
[23] Loh K H 2017 The relationship between academic self efficacy and learning performance in problem based learning environment *WASJ* 35(8) 1568-1573

Journal of Physics: Conference Series **1397** (2019) 012034 doi:10.1088/1742-6596/1397/1/012034

Development of Media Three-dimensional (3D) Visualization using Virtual Reality on Chemistry Education

M Suleman [1,a)], K H Sugiyarto, and J Ikhsan[2]
[1]Study Program of Chemistry Education, Graduate School
[2]Department of Chemistry Education, Faculty of Mathematics and Natural Sciences
Yogyakarta State University, Indonesia

[a]*Corresponding author:* mirnawatisuleman@gmail.com

Abstract. This study aims to develop and test feasibility three-dimensional (3D) visualization of media using virtual reality in chemistry learning . The development model used in this study is Research and Development (R&D) model. This study implemented the stages of planning, preliminary development of the product, preliminary field testing, main product revision adapted from Borg and Gall. The development of the learning media in this study was validated by media expert, material expert, peer reviewers, and subject teachers, and was tested to 23 students. The results of this study revealed that three-dimensional (3D) media using virtual reality is feasible to use as learning media for high school students. The percentages of validation results are 74% by media and material experts, 76% by chemistry teachers, 84% by peer reviewers, and 77% by students. Overall, the average total percentage score of feasibility of the three-dimensional media is 78% categorized to 'good'. In conclusion, the developed three-dimensional (3D) learning media using virtual reality is feasible to use in the chemistry learning process for the material of reaction rate.

1. Introduction

Learning in the education field should be interactive, inspiring, challenging, and improving the quality and motivation in learning process [1]. The use of technology can assist in the improvement of knowledge called the 'information super highway' [2]. The development of technology use also makes new innovations appear that they can be developed in learning process. In chemistry learning, students are expected to have scientific attitude, and doing the chemistry experiment helps students have scientific experience, understand the advantage and disadvantage of chemistry learning, and understand chemistry concepts to solve problems in daily life [3]. Moreover, chemistry is considered difficult because it discusses more abstract concepts, such as the concept of atoms, oxidation numbers, chemical equation, and energy [4]. That one of the most efficient ways in learning chemistry is through laboratory [5].

A laboratory is very important to have chemistry learning be easy to understand for students[6].The function of laboratory is to complete the theoretical materials that were received, to gather a courage to discover scientific fact from an environmental and social object, and to add skills in utilizing the tools [7]. One of chemistry materials that need a practicum is reaction rate. This concept cannot be only explained in speech method. There needs a practicum that helps in explaining the concepts which ease students in constructing the concepts, because reaction rate is a change that can be observed occasionally but the concentration change in reaction rate cannot be observed directly [8]. However, not all schools have laboratory that facilitates the process in laboratory.

Based on the observation conducted by seven schools in Ternate, there are many schools that have minimum facilities, such as the tools and materials provided in the laboratory, expired materials, and insufficient laboratory. To solve the problem on the minimum facilities in practicum, technology can be utilized, namely virtual laboratory with several facilities done in a computer-based interactive multimedia software [9]. One of recent technologies is the use of three-dimensional (3D) using virtual reality (VR). Virtual reality is a sophisticated technology that integrates multi-sensory, such as sound tracking and gesture as if the students were in the environment and interacts actively. This virtual reality is able to use to ease students in understanding the chemistry learning.

In the learning, virtual reality can make the students be more active because it provides more interactive environment [10]. Virtual reality can created an environment which is abstract and can be stimulated by computer so that students can interact to the simulated environment [11]. Virtual reality can also assist teacher in teaching because the use of virtual reality can create an activity which makes human's brain and the function of senses are able to work together with computer's help [12]. The development of virtual reality does not only include in education field but has been developed for extent needs [13]. Virtual reality also cannot be separated from visual display in which the visual is in the form of three-dimensional (3D) that the category of each visual display is different [14]. The visual display used in this study is head-mounted display (HMD).

The use of virtual reality in the education field, particularly in chemistry education, is necessary for schools that still lack of facilities in their laboratory, and the use of virtual reality is able to minimalize the danger in conducting the practicum [15]. The use of virtual reality can also improve students' interest in learning [16]. The development of virtual reality is expected to assist teacher in solving the problem, particularly chemistry learning that requires other learning media to explain chemistry material. In general, the implementation of activities in the laboratory is very important in chemistry learning, but for some reasons there are activities that require to be conducted in the laboratory cannot be carried out well. One of the ways that helps students in understanding chemistry material about reaction rate is through media learning. The media is three-dimensional (3D) technology. The technology of three-dimensional (3D) visualization is introduced with visual reality (VR). Three-dimensional technology using virtual reality can help students solve problems, such as the lack of facilities in the school that has been observed because the school has used form of software for practicum [14]. This study aims to develop and test the feasibility of using virtual reality media which will be applied to high school students in the chemistry learning process.

2. Research Methods

This study is descriptive quantitative research which describe about feasibility study of three-dimensional (3D) visualization media using virtual reality. The three-dimensional of learning media using virtual reality was developed by adapting a development model by Borg and Gall (1989), namely collecting information, planning, preliminary development of the product, preliminary field testing, and main product revision. Research and development of education is one method that aims to produce desired products in the field of education both in the form of media,test the validity, effectiveness and feasibility of the products used.

The first stage is collecting information. This stage is need analysis phase of this study, namely field observation to find out information about the implementation of the learning process in classroom, to obtain relevant sources to the study about development of three-dimensional media using virtual reality by literature review and need analysis. The second stage is planning, namely collecting material and reference related to three-dimensional media development using virtual reality, including designing the product and designing the learning. The third stage is preliminary development of the product, namely merging the design of product and learning. The product that has been finished was validated by material and media expert. Media expert is a lecturer who is expert at developing the learning media.

Journal of Physics: Conference Series 1397 (2019) 012034 doi:10.1088/1742-6596/1397/1/012034

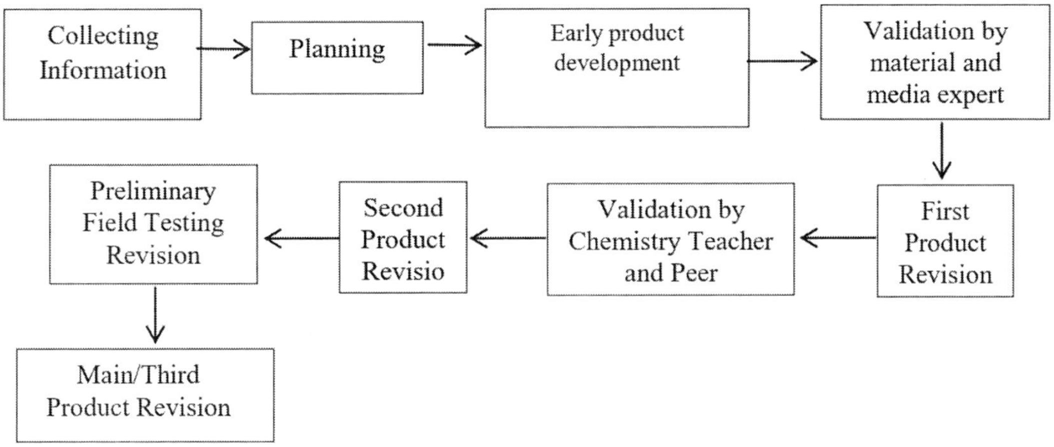

Figure 1. Data collection from five stages of development study

The validation conducted by material expert, media expert, chemistry teacher and peer reviewer using five-scale questionare was modified from Likert-scale. The learning media using virtual reality were validated by two judgment expert from content expert who profecient on teaching-learning chemisty and media expert who proficient on media development. Judgement content expert assessed both learning and material aspects. While, judgment media expert assessed both performance and opwrational aspects. Moreover, media assessment also validated by six chemistry teacher and ten peer reviewer. Validation of media feasibility by teacher and peer reviewer consider on four aspects i.e. audio-visual, software engineering, learning, and material. The instrument of virtual reality media were developed from Squires & Preece (1996), Nesbit & Leacock, (2009) Oyelekan & Olurundare (2009) dan Kustandi & Sutjipto (2011). The aspects developed are in the form of learning, material, appearance and operational aspects. There are four aspects were developed namely learning aspect, content aspect, performance aspect and operational aspect. The indicators of assessment can be seen in Table 1.

Table 1. The Indicators of media assessment

Aspects	Indicators
Learning	1. The suitability of indicators with material
	2. The suitability put in the learning media
	3. The clarity of the user target
	4. The ease of media access in learning
	5. Interactive subject of the learning with learning media
Material	6. The provision of material explanation
	7. The suitability of the composition of colors, shapes, and steps
	8. The clarity of practicum simulation with virtual reality media
	9. The conspicuousness of material in motivating the user
	10. The suitability in choosing the background color
Audio Visual	11. Font suitability
	12. Font size suitability
	13. Font color suitability
	14. Layout suitability
	15. Picture size suitability
	16. The effectiveness and efficiency in media development and use
Software	17. Ease of operation
Engineering	18. The creativeness and innovation in learning media

Aspects	Indicators
	19. The accuracy in the choices of operation type with the program
	20. The opportunity of media development towards science and technology development

The fourth stage is preliminary field testing, namely the readability of media. In this stage, the media were tested to 23 students to obtain the analysis and the consideration for main product revision. The fifth stage is main product revision to complete the learning media based on the feedbacks and the results obtained from when the media were tested.

The study was conducted using three-dimensional visualization of media on chemistry material of reaction rate which contains simulation of practicum. Data obtained from material expert, media expert, teachers and students were analyzed by calculating the formula and were suit to the criteria and aspects of media feasibility [17].

Table 2. Category of Media Quality

No	Rentang Skor (i)	Kategori
1	$Xi + 1{,}8\ SBi < \bar{x}$	Very Good
2	$Xi + 0{,}6\ SBi < \bar{x} \leq Xi + 1{,}8\ SBi$	Good
3	$Xi - 0{,}6\ SBi < \bar{x} \leq Xi + 0{,}6\ SBi$	Adequate
4	$Xi - 1{,}8\ SBi < \bar{x} \leq Xi - 0{,}6\ SBi$	Poor
5	$\bar{x} \leq Xi - 1{,}8\ SBi$	Very Poo

Note: X= average score; Xi= 1/2 (ideal maximum score + ideal minimum score); SBi = 1/6 (ideal maximum score + ideal minimum score)

3. Results and Discussion

Three-dimensional (3D) visualization of learning media using virtual reality on the reaction rate material was designed in an Android-based application. Visual display is used to provide cyberspace in three-dimensional display. The visual display developed for projecting image to the screen is active stereos-copy that needs shutter glasses. Every eye receives different images because every filter of eyeglass lens come out a set of images. The type of passive solution cannot be used on desktop display [18]. The illustration of shutter glasses is seen in Figure 2.

Figure 2. Shutter Glasses

The start display of three-dimensional visualization of learning media using virtual reality can be used as substitute of experiment activity in laboratory. Its illustration is seen in Figure 3.

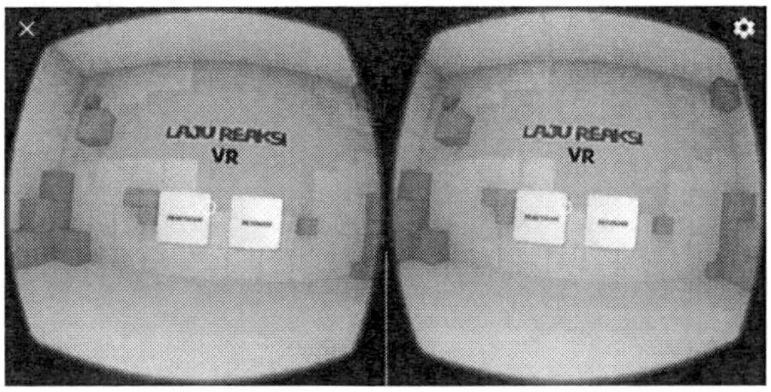

Figure 3. Homepage Display

The experiment conducted in laboratory needs tools, materials, and procedures in trying the display. The experiment display can be seen in Figure 4.

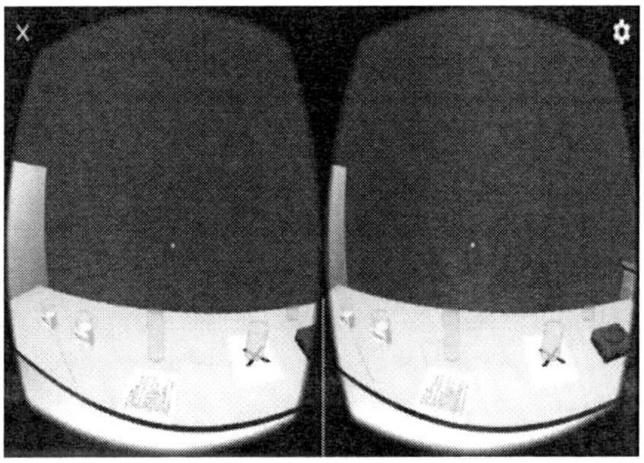

Figure 4. Display of Experiment

In testing the developed media, the data were general testing about feasibility media. The result data of three-dimensional visualization of media using virtual reality by the experts are seen in the following table.

Table 3. Media Validation Results by Experts

No	Aspects	Score	Category
1	Learning Aspect	3.6	Good
2	Material Aspect	3.8	Good
3	Audio-Visual Aspect	3.5	Good
4	Operational Aspect	4.0	Very Good

The validation results by material expert were focused on the learning aspect and material aspect while the validation results by media expert were focused on audio-visual aspect and software engineering aspect. The media expert assessed audio-visual aspect with the score 3.5 and its percentage score is 70% which belongs to 'good' criteria, while the software engineering aspect score is 4.0 with the percentage score 80% categorized to 'very good' criteria. Then, the material expert assessed learning aspect with the score 3.6 and its percentage score is 72% categorized to 'good' criteria, while the material aspect score is 3.8 with the percentage score 75% and belongs to 'good' criteria. Therefore, the results of validation by the experts showed that the development media of three-dimensional visualization using virtual reality is feasible to use as chemistry learning for the material of reaction rate. However, there are feed-backs that the description the tool names should be clearer and the tool choices that the students will use should have different color.

Table 4. Media Validation Results by Teachers and Peer Reviewer

No	Aspects	Chemistry Teachers	Peer Reviewers	Category
1	Learning Aspect	3.8	3.9	Good
2	Material Aspect	3.8	4.0	Good
3	Audio-Visual Aspect	3.7	4.5	Very Good
4	Software Engineering	4.5	4.3	Very Good

Table 4 showed the media validation conducted by chemical teachers and peer reviewers. According to five chemical teachers who have given the feasibility score for the media, the learning and material aspect has each 3.8 with the percentage score 76 % while the audio-visual aspect score is 3.7 with the percentage score 74% and the software engineering aspect is 4.5 with the percentage score 80%. Ten peer reviewers also assessed the developed media in this study. The validation conducted by peer reviewers showed that the learning aspect score is 3.9 with the percentage score 78% while the material aspect score is 4.0 with the percentage score 80%. Then, the result score for audio-visual aspect is 4.5 with the percentage score 90% and the result score for software engineering is 4.1 with the percentage score 87.50%. Based on the assessment carried out by the chemistry teacher and Peer Reviewer it was stated that this shows that virtual realty is very feasible to use but there are some inputs from teachers and Peer Reviewers that the media is made easier in choosing the desired object.

After the media were assessed by the experts, subject teachers, and peer reviewers, the students also assess the media, namely the aspect of learning and material and the aspect of media appearance and operation. On stages to get input and advice given by students on media readability. There are 23 students involved in using three-dimensional visualization media using virtual reality and assessed that media. The results of the media by students can be seen in table 5.

Table 5. Media Validation Results by Students

No	Aspects	Score	Category
1	Learning and Material Aspect	3.8	Good
2	Display and Operational Aspect	3.9	Good

Based on the results of media readability by students, the learning and material aspect score is 3.8 with the percentage score 76% while display and operational aspect score is 3.9 with the percentage score 78%. Then, the average total percentage score of students' readability of media is 77 %. Based on the feasibility table, this media categorized as good cause of it was developed a series of laboratory tools in the form of software, which is operated by computer and can simulate the activities in the laboratory as if the user were in an actual laboratory, cause of the media of virtual reality has never been used before by students in the learning process, it can increase students 'motivation in the learning process due to the students' interest in virtual reality media (Shim et al. 2003). Virtual strengths are more economical, easy to use for students while learning process both in the classroom and independent learning. Furthermore, it can increase the students' understanding because they can be repeated to practice if they do not understand about the material.And also, student are safe to carry out because of minimum occupational health safety [19]. In other words, the use of three-dimensional visualization media using virtual reality is categorized as a 'good' criterion so that it can be applied in learning to facilitate students in understanding and memorizing material given to them [20].

4. Conclusion

This study focuses on developing the product which is a three-dimensional visualization of media using virtual reality. Media expert, material expert, chemical teachers, peer reviewers and students assessed the media and the total ideal percentage is 78% categorized to 'good' criteria. The results of this study revealed that the media of three-dimensional visualization using virtual reality that is developed can be used as one of learning media for the material of reaction rate at senior high school, and the use of technology in the learning process can influence learning achievement

The media of three-dimensional visualization using virtual reality can assist teacher in doing the practicum without using the tools, materials, and the laboratory itself. It solves teacher's problem about the lack of tools and materials. This technology can increase the understanding on the learning content [21]. The developed technology using virtual reality can be used as the effective learning media to enhance students' involvement in order to motivate them in learning. The media of three-dimensional visualization using virtual reality is also able to help students who are usually passive in the real laboratory and it becomes an opportunity for the chemical learning media in the future.

5. References

[1]. Mendikbud 2013a Perubahan atas Peraturan Pemerintah Nomor 19 Tahun 2005 Tentang Dasar Nasional Pendidikan

[2]. Gates B, Myhrvold N, and Rinearson P1996 *The road ahead, penguin Books* (ISBN 978-0-14-026040-3)

[3]. Mendiknas 2006 Permendiknas Nomor 22 Tahun 2006 *Tentang Standar Isi untuk satuan*

[4]. Ristiyani E and Bahriah E S 2016 Analisis kesulitan belajar kimia siswa di SMAN X Kota Tangerang Selatan *Jurnal Penelitian dan Pembelajaran IPA 2*(1) 18-29 Retrieved from http://jurnal.untirta.ac.id/index.php/JPPI/article/view/431/573.

[5]. Tatli Z and Ayas A 2013 Effect of a virtual chemistry laboratory on students' achievement. *Educational Technology & Society **16***(1) 159-170 Retrieved from https://www.j-ets.net/ETS/journals/16_1/14.pdf.

[6]. Altun and Eralp 2009 Developing an Interactuve virtual chemistry Laboratory Enrich with constructivist Learning Activities for secondary School *Science Direct* (1) 1895-1898

[7]. Jahro I S 2009 Analisis penerapan metode praktikum pada pembelajaran ilmu kimia di sekolah menengah atas *Jurnal Pendidikan Kimia* **1**(1) 20 – 26.

[8]. Tusiyam S N H 2011 *Pemanfaatan laboratorium biologi untuk mencapai standar kompetensi pembelajaran biologi kelas XI IPA semester I di MAN Kendal* (Doctoral dissertation, IAIN Walisongo).

[9]. Manitoba 2013 *Grade 12 Chemistry: A Foundation for Implementation* (Winnipeg: Manitoba Education School Programs Division)

[10].Herga N R, Čagran B and Dinevski D 2016 Virtual laboratory in the role of dynamic visualisation for better understanding of chemistry in primary school *EURASIA J. Math., Sci Tech. **12***(3), 593-608

[11].Kim J H, Park S T, Lee H, Yuk K C, and Lee H 2001 Virtual reality simulations in physics education *IMEJCL 3*(2)

[12].Sihite B, Samopa F, and Sani N A 2013 Pembuatan Aplikasi 3D Viewer Mobile dengan Menggunakan Teknologi Virtual Reality (Studi Kasus: Perobekan Bendera Belanda di Hotel Majapahit) *Jurnal Teknik ITS 2*(2) A397-A400 12

[13].Lu S Y, Shpitalni M, and Gadh R 1999 Virtual and augmented reality technologies for product realization *CIRP Annals **48***(2) 471-495

[14].Bahar Y N, Landrieu J, Pere C, and Nicolle C 2013 CAD Data Workflow toward the Thermal Simulation and Visualization in Virtual Reality *IJIDeM*

[15].Shneiderman B 1998 Designing the User Interface: Strategies for Effective Human-Computer Interaction 3rd Edition (Reading M A: AddisonWesley)

[16].Herga N R, Cagran B, and Dinevski D 2015 Virtual laboratory in the role of dynamic visualisation for better understanding of chemistry in primary school *EURASIA J. Math., Sci Tech.* **12**(3) 593-608.
[17].John N W and Leng J 2001 *Scientific Applications of Visualization, Virtual Reality and High Performance Visualization Computers, Briefing paper* (UKHEC – UK High Performance Computing: Manchester Research Centre for Computational Science) pp.1-9.
[18].Shim K C, Park J S, Kim H S, Kim J H, Park Y C, and Ryu H I 2003 Application of virtual reality technology in biology education *Journal of Biological Education 37*(2) 71-74
[19].Wisudawati AW and Eka S 2015 *Metodologi pembelajaran IPA* (Jakarta: Bumi Aksara)
[20].Azwar S 2015 *Penyusunan skala psikologi* (Yogyakarta: Pustaka Pelajar)
[21].Ambusaidi A, Musawi A A, Al-Balushi S, and Al-Balushi K 2018 The impact of virtual Lab learning experiences on 9th grade students' achievement and their attitudes towards science and learning by virtual Lab *Journal of Turkish Science Education* **15**(2) 13-29
[22].Borg W R and Gall M D 1983 *Educational research an introduction* (New York: Longman Inc.)
[23].Squires D and Preece J 1999 Predicting quality in educational software *Interacting with computers 11*(5) 467-483 **DOI:** 10.1016/S0953-5438(98)00063-0.
[24].Nesbit J C and Leacock T L 2009 Collaborative argumentation in learning resource evaluation. In *Handbook of Research on Learning Design and Learning Objects: Issues, Applications, and Technologies* pp. 574-588
[25].Oyelekan O S and Olorundare A S 2009 Development and validation of a computer instructional package on electrochemistry for secondary schools in Nigeria 5(2) 88-104 *Retrieved from*: http://ijedict.dec.uwi.edu/viewarticle.php?id=677&layout=html.
[26].Kustandi C and Sutjipto B 2011 *Media pembelajaran manual dan digital* (Bogor: Ghalia Indonesia)

Students' Chemical Literacy on Context-Based Learning: A Case of Equilibrium Topic

Hestu Anggrah Eny[1,a)], **Antuni Wiyarsi**[2]
[1]Graduate Program of Chemistry Education, Universitas Negeri Yogyakarta, DIY 55281, Indonesia
[2]Chemistry Education Department, Faculty of Mathematics and Natural Science, Universitas Negeri Yogyakarta, DIY 55281, Indonesia

[a)] Corresponding author e-mail: hestuanggraheny@gmail.com

Abstract. This study was aimed to analyze the students' chemical literacy on context based chemical equilibrium learning. This research used quantitative approach with survey method. A total 92 students of 11th grade from two public high schools in Special Region of Yogyakarta, Indonesia were selected using the purposive sampling technique. The data were collected using open-ended questions namely Chemical Equilibrium Literacy Test (CELT) to analyze the students' chemical literacy ability in the topics of coral reef, hypoxia, teeth enamel, and ammonia production. CELT used 3 indicators of Higher-Order Learning Skills (HOLS) on chemical literacy, namely identifying the information, connecting the information and analyzing information. Data from CELT was analyzed using ideal rating category. The result of this study showed that only one third of the sample that has a good chemical literacy ability on the context-based chemical equilibrium learning. This study suggests that the context-based learning process should be always applied to school in order to improve the students' chemical literacy ability better.

1. Introduction

Scientific literacy is defined by the Programme for International Student Assessment (PISA) as the ability to engage with science-related issues and the ideas of science as a reflective citizen [1]. The PISA result towards the students' scientific literacy that were conducted on 2015 has revealed that the students' scientific literacy in Indonesia were still below the average score, which is 403 out of the average score 493 [1]. Meanwhile, chemistry is a part of science. The low of scientific literacy in Indonesia may indicate that the students' chemical literacy also not good. This was supported by the research conducted by [2] which state that students' chemical literacy is still in medium category. The results of this study were in line with [3] that concluded only some students achieved high levels of chemical literacy. The students aren't be able to relate the chemical learning that they received in the school into daily life activities. For the students, science learning that applied in the school is the separate lessons from life. This caused the students couldn't be able to relate and make a use of the scientific concepts that has been studied to encounter with the problems in their life [4].

Therefore, to overcome that problems, government applied curriculum 2013 as an effort to improve the education quality in Indonesia. The low score of PISA result is one of the rational foundation for curriculum 2013 development. The curriculum 2013 also has a principle that education is not allowed to separate the learner from their environment and the curriculum development should be based on the principle of relevance of the education that suitable to the needs and the environment of the learner [5]. In other words, curriculum gives another opportunity to the students to learn about the problems in their community environment as the curriculum content and an opportunity to apply the lessons they have studied in the classroom into the daily life. This principle is very closely related to chemical literacy.

According to [6], chemical literacy consisted of 4 domains which are the chemical content of knowledge, chemistry in the contexts, Higher Order Learning Skill (HOLS) and affective aspect. The student who understands the chemistry literacy should be able to comprehend the general ideas of science and the characteristic of chemistry. Contextually, the students who are chemically illiterate should be able to acknowledge the importance of chemical knowledge to explain the phenomenon of daily life and use his understanding in daily life. They also should be able to ask a question, look for the information and explain those phenomenon using the scientific evidences also be able to evaluate the pros and cons of an argument. The students who are chemically illiterate has a fair and rational perspective towards chemistry and showed the interest towards the chemical related issues especially in the non-formal environment like pers.

The students who only taught about the concept usually tend to memorize and comprehend the lesson they've learnt. One of the learning approach that can help the students to relate the learning content and the daily life is the context-based learning. Context-based learning has a purpose to cope with the students less interest in learning science at school and to provide the learning environment that could help the students to understand the relevance of scientific content towards individual life, society or occupation in the future [7]. Context-based learning will be more beneficial for the students since the contents on their learning can be directly related and be applied in daily life. Moreover, in the terms of affective, context-based learning can help to improve the students' positive responses on chemistry [1]. The result of [8] research showed that learning using the context-based approach is better compared to traditional learning to improve the students' chemical literacy on the thermochemical and thermodynamics concepts.

One of the chemistry concept that has a lot of context in daily life is the chemical equilibrium. Fundamentally, the chemical equilibrium is important to comprehend the basic concept of chemistry (the solubility equilibrium and acid-base equilibrium). Chemical equilibrium is a concept learned in high school and present at the national examination. However, compared to the other topics, the chemical equilibrium subject is one of the most difficult topics to learn [9]. There are a lot of contexts that can be used on chemical equilibrium concept, for example the coral reef, the oxygen binding in the blood, the teeth enamel damage and others. The results of the study of [10] revealed that context-based chemical equilibrium learning had a positive influence on student achievement, motivation in learning chemistry and a constructive learning environment. The education for the chemical literacy has been a general purpose to understand the chemistry from any range of age [11]. Hence, the objective of the research is to analyze the students' chemical literacy ability on the context-based chemical equilibrium learning.

2. Research Methods

2. 1. Research design and procedure

This research used quantitative approach with survey method. In this research, the survey method used to analyze the students chemical literacy profile on the chemical equilibrium which consists of various topics of chemical equilibrium in daily life. The test was given after the implementation of the context-based chemical equilibrium learning process. Samples in this study were chosen from two public high schools in the Special Region of Yogyakarta, Indonesia using the convenience sampling. The school that was chosen as the samples have same characteristic i.e have a very good school accreditation and on the middle-categorized for the national chemistry test scores. A total 92 students were randomly selected. They are 16- 17 years old.

2.2. Data collection and research instrument

The data used in this study were obtained from the Chemical Equilibrium Literacy Test (CELT). CELT was open-ended questions which consisted of 4 chemical equilibrium topics (coral reef, hypoxia, teeth enamel, and ammonia production). The researcher developed the chemical literacy ability test by synthesizing the aspects of chemical literacy according to [12] and [1]. There are 3 indicators of HOLS in the chemical literacy ability which are identifying the information, connecting the information and analyzing the information that were used in CELT. Other than that, the students were asked to give an argument on the issues for every topic in the affective aspect questions but discuss in other papers. CELT also adjusts chemical equilibrium concept that was present on syllabus of high school chemistry course. The grid of CELT could be seen on the Table 1.

Table 1. The grid of CELT

Chemical in the context	Chemical content knowledge	HOLS	No
Coral reef (environment)	The dynamic equilibrium	Identifying the information	1a
	The reaction equation of chemical equilibrium	Identifying the information	1b
	Homogeneous and heterogeneous equilibrium	Analyzing the information	1c
	The factors that affecting the friction of equilibrium	Analyzing the information	1d
Hypoxia (health)	The reaction equation of chemical equilibrium	Identifying the information	2a
	The equilibrium constant of Kc and Kp	Identifying the information	2b
	The determination of Kc value	Connecting the information	2c
	The connection between Kp and Kc	Analyzing the information	2d
Teeth enamel (health)	The factor that affecting the friction of chemical equilibrium	Connecting the information	3a
		Connecting the information	3b
Ammonia production (industry)	The reaction equation of chemical equilibrium	Identifying the information	4a
	The determination of Kc value	Analyzing the information	4b
	The factor that affecting the friction of chemical equilibrium	Connecting the information	4c
		Connecting the information	4d

CELT has been tested for its validity and reliability. The validation has been done theoretically and empirically. The theoretical validation has been done by asking for the expert judgement namely 2 expert lecturers who are majoring in chemistry. According to the expert judgement, the instrumental was stated as the decent instrumental materially, constructively and by the language.

Furthermore, CELT has been empirically tested to 110 students. Based on the result of the empirical validity analysis, there were 5 items that were not accordance with the Partial Credit Model (PCM) out of the 19 item questions that has been tested. This showed that there were 14 item questions (consists of 4 topics) that can be used on CELT. Cronbach Alpha reliability from CELT was 0,92. It was concluded that the CELT can be used to collect the data.

2.3. Data analysis

The collected data from CELT was analyzed using the descriptive quantitative method with categorization. The score was categorized based on the ideal rating category suggestion by [13] with ideal maximal score is 34 and ideal minimal score is 14. It divided into 5 categories, namely very good, good, quite good, less good, and bad. The score of the ideal rating category can be seen in Table 2.

Table 2. Ideal rating category

No.	Score Range	Category
1	X > 30.50	Very good
2	26.83 < X ≤ 30.50	Good
3	23.17 < X ≤ 26.83	Quite good
4	19.5 < X ≤ 23.17	Less good
5	X ≤ 19.5	Bad

3. Result and Discussion

3. 1. The level of students' chemical literacy

The chemical literacy profile data was obtained from CELT that has been done on the 92 samples. The result of the chemical literacy profile that has been analyzed based on the ideal rating category on this research could be seen on the Figure 1.

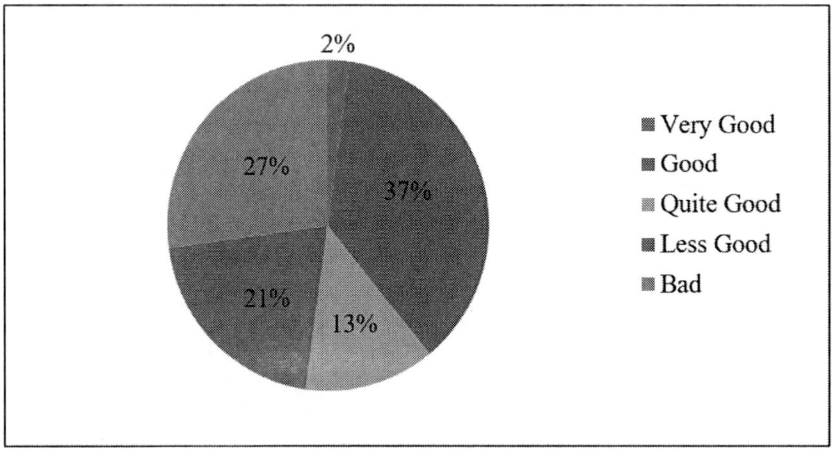

Figure 1. The level of students' chemical literacy.

Based on the Figure 1, it could be seen that the students' chemical literacy ability in good category was highest (37%). However, students' chemical literacy in bad (21%) and less good (27%) also had a high percentage. The students with chemical literacy category were very good only 2%. It can be concluded that there were many students who had low chemical literacy. This is supported by [14] research which stated that students' chemical literacy levels were low, students also lacked high order thinking skills in analytical thinking and reasoning to applied their knowledge for solving everyday life problems related to chemistry. Students' chemical literacy was still low because students were not familiar with context-based learning. Students were less active and less collaborative when context-based chemical equilibrium learning was carried out. However, some of students were interested with context-based chemical equilibrium learning because they could learn chemistry based on real life.

According to the students' chemical literacy profile on the context-based chemical equilibrium learning, there should be a reconstruction that can improve the chemical literacy ability. Learning concept with the daily life problems is very necessary. The chemical learning was introduced to the students with the chemical aspects that relatively restricted and unnecessary theoretical concept that can't be used in their future life [3]. The research that conducted by [8] revealed that learning using the context-based approach is better to improve the students' chemical literacy than traditional learning on the thermochemical concept. This learning allows the teacher to connecting the context that is used and the other relevant contexts in order to stimulate the students' way of thinking. Furthermore, context-based learning can be integrated with multiple representation approach that facilitate students' creative thinking skills [15]. It will give more chance to improve students' chemical literacy.

3. 2. Students' chemical literacy on each topic

The analysis of chemical literacy profile has been done by counting equally each scores topic on CELT which consisted of coral reef, hypoxia, teeth enamel, and the industry of ammonia production which is categorized into the very good, good, quite good, less good, and bad. The analysis result of the students' chemical literacy profile on the chemical equilibrium concept showed that there were still a lot of students on the less good and bad categories.

3.2.1. Topic 1: Coral reef (environment). The students' chemical literacy profile was reviewed based on the category percentage on the topic 1 could be seen on the Figure 2.

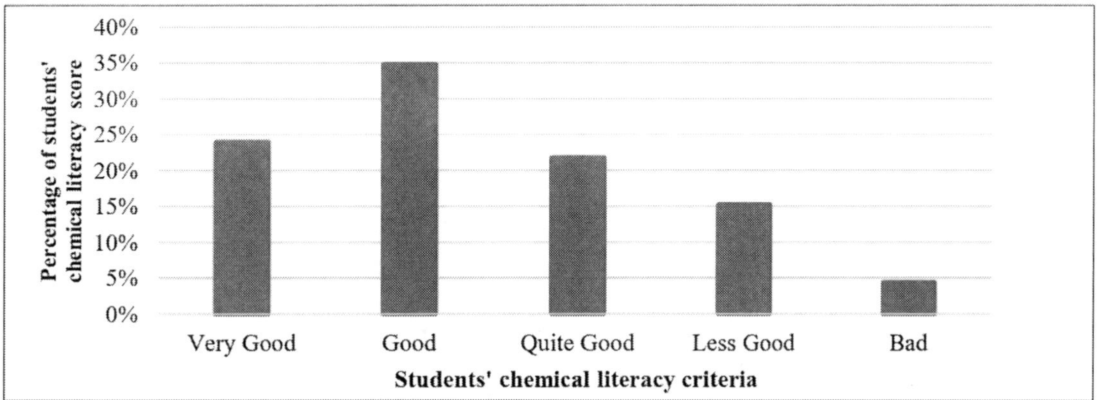

Figure 2. The students' chemical literacy percentage on the topic 1

Based on the Figure 1, it could be concluded that the most dominant students have the good criteria of the chemical literacy (34.78%). Meanwhile, the percentage of the students with very bad category was the very low (4.35%). This showed that mainly the students on the topic 1 already has the good chemical literacy ability.

This topic discussed about the coral reef which is one of the examples of the dynamic equilibrium in daily life. The students were asked to identify the definition of chemical equilibrium based on the coral reef concept that has been explained on the question. Mainly the students were only able to identify the equilibrium dynamic concept generally without related it to the dynamic equilibrium that happened on the coral reef. [16] stated that the equilibrium on the closed system usually is explained rapidly. The teacher should use some relatable examples in order to help the students understand the concept easily. On this topic, the students also were asked to make a chemical equilibrium reaction that happened on the coral reef and also tried to differentiate homogenous and heterogenous equilibrium. On this part, some of the students were already able to distinguish it very well. Before this test, students have been taught with context-based chemical equilibrium learning, so students are accustomed to associating the concept of chemical equilibrium with everyday life. Most students could answer questions well by identifying information and linking information. Based on the test results on topic 1, most students have reached HOLS.

3.2.2. Topic 2: hypoxia (health). The students' chemical literacy profile on the chemical equilibrium context-based learning on the topic 2 (hypoxia) could be seen on the Figure 3.

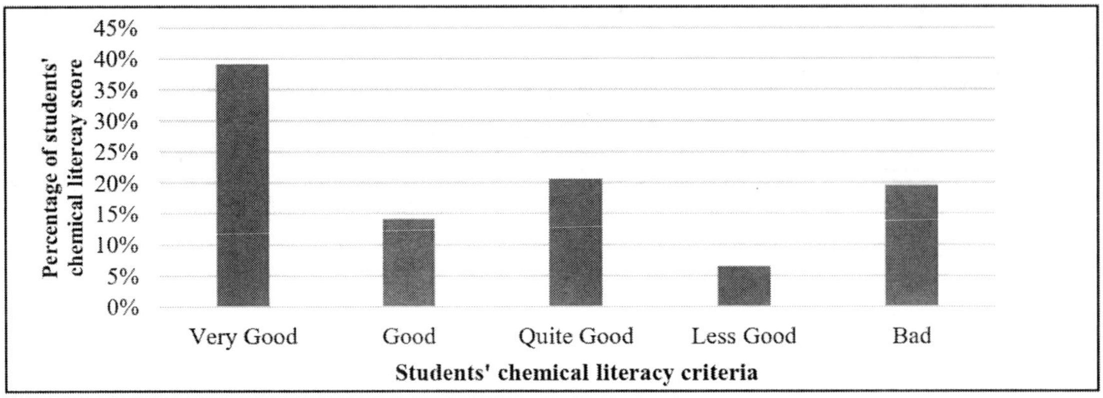

Figure 3. The students' chemical literacy percentage on the topic 2

Based on the Figure 3, it could be seen that the students' chemical literacy percentage on the topic 2 with very good category has the biggest percentage (39.13%) and the smallest percentage (6.52%) on the bad category. This could be concluded that on this topic, some of the students had already has the good chemical literacy ability.

This topic discussed about the hypoxia phenomenon that usually happened when someone climbed the mountain. Almost all the students could wrote down the right chemical equilibrium reaction that was happened between the hemoglobin and the oxygen on blood. However, some students still face the difficulty to determine the value of Kc when the blood pressure was known on the connected part of Kp and Kc. There were still lot of the students that made a wrong analysis of the questions that was why they answered it wrong too. [17] on his research showed that the students still has the limited ability when being asked to count the equilibrium constants. On this topic, the students were also asked to give an argument about the necessary preparation to do in order to avoid experiencing hypoxia. Some of the students were already gave the right arguments even though the others were still confused about the answer. This was because in the some areas around Yogyakarta, there are lot of mountains that could be climbed, so most of the students has done climbing the mountain and had some good experience about hypoxia phenomenon.

3.2.3. Topic 3: teeth enamel (health). The students' chemical literacy profile on the context-based chemical equilibrium learning on the topic 3 (teeth enamel) could be seen on the Figure 4.

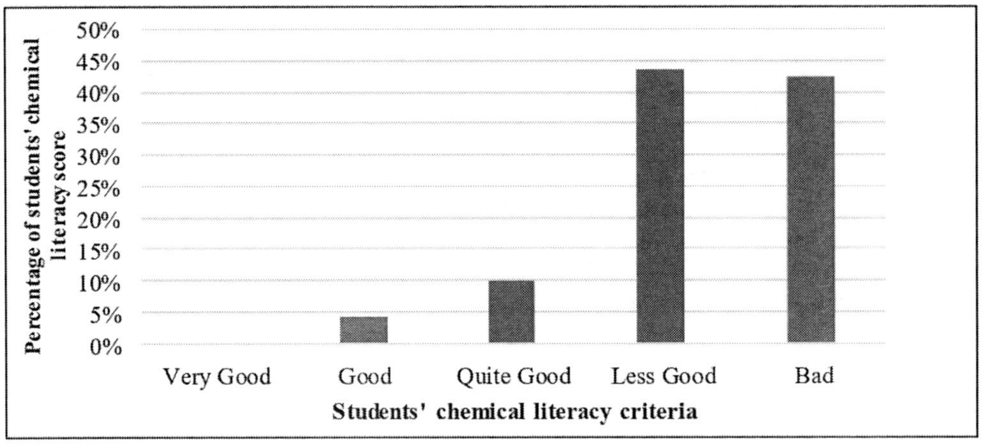

Figure 4. The students' chemical literacy percentage on the topic 3.

IOP Publishing
1397 (2019) 012035 doi:10.1088/1742-6596/1397/1/012035

Based on the Figure 4, it could be seen that the students' chemical literacy percentage on the less good category (43.48%) and bad category (42.39%) were the biggest. The students on the good chemical literacy category was very small (4.35%). In short, the students' chemical literacy on the topic 3 was still low.

This topic discussed about the chemical equilibrium on the teeth enamel. The students were asked to determine the friction of the chemical equilibrium that happened on the teeth enamel based on some familiar factors, like the used of toothpaste that contains calcium and consuming the soft drinks. Most of the students were able to answer the friction question correctly, but some of the students were still confused about how could the friction happened. [18] on his research revealed that on the concentration change factor which could cause the friction in chemical equilibrium obtained the smallest average score. This was because of some students were stuck with the question and it made them solving the question inaccurately. This also supported by the result of [19] research that stated, the most common misconception which is about the interrupted equilibrium system by an increase in concentration has been found.

3.2.4. Topic 4: the industry of ammonia production (industry). The students' chemical literacy profile on the context-based chemical equilibrium learning on topic 4 could be seen on the Figure 5.

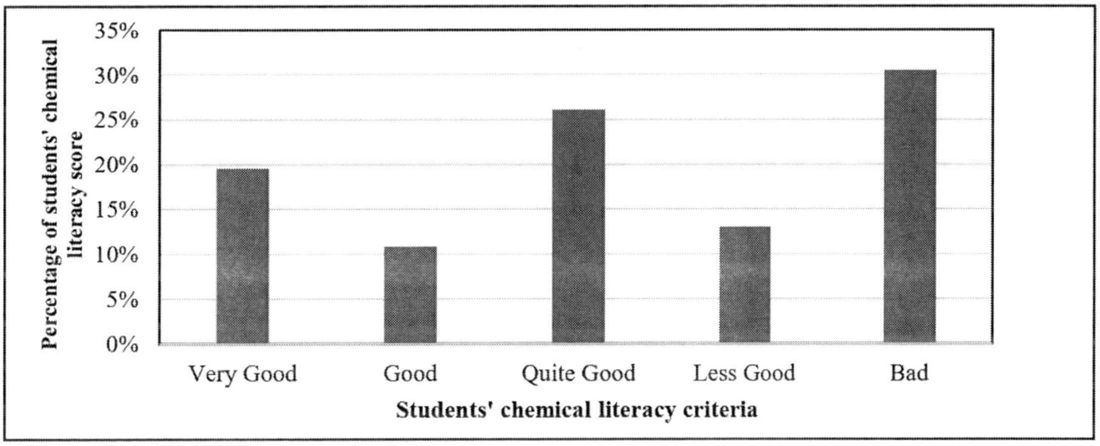

Figure 5. The students' chemical literacy percentage on topic 4.

Based on the Figure 5, it could be concluded that the most dominant of the students' chemical literacy on topic 4 is on the quite good category (26.09%) and bad category (30.43%). This showed that the students' chemical literacy on this topic are low.

Topic 4 discussed about the industry of ammonia production where the students were asked to determine the value of Kc on the ammonia equilibrium reaction and to predict the most on advantage ammonia production based on the factors that affecting the friction of the chemical equilibrium. However, most of the students were still having difficulty on answering the questions. [16] based on the result of his research revealed that not all the students were succeed on making a prediction on the Le Chatelier principle or the equilibrium law. Therefore, the teacher should be more focus on explaining to the students about the changing concentration and temperature against the friction of the equilibrium because those two factors are the main factors. [20] also revealed that the student difficulty on comprehending the chemistry concept related to the learning method that used by the teacher. The topic of ammonia industry was a difficult topic for students because this concept is not close to their lives.

4. Conclusion

Based on the obtained data on analysis of students' chemical literacy in context-based chemical equilibrium learning showed that only one third of students have good categories (2% very good and 37% good). That means there were many students (61%) who had chemical literacy skills in the low category. Context-based learning must often be applied in schools in order to improve students' chemical literacy skills.

References

[1] OECD. 2016. *PISA 2015 Assessment and Analytical Framework: Science, Reading, Mathematic and Financial Literacy*. (Paris: OECD Publishing).

[2] Imansari, M., Sudarmin, & Sumarni, W. 2018. Analysis of students' chemical literacy through guided inquiry learning with ethnoscience. *Jurnal Inovasi Pendidikan Kimia, 12*(2), 2201-2211. Retrieved from https://journal.unnes.ac.id/nju/index.php/JIPK/article/view/15480

[3] Shwartz, Y., Ben-Zvi, R., & Hofstein, A. 2006. The use of scientific taxonomy for assessing the development of chemical literacy among high-school students. *Chemistry Education Research and Practice, 7* **4** 203-225

[4] Holbrook, J., & Rannikmae. 2007. The nature of science education for enhancing scientific literacy. *Int. J. Sci. Educ., 29* **11** 1347-1362

[5] Kemendikbud. 2012. *2013 Curriculum Documents*. Retrieved from http://yogyakarta.kemenag.go.id/file/file/dikmad/ldue1388737894.pdf

[6] Shwartz, Y., Ben-Zvi, R., & Hofstein, A. 2006. Chemical literacy: what it means to scientists and school teachers?. *J. Chem. Educ., 83*(10), 1557-1561. doi: 10.1021/ed083p1557

[7] Stuckey, M., Hofstein, A., Mamlok-Naaman, R., & Eilks, I. (2013). The meaning of 'relevance' in science education and its implications for the science curriculum. *Studies in Science Education,* 49 **1** 1-34.

[8] Cigdemoglu, C., & Geban, O. 2015 Improving students' chemical literacy level on thermochemical and thermodynamics concepts through context-based approach. *Chem. Educ. Res. Pract, 16* **2** 302-317.

[9] Quilez-Pardo, J., & Solaz-Portoles, J., 1995 Students' and teachers' misapplications of le chatelier's principle: implications for the teaching of chemical equilibrium. *J Res Sci* Teach *32* **9**, 939-957.

[10] Ilhan, N., Yildirim, A., & Yilmaz, S. S. 2016. The effect of context-based chemical equilibrium on grade 11 students' learning, motivation and construtivistic learning environment. *International Journal of Environmental & Science Education, 11* **9** 3117-3137

[11] Lin, S. 2009. Chemical literacy and learning sources of non-science major undergraduates on understanding of environmental issues. *Chemical Educational Journal, 13* **1** Retrieved from http://chem.sci.utsunomiya-u.ac.jp/cejrnlE.html

[12] Shwartz, Y., Ben-Zvi, R., & Hofstein, A. 2005. The importance of involving high-school chemistry teachers in the process of defining the operational meaning of 'chemical literacy'. *International Journal of Science Education, 27* **3** 323-344.

[13] Stiggins, R. J. 1994. *Student-Centered Classroom Assessment*. (Boston: Merrill Publishing Company).

[14] Thummathong, R., & Thatong, K. 2018. Chemical literacy levels of engineering students in northeaster Thailand. *Kasetsart Journal of Social Science, 3* **3** 478-487.

[15] Wiyarsi, A., Sutrisno, H., & Rohaeti, E. 2018. The effect of multiple representation approach on students' creative thinking skills: a case of 'rate of reaction' topic. *J of Phys: Conf. Ser., 1097,* 012054.

[16] Tyson, L., Treagust, D. F., & Bucat, R. B. 1999. The complexity of teaching and learning chemical equilibrium. *J. Chem. Educ., 76* **4** 554-558.

[17] Karpudewan, M., Treagust, D. F., Mocerino, M., Won, M., & Chandrasegaran, A., L. 2015. Investigating high school students' understanding of chemical equilibrium concepts. *International Journal of Environment & Science Education, 10* **6** 845-863.

[18] Rafiuddin. 2016. Application of hypothesis deductive cycle learning model in the matter of chemical equilibrium to improve critical thinking skills student high school. *International Journal of Education and Research, 4* **6** 249-262. Retrieved from https://www.ijern.com/June-2016.php

[19] Canpolat, N., Pinarbasi, T., Bayrakceken, S., & Geban, O. (2006). The conceptual change approach to teaching chemical equilibrium. *Res Sci Technol Educ 24* **2** 217-235

[20] Juliao, M. S. S., Rodrigues, S. H. B. G., Andrade, L. B. S., & Melo, L. C. (2018). Teaching chemical equilibria: a contextualized scientific method and forensic chemistry class. J. Lab. Chem. Educ., *6* **5** 148-155.

Journal of Physics: Conference Series **1397** (2019) 012036 doi:10.1088/1742-6596/1397/1/012036

Students' chemical literacy: A study in chemical bonding

Dessy Lusyana Yustin[1, a)] and Antuni Wiyarsi[2]
[1]Graduate Program of Chemistry Education, Yogyakarta State University
[2]Chemistry Education Deparment, Faculty of Mathematics and Natural Sciences, Yogyakarta State University

Corresponding author e-mail: [a)]dessylusyana@gmail.com

Abstract. Scientific literacy is the ability to understand science, communicate science, and apply science skills to solve a problem. Scientific literacy skills, especially chemical literacy are things that must be possessed by students in the 21st century. This study aims to analysis students' chemical literacy abilities in chemical bonding material. The research use quantitative approach by survey method. A total of 199 first year students from four Public High Schools were randomly selected in the school category, they have very good accreditation in the Special Region of Yogyakarta, Indonesia. Data were collected by test techniques with Chemical Bonding Literacy Test (CBLT). The instrument consist of chemical literacy ability open-ended questions with three aspects namely the context, chemical knowledge content, and Higher-Order Learning Skills (HOLS) in the form of 21 questions about chemical bond material. The analysis of students' chemical literacy ability is done by categorizing the scores obtained from the test into the ideal ranking category. The results of this study indicate that students' chemical literacy ability are still very low as evidenced by the results of the analysis students entered into the bad category. This study suggest that students need to be extra in develop their chemical literacy skills, especially in the aspects of connecting and analyzing scientific information.

1. Introduction

The 21st century was marked by changes in the rapid development of science and technology in people's lives. Therefore education are faced with increasing challenges by delivering human resources who have the ability and skills to face the challenges of life in this century. Education today is directed to prepare students to be successful life in the 21st century by encouraging learners to have a base of knowledge and a deep understanding to become life-long learners [1]. Because education is a crucial aspect of advance or retreat of a life [2].

Students need to be equipped with skills to deal with the challenges of the 21st century in order to be able to compete in the era of globalization. The rapid development of the globalization era has made education required to be able to shape scientific literacy that has the ability to think scientifically to solve individual problems and community issues so that it can play a role as a qualified human resource as shown by the scientific literacy. Previous research [3] stated that chemistry education provide a great potential and strategic role in preparing qualified human resources to face the era of industrialization and globalization. Therefore, it is important to incorporate 21st century skills in science education.

One of the skills needed in the 21st century is scientific literacy [4]. Scientific literacy is defined from the Program for International Student Assessment (PISA) as the ablility to apply scientific knowledge, identify issues, describe scientific phenomena, draw conclusions based on evidence, and the willingness to reflect on and engage with scientific ideas and subjects [5]. Scientific literacy embodies scientific ideas, concepts and concepts in practice in and in many disciplines [6]. Chemistry, which is one of the branches of science, also has educational goals that are in line with science, that is, people who are chemically illiterate can later form a society of literary science.

Indonesia is being reviewed from the Indonesian Education quality and access mapping results in 2013-2014. Indonesia is 40 from 40 countries on the mapping of the learning curve-person [7]. In a report by OECD for scientific literacy in 2015, Indonesia was 62 from 70 countries with a score of 403 [8]. The results of this survey of scientific literacy abilities led Indonesia to be categorized as a country with low scientific literacy skills. Obtaining a lower score implies that the Indonesian students have limited knowledge of science and chemistry learning achievement also reflects the low level anyway. The low level of this chemistry literacy can be influenced by several factors including the learning process, teacher competence, curriculum used, and learning resources. Lack of students' scientific literacy skills is a reason that undertook the Indonesian Government to perform the curriculum's revision of the 2006 to 2013. Gilbert, Bulte, and Pilot [9] categorizes problems in the chemistry curriculum in five different groups; i.e. overload, separate facts, lack of transfer, lack of relevance, and a lack of emphasis in learning.

Chemical literacy describe [6] includes four components: first is the chemical content of knowledge that explain how a chemically-literate student should understand a common chemical idea, as well as using knowledge to explain a phenomenon and understand the characteristics (key ideas) of chemistry in explaining the process, the reaction in chemistry. The second component is chemistry in context stating that a chemically-literate student should be able to use chemical knowledge to explain chemistry in explain the chemistry in everyday life in providing social arguments regarding problems and innovations in chemistry. The third component is about Higher-Order Learning Skills (HOLS) that referring to the questions and investigate any relevant information pertaining to the chemical/chemistry. The last component is an affective aspects which explain that a chemically-literate person must have a chemical perspective and can apply it and must demonstrate an interest in chemical problems in everday life.

Chemistry is a subject that most students consider difficult. For students the concept of chemistry is difficult to learn and for them chemistry is just a field that is less popular in the future to learn [10]. In addition, they assume many calculations in chemistry and eventually they have learning difficulties [11]. For example in chemical bonding material, Students can not understand the concept of the types of chemical bonds so that difficulties in distinguishing between these chemical bonds. Even though chemical bonding material is an important sub-material for high school students to study. Previous research [12] also confirmed that the chemical bonding material is more abstract, it is usually more difficult to understand, the linkage with the understanding of the previous material relatively much like the structure of atoms, the properties of the element periodicity makes students less motivated to learn.

Learning chemistry includes the discussion of the material in submakroskpik, submicroscopic and symbolic, sometimes referred to the three tiers of representative chemical [13] Previous research results [10] stated that the methods used in learning are lacking in connecting chemical concepts with real life, so that they are not sufficient to improve students' chemical literacy. Chemical literacy is very important, this is related to how students are able to appreciate nature by utilizing the science and technology they have mastered [14]. Students who have chemical literacy must understand the basic concepts of science/chemistry [6]. Therefore, chemical literacy of students is very important to analyze in order to improve student performance in the process of learning chemistry.

2. Methods

2.1 Research design and samples

This research is a quantitative approach by survey method. Participants in this study came from four Public High Schools in the Special Region of Yogyakarta, Indonesia. This study used a convienence sampling. A total of 199 10th grade high school students with an average age of 16 years consisting of 99 female and 100 male students were participants in this study.

2.2 Techniques and data collection instruments

Chemistry literacy data are collected by test techniques. The instrument used in this study is called a Chemical Bonding Literacy Test (CBLT) consist of chemical literacy ability open-ended questions. The CBLT was developed based on the definition of chemical literacy [6] and [15] in measuring chemical literacy. Chemical bonding materials in the chemical curriculum in Indonesia include Lewis structure, ionic bonds, covalent bonds, coordination covalent bonds, compound polarity, metal bonds, intermolecular forces and physical properties of compounds. The question sheet consists of 21 description questions which are used as post-learning tests which are broken down into 4 topics which are adjusted to the sub-material of chemical bonds namely ionic bonds, covalent bonds, compound polarity and metal bonds. Each sub-subject of chemical bonding includes 3 aspects of chemical literacy ability, namely the chemical aspects in the context, aspects of chemical knowledge content, aspects of Higher-Order Learning Skills (HOLS). HOLS consists of identifying the information needed, linking scientific information, analyzing scientific information, and the ability to argue. The following is a summary of the grids of chemical literacy instruments can be seen in Table 1.

Table 1. Chemical literacy instruments grids.

Topic 1. Ion bonds: chemistry in kitchen salt (NaCl)		
Contexts	Chemistry Sciences Content	HOLS
The dangers of excessive NaCl kitchen salt consumption.	Understand and disclose the process of formation of an ionic bond.	Connecting scientific information.
	Understand and disclose ion bonds using their structures and processes.	Analyzing scientific information.
	Understand and disclose the properties of ion compounds using their structures and processes.	Analyzing scientific information.
	Providing chemical/chemistry knowledge to explain phenomena in the health field.	Identifying required information.
		Argue.
Topic 2. Covalent bonds: chemistry in bleach (NaClO) and ammonia (NH3)		
Contexts	Chemistry Sciences Content	HOLS
The danger of mixing two cleaning materials to clean obstinate stains.	Understand and disclose covalent bonding formation processes.	Connecting scientific information.
	Understand and disclose the process of forming mixed chemical bonds (ionic bonds and covalent bonds).	Connecting scientific information.
	Understand and disclose covalent bonds using the structure and process.	Analyzing scientific information.
	Understand and disclose mixed chemical bonds (ion bonds and covalent bonds) using the structure and process.	Analyzing scientific information.
	Understand and disclose PEB and PEI using the structure and process.	Identifying required information.
	Understand and disclose chemical bonds using structures and processes.	Analyzing scientific information.
	Providing chemical/chemistry knowledge to explain phenomena in the health field.	Argue.
Topic 3. Compound polarity: chemical bonding in the used cooking oil		
Contexts	Chemistry Sciences Content	HOLS
Used cooking oil is B3 waste that has the potential	Providing chemical/chemistry knowledge to explain phenomena in the enviroment field.	Identifying required information.
	Understand and explain the polence of compounds that are	Analyzing scientific

to damage the environment.	reviewed from its properties.	information.
	Understand and explain polar and non-polar compounds using their structures and processes.	Analyzing scientific information.
	Providing chemical/chemistry knowledge to explain phenomena in the enviroment field.	Argue.
Topic 4. Metal bonds: chemical bonding in aluminum foil		
Contexts	Chemistry Sciences Content	HOLS
The danger of consuming foods that are heated with aluminum foil.	Understand and explain the bonding of metals using the structures and processes.	Connecting scientific information.
	Understand and explain the properties of the metal bonding using the structure and process.	Analyzing scientific information.
	Providing chemical/chemistry knowledge to explain phenomena in the health field.	Identifying required information.
		Argue
		Identifying required information.

Analysis instrument chemistry literacy tests conducted by theoretical and empirical validity and reliability. Theoretical validity is done by asking for improvement and expert advice in terms of material, construction, and language (item readability) in the form of an assessment of the feasibility of the items of the instrument to two experts from the Department of Chemical Education. The assessment by experts is qualitative. After getting feedback from the validator such as suggestions and improvements for the instruments used in this study, revisions were made and then validation continues with empirical validity.

Empirical validity was conducted by administrate the CBLT to other students who are not used as research samples. Students used to test chemical literacy test instruments empirically have certain criteria. The criteria in this case include having an average age of 16 years and have studied the subject of chemical bonding. A total of 134 students obtained for testing chemical literacy test instruments that have been developed. Based on the results of empirical validity analysis there is 1 item from 22 items the CBLT that is not in accordance with the Partial Credit Model (PCM). This means that as many as 21 items of CBLT are in accordance with the PCM model. These 21 items were used as instruments to measure students' chemical literacy abilities. In addition, the results of the analysis showed an estimate of Cronbach's Alpha reliability of 0.78 which was included in the good category. It means that the CBLT can be used to collect the data.

2.3 Data analysis technique

The data analysis technique used to determine the profile of chemical literacy abilities is based on the scores of chemical literacy tests on each topic in a quantitative descriptive analysis that has steps namely: (1) recap the score of each topic from the instrument data used, which is the test chemical literacy ability; (2) make the ideal score range and assessment category; (3) categorizing the scores obtained according to the ideal rating category.

Chemical literacy profile categories are divided into 5 categories, namely very good, good, sufficient, less good, bad. Determination of ideal assessment categories refers to the suggestion by [16] with the highest ideal score of 42 and the lowest ideal score of 21. Ideal rating categories as afollows.

Table 2. Ideal rating categories.

Category	Score Range
Very Good	$36,75 < X$
Good	$35 < X \leq 36,75$
Sufficient	$29,75 < X \leq 35$
Less Good	$26,25 < X \leq 29,75$
Bad	$X \leq 26,25$

3. Results and discussion

The results of chemical analysis of chemical literacy ability on chemical bonding materials as follows.

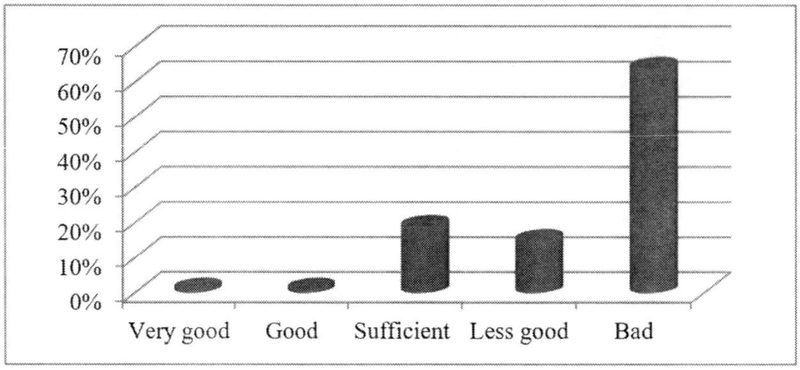

Figure 1. Profile of Chemical Literacy Ability on Chemical Bonding Materials

Figure 1. shows that the largest number results of students' chemical literacy abilities in chemical bonding material are included in the bad category. This portray that the school learning has not yet fully trained the students' chemical literacy ability. Students are not familiar with chemistry learning associated with the context of the life around them, so the concepts of chemical bonding materials that can not be mastered by the student and the student having difficulty in resolving the problem in these contexts. Students' misconceptions about these concepts are based on the fact that they are only able to understand lecture at the macroscopic level and the difficulty follows a shift between the macroscopic and sub-macroscopic levels [17]. Most students also have a lot of mistakes in identifying what is require from the problem. If students are not accustomed to complete the test, the students' scientific literacy skills will tend to lower [18].

Most students cannot connect and analyze scientific information from reading texts from the question of literacy. Especially in covalent and metal bonding materials, many students are trapped so that they cannot answer what is asked of the problem. This is because the methods used in learning are not suitable to make students accustomed to working on literacy-based questions. And also educators must have good abilities in teaching students so students have chemical literacy ability. Science literacy provides aspirations for curriculum development, teaching materials and assessment practices, so that if the material and learning science are facilitated with the aforementioned competencies, scientific literacy of students will develop [19] [20].

3.1 Topic 1. Ion bonds: chemistry in kitchen salt (NaCl)

Results of chemical analysis of literacy skills in ionic bonding material on the topic of chemical bonds in kitchen salt (NaCl) which is included in the health field is presented in Figure 2.

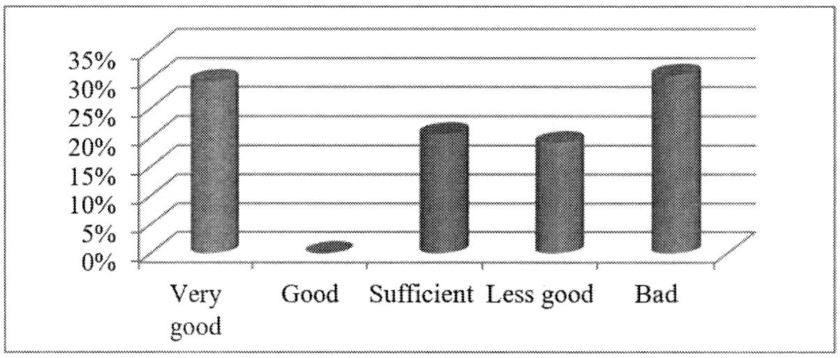

Figure 2. Profile of Chemical Literacy Ability on Topic 1.

Topic 1 contains the context of the dangers of excessive consumption of NaCl salt. The results of the analysis show that the categories of students' chemical literacy abilities with the highest percentage are in the very good and bad categories. On this topic students can connect scientific information in determining the type of bonding in a salt compound. Students can identify information on chemical knowledge in explaining the hazardous context of excessive consumption of NaCl salt and can provide arguments to the phenomenon in the health sector. This portray that students are able to connect the context in the problem with the chemical concepts they have. A learning that uses students 'daily experiences combined with chemical content will produce chemical concepts that will be left behind in students' thinking, so students can use that knowledge to solve problems and students have new knowledge and experience [21]. The relevance of science in everyday events can improve students' ability to transfer knowledge and make students understand why they must learn science so learning becomes meaningful [9].

However some students are weak in explaining the process of forming ion bonds in NaCl compounds. The students' mistakes in describing the ionic bonds in reaction and Lewis structure are, among other things, are incorrectly determining the number of electrons released and received by NaCl constituent elements, writing symbols in ion bond formation reactions, and depicting the number of electron electrons. Students have a problem in understanding the ionic bonds that occur between the atoms of Na and Cl, in terms of the images of students showing that chemical bonds are formed between the atoms of Na and Cl through the division of electrons instead of transferring electrons from one atom to another [22].

3.2 Topic 2. Covalent bonds: chemistry in bleach (NaClO) and ammonia (NH3)

Topic 2 contains about the dangers of mixing the two cleaning agents to clean obstinate stains. The results of the analysis of the ability of chemical literacy on the material of covalent bonds on topic 2 are presented in Figure 3.

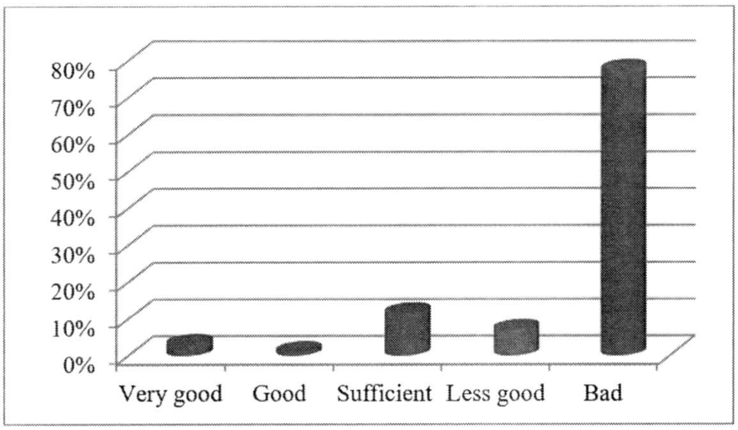

Figure 3. Profile of Chemical Literacy Ability on Topic 2.

The analysis illustrated that the covalent bonding material is the most difficult topics when compared to three other chemical bonding material, as it contains issues that most of the students are not able to finish. More than half of all students fall into the bad category because many students cannot connect and analyze scientific information in understanding and disclosing the process of forming mixed chemical bonds (ionic bonds and covalent bonds) using their structures and processes. The results of a similar study [23] reveals there are many misconceptions concept of covalent bonds even though the students were able to recite the definition of the covalent bond.

This has an effect in determining the number of Free Electron Pairs (PEB) and Bond Electrons (PEI) in compounds containing ionic and covalent bonds. In the terms of the students' answers, most of them are misconceptions about the mixed chemical bonds. Misconceptions occur in determining and disclosing the process of the formation of NaClO compounds. Students are correct when answered that there is an ionic bond between Na^+ and ClO^-, but many students have misconceptions about the bonds that occur in ClO^-. The answers of some students indicate that the bond that occurs in ClO^- is a coordination covalent bond, whereas the answer should be a single covalent bond. Chemical bonds are basically taught by emphasizing the dichotomy of molecules and compounds that will be classified as ionic or covalent [24]. If the students have a different concept of the concept of science is what is meant by misconceptions [25]. Overall on this topic students are not quite mastered the concept of chemical bond materials. It is characterized by students can not use the concepts he already has in solving the existing problems in the question, contrary to the statement [26] about the possibility that occurs when the student has mastered a concept that has the characteristics of students can use these concepts to problem solving and mastery of concepts make it easier for students to learn other concepts .

3.3 Topic 3. Compound polarity: chemical bonding in the used cooking oil

Figure 4 shows the results of chemical analysis of literacy skills in the context of the B3 waste cooking oil that could potentially damage the environment.

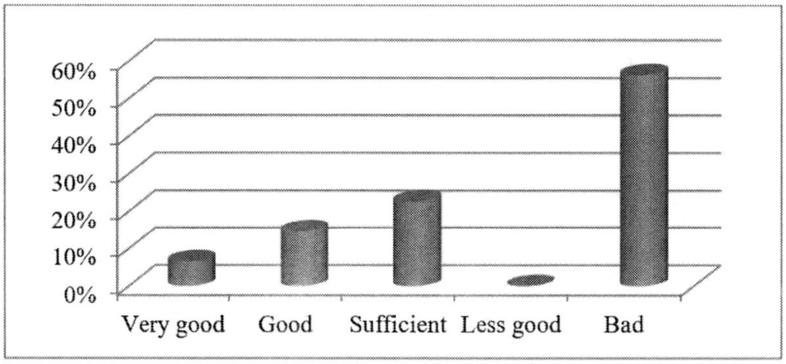

Figure 4. Profile of Chemical Literacy Ability on Topic 3.

Topic 3 contains the context of the B3 waste cooking oil that could potentially damage the environment. The results of the analysis show that the average student falls into the bad category. Students are weak in analyzing scientific information about understanding and disclosing the compound polarity using their structures and processes. Some students have a limited understanding of the concepts of polar covalent bonds and polarity, specifically students' understanding of the concept of electronegativity. This is in line with the research conducted by [27] which shows that many students have difficulty making meaningful connections between the concepts of electronegativity and polar covalent bonds.

Concept of students' understanding of polar covalent bonds and related concepts such as polarity and other electronegativitare y difficult to understand because they are blurry [28]. But some other students can distinguish whether the compound is polar or nonpolar based on the readings in the question. Students have an understanding that the oil cannot unite with water so that used cooking oil can become B3 waste (Hazardous and Toxic Material) which can potentially damage the environment. This is because in the process of learning educators has presented real-world situations into learning about the properties of the polarity of a compound. Learning that connects the concept of chemistry with problems in everyday life through learning approaches such as contextual approaches will produce meaningful chemistry learning [21]. Corresponds [29] statement that in developing the chemical literacy ability chemistry can be done by implementing learning-oriented material concepts to intra- problem students in their daily lives.

3.4 Topic 4. Chemical bonding in aluminum foil (metal bonds)

Topic 4 contains the context of the dangers of consuming food that is heated with aluminum foil. The results of the analysis of topic 4 chemical literacy ability are presented in Figure 5.

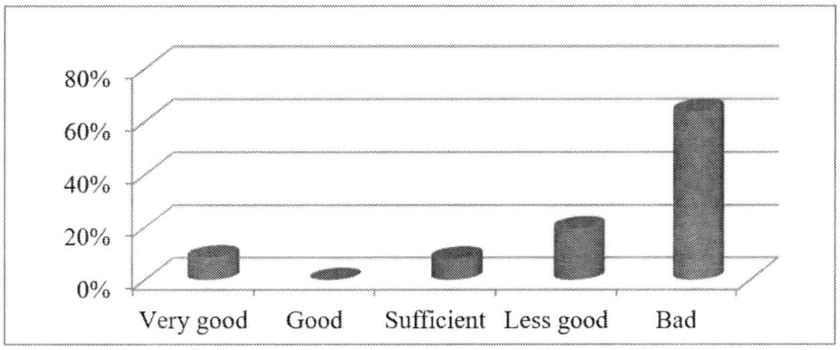

Figure 5. Profile of Chemical Literacy Ability on Topic 4.

The results of the analysis portray the results that are not much different from the previous 3 topics that the average student falls into the bad category. From the results of student answer analysis, only a few students can disclose the process of forming metal bonds appropriately. Some students can only express that metal bonds occur between metal elements, but cannot explain the process of forming metal bonds. This causes the students can not work on the next question on the nature of the metallic bond, which is the difference properties of metal compounds that are not easily broken compared to ionic compounds which are fragile. But other responses analysis results found that students could give other examples of metals that can be found in their daily lives. This proves that students' knowledge of metal ties in everyday life can provide meaningful learning in everyday life for students. In PISA 2015 it was described that scientific literacy skills of students would be more visible if they applied scientific knowledge in daily life.

This attainment of students' low chemical literacy ability proves that students are not familiar with chemical literacy-based questions where students are required to be too high in solving problems in the problem in questions so that students have difficulty in transferring the use of various chemical bond material concepts. One reason for the difficulties students experience in understanding the nature of matter is related to the various levels of representation used in chemistry learning to describe and explain chemical phenomena [19] [30]. Therefore teachers need more references in a learning to improve students' chemical literacy abilities. Chemical literacy can be developed by applying learning approaches and models. One of the learning models that can develop chemical literacy is a concept attainment model with a multiple representation approach in which this model can facilitate students' creative thinking which is one of the bases for developing chemical literacy [31].

4. Conclusion

Based on the results of this study, it can be concluded that the profile of chemical literacy abilities of 10th grade high school students on chemical bonding materials is still definetely low which falls into the bad category at 64%, indicated by many students are weak on the aspect of connecting and analyzing scientific information.

References

[1] NCREL & Metiri Group. 2003. *enGauge 21st century skills: digital literacy for digital age.* (Los Angeles : Napierville IL).

[2] Bahriah, E. S. (2013). Increasing Student's Literacy in the Aspects of the Science Process through Interactive Multimedia Based Learning. *Proc of the National Seminar on Science Education - Professional Development of Science Teachers through Research and Technology Work in Accordance with Curriculum Requirements.* Available in

http://evisapinatulbahriah.wordpres.
[3] Hernani, M., Mudzakir, A., & Aisyah, A. 2009. Teaching Science-Chemistry Concepts from Social Perspectives to Improve Junior High School Student Literacy Literacy. *Journal of Mathematics and Natural Sciences Teaching, 13* **1** 71-93.
[4] Liu, X. 2009. Special Issue on Scientific Literacy. *Int J Env Sci Ed, 4* **3** 301-311.
[5] OECD. 2010. *PISA 2009 Results: Executive summary*. New York: Columbia University.
[6] Shwartz, Y., Ben-Zvi, R., & Holstein, A. 2006. The Use of Scientific Literacy Taxonomy for Assessing the Development of Chemical Literacy among High-School Students. *Chem. Educ. Res. Pract, 7* **4** 203-225.
[7] Baswedan, A.R. 2014. Emergency Department of Education in Indonesia. *Hospitality Gathering with the Head of Service*. (Jakarta: Kemendikbud).
[8] OECD. 2016. *PISA 2015 Assessment and Analytical Framework: Science, Reading, Mathematic and Financial Literacy*. (Paris: OECD Publishing).
[9] Gilbert, J. K., Bulte, A. M. W., & Pilot, A. 2011. Concept Development and Transfer in Context-Based Science Education. *Int. J. Sci. Educ. 33* **6** 817–837.
[10] Osborne, J. F., & Dillon, J. 2016. Teacher Professional Development in Outdoor and Open Learning Environments: A Research Based Model. *Creative Education*, *7* **10** 1392-1403.
[11] Febrianto, Wiyarsi, A., Partana, C. F., & Sulistyo, B. 2019. Chemistry in context: The development of hydrocarbon chemistry and petroleum module based on vehicle case *j. phys. conf. ser. 1156*, 012021.
[12] Nurbaity, & Mustikasari, I. 2012. Mastery Analysis of Chemical Bonding Concepts in Organic Chemistry Subjects through Two Tier Instruments. *Journal of Chemistry Education Research, 2* **1**, 99-106.
[13] Johnson, N. 2014. Teacher's And Student's Perceptions Of Problem Solving Difficulties In Physics. *International Multidisciplinary e-Journal, 1* **5** 97-101.
[14] Nisa, A., Sudarmin, & Samini. 2015. Effectiveness of the Use of Integrated Ethnoscience Modules in Problem-Based Learning to Improve Student's Literacy in Science. *Unnes Science Education Journal, 4* **3** 1049-1056.
[15] Cigdemoglu, C., Arslan, H. O., & Cam, A. 2017. Argumentation to foster pre-service science teachers' knowledge, competency, and attitude on the domains of chemical literacy of acids and bases. *Chem. Educ. Res. Pract, 18* **2** 288–303.
[16] Stiggins, R. J. 1994. *Student-Centered Classroom Assessment*. (Boston: Merrill Publishing Company).
[17] Harrison, A. G., & Treagust, D. F. 2000. Learning about atoms, molecules, and chemical bonds: A case study of multiple-model use in grade 11 chemistry. *Science Education, 84* **3** 352–381.
[18] Odja, A. H., & Payu, C. S. 2014. Analysis of Students' Early Literacy Ability Capabilities on Science Concepts. *Proceedings of the National Chemistry Seminar,* Chemistry FMIPA Department, Surabaya State University.
[19] Shwartz, Y., Ben-Zvi, R., & Hofstein, A. 2005. The importance of involving high-school chemistry teachers in the process of defining the operational meaning of "chemical literacy". *International Journal of Science Education, 27* **3** 323–344.
[20] Roberts, D. 2007. *Scientific literacy/science literacy: threats and opportunities*. in Abell S. K. and Lederman N. G. (ed.), Handbook of research on science education, Mahwah (New Jersey: Lawrence Erlbaum Associates) pp. 729–780.
[21] Wiyasi, A., Pratomo, H., Priyambodo, E., Marfuatun & Kusumaningtyas, H. 2019. Chemistry enrichment in tourism vocational school: The development and validation of food additives module. *J. of Physics: Conf. Ser. 1156*, 012015.
[22] Karacop, A., & Doymus, K. 2012. Effects of Jigsaw Cooperative Learning and Animation Techniques on Students' Understanding of Chemical Bonding and Their Conceptions of the Particulate Nature of Matter. *J Sci Educ Tech 22* **2** 186–203.

[23] Luxford, C. J., & Bretz, S. L. 2013. Moving beyond definitions: what student-generated models reveal about their understanding of covalent bonding and ionic bonding. *Chem. Educ. Res. Pract., 14* **2** 214–222.

[24] Nahum T. L., Mamlok-Naaman R., Hofstein A., & Taber K. S. 2010. Teaching and learning the concept of chemical bonding. *Studies in Science Education, 46* **2** 179–207.

[25] Driver, R. 1989. 'Changing conceptions', in P.Adey (ed.). *Adolescent development and school science* (London: Falmer Press) 79-103.

[26] Slameto. 2010. *Learning and factors that influence*. (Jakarta: Rineka)

[27] Burrows, N. L., & Mooring, S. R. 2015. Using concept mapping to uncover students' knowledge structures of chemical bonding concepts. Chem. Educ. Res. Pract, *16* **1** 53–66.

[28] Peterson, R. F., & Treagust, D. F. 1989. Grade-12 students' misconceptions of covalent bonding and structure. *J. Chem. Educ., 66* **6** 459.

[29] Yuliati, Y. 2017. Literasi Sains dalam Pembelajaran IPA. *Jurnal Cakrawala Pendas*, *3*(2), 21-28.

[30] Othman, J., Treagust, D. F., & Chandrasegaran, A. L. 2008. An Investigation into the Relationship between Students' Conceptions of the Particulate Nature of Matter and their Understanding of Chemical Bonding. *Int. J. Sci. Educ., 30* **11** 1531–1550.

[31] Wiyarsi, A., Sutrisno, H., & Rohaeti, E. 2018. The effect of multiple representation approach on students' creative thinking skills: A case of 'Rate of Reaction' topic. *J. of Phys: Conf Ser., 1097*, 012054.

The effect of problem-based learning model toward students' conceptual understanding and verbal communication skills in reaction rate learning

Marsianus Satrio Nangku[1] and Eli Rohaeti[2]
[+]Chemistry Education-Master Program, Universitas Negeri Yogyakarta, Yogyakarta, Indonesia, 55281
[2]Department of Chemistry Education, FMIPA, Universitas Negeri Yogyakarta, Indonesia, 55281

Corresponding author e-mail: Marsianusnangku.2017@student.uny.ac.id.

Abstract. This quasi-experimental study aims to determine the effect of applying Problem Based Learning (PBL) model toward students' conceptual understanding and verbal skills in learning reaction rate materials. It is designed using posttest only control group design subjecting the students of SMAN 1 Poco Ranaka. The PBL model was applied to the experimental class while 5E Learning Cycle model was treated for the control class. The data was collected using the observation sheet of verbal communication skills, posttest of students' conceptual understanding, and the questionnaire of students' response to the application PBL model. The data of observation and posttest, analyzed using MANOVA, revealed the significant effect of applying PBL model toward students' conceptual understanding and verbal communication skills with $P = 0.000 < 0.05$ of significance level. The percentage of contribution given by PBL model to the improvement of students' conceptual understanding and verbal communication skills was also analyzed. The percentage of contribution of PBL model to the improvement of students' conceptual understanding and verbal communication skills was, respectively, 22.26% and 33.17%, both with $P = 0.000 < 0.05$ of significance level. This shows the positive effect of applying PBL model on the two variables, though still in low and medium category of contribution. The result of questionnaire analysis using quantitative descriptive, students give positive response to the application of PBL model. Thus, it is concluded that the application of PBL model effects positively to the student's conceptual understanding as well as verbal communication skills.

1. Introduction

Learning is essentially to move forward to the state of more knowing, understanding and/or applying something. To achieve such goal, learning should be considered as a process to ship students from the state of their current knowledge to the proximate ideal target knowledge. As in other countries, Indonesian government have made change of the curriculum, the major tool of that students' knowledge shifting process, by taking into consideration the distance between these two knowledge edges (the current knowledge and the target knowledge). It is done for the final desired purpose which is to improve the quality of national education. In line with this, the Ministerial Ordinance of National Education Ministry No. 41 year 2007 [1] have ruled the process standard of education in which the role of teachers is

 1397 (2019) 012037 doi:10.1088/1742-6596/1397/1/012037

important to plan and execute well-designed learning, in line with the set curriculum, to achieve national education purpose [2].

Realizing that learning should be designed to the enhancement of students' knowledge end, learning practice nowadays has turned out attention to the importance of having students' active engagement and involvement in constructing their knowledge. This perspective, in more or less portion, is influenced by the constructivism theory of cognitive psychology proposed by Piaget and Vigotsky where students are treated to be the main actor of constructing their own knowledge while teachers act to scaffold or facilitate the construction [3]. This view of learning may give positive contribution to students' learning achievement in as an abstract subject as chemistry. It is due to the fact that the effort of shifting students' knowledge on this typical subject might be much more challenging, resulting to high demand of teachers' ability to provide learning in an innovative and meaningful way.

One fruitful learning model being able to trigger students' knowledge construction and to be usefully applied in chemistry class is Problem Based Learning (PBL) model. PBL model, the strategies of which may include posing students to the contextual problem, searching for the problem-related concept, and articulating the problem solving, is believed to abundantly promote both students' conceptual understanding and verbal communication skills. In its very essence, PBL model is naturally reduced, constricting the learning of concepts for the purpose of problem solving. It will enable students to focus more on learning the specific concepts, deepening their understanding of those concepts, while simultaneously thinking about the way they could apply such concepts to solve the given problem. PBL model facilitates students to learn, find out, and apply the knowledge they learn from any sources such as from online, library, and experts [4] which consequently results to students' achievement of conceptual understanding. Aside from being able to promote students' understanding to problem-related concepts, applying PBL model is also beneficial for the development of students' verbal communication skills. For its problem solving purposes, PBL model offers encouragement of students' development of verbal communication skills on at least its three learning stages. First, on the stage of problem investigation, students are encouraged to articulate the known concepts in searching for understating toward the assumptions and hypotheses of the investigated problem [5]. Second, on the stage of problem solving development and result presentation, students are fostered to present and articulate in a proper manner the result of their working on problem by which they are potentially facilitated to develop their independent learning ability [6]. Third, on the stage of analysis and evaluation, the students are encouraged to share their experience of problem investigation process and cooperatively construct the evaluation and conclusion regarding the learning content with peers and the teacher. Thus, by its ability to have students being active in each learning stage, PBL aids students to develop their skills in communicating, arguing, and investigating the problem and the problem-solving process [7].

Regardless those abundant potentialities of PBL model, the theoretical and the empirical problems have been revealed in terms of students' conceptual understanding and verbal communication skills, particularly in chemistrylearning. In his study, Setiyono [8] reports that in learning the abstract materials in chemistry, students tend to memorize and perform the direct application of the learnt concept and take less effort to understand the concept which could be stored in students' long-term memory. Memorizing the concepts, however, only rests in students' short-term memory, gives no contribution to solve analytical problems, and will not be useful to fulfill students' future needs. As the result of having no well-structured conceptual understanding, the students will in turn find difficulties in communicating the learnt concepts. The empirical problem related to these issues correspondingly echoes and emphasizes the same result. From the preliminary interview to some chemistry teachers in several schools at ManggaraiTimur Regency of Nusa Tenggara Timur Province, it is found that students have problems both in conceptual understanding and verbal communication skills in a reciprocal way. In one hand, the shallowness of students' conceptual understanding, due to their memorizing learning habits, effects to their inability to establish and articulate their understanding using appropriate verbal communication when they are assigned to present, argue or clarify their learning results. On the other hand, since the learning models applied by the teachers in their classrooms do not give enough opportunity for students to practice their verbal communication skills, the students tend to be defensive, relying their learning highly on their teachers without being motivated to search for other learning resources to deepen their understanding.

In respect to these problems while glancing on the theoretical preeminence of PBL model, this study is conducted to investigate the effect of applying PBL model toward students' conceptual understanding and verbal communication skills in learning reaction rate materials. The learning topic was chosen based on the assumption that reaction rate material is one of learning topics in chemistry that would trigger students to relate their learning content to the contextual problems they can identify in daily lifeexperience. In line with its daily identifiable and observable problems, the students will be indirectly posed to the abstract concepts of reaction rate learning material but be prepared for the concepts they engage every day. To do so, this study is conducted with the objectives formulated as follows.

1) Investigating the effect of applying Problem Based Learning model to students' conceptual understanding and verbal communication skills.
2) Analyizing the contribution of Problem Based Learning model to students' conceptual understanding.
3) Analyizing the contribution of Problem Based Learning model to students' verbal communication skills.

2. Methods

This study was quasi experimental using posttest only control group design [9]. The research subject was the first semester of XI class in the academic year 2018/2019. The population was the senior high schools in Manggarai Timur regency which were equal to or parallel with the first public senior high school (SMAN I) of Poco Ranaka. Using random sampling technique, the sample of this research was the 63 students of SMAN I Poco Ranaka, divided into two groups: experimental group (31 students) and control group (32 students). The experimental group was taught by applying PBL model and the control group was taught using 5E Learning Cycle model. The learning topic chosen for both groups was reaction rate learning materials.

Data were collected by conducting the observation, assigning the posttest, and distributing the questionnaire. The observation was done to observe students' verbal communication skills in the learning process of those two classes. The posttest was given to investigate students' conceptual understanding on reaction rate materials after the application of the aforementioned models in each researched class. The questionnaire, in addition, was distributed only for the experimental group students to identify their response after the treatment using PBL model.

Before used for data collection, the observation sheet, the test, and the questionnaire were tested to measure the validity and the reliability level. In validation process, the observation sheet and the questionnaire were validated using logic validity while the validity of posttest instrument was measured using empirical validity. The score of logic validity was gathered by having the instruments assessed by the experts for their content and construction validity. It was then counted that all 13 items of observation sheet and all 10 question items of the questionnaire were valid. The score of empirical validity was determined by piloting the instruments. Before piloting the posttest, there were 14 question items hich covered these indicators: determining the reactant concentration (5 items), explaining the reaction rate equations (3 items), determining the orders of reaction (1 item), identifying the reaction rate (1 item), and analyzing factors of reaction rate (4 items). The pilot testing result of the posttest question items revealed that out of those question items, there were 2 invalid question items, both of which were in te indicator of determining the reactant concentration. This confirmed that the posttest consisted of the 12 valid question items. Whereas, the reliability statistics of the instruments was measured by conducting α Cronbach reliability testing assisted by SPSS 22.

Furthermore, the collected data was analyzed using two data analysis techniques. First, the data of observation and posttest were analyzed using MANOVA inferential statistics with the procedures consisted of: 1) formulating the hypothesis, 2) testing of assumptions, and 3) hypothesis testing. The analysis of MANOVA was assisted by SPSS 22. Second, the data of questionnaire was analyzed using descriptive statistics which was assisted by using Microsoft Excel program. The procedures of it included: 1) data tabulation, and 2) percentage computation.

3. Results and discussion

3.1 The effect of applying Problem Based Learning (PBL) model toward students' conceptual understanding and verbal communication skills

Before taking into the statistical computation, the hypothesis of statistics is formulated as follows:

H_0 : There is no significant difference of students' conceptual understanding and verbal communication skills between those treated using Problem Based Learning (PBL) model and those treated using 5E Learning Cycle model.

H_a : There is significant difference of students' conceptual understanding and verbal communication skills between those treated using Problem Based Learning (PBL) model and those treated using 5E Learning Cycle model.

The testing of MANOVA applied to test the hypothesis was Hotteling's Trace testing. The result of analysis is displayed in table 1.

Table 1. Hotteling's Trace Testing Result

	Value	F	*Error df*	Sig.	*Partial Eta Squared*
Pillai's Trace	0.717	75.935^b	60.000	0.000	0.717
Wilks' Lambda	0.283	75.935^b	60.000	0.000	0.717
Hotteling's Trace	2.351	75.935^b	60.000	0.000	0.717
Roy's Largest Root	2.351	75.935^b	60.000	0.000	0.717

Based on the table, the result of Hotteling's Trace testing was significantly different in which the significance level (P) was 0.000 < 0.05. The result of statistical analysis rejectsthe null hypothesis (H_0) and confirms the alternative hypothesis (H_a). Therefore, the application of applying Problem Based Learning (PBL) model effects positively toward conceptual understanding and verbal communication of senior high school students.

Problem based learning encourage students to make meaning of real-life problem, engage higher order thinking skills, promote problem solving skills, process interdisciplinary learning, develop independent learning, facilitate learning to search for information, apply cooperative learning, and enhance learning of communicative skills [10]. Creative application of PBL aids students to identify, learn, and apply the known concepts from various learning resources and promote self-directed learning which is to be useful for their future needs[11] and contributes to students' self-actualization.

3.2 The contribution given by PBL model in improving students' conceptual understanding

To identify the percentage of contribution given by PBL in improving students' conceptual understanding in reaction rate material, univariate significance testing was conducted. The statistical hypothesis is formulated as follows:

H_0 : There is no contribution of Problem Based Learning (PBL) model to improve students' conceptual understanding.

H_a : There is contribution of Problem Based Learning (PBL) model to improve students' conceptual understanding.

Table 2. The Result of Univariate Testing of Students' Conceptual Understanding (CU)

Variabel Dependen	*Sum of Squares*	Df	*Mean Square*	F	Sig.	*Partial Eta Squared*
CU	417.330	1	417.330	55.383	0.000	0.476

Table 2 that presents the result of univariate testing of students' conceptual understanding contains the value ofpartial eta squared. The partial eta squared value isthen computed to find out the eta squared which is used to determine the percentage of contribution of the PBL model to improve the conceptual understanding of senior high school students. The percentage of contribution is the quadratic value of partial eta squared multiplied by 100% = $(0.476)^2 x\, 100\%$. It was counted that the percentage of the contribution of PBL model in improving students' conceptual understanding is 22.66% in which the rest percentage of contribution was given by other factors that also take effect to the improvement. This result reveals that PBL effects to students' conceptual understanding improvement in some way but still in low category of contribution. The low contribution of PBL model is assumedly caused by the fact that the students are not yet accustomed to the application of PBL model in their chemistry class. Further, to test the statistical hypothesis, the univariate testing shows F = 55.383, df = 1, and P = 0.000 <0.05. It indicates the significant difference of students' conceptual understanding before and after the application of PBL model. It is concluded that the result of statistical analysis confirms the alternative hypothesis (H_a) stating the positive contribution of PBL model to improve students' conceptual understanding and rejects the null hypothesis (H_0).

PBL, which the learning process is conducted by presenting the problem, posing questions related to the problem, facilitating investigation, and opening dialogic working on the problem, demands students' active engagement to participate. Students are also motivated to be self-directed in constructing their problem-related knowledge for problem solving purpose. The encouragement of students' active participation on each stage of PBL will affect to the improvement of students' conceptual understanding. Conceptual understanding refers to students' abilities to identify the known meaning or concept, situations, and facts [12]. In PBL, such understanding becomes the embodiment of self-constructive, self-explorative, and collaborative process of meaning making which incorporates teacher to facilitate and scaffold learning. Encouraging students to identify problem and the process of problem solving also determines students' conceptual understanding [13]. PBL model which offers various problem identification process, triggers students to activate their prior knowledge to work on the problem [14] that may leave the students to achieve their proximate development at the end of learning process.

3.3 The contribution given by PBL model in improving students' verbal communication skills

To identify the percentage of contribution given by PBL in improving students' verbal communication skills in learning reaction rate material, univariate significance testing was conducted. The statistical hypothesis of this research problem is formulated as the following:

H_0 : There is no contribution of Problem Based Learning (PBL) model to improve students' verbal communication skills.

H_a : There is contribution of Problem Based Learning (PBL) model to improve students' verbal communication skills.

Table 3. The Result of Univariate Testing of Students' Verbal Communication Skills (VCS)

Variabel	Sum of	Df	Mean	F	Sig.	Partial Eta

Dependen	Squares		Square			Squared
VCS	1311.393	1	1311.393	82.837	0.000	0.576

Table 3 presents the partial eta squared value (0.576). The percentage of contribution is the quadratic value of partial eta squared multiplied by 100% = $(0.576)^2 \, x \, 100\%$. It was then counted that the percentage of the contribution of PBL model in improving students' verbal communication skill is 33.17%. The rest of the percentage of contribution was given by other factors that also take effect to the improvement of students' verbal communication skills. This result reveals that PBL effects to students' verbal communication improvement but still in mid category of contribution. This is also assumedly caused by the fact that the students are not yet accustomed to the application of PBL model in their chemistry class that requires them to articulate their understanding through acceptable verbal communication skills. Afterwards, to test the statistical hypothesis, the univariate testing shows F = 82.837, df = 1, and P = 0.000 <0.05. It indicates the significant difference of students' verbal communication skills before and after the application of PBL model in their learning of reaction rate material. In conclusion, the result of statistical analysis confirms the alternative hypothesis (H_a) stating the positive contribution of PBL model to students' verbal communication skills and rejectsthe null hypothesis (H_0).

Verbal communication refers to one's ability to express the ideas using acceptable words and language function, in both written and spoken form [15]. Being skilled in verbal communication is then emphasized by Sullivan dan Mousley[16] as to not only have the ability to state the ideas in written form, but also in a broader view to articulate, explain, describe, clarify, listen to, question, share, and report the ideas one has got from a learning process. In its process, PBL model has advantageous aspects regarding the enhancement of students' verbal communication skills. As the students are engaged in a very effective way to work on the problem, students' verbal communication skills will be developed when they are trying to investigate the problem, arguing theories related to the problem, theorizing their identification of the problem, proposing problem solving, analyzing and evaluating the process of problem solving, and sharing experience related to the process. PBL model, which is able to promote students' conceptual understanding, will in turn improve their verbal communication skills as on each stage of PBL, the students will posit students to produce various kinds of verbal communication to enunciate their understanding related to the learnt concepts. Other than that, the application PBL model will somehow improve students' motivation and self-confidence in processing verbal communication. It is, on the very least, due to the fact that the knowledge attained by students at the end of a learning process is their own genuine knowledge contributed by self-directed process of knowledge construction and exploration in PBL model. Having the *weapon* (the attained knowledge) on their own, the students will obviously be more motivated to process and articulate their understanding using well-grounded verbal communication skills.

3.4 Students' response on the application of PBL model

To identify students' response on the application of PBL model, the Likert scale questionnaire was analyzed using descriptive statistics, assisted by Microsoft Excel program. The questionnaire was distributed to the students in experimental group to reveal students' response related to PBL model in two main aspects: 1) students' response related to learning strategies in PBL model and 2) students' response regarding the efficacy of applying PBL model to solve problem in the classroom. The result of the analysis of the questionnaire is displayed in Figure 1.

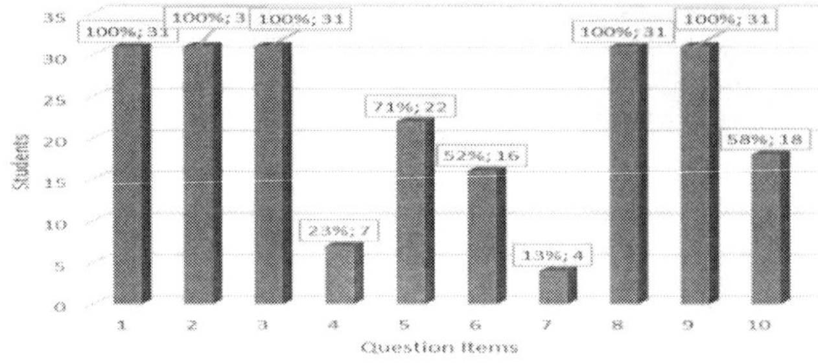

Figure 1. Students' Response on the application of PBL model

The data analysis of the questionnaire shows that most of the students in experimental group have positive response to the application of PBL model in the classroom. In term of learning strategies of PBL model, the result of data analysis shows students' high enthusiasm to engage actively in PBL. This finding endorses the theory saying that PBL model turns the learning approach in the classroom from teacher-centered to student-centered learning [17] and results positively to students' active and independent learning [18]. Dealing with the efficacy of PBL to solve students' classroom problem, the data analysis also reveals students' high belief on PBL model. The students have high belief on PBL model that can encourage them to be involved in learning real-life problems, promoting higher order thinking skills, enriching problem-solving skills, encouraging multidisciplinary learning, fostering independent learning, facilitating learning to search for information, supporting cooperative learning, and enhancing learning of verbal communication skills.

4. Conclusion

The quality of learning process is characterized by its ability to bridge students' knowledge gap, treating students to involve actively in its process. Problem Based Learning (PBL) model, in which the learning is centrally posited to activate and construct students' knowledge by problem posing and problem-solving activities, contains, in its nature, the promising effect to students' improvement of conceptual understanding and verbal communication skills in chemistry classes. This study confirms that challenging and engaging students' self-directed learning of problem-related concepts is able to accelerate students' conceptual understanding as well as verbal communication skills. It is identified from the findings of this study in which the application PBL model effects positively to students' conceptual understanding and verbal communication skills ($P < 0.05$), with the contribution of PBL model to the improvement both variables are, concecutively, 22.66% and 33.17%. However students' conceptual understanding and verbal communication skill may improve gradually as the result of the precise application and the frequent use of PBL model in classrooms. Other than its advantageous effect for students' individual development, getting students accustomed to problem-based learning situation will promote students' cooperative skills in problem solving which is highly prominent in students' future profesional.

5. References

[1] Kemendikbud. 2007. Peraturan Menteri Pendidikan Nasional Nomor 41 Tahun 2007, tentang Standar Proses.

[2] Slameto. 2013. *Belajar & Faktor-Faktor yang Mempengaruhi*. (Jakarta: Rineka Cipta).

[3] Sani, A. 2014. The effects of problem based learning on mathematics performance and affective attributes in learning statistics at form four secondary level. *Procedia Social and Behavioral Science.*

[4] Kek & Huijer. 2017. A Critical Use of Foucault's Art of Living. *Foundations of Science.*

[5] Fang, N. 2011. Students'perceptions of dynamics concept pairs and correlation with their problem – solving performance. *j. of. sci. technol.*
[6] Arnock& Muhammadi. 2016. Language and Individual Identities. *Language in India.*
[7] Wilder. 2015. Impact of problem-based learning on academic achievement in high school: a systematic review. *Educational Review.*
[8] Witte& Rogge. 2016. Problem-based learning in secondary education: evaluation by an experiment. *Education Economics.*
[9] Setiyono. 2011. Pengembangan Perangkat Pembelajaran Kimia Kelarutan Dan Hasil Kali Kelarutan (Ksp) Dengan Pendekatan Sets Untuk Meningkatkan Kemampuan Berpikir Kritis Dan Kreatif Siswa. *Jurnal PP*
[10] Creswell, J. 2015. *Riset Pendidikan Perencanaan, Pelaksanaan, dan Evaluasi Riset Kualitatif dan Kuantitatif.* (New Jersey : Pearson Education, Inc).
[11] Campbell, D. T & Stanley, J. C. 1963. *Experimental and Quasi-Experimen Designs for Research.* (USA: Houghton Mifflin Company).
[12] Tosun & Taskesenligil. 2013.The effect of problem-based learning on undergraduate students' learning about solutions and their physical properties and scientific processing skills. Chem. Educ. Res. Pract., 2013, **14**, 36-50
[13] Yoon, et al. 2016. Professional Development Workshops—What Do Teachers Learn and Use?. *J Teach Phys Educ 27* **1** *116 - 126*
[14] Tatar & Oktai. 2011.The effectiveness of problem-based learning on teaching the first law of thermodynamics. *Res Sci Technol Educ* 29 **3**
[15] Sungur & Tekkaya 2006. Improving achievement through problem-based learning. *J. Biol. Educ.* 40 155-160
[16] Schunk. 2014. Handbook of Self-Regulation of Learning and Performance. (New York: Routledge).
[17] Kelly, O. C. & Finlayson, O. E. 2007. Providing solutions through problem – based learning for the undergraduate 1 year chemistry laboratory. Chem. Educ. Res. Pract **8** 347 – 361.
[18] Kaya, E. 2013. Argumentation practices in classroom: pre-service teachers'conceptual understanding of chemical equilibrium. *Int. J. Sci. Educ* **35** 1139 – 1158
[19] Tarhan & Ayyildiz. (2018). Problem-based learning in teaching chemistry: enthalpy changes in systems. *Res Sci Technol Educ.* 36 **1** 35-55.

Exploration of pre-service chemistry teacher's ability in constructing context-based content representation on electrochemistry topic

A Wiyarsi*, H Sutrisno and E Rohaeti
Chemistry Education Department, Faculty of Mathematics and Natural Sciences, Universitas Negeri Yogyakarta, Indonesia, 55281

*Corresponding author email: antuni_w@uny.ac.id

Abstract. The study aims to explore the pre-service chemistry teacher ability in constructing context-based Content Representation on Electrochemistry topic (Cb-CoRe-E). Due to the exploration of pre-service chemistry teacher ability in constructing Cb-CoRe-E, this research adopted a descriptive study with quantitative approach. A total of 20 pre-service chemistry teacher enrolled on Curriculum course were the subject of this research. The data of pre-service chemistry teacher ability in constructing Cb-CoRe-E were collected according to the document which consist a questions on pedagogical aspect. As many as 9 indicators were used as a references to be explored the pre-service chemistry teacher ability in constructing Cb-CoRe-E. The data of pre-service chemistry teacher ability were analysed according to descriptive quantitative technique by classifying the mean score obtained into five categories from very good to bad. The result of this study showed that the pre-service chemistry teacher ability in constructing Cb-CoRe-E was good. It can be concluded that pre-service chemistry teacher were ready to conducted a good chemistry learning. The pre-service chemistry teacher in the future is expected applying CoRe in every learning to promote students'chemistry meaningful learning.

2.1. Introduction

The essence of chemistry education lies in the quality of learning in the classroom. Studies shown that teacher quality becoming the most contributing factor in successfully of chemistry learning. It means that teachers are considered as the most important factor that determines the quality of chemistry education [1-5]. Chemistry teachers who are competent and highly committed to quality are demands for the realization of high quality chemistry learning.

Professional and pedagogic competencesare the qualifications that should be possessed by chemistry teachers in Indonesia. Pre-service chemistry teachers are required to have a good understanding of chemistry as the subject matter that must be taught and how to teach it. Some researchers use the term content knowledge to conceptualize the professionalism of teachers in mastering the subject matter in their classroom. A good content knowledge is a strong foundation for pre-service teachers to be able to teach chemistry effectively. Some studies show the importance of teacher's content knowledge in developing quality chemistry learning. Mastery of content knowledge has a positive influence on teaching effectiveness, being the main key on developing teacher professionalism and correlating strongly with the ability of how teachers teach appropriate chemistry content [5-9].

The way teachers delivering content influences the things students learn, so that the teacher's teaching methods cannot be separated from related content [10]. In the other words, teacher should have pedagogical knowledge that could be implemented appropriately on the certain chemistry content

[11].state that content knowledge and pedagogical knowledge should be integrated into learning process to create a new knowledge, called Pedagogical Content Knowledge (PCK). PCK is part of the slice between pedagogical knowledge and content that helps teachers understand how to teach content. The previous study [12] developed the Content Representation (CoRe) model as one of the chemistry teacher's PCK representations. CoRe is a document that presents an overview of important ideas or concepts from a content that will be taught and developed based on the questions of pedagogy aspects[12]. CoRe development could help pre-service teachers to study the content critically and explore the difficulties experienced in understanding the chemistry content that will be taught.

PCK as a specific new knowledge is needed by pre-service chemistry teachers as a provision to prepare themselves to become professional teachers. One study[13] conducted a research of the constructingCoRe as astrategy to buildpre-service chemistry teachers'PCK. The study was conducted with a case study into nine chemistry education students. The research result indicated that the constructing CoRe exercise followed by appropriate scaffolding during the constructing CoRe process enables the development of PCK for beginner teachers. The ability of pre-service teachers in constructing CoRe is very important as teacher readiness in preparing quality chemistry learning.

The trend of chemistry curriculum changing occurs with the development and introduction of contextbased education [14-15]. Context-based chemistry education adopts the view that chemistry content must be associated with reality, evolving, flexible and not merely a set of rules and principles to remember. Context is used as the starting point in learning. This contextbased chemistry curriculum has implications for a new role of a teacher in the relation with students for conditioning content in context settings. The ability of pre-service chemistry teachers in constructing CoRe for chemistry learning in vocational contexts is still low [16]. The main difficulties encountered by pre-service teachers were predicting student's difficulties in learning content and ensuring their understanding. The lack of understanding and experience of pre-service teachers regarding chemistry learning process in vocational schools becomethe main obstacle.

The implementation of context based chemistry learning is broader for the context in high school. Pre-service chemistry teachers still have weaknesses in constructing CoRe in the components of curriculum knowledge, learning strategies, assessment, and mastery thechemistry content [17]. Context based chemistry learning can actually facilitate students in linking experiences, increasing learning motivations and scientific attitudes,also training student collaboration in solving the [18-20]. The analysis of pre-service chemistry teachers' ability related to the constructing of chemistry CoRe in this broader context needs to be assessed as an early indication of teacher readiness in carrying out quality chemistry learning.

Electrochemistry is classified as hard material in chemistry subject matter for high school students. Previous study [21] reviewed a number of articles then summarized some of the main alternative conceptions faced in electrochemistry as follow: (1) the cathode is a negative electrode, the oxidation reaction of half cells releases electrons, and reduces mass over time, (2) the anode is a positive electrode, the reduction reaction of half the cell accepts the electron and increases the mass over time, (3) the salt bridge allows electrons to move from the anode to the cathode, supplies the ions needed to move from the cathode to the anode, allows the cation to migrate towards anode electrode, while the anions migrate towards the cathode electrode, and (4) lack of ability to report cell reactions correctly. The teacher should plan learning well to help students understand electrochemistry topic. Developing CoRe is one way to develop the ability of pre-service teachers in designing proper electrochemistry learning. Therefore, this study aims to examine the ability of pre-service chemistry teachers in constructing Context-based Content Representation on Electrochemistry topic (Cb-CoRe-E).

2.2. Methods

2.3. Research design and samples

This study adopted descriptive research with quantitative approach. A total of 20 pre-service chemistry teachers involved in the Curriculum Review of Chemistry Education Course in Chemistry Education Study Program, Yogyakarta State University became the subject of this research. The samples were selected by convenience sampling.

2.4. Data collection technique and instrument

Data on the ability of pre-service chemistry teachers in constructing Cb-CoRe-E was measured through questions in pedagogical aspects [12]. The questions in the pedagogical aspects used in this study were based on the context used as the starting point in chemistry learning. The starting point of learning is one of the characteristic of context based learning [12, 22]. A total of 9 indicators were used to arrange the questions in pedagogical aspects to measure the constructing CoRe ability of pre-service chemistry teachers. The 9 indicators can be seen in Table 1.Beside the data on the ability of pre-service chemistry teachers in constructing Cb-CoRe-E, data from ideas or notions related to chemistry content formulated by pre-service chemistry teachers was also obtained.

Table 1. The Indicator of ability in constructing Cb-CoRe-E

No	Indicator
1	Determining the context used to initiate learning process
2	Knowing what pre-service chemistry teachers want students to learn from the developed ideas
3	Analyzing the importance of learning the developed content
4	Knowing what ideas/ content that the pre-service chemistry teacher knows but not the time yet for students to know
5	Knowing the difficulties or limitations related with the method on teaching the developed content
6	Knowing students' thoughts that might influence the pre-service chemistry teachers in constructing on the developedcontent
7	Knowing other factors that might influence the way of pre-service chemistry teachers construct on the content that has been developed
8	Knowing the teaching procedure and the specific reasons to apply it
9	Knowing whether to specifically ensure students' understanding or confusing related with the developed content

2.5. Data Analysis

The data obtained in this study were the ability of pre-service chemistry teachers in constructing Cb-CoRe-E then analyzed using quantitative descriptive techniques. There were 9 indicators that used as benchmarks in measuring the ability of pre-service chemistry teachers in constructing CoRe. Each indicator was given 0 to 5 score. The scores that obtained by pre-service chemistry teachers on each CoRe indicators then counted to obtain the mean value. The average score of each pre-service chemistry teacher in answering 9 CoRe indicators was also counted. The average ability of pre-service chemistry teachers in constructing Cb-CoRe-E based on predetermined indicators and the average of overall chemistry teachers ability in each indicator then classified into 5 categories ranging from very good to bad categories based on Sturgess formula. The average ability of pre-service chemistry teachers in each indicator are: having a bad category if the mean score is between the range 0.00 - 1.00; a poor category if 1.01 - 2.00; a sufficient category if 2.01 - 3.00; a good category if 3.01 - 4.00; and a very good category if 4.01 - 5.00.

2.6. Result and Discussion

The results of the analysis of the Cb-CoRe-E document showed that the mean score of the ability of the pre-service chemistry teacher in constructing Cb-CoRe-E was 3.4 and in the good category. The highest score achieved was 4.6 with very good categories, while the lowest score was 1.7 with poor categories. The ability of most pre-service chemistry teachers in constructing Cb-CoRe-E was included in the good category (35%). Figure 1 showed the distribution of teacher candidates' ability categories. One thing that needs to be considered is that there is still one prospective teacher who has the poor ability in constructing Cb-CoRe-E documents.

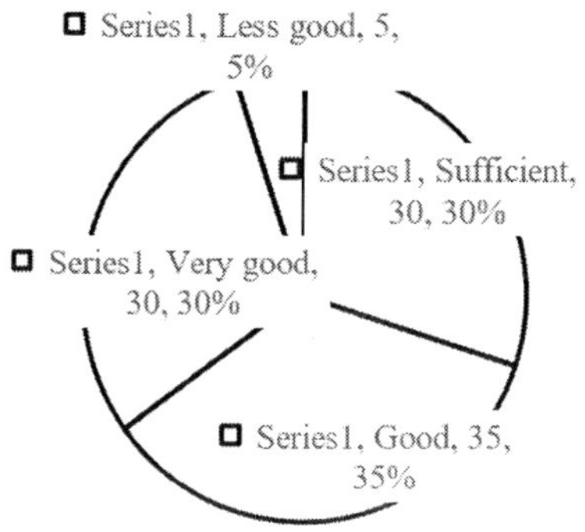

Figure 1. Distribution of categories of the ability of pre-service chemistry teachers in constructing Cb-CoRe-E

The result showed that the ability of pre-service chemistry teachers in constructing Cb-CoRe-E was good. Eventhough, there were some things that still need to be improved. For a deeper understanding, the ability of prospective chemistry teachers is analyzed based on nine questions in preparing Cb-CoRe-E.

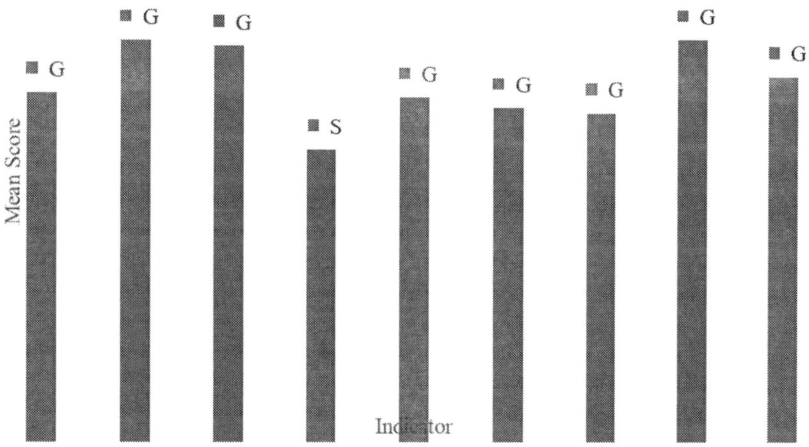

Figure 2. Mean score of pre-service chemistry teachers' ability for each indicator

Figure 2 showed that, of the nine Cb-CoRe-E constructing indicators, eight (Indicators 1-3 and 5-9) are achieved in good categories and others in sufficient categories (Indicator 4). One of the skills that should be improved is the pre-service chemistry teacher who knows about the content that they know but is not the time yet for students to know. Pre-service chemistry teachers have not fully mastered the concepts related to electrochemistry in depth so that it is difficult to describe their knowledge. On the other hand, there are limitations to prospective teachers in translating chemical concepts that students must know based on the formulation of competencies in the curriculum. This knowledge includes basic knowledge and core knowledge of electrochemical topics. As for the eight indicators that have been achieved in the good category, there are two that still need to be improved harder. First, indicator 7 related to the knowledge about other ways in influencing the way in teaching the topics.

Pre-service chemistry teachers need to be trained in explaining in detail the obstacles encountered when teaching content for each topic, not globally only stating because of limited facilities and infrastructure. Thus the pre-service chemistry teacher in determining the best learning according to the characteristics of each concept and conditions that exist. One other indicator (6) is related to the thoughts that students have and will influence pre-service chemistry teachers in teaching-related topics. Alternative knowledge possessed by students, for example about the anode which is considered always a positive pole becomes the basis of prospective teachers to find the best strategy. Otherwise, Cb-CoRe-E development help pre-service chemistry teachers to be better in preparing of chemistry learning. Through Cb-CoRe-E, pre-service chemistry teachers study chemistry content critically so they understand chemistry content deeply. The ability of pre-service teachers in constructingCb-CoRe-E is very important to be reviewed as the teacher's effort to prepare meaningful chemistry learning.

The overall ability of prospective teachers is also influenced by the ability to identify subtopics that must be taught in electrochemical content. Based on curriculum analysis, there are eight subtopics that should be taught including equalizing redox reactions, potential of reduction, voltaic cell, application of voltaic cell, corrosion, electrolysis, Faraday' Laws, and Application of electrolysis cell. However, the results of the study indicate that not all prospective teachers can bring up subtopics (see Figure 3).

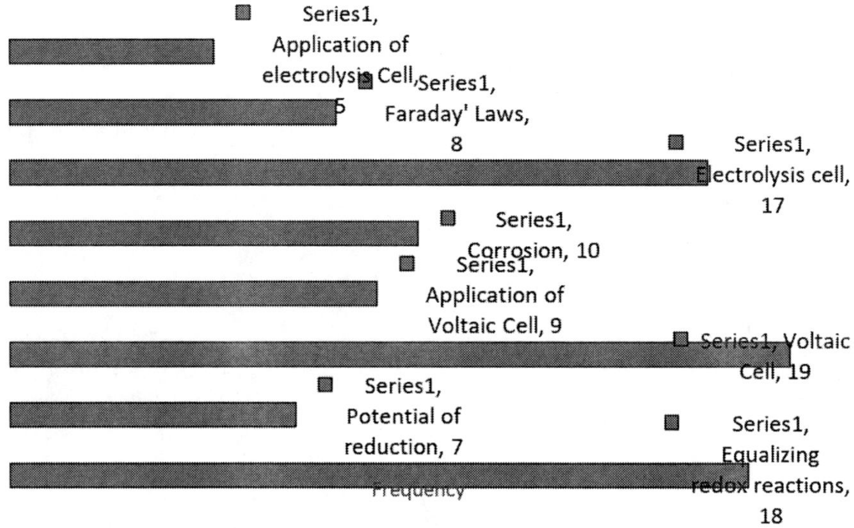

Figure 3. Frequency of appearance of subtopics in Cb-CoRe-E

Among the eight subtopics, only one subtopic is the voltaic cell that is raised by almost all prospective chemistry teachers. One prospective teacher who does not bring up this subtopic immediately displays it in the application subtopic. Most of the pre-service chemistry teachers have not separated the importance of separating the study of basic concepts and applications for both voltaic and electrolysis cells. This is very important so that pre-service teachers can provide a more thorough explanation, especially in relating context. Thus in teaching teachers not only gives an example of the use of concepts in general but trains students to overcome the problems of daily life related to the use of concepts.

In other hand, relating to integrate context, most of students can give simple example whether the daily process that can be used as context in electrochemistry learning. They propose to use some cases such as photosynthesis, painting fence of house, coating of vehicle component, gold bracelet, battery, and accumulator. Even though, pre-service chemistry teachers have not depict the case clearly why these cases can be used as starting point in learning. It is should be improved since this is the emphasis in context-based learning so as to facilitate the development of high-level thinking skills for students. These results show the importance of equipping pre-service chemistry teachers on how to analyze curriculum and mastery of chemistry content and integrate it with the context properly. The results of this study are in line with the research [23] which states that the ability of pre-service chemistry teachers to discuss electrochemistry uses in daily life is an important basis for the success of developing good CoRe. The development of Cb-CoRe-E is one way to train the ability of pre-service teachers to prepare meaningful and quality chemistry learning.

2.7.Conclusion

The ability of pre-service chemistry teachers in constructing Cb-CoRe-E was good. It mean that pre-service chemistry teachers were have a good knowledge both of pedagogical and content. Even though, they should improve their understanding of enrichment chemistry content knowledge and skill in integrating context for the learning effectiveness. They ready to start for conduct qualified chemistry learning. In the future, pre-service chemistry teachers can apply Cb-CoRe in each chemistry topic in order to hold meaningful learning for students.

2.8. References

[1] Khasawneh S A, Olimat Q and Abu T 2008 *International Journal of Applied Educational Studies* **2** 24-37
[2] Rohaan E J, Taconis R and Jochems W M G 2009 *Res Sci Technol Educ* **2** 727-338
[3] AdodoSO andGbore L O 2012 *Int. J. Psychol. Couns.* **4** 68-72
[4] Karaman A 2012 *Atlas Journal of Science Education* **2** 56-60.
[5] Ghazi S R, Shahzada G, Shah M T and Shauib M 2013 *Journal of Educational and Social Research* **3** 453-460
[6] Robinson J B 2005 *Chem. Educ. Res. Pract* **6** 83-103.
[7] JustiR dan van Driel J 2005 *Research in Science Education* **35** 197–219.
[8] Ozden M 2008 *Educational Science: Theory & Practice* **8** 633-645.
[9] Karisan D, Senay A and Ubuz B 2013 *Journal of Educational and Instructional Studies* **3** 22-31.
[10] NRC 1996 *National science education standards* (Washington: National Academy Press).
[11] Shulman LS 1986 *Educational Researcher* **15** 4-14.
[12] Loughran J, Berry A and Mulhall P 2006 *Understanding and developing science teachers' pedagogical content knowledge* (Rotterdam: Sense Publishers).
[13] Hume A and Berry A 2011 *Research in Science Education* **41** 341–355.
[14] Coenders F, Terlouw C, DijkstraS and Pieters J 2010 *Journal Science Teacher Education* **21** 535–557
[15] Dolfing R, BulteMW, Pilot A andVermunt J D 2011 *Research of Science Education* **42** 1-22
[16] Wiyarsi A, Hendayana S and Anwar S 2017 Journal of Science Education **18** 50-53
[17] Kondakci E U, Demirdogen,B, Akın F M,Tarkin A and Gunbatar S A 2017 *Chem. Educ. Res. Pract* **18** 250—270
[18] Vos M A J, Taconis R, Jochems W M and Pilot A 2010 *Chem. Educ. Res. Pract* **11** 193-206.
[19] Ilhan N, Dogan and Cicek O 2015 *Bartin University Journal of Faculty of Education* **4** 666–681.
[20]Ilhan N, Yildirim A and YilmazS D, 2016 *Int J Env Sci Ed* **11** 3117-3137.
[21] Karsli F and Çalik M 2012 *Asian J. Chem.* **24** 485–491.
[22] Bulte A M W, Westbroek H B, de Jong O and Pilot A 2006 *Int. J. Sci. Educ* **28** 1063-1086
[23] Aydin S, Friedrichsen P M, Boz Y and Hanuscin D L 2014 *Chem. Educ. Res. Pract* **15** 658—674

Acknowledgments
Thanks to the UNY graduate school that has facilitated the funding of this research and for the reviewers for the suggestions and feedback.

The Effect of a Macromedia Flash-based Guided Inquiry on Students' Critical Thinking Skill and Self-Regulated Learning

Srireski Jainal[1*] and Isana Supiah Yosephine Louise[2]
[1]Postgraduate of Chemistry Education, Yogyakarta State University, Indonesia
[2]Department of Chemistry Education, Faculty of Mathematics and Natural Science, Yogyakarta State University, Indonesia

* Corresponding author email: srireski.2017@student.uny.ac.id

Abstract. This study aims to determine the effect of a macromedia flash-based guided inquiry model on students' critical thinking skill and self-regulated learning. This study was a quasi-experiment, with a post-test only group design. The subjects were 96 eleventh-graders of divided into two groups. Fifty students in the experimental group used macromedia flash-based guided inquiry model and 46 students in the control group used direct instruction model. Data collection techniques were a test of critical thinking skill consisting seven essay questions to see students' critical thinking skill and a questionnaire of self-regulated learning consisting 33 statements to see students' self-regulated learning. The data were analyzed using MANOVA. The results showed that students taught with macromedia flash-based guided inquiry model have a statistically difference on their critical thinking skill and self-regulated learning at significance level 0.05 while students taught with direct instruction model showed critical thinking at lower level. In conclusion, a macromedia flash-based guided inquiry model can improve students' critical thinking skill and students' self-regulated learning.

1. Introduction

Teacher has an important role in learning. Learning strategies including learning model, method, and media used in the learning activity determine students' thinking skill and students' independence [1]. As stated in the basic competence of the 2013 curriculum, students are expected to show logical, critical, consistent and thorough, responsible, and responsive attitudes, and the students are also expected not to give up easily in solving problems.Based on the results of the survey conducted by PISA (Programme for International Students Assessment), Indonesia is still far below the international average in that Indonesia ranks the third from the bottom in the science performance if viewed from the average score obtained in the division of men and women [2]. There are several factors influencing the low level of students' critical thinking skill, namely theteachers who still use conventional learning in the class [3], the one-sided learning where the teacher prefers to focus on one student rather than on most students [4], and teacher's domination in the learning that causes students to have less practice in developing the students' thinking on problem solving and in applying the concept that they have learned in the class to real life [5].Teacher's conventional learning is very familiar with transferring knowledge and mostly encouraging students to have mathematical learning exercises.This learning can foster the learning with memorizing method that facilitates less learning than deep understanding [6].

Science plays an important role in the education field. At high school level, science lessons are divided into Physics, Chemistry, and Biology. Particularly, chemistry has the scope of chemical materials which mostly consist of abstract concepts [7].This abstract material brings students to have difficulties in learning chemistry [8-9].

In order to help students in increasing their critical thinking skill, the learning process should be interactive, innovative, interesting and motivating to students so that they can actively involve in the learning. The learning process should also provide enough chance for students' independence.To achieve learning activity well, the learning model and media are needed as tools in the learning process. Importantly, the students will succeed in their academic achievement if they have self-regulated learning or independent learning [10].

The learning model and media that support the statements above are guided inquiry and macromedia flash.Guided inquiry model is a learning model that is centred on constructivist theory[11]. Constructivist teaching leads students to develop knowledge and meaning from individual's experience[12-13].According [14], teacher's role are to ensure that the students can understand the concept which has learned and to guide students in dealing with the concept.The steps of guided inquiry learning model are (1) identifying problems; (2) suggesting hypothesis; (3)collecting data; (4) testing hypothesis; (5) suggesting temporary conclusion; (6) replicating; (7)drawing final conclusion[15].

In the learning process, guided inquiry needs suitable media. One of media that supports guided inquiry is macromedia flash. According to [16] macromedia flash is a vector-based interactive animation program that has action script facilities. The vector-based interactive program is widely used to create games, cartoons, presentations, and interactive learning models [17].User can include text, graphic, animation, simulation, audio and video into macromedia flash application. Moreover, media that can display animation is a medium that can reduce students' difficulties in receiving learning material. Through animation, analogies can be created to visualize abstract materials.Conceptual visualization in the form of animation and analogies is very fruitful for students in understanding chemistry [18].

Based on the elucidation above, this study was aimed to determine the effect of macromedia flash-based guided inquiry learning model on students' critical thinking skill and students' self-regulated learning in the material of chemical equilibrium concept.

2. Methods

This study was a quasi-experiment with post-test control group design. The population of this study were eleventh-grade students. Ninety six students were randomly selected as samples of this study and were grouped into experimental group and control group. Macromedia flash-based guided inquiry learning model was implemented in the experimental group while direct instruction model was implemented in the control group. The study was conducted by taking the material of chemical equilibrium concept.The sample size of the study is presented in the Table 1.

Table 1. Sample size of the study

Group	Number of students
Experimental	50
Control	46
Total	96

2.1 Research instruments

This study used the following instruments:

2.1.1 Validation test of media consisting of four aspects. The media was validated by material and media experts. After validated by experts, the score 3.84 was obtained.

$$M_i + 0.5\ SB_i < \overline{X} \leq M_i + 1.5\ SBi \qquad \text{(i)}$$

The score range showed that the media is categorized to good and feasible to use.

2.1.2 Seven items of critical thinking test on the concept of chemical equilibrium. The type is a test with essay question. Before the test is used, the test was validated theoretically and empirically. The instrument was tested using QUEST with difficulty index of the test between -2.90 and 4.26

and reliability 0.87. Reliability coefficient is above 0.7 which is acceptable. These indicated that there is a good internal constituency of the items [19]

2.1.3 Questionnaire responses about students' self-regulated learning in the chemistry learning consist of 33 items using a 4-Likert scale. Before the questionnaire is distributed to students, the questionnaire was theoretically validated by experts.

2.2 Teaching intervention

Learning material taught to students was chemistry equilibrium. The experimental group was taught using macromedia flash-based guided inquiry model and the control group was taught using direct instruction model. Each group had five meetings for learning and one meeting for a post-test. The test, namely critical thinking skill test, was given to students after the learning as well as the questionnaire about self-regulated learning in the chemistry learning. Because the activities in the guided inquiry learning optimally involve the students' skills to discover and investigate systematically, critically, logically, and analytically, the students are able to find out the findings by themselves with confidence [20]. The guided inquiry learning emphasizes the thinking process that relies on the learning process and learning outcomes, and it does not only develop students' intellectual skills. Knowledge will be embedded longer if the students are involved directly in the learning process. This is also due to the advantages of guided inquiry learning for students. The advantages are that the students are able to develop their own understanding, gain their independence in the learning, and improve high motivation and participation in the learning [21].

2.3 Data analysis

To compare the score obtained from students' critical thinking and self-regulated learning test, one-way MANOVA was analyzed using SPSS version 21 (a software package used for statistical analysis).

3. Results and discussion

This study was conducted to see the different influence of students' critical thinking skill and students' self-regulated learning. The experimental group used macromedia flash-based guided inquiry while the control group used direct instruction in the learning process. Macromedia flash was validated before it was used in the learning. The validation result carried out by media expert revealed that the mean score of media quality was 3.84.

Assessment of students' critical thinking skill was taken from the results of post-test and assessment of students' self-regulated learning was taken from students' questionnaire response. Based on the result of one-way MANOVA, there is a difference of students' critical thinking skill between experimental group and control group in the material of chemical equilibrium concept with significance level 0.05. The results of one-way MANOVA can be seen in the Table 2.

Table 2. MANOVA results of critical thinking skill and self-regulated learning.

	Value	F	Hypothesis df	Sig
Model	0.142	6.609[b]	2.000	0.002

[b]*Exact statistic.*

The results revealed that the significance level is 0.002 which is lower than 0.05. It showed that there is a difference of students' critical thinking skill and self-regulated learning between experimental group and control group. In addition, the following table is a difference test result of students' critical thinking skill between experimental group and control group.

Table 3. Result of between-subject effect test for students' critical thinking skill

Dependent variable	df	Mean square	F	Sig

Class	Critical thinking skill	1	327.219	5.046	0.027

Based on the significance value 0.015<0.05, H_0 was rejected so that there is a difference of students' critical thinking skill. In other words, students who learned using macromedia flash-based guided inquiry model have higher critical thinking skill. Guided inquiry enables students to be actively involved in the learning process. The students are encouraged to be active in recognizing problems, making hypothesis, analyzing and interpreting data, finding answers and discussing the results until they make final conclusion. In line with results of the study by [23-26], they demonstrated that guided inquiry contributes to improving critical thinking skill.

In addition to learning model, media also effect students' critical thinking skill. Media is one of components of delivery system. The delivery system is the delivery of messages from teacher to students, and the message is information or knowledge in the learning [27].The media used in this study is macromedia flash. Macromedia flash is designed attractive so it can work maximally. Moreover, the abstract learning material was visualized to assist students to easily understand the learning. Visual media has important role in the learning since it can help the students understand and strengthen memories [28]. Hence, the macromedia flash in this study helps students increase their critical thinking skill.

Furthermore, macromedia flash-based guided inquiry also influences on the students' self-regulated learning.The result of difference test for students' self-regulated learning is shown in the Table 4.

Table 4. Result of between-subject effect test for students' self-regulated learning

	Dependent variable	df	Mean square	F	Sig
Class	Self-regulated learning	1	893.508	6.197	0.015

Based on the table above, the significance level is $0.015 < 0.05$. H_0 was rejected so it can be concluded that there is a difference of students' self-regulated learning between using macromedia flash-based guided inquiry and using direct instruction.

The results of this study asserted that students' self-regulated learning contributes significantly. Self-regulated learning allows someone not to continuously need assistance from others [29]. Students who have self-regulated learning belong to personality type of autonomy. Autonomous personality is a type of personality with the desire to do something independently, to act and to take initiative by themselves [30]. In this situation, the students do not continuously need the help of the teacher in the learning process because the role of the teacher is as a facilitator. Thus, it is obvious that the presence of learning media can help students have self-regulated learning or independence as a means of learning on their own [31].

4. Conclusion

Based on the results and discussion, the study concludes that there is a difference between students' critical thinking skill and students' self-regulated learning using macromedia flash-based guided inquiry learning model in the material of chemical equilibrium concept in the chemical learning. The students who are taught using macromedia flash-based guided inquiry learning model have higher critical thinking skill than the students who are taught using direct instruction. Guided inquiry learning enables students to have the opportunity in discovering their knowledge, to assist students in understanding the concept well, to be independent and to be active in the learning. Moreover, the media also assist students in understanding and receiving the chemical learning material. Therefore, macromedia flash-based guided inquiry learning model can be used to improve students' critical thinking skill and students' independent learning or self-regulated learning.

References

[1] Setiani I., Dafik., & Darojat, O. 2015. Pengembangan perangkat pembelajaran berbasis pendekatan saintifik dengan teknik whole brain teaching materi bangun ruang sisi lengkung pada siswa kelas IX. *Jurnal Pancaran, 4* **1** , 193-210.

[2] Organisation for Economic Cooperation and Development (OECD). 2016. *OECD Database.* OECD.

[3] Priyadi, R. 2018. Analisis kemampuan berpikir kritis siswa SMA kelas X MIPA dalam pembelajaran fisika. *Jurnal Pendidikan Fisika Tadulako (JPFT), 6* **1** 53-55

[4] Abdi, A. 2014. The effect of inquiry-based learning method on students' academic achievement in science course. *Univers. J. Educ. Res., 2* **1** 37-41.

[5] Subhan, S. 2018. Pengaruh media animasi dalam model pembelajaran inkuiri terbimbing terhadap keterampilan berpikir kritis dan aktivitas belajar peserta didik pada materi kesetimbangan kimia. *Chemistry Education Review 1 2* 125-141.

[6] Sandi, G. 2012. Pengaruh blended learning terhadap hasil belajar kimia ditinjau dari kemandriian siswa. *Jurnal Pendidikan dan Pengajaran, 45 3* 241-241.

[7] Kean, E., & Middlecamp, C. 1985) Panduan belajar kimia dasar. (Jakarta: PT. Gramedia).

[8] Djamarah & Zain 2006. *Strategi belajar mengajar.* (Jakarta: Rineka Cipta).

[9] MPBPTIK. 2010. *Membangun LSM Berbasis WEB dengan Aplikasi Moodle* (Kementrian Pendidikan Nasional Direktorat Jendral Manajemen Pendidikan Dasar Menengah, Direktorat Pembina SMA)

[10] Zimmerman, B.,J. 2002. Becoming a self-regulated learner: an overview. *Theory Into Practice, 41 2,* 65-70.

[11] Maikristina, N. I., Dasna, W., & Sulistina, O. 2013. Pengaruh penggunaan model pembelajaran inkuiri terbimbing terhadap hasil belajar dan keterampilan proses sains siswa kelas XI IPA SMAN 3 Malang pada materi hidrolisis garam. *Jurnal Online Universitas Negeri Malang, 2* **2**

[12] Jonassen, D. 1991. Objectivism versus constructivism: Do We Need a New Philosophical Paradigm?. *Educ Technol Res Dev,* 39 **3**, 5-14.

[13] Olusegun, S., & BADA. 2015. Constructivism learning theory: a paradigm for teaching and learning. *IOSR Journal of Research & Method in Education 5* **6** 66-70.

[14] Oliver, R., & Herrington, J. 2000. An instructional design framework for authentic learning environments. *Educ Technol Res Dev, 48* **3** 23-48

[15] Orlich, D. C., Harder, R. J., Callahan, R. C., Trevisan, M. S., & Brawn, A. 2010 *Teaching strategis a guide to effective instruction (8th ed).* (Boston New York: Wadsworth Cengage Learning).

[16] Dikes, I.W. 2010. *Animasi dengan flash 8.* (Yogyakarta: Graha Ilmu).

[17] Madcoms. 2013. *Adobe flash CS6.* (Yogyakarta: Andi Offset).

[18] Kirna, I. M. 2010. Determinasi proposisi pembelajaran pemahaman konsep kimia melalui implementasi pembelajaran sinkronisasi kajian makroskopis dan submikroskopis *Jurnal Pendidikan dan Pengajaran, 43* **3**, 185-191

[19] Fraenkel, R. J., & Wallen, E. N. 2000 *How to design and evaluate research in education (4thed).* (San Fransisco: McGraw-Hill).

[20] Azwar, S. 2015. *Penyusunan skala psikologi.* (Yogyakarta: Pustaka Pelajar).

[21] Gulo, W. 2002. *Strategi belajar-mengajar.* (Jakarta: PT. Grasindo).

[22] Kuhlthau, C., Maniotes, L., & Caspari, A. 2007. *Guided inquiry: Learning in the 21st century.* (Westport, CT: Libraries Unlimited).

[23] Azizmalayeri. K., Mirshahjafari, E., Sharif, M., Asgari, M., & Omidi, M. 2012. The impact of guided inquiry methods of teaching on the critical thinking of high school students. *Journal of Education and Practice, 3* **10** 1-7.

[24] Fuad, N. M., Zubaidah, S., Mahanal, S., & Suarsini, E. 2017. Improving Junior High Schools' critical thinking skills based on test three different models of learning. *International Journal of Instruction,10* **1** 101-116.

[25] Nisa, E. K., Koestiari, T., Habibbulloh, M., & Jatmiko, B. 2018. Effectiveness of guided inquiry learning model to improve students' critical thinking skills at senior high school. *Journal of Physics: Conderence Series, 997.*

[26] Priono, A. 2015. *Penerapan Model Pembelajaran inkuiri terbimbing untuk meningkatkan kemampuan berpikir kritis siswa kelas XI SMA Negeri 3 Lubuklinggau. Skripsi.* Lubuklinggau: STKIP-PGRI.

[27] Gagne, R. M., & Leslie, J. B. 1979. *Principles of instructional design.* (United States of America: Holt, Rinehart and Winston).

[28] Arsyad & Azhar 2011. *Media pembelajaran.* (Jakarta: PT Raja Grafindo).

[29] Covey, S. R. 1994. *The seven habits of highly effective people.* Jakarta: Binarupa Aksara.

[30] Rohman & Arif. 2009 *Memahami pendidikan dan ilmu pengetahuan* .Yogyakarta: Laks Bang Mediatama.

[31] Oktavera, S. 2015. Pengaruh media pembelajaran dan kemandirian belajar terhadap hasil belajar IPA siswa kelas VI Sekolah Dasar. *Jurnal Pendidikan Dasar, 6* **2** 327-338.

Journal of Physics: Conference Series **1397** (2019) 012040 doi:10.1088/1742-6596/1397/1/012040

Effects of use 3D visualization virtual reality to increase scientific attitudes and cognitive learning achievement

Ika Setiawati A Bakar[1], **K H Sugiyarto**[2] **and J Ikhsan**[2]

[1] Chemistry Education Program, Graduate School of Yogyakarta State University, Indonesia
[2] Department of Chemistry Education, Faculty of Mathematics and Natural Sciences, Yogyakarta State University, Indonesia

*Corresponding author e-mail: setiawatyika68@gmail.com

Abstract. This study was conducted to analyze the effect of the use of 3D visualization media with virtual reality on the achievement of cognitive learning and scientific attitudes of students on chemical equilibrium topic. This study was a quasi-experimental with only posttest design with the subject of research in class XI of State High School of Ternate consisting of 31 students for the experimental class and 27 students for the control class. Collecting cognitive learning achievement data with test and scientific attitudes using questionnaires. The statistical analysis used was MANOVA to measure differences in cognitive learning achievement and scientific attitudes of experimental and control class students. The results showed a significant difference in cognitive learning achievement and scientific attitudes between classes using virtual 3D visualization media and discovery learning models with classes that only use discovery learning models. The implementation of practicum using 3D visualization media with virtual reality can provide opportunities to students to learn about the implementation of the practicum that they design by themselves.

1. Introduction

Growing information technology presents numerous innovations in various fields. Likewise in the education system that gets the impact of the development of information technology. In the world of education, forms of learning activities using information technology are arranged to assist in constructing concepts in learning. The education also began to prepare the delivery of material to students by using information technology capabilities, with the hope that the material delivered is capable to be easily understood by students and also can advance the quality of education especially in the learning of chemistry.

Chemistry topics that requires expertise in calculating, memorizing, and conducting experiment, are often difficult for students to learn [1]. Chemistry knowledge also has levels, namely submicroscopic, microscopic, and symbolic. These three levels are very important in the chemistry learning. The learning process chemistry is not confined to theories or concepts, but also supported by laboratory experiments to prove and develop the theory learned. The function of the laboratory is to provide completeness for accepted theoretical lessons, provide and foster the courage to seek the nature of scientific truths from an object in the natural and social environment, adding skills in using tools [2]. Laboratories are very important to make abstract chemistry concepts become concrete and make chemical material easier for students to understand [3].

One of the most efficient ways of learning chemistry is through laboratories [4]. The practicum can also provide better results for increasing the understanding of students in the practiced material. In addition to improvement in learning achievement, practicum can also improve the scientific attitude of students. However, learning at school still does not utilize the laboratory due to a lack of tools and materials in the laboratory. Chemicals also require large costs and some trials are at risk for safety.

In general it can be said that the implementation of activities in the laboratory is very important in chemistry learning, but for some reasons this activity cannot be carried out properly. So, we need an alternative laboratory or media that can assist learners in doing practical work they need. One way to help students is through learning media. Media that can be used are embodied learning media technology with three-dimensional (3D). 3D visualization technology was introduced using virtual reality (VR) which is used for chemistry learning.

The VR concept refers to the system of principles, methods, and techniques used to design and create software products to use by several multimedia computer system with particular device system [5]. Virtual laboratory brings advantages that practitioners can conduct dangerous experiment without endangering themselves or others, the simulation is affordable, and after the virtual laboratory is developed, and the practitioners are able to conduct the experiment without great expense [6]. The purpose of this study was to determine the differences between the scientific attitudes of high school students and cognitive learning achievement in the use of three-dimensional visualization media with virtual reality.

2. Methods

This study uses a quasi-experimental method with a posttest only control group design. The sample in this study using two classes of experimental classes and control classes. The experimental class uses discovery learning models with 3D visualization media using virtual reality and control classes using discovery learning models. The trial design can be seen in Table 1.

Tabel 1. Posttest Only Control-Group

No	Group	Method	*Posttest*
1	Experimental	Y1	X1,X2
2	Control	Y2	X1,X2

Note:

Y1: students use 3D visualization media with virtual reality with discovery learning models;

Y2: students using discovery learning models;

X1: scientific attitude questionnaire;

X2: test cognitive learning achievement.

In this study using, a sample of 58 consisting of two classes, the experimental class with 31 students and the control class with a total of 27 students taken from one school in Ternate. The sample used is random sampling technique. The experimental class was treated with the implementation of 3D visualization media with virtual reality with discovery learning while for the control class using discovery learning.

2.1 Instruments

The instrument used in this study is a question of cognitive learning achievement test consisting of 10 essay questions and scientific attitude questionnaire. The question of cognitive learning achievement test before use is done by expert validation and empirical validation. Empirical validation is done by testing the test of learning achievement questions for high school students. Analysis of the results of empirical validation using the QUEST program to find out the number of fit questions that meet the requirements is used based on the difference in power and level of difficulty of the questions.

2.2 Data Analysis
Differences scientific attitude and cognitive learning achievement were statistically analyzed using Multivariate Analysis of Variance (MANOVA) using SPSS 21. Analysis Manova has carried out in this study via procedures.

a. Interval data and ratio
b. Independent variable consists of two or more categories
c. The observation is independent
d. The number of samples must be adequate
e. There is no univariate or multivariate outliers
f. There is multivariate normality
   The normality test is used to find out whether the data is normally distribution or not. Data normality test used in the study was based on ratio of the cost of a Mahalanobis distance and the chi square value using SPSS software. Data can be said to be normally distributed if the significance is greather than 0.05 or p-value > 0.05 at a significance level of 5%
g. There is a linear relationship between each pair of dependent variables and each independent variable
h. It has covariant metric homogeneity
   Homogeneity tests were conducted to determine whether the sample came from a homogeneous population or not. The homogeneity test of the covariance matrix in this study uses the Box's M. test. The covariant variance matrix is said to be homogeneous if the value of sig. > α (0.05).
i. There is no multicolinerity.
j. Correlation Test
   Correlation test is useful for knowing the relationship between one variable and another. This study uses Pearson test with the help of SPSS software. The data has correlation if the significance value was less than 0.05 or p <0.05 at a significance level of 5%. If the hypothesis prerequisite test is fulfilled then MANOVA analysis is performed using SPSS to determine the difference between the experimental group and the control group.

3. Results and Discussion
In this study, the learning achievement data in the form of posttest results are given to the students after receiving treatment with media using virtual reality 3D visualization. The results showed that the use of virtual reality 3D visualization media with discovery learning learning model produces a difference of learning achievement and scientific attitude with class without the use of media. Virtual chemistry laboratory has a positive effect on achievement and scientific attitude of learners [7].

Prerequisite analysis carried out is the normality test and homogeneity test. The results of the normality test can be seen in Table 2.

Table 2. Normality Test Results

Variable	Class	Sig
Scientific Attitude	Experiment	0.858
	Control	0.744
Cognitive Learning Achievement	Experiment	0.193
	Control	0.152

Based on Table 2. it can be seen that the significance value of scientific attitudes and cognitive learning achievement of students from the experimental class and control class is greater than 0.05 so that it can be concluded that the data is normally distributed. The homogeneity of the variance-covariance matrix test was performed using the Box's-M Test. The results of the homogeneity test obtained Box's M value of 0.230 with an F value of 0.074 and a significance value of 0.974. The significance value obtained is greater than 0.05 so that the covariance matrix is homogeneous.

After the prerequisite test of normality and homogeneity done, proceed to test the hypothesis, MANOVA test was used to test whether there are differences in the scientific attitude and cognitive learning achievement of learners using instructional media 3D virtual reality visualization and learners who do not use the media. A summary of the results of the manova test can be seen in Table 3.

Tabel 3. Result of Manova Test

	Effect	Value	F	Hypothesis df	Sig.
Class	Pillai's Trace	0.116	3,619[b]	2,000	0.033
	Wilks' Lamda	0.884	3,619[b]	2,000	0.033
	Hotelling's Trace	0.132	3,619[b]	2,000	0.033
	Roy's Largest Root	0.132	3,619[b]	2,000	0.033

Based on the manova test, the significance value of 0.03 is smaller than 0.05, indicating that there are differences in scientific attitudes and cognitive learning achievements of students who use 3D visualization media with virtual reality and those who do not use the media. There are several factors that cause the learning in the experimental class higher than in the control class, namely the learning using 3D visualization media with virtual reality that makes the students be more interested in learning. The learning process with virtual laboratory can also be carried out independently by the students so that they become more focused on conducting a practicum. Virtual media let students to conduct the experiment easily and to help students understand the concept. The use of 3D visualization media with virtual reality can help students understand the practicum results because the students can repeat the practicum that has been done in the learning so that it can deepen their experience and material that has been learned [8].

In the experimental class, aspect of curiosity has the highest score because of the learning media use. The existence of virtual laboratory is a new thing for students, so that it creates their curiosity to ask and investigate the things that the students do not know yet and to attract the students to learn. The advantages in general are that the learning process becomes more interesting, interactive and appropriate to stimulate students' curiosity. Implementation of media is able to increase students' science attitude. Demonstrated that the use of virtual laboratory can give an experience and improve students' learning motivation [4]. The dominant science attitude in the control class are open mindedness and cooperation, namely the indicator on showing respect for other opinions and findings. In the control class, the students are accustomed to work in groups so that they more appreciate the other students' opinions and findings. Students' good attitudes and interests on the use of virtual media will have a direct effect on the improvement of their learning outcomes [9].

This study does not claim that laboratories with virtual reality are more effective than real laboratories, but laboratories with virtual reality can be an alternative for reasons such as the dangers of chemical reactions, time problems, lack of tools and materials in real laboratories. The use of laboratory media has become an increasing issue [10]. Learning using a virtual laboratory is a simulation, whereas, a real laboratory is a conventional practicum learning in the real situation. In this study, the laboratory using virtual reality can also be utilized as an alternative for the lack of practical equipment and learning media that support students to better understand the concept of chemical equilibrium. Media in the virtual laboratory is a medium used to help students understand a subject and can provide a solution to the lack and absence of laboratory tools [11].

Learning with 3D visualization media with virtual reality can be used for self-directed learning if the students want to repeat the practice. Learning with virtual laboratory can provide flexibility in time and place and can overcome other obstacles in the school's laboratory [12]. The use of instructional media technology 3D visualizations using virtual reality to make the participants more interested in learning. Moreover, it can improve learning and make learning more effective, adding to the experience of learners and also provides practical needs of learners. 3D Visualization using virtual reality make learners feel as though I was in a real laboratory. Learners are more motivated in learning using a virtual 3D environment. The use of information, communication and technology can be a good opportunity to create chemical learning programs with effective tools to develop new methods and techniques in educational programs [13].

4. Conclusion

The conclusion of this study is that there are significant differences in cognitive learning achievement and scientific attitudes between classes using virtual 3D visualization media and discovery learning learning models with classes that only use discovery learning models for chemical equilibrium subjects. The implementation of practicum using 3D visualization media with virtual reality can provide opportunities to students to learn about the implementation of the practicum that they design by themselves. As a result, the students become active and creative in learning chemistry and the use of technology-based media is able to make the learning process interesting for students.

5. References

[1] Chittleborough, G., & Treagust, D.F. 2007. The modelling ability of non-major chemistry students and their understanding of the sub-microscopic level. *Chem. Educ. Res. Pract* , *8* **3** 274-292.

[2] Tusiyam, S.N.H. 2011. Pemanfaatan laboratorium biologi untuk mencapai standar kompetensi pembelajaran biologi kelas XI IPA semester I di MAN Kendal, (Semarang: Institut Agama Islam Negeri Walisongo)

[3] Altun, E., Demirdag, B., Feyzioglu, B., Ates, A., & Cobanoglu, I. 2009. Developing an Interactive Virtual Chemistry Laboratory Enrich with Constructivist Learning Activities for Secondary School. Science Direct, I **1** 1895-1898.

[4] Tatli, Z. dan Ayas, A. 2012. Virtual chemistry laboratory: Effect of constructivist learning environment. *Turkish Online J. Distance Educ.*. *13* **1** 183 – 199.

[5] Lacrama, D. 2007. Virtual reality. *Journal Anale Seria Informatica*, *5* **1** 137-144.

[6] Herga, N.R., & Dinevski, D. 2012. Virtual laboratory in chemistry-experimental study of understanding, reproduction and application of acquired knowledge of subject's chemical content. *Organizacija*, *45* **3**

[7] Tuysuz, C. 2010. The effect of the virtual laboratory on students' achievement and attitude in chemistry. *Int. Online J. Educ. Sci.*, *2* **1** 37-53.

[8] Ardac, D., & Akaygun, S. 2004. Effectiveness of multimedia-based instruction that emphasizes molecular representations on students' understanding of chemical change. *J Res Sci* Teach, *41* **4**, 317–337.

[9] Garrison, D. R., & Akyol, Z. 2009. Role of instructional technology in the transformation of higher education. *J. Comput. High. Educ*, *21* **1** 19–30.

[10] Nana.S. 2010. Dasar-Dasar Proses Belajar Mengajar. (Bandung: Sinar Baru Algesindo).

[11] Razi, P. 2013. Hubungan motivasi dengan kerja ilmiah siswa dalam pembelajaran fisika menggunakan virtual laboratorium di kelas X SMAN kota Padang. *Jurnal Teknologi Informasi & Pendidikan*, 119-124.

[12] Carnevale, D. 2003. The Virtual Lab Experiment. *The Chronicle of Higher Education*. *1* **1** 1-5

[13] Pekdag, B. 2010. Alternative methods in learning chemistry: learning with animation, simulation, video and multimedia. *J. Turk. Sci. Educ.*,*7* *2* 111-118.

Developing an android-based application (AVALIMA) for chemical literacy evaluation

Erfan Priyambodo*, Anggiyani Ratnaningtyas Eka Nugraheni, Antuni Wiyarsi, Dina, Dini Lathifa Husna
Department of Chemistry Education, Faculty of Mathematics and Natural Sciences, Universitas Negeri Yogyakarta, Jl. Colombo No.1 Yogyakarta 55281, Indonesia

*Corresponding author email: erfan@uny.ac.id

Abstract. The aim of this research is to determine the feasibility and perceptions of teachers and students about AVALIMA, a based on android application, as a chemical literacy evaluation media. AVALIMA was developed by taking the first stage of 4D model, that is define, design, and develop without disseminate. The initial product of AVALIMA was reviewers by material expert, media expert, and peer reviewers. Furthermore, AVALIMA was also assessed by five chemistry teachers by regarding to six aspects, namely material, science literacy, construction, language, visual appearance, and program performance. AVALIMA was also tried out to ten students in the eleventh grade. The result of the analysis shows that AVALIMA is a good android-based media that can be used to evaluate chemical literacy. According to the reviewers, the application is 88.2% appropriate to be used to evaluate chemical literacy. Meanwhile, according to the students' response in using this media, it is 86.31% appropriate to be used as media to evaluate their chemical literacy. The teacher's and the students' perception toward the application is also very good.

1. Introduction

The influence of technological development in many aspects has grown rapidly these days, including education. The use of information and communications technology (ICT) in education can be found in the form of the media used, for example audio-visuals, computers, laptops, smartphones, and others. Compared to all the technology devices mentioned, smartphone always seems to outshine the others. It is merely because it has more capabilities in resolution, features, and computing, including the operating system [1]. One of operating systems on smartphone is Android. Android is a Linux operating system for smartphone which is free and uses open source code that offers rapid processing and innovative features [2]. Based on the research of Stat Counter [3], the use of Android as a developing operating system in Indonesia has grown up to 67% and is now dominating the smartphone market. Many consumers choose to use Android as an operating system because it is easier to use and provides access to install multiple applications as desired for free. The operation of the Android system on smartphone is also easy. Hence, it is also popularly used as a learning media.

In accordance with the development of ICT, teachers are expected to be able to develop a technology-based learning media as an innovation. Learning media is a tool to convey message information and as everything that can affect student learning [4]. One of the learning media innovations is technology-based practice questions that can be used by students to evaluate or measure learning outcomes.

Learning evaluation was defined as a process to evaluate activities or correct things that have happened or been implemented during the learning that has taken place [5]. To obtain good evaluation results, evaluation activities must have the principles of (a) continuity, (b) comprehensive, (c) fair and objective, (d) cooperative, and (e) practical. According to Sukardi [5] the principles of evaluation (a) the evaluation must still be in the working lattice of the intended purpose; (b) evaluation should be carried out comprehensively; (c) evaluations are held in a cooperative process between the teacher and students; (d) valuations are carried out in a continuous process; (e) evaluation must be concerned and consider the values that apply.

The common evaluation in teaching and learning through ICT based learning is objective test. There are many advantages in developing of this type of evaluation [6], i.e. (a) multiple choice tests have flexible properties to measure student learning outcomes; (b) test items that are intensively constructed can cover all learning materials; (c) multiple choice test items can measure students' knowledge, attitudes and skills; (d) students' answers can be corrected easily by using answer keys that have been prepared separately; (e) the results of student answers can be corrected together with a more conducive situation.

The outcomes of students' learning in chemistry subject are considered to be not yet satisfying. There are many students who still experience some difficulties in studying chemistry subject at school [7]. They view chemistry as a difficult, complex and abstract subject that requires special intellectual talent and a strong effort to understand it [8]. Students' difficulties in understanding the concept of chemistry, or science in general, are revealed in the 2015 PISA (Program for International Student Assessment) results. In the field of science, Indonesian students are in the rank of 64 out of 70 surveyed countries [9]. PISA tests are usually in the form of open questions based on real life contexts. Students are required to use their abilities to understand and interpret the problem. This kind of process is a part of chemical literacy.

Shwartz, Ben-Zvi, & Hofstein [10] define chemical literacy in four domains: content, context, skills, and affective. Chemical literacy in the content domain of chemistry deals with macroscopic phenomena in terms of microscopic structures of matter. It is aimed to understand and explain the chemical structure, investigate the dynamics of processes and reactions, and change in energy during chemical reactions. Chemical literacy in the context domain deals with chemical knowledge that explains daily phenomena, understands the relationship between innovation in chemistry and sociological processes. Chemical literacy in the skill domain deals with the ability of individuals to ask questions and find information that is interconnected. Chemical literacy in the affective domain deals with how chemically educated individuals deliver their chemical views and can express interest in chemical issues and topics. If students have the skills in using the information to solve chemical problems (valid arguments with pro and contra ideas), it means that they have a high level of chemical literacy [11].

Responding to the need of technology-based learning media as an innovation in education, an Android-based smartphone application is developed. A smartphone equipped with an Android operating system is possible to be used as an online learning media or currently better known as M-Learning [12]. The application developed in this research is aimed to provide a flexible and practical chemical literacy evaluation tool for students in grade 11. The application is called AVALIMA. This application provides chemical literacy-based chemistry questions, like the ones appear in PISA, but have been adjusted to the level of competence of the eleventh-grade students.

The purpose of developing AVALIMA media in this study is to know the feasibility and the perception of teachers and students towards AVALIMA media as an android-based application to evaluate chemical literacy in grade eleven. In this case, the researchers assume that all the reviewers and the students have Android-based smartphones. For the parameter of success in this study, AVALIMA media is assumed to be suitable to be used as a chemical literacy evaluation tool for grade eleven if the results of the review given by the reviewers is at least B.

2. Methods

2.1. Reasearch Procedure

The procedure of this study was carried out by adapting the research methods used by Thiagarajan, Semmel, and Semmel [13] which takes only the first 3-D from the 4-D model. This model is a simple instructional model that helps the researcher in producing the android based media to evaluate students' chemical literacy. This 3-D model only covers define, design, and develop; excluding disseminate. The disseminate stage could not be conducted in this study due to the fact that the steps done in the development level was not yet perfect. The stages done in the development procedure are presented in Figure 1.

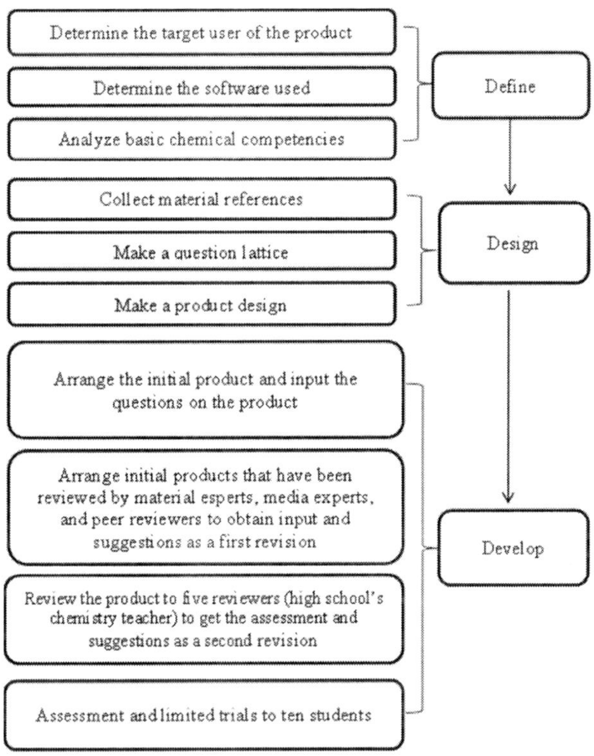

Figure 1. AVALIMA media development procedure diagram

2.2. Data Collection

There were two instruments used in this study. The first instrument was an open questionnaire for material experts and media experts to assess the android based media. The second instrument was questionnaire related to the product trial for eleventh grade students who chosen as the respondents.

The questionnaire employed in this study used Likert scale and was adapted from Wahono [14] and OECD [9] which covers the aspects to be assessed from a technology-based learning media and chemical literacy. These aspects were then further developed into 20 assessment indicators. The Likert scale used in this questionnaire was ranging from very good, good, fair, less and poor. Table 1 and Table 2 show the blueprint of the questionnaires developed for the reviewers and the students.

2.4. Data Analysis Technique

2.4.1. Data of Product Development Process

The data collected for the product development process was in the form of input and suggestions from material experts and media experts. This data was qualitative data which used to revise the android based media.

Table 1. The Blueprint of the Questionnaire Developed for the Reviewers

No.	Aspect	Sum of Indicator
A	Material	4
B	Science Literacy	3
C	Construction	3
D	Language	3
E	Visual Appearance	3
F	Program Performance	4

Table 2. The Blueprint of the Questionnaire Developed for the Students

No.	Aspect	Sum of Indicator
1.	Material	3
2.	Language	3
3.	Visual Appearance	3
4.	Program Performance	4

2.4.2. Data of Product Quality
The data of product quality was obtained from the questionnaire filled by the reviewers and the students. Then, it was further analyzed by using descriptive data analysis by employing the following steps:

2.4.2.1. Converting the feedbacks given by the reviewers and the students into quantitative data by using a Likert scale. Scoring scale of data was: very good scored by 5, good by 4, fair by 3, less by 2 and poor by 1.

2.4.2.2. Calculating the average score of each aspect and also the overall aspects. The following is the formula used to calculate the average score:

$$\bar{X} = \frac{\Sigma x}{n}$$

Note:

$\bar{X}$ = average score for overall aspects/ average score for each aspect
Σx = the total score of overall aspects/the total score of each aspect
n = the total number of the reviewers/the total number of the students

2.4.2.3. Converting the average score obtained into qualitative data in order to determine the quality of the Android-based Chemical Literacy Evaluation Application (AVALIMA) by using the ideal assessment criteria proposed by Widoyoko [15].

2.4.2.4. Determining the quality of the Android-based Chemical Literacy Evaluation Application (AVALIMA) by referring the average scores obtained to the ideal assessment criteria.

3. Result and Discussion
The results of the development research carried out consisted of two data, namely the development process data and the product quality data. Data on the development process of Android-based AVALIMA media was obtained from conducting define, design, and develop stages. Meanwhile, the product quality data was obtained from the feedbacks by the reviewers and the ten eleventh-grade students chosen as samples.

The reviewers involved to give feedbacks toward the AVALIMA's media quality were five chemistry teachers from SMAN 7 Yogyakarta, SMAN 2 Klaten, SMAN 1 Prambanan Klaten, SMAN 1 Cilacap, and MAN 1 Sleman. Meanwhile the reviewers asked to give feedbacks for the product trial were ten eleventh-grade students of MAN 3 Sleman. The results of the AVALIMA media quality feedbacks by the reviewers can be seen in Table 5.

Table 3. The Results of AVALIMA Media Quality Assessment by Reviewers

Reviewer	Aspect						Σ
	A	B	C	D	E	F	
1	16	10	12	12	12	16	78
2	16	12	13	12	12	20	85
3	19	14	14	15	15	20	97
4	16	12	12	12	15	17	84
5	20	14	15	14	15	19	97
$\bar{X}$	17.4	12.4	13.2	13.0	13.8	18.4	88.2
Percentage							88.2
Quality Criteria	very good	good	very good	very good	very good	very good	very good

Information:
A : Material Aspect
B : Science Literacy Aspect
C : Construction Aspect
D : Language Aspect
E : Visual Appearance Aspect
F : Program Performance Aspect

The following is the comparative percentage diagram of the six aspects based on the feedbacks given by the reviewers.

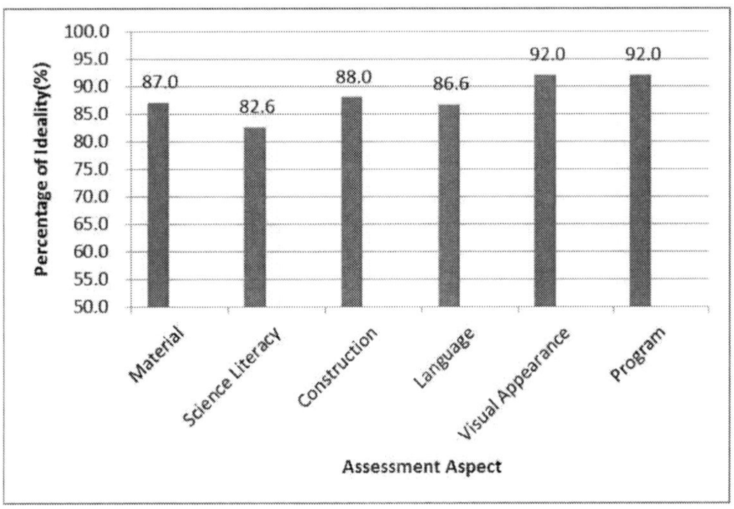

Figure 2. Comparative percentage diagram based on the reviewers' feedbacks on the six aspects measured

The questionnaire distributed to the students only consisted of four aspects: material, language, visual appearance, and program performance. The results of the students' feedbacks on the four aspects while trying out the AVALIMA media can be seen in Table 6.

Table 4. The results of AVALIMA media trial by the students

Value	Aspect				Σ
	A	B	C	D	
$\bar{X}$	18.0	8.3	12.3	17.5	56.1
Percentage					86.31
Quality Criteria	very good	good	good	very good	very good

Information:
A : Material Aspect
B : Language Aspect
C : Visual Appearance Aspect
D : Program Performance Aspect

The following is the comparative percentage diagram of the four aspects based on the feedbacks given by the students regarding to the media trial they had

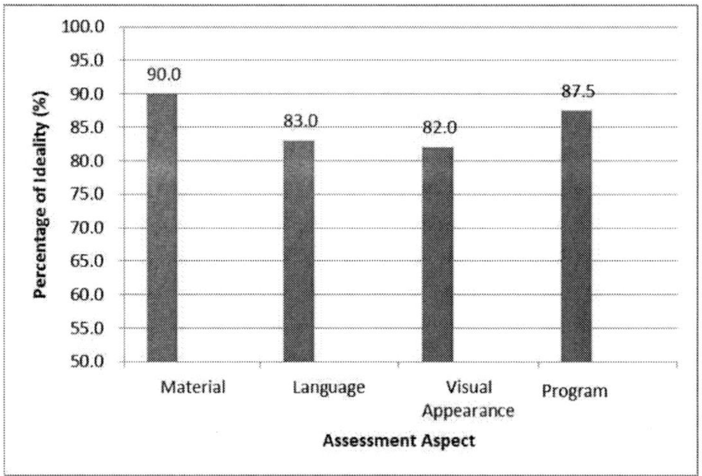

Figure 3. Comparative percentage diagram based on the students' feedbacks on the four aspects measured

Based on the result of the feedbacks given in material aspect, it can be indicated that the materials provided in AVALIMA media are in accordance with the basic competences (KD) stated in Curriculum 2013, particularly the ones that should be achieved by the eleventh-grade students while learning Chemistry subject. The questions given are also in accordance with the eleventh-grade students' level of competence. The questions compiled are considered to cover chemical concepts and theories related to scientific phenomena in daily life. According to Fives, Huebner, Birnbaum, & Nicolich [16] scientific literacy is an ability to understand scientific processes and scientific information in everyday life meaningfully. Thus, it can be said that the questions in AVALIMA media convey the chemical literacy.

Based on the result of the feedbacks given in the chemical literacy aspect, it can be indicated that the problem formulated has conveyed chemical literacy. The chemical literacy here is further broken down into content and context. The content is said to convey chemical literacy if the questions and materials given contain chemical materials, such as processes and reactions, changes in chemical energy, and chemical structure. Then, the context is said to convey chemical literacy if it deals with anything happens in everyday life or phenomena related to chemistry and the relationship between chemistry and other sociological processes.

Chemical literacy is related to other sciences. So chemical literacy does not only concern about chemistry, but there can also be integrated biology, physics, mathematics, and other sociological subjects. Most of the studies on the identification of chemical literacy are based on studies of scientific literacy, just as scientific chemical science measurements are also based largely on the same study of scientific literacy [17].

The results of the feedbacks given in the construction aspect can be used as a reflection of the construction of the questions that has been formulated. One set of multiple choices question consists of stem, option, key, and distractor. In this type of question, there is only one right answer; there is no question that depends on other questions; and some questions may have to be solved through several steps.

Based on the result of the feedbacks given in the linguistic aspects, it can be indicated that the products have followed the Indonesian rules stated in EYD. Besides, the questions and discussions provided have also used communicative language, are easy to understand, and do not cause multiple interpretations. The length of the sentences or paragraphs used to convey chemical literacy is also appropriate to the students' level of competence.

Based on the result of the feedbacks given in the visual appearance aspect, it can be indicated that the type and size of the font (font), background, and display color are appropriate. The AVALIMA media uses Calibri font, 12 points. The background used is designed by using Photoshop CS6. The color chosen for AVALIMA media is green. The students could give better feedbacks on the aspect of visual appearance because the appearance of the application attracts the eyes.

The results of the feedbacks given in the AVALIMA media program performance aspect show that the product has compatible properties, easy navigation, user friendly, and easy to install. Compatible means that the product can be used on all smartphones with android platforms (at least Jellybean) smoothly, stably, and accurately. Easy navigation means that the product is easy to operate, not confusing, interactive, and form and its location is consistent. User friendly means that the product or media is easy to use and simple in operation [13]. The product is claimed to provide easy installation since it can be installed by only downloading it or sharing it by using Bluetooth, WhatsApp, email, etc. For future researcher who wants to conduct development research in the same area of interest, it is recommended that the technology projects to be developed is not only compatible for Android phones, but also other mobile applications or smartphones by testing the effectiveness and efficiency of a group of users [18].

4. Conclusion

Based on the findings and the discussion of this study, it can be concluded that (1) AVALIMA media is feasible to be developed as a media for chemical literacy evaluation for the eleventh-grade students since it is found that the ideal percentage calculated from the feedbacks given by the reviewers and the students is high, that is 88.2% and 86.31%, (2) The teachers' and the students' perceptions toward this Android-based AVALIMA media as an application to evaluate chemical literacy of the eleventh-grade students is very good.

Based on the results of this study, there some suggestions could be highlighted as the following:
4.2.1. Future researchers need to make sure if the the level of difficulty of the questions given and the number of options provided in multiple- choice items is appropriate for the eleventh-grade students.
4.2.2. AVALIMA media should be used as a teacher's evaluation tool and students' independent learning media in understanding and recognizing chemical literacy questions.

5. References

[1] Yamamoto N 2014 An interactive learning system using smartphone:Improving students' learning motivation and self-learning *Ninth Internatiional Conference on Broadband and Wireless Computing* pp 428-431
[2] Andrus J and Nieh J 2012 Teaching operating systems using android *Proceedings of the 43rd ACM technical symposium on computer science education* pp 613-618
[3] Statcounter, GlobalStats. *Operating System Market Share Indonesia Mar 2017-Mar 2018.* http://gs.statcounter.com/os-market-share/all/indonesia.
[4] Kozma R B 1991 Learning with media *Review of educational research* **61** pp 179-211
[5] Boyan J and Moore A W 2000 Learing evaluation functions to improve optimization by local search *Journal of Machie learninng Research* **1** pp 77-112
[6] Priyambodo E 2016 Validity and Realibility of chemistry systemic multiple choices questions (CSMCQs) *International Journal of Evaluation and Research in Education* **5** pp 306-309
[7] Sirhan G 2007 Learning Difficulties in Chemistry: An Overview *Journal of Turkish Science Education* **4** pp 1-20
[8] Cardellini L 2012 Chemistry: Why the Subject is Difficult? *Educacion Quimica* pp 1-6
[9] Organisation for Economic Co-operation and Development (OECD). 2016.
[10] Shwartz Y, Ben-Zv R and Hofstein A 2006 Chemical Literacy: What Does This Mean to Scientists and School Teachers? *J. Chem. Educ.,* **83** pp 1557-1561.
[11] Cigdemoglu C and Geban O 2015 Improving Student' Chemical Literacy Level on Thermochemical and Thermodynamics Concepts through Context-Based Approach *Chem. Educ. Res. Pract* **16** pp 302-317.
[12] Lu'mu 2017 Learning Media Of Applications Design Based Android Mobile Smartphone *Int. J. Appl. Eng* **12** pp 6576-6585.
[13] Thiagarajan S, Semmel D S and Semmel M I 1974 *Instructional development for training teachers of exceptional children* (Minneosota: University of Minnesota)
[14] Wahono RS *Aspek dan Kriteria Penilaian Media Pembelajaran.* http://romisatriawahono.net/2006/06/21/aspek-dan-kriteria-penilaian-media-pembelajaran
[15] Widoyoko, E P 2009 *Evaluasi Program Pembelajaran Panduan Praktis bagi Pendidik dan Calon Pendidik* (Yogyakarta: Pustaka Pelajar)
[16] Fives H, Huebner W, Birnbaum A S and Nicolich M 2014 Developing a Measure of Scientific Literacy for Middle School Students *J. Sci. Educ* **98** *4* 549-580.
[17] Celik S 2014 Chemical Literacy Levels of Science and Mathematics Teacher Candidates *Australian Journal of Teacher Education,* **39** 1 1-15
[18] Calimag J N, Miguel P A and Conde R S 2014 Ubiquitous Learning Environment Using Android Mobile Application *International Journal of Research in Engineering & Technology,* **2** 2 119-128

Molecular Identification of Several Orchid Species Based on OPA10 and OPA18 RAPD Marker

Junita Kurniawati[1], Lili Sugiyarto[1], Evy Yulianti[1], Heru Nurcahyo[1], Ixora S. Mercuriani[1,a)]

[1]Biology Education Department, Faculty of Mathematics and Science, Universitas Negeri Yogyakarta
Jalan Colombo No.1, Sleman, Yogyakarta, 55281, Indonesia

[a)] corresponding author: ixomerc@uny.ac.id

Abstract. The purpose of this study was to determine the genetic characteristics of nine orchids plants collected by Biology Gardens of Mathematics and Science Faculty of UNY based on OPA10 and OPA18 markers, to detect their genetic relationship, and determining the exact RAPD markers for identification of intraspecies and interspecies orchids. The analysis was carried out by isolating DNA from the leaves of 9 species of orchids (*Rhynchostylis retusa, Vanda tricolor, Dendrobium aphyllum, Dendrobium fimbriatum, Dendrobium moschatum, Dendrobium crumenatum, Dendrobium antennatum, Dendrobium anosmum* Gigantea Alba and *Dendrobium anosmum* Gigantea) using TIANGEN DNA healing plant kit. Molecular identification was analysed based on the PCR-RAPD technique using OPA 10 and OPA18 primers. Electrophoregraph of PCR results are converted into binary data and analysed with NTSYSpc 2.02 software. The similarity index (kinship level) was analysed based on UPGMA dendrogram. The results of the analysis showed that there was genetic diversity in 9 orchid plants tested. Intra-species orchids show higher similarity than inter-species. Primer OPA 10, OPA18, and a combination of both shows polymorphism in intra-species orchids.

1. Introduction

Indonesia has around 5000 of the 25,000 orchid species in the world that are scattered throughout the island [1]. High commercial values in the fields of medicine, food, ornamental plants, and cut flowers, make orchids as one of the flower plants with the highest trade rank globally. CITES (Convention on the International Trade in Endangered Species of Wild Fauna and Flora) reports that the world orchid trade is dominated by orchids that have beautiful, fragrant, long-lasting flower colors, shapes and patterns. Some genera found in the market are Cymbidium, Dendrobium, and Phalaenopsis [2].

At present 736 genera are recorded in Orchidaceae. New discoveries continue to occur every year and the orchid genus is described as reaching 13 per year [3]. Updates to the classification of plants can occur due to new findings. Morphological identification, in general, is an accepted basis for describing new taxa. At present, there are many descriptions of new genera that involve molecular analysis (DNA). The description of the new species currently also involves molecular DNA studies to show its peculiarities [4]. The development of molecular techniques for genetic analysis in the past decade has increased information on orchid genetic diversity. Molecular techniques have been used to study variations in DNA sequences among orchid species and cultivars [5]. According to [6] descriptions of orchid taxa at any level must include molecular studies and morphological studies.

PCR-based molecular markers are more widely used because they require low concentrations of DNA, are able to amplify DNA sequences from stored tissue, are simple, and relatively inexpensive [7]. Among the molecular markers that have been found, RAPD is a molecular marker that is widely used to identify genetic characters in plants. The RAPD technique is based on DNA amplification with a single primer that targets random sequences. Many current research publications use the RAPD technique to determine genetic equations in orchid plants such as *Vanda tricolor* which uses OPA10 primers to detect soma clonal variation in propagation through in vitro culture [8].

At this time the Biology Garden of Faculty of mathematics and science of UNY collects various types of orchids for conservation purposes as well as germplasm sources in the development of varieties. Morphological identification has been carried out, but molecular identification has not been done. The genetic characteristics of these orchids have not been identified so that the research is important to support data on a collection of orchid species (*Rhynchostylis retusa, Vanda tricolor, Dendrobium aphyllum, Dendrobium fimbriatum, Dendrobium moschatum, Dendrobium crumenatum, Dendrobium antennatum, Dendrobium anosmum* Gigantea Alba and *Dendrobium anosmum* Gigantea).

2. Method

Nine orchids (*Rhynchostylis retusa, Vanda tricolor, Dendrobium aphyllum, Dendrobium fimbriatum, Dendrobium moschatum, Dendrobium crumenatum, Dendrobium antennatum, Dendrobium anosmum* Gigantea Alba and *Dendrobium anosmum* Gigantea) were collected from Biology Gardens Faculty of Mathematics and Natural Science UNY. DNA isolation was carried out using TIANGEN DNA isolation plant kit. The source of DNA comes from the leaves of 9 orchid plants analysed. The subsequent DNA isolation is visualized through electrophoresis. The concentration that is used is 0.8% which is added with 5µl DNA staining (fluorosafe). The electrophoresis process is carried out at a voltage of 100 V for 20 minutes. The reading and documentation of the visualization results are done using the 1012 Series Gel Documentation System. The isolated DNA was amplified using the OPA10 (5`CTGATCGCAG3`) and OPA18 (5`AGGTGACCGT3'). The amplification process was carried out using the Variti Thermal Cycler tool. The results of the amplification process were then migrated on agarose gel with a concentration of 0.8%. Migration is carried out on electrophoresis devices with a voltage of 50 V for 60 minutes. Electrophoresis was continued by visualization using the 1012 Series Gel Documentation System.

Analysis of RAPD data was done by scoring the amplification results. The bands that appear are given a score of 1 and those that do not appear are given a score of 0. Subsequent data is applied to the NTSYSpc2.0.2 program to calculate the equation coefficients and make a dendrogram using the Unweighted Pair-Group Method using Arithmetic Average (UPGMA).

3. Result and Discussion

3.1. Result

The results of genomic DNA amplification from nine orchids plants tested using RAPD OPA10 and OPA18 markers showed a diversity of genetic (polymorphism) (Figure 1). The diversity is indicated by the number and variation in the size of the fragments produced (Table 1).

Table 1. Number and measure of DNA fragmen

Sample	OPA10		OPA18	
	Σ fragmen	Measurement (bp)	Σ frgamen	Measurement (bp)

1	5	1531, 1052, 839, 626, 200	7	1739, 1397, 1282, 1168, 1026, 826, 541
2	4	998, 839, 572, 359	4	997, 826, 427, 228
3	4	1078, 918, 492, 359	3	1054, 826, 456
4	3	918, 492, 359	3	1054, 836, 427
5	3	1584, 1344, 439	5	1379, 1168, 940, 826, 541
6	4	1424, 1238, 1078, 413	1	541
7	3	1078, 839, 705	3	627, 456, 313
8	4	142, 1131, 918, 519	4	1282, 1111, 997, 714
9	5	1398, 1078, 998, 839, 679	6	1739, 1454, 1311, 1026, 855, 541

Note: (1) *R. retusa*, (2) *V. tricolor*, (3) *D. anosmum* Gigantea Alba, (4) *D. anosmum* Gigantea, (5) *D. aphyllum*, (6) *D. fimbriatum*, (7) *D. moschatum*, (8) *D. crumenatum*, (9) *D. antennatum*

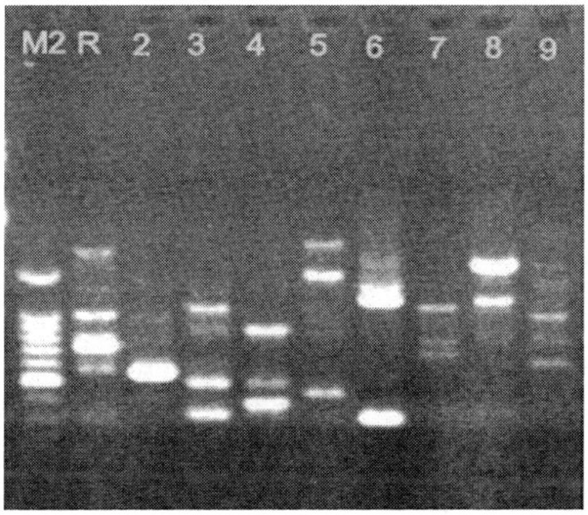

Figure 1.a) Electrophoregram of nine orchids using RAPD OPA10

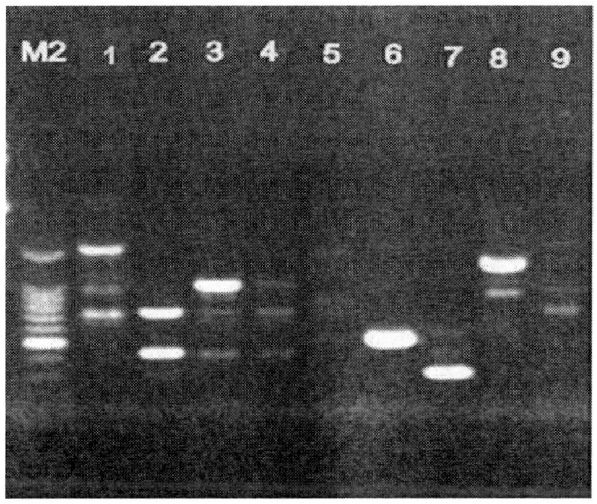

Figure 1.b) Electrophoregram of nine orchids using RAPD OPA18

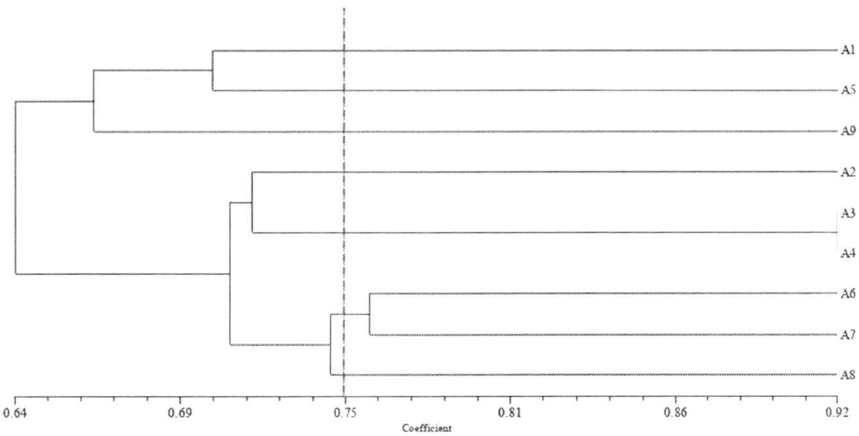

Figure 2. UPGMA dendrogram of OPA10 and OPA18

3.2. Discussion

OPA10 RAPD markers produce more than 2 fragments for all samples. The highest number of fragments (5 fragments) was produced by *R. retusa* (sample 1) while the least number of fragments (3 fragments) was produced by *D. anosmum* Gigantea (sample 4), *D. aphyllum* (sample 5), *and D. moschatum* (sample 7). Otherwise, OPA18 produce the highest number of fragments (7) by *R. retusa* (sample 1) and the least number of fragmen was produced by *D. fimbriatum* (sample 6). The diversity of DNA migration patterns produced through amplification using RAPD is widely used for the identification of organisms based on plant genetic characters.

Several RAPD-based studies to analyse diversity and genetic characters have been carried out. The research on *Gosypium hirsutum* L. germplasms aimed at analysing diversity and genetic characters using OPA10 RAPD markers showed high polymorphism with a total of 44 fragments with fragments ranging in size from 500 - 2400 bp. Similar research was also conducted on *Oryza sativa* L. which showed 32 polymorphism fragments from 34 fragments [9]. The number and intensity of the DNA fragments produced are very dependent on the primer complementation component of the DNA template used [10]. DNA polymorphism is detected based on the appearance and absence of fragments. The absence of fragments can be caused by a primary failure in the process of annealing a site in some individuals due to differences in nucleotide sequences or due to insertion or deletion at the primary site [11].

D. anosmum Gigantea Alba (sample 3) and *D. anosmum* Gigantea (sample 4) are intraspecies orchids. The morphology of the two types of orchids has similar shapes, sizes, and colours of leaves, stems, roots. The difference between the two plants lies in the colour of the flower. Based on the results of the RAPD using the RAPD OPA10 and OPA18 the two plants showed different genetic characters but several DNA fragments showed the same size (i.e. fragments measuring 918 bp and 492 bp). In interspecies orchids (*D. fimbriatum* and *D. crumenatum*) show the same number of fragments (4 fragments) and have one fragment which is the same size (1424 bp). Based on the classification according to morphological characters, *D. anosmum, D. aphyllum, D. fimbriatum*, and *D. moschatum* in the same section (Dendrobium). Typical morphological characters in this section are the shape of the pseudo bulb with medium length. The leaves are in the upper 2/3 of the stem. The leaves change after a year. Have a leaf midrib. Flowers appear from the apical stem with a long or short flower arrangement during the dry season. The size of the flower is usually large and attracts attention [12].

The results of the relationship analysis based on the RAPD OPA10 and OPA18 markings contained in the UPGMA dendrogram indicate the genetic relationship of orchid plants based on the genetic similarity index (Figure 2). At the 75% similarity level, the genotype analysed is divided into seven clusters. Just one cluster (Cluster 5) showed genetic similarities in intraspecies orchids (*D. anosmum* Gigantea and *D. anosmum* Gigantea Alba) with level of similarity 92%. *D. moschatum* and *D. fimbriatum*) showed genetic similarities in interspecies orchids with level 76%.

4. Conclusion

Nine orchid plants showed genetic diversity based on the RAPD OPA 10 and OPA 18 markers, each producing different number and size of fragments.The results of genetic analysis based on OPA10 and OPA18 RAPD markers showed that there was genetic diversity in 9 orchid plants. The similarity level of intraspecies orchids (*D. anosmum* Gigantea and *D. anosmun* Gigantea Alba) were higher than interspecies orchids (*D. aphyllum, D. fimbriatum, D moschatum, D. crumenatum, D. antennatum*). The amplification results using RAPD OPA 10, OPA 18 markers, and a combination of the two markers, genotypically showing orchids *D. anosmun* Gigantea and *D. anosmum* Gigantea Alba is in the same group, OPA10 and OPA18 can be used as molecular markers for classification intraspesies orchid.

References

[1] Irawati. (2002). Konservasi Anggrek Slpesies di Indonesia. *Prosiding Seminar Anggrek Indonesia 2002*. Yogyakarta. Hal : 46-56.

[2] Hinsley, A., H.J. De Boer, M.F. Fay, S.W. Gale, L.M. Gardiner, R.S. Gunasekara, P. Kumar, S. Masters, D. Metusala, D.L. Roberts, S. Veldman, S. Wong, J. Phelps. (2018). A review of the trade in orchids and its impications for conservation. *Botanical Journal of the Linnean Society*. 186. Pp : 435-455.

[3] Scuiteman, A. (2004). *Devogelia* (Orchidaceae), a new genus from the Moluccas and New Guinea. *Blumea*. 49 : 245-257.

[4] Zhai, J.W., Zhang C.Q., Chen L.J., Xiao X.J., Liu K.W., Tsai W.C., Hsiao Y.Y, Tian H.Z., Zhu J.Q., Wang F.G., Xing F.W., Liu Z.J. (2013). A new orchid genus, Danxiaorchids, and phylogenetic analysis of the tribe Calypsoeae. *PloS ONE*. 8: e60371.

[5] Wang LX, RX Guan, YH Li, FY Lin, WJ Luan, W Li, YS Ma, ZX Liu, RZ Chang & LJ Qiu (2009). Genetic diversity of Chinese spring soybean germplasm revealed by SSR markers. *Plant Breed.* 127: 56-61.

[6] Chase, M.W., Kenneth M., Cameron, J.V. Freudenstein, A. M. Pridgeon, G. Salazar, C. Van Den Berg, A. Schuiteman. (2015). An update classification of Orchidaceae. *Botanical Journal of the Linnean Society*. 177: 151-174.

[7] Putranto, RA and Priyono. (2014). Molecular markers and their aplication for DNA fingerprinting and genetic diversity studies in *Coffea* species. *Menara Perkebunan*. 28(1): 39-50.

[8] Dwiyani, Rindang. (2016). Induksi Organogenesis pada Mikropropagasi Anggrek *Vanda tricolor* Lindl. serta Deteksi Variasi Genetik Hasil Perbanyakan dengan RAPD. *AGROTOP*. 6 (2): 161-170.

[9] Kibria, K., F. Nur, S.N. Begum, M.M. Islam, S.K. Paul, K.S. Rahman, M.M. Azam. (2009). Molecular marker based genetic diversity analysis in aromatic rice genotypes using SSR and RAPD markers. *International Journal Sustain Corp Product.* 4 (1): 23-34.

[10] Tingey, S.V., Rafalaski, J.A., Hanafey, M.K. (1994). Genetic analysis with RAPD markers. Di dalam: Puigdomenech, P. & Coruzzi, G *(eds)*. *Plant Molecular Biology*. Berlin: Springer-Velag.

[11] Clark, A.G. and Langigan, C.M.S. (1993). Prospects for esimating nucleotide divergence with RAPDs. *Journal of Molecular Evolution*. 10 : 1096-1111.

[12] Lavarack, B., Wayne Harris, Geoff Stocker. (2006). *Dendrobium and its relatives*. Portland : Timber Press.

Isolation And Characterization Of Lactic Acid Bacteria (Lab) From Small Intestine Content Of Duck (*Anas Sp.*) As A Probiotic Candidate

Heru Nurcahyo[1], Suyanta[2], Abdollah Dale[3], and Fadholi Yudha Alif Furqon[4]

[1]*Lecturer of Biology Education Department, Mathematics and ScienceFfaculty, YSU*
[2]*Lecturer of Chemistry Education Department, Mathematics and Science Faculty, YSU*
[3]*Lecturer of Biology Education Department, Mathematics and Science Faculty, YRU*
[4]*Student of Biology Education Department, Mathematics and Sciences Faculty, Yogyakarta State University jl. Colombo No 1, Yogyakarta. 55281. Indonesia*

Abstract. A group of lactic acid bacteria (LAB) as a probiotic agent have been known to have a beneficial effect on health chicken. The objectives of this study were to isolate, characterize and determine of the LAB which originated from the small intestine content of domestic duck (*Anas sp.*) for candidate of probiotic. Isolation, selection and identification of bacteria was done using de Man Rogosa Sharpe (MRS) agar + 1% bromocresol purple. Characterization of bacteria based on morphological by Gram staining methods, and identification of biochemical characters using catalase activity, and Triple Sugar Iron Agar (TSIA). The data of isolate confirmed to *Bergey's Manual of Systematic Bacteriology*. This research carried out on the Microbiology Laboratory, Microbiology Study Program, Yala Radjabhat University (YRU), Faculty of Science and Agriculture.
The results showed that according to physiological and biochemical characteristics in various media, the bacteria had Gram +, catalase -, coccus in shape, and were able to ferment glucose to be lactic acid. Based on the result, it is concluded that the LAB isolate shows a similar characteristic to the member of genus *Lactobacillus sp*. The isolated bacteria shows characteristic resistance to acidic pH and bile salts so the bacteria were fulfilled as a probiotic candidate.
Keywords: Isolation of LAB, duck (*Anas sp.*), probiotic candidate

1. Introduction

Naturally there are enormous bacteria which live in various environments such as soil, stale food, water, and even in the intestine pathways in humans or animals. They are many types of bacteria that can produce lactic acid, antimicrobials and other metabolic results that have a positive effect on the host. One type of the useful bacteria is lactic acid bacteria (LAB). They can be used as a probiotics which can improve lactose digestibility, control pathogenic bacteria in the digestive tract, produce antibacterial and inactivation of various toxic compounds. Åsa Ljungh & Torkel Wadström (2004) said that a number of *Lactobacillus* species, *Bifidobacterium* sp, *Saccharomyces boulardii,* and some other microbes have been proposed as and used as probiotic. Probiotics are defined as microbial dietary adjuvants that beneficially affect the host physiology by modulating mucosal and systemic immunity, as well as improving nutritional and microbial balance in the intestinal tract [23]. The concept of probiotics has been defined by Fuller [9] which means "food supplement in the form of microbial life that has beneficial effects for the host animal by improving intestinal microbial balance". Several modes of probiotic action have

been proposed in the literature such as controlling the pathogenic bacteria, modulating immune responses, competing for adhesion receptors in the gut epithelium with toxin-producing bacteria [4, 5] and altering metabolism by increasing digestive enzyme activity [7].
Most of probiotic microorganisms belong to Lactic Acid Bacteria (LAB), such as *Lactobacillus* sp, *Bifidobacterium* sp and *Enterococcus* sp (Klein *et al*., 1998) [8]. Gabriela Perdigón *et al*., (2001) said that Lactic acid bacteria (LAB) are present in the intestine of most animals. Herich, and Levkut (2002), said that probiotic microorganisms including lactic acid bacteria (LAB) positively influence the composition of the gut microflora, they stimulate the production of secretory IgA, they affect the targeted transportation of the luminal antigens to Peyer's patches and they increase the production of IFN-γ.
Lactic acid bacteria in this study isolate from the intestine contents of ducks (*Anas sp.*), because ducks are more resistant to poultry diseases. These scientific evidences have been a motivating factor to choose a domestic duck (*Anas sp*) which could further confirm the results of this study. The objectives of this research was to determine the LAB to isolate, characterize and determine of the LAB which originated from the small intestine content of domestic duck (*Anas sp*) for candidate of probiotic.

2. Research Methods

This research design was to isolate, characterize, and further select the most suitable strain of LAB from small intestine contents of local domestic duck (*Anas sp*). Preparation of isolation, selection, and characterization of LAB was carried out in the Microbiology Laboratory, Microbiology Study Program, Yala Radjabhat University Faculty of Science and Agriculture.
Materials and Chemicals: The main materials used in this study were duck intestinum. All chemicals used in this study were of analytical grade, mostly purchased from Sigma Chemical Co. (St. Louis, MO). De Man, Rogosa, and Sharpe (MRS) and nutrient broth media were purchased from Merck (Darmstadt, Germany) and Oxoid Ltd. (Basingstoke, UK), Bromocresol Purple, Paint Gram Kit, HCl, NaOH, NaCl, Bile Salt.

Methods to Isolation of Bacterial Strains

The sample used in this research is taken from the contents of the duck intestine (*Anas sp*) aged 8 week, body weigh 1 kg, and purchased from a local farm ducks are taken from a farm in Pattani Province, Yala, Thailand. The section of 10-cm yeyunum, ileum and caecum of each ducks was removed and aseptically minced with contents. The resulting homogenates were serially diluted in phosphate buffer (pH 6) and plated onto de Man Rogosa Sharpe (MRS) agar + Bromocresol purple (BP) and then incubated in anaerobic jars at 37°C for 48 hours. Based on the halo zone surrounding each colony appearance as a lactic acid bacteria.

Gram Staining

The isolates were characterized by standardized methods using Gram staining. One batch of lactic acid bacteria is taken and mixed with distilled water. The mixture was flattened so that the bacterial cell is not too dense, and is passed over the fire to dry. Violet crystals are added above the isolate, wait 1 minute then rinse with running water. Iodine was added to the isolate, awaited 1 minute and then rinsed with alcohol. Safranin coloring is added, awaited 30 seconds then rinsed with running water.

Screening for Catalase Activity

Methods for screening a number of bacteria as a include catalase activity was measured by using hydrogen peroxide (H_2O_2).

Screening for Fermentation Activity

The assay of the bacteria by fermentation and lactic acid test to know homo-fermentative bacteria or heterofermentative bacteria Finally the lactic acid assays was done to all bacteria by measuring the lactic acid production. Lactic acid production activities of the selected bacteria were examined. One strains obtained from avian intestine origin was indentified.

Screening for TSIA Fermentation Activity

Profile of *Lactobacillus* were identified based on biochemical characters using Triple Sugar Iron Agar (TSIA). For fermentation activity, all strains were subcultured on modified MRS (MRS with 0.25% starch instead of whole glucose) broth.

Screening for Candidate of Probiotic

For screening probiotic activity, all strains were subcultured on modified MRS broth. For detection of acid activity, the strains were cultured on MRS broth and after anaerobic incubation for 24 h at 37°C, 30 µL of culture supernatant was transferred onto a disc placed over a medium consisting of HCl (1.5%).

For screening probiotic for bile salt activity, 1 ose of bacteria was isolated into a tube containing 5 ml MRS + Bile Salt 0.3; 0.5; 1. then isolated into an incubator at 37°C for 20 hours. Isolates that are able to grow in bile salt conditions can be seen from the turbidity of the media and the presence of deposits and then 1 ml of isolates after 20 hours incubation was isolated into 10 ml MRS + Bile Salt then 1 ml of isolates were tested for OD in a spectrophotometer each stages are used for measuring 0 hours, and tests for the 1st, 2nd and 3rd hours are made.

Inhibition test with *S.aureus* and *E.Coli*

A total of 100 µl of isolates were isolated into 5 ml MRSB. Isolates were incubated at an incubator at 37°C for 24 hours. The incubation results were centrifuged at 6000 rpm for 15 minutes. Then the suspension of pathogenic bacteria was made i.e, *S. aureus* and *E. coli* were isolated into 10 ml of 0.85% NaCl solution and matched with Mc Farland index 0.5. After that the pathogenic bacterial suspension was isolated on the MHA media by using cotton swabs to cover the entire surface of the MHA media. Holes with 1 cm diameter are made on MHA media. Then 100 µl of the supernatant from the centrifuge was isolated at the hole in the MHA media. Incubated in an incubator at 37°C for 1x24 hours.

Data Analysis

The data of research were statistically analyzed by one-way analysis of variant (ANOVA) using procedure of statistical analysis software (SAS). When significant difference present then data analysis continously by *Duncan Multiple Range Test* (DMRT) was used for comparison for each group. A value of $p<0.05$ was considered statistically significant.

3. Results And Discussion

A. Results

1. Isolation of Lactic Acid Bacteria (LAB)

Isolation of Lactic Acid Bacteria (LAB) from the small intestine contents of the Ducks (*Anas sp*) using De Man, Rogosa, and Sharpe agar (MRSA) media contains Bromocresol Purple (BP) and spread plate method was obtained some colonies of LAB are shown in the table 1 as follow:

Table 1: Isolated colonies of LAB

Source	Type of Incubation	Sample	Number		
			Dilution	Colony	Subculture
Duck Intestine	Aerobic / (A)	3	10 (-6)	21	5
			10 (-7)	(-)	-
	Anaerobic / (N)	3	10 (-6)	3	3
			10 (-7)	(-)	-
Total Colony Count				24	8

Notes: (-): Not growing, A: Aerobic Isolate, N: Anaerobic Isolates.

Isolation results have obtained 8 colonies of LAB based on the presence or absence of clear zones or discoloration in the MRSA media contain Bromocresol Purple (BP) its means that bacteria can produce acid or can make the environmental conditions acidic. Lactic acid bacteria (LAB) is a group of bacteria that produce lactic acid as one of the main fermentation products of carbohydrate metabolism [7].

2. Morphological Characterization of LAB

All isolates that had been subcultured were 8 isolates then morphological characterization of the colonies. Observations result on colony morphology include shape, edges, elevation and color of colonies are shown in the table 2 as follow:

Table 2. Characterisation of Morphology Isolate LAB

Karakteristik BAL		Kode Isolat Bakteri Asam Laktat							
		A1	A2	A3	A4	A5	N1	N2	N3
Morfologi Koloni	Warna	Putih	Putih	Putih	Putih	Putih	Putih	Putih	Putih
	Configuration	Round	Round	Round	Round	Round	Round	Round	Irregular
	Margin	Entire	Entire	Entire	Entire	Entire	Entire	Entire	Lobate
	Elevation	Flat	Flat	Flat	Flat	Flat	Flat	Covex	Flat
Morfologi Sel	Gram	Gram +	Gram +	Gram +	Gram +	Gram +	Gram +	Gram +	Gram +
	Bentuk	Coccus	Coccus	Coccus	Coccus	Coccus	Coccus	Bacil	Bacil

Colonies obtained were white, had a round configuration and only N3 isolates had irregular configurations, the shape of the entire colony edge except for N3 lobate isolates and mostly flat elevations and only N2 isolates had convex elevation.

All subculture isolates were then characterized by Gram staining. Lactic Acid Bacteria generally have gram-positive characteristics (Seely *et al.*, 2001: 145-146). As a result, all 8 isolates of LAB obtained from the duck intestine (*Anas sp.*) were gram-positive bacteria.

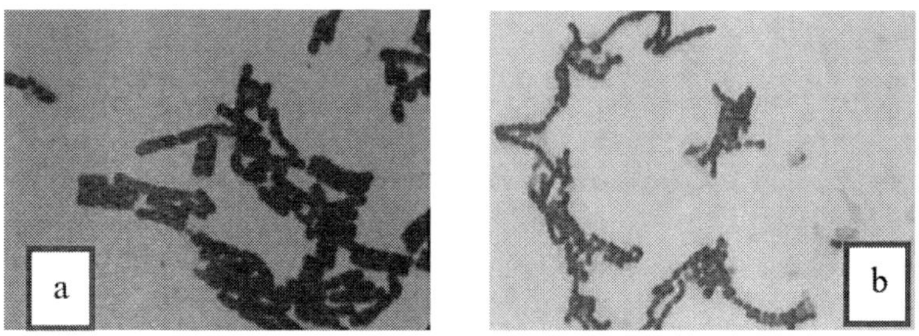

Figure 1. a) LAB Gram positif, bacillus, b) Gram positif, coccus

Gram-positive bacteria have a special element, namely teichoic acid as much as 50% of the dry weight of cell walls. This element has a function to maintain ion transport, cell wall integrity and others so that it is resistant to autolysis and maintains external permeability. This ability is the basis for choosing gram-positive bacteria as probiotics because morphology and biochemistry are better able to survive in the digestive tract. All isolates obtained showed gram positive so that they could be used as probiotics.

3. Biochemical Test of Isolates of Lactic Acid Bacteria

Catalase test

The results of the catalase test on 8 bacterial isolates showed a negative result indicated by the absence of gas bubbles containing oxygen when the isolates were dripped with H_2O_2 solution. This is in accordance with the research of Stamer *et al.*, (1979) which said that lactic acid bacteria including bacteria with negative catalase. Lactic acid bacteria generally have negative catalase characteristics (Seely *et al.*, 2001).

Temperature Resistance

From the results of the temperature resistance test, all LAB isolates from the duck intestine are a type of mesophilic bacteria. This can be seen in table 3, where there are no isolates which can grow in temperatures of 45°C and 50°C. These results are in accordance with Anastiawan (2014: 60) Lactic acid bacteria isolated from the intestines of ducks can grow at a temperature of 15°C, 37°C, and 45°C. But growth looks better or maximum at 37°C.

Table 3. Result of temperature test

Temp	Code of Isolates							
	A1	A2	A3	A4	A5	N1	N2	N3
20°C	✓	✓	✓	✓	✓	-	-	-
30°C	✓	✓	✓	✓	✓	-	-	-
37°C	✓	✓	✓	✓	✓	✓	✓	✓
45°C	-	-	-	-	-	-	-	-
50°C	-	-	-	-	-	-	-	-

Notes :
✓ = growth
- = not growth

PH conditions

The pH test results in Broth MRS can be seen from the turbidity of the medium and the presence of deposits. The pH test results are in table 4 as follows:

Table 4. Result of pH test

Ph	Kode							
	A1	A2	A3	A4	A5	N1	N2	N3
3	+	+	+	+	++	+	++	++
3,5	+	+	+	+	++	+	+	+
4	+	+	+	+	+	-	-	+
5	+++	+++	+++	+++	+++	-	-	+
7	+++	+++	+++	+++	+++	+++	++	++
9	++	++	++	++	++	-	-	-

Notes:
\+ = little
++ = bit little
+++ = many sediments

The pH test results (Fig 5) showed that 8 isolates were able to survive at low pH conditions of 3 and 3.5. But the most effective growth is at a normal pH between 6-7. Resistance to low pH is one of the main characteristics for probiotic bacteria (Saarela et al., 2000). The pH value in the digestive tract of poultry in each part is: cached (4.5), proventriculus (4.4), gizzard (2.6), duodenum (5.7- 6.0), jejunum (5.8), ileum (6.3), colon (6.3), ceca (5.7), and bile (5.9) (Sun, 2004).

NaCl Test

Digestive conditions of chickens have different NaCl content, depending on the content of NaCl contained in rations or chicken feed. NaCl (3, 4, 6.5, 10). The results of the NaCl test in MRS Broth media with isolation for 2x24 hours in 37°C temperatures are found in table 5 below:

Table 5. Result of LAB in various concentration of NaCl

NaCl	Kode							
	A1	A2	A3	A4	A5	N1	N2	N3
3	+	+	+	+	+	++	+	+
4	+	+	+	+	+	+	+	+
6,5	+	+	+	+	+	+	+	++
10	+	+	+	+	+	+	+	+
K	+++	+++	+++	+++	+++	+++	+++	+++

Notes:
\+ = little
++ = bit little
+++ = many sediments

From the results of the test of resistance to NaCl the results are positive if turbidity and the presence of sediment arise. All lactic acid bacterial isolates from duck intestines (Anas sp) (Fig 6) can still survive in NaCl levels from 3% to 10%. LAB will tolerate high salt levels to start the process of metabolism to produce acids that can inhibit the growth of undesirable bacteria.

According to Axelsson (2004), LAB which can live at a salt concentration of 6.5% if it is rod-shaped is included in the genus Lactobacillus. Certain lactic acid bacteria such as *Pediococcus acidilactici* able to grow at 5-10% NaCl levels (Romadhon, *et al.*, 2012: 5). According to Buckle et al., (1987), in the fermentation process of fish using high salinity, it is estimated that the type of LAB that is able to grow and develop is from the genera *Lactobacillus sp, Pediococcus sp, and Leuconostoc sp.*

a. Tolerance to Bile Salt

The results of the preliminary test of resistance in the various concentration of bile salt on MRS Broth media for 24 hours in temperatures 37°C are found in table 6 below:

Table 6. Result of Bile Salt Test

Bile Salt	Kode							
	A1	A2	A3	A4	A5	N1	N2	N3
0,3	+	+	++	+	+++	-	+++	+++
0,5	+	+	++	+	+++	-	+++	+++
1	+	+	++	+	+++	-	+++	+++
K	+++	+++	+++	+++	+++	+	+++	+++

Notes:
\+ = little
++ = bit little
+++ = many sediments

From the results of the preliminary test all isolates had 3 isolates which showed that the isolate was able to continue to grow effectively in 0.1 to 1% bile salt content (Figure 7) which was seen from the change in color of the MRS medium broth to become cloudy and there were many deposits. The isolates are isolates A1, N2, and N3. Bezkorovainy (2001: 401) states that bile in the small intestine inhibits the growth of existing microbes, therefore LAB, especially Streptococcus which will be used as a probiotic must be able to survive against bile salts in order to live and perform its role when in the chicken intestine. The concentration of bile salts in the digestive tract averages around 0.3%, and can range up to an extreme of 2.0% during the first hours of digestion (Gotcheva et al., 2002).
The three isolates were further tested by conducting a test using the plate count agar method. To ascertain whether within the first 3 hours after isolation in the MRS broth media added Bile Salt isolates still grow and live also to count the number of bacterial colonies. Further spectro test results of the 3 selected LAB isolates can be seen in table 7 as follows:

Table 7. Bile salt test result of 3 isolates choosed

Code	hour ke-	Bile salt (%)			
		K	0,3	0,5	1
A5	0	0,291	0,240	0,197	0,183
	1	0,495	0,462	0,322	0,162
	2	0,669	0,665	0,552	0,131
	3	0,130	0,129	0,112	0,103
N2	0	0,243	0,232	0,197	0,207
	1	0,824	0,668	0,815	0,706
	2	0,266	0,252	0,271	0,325
	3	0,290	0,255	0,314	0,365
N3	0	0,087	0,111	0,072	0,124
	1	0,082	0,414	0,136	0,518
	2	0,090	0,137	0,431	0,173
	3	0,154	0,179	0,142	0,174

According to Siti Umniyati (2009: 171) Lactic acid bacteria genus Streptococcus when tested in a concentration of bile salt (bile salt) 0% to 1% for 20 hours still able to grow and live. Where in the growth phase there is a slow phase that occurs at the 0th hour to the 4th hour because the initial hours of incubation density have not increased significantly ($P > 0.05$). Density starts to increase significantly ($P < 0.05$) from the 5th to the 11th hour, this phase is then called the exponential phase. The next phase is the stationary phase which lasts from the 12th hour to the 20th hour (end of observation). This can be proven in the results of further tests on 3 selected isolates (Table 7) where the spectro results show that until the third hour there is still growth of lactic acid bacteria. This growth does not increase significantly according to the statement of Siti Umniyati (2009: 171) where in the 3rd hour it is still included in the slow phase so that growth does not increase significantly. In addition, colony counts are also carried out using the plate count agar method. The three isolates selected were isolated in 10ml MRS broth + Bile Salt media and isolated for 3 hours. 1ml is taken every 1 hour to see the OD values in the spectrophotometer and 1ml taken for serial dilution to 10-3. A total of 0.1 ml of 10-3 dilution results were isolated in Nutrient Agar medium with 2x repetitions and isolated for 1x24 hours at 37°C. Calculation of the number of colonies needs to be done because the results of the OD values in the spectrophotometer only show the turbidity level of the MRS broth media, without knowing the condition of the bacteria in the media is still alive or not. From the calculation of the number of colonies 3 isolates in all levels of bile salt cannot count the number of colonies, because the colonies that grow in all plates show a number of 300 < (Fig. 8). But these results still show that 3 selected LAB isolates are probiotic candidates because they can survive and grow in 1% bile salt content.

Anti Pathogenic Bacteria

Table 8. Diameter inhibition of pathogenic bacteria *S. aureus* and *E. coli*

Patogens	Sample	Code		
		A5	N2	N3
	1	0,605	0,205	0,025
S. aureus	2	0,495	0,015	0,02
	3	0,235	0,01	0,025
average		0,445	0,076	0,023
	1	0,434	0,105	0,05
E. coli	2	0,38	0,075	0,015
	3	0,33	0,115	0
average		0,38	0,098	0,022

These isolates showed resistance to stomach pH (pH 3,0), tolerance against 0,3% bile concentration. Three of the isolates were observed as potential probiotic. Two of them are bacilli and the other is cocci. Two lactobacilli were identified as *Lactobacillus sp.* In the light of this study, it is observed that, small intestine of duck are a source of potential probiotic strains.

4. Conclusion

The results of this study suggest that the isolated bacteria have been isolated from small intestine of domestic duck (*Anas sp.*) having the character of Gram positive, catalase negative, rod form, respectively, and therefore the bacteria are genus *Lactobacillus spp.* Based on the result, that the bacteria could survive on the acidic pH and bile salts is prerequisite for choosing as a probiotic candidate.

References

[1] Barrow, P.A., (1992). Probiotics for chickens. *Probiotics The Scientific Basis*. Chapman & Hall: London.
[2] Fuller, R. 1989. Probiotic in Man and Animal. *J. Appl. Bacterial*. 66: 365-378.
[3] --------, 1992. History and development of probiotics. *Probiotics The Scientific Basis*. Chapman & Hall, London.
[4] --------, 1999. Probiotics for farm animal. *Probiotics A Critical Review*. Horizon Scientific Press, Wymondham, U.K.

[5] Gabriela Perdigón, Roy Fuller, and Raúl Raya. (2001). *Lactic Acid Bacteria and their Effect on the Immune System.* For correspondence. Email perdigon@cerela.org.ar; Tel. 54-381-4310465; Fax. 54-381-4311720. Argentina, 359 Ryeish Green, Three Mile Cross, Reading RG7 1ES, UK

[6] Janeway,CA and Travers,P. 1994. *Imunnobiology, the imune System in health and disease.* Blackwell Scientific publ. Oxford,9:11- 9:12.

[7] Kiyoshi Takeda & Shizuo Akira, 2003. TLR Signaling Pathways. Elsevier

[8] Marcinakova, M., Simonova, M. dan Laukova, A. 2004. *Probiotic Properties of* Enterococcus *faecium EF9296 Strain Isolated from Silage.* (Diakses Minggu, tanggal 20 Februari 2011 pukul 20.14 WIB). http://actavet.vfu.cz/pdf/200473040513.pdf.

[9] Miettinen M., Sareneva T., Julkunen I. dan Matikainen S., 2001. *IFNs Activate Toll-Like Receptor Gene Expression In Viral Infections.* (Diakses Minggu, tanggal 20 Februari 2011 pukul 20.44 WIB). http://www.ncbi.nlm.nih.gov/pubmed/11607792.

[10] Salminen, S., Ouwehand, A., Benno, Y., and Lee, Y. K. 1999. Probiotics: *How should they be defined Trends Food Sci.* Technol. 10: 107–110.

[11] Salminen, S., A.V. Wright & A.C. Ouwehand (eds). *Lactic acid Bacteria, Mikrobiology and Functional Aspects*. Third Edition. Marcel Dekker, Inc. New York-Basel.

[12] Tadao Saito, 2004. Search for excellent probiotic lactic acid bacteria and their applications.

[13] Taheri, H.R., Moravej, H., Tabandeh, F., Zaghari, M., and M. Shivazad. 2009. Screening of lactic acid bacteria toward their selection as a source of chicken probiotic. Poultry Science 88: 1586–1593

Detection *Staphylococcus aureus* Producing Enterotoxin A on the Skewers Meatballs Product in Yogyakarta City Indonesia

T Y Budiarso [a)], G Prihatmo, R Restiani, S Pakpahan, L Sari

Biology Department, Faculty of Biotechnology, Universitas Kristen Duta Wacana. Jl Dr.Wahidin Sudirohusodo 5-25 Yogyakarta.55224.

[a)]yahya@staff.ukdw.ac.id

Abstract. Meatballs are snacks that are sold by street vendors in the school area and some public places in Yogyakarta city. Based on the materials used and the traditional producing process, it is possible that the hearty snacks can be contaminated by enterotoxin-producing *Staphylococcus* bacteria. This study aims to detect the presence of enterotoxin-producing *Staphylococcus* on skewers meatballs. Skewed meatball samples were taken from ten different locations in Yogyakarta city. The samples were grown on the *Staphylococcus* sp test standard medium, Baird Parker Agar (BPA). The colony candidates were then selected into Mannitol Salt Agar (MSA), carbohydrate fermentation medium, gram staining and then confirmed using the API Staph. Molecular characterization by detecting *nuc* genes and *sea* genes as encode enterotoxin A producers. The result of this research showed that the total bacteria ranged from $1.7x10^4$ - $2.0x10^9$ CFU/g. Identification results using API Staph were found as *S. aureus*, *S. cohni*, *S. xylosus*, *S. lentus*, *S. warneri*, and *S. sciuri* while *nuc* and *sea* genes were only found in *S. aureus*.

1. Introduction

The *Staphylococcus* genus currently has more than 50 species and 28 sub-species i.e. gram-positive bacteria, with a diameter of 0.5-1.5 µm, its cocci form is in grape like clusters, commonly living in colony on the skin or mucous membrane of human and other animals. This bacteria is naturally found as normal flora on animal skin and humans [9][10]. Most *Staphylococcus* sp bacteria are pathogenic and capable of producing toxins that can cause health problems for humans. Based on the findings of previous studies, this member of *Staphylococcus* sp genus that produces the most toxins and has caused health problems is *Staphylococcus aureus*. This bacteria mostly contaminates food, mainly due to traditional processing or dirty hand-contact. Contamination may also be caused by raw materials, equipment, containers, and poor environmental sanitation [7].

The previous studies had found *Staphylococcus* sp isolates in fresh milk that were sold by street vendors and cafes in Yogyakarta [12]. Evidently, *S. aureus*, *S. epidermidis*, and *S. hemoliticus* were also found in infant formula [19]. In addition to milk and its products, it is also found in home-scale snacks, i.e. the skewed meatballs in Bantul, Yogyakarta. The results of isolation and identification during the production process of skewed meatballs from raw materials preparation to the presentation were obtained by various kinds of *Staphylococcus* sp. Based on the confirmation results using API-STAP, several types of *Staphylococcus* were identified with the following determination index (% ID):

from the raw material, *S. epidermidis* (97.6%) and *S. aureus* (97.9%) were found, in the mixing process *S. xylosus* (92.6 %) was found; in kneading process isolates *S. epidermidis* (97.1%) and *S. lentus* (99.8%) were found, and in the presentation process *S. lugdunensis* (74.7%) and *S. epidermidis* (97.1%) were found [21]. *Staphylococcus aureus*, *S. epidermidis*, *S. lentus*, and *S. xylosus* isolates were also found in fruit and vegetable salads [22]. Skewers meatballs is the one of favorite snacks for many people in Yogyakarta. These meatballs are made using workers' hands to allow staphylococcus contamination. This study aims to detect enterotoxin-producing *S. aureus* in skewers meatballs sold in the school area and several public places in Yogyakarta and its surrounding areas.

2. Experimental

2.1. Samples

Skewer meatball samples were taken from 10 different locations which were often visited by the students and some public places which were crowded by costumers in Yogyakarta city and its surrounding areas. The samples were then transported to the Microbiology Laboratory of Universitas Kristen Duta Wacana for testing according to the standard method for the detection of *Staphylococcus* sp.

2.2. Isolation and selection of Staphylococcus

A sample of 10 grams was then mashed aseptically. It aimed to suspend all bacteria in 90 ml of 1% peptone water medium. The cell cultures were grown into Baird Parker Agar (BPA) medium, incubated for 48 hours at 37 °C. The suspected colonies of *Staphylococcus aureus* will give a blackish-gray appearance and have an opaque zone around the colony. The suspected colonies were then selected by streaking them into the Mannitol Salt Agar (MSA) medium and then incubating at 37°C for a maximum of 24 hours[1] [5][13]. The growth of *S. aureus* colonies was marked by the color change of the colony to yellow in less than 24 hours. The colonies that turned yellow after 24 hours were non-aureus *Staphylococcus* group. The yellow colonies were then isolated and grown into the Brain Heart Infusion (BHI) medium as a culture stock for the following tests, that being gram staining and carbohydrates fermentation. The biochemical test culture tubes were incubated at 37C for 48 hours. The color change of the red medium to yellow indicated a positive result. Colonies that have been indicated to certain species of a *Staphylococcus* sp genus were then separated to be subcultured into BHI Agar medium for 18-24 hours. It was then used as culture stock for the *Staphylococcus* confirmation test using the API-Staph [17][18][24].

2.3. Biochemical identification of Staphylococcus

The isolates were grown into the API-Staph medium, then homogenized and its turbidity levels were standardized using a 0.5% McFarland standard solution. Furthermore, the cultures were dropped on biochemical test wells found in the Biomeriux Staph-API test kit. Mineral oil was then added into the wells used for ADH and UREA test to create an anaerobic condition. The incubation process was carried out at 37°C for 18-24 hours. After the incubation process, VP reagents (Voges Proskauer), PAL, NIT (nitrate) were added into wells used for the NIT, ZYM, and VP tests. Positive results of *Staphylococcus* sp were confirmed using a positive control table and API web software (bioMerieux).

2.4. Molecular detection of nuc and sea genes

Each of the confirmed *S.aureus* and non-aureus isolates from the API Staph test was taken for molecular testing by detecting *nuc* and *sea genes*. Molecular detection initiated with DNA isolation phase. The DNA isolates were tested using electrophoresis. The *nuc* genes were amplified using reference primers [3][4][20]. It was performed using the following oligonucleotide sequences:

fw5'GCGATT GATGGTGATACGGTT-3' and rv3' GCCAAGCCTTGACGAACTAAAGC-5'. The reaction consisted of 1 µl ddH_2O, 12.5 µl Mix PCR Promega, 1 µl of primer F, 1 µl of primer R, 1 µl DNA. The PCR phase of the *nuc* gene was amplified at 270 bp, pre-denatured 95^0C for 2 minutes; 35 cycles of amplification (denaturation at 95^0C for 60 seconds, annealing at 58^0C for 30 seconds, and extension at 72^0C for 60 seconds) and final extension at 72^0C for 5 minutes. *Sea* gene was amplified using the following primers: (sea) - F5'-TTGGAAACGGTTAAAACGAA-3' and (sea) R5'GAACCTTCCCATCAAAAACA-3' [14]. The reaction consisted of 1 µl of ddH2O, 12.5 µl of Mix PCR Promega, 1 µl of F Primer, 1 µl and R Primer, 1 µl of DNA. In PCR phase, *sea* genes were amplified at 120 bp, pre-denaturation was carried out for 3 minutes at 95°C, denaturation for 2 minutes at 94°C, annealing for 35 cycles for 1 minute 30 seconds at 58°C, and final extension for 5 minutes at 72°C.

3. Results and Discussion

3.1. Isolation and selection of Staphylococcus bacteria

The results of the growth of colonies of *aureus* and *non-aureus Staphylococcus* from the samples studied are shown in Figure 1. Colonies of *S. aureus* have a type of the selective medium of BPA. The BPA medium is a differential selective medium for the isolation and detection of *Staphylococcus* sp and *S. aureus* bacteria groups. BPA medium was chosen for the isolation of *Staphylococcus* sp in this study because the medium has been recommended by The International Organization for Standards, the Association of Official Analytical Chemists International (AOAC), Bacteriological Analytical Manual (BAM), American Public Health Association (APHA) and International Dairy Federation (IDF) [5].

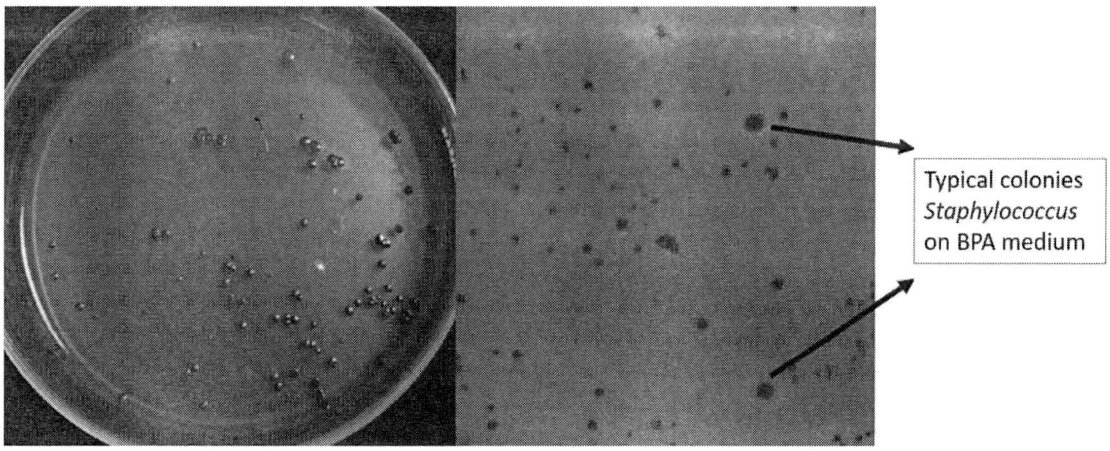

Figure 1. Growing of *Staphylococcus* sp colonies on BPA medium.

The medium of BPA contained sodium pyruvate which played a role in stimulating the growth of *S. aureus*. The suspected *S. aureus* colonies that grew on the BPA medium have typical grayish black colonies due to tellurite reduction to tellurium. In addition the colonies also have haloed zones around the colony (opaque halo) due to lecithinase activity. The occurrence of lipolytic and egg yolk proteolytic formed a clear zone to appear around the colony. The opaque zone in the clear zone would only appear after 48 hours of incubation. In addition, the growth of *S. aureus* and S. *xylosus* in the BPA medium was also characterized by the presence of a clear zone that could be clearly seen in the medium, grayish black colonies. The BPA medium also contained lithium chloride and tellurite which functioned to inhibit the growth of other microorganisms, whereas pyruvate and glycine selectively stimulated the growth of the *Staphylococcus* sp group [5][13][25].

3.2. Biochemical characteristics of Staphylococcus

At this stage, the suspected *S. aureus* colonies were selected into the MSA medium to confirm the suspicion because of their resistance to salinity. MSA medium contained beef extract and protease peptone so that it strongly supported the growth of *Staphylococcus* sp and other bacterial growths which would be inhibited by high salt content of 7.5%. *Staphylococcus aureus* would grow well on MSA medium because of its ability to ferment mannitol and resistance to salt which produced yellow colonies in less than 24 hours incubation. The group of coagulase-negative *Staphylococcus* and *Micrococcus* sp bacteria could not ferment mannitol, thus *S. aureus* has a very good opportunity to grow and it is easily separated from other bacteria. Yellow colonies isolated from the MSA medium were selected through gram staining [6] and carbohydrate fermentation tests. This test aimed to determine the ability of the suspected *S. aureus* isolates in fermenting sugar substrates in order to get the suitability of *S.aureus* properties based on identification keys of *Staphylococcus* sp (Bergey's Manual of Systematic Bacteriology).

Staphylococcus aureus was able to change the color of the carbohydrate fermentation medium on the substrate maltose, trehalose, and sucrose while on the xylose substrate, the medium color did not change. Of all the samples were taken from 10 different locations, 25 isolates leading to *S. aureus* were discovered based on the appearance of gram staining which had the shape of cocci, purple-colored, and clustered. From the carbohydrate fermentation test, 10 suspected *S. aureus* isolates, 8 suspected *S. xylosus*, 2 isolates of *S. lentus*, and 3 isolates of *S. epidermidis* were found. All of which were confirmed biochemically using the API Staph (bioMerieux). The isolates were inoculated into the medium contained in the kit. They were then homogenized so that they had turbidity equivalent to 0.5 McFarland, then cell cultures were inoculated into the API Staph and incubated at 37^{0}C for 24 hours. The biochemical test results indicated a change in the color of the reagent of the API Staph kit according to the bioMerieux procedure which gave positive and negative results. The test results were then inputted into the bioMerieux web API to find out the identification results of the bacteria tested. Of the 10 isolates tested, only 8 isolates had %ID≥70%, that being *S. xylosus* with 2 isolates, *S. warneri* with 1 isolate, *S. sciuri* with 1 isolate, *S.cohni* with 1 isolate and *S. aureus* with 2 isolates (Figure 2).

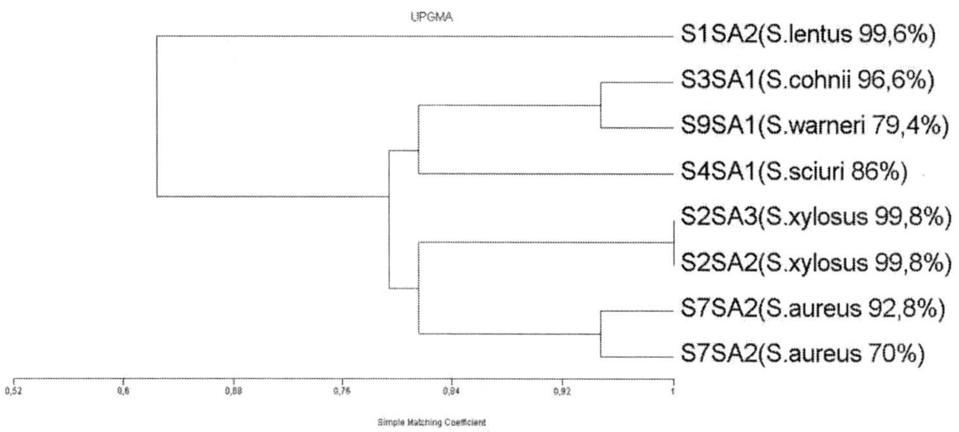

Figure 2. The diversity of *Staphylococcus* sp contaminant on skewed meatballs.

3.3. Molecular detection of nuc and sea genes of Staphylococcus

All isolates identified as *S. aureus* were then molecularly characterized using *nuc* gene markers. The *nuc* gene is a marker for the thermonuclease enzyme that established to molecular identification of *S. aureus* [4], while the *sea* gene marker was used to determine the nature of enterotoxin. Detection of *nuc* and *sea* genes in *S. aureus* suspected isolates used *nuc1*, *nuc2* and *sea1*, *sea2* primers. The results of *nuc* and *sea* gene detection in *S. aureus* suspected isolates are shown in Figure 3.

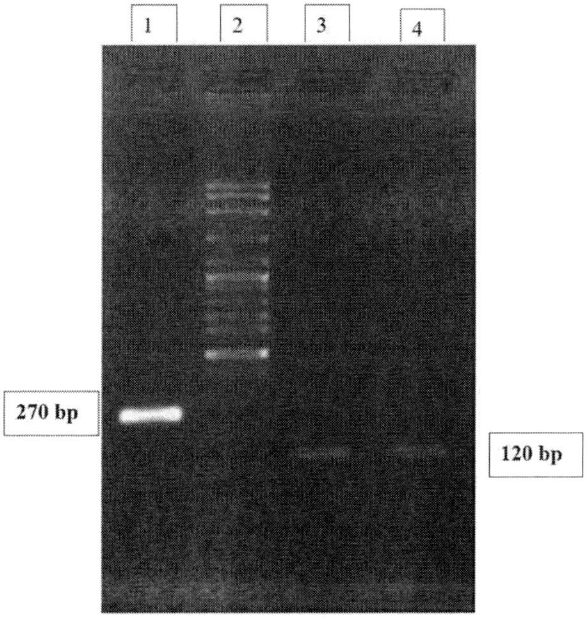

Figure 3. Detection *nuc* and *sea* genes of *S. aureus* and *S. lentus*.

Line 1.Amplified fragment using primer for nuc gene (270 bp) of S. aureus; 2, Marker; 3, Amplified fragment using primer for sea gene (120 bp) of *S.aureus* and 4, *S. lentus*.

Figure 3 shows that the DNA amplification process of *S. aureus* isolates was successfully carried out using *nucl, nuc2* and *seal, sea2* primers. The findings shows that the isolates identified as *S.aureus* based on the confirmation results using the API Staph with% ID 70% have *nuc* and *sea* genes. While the isolates identified as S. *lentus* showed no *nuc* gene, but had the *sea* gene. The presence of the *nuc* and *sea* genes in *S. aureus* indicates that these bacteria are high-heat resistant and able to produce enterotoxin A which may pose risk for human health [23]. *Staphylococcus aureus* strains that grow under different stress conditions have a higher resistance to heat than those grown in a physiological environment suitable for their growth [8]. In this study, it was found that not only *S. aureus* bacteria were contaminant bacteria in skewed meatballs, but also other *Staphylococcus* group bacteria such as *S. cohni*, *S. warneri*, *S. lentus*, *S. xylosus,* and *S. sciuri*. *Staphylococcus cohni*, *S. warneri*, *S. xylosus* is a group of *Staphylococcus* sp bacteria that also have opportunistic pathogenic properties [2]. *Staphylococcus lentus* is a commensal bacteria whose habitat is on the skin of various animal species, such as poultry and dairy animals. In addition workers on farms are also at risk as hosts of *S. lentus* bacteria [11]. Cases of infection caused by *S. lentus* bacteria are very rare in humans [15]. *Staphylococcus sciuri* is a bacteria that can be found in pets and domestic animals, both healthy and sick. This bacteria has a characteristic of surviving using inorganic salt as the only source of nitrogen. Similarly, *S. lentus* and *S. sciuri* also rarely causes infection in humans [16].

4. Conclusion

Meatball samples obtained from 10 different locations in Yogyakarta contained contaminants of *Staphylococcus* bacteria between 1.7 x 10^4 to 2.0 x 10^9 colonies/gram. The biochemical identification findings using the Staph API indicate that each skewed meatball sample has been contaminated by *S. aureus* bacteria with %ID of 70% and 92%. In addition, other types of *Staphylococcus* were also found, that being *S. lentus* with% ID of 99.6%, *S. xylosus* with %ID of 99.8%, *S. cohni* with %ID of 96.6%, *S. sciuri* with %ID of 86% and *S. warneri* with %ID of 79.4%. Molecular identification findings on isolates identified as *S. aureus* with %ID of 70% in the Staph API confirmation test shows a positive result of having a *nuc* gene at 270 bp and *sea* gene at 120 bp. In addition to other *Staphylococcus*, *S. lentus* only has *sea* genes that have the potential to produce enterotoxins A.

Acknowledgments
This paper is sponsored by the Ministry of Research, Technology and Higher Education of the Republic of Indonesia. The project number: 227/SP2H/LT/DRPM/2019.

References
[1] Bennett R W and Lancette G A 2016 Bacteriological Analytical Manual **12** 1-6
[2] Bannerman TL 2003 *Manual of Clinical Microbiology* vol 8, ed P R Murray, E J Baron, J H Jorgensen, M A Pfaller, R H Yolken (Washington: American Society Microbiology) pp 384-404
[3] Becker K, von Eiff C, Keller B, Brück M, Etienne J, Peters G 2005 *Diagnostic Microbiology and Infectious Disease* **51** 237–244
[4] Brakstad O G, Aasbakk K, Maeland J A 1992 *Journal of Clinical Microbiology* **30** 1654–1660
[5] Capita R, Allonso-Calleja C, Moreno B, and Garcia-Fernandez MC 2001. The Journal of Microbiology **39** 321-325
[6] Cappucino, James G, and Natalia S 2017 Microbiology: A Laboratory Manual **11** 171-176
[7] de Oliveira LP, e Barros L S, S., Silva C V, Cirqueira M G 2011) *Journal of Food Processing & Technology* **2** 2-6
[8] El-Banna A A, Hurst A 1983. *Canadian Journal Microbiology* **29** 297-302
[9] Fetsch A, Johler S 2018 Current Clinical Microbiology Reports **5** 88-96
[10] Gotz F, Bannerman T, Schleifer K H. 2006.The Prokaryotes: A handbook on the Biology of Bacteria vol 4, ed Dworkin M, Falkow S, Rosenberg E, Scheleifer K-H, Stackebrandt E (New York: Springer) pp 5-75
[11] Huber H, Ziegler D, Pfluger V, Vogel G, Zweifel C, Stephan R 2011. *BMC Veterinary Research* **7** 6.
[12] Irwanto M I and Budiarso T Y 2015. Proceeding 1st International Seminar on "Natural Resources Biotechnology: from Local to Global **1** 290-295
[13] Jamali H, Paydar M, Radmehr B, Ismail S 2015 *Food Control* **54** 383–388
[14] Johnson W M, Tyler S D, Ewan E P, Ashton F E, Pollard D R, Rozee K R 1991 *Journal of Clinical Microbiology* **29** 426-430
[15] Karachalios G N, Michelis F V, Kanakis K V, Karachaliou L, Koutri R, Zacharof A K 2006 Scandinavian Journal of Infectious Diseases **38** 708 –710
[16] Kloos W E, Ballard D N, Webster J A, Hubner R J, Tomasz A, Couto I, Sloan G, Dehart H P, Fiedler F, Schubert K, de Lencastre H, Sanches I S, Heath H E, Leblanc P A, Ljungh Å 1997
[17] Koop G, de Visscher A, Collar C A, Bacon D A C, Maga E A, Murray J D, van Werven T 2012 *Journal of Dairy Science* **95** 7200–7205
[18] Langlois B E, Harmon R J, Akers K 1984 *Journal of Clinical Microbiology* **20** 227–230
[19] Palilu PT and Budiarso T Y 2017 AIP Conference Proceeding **1844** 010001
[20] Sahebnasagh R, Saderi H, Owlia P 2014 *Iran J. Public Health* **43** 84–92
[21] Salle G E, and Budiarso T Y 2017 Prosiding Seminar Biodiversitas **6** 1-5
[22] Saputra T N and Amarantini C 2017 Prosiding Seminar Biodiversitas **6** 6-9

Journal of Physics: Conference Series **1397** (2019) 012044 doi:10.1088/1742-6596/1397/1/012044

[23] Tucker P W, Hazen E E, Cotton F A 1987 *Molecular and Cellular Biochemistry* **22** 67-77
[24] Vanderhaeghen W, Piepers S, Leroy F, van Coillie E, Haesebrouck F, de Vliegher S 2015 *Veterinary Journal* **203** 44–51
[25] Vermassen E, Fissore R A, Nadif Kasri N, Vanderheyden V, Callewaert G, Missiaen L, Parys J B, De Smedt H 2004 Biochemical and Biophysical Research Communications **319** 888–893

Screening of antimicrobial-producing lactic acid bacteria isolated from traditional fish fermentation against pathogenic bacteria

C Amarantini[a)], D Satwika, T Y Budiarso, E R Yunita, E A Laheba

Biology Department, Faculty of Biotechnology, Universitas Kristen Duta Wacana.
Jl dr Wahidin Sudirohusodo 5-25 Yogyakarta.55224.

[a)]charis@staff.ukdw.ac.id

Abstract. This study aims to obtain lactic acid bacteria (LAB) which have probiotic properties and inhibitory effects against pathogens. Peda fish fermented in 20% salt solution used as samples. Antimicrobial properties of LAB isolates against gram-positive and gram-negative pathogens (*Salmonella typhi* BPE 127.1.MC, *Salmonella typhi* BPE 122.4.CCA, *Salmonella typhi* NCTC 786, *Salmonella typhimurium* FNCC 0050, *Pseudomonas putida* FNCC 0071, *Bacillus subtilis* ATCC 6633 and *Staphylococcus aureus* ATCC 25923) were tested using agar well diffusion method. To exclude acid production, the cell-free culture supernatant (CFCS) solution was neutralized using 1 N NaOH before antimicrobial activity was tested. Likewise with bacteriocin tests, CFCS solutions that have been neutralized, were treated first by heating at 100°C for 10 minutes in order to enzyme inactivation. A total of 26 out of 50 LAB isolates showed clear zones in the MRS-CaCO3 agar medium, gram positive, and negative catalase as the main characteristics of LAB. Sixteen of the 26 LAB isolates were known to be tolerant to pH 2 and 1% of bile salts. In addition, 92.3% of LAB isolates (15 of 16 LAB isolates) were homofermentative. The antimicrobial test results found that two of 26 LAB isolates (i.e. LAB Pr.3.4L and Pi.5.8 isolates) were shown to have very strong inhibitory effects against gram-positive pathogens *S. aureus* ATCC 25923 compared to *S. typhi* BPE 122.4.CCA. Thus the two strains of LAB are indicated as bacteriocin-producing probiotic strains, and need to be pursued further for identification and optimization of production.

1. Introduction

Lactic acid bacteria (LAB) isolated from traditional fermentation processes are known as antibacterial producers which are active against both gram positive and negative bacteria. This broad antibacterial spectrum causes LAB as potential natural preservative in various food products [1]. As reported elsewhere, many LAB could be recovered from traditional Indonesian fermented foods, including species from *Lactobacillus, Pediococcus, Enterococcus, Weisella* and *Leuconostoc*, as well as some fungi [2]. On that account, Indonesian fermented foods are a potential source of probiotics that benefit human health.

Studies on the antimicrobial potential of lactic acid bacteria in vitro have been carried out to assess the potential of LAB probiotics from fermented foods. The antimicrobial function of lactic acid bacteria (LAB) in fermented foods is mainly served by lactic acid and acetic acid. The production of organic acids is beneficial for nutritional quality, organoleptic properties, and shelf life characteristics.

Traditionally, lactic acid bacteria develop spontaneously during fermentation process, or in their industrial application, they are added intentionally in the food matrix as a natural preservative. Lactic acid bacteria also produce acetic acid, ethanol, aroma compounds, exopolysaccharide, and several important enzymes. In addition, antimicrobial production by probiotic LAB plays a role during in-vivo interactions in the gastrointestinal tract, thus contributing to the intestinal health. Many LAB strains also produce bacteriocins that show antibacterial activity. Some LABs are also capable of synthesizing other antimicrobial peptides that also contribute to food preservation and safety [3]. Thus, compounds such as bacteriocin and antifungal peptides produced by LAB have the potential for in depth studies as new antimicrobials that will be relevant in ensuring food quality and safety in the future. Considering this background, LAB producing bacteriocin, known as bacteriocinogenic strain, deserve further research for the improvement of food quality and its safety. It is then the purpose of this study to screened LAB strain(s) with the ability to produce bacteriocins having bacteriocin activity able to inhibit other bacteria.

2. Experimental

2.1. Samples

Twelve samples of Peda fish (*Rastrelliger* sp.) were purchased from local markets namely Demangan and Kranggan in Yogyakarta, Indonesia. The samples were divided into two types. The first type is dried peda fish. For this category, each sample purchased is immediately taken to the laboratory for isolation based on the established procedures. The second type was samples taken from a solution of dried peda fish soaked in 20% salt solution for 2 weeks.

2.2. Pathogenic strain

The pathogenic strains that represent gram negative bacteria used in this study were *Salmonella typhi* BPE 127.1.MC and *Salmonella typhi* BPE 122.4.CCA obtained from previous research collections [4][5]; *Salmonella typhi* NCTC 786 obtained from PT Biofarma; and *Salmonella typhimurium* FNCC 0050 and *Pseudomonas putida* FNCC 0071 obtained from the Food and Nutrition Culture Collection (FNCC), Center for Food and Nutrition Studies, Gadjah Mada University, Yogyakarta. Whereas, for gram-positive bacteria, the indicator strains *Bacillus subtilis* ATCC 6633 and *Staphylococcus aureus* ATCC 25923 were used.

2.3. Isolation and selection of lactic acid bacteria isolates

Fifty grams of samples were cultured in 500ml Buffered Peptone Water (BPW) 0.1% and incubated at 37°C for 24 hours. Then, 25 ml of the culture was taken to be grown in 225 ml of de Man-Rogosa-Sharpe (MRS) broth medium and re-incubated at 37°C for 24 hours. Through a series of dilutions using 0.1% peptone solution, the culture was grown in an MRS-$CaCO_3$ agar medium using the spread plate method, and incubated at 37°C for 24–48 hours. Based on the characteristics of the LAB colonies, LAB isolates were then selected by picking creamy-white colonies with clear zones formed around it. The obtained isolates were collected as LAB candidates that are potential to produce antimicrobials. To ensure that these isolates are gram-positive bacteria as the main characteristic of LAB, the LAB isolates were observed microscopically for determining cell shape and gram properties to. All of the collections of LAB isolates were maintained routinely and stored in MRS agar medium containing 20% glycerol at -20°C.

2.4. Screening and characterization of antimicrobial-producing lactic acid bacteria.

Several tests were done to identify the antimicrobial-producing LAB strains by observing its morphology, gram staining, catalase formation, growth on MRS broth at different pH, the ability to grow on different concentrations of bile salts, and homofermentative or heterofermentative test.

Growth of LAB strains at various pH. To ensure the growth of LAB isolates at different pH, the MRS broth was adjusted at pH 2, pH 3, and pH 4 using 1 N HCL. The tested culture was prepared by growing it in an MRS broth for 48 hours. Cells were harvested by centrifugation at 13,500 rpm for 15 minutes. Cells were then washed twice using 100 µl Phosphate Buffer Saline (PBS) solution, after which the cell pellets were inoculated into 5 ml of MRS broth and incubated at 37 °C for 4 hours. Their growth was confirmed by plating the culture in the MRS agar [6].

Growth of LAB isolates at different concentrations of bile salts. To determine the growth of LAB isolates in various concentrations of bile salts, 1 ml of the test culture were grown in 5 ml MRS broth containing bile salts at concentrations of 0.3%, 0.5%, and 1%. Using the spread plate method, culture was grown in MRS agar and incubated at 37°C for 24 hours. Growth in the MRSA medium was determined by observing the growth of colonies in the medium [6].

2.5. Determining antimicrobial activity

To determine the antimicrobial activity of the obtained LAB strains, the indicator strain was tested against the LAB strains using well-diffusion agar method.

Cell-free culture supernatant (CFCS) *preparation.* The antimicrobial LAB isolates were grown in MRS broth at 37 °C for 18 hours. The supernatant was neutralized at pH 6.5 with 1 N NaOH solution before used for antimicrobial testing. The cells were then harvested through a 10,000-rpm centrifugation process for 10 minutes. The supernatant was collected and stored in the refrigerator until it would be used [7].

Preparation of bacteriocin-like substances. The antimicrobial-producing LAB isolates were grown in a MRS broth at 37 °C for 18 hours. The supernatant was neutralized at pH 6.5 with 1 N NaOH solution before used for antimicrobial testing. The cells were then harvested through a 10,000-rpm centrifugation process for 10 minutes. The supernatant is heated at 100° C for 10 minutes, collected, and stored in the refrigerator until it would be used [8][9].

Antimicrobial test. The antimicrobial activity was tested visually using well diffusion agar method. Mueller-Hinton Agar (MHA) is used as a standard medium recommended in the guideline [10][11]. Indicator strains were grown in Brain Heart Infusion (BHI) broth at 37° C for 16–18 hours. Then, sterile cotton swabs are dipped in the standardized bacterial suspension. The excess inoculum fluid was removed by pressing a little swab on the tube wall above the liquid. The swab that contains a stained inoculum on the MHA medium was inoculated. This swab was repeated twice so that the test culture was evenly distributed on the surface of the MHA medium. A 8 mm-diameter well was cut into a compacted agar using a sterile blue tip and filled with 150µl of supernatant culture or a solution containing a bacteriocin-like substance. The plates were incubated at 37° C for 24 hours, after which the zone of the inhibition was measured [7][9][12].

3. Results and Discussion

3.1. Isolation of antimicrobial-producing lactic acid bacteria

As seen from the colonies that grew on the MRS-CaCO3 medium, it was found that the colonies of lactic acid bacteria could be single or group colonies with neat edges, smooth surfaces, and transparent or creamy yellow colors (Figure 1). The diameter of the colony varies from 1 mm to 5 mm. From each petri dish, 5 colonies were taken to be tested further and there was certainly no contamination. Fifty candidate isolates of lactic acid bacteria from peda fish that showed clear zones were selected from MRS-CaCO3 medium. Among the 50 isolates, 26 isolates showed the characteristics of gram-positive and were negative catalase in nature. Based on the results of gram staining, under the microscope purple cells appeared in the form of basil without spores. These data follow the characteristics of lactic acid bacteria found in various fermented foods [6][8][13][14][15]. Thus, based on these characteristics, 26 isolates of lactic acid bacteria were obtained from peda fish and were potential to produce antimicrobial compounds (Table 1).

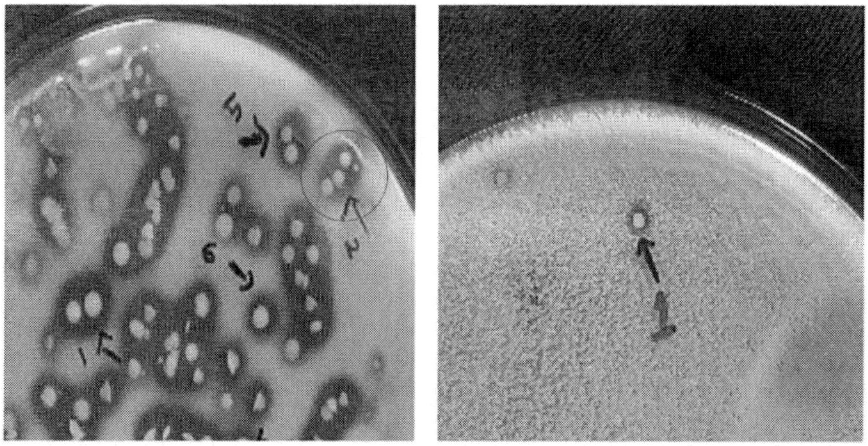

Figure 1. Colonies of lactic acid bacteria that grow on the MRS-$CaCO_3$ medium.

From microscopic observations, all of these isolates were found to be in the form of bacilli. Interestingly, most of the isolates obtained were known to be resistant to bile salts; 19 of the 26 isolates selected were even known to grow in 1% of bile salts. In addition, 16 of the 26 isolates were able to grow at pH = 2 and almost all (92.3%) were homofermentative (Table 1). These characteristics are important for assessing their potential as probiotics and as a producer of antimicrobial compounds.

To ensure that the lactic acid bacteria isolates obtained benefit human health, the bacteria must be able to reach the digestive tract alive. Although they only temporarily colonize the digestive tract, their ability to affect the mucous layer and the extracellular matrix components in the digestive tract are the main functions of LAB that give healthy effects, in addition to their role in terms of antibacterial activity and immunomudulators. In this case, lactic acid bacteria that act as probiotics play an important role in protecting host defenses against pathogenic infections in the intestinal lumen. With the fast formation of microbial communities in the digestive tract, pH will reduce, and adhesion competition will occur against pathogenic bacteria, preventing pathogenic from colonizing. In addition to competing in terms of adhesion, the production of antibacterial compounds by lactic acid bacteria such as acetic acid and bacteriocin also prevents the growth of pathogenic bacteria [6].

Table 1. Characteristics of morphology and physiology of lactic acid bacteria isolates as a result of screening from peda fish samples.

Isolate code	Gram stain	Cell shape	Ability to grow in acidic conditions			Ability to grow in bile salts (%)			Homo-/Hetero-fermentative
			pH 2	pH 3	pH 4	0,3	0,5	1	
Solution of dried peda fish:									
Pr.3.1 L	+	basil	-	+	+	-	+	+	Homofermentative
Pr.4.3 L	+	basil	+	+	+	+	+	+	Homofermentative
Pr.4.2.	+	basil	-	+	+	-	-	-	Homofermentative
Pr.4.3.	+	basil	-	+	+	-	-	-	Homofermentative
Pr.4.5.	+	basil	+	+	+	+	-	-	Homofermentative
Pr.4.6.	+	basil	+	+	+	+	+	+	Homofermentative
Pr.4.8.	+	basil	-	+	+	-	-	-	Homofermentative
Peda fish:									
Pi.5.1.	+	basil	+	+	+	+	+	+	Homofermentative
Pi.5.2.	+	basil	+	+	+	+	+	+	Homofermentative
Pi.5.3.	+	basil	+	+	+	+	+	+	Homofermentative
Pi.5.4.	+	basil	+	+	+	+	+	+	Homofermentative
Pi.5.5.	+	basil	+	+	+	+	+	+	Homofermentative
Pi.5.6.	+	basil	+	+	+	+	+	+	Heterofermentative
Pi.5.7.	+	basil	+	+	+	+	+	+	Homofermentative
Pi.5.8	+	basil	+	+	+	+	+	+	Homofermentative
Pi.5.9	+	basil	+	+	+	+	+	+	Homofermentative
Pi.5.10	+	basil	-	+	+	+	+	+	Homofermentative
Pi.6.1.	+	basil	-	+	+	+	+	+	Homofermentative
Pi.6.2.	+	basil	+	+	+	+	+	+	Homofermentative
Pi.6.3.	+	basil	-	-	+	+	+	+	Homofermentative
Pi.6.4.	+	basil	-	+	+	+	+	+	Homofermentative
Pi.6.5.	+	basil	+	+	+	+	+	+	Homofermentative
Pi.6.6.	+	basil	-	+	+	+	+	+	Homofermentative
Pi.6.8.	+	basil	-	+	+	+	+	+	Homofermentative
Pi.6.9	+	basil	+	-	-	+	-	-	Homofermentative
Pi.6.10	+	basil	+	+	+	+	+	+	Homofermentative

3.2. Screening for antimicrobial-producing lactic acid bacteria

At this stage, tests were performed to measure the ability of lactic acid bacteria in inhibiting pathogenic bacteria. These included testing the antimicrobial activity using the well diffusion assay method. To exclude the effects of acid production, antimicrobial activity was tested after the supernatant was neutralized with NAOH 1 N [16]. A total of 26 BAL isolates obtained from the isolation stage were tested for their ability as antimicrobial producers (Table 2). The results show that not all candidates for lactic acid bacteria have inhibitory effects against gram negative bacteria and against gram-positive bacteria at the same time. Overall, it was found that 21 of the 26 LAB isolates showed inhibitory effects on both strains of gram negative bacteria and gram-positive bacteria. Five LAB isolates (Pr.4.8, Pi.5.8, Pi.5.9, Pi.6.2, Pi.6.4) showed inhibitory effects only on gram-positive pathogens. *Bacillus subtilis* ATCC 6633 was found to be the most sensitive strain on the CFCS inhibitory effect of LAB isolates. The biggest inhibitory effect was seen in LAB Pr.4.3.L isolates against *Bacillus subtilis* ATCC 6633 and LAB Pi.5.8 isolates against *Staphylococcus aureus* ATCC 25923 (Figure 2). Both LAB isolates (Pr.4.3.L and Pi.5.8) also showed inhibitory effects on gram negative pathogens (Figure 3).

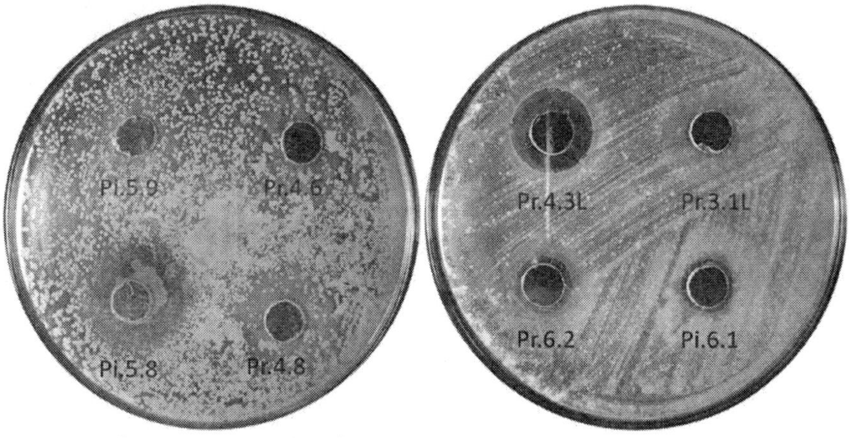

Staphylococcus aureus ATCC 25923 *Bacillus subtilis* ATCC 6633

Figure 2. The inhibitory effect of antimicrobial compounds produced from LAB Pi.5.8 isolates against *Staphylococcus aureus* ATCC 25923 and LAB Pr.4.3.L isolates on *Bacillus subtilis* ATCC 6633.

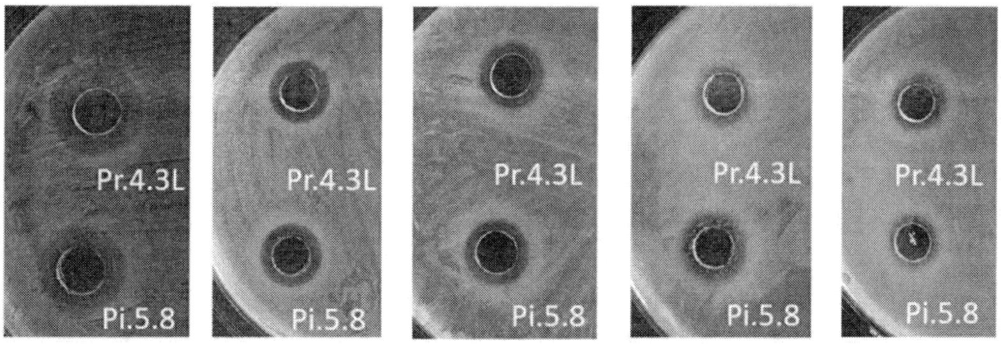

Salmonella typhi BPE 122.4.CCA | *Salmonella typhi* BPE 127.1.MC | *Pseudomonas putida* FNCC 0071 | *Salmonella typhi* NCTC 786 | *Salmonella typhimurium* FNCC 0050

Figure 3. The inhibitory effect of antimicrobial compounds produced from LAB Pr.4.3.L and LAB Pi.5.8 isolates against pathogen gram-negative bacteria.

The results of screening as shown in in Table 2, Figure 1, and Figure 2 indicate that only 2 isolates were proved to be able to inhibit gram negative and gram positive pathogenic bacteria, namely LAB Pr.4.3L and LAB Pi.5.8. Furthermore, these 2 isolates were selected for testing to detect bacteriocin activity against gram negative and gram positive pathogens (Table 3).

Table 3 shows that for the group of gram negative pathogens, both LAB Pr.4.3L and Pi.5.8 isolates were able to inhibit all gram negative indicator strains. The biggest inhibition zone (19.5 ± 0.3 cm) was detected in *Salmonella typhi* BPE 122.4.CCA. The two LAB isolates also showed inhibition against *Staphylococcus aureus* ATCC 25923 and *Bacillus subtilis* ATCC 6633. Cytoplasm membranes of gram negative bacteria were easily damaged by organic acids, making gram negative bacteria sensitive to organic acids [14]. In this study, the inhibitory effect of acid production had been neutralized with NaOH solution so that the pH of the CFCS solution becomes neutral. Therefore, it can be concluded that CFCS solutions from LAB Pr.4.3L and Pi.5.8 isolates contain antimicrobial compounds.

Table 2. The inhibitory effects of antimicrobial compounds produced by LAB isolates against indicator bacteria.

Isolate code	Antimicrobial activity of LAB isolates against indicator bacteria						
	Salmonella typhi NCTC 786	*Salmonella typhi* BPE 127.1MC	*Salmonella typhi* BPE 122.4CCA	*Salmonella typhimurium* FNCC 0050	*Pseudomonas putida* FNCC 0071	*Bacillus subtilis* ATCC 6633	*Staphylococcus aureus* ATCC 25923
Pr.4.3 L	+	+	+	-	+	+	-
Pr.4.3.	-	-	-	-	+	+	-
Pr.4.5.	-	+	+	-	-	+	-
Pr.4.8.	-	-	-	-	-	+	-
Pi.5.3.	-	-	-	+	+	+	-
Pi.5.4.	+	-	-	+	+	+	+
Pi.5.5.	+	-	-	+	+	+	-
Pi.5.6.	-	-	-	+	+	-	-
Pi.5.8	-	-	-	-	-	+	+
Pi.5.9	-	-	-	-	-	+	-
Pi.5.10	-	-	+	-	-	+	-
Pi.6.1.	-	+	-	-	+	+	-
Pi.6.2.	-	-	-	-	+	+	-
Pi.6.4.	-	-	-	-	-	+	-
Pi.6.8.	-	+	-	-	-	+	-
Pi.6.10	-	+	+	-	-	+	-

Note: + (presence) - (absence)

Table 3. Diameter of inhibitory zone (mm) of LAB strains Pr.4.3L and Pi.5.8, and control (without LAB) against pathogenic bacteria based on antimicrobial activity test and bacteriocin activity test with well diameter of 8 mm.

Pathogenic Strains	Clear zone (mm) based on antimicrobial activity test		Clear zone (mm) based on bacteriocin-like substance activity test		Clear zone (mm) of control (without LAB)	
	Pr.4.3L	Pi.5.8	Pr.4.3L	Pi.5.8	Pr.4.3L	Pi.5.8
Salmonella typhi NCTC 786	13.3 ± 0.8	14.1 ± 0.2	13.6 ± 0.4	14.1 ± 0.2	0 ± 0.0	0 ± 0.0
Salmonella typhi BPE 127.1MC	14.0 ± 0.0	13.0 ± 0.7	14.8 ± 0.8	14.0 ± 1.3	0 ± 0.0	0 ± 0.0
Salmonella typhi BPE 122.4CCA	**19.5 ± 0.3**	**19.0 ± 0.7**	**17.3 ± 0.9**	**17.7 ± 1.1**	0 ± 0.0	0 ± 0.0
Salmonella typhimurium FNCC 0050	13.7 ± 0.3	13.3 ± 1.1	14.8 ± 0.8	13.6 ± 0.4	0 ± 0.0	0 ± 0.0
Pseudomonas putida FNCC 0071	14.7 ± 0.9	14.3 ± 0.4	15.3 ± 0.8	15.7 ± 1.1	0 ± 0.0	0 ± 0.0
Bacillus subtilis	14.7 ± 0.4	13.0 ± 0.3	14.3 ± 1.7	13.3 ± 0.9	0 ± 0.0	0 ± 0.0

ATCC 6633						
Staphylococcus aureus ATCC 25923	**18.7 ± 2.1**	**18.7 ± 1.8**	**18.3 ± 0.4**	**18.0 ± 2.7**	0 ± 0.0	0 ± 0.0

Bacteriocin is a protein compound synthesized by bacteria. It can be a peptide that gives a bactericidal effect. This antimicrobial peptide is acid resistant, heat-stable, but sensitive to the action of protease enzymes and inactivated by digestive enzymes [14][17]. Therefore, to observe the effect of heat treatment on the CFC S inhibitory effect on the indicator strain, the compounds were heated at a temperature of 100°C for 10 minutes for inactivating the enzyme. Overall, it was found that the bactericidal compounds contained in CFCS were resistant to heat, as evidenced by the inhibitory effect of bacteriocin compounds on gram positive and gram negative pathogens (Table 3).

Bacteriocin has especially caught the attention of microbiologists for its ability to control pathogenic bacteria. Bacteriocin from *Lactobacillus fermentum* was known to have antimicrobial potential against the methicillin-resistant *Staphylococcus aureus*. Likewise, it was also found that bacteriocin from *L. acidophilus* inhibits *Salmonella* and *Escherichia*, although it is not unusual to find that gram negative bacteria are sensitive to bacteriocins. Another important finding was the inhibition zone produced by *L. fermentum* that was greater than *L. acidophilus* [18]. This finding supports the conclusion that bacteriocin antimicrobial activity has a broad spectrum. Table 3 shows that the strongest inhibitory effect was known against *Staphylococcus aureus* ATCC 25923 compared to *Salmonella typhi* BPE 122.4CCA.

The Lactobacillus group is known for its use as probiotics and food preservatives. They can be used as food preservative due to the production of inhibiting compounds such as organic acids, hydrogen peroxide, diacetyl, and bacteriocin. Bakteriocin is a cationic protein produced by ribosomes that block other bacteria that live in the same ecological niche. This ability to inhibit other bacteria living in the same ecological niche generally causes bacteriocin to have a narrow inhibitive spectrum activity or, in other words, only inhibit closely related species [19]. However, the terminology still causes debate and by far has not been precisely defined. It has also been reported that bacteriocins that have broad spectrum have been found to have inhibitory effects on gram positive and gram negative inhibitory, and give antifungal activity [20][21]. Bacteriocin-producing strains isolated from various food fermentation processes such as yogurt, milk, sour milk, and milk cream are proved to have a very strong inhibition zone against *Bacillus subtilis* [13]. LAB strains isolated from meat are known to have inhibitory effects on gram-positive bacteria such as *Staphylococcus aureus, Listeria monocytogenes, Clostridium perfringens, C. botulinum, Enterococcus faecalis*, and *Bacillus*, in addition to LAB species. The inhibiting mechanism occurs by forming pores on the cytoplamic membrane [14][22]. This opinion affirms the findings in this study where the LAB strain obtained are found to have a stronger inhibitory effect on gram-positive pathogens than that on gram-negative pathogens and is indicated as bacteriocin because of its resistance to pH and heat.

4. Conclusion

All isolates obtained from the screening in MRS-CaCO3 medium were lactic acid bacteria (LAB) based on its characteristic as gram-positive, did not have catalase activity and did not form spores, with basil or cocci form. Two out of twenty six (26) isolates have the potential as antimicrobial producers, namely LAB Pr.4.3L and LAB Pi.5.8 isolates. Both isolates were shown to have broad-spectrum inhibitory effects, with stronger inhibitory effect on gram-positive pathogenic strains. The CFCS solution used for the antimicrobial test has been neutralized, therefore the inhibitory effect on pathogenic indicator strains is caused by antimicrobial compounds produced by the LAB. The antimicrobial compound was stable after heat treatment, giving the strongest inhibitory effect against *Staphylococcus aureus* ATCC 25923 compared to *Salmonella typhi* BPE 122.4CCA. Based on this finding, these antimicrobial compounds are assumed to be bacteriocin. Further research is still need to be done to optimize bacteriocin production and to identify it.

Acknowledgements
This research was funded by a grant from Kementerian Riset, Teknologi, dan Pendidikan Tinggi Republik Indonesia. The project number are 227/SP2H/LT/DRPM/2019.

References
[1] Mokoena M P 2017 *Molecules* **22** 1-13
[2] Nuraida L 2015 *Food Science and Human Wellness* **4** 47–55
[3] Leroy L De V F L 2007 *J Mol Microbiol Biotechnol* **13** 194–9
[4] Amarantini C, Sembiring L, Kushadiwijaya H and Asmara W 2011 *Biodiversitas* **12** 1-6
[5] Amarantini C and Satwika D 2015 *Microbiology Indonesia* **9** 1-8
[6] Shi T, Nishiyama K, Nakamata K, Aryantini N P D, Mikumo D, Oda Y, Yamamoto Y, Mukai T, Sujaya I N, Urashima T and Fukuda K 2012 *Bioscience, Biotechnology, and Biochemistry* **76** 1897-1903
[7] Baloch M N, Siddiqi R, Erum H and Zia M 2015 *International Journal of Current Research* **7** 13431-36
[8] Pramono H, Suciati P and Andriyono S 2015 *Ilmu Kelautan* **20** 33-7
[9] Fernandez B, Le Lay C, Jean J and Fliss I 2013 *Journal of Applied Microbiology* **114** 877-85
[10] Balouiri M, Sadiki M, Ibnsouda S K 2015 *Journal of Pharmaceutical Analysis* **6** 71–9
[11] Nassar M S M, Hazzah W A and Bakr W M K 2019 *Journal of the Egyptian Public Health Association* **94** 1-5
[12] Klayraung S and Okonogi S 2009 *Brazilian Journal of Microbiology* **40** 757-766
[13] El-Shafei H A, El-Sabour H A, Ibrahim N, Mostafa Y A 2000 *Microbiol. Res.* **154** 321-331
[14] Putra T F, Suprapto H, Tjahjaningsih W and Pramono H 2018 *IOP Conf. Series: Earth and Environmental Science* **137** 012060-7
[15] Ren L and SUO H 2017 Advances in Computer Science Research (ACSR) **76** 1248-56
[16] Crispim S M, Nascimento A M A, Costa P S, Moreira J L S and Nunes A C, Nicoli J R, Lima F L, Mota V T and Nardi R M D 2013 *Brazilian Journal of Microbiology* **44** 15-21
[17] Hwanhlem N, Chobert Jean-Marc and H-Kittikun A 2014 *Food Control* **41** 202-11
[18] Riaz S, Nawaz S K and Shahida H 2010 *Brazilian Journal of Microbiology* **41** 643-648
[19] Hun-Joo L, Park C, Joo Y, Kim S, Yoon J, Park Y, Hwang I, Ahn J and Mheen T 1999 *J. Microbiol. Biotechnol.* **9** 282-91
[20] Ocaña V S and Nader-Macías M El 2004 *Public Health Microbiology: Methods and Protocols* vol 268, ed J F T Spencer and A L Ragout de Spencer (Totowa NJ: Humana Press Inc.) pp 347-53
[21] Agaliya P J and Jeevaratnam 2013 *Brazilian Journal of Microbiology* **44** 1199-1206
[22] Ammor M S and Mayo B 2007 *Meat Science* **76** 138-46

Journal of Physics: Conference Series **1397** (2019) 012046 doi:10.1088/1742-6596/1397/1/012046

Isolation And Characterization Of Fermenting Yeast From Traditional Ethanol Production

Dhira Satwika*, Suhardi Djojoatmodjo, Gustin Finnegan, Dhira Puttajaya, Jovita Ivana

Faculty of Biotechnology, Duta Wacana Christian University Yogyakarta 55224, Indonesia

*Corresponding author: dhira@staff.ukdw.ac.id

Abstract. Ethanol fermentation still plays a significant role in the industry, as it has a huge economical value. *Saccharomyces cerevisiae* is the most often culture being used for this fermentation because its superiority in producing ethanol. However, this yeast also showed its limitation because of the toxic effects of ethanol to the cells. This research was done to isolate new fermenting yeasts from traditional fermentation process in Solo, Central Java, with the hope to obtained more potential strains. A series of isolation and purification steps were done and two potential yeasts isolates obtained. They are identified as *Schizosaccharomyces pombe* and *Zygosaccharomyces parabailii*, respectively. At least one isolates showed comparable ethanol fermentation capacity compared to *S. cerevisiae* and deserve further research.

1. Introduction

Ethanol fermentation could be considered one of the oldest cultivation method in man's history. Researches, reports, as well as reviews have been produced related with this fermentation. *Saccharomyces cerevisiae* is the most studied fermenting yeast which perform anaerobic facultative process [10], it is also the most common strain used for ethanol fermentation as it high osmotic tolerance, as well as higher sugar conversion into ethanol [4].

Despite this rapid ethanol conversion, it is reported elsewhere that ethanol itself has deleterious effect on the producing yeast [5], as the fermentation is influenced by factors such as type of substrate, final ethanol concentration, oxygen contents, and culture being used [3]. [7] reported the denaturation of glycolytic enzymes due to the high level of ethanol resulting in the reduction of biomass growth and glycolytic activity. It is already observed the organel damaged that cause in the reduction up to growth termination due to inhibition effect of ethanol [1]; [3]; [7]. The higher ethanol content in the fermentation medium, the more severe damage observed to the cells; 5-7% ethanol content may result in cell growth reduction, while 10% ethanol may cause protein denaturation which in turn will resulting in cell mortality [8].

Researches have been done to overcome the problem, such as the use of specific fermenter tank to continuously removed the ethanol content from the fermentation medium [11], which in turn could enhanced ethanol yield up to 31.3%. Another approach have been done by [9] who basically destroy the autophagy system, especially the mitophagy in *S. cerevisiae*. Mitophagy is known as selective form of autophagy that degrade mitochondriae of *S. cerevisiae,* which resulting in the improvement of ethanol yield up to 2.12%.

As the ethanol effects on the fermenting yeasts are still need to be overcome, however, another approach could be done to improve the ethanol fermentation efficiency. It is reported that the laboratory strains behave differently as compared to its wild-type. As reported by [6] that working culture which are used in the laboratory may have not represent the optimum potency as compared to the wild-type *S. cerevisiae*.

Based on the above mentioned background, it is the aim of this research to isolate yeast strain(s) from traditional ethanol fermentation, and characterize its fermentation kinetics in attempt to improve ethanol yield from sugar fermentation..

2. Research Methods

2.1. Isolation of yeasts from traditional ethanol fermentation
Isolation was done by collecting "laru" samples of "Ciu" production from Bekonang, Solo, Central Java. A series of dilution was done and continued with plating on PGY with or without antibiotics addition, *i.e.* chloramphenicol and tetracycline with the concentration of 75 and 100 ppm, respectively.

2.2. Yeast regeneration and maintenance
Cells of *S. cerevisiae* D-01, considered as working culture, and isolated yeasts, as wild-type, are maintain and grown on PGY medium.

2.3. Characterization
The obtained isolates were characterized by observing their macroscopic and microscopic morphology, for examples the colonies colour, cell size, cell shape and budding type.

2.4. Antagonistic assay
The assay was basically done by streaking the obtained isolates and observed for the inhibition zone against the test organisms.

2.5. Fermentation
Fermentation were done by inoculating 10% starter to the fermentation media of 10-15% glucose, placed on a rotary shaker for 4 days at room temperature. Cell biomass were basically measured as cell number (CFU/ml) by plating 1 ml samples continued by serial dilution, and plating on PGY medium. Sugar content was measured by employing DNSA method by collecting 0.3 ml samples, added with 0.3 ml DNSA reagent, added with 2.4 ml water and heated for 10 min before taking OD measurement at 540 nm. Ethanol concentration was determined by GC analysis by sending the samples to UII, Yogyakarta. 30 ml samples were distilled until it reached 10 ml, placed in an eppendorf and tightly sealed before being sent for measurement.

2.6. Molecular identification of obtained isolates
Obtained isolates were further characterized by conducting DNA barcoding, *i.e.* by sending the samples to be sequenced. The isolates were send to Macrogen, Korea and sequenced for 16S rRNA. After obtaining the sequences, data were analysed by conducting some bioinformatics analysis using MEGA software and comparing the sequences with the online database, *i.e.* GenBank.

3. Results and Discussion
Isolation of yeast(s) from laru obtained from Bekonang, Solo, Central Java resulting in several different yeast colonies based on their morphological appearance, and the addition of antibiotics effectively prevent the occurrence of bacteria. Based on this observation, 17 different yeasts were obtained. Figure 1 shows typical colonies obtained on media either with or without antibiotics addition. It is worth to mention here that the strains in Fig.1 are the most often and dominant colonies obtained in isolation steps as described in Materials and Methods.

After following several purification steps, a single colony could be recovered and further examination be performed. The resulting colonies are presented in Figure 2, shows that the small colony are more whitish in colour compared to the bigger colony.

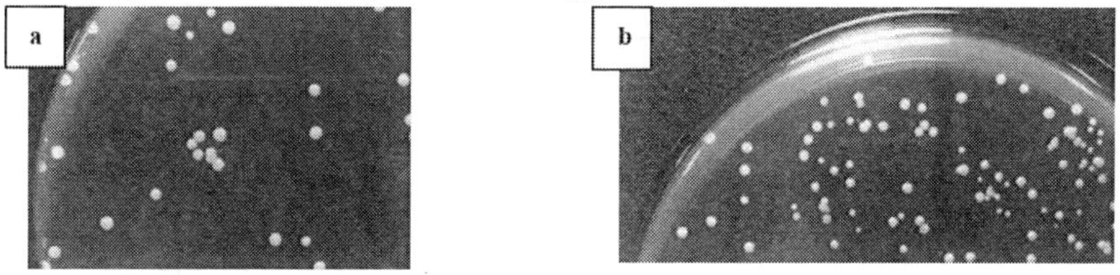

Figure 1. Yeasts (a) big and (b) small colonies, obtained from isolation steps grown on PGY media

Subsequently, a microscopic examination was done to further characterize both isolates obtained. The resulting observation is presented in Figure 3 and Table 1, showing that the two colonies are differ in cell shape, colour and its budding type.

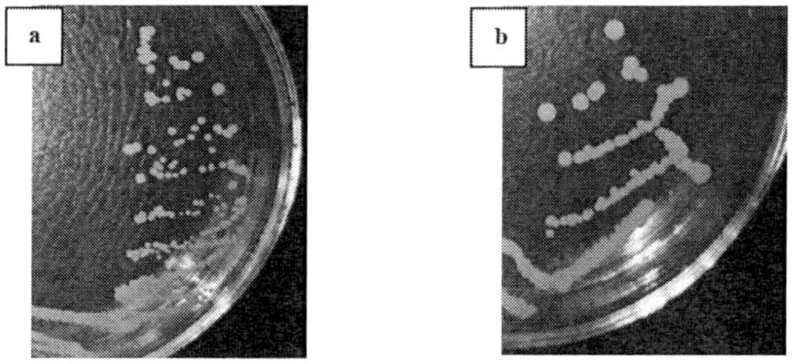

Figure 2. Single colonies from both (a) small and (b) big colony grown on PGY medium

As can be seen in Fig.3 and Table 1, the small colony has an ellipse shape, lighter in colour and with bipolar budding type, as compared to the bigger colony which has a smaller cell size, round-ellipse cell shape and darker, and unipolar budding type. Based on the observed data, it is most probably that the two isolates may belong to the different species/strain, regardless the difficulty to separate these two isolates during the isolation and purification steps. The next step will be molecularly identify the obtained isolates.

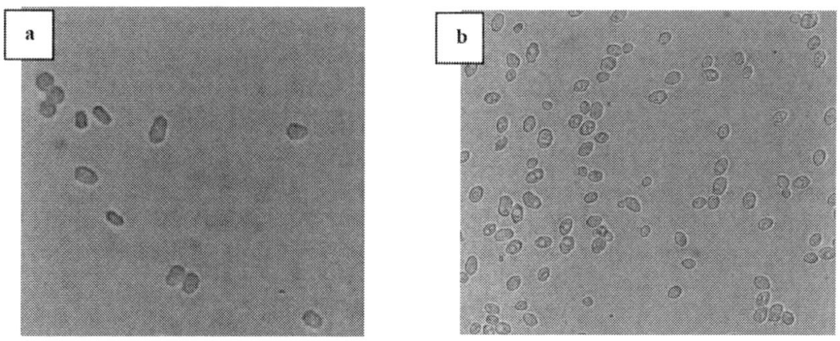

Figure 3. Microscopic properties of (a) bipolar budding of the small colony and (b) monopolar budding for the big colony (magnification: 400x)

Table 1. Selected macroscopic and microscopic properties of obtained isolates

Colony	Shape	Cell colour	Budding
Small	Ellipse	Transparent	Bipolar
Big	Round-ellipse	Dark transparent	Monopolar

In parallel to the molecular identification, a series of antagonistic assays were done to further characterize whether these isolates have the ability to inhibit the growth of other organisms. The results could be seen in Figure 4, showing that there are no antagonistic activity from both of the isolates.

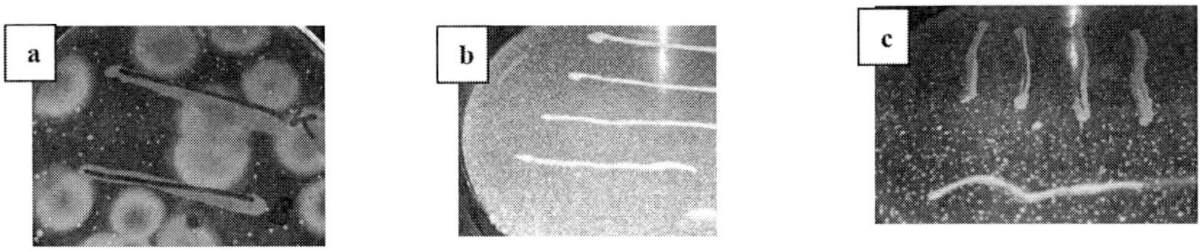

Figure 4. A typical antagonistic assay of the isolates obtained showing no antagonistic activity; (a) both colonies are grown on plate spread with laru; (b) small colony grown on lawn of big colony, and (c) big colony grown on lawn of small colony

Molecular identification of both isolates based on their 16S rRNA resulting in the identification of two different species, as presented in Table 2, namely *Schizosaccharomyces pombe* and *Zygosaccharomyces parabailii*, respectively. It was reported that *S. pombe* was also recognized from laru of Bekonang, Solo, and considered one of the most active fermenting yeast resulting in high ethanol content compared to other microorganisms found, either yeasts, fungi and bacteria [2].

Table 2. Molecular identification of obtained isolates

No.	Colony	Species	Similarities	Acc. No.
1.	Small	*Schizosaccharomyces pombe* strain BY-10	95%	KJ562356.1
2.	Big	*Zygosaccharomyces parabailii* ATCC 56075	95%	JX458094.1

To test the finding, we performed fermentation with 12.5% glucose by using *S. cerevisiae* D-01 which is considered as working culture, as it is normally being used in ethanol fermentation in the laboratory, and compared it with the two isolates obtained in this research. The results are presented in the Figure 5 and Figure 6 below.

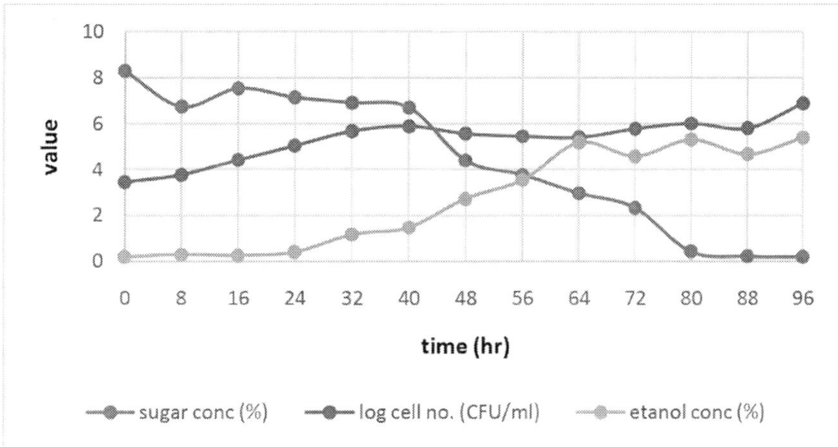

Figure 5. Fermentation performance of *S. cerevisiae* D-01 for 4 days on PGY medium containing 12.5% glucose

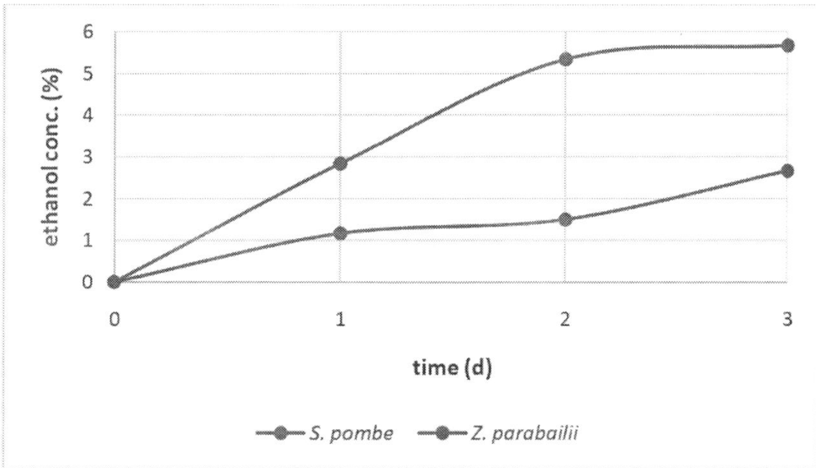

Figure 6. Fermentation performance of isolates obtained on 12.5% glucose

The pictures on Fig.5 and Fig.6 clearly shows that at least the small colony obtained from laru and identified as *S. pombe* give a comparable amount of ethanol yield compared to *S. cerevisiae* D-01. Further researches are still going on to better characterize the isolates obtained, as well as to test their ability to produce ethanol compared to laboratory strains. More data are needed to better analyze the fermentation kinetics of the isolates being used.

4. Conclusion

Based on our findings, two yeasts species identified as *S. pombe* and *Z. parabailii*, respectively, are recovered from laru. The isolates showed comparable fermentation capability compared to working culture, based on the ethanol yield.

References

[1] Aguilar-Uscanga, M. G., Alvarado, Y. G., Rodriguez, J. G., Phister, T., Delia, M. L., Strehaino, P. (2011). Modelling the Growth and Ethanol Production of *Brettanomyces bruxellensis* at Different Glucose Concentrations. *Appl. Microbiol.*, *53*, 141-149.

https://doi.org/10.1111/j.1472-765X.2011.03081/x

[2] Agustini, L. (2019). Biodiversitas Mikroorganisme yang Diisolasi dari Proses Pembuatan Minuman Beralkohol 'Ciu' di Jawa Tengah. *Seminar Nasional Biologi dan Pendidikan Biologi 2019: Inovasi dalam Penelitian dan Pembelajaran Biologi* [Indonesia]

[3] Amenaghawon, N. A., Okieimen, C. O., Ogbeide, S. E. (2012). Kinetic Modelling of Ethanol Inhibition during Alcohol Fermentation of Corn Stover Using *Saccharomyces cerevisiae. International Journal of Engineering Research and Application (IJERA), 2*(4). 798-803.

[4] Chen, Y., Wu, Y., Zhu, B., Zhang, G., Wei, N. (2018). Co-Fermentation of Cellobiose and Xylose by Mixed Culture of Recombinant *Saccharomyces cerevisiae* and Kinetic Modeling. *PLoS ONE, 13*(6), 1-18. https://doi.org/10.1371/journal.pone.0199104

[5] Jin, H., Liu, R., He, Y. (2014). Kinetics of Batch Fermentation for Ethanol Production with Immobilized *Saccharomyces cerevisiae* Growing on Sweet Sorghum Stalk Juice. *Proc. Environ. Sci.*, 12, 137-145. https://doi.org/10.1016/j.proenv.2012.01.258

[6] Jouhten, P., Ponomarova, O., Gonzales, R., & Pati, K. R. 2016. *Saccharomyces cerevisiae* metabolism in Ecological Context. FEMS Yeast Research. doi: 10.1093/femsyr/fow080

[7] Stanley, D., Bandara, A., Fraser, S., Chambers, P. J., Stanley, G. A. (2010). The Ethanol Stress Response and Ethanol Tolerance of *Saccharomyces cerevisiae. J. Appl. Microbiol., 109*(1), 13-24. https://doi.org/10.1111/j.1365-2672.2009.04657.x

[8] Schiavone, M., Formosa-Dague, C., Elsztein, C., Teste, M. A., Martin-Yken, H., Morais, M. A. D., Dague, E., Francois, J. M. (2016). Evidence for a Role for the Plasma Membrane in the Nanomechanical Properties of the Cell Wall as Revealed by an Atomic Force Mircoscopy Study of the Response of *Saccharomyces cerevisiae* to Ethanol Stress. *Applied and Environmental Microbiology, 82*(15), 4789-4801

[9] Shiroma, S., Jayakody, L. N., Horie, K., Okamoto, K., Kitagaki, H. (2014). Enhancement of Ethanol Fermentation in *Saccharomyces cerevisiae* Sake Yeast by Disrupting Mitophagy Function. *Applied and Environmental Microbiology, 80*(3), 1002-1012.

[10] Walker, G. M. & Stewart, G. G. (2016). *Saccharomyces cerevisiae* in the Production of Fermented Beverages. *Beverages, 2*(30), 1-12. https://doi.org/10.3390/beverages2040038

[11] Tavares, B., Felipe, M. G. A., Santos, J. C., Pereira, F. M., Gomes, S. D., Sene, L. (2019). An Experimental and Modeling Approach for Ethanol Production by *Kluyveromyces marxianus* in Stirred Tank Bioreactor Using Vacuum Extraction as a Strategy to Overcome Product Inhibition. *Renewable Energy, 131*, 251-267. https://doi.org/10.1016/j.renene.2018.07.030

Journal of Physics: Conference Series **1397** (2019) 012047 doi:10.1088/1742-6596/1397/1/012047

The Effectiveness of Problem Based Learning in Biology with Fishbone Diagram on Critical Thinking Skill of Senior High School Students

A. A Priyadi[1*] and S. Suyanto[2]
[1]Biology Education, Post-graduate Program, Yogyakarta State University, Yogyakarta, Indonesia
[2]Biology Education Department, Yogyakarta State University, Yogyakarta, Indonesia

[*]Corresponding author e-mail: annaatiqap@gmail.com

Abstract. This research aims at investigating the effectiveness of Problem Based Learning with Fishbone Diagram on the students' critical thinking skill in the Biology learning. This research adopted a quasi-experimental design notably pretest-posttest nonequivalent control group design. Population of this research included 73 students of grade X of the State Senior High School 1 of Pundong. The population was taken by using target population, and 49 students were determined as the sample by using random sampling technique. This research adopted critical thinking tests that were administered before and after the learning process took place. The instrument to measure critical thinking skill was a test consisting of 11 items. The instrument was initially validated the expert lecturer (expert judgment) to obtain its content validity and construct validity. Subsequently, empirical validity and instrument reliability were done by using Alpha Cronbach. The data of critical thinking skill were analyzed by using independent sample t-test at the significance level of (α) = 0.05. The result showed the significance value of 0.00<0.05. This study suggests that PBL with FD was effective in developing the students' critical thinking skill.

1. Introduction

The world has experienced fast and unpredictable changes in the 21st century that involved voluminous aspects of life such as technology, communication, information, economy, transportation, etc. Schools need to anticipate these changes by developing the 21st century skills [21]. There are some competencies that must be developed in living in the 21st century. They are critical thinking skill, problem solving, creativity and innovation, communication and collaboration, social skill and cross culture, and mastery of information [17].

The change of curriculum must be done to anticipate the needs of the 21st century. In Indonesia, the alteration and revision of curriculum have been performed by the government. The recent curriculum – as a result of the alteration and revision of the previous curriculum – is called the Curriculum of 2013. This curriculum has been designed to develop the 21st century skills, either viewed from the standard of content, process, or evaluation. In the process standard, for example, the curriculum of 2013 adopts scientific approach in which a teacher functions as the facilitator or motivator, no longer as a solitary learning source. However, most of the learning processes remain teacher-centered. The learning process should not be a transfer of knowledge in which the knowledge is simply transferred from teachers to students, yet it should be a stimulus to students to develop their thinking skill. Students are expected to be a critical thinker and problem solver [19]. The teacher-centered learning makes students passive during the learning process, therefore they are unable to critical thinking skill excellently.

Critical thinking skill of Indonesian students is still low according to a survey conducted by PISA (Program for International Student Assessment) and TIMSS (Trends in International Mathematics and Science Study) in 2015. The survey revealed that Indonesia was at the bottom rank compared to other countries. Of all the tests, Indonesian students were only able to solve the second difficulty level (understanding), whereas higher order thinking skill (HOTS) were categorized into the fourth up to sixth difficulty level [21]. The learning process should be directed to develop the students' critical thinking skill. Critical thinking skill is not only essential in the learning process to achieve a high score, but also it is used to help students solve problems in daily life and career [4], [9].

The students' critical thinking skill can be trained by applying a learning model namely Problem Based Learning (PBL). PBL in Biology is a learning model in which students learn by solving realistic, general, crucial, complex, and open-ended problems about Biology. Based on the literature study, Biology problems in PBL include ecosystems, environmental changes, organ systems in humans, fungi, bacteria, metabolism (photosynthesis), Spermatophyte, biotechnology, invertebrates, and classification of living creatures. In this study, the Biology problems are only limited to the problem of environmental change. The aforesaid problems are the real-life problems that must be resolved by applying some concepts and principles [23], [1]. The real problems adopted in PBL stimulate students to vigorously construct new knowledge that relates strongly to their prior knowledge; therefore, learning is a constructive and contextual process [7]. According to Surya et al. [25], PBL develops the students' critical thinking skill since students are directed to find a problem in the Biology learning, identify any possible solutions, decide the best solution, and discuss and evaluate the applied solution. Thus, students are truly trained to develop their critical thinking skill. This is in line with Orozco & Yangco [16] arguing that PBL improves the students' critical thinking skill and creativity in the Biology learning.

Before solving a problem, one must identify and understand the causes of the problem, thus problem solving can be more straightforward. One of the tools of Root Cause Analysis (RCA) to help a group or individual identify the causes of a certain problem is Fishbone Diagram (FD) [6]. Fishbone Diagram – known as the Cause and Effect or Ishikawa Diagram – was firstly introduced by Kaoru Ishikawa, a Japanese scientist. Fishbone Diagram is an analysis tool to see and find the causes of a problem contributing to it systematically [27]. In reference to [13], Fishbone Diagram is a diagram that is used to investigate the real causes or potentials of a problem. Meanwhile, [10] said that Fishbone Diagram is created to identify and classify the causes of a problem. According to [15], the implementation of FD in PBL fostered students in analyzing; finding and identifying factors affecting a problem more deeply; determining the solution more straightforwardly due to a meaningful and systematic process; building their own knowledge; and expressing and accepting other people's opinion.

Most of the researchers suggest adopting FD as an assistant tool to solve problems. According to Wong [28], FD was utilized to show the connection between a clinical problem and its potential causes. FD enhanced memory skills and helped someone take the relevant medical cases as well as literature. FD can be used in PBL model either in a group or individual learning. Based on a study by Ilie & Ciocoiu [10], it was revealed that FD was a tool to determine the risks of a case with any relevant causes. FD was also straightforward in its practice. A literature study also indicated that PBL enhanced the students' achievements, whereas FD assisted in resolving problems [11].

Several studies have confirmed the effectiveness of PBL with FD in learning. Meylani et al. [15] showed that PBL with FD developed the students' science process skill in the Biology learning of grade X students, notably in the topic of environmental contamination. This research used a true experimental design. Another study done by Prasasti [20] indicated that PBL with FD was effective in enhancing the student's analyzing skill in the environmental contamination material. The type of this research was a quasi-experimental one group pretest posttest design. Istikomah [11] also investigated the effect of PBL with FD on the critical thinking skill of Vocational High School students majoring in

Computer Engineering. The material used in this research was problem diagnosis of computer network. A quasi-experimental pretest-posttest non-equivalent control group design was adopted in this research. The result demonstrated that the implementation of PBL with FD was effective on the students' critical thinking skill.

Referring to the previous studies, the researchers want to investigate the effectiveness of PBL with FD in the Biology learning. The similarity of the present study with Meylani et al. [15] and Prasasti [20] is that the material used – either the present study or the previous studies focus on the topic of environmental change for Senior High School students majoring in Mathematics and Science. The present study, however, differs from the previous studies by Meylani et al. [15] and Prasasti [20] in terms of the dependent variable and research design. In addition, comparing to Istikomah [11], the present study embodies both the similarity and difference. The similarity was apparent in the dependent variable and research design, yet this present study is different in terms of the learning materials as the focus of this study. The novelty of this study lies in the effectiveness of PBL in Biology with FD on the students' critical thinking skills in State Senior High School 1 of Pundong, which was conducted by adopting quasi-experimental non-equivalent group design. This research is crucial to provide novel information about the effectiveness of Problem Based Learning in Biology with Fishbone Diagrams on the students' critical thinking skills in State Senior High School 1 of Pundong.

Based on the elaboration above, the effectiveness of PBL with FD on the students' critical thinking skill is essential to be explored. This research aims to investigate the effectiveness of PBL with FD on the students' critical thinking skill in the Biology learning about environmental change di State Senior High School 1 of Pundong.

2. Method

This research used a quasi-experimental design notably pretest-posttest nonequivalent control group design. This design is called nonequivalent because the experimental and control groups are not randomly selected. This research applied experimental group and control group that were given different treatments. The experimental group received the treatment of Problem Based Learning (PBL) with Fishbone Diagram (FD), whereas the control group obtained 5M scientific approach as the treatment in the learning process of Biology focusing on environmental change. This research was carried out in the State Senior High School 1 of Pundong. The subjects were 49 students of grade X in the academic year of 2018/2019. The population in this research used target population, and the sample was determined by using random sampling technique. To collect the data, the researchers used testing technique. It adopted 11 essay questions to measure the students' critical thinking skill. Before the instrument was used, the validation of instrument was performed to identify the content validity and construct validity. This validation process involved the expert lecturers (*the expert judgment*). Subsequently, it was performed the empirical validity and instrument reliability tests by using Alpha Cronbach. The data of critical thinking test were then analyzed by using independent sample t-test from SPSS 22 when the result of prerequisite tests were met, that is, the data were normally distributed and homogeneous.

3. Result and Discussion

Before the instrument was applied, the expert validation was undertaken by two expert lecturers (expert judgment). The quantitative data were transformed into the mean score, and converted to scores ranging from scale 1-5. Table 1 displays the conversion from quantitative data to qualitative data adopted from [22].

Table 1. The Conversion of Five-Point Score

Score Interval		Category
$X > X_i + (1.80.\ Sb_i)$	$X > 4.2$	Very Good (A)
$X_i + (0.60.Sb_i) < X \leq X_i + (1.80.Sb_i)$	$3.4 < X \leq 4.2$	Good (B)
$X_i - (0.60.Sbi) < X \leq X_i + (0.60.Sb_i)$	$2.6 < X \leq 3.4$	Fair (C)
$X_i - (1.80.Sbi) < X \leq X_i - (0.60.Sb_i)$	$1.8 < X \leq 2.6$	Worse (D)
$X < X_i - (1.80.Sb_i)$	$X \leq 1.8$	Poor (E)

The result of validity test obtained the mean of 4.95 from the scale of 1-5; thus, it belonged to a very good category. The validation result by two validators showed that the critical thinking test was decent to be used without any revision.

The empirical try-out of the test was conducted in State Senior High School 1 Pundong to know which critical thinking questions were valid. An empirical try-out was conducted on a sample of the students of grade XI at the school. The assumption was that the students have studied the subject of Environmental Change. The validity test was done by benefiting SPSS 22. Referring to the result of analysis, it was obtained 11 valid questions of 17 questions in total. The questions were valid if r obtained > r table (0.3673) for n = 29.

In this research, the reliability test was performed by using Alpha Cronbach formula. SPSS 22 was utilized in the reliability test. The result of reliability test showed a reliability coefficient of 0.666 > 0.6 for the variable of critical thinking skill. The reliability degree of Alpha Cronbach of 0.6 has been declared as the minimum level of acceptance [12]. Based on the reliability test, it can be concluded that the instrument was reliable.

The data of the students' critical thinking skill comprised of pretest and posttest scores from the experimental and control groups. The students reached the minimum criteria of success if they attained the score of 68. Table 2 depicts the descriptive analysis of pretest and posttest of the students' critical thinking skill.

Table 2. The Result of Descriptive Analysis of the Students' Critical Thinking Skill

Analysis of	PBL+FD Class		Control	
	Pretest	*Posttest*	*Pretest*	*Posttest*
Mean	42.04	78.16	37.92	59.33
The highest score	67	91	67	79
The lowest score	15	64	15	45
Deviation standard	13.268	7.116	13.161	9.635
N	25	25	24	24

As Table 2 shows, the results of pretest of experimental group and control group were nearly similar. The difference between the two groups was 4.12. The mean difference was also tested by using SPSS program resulting in the significance value of 0.280 ($p > 0.05$). Thus, it can be said that H_0 was rejected and H_a was accepted. It indicated that the students' critical thinking skill in either experimental group or control group was alike before the treatment was given. Therefore, both groups met the criteria to participate in this research. The difference of posttest result regarding the students' critical thinking skill was 19.43. The mean obtained by the experimental group was higher than that of the control group. Based on Table 2, it is known that the mean of posttest of PBL with FD class FD, 78.16, has reached the Minimum Completion Criteria (KKM), that is, 68 (mean 78.16>68), whereas the mean of posttest of the control class, 59.33, has yet to reach KKM (mean 59.33<68). The mean of posttest was then interpreted in the form of qualitative data [3]. The criteria of critical thinking skill are presented in Table 3.

Table 3. Criteria of Critical Thinking Skill

Interval Score		Criteria
$X > Mi + 1.5\ Si$	$X > 75$	Very Good (A)
$Mi + 0.5\ Si < X \leq Mi + 1.5\ Si$	$58.33 < X \leq 75$	Good (B)
$Mi - 0.5\ Si < X \leq Mi + 0.5\ Si$	$41.67 < X \leq 58.33$	Fair (C)
$Mi - 1.5\ Si < X \leq Mi - 0.5\ Si$	$25 < X \leq 41.67$	Worse (D)
$X \leq Mi - 1.5\ Si$	$X \leq 25$	Poor (E)

Explanation: Mi : Ideal mean; Si : Ideal standard deviation, where

$$\mathrm{Mi} = \frac{1}{2} \times (ideal\ maximum\ score + ideal\ minimum\ score) = \frac{1}{2} \times (100 + 0) = 50$$

$$\mathrm{Si} = \frac{1}{6}\ (ideal\ maximum\ score - ideal\ minimum\ score\ )\ = \frac{1}{6} \times (100 - 0) = 16.67$$

As depicted in Table 3, the result of posttest of PBL with FD class was very good (A), while the posttest result of the control class was good (B). Although the experimental results were classified into a very good category, the mean score was far from the maximum score of 100. This is due to the students' unfamiliarity with PBL. Therefore, the application of PBL needs to be carried out continuously by teachers in schools so that the results can be more optimum. The high standard deviation in the experimental and control groups was owing to the varied students' thinking skills; some were low, and some others were high. Data that are far from the mean will produce deviation greater than that of close to the mean. After being given treatment, the results of the students' critical thinking skills in the experimental and control groups also varied.

Once the descriptive analysis had been performed, the prerequisite tests including normality test and homogeneity test were carried out before analyzing the data by using independent sample t-test. The data analysis adopted SPSS 22 to test the normality and homogeneity. The normality test of the data was performed by adopting Saphiro-Wilk test. Table 3 shows the result of normality test of the students' critical thinking skill.

Table 4. The Normality of the Data of the Students' Critical Thinking Skill

Groups	Data	Significance	Conclusion of Data Distribution
PBL+FD	Pretest	0.335	Normal
	Posttest	0.364	Normal
CONTROL	Pretest	0.442	Normal
	Posttest	0.415	Normal

As depicted in Table 4, the result of normality test of the students' critical thinking skill resulted in the significance value of > α (0.05); thus, H_0 was rejected. This finding indicated that the data of the students' critical thinking skill was normally distributed.

The homogeneity of data variance of the students' critical thinking skill in both the experimental group and control group was analyzed by using Levene test. Table 5 demonstrates the result of homogeneity test of data variance of the students' critical thinking skill.

Table 5. The Result of Homogeneity Test of Data Variance of the Students' Critical Thinking Skill

The Data of the Students' Critical Thinking Skill	Significance	Conclusion of Data Variance
Pretest	0.916	Homogeneous
Posttest	0.218	Homogeneous

Referring to Table 5, the result of homogeneity test of data variance of the students' critical thinking skill revealed sig. *based on mean* pretest of 0.916> α (0.05) and posttest 0.218> α (0.05); therefore, H_0 was rejected. This finding suggests that the data of the students' critical thinking skill in pretest and posttest came from the homogeneous population – in other words, the data had the same variance.

Once the prerequisite tests were accomplished, hypothesis testing was then performed by using independent sample t-test. The hypothesis testing was done to figure out if there was any effect of PBL with FD on the development of the students' critical thinking skill. Table 6 displays the result of independent sample t-test of the students' critical thinking skill.

Table 6. The Result of Independent Sample T-Test of the Students' Critical Thinking Skill

Type of Data	t	Df	Sig. (2-tailed)	A	Decision	Detail
Posttest	7.802	47	0.000	0.05	H_a was accepted	Significantly different

As showed in Table 6, the value of sig. (2-tailed) was $0.000 < 0.05$; hence, it can be concluded that H_0 was rejected while H_a was accepted. This finding indicated the significant difference of the mean score of the students' critical thinking skill in the experimental and control groups. The result showed that there was a significant effect of Problem Based Learning (PBL) with Fishbone Diagram (FD) on the students' critical thinking skill.

The students in the experimental group were more active comparing to those involved in the control group. The students were more attracted to make FD with their groups [11]. The real problems in PBL stimulate the students to actively construct new knowledge that strongly allied with their prior knowledge. The learning process, therefore, can be constructive and contextual [7]. In addition, the complex problems in PBL boost the students' critical thinking since the students must solve the problems by evaluating some valid solutions [5].

According to [6], FD is an assistant tool to solve problems. FD identifies factors causing a problem in a structured way by using a technique focusing on the identification and completion of the problem. FD makes the students more straightforward in finding, identifying and analyzing the root causes that relate to the materials, and identifying which parts have the interrelated problems [13]. Each student involved in the learning process adopting FD can participate in providing solutions concerning on the potential causes of a problem or the core of the problem. Active learning in groups can enhance the students' critical thinking skill since the students can be open-minded, and they can communicate effectively with other students [8], [18].

The use of Fishbone Diagram as a graphic organizer can help the students understand the entire information, in addition to enhancing the students' creativity. Besides, Fishbone Diagram assists the students in clarifying and organizing the relationship between concepts and ideas; helps the students focus on the materials; facilitates the students in doing brainstorming of the possible causes of a problem; aids the students in solving a problem, making decision or action, and helping the students enhance memory skill and understanding; and guides the students in communicating their thinking process in either spoken or written form [26], [14].

The implementation of PBL with FD in the form of fishbone makes the discussion more directed to a problem and its causes, whereas in the implementation of PBL without FD, it was not found the same finding. The students in the experimental group were more enthusiastic in learning comparing to the control group. This is seen from the students' behavior that was more active, careful, and collaborative during the learning process. Not to mention, the students performed better in making and answering questions in presentations [15].

PBL with FD facilitates the students in understanding the root causes of a problem. This way, they can be more straightforward in determining the solutions. When the students were accustomed to analyzing, organizing, evaluating, and solving problems, their critical thinking skill will improve. Based on a research done by [20], PBL with FD enhanced the students' critical thinking skill notably analyzing skill because the students were encouraged to analyze the root of a problem related to the materials. Besides, the students were actively urged to develop analyzing skill through experimentation or direct observation. The development of the students' critical thinking skill can be seen from the students' willingness to ask questions, respond to questions with unpredictable answers, propose varied solutions, and numerous alternatives of correct answer the students have.

The finding of this study is in line with [2] stating that PBL enhanced the students' achievements. In their experimental study, the experimental group performed better than the control group. This study is also consistent with [11] revealing that PBL with FD was effectual on the students' critical thinking skill since PBL trained the students to solve a problem, and FD helped the students undertake the problem.

4. Conclusions

This research suggests that Problem Based Learning with Fishbone Diagram was effective in developing the students' critical thinking skill. Based on the mean of posttest, the critical thinking skill of the experimental class has reached the KKM (Mean 78.16> 68), while the control class has yet to reach the KKM (mean 59.33 <68). After converting quantitative data to qualitative data, the mean of posttest of the critical thinking skill in the experimental class belonged to a very good criteria (A), whereas the control class was classified into a good criteria (B). Although the experimental result was very good, the mean score was still far from the maximum score of 100. It happened since students were unfamiliar with PBL. Therefore, teachers should continually apply PBL so that the result can be more optimal. The high standard deviation was due to the diverse students' thinking skills; some were low, and some were high. After the treatment was given, the result of the students' critical thinking skills in the experimental class also varied. Although the pretest score between the experimental group and control group was the same, there was a significant difference in posttest in which the experimental group achieved a higher score comparing to that of the control group. The result of independent sample t-test showed the significance value of $0.00 < 0.05$.

The researchers expect this study can be a source of information for teachers in applying Problem Based Learning with Fishbone Diagram in schools, thus students can construct their knowledge. The implementation of this model should be done continuously to improve the students' critical thinking skill. Before the class begins, teachers should prepare everything well mainly the selection of real problems appropriate with the materials. Therefore, students can analyze and solve problems accordingly. Teachers also need to manage their time to better implement Problem Based Learning with Fishbone Diagram. Further researchers can conduct the similar study focusing on other materials, and even other fields of study.

References

[1] Ali, S S 2019 *English Language Teaching* **12** (5) 73
[2] Ayyildiz, Y & L Tarhan 2018 *Research in Science and Technological Education*, **36**(1), 35–54.
[3] Azwar, S 2015 *Tes Prestasi* (Yogyakarta : Pustaka Pelajar)
[4] Brookfield, Stephen D 2012 *Teaching for Critical Thinking: Tools and Techniques to Help Student Question Their Assumptions* (San Francisco: Jossey-Bass)
[5] Burrow, A K 2018 *The Bulletin of the Ecological Society of America* **99** (1) 137–150
[6] Doggett, A M 2004 *Journal of Industrial Technology* **20** (2) 1–9
[7] Dolmans, D H J M., De Grave, W, Wolfhagen, I H A P, & Van Der Vleuten, C P M 2005 *Medical Education* **39** (7) 732–741
[8] Espey, M 2018 *Higher Education Research and Development* **37** (1) 15–29
[9] Gormley, William T 2017 The critical advantage : developing critical thinking skills in school (Cambridge, Massachusetts: Harvard Education Press)
[10] Ilie, G & Ciocoiu, C N 2010 *Management Research and Practice* **2** (1) 1–20
[11] Istikomah 2017 *Indonesian Journal of Informatics Education R* **1** (2) 83–91
[12] J C Nunnally and I H Bernstein 1994 *Psychometric Theory*, (3 ed) (New York: McGraw-Hill)
[13] Kang, Chang W, & Paul H Kvam 2011 *Basic Statistical Tools for Improving Quality* (New Jersey: Hoboken)

[14] Krasnic, T 2011 *How to study with mind maps: The concise learning method.* (Concise Books Publishing)
[15] Meylani, Vita, Purwati Kuswarini, and Nurhidayah 2018 *EKSAKTA : Jurnal Penelitian dan Pembelajaran MIPA,* **3** (2) 11–18
[16] Orozco, J A & Yangco, R T 2016 *Asian Journal of Biology Education* **9** 1–10
[17] Partnership for 21st Century Skills (P21) 2015 *Framework for 21st century learning.*
[18] Paul, R, & Elder, L 2006 *The miniature guide to critical thinking: Concepts and tools, fourth edition* (Dillon Beach, CA: The Foundation for Critical Thinking)
[19] Permendikbud No 81a 2013 Tentang Implementasi Kurikulum
[20] Prasasti, P A T 2015 *Premiere Educandum* **5** (2) 223-238
[21] Redhana, I Wayan 2019 *Jurnal Inovasi Pendidikan Kimia* **13** (1) 2239-53
[22] Sani, R A 2016 *Penilaian Autentik* (Authentic Assessment). Jakarta: Bumi Akasara.
[23] Sani, R A 2017 *Pembelajaran Saintifik untuk Implementasi Kurikulum 2013* (Jakarta: Bumi Aksara)
[24] Sukardjo 2009 *Landasan Pendidikan Konsep dan Aplikasinya* (Jakarta: Rajawali Press)
[25] Surya, Erdi, Khairil, & Razali 2014 *Jurnal Edubio Tropika,* **2** (1) 121-186
[26] USAID 2014 Buku Sumber untuk Dosen LPTK: Pembelajaran Literasi Kelas Awal di LPTK (Jakarta: USAID)
[27] Watson, G 2004 *Quality Progress* **37** (4) 54-47
[28] Wong, K C 2011 *Journal of Medical Case Reports* **5** (120) 1–3

Journal of Physics: Conference Series **1397** (2019) 012048 doi:10.1088/1742-6596/1397/1/012048

Comparison of Real Object and Blended Learning towards Students' Metacognitive Knowledge

A. Musyaddad* and S. Suyanto

Department of Biology Education ,Graduate School of Yogyakarta State University Colombo Street No 1, Karangmalang, Depok, Sleman, Yogyakarta, Indonesia

*Corresponding author e-mail: anwarmusyaddad10.2017@student.uny.ac.id;

Abstract. This study aims to compare the effectiveness of real object-based learning and blended learning towards students' metacognitive knowledge in ecosystem material. This research is a quasi-experiment using a pretest-posttest comparison group design. The population of this study was all students of class X in Bantul Public High School 1 and Bantul Public High School 2. The sample of this study was class X MIA 3 students as experimental group 1 and class X MIA 6 students as experimental group 2 in Bantul Public High School 1 and class X MIA 4 students as experimental group 1 and class X students MIA 6 as experimental group 2 in Bantul Public High School 2 which is determined through cluster random sampling. Data collection techniques in this study used tests and questionnaires. The research data analysis technique uses One Way ANOVA tests to determine the effect of real objects and blended learning on students' metacognitive knowledge. The results showed that the two learning methods did not have a significant effect on students' metacognitive knowledge, but blended learning had a higher average increase in metacognitive knowledge than real object-based learning.

1. Introduction

The curriculum has drawn attention to be developed in education including Indonesia's stakeholder does. A refined purpose to accomplish the national education target is curriculum 2013 which concern students' competency and ability. In the matter of global areas, they are expected to compete fairly as much as other countries do so, so that they are involved to learn for having factual, conceptual, procedural and metacognitive knowledge.

The learning process in schools must be designed in such a way as to develop metacognitive knowledge as an effort to train students' skills in various fields, especially science. Metacognitive knowledge is the ability of students to control cognitive components in understanding assignments or problems and can complete tasks or problems properly and correctly [9]. Metacognitive knowledge is still very rarely developed, students must be accustomed to being trained using it on an ongoing basis because it can foster students' confidence in learning and improve mastery of their competencies [8]. Metacognitive knowledge must be continuously developed in the learning process. Some experts recommend implementing more activities that can improve metacognitive knowledge in the learning environment. Teachers must be able to provide stimuli to students to build students' meta-cognitive knowledge through innovative and fun learning activities [4]. For that reason, we need a model, strategy or method that is appropriate in the learning process so that students can comprehend better.

The application of the model, strategy or method in learning is very important to do because by using the model, strategy or method can help the teacher in delivering learning material and the purpose of learning. However, not all models, strategies and methods are suitable for use in all learning materials. For this reason, the teacher must be able to choose the right model, strategy or method before being used for learning. Models, strategies or methods in learning are based offline and online. Internet or online learning has begun to be developed, one of which is using blended learning. Blended learning is a combination of online learning and face-to-face or offline learning.

Blended learning is learning that combines online learning with offline learning or face-to-face [7]. Blended learning as an imitation of active, individual learning and learner-centered strategies [12]. Blended learning is claimed to overcome students' boredom in learning, this occurs because during the learning process students will not experience such routine. In addition, students will also participate more in the learning process and can get more learning resources, because they can connect directly to the internet which provides abundant learning materials. Technological advances enhance students to experience rapid development and often make teachers difficult to deal with the learning process. Teachers should be more open and more understand technological developments because students in the 21st century are active users of technology [5]. Therefore blended learning will be a way for teachers to keep up with student technology and skills [13].

Blended learning can provide students with a lot of knowledge because it can explore a variety of learning resources. Those who use interactive media based online will get the opportunity to gain new knowledge faster [2]. In addition, digital learning can also practice students-centered. Even though online learning is considered to have many benefits, learning will be more meaningful when students can interact directly with what they are learning, one of them is learning real objects, in example all objects that are still in their original state, or naturally in a living or in a dead state in the form of specimens representing specimens in their original location. The use of real objects in the learning process will make it easier for students to carry out observations and conduct investigations about what they are learning [10]. Learning by using a real object does not only bring learning objects into the classroom but also through direct observation in the student's learning environment. The purposes of this research are: (1) to find out comparison of the effectiveness of real object-based learning and blended learning towards students' metacognitive knowledge.

2. Research Methods

This research is a quasi-experimental study using a pretest-posttest comparison group design. The study was conducted at Bantul Public High School 1 and Bantul Public High School 2 2018/2019 academic year in the second semester. The population in this study were all students of class X MIA in Bantul Public High School 1 and Bantul Public High School 2 which totaled 378 students. The sample in this study amounted to 130 students selected using cluster random sampling technique consisting of 31 students of class X MIA 3 as experimental class 1, and 31 students of class X MIA 2 as experimental class 2 from Bantul Public High School 1. Then for the Bantul Public High School 2 as many as 34 students of class X MIA 4 as the experimental class 1, and 34 students of class X MIA 6 as experiment 2. Data collection techniques in this study used tests and questionnaires. The research instrument consisted of metacognitive tests to see students' metacognitive knowledge levels and questionnaires on metacognitive knowledge to find out the increase in students' metacognitive knowledge after the learning process. The research data analysis technique uses One Way ANOVA with the SPSS 24 for Windows application.

After all data has been collected, the validity and reliability test is carried out with the Quest program. The results of the validation obtained by infit mean MNSQ 1.04 and a standard deviation value of 0.43 so that it can be concluded that the test fit with the Rash Model so that all test items are suitable to be used in the study. The reliability result according to Alpha Chronbach is 0.80, so it can be concluded that the reliability of the instrument is a very high category. Because the instrument is valid and reliable, so it can be used as a tool to get data about metacognitive knowledge.

3. Finding and Discussion

The comparative study aims to compare the results of a treatment that has been done to a selected sample. In this research, a comparative study is done to determine the effectiveness of real object-based learning and blended learning in ecosystem topic on students' metacognitive knowledge. The activity of real object-based learning conducted by the experimental class 1 is to observe the ecosystem directly in the environment around the school with various activities that have been compiled in the student activity sheet, and blended conducted by the experimental class 3 is learning activities using quipper school, they do learning activities online on it apart from formal schedule that they have, furthermore when students in the classroom they do offline learning activities, discussing the material they have learned then the teacher guides the discussion process and gives confirmation.

The data obtained from the results of the study were then analyzed using the SPSS software and the following results were obtained: Table 1 shows the results of the comparison of descriptive values between real objects and blended learning towards students' metacognitive knowledge.

Table 1. Descriptive Results of Student Metacognitive Knowledge

	Real object		Blended learning	
	Pretest	Posttest	Pretest	Posttest
N	65	65	65	65
Range	26,00	32,00	35,00	37,00
Minimal Score	48,00	55,00	39,00	49,00
Maximal Score	74,00	87,00	74,00	86,00
Mean	60,07	69,21	54,46	70,07
Std. Deviation	6,82	8,36	8,25	7,47
Variance	46,54	69,98	68,06	55,88

The data in Table 1 shows the results of the descriptive of students' metacognitive knowledge, the data shows that the class given real object learning treatment from 65 students who took pre-test, obtained a range of 26 with a minimum value of 48 and a maximum value of 74, with a standard the deviation is 6.82209 and the variance value is 46.541 with an average value of 60.07. While for the class given treatment for blended learning based on 65 students who took the pre-test, the ranged value was 35 with a minimum value of 39.00 and a maximum value of 74, with a standard deviation of 8.25015 and a variance value of 68.065 with a value an average of 54.46.

After treatment, then data is collected by giving a post-test. In the class given real object-based learning treatment of 65 students who took post-test the range value was 32 with a minimum value of 55 and a maximum value of 87, with a standard deviation of 8.36565 and a variance value of 69,984 with an average value amounting to 69.21. While the class given the treatment of blended learning based learning from 65 students who took post-test found a range value of 37 with a minimum value of 49 and a maximum value of 86, with a standard deviation of 7.47560 and a variance value of 55.885 with an average value average of 70.07.

Based on the descriptive results of students 'metacognitive knowledge, there are differences in the increase in students' metacognitive knowledge between classes given the treatment of real objects and blended learning. This can be seen by comparing the minimum and maximum values before being given treatment or after being given treatment. These data show blended learning is better than real objects towards increasing students' metacognitive knowledge. Furthermore, the prerequisite test was carried out, Table 2 shows the results of tests on the normality of students' metacognitive knowledge.

Table 2. Normality Test Results of Student Metacognitive Knowledge

	Kolmogorov-Smirnov(a)			
	Statistic	Df	Sig.	Keterangan
Real object	,103	65	,086	Normal
Blended	,080	65	,200*	Normal

The data in Table 2 show the results of the one-sample Kolgomorov-Smirnov test to determine the level of normality. Based on the tests that have been done, the results of the Real object normality test are 0.86 and the Blended learning normality is 0.200. Because the normality test value is above 0.05, it can be assumed that the data is normally distributed. then carried out homogeneity test and obtained results:

Table 3. Results of Homogeneity Tests

		Levene Statistic	df1	df2	Sig.
Metacognitive	Based on Mean	1,238	1	128	,268
	Based on Median	1,323	1	128	,252
	Based on Median and with adjusted df	1,323	1	127,947	,252
	Based on trimmed mean	1,194	1	128	,277

The data in Table 3 show the results of homogeneity tests, from these results it can be seen that the significance values are above 0.05, so it can be concluded that the similarity of population variance based on the applied learning methods can be met. Furthermore, One Way ANOVA test was conducted to find out which method had an effect on students' metacognitive knowledge and the results were obtained:

Tabel 4. Results of One Way Anova Test

			Sum of Squares	df	Mean Square	F	Sig.
Between Groups	(Combined)		24,123	1	24,123	,383	,537
	Linear Term	Contrast	24,123	1	24,123	,383	,537
Within Groups			8055,600	128	62,934		
Total			8079,723	129			

Based on the data in table 4, it is obtained a significance value of 0.537. Because the significance value obtained is more than 0.05, it can be concluded that the two teaching methods have no significant effect on students' metacognitive knowledge. Measurement of metacognitive knowledge in this study in addition to using questionnaires, also using essay tests and obtained results as follow.

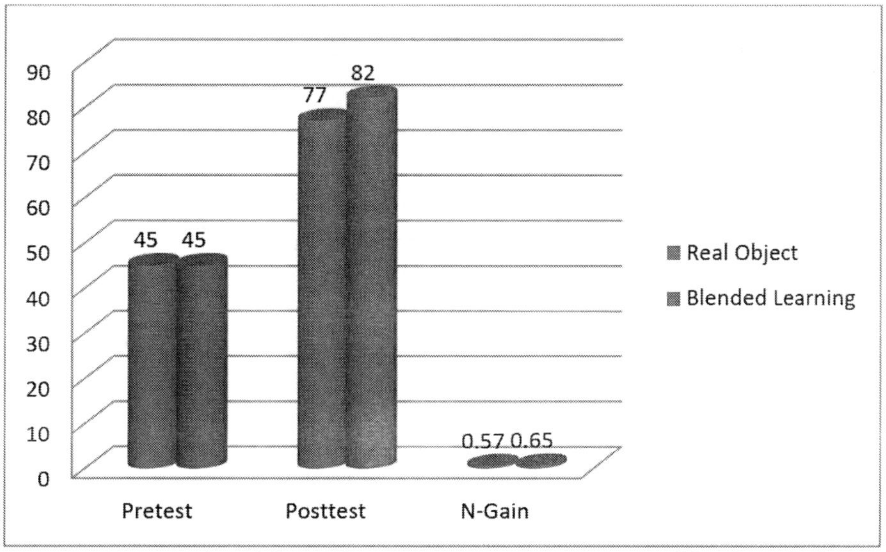

Figure 1. Results of the students' pretest-posttest metacognitive knowledge

Figure 1 shows the results of the metacognitive measurement of students with essay tests. Based on these data, blended learning obtained a pretest value of 45 and a posttest value of 82, while real object-based learning obtained a pretest value of 45 and a posttest value of 77. Based on these data it can be concluded that blended learning gained the highest average increase compared to real object-based learning.

This study was conducted to compare real object-based learning, and blended learning to students' metacognitive knowledge. Classes that get real object-based learning will directly observe the ecosystem around the school environment, while the class that gets treatment based on blended learning, students will learn using Quipper School. After the learning process, it was found that according to the data in table 4, real object-based learning and blended learning did not significantly influence students' metacognitive knowledge. Real object-based learning makes students engage and interact directly with learning objects so that they can explore knowledge directly through the objects they learn. Using realia can make the learning process more meaningful and more memorable for students because they can connect it to their real-life [1]. Therefore, real object-based learning will be more attractive to students so that they can develop various abilities they have. Students who are taught using human skeletal media show better learning outcomes than students who study with media images [3].

Meanwhile, web-based learning is already familiar to students, they have already used their smartphones. Web-based learning makes students free to find resources to learn about what they learn from various sources on the internet. In addition, students can have convenience, because they can access it whenever and wherever, not necessarily at school, so they can improve their learning achievement. Students' achievement and cognitive skills which are web-based learning methods are improved and developed [6]. In the digital era, web-based learning is indeed properly implemented by teachers in schools, besides being able to make it easier for them, web-based learning will also provide youthfulness to students. The advantage of web-based learning is pages from the web contain many hyperlinks from various other webs so that students have more information [14] [15].

Web-based learning in this research is by using blended learning based learning, which means there is a combination of online and offline learning. Based on the data in table 3, there is no significant influence on the dimensions of student knowledge. This is due to the lack of maximum learning when offline, students rarely open the quipper school application that is used for blended learning based learning. So when learning face-to-face students don't get the initial knowledge they should get when opening the quipper school application, because the application has provided a variety of materials that they will learn when meeting face-to-face in class. In addition, other factors that might influence students who are given blended learning are constrained by networks and internet quota and time management [11]. Because to open Quipper School requires a stable internet network and large internet quota.

4. Conclusion

The research that has been done shows the results of the two learning methods have no effect on students' metacognitive knowledge. however blended learning-based learning gets a higher increase in the average metacognitive students than real object-based learning. The results of this study hopefully can help teachers to prepare the right learning method when going to deliver learning in class. So that it can convey the learning objectives well, and hopefully the teacher will be more creative so that they are able to adapt the learning method to the material taught in the class.

Acknowledgement

The authors thank students and teachers of SMAN 1 Bantul and SMAN 2 Bantul for helping this research and special thanks to Department of Biology Education Graduate School of Yogyakarta State University, due to supporting this work.

References

[1] Bably, T. & Nusrat, D. (2017). Using realia as an effective pedagogical tool. *Journal Of Humanities And Social Science (IOSR-JHSS)*, *22*(11), 1–7. https://doi.org/10.9790/0837-2211040107

[2] Gurubatham, M. R. (2013). Blended action learning involving metacognition and active discussion internationally. *Procedia-Social and Behavioral Sciences*, *93*, 2157–2172. https://doi.org/10.1016/j.sbspro.2013.10.182

[3] Humaira, Sardinah, & Yusuf, M. N. (2015). Perbandingan hasil belajar siswa pada materi kerangka manusia melalui media kerangka manusia dan media gambar siswa kelas IV SDN Lampeuneurut, Aceh Besar. *Jurnal Pesona Dasar*, *3*(3), 60–72. Retrieved from http://www.jurnal.unsyiah.ac.id/PEAR/article/view/7515/6186

[4] Husamah. (2015). Blended project based learning: Metacognitive awareness of biology education new students. *Journal of Education and Learning*, *9*(4), 274–281. https://doi.org/http://dx.doi.org/10.11591/edulearn.v9i4.2121

[5] Jayawardana, H. B. (2017). Paradigma pembelajaran biologi di era digital. *Jurnal Bioedukatika*, *V*(1), 12–17. Retrieved from http://journal.uad.ac.id/index.php/Bioedukatika

[6] Lakonpol, T., Ruangsuwan, C. & Terdtoon, P. (2015). Development of web-based learning environment model to enhance cognitive skills for undergraduate students in the field of electrical engineering. *Academic Journals*, *10*(21), 2806–2813. https://doi.org/10.5897/ERR2015.2470

[7] Lalima, & Lata Dangwal, K. (2017). Blended learning : An innovative approach. *Universal Journal of Educational Research*, *5*(1), 129–136. https://doi.org/10.13189/ujer.2017.050116

[8] Lukitasari, M., Hasan, R. & Murtafiah, W. (2019). Using critical analysis to develop metacognitive ability and critical thinking skills in biology. *JPBI (Jurnal Pendidikan Biologi Indonesia)*, *5*(1), 151–158. https://doi.org/https://doi.org/10.22219/jpbi. v5i1.7262

[9] Paidi. Wibowo, Y. & Rachmawati, A. (2009). Analisis tingkat kemampuan metakognitif mahasiswa jurusan pendidikan biologi, FMIPA UNY. *FMIPA, UNY*, 1–16. Retrieved from http://staffnew.uny.ac.id/upload/132048519/penelitian/artikel-metakognitif-makalah-semnas-bio-2013-paidi.pdf

[10] Rivai, A. & Sudjana, N. (2017). *Media pengajaran*. Bandung: Sinar Baru Algensindo

[11] Sayed, M. (2013). Blended learning environments : The effectiveness in developing concepts and thinking skills. *Journal of Education and Practice*, *4*(25), 12–18. Retrieved from www.iiste.org

[12] Suana, W., Maharta, N., Nyeneng, I. D. P. & Wahyuni, S. (2017). Design and implementation of schoology-based blended learning media for basic physics 1 course. *Jurnal Pendidikan IPA Indonesia*, *6*(1), 170–178. https://doi.org/10.15294/jpii.v6i1.7205

[13] Suyanto, S. (2018). The implementation of the scientific approach through 5Ms of the revised curriculum 2013 in Indonesia. *Cakrawala Pendidikan*, *XXXVII*(1), 22–29. https://doi.org/https://doi.org/10.21831/cp.v37i1.18719

[14] The Dung, N. & Fatmawati, D. (2018). General informatics teaching with b-learning teaching model. *Jurnal Pendidikan Biologi Indonesia*, *4*(1), 85–94. https://doi.org/doi.org/10.22219/jpbi.v4i1.5312

[15] Wasim, J., Sharma, S. K., Khan, I. A., & Siddiqui, J. (2014). Web based learning. *International Journal of Computer Science and Information Technologies*, *5*(1), 446–449. Retrieved from www.ijcsit.com

Developing Chamilo-Based E-Learning in Environmental Change Material to Enhance Students' Scientific Literacy Skills

F. Astriawati[1*] and Djukri[2]

[1]Student of Biology Education, Graduate Program, Yogyakarta State University Colombo Street No 1, Karangmalang, Depok, Sleman, Yogyakarta, Indonesia
[2] Department of Biology, Graduate Program, Yogyakarta State University

[*]Corresponding author e-mail: fitriastriawati@gmail.com

Abstract. The main goal of learning science today is the realization of learners who are literate in science. Unfortunately, Indonesian scientific literacy skill is still low based on the assessment of OECD for three evaluation periods. Therefore this study aims to produce an electronic-based learning system in learning science using LMS Chamilo which is intended to enhance students' scientific literacy skills. This study is research and development (R & D) using the ADDIE model. Research findings indicated that (1) the Chamilo-based e-learning was feasible to implement in biology teaching based on experts judgment with a total score of 3.83 (very good) from environmental material expert and 3.22 (good) from learning media expert; (2) the Chamilo-based e-learning was accepted very well by users with a total score of 3.78 (very good) from biology teachers and 3.36 (very good) from students; (3) the Chamilo-based e-learning was effective in enhancing students' scientific literacy skills with n gain score of 0.74 (high criteria). Based on the results, it can be concluded that Chamilo-based e-learning is recommended to implement and to use in the biology learning process.

1. Introduction

The main goal of learning science today is the realization of a society that literate in science through understanding mathematics, physics, chemistry, biology, and technology. The demand for mastering scientific literacy either in national level or international level arises because all of the people are required to participate in solving real-world problems such as environmental pollution, climate change, natural disaster, and health [1].

The term scientific literacy was first revealed by Paul DeHart Hurd in 1958 which was used to describe the understanding of science and its applications in society. Scientifically literate people can engage with science issues and science ideas as reflective citizens based on the ability to explain phenomena scientifically, evaluate and design scientific inquiry, and interpret data and evidence scientifically [2]. It means that the students who are literate in science are those who have scientific knowledge and use it to explain the natural phenomena, get new knowledge through scientific inquiry, and draw conclusions based on scientific evidence.

Scientific literacy skills are evaluated internationally by the OECD once in three years and show that Indonesia is still below the average score of OECD during three evaluation periods (2009, 2012 and 2015), it causes Indonesia categorized as a country with low scientific literacy skills. This background then leads a demand to make the efforts of improving science learning quality in schools gradually and continuously.

The rapid progress of information, communication, and technology (ICT) offers a new learning innovation through the utilization of the internet with world-web-technology what it is called electronic-based learning (e-learning). The term "e-learning" is defined as learning without using printed teaching materials [3], the delivery of learning materials is conducted using an electronic course [4]. The use of e-learning offers the interactions between students and teachers from unlimited time and space through the asynchronous and synchronous learning network model [5] where the traditional instructional delivery system in the schools are limited to the classroom with the teacher giving speeches to students and students listening and taking notes [6]. With the progress of information and communication technology development, e-learning is emerging as the paradigm of modern education.

However, further development of e-learning is not just learning using an electronic course, but it necessary to build a rule and management of good teaching by using information technology [7], it is what is often referred to as a Learning Management System (commonly abbreviated to "LMS"). Through LMS, the teachers not only utilize the internet to give interesting information to the students but also can manage the learning activity through the served features inside it. One kind of LMS providing platform for e-learning and collaboration is Chamilo. It is free software focused on building e-learning portals easily, quickly and has attractive interfaces [8]. Chamilo is included in the 10 best open sources of LMS and takes place in the first rank in the course of development features, collaboration features, and instruction methods. Chamilo also takes place in the second rank in the administration features instead of the other open sources of LMS such as Moodle, LRN, eFront, Dokeos, Sakai, Latitude Learning, Canvas, OLAT and Totara [9].

Through the features available in Chamilo, the teachers can present distant biological objects like pollution and environmental damage occurring in various regions or hard-observed biological phenomena like global warming, polar ice melt, depletion of the ozone layer, and the acid rain process. These objects and phenomena can be presented in the form of pictures, animation, or video. These advantages will make easier for students to understand biological concepts and relate them to various phenomena and current scientific issues, draw conclusions based on evidence, make decisions, and apply them in their daily life. Moreover, the teachers can manage in distributing learning materials, building communication, giving the task, doing the test, monitoring student's learning progress and other aspects of learning management. As it provides a pleasant learning experience, the students will be more interested in learning biology. The learning can also give positive effects related to their scientific literacy skills.

Based on the explanation above, this study aims to develop Chamilo-based e-learning in environmental change material to enhance students' scientific literacy skills.

2. Method

2.1. Research Type and Design
This study is Research and Development (R&D) with the ADDIE model. Research and Development is the systematic study of designing, developing and evaluating instructional programs, processes, and products that must meet the criteria of internal consistency and effectiveness [10]. The systematic process in this model was represented in the ADDIE (Analysis, Design, Development, Implementation, and Evaluation) which stood for the important components in the process of creating the instructional design [11]. The ADDIE model is an iterative instructional design process where the results of the formative evaluation of each phase might lead the instructional designer back to any previous phase. The end of one phase was the starting product of the next phase [12].

2.2. Data Collection and Analysis Technique
The developed e-learning was evaluated by the experts and the users and it was implemented to the students to find out its effectiveness in enhancing scientific literacy skills. The evaluation by experts and users were measured using feasibility assessment sheets. The data of evaluation was calculated using a

Likert scale 1-4 and analyzed based on the scoring system and the criteria in Table 1[13] to determine the level of feasibility of e-learning devices.

Table 1. Score interpretation (conversion quantitative data to qualitative)

Formula	Interval Score	Criteria
$Mi + 1.5SDi \leq M \leq Mi + 3.0SDi$	$3.25 \leq M \leq 4.00$	Very Good
$Mi + 0SDi \leq M \leq Mi + 1.5SDi$	$2.50 \leq M \leq 3.25$	Good
$Mi - 1.5SDi \leq M \leq Mi + 0SDi$	$1.75 \leq M \leq 2.50$	Not Good
$Mi - 3.0SDi \leq M \leq Mi - 1.5SDi$	$1.00 \leq M \leq 1.75$	Very Not Good

Scientific literacy skills were measured using an essay instrument formulated following the indicators of the *Program for International Student Assessment* (PISA) 2015. Indicators of scientific literacy are developed based on the science competency aspect, including (a) explain phenomena scientifically - recognizing, offering and evaluating the explanations for various natural things and technological phenomena; (b) evaluate and design scientific inquiry - describing and evaluating scientific investigations and proposing ways to answer questions scientifically; (c) interpret data and evidence scientifically - analyzing and evaluating data, claims, and arguments in various representations and drawing appropriate scientific conclusions. The test was carried out using *one group pretest-posttest design* and the data of test were calculated using the formula of the n-gain score [14]:

$$< g > = \frac{S_f - S_i}{S_{max} - S_i}$$

Noted:
S_f = Post-test score
S_i = Pre-test score
S_{max} = maximum score

The acquisition of n-gain scores was analyzed based on the criteria system in table 2 to determine the category of enhancement in students' scientific literacy skills.

Table 2. The n-gain interpretation score

Range	Criteria
$(<g>) < 0.3$	Low
$0.3 \leq (<g>) < 0.7$	Medium
$(<g>) \geq 0.7$	High

3. Result and Discussion

The produced product in this research and development is e-learning in environmental change material. E-learning devices are developed using LMS Chamilo version 1.11.8 for PHP 7.0.x or later. Chamilo is free software providing a platform for e-learning and collaboration. It is what is often referred to as a "Learning Management System", a popular term used to describe any kind of software platform designed to facilitate the management and delivery of courses and the monitoring of students' progress [15].

The developed e-learning consists of four main components including homepage, courses catalog, my courses, and course homepage. There are two types of homepage. The first is homepage before login that displays login navigator, usage instructions, learning resources (containing learning videos), list of contents, and developer biography. The second is homepage after login that displays the most popular courses, user profiles, personal agenda, progress, social networks, and the number of users online.

The courses catalog displays all content in e-learning which is classified into four categories, namely learning materials, practical worksheets, articles, and tests. All content in e-learning is related to environmental change material. My courses displays a list of courses that have been subscribed by students. To access a course, click on the course name and the course home will appear. The course homepage displays menus in the form of course descriptions, agendas, learning paths, announcements, users, chat, assignments and glossary. E-learning users simply click on one of the menus as needed. In detail, the developed e-learning can be accessed on the website *chamilogi-learning.com*.

The development of Chamilo-based e-learning involves the experts and users to provide an evaluation of the developed e-learning devices. Furthermore, e-learning is implemented for students to find out its effectiveness in enhancing scientific literacy skills. The assessment results are shown in the following explanation.

3.1. Evaluation Result of Experts

3.1.1. Evaluation Result of Environmental Expert. The evaluation of the material aspect was carried out by Dr. Suhartini, M.S., an environmental expert in Biology Education, Graduate Program, Yogyakarta State University. The material evaluation is reviewed based on four aspects, including the suitability of material, material presentation techniques, language and suitability of evaluation form. The evaluation result of an environmental expert can be seen in table 3.

Table 3. The evaluation result of environmental expert

No	Assessment Aspect	Score	Criteria
1	Suitability of material	3.86	Very Good
2	Material presentation techniques	3.80	Very Good
3	Language	4.00	Very good
4	Suitability of evaluation form	3.67	Very Good
Average		3.83	Very Good

The evaluation result by environmental expert states that the material concept is correct and following the core competencies (KI) and basic competencies (KD). It is presented clearly, systematically and interestingly. The language chosen in explaining the concept is communicative, helps the students to understand learning material easily. The type of test used is also under the material presented. However, there are some typing errors needed to be corrected and some environmental laws that must be updated. Based on table 3, the average score of all aspects given by material experts is 3.83. It can be concluded that the presented material in e-learning is feasible to use in biology teaching with very good criteria.

3.1.2. Evaluation Result of Learning Media Expert. The evaluation of the media aspect is conducted by Dr. Slamet Suyanto, M.Ed., an education technology expert in Biology Education, Graduate Program, Yogyakarta State University. The media evaluation is reviewed based on eight aspects, including quality of content, quality of pictures, videos, and animations, creativity in expressing the idea, display, language, interaction, usability, and compatibility. The evaluation result of learning media expert can be seen in table 4.

Table 4. The evaluation result of learning media expert

No	Assessment Aspect	Score	Criteria
1	Quality of content	3.33	Good
2	Picture, video, animation	3.00	Good
3	Creativity in expressing idea	4.00	Very Good
4	Display	3.50	Very good
5	Language	3.00	Good
6	Interaction	3.00	Good
7	Usability	3.00	Good
8	Compatibility	3.00	Good
Average		3.22	Good

The evaluation result of learning media expert states that e-learning has functioned well. It has an attractive interface and uses communicative language. Various kinds of e-learning content such as learning materials, learning videos, environmental articles, competency tests, and discussion forums are presented creatively. Moreover, e-learning compatibility is good; it can be accessed via personal computers and smartphones, making it easier for students to use. Based on table 4, the average score of all aspects given by learning media expert is 3.22. From this, the researcher concludes that e-learning media is feasible to use in biology teaching with good criteria.

However, there is a feedback from learning media expert. The expert says that it is necessary to add the instructions of use, so the new students can be easier to access the material in e-learning.

3.2. Evaluation Result of Users

3.2.1. Evaluation Result of Biology Teachers. The Chamilo-Based e-learning was evaluated by Mr. Kusnardi S.Pd and Mrs. Lilik Tri Utami S.Pd as biology teachers of SMA N 1 Wates, Yogyakarta. The aspects evaluated by biology teachers include the material/content, media, language, and learning. The evaluation results of biology teachers can be seen in table 5.

Table 5. The evaluation result of biology teachers

No	Assessment Aspect	Score	Criteria
1	Material/content	4.00	Very Good
2	Media	3.75	Very Good
3	Language	3.67	Very Good
4	Learning	3.71	Very Good
Average		3.78	Very Good

The evaluation result of biology teachers states that e-learning content is interesting to learn. The material is arranged systematically and equipped with images and videos, support students in understanding learning materials. E-learning design is good and easy to operate via personal computers and smartphones. Moreover, e-learning can help teachers to convey more interesting learning materials to students. Through e-learning, the teachers can manage the learning materials and distribute them to the students. It also helps the teachers to monitor students' progress in learning biology. Lastly, e-learning helps the students to learn the materials independently; students can access the learning material anytime anywhere making the learning process more flexible. Overall, it can be concluded that e-learning can improve the quality of learning.

Based on table 5, the average score of all aspects scored by biology teachers is 3.78 (very good), meaning that e-learning media is feasible to use in biology teaching.

3.2.2. Evaluation Result of Students' Responses. The evaluation of students' responses to Chamilo-based e-learning is conducted by giving a questionnaire to thirty-two students of the tenth grade of SMA N 1 Wates, Yogyakarta. The questionnaire contains the statements categorized into seven aspects; including interest of content, interest of media, language, usability, compatibility, flexibility, and interaction. The evaluation result of students' responses can be seen in table 6.

Table 6. The evaluation result of students' responses

No	Assessment Aspect	Score	Criteria
1	Interest of content	3.55	Very Good
2	Interest of media	3.54	Very Good
3	Language	3.16	Good
4	Usability (simple to operate)	2.98	Good
5	Compatibility (easy to run on PC/tablet/ smartphone)	3.13	Good
6	Flexibility (can be accessed anytime and anywhere)	3.78	Very Good
7	Interaction (chat, online discuss)	3.34	Very Good
Average		3.36	Very Good

The evaluation result of students' responses state that the design of environmental change material presented in e-learning is very good. The support of images and videos on the materials makes it easier to understand. Moreover, e-learning has an attractive interface. It can be accessed not only through personal computers but also through smartphones. It makes students easier to learn; they do not have to bring a lot of books. Only with a smartphone, they can still access the materials anytime anywhere. Through e-learning, they can create a group of discussion to discuss assignments from the teacher and can also ask the teacher through a private chat when they cannot meet the teacher face to face. In other words, the use of e-learning provides new learning experiences for students.

Based on table 6, the average score given by students is 3.36, meaning that the development of Chamilo-based e-learning can be accepted by students very well. However, some students find some difficulties in using e-learning, especially in accessing the material and uploading the assignments. This case can be a reason why the developers need to improve e-learning more easily or at least add the instructions of use on each page.

3.3. The Effectiveness of Chamilo-Based E-Learning in Enhancing Students' Scientific Literacy Skills

The Chamilo-based e-learning is implemented to 32 students of the tenth grade of SMA N 1 Wates Yogyakarta to find out its effectiveness in enhancing students' scientific literacy skills. The scientific literacy skills are measured based on the pre-test and post-test results using n-gain score analysis. The measurement results of scientific literacy skills can be seen in Table 7.

Table 7. The review of pre-test and post-test results

No	Explanation	Pre-test	Post-test
1	Total number of students	32	32
2	Lowest score	38.25	75.00
3	Highest score	65.00	100.00
4	Group's average	52.08	87.52
n-gain score		0.74	
The category of *n-gain* score		High	

Based on Table 7, the n-gain score gets 0.74. It means that there is an improvement of students' scientific literacy skills after using Chamilo-based e-learning in biology teaching with the high category. The spread of categories of scientific literacy skill improvement in 32 students is represented in the pie chart below (figure 1).

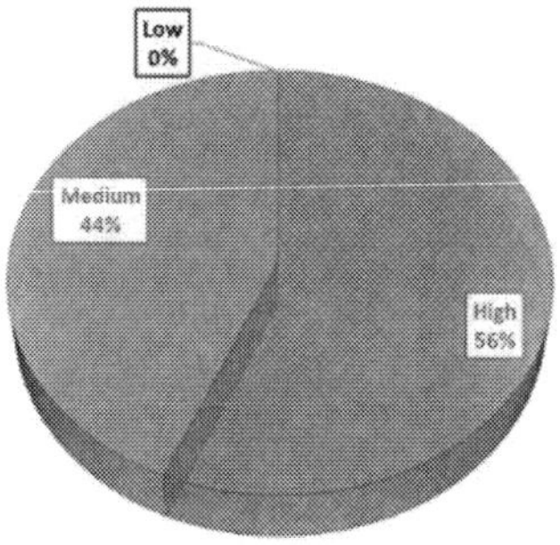

Figure 1. The spread of categories of students' scientific literacy skills improvement after using Chamilo-based e-learning in the learning process.

Based on Figure 1, it can be explained that 56% of the students (18 students) enhance their scientific literacy skills with the high category and 44% (16 students) enhance their scientific literacy skills with the medium category. In other words, there is a big difference in the range of students' scientific literacy skills. The difference in students' scientific literacy skills enhancement can occur because of some reasons such as students' potential reasons, the characteristics of social areas, and the various cultures of the students in Indonesia. These reasons give an effect on the aspects of learning and the students' scientific literacy skills as well. Based on the results, there is the enhancement in each scientific literacy competency after using Chamilo-based e-learning in biology teaching represented in the chart below (figure 2).

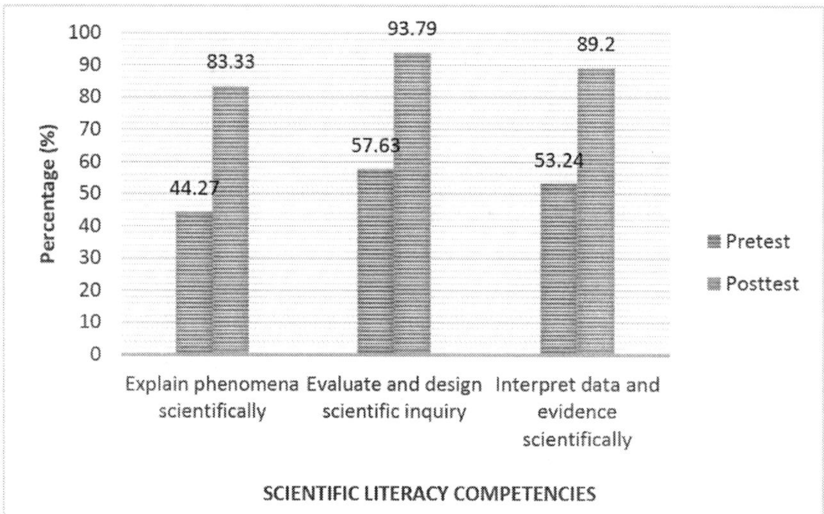

Figure 2. The chart of the enhancement of the scientific literacy competencies

The chart in figure 2 shows that the enhancement of Explain Phenomena Scientifically aspect is 39.06%, the enhancement of Evaluate and Design Scientific Inquiry aspect is 36.16%, and the enhancement of Interpret Data and Evidence Scientifically aspect is 35.96%. The increase of these three aspects is stimulated by the use of text, picture, video, and animation presented in the interactive learning media like e-learning. It can help the students in understanding the content learned [16]. Besides, Chamilo-based e-learning provides the environmental pollution problems happened in the society, so the students can understand the situations of life in which it does not only focus on the school life, but also focus on daily life, so this kind of e-learning will stimulate the students to relate their learned knowledge with the problems existed in the society.

The Chamilo-based e-learning also presents various scientific phenomena such as a mass of fish death in the river, acid rain, depletion of the ozone layer, hotter weather, explosion of the particular animal population like rats, and other phenomena. Presenting the scientific phenomena will train the students to collect various information, data, and evidence from many resources such as news, journals, or the internet to explain these scientific phenomena based on the correct data and evidence. Moreover, they can also interpret the data and evidence they've obtained to strengthen their arguments. Chamilo-based e-learning also provides a worksheet to train the students in arranging and using the correct procedure in conducting some investigations scientifically.

Fundamentally, the development of Chamilo-based e-learning is not only intended to improve students' scientific literacy skills but also to help teachers in collecting and organizing the learning materials online. The e-learning is also helpful in sharing the materials to the students, so they do not have to carry a pile of books to school. The use of e-learning can offer much time-saving by taking over a range of repetitive administrative tasks and allowing teachers to focus on supporting students. The spare time, then, can be used to create many types of teaching materials. More importantly, it can provide students with a range of motivational tools supporting the effectiveness of independent study at their own space and as a purpose to interact more with their teachers and fellow students.

The use of e-learning in biology teaching has advantages and disadvantages. The advantages of using e-learning are: (a) providing complete learning materials accompanied by practice questions with attractive displays so they can easily be understood by students; (b) facilitating teachers to monitor the students' learning progress because it is equipped with students' developmental data analysis; (c) facilitating the students to review the learning materials anytime anywhere; (d) facilitating the teachers and the students to interact each other through the forums, so it makes easier for students to ask teachers when they cannot meet face to face, (e) being more efficient in time, place, and cost.

In conclusion, the use of e-learning is intended to improve the quality of learning. The results of some studies state that e-learning can increase student motivation [17] [18]; give positive impact to academic performance [19]; improve students' academic achievement and creativity [20]; Improve students' knowledge and skills [21]; improve the quality of practical training; and provide a better understanding of learning materials [22]. Indeed, e-learning is not intended to replace conventional classroom settings, but it can be a complementary learning process related to learning materials possibly can be accessed online [23].

The disadvantages of e-learning are: (a) requiring additional devices such as personal computers, smartphones, or tablets; (b) only be used if devices definitely connect to internet with stable networks; (c) having negative effect on student communication skills (students may have excellent knowledge in academics but they may not have good skills to convey their knowledge to others); (d) lacking of effectiveness for practical learning materials like observing microorganism directly using a microscope; (5) being less effective for the teachers and the students who are not familiar with or lacking the knowledge of internet.

4. Conclusions

Based on the result of the research and development, it could be concluded that: (1) the Chamilo-based e-learning was feasible to implement in biology teaching based on experts judgment with a total score 3.83 (very good) from environmental material expert and 3.22 (good) from learning media expert; (2) the Chamilo-based e-learning was accepted well based on users' judgment with a total score 3.78 (very good) from biology teachers and 3.36 (very good) from senior high school students; (3) the Chamilo-based e-learning was effective in enhancing students' scientific literacy skills with n gain score of 0.74 (high category).

Acknowledgment
The researcher thanked to Prof. Dr. Djukri, M.S. who had assisted in developing the Chamilo-based e-learning and writing this article. Also, many thanks to material experts, media experts, biology teachers, and students who had provided the evaluation of e-learning devices developed.

References
[1] Cardwell V B 2005 *J. of Natural Resources and Life Science Education* **34** 112-117
[2] OECD 2016 PISA 2015 *Assessment and Analytical Framework: Mathematics, Reading, Science, and Financial Literacy* (Paris: OECD Pubishing)
[3] Goyal S 2012 *J. of Education and Learning* **6** 239-242
[4] Cheng K 2006 *J. of American Academy of Business* **8** 265-270
[5] Sun P C 2008 *Computers & Education* **50** 1183-1202
[6] Harandi S R 2015 *Procedia-Social and Behavioral Science* **181** 423-430
[7] Istambul M R 2016 *Int. J. of Information and Education Technology* **6** 423-429
[8] Torreblanca A, Steve M, Yannik W 2015 *Chamilo LMS 1.9 Starter Guide. A brief guide to a great e-learning platform* (Lean Publishing)
[9] Elabnody M R 2016 *Int. J. of Scientific & Technology Research* **5** 7-11
[10] Richey R C and Klein J D 2005 *J. of Computing in Higer Education* **16** 23-38
[11] Aldoobie, N 2015 *American Int. J. of Contemporary Research* **6** 68-72
[12] McGriff S J 2000 *Instructional System* (Philadelphia: Penn State University)
[13] Direktorat Pembinaan SMA 2010 *Juknis penyusunan perangkat penilaian afektif di SMA* (Jakarta: Direktoran Pembinaan SMA)
[14] Hake R 1999 *Analyzing Change/Gain Scores*. American Education Research Association's Devision. Measurement and Reasearch Methodology
[15] Torreblanca A, Steve M, Yannik W 2015 *Chamilo LMS 1.9 Starter Guide. A brief guide to a great e-learning platform* (Lean Publishing)
[16] Mayer R E 2015 *Appl. Cognit. Psychol* **29** 614-620
[17] El-Seoud M S A, Islam. A T F, Naglaa S, Mahmoud M E, Ann N 2014 *Int. J. of Engineering & Technology* **9** 20-26
[18] Harandi S R 2015 *Procedia-Social and Behavioral Science* **181** 423-430
[19] Suresh M., Visnhu P, Gayathri 2018 *Research Article Drug Invention Today* **10** 1797-1800.
[20] Zare M., Sarikhani R., Salari M., Mansouri V 2016 *J. of Technical Education and Training* **8** 25-33
[21] Istambul M R 2016 *Int. J. of Information and Education Technology* **6** 423-429
[22] Yanuschik O V, Elena G P, Khongorzul B 2015 *Procedia-Social and Behavioral Science* **215** 147-155
[23] Suyanto, S. 2017. A Reflection on the Implementation of a New Curriculum in Indonesia: A crucial Problem on School Readiness. *AIP Conference Proceedings* **1868** 100008-1–100008-9

Journal of Physics: Conference Series **1397** (2019) 012050 doi:10.1088/1742-6596/1397/1/012050

TPACK Mastery of Biology Teachers: A Study Based on Teacher Gender

F. E. C. Astuti[1*], Paidi[2], B. Subali[2], N. Hapsari[1], S. P. Pradana[1], and M. K. Antony[1]

[1]Graduate Student of Biology Education, Universitas Negeri Yogyakarta, Indonesia
[2]Graduate Program, Universitas Negeri Yogyakarta, Indonesia

*Corresponding author e-mail: fitriaeka.2017@student.uny.ac.id

Abstract. TPACK is knowledge that needs to be mastered by teachers to be able to integrate technology in the learning process. The success of integrating ICT can be influenced by teacher gender. Differences in character between male and female teachers can influence teacher's steps and ways to teach to integrate technology in learning. This study aimed to know TPACK mastery of biology teacher and how it differs between male and female biology teachers. This study involved 29 biology teacher respondents who were taken using the convenience sampling of the hypothetical population. Data collection is conducted using a test instrument consisting of 40 items. Instruments meet the logical validity test based on expert judgment. The empirical validity showed infit MNSQ 1.00 invitations and standard deviation of 0.16 which means all items were valid. Test instrument reliability results of 0.57 means that test was good enough to provide consistent measurement results. The data were analysed descriptively and inferentially through the independent sample t-test. The results of this study showed teacher TPACK mastery was good with a score of 59.15 but TK domain with score of 46.67 and TPACK domain with score of 45.83 were still fair. TPACK mastery of biology teacher was not differentiated by teacher gender with $p>0.05$.

1. Introduction

The 21st century was marked by the rapid development of information and communication technology (ICT). The development of technology has an impact on all sectors of life, one of which is the education sector. Teachers as key in education need to pay attention to this development. Professional teachers will always develop themselves along with the development of the times in accordance with the 2005 Constitution No. 14 concerning Teachers and Lecturers stating that teachers are obliged to improve and develop academic qualifications and competencies on an ongoing basis. This is in line with the development of science, technology, and art. Therefore, teachers need to master ICT therefore that learning is in line with current developments. ICT mastery is needed by teachers to be able to integrate technology in learning according to the demands of 21st-century education.

To be able to integrate ICT in the learning process, teachers must master the knowledge of technology that is integrated with pedagogical and content knowledge. The integration of pedagogic, content and technology knowledge in order to realize ICT-integrated learning is called Technological Pedagogical and Content Knowledge (TPACK) [1]. Good teaching with technology requires at least three domains, namely Pedagogical Knowledge (PK), Content Knowledge (CK), and Technological Knowledge (TK) as well as relationships between these domains. The three are interconnected to form the Pedagogical Content Knowledge (PCK), Technological Pedagogical Knowledge (TPK),

Technological Content Knowledge (TCK), and Technological Pedagogical and Content Knowledge (TPACK).

The TPACK mastery of teacher need to be aware to see the teacher's readiness in integrating ICT in learning. Teachers who master TPACK will have a great potential to successfully integrate ICT in the learning process. Teachers could not integrate ICT if teacher did not have a good enough knowledge on integrating ICT in the learning process. TPACK's mastery of biology teachers has not been measured especially in Yogyakarta city. Yogyakarta City as a barometer for other areas of education quality issues. Thus the mastery of TPACK of biology teachers in Yogyakarta becomes more urgent, because it will be able to provide reference information for biological learning in other regions in Indonesia. The teacher assessment that has been done is that UKG (Uji Kompetensi Guru) and PKG (Penilaian Kinerja Guru). This assessment measure teacher pedagogic and professional knowledge but have not measured the teacher's knowledge of technology and its integration in pedagogic and content. Some research has been conducted measuring the TPACK mastery using an instrument with attitude scale. Whereas TPACK is a knowledge therefore the appropriate measuring instruments use test. Therefore, research surveys need to be conducted to know the mastery of TPACK biology teachers using test instruments.

Teacher TPACK are very complex because they are influenced by many factors. Each teacher has different characteristics that influence the learning process that will be carried out. The teacher's personality can influence teacher teaching methods including gender, age, teaching experience, character, and beliefs [2]. Gender was one of the factors that could influence the teacher's TPACK mastery. Differences in characteristics, attitudes, and abilities by male and female teachers could influence teacher mastery of teacher TPACK. Research into the influence of gender on TPACK has been carried out in several countries and shows diverse results. Some studies show gender was a factor that influences teacher TPACK and there were significant differences in TPACK towards teacher gender where male have higher TPACK than female [3, 4, 5, 6]. This is due to the male's knowledge of ICT better than female [7]. Different ICT knowledge between male and female can affect the mastery of TPACK. This difference in results showed gender influences and the TPACK teachers need to be further examined. Some studies show the opposite results where there was no significant gender influence on TPACK [8, 9, 10, 11]. The difference in results caused the influence of gender to be studied more deeply. results of the study showed that female teachers were less interested in using ICT [12], male were more experienced with ICT [13, 14]. The research above showed differences in the characteristics of male and female teachers could influence the integration of ICT in learning. Gender influences the adoption of ICT in learning activities [15]. The level of success of the integration of ICT in learning could be reflected in TPACK therefore that if the teacher was able to apply ICT in learning then had a high TPACK.

Male teacher were more confident in ICT but female teachers were more confident in the pedagogical knowledge [16]. Male and female teacher had different type in teaching. Some research results show there are differences in the teaching characteristics of female and male teachers. Female teachers are more interactive, providing support and motivation to learners and expressive [17], more guiding [18], more open to learners [19], more interaction and collaboration with learners [20], and using more flexible learning methods [19]. Male teachers tend to dominate more in teaching [19], more emphasizing group learning and structured activities [19], actively asking short questions to student [17], and more authoritarian teaching styles [20].

According mastery of biological content, some research shows female have more mastery biology content than male [21, 22]. Female have a higher recall ability than male because female have a wider hippocampus therefore as to potentially improve better long-term storage memory than male [23]. Biology were many concepts that must be memorized very supportive for female.

In Indonesia, technology is still seen as a task of male. Male were more trusted in technical tasks related to technology while female were still underestimated [24]. The results of other studies by the State Ministry of Empowerment of Indonesian Women, in the field of technology, especially ICT, obtained results that information and communication technology was still very close to male's identities. One of the factors that influence this condition was value and culture. In Indonesia, the existence of a patriarchal culture that marginalizes female had an impact on the use of technology. Technology was not free of values or culture, meaning the level of use of technology was influenced by culture and user value [25]. The value of technology tended to be more masculine than feminine which was one of the causes of technological inequality between male and female. The cause of the low involvement of female in the ICT field was the socio-cultural construction which emphasizes the ICT field was the domain of male [26, 27]. In addition, variables that cause ICT integration were slow in the curriculum, namely teacher factors, one of which was due to the influence of gender aspects [20]. Therefore, this research was important to determine whether there was an influence of gender on the TPACK mastery of biology teacher.

2. Method

This type of research was survey research. This research was conducted from September 2018 – March 2019 in 20 high schools in Yogyakarta City, Indonesia. The population of this research was hypothetical population of all biology teacher in public and private high school in Yogyakarta city. The sample of this study amounted to 29 biology teachers who were selected using the convenience sampling technique. Characteristics of respondents dominated by teachers who have been teaching more than 15 years, certified, and bachelor degree of biological education. This research is not conducted by strict variable control. Data collection instruments used tests. Test instruments in the form of multiple-choice test amounted to 40 items. The indicator of teacher TPACK mastery were presented as follows.

Table 1. Indicator of TPACK Mastery

TPACK Domain	Indicator	Items
PK	Teacher understand students' character, learning theory, learning strategies, and learning planning	6
CK	Teacher understand biology subject (i.e. bacterial, plantae, digestive system, respiration system, excretion system, immune system, metabolism, genetics, and biotechnology).	10
TK	Teacher understand various technological elements include ICT-based media, blogs, email, educational websites, and Internet-based search engines	4
PCK	Teacher understand how to adjust biology subject with learning approaches, models, methods, learning media, student characteristics, and assessment	6
TCK	Teacher understand how to adjust biology subject with ICT	5
TPK	Teacher understand how to adjust ICT with approaches, models, methods, and learning media	5
TPACK	Teachers ' understanding of ICT, learning strategies, and learning materials are adjusted to the students ' characters	4
	Total	40

The instrument used had fulfilled validity and reliability. The logical validity of the test instrument was done through expert judgment and empirical validity obtained from the results of the quest test with the results of all items fit with MNSQ 1.00 invitations and standard deviation of 0.16, meaning all items were in accordance with the Rasch model. Test instrument reliability results of 0.57 means that the reliability of the test is good enough to provide consistent measurement results.

Data analysis of this study was carried out descriptively and inferentially through the independent sample t-test. The results of the scores obtained are converted into values using the following formula.

$$P = \frac{f}{N} x\, 100$$

Description:
P = values
F = obtained score
N = maximum score

The values gained was then compared to the TPACK criteria [28]. The criteria of TPACK mastery presented as follows.

Table 2. Biology Teacher's TPACK Mastery

Values	Criteria
84-100	Excellent
68-83	Very good
52-67	Good
36-51	Fair
$\leq$ 35	Poor

3. Result

Teacher TPACK were obtained through test results. Data obtained were profile of TPACK mastery of biology teachers in Yogyakarta city, Indonesia and TPACK mastery difference by teacher gender. Gender teachers were consisted of male and female teachers. The data of research were presented as follows.

3. 1. Profile of TPACK Biology Teacher

TPACK mastery consist pedagogical knowledge (PK), content knowledge (CK), technological knowledge (TK), pedagogical content knowledge (PCK), technological content knowledge (TCK), technological pedagogical knowledge (TPK), and technological pedagogical and content knowledge (TPACK). Profile of TPACK mastery of biology teacher in Yogyakarta city was showed as follows.

Table 3. Biology Teacher's TPACK Mastery

TPACK Domain	Mean	Standard Deviation	Criteria
PK	72.22	18.22	Very good
CK	72.67	13.48	Very good
TK	46.67	22.49	Fair
PCK	53.33	15.41	Good
TCK	54.67	18.14	Good
TPK	68.67	22.09	Very good
TPACK	45.83	26.33	Fair
Total Average	59.15	19.45	Good

Based on the test results, the average mastery of TPACK teachers obtained a score of 59.15 with good criteria. Teachers of biology have mastered the knowledge of PK, CK, PCK, TCK, and TPK well but the domain of TK with score of 46.67 and TPACK with score of 45.83 was still fair. These results showed that teachers still lack knowledge of ICT, especially in relation to websites, blogs, and email as well as ICT-based media. Teachers also have not mastered the knowledge of ICT, learning strategies, and learning materials adjusted to the student's character. Nevertheless, the teacher has been quite well mastered the knowledge of ICT to teach biology subject and knowledge about the proper ICT selection to support approaches, models, methods, and learning media.

3. 2. TPACK Mastery of Biology Teachers Based on Gender

Mastery of teacher TPACK is examined based on teacher gender. TPACK data compared between male and female teachers with independent sample t test. The results of the teachers TPACK mastery analysis based on gender are presented in the following table.

Table 4. TPACK Mastery of Biology Teacher Difference by Teacher Gender

TPACK Domain	Male			Female			t	P (two-tailed)
	Mean	SD	Criteria	Mean	SD	Criteria		
PK	64.28	24.28	Good	74.63	15.78	Very good	-1.334	0.193
CK	71.42	15.73	Very good	73.04	10.63	Very good	-0.314	0.756
TK	46.42	22.49	Fair	46.73	22.99	Fair	-0.031	0.975
PCK	45.23	15.85	Fair	55.79	14.74	Good	-1.632	0.114
TCK	48.57	19.51	Fair	56.52	17.73	Good	-1.016	0.318
TPK	68.57	25.44	Very good	68.69	21.59	Very good	-0.013	0.990
TPACK	39.28	37.79	Fair	47.82	22.50	Fair	-0.746	0.462

Based on the results of the test, male teachers have a score of PK, CK, and TPK in very good category while the mastery of TK, PCK, TCK, and TPACK was still fair. The female teacher has a score of PK, CK, PCK, TCK, and TPK in good category but the TK and TPACK domains were still lacking. Furthermore, a different test was conducted which showed that there was no significant difference in the average TPACK between male and female teachers with a value of $P > 0.05$, meaning that mastery of teacher TPACK was not differentiated by teacher gender.

4. Discussion

Teacher's TPACK mastery was reviewed based on test. The result show that the average mastery of TPACK teachers was good with score of 59.15. Teacher has mastered the knowledge of pedagogic (PK), content (CK), pedagogical content (PCK) well. This results in accordance with the results of the UKG 2015, the biology teacher in the city of Yogyakarta were 63.17 in pedagogic, 74.22 in professional, and a total of 70.90. The professional score of teachers in UKG reflects the teacher's mastery of subject matter knowledge. Overall, the results of the UKG obtained has reached a minimum score limit of 55.00, meaning that the mastery of biological teachers in Yogyakarta has exceeded the minimum limit but the assessment based on UKG has not revealed teacher mastery about technology knowledge.

The results of this research showed that the teacher TK and TPACK domain was still fair. This result means that the teacher was still weak in the knowledge of technology and its integration in pedagogic and subject matter. This outcome in line with the research shows the biology teacher was still weak in the knowledge technology (TK) and best on pedagogical content knowledge (PCK) [29]. Nevertheless, the teacher has well mastered the knowledge of technology to present subject matter as well as technological knowledge to support the application of learning strategies. The lack of knowledge of ICT was the main cause teachers did not adopt the ICT in learning process [30]. Teachers who lack ICT knowledge tend to feel anxious if they have to use the ICT in class because the students were smarter in ICT [31]. Therefore, it is necessary to do a workshop to improve the knowledge of ICT teachers and how to apply it in the learning process.

TPACK mastery were measured based on mastering the knowledge. These results indicated that teacher TPACK wasn't influenced by gender. This results in accordance with some research that showed no significant gender influence on PK, CK, TK, PCK, TCK, TPK, and TPACK [8, 9, 10, 11]. The difference TPACK mastery may be due to a technological gap between male and female getting reduced. Imhof, Vollmeyer, & Beierlein reported that in the year 1990, males using ICT were more effective than female but these percentages decreased over time due to increased chances of accessing technology [32]. This condition indicates that the ease of internet access and the increasing number of female of ICT Gap due to gender is decreasing therefore that the male and female levels in the use of ICT become equal. Gender influenced attitudes of ICT where the results of several studies had found that male had a more confident, positive attitude and relatively more experience with ICT [13, 14, 33, 34, 35] but were not affect the ability of ICT's mastery between male and female [14]. These results are backed research results showing no significant different between gender and ICT [36, 37]. The results of the study also showed no significant technological and pedagogical mastery differences between male and female teachers [38]. Therefore, teacher mastery of pedagogy, biology content, and technology between male and female teacher were same.

Other research showed obstacles in the integration of ICT in learning, namely the lack of teacher ICT ability, the lack of confidence in teachers being exposed to ICT, and lack of ICT training [39]. Lack of ICT knowledge was the main reason teachers lack confidence in adopting and integrating ICT in learning [31]. This is due to the fact that teachers who lack the knowledge of ICT would feel anxious if they have to use ICT in front of students who are smarter in using ICT [32, 40]. Teachers tend not to use technology in learning if they do not master technology knowledge.

5. Conclusion

TPACK mastery of biology teachers was good but still low in TK and TPACK domain. Teacher's TPACK mastery were not influenced by gender. The TPACK gap between male and female teachers does not seem even therefore the mastery of TK and TPACK between male and female teachers was equally fair. Therefore, the results of this research could provide recommendations for school principals and educational policy makers to provide teacher training to improve knowledge about technology and its integration to support pedagogic and content. In addition, it is hoped that teachers could apply their knowledge about TPACK in the learning process therefore that integrated ICT learning can be realized in accordance with the education in the 21st century.

References

[1] Koehler, M. J., & Mishra, P. 2006 Technological Pedagogical Content Knowledge: A Framework For Teacher Knowledge *Teachers College Record* **108** 1017-1054

[2] Cruickshank, D. R., Jenkins, D. B., & Metcalf, K. K. 2012 The Act Of Teaching: Sixth Edition (New York: Mc Graw Hill) 3-14

[3] Lin et al, 2013; Lin, T. C., Tsai, C. C., Chai, C. S., & Lee, M.H 2013 Identifying science teachers' perceptions of technological pedagogical and content knowledge (TPACK). *Journal Science Education Technology* **22** 325-336

[4] Koh, J.H.L, Chai, C. S., & Tsait, C.C 2010 Examining The Technological Pedagogical Content Knowledge Of Singapore Pre-Service Teachers With A Large-Scale Survey *Journal of Computer Assisted Learning* **26** 563-573

[5] Jang, S. J., & Tsai, M. F 2013 Exploring The TPACK Of Taiwanese Secondary School Science Teachers Using A New Contextualized TPACK model *Australasian Journal of Educational Tech* **29** 566-580

[6] Liu, Q., Zhang, S., & Wang, Q 2015 Surveying Chinese In-Service K12 Teachers' Technology, Pedagogy, And Content Knowledge *Journal of Educational Computing Research* **0** 1-20

[7] Berber, D. C., & Erdem, A.R 2015 An Investigation Of Turkish Pre-Service Teachers' Technological, Pedagogical And Content Knowledge *Computers* **4** 234-250

[8] Koh, J.H.L., & Chai C. S 2011 Modeling Pre-Service Teachers' Technological Pedagogical Content Knowledge (TPACK) Perceptions: The Influence Of Demographic Factors And TPACK Constructs Changing Demands, Changing Directions *Proceedings ascilite Hobart 2011* 735-746

[9] Ay, Y., Karadag, E., & Acat, M. B 2016 ICT Integration Of Turkish Teachers: An Analysis Within TPACK-Practical Model International Journal of Progressive Education 12 148-165

[10] Ersoy, Mehmet., Yurdakul, I. K., & Ceylan, B 2016 Investigating Preservice Teachers' TPACK Competencies Through The Lenses of ICT Skills: An Experimental Study *Education and*

Science **41** 119-135
[11] Karataş, F.İ., & Tutak, F. A 2015 An examination of secondary mathematics teachers technological pedagogical content knowledge [Online]
[12] Volman, M., Van Eck, E., Heemskerk, I., & Kuiper, E 2005 New Technologies; New Differences, Gender And Ethnic Differences In Pupils' Use Of ICT In Primary And Secondary Education *Computers & Education* **45** 35-55
[13] Ertl, B., & Helling, K 2011 Promoting Gender Equality In Digital Literacy *Journal Educational Computing Research* **45** 477-503
[14] Varank, I 2007 Effectiveness of quantitative skills, qualitative skills, and gender in determining computer skills and attitudes: a causal analysis. *The Clearing House: A Journal of Educational Strategies, Issues and Ideas* **81** 71-80
[15] Gode, V. O., Obegi, F.M., & Macharia, A 2014 Factors Influencing Integration of Information and Communication Technologies In Public Primary Teacher Training Colleges In Central Region of Kenya *Journal of Emerging Trends in Computing and Information Sciences* **5** 968-974
[16] Lin, T. C., Tsai, C. C., Chai, C. S., & Lee, M.H 2013 Identifying Science Teachers' Perceptions of Technological Pedagogical and Content Knowledge (TPACK) *Journal Science Education Technology* **22** 325-336
[17] Rashidi, N., & Naderi, S 2012 The Effect of Gender on The Patterns of Classroom Interaction. *Education* 2 30-36
[18] Wood, T. D 2012 Teacher perceptions of gender-based differences among elementary school teachers *International Electronic Journal of Elementary Education* 4 317-345.
[19] Lacey, Saleh, & Gorman, 1998:5 Lacey, C.H., Saleh, A., & Gorman, R 1998 Teaching Nine to Five: A Study of The Teaching Styles of Male and Female Professors [Online]
[20] Chen, Y. M 2000 Feminization In Writing Pedagogy A Study Of Teacher's Gender At EFL University Composition Classrooms [Online].
[21] Spall, K., Barrett, S., Martin Stanisstreet, Dickson, D., & Boyes, E 2003 Undergraduates' views about biology and physics *Research in Science & Technological Education* **21** 193-208.
[22] Tsabari, A. B., & Yarden, A 2008 Girls' Biology, Boys' Physics: Evidence From Free-Choice Science Learning Settings *Research in Science & Technological Education* **26** 75-92
[23] Sasser, L 2010 Gender Differences In Learning [Online]
[24] Lestari, R.B 2011 Pemberdayaan wanita melalui teknologi informasi. *Makalah* disajikan dalam Seminar Nasional Aplikasi Teknologi Informasi di Yogyakarta
[25] Dholakia & Kshetri 2003 Gender And Internet Usage: The Internet Encyclopedia (New York: Wiley) 18-19
[26] Trauth, E. M, Jeria L. Quesenberry, & Haiyan Huang 2006 Cross-Cultural Influences On Women In The It Workforce *SIGMIS CPR '06 Proceedings of the 2006 ACM SIGMIS CPR conference on computer personnel research: Forty four years of computer personnel research: achievements, challenges & the future* 12-19
[27] Buskens, I. & Webb, A 2009 African Women & ICTs: Investigating Technology, Gender, And Empowerment (London: Zed Books) 177
[28] Rather, S. 2014 Secondary School Principals' And Teacher Attitude Towards ICT And ITS Implementation In Uttar Pradesh (U.P) and Jammu and Kashmir (J&K) *Disertasi,* unpublished, Aligarth Muslim University, India
[29] Arikunto, S 2011 *Dasar-dasar evaluasi*. Jakarta: Bumi Aksara.
[30] Suryawati, E., Firdaus, L.N & Hernandez, Y 2014 Analisis keterampilan technological pedagogical content knowledge (TPCK) guru biologi SMA negeri kota Pekanbaru *Jurnal Biogenesis* **11** 67-72
[31] Balanskat, A., Blamire, R., & Kefala, S 2006 The ICT Impact Report A Review of Studies of ICT Impact on Schools In Europe [Online]
[32] Jones, A 2004 A Review of The Research Literature on Barriers to The Uptake of ICT by

Teachers [Online]
[33] Imhof, M., Vollmeyer, R., & Beierlein, C. (2007). Computer use and the gender gap: the issue of access, use, motivation, and performance. *Computers in Human Behavior*, 23, 2823-2837
[34] Baturay, M. H., Gokcearslan, S., & Şahin, S 2017 Associations among teachers' attitudes towards computer-assisted education and TPACK competencies *Informatics in Education* **16** 1-23
[35] Jimoyiannis, A., & Komis, V 2007 Examining Teachers' Beliefs About ICT in Education: Implications Of A Teacher Preparation Programme *Teacher Development* **11** 149-173
[36] Sweeney, T & Drummond, A 2013 How Prepared Are Our Pre-Service Teachers To Integrate Technology? A Pilot Study *Australian Educational Computing* **27** 117-123
[37] Teo,T 2008 Pre-service teachers' attitudes towards computer use: a singapore survey *Australasian Journal of Educational Technology* **24** 413-424
[38] Yildiz, A 2017 The Factors Affecting Techno-Pedagogical Competencies and Critical Thinking Skills of Preservice Mathematics Teachers *Malaysian Online Journal of Educational Sciences* **5** 56-81
[39] Semerci, A. M., & Aydin, K 2018 Examining high school teachers' attitudes towards ICT use in education *International Journal of Progressive Education* **14** 93-105
[40] Suyanto, S 2017 A Reflection on the Implementation of a New Curriculum in Indonesia: A crucial Problem on School Readiness *AIP Conference Proceedings* **1868** 1-9

Developing a Learning Continuum of the Pedagogic Materials of Genetics Aspects from Elementary School to Senior High School Level Based on the Opinions of Biology Education Experts

L. Orizasativa*, B. Subali, and Paidi
Biology Education Program, Graduate School of Yogyakarta State University,Yogyakarta, Indonesia

*Corresponding author e-mail: liandraorizasativa@gmail.com

Abstract. Learning continuum is a set of competencies mutually related from one concept to the others. The competencies are ordered from the easy ones to the difficult and specific ones at all levels sustainably from the elementary to the secondary level. This study aims to reveal the opinions of biology education experts on the learning continuum of genetics materials viewed from the competency level and material characteristics and also to reveal the learning continuum of the specific pedagogic materials of the genetics aspects according to experts. The expert respondents were selected by considering their educational background, namely their expertise in the field of biology education, with regard to their bachelor, masters, and doctoral programs. This study is a survey with a sample of 85 biology education experts, established using the convenience technique, i.e. those who are easy to find. The result of this study shows that 25% of the respondents agree that genetics is taught to grade 9 students at the cognitive level of C2 (comprehension), on the sub-aspect of character inheritance, while 75% have the opinion that genetics is taught to grade 12 students at the cognitive level of C2 (comprehension). However, 100% of the participants of the Forum Grup Discussion have the opinion that genetics is taught to grade 12 students at the cognitive levels of C2 and C3. Therefore, it can be concluded that the genetics aspect is taught to grade 9 students of junior high school and it is taught again to high school students for the different cognitive level and the learning materials are different and more complex

1. Introduction

A good teaching and learning system is the one that refers to a current curriculum, and of course the curriculum becomes a very important consideration. A curriculum is a set of plans and arrangement containing instructional goals[1]. The structure of a curriculum is determined based on the National Standard of Education, which aims to manage every competency aspect which contains teaching materials in an educational program. The structure of a curriculum consists the goal, content (teaching materials), strategies, learning resources, time allocation, and infrastructure, all of which support the teaching and learning activities[2]. A curriculum is improved in order to adapt to the current development. Developing a curriculum is not an easy matter, because it has to refer to a various types of needs analyses, one of which is the content aspect (teaching materials)[3].

Teaching materials (*Content Knowledge*) are specific pedagogic materials which, if combined with competencies and teaching strategies, will enable to determine the curriculum well and will determine the materials and cognitive levels that have to be developed in the students. However, in the year-to-year curriculum development, there has not been an explanation concerning the description which shows the degrees of complexity and abstraction of *Content Knowledge* or specific pedagogic materials (SPM) from elementary school level to high school level, and therefore, there needs to be orderly graded competencies which then become a *Learning Continuum* (LC) or *Learning Trajectory* (LT) for a good curriculum, one of which is that the curriculum is in line with current advancement and with students' needs.

The principal use of a *learning continuum* is to facilitate the teachers' understanding of general and specific instructional objectives, teachers' references in teaching easy materials to difficult materials, monitoring different students' learning achievement, and teachers' references in using teaching strategies suitable with the students' characteristics[4]. Determining a continuum in one aspect refers to the difficulty level of learning materials and the student cognitive development[5]. According to Knowles learning continuum can be revealed through the teaching materials taught to the students, including simple (easy) concepts to complex (difficult) ones[6]. This is in line with the dimension of both knowledge and skill which are developed and it is adjusted to the students' level of cognitive development. For example, in elementary school (at the ages of 7 to 11 years old) the students learn simple concepts with the cognitive domain to achieve being to remember and to understand[7].

One of the reasons for using a learning continuum is that the materials taught to the students are in accordance with their competence and cognitive development. A learning continuum is highly related to the curriculum, while the curriculum used in schools and practised by teachers contains overlapping materials and is not in line with their students' cognitive development, i.e. it is not developed from the easy competency, and it makes teaching ineffective. A curriculum which is developed and used by referring to a learning continuum will be better, because it has really been adjusted to the students' competence and cognitive development. That is why a learning continuum needs to be used in the education in Indonesia.

Genetics is one of the aspects in biology materials which is quite important in the analysis of the content (teaching materials). This is due to the fact that genetics is an essential part of biology materials, and that there have not been studies on the learning continuum which deals with the genetics aspect up to the level of experts' opinions.

The study of genetics aspect materials involves practitioners consisting of supervisors, teachers, and school principals. Practitioners (supervisors, teachers, and school principals), the people who are very close to students, give different opinions about the content of genetics material aspects. Most teachers refer only to the available curriculum[8],[9],[10],[11]. The problem is, if the materials of the genetics aspect are not suitable with the students' cognitive development, the students do not develop following the continuum in their study.

2. Method

This research is a descriptive study using the survey method to reveal the data on the biology education experts' opinions about the students' competence level and specific pedagogic materials of the genetics aspects for elementary and secondary school students. It was conducted in five provinces, including West Java, Jakarta, Central Java, and West Java Provinces, and Yogyakarta Special Territory, from early 2018 to end of 2019. The biology education experts are lecturers of biology education who became the sample respondents from the afore-mentioned provinces. The number of the respondents is 85 lecturers spread in universities – state and private universities. They were selected by considering their educational background, i.e. expertise in biology education in their bachelor, masters, and doctoral programs. The sample was established using the convenience sampling technique, i.e. those who are easy to find[12]. The research variables include the percentage of learning continuum of specific pedagogic materials and competency level. This research uses a non-test instrument in the form of a questionnaire. The data were collected using the questionnaire on students' competency level and specific pedagogic materials of *the Aspect of Life in Nature, Science and Biology Subject* written by Bambang Subali and Team in 2018. The questionnaire was validated through a forum grup discussion (FGD) of a team of validators consisting of three lecturers of UNY and five lecturers of UNS[13]. In addition to the experts' opinions, the result of the FGD was used as the supporting data in mapping out the grid of the learning continuum of the genetics aspects.

In the FGD, the lecturers were asked to decide whether or not the aspects were logical to teach in the classroom and to decide the cognitive level chosen by the experts. If the experts' opinions are illogical, then in the FGD the lecturers had to decide the material aspect suitable to teach in class and the appropriate cognitive level according to the analysis of a variety of considerations.

The result of the FGD on experts' opinions is used to consider a number of things, including the difficulty level of materials, depth of materials, cognitive levels, and students' cognitive development. The calculation of scores in the process of obtaining percentages is as follows: if an expert chooses a certain sub-topic to be taught, for example to grade 10 students at the cognitive level of C2, then he/she is given a score of 1, but if he/she does not suggest that it is taught to grade 10 students, for example: kelas X; cognitive level C2, then he/she is given a score of 0. The percentage is obtained by summing up the scores and then conversing them in the form of percentage.

Each genetics aspect has a variety of sub-aspects, and thus, the percentage is presented in a range of percentages, for example, in the crossing aspect, the percentage of experts' opinions ranges from 28% to 36%, which means that for the available material sub-aspect in the crossing aspect, 28 to 36% of the experts suggest that the materials should be taught to grade 12 students at the cognitive level of C2 (comprehension). The collected opinions of the experts were analysed in FGD in order to determine the straight line of the learning continuum of specific-pedagogic materials of the genetics aspects. The learning continuum is chosen by analysing every opinion of each expert by considering the cognitive level, scope, depth of specific pedagogic materials of the genetics aspect, and the students' cognitive development.

3. Result and Discussion

Based on the scope of BSCS Biology A Molecular Approach 9th Edition, genetics materials are taught starting from the easy to the difficult materials, from a concrete objects including animals and vegetation with the level of biome organization up to individuals and then followed by the abstract objects and the level of organ system organization up to abstract molecules[14]. The experts' opinions vary. The opinion most frequently appears will become the mode of each genetics sub-aspect. The result shows that the mode of each genetics sub-aspects varies.

As many as 44 – 45% of the experts' opinions are in the mode of character inheritance through fission and genetics materials taught to grade 12 students at the cognitive level of C2. As many as 59% of the experts suggest that the genetics materials of the sub-aspect of DNA replication should be taught to grade 12 students at the cognitive level of C2. As many as 42% of the experts suggest that the genetics materials of the sub-aspect of genotype and phenotype should be taught to grade 12 students at the cognitive level of C2. As many as 41% of the experts suggest that the genetics materials of the sub-aspect of gene characteristics: dominant, recessive, and intermediate, should be taught to grade 12 students at the cognitive level of C2. As many as 53% of the experts suggest that the genetics materials of the sub-aspect of allele and multiple allele should be taught to grade 12 students at the cognitive level of C2. For sub-aspect 1 up to sub-aspect 5, the result of the FGD shows that the experts have the opinions that genetics materials are suitable/logical to be taught to grade 12 students at the cognitive level of C2 (understanding) because they have considered the cognitive development of the students. The following are the experts' modes and the result of the FGD.

Table 1. The percentages of biology education experts' opinions and the result of Forum Grup Discussion.

Genetics Aspects	Biology Education Experts' Opinions		FGD	
	Modes	%	Logical	Illogical
Character inheritance through fission and genetics materials	XII/C2	44-45	√	-
DNA replication	XII/C2	59	√	-
Genotype and phenotype	XII/C2	42	√	-
Gene characteristics: dominant, recessive, and intermediate	XII/C2	41	√	-
Allele and multiple allele	XII/C2	53	√	-
Crossing	XII/C2	28-36	-	√ (replaced by XII/C3)
Law of character inheritance	XII/C2	28-33	-	√ (replaced by XII/C3)
Character inheritance in human, animal and vegetation	IX/C2; XII/C2	25-44	-	√ (replaced by XII/C3)
Mutation	XII/C2	51-66	√	

As many as 28 – 36% of the experts suggest that the genetics materials of the sub-aspect of crossing should be taught to grade 12 students at the cognitive level of C2. As many as 28 – 33% of the experts suggest that the genetics materials of the sub-aspect of the law of character inheritance should be taught to grade 12 students at the cognitive level of C2. As many as 25 – 44% of the experts suggest that the genetics materials of the sub-aspect of the character inheritance in human, animal and vegetation should be taught to grade 12 students at the cognitive level of C2. The result the FGD for sub-aspects 6 and 8 shows that the experts think that it is illogical that those materials are taught to students of grades 9 and 12 at the cognitive level of C2 (comprehension) but it will be logical that the materials are taught to grade 12 students at the cognitive level of C3 (application). This is due to the fact that the students are expected not only to understand the materials but also to do exercises and solve problems related to the materials so that the cognitive level is raised to C3 (application). For the sub-aspect of mutation, 51-66% of the experts suggest that the materials should be taught to grade 12 students at the cognitive level of C2. The result of the FGD on the experts opinions is logical that the materials are taught to grade 12 students at the cognitive level of C2 (comprehension) because it has considered the students' cognitive development.

Viewed holistically, all sub-aspects, from the sub-aspect of characteristic inheritance to the sub-aspect of mutation, the mode of biology education experts' opinions shows that the genetics materials are taught to grade 9 or grade 12 students. The differences in the biology education experts' opinions are affected by the factors of ideology, competence, and teaching experience as well as insight into the curriculum[8]. Some of the experts suggest that genetics materials should be taught to grade 9 students at the cognitive level of C2 (comprehension). This is in line with the Decree of the Minister of Education and Culture of the Republic of Indonesia, No. 37, Year 2018 that in junior high school, genetics materials begin to be taught to grade 9 students[15]. The result of the FGD (*Forum Grup Discussion*) was obtained from observing biology education experts' opinions by considering students' cognitive development.

Indonesia has applied the spiral model curriculum, in which teaching materials in the lower level educational institution will be continued in the next level[8],[16]. Burner states "If the understanding of amount, size, and probability is considered very important in science, then instruction in school subjects must begin intellectually with honesty and as early as possible consistent with students' way of thinking. Let it be developed in the following class"[17]. Of course, this also applies to the biology subject of the genetics aspect, in which if genetics materials are considered important then they are introduced as early as possible to the students by consistently considering the students' cognitive development. In delivering the teaching materials, the teacher has to fully understand which materials can be understood and which materials cannot be understood by the students. Of course, the teacher has to understand his/her students' cognitive development. The genetics materials are taught starting from the simple to the complex, from concrete to abstract, and of course it develops in accordance with the education level. The higher the education level, the wider the scope and depth of the materials, and the concepts developed by the students also improve because the initial materials become the basis for the students in finding the following concept[18],[19].

Table 1 shows that all sub-topics of genetics materials, from character inheritance to mutation are suggested to be taught to grade 12 students at the cognitive level of C2 (comprehension) and up to C3 (application). This is in line with the Decree of the Minister of Education and Culture of the Republic of Indonesia, No. 37, Year 2018 that in senior high school, genetics materials begin to be taught to grade 12 students[15]. The genetics materials for grades 9 and 12 are basically repeated materials, with a deeper and higher complexity in grade 12[20],[21], [22], [23]. In its implementation, the application of a learning contiuum in genetics materials is a solution to the making of students' more focus on understanding the lesson. The following is the table of learning continuum of the genetics aspect.

Table 2. Grid of learning contiuum of genetics aspect materials

Genetics aspects	SD	SMP	SMA			Cognitive competence levels					
	(I-VI)	(VII-IX)	X	XI	XII	C1	C2	C3	C4	C5	C6
Character inheritance through fission and genetics materials	-	-	-	-	√	-	√	-	-	-	-
DNA replication	-	-	-	-	√	-	√	-	-	-	-
Genotype and phenotype	-	-	-	-	√	-	√	-	-	-	-
Genetic characteristics: dominant, recessive, and intermediate	-	-	-	-	√	-	√	-	-	-	-
Allele and multiple allele	-	-	-	-	√	-	√		-	-	-
Crossing	-	-	-	-	√	-	-	√	-	-	-
Law of character inheritance	-	-	-	-	√	-	-	√	-	-	-
Character inheritance in human, animal and vegetation	-	-	-	-	√	-	-	√	-	-	-
Mutation	-	-	-	-	√	-	√	-	-	-	-

In accordance with Table 2 above, a line of learning contiuum can be drawn based on biology education experts' opinions and the result of the FGD. A teacher as one of the practitioners in the field can apply the grid of learning continuum in teaching specific pedagogic materials of the genetics aspect starting from character inheritance to mutation. In the aspect of genetics materials, the students' cognitive process achieves the level of C3 or application. This does not necessarily mean that all materials presented in the lesson focus only on the cognitive process of C3 (application); they also include the cognitive process of C1 (knowledge) and C2 (comprehension). When a student has achieved a high cognitive level, automatically he/she has achieved the cognitive level below it. For example for the sub-aspect of crossing material which has been taught to grade 12 students at the cognitive level of C3 (application), the students who have learned it at the cognitive level of C3 (application) automatically have also been able to understand and to do the work on it at the cognitive level of C2 (to understand) and C1 (to remember). Overall, the result of this study is a grid of learning continuum which is formed using the biology education experts' opinions expressed in focus group discussions. As practitioners, teachers can apply the developed grid of *learning continuum* in the field for try-outs in the form of classroom action research, so that they can see the differences in their student's mastery of genetics materials.

4. Conclusion

The learning continuum of specific-pedagogic materials of the genetics aspect is the end result of this study. Based on the entire discussion of biology education experts, the grid of the learning continuum containing the aspect of genetics materials which is taught to grade 9 students of junior high schools at the cognitive level of C2 (comprehension), and C3 (application). The higher the students' education level, the higher the cognitive level they have to reach.

Acknowledgments

I would like to thank my two colleagues; 85 biology education lecturers throughout the provinces of West Java, Jakarta, Central Java, and West Java, and Yogyakarta Special Territory; and my academic supervisor, who have helped me complete this article.

References

[1] Republik Indonesia 2012 Undang Undang Republik Indonesia Nomor 20 Tahun 2003 tentang Sistem Pendidikan Nasional.pdf (Bandung: Citra Umbara)
[2] Andriani A E and Subali B 2017 Teachers' opinion about learning continuum based on student's level of competence and specific pedagogical material in classification topics AIP Conf. Proc. 1868
[3] Situmorang R P 2016 Analisis Learning Continuum Tingkat SD sampai SMP pada Tema Sistem Pencernaan Manusia Scholaria 6 1–13
[4] Association N E and Idaho 2003 Idaho State Aligned Learning Continuum
[5] Subali B, Aminah N S and Sumintono B 2019 Student Achievement Based on The use of Scientific Method in The Natural Science Subject in Elementary School J. Pendidik. IPA Indones. 8 41–53
[6] Knowles M S 1980 The Modern Pratice of Adult Education From Pedagogy to Andragogy (New York: Cambridge The Adulth Education Company)
[7] Krathwohl L W A and D R 2001 A Taxonomy for Learning, Teaching, and Assessing ed L W A and david R Krathwohl (New York: Longman)
[8] Juniati E and Subali B 2017 Teacher's opinion about learning continuum of genetics based on student's level of competence AIP Conf. Proc. 1868 1–11
[9] Pramesti I C and Subali B 2017 The learning continuum of ecology based on teachers' opinion about student's level of competence and specific pedagogical learning material AIP Conf. Proc. 1868
[10] Fitriyani R and Subali B 2017 The Learning Continuum based on Student ' s Level of Competence and Specific Pedagogical Learning Material on Physiological Aspects from Teachers ' s Opinions 100006
[11] Astuti L D and Subali B 2017 Teacher's opinions about learning continuum based on the student's level of competence and specific pedagogical materials on anatomical aspects Am. Inst. Phys. 100005 1–13

[12] Daniel W W 1989 Statistik Nonparametrik Terapan (Jakarta: Gramedia)
[13] Subali, B., Kumaidi, & Aminah N S 2018 Developing a Scientific Learning Continuum of Natural Science Subjects at Grades 1–4 J. Turkish Sci. Educ. 15 66–81
[14] BSCS 2006 BSCS Biology A Molecular Approach (Columbus Ohio: Glencoe/McGraw-Hill Companies, Inc.)
[15] Kemendikbud 2018 Undang-Undang RI Nomor 37 Tahun 2018 Tentang Kompetensi Inti dan Kompetensi Dasar Pelajaran pada Kurikulum 2013 pada Pendidikan Dasar dan Pendidikan Menengah
[16] Farisi M I 2013 Kurikulum Rekontruksionis dan Implikasinya Terhadap Analisis Dokumen Kurikulum 2013 PAEDAGOGIA 16 143–64
[17] Yumuşak G K 2016 An Analysis of the Science Curricula in Turkey with Respect to Spiral Curriculum Approach J. Educ. Pract. 7 6–9
[18] Xuan Jiang K P 2013 A Conceptual Paper on the Application of the Picture Word Inductive Model Using Bruner â€TM s Constructivist View of Learning and the Cognitive Load Theory Interdiscip. J. Teach. Learn. 3 8–17
[19] Efland A D 1995 The Spiral and The Lattice: Changes in Cognitive Learning Theory with Implications for Art Education Stud. Art Educ. A J. Issues Res. 36 134–53
[20] R.M. Harden dan N. Stamper 1999 What is a spiral curriculum ? Med. Teach. 21 141–3
[21] Snider D B C and V E 2019 Effective Mathematics Instruction The Importance of Curriculum Educ. Treat. Child. 23 122–42
[22] Dibiasio D, Clark W M, Dixon A G, Comparini L and Connor K O 1999 Evaluation of a Spiral Curriculum for Engineering 29'h ASEE/IEEE Front. Educ. Conf. 15–8
[23] Yanto B. E S . & S S 2019 Measurement Instrument of Scientific Reasoning Test for Biology Education Students Int. J. Instr. 12 1383–98

Developing A Learning Continuum of Biological Resources Management Aspect from Elementary School to Senior High School Based on The Experts' Opinions

M. U. Kusumawati*, B. Subali, and Paidi
Biology Education, Postgraduate Program, Yogyakarta State University

*Corresponding author e-mail: megautamik.2017@student.uny.ac.id

Abstract. This study aims to determine the experts' opinions and the design of the learning continuum on aspects of biological resource management reviewed at the level of competence and characteristics of specific pedagogical material from elementary to high school. The method used in this research was survey by giving questionnaire to the experts. The sample of this study is convenience sample, which is 85 lecturers. This research was conducted at several State and Private Universities in the Special Region of Yogyakarta, Central Java, East Java, West Java, Jakarta. Results of this study indicate that the experts believe that the aspect of biological resource management has been taught in the tenth grade of senior high school (SHS) with cognitive levels (CL) "understanding" (C2) and "applying" (C3), and in seventh grade junior high school (JHS) with C2 CL. The design of learning continuum aspects of the management of biological resources from the FGD results based on the opinions of experts is that most of the specific pedagogical materials have begun to be taught at the JHS level with C2 CL, then continued at the SHS level with C3 CL, and there are also some specific pedagogic materials that have begun to be taught at the SHS level with C3 CL.

1. Introduction

Curriculum is a set of plans and rules about the objectives, contents, and learning materials and the techniques that are used as the manual in the implementation of teaching and learning activities to achieve certain education objective. [1] In the curriculum substance there is inconsistence of the content aspect in the materials that becomes main problem since there were many basic competences in the materials that are overlap. [2] Therefore, the characteristics, broadness, and depth of the materials must be adapted with the learning continuum. If there is no attention from the development of curriculum about the importance of the continuity materials in each level, it will cause the learning process be ineffective. [3] The efficiency of a learning program can be seen from the program that is sustainability and continuity. The implementation process of learning must be able to support the achieved purposes in the education, for example its implementation in the elementary school (ES), junior high school (JHS), or senior high school (SHS) level. [4] The learning process in the level of education, however, can work well if it can be sustainable and appropriate with the students' development. [5]

The ability in human being based on the Taxonomy Bloom is divided into three domains that are cognitive domain, affective domain, and psychomotor domain. One of the competencies that must be had by the students for learning science or especially biology is cognitive competency. [6] Piaget theory stated that child actively build an understanding about the world through four stages of cognitive development. [7] The cognitive development is a learning process that follows on the thought and the way of working. [8] Biology/science learning places the students as subject of assessment where the assessment that focuses on cognitive level actually must consider about the achievement levels and stages. Based on Biological Science Curriculum Study (BSCS) biology has scientific scope that consists of biology objects, biology problems, and organizational level of life. [9] Essentially, biology has various aspects of life that are taught to the students in the middle school (MS) either in JHS or SHS from morphology aspects of living things, anatomy of living things, physiology of living things, diversity of living things, biological resources management, the classification of living things, ecology, genetic, evolution, and physic and chemistry aspects. [10] Biology is a science that has a very broad scope in studying the science of living things and their interactions with surrounding environment. [11] It can be said that aspects of management of biological resources are within the scope of learning biology/natural science subjects.

One of the basic in arranging the curriculum is considerations that are used to arrange the materials (content) so that they become the learning materials (content knowledge) in learning activities to support in achieving the competency targeted. The unification between content knowledge as the specific pedagogical materials and the competencies and the learning strategy will be able to determine the curriculum. This explanation cannot be found in curriculum development of biology in SHS whether in biology aspects in natural science at ES or JHS. It must be a presentation course that shows the gradation of complexity and abstract about the content knowledge from ES to SHS. Since there is no presentation about it, it need to arrange the learning continuum.

Learning continuum is learning where specific pedagogical material is arranged on a continuum from one level to another higher level that is adjusted to the development of students (metal and age). The basis for developing or developing a learning continuum is the preparation of specific pedagogic material from concrete to abstract, or easy to difficult, or simple to complex. With the learning continuum, learning will be more meaningful and effective. According to Subali, the students are expected to learn the material from basic level to complex level. To be able to guarantee the sustainability between levels, it is needed a concrete standard about the learning materials that will be taught to the students in each level or degree. This standard is started from basic education level to MS level. So that, it is also needed the references in arranging the grillwork such as learning continuum that shows the abstract continuum from all competencies that are targeted from simple level to the complex level. [3] The teachers', principals', and school supervisors' views could be used for developing a learning continuum of "natural science". [12]

However, most of the opinions of teachers (science teachers in junior high school and Biology teachers in senior high school) regarding the continuity of learning aspects of biology based on the level of student competence and specific pedagogical material still refer to the current syllabus and curriculum. Therefore, the teachers do not have personal opinions about the continuum of learning specific pedagogical material that must be taught at school at each level.[13] [14] [15] [16] [17] Based on these ideas, the researcher is interested in conducting a research to formulate a learning continuum on aspects of biological resource management that are taught from elementary to high schools based on the opinion of experts in Biology Education. The basis for the preparation of a learning continuum is the abstractness of concepts (from the concrete to the abstract), the ease of the concept (from the very easy to the most difficult), and the complexity of the concept (from the simplest to the most complex). By this research, it is expected that the specific pedagogical material and competency levels targeted at elementary to high school students will be organized according to class levels and students' cognitive development. It is hoped that from the development of learning continuum which makes specific pedagogical material no longer overlaps. The difference between this study and other studies is that the development of learning continuum in this study is based on the opinions of experts so that the design of the pedagogical learning continuum produced specific aspects of the management of biological resources.

2. Methods

This research is descriptive research by using survey method to gather the data of experts' opinions about the competency level and specific pedagogical materials toward the aspects of biological resources management. This research is a follow-up study conducted by Bambang Subali about the learning continuum aspects of biology in 2016 (at ES) and 2017 (at JHS and SHS). The basis of the preparation of learning continuum aspects of natural resource management in terms of competency levels and specific pedagogical material characteristics are concrete to abstract, easy to difficult, and simple to complex.

This research is the descriptive research by using survey method to gather the data of the experts. Gathering the experts' opinion was already done in the previous research by Bambang Subali in 2016 and 2017. This research is done in the private and state university in Yogyakarta province, Central Java, East Java, West Java, and Jakarta. Gathering the population and sample follows Daniel terminology. In this case, the hypothetic of the population is the biology education lecturers that teach in the private and state university in Indonesia. The sample is convenience sample that consists of 85 respondents of the biology education lecturers from 9 state university and 9 private university. The collecting data technique in this research uses instrument of questionnaire that is given to the respondent (expert). The collecting data is done by using questionnaire instrument of learning continuum-based confirmatory assessment model about the aspects of biological resources management. The validity of the questionnaire is done through Focus Group Discussion (FGD) that is done by Bambang Subali in 2016. The discussion consists of 3 lecturers of Biology Education at Yogyakarta State University and 5 lecturers of Biology Education at UNS.

The data of experts' opinions collected through questionnaires were analyzed by searching for a mode, where the mode illustrates the agreement between respondents. For example, from the overall opinion of the experts, the highest category is 55% of respondents who think that the material is taught in class X with the cognitive domain of "*applying*" (C3). This opinion can be used as a mode of describing the agreement of opinion between respondents. Data from the analysis of the questionnaire were then discussed through the FGD by lecturers with Biology Education expertise or their allies (experts) at Yogyakarta State University to compile a learning continuum on specific pedagogic material for the management of biological resources. This FGD determines the logical or illogical opinion of experts. The logical or illogical determination of experts' opinion depends on the level of mental development of students with specific pedagogical material aspects of the management of biological natural resources.

3. Results and Discussion

The percentage data of the experts' opinions (the lecturers of biology education) about the learning of the aspects of biological resources management that must be taught in the school is gathered from 85 respondents that spread out from 18 colleges in Yogyakarta, Central Java, East Java, West Java, and Jakarta. The data will be shown as follows. The result of data analysis that is done shows the experts' opinions about the teaching of the aspects of biological resources management in the school are various that contain from the class level to the competency level (cognitive scope) that are targeted to the students in ES level to SHS level.

Table 1. The percentage percentage of expert opinion data and results of FGD study about the learning of the aspects of biological resources management that must be taught in the school

The aspects of biological resources management		The experts' opinions of biology lecturers (N[a]=85)		Results of the FGD Study
		Modus	Cls[b]/CL[c]	
Sub-aspects 1	**The benefit of biological resources**	26 %	X/C2[d]	Illogical, replaced VII/C2
Sub-aspects 2	**The kinds of biological resources**	27 %	X/C2	Illogical, replaced VII/C2
Sub-aspects 3	**The balance of nature**			
	a. The harmony of nature	25 %	X/C3[e]	Logical
	b. The conservation and preservation	26 %	X/C3	Logical
Sub-aspects 4	**The conservation**			
	a. The conservation with general purposes			
	1) The tight conservation	24 %	X/C3	Logical
	2) The guided conservation	24 %	X/C3	Logical
	b. The conservation with certain purposes			
	1) Geological protection	43 %	X/C2	Illogical, replaced VII/C2
	2) Botany conservation	44 %	X/C2	Illogical, replaced VII/C2
	3) Zoology conservation	41 %	X/C2	Illogical, replaced VII/C2
	4) Anthropology protection	40 %	X/C2	Illogical, replaced VII/C2
Sub-aspects 5	**Conservation efforts of biological resources**			
	a. Conservation efforts of biological resources artificially	27 %	X/C2	Illogical, replaced VII/C2
	b. Conservation efforts of biological resources naturally	26 %	X/C2	Illogical, replaced VII/C2
	c. In-situ maintenance	24 %	VII/C2	Logical
	d. Ex-situ maintenance	24 %	VII/C2	Logical
Sub-aspects 6	**Forms of nature conservation**			
	a. National park	26 %	X/C2	Illogical, replaced VII/C2
	b. Sea park	26 %	X/C2	Illogical, replaced VII/C2
	c. Nature preserve	26 %	X/C2	Illogical, replaced VII/C2
	d. Wildlife sanctuary	26 %	X/C2	Illogical, replaced VII/C2
	e. Botanical garden	25 %	X/C2	Illogical, replaced VII/C2

[a] total respondent
[b] class
[c] competency level (cognitive scope)
[d] understand
[e] apply

Finding out the agreement of the experts in this research, it needs to look for the highest opinions. The modus is the experts' opinions that grouped together with the highest number, so the modus can draw the agreement of the experts' opinions. These opinions relate to the sub-aspects 1 and sub-aspects 2 that are taught in the tenth grade with "*understand*" (C2) cognitive level. Based on the FGD activity, they are illogical since these aspects must be taught in the seventh grade with C2 cognitive level. In the sub-aspects 3 that is taught in the tenth grade with "*apply*" (C3) cognitive level, the experts stated that it is logic whereas they also stated that sub-aspects 4 about the strict conservation and long guided protection is logically taught in the tenth grade with C3 cognitive level. Thus, the sub-aspects 4 about the geology, botany, zoology, and anthropology conservation are illogically taught in the tenth grade with C2 cognitive level based on the FGD results since they have to be taught in the seventh grade with C2 cognitive level. Moreover, the sub-aspects 5 about the conservation efforts of biological resources artificially and naturally are illogical if they are taught in the tenth grade with C2 cognitive level based on the FGD results since they have to be taught in the seventh grade with C2 cognitive level. Thus, the sub-aspects 5 about in-situ and ex-situ maintenance are logically taught in the seventh grade with C2 level based on the FGD results. Finally, the sub-aspects 6 are illogically taught in the tenth grade with C2 cognitive level based on the FGD results since they have to be taught in the seventh grade with C2 cognitive level.

The deciding of logical or illogical statement of the experts follows the students' mental development level with the specific pedagogic materials of biological resources management. The basis of the preparation of learning continuum aspects of natural resource management in terms of competency levels and specific pedagogical material characteristics are concrete to abstract, easy to difficult, and simple to complex. Based on Subali and Suyata, "the students' competency level in the elementary level is "*remember*" (C1) and "*understand*" (C2), for the students of JHS placed in "*understand*" (C2), "*apply*" (C3), and "*analyze*" (C4), and the students of SHS placed in "*apply*" (C3), "*analyze*" (C4), "*evaluate*" (C5), and "*create*" (C6). According to the FGD activity, the opinion that is illogical caused by the sub-aspects that are easy to be learned, so they can be taught in the beginning level." [18] On the other hand, according to Santrock explains that, based on the theory of cognitive development by Piaget, the students in the ES level from the 7 to 11 ages are in the stage of concrete operational development, so they must be able to think logically and be able to implement their reasoning and cognitive skill to the concrete examples specifically in the cognitive scope of C1 and C2. [19] Anderson and Krathwohl stated that the students can reach the C2 level when they are able to build the meaning from instructional message (oral communication, written, and graphics). The cognitive processes in the C2 level are interpreting, summarizing, giving example, classifying, explaining, comparing, and concluding. This cognitive level of C2 is in the lower order thinking skills. [6] Moreover, according to Subali and Suyata in their research stated that almost all sub-aspects of biology are taught in the competency level of remembering and understanding. These competencies are easier and simpler than other skills such as applying, analyzing, evaluating, and creating. On the JHS and SHS level, the students with 12 to 17 ages are in the formal development stage. In this stage, the students are already be able to do reasoning in the way that is more abstract, idealism, and logic. Thus, the students in the JHS level are in the thinking skill level of C2, C3, and C4. In the SHS level, the students are expected to think in the high order thinking skill that is C3, C4, C5, and C6. [18] Determining the level of the cognitive scope is important in teaching biology since it can help the students to understand the materials. It is also stated by Powell and Kalina that the cognitive development of the students is needed to be the consideration to arrange he pedagogic materials and its level of competencies that refers to the cognitive development theory by Piaget. [20]

The presentation data of learning continuum framework toward the aspects of biological resources management is gotten from the FGD activity based on the experts' opinions, and it is considered from competency level and the characteristics of specific pedagogical materials from ES to SHS. The FGD activity actually focuses to justify and analyze the experts' opinions that are discussed or be an analysis substance. Thus, the data of learning continuum framework toward the aspects of biological resources management based on the experts' opinions that is considered from competency level and the characteristics of specific pedagogical materials from ES to SHS will be shown further.

Table 2. Learning continuum framework toward the aspects of biological resources management based on the experts' opinions

The aspects of biological resources management	The experts' opinions of biology lecturers					
	ES[a]		JHS[b]		SHS[c]	
	Cls[d]	CL[e]	Cls	CL	Cls	CL
1. The benefit of biological resources	-	-	VII	C2[f]	X	C3[g]
2. The kinds of biological resources	-	-	VII	C2	X	C3
3. The balance of nature						
a. The harmony of nature	-	-	-	-	X	C3
b. The conservation and preservation	-	-	-	-	X	C3
4. The conservation						
a. The conservation with general purposes						
1) The tight conservation	-	-	-	-	X	C3
2) The guided conservation	-	-	-	-	X	C3
b. The conservation with certain purposes						
1) Geological protection	-	-	VII	C2	X	C3
2) Botany conservation	-	-	VII	C2	X	C3
3) Zoology conservation	-	-	VII	C2	X	C3
4) Anthropology protection	-	-	VII	C2	X	C3
5. Conservation efforts of biological resources						
a. Conservation efforts of biological resources artificially	-	-	VII	C2	X	C3
b. Conservation efforts of biological resources naturally	-	-	VII	C2	X	C3
c. In-situ maintenance	-	-	VII	C2	X	C3
d. Ex-situ maintenance	-	-	VII	C2	X	C3
6. Forms of nature conservation						
a) National park	-	-	VII	C2	X	C3
b) Sea park	-	-	VII	C2	X	C3
c) Nature preserve	-	-	VII	C2	X	C3
d) Wildlife sanctuary	-	-	VII	C2	X	C3
e) Botanical garden	-	-	VII	C2	X	C3

[a] elementary school
[b] junior high school
[c] senior high school
[d] class
[e] competency level (cognitive scope)
[f] understand
[g] apply

Table 2 is a learning continuum framework toward the aspects of biological resources management based on the experts' opinions that is considered from competency level and the characteristics of specific pedagogical materials from ES to SHS. The research result of learning continuum framework toward the aspects of biological resources management from FGD activity based on the experts' opinions that is considered from competency level and the characteristics of specific pedagogical materials from ES to SHS shows that almost all sub-aspects and specific pedagogical materials of the aspects of biological resources management are started to be taught in the JHS level with the cognitive scope of "*understand*" (C2). Thus, it is continued in the SHS level with the cognitive scope of "*apply*" (C3). Moreover, there are several sub-aspects and specific pedagogical materials of the aspects of biological resources management that is started to be taught in the SHS level with the cognitive scope of C3. In the sub-aspects of the benefit of biological resources, sub-aspects of the kinds of biological resources, sub-aspects of the forms of conservation, and the specific pedagogical materials of conservation with certain purposes (geological conservation, botany conservation, zoology conservation, and anthropology conservation) are started to be taught in the JHS level at the seventh grade with the cognitive level of C2, then, they are continued to be taught in the tenth grade of SHS with the cognitive scope of C3. Afterward, on the sub-aspects of the balance of the nature and the specific pedagogical materials of conservation with general purposes (the tight conservation and the guided conservation) are started to be taught in the tenth grade of SHS with the cognitive scope of C3.

The learning continuum design shows that the aspects of managing biological resources have been begun to be taught in the seventh grade and then later on in the tenth grade. According to Subali, it is more difficult to teach several aspects of the scientific method (both basic and process skills) to students in grades 4 to grade 6 in elementary school. [21]

Determining the cognitive scopes (competency level) that are adapted with the education level is in line with Subali and Suyata statement where the competency levels for the ES students are "*remember*" (C1) and "*understand*" (C2), for JHS students are "*understand*" (C2); "*apply*" (C3); and "*analysis*" (C4) whereas for SHS students are "*apply*" (C3), "*analysis*" (C4), "*evaluate*" (C5), and "*create*" (C6). [18] Moreover, theory of the cognitive development based on Piaget (Piaget theory) can show the students' development based on their age and mental. [7] The teacher roles, here, are very important. According to Feiman-Nemser to correlate the students and the materials that are suitable with their age and with the correct ways then the teachers must develop the pedagogical attitude that is formed from the knowledge about the progress and learning of the children. [22] Learning continuum can affect the way of the teacher in teaching to be more innovative in teaching the materials through the concepts that are already analyzed. [23] So that, the arrangement of learning continuum in the aspects of biological resources management is important to support the learning that is done by the teachers to give the learning outcome that is suitable with the learning objectives.

Learning continuum takes place as an abstract continuum of measurement that contains of competency aspects of the lessons that are related each other, and it is a basic of developing the learning assessment instrument of the lessons. [24] The learning process in the education level will work well if it can be continuity and suitable with the students' progress. [5] The development of learning continuum must first pay attention to the aspects of targeted cognitive abilities starting from the simplest/easiest abilities to the most difficult aspects. If referring to the level/level of the cognitive domain, the ability to remember will be easier/simpler than the ability to understand, and the ability to understand will be easier than the ability to apply. The second thing to note is the complexity of teaching material (subject matter specific pedagogic). It depends on the breadth and depth of a piece of science that is used as material and as a tool to master the targeted competencies. [25] The presence of curriculum and basic competence in the materials of learning continuum will give many benefits. The learning continuum-based in the teaching and learning process has several benefits, such as: 1) the learning based on learning continuum where the teachers can find out the materials that are suitable with the students. Because the students have to get the materials that are appropriate with their needs, the teachers must be able to develop the students' competency level in the classroom by selecting the materials; 2) the learning strategies that are used will be different when the demand of competency in the learning process increases; 3) the planning for learning curriculum based on learning continuum can be the affordable resources during the development of curriculum or the revising program of curriculum; 4) the control toward knowledge had by the students through learning continuum can be a big potency to increase the students' quality of knowledge in cognitive, affective and psychomotor scope. Because the students' knowledge relatively depends on the learning continuum presentation that is ongoing in education, it will be better to revise the school planning; 5) the continuity of education planning for the individual learning can focus on the students. [26]

The development of curriculum must show the relation between the subject matters that are needed in the further learning process for the higher level of education, and these matters actually must be taught in the previous level / stage or below it. Moreover, the learning materials that are taught in the lower education level must not be taught again in the next level, so it can avoid the overlapping of the materials' setting in teaching and learning process. Besides, it is also needed the standard or curriculum that pays attention to the essence of learning continuum and in developing it. [27] [28]

4. Conclusion

Based on the analysis data result of this research, it can be concluded in several ways, as follows: (i) Results of this study indicate that the experts believe that the aspect of biological resource management has been taught in the tenth grade of senior high school with cognitive levels "*understanding*" (C2) and "*applying*" (C3), and in seventh grade junior high school with cognitive level C2. (ii) The design of learning continuum aspects of the management of biological resources from the FGD results based on the opinions of experts is that most of the specific pedagogical materials have begun to be taught at the JHS level with cognitive domain C2, then continued at the SHS level with C3 cognitive domain, and there are also some specific pedagogic materials that have begun to be taught at the SHS level with cognitive domain C3.

Acknowledgments

The deepest gratitude goes to the lecturers of Biology Education Study Program in Yogyakarta, Central Java, East Java, West Java & Jakarta who have been become the respondents in this research and have provided assistance, Department of Biology of Postgraduate Program of Yogyakarta State University who has given research permission, and the Directorate of Research and Community Service from the Ministry of Research, Technology and Higher Education, which has funded this research so that the research can be conducted.

References

[1] Undang-Undang Republik Indonesia Nomor 20 Tahun 2003 tentang Sistem Pendidikan Nasional 1–33
[2] Situmorang R P 2016 Analisis Learning Continuum Tingkat SD sampai SMP pada Tema Sistem Pencernaan Manusia *Scholaria* **6** 1–13
[3] Subali B 2009 Pengembangan Tes Pengukur Keterampilan Proses Sains Pola Divergen Mata Pelajaran Biologi SMA *Makalah disajikan dalam Seminar Nasional Biologi, Lingkungan dan Pembelajarannya, Jurdik Biologi, FMIPA, Universitas Negeri Yogyakarta, Yogyakarta* pp 581–93
[4] Subali B 2014 Evaluasi Pembelajaran (Proses dan Produk) *Makalah disajikan dalam Workshop Evaluasi program Pembelajaran (Proses & Produk) bagi Dosen* pp 1–32
[5] Sugihartono, Fathiyah K N, Setiawati F A, Harahap F and Nurhayati S R 2007 *Psikologi Pendidikan* (Yogyakarta: UNY Press)
[6] Anderson L W and Krathwohl D R 2001 *A Taxonomy for Learning, Teaching and Assessing : A Revision of Bloom's Taxonomy of Educational Objectives* (New York: Addison Wesley Longman, Inc)
[7] Shaffer D R and Kipp K 2010 *Developmental Psychology Childhood & Adolescence* (Georgia: Cengage Learing)
[8] Dodge D T, Colker L J and Heroman C 2004 *The Creative Curriculum for Preschool Fourth Edition* (Washington, D.C: Teaching Strategies, Inc)
[9] BSCS 1987 *Biologi Science : An Ecological Approach BSCS Green Version* (Colorado: The Colorado Collage)
[10] Agustiana I G A T 2014 *Konsep Dasar IPA Aspek Biologi* (Yogyakarta: Ombak)
[11] Campbell N A, Reece J B, Urry L A, Cain M L, Wasserman S A, Minorsky P V. and Jackson R B 2010 *Biologi* ed W Hardani (Jakarta: Erlangga)
[12] Subali B, Kumaidi and Aminah N S 2018 Developing a Scientific Learning Continuum of Natural Science Subjects at Grades 1 – 4 *J. Turkish Sci. Educ.* **15** 66–81
[13] Andriani A E and Subali B 2017 Teacher's Opinion about Learning Continuum based on

Student's Level of Competence and Specific Pedagogical Material in Classification Topics *AIP Conference Proceedings* vol 1868 pp 1–10
[14] Juniati E and Subali B 2017 Teacher's Opinion about Learning Continuum of Genetics based on Student's Level of Competence *AIP Conference Proceedings* vol 1868 pp 1–11
[15] Pramesti I C and Subali B 2017 The Learning Continuum of Ecology based on Teachers' Opinion about Student's Level of Competence and Specific Pedagogical Learning Material *AIP Conference Proceedings* vol 1868 pp 1–10
[16] Astuti L D and Subali B 2017 Teacher's Opinion about Learning Continuum based on The Student's Level of Competence and Specific Pedagogical Materials on Anatomical Aspects *AIP Conference Proceedings* vol 1868 pp 1–12
[17] Hadi R F and Subali B 2017 The Learning Continuum based on The Student's Level of Competence and Specific Pedagogical Learning Material on Physiological Aspects from Teachers's Opinions *AIP Conference Proceedings* vol 1868 pp 1–11
[18] Subali B and Suyata P 2012 *Pengembangan Item Tes Konvergen dan Divergen dan Penyelidikan Validitasnya secara Empiris* (Yogyakarta: Diandra)
[19] Santrock J W 2007 *Masa Perkembangan Anak* (Jakarta: Salemba Humanika)
[20] Powell K C and Kalina C J 2009 Cognitive and Social Constructivism : Developing Tools for An Effective Classroom *Education* **130** 241–50
[21] Subali B, Kumaidi, Aminah N S and Sumintono B 2019 Student achievement based on the use of scientific method in the natural science subject in elementary school *J. Pendidik. IPA Indones.* **8** 41–53
[22] Feiman-Nemser S 2001 From Preparation to Practice : Designing a Continuum to Strengthen and Sustain Teaching *Teach. Coll. Rec.* **103** 1013–55
[23] Torre D M 2015 Learning Throughout a Continuum , an Innovative and Feasible Educational Agenda ! *Mil. Med.* **180** 163
[24] Subali B and Suyata P 2013 The Standardization of School Based Assessment *J. Penelit. dan Eval. Pendidik.* **17** 1–18
[25] Subali B, Kumaidi and Aminah N S 2018 *Pengembangan Model Asesmen Konfirmatori Penguasaan Metode Ilmiah pada Mata Pelajaran IPA di Sekolah Dasar* (Yogyakarta: UNY Press)
[26] North West Evaluation Association 2003 *Primary Use of the Learning Continuum* (Pocatello: Idaho State University)
[27] Idi A 2007 *Pengembangan Kurikulum Teori & Praktik* (Yogyakarta: Ar-Ruzz)
[28]Suyanto S 2018 The Implementation of The Scientific Approach through 5Ms of The New Curriculum 2013 in Indonesia *Jurnal Cakrawala Pendidikan No. 1 Th.2018* 22–29 ISSN:0216-1370 online ISSN 24428620

Developing A Learning Continuum on Ecological Aspect from Elementary to Senior High School Based on The Opinions of Biology Education Experts

Mendala*, B. Subali and Paidi

Biology Education, Graduate School, Yogyakarta State University, Indonesia

*Corresponding author e-mail: mendala95@gmail.com

Abstract. This research uses survey method aiming to investigate the opinions of biology education experts toward the teaching and the assessment of ecological aspects previewed from the level of competency and characteristic of specific pedagogical materials in order to develop the learning continuum of ecological aspects from elementary to senior high schools was conducted in 2018-2019. The population used in this research was hypothetical population which consisted of the all experts of biology education who taught in public and private universities in Indonesia. The sample used in this research was convenience sample consisted of 85 experts of biology education who taught at 18 public and private universities in Yogyakarta, Central Java, East Java, West Java, and Jakarta determined by using purposive technique. The instrument used was confirmatory assessment questionnaire learning continuum based on mastery of ecological aspects which has been arranged by Subali et al. (2016) and the validty was well examined through the focus group discussion activity by 3 lecturers of UNY and 5 lecturers of UNS. The result of this research showed that learning continuum of ecological aspects according to the biology education experts' opinion, start be taught and assessed in the fourth and fifth grade of elementary school with the competency level C1 and C2. Then, it will be continued to the higher level in seventh grade of junior high school with the competency level C2 and tenth grade of senior high school with the competency level C3.

1. Introduction

Learning continuum (LC) is a logical order which shows the continuity between the competency level and characteristics of specific pedagogical materials vertically in every level of education [1]. The basic concept of LC is implemented by doing some mappings toward some competencies which have correlation in each level starting from the simples to the most complex and also composing the characteristic of pedagogical material from the easiest one until the most difficult one and from the most concrete until the most abstract thing [2]. Therefore, the learning process follows the rules of LC so it can give some descriptions toward the developmental phase of students and makes the learning process becoming more directed, efficient, and effective[3]. The most important thing to be considered is that the mental and cognitive development of students in every level of education are different[4].

LC has many benefits in the learning process, some of those benefits are LC can be made as the basic in selecting the learning materials and learning sources which are appropriate with the variety level of students, LC can be utilized by teachers to develop individuals' learning and the further lesson plans, as the basic to compose, to develop, and to fix the curriculum[5],[6].

LC can be implemented in various aspects of knowledge fields, one of them is in ecological aspect. Ecological aspect has the various scopes of learning material from the level of organism until biosphere, having correlation with the other aspects of Biology, and when studying about curriculum in Indonesia the material about ecology has been taught to students from the early level of education in elementary school until in the level of high school[7],[8],[9]. Therefore, the order which shows the appropriate gradation with LC so students can easily learn and understand the materials is urgently needed.

Unfortunately, the fundamental problem faced until now is the serving of competency level and characteristics of pedagogical materials of ecological aspect in the curriculum which is still overlapping, whether in elementary or igh school[10]. Besides that, the result of the previous research taken from the opinion of practitioners still shows the grouping of opinion and tends to follow the serving in curriculum. Based on the description explained above, the further research about this topic is highly recommended by asking the opinions from the experts of biology education about LC of ecological aspects from elementary to senior high school. The modus of practitioners' opinion from the previous research is served in the research questionnaire which is purposed to become considerations from the experts when giving opinions.

2. Method

This research uses survey method aiming to investigate the opinions of some experts of biology education toward the teaching and the assessment of ecological aspects previewed by the level of competency and characteristic of specific pedagogical materials in order to develop the LC of ecological aspects from elementary to senior high schools was conducted in 2018-2019. The population used in this research was hypothetic population which consisted of the all experts of biology education who taught in public and private universities in Indonesia. The sample used in this research was convenience sample consisted of 85 experts of biology education who taught at 18 public and private universities in Yogyakarta, Central Java, East Java, West Java, and Jakarta that become the respondents in this research determined by using purposive technique. The instrument used was confirmatory assessment questionnaire LC based on mastery of ecological aspects which has been arranged by Subali et al. (2016) and the validty was well examined through the Focus Group Discussion (FGD) activity by 3 lecturers of Universitas Negeri Yogyakarta (UNY) and 5 lecturers of Universitas Sebelas Maret (UNS). These aspects are presented in the questionnaire that contain of 10 sub-aspects that are characteristics of biome, characteristics of ecosystem, characteristics of ecosystem community, characteristics of the arrangement population of the ecosystem community, characteristics of the functions happened in the ecosystem, interaction interspecific inter population, interaction interspecific inter individual in a population, functions of the biotic components in the ecosystem, functions of abiotic components in the ecosystem, and interaction between biotic components and abiotic components in the ecosystem. Thus, the competency level in this research is the competency level of cognitive area in taxonomy Bloom by Anderson and Krathwohl that contain of six levels that are remembering (C1), understanding (C2), applying (C3), analyzing (C4), evaluating (C5), and creating (C6).

The data analysis in this research used descriptive statistic which served the modus data of respondents' opinion through the use of questionnaire. The serving of modus aims to find the agreement of opinion among respondents. It means that, the respondents who have same opinion will be on the one group of modus. The modus of respondents previewed furthermore by the group of biology education expert through the activity of FGD to decide whether the opinion of respondents is logic or not. Its decision follows the basic concepts of the LC arrangement that are competency level and the pedagogic materials, and they are arranged from the abstraction of the concepts (concrete to abstract), the easiness of the concepts (easy to difficult), and the complexity of the concepts (simple to complex). Besides, they also consider about the appropriateness with the students' cognitive development step in every level of education starting from the elementary school to senior high school.

3. Result and Discussion

The result of this research shows the modus of respondents' opinions then it is analyzed further by the group of experts of biology education through FGD activity that can be seen in Table 1 to 10. The analysis result can become a reference to arrange the grid LC of ecological aspects in elementary to senior high school (from the first grade to the twelfth grade) that will be shown in Table 11. The description of the research result further can be seen in the table below:

Table 1. The Opinion of Biology Education Experts on LC Ecological Aspects of Sub-aspect 1

Ecology Aspects	Modus		FGD
	%	G/CL	
Sub-aspect 1: Biome characteristics with the indicators, as follows:			
a. The characteristics of land biome based on its structure (its arranging components)	36	VII/C2	L
b. The characteristics of water biome based on its structure (its arranging components)	34	VII/C2	L
c. The growth of land biome and the factors that influence it	31	VII/C2	L
d. The growth of water biome and the factors that influence it	31	VII/C2	L
e. The evolution of land biome and the factors that influence it	31	VII/C2	L
f. The evolution of water biome and the factors that influence it	31	VII/C2	L

Note: %= Percentage, G= Class, CL= Level of competency, L= Logic

Table 1 shows that between the modus of respondents' opinions and the analysis result of FGD have the similarity. It means that the sub-aspect 1 in ecology materials that is biome characteristics shows that all indicators from indicator **a** to **f** are logically started to be taught in the seventh grade of junior high school with the competency level C2. The description of the respondents' modus with the analysis result of FGD in the sub-aspect 2 can be seen in Table 2, as follows:

Table 2. The Opinion of Biology Education Experts on LC Ecological Aspects of Sub-aspect 2

Ecology Aspects	Modus		FGD
	%	G/CL	
Sub-aspect 2: The characteristics of ecosystem with the indicators, as follow:			
a. The characteristics of natural ecosystem and artificial ecosystem	26	VII/C2	TL, V/C1
b. The characteristics of land ecosystem based on its structure (its arranging components)	22	VII/C2	TL, V/C1
c. The characteristics of water ecosystem based on its structure (its arranging components)	25	VII/C2	TL, V/C1
d. The growth of land ecosystem and the factors that influence it	25	VII/C2	L
e. The growth of water ecosystem and the factors that influence it	25	VII/C2	L
f. The evolution of land ecosystem and the factors that influence it	25	VII/C2	L
g. The evolution of water ecosystem and the factors that influence it	25	VII/C2	L

Note: %= Percentage, G= Class, CL= Level of competency, L= Logic, TL= illogical

Table 2 shows that the indicator **a**, **b**, and **c** based on the FGD result are illogical if they are started to be taught in the seventh grade of junior high school because the three indicators have the characteristics and the scope of materials that are easy and simple. So that, the FGD result suggests to teach them in the fifth grade of elementary school with the competency level C1 whereas the indicator **d**, **e**, **f**, and **g** between the modus of respondents' opinions and the analysis result of FGD agree if these indicators start to be taught in the seventh grade of junior high school with competency level C2. The description of the respondents' modus with the analysis result of FGD in the sub-aspect 3 can be seen in Table 3, as follows:

Table 3. The Opinion of Biology Education Experts on LC Ecological Aspects of Sub-aspect 3

Ecology Aspects	Modus		FGD
	%	G/CL	
Sub-aspect 3: The characteristics of ecosystem community with the indicators, as follows:			
a. The characteristics of land ecosystem community based on its structure (its arranging components)	28	VII/C2	TL, V/C1
b. The characteristics of water ecosystem community based on its structure	28	VII/C2	TL, V/C1
c. The growth of land ecosystem community and the factors that influence it	24	VII/C2	L
d. The growth of water ecosystem community and the factors that influence it	24	VII/C2	L
e. The evolution of land ecosystem community and the factors that influence it	24	VII/C2	L
f. The evolution of water ecosystem community and the factors that influence it	24	VII/C2	L

Note: %= Percentage, G= Class, CL= Level of competency, L= Logic, TL= illogical

Table 3 shows that the indicator **a** and **b** based on the FGD result are illogical if they are started to be taught in the seventh grade of junior high school because the three indicators have the characteristics and the scope of materials that are easy and simple. So that, the FGD result suggests to teach them in the fifth grade of elementary school with competency level C1 whereas the indicator **c**, **d**, **e**, and **f** between the modus of respondents' opinions and the analysis result of FGD agree if these indicators start to be taught in the seventh grade of junior high school with competency level C2. The description of the respondents' modus with the analysis result of FGD in the sub-aspect 4 can be seen in Table 4, as follows:

Table 4. The Opinion of Biology Education Experts on LC Ecological Aspects of Sub-aspect 4

Ecology Aspects	Modus		FGD
	%	G/CL	
Sub-aspect 4: The characteristics of the arrangement population of the ecosystem community with the indicators, as follows:			
a. The characteristics of the arrangement population of the land ecosystem community	34	VII/C2	TL, V/C1
b. The characteristics of the arrangement population of the water ecosystem community	34	VII/C2	TL, V/C1
c. The growth of the arrangement population of the land ecosystem community and the factors that influence it	33	VII/C2	L
d. The growth of the arrangement population of the water ecosystem community and the factors that influence it	33	VII/C2	L
e. The evolution of the arrangement population of the land ecosystem community and the factors that influence it	34	VII/C2	L
f. The evolution of the arrangement population of the water ecosystem community and the factors that influence it	35	VII/C2	L

Note: %= Percentage, G= Class, CL= Level of competency, L= Logic, TL= illogical

Table 4 shows that the indicator **a** and **b** based on the FGD result are illogical if they are started to be taught in the seventh grade of junior high school because the three indicators have the characteristics and the scope of materials that are easy and simple. So that, the FGD result suggests to teach them in the fifth grade of elementary school with competency level C1. Whereas, the indicator **c**, **d**, **e**, and **f** between the modus of respondents' opinions and the analysis result of FGD agree if these indicators start to be taught in the seventh grade of junior high school with competency level C2. The description of the respondents' modus with the analysis result of FGD in the sub-aspect 5 can be seen in Table 5, as follows:

Table 5. The Opinion of Biology Education Experts on LC Ecological Aspects of Sub-aspect 5

Ecology Aspects	Modus		FGD
	%	G/CL	
Sub-aspect 5: The characteristics of the functions happened in the ecosystem with the indicators, as follows:			
a. Food chain	26	IV/C2	TL, IV/C1
b. Food webs	26	IV/C2	TL, IV/C1
c. Energy flow	29	IV/C2	TL, IV/C1
d. Pyramid ecology (total number, biomass, and energy)	41	VII/C2	L
e. Biogeochemical cycle (carbon, nitrogen, phosphor, sulfur, and other mineral)	34	X/C3	L

Note: %= Percentage, G= Class, CL= Level of competency, L= Logic, TL= illogical

Table 5 shows that the indicators **a**, **b**, and **c** according to the FGD results are illogical when they have to be taught in the fourth grade of elementary school with competency level C2. It caused by the three indicators that are taught in elementary school only in introducing level, so the FGD results suggest to teach them in the fourth grade of elementary school with competency level C1. Thus, the indicator **d** starts to be taught in the seventh grade of junior high school with competency level C2 whereas the indicator **e** starts to be taught in the tenth grade of senior high school with C3 competency level. The description of the respondents' modus with the analysis result of FGD in the sub-aspect 6 can be seen in Table 6, as follows:

Table 6. The Opinion of Biology Education Experts on LC Ecological Aspects of Sub-aspect 6

Ecology Aspects	Modus		FGD
	%	G/CL	
Sub-aspect 6: The interaction interspecific inter population with the indicators, as follows:			
a. Symbiosis (mutualism, parasitism, and commensalism)	27	IV/C2	TL, VII/C2
b. Neutralism	28	VII/C2	L
c. Predation	27	VII/C2	L
d. Competition	26	VII/C2	L
e. Cooperation	26	VII/C2	L

Note: %= Percentage, G= Class, CL= Level of competency, L= Logic, TL= illogical

Table 6 shows that the indicator **a** according to the FGD result is illogical if it is taught in the fourth grade of elementary school with competency level C2. The indicator has the characteristics and scope of the materials that are difficult, abstract, and complex. So that, the FGD results suggest the indicator of symbiosis logically to be taught in the seventh grade of junior high school with competency level C2. Otherwise, the indicators **b, c, d**, and **e** based on the respondents' opinions and discussion results that agree will be taught in the seventh grade of junior high school with competency level C2. The description of the respondents' modus with the analysis result of FGD in the sub-aspect 7 can be seen in Table 7, as follows:

Table 7. The Opinion of Biology Education Experts on LC Ecological Aspects of Sub-aspect 7

Ecology Aspects	Modus		FGD
	%	G/CL	
Sub-aspect 7: The interaction interspecific inter individual in a population with the indicators, as			

follows:			
a. Neutralism	32	X/C2	TL, VII/C2
b. Predation	32	X/C2	TL, VII/C2
c. Competition	32	X/C2	TL, VII/C2
d. Cooperation	32	X/C2	TL, VII/C2

Note: %= Percentage, G= Class, CL= Level of competency, TL= illogical

Table 7 shows that the indicators **a, b, c,** and **d** based on the FGD results are illogical if these indicators are taught in the tenth grade of senior high school with competency level C2 since the characteristics and the scope of the materials are easy and simple. So that, the FGD results suggest these indicators logically to be taught in the seventh grade of junior high school with competency level C2. The description of the respondents' modus with the analysis result of FGD in the sub-aspect 8 can be seen in Table 8, as follows:

Table 8. The Opinion of Biology Education Experts on LC Ecological Aspects of Sub-aspect 8

Ecology Aspects	Modus		FGD
	%	G/CL	
Sub-aspect 8: The functions of the biotic components in the ecosystem with the indicators, as follows:			
a. Functional status (*nisia*) in the food chain and food webs (producer, consumer, garbage eater, and decoder)	26	IV/C2	L
b. Functional status in the food sources (autotroph, heterotroph, and saprotrophic)	38	VII/C2	L

Note: %= Percentage, G= Class, CL= Level of competency, L= Logic

Table 8 shows that between the respondents' opinions and the FGD results have the similarities. It means that the indicator **a** is logically started to be taught in the fourth grade of Elementary school with competency level C2. Thus, the indicator **b** is logically started to be taught in the seventh grade of Junior High School with competency level C2. The description of the respondents' modus with the analysis result of FGD in the sub-aspect 9 can be seen in Table 9, as follows:

Table 9. The Opinion of Biology Education Experts on LC Ecological Aspects of Sub-aspect 9

Ecology Aspects	Modus		FGD
	%	G/CL	
Sub-aspect 9: The functions of abiotic components in the ecosystem with the indicators, as follows:			
a. The functions of land in the ecosystem based on the physical characteristics	31	X/C2	TL, V/C2
b. The functions of land in the ecosystem based on the chemical characteristics	32	X/C2	TL, V/C2
c. The functions of atmospheric components in the ecosystem based on the physical characteristics	32	X/C2	TL, V/C2
d. The functions of atmospheric components in the ecosystem based on the chemical characteristics	32	X/C2	TL, V/C2
e. The functions of hydraulic in the ecosystem	25	X/C2	TL, V/C2

Note: %= Percentage, G= Class, CL= Level of competency, TL= illogical

Table 9 shows that all indicators are illogical to be taught in the tenth grade of senior high school with competency level C2 because the characteristics and the scope of the materials are still easy and simple. So that, the FGD suggests these indicators logically to be taught in the seventh grade of

Elementary School with competency level C2. The description of the respondents' modus with the analysis result of FGD in the sub-aspect 10 can be seen in Table 10, as follows:

Table 10. The Opinion of Biology Education Experts on LC Ecological Aspects of Sub-aspect 10

Ecology Aspects	Modus		FGD
	%	G/CL	
Sub-aspect 10: The interaction between biotic components and abiotic components in the ecosystem with the indicators, as follows:			
a. The interaction between biotic components and land/edaphic components	26	X/C2	TL, V/C2
b. The interaction between biotic components and air/ atmospheric components	25	X/C2	TL, V/C2
c. The interaction between biotic components and hydraulic components	25	X/C2	TL, V/C2

Note: %= Percentage, G= Class, CL= Level of competency, TL= illogical

Table 10 shows that all indicators are illogical to be taught in the tenth grade of Senior high school with the competency level C2 because the characteristics and the scope of the materials are still easy and simple. So that, the FGD suggests these indicators logically to be taught in the seventh grade of Elementary School with the competency level C2. From Table 1 to Table 10 that are explained previously can be concluded in the grid LC of ecological aspects in elementary and senior high school, as follows:

Table 11. The Grid LC of Ecological Aspects in Elementary and Senior High School

Ecology Aspect	Education level/ Competency level (C1-C6)											
	I	II	III	IV	V	VI	VII	VIII	IX	X	XI	XII
Characteristics of biome	-	-	-	-	-	-	C2	-	-	-	-	-
Characteristics of ecosystem	-	-	-	-	C1	-	C2	-	-	-	-	-
Characteristics of ecosystem community	-	-	-	-	C1	-	C2	-	-	-	-	-
Characteristics of the arrangement population of the ecosystem community	-	-	-	-	C1	-	C2	-	-	-	-	-
Characteristics of the functions happened in the ecosystem	-	-	-	C1	-	-	C2	-	-	C3	-	-
Interaction interspecific inter population	-	-	-	-	-	-	C2	-	-	-	-	-
Interaction interspecific inter individual in a population	-	-	-	-	-	-	C2	-	-	-	-	-
Functions of the biotic components in the ecosystem	-	-	-	C2	-	-	C2	-	-	-	-	-
Functions of abiotic components in the ecosystem	-	-	-	-	C2	-	-	-	-	-	-	-
Interaction between biotic components and abiotic components in the ecosystem	-	-	-	-	C2	-	-	-	-	-	-	-

Note: C= Competency level, I-VI= Elementary school level, VII-XII= Junior and Senior high school

Table 11 shows that, according to the biology education experts' opinions, the ecological aspects start to be taught and assessed in the fourth grade of elementary school with C1 and C2 competency level and in the fifth grade with the competency level C1 and C2. Then, it will be continued to the higher level in the seventh grade of junior high school with the competency level C2 and in senior high school with the competency level C3.

Based on the result explained previously, there are some sub-aspects from the modus of respondents' opinions (the biology education experts that are given the questionnaire) which are illogical according to the FGD results. In this case, it can be happened because they are not suitable with the basic concepts of the LC arrangement that are competency level and the characteristics of pedagogical materials. These things are not arranged continually from the simple to complex, easy to difficult, and concrete to abstract in every level of education and the development stage of the students[2],[10],[11]. For example, the modus from the respondents' opinions stated that there are several indicators that have the characteristics and the scope of the materials which are easy and simple. Besides, they become the basic materials and are still taught in the higher level of education. It can be seen in the Table 2 (indicator **a**, **b**, and **c**), 3 (indicator **a** dan **b**), 4 (indicators **a** and **b**), 5 (indicator **a**, **b**, and **c**), 7 (indicator **a** to **d**), 9 (indicator **a** to **e**), dan 10 (indicator **a** to **c**). Otherwise, in the Table 6, indicator **a** has the characteristics which is more difficult, complex, and abstract, so it is illogically taught in the elementary school.

According to the competency level in the cognitive area of taxonomy Bloom by Anderson and Krathwohl that contains of six levels [12]. If they will be arranged from the simple one to the complex, so that the competency level of remembering (C1) is more simple that the competency level of C2 understanding (C2), then, the competency level of applying (C3), analyzing (C4), evaluating (C5), and creating (C6)[13]. Besides, the essential thing in the LC arrangement especially for developing the curriculum is the continuity of the pedagogic materials specifically in every level of education. The materials in the elementary school as the basic level must learn the basic materials to build the concepts of the next materials [14],[15], for example in Table 5 that shows the sub-aspects of the functional characteristics that happen in the ecosystem that consist of the food chain, food webs, and the energy flow are taught in the elementary school as the reference to learn the pyramid of ecology and biogeochemical cycle in the Junior High School.

The development stages of the students by Piaget theory consists of four stages that are sensorimotor (age 0-2 years), pre-operational (age 2-7 years), concrete operational age 7-11 years), and formal operational (age 11 to upward) [16],[17]. If they are connected with the students' age in every level of education, the students will be in elementary school at the age from 7 to 11 years old. It means that they are in the formal operational stage where, in this stage, they must be able to think logically and be able to apply their way of thinking to the concrete and specific example with remembering (C1) and understanding (C2) competency level[18],[19].

The students that are in the junior and senior high school are about 12 to 17 years old. It means that they are in the formal operational stage where, in this stage, they must be able to think abstractly, ideally, and logically [20], [21]. If they are linked with the competency level, the students of junior and senior high school are in the level of understanding (C2), applying (C3), and analysing (C4) [11]. Ideally, the materials that are presented in the curriculum between the competency level and the pedagogic materials taught by the teachers must be suitable with the sub-aspects or the learning materials and the students' development [22], [23], [24].

4. Conclusion

By considering the result of the research, it can be concluded that LC of ecological aspects according to the opinions from some experts of biology education should be taught earlier in the fourth grade of elementary school in sub-aspect characteristic of function happened in the the ecosystem (indicator **a**, **b**, and **c**) with the competency level C1 and sub-aspect of the function of biotic components in the ecosystem (indicator **a**) with the competency level C2. The fifth Grade of elementary school in sub-aspect characteristics of ecosystem (indicator **a**, **b**, and **c**), sub-aspect characteristics of ecosystem community (indicator **a** and **b**), sub-aspect characteristics of the arrangement population of the ecosystem community (indicator **a** and **b**) with competency level C1 and sub-aspect functions of abiotic components in the ecosystem (**a**, **b**, **c**, **d**, and **e**), sub-aspect interaction between biotic components and abiotic components in the ecosystem (indicator **a**, **b**, and **c**) with competency level C2. Furthermore, in the level of junior high school in seventh grade sub-aspect characteristics of the biome (indicator **a**, **b**, **c**, **d**, **e**, and **f**), sub-aspect characteristics of ecosystem (indicator **d**, **e**, **f**, and **g**), sub-aspect characteristics of ecosystem community (indicator **c**, **d**, **e**, and **f**), sub-aspect characteristics of the the arrangement population of the ecosystem community (indicator **c**, **d**, **e**, and **f**), characteristics of the functions happened in the ecosystem (indicator **d**), sub-aspect interaction of interspecific among populations (indicator **a**, **b**, **c**, **d**, and **e**), sub-aspect interaction of intraspecific among individuals in a population (indicator **a**, **b**, **c**, and **d**), sub-aspect functions of biotic components in the ecosystem (indicator **b**) with competency level C2 and in the level of senior high school tenth grade sub-aspect characteristic of the functions happened in the ecosystem with the competency level C3.

Acknowledgments
The researchers give thanks to the Ministry of Research, Technology and Higher Education in Indonesia that already helps this research fee.

References

[1] Prihatni Y, Kumaidi & Mundilarto 2016 Pengembangan Instrumen Diagnostik Kognitif pada Mata Pelajaran IPA di SMP *J. Penelitian & Evaluasi Pendidikan.* **20** 111–25
[2] Subali B & Suyata P 2012 *Pengembangan Item Tes Konvergen dan Divergen* (Yogyakarta: Diandra Pustaka Indonesia)
[3] Dodge D T, Heroman C, Charles J, & Maiorca J 2004 Beyond Outcomes: How Onging Assessment Supports Children's Learning and Lead to Meaningful Curriculum *YC Young Child.* **59** 20–8
[4] Pramesti I C & Subali B 2017 The Learning Continuum of Ecology Based on Teachers' Opinion About Student's Level of Competence and Specific Pedagogical Learning Material *Proceedings The 4th International Conference on Research, Implementation, and Education of Mathematics and Science (4th ICRIEMS)*
[5] Northwest Evaluation Association 2003 *Primary Use of The Learning Continuum* (Pocattello: Idaho State University)
[6] Department of Education and Communities 2013 *An Overview of the Literacy Continuum K-10* (Sydney: State of New South Wales)
[7] Solomon E P, Berg L R, & Martin D W 2008 *Biology 8th edidtion* (Belmont: Thomson Brooks/Cole)
[8] Ministry of National Education (2016) *The Decree of National Education Minister No. 21 of 2016 about Standard Contenct in Primary and Secondry Education*
[9] Odum E P & Barret G W 2005 *Fundamantal of Ecology 5th edition* (Belmont, CA: Cengage Learning)
[10] Situmorang R P 2016 Analisis Learning Continuum Tingkat SD sampai SMP pada Tema Sistem Pencernaan Manusia *Scholaria* **6** 1–13
[11] Subali B, Kumaidi, & Aminah N S 2018 Developing a Scientific Learning Continuum of Natural Science Subjects at Grades 1–4 *J. Turkish Science Education.* **15** 66–81
[12] Alavinia P 2013 The Overriding Role of Critical Thinking in Nahjolbalaqa 'Letters': Insights from Bloom's Cognitive Model *J. Language Teaching Research.* **4** 97–104
[13] Anderson L W & Krathwohl D R 2001 *A Taxonomy for Learning, Teaching, and Assessing: A Revision of Bloom's Taxonomy of Educational Objectives.* (New York: Addision Wesley Longman)
[14] Schunk D H 2012 *Learning Theories 6th edition* (Boston: Pearson Education Inc)
[15] Knowles M S 1997 The Modern Practice of Adult Education from Pedagogy to Andragogy (New York: Cambridge The Adult Education Company)

[16] Bucler S & Castle P 2014 *Psychology of Teachers* (London: Sage Publications)
[17] Santrock J W 2014 *Educational Pschology 5th edition* (Asia: McGraw-Hill Education)
[18] Hughes A G & Hughes E H 2012 *Learning and Teaching* (New Delhi: Sonali Publication)
[19] Febrini D 2017 *Psikologi Pembelajaran* (Yogyakarta: Putaka Pelajar)
[20] Desmita 2009 *Psikologi Perkembangan* (Bandung: Remaja Rosdakarya)
[21] Henson K T 2007 *Curriculum Planning 4th edition* (United States of America: Waveland Press)
[22] Shawer S F 2017 Teacher-Driven Curriculum Development at the Classroom Level: Implications for Curriculum, Pedagogy and Teacher Training *J. Teaching & Teacher Education.* **63** 296–313
[23] Lu, Q and Liu E 2012 Alignment between High School Biology Curriculum Standard and the standardised tests of four provinces in China *J. Biological Education.* **46** 149–64
[24] Suyanto, S. & Wibowo,Y. *2017. Curriculum review of teacher professional development program based on biology teacher profile in technological pedagogical and content knowledge*. https://iopscience.iop.org/article/10.1088/1742-6596/1097/1/012042/pdf

Teacher's TPACK Profile: The Affect of Teacher Qualification and Teaching Experience

M. K. Antony*, Paidi, B. Subali, S. P. Pradana, N. Hapsari, F. E. C. Astuti

Biology Education, Graduate School, Yogyakarta State University, Yogyakarta, Indonesia

*Corresponding author e-mail: m.khoirulantony10@gmail.com

Abstract. The 21st century is the era of technology and digital. The 21st century affects all aspects of life, regardless education. This is clearly the teacher must improve their ability to master ICT in learning. The aim of this study was to determine the affect of teacher qualifications and teaching experience on biology teacher's TPACK in the Magelang City. This was a survey research. The data collection technique uses TPACK test instruments from Paidi & Subali (2018). This research was conducted in all public secondary schools in the Magelang City, Central Java. Population in this study was hypothetico population. The sample in this study were 14 high school biology teachers in Magelag City. The analysis of data was two-way ANOVA, with provisions $p < 0.05$, which means that there is a significant variable influence on TPACK implementation. This finding reveals that teacher qualifications and teacher teaching experience have Sig. score $p < 0.05$. In conclusion, there was a significant affect of teacher qualifications and teaching experience on biology teacher' TPACK.

1. Introduction

The 21st century is a technological era. Technology has penetrated almost all segments of life, including education. The learning the era of globalization aims to prepare a generation that is ready for the advancement of ICT that develops so quickly and has an impact on various aspects of life, including in the teaching and learning process [14]. Technology has become an integral part of educators and students' lives, it changed interaction of teachers and students how to learn in technology-rich environments [6][26]. The conclusion from [17], the use of technology in the classroom is described as a great support tool for access to information, student academic support, and for individualization and personalization.

According to [20], along with digital technology, technical competencies have also emerged. This requires knowledge and skills in using technology to provide comprehensive learning. This confirms that the task of the teacher as an educator will increase. According to Shulman (1987), there is a teacher's ability to manage classes, namely mastery of Content Knowledge (CK) [15], and the importance of mastering Paedagogical Knowledge (PK) [10][25] which was integrated into PCK. Koehler & Mishra integrated TK in the Shulman PCK concept (1987) so as to form new types of knowledge, TCK, TPK, and TPACK. The related literature review shows that school-based determinants that was important to support education teacher quality [1] [11] because the teacher determines Adeyemi students' academic performance (2010). Therefore, there was a need to assess secondary school teachers's characteristics in terms of qualifications and teaching experience to ensure the quality of education. Teacher qualification is described that skill or type of experience or certain knowledge that a person has to make or is suitable for teaching. Accordingly, teacher qualifications can mean all the skills teachers need to teach effectively. In this study, teaching qualifications focused on academic degrees that a teacher must have to fulfill his teaching requirements. In Nigeria, this includes Bachelor, Master, Doctor, other certificates, and inclusive diplomas. However, only degree holders are eligible to teach in high school. Some scholars believe that higher degree shows a positive correlation with student academic achievement [9][29]. The teachers who have more training produce students who have better achievements. Years of teacher experience is one indicator of teacher qualifications which is believed to be a significant determinant of student academic performance [8]. A greater teaching experience will result in students with higher achievement [6]. Research has shown that inexperienced teachers are usually less efficient than experienced teachers [7]. Studies have found a positive relationship between teacher effectiveness and their experience over the years and efficient teachers positively influence student academic achievement of [2].

2. Research Methods

This was survey research. The study was conducted in all Senior High Schools in Magelang City, Central Java, which amounted to 5 Public High Schools. Populotion in this studied was hypothetico population with konvenien sample, that is 24 biology teachers in Magelang City. The instrument in this research is the TPACK biology teacher test instrument by Paidi & Subali (2018). Empirical validity of test instruments obtained MNSQ infit value of 1.00 with a standard deviation of 0.16. In this case, it shows that all test items that have been prepared fit the Rasch model because the MNSQ infit score obtained is at the acceptance limit of ≥ 0.77 to ≤ 1.30. . The instrument contains 40 multiple choice questions, each TPACK component consists of types of conceptual, factual, and procedural questions. The study was conducted from January to February 2019. The independent variables measured in this study were teacher qualifications and teaching experience. Qualifications teachers in this studied were teachers with master's degrees (S2 teachers) and bachelor degree graduates (S1 teachers). Teaching experience were teachers had taught more than 15 years (senior teachers) and teachers had taught less than 15 years (junior teachers). The dependent variable (y) was the teacher's kowledge about seven components of TPACK. The data analysis used the multivariat analysis of variance with the Sig. provisions <0.05, which means that there is a significant effect on teacher qualifications and teaching experience on teacher's TPACK.

3. Findings and Discussion

Table 1. The result of the Effect of Experienced Teaching on the implement TPACK

No	TPACK's components	Teaching Experience	Teacher Qualification	Means	Probability
1	CK	Senior	S1	7.5000	0,016*
			S2	8.7500	
		Junior	S1	6.6667	
			S2	8.3333	
2	PK	Senior	S1	4.2500	0,033*
			S2	5.3333	
		Junior	S1	3.6667	
			S2	5.0000	
3	TK	Senior	S1	1.0000	0,012*
			S2	2.3333	
		Junior	S1	2.6667	
			S2	2.7500	

4	PCK	Senior	S1	4.5000	0,048*
			S2	5.6667	
		Junior	S1	3.3333	
			S2	4.0000	
5	TCK	Senior	S1	2.7500	0,010*
			S2	3.6667	
		Junior	S1	3.6667	
			S2	4.5000	
6	TPK	Senior	S1	2.7500	0,041*
			S2	3.6667	
		Junior	S1	3.6667	
			S2	4.5000	
7	TPACK	Senior	S1	1.7500	0,010*
			S2	3.0000	
		Junior	S1	3.0000	
			S2	3.2500	

Based on the results of the study, 100% were certified teachers with 21.4% of men and 78.6% of women. Table 1 showed that the TPACK by postgraduate graduate (S2) teachers were much higher than those of Diploma / Bachelor graduates (S1). This study revealed that 42.8% of participants served more than fifteen years as teachers and 57.2% of teachers had taught less than 15 years. Table 3 shows that there are significant difference of seven TPACK components between S2 teachers and S1 both teachers who have more than 15 years of teaching experience and those who are less than 15 years old in Magelang City, Central Java, $p< 0.05$. This means that teacher qualifications and teaching experience contributed significant to the teacher's TPACK in the Magelang City.

4. Discussion

The purpose of this study was to see the extent to which teacher qualifications and teacher teaching experience were related to to the ability of seven components of the teacher's TPACK on biology subjects. The findings of this study indicated that 42.8% were teachers with master's degrees (S2 teachers) while 57.2% have Bachelor degree graduates (S1 teachers). Further research revealed that the teachers taught Biology subjects where they were trained. This implies that the teachers meet the requirements so that they are expected to be able to integrate TPACK's abilities in the learning process. These findings agreed with the views of Rivkin et al. (2005) who found no convincing evidence that master's degrees increased the effectiveness of teachers at the secondary school level. Further findings agreed with [13] which stated that additional professional qualifications outside the first level did not always lead to increased teaching competencies at the secondary school level. However, the findings of this study contradict the findings of research conducted by other scholars who found that teacher qualifications and experience played an important role in improving teacher abilities [3][4][16][18] (Yala and Wanjohi, 2011; [19]. The research findings showed that a low percentage (42.8%) of biology teachers who have had many years of experience in the field, these findings revealed that there was no significant influence from teaching experience. This confirmed that the teacher experience did not significantly influence the ability to implement 4 components of TPACK (Table 1). Regarding the development of the teacher's career over the years, this study determined that the majority of teachers (42.8%) had taught for more than fifteen years, 57.2% of the participants had taught less than fifteen years.

Finding showed that teacher qualification, S1 and S2 teachers have significant differences in the mastery of the seven components of TPACK principle. This is possible because of seven components of TPACK are competency in the form of knowledge about science disciplines that must be possessed by teachers, therefore both senior and junior teachers have taken education to obtain this knowledge, but teachers with S2 graduates have better knowledge than teachers with S1 graduates. In line with Musou and Migasi (2015), the average knowledge of students taught by S2 graduates is higher than students taught by S1 graduates, because the mastery ability of master graduates is better than S1 graduates. This finfing showed that senior teachers have PK, CK, and PCK better than junior teachers. This is because more teaching experience causes flexibility in teaching. According to [22] reveals that learning is a long-term change in behavior or inability to behave in a certain way, resulting from practice or other forms of experience. This is in line with the opinion of [12] "definitions of organizational effectiveness are constantly changing because they are influenced by context preferences, which in turn reflect social change." Learning occurs when experience (including exercise) results in a change in behavior and a person's knowledge that is stable. This emphasizes that the more an experienced teacher, the better the way to teach. The contrast finding that, junior teachers have TK, TCK, TPK, and TPACK better than senior teachers. The same thing was also proven based on [17] that as teachers age, senior factors are challenges to the application of technology. Thus, teacher qualification and teaching experience have significant affect to teacher's TPACK. and [16] suggest that teacher qualification and teaching experience have an important role in the self-development of a teacher to master the competencies that demand the development of the times[3].

5. Conclusion

The results of the study showed that teacher qualifications and teaching experience have a significant affect on biology teachers's TPACK in Magelang City, the significance value of the two way annova test on seven TPACK components were $p < 0.05$.

Aknowledgement

The results of this study will provide information regarding mastery of teacher TPACK so that it can be used to improve the quality of learning. In addition, the results of the study also provide consideration in determining education policies, related to the number of basic competencies that must be mastered by students, learning experiences or abilities that need to be developed in the curriculum, as well as educational or learning methods or technologies that need to be accommodated. The researcher gratitude the senior high school headmaster, biology teacher, head of MGMP in Magelang City, and all those who had helped in this research.

Refferences

[1] Aaronson D, Barrow L, Sander W 2007 Teachers and Student Achievement in the Chicago Public High Schools. J. Labor Econ. 25(1):95-135.

[2] Agharuwhe, A. A. 2013 Effects of Teachers' Effectiveness on Students' Academic Performance in Public Secondary Schools; Delta State – Nigeria, *Journal of Educational and Social Research, 3*(3), 105-111.

[3] Ankomah Y, Koomson J, Bosu R, Oduro GK 2005 Implementing Quality Education in Low Income Countries. Institute for Educational Planning & Administration (IEPA) University of Cape Coast Ghana.

[4] Asikhia OA 2010 Students and Teachers' Perception of the Causes of Poor Academic performance in Ogun State SecondarySchools: Euro. J. Soc. Sci. 13(2):229-242.

[5] Bilici, S.C., Yamak, H., Kavak, N., & Guzey, S.S. 2013 Technological Pedagogical Content Knowledge Self-Efficacy Scale (TPACK-SeS) for pre-service science teachers: construction, validation, and reliability.

[6] Boyd, D., Landford, H., Loeb, S., Rockoff, J., &Wyckoff, J. 2008 The Narrowing Gap in New York City Teacher Qualifications and Its Implications for Student Achievement in High-Poverty Schools.*Journal of Policy Analysis and Management, 27*(4), 793–818.

[7] Darling-Hammond, L. 2000 Teacher quality and student achievement: A review of the state policy evidence. *Education Policy Analysis Archives, 8*(1), 1-44.
[8] Darling-Hammond, L., Berry, B. & Thoreson, A. 2001 Does teacher certification matter? Evaluating the evidence. *Educational Evaluation and Policy Analysis*, *23*(1), 57-77.
[9] Goldhaber, D. D., & Brewer, D. J. 2000 Does teacher certification matter? High school teacher certification status and student achievement. *Educational Evaluation and Policy Analysis, 22* (2), 129–145.
[10] Gore, J.M., Griffiths, T., & Ladwig, J.G. 2004 Towards better teaching: productive pedagogy as a framework for teacher education. *Teaching and Teacher Education,* 20:375- 387.
[11] Harris DN, Sass TR. 2008 Teacher Training, Teacher Quality and Student Achievement." CALDER Working Paper 3. Washington, D.C.: The Urban Institute.
[12] Hoy, W.K. & Miskel, C.G. 2008 *Educational Administration 9th Edition*. New York: McGraw-Hill. ISBN: 978-0-07-131506-7.
[13] Kimani GN, Kara AM, Njagi LW. 2013 Teacher Factors Influencing Students'Academic Achievement in Secondary schools in Nyandarua County, Kenya. Int. J. Educ. Res. 1(3):1-14.
[14] Kose, N.K. 2016 Technological Pedagogical Content Knowledge (TPACK) of English language instructors. *Journal of Educational and Instructional Studies in the World.* May 2016, Volume: 6 Issue: 2 Article: 02 ISSN: 2146-7463. Retrived by http://www.wjeis.org/FileUpload/ds217232/File/02.naran_kayacan_kose.pdf
[15] Liversidge, T., Cochrane, M., Kerfoot, B., & Thomas, J. 2009 Teaching science: developing as a reflective secondary teacher. London: SAGE Publications Ltd.
[16] Maguswi BV 2011 Factors contributing to under achievement of Zambian female Students in O-level Physics examinations. A case of selected high Schools in Central Province. A Masters Thesis, University of Zambia.
[17] Morelock, J.S. 2015 Effective technology implementation in schools: differing perceptions of teachers, administrators, and technology staff". *Dissertations and Theses.* Paper 2626. DOI: 10.15760/etd.2622
[18] Njeru EHN, Orodho JA 2003 Access and Participation in Kenya. Nairobi. Institute of Policy Analysis and Research.
[19] Olaleye FO 2011 Teacher characteristics As Predictor of Academic Performance of Students in Secondary Schools in Osun State – Nigeria. Euro. J. Educ. Stud. 3(3):505 -511.
[20] Qasem, A.A. & Viswanathappa, G. 2016 Blended learning approach to develop the teachers' TPACK. *Contemporary Educational Technology*, 2016, 7(3), 264-276. Retrived by http://www.cedtech.net/articles/73/735.pdf
[21] Ritzhaupt, A.D., Dawson, K., & Cavanaugh, C. 2012 An investigation of factors influencing student use of technology in k-12 classrooms using path analysis. *J. Educational Computing Research*, Vol. 46(3) 229-254. DOI: http://dx.doi.org/10.2190/EC.46.3.b
[22] Schunk, D.H. 2012 *Learning Theories: An Educational Perspective 6th Edition. Teori-Teori Pembelajaran: Perspektif Pendidikan Edisi Keenam*. Yogyakarta: Pustaka Belajar. ISBN: 978-602-229-040-7
[23] Stošić, L. 2015 The importance of educational technology in teaching. *IJCRSEE* Vol. 3, No.1. Retrived by https://www.researchgate.net/publication/278848636_The_importance_of_educational_technology_in_teaching.
[24] Unser, C.E. 2017 A Study on the positives and negatives of using technology in the classroom. *Undergraduate Honors College Theses*. 18. Retrived by http://digitalcommons.liu.edu/post_honors_theses/18
[25] Wellington, J. & Ireson, G. 2012 Science learning, science teaching 3rd edition. New York:

Routledge ISBN 978-0-415-61971-4

[26] Whitehead, B.M., Jensen, D.F.N., & Boschee, F. 2013 Planning for technology second edition. USA: SAGE Company. ISBBN 9781452268262

[27] Wilson, S.M., Floden, R.E., & Ferrini-Mundy, J. 2001 *Teacher Preparation Research: Current Knowledge, Gaps, and Recommendations*. Center for the Study of Teaching and Policy, Michigan State University.

[28] Zimlich, S. L. 2015 Using technology in gifted and talented education classrooms: The teachers' perspective. *Journal of Information Technology Education: Innovations in Practice, 14,* 101-124. Retrieved from http://www.jite.org/documents/Vol14/JITEv14IIPp101-124Zimlich0846.pdf

[29] Zuzovsky, R. 2009 *Teachers' qualifications and their impact on students' achievement findings from TIMMS-2003 data in Israel.* IERI Monograph Series: Issues and Methodologies in large scale Assessment, 2, 37-62.

Journal of Physics: Conference Series **1397** (2019) 012055 doi:10.1088/1742-6596/1397/1/012055

The TPACK Profile of Biology Teacher Based on Certification Status: A Case Study in Bantul Regency

N. Hapsari*, Paidi, B. Subali, F. E. C. Astuti, S. P. Pradana, M. K. Antony
Biology Education, Post-graduate, Universitas Negeri Yogyakarta, Indonesia

*Corresponding author e-mail: novina07@gmail.com

Abstract. This study aims to find out the differences TPACK profile of biology teacher based on certification status. TPACK is a framework that must be possessed by teachers in utilizing technology to deliver the subject matter to students. The sample used in this study was 38 biology teachers in Bantul. The selection of samples in this study used a convenience sampling technique. This study used observation techniques in data collection. The data was then analyzed descriptively and inferentially. The results of the study showed that the PCK and TCK profile is no different between certified biology teachers and un-certified biology teachers, certification status has more influence on the mastery of TPK and TPACK.

1. Introduction

The teacher is one of the important components of education system. Law number 14 of 2005 concerning teachers and lecturers states that teachers are professions that have the main task of educating, teaching, guiding, directing, training, evaluating, and evaluating students[36]. The teacher must have competence in carrying out his professional duties. Competencies possessed by the teachers influence the achievement of learning objectives. The competencies are pedagogical competence, personality competence, social competence, and professional competence, according to the mandated the law number 14 of 2005 concerning teachers and lecturers[36]. Competent teachers must have good communication skills, ability to motivate students, and good material content knowledge[19].

In case of implementing 21st-century learning, competent teachers hold significant roles. The 21st-century learning focuses on the importance of developing students' skill in confronting the challenges of 21st-century life. The skills of 21st-century students are critical thinking, creativity, communication, and collaboration[30]. 21st-century learning allows students to have new knowledge and skills, the ability to connect their knowledge with new information, analyze, and cooperate with others. Media that supports 21st-century learning are information and communication technology (ICT), such as computers, networks, and other digital and non-digital computing technologies, as well as audio, video, and others[7]. Effective use of ICT in learning accommodates students to work collaboratively and solve problems as a form of 21st-century learning[16]. The use of ICT in learning activities also positively correlates with student performance in the class[4]. Various benefits of using technology, especially ICT in improving the quality of students, underlie the importance of ICT utilization by teachers.

As stated in the previous sentences that ICT are seriously beneficial and essential in the scope of classroom learning, Mishra & Koehler[22] try to develop Technological Pedagogical and Content Knowledge (TPACK) framework. This framework is a form of development of Shulman's

Pedagogical Content Knowledge (PCK) design[37]. Integrating technology, pedagogics, and subject matter content in the learning process in the classroom are seen as important[15]. A good learning process has 3 main components of knowledge, namely Pedagogical Knowledge (PK), Content Knowledge (CK), and Technological Knowledge (TK), where the interactions/relationships of these three components are important in learning[23]. The interaction of the 3 main components in TPACK resulted in 4 other knowledge components, namely PCK (Pedagogical Content Knowledge), TCK (Technological Content Knowledge) TPK (Technological Pedagogical Knowledge), and TPACK (Technological Pedagogical and Content Knowledge)[12].

The teacher must have TPACK because it attracts students' learning interest in the subject matter delivered[26]. Knowledge about TPACK will make it easier for teachers to teach science material clearly to students[21]. Biology is one part of science, in biology learning students are required to memorize various scientific names and understand fact material that students cannot easily see. Facts that are not seen directly by students can cause students to have difficulty understanding biology subject matter. The main topics that are difficult for students to learn include the material cycle, endocrine systems and hormones, aerobic respiration, cell division, and genetics[6]. Combining ICT in learning especially biology learning will make it easier for students to understand the material presented by the teacher[34]. In addition, students and teachers will find it easier to get new information related to biology subject matter by utilizing ICT. Therefore, it has become an obligation for teachers to mastery TPACK, characterized by use of technology in biology learning process.

Various factors influence teacher performance, both internal and external factors. One factor that influences the performance of biology teachers is teacher salaries [28] [10]. In order to improve teacher performance and the quality of education, the Indonesian government has provided salary increases for professional teachers who meet the qualifications. This is stated in the law number 14 of 2005 concerning teachers and lecturers that the teacher has the right to receive a professional allowance of one time the basic salary if has an educator certificate. The teacher's certificate can be obtained by the teacher by participating in the certification program[36].

Teacher certification is the process of giving recognition in the form of a certificate of an educator to someone who has the competence to do his job after passing the competency test held by the certification institution [36] [24] [40]. The implementation of teacher certification program is believed to determine the quality of teachers in carrying out their duties as professional teachers, improve learning processes and outcomes, improve teacher welfare and teacher dignity[18]. Teachers who have gone through certification programs are more competent in teaching practices and make it easier for students to understand the material taught[5][38]. Biology teachers who already have teacher certificate are expected to be able to mastery TPACK better than un-certified.

Based on the description that has been stated, needed to analyse TPACK profile of biology teacher. It was also important to find out the difference TPACK profile of certified biology teachers and un-certified biology teachers.

2. Methodology

2.1. Research Design

This research was a descriptive study with a quantitative approach. Data collection was done using a survey method.

2.2. Research Sample

This research was conducted in Bantul Regency, one of the regency in the province of Yogyakarta Special Region. The population in this study were all high school biology teachers in Bantul who were hypothetical. The sample in this study were 38 biology teachers from 23 high schools taken based on convenience sampling techniques.

2.3. Research Instrument

The instrument used in this study was the observation sheet. The observation sheet instrument consisted of 46 items, including PCK components (18 items), TCK (10 items), TPK (11 items) and TPACK (7 items). The empirical validity of the observation sheet instrument is known from the Biserial Correlation value, each statement shows R count> R table (0.602). The reliability of the observation sheet instrument is known from the Intraclass Correlation Coefficient (ICC) value, which is 0.949 (PCK); 0.921 (TCK); 0,968 (TPK); 1,000 (TPACK). The instrument of observation sheet uses Guttman scale completed with 'yes -no' as its alternative answer. Furthermore, 1 is a score given to the 'yes' answer and 0 is a score given to the 'no' answer.

2.4. Data collection

Data collection was conducted from October 2018 to March 2019. This study used observation techniques in obtaining quantitative data regarding TPACK profile in the learning process. The observation is done in one time during the biology learning process.

2.5. Data Analysis

Data of TPACK is analyzed by descriptive and inferential. The descriptive analysis aimed to determine TPACK profile of biology teacher in the learning process. The inferential analysis aimed to determine differences in the TPACK profile of biology teachers based on certification status.

Descriptive analysis was taken by finding the minimum value, maximum value, mean, standard deviation, and criteria. The score for each TPACK component was changed to a value form, provided that:

$$\text{value per component} = \frac{\text{score obtained}}{\text{maximum score for each component}} \text{x } 100$$

The values obtained were then analyzed to find out how the TPACK profile of the biology teacher. The criteria for the TPACK profile were based on the criteria according to Arikunto[2] which are presented in Table 1.

Table 1. Criteria for Profile of Teacher's TPACK

Nilai	Criteria
84 – 100	Very Good
68 – 83	Good
52 – 67	Fair
36 – 51	Poor
≤ 35	Bad

The inferential analysis was using Independent sample t-test. If the probability score was greater than 0.05 ($p <$ Sig. 0.05), there was a significant difference between the certified teacher and un-certified in terms of the mastery TPACK.

3. Result and Discussion

3.1. Result

TPACK profiles of high school biology teachers in Bantul Regency are obtained from the observation of the biology learning process. TPACK profile of biology teacher was emphasized the integration of PK, CK, and TK components. Therefore, the TPACK profile of biology teacher is only viewed from 4 components, i.e PCK, TCK, TPK, and TPACK components. The TPACK profile of high school biology teachers in Bantul Regency can be seen in Table 2.

Table 2. TPACK Profile of Biology Teacher in Bantul Regency

Component	N	Minimum	Maximum	Mean	SD	Criteria
PCK	38	33.33	100	72.66	17.39	Good
TCK	38	0	100	22.89	20.91	Bad
TPK	38	0	100	33.49	27.08	Bad
TPACK	38	0	100	34.21	26.89	Bad
Average	38	-	-	40.81	23.07	Poor

The results showed that the PCK profile of biology teacher were in good criteria, nevertheless the TCK, TPK, and TPACK components were in bad criteria. Based on the average of all TPACK components (including 4 components), biology teachers are poor at mastering TPACK with a value of 40.81.

The status of teacher certification was divided into certified teachers and un-certified teachers. The TPACK profile of high school biology teachers in Bantul Regency with differences in certification status can be seen in Table 3.

Table 3. TPACK Profile of Biology Teacher with Differences in Certification Status

Component	Group	N	Mean	SD	Criteria	t	P (two tailed)
PCK	Certified	32	73.78	18.41	Good	0.918	0.365
	Un-certified	6	66.67	9.29	Fair		
TCK	Certified	32	25.63	20.78	Bad	1.925	0.062
	Un-certified	6	8.33	16.02	Bad		
TPK	Certified	32	37.22	26.47	Poor	2.039	0.049*
	Un-certified	6	13.64	22.82	Bad		
TPACK	Certified	32	38,39	26.27	Poor	2.345	0.025*
	Un-certified	6	11.91	18.99	Bad		
Average	Certified	32	43.76	20.34	Poor	2.157	0.038*
	Un-certified	6	25.14	11.99	Bad		

Judging from the mean value, certified biology teachers were better at mastering the components of PCK, TCK, TPK, and TPACK than un-certified biology teachers. The inferential analysis results showed that there was no difference in mastery PCK and TCK between certified biology teachers and un-certified biology teachers because the probability value was greater than 0.05 ($p > 0.05$). Whereas the TPK and TPACK components obtained a probability value smaller than 0.05 ($p < 0.05$), meaning that the mastery of TPK and TPACK was significantly different between certified biology teachers and un-certified biology teachers. The mastery of TPACK as a whole (including 4 components) showed a significant difference between certified biology teachers and un-certified biology teachers, with $p = 0.038$ ($p < 0.05$).

3.2. Discussion

3.2.1. TPACK Profile

The TPACK profile of biology teacher in Bantul Regency showed that the PCK component of the biology teacher was in good criteria. This indicated that high school biology teachers in Bantul Regency were able to integrate pedagogical knowledge with biology subject matter. The mastery of PCK could be seen from the number of teachers who were able to: (1) adjust material with approaches, models, and learning methods, (2) adjust material with learning media, (3) adjust material to student character, (4) adjust material with assessment, and (5) teacher and student interactions were seen during the learning process. PCK was the teacher's knowledge framework on how to deliver material effectively and efficiently in the learning process[37].

It is known that in the learning process the teacher will face various students with various characteristics, abilities, and needs, therefore the teacher must have good teaching skills. Teaching is not only limited to the transfer of knowledge to students, nevertheless the teacher must also make students understand well the knowledge that shared by the teacher, also instill a variety of values and characters that are good for the future of students. Effective learning processes require content knowledge and in-depth understanding of how to facilitate students to learn the content of knowledge delivered[3]. Effective and efficient learning can be created by mastering PCK.

PCK is the right predictor of what the teacher knows and what the teacher actually does in the classroom[32]. Complex interactions between knowledge of content, teaching, and students in the learning process are depend on the mastery of PCK that the teacher has[8]. PCK that the teacher has affects the learning outcomes and student motivation in the learning process[8] [14]. Based on this, it is very important for biology teachers to develop their PCK in teaching practices therefore the material presented is easily accepted by students and easily organizes classroom situations.

The TCK, TPK, and TPACK profile of biology teachers was not good. This fact showed that most high school biology teachers in Bantul had not been able to integrate the use of ICT in supporting the teaching and learning process. Seen in the learning process observation, many teachers had not been able to: (1) adjust material with ICT, (2) adjust ICT with approaches, models, methods and learning media, and (3) adjust ICT, learning materials, approaches, models, methods and learning media with student characteristics. The fact that the integration of ICT in learning had not been maximally implemented could lead to a lack of information development and knowledge received by students. ICT can facilitate a learning process in obtaining true, accurate, and transparent information therefore Indonesian education has competitiveness with other countries[25].

ICT that is exemplified by the internet and interactive multimedia clearly need to be effectively integrated into the learning process by teachers[35]. Actually the use of ICT in the learning process makes it easy for teachers to convey the content of biology material attractively therefore student's' interest in learning biology increases. ICT is currently very important in the learning and teaching process, students can access various information, images and videos through electronic gadgets, while teachers can upload articles, pictures and videos using ICT-based tools[33]. The success of the integration of ICT in the teaching and learning process depends on the abilities, skills, and creativity of the teacher in utilizing ICT[1].

Ineffective use of ICT can be caused by a lack of teacher knowledge about ICT-based learning media. Based on the results of the interview, it was revealed that some teachers did not understand the technology and its use. Basically, teachers have been able to use the internet to communicate and obtain information nevertheless had not been able to use it in learning. Meenakshi [20] revealed that many teachers were reluctant to use ICT, especially computers and the internet, the reasons being among others, teachers lacked trust in the effectiveness of using ICT in improving learning outcomes, lack of school support, the need for time and effort to learn how to use ICT in learning, and fear of losing authority because it was more student-centered.

The results of observations in the field illustrated that many teachers who use technology by teachers were limited to using power points and videos to support the learning process. Although there were some teachers who utilized internet facilities, it seemed that teachers were less creative in using them. The use of the internet in learning was mostly done by students, the teacher only asked students to use the internet for additional information in completing the tasks given by the teacher.

The poor ability of biology teachers to integrate ICT in teaching practices was very unfortunate. Teachers should be able to meet the demands of the development era with the use of technology in accommodating student learning activities. The importance of the use of ICT in the world of education and the demands of the digital era should alert teachers to develop their TPACK. The poor mastery of TCK, TPK, and TPACK of high school biology teachers in Bantul should be an evaluation material both by teachers, schools and the government.

3.2.2. TPACK Profile Based on Certification Status

Certified biology teachers are better at mastering the components of PCK, TCK, TPK, and TPACK than un-certified biology teachers. In line with this, Kartowagiran[13], Palupi[31], Nuraeni & Retnawati[27] revealed that certified teachers were good at carrying out classroom learning. The inferential analysis results showed that the mastery of PCK and TPK was not significantly different between certified biology teachers with un-certified biology teachers. The mastery of PCK that did not differ significantly in the two groups of teachers could be caused of both groups of teachers were almost the same in choosing learning strategies, teaching materials, and learning media. Discussing learning strategies, approaches and models that were widely used by both groups of teachers was the scientific approach with the model of Discovery Learning and Problem Based Learning (PBL). Regarding the learning method, both certified biology teachers and un-certified biology teachers often used the method of discussion and presentation. Teaching materials and learning media that were often used by both groups of teachers were textbooks, internet sources, and power point media.

Widodo[39] revealed the fact that most certified teachers had the ability to open lessons with sufficient criteria, nevertheless the ability to implement learning strategies, teach materials, and encourage active students belong to bad criteria. Although the mastery of PCK and TPK did not show a difference between certified teachers and un-certified teachers, Helding & Fraser[11] proved that rather than certified teachers, certified teachers had a good attitude and were able to create a learning environment students were better.

The mastery of TCK and TPACK had a probability value smaller than 0.05 therefore it was concluded that the mastery of TCK and TPACK was significantly different between certified biology teachers with un-certified biology teachers. Certified biology teachers had the mastery of TCK and TPACK better with mean value 37.22 and 38.39, while un-certified biology teachers obtained mean value 13.64 and 11.91. Overall (including 4 components), the mastery of TPACK of certified biology teachers differed significantly from un-certified biology teachers. These results indicated that certified biology teachers were able to use ICT well in learning compared to un-certified biology teachers. This result was in line with Farida & Pratiwi[9] that there was a difference in performance between certified teachers and un-certified teachers in the learning process.

The difference in the mastery of TPACK between a certified biology teacher and a un-certified biology teacher can be caused the certified teacher has responsibility and demands for his title as professional teacher. Certified teachers are required to have the ability to use a variety of learning strategies, maximize educational facilities, and innovate in the learning process therefore students are interested in learning[17]. Teacher certification required teachers to have innovation in creating learning methods that were in accordance with student characteristics[9]. Teacher certification could also improve teacher welfare, dignity, discipline and pedagogical competence[29].

4. Conclusion

Based on certification status, certified biology teachers were better at mastering the components of PCK, TCK, TPK, and TPACK than un-certified biology teachers. However the result of inferential analysis show that the mastery of PCK and TCK is no different between certified biology teachers and un-certified biology teachers, certification status has more influence on the mastery of TPK and TPACK.

References

[1] Anshori, S 2017 Pemanfaatan TIK sebagai sumber dan media pembelajaran di sekolah. *civic-culture: Jurnal Ilmu Pendidikan PKN dan Sosial Budaya*, *1*(1), 10-20.

[2] Arikunto, S 2011 *Dasar-dasar evaluasi*. Jakarta: Bumi Aksara.

[3] Bertram, A., & Loughran, J 2012 Science teachers' views on CoRes and PaP-eRs as a framework for articulating and developing pedagogical content knowledge. *Research in Science Education*, *42*(6), 1027-1047.

[4] Biagi, F., & Loi, M 2012 ICT and Learning: Results from PISA 2009. *Luxemburgo: Joint*

Research Centre, Instituto para la Protección y Seguridad del Ciudadano [Comisión Europea].

[5] Bond, L., Smith, T., Baker, W. K., & Hattie, J. A.2000 The certification system o f the National Board fo r Professional Teaching Standards: A construct and consequential validity study. Greensboro, NC: Center for Educational Research and Evaluation, The University of North Carolina at Greensboro, p. 140.

[6] Cimer, A 2012 What makes biology learning difficult and effective: students' views. *Educational Research and Reviews*, *7*(3), 61.

[7] Crane, T., Wilson, J., Maurizio, A., Bealkowski, S., Bruett, K., & Couch, J 2003 Learning for the 21st century: A report and mile guide for 21st century skills.

[8] Damawati, N. A. C. 2015 Pentingnya pemahaman terhadap PCK untuk meningkatkan profesionalisme guru. *Seminar Nasional ALFA IV*, 1-5.

[9] Farida, U & Pratiwi, Y. 2018 Peran sertifikasi terhadap kinerja guru di SMA negeri 1 Ponorogo. In *Seminar Nasional dan Call for Paper III Fakultas Ekonomi* (pp. 504-512).

[10] Hasbay, D., & Altındag, E. 2018 Factors that affect the performance of teachers working in secondary-level education. *Academy of Educational Leadership Journal*, *22*(1).

[11] Helding, K. A., & Fraser, B. J. 2012 Effectiveness of National Board Certified (NBC) teachers in terms of classroom environment, attitudes and achievement among secondary science students. *Learning Environments Research, 16(1), 1–21.*

[12] Herring, M. C., Koehler, M. J., & Mishra, P. (Eds.). 2016 Handbook of technological pedagogical content knowledge (TPACK) for educators. Routledge.

[13] Kartowagiran, B. 2011 Kinerja guru profesional (Guru pasca sertifikasi). *Jurnal Cakrawala Pendidikan*, *3*(3).

[14] Keller, M. M., Neumann, K., & Fischer, H. E. 2017 The impact of physics teachers' pedagogical content knowledge and motivation on students' achievement and interest. *Journal of Research in Science Teaching*, *54*(5), 586-614.

[15] Koehler, M., & Mishra, P. 2009 What is technological pedagogical content knowledge (TPACK)?. *Contemporary issues in technology and teacher education*, *9*(1), 60-70.

[16] Koh, J. H. L., Chai, C. S., & Lim, W. Y. 2016 *Teacher professional development for TPACK-21CL. Journal of Educational Computing Research,* 55(2), 172–196.

[17] Kurniawan, B. D. 2011 Implementasi kebijakan sertifikasi guru dalam rangka meningkatkan profesionalitas guru di Kota Yogyakarta. *Jurnal Studi Pemerintahan*, *2*(2).

[18] Ma`arif, Syamsul. 2011 *Guru profesional: harapan dan kenyataan.* Semarang: Walisongo Press.

[19] Malik, A. N., & Bashir, S. 2015 Good teacher. *The Professional Medical Journal*, *22*(05), 670-673.

[20] Meenakshi, K. 2013 Importance of ICT in education. *Journal of Research and Method in Education (IOSR-JRME)*, *1*(4), 03-08.

[21] Mercado, N. L., Panganiban, J. M., & Ramos, M. I. 2019 Technology integration in teaching science using TPACK among pre-service science teachers of St. Bridget College, Batangas City, Philippines. *International Multidisciplinary Research Journal, 1* (1), pp. 67-75.

[22] Mishra, P., & Koehler, M. J. 2006 Technological pedagogical content knowledge: A framework for teacher knowledge. *Teachers college record*, *108*(6), 1017-1054.

[23] Mishra, P., & Koehler, M. J. 2008 Introducing technological pedagogical content knowledge. In *annual meeting of the American Educational Research Association* (pp. 1-16).

[24] Mulyasa 2009 *Standar kompetensi dan sertifikasi Guru*. Bandung: PT. Remaja Rosda Karya.

[25] Munir 2009 Kontribusi teknologi informasi dan komunikasi (TIK) dalam pendidikan di era globalisasi pendidikan Indonesia. *Jurnal Pendidikan Teknologi Informasi dan Komunikasi (PTIK)*, *2*(2), 1-4.

[26] Naziri, F., Rasul, M. S., & Affandi, H. M 2019 Importance of technological pedagogical and content knowledge (TPACK) in design and technology Subject. *International Journal Of*

Academic Research In Business And Social Sciences, *9*(1).

[27] Nuraeni, Z., & Retnawati, H 2016 The post-certification performance of mathematics teachers. *The Online Journal of New Horizons in Education*, *6*(2), 130-142. Retrived from

[28] Nurhayati, B 2006 Faktor-Faktor yang mempengaruhi profesionalisme dan kinerja guru Biologi di SMAN Kota Makassar Sulawesi Selatan. *Mimbar Pendidikan*, *4*(25), 64-70.

[29] Nusantara, S. B. U. V. B 2012 Kinerja guru yang telah lulus sertifikasi guru dalam jabatan. *Jurnal Manajemen Pendidikan (JMP)*, *1*(2).

[30] P21 2019a Framework for 21st century learning definitions. *Battelle for Kids*.

[31] Palupi, Retno 2011 *Kinerja Guru Rumpun IPA SMA (Biologi, Kimia, dan Fisika) se- Propinsi Daerah Istimewa Yogyakarta Pascasertifikasi.* S2 thesis, UNY.

[32] Park, S., Jang, J. Y., Chen, Y. C., & Jung, J 2010 Is pedagogical content knowledge (PCK) necessary for reformed science teaching?: Evidence from an empirical study. *Research in Science Education*, *41*(2), 245-260.

[33] Pathak, Kaushlendra & Manoj, Nawal Kishor 2018 ICT in educational institution: need, role and importance. *IOSR Journal of Humanities and Social Science (IOSR-JHSS) Volume 23, Issue 1, Ver. 8 (January. 2018)* PP 42-46.

[34] Raja, R & C. Nagasubramani, P 2018 Impact of modern technology in education. *Journal of Applied and Advanced Research*, 3(1), 33–35.

[35] Ratheeswari, K 2018 Information communication technology in education. *Journal of Applied and Advanced Research*, 3 (Suppl. 1) S45–S47.

[36] Republik Indonesia 2005 *Undang-Undang Nomor 14 Tahun 2005, pasal 10 ayat 1 tentang Guru dan Dosen.*

[37] Shulman, L 1986 Those who understand: knowledge growth in teaching. *Educational Researcher, 15*(2), 4-14.

[38] Suyanto, S. & Wibowo,Y 2018 Curriculum review of teacher professional development program based on biology teacher profile in technological pedagogical and content knowledge. In *Journal of Physics: Conference Series* (Vol. 1097, No. 1, p. 012042). IOP Publishing. https://iopscience.iop.org/article/10.1088/1742-6596/1097/1/012042/pdf

[39] Widodo, S. 2016 The evaluation of teacher certified performance in Tulang Bawang Lampung. In *lst International Cohference on Geography and Education (ICGE 2016)*. Atlantis Press.

[40] Yamin, Martinis 2006 *Sertifikasi profesi keguruan di Indonesia*. Jakarta: Gaung Persada Press.

The TPACK Analysis of High School Biology Teachers in Sragen Regency, Central Java Based on Teacher Certification Status

S. P. Pradana*, Paidi, B. Subali, M. K. Antony, N. Hapsari, F. E. C. Astuti

Biology Education, Graduate School, Yogyakarta State University, Yogyakarta, Indonesia

*Corresponding author e-mail: sendypradana368@yahoo.com

Abstract. Technological Pedagogical and Content Knowledge (TPACK) is the combination of knowledge about PK, CK and TK. The purposes of this study were to find out how the TPACK of high school biology teachers in Sragen Regency and to find out whether there were differences in the TPACK between certified biology teachers and those who were not certified. The type of this research was descriptive research with survey methods. The population in this study was a hypothetical population (the forecast of the population, what the population is expected to be) and the sample representing the population was taken using convenience sampling techniques which included 17 biology teachers spread across 13 high schools in Sragen Regency. Based on the research that has been carried out, the results showed that the knowledge of PCK is very good (85.0%), TCK is poor (47.6%), TPK is fair (64.2%), and TPACK teachers were in a fairly poor category (51.3%). The results also showed that there was no significant difference between the TPACK of certified biology teachers and non-certified biology teachers with a significance value greater than 0.05.

1. Introduction

The 2005 Constitution No. 14 concerning teachers and lecturers stated that teachers are professional educators with the main task aucating, teaching, guiding, directing, training, assessing and evaluating students, in early childhood education, formal education, basic education, and secondary education. The government has also formulated four types of teacher competencies in the perspective of national education policy as stated in the Explanation of Government Regulation No. 19 of 2005 concerning National Education Standards, namely pedagogical competencies, personality competencies, social competencies, and professional competencies. Competence is basically a description of what a person can do at work, and what forms of work can be shown. The teacher is required to be an agent of change (the agent of change), so that the teacher must be able to develop the teaching process in the class and continue to explore the ability to make a design or lesson plan, one of them is by understanding Pedagogical Content Knowledge (PCK). To teach biology is not enough to understand the content or teaching material (knowing science), but also have to think about how to teach well (how to teach)[1]. The main element that can be used as the basis of a teacher in carrying out the learning process is to master CK and PK, where the two components will become a unit called PCK[2].

Technological developments in the 21st century make teachers always have to develop their teaching abilities by not only mastering PCK, but teachers must be able to teach learning materials by

Journal of Physics: Conference Series **1397** (2019) 012056 doi:10.1088/1742-6596/1397/1/012056

utilizing technology (ICT), which components will be integrated with each other and related to each other in Technological Pedagogical Content Knowledge (TPACK). Integrating technology with pedagogical knowledge and learning content has become important from educational programs[3]. Integration of technological elements in the learning process in the 21st century is indeed necessary because learning activities require the application of technology both for teachers and students because the use of technology is expected to produce innovative learning processes.

The TPACK concept was taken from the description of PCK by Shulman, which was integrated to find out how the interaction between technology and PCK in producing an effective learning process. TPACK became a design that integrates technology in the learning process. TPACK is an intact combination of three basic knowledge, there are PK (Pedagogical Knowledge), CK (Content Knowledge), and TK (Technological Knowledge) that form the balance of PCK (Pedagogical Content Knowledge), TCK (Technological Content Knowledge), TPK (Technological Pedagogical Knowledge), and TPACK (Technological Pedagogical and Content Knowledge)[4]. Mastery of TPACK is considered very important in answering the challenges of education in the 21st century because technology is basically helping teachers to continue to hone and develop their professionalism. For example, by using the internet, teachers can access various kinds of information about educational issues, teaching strategies, as well as being able to keep up with the progress of science that continues. TPACK can also help teachers build good competencies in schools because TPACK explains how to think new for teachers in dealing with problems in the class[5].

TPACK is indeed supposed to be understood by the teacher because in the standards of the Indonesian National Qualifications Framework, the teacher is at level 7. In this level, the teacher is required to be able to manage resources under their responsibility and evaluate comprehensively their work by utilizing science, technology, and art to produce strategic development steps for the organization. In improving the ability to use technology in the learning process, teachers must take part in teacher education programs. Although the teacher education program aimed to equip teachers in utilizing technology in learning, in reality, teacher training programs were still not able to improve the mastery of technology in learning[6].

Teacher's TPACK is influenced by many factors, both the factors within the teacher and external factors such as the carrying capacity of the teaching school to develop the teacher's TPACK. The factors that are thought to influence teacher's TPACK include the status of teacher certification. Teacher certification is the provision of real recognition for a teacher who already has the competence to carry out educational services to certain educational units after passing the competency test held by the certification body[7]. In the 2005 Constitution of the Republic of Indonesia, Number 14, Article 8 explains that certification teachers must have academic qualifications, competence, physical and spiritual health, and have the ability to realize national education goals.

Teacher certification programs can improve the quality of education, but the size of the effect of this increase differs for each school category [8]. In Indonesia, teacher certification programs began in 2006. Furthermore, teachers can be assumed to be responsible for the success or failure of student performance, therefore the Indonesian government has sought to improve the qualifications and skills of Indonesian teachers through teacher certification programs [9]. Teacher certification assumes that the work done by the teacher is professional work therefore that the teacher needs to be certified professionally according to the standards of professional work [10]. Based on the statement, it can be concluded that the teacher with the title of certification must possess and master a number of competencies in terms of learning, one of which is to be able to understand TPACK.

Understanding of technology, material (content) and learning strategies is an important thing that must be mastered by teachers to become professional teachers. However, not only being a professional teacher, but the teacher must also be able to provide effective learning, with special abilities to integrate knowledge about technology, materials, learning processes, and students, including the biology teachers in State High School in Sragen Regency. The information on the government administration report at the end of the 2016 Fiscal Year of Sragen Regency explained that the percentage of Transition Numbers (AT) of students from junior high to senior high school has increased from 87.4% to 94.22% (an increase of 6.74%). The increase in numbers was inseparable from the increasingly advanced human resources in Sragen Regency. This must be addressed by high school teachers to constantly develop the ability to teach in the classroom and sharpen their abilities in making learning designs, one of them is by understanding Technological Pedagogical Content Knowledge (TPACK). Based on the description related to the TPACK that teachers need to have in learning, then a study was conducted on the TPACK possessed by the teacher with the title "The TPACK Analysis of high school biology teachers in Sragen Regency, Central Java based on teacher certification status".

2. Methods

2.1 Research Design

This type of research was a descriptive study with a survey method. This study aimed to find out how teacher's TPACK in the learning process and to know the difference in teacher's TPACK between certified biology teachers and non-certified biology teachers. The population in this study was a hypothetical population (the forecast of the population, what the population is expected to be), which included all biology teachers who taught at high schools in Sragen that once existed, which now exist, or which would exist. The sample that representing the population was taken using convenience sampling techniques or samples that are now available and easy to observe, in this case, the sample included 17 biology teachers spread from 13 high schools in Sragen Regency.

2.2 Research Instrument

The data in this study were the TPACK of the high school biology teachers. Data obtained through observation of learning activities carried out by the teacher by using instruments in the form of observation sheets. The data obtained then be collected and included in the observation sheet table which was then analyzed from the data that had been obtained. This method was chosen because it was more practical and simple. The instruments used were first tested for validity and reliability. After being declared valid and reliable, the instrument can be used to retrieve the TPACK of biology teachers.

2.3 Data Analysis

The data in this study were the TPACK of the high school biology teachers in Sragen Regency as seen from the learning process. The data was tabulated in the form of tables using instruments that had been made. The knowledge of TPACK was analyzed using descriptive analysis, while the influence of the variables on the teacher's TPACK was analyzed using the independent sample t-test. To facilitate the grouping of teacher's TPACK, several criteria were made. The criteria are: 84 % - 100 % (Very Good); 68 % - 83 % (Good); 52 % - 67 % (Fair); 36 % - 51 % (Poor); 0 % - 35 % (Bad)[11].

3. Results and Discussion

3.1. Result

3.1.1 Descriptive Analysis of teacher's TPACK

Descriptive analysis of TPACK can be seen in the following table:

Table 1. Descriptive Analysis of TPACK

Aspect	Statistic	Score	Note
	Mean	85,0	
PCK	Maximum	94,4	Very Good
	Minimum	44,4	
	Mean	47,6	
TCK	Maximum	60,0	Poor
	Minimum	40,0	
	Mean	64,2	
TPK	Maximum	72,7	Fair
	Minimum	36,4	
	Mean	51,3	
TPACK	Maximum	57,1	Fair
	Minimum	42,9	

Table 1 explains how TPACK of high school biology teachers in Sragen Regency. The teacher's TPACK is divided into four main components, namely PCK, TCK, TPK, and TPACK. Each component has a different level of knowledge. The results listed in table 1 are results that are still comprehensive and not specific. It means that these results still cannot distinguish how the descriptions of teacher's TPACK with different certification statuses. These results are still in the form of overall results from the sample used. In this case, the status of teacher certification is divided in two groups, namely certified teachers and non-certified teachers. To see the description of the biology teacher's TPACK based on certification status can be seen in table 2.

3.1.2 Description of the tendency of differences in the knowledge of TPACK

The tendency of differences TPACK of Biology Teacher can be seen in the following Table:

Table 2. The tendency of differences in Teacher's TPACK

Aspect	Teacher Certification Status	
	Certified	Not Certified
PCK	88,4	76,6
TCK	48,3	46,0
TPK	65,1	61,8
TPACK	51,1	51,4
Mean	**63,23**	**58,95**

Based on table 2, it can be seen that TPACK knowledge of certified teachers is better when compared to teachers who are not certified. But, actually these results haven't been able to reveal significantly how different the knowledge of TPACK biology teachers. These results are still in the form of descriptive results that lead to the tendency of biology teachers to understand TPACK based on certification status.

To find out how the significance level of differences in knowledge of TPACK biology teachers, different tests were carried out using Independent Sample t-Test with results that can be seen in table 3. The Independent Sample t-Test compares the means of two independent groups in order to determine whether there is statistical evidence that the associated population means are significantly different. The Independent Samples t-Test is a parametric test. With this analysis, the significance of differences in biology teacher tpack based on certification status will be known.

3.1.3 t-Test

The data was analyzed using the Independent Sample t-Test with the following results in table 3.

Table 3. t-Test

Aspect	Certification	Sig
PCK	Certified Not Certified	0,224
TCK	Certified Not Certified	0,453
TPK	Certified Not Certified	0,599
TPACK	Certified Not Certified	0,951

Table 3 explains that there is no difference in teacher's PCK, TCK, TPK, and TPACK between certified teachers and non-certified teachers. This can be seen in the significance value or in the table stated in Asymp. Sig. (2-tailed) which has a value of more than 0.05, it can be said that there is no significant difference in teacher's PCK, TCK, TPK, and TPACK between certified biology teachers and non-certified biology teachers.

3.2. Discussion

Based on the results obtained, it could be said that the biology teacher's PCK was very good (85%), TCK was poor (47.6%), TPK was fair (64.2%), and TPACK was in the fair category (51.3%). The knowledge of PCK in the learning process was in a very good category with a percentage of 85.0%. In this case, the teacher had been able to choose the right model, method, and learning media in accordance with the teaching material to be delivered to the students. The selection of learning models, methods and media were appropriate because when the learning process took place, students could follow the whole set of learning well. Active students asked questions and were also active in expressing opinions, and when the teacher gave questions, students were actively brave enough to answer these questions. When students expressed their opinions and also actively asked questions, the teacher swiftly made a small note to give an assessment to the student, both in the form of cognitive, affective, and psychomotor assessment.

The knowledge of TCK in the learning process fell into the poor category with a percentage of 47.6%. TCK is a unity of the two main elements that are important in learning, namely TK and CK. When observing, teachers were still limited in utilizing technology to be adapted to teaching material. In this case, the teachers had not been able to provide new information related to the material to the students, the teachers only used technology facilities (internet) to collect material to be delivered in the form of power points, where the material collected by the teacher was actually already in the student handbook. The importance of utilizing technology facilities to support the achievement of learning objectives was delivered by [12], the maximum use of technology will be able to make abstract material easier for students to learn. In this era of globalization, the use of technological resources as learning materials for students will make students have a high collaborative spirit, be flexible in exchanging information and knowledge, and constantly hone their abilities by continuing to learn.

TPK knowledge in the learning process was included in a fair category with a percentage of 64.2%. Knowledge about TPK refers to the suitability of technological elements for different learning activities [13]. In this case, the teacher's TPK can be seen from how the teachers adjusted the use of technology (TK) with models, methods, and also learning media and the use of technology that was tailored to the characteristics of students (PK). From the observation results obtained by the fact that the teachers had not been fully able to use ICT-based learning media for the learning process, the teachers were limited to the use of power points that contain many texts which were felt not enough to make students understand the material being taught.

The knowledge of TPACK in the learning process was included in a fair / sufficient category with a percentage of 51.3%. The application of ICT (Information and Communication Technologies) in learning was still not maximal. Based on the results of interviews from several teachers, information was obtained that most of the teachers claimed that they did not really understand ICT because they did not use technology facilities too often in their daily lives. The use of technology facilities was limited to the use of the internet to gather information and use social media in the form of WhatsApp to communicate. The lack of maximum teachers in knowing TPACK in the learning process was also triggered by limited infrastructure facilities from the school. Some schools had wifi (hotspot area) facilities, but the area was only at one location point and the signal from the wifi did not reach the classroom. The problems in teacher performance were influenced by several factors, one of which was the low use of ICT due to the lack of teachers' ability to use it, and there was no obligation on the part of the school to use ICT during the learning process[14]. TPACK is a central element in achieving learning objectives because if TPACK is implemented correctly, students will get more meaningful learning. TPACK can be applied by collaborating three main elements, namely TK, PK, and CK. The importance of implementing TPACK is expressed by [15] which states that the implementation of TPACK strategies can improve mastery of concepts and learners' active learning. From the results of the study, it could be seen that the teacher's knowledge about TPACK had not fully fulfilled the main elements. But in this case, the teacher had been able to compile the material using modern learning media such as powerpoint, some of which already contained images, videos, and animations taken from the internet. Technology must be utilized properly to package a teaching material or certain material into a good media such as animation, simulation, and virtual lab so that students will be easier understand the material[16].

From the results of the study also found that there were no significant differences regarding the knowledge of PCK, TCK, TPK, and TPACK in the learning process between the certified and non-certified teachers, with the results of the analysis in the form of significance values above 0.05. There was no significant difference because overall, the way to teach teachers in the classroom was not much different. The average teacher used discovery learning and problem-based learning models, and the methods used in teaching were lecture and discussion methods, the use of technological elements was limited to finding information related to learning material. Learning media was still limited to the use of laptops and the presentation of material in the form of power points which were dominated by the amount of text. Certified teachers had the understanding that the teacher had the competence to carry out educational services to certain educational units after passing the competency test held by the certification body[8]. But based on the results obtained, it was said that there was no significant difference between certified and non-certified teachers in knowing TPACK. To be able to master TPACK, of course, the teacher first mastered the PCK, TCK, and TPK. The absence of differences in the knowledge of TPACK is also due to the ability of teachers to master content, pedagogics, and technology is still not optimal. Even though, the NBC process (National Board Certification) or in this case the certification process affected five aspects of teaching practices related to PCK, including reflection on teaching practices, implementation of new teaching strategies and / or innovative, inquiry-oriented instruction, student learning assessment, and student understanding[17].

The absence of a significant difference in the TPACK between certified and non-certified teachers is clearly a problem. Teachers who get a certification degree should have good competence in the world of teaching. Whatever form of program to improve teacher quality and education, follow-up evaluations that maintain the sustainability of the program are very necessary. Status of certification for a teacher should not be permanent as in Indonesia, certification must have expired. When teacher certification has reached its final year, teachers need to be retested, and those who fail the certification test will not get the certification title. By applying this evaluation, certified teacher quality can be maintained[18]. Certified teachers are closely related to professional teachers, where teacher professionalism can be seen from a number of competencies they have. Teachers who have qualified competencies are proven by obtaining a teacher certification degree[19]. Based on this explanation, as a teacher with a certification title, it should have a number of competencies in an adequate teaching field, one of which is being able to understand TPACK. Because, whether a teacher is good or not in providing learning to students will determine the achievement of learning goals.

Teachers must continue to learn to be able to design and develop technology to foster the success of students in this advanced and modern learning environment because applying technology will produce meaningful learning[20](22). TPACK's poor knowledge illustrates that teachers still tend not to be able to properly implement technological, pedagogical and content elements. The teacher's lack of knowledge about TPACK is very unfortunate because TPACK is an important thing in addressing technological advances in the 21st century. Good knowledge about TPACK makes it easy for teachers to find solutions to deal with existing problems in class, TPACK can also be used to build teacher competency in schools[5].

Based on the results of the research obtained, it can be said that indeed the teacher certification status did not have a significant impact on the teacher's TPACK. As a teacher, both certified and non-certified teachers should have the skills to master everything related to their field of teaching. The teacher's profession is obtained by various conditions, one of which is the prospective teacher must take the level of teacher education which means that after graduating from teacher training, a teacher already has qualified capital to be applied in the world of teaching. Certification candidates must have completed a bachelor's degree, completed at least three full years of teaching or counseling before earning a degree, and have valid teaching or counseling licenses during that period[21].

4. Conclusion

Based on the research that had been carried out, the results showed that the teacher's PCK was very good (85.0%), TCK was poor (47.6%), TPK was fair (64.2%), and TPACK teachers were in a fairly fair category (51.3%). The results also showed that there was no significant difference in TPACK between certified biology teachers and non-certified biology teachers with a significance value above 0.05.

References

[1] Susilowati., Widhy, P 2015 Analisis *Pedagogical Content Knowledge* Guru IPA SMP Kelas VIII dalam Implementasi Kurikulum 2013 *Jurnal Pendidikan Matematika dan Sains Tahun III,* 1(1) 72-78

[2] Etkina, E 2010 Pedagogical Content Knowledge and Preparation of High School Physics Teachers *Physical Review Special Topics-Physiscs Education Research*

[3] Lestari, S 2015 Analisis Kemampuan Technological Pedagogical Content Knowledge (TPACK) pada Guru Biologi SMA dalam Materi Sistem Saraf *Seminar Nasional XII Pendidikan Biologi FKIP UNS*

[4] Koehler, M. J., and Mishra, P 2005 What happens when teachers design edu-cational technology? The development of technological pedagogical content knowledge *Journal of Education Computing Research* 32(2) 131 – 152

[5] Chai, C. S., Koh, J. H., and Tsai, C. C 2013 A Review of Technological Pedagogical Content knowledge *Educational Technological & Society* 31-51

[6] Hosseini, Z., and Kamal, A 2013 A Survey on Pre-service and In-service Teachers' Perceptions of Technological Pedagogical Content Knowledge (TPACK) *Educational Technology* 1-7.

[7] Mulyasa, E 2007 *Standar Kompetensi dan Sertifikasi Guru (*Bandung : Remaja Rosda Karya)

[8] Wahyudi, K. E., Supranoto., and Suji 2012 Measuring performance of teacher certification program *International Journal of Administrative Science & Organization* 19(3) 153 – 161

[9] Utami, A. L. P 2015 Teacher certification program in indonesia: problems and recommendation for the betterment of the program *International Journal of English and Education* 4(2) 471 – 481.

[10] Hakim, L., Tunas, B., and Rubini, B 2018 Evaluation of the professionalism of teachers accomplishing certification program: evidence from bekasi city, Indonesia *International Journal of Multidisciplinary Education and Research* 3(4) 32 – 35.
[11] Arikunto, S 2011 *Dasar-dasar Evaluasi Pendidikan* (Jakarta: Bumi Aksara)
[12] Mairisiska, T., Sutrisno, and Asrial 2014 Pengembangan perangkat pembelajaran berbasis tpack pada materi sifat koligatif larutan untuk meningkatkan ketrampilan berfikir kritis siswa *Edu-Sains* 3(1) 28 - 37
[13] Puspitarini, E. W., Sunaryo, S., dan Suryani, E 2013 Pemodelan technlogical pedagogical content knowledge (tpack) berbasis teknologi informasi dan komunikasi (tik) dengan pendekatan structural equation modeling (sem) *Prosiding Seminar Nasional Manajemen Teknologi XVIII* (pp. 1-8)
[14] Fuad, N 2017 Pengaruh Sertifikasi Guru Terhadap Peningkatan Kinerja Guru PAI di SMP dan MTs *Jurnal Manajemen Pendidikan* 8(1) 23 – 32.
[15] Putriani, E. D., dan Sarwi 2014 Implementasi Strategi TPCK dengan Media Simulasi Berbasis Inkuiri Terbimbing Pada Konsep Getaran dan Gelombang *Unnes Physics Education Journal* 3(2) 34 – 41.
[16] Hayati, D. K., Sutrisno, dan Lukman, A 2014 Pengembangan Kerangka Kerja TPACK pada Materi Koloid untuk Meningkatkan Aktivitas Pembelajaran dalam Mencapai HOTS Siswa *Edu-Sains* 3: 53-61.
[17] Park, S., and Oliver, J. S 2009 National Board Certification (NBC) as a catalyst for teacher's learning about teaching: The effects of the NBC process on candidate *Journal of Research in Science Teaching* 45(7) 812 – 834
[18] Utami, A. L. P. 2015 Teacher certification program in indonesia: problems and recommendation for the betterment of the program. *International Journal of English and Education* 4(2) 471 – 481
[19] Dewi, R. S., Kurniatun, T. C., dan Abubakar 2018 Kemampuan Profesional Guru dan Motivasi Kerja Terhadap Kinerja Mengajar Guru Sekolah Dasar *Jurnal Administrasi Pendidikan* 25(1) 150 – 158
[20] Keengwe, J., Onchwari, G., and Onchwari, J 2009 Technology and student learning: toward a learnercentered teaching model *Associating for the Advancement of Computing in Education* 17(1) 11-22
[21] Elliott, S. W., Koenig, J. A., and Hakel, M. D 2008 *Assessing accomplished teaching: Advanced-level certification programs* (Wahington DC: National Academies Press 500 Fifth Street NW)
[22[Suyanto, S, and Wibowo, Y 2017 Curriculum Review of Teacher Professional Development Program Based on Biology Teacher Profile in Technological Pedagogical and Content Knowledge *Journal of Physic Conference Series* 1097 012042

Combination of CPS and SAVI on Biology Learning Media to Improve Student's Understanding and Creativity on Environmental Change Topic

S. G. N. Sagita[1*] and T. Aminatun[2]

[1]Biology Education, Graduate Program, Yogyakarta State University, Colombo St. No.1, Karangmalang, Depok, Sleman, Yogyakarta, Indonesia
[2]Biology, Faculty of Mathematics and Natural Sciences, Yogyakarta State University, Colombo St. No.1, Karangmalang, Depok, Sleman, Yogyakarta, Indonesia

*Coressponding author e-mail: shintyagaluh94@gmail.com

Abstract. The purpose of this research is to determine the feasibility and effectiveness of biology learning media with a combination of CPS and SAVI to improve students' understanding and creativity. This research was conducted on students of grade X MIPA program in SMA Negeri 7 Purworejo, the subject was selected using Purposive Sampling method. The research method used is Research and Development (R & D) by Dick and Carey's ADDIE (Analyze, Design, Development, Implementation, and Evaluation) model research. The instruments used were expert judgment feasibility questionnaires, multiple choice questions and student worksheets. Data of the feasibility was analyzed by using percentage. The results indicated that Biology Learning Media with a combination of CPS and SAVI is effective to use shown on (1) expert's validity percentage which above 75 means that learning media is feasible to use, (2) user response on usability showed that the percentage is 89,02 means that the learning media is feasible to use, (3) effectiveness of media showed that the N-Gain result is 0,52 which included on middle category of improving students' ability, and the results of the student worksheet assessment is 84,16 which categorized high on students'.

1. Introduction

The world of education this era has developed very rapidly, the use of science and technology is applied to the world of education as the times developed. Various types of media, models, and learning approaches have been used to improve students' abilities, both affective, cognitive, and psychomotor. The selection of learning strategies is not easy, it must be adapted to the students needs. As time goes on globalization occurs, too many problems arises and students are required to have the ability to understand and be creative in their lives to solved problems and to survive, whether it is related to environmental issue or many other issues. This ability is included in the cognitive domain which can be improved through a learning system. From PISA (Programme for International Students Assessment) rank that published by OECD (Organisation for Economic Co-operation and Development) in 2015, shows that Indonesia is placed on 62 from 70 countries. This subjects of assessment is foccused on science, reading, and mathematics, it's surveyed on 15 years old students around the world. Most importantly, Indonesia is not ranked on Collaborative Problem Solving assessment aspect (OECD, 2016). To improve this ability, using an appropriate learning media with learning model and learning approach is highly needed. Educators must find the best solution to solves the problems of students based on student's characteristics and other things to adjust the appropriate learning strategy.

Form of interactions between students, teachers, learning tools, and learning environments is called learning system. The success of the learning process is influenced by the methods and learning media used. The process of learning consists of five components: goals, material, methods, media and evaluation (Wina Sanjaya, 2006). Learning objectives are needed to find out how far students have learned what has been taught, this goal can be seen from the results of cognitive and knowledge tests, there are four dimensions of knowledge, it is factual knowledge; conceptual knowledge; procedural knowledge; and metacognitive knowledge. And there are six dimentions of cognitive, it is remember; understand; apply; analyze; evaluate; create (Anderson, 2001)(Krathwohl. 2002).

The basic cognitive ability is understand, one of cognitive dimentions as mentioned on Bloom's Taxonomy (krathwohl, 2002). Understand means that someone already knows and absorbs the meaning of a thing or material that has been learned, on Cambridge dictionary, understand has meaning "know" or "realize". Someone can be said understand if they know or realize about how and why something had happened. Understand has been categorized into three: 1) the lowest level is understanding translation, starting from translating into the true meaning, interpreting and applying principles; 2) the second level is the interpretation understanding, which can link the relationship between or from the lowest level to the next level, or connect parts of the graph with a real event, distinguish between the principal and the non-principal; 3) the third level is to interpret extrapolation (Sudjana, 2013).

After know or realize about how and why something had happened, the next level is how to solve problems using creativity. Creativity comes from the word "to create" which means "to make". It is the ability to make things, whether in the form of ideas, steps, or products (Rahmat, 2014). Someone who is creative is someone who can think synthetically, who can see relationships where others cannot see, and can analyze his own ideas and evaluate the quality of his personal work, able to translate theories and also abstract things into practical ideas to be able to convince others of the ideas that are going to implement (Sternberg, 2003). Someone categorized creative if they have ability to developing their original ideas that useful, different, new and unsual (Simonton, 2012). Creativity is one part of meta knowledge. Meta Knowledge it self is knowledge about cognition or thoughts in general, especially self-cognition (Anderson, 2001). Metacognition, self-efficacy and self-regulation are the three main elements that can help student's learning activities (Cera, 2013). Meta cognition has a relationship with Higher Order Thinking (HOT), which deals with cognitive processes in learning activities (Aguayo, 2010).

Difference with in general, creativity in science is called scientific creativity, because it is concerned with creative science experiment (Simonton, 2012). Scientific creativity is an ability to study science and to solve scientific problems (Wang & Yu, 2011). To study more about scientific creativity, Hu & Adey (2002) proposed Scientific Structure Creativity Model (SSCM). SSCM consist three dimensions, the process dimension, the trait dimension, and the product dimensions. The process dimension has two sub-dimension, that is thinking and imagination. The trait dimension has three sub-dimension, that is fluency; flexibility; and originality. The last is product dimension, has four sub-dimension that is technical product; science knowledge; science phenomena, science problem (Hu & Adey, 2002). The three-dimensional Scientific Structure Creativity Model (SSCM) has 24 cells contains components of scientific creativity, shown in Figure 1.

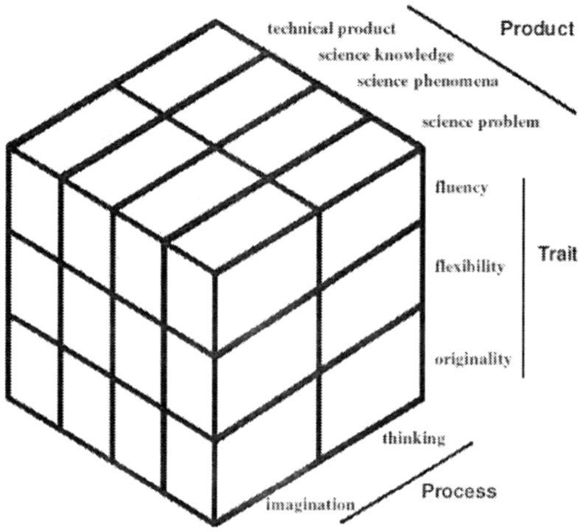

Figure 1. The Scientific Structure Creativity Model (SSCM) (Hu & Adey, 2002)

Biology is branch of science that study about living thing. Biology is study about the origins; histories; structures; functions and interaction between living things (Alton Bigs, 2004). Included in domain of biology is Ecosystem, in Ecosystem there is sub-domain namely Environmental Change, it is study about how Environment in Ecosystem has changed, what the consequence if it changes, what could we do to prevent the worst consequence, how to conserve the environment, and raise awarness about environmental change (Baharudin, 2009).

One of the method that can be use to improve understanding and crativity on environmental change is using Creative Problem Solving (CPS) learning model combined with Somatic Auditory Visual Intelectual (SAVI). CPS is learning model based on problem solving with systematic technique to provide creativity when solving the problems (Isaksen, 1995). Because someone creative is who could become good problem solver when solving the problem use different way more than using his knowledge only. Treffinger (2003) describe that "CPS is a model to help you solve problems and manage change creatively. It gives you a set of easy to use tools to help translate your goals and dreams into reality", it means that CPS is a method that solve problem creatively and developing new result that useful to solve other problems. There is four steps of CPS learning model described by Treffinger 1). Understanding the challenge; 2). Generating ideas; 3). Preparing for actions; 4). Planning your approach (Treffinger, 2003).

Not only using learning model, but also using learning approach is higly needed. SAVI (Somatic Auditory Visual Intelectual) is a theory by Dave Meier from Accelerated Learning concept on his book "The Accelerated Learning Handbook". SAVI learning approach is part of the accelerated learning theory, left/right brain theory; visual, auditorial, and kinesthetic; double intelectual theory; a thorough education; learning by experience; and learning by symbols. Four aspecs in SAVI is: 1) Learning Somatic, is learning by moving and doing and learning process is involving the senses of touch, kinetic, physical, and move all the body; 2) Learning Auditory, is learning by talking and hearing; 3) Learning Visual, is learning by observing and picturing; 4)Learning intelectual, is learning by Problem Solving and reflecting. Learning process is optimal if four aspects in SAVI is used all together (Meier, 2000).

CPS and SAVI is combined in Learning Media based on Adobe Flash application on computer. If learning media designed creatively, it could make students learn much, increase students performance, and more understand about learning object (Nuryani, 2005). Using learning media based on Adobe Flash is caused by development of Information Communication and Technology (ICT), on 21st century, its development is greatly influecing the world of education, and the implementation on curriculum should encourage learning activites by using the technology (Linder, 2010). By combining

CPS and SAVI on learning media based on Adobeflash, the student has expected to improve their understanding and their creativity about Environmental Change topic on Biology.

2. Methodology

The type of this research is Research and Development (R&D), by using ADDIE model, ADDIE model is a systematic instructional design model consisting of five phases: 1) Analysis; 2) Design; 3) Development; 4) Implementation; and 5) Evaluation (Dick and Carey, 1996). The implementation of this research is performed on students in SMA Negeri 7 Purworejo. The subject selection technique in this research was random sampling to decide the control-experiment group, one group of control group and one group of experiment group, on each group has 31 students. The treatment of experiment class is using Learning Media based on Adobe Flash with combination of CPS and SAVI, and the treatment of control class is using conventional Learning Media and PBL learning model which commonly used in Indonesia. The learning media is validated by experts on each fields: instrumens expert, medias expert, subjects expert, and user response. Before and after going through the improvement process in the implementation phase, pretest and postest is given to know the effectivity of learning media, the effectiveness can be seen on the result of the test using normalized gain (N-gain) score.

3. Result and Discussion

The following steps of developing the understanding and creativity of students on Environmental Change is according to the ADDIE model by Dick and Carey's the Systematic Design of Instruction (Dick and Carey, 1996). The result of this reseach are:

3.1. Expert Validity

Expert validity is one of the ADDIE steps on the developement stage (Dick and Carey, 2015). Development phase is the actual creation of the content and learning materials based on the Design phase. Validation is carried out by one material expert, one media expert, one instrument expert, and one biology teacher. Revisions stage are done before implementation is carried out based on advice from experts (Dick and Carey, 2015).

All the Instrumen of validation was the result of adapting and modifiying the research instrument by Kurniawan Aji Sutardi (2012) with the title "Pengembangan Multimedia Pembelajaran Interaktif (MPI) IPA dengan Tema Air Menggunakan Metode Webbed Berbasis Adobe Flash CS4 Kelas VII SMP RSBI" and Azhar Arsyad (2011). Instrument used to measure validity using rating scale with a scale of 0 to 4 per item, the final result is showed on percentage for each aspect. Percentage category division based on: 0-25 is very unfeasible, 25-50 is not feasible, 50-75 is feasible, 75-100 is very feasible (Azhar, 2011).

Table 1. The Media Expert's Validation Result

Assessment Aspect	Assessment Results (%)	Category Feasibility
Aspect of Concept Truth	90	Very Feasible
Language Accuracy Aspects	81,25	Very Feasible
Display Aspects	85	Very Feasible
Software Aspects	100	Very Feasible
Interaction Aspects	81,25	Very Feasible
Implementation aspects	100	Very Feasible
Average Percentage	87,4	Very Feasible

The validity result of Media Expert shows that all the aspect is very feasible, and also can be seen on the average of all the percentage aspects is 87,4%, above 75% means that the learning media is very feasible based on media expert validation.

Table 2. The Material Expert's Validation Result

Assessment Aspect	**Assessment Results (%)**	**Category Feasibility**
Aspect of Concept Truth	85	Very Feasible
Language Accuracy Aspects	81,25	Very Feasible
Display Aspects	79,54	Very Feasible
Average Percentage	82	Very Feasible

The validity result of Material Expert shows that all the aspect is very feasible, and also can be seen on the average of all the percentage aspects is 82%, above 75% means that the learning media is very feasible based on material expert validation.

3.2. User Response on Usability

The user response on usability is done by Biology Teacher, this response is one of the result on development stage (Dick and Carey, 1996). The teacher is the one who teaches Biology more than five years and be able to use multimedia learning

Table 3. The Biology Teacher's Validation Result

Assessment Aspect	**Assessment Results (%)**	**Category Feasibility**
Aspect of Concept Truth	90	Very Feasible
Language Accuracy Aspects	75	Very Feasible
Completeness of Material Aspects	87,5	Very Feasible
Technical aspects of material presentation	93,75	Very Feasible
Display Aspects	82,5	Very Feasible
Interaction Aspects	100	Very Feasible
Implementation aspects	100	Very Feasible
Average Percentage	89,02	Very Feasible

The validity result of Biology Teacher shows that all the aspect is very feasible, and also can be seen on the average of all the percentage aspects is 82%, above 75% means that the learning media is very feasible based on biology teacher validation.

3.3. Effectiveness of Biology Learning Media

Biology learning media about Environmental Change material are used in learning activities, the combination of CPS and SAVI learning models are used to improve students' understanding and creativity. The implementation was carried out in SMA Negeri 7 Purworejo, using the control class and experimental class to compare the final results at the evaluation stage. The control class is consisting of 31 students, and the experimental class in this study consisting of 31 students.

Combination was carried out at each stage of the learning model, the SAVI approach was applied at each stage of the CPS Treffinger (2003) model: 1). Understanding the challenge; 2). Generating ideas; 3). Preparing for actions; 4). Planning your approach. Each CPS stage has its sub-stages, the sub-stage if sorted again will be 6 stages and then combined with the SAVI approach becomes: 1). *Constructing Opportunities* (SAVI-Somatic, Audio, Visual, Intelectual); 2). *Exploring Data* (SAVI-Somatic, Audio, Visual, Intelectual); 3). *Framing Problem* (SAVI-Somatic, Audio, Visual, Intelectual); 4). *Generating Ideas* (I-Intelectual); 5). *Developing Solution* (SVI-Somatic, Visual, Intelectual); 6). *Building Acceptance* (SAVI-Somatic, Audio, Visual, Intelectual).

3.4. Improvement of Students' Understanding

To determine the effectiveness of learning media on student's understanding and creativity, students are given a pretest and posttest to compare students' abilities before and after being given treatment. The results of the pretest and posttest were calculated using the Normalized Gain Score (N-gain) to determine the increase in students' abilities of undestranding. The criteria for increasing N-gain are shown in the following table

Table 4. N-Gain Improvement Criteria (Hake, 1999)

Quantitative Value	**Qualitative Value**
$g > 0.70$	High
$0.30 < g \leq 0.70$	Medium
$g \leq 0.30$	Low

From the results of N-gain we can see the difference in the increase of students' ability in each treatment for the control class and the experimental class. The results of the pretest and posttest in the control class and experimental class were calculated using N-Gain with the results shown in the following table.

Table 5. N-Gain Calculation Result

Criteria	Control Class		Experimental Class 1	
	Pretest	Posttest	Pretest	Posttest
Min	46.60	53.30	46.40	66.60
Max	86.60	93.30	86.60	100.00
SD	8.68	8.68	1.04	7.31
Average	66.42	75.46	67.49	85.56
N-Gain	0.26		0.52	

Based on the Table 5. The average of Normalized Gain Score on the control class is 0.26 categorized Low, means that the improvement of control class students' understanding and creativity using conventional learning media was in low effectiveness. The average of experimental class' Normalized Gain Score is 0.52 categorized medium, means that using learning media based on Adobe Flash with combination CPS and SAVI is improving the students' understanding and creativity in medium effectiveness (Hake, 1999).

3.5. Improvement of Students' Creativity

To strengthen the effectiveness of learning media on student creativity, students are given a project using combination of CPS and SAVI on learning media. The students' scientific creativity is categorized using SSCM by Hu and Adey (2002), the 24 cells of SSCM is simplified by Hu and Adey (2002) into seven category for aspect of the assesment. The results of the simplification is problem finding, problem solving, science experiment, creative imagination, unusual uses, product improvement, and product design. Comparisons were made between the control class and the experimental class to determine differences in creativity after the use of learning media, these results are shown in the following table

Table 6. The Result of Students' Creavity

Aspect	Control Class	Experimental Class
Problem Finding	2,13	3
Problem Solving	2,65	2,87
Science Experiment	2,65	2,48
Creative Imagination	2,16	2,87
Unusual Uses	1,10	1,97
Product Improvement	1,90	2,48
Product Design	1,90	2,55
Score All Aspect	14,48	18,23
Mean	2,07	2,60
Percentage	66,88	84,16

From the Table 8. The students' creativity of experimental class is higher than the control class, can be seen based on the mean of experimental class is higher than the control class, the mean of the control class is 2,07 and the mean of the experimental class is 2,60, the percentage of control class is 66,88 and experimental class is 84,16. From the result can be concluded that the use of Learning Media with combination CPS and SAVI is increasing the students creativity. This conclusion strengthened by the result of Independent Sample T-Test, the value of sig. (2-Tailed) is 0,000. The value of sig. (2-Tailed) < 0.05 is Ho rejected and Ha accepted, could be interpretated that there is a difference mean of learning result on control class and experimental class.

4. Conclusion

From the results of this research, can be concluded the learning media with a combination of CPS and SAVI are feasible and effective to be used to improve students' understanding and creativity about environmental change topic by the result of (1) the expert judgement showed that the percentage is more than 75 it means the learning media is feasible to use, (2) the user response on usability showed that the percentage is more than 75 it means the learning media is feasible to use, (3) effectiveness of media showed that the N-Gain result is 0,52 which included on middle category of improving students'understanding, and the results of the student worksheet assessment is 84,16 which categorized high on students' creativity.

Acknowledgement

The author would like to thank to Dr. Tien Aminatun who has provided advice, input, and motivation in preparing this article. Also thanks to the Biology Education study programs of Graduate Program Yogyakarta State University that has provided support in the preparation of this article.

References

[1] Aguayo, R. (2010). *The Metaknowledge Advantage*. New York: Free Pass
[2] Anderson, L. W & Krathwohl, D. R. (2001). *A taxonomy of Bloom's Taxonomy of Education Objectives*. New York: Addison Wesley Longman
[3] Arsyad, Azhar. (2011). *Media Pembelajaran*. Jakarta: Erlangga
[4] Aziz, Rahmat. (2014). Psikologi Pendidikan: Model Pengembangan Kreativitas dalam Praktik Pembelajaran. Malang: UIN-MALIKI PRESS.
[5] Baharuddin, & I, Taskiawati. 2009. *Hasil Hutan Bukan Kayu*. Makasar: Universitas Hasanuddin.
[6] Biggs, Alton. (2004). Biology: The Dynamics of Lyfe
[7] Budiningsih, C. A. (2004). Karakteristik Siswa: Sebagai Pijakan Pembelajaran. Yogyakarta: FIP UNY
[8] Cera, R., Mancini, M. & Antonietti, A. (2013). *Relationship Between Metacognition, Self*

Efficacy, and Self Regulation in Learning. ECPS Journal. Vol. 7, pp. 115-141. Retrieved from: http://www.ledonline.it/index.php/ECPS-Journal/article/view/511
[9] Dick, W., & Carey, L. (1996). *The Systematic Design of Instruction, 4th Edition*. New York: Harper Collin.
[10] Dick, W., & Carey, L. (2015). The Systematic Design of Instruction, 8th Edition. USA: Pearson
[11] Hake, R. R. (1999) Analyzing Change/Gain Score. *Unpublished* [online]. Retrieved from http://www.physics.indiana.edu/~sdi/AnalyzingChange-Gain.pdf
[12] Hu, W., & Adey, P. (2002). A Scientific Creativity Test for Secondary School Students. *International Journal of Science Education*, vol. 21(4), pp. 389-403. https://doi.org/10.1080/09500690110098912
[13] Isaksen, S. G. (1995). Creativity and Innovation Management 4. On the Conceptual Foundations of Creative Problem Solvng: a Response to Magyari-Beck. USA: Basil Blackwell.
[14] Krathwohl, D. R. (2002). A Revision of Bloom's Taxonomy: An Overview. *Theory into Practice*, vol. 41, Issue 4. https://doi.org/10.1207/s15430421tip4104_2
[15] Linder, C.; Ostman, L. & Roberts, D. A. (2010). Exploring the landscape of Scientific Literacy. New York: Routledge.
[16] Meier, Dave. (2000). The Accelerated Learning Handbook: A Creative Guide to Designing and Delivering Faster, More Effective Training Programs. US: McGraw Hill Professional.
[17] Nuryani, R., et. al. (2005). Trategi Belajar Mengajar Biologi. Malang: UM Press.
[18] OECD. (2016). PISA 2015 Results (Volume I): Excellence and Equity in Education, PISA. Paris: OECD Publishing. http://dx.doi.org/10.1787/9789264266490-en
[19] OECD. (2017). PISA 2015 Results (Volume V): Collaborative Problem Solving, PISA. Paris: OECD Publishing. http://dx.doi.org/10.1787/9789264285521-en
[20] Sanjaya, Wina. (2011). Strategi pembelajaran berorientasi standar proses pendidikan. Jakarta : kencana
[21] Simonton, D. K. (2012). Assessing Science Creativity: Conceptual Analyses of Assessment Comlpexities. National Academy of Sciences. *Science of Science and Innovation Policy Conference* [Commissioned paper], . 1-24. Retrieved from https://psychology.ucdavis.edu/Simonton/SimontonSciSIPFinalDraft.pdf
[22] Sternberg, R. J., Kaufman, J. C., & Pretz, J. E. (2003). A propulsion model of creative leadership. *Leadership Quarterly*, 14, 455–473. https://doi.org/10.1111/j.0963-1690.2004.00304
[23] Sudjana, Nana. (2013). *Penilaian Hasil Proses Belajar Mengajar*. Bandung: PT. Remaja Rosdakarya Offset
[24] Sutardi, Kurniawan Aji. (2012). Pengembangan Multimedia Pembelajaran Interaktif (MPI) IPA dengan Tema Air Menggunakan Metode Webbed Berbasis Adobe Flash CS4 Kelas VII SMP RSBI [skripsi]. Yogyakarta: Universitas Negeri Yogyakarta.
[25] Treffinger, D. J. (2002). *Asessing Creativity: a guide for educators*. Florida: Center for Creative Learning
[26] Treffinger, D. J.; Isaksen, S. G & Dorval, K. B. (2003). Creative Problem Solving (CPS version 6.1) a Contemporary Framework for Managing Change. Retrieved from https://www.cpsb.com
[27] Wang, Jun & Yu, Jiayuan. (2011). Scientific Creativity Research Based on Generalizability Theory and BP_Adaboost RT. *Procedia Engineering*, vol. 15, pp. 4178-4182. https://doi.org/10.1016/j.proeng.2011.08.784

The Effectiveness of Student Worksheet Development Based on Problem-Based Learning in Respiratory System Material to Improve High School Students' Quantitative Literacy

W. Apriyani[1]* and Suhartini[2]

[1]Biology Education, Graduate Program, Yogyakarta State University, Colombo St. No 1, Karangmalang, Depok, Sleman, Yogyakarta, Indonesia
[2]Department of Biology, Faculty of Mathematics and Natural Sciences, Yogyakarta State University, Colombo St. No 1, Karangmalang, Depok, Sleman, Yogyakarta, Indonesia

*Corresponding author e-mail: apriyaniwinda04@gmail.com

Abstract. This study aims to produce teaching materials in the form of student worksheet based on problem-based learning (PBL) in respiratory system material to improve quantitative literacy of eleventh grade students of senior high school using ADDIE development model method (Analysis-Design-Development-Implement-Evaluation). The instruments of this study were validation sheets for the teaching materials and pretest and posttest for quantitative literacy. The validation sheets were validated by an expert at teaching material, an expert at material, two biology teachers, four colleagues and fifteen students. Pretest and posttest to measure students' quantitative literacy were validated by two experts and were tested to 29 eleventh grade students of senior high school. The results revealed that the teaching materials were theoretically feasible to use in that the feasibility score of the materials was 3.50 (very good), the feasibility score of the teaching materials was 3.59 (very good), and the feasibility score of the biology teachers and colleagues was 3.52 (very good) and the students' response score was 3.48 (very good). Moreover, the results of quantitative literacy showed that there are nine students categorized to high criteria and twenty students categorized to moderate criteria. The results indicated that the PBL-based student worksheet as a teaching material is feasible and recommended to be implemented in the learning process.

1. Introduction

The implementation of 2013 curriculum is a process of "student-centered learning" in order to improve students' skill and proficiency and to encourage students to develop quantitative literacy. The 2013 curriculum directs the learning using the data related to the observation which has quantitative characteristics, namely suggesting hypothesis, processing data, calculating and presenting the data systematically and attractively, interpreting, drawing a conclusion based on a gained fact, and scientific communicating to the results of experiment in oral and written. Starting from 21th century, biology learning is expected to develop the learning that tends to science which is more quantitative because a quantitative literacy is an ability to analyze numbers and to criticize in solving the problem in daily life [1]. Improving students' quantitative literacy can be assisted with the development of learning material, namely student worksheet. Several students' literacy skills have not trained yet, and it is evidenced from the teachers' interview and the students' result score of the pretest on the respiratory system. Thus, teachers are able to prepare teaching materials that are made interesting so that the learning process can be interesting as well. The monotonous learning activity can give an impact on the students who are not accustomed to learning through discovery or experiment, as a result, teaching materials are needed. Majid [2] stated that the teaching materials are all forms of materials used to assist teachers or instructors in carrying out the learning activities. The teaching materials or learning media mostly used are modules, magazines, journals and student worksheet [3]. This study develops a teaching material namely student worksheet, which is a sheet containing the student's learning activities to be carried out in which the students learn the real objects and collect them to become a book. According to Prastowo [4] student worksheet as a teaching material helps the students direct and ease the learning activities because it contains materials, summaries and assignment's instructions for the students to work in the form of theory and practice that refer to the basic competencies. The student's understanding of the material to be delivered can be maximized according to the predetermined achievement indicators [5]. The teacher can develop the student worksheet independently by analyzing the curriculum, including core competencies, basic competencies, indicators and learning materials.

The development of student worksheet which is independently carried out by the teacher should have an innovation. The innovation is in the form of student worksheet that is based on problem-based learning (PBL) and is provided with the videos that can be accessed with a barcode. The problem-based learning is not intended to convey a large amount of knowledge for the students, but to focus on encouraging the ability to solve the problems and to actively build their own knowledge [6]. The use of PBL-based student worksheet in science learning, especially biology as an investigation and resolution of real-world problems requires a guidance for the activities in order to make the learning directed and systematic based on the scientific method. It is similar to the research conducted by Wulansari [7] that the implementation of problem-based learning on student worksheet can improve students' cognitive and psychomotor aspects. Problem-based learning has stages that encourage students to be active in the learning activities and to help students improve their quantitative literacy, such as interpretation, representation, calculation, assumption, application/ analyzation, and communication. Besides, the learning using the student worksheet has videos that can be accessed with a barcode connected to the address site of the videos. The student worksheet provided with videos will help the students better understand the material being learned. Similar to Megahati's opinion [8] the students' understanding can be increased using the student worksheet during the learning process. The students who use the student worksheet during the learning activities will know the steps of how the students should learn because they will learn to be responsible for completing the work that they are learning [9]. The development of PBL-based student worksheet will give a benefit for the students because Sasmaz Oren [10] suggested that there are various objectives in the use of student worksheet and their effects impact on the student's success, information retention and motivation while participating the learning activities in the classroom. The use of PBL-based student worksheet in the classroom is expected to provide a positive effect on the student's learning outcomes.

2. Research Methodology

To determine the quality of a teaching material developed from the aspects of validity, practicality and effectiveness, the Research and Development (R & D) method was used. This research method aims to produce a certain product by analyzing the results of the lack of the previous products and examining the effectiveness of the new product [11]. Through the various developments toward the product improvement (development), the product development in this study used the ADDIE development model which consists of five stages, namely Analysis, Design, Development, Implementation and Evaluation [12]. This study was conducted in February 2019 and the place of the study was at State Senior High School 6 Yogyakarta.

The instruments used in this study were the validity sheets of student worksheet development by a lecturer who is expert at teaching material, a lecturer who is expert at material, two biology teachers, and four colleagues, and the response sheet by fifteen students. The validation results were calculated using Likert scale with the lowest scale 1 and the highest scale 4 and were analyzed using quantitative and qualitative method. The quantitative data given by the validators were converted into qualitative data through descriptive analysis. The conversion of quantitative data to qualitative data is presented in Table 1 below [13].

Table 1. The conversion of quantitative data to qualitative

Interval	Interval	Criteria
Mi + 1.5 SDi ≤ M ≤ Mi + 3 Sdi	3,25 ≤ M ≤ 4	Very good
Mi + 0 SDi ≤ M ≤ Mi + 1.5 Sdi	2,5 ≤ M ≤ 3,25	Good
Mi – 1.5 SDi ≤ M ≤ Mi + 0 Sdi	1,75 ≤ M ≤ 2,5	Fair
Mi – 0 SDi ≤ M ≤ Mi – 1.5 Sdi	1≤ M≤ 1,75	Poor

In addition, the instruments used to examine students' quantitative literacy skill are pre-test and post-test sheets. The data analysis of quantitative literacy was through descriptive quantitative analysis. The improvement skill of students' quantitative literacy was calculated using the n-gain formula [14]:

$$< g \geq = \frac{< Spost > \; - < Spre >}{Smax- < Spre >}$$

Then, the n-gain was interpreted as follows:

Tabel 2. Criteria conversion of quatitative data to qualitative

Range	**Criteria**
$g > 0{,}7$	High
$0{,}3 \leq g \leq 0{,}7$	Moderate
$g < 0{,}3$	Low

3. Results and Discussion

Student worksheet should be developed as creatively as possible in order to meet the criteria for implementing the learning activities effectively because the main function of the student worksheet is to encourage active students to participate in the learning activities through a designed series of activities based on the student worksheet and the applied learning models, such as discussions, observations and experiments. According to Suyanto, Paidi and Wilujeng [16] the student worksheet has a function as a guide for the students to conduct an investigation, an experiment and a guideline in the discussions for the conceptualization. A good student worksheet helps the students find, apply and integrate a concept that has been found. The previous research conducted by Dyah [17] found out three aspects that must be fulfilled in the development of PBL-based student worksheet, namely didactic aspect, constructive aspect and technical aspect. The research on the development of PBL-based student worksheet involves an expert at materials, an expert at teaching materials, biology teachers, colleagues and students' responses to provide an evaluation of the developed teaching materials. The research conducted by Firdaus and Wilujeng [18] also involved material expert, teaching material expert, biology teachers, colleagues and students' responses. Based on the function of student worksheet, therefore, the use of PBL-based student worksheet is able to give an impact on students in improving their quantitative literacy. The quantitative literacy is a knowledge, a skill to solve a problem, a thinking method, a communication and a capability in the education field [19].

3.1Validation Results of Material Expert

The PBL-based student worksheet was validated by an expert at material, namely Dr.drh. Heru Nurcahyo, M.S, a lecturer in Biology Education of Postgraduate Programs, Yogyakarta State University. The feasibility evaluation of PBL-based student worksheet includes didactic aspects and language aspects with fifteen statement items. The results of the feasibility validation of PBL-based student worksheet from the material expert can be seen in Table 3 below.

Table 3. Validation Result of Material Expert

No	Evaluation Aspects	Score
	Didactic Aspects	
1	The presented material concept is suitable with core competencies, basic competencies, indicators, and the learning objectives about respiratory system which are complete and easy to understand.	3
2	The materials presented thoroughly represent the scope of the respiratory system materials which are complete and suitable with the education level of eleventh grade of senior high school.	4
3	The learning activities in the PBL-based student worksheet aim to improve quantitative literacy.	4
4	The accuracy of the materials or the accuracy of the presented discourses is appropriate and easy to understand.	3
5	The depth of the materials is suitable with the student's competencies in the 2013 curriculum.	4
6	The development of the concepts is related to daily life.	4
7	The concept of the respiratory system is complete.	3
8	The materials are suitable with the learning indicators.	3
9	The respiratory system materials are up-to-date, following the development of science.	4
10	The presented respiratory system materials and problems are relevant between the presented fact and concept.	4
Total		**36**
Average		**3.60**
Criteria		**Very good**
	Language Aspects	
11	The language use is adjusted to the intellectual level of the students.	3
12	The accuracy of the sentence use is simple and easy to understand by the students.	3
13	The accuracy of the language use in the instructions of the learning activities can be clearly understood.	3
14	The use of communicative language is accurate.	4
15	The accuracy of the language use does not give a dual meaning (ambiguous).	4
Total		**17**
Average		**3.40**
Criteria		**Very good**

The average score given by the material expert in the didactic and language aspects was 3.60 and 3.40, and the average score of the two aspects was 3.5. These aspects are categorized to very good criteria. These results indicated that the teaching materials in the form of PBL-based student worksheet are feasible to be implemented and used as a teaching material that supports the student's learning activities although there are several language uses used in the student worksheet that must be improved.

3.2 Validation Results of Teaching Material Expert

The teaching materials were validated by Dr. Paidi, M.Sc, a lecturer in Biology Education of Postgraduate Program, Yogyakarta State University. The feasibility evaluation of PBL-based student worksheet includes technical, language and constructive aspects with thirty statement items. The validation results of the feasibility of PBL-based student worksheet from a teaching material expert can be seen in Table 4 below.

Table 4. Validation Result of Teaching Material

No	Evaluation Aspects	Score
	Technical Aspects	
1	Interesting cover design of PBL-based student worksheet	4
2	The use of font and text arrangement	4
3	Attractive PBL-based student worksheet display	3
4	Well-matched writing layout and picture	3
5	Effectiveness of images that can convey messages related to the materials	3
6	Interesting design of each page and the layout of PBL-based student worksheets attracting readers' interest	4
7	Typography of the content	4
Total		**25**
Average		**3.57**
Criteria		**Very good**
	Language Aspects	
8	The language use is adjusted to the intellectual level of the students.	3
9	The accuracy of the sentence use is simple and easy to understand by the students.	4
10	The accuracy of the language use in the instructions of the learning activities can be clearly understood.	3
11	The use of communicative language is accurate.	4
12	The accuracy of the language use does not give a dual meaning (ambiguous).	4
Total		**18**
Average		**3.6**
Criteria		**Very good**
	Constructive Aspects	
13	Respiratory system material in student worksheet has clear syntax.	4
14	Compilation of the materials in each chapter in PBL-based student worksheet is correct.	3
15	The questions in PBL-based student worksheet can be answered by processing the information.	3
16	Provided space for writing or drawing is enough to give a freedom to the students.	3
17	Clear and useful goals for students are listed.	4
18	Student worksheet contains the student's identity column.	4
19	The presence of images or video barcodes can convey messages related to the materials.	4
20	Various ways of presenting the information in the form of discourse news or video are developed.	4
Total		**29**
Average		**3.62**
Criteria		**Very good**

The average score given by the experts on the technical aspects of the media was 3.57, the average score of the language aspects was 3.6 and the average score of the constructive aspects was 3.62. From these results, the average total score of the three aspects was 3.59 and it is categorized to very good criteria. The results of this evaluation indicated that PBL-based teaching material is feasible to be implemented and used as a learning material in the school. However, there is a feedback from teaching material expert about selecting the more appropriate videos so that they can support the learning material.

3.3 Validation Results of External Validators

The feasibility of PBL-based student worksheet was also validated by external validators, namely biology teachers and colleagues. The external validators evaluated didactic aspects, language aspects, technical aspects and constructive aspects. Table 5 below shows the validation results from two biology teachers of State Senior High School 6 Yogyakarta and four colleagues who have conducted or are conducting a research about learning development, media or learning material.

Table 5. Validation Results by External Validators

No	Aspects	Biology Teachers		Colleagues				Average	Category
		V1	V2	V3	V4	V5	V6		
1	Didactic	35	33	35	34	37	37	3.51	Very good
2	Language	18	18	19	18	16	18	3.56	Very good
3	Technical	23	23	25	25	26	26	3.52	Very good
4	Constructive	27	25	28	29	30	28	3.52	Very good
Total score		103	99	107	106	109	109	3.52	Very good

Table 4 shows the feasibility evaluation by biology teachers and colleagues. The average score of the didactic aspects was 3.51, the average score of the language aspects was 3.56, the average score of the technical aspects was 3.52 and the average score of the constructive aspects was 3.52. Then, the average total score of the four aspects was 3.52, which means that the four aspects are categorized to very good criteria. Thus, the PBL-based student worksheet is feasible to be implemented and used as a biology teaching material.

3.4 Evaluation Results of Students' Responses

The evaluation of the responses to PBL-based student worksheet in a small-scale feasibility test carried out by fifteen students. The evaluation includes learning aspects, language aspects and constructive aspects. The evaluation results of the responses to the PBL-based student worksheet by twelfth grade of science students of State Senior High School 6 Yogyakarta are presented in the following table.

Table 6. Evaluation Results of Students' Responses

NO	**Total Score**			**Average**	**Category**
	Learning Aspects	**Language Aspects**	**Constructive Aspects**		
1	25	10	15	3.57	Very good
2	21	10	12	3.07	Good
3	20	9	13	3.00	Good
4	21	11	13	3.21	Good
5	24	11	14	3.50	Very good
6	28	12	16	4.00	Very good
7	25	1 2	16	3.79	Very good
8	25	10	14	3.50	Very good
9	26	12	14	3.71	Very good

NO	Total Score				Category
	Learning Aspects	**Language Aspects**	**Constructive Aspects**	**Average**	
10	27	11	16	3.86	Very good
11	23	10	14	3.36	Very good
12	20	10	11	2.93	Good
13	24	12	16	3.71	Very good
14	25	11	14	3.57	Very good
15	21	10	13	3.14	Good
Total Score	355	161	211	51.92	Very good
Average	3.38	3.57	3.51	3.46	Very good

The results presented in Table 5 were obtained from the evaluation by fifteen students' giving the responses that showed the average score of 3.38 for the learning aspects. The average score of the language aspects was 3.57 and the average score of the constructive aspects was 3.51. As a result, the average total score of the three aspects was 3.48 and the three aspects are categorized to very good criteria. From these results, it can be declared that the PBL-based student worksheet is feasible to be implemented and used by the students as biology teaching materials that support more meaningful learning. This is in line with a research conducted by Chandra [20] and scientific approach [21] that the activities designed using the PBL-based student worksheet are able to make the learning activities interesting and fun for the students.

3.5 N-gain Value of Students' Quantitative Literacy

To measure students' quantitative literacy, pre-test and post-test regarding respiratory system material were given to 29 science students of eleventh grade of State Senior High School 6 Yogyakarta. The illustration of n-gain value is seen in Figure 1 below.

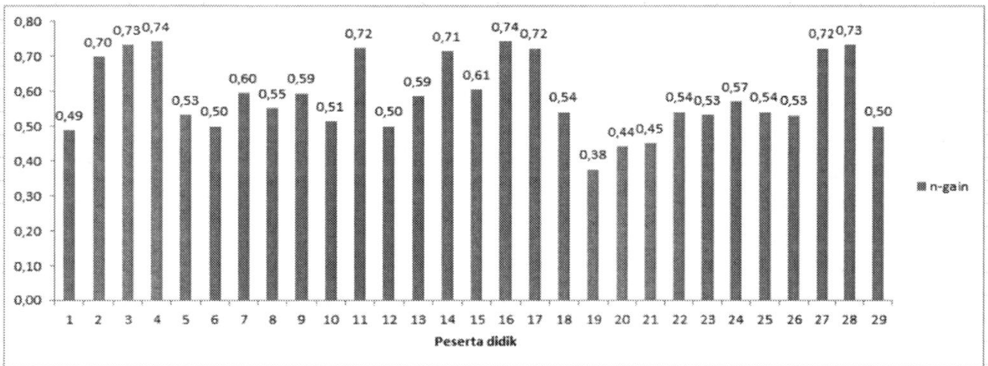

Figure 1. N-gain Value of Students' Quantitative Literacy

Figure 1 presented the n-gain value from 29 students. Based on the n-gain value, there are nine students who have the n-gain higher than 0.70, so that it is categorized to high criteria. In other words, the PBL-based student worksheet gives an impact on students in improving their quantitative literacy. In addition, there are twenty students in the n-gain range between 0.38 and 0.61, and these are categorized to moderate criteria.

4. Conclusion

The PBL-based student worksheet in the respiratory system material of eleventh grade science students of State Senior High School 6 Yogyakarta is feasible to be used as a teaching material for the students as an effort in assisting the students to improve their quantitative literacy. This is evidenced by the results of the validation scores from the validators and the students. The total score of the material expert was 3.50 (very good), the total score of the teaching material expert was 3.59 (very good), the total score of the external experts was 3.52 (very good) and the total score of the students' responses was 3.48 (very good). Moreover, the calculation results of students' quantitative literacy skill were found out that there are nine students with high n-gain value and twenty students with moderate n-gain value. Thus, the PBL-based student worksheet as a teaching material can be implemented in biology learning.

Acknowledgment

The researcher would like to thank Dr. Suhartini, MS who has assisted the researcher in writing this article. The researcher also thanks all the validators, namely a lecturer who is an expert at material, a lecturer who is an expert at teaching material, biology teachers, colleagues and students who have been involved in providing the evaluation.

References

[1] Speth E B, Momsen J L M, G A, Ebert M D. Long T M, Wyse L & Linton, D 2010 *CBE life Science Education* Vol. 9: 323-332
[2] Majid A 2014 *Strategi Pembelajaran* (Bandung: PT Remaja Rosdakarya)
[3] Cahyani D N 2017 Jurnal Inovasi Pendidikan IPA 3(1) 30-41
[4] Prastowo A 2011 *Panduan Kreatif Membuat Bahan Ajar Inovatif Menciptakan Metode Pembelajaran yang Menarik dan Menyenakan* (Yogyakarta: Diva Press)
[5] Trianto 2010 *Mendesain Model Pembelajaran Inovatif Progresif* (Jakarta: Kencana Predana Media Group)
[6] Hosnan 2014 *Pendekatan Saintifik dan Kontekstual dalam Pembelajaran Abad 21*(Bogor: Ghalia Indonesia)
[7] Wulansari N T, I Putu G S, Ni Wayan K D 2018 *SHS Web Of Conferences*. Les Ulis Vol.42 (Les Ulis: EDP Sciences)
[8] Megahati R R P 2017 *Barat Asia-Pacific Forum on Science Learning and Teaching*. Hong Kong Vol. 18, Iss 2
[9] Yildrim N, Kurt, S. & Ayas, A 2011 *Journal of Turkish Science Education* Vol 8 Issue 3
[10] Sasmaz O, Fatma & Ormanci U 2012 *Educational Scineces: Theory & Practic*. Istanbul Vol. 12 Iss. 1
[11] Sugiyono 2012 *Metode Penelitian Pendidikan Pendekatan Kuantitatif, Kualitatif dan R&D* (Bandung: Alfabet)
[12] Aldoobie N 2015 American *Int. J. of Contemporary Research* **6** 68-72
[13] Direktorat Pembinaan SMA 2010 *Juknis Penyusunan Perangkat Penilaian Afektif di SMA*. Jakarta
[14] Meltzer 2002 *American Journal of Physics* 70 (12): 1259-1268
[15] Chiappetta E L & Koballa T R Jr 2010 *Science Intruction In The Middle And Secondary School: Developing Fundamental Knowledge And Skill (*New York: Allyn &Bacon)
[16] Suyanto S, Paidi & Wilujeng I 2011 *Lembar Kerja Siswa* (Universitas Negeri Yogyakarta)
[17] Dyah K A P 2016 *Jurnal Pendidikan Ilmu Pengetahuan Alam* (Universitas Negeri Yogyakarta) Vol 5, No.1
[18] Firdaus M & Wilujeng I 2018 *Jurnal Inovasi Pendidikan IPA* (Universitas Negeri Yogyakarta) Vol 4, No1: April 2018

[19] Saputra I D, Sri A, Supriatno B 2016 *Proceeding Biology Education Conference* (Bandung: Sekolah Pascasarjana UPI) ISSN:2528-5742 Vol 131
[20] Chandra M W 2016 *Jurnal Pendidikan Matematika dan Sains* (Universitas Negeri Yogyakarta) Vol 4, No.2: December 2016
[21] Suyanto S 2018 *Jurnal Crakrawala Pendidikan,* ISSN:0216-1370 Online ISSN 24428620. https://doi.org/10.21831/cp.v37i1.18719

Dual Channel Supply Chain Model with Delivery Lead Time on The Imperfect Production Process by Notice Into Carbon Emission Capacity Regulation

A R Yanuarsih[1], R Setiyowati, Pangadi, Sutanto

Department of Mathematics, Faculty of Mathematics and Natural Sciences, Universitas Sebelas Maret, Surakarta, Indonesia

[1]rismaadhieka@gmail.com

Abstract. The development of the internet, adds flexibility of supply chain with an online channel for the product distribution. In this research, the dual channel supply chain model is used on the imperfect production process by notice into carbon capacity regulation and also delivery lead time. We construct the model to maximize the system profit of one manufacturer and one retailer by considering a carbon emissions capacity constrain from the government. Furthermore, we determine the optimal solution with Karush Kuhn Tucker condition. Based on sensitivity analysis maximal profit is obtained when probability of defective product, delivery lead time sensitivity, and carbon emissions are minimal.

1. Introduction

Supply chain is the system that includes processes from production to product distribution to consumer [5]. Production processes and products distribution on supply chain management can cause carbon emissions which contributes to global warming. To solve the problem, the government issued carbon emission regulations in industrial sector. We can construct the supply chain management mathematically through supply chain inventory model. The development of information and communication technology, resulting in the development on supply chain inventory model with products distribution through online and offline channel. In the process of products distribution through online channel, we need to notice the delivery lead time caused by shipping process. Supply chain with distribution process through online and offline channel called dual channel supply chain.

Dual channel supply chain's related research conducted by reference [15] with consider to carbon emission capacity regulation, consists of one supplier and one retailer. Supplier buys products to manufacturer, so there is no production process. Reference [3] did a research about production process with defective product that are sold at a discounted price. The next research about rework process for defective products [11]. The defective and non defective products are classify through the inspection process. In fact, errors occur in the inspection process. There are two types of inspection errors. Type I of inspection error occur when classify perfect products into defects and type II of inspection error occur when classify defect products to perfect products [13].

In this research, we construct dual channel supply chain model with delivery lead time on the imperfect production process by notice into carbon emission capacity regulations. The defective product which produced from the imperfect production process, selected through the inspection process that notice to inspection error, then repaired through the rework process. In addition, we also pay attention to delivery lead time because of online channel. The purpose of this research is to maximize the system profit that consist of one manufacturer and one retailer. We determine the optimal selling price through online and offline channels, so the system total profit is maximum.

2. Literature Review

In the dual channel supply chain research, the manufacturer makes green products for the environmental conscious. The research discussed pricing and greening strategies in centralized and decentralized cases, then compare the results [6]. Next research is studied about single channel and dual channel strategy to examine the effect of different return policy. That paper of supply chain system comprised of production, refurbishing, collection, and waste disposal processes [2]. Other research studied optimal product distribution strategy for symmetric manufacturers that use dual channel supply chains. Including an asymmetric distribution policy, where one manufacturer distributes products only through the direct channel, while the other manufacturer distributes through both the direct and retail channel [7]. Pricing and channel priority strategies of dual channel supply chain in the presence of supply shortage caused by random yields [14]. Pricing and service decisions for complementary products in a dual channel supply chain are determined by two manufacturers and one retailer. One of two manufacturers distributes products through the online and traditional retail channel [12]. Then, the research about dual channel supply chain for standard and customize products in centralized system. The manufacturer offers customized products through a direct online channel and standard product through the traditional retail channel [1].

Reference [16] developed the decision making models of the centralized and decentralized dual channel supply chain, which consist of one manufacturer and one retailer from game theoretical perspective. The research is design an improved revenue sharing contract to effectively coordinate the manufacturer and retailer and suggest government make cap and trade regulation to reduce carbon emission efficiently. The research from [15] considered the coordination of a dual-channel supply chain under carbon emission capacity regulation in centralized and decentralized system which consist of one supplier and retailer. Supplier sell products through online channel to brand loyal and offline channel to retailer. Retailer only sell products through offline channel to brand loyal and retailer loyal. Dual channel supply chain in [9] studied about optimal pricing policies for a centralized and decentralized dual channel retailer with same and cross channel returns. Retailer are adopting a dual channel retailing strategy in which products are offered through two channels, there are physical stores and online stores. The demand disruption management in a dual channel supply chain producing and selling green products, composed of one manufacturer and one retailer in centralized and decentralized cases are taken from [10]. The research examines pricing and greening issues for a dual channel green supply chain when the market demand is disrupted. The study from [8] examined about a dual channel supply chain under price and delivery time dependent stochastic customer demand. The research analyze centralized and decentralized systems for unknown distribution function of the random variables through a distribution free approach.

Based on the related research above, the supplier in the dual channel supply chain model from reference [15] replaced by manufacturer. Manufacturer did imperfection production process, and the defective product fix through rework process. The inspection process in manufacturer, also are noticed into inspection error type II. The aim of this research is to maximize the system profit under carbon emission capacity regulation.

3. Model Construction

3.1. Demand Function

Customers are divided into two types, there are brand loyal and retailer loyal, which have initial demand d_m and d_r. The retailer loyal customers purchase products only from the retailer. The brand loyal customers can purchase the products from both the manufacturer and retailer. Demand on online channel just done by brand loyal with initial proportion θ. Demand on offline channel is done by brand loyal with initial proportion $(1-\theta)$ and retailer loyal. All initial demand is reduced when the price rises. Changes in demand caused by the increase of price describe with elasticity selling price parameter β. When the selling prices of the two channels differ, customers will change their priority to the channel that has lower price. The priority of consumer purchases is described with η. The delivery lead time on online channel (l_m) affects to the demand of both channel. The longer delivery lead time in online channel makes consumers change their priority to offline channel. Priority of consumer purchases due to delivery lead time is described with α. Then, the demand functions of the online, offline, and both channels are

$$D_m(p_m,p_r) = d_m(\theta(1-\beta p_m) - \eta(p_m - p_r) - \alpha l_m), \tag{1}$$

$$D_r(p_m,p_r) = d_r(1-\beta p_r) + d_m((1-\theta)(1-\beta p_r) + \eta(p_m - p_r) - \alpha l_m), \tag{2}$$

$$D_c(p_m,p_r) = D_m(p_m,p_r) + D_r(p_m,p_r). \tag{3}$$

3.2. *Manufacturer Profit Function*

The profit function is obtained from total revenue minus total cost. Manufacturer issued manufacturing, inspection, post sale failure, and rework cost, with cost per unit product are c_{mp}, c_{imp}, c_{pfp}, and c_{rwp}. Manufacturer follows the lot for lot policy, so manufacturer produces products according to consumer demand on equation (3). Manufacturer does rework process to fix the defective products with probability γ and inspections error type II with probability f_2. That error makes retailer return the product, then manufacturer does the inspection and rework process. Retailer receives defective product with proportion δ and retailer doesn't make inspection errors, so retailer return the defective products to manufacturer. Manufacturer cost is

$$TC_M(p_m,p_r) = \left(c_{mp} + c_{imp}(1+\gamma f_2\delta) + c_{pfp}\,\gamma f_2\delta + c_{rwp}\gamma(1-f_2(1-\delta))\right)D_c. \tag{4}$$

Manufacturer sells the products through online channel to brand loyal with price p_m and through offline channel to retailer with wholesale price w. Manufacturer revenue on online and offline channel are $p_m D_m$ and wD_r. Then, manufacturer revenue is

$$TR_M(p_m,p_r) = p_m D_m + w D_r. \tag{5}$$

Manufacturer sells products through online and offline channel, so the profit function of manufacturer is the sum of online and offline profit. The profit function of manufacturer is

$$\Pi_M(p_m,p_r) = (p_m - C_M)D_m + (w - C_M) D_r, \tag{6}$$

with $C_M = c_{mp} + c_{imp}(1+\gamma f_2\delta) + c_{pfp}\,\gamma f_2\delta + c_{rwp}\gamma(1-f_2(1-\delta))$.

3.3. *Retailer Profit Function*

Retailer issued product purchases, inspection, and operational cost, with cost per unit product are w, c_{irp}, and c_{orp}. Retailer total cost is cost per unit multiplied by offline channel demand on equation (2). So, retailer total cost is

$$TC_R(p_m,p_r) = (w + c_{irp} + c_{orp})D_r. \tag{7}$$

Retailer just sells the products through offline channel to brand loyal and retailer loyal with price p_r. Then, retailer revenue is

$$TR_R(p_m,p_r) = p_r D_r. \tag{8}$$

The profit function of retailer is total revenue on equation (8) minus total cost on equation (7). Retailer profit function is

$$\Pi_R(p_m,p_r) = (p_r - C_R)D_R, \tag{9}$$

with $C_R = w + c_{irp} + c_{orp}$.

3.4. Dual Channel Supply Chain Model Manufacturer-Retailer

In centralized system, manufacturer and retailer determine the decisions about price together. The system profit is sum of manufacturer and retailer profit on equation (6) and (9). So, the system profit is

$$\Pi_c(p_m, p_r) = (p_m - C_M)D_m + \left(p_r - (C_M + c_{irp} + c_{orp})\right)D_r. \tag{10}$$

The government determines carbon emission capacity K_c. The maximal profit should not exceed the K_c limit. The main carbon emissions are generated from production process (e), and product distribution in online and offline channel (e_m and e_r). The total carbon emissions is

$$E_c(p_m, p_r) = D_m e_1 + D_r e_2 \tag{11}$$

with $e_1 = e + e_m$ and $e_2 = e + e_r$. So, we can formulate the mathematical model as

$$\begin{cases} max, & \Pi_c(p_m, p_r) = (p_m - C_M)D_m + \left(p_r - (C_M + c_{irp} + c_{orp})\right)D_r, \\ s.t., & E_c(p_m, p_r) \leq K_c. \end{cases} \tag{12}$$

4. Optimal Solution

The total profit of the dual channel supply chain system of $\Pi_c(p_m, p_r)$ is strictly concave function. So, there is only one value of (p_m, p_r) which is resulted in the maximal value of $\Pi_c(p_m, p_r)$. The proof of concavity is given in Appendix A. The mathematical model on equation (12) is optimization problem with constraints. The optimal solution of the model is determined by completing Karush Kuhn Tucker conditions. The proof of optimal solutions is given in Appendix B. The optimal value of p_m, p_r, and λ are

$$\begin{aligned} p_m^* = {} & \left((d_r\beta + d_m(\eta + (1-\theta)\beta))(A + c_{irp} + c_{orp}) + \eta(d_m(1 - \theta + l_m\alpha) + d_r) + \right. \\ & \left(d_r e_2\beta + d_m(-e_1\eta + e_2(\beta(1-\theta) + \eta))\right)\lambda^* + \beta(d_m(1-\theta) + d_r)C_M)/2B, \end{aligned} \tag{13}$$

$$\begin{aligned} p_r^* = {} & -\frac{1}{2\eta}((\mathrm{A} - 1/\mathrm{B}(\eta + \beta\theta)((d_r\beta + d_m(\eta + (1-\theta)\beta))(A + \eta(d_m(1 - \theta + l_m\alpha) \\ & +d_r + \left(d_r e_2\beta + d_m(-e_1\eta + e_2(\beta(1-\theta) + \eta))\right)\lambda^* \\ & + \left(d_r\beta + d_m((1-\theta)\beta + \eta)\right) \\ & (c_{irp} + c_{orp}) + \beta(d_m(1-\theta) + d_r)C_M)))), \end{aligned} \tag{14}$$

$$\begin{aligned} \lambda^* = {} & -(-e_2 d_c + d_m(e_1 - e_2)(l_m\alpha - \theta) + \beta\left(d_r e_2 + d_m(e_2(1-\theta) + e_1\theta)\right)(1 + f_2\delta\gamma) \\ & c_{imp} + \left(d_r e_2\beta + d_m\left(-e_1\eta + e_2((1-\theta)\beta + \eta)\right)\right)c_{irp} \\ & + (e_2\beta d_c + d_m\beta\theta(e_1 - e_2)) \\ & \left(C_M - c_{imp}(1 + \gamma f_2\delta)\right) + e_2\beta c_{orp}(d_m(1-\theta) + d_r) - d_m\eta c_{orp}(e_1 - e_2) + 2K_c) \\ & /(e_2^2\beta d_c + d_m e_1(-2e_2\eta + e_1(\eta + \beta\theta))), \end{aligned} \tag{15}$$

with $A = -l_m\alpha + \theta - \eta\lambda^*(e_1 - e_2) + \beta\theta(e_1\lambda^* + C_M) - \eta(c_{orp} + c_{irp})$ and $B = \beta\left(d_r(\eta + \beta\theta) + d_m(\eta - \beta(\theta - 1)\theta)\right)$.

5. Numerical Example

The parameters values are taken from the research of [4, 11, 15]. There are $d_c = 6500$ unit, $\theta = 0.5$, $\beta = 0.01/\$$, $\eta = 0.05/\$$, $l_m = 4$ days, $\alpha = 0.025/day$, $c_{mp} = 25\$$, $c_{imp} = 1\$$, $c_{pfp} = 3\$$, $c_{rwp} = 25\$$, $\gamma = 0.02$, $f_2 = 0.05$, $\delta = 0.5$, $c_{irp} = 1\$/unit$, $c_{orp} = 3\$/unit$, $K_c = 13960$, $e = 3$, $e_m = 4$, and $e_r = 5$.

There are three possible demands from the brand loyal and retailer loyal, $d_m < d_r$, $d_m = d_r$, and $d_m > d_r$. These optimal solutions of three possible demands are in Table 1.

Table 1. Optimal solution of DCSC model

d_m (unit)	d_r (unit)	λ^*	p_m^* (\$)	p_r^* (\$)	Π_c^* (\$)
2500	4000	1.81	68.67	72.7	77026.4
3250	3250	1.77	68.75	72.56	77790.2
4000	2500	1.72	68.63	72.42	78549.6

From Table 1, for every possible demand we obtain the value of $p_m^* < p_r^*$ and $\lambda^* > 0$, and the profit system is maximal when the value of $d_m = 4000$ unit and $d_r = 2500$ unit. Manufacturer sells products in online channel at 68.63\$ and retailer sell products in offline channel at 72.42\$. The optimal price obtained when the value of Lagrange multiplier is 1.72. With that optimal solution, we obtain the maximal profit is 78549.6\$. Furthermore, we did sensitivity analysis to know the optimal solution when we changed parameter values.

6. Sensitivity Analysis
We did sensitivity analysis for $\beta, \eta, \theta, \alpha, e_1, e_2, \gamma, \delta$, and f_2. The result of the sensitivity analysis of parameters is in Table 2.

Table 2. Result of sensitivity analysis

Parameter	β	η	θ	α	e_1
	0.01	0.01	1	0.01	3.26
Π_c^* (\$)	78549.6	77741.6	81894.7	79075.7	81562.8
Parameter	e_2	γ	δ	f_2	$f_2(\delta = 0.87)$
	5.44	0.01	0	1	0.01
Π_c^* (\$)	81562.7	79006.9	78576.7	78922.4	78529.9

From Table 2, the maximal profit is 81894.7\$, achieved when brand loyal only choose online channel. The minimal profit is 77741.6\$, acquired when the price sensitivity is 0.01. When probability of defective product approach to zero and carbon emissions in online and offline channel are minimal, the profit are maximal. Change in value of type II inspection error in the range $0 \leq \delta < 0.87$ resulted the maximal profit when the inspection for all defective products are wrong. When $0.87 \leq \delta \leq 1$, the maximal profit achieved when the probability of type II inspection error approach to zero. The result of the probability of type II inspection error in this paper similar to the result of research investigate by [13].

7. Conclusion
This paper explains how the manufacturer and retailer determine the optimal selling prices to maximize their system profit by notice into carbon emission capacity regulation. The maximal profit in centralized system can be obtained when probability of defective product, delivery lead time sensitivity, and carbon emissions are minimal. This paper only focus on manufacturing, inspection, and rework process. Future research could explore a dual channel supply chain model with additional to other production processes.

Appendix A
To determine if the total profit of the dual channel supply chain strategy is a strictly concave function, we must examine the principal minor determinant from Hessian matrix. The Hessian matrix of $\Pi_c(p_m, p_r)$ is

$$\boldsymbol{H} = \begin{pmatrix} -2d_m(\eta + \beta\theta) & 2d_m\eta \\ 2d_m\eta & -2(d_r\beta + d_m((1-\theta)\beta + \eta)) \end{pmatrix}. \quad (A.1)$$

From (A.1) we get

$$\boldsymbol{H}_{11} = -2d_m(\eta + \beta\theta) < 0, \quad (A.2)$$

$$\boldsymbol{H}_{22} = 4d_m\beta\big(d_r(\eta + \beta\theta) + d_m(\eta - \beta(\theta - 1)\theta)\big) > 0. \quad (A.3)$$

Sylvester Criterion says if the principal minor determinant from Hessian matrix a function is $(-1)^i$, with $i = 1,2,3,\dots,n$ then the function is strictly concave function. From (A.2) and (A.3) we can conclude that $\Pi_c(p_m, p_r)$ is a strictly concave function.

Appendix B

Necessary conditions of KKT for maximization problem is

$$\begin{gathered} \vec{\boldsymbol{\lambda}} \geq \mathbf{0}, \\ \boldsymbol{\nabla} f(\boldsymbol{X}) - \vec{\boldsymbol{\lambda}}\boldsymbol{\nabla} g(\boldsymbol{X}) = \mathbf{0}, \\ \lambda_i g_i(X) = 0 \\ \boldsymbol{g}(\boldsymbol{X}) \leq \mathbf{0}. \end{gathered} \quad (A.4)$$

i. Let $\frac{\partial(p_m,p_r,\lambda)}{\partial p_m} = 0$ and $\frac{\partial(p_m,p_r,\lambda)}{\partial p_r} = 0$, we can get the optimal selling prices p_m^* and p_r^* in the centralized system as

$$\begin{aligned} p_m^* &= \big((d_r\beta + d_m(\eta + (1-\theta)\beta))\big(A + c_{irp} + c_{orp}\big) + \eta(d_m(1 - \theta + l_m\alpha) + d_r) + \\ &\quad \Big(d_r e_2\beta + d_m\big(-e_1\eta + e_2(\beta(1-\theta) + \eta)\big)\Big)\lambda^* + \beta(d_m(1-\theta) + d_r)C_M)/2B, \\ p_r^* &= -\frac{1}{2\eta}((\mathrm{A} - 1/\mathrm{B}(\eta + \beta\theta)((d_r\beta + d_m(\eta + (1-\theta)\beta))(A + \eta(d_m(1 - \theta + l_m\alpha) \\ &\quad + d_r + \Big(d_r e_2\beta + d_m\big(-e_1\eta + e_2(\beta(1-\theta) + \eta)\big)\Big)\lambda^* + \Big(d_r\beta + d_m\big((1-\theta)\beta + \eta\big)\Big) \\ &\quad \big(c_{irp} + c_{orp}\big) + \beta(d_m(1-\theta) + d_r)C_M)))). \end{aligned}$$

ii. And from the other necessary condition we get the optimal λ is

$$\begin{aligned} \lambda^* = &-(-e_2 d_c + d_m(e_1 - e_2)(l_m\alpha - \theta) + \beta\big(d_r e_2 + d_m(e_2(1-\theta) + e_1\theta)\big)(1 + f_2\delta\gamma) \\ &c_{imp} + \Big(d_r e_2\beta + d_m\Big(-e_1\eta + e_2\big((1-\theta)\beta + \eta\big)\Big)\Big)c_{irp} + \big(e_2\beta d_c + d_m\beta\theta(e_1 - e_2)\big) \\ &\Big(C_M - c_{imp}(1 + \gamma f_2\delta)\Big) + e_2\beta c_{orp}(d_m(1-\theta) + d_r) - d_m\eta c_{orp}(e_1 - e_2) + 2K_c) \\ &/(e_2^2\beta d_c + d_m e_1\big(-2e_2\eta + e_1(\eta + \beta\theta)\big)). \end{aligned}$$

References

[1] Batarfi R, Jaber M Y and Aljazzar S M 2017 *Comput. Indus. Eng.* **106** 58
[2] Batarfi R, Jaber M Y and Zanoni S 2016 *Appl. Math. Model.* **40** 9454
[3] Goyal S K, Huang C and Chen K 2003 *J. Prod. Plan. Cont.* **14** 596
[4] Hua G, Wang S and Cheng T C E 2010 *Eur. J. Oper. Res.* **205** 113
[5] Kumar N R and Kumar R M S 2013 *J. Eng. Res. Technol.* **6** 455
[6] Li B, Zhu M, Jiang Y and Li Z 2016 *J. Clean. Prod.* **112** 2029
[7] Matsui K 2016 *Eur. J. Oper. Res.* **248** 646
[8] Modak N M and Kelle P 2019 *Eur. J. Oper. Res.* **272** 147
[9] Radhi M and Zhang G 2018 *Comput. Indus. Eng.* **119** 63
[10] Rahmani K and Yavari M 2019 *Comput. Indus. Eng.* **127** 493
[11] Taheri-Tolgari J, Mirzazadeh A and Jolai F 2012 *Comput. Math. Appl.* **63** 1007
[12] Wang L, Song H and Wang Y 2017 *Comput. Indus. Eng.* **105** 223
[13] Widianto I P, Jauhari W A and Rosyidi C N 2014 *J. Mult. Eng. Sci. Technol.* **2** 106
[14] Xiao T and Shi J 2016 *Eur. J. Oper. Res.* **254** 813
[15] Xu J, Qi Q and Bai Q 2018 *Appl. Math. Model.* **56** 449

[16] Xu L, Wang C and Zhao J 2018 *J. Clean. Prod.* **197** 551

Journal of Physics: Conference Series **1397** (2019) 012060 doi:10.1088/1742-6596/1397/1/012060

Estimation of type I censored exponential distribution parameters using objective bayesian and bootstrap methods (case study of chronic kidney failure patients)

A Wiranto, A Kurniawan[1], D A Fitria, Suliyanto, N Chamidah

Department of Mathematics, Faculty of Sciences and Technology, Airlangga University, Surabaya, Indonesia

[1]Corresponding Author : ardi-k@fst.unair.ac.id

Abstract. Bayesian point estimation is an estimation method based on prior selection and loss function. In Objective Bayesian estimation are chosen prior to Jeffrey and used intrinsic discrepancy loss functions based on the Kullback-Leibler divergence equation which will have a minimum effect of data on the posterior distribution. The objective Bayesian point estimator provides estimates of population parameters based solely on the assumed population distribution and data. The goal of this paper is to estimate parameters from the exponential distribution on type II censored data using the objective Bayesian and bootstrap methods. The bootstrap method is used to resampling and built a confidence intervals for parameters whhich will be estimated. The methods were applied on the life-time data of 63 patients of chronic renal failure and the initial diagnosis was non-diabetic disease with bootstrap methods using 10,100, and 1000 times used in this study. So that the bigger bootstrap samples rendered the estimated value $\hat{\theta}$ will be better and the result confidence interval ranges narrower.

1. Introduction

Survival data analysis is an analysis that discusses the survival of an object or individual in certain operational conditions. This analysis is usually applied to observe the survival of a product up to observations in the field of human disease that can be used to improve product quality and develop modern medical methods [1]. In observing survival data, life time data can be divided into censored and uncensored data. Censored observations occur if the survival data of the observed individuals is unknown [7].

One type of censorship is type I censorship, which occurs when all the objects studied are tested at the same time and the test is stopped after the specified observation time limit [6]. Observation with censorship is often have disadvantages, which one is the limited sample observed. So, in the analysis a method called bootstrap method used to solve the disadvantages. The bootstrap approach uses a recurring sampling method (resample). Based on the characteristics of the Bootstrap method which is a data-based simulation method for the certain simpler statistical inferential will require help from modern computers [4].

Exponential Distribution is a random variable distribution that is often used for modeling purposes. Exponential distribution is widely used as a lifetime model based on research that includes the durability of survival time. One of the inferences carried out in statistical modeling is parameter estimation [5]. To do estimations, the objective of this research is to use the Bayesian method which is an estimation method using a reference prior so that the completion of the estimation depends only on the presumption model and observation data [2]. The objective Bayesian method uses the intrinsic discrepancy loss function, which is the minimum function adapted from the Kullback-Leibler discriminant function. The principle of this function is to represent the parameters of a distribution with other parameters which still have the same information as the initial parameters, the parameter parameters are called proxies [3].

Several researches related to the use of bootstrap method in the field of life time analysis have been carried out. Bootstrap method can be used as an alternative method for predicting confidence intervals for geometric Weibull process model parameters in life test analysis for type II censored data [8]. Then another research concerning the Objective Bayesian estimation in uniform distribution. The study was able to produce intrinsic estimator values for θ and the intrinsic estimator obtained was close to the minimum value of the expected losses [3]. Based on the description above, this paper contains research on the application of parameter estimation Exponential distribution for type I censored data with an Objective Bayesian and bootstrap method. The result of those method used to modelling using the survival time data of patients with chronic renal failure.

2. Data and Research Method

The data used in this research based on survival time of patients with chronic renal failure undergoing Hemodialysis therapy diagnosed cause is non-diabetic disease in the hospital Dr. Sosodoro Djatikoesoemo Bojonegoro in 2014 was 63 patients. This paper describes the steps in estimating parameters, the confidence interval with the bootstrap method, and the application of theory to data on chronic kidney failure.

2.1 Estimation of Type I Censored Exponential Distribution Parameters with Objective Bayesian Method

The methods that used to obtain the estimation of type I censored exponential distribution parameters with objective bayesian methods

a. Take random sample $\boldsymbol{t_1, t_2, \dots, t_n}$ from Exponential distribution $(\boldsymbol{\theta})$

b. Determine the intrinsic discrepancy of Exponential distribution using the formula
$$\delta(\theta_0,\theta) = n\min\{K(\theta_0|\theta),K(\theta|\theta_0)\} \text{ dengan } K(\theta_i|\theta_j) = \int_t^{\infty} f(t|\theta_j)\ln\left[\frac{f(t|\theta_{ji})}{f(t|\theta_i)}\right]dt$$

c. Determine the censored time limit as $\boldsymbol{L_i}$ and the lifetime $\boldsymbol{T_i}$ censored or uncensored with $\boldsymbol{d_i = 1}$ if $\boldsymbol{t_i = T_i}$ or $\boldsymbol{T_i < L_i}$ and $\boldsymbol{d_i = 0}$ if $\boldsymbol{t_i = L_i}$ or $\boldsymbol{T_i \geq L_i}$

d. Determine the likelihood function $\boldsymbol{\ell(\theta|t)}$ from Exponential distribution of type I censored data
$$L(\theta|t) = \prod_{i=1}^{n}\left(f(t_i)^{d_i} S(L_i)^{1-d_i}\right)$$

e. Define the prior distribution use *Jeffrey's* method $\pi(\theta) \propto \sqrt{I(\theta)}$

Where, $I(\theta)$ is Information of Fisher's from θ .

f. Define posterior distribution $\pi(\theta|t) = \dfrac{L(t|\theta)\pi(\theta)}{\int_{-\infty}^{\infty} L(t|\theta)\pi(\theta)d\theta}$ s

g. Define intrinsic statistics $\boldsymbol{d(\theta_0|t) = \int_\theta \delta(\theta|\theta_0)\pi(\theta|t)\,d\theta}$

h. Obtain objective Bayesian estimation by $\boldsymbol{\arg\min_{\theta_0 \epsilon \Theta} d(\theta_0|t)}$

2.2 Determination Of Confidence Interval With Bootstrap Method

In this section will be explained how to acquire Mean Square Error from the parameter, and also obtain the lower bound and upper bound according to bootstrap sample size

a. Suppose $\boldsymbol{t_1, t_2, \ldots, t_n}$ Exponential Distribution
b. Make a withdrawal by returning data at step a one by one 63 times
c. Perform parameter estimation as in step 1 and obtained $\widehat{\boldsymbol{\theta}}^*$ (1)
d. Repeat steps b and c as much as B = 10 times so that an estimator is obtained $\widehat{\theta}^*$ (1), $\widehat{\theta}^*$ (3), ..., $\widehat{\theta}^*$ (10)
e. Calculate the average of $\widehat{\theta} = \sum_{i=1}^{10} \widehat{\theta}^*(i)$
f. Calculate the MSE by $MSE(\widehat{\theta}) = \frac{\left(\sum_{i=1}^{B} \bar{\theta} - \widehat{\theta}^*(i)\right)^2}{B}$
g. Sort the value of $\widehat{\theta}^*$ (i) from smaller to bigger
h. Determine confidence interval of 95% for $\widehat{\theta}^*$ with 2.5% quantile for lower bound and 97.5% quantile for upper bound.

2.3 Apply the theory obtained from steps 2.1 and 2.2 to the survival data of patients with chronic renal failure.

3. Result and Analysis

3.1 Estimation of Exponential Distribution Parameters in Type I Censored Data with Objective Bayesian Methods

If $t_1, t_2, \ldots, t_n$ is a random sample with Exponential distribution with scale parameters is θ, then it is known that the form of probability distribution function of Exponential distribution functions as follows:

$$f(t) = \frac{1}{\theta}\exp\left(-\frac{t}{\theta}\right), \theta > 0, t > 0 \tag{1}$$

The first step to estimate using the Objective Bayesian method is determine intrinsic discrepancy as follows:

a. Intrinsic Discrepancy for Type I Censored Data of Exponential Distribution

In general *Kullback-Leibler* equation based on $f(t|\theta)$ are $K(\theta_0|\theta)$ and $K(\theta|\theta_0)$ as follows :

$$K(\theta_0|\theta) = \frac{\theta}{\theta_0} - 1 - \ln\left(\frac{\theta}{\theta_0}\right) \text{ and } K(\theta|\theta_0) = \frac{\theta_0}{\theta} - 1 - \ln\left(\frac{\theta_0}{\theta}\right) \tag{2}$$

Furthermore, the intrinsic discrepancy will be determined $\delta(\theta_0, \theta)$ as follows:

$$\delta(\theta_0, \theta) = n\min\left\{\frac{\theta}{\theta_0} - 1 - \ln\left(\frac{\theta}{\theta_0}\right), \frac{\theta_0}{\theta} - 1 - \ln\left(\frac{\theta_0}{\theta}\right)\right\} \tag{3}$$

Mathematically, we can determine the minimum Kullback-Leibler equation with the conditions $\theta_0 > 0$ and $\theta > 0$. With the θ_0 is a proxy from θ which the value cannot be determined, then Mathematica Software is used to be solved numerically. After determining the intrinsic discrepancy, the next step is to determine the survival function in type I censored samples.

b. Determine Jeffrey Prior's for Type I Censored Exponential Distribution

The form of the survival function for exponential distribution is:

$$S(t_i) = \exp\left(-\frac{t_i}{\theta}\right) \tag{4}$$

Type I censored sampling is by determining the number of observation samples as much as n and the censor time as L_i. In taking type I censored samples, researchers can subject the censor deadline subjectively according to the research needs. The research stopped according to the predetermined time, if it exceeds the time limit, it is classified as censored observation and vice versa. Following is the form of the likelihood function of type I censored Exponential distribution:

$$L(\theta \mid t)=\frac{1}{\theta^{r}}\exp\left(-\frac{\sum_{i=1}^{n} t_i}{\theta}\right) \text{ with } r=\sum_{i=1}^{n} d_i \tag{5}$$

Where, the d_i shows that the lifetime of T_i Censored or Not Censored

The form of prior distribution $\pi(\theta)$ can be determined using the Jeffrey's method. In Jeffrey's method, it is known that the proportional to the square root of Fisher's information $I(\theta)$ is obtained by looking for the second derivative of the function $\ln(f(t \mid \theta))$. The Fisher's information formula is given below.

$$I(\theta)=-E\left[\frac{1}{\theta^2}-\frac{2t}{\theta^3}\right]=\frac{1}{\theta^2} \tag{6}$$

Using the Jeffrey's method, the prior distribution is obtained as follows:

$$\pi(\theta)\propto\sqrt{\frac{1}{\theta^2}}=\frac{1}{\theta} \tag{7}$$

c. Determine Posterior Distribution for Type I Censored Exponential Distribution

Posterior distribution is obtained from the result of a combination of likelihood functions with prior distributions that have been previously known.

$$\pi(\theta)=\frac{\left(\sum_{i=1}^{n} t_i\right)^{r}\exp\left(-\frac{\sum_{i=1}^{n} t_i}{\theta}\right)}{\theta^{r+1}\Gamma(r)} \tag{8}$$

d. Parameter Estimation Results θ with Objective Bayesian Method

The last step to estimate with the Objective Bayesian method is to determine the intrinsic statistics and $\hat{\theta}^*$ based on 2.1 as follows

$$d(\theta_0 \mid t)=\frac{n\left(\sum_{i=1}^{n} t_i\right)^{r}}{\Gamma(r)}\int_0^{\infty}\left(\frac{\delta(\theta_0 \mid t)\exp\left(-\frac{\sum_{i=1}^{n} t_i}{\theta}\right)}{\theta^{r+1}}\right)d\theta \tag{9}$$

$$\hat{\theta}^* = \underset{\theta_0 \in \Theta}{\arg\min}\left(\frac{n\left(\sum_{i=1}^{n} t_i\right)^{r}}{\Gamma(r)}\int_0^{\infty}\left(\frac{\delta(\theta_0 \mid t)\exp\left(-\frac{\sum_{i=1}^{n} t_i}{\theta}\right)}{\theta^{r+1}}\right)d\theta\right) \tag{10}$$

Equation (10) will determine the estimated value of the parameter for θ. The solution for Equation (10) is not closed form, so we use a Mathematica software to run the estimation.

3.2 Application Of Exponential Distribution Parameter Estimation Results On Type I Censored Data Of The Lifetime For Patients With Chronic Renal Failure

The data used in this research are data on patients with chronic renal failure undergoing Hemodialysis therapy with the initial cause is non-diabetic disease. In this research 63 samples were observed together, after that built the samples with bootstrap value consists of 10, 100, and 1000 times and the observation was stopped until the 100th day. The results of the parameter estimation are as follows:

Table 1. Parameter Estimation Results Of $\hat{\theta}^*$

Bootstrap	$\bar{\hat{\theta}}$ Estimate	$\overline{MSE}$
10	165.358	1841.303
100	166.825	951.5675
1000	162.280	922.2375

Table 1 shows the results of parameter estimation when bootstrapping is 10, 100, and 1000 times. It can be said that estimated $\bar{\hat{\theta}}$ values obtained for each bootstrap sample size are not much different, but for th $\overline{MSE}$ values can conclude that the more bootstrap simulation performed the smaller $\overline{MSE}$ value will produced. Furthermore, the confidence interval for the parameter $\hat{\theta}^*$ is calculated using the bootstrap method. So that the lower and upper limits are obtained as in Table 2, and when presented in the form of a histogram can be seen in Figure 1.

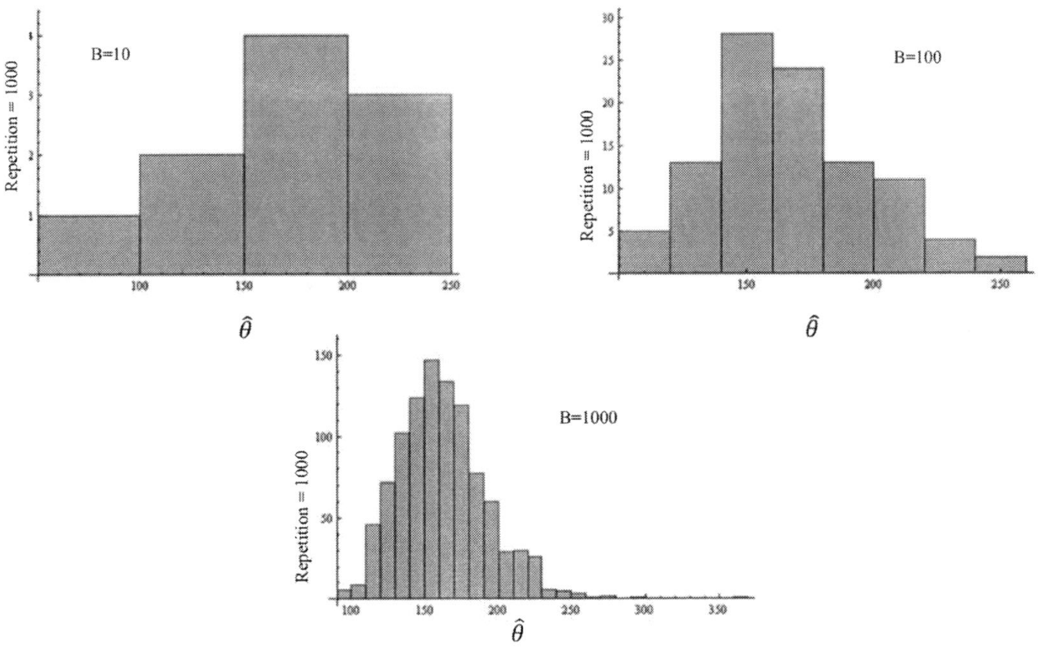

Figure 1. Histogram for Confidence Interval Parameters $\hat{\theta}^*$

Journal of Physics: Conference Series **1397** (2019) 012060 doi:10.1088/1742-6596/1397/1/012060

Table 2. Confidence Intervals for Each Bootstrap

Bootstrap Sample Size	Lower Bound	Upper Bound
10	98.8478	224.571
100	117.318	235.893
1000	112.857	225.69

Based on table 2 that obtain from smallest value and biggest value from estimated parameter value obtained. With a 95% confidence interval seen for the bootstrap, the bigger the confidence interval produced will be narrower. This is in accordance with the illustration in Figure 2, for bootstrap observations as much as 10 times shows that the beam tends to be wide. When bootstrap is done 100 times, the beam size will be narrower, and so on when bootstrap is done 1000 times, the beam size will narrow. This shows that the more bootstrap is carried out, the estimation results for each test will be conical. So that the confidence interval is getting shorter.

4. Conclusion

The application of estimation results to the data on chronic renal failure patients shows that the more bootstrap simulations performed can produce smaller $\overline{MSE}$ values. The bigger the bootstrap the better the estimation $\hat{\theta}$ is generated. Thus, the best fit model is $f\ (t) = \frac{1}{112.857}\exp(-\frac{t}{112.857})$, with the t are consists of lifetime With a 95% confidence interval seen for the bootstrap where, the more bootstrap sample size performed then affect the confidence interval produced will be even shorter.

References

[1] Barlow R E, and Proschan F 1996 *Mathematical Theory of Reliability Society for Industrial and Applied Mathematics* (Philadelphia)

[2] Bernardo J M 2010 *Integrated Objective Bayesian Estimation and Hypothesis Testing* (Oxford: Oxford University Press)

[3] Bernardo J M, and Juarez M A 2010 *Intrinsic Estimation (Bayesian Statistics vol 7)* (Oxford: Oxford University Press) pp 465-476

[4] Cox D R 1953 *J.R. Stat.Soc. B* **21** 411-421

[5] Efron B, and Tibshirani R J 1993, *An Introduction to the Bootstrap*, Chapman and Hall (New York)

[6] Harinaldi 2006 *Statistics Principles for Engineering and Science* Erlangga (Jakarta)

[7] Lee E T 1992 *Statistical Methods for Survival Data Analysis* John Willey & Sons, Inc. (Canada).

[8] Lee E T, and Wang J W 2003 *Statistical Methods for Survival Data Analysis*, 3th Edition (Oklahoma: University of Oklahoma Health Sciences Centre)

[9] Shanker R, Fesshaye H, and Selvaraj S, On *Modelling of Lifetimes Data Using Exponential and Lindley Distribution*, *Biom Biostat Int*, 2(5).

[10] Solih A., Cahyandari R, and Tarkinih 2017 *Estimation of Confidence Intervals Parameters of Weibull Geometric Process Models in Life Test Analysis for Censored Type Data II* ISTEK 10(1).

Fuzzy Multi Attribute Decision Making (FMADM) Implementation for Classifying Student's Single Tuition Fee (UKT) Based on Android Applications

A W Sugiyarto[1*], R Pamungkas[1], A R Rasjava[2], A M Abadi[1]

[1] Mathematics Department, Universitas Negeri Yogyakarta
[2] Chemistry Department, Universitas Negeri Yogyakarta

*aditya.wisnugraha2016@student.uny.ac.id

Abstract. Universitas Negeri Yogyakarta is one of the state universities in Indonesia that assigned Student's Single Tuition Fee (UKT) policy. UKT is an Indonesian government policy of tuition fees charged to every student with nominal based on the student's financial capabilities. Although the implementation of UKT nominal has been arranged in such ways, there are problems occurred, such as inconsistency and inaccuracy of UKT nominal classification. This causes so many dissatisfactions between students to the UKT nominal that did not meet the expectations and raises questions about the transparency of the determination of student's UKT nominal. Therefore, the application of Fuzzy Multi Attribute Decision Making (FMADM) can improve the accuracy and the effectivity of the determination of student's UKT nominal also answering the student's questions and giving clarifications about student's UKT nominal. The result of this research was an Android-based application that can be used for determining student's UKT nominal and determining what group that nominal belongs to with ease for under 5 minutes fast. With this android application, determining and calculating student's UKT nominal will be much faster and more accurate when groupping the UKT nominal than the conventional method used before.

1. Introduction

Universitas Negeri Yogyakarta is one of the state universities in Indonesia that assigned Student's Single Tuition Fee (UKT) policy. UKT policy was set based on circular letter from Dirjen Dikti Number 97/E/KU/2013 about UKT that instructs all state universities in Indonesia to write off base fees while setting and implementing UKT policy for the new undergraduate and diploma regular students starting from academic year of 2013/2014 [1][2].

UKT is an Indonesian government policy of tuition fees charged to every student with nominal based on the student's financial capabilities [3][4]. UKT included in the Overall Tuition Fee (BKT). BKT is the total of academic operational costs per student per semester [5]. The application of UKT nominal was set corresponding to the university operational costs by the constitution of UU Number 12 Year 2012 about higher education on State University Operational Subsidies, BKT, and UKT [4] and Regulation of Ministry of Education and Culture of Republic Indonesia Number 55 Year 2013 about BKT and UKT on State University under Ministry of Education and Culture [6].

UKT nominal affected by some criteria that affecting student's financial capabilities [3], such as parent's income, building and land area, and number of family dependents [7]. UKT nominal in the

Universitas Negeri Yogyakarta on 2018 was divided from group I to group VII and has been set by the Decision of Ministry of Research, Technology, and Higher Education Number 91/M/KPT/2018 [8].

According to Karim et al. (2011) [6] although the implementation of UKT nominal has been arranged in such ways, there are problems occurred, such as inconsistency and inaccuracy of UKT nominal classification, for example there are students with lower financial capabilities got higher UKT nominal than the students with higher financial capabilities. This causes so many dissatisfactions between students to the UKT nominal that did not meet the expectations and raises questions about the transparency of the determination of student's UKT nominal. Therefore, the application of Fuzzy Multi Attribute Decision Making (FMADM) can improve the accuracy and the effectivity of the determination of student's UKT nominal also answering the student's questions and giving clarifications about student's UKT nominal.

In fuzzy systems there are many algorithms that can be used for classification such as fuzzy inference system (FIS), fuzzy C-means (FCM), and FMADM [9]. FMADM is a method to find the alternative from the alternatives by some specific criteria [10]. FMADM involving two main processes, the rating and ranking of the alternatives [11]. The researchers chose to use this method because it can determine the best alternative from the alternatives using some specific criteria [12]. FMADM method is used for determining the UKT classification combined with Technique for Order Preference by Similarity to Ideal Solution (TOPSIS) method for ranking the alternatives [13]. The solution of the Multi Attribute Decision Making (MADM) case can be done by considering problem representation, evaluation of fuzzy sets, optimal alternative selection, also data sources and types [14].

The objective of this research was to create an Android-based application that can be used for determining student's UKT nominal and determining what group that nominal belongs to. With this android application, determining and calculating student's UKT nominal will be much faster and more accurate when groupping the UKT nominal than the conventional method used before.

2. Method of Research

2.1. Data Processing Method

In general, data processing method starts as the data become problems representation, then Fuzzi Sets began the evaluation producting optimal alternatives selection from the data as the result. The method for data processing as shown at figure 1.

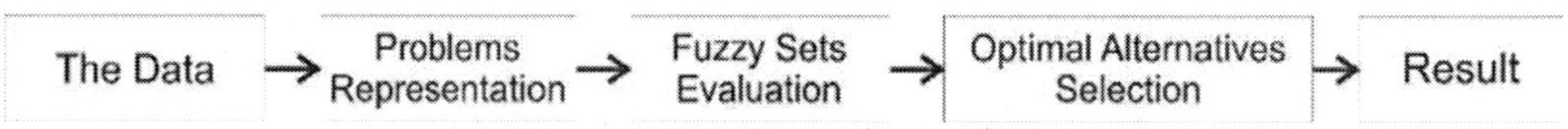

Figure 1. Data Processing Method

In this research, FMADM and Technique for Order Preference by Similarity to Ideal Solution (TOPSIS) were used for determining student's UKT nominal group. According to [14] steps for solving this MADM case are problem representation, fuzzy sets evaluation, and optimal alternatives selection.

In the problem representation there are 3 things to do: research objectives identification and groups of alternatives decision, criteria identification, and building structural hierarchy of the problems based on specific considerations. After performing a problem representation process, a fuzzy set evaluation process is carried out. TOPSIS method was used by the researchers on this step for evaluating the fuzzy sets on this MADM case. TOPSIS is one of the multicriteria decision making method based on the concept that the best-chosen alternative did not only have the shortest length from the ideal positive solution, but also have the longest length from the ideal negative solution. On the TOPSIS method, there are some steps as follows: Determining fuzzy sets from each criteria, calculating normalized decision matrix, building weighted normalized decision matrix, determining positive ideal solution matrix and negative ideal solution matrix, determining range between each alternative value with the positive ideal solution matrix and negative ideal solution matrix, determining preference score for each alternatives [13]. This research did not stop at the end of the last TOPSIS process, but it continues to rank the

Journal of Physics: Conference Series **1397** (2019) 012061 doi:10.1088/1742-6596/1397/1/012061

preference score from each alternative from the biggest to the smallest. In the last step to process the data, we must select the best solution from some of the optimal solution variable. On this step, the results from the fuzzy sets evaluation step will be ranked based on the alternatives preferences and will be grouped based on the percentage of the data for determining every UKT groups. On this research, there are 7 UKT groups, the group I for the lowest group, group II, III, IV, V VI, and group VII as highest group.

2.2. *Data Sources and Types*
Data source for this research comes from the 50 students input. Students will fill out the online form including 26 questions. From the 26 questions, 13 points will be taken as main criteria to describe financial capabilities of the new student of Universitas Negeri Yogyakarta, these 13 points are:

a. Number of Family Dependents (person) (C1)
b. Number of Family Dependents on Education (person) (C2)
c. Latest Electric Bill (Rp.) (C3)
d. Latest Land and Building Tax (Rp.) (C4)
e. Building Ownership Status (C5)
f. Land Ownership Status (C6)
g. Overall Area of Building (m^2) (C7)
h. Motorcycle Price (Rp.) (C8)
i. Car Price (Rp.) (C9)
j. Overall Gold Value (Rp.) (C10)
k. Monthly Father's Gross Income (Rp.) (C11)
l. Monthly Mother's Gross Income (Rp.) (C12)
m. Overall Monthly Family Income (Rp.) (C13)

3. Result and Discussion
On this research, economic data from the new student of Universitas Negeri Yogyakarta will be ranked. There are 13 criteria used for determining student's UKT nominal on this research: C1 = Number of Family Dependents, C2 = Number of Family Dependents on Education, C3 = Latest Electric Bill, C4 = Latest Land and Building Tax, C5 = Building Ownership Status, C6 = Land Ownership Status, C7 = Overall Area of Building, C8 = Motorcycle Price, C9 = Car Price, C10 = Overall Gold Value, C11 = Monthly Father's Gross Income, C12 = Monthly Mother's Gross Income, dan C13 = Overall Monthly Family Income.

The researchers using TOPSIS method for solving this MADM case. These are steps:

Step 1: Problems Representation

a. The objective of this decision was to achieve the determination for grouping student's UKT nominal starting from group I to group VII based on the data obtained from the new students of Universitas Negeri Yogyakarta.
b. There are 13 decision criteria: C = {C1,C2,C3,C4,C5,C6,C7,C8,C9,C10,C11,C12,C13}.
c. Structural hierarchy of the problems as shown in figure 2.

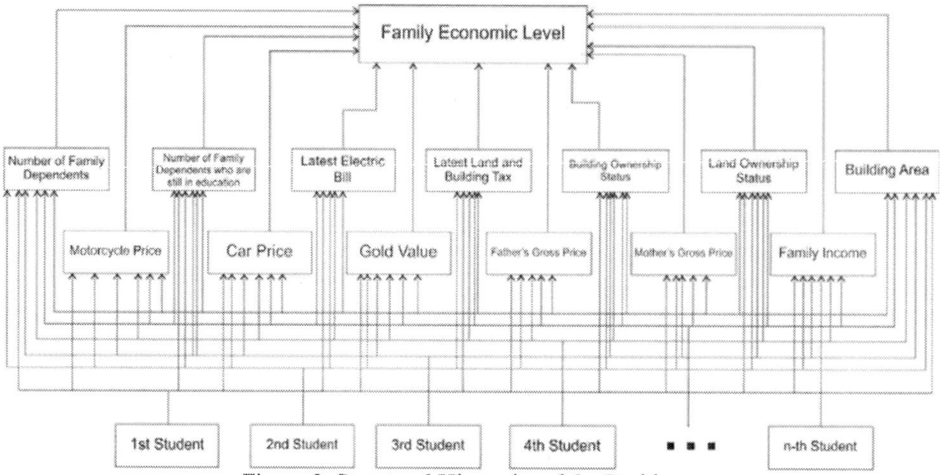

Figure 2. Structural Hierarchy of the Problems

Step 2: Fuzzy Sets Evaluation

On this step, researchers used TOPSIS method explained as follows:

a. Determining Fuzzy Sets from Each Criteria

The first thing to do is to form fuzzy sets for each criteria, in this research using discrete data so that fuzzy sets from each criteria on this research as shown in figure 3 for criteria C1, figure 4 for criteria C2, figure 5 for criteria C3, figure 6 for criteria C4, figure 7 for criteria C5, figure 8 for criteria C6, figure 9 for criteria C7, figure 10 for criteria C8, figure 11 for criteria C9, figure 12 for criteria C10, figure 13 for criteria C11, figure 14 for criteria C12, dan figure 15 for criteria C13.

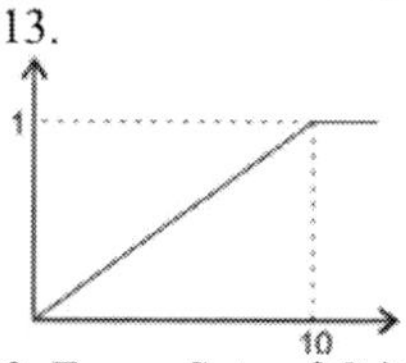

Figure 3. Fuzzy Sets of Criteria C1

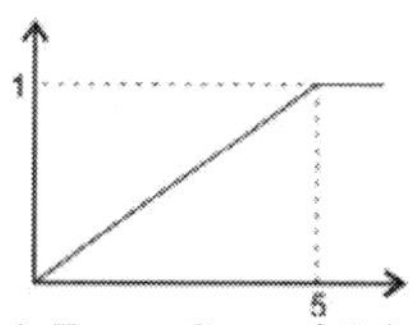

Figure 4. Fuzzy Sets of Criteria C2

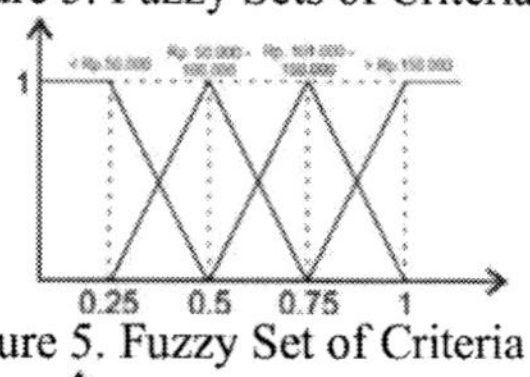

Figure 5. Fuzzy Set of Criteria C3

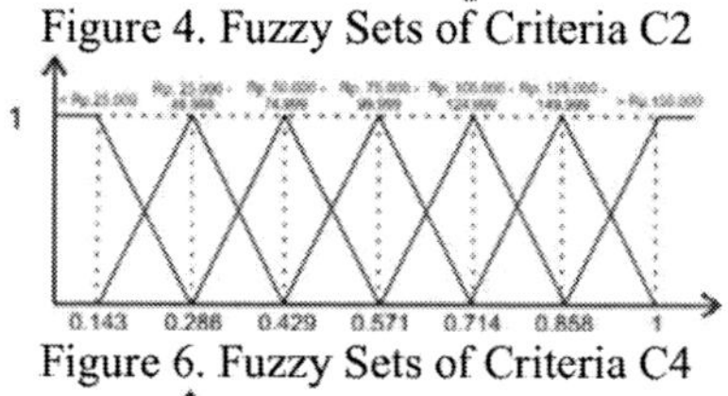

Figure 6. Fuzzy Sets of Criteria C4

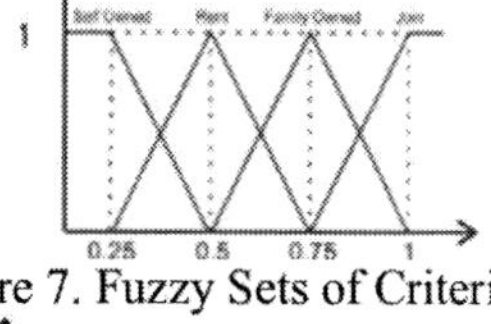

Figure 7. Fuzzy Sets of Criteria C5

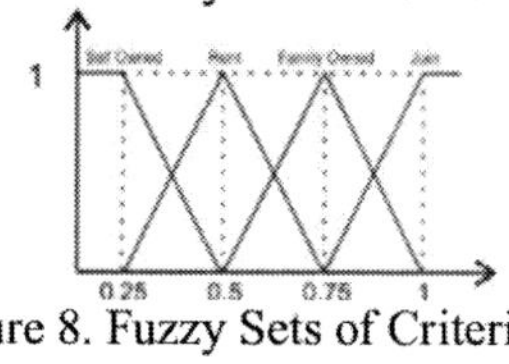

Figure 8. Fuzzy Sets of Criteria C6

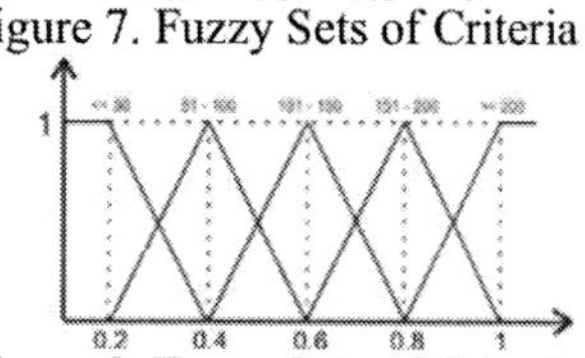

Figure 9. Fuzzy Sets of Criteria C7

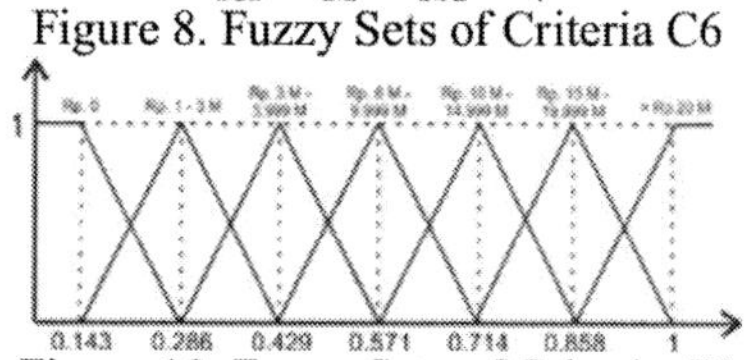

Figure 10. Fuzzy Sets of Criteria C8

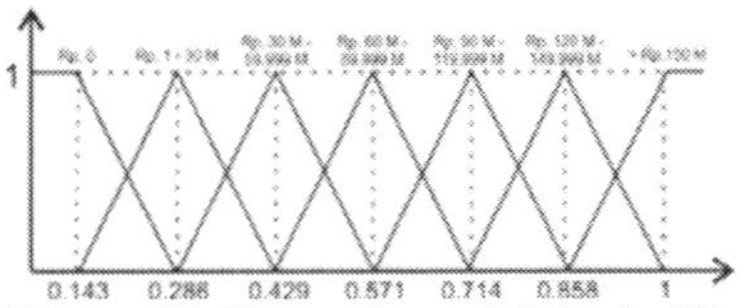

Figure 11. Fuzzy Sets of Criteria C9

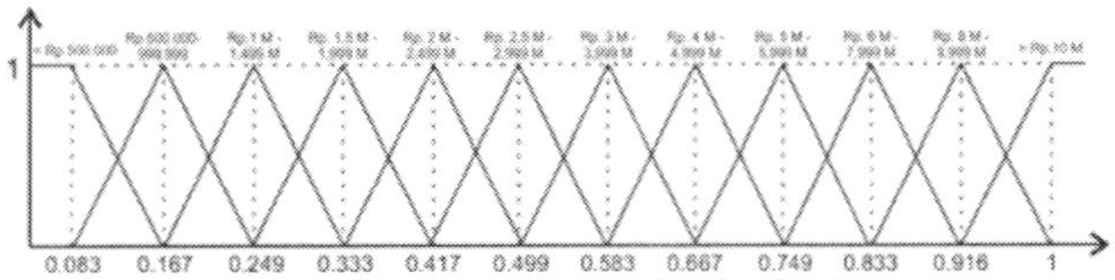

Figure 12. Fuzzy Sets of Criteria C10

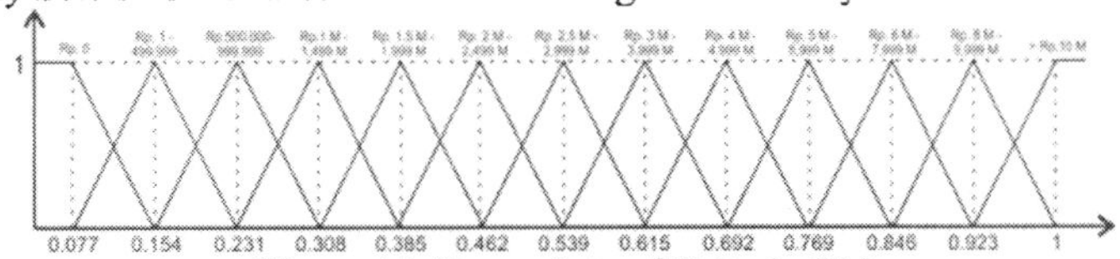

Figure 13. Fuzzy Sets of Criteria C11

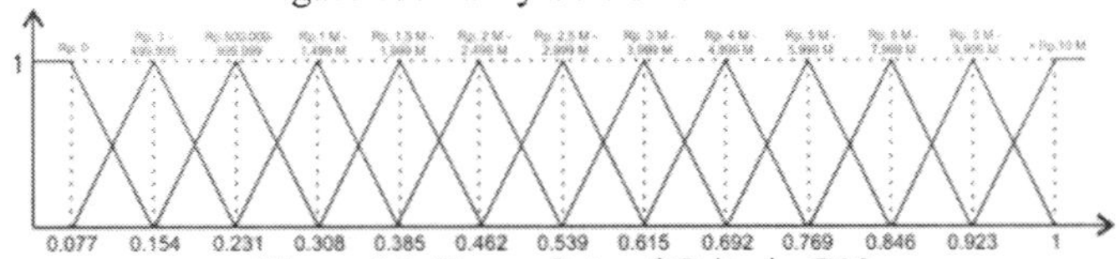

Figure 14. Fuzzy Sets of Criteria C12

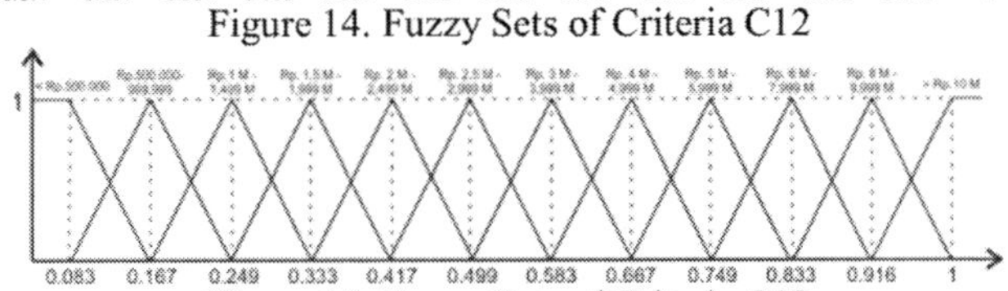

Figure 15. Fuzzy Sets of Criteria C13

b. Calculating Normalized Decision Matrix
Normalized decision matrix calculation can be done using equation 1.

$$r_{ij} = \frac{x_{ij}}{\sqrt{\sum_{i=1}^{m} x_{ij}^2}}; where\ i = 1,2,\dots,m; and\ j = 1,2,\dots,n \quad (1)$$

Where, x_{ij} is the i-th alternative performance rating of the j-th criteria.

c. Building Weighted Normalized Decision Matrix
A^+ positive ideal solution and A^- negative ideal solution can be determined by normalized score rating (Y_{ij}) as shown on the equation 2.

$$y_{ij} = w_i r_{ij}; where\ i = 1,2,\dots,m; and\ j = 1,2,\dots,n \quad (2)$$

Where, w_i is the weight value of the i-th criteria chosen in accordance with the policies in force at each university.

d. Determining Positive Ideal Solution Matrix and Negative Ideal Solution Matrix
Positive ideal solution (A^+) calculated using equation 3.

$$A^+ = (y_1^+, y_2^+, y_3^+, \dots, y_n^+) \quad (3)$$

by,

$$y_j^+ = \begin{cases} \max y_{ij}; if\ j\ is\ a\ profit\ attribute \\ \min y_{ij}; if\ j\ is\ a\ cost\ attribute \end{cases}$$

On this research, the researchers using criteria C1, C2, C5, and C6 as profit attributes while using criteria C3, C4, C7, C8, C9, C10, C11, C12, and C13 as cost attributes in the y_j^+ calculation.
Negative ideal solution (A^-) calculated using equation 4.

$$A^- = (y_1^-, y_2^-, y_3^-, \dots, y_n^-) \quad (4)$$

by,

$$y_j^- = \begin{cases} \min y_{ij}; if\ j\ is\ a\ profit\ attribute \\ \max y_{ij}; if\ j\ is\ a\ cost\ attribute \end{cases}$$

On this research, the researchers using criteria C3, C4, C7, C8, C9, C10, C11, C12, and C13 as margin attributes while using criteria C1, C2, C5, and C6 as cost attributes in the y_j^- calculation.

e. Determining Range Between Each Alternative Value with the Positive Ideal Solution Matrix and Negative Ideal Solution Matrix
Range between alternative A_i with positive ideal solution formulated on the equation 5.

$$D_i^+ = \sqrt{\sum_{j=1}^{n}(y_i^+ - y_{ij})^2}\, ; i = 1,2,\dots,m \qquad (5)$$

Range between alternative A_i with negative ideal solution formulated on the equation 6.

$$D_i^- = \sqrt{\sum_{j=1}^{n}(y_{ij} - y_i^-)^2}\, ; i = 1,2,\dots,m \qquad (6)$$

Step 3: Optimal Alternative Selection

There are processes on this step as follows:

a. Determining Preference Score for Each Alternatives
Proximity of each alternatives to the ideal solution calculated using equation 7.

$$V_i = \frac{D_i^-}{D_i^- + D_i^+}; i = 1,2,\dots,m \qquad (7)$$

b. Data Ranking based on Preference Score
After each new students' family economic preference score obtained, process carried out by ranking the data from the biggest to the smallest. From the sequence taken 5% data from the first sequence as UKT group, next 5% as UKT group II, next 10% as UKT group III, next 15% as UKT group IV, next 20% as UKT group V, next 20% as UKT group VI, and the rest as UKT group VII.

On this research, the result from the TOPSIS method in this MADM case compiled in an android application so it can be used directly. Figure 15 shows the input page of student's economic data. Meanwhile, figure 16 shows the configuration page where settings related to the score of each variable are located.

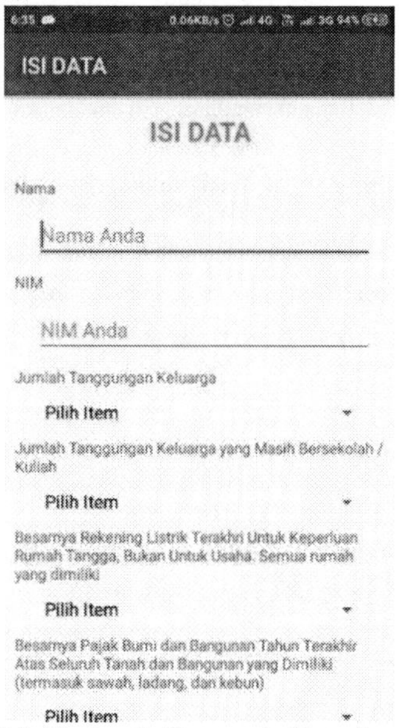

Figure 15. Student's Economic Data Input Page

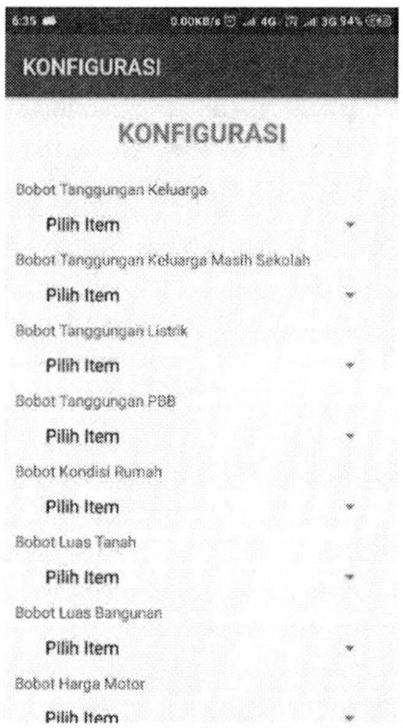

Figure 16. Criteria Score Configuration

After the student input their family economic data and the admin input score configuration of each criteria, the system will perform calculations using TOPSIS method that has been applied in the system. The implementation process is finding the model first after that the coding process is carried out to form an android application based on the model. In designing this application, the database system used is the Firebase online database so the calculation process is based online. Figure 17 shows UKT nominal calculation page based on the economic data input and TOPSIS method.

Figure 17. UKT Nominal Calculation Page

On the nominal calculation page of the UKT that has been shown in figure 17 it is shown that after the data is entered into the application system, the UKT group is generated from each data. The classification of UKT from each data is based on the order of the economic level taking into account the proportion of each class of UKT that has been previously determined. The results was 7 category or group, group I UKT is the lowest group, with UKT nominal as Rp500.000,00 then group II as Rp1.000.000,00 group III as Rp2.400.000,00 group IV as Rp3.630.000,00 group V as Rp4.235.000,00 group VI as Rp4.840.000,00 and group VII as the highest UKT group with UKT nominal as much as Rp5.645.000,00.

4. Conclusion and Recommendation

From this research result, we can conclude that by using Fuzzy Multi Attribute Decision Making, we can classify UKT groups for the new students of Universitas Negeri Yogyakarta starting from group I UKT to group VII UKT faster and easier. By using the TOPSIS method, we can improve the decision-making process for MADM case for better results. Process needed for determining new student's UKT group will be more effective and accurate by using the android application compared to conventional method used before because it can prevent human error when groupping the UKT nominal and can determine what group that nominal belongs to with ease for under 5 minutes fast. The process become more effective because the system was created as an application based on android. Users only needed to input the data then the application will process it, so users did not need to manually process the data. This application will do the calculation with better accuration because in the calculating process was more clearly and objective than manual calculation. In this study only limited to implementing calculations from the TOPSIS technique on FMADM problems so that even though the data has the same type it can always classified into 7 UKT groups. The recommendation for further research is to consider the condition if the data has the same type must produce a UKT group or not depend entirely on the proportion distribution for each UKT group.

References

[1] Rambing P N, Rorong A J and Palar N R A 2014 *Sam Ratulangi University Students Perceptions about Single Student's Tuition Fee Policy*

[2] Retnoningsih Y D and Marom A. *Analysis of the Policy for the Implementation of Student's Single Tuition Fee-Based Education for State Universities of the Faculty of Social and Political Sciences, Diponegoro University, Semarang, Central Java* pp 1–15

[3] Nur M, Utomo Y, Permanasari A E, Tungadi E and Syamsuddin I 2017 *Education In Indonesia : A Comparative Analysis Of Data Mining Classification Algorithms* pp 113–117

[4] Muchsin A K and Sudarma M 2015 *Application of Fuzzy C-Means for Determining the Student's Single Student Tuition Fee* 6(3) pp 175–84

[5] Rokhman S, Rozi I F and Asmara R A 2017 *Development of Decision Support Systems for Determining Student UKT Using the Moora Method in Malang State Polytechnic CaseStudy. J Inform Polinema* 3(4) pp 36–42

[6] Karim B, Sentinuwo S R and Sambul A M 2011 *Determination of the Amount of Student's Single Tuition Fee for New Students at Sam Ratulangi University Using Data Mining* 11(1)

[7] Wardi Y, Abror and Trinanda O 2018 *The Marketing of Higher Education: Managing Student Loyalti Based on Tuition Fee Policy and Service Quality Journal of Business dan Management* 19(2) pp 101–8

[8] UNY 2018 *KEPMEN RISTEKDIKTI UKT & BKT UNY* (Yogyakarta: Universitas Negeri Yogyakarta)

[9] Permatasari D, Azizah I N, Hadiat H L and Abadi A M 2017 *Classification of Toodler Nutritional Status Using Fuzzy Inference System (FIS)* AIP Conference Proceedings 1868

[10] Fatoni, Kurniawan and Munandar W 2014 *Fuzzy Multiple Attribute Decision Making (FMADM) Method of Assessment of Prospective Beneficiaries* J Ilm MATRIX 16(4) pp 45-58

[11] Ribeiro R A 1996 *Fuzzy multiple attribute decision making : A review and new preference elicitation techniques Fuzzy Multiple Attribute Decision Making : A Review and New Preference Elicitation Techniques* Fuzzy Sets Syst 0114

[12] Wahyuningsih D 2016 *System for Measuring Lecturer Performance with Fuzzy Multiple Attribute Decision Making (FMADM)* 17(2) pp 227–36

[13] Rizal M, Nurkifli E H and Nurinasari B *Hybrid Fuzzy Multiple Attribute Decision Making in Selection of Elective Courses*

[14] Kusumadewi S, Hartanti S, Harjoko A and Wardoyo R 2006 *Fuzzy Multi-Attribute Decision Making (FUZZY MADM)*. (Yogyakarta: Penerbit Graha Ilmu)

Multivariate Adaptive Generalized Poisson Regression Spline (MAGPRS) on the number of acute respiratory infection infants

B W Otok[1*], S Hidayati, and Purhadi

Departement of Statistics, Institut Teknologi Sepuluh Nopember, Surabaya

*Corresponding author e-mail: dr.otok.bw@gmail.com

Abstract. Acute Respiratory Infection (ARI) is an infectious disease of the respiratory tract that affects the structure of the respiratory tract. The ARI is a health problem that should not be ignored because it causes high infant mortality. Therefore, it is important to know the factors that influence the ARI. Generalized Poisson Regression (GPR) is one of method that can handle cases of overdispersion (variance is greater than the mean) or underdispersion (variance is less than the mean). Multivariate Adaptive Regression Spline as a statistical method for fitting the relationship between a set of input variables and dependent variables. This research is the development of the MARS method and GPR namely MAGPRS. The application of the MAGPRS model was carried out in the case of the number of ARI in infants from surabaya health departement 2017. The results showed that the importance of predictor variables in MAGPRS, the variables affecting the number of ARI patients in infants are the percentage of low birth weight (X_2), the percentage of unhealthy houses (X_5), and the percentage given non-exclusive breastfeeding to infants (X_1).

1. Introduction

Acute Respiratory Infection (ARI) is a disturbing infection respiratory processes caused by viruses or bacteria that attacks nose, respiratory tube, sinuses, and pharynx larynx. Channel infections acute breathing is recorded as the most common disease the community, especially the babies and children who usually experience the symptoms of a cold cough are at least three to six times a year. ARI is classified as an airborne disease, contact with people who are infected and can be contaminated by carrying a virus and bacteria. If ARI is not immediately there will be special treatment severe illness because it enters the lung tissue and will cause it pneumonia, so that it affects death to people who experience it especially the age of a baby child. ARI is a disease caused by viruses, bacteria and rickets and fungi which are transmitted through the air (water borne disease). Causative factor which can increase the incidence of ARI, namely environmental factors, factors babies, behavioral factors, family factors, and health services [1].

A generalized poisson regression is an alternative the model for the data in the form of count where there is an assumption violation in the poisson distribution, the mean and variance have the same value [3]. One of the based model that can overcome overdispersion or underdispersion cases has been developed is MARS. Multivariate Adaptive Regression Spline as a statistical method for fitting the relationship between a set of input variables and dependent variables [4].

Multivariate Adaptive Generalized Poisson Regression Spline (MAGPRS) model is used as the development of the MARS and Generalized Poisson Regression (GPR). The purpose of this study is the first parameter estimation of MAGPRS model using Weighted Least Squares (WLS) method. The next do the test statistics from the MAGPRS model using Maximum Likelihood Ratio Test (MLRT). The application of MAGPRS model was carried out in the case of the number of Acute Respiratory Infection in babies from surabaya health departement 2017.

2. Literature review

2.1. Generalized Poisson Regression (GPR)

Poisson regression cannot be used in overdispersion or underdispersion data. Let $Y_i \sim GP(\mu,\theta)$, $i=1,2,...,n$ then Y_i has a probability function [5]

$$f(y_i;\mu,\theta)=\left(\frac{\mu}{1+\theta\mu}\right)^{y_i}\frac{(1+\theta y_i)^{y_i-1}}{y_i!}\exp\left(\frac{-\mu(1+\theta y_i)}{1+\theta\mu}\right) \tag{1}$$

where mean of Y_i is given by $E(Y_i)=\mu$ and the variance of Y_i is given by $Var(Y_i)=\mu(1+\theta\mu)^2$. The parameter θ measures the dispersion. When $\theta=0$, the model reduces to the Poisson regression (PR) model. When $\theta>0$, the model represents count data with over-dispersion and when $\theta<0$, the GPR model represents count data with under-dispersion.

2.2. Multivariate Adaptive Regression Spline (MARS)

MARS use a set of special function to capture the nonlinearity of data, which is approximated with different regression slope in the corresponding interval of each predictor. The intervals are closed and non-overlapping expect for the boundaries, which are called knots. MARS model which is a two-stage process; forward and backward. In the first stage, MARS constructs a model with an extra large number of Basis Function (BF), which deliberately overfits the data. Then, some of the BFs that contribute least to the overall performance are removed. Therefore, the forward construction may initially include insignificant model terms. In the backward pruning step, these terms are excluded. Thus, the backward step reduces the complexity of the model without degrading the fit to the data.

The MARS model is formulated as follows:

$$f(\mathbf{x}_i)=\alpha_0+\sum_{m=1}^{M}\alpha_m\prod_{k=1}^{K_m}\left[s_{km}(x_{v(k,m)\mathrm{i}}-t_{km}\right]_+ +\varepsilon_i=\alpha_0+\sum_{m=1}^{M}\alpha_m B_{mi}(\mathbf{x}_i)+\varepsilon_i \tag{2}$$

where,

α_0 : intercept of the MARS model

α_m : regression coefficient on the m th basis function

$B_{mi}(\mathbf{x}_i)$: the m th basis function in MARS

t_{km} : knot

s_{km} : the basic function sign is if +1 knot is located on the right or -1 if the knot is on the left

ε_i : error in observation to i

$x_{v(k,m)\mathrm{i}}$: independent variable

At the end of the forward stage of the MARS algorithm, the largest model that overfits the data is obtained. In the backward stage, in each step, the least contributing BF that causes the smallest increase in the residual sum of squares is deleted from the model set. In order to estimate the optimal value of the model size, MARS again uses the GCV, defined as follows:

$$GCV = \frac{MSE}{\left[1-\frac{C(\tilde{M})}{n}\right]^2} = \frac{n^{-1}\left[\left(\mathbf{y}-\hat{\mathbf{f}}(\mathbf{x})\right)^T\left(\mathbf{y}-\hat{\mathbf{f}}(\mathbf{x})\right)\right]}{\left[1-\frac{C(\tilde{M})}{n}\right]^2} \tag{3}$$

where,

$C(\tilde{M})$: penalty function of model complexity

$\hat{\mathbf{f}}(\mathbf{x})$: estimated function in MARS

GCV : generalized cross validation statistic

2.3. *Multivariate Adaptive Generalized Poisson Regression Spline (MAGPRS)*

The Multivariate Adaptive Generalized Poisson Regression Spline model is a development from MARS and GPR. The response variable is generalized poisson distribution $Y \sim GP(\mu,\theta)$ with $E(Y)=\mu$, and $Var(Y)=\mu(1+\theta\mu)^2$ [10]. The assumptions can be written as,

$$\underset{\sim}{y} = \exp(\mathbf{B}\underset{\sim}{\alpha}) + \underset{\sim}{\varepsilon} \tag{4}$$

where,

$$\underset{\sim}{y} = (y_1, y_2, ..., y_n)^T, \underset{\sim}{\alpha} = (\alpha_0, \alpha_1, ..., \alpha_M)^T, \underset{\sim}{\varepsilon} = (\varepsilon_1, \varepsilon_2, ..., \varepsilon_n)^T \tag{5}$$

Probability density function is said to be an exponential family if the function can be transformed into an exponential family function. Recall that the single-parameter exponential family is expressed as:

$$P(x|\eta) = h(x)\exp\{\eta^T T(x) - A(\eta)\} \tag{6}$$

where η is the natural parameter, $T(x)$ is the sufficient statistics, $A(\eta)$: log partition function and μ is the mean parameter. We can write out the Generalized Poisson distribution in the exponential family form by applying the exp(log(.)) function:

$$\begin{aligned} P(x|\eta) &= \exp\left\{\log\left(\left(\frac{\mu}{1+\theta\mu}\right)^x \frac{(1+\theta x)^{x-1}}{x!}\exp\left(\frac{-\mu(1+\theta x)}{1+\theta x}\right)\right)\right\} \\ &= \exp\left\{\log\left(\frac{\mu}{1+\theta\mu}\right)^x + \log\frac{(1+\theta x)^{x-1}}{x!} + \log\left(\exp\left(-\frac{\mu(1+\theta x)}{1+\theta x}\right)\right)\right\} \\ &= \frac{1}{x!}\frac{(1+\theta x)^{x-1}}{(1+\theta\mu)^x}\exp\left\{x\log\mu - \frac{\mu(1+\theta x)}{1+\theta x}\right\} \end{aligned} \tag{7}$$

where $\mu = \exp\{\eta\}$ and $\eta = \log\mu$, $\eta = \mathbf{B}\underset{\sim}{\alpha}$

The MAGPRS model as follows,

$$\begin{aligned} \mu &= \exp\{\eta\} \\ \mu &= \exp(\mathbf{B}\underset{\sim}{\alpha}) \\ \mu &= \exp\left(\alpha_0 + \sum_{m=1}^{M}\alpha_m B_{mi}(\mathbf{x}_i)\right) \end{aligned} \tag{8}$$

the parameter estimation Multivariate Adaptive Generalized Poisson Regression Spline use Weighted Least Squares method. The parameter estimation of the MAGPRS model as follows [10],

$$\hat{f}(\underset{\sim}{x}) = \exp(\mathbf{B}\hat{\underset{\sim}{\alpha}}) = \exp\left(\mathbf{B}(\mathbf{B})^{-1}\ln\left(\left(\mathbf{B}^T\exp\left(\hat{\underset{\sim}{\alpha}}^T\mathbf{B}^T\right)W\right)^{-1}\mathbf{B}^T\exp\left(\hat{\underset{\sim}{\alpha}}^T\mathbf{B}^T\right)W\underset{\sim}{y}\right)\right)$$
$$\hat{f}(\underset{\sim}{x}) = \left(\mathbf{B}^T\exp\left(\hat{\underset{\sim}{\alpha}}^T\mathbf{B}^T\right)W\right)^{-1}\mathbf{B}^T\exp\left(\hat{\underset{\sim}{\alpha}}^T\mathbf{B}^T\right)W\underset{\sim}{y} \quad (9)$$

2.4. *Statistics Test of Multivariate Adaptive Generalized Poisson Regression Spline (MAGPRS)*

The parameter estimation in Multivariate Adaptive Generalized Poisson Regression Spline (MAGPRS), carried out testing simultaneously use the Maximum Likelihood Ratio Test (MLRT) method. The hypothesis of simultaneously testing is $H_0 : \alpha_1 = ... = \alpha_m = 0$ and H_1: at least have one $\alpha_j \neq 0, j = 1,....m$. The set of parameters that under the population $\Omega = \{\theta, \alpha_1,.....,\alpha_m\}$, and set of parameters that under H_0 is $\omega = \{\theta\}$. Then, the test statistics with equation below $G^2 = -2\ln\Lambda = 2\left(\ln L(\hat{\Omega}) - \ln L(\hat{\omega})\right)$ reject H_0 if $G^2 > \chi^2_{\alpha,v}$ when $v = n(\Omega) - n(\omega)$ [10].

3. Research methodology

The application of MAGPRS was carried out in the case of the number of Acute Respiratory Infection in babies in Surabaya, East Java Province 2017. Data sourced from surabaya health departement. The dependent variable of this study is the number of ARI infants (Y). The seven independent variables which influence the response variables are the percentage of babies given non-exclusive breastfeeding (X_1), the percentage of birth weight is low (X_2), the percentage of basic immunization is incomplete (X_3), the percentage of unclean and unhealthy life behavior (X_4), the percentage of unhealthy houses (X_5), the percentage of babies not getting vitamin A (X_6), the percentage of not getting coverage of infant health services (X_7). Step of analyses in this study are as follows:

- Describing the characteristics of the number of ARI infants in babies as research object
- Carrying out overdispersion and underdispersion testing
- Estimating the parameters of the MAGPRS model with the Weighted Least Squares (WLS)
- Conducting hypothesis testing simultaneously and partially with Maximum Likelihood Ratio Test (MLRT)
- Drawing conclusions from the analysis of the MAGPRS model

4. Analysis and discussion

The first step of analysis is to detect overdispersion or underdispersion of the number of ARI in infants. In the Poisson regression, the assumption of equidispersion between a mean and a variance of the response variable must be satisfied. However, this assumption is often violated in real cases. The variance of the response is often greater than the mean resulting an overdispersion problem. To detect the overdispersion, we use a value of deviance/df or pearson/df. If the deviance/df or pearson/df is greater than 1, then the overdispersion occurs.

Table 1. Deviance and Pearson values from the Poisson regression model

Variable	Criteria	Value/df
The number of ARI	Deviance	**14.44**
	Pearson chi-square	16.53

$$H_0 : \frac{Var(Y)}{E(Y)} = 1$$
$$H_1 : \frac{Var(Y)}{E(Y)} > 1$$

$$D = 2\sum_{i=1}^{n}\left(Y_i \ln\frac{Y_i}{\hat{\mu}_i} - (Y_i - \hat{\mu}_i)\right)$$

Table 1 shows that the Deviance/df value of 14.44 is greater than 1. Thus, the null hyphothesis is rejected and concluded that the number of ARI in infants has an overdispersion.

The number of predictors in this study is 7 so the appropriate numer of BF includes 14, 21, and 28. Maximum Interaction (MI) is the number of interactions that can occur in the model. The maximum interaction used in this study is 1, 2, and 3 because according to Friedman (1991) if the maximum interaction used is more than 3, the value of GCV will increase and the model used will be increasingly complex. Minimum Observation (MO) is the minimum number of observations between knots. In this study the MO used is 0,1,2,3,4, and 5 because the GCV value above will increase. The number of possible models based on the combination are as many as 54 models. Based on the empirical using Multivariate Adaptive Generalized Poisson Regression Spline (MAGPRS) of trial and error combination BF, MI, and MO, the combination of which produces minimum GCV value is combination of 28, 3, 1 with a value of GCV= 6.08E-05 with $R^2 = 0.878131$. Based on the results of this combination, it is known MAGPRS models produced are as follows:

$$\hat{\mu} = \exp\left(\hat{\alpha}_0 + \sum_{m=1}^{M} \hat{\alpha}_m \prod_{k=1}^{K_m} \left[s_{km} (x_{v(k,m)i} - t_{km} \right]_+ \right)$$

$$\hat{\mu} = \begin{bmatrix} 2.05874+1.27478h(x_2-0.12)+1.18897h(96.64-x_5)+3.19983h(x_5-96.64)+0.34207h(x_2-0.12) \\ \times h(x_5-96.47)+0.89811h(x_2-0.12)\times h(96.47-x_5)+0.99921h(x_1-85.17)\times h(x_2-0.12) \\ \times h(96.47-x_5)+1.00446h(85.71-x_1)\times h(x_2-0.12)\times h(96.47-x_5)+1.00356h(x_1-72.3) \\ \times h(96.64-x_5)+0.99399h(72.3-x_1)\times h(96.64-x_5)+1.00450h(x_1-72.3)\times h(x_2-1.38)\times h(96.64-x_5) \\ +0.97645h(x_1-72.3)\times h(1.38-x_2)\times h(96.64-x_5)+1.00043h(x_1-72.3)\times h(x_4-91.31)\times h(96.64-x_5) \\ +1.00018h(x_1-72.3)\times h(91.31-x_4)\times h(96.64-x_5)+0.99099h(72.3-x_1)\times h(96.64-x_5)\times h(x_7-99.88) \\ +0.99908h(72.3-x_1)\times h(96.64-x_5)\times h(99.88-x_7)+0.99999h(x_1-72.3)\times h(96.64-x_5) \\ \times h(x_6-83.15)+0.99559h(x_1-72.3)\times h(96.64-x_5)\times h(83.15-x_6)+0.99402h(x_3-92.07) \\ \times h(96.64-x_5)+1.00260h(92.07-x_3)\times h(96.64-x_5)+1.05517h(x_1-78.36)\times h(x_2-0.12) \\ \times h(x_5-96.47)+1.0619h(78.36-x_1)\times h(x_2-0.12)\times h(x_5-96.47)+1.00114h(x_3-92.07)\times h(x_4-70.2) \\ \times h(96.64-x_5)+1.00047h(x_3-92.07)\times h(70.2-x_4)\times h(96.64-x_5)+0.99972h(72.3-x_1) \\ \times h(x_4-59.56)\times h(96.64-x_5)+0.99991h(72.3-x_1)\times h(59.56-x_4)\times h(96.64-x_5) \end{bmatrix}$$

Simultaneous hypothesis testing of parameters was carried out to find out predictor variables together to influence the response variable. The testing hypothesis of the MAGPRS parameters simultaneously as follows,

$$H_0 : \alpha_1 = ... = \alpha_M = 0$$
$$H_1 : \textit{there is at least one } \alpha_j \neq 0, j = 1, ..., \mathrm{M}$$

The test statistics of G^2 is 419.45 which is greater than $\chi^2_{(0,05;25)}$ of 37.65. So that the decision to reject H_0 is obtained which means that at least one basis function selected in the MAGPRS model has an influence on the response variable. Therefore it is necessary to conduct partial test as presented in Table 2.

$$H_0 : \alpha_j = 0$$
$$H_1 : \alpha_j \neq 0, j = 1, ..., \mathrm{M}$$

Table 2. Estimation value and testing the partial hypothesis of the parameter MAGPRS

Parameter	Estimation value	Standard Error (SE)	t value	P-Value
(Intercept)	7,22E-01	3.09E-16	2.341E+15	< 2E-16
$h(X_2-0.12)$	2.43E-01	1.14E-16	2.12E+15	< 2E-16
$h(X_2-0.12)*h(X_5-96.47)$	-1.07E+00	4.75E-16	-2.26E+15	< 2E-16
$h(X_2-0.12)*h(96.47-X_5)$	-1.08E-01	2.93E-17	-3.67E+15	< 2E-16
$h(X_5-96.64)$	1.16E+00	3.37E-16	3.45E+15	< 2E-16
$h(96.64-X_5)$	1.73E-01	4.44E-17	3.90E+15	< 2E-16
$h(X_1-85.71)*h(X_2-0.12)*h(96.47-X_5)$	-7.90E-04	7.07E-18	-1.12E+14	< 2E-16
$h(85.71-X_1)*h(X_2-0.12)*h(96.47-X_5)$	4.46E-03	1.52E-18	2.94E+15	< 2E-16
$h(X_1-72.3)*h(96.64-X_5)$	-3.6E-03	4.55E-18	-7.82E+14	< 2E-16
$h(72.3-X_1)*h(96.64-X_5)$	-6.02E-03	2.82E-18	-2.13E+15	< 2E-16
$h(X_1-72.3)*h(X_2-1.38)*h(96.64-X_5)$	4.49E-03	4.01E-18	1.12E+15	< 2E-16
$h(X_1-72.3)*h(1.38-X_1)*h(96.64-X_5)$	-2.38E-02	8.66E-18	-2.75E+15	< 2E-16
$h(X_1-72.3)*h(X_4-91.31)*h(96.64-X_5)$	4.31E-04	1.41E-18	3.05E+14	< 2E-16
$h(X_1-72.3)*h(91.31-X_4)*h(96.64-X_5)$	1.82E-04	2.22E-19	8.21E+14	< 2E-16
$h(72.3-X_1)*h(96.64-X_5)*h(X_7-99.88)$	-9.05E-03	2.94E-18	-3.07E+15	< 2E-16
$h(72.3-X_1)*h(96.64-X_5)*h(99.88-X_7)$	-9.18E-04	1.82E-19	-5.03E+15	< 2E-16
$h(X_1-72.3)*h(96.64-X_5)*h(X_6-83.15)$	-8.31E-07	4.57E-19	-1.82E+12	< 2E-16
$h(X_1-72.3)*h(96.64-X_5)*h(83.15-X_6)$	-4.41E-03	4.52E-18	-9.77E+14	< 2E-16
$h(X_3-92.07)*h(96.64-X_5)$	-5.59E-03	5.82E-18	-1.03E+15	< 2E-16
$h(92.07-X_3)*h(96.64-X_5)$	2.60E-06	9.83E-19	2.65E+15	< 2E-16
$h(X_1-78.36)*h(X_2-0.12)*h(X_5-96.47)$	5.37E-02	3.44E-17	1.56E+15	< 2E-16
$h(78.36-X_1)*h(X_2-0.12)*h(X_5-96.47)$	6.01E-02	3.62E-17	1.66E+15	< 2E-16
$h(X_3-92.07)*h(X_4-70.2)*h(96.64-X_5)$	1.14E-03	5.05E-19	2.27E+15	< 2E-16
$h(X_3-92.07)*h(70.2-X_4)*h(96.64-X_5)$	4.79E-04	4.33E-19	1.11E+15	< 2E-16
$h(72.3-X_1)*h(X_4-59.56)*h(96.64-X_5)$	-2.76E-04	1.54E-19	-1.80E+15	< 2E-16
$h(72.3-X_1)*h(59.56-X_4)*h(96.64-X_5)$	-8.61E-05	4.07E-19	-2.11E+14	< 2E-16
θ	-0.080326	0.005364267	-1495687	0.000

This partial test uses a test $|t_{value}|$ compared to t_{table} by using a significance level 5% obtained $t_{(0.05/2;35)} = 2.030$. Table 2 show that $|t_{value}|$ from each predictor variables greater than 2.030 so the decision reject H_0 so that all basis functions selected in the MAGPRS model have an influence on the number of ARI infants in Surabaya 2017.

The value of the importance of predictor variables in the grouping function with increasing GCV values in Table 3. It can be seen that the percentage of low birth weight (X_2) and the percentage of unhealthy houses (X_5) are the most important variables in the MAGPRS model with 100 importance. Then followed by the percentage given non-exclusive breastfeeding to infants (X_1) with an importance of 81.4. In the fourth order, the variable the percentage of basic immunization is incomplete (X_3) and the percentage of unclean and unhealthy behavior (X_4) with a level of importance of 63.6. The last two variables are the the percentage of not getting coverage of infant health services (X_7) with a value of importance of 50.6 and the percentage of babies not getting vitamin A (X_6) has the smallest value for the variable importance level of 16.

Table 3. The importance of the predictor variable

variable	nsubsets	gcv	rss
X_2	25	100	100
X_5	25	100	100
X_1	21	81.4	81.4
X_3	18	63.6	63.6
X_4	18	63.6	63.6
X_7	16	50.6	50.6
X_6	4	16	16

Based on the value importance of predictor variables in MAGPRS, the variables that affect the number of ARI patients in infants are the percentage of low birth weight (X_2), the percentage of unhealthy houses (X_5), and the percentage given non-exclusive breastfeeding to infants (X_1).

5. Conclusion

The results showed that the importance of predictor variables in Multivariate Adaptive Generalized Poisson Regression Splines (MAGPRS), the variables affecting the number of ARI patients in infants are the percentage of low birth weight (X_2), the percentage of unhealthy houses (X_5), and the percentage given non-exclusive breastfeeding to infants (X_1).

References

[1] Notoatmodjo S 2010 *Health Research Methodology* Jakarta Rineka Cipta

[2] Consul P and Jain G 1973 *A Generalization of the Poisson Distribution* Technometrics **15** 791-799

[3] Wang W And Famoye F 1997 *Modeling Household Fertility Decisions with Generalized Poisson Regression* Joural of Population Economics **10** 273-283

[4] Friedman J H 1991 *Multivariate Adaptive Regression Splines (with discussion)* Annual Statistics 191-141

[5] Famoye F and Wang W 2004 *Censored Generalized Poisson Regression Model* Computational Statistics and Data Analysis **46** 547-560

[6] Friedman J H 1990a *Estimating Functions of Mixed Ordinal and Categorical Variables Using Multivariate Adaptive Regression Splines* Technical Report LCS 107 Statistics Departement Stanford University

[7] Friedman J H 1990b *Some Comments on Multivariate Adaptive Regression Splines* Seattle WA 98195

[8] Friedman J H 1991 *Multivariate Adaptive Regression Splines (with discussion)* Annual Statistics 191-141

[9] Cameron A C and Trivedi P K 1998 *Regression Analysis of Count Data* Cambridge University Press Cambridge

[10] Hidayati S 2019 *Multivariate Adaptive Generalized Poisson Regression Splines* (MAGPRS) Thesis Department of Statistics FMKSD ITS Surabaya

[11] Ampulembang A P Otok B W Rumiati A T Budiasih 2015 *Biresponses Nonparametric Regression Model Using MARS and its Properties* Applied Mathematical Sciences **9** 1417-1427

Journal of Physics: Conference Series **1397** (2019) 012063 doi:10.1088/1742-6596/1397/1/012063

Modeling and hypothesis testing for the factors affecting infant's diarrhea using Generalized Poisson Regression

B W Otok [1*], C B G Allo[1], and Purhadi[1]

[1]Institut Teknologi Sepuluh Nopember (ITS), Surabaya 60111, Indonesia

*Corresponding author's email: dr.otok.bw@gmail.com

Abstract. Infants are weak individuals. Number of infants with Diarrhea is count data. Count data can be modeled using Poisson Regression. Poisson Regression has assumption that must be met. In the real case, overdispersion or underdispersion often occurs in data. This condition causes Poisson Regression cannot be used to model the data. Another alternative used to model the data with violation of assumption in Poisson Regression is Generalized Poisson Regression. This article will estimate the parameters of Generalized Poisson Regression using Generalized Poisson Regression. After getting the estimate parameters, parameters hypothesis testing simultaneously is done using Maximum Likelihood Ratio Test. There are three independent variables. They are percentage of infants who get exclusive breastfeeding, percentage of infants who get complete basic immunization, and percentage of households who have healthy living behavior. Significant parameter used to build the model. So, model for the factors affecting Diarrhea in infants in Pasuruan Regency is a model consisting complete basic immunization and healthy living behavior in the model.

1. Introduction

Poisson Regression is a standard model for count data. The assumption in Poisson Regression, equidispersion, is a condition that the variance is equal with the average of the response variable. If the variance of variable response is greater than the mean, then it is called overdispersion. If the variance of variable response is smaller than the mean, then it is called underdispersion. The development of the Poisson Regression model has been done by many researchers. One of the developments is parameter estimation of Geographically Weighted Multivariate Poisson Regression [1].

Poisson Regression model cannot be used if the equidispersion assumption is not fulfilled. Generalized Poisson Regression have been developed by many researchers. Generalized Poisson Regression can be used when equidispersion assumption is not fulfilled. Poisson Regression (PR) and Generalized Poisson Regression (GPR) are applied in motor insurance claim data in Malaysia [2]. There are two types of data used and one of them is not fulfill equidispersion assumption. The performance of Poisson, Negative Binomial, and GPR model are compared in the prediction of antenatal care visits in Nigeria [3].

Parameter estimation method is used to get estimator from regression model. There are many parameter estimation methods. There are many parameter estimation methods, like Ordinary Least Square (OLS), Maximum Likelihood Estimation (MLE), Method of Moments (MM), and Generalized Method of Moment (GMM). Maximum Likelihood Estimation (MLE) is used to get estimation of Bivariate Generalized Poisson Regression (BGPR) [4]. They also used Maximum Likelihood Ratio

Test (MLRT) to get hypothesis testing of BGPR. The result of estimate parameter of PR model using MLE and GMM are compared using Acute Respiratory Tract Infection data [5].

Diarrhea is a disease that is familiar to the public. Diarrhea is infectious disease. Diarrhea can be transmitted through food. Children specifically infants affected by Diarrhea need special attention because Diarrhea can cause death. Prevalance of Diarrhea is 8 percent in Indonesia and East Java is in the nineteen position of thirtythree province [6]. Diarrhea is both preventable and treatable. So, it is important to reduce incident of diarrhea in infants. Parameters of GPR are estimated by GMM because the number of moment condition is greater than the parameters. Data of infants affected Diarrhea in Pasuruan Regency is used as application of GPR.

2. Literature review

2.1. *Poisson Regression* (PR)

A random variable Y is said to have a Poisson probability distribution if only if

$$f(y)=\frac{e^{-\mu}\mu^{y}}{y!} \qquad y=0,1,2,... \tag{1}$$

with $\mu>0$ [7]. The mean and variance of Y can be seen in (2).

$$E(Y)=V(Y)=\mu \tag{2}$$

The Poisson Regression (PR) model can be seen in (3).

$$\mu_i=\exp(\beta_0+\beta_1x_{1i}+\beta_2x_{2i}+...+\beta_kx_{ki}) \tag{3}$$

2.2. *Generalized Poisson Regression* (GPR)

Let $Y_j \sim GP(\mu_j,\phi)$ where $j=1,2,...,n$ then Y_j has a probability function:

$$f(y_j\mid\mu_j,\phi)=\left[\frac{\mu_j}{1+\phi\mu_j}\right]^{y_j}\frac{(1+\phi y_j)^{y_j-1}}{y_j!}\exp\left(-\frac{\mu_j(1+\phi y_j)}{1+\phi\mu_j}\right) \tag{4}$$

where:

$\mu_j=e^{\mathbf{x}_j^T\boldsymbol{\beta}}$

$\mathbf{x}_j=\begin{bmatrix}1 & x_{1j} & x_{2j} & \cdots & x_{kj}\end{bmatrix}^T$

$\boldsymbol{\beta}=[\beta_0 \;\; \beta_1 \;\; \beta_2 \;\; \cdots \;\; \beta_k]^T$.

If $\phi=0$ then equidispersion. It means equation (4) will same with equation (1). If $\phi>0$ then overdispersion. If $\phi<0$ then underdispersion [8]. The mean and variance of GP can be seen in (5) and (6). There is no difference between Poisson Regression model and Generalized Poisson Regression model. So, the GPR model can be seen in (3).

$$E(Y_j)=\mu_j \tag{5}$$

$$V(Y_j)=\mu_j(1+\phi\mu_j)^2 \tag{6}$$

2.3. *Method of Moments* (MM)

Important thing in Method of Moments is choose the moment condition of model. The moment conditions can be seen in (7) [9]:

$$E(\mathbf{h}(y_j,\boldsymbol{\theta}_0))=\mathbf{0}. \tag{7}$$

After getting moment conditions, the sample moment conditions can be searched by (8).

$$\mathbf{h}_{\mathbf{n}}(\boldsymbol{\theta})=\frac{1}{n}\sum_{j=1}^{n}\mathbf{h}(y_j,\boldsymbol{\theta}) \tag{8}$$

So, the $\hat{\boldsymbol{\theta}}$ by Method of Moments can be obtained by solve the equation in (9). The method of moments only applies when the number of moment conditions is equal with the number of parameters.

$$\mathbf{h}_{\mathbf{n}}(\boldsymbol{\theta})=\mathbf{0} \tag{9}$$

2.4. Generalized Method of Moments (GMM)

Another condition is when the number of moment conditions is greater than the number of parameters. So, we cannot use Method of Moments (MM). First time, GMM is proposed by Hansen (1982) [10]. The best way to get solution for (9) is by the quadratic function $\mathbf{h}_{\mathbf{n}}(\boldsymbol{\theta})^T\mathbf{W}\mathbf{h}_{\mathbf{n}}(\boldsymbol{\theta})$, where $\mathbf{W}$ is a positive definit and symmetric matrix. The optimal matrix can be estimated by an heteroskedasticity and autocorrelation consistent (HAC) matrix like the one proposed by Andrews (1991) [11]. Estimation of weighted matrix as follows:

$$\hat{\mathbf{W}}(\boldsymbol{\theta}^*)=\frac{n}{n-p}\sum_{s=-(n-1)}^{n-1}t\left(\frac{s}{b}\right)\hat{\boldsymbol{\Gamma}}(s) \tag{10}$$

where:

n : number of observations

p : number of parameters

t : kernel function

b : bandwidth

$$\hat{\boldsymbol{\Gamma}}(s)=\begin{cases}\frac{1}{n}\sum_{j=s+1}^{n}\mathbf{h}(\boldsymbol{\theta}^*,y_j,\mathbf{x}_j)\mathbf{h}(\boldsymbol{\theta}^*,y_j,\mathbf{x}_{j-s})^T & \text{if } s\geq 0\\ \frac{1}{n}\sum_{j=-s+1}^{n}\mathbf{h}(\boldsymbol{\theta}^*,y_j,\mathbf{x}_{j+s})\mathbf{h}(\boldsymbol{\theta}^*,y_j,\mathbf{x}_j)^T & \text{if } s<0\end{cases}.$$

There are five types of kernel functions. They are Truncated, Bartlett, Parzen, Tukey-Hanning, and Quadratic Spectral.Choice of kernel functions does not affect the asymptotic properties of GMM, very little is known about the impacts in finite samples [12]. The best bandwidth can be obtained by minimizing the MSE.

Iterative GMM (ITGMM) proposed by Hansen et al (1996) is development of two stage GMM (2GMM) proposed by Hansen (1982) [13]. The algorithm is as follows [12]:

- Compute $\boldsymbol{\theta}^{(0)}=\arg\min_{\theta}\mathbf{h}_{\mathbf{n}}(\boldsymbol{\theta})^T\mathbf{h}_{\mathbf{n}}(\boldsymbol{\theta})$
- Compute the weighted matrix, $\hat{\mathbf{W}}(\boldsymbol{\theta}^{(c-1)})$. $\hat{\mathbf{W}}(\boldsymbol{\theta}^{(c-1)})$ is computed by (10).
- Compute $\boldsymbol{\theta}^{(c)}=\arg\min_{\theta}\mathbf{h}_{\mathbf{n}}(\boldsymbol{\theta})^T\hat{\mathbf{W}}(\boldsymbol{\theta}^{(c-1)})^{-1}\mathbf{h}_{\mathbf{n}}(\boldsymbol{\theta})$
- If $\left\|\boldsymbol{\theta}^{(c-1)}-\boldsymbol{\theta}^{(c)}\right\|<\varepsilon_{IT}$ stops where *tol* as small as we want, else $\boldsymbol{\theta}^{(c-1)}=\boldsymbol{\theta}^{(c)}$ and go to 2 where $c=1,2,\cdots,C$ and C is number of iteration.
- Define the ITGMM estimator $\hat{\boldsymbol{\theta}}$ as $\boldsymbol{\theta}^{(C)}$.

2.5. Maximum Likelohood Ratio Test (MLRT)

Maximum Likelihood Ratio Test (MLRT) is used to get statistics test in hypothesis testing simultaneously. There are several steps in MLRT. The steps for GPR model are as follows:

- Making the hyphotesis
  $H_0 : \beta_1 = \beta_2 = \cdots = \beta_k = 0$
  H_1 : There is at least one $\beta_l \neq 0$ where $l = 1,2,\ldots,k$.
- Determine the parameters set under population
  $\Omega = \{\beta_0, \beta_1, \ldots, \beta_k, \phi\}$ where $n(\Omega) = k+2$
- Find $\hat{\Omega}$
- Determine the parameters set under null hypothesis
  $\omega = \{\beta_0, \phi\}$ where $n(\omega) = 2$
- Find $\hat{\omega}$
- Statistics Test finds the distribution of $G^2 < c$, where $G^2 = \dfrac{L(\hat{\omega})}{L(\hat{\Omega})}$. H_0 is rejected if only if $G^2 > \chi^2_{\alpha,v}$, where $v = n(\Omega) - n(\omega)$.

3. Methodology

The data is secondary data obtained from thirtythree health center in Pasuruan Regency, East Java in 2017. The variables used in this study are same with research of Allo *et al* (2019) [14]. Step analysis is as follows:

Step 1. Check the equidipersion. If equidispersion occur in the data then use Poisson Regression, else go to step 2.

Step 2. Check the multicollinearity between independent variables. If there is multicollinearity between independent variables the handle the multicollinearity, else go to step 3.

Step 3. Estimate the parameters of GPR using GMM.

Step 4. Testing the estimated parameters.

4. Result and discussion

Step 1 check the assumption of PR. Overdispersion checking needs to be checked. Value of deviance divided by degree of freedom of Infants affected Diarrhea in Pasuruan Regency is 143.15. Value of pearson chi-square divided by degree of freedom of Infants affected Diarrhea is 160.81. This result shows that overdispersion occurs in the data because the value of deviance divided by degree of freedom of Infants affected Diarrhea and value of pearson chi-square divided by degree of freedom of Infants affected Diarrhea are greater than one. Step 2 check the multicollinearity using Variance Inflation Factor (VIF). If VIF value is more than 10 then there is multicollinearity between independent variables. Every VIF value in Table 1 is not more than 10.

Table 1. VIF value

Variable	VIF
X_1	1.088
X_2	1.049
X_3	1.038

Step 3 estimate the parameter of GPR using GMM. The moment conditions must satisfy equation (7). So, the moment conditions for GPR model with the data is as follows:

$$\mathbf{h}\left(y_j,\mathbf{x}_j,\boldsymbol{\theta}\right)=\begin{bmatrix} y_j-e^{\mathbf{x}_j^T\boldsymbol{\beta}} \\ x_{1j}\left(y_j-e^{\mathbf{x}_j^T\boldsymbol{\beta}}\right) \\ x_{2j}\left(y_j-e^{\mathbf{x}_j^T\boldsymbol{\beta}}\right) \\ x_{3j}\left(y_j-e^{\mathbf{x}_j^T\boldsymbol{\beta}}\right) \\ \left(y_j-e^{\mathbf{x}_j^T\boldsymbol{\beta}}\right)^2-\left(e^{\mathbf{x}_j^T\boldsymbol{\beta}}\left(1+\phi e^{\mathbf{x}_j^T\boldsymbol{\beta}}\right)^2\right) \\ x_{1j}\left(\left(y_j-e^{\mathbf{x}_j^T\boldsymbol{\beta}}\right)^2-\left(e^{\mathbf{x}_j^T\boldsymbol{\beta}}\left(1+\phi e^{\mathbf{x}_j^T\boldsymbol{\beta}}\right)^2\right)\right) \\ x_{2j}\left(\left(y_j-e^{\mathbf{x}_j^T\boldsymbol{\beta}}\right)^2-\left(e^{\mathbf{x}_j^T\boldsymbol{\beta}}\left(1+\phi e^{\mathbf{x}_j^T\boldsymbol{\beta}}\right)^2\right)\right) \\ x_{3j}\left(\left(y_j-e^{\mathbf{x}_j^T\boldsymbol{\beta}}\right)^2-\left(e^{\mathbf{x}_j^T\boldsymbol{\beta}}\left(1+\phi e^{\mathbf{x}_j^T\boldsymbol{\beta}}\right)^2\right)\right) \end{bmatrix} \tag{11}$$

Follow equation (8), we can get sample moment conditions where $\mathbf{x}_j=\begin{bmatrix}1 & x_{1j} & x_{2j} & x_{3j}\end{bmatrix}^T$ and $\boldsymbol{\beta}=\begin{bmatrix}\beta_0 & \beta_1 & \beta_2 & \beta_3\end{bmatrix}^T$. To get the estimator, we need numerical iteration. This paper uses Nelder Mead iteration because function does not need to be differentiable. This paper also uses Quadratic Spectral kernel because the choice of kernel selection does not affect the asymptotic properties of GMM [12] and initial value used is the estimation coefficient from PR with $\varepsilon_{NM}=\varepsilon_{IT}=10^{-30}$. The result can be seen in Table 2. Every independent variable gives negative impact to dependent variable.

Table 2. Estimation result of GPR

Parameter	Estimator	SE
ϕ	0.0986	0.0148
β_0	6.5535	0.4851
β_1	-9.5227e-05	0.0036
β_2	-0.008	0.0038
β_3	-0.0223	0.0081

Step 4 is testing the hypoyhesis. The hypothesis for this case is as follows:
$H_0:\beta_1=\beta_2=\beta_3=0$
H_1: There is at least one $\beta_l\neq 0$ where $l=1,2,3$.
The statistics test can be seen in (12).

$$G^2=-2\ln\left(\frac{L(\hat{\omega})}{L(\hat{\Omega})}\right)=2\left[\ln L(\hat{\Omega})-\ln L(\hat{\omega})\right] \tag{12}$$

G^2 follows Chi Square dsitribution with degree of freedom v, where $v=n(\Omega)-n(\omega)$. The null hypothesis is rejected if $G^2>\chi^2_{\alpha,v}$ [15]. For this case, G^2 is 235.846. If G^2 is compared with $\chi^2_{0,05;3}$ then $G^2>\chi^2_{0.05;3}$. So, At least one independent variable is different from zero. On the other hand, the GMM estimator converges as n goes to infinity to the following distribution [12]:

$$\sqrt{n}\left(\hat{\boldsymbol{\theta}}-\boldsymbol{\theta}_0\right)\xrightarrow{L}N(0,V) \tag{13}$$

where:

$$V=E\left(\frac{\partial \mathbf{h}(x_i,\boldsymbol{\theta}_0)}{\partial \boldsymbol{\theta}}\right)^T \mathbf{W}(\boldsymbol{\theta}_0)^{-1} E\left(\frac{\partial \mathbf{h}(x_i,\boldsymbol{\theta}_0)}{\partial \boldsymbol{\theta}}\right)$$ and $\boldsymbol{\theta}_0$ is true value.

So, $\hat{\boldsymbol{\theta}}$ approximately distributed as $N\left(\boldsymbol{\theta}_0,\frac{\hat{V}}{n}\right)$. The result for hypothesis testing partially can be seen in Table 3. Table 3 shows the significant parameters are ϕ, β_0, β_2, and β_3 with $\alpha=0.05$.

Table 3. p-value for every parameter

Parameter	p-value
ϕ	0.00000
β_0	0.00000
β_1	0.48955
β_2	0.01955
β_3	0.00286

Only significant variables are interpreted in the model. The model for factors affecting Diarrhea in infants in Pasuruan regency can be seen in (14).

$$\hat{\mu}=\exp(6.5535-0.4895X_1-0.008X_2-0.0223X_3) \tag{14}$$

Each there is increasing in the X_2, then the average number of infants affected by diarrhea will decrease one with all other variables are constant. Each there is increasing in the X_3, then then the average number of infants affected by diarrhea will decrease 1.02 with all other variables are constant. The result shows how basic immunization and households who healthy living behavior are important. Table 3 shows X_1 is not significant, but exclusive breastfeeding for infants is very important. So, infants can get good immune.

5. Conclusion

Overdispersion occurs in infant's Diarrhea data in Pasuruan regency. So, Poisson Regression (PR) cannot be used in this data. Generalized Poisson Regression (GPR) can handle the overdispersion in data. In GPR model, the number of moment conditions is greater than the number of parameters. Parameters of GPR are estimated using Generalized Method of Moments. Based on hypothesis testing simultaneously using Maximum Likelihood Ratio Test (MLRT), there are independent variables that significantly affecting Diarrhea in infants. Every coefficient of independent variable is negative. It means all independent variables can reduce Diarrhea in infants. The percentage of infants who get exclusive breastfeeding is not significantly affecting Diarrhea in infants. It can be caused by the high awareness of mothers to give exclusive breastfeeding to their infant. Infants who get complete basic immunization and households who have healthy living behavior can reduce the Diarrhea in infants.

References

[1] Triyanto, Purhadi, Otok B W and Purnami S P 2015 Parameter Estimation of Geographically Weighted Multivariate Poisson Regression *Applied Mathematical Sciences* **9** 4081-93.

[2] Noriszura I and Jemain A A 2005 Generalized Poisson Regresion: An Alternative for Risk Classification *Jurnal Teknologi* **43** 39-54.

[3] Yusuf O B and Ugalahi L O 2015 On the Performance of the Poisson, Negative Binomial and Generalized Poisson Regression Models in the Prediction of Antenatal Care Visits in Nigeria *American Journal of Mathematics and Statistics* **5** 128-36.

[4] Wardani D K, Purhadi and Wibowo W 2016 Parameter Estimation and Hypothesis Testing on Bivariate Generalized Poisson Regression *Yogyakarta. The 3RDICRIEMS.*
[5] Mahpolah, Suharto, Wibowo A and Otok B W 2018 The Estimation of Generalized Method Moment Poisson Regression Model on The Prevalence of Acute Respiratory Tract Infection (RTI) in South Kalimantan *CAUCHY-Jurnal Matematika Murni dan Aplikasi* **5**(3) 161-8.
[6] Kementerian Kesehatan Republik Indonesia 2019 *Riset Kesehatan Dara Tahun 2018* (Jakarta: Kementerian Kesehatan Republik Indonesia).
[7] Wackerly D D, Mendenhall W I and Scheaffer R L 2008 *Mathematical Statistics with Applications* (United States: Thomson Brooks/Cole).
[8] Famoye F, Wulu J T and Singh K P 2004 On The Generalized Poisson Regression Model with An Application to Accident Data *Journal of Data Science* **2** 287-95.
[9] Matyas L 1999 *Generalized Method of Moments Estimation* (New York: Cambridge University Press).
[10] Hansen L P 1982 Large Sample Properties of Generalized Method of Moments Estimators *Econometrica* **50** 1029-54.
[11] Andrews D W K 1991 Heteroskedasticity and Autocorrelation Consistent Covariance Matrix Estimation *Econometrica* **59**(3) 817-58.
[12] Chausse P 2010 Computing Generalized Method of Moments and Generalized Empirical Likelihood with R *Journal of Statistical Software* **34**(11) 1-35.
[13] Hansen L P, Heaton J and Yaron A 1996 Finite-Sample Properties of Some Alternative GMM Estimators *Journal of Business and Economic Statistics* **14** 262-80.
[14] Caecilia B G Allo, B W Otok and Purhadi 2019 Estimation Parameter of Generalized Poisson Regression Model Using Generalized Method of Moments and Its Application *IOP Conference Series: Materials Science and Engineering* **546** 1-7.
[15] Agresti A 2002 *Categorical Data Analysis* (New York: John Wiley & Sons).

Journal of Physics: Conference Series **1397** (2019) 012064 doi:10.1088/1742-6596/1397/1/012064

Smoothing parameter selection method for multiresponse nonparametric regression model using smoothing spline and Kernel estimators approaches

B Lestari[1], Fatmawati[2], I N Budiantara[3], and N Chamidah[2]

[1]Faculty of Mathematics and Natural Sciences, The University of Jember, 37 Kalimantan Street, Campus of Tegal Boto, Jember 68121, Indonesia.
[2]Mathematics Department, Faculty of Science and Technology, Airlangga University, Mulyorejo Street, Campus C UNAIR, Surabaya 60115, Indonesia.
[3]Statistics Department, Sepuluh Nopember Institute of Technology, Campus of Sukolilo, Surabaya 60111, Indonesia.

Corresponding author's e-mail: fatmawati@fst.unair.ac.id

Abstract. The principle problem in multiresponse nonparametric regression model is how we estimate the regression functions which draw association between some dependent (response) variables and some independent (predictor) variables where there are correlations between responses. There are many techniques used to estimate the regression function. Two of them are spline and kernel smoothing techniques. Speaking about smoothing techniques, not only in uniresponse spline and kernel nonparametric regression models but also in multiresponse spline and kernel nonparametric regression models, the estimations of regression functions depend on smoothing parameters. In the privious researches the covariance matrices were assumed to be known. Matrix of covariance is not assumed known in this research. The goals of this research are selecting of optimal smoothing parameters for the model we consider through spline and kernel smoothing techniques. Optimal smoothing parameters can be obtained by taking the solution to generalized cross validation (GCV) optimization problem. The obtained results of this research are the optimal smoothing parameter for smoothing spline estimator approach and the optimal smoothing parameter namely optimal bandwidth for kernel estimator approach.

1. Introduction

In statistical analysis we frequently involve a building mathematical model which examines the functional association between one or more than one dependent (response) and one or more than one independent (predictor) variables. Discussing about association between some dependent (response) variables and some independent (predictor) variables which is drawn by a function, we may not neglect an usual model namely a regression model. In analysis of regression which uses parametric and nonparametric regression models approaches we often meet to principle problem, i.e., estimate functions of regression. For approaching model of nonparametric regression where it has no specific pattern of the regression function, we can use some estimators to estimate function of regression in the model, for example spline, kernel, local linear, etc. In this approach, spline estimator is more flexible to estimate regression function than other estimators. Researchers who have used spline estimator in various cases of models of nonparametric regression with single response are [1-21]. Next, [22-28] used this estimator to estimate functions of regressions in models of nonparametric regression with two responses and more than two responses. On the other hand, [29-35] have proposed kernel estimator to estimate models of nonparametric regression with single response. Furthermore, [36-38] have studied these estimators, i.e., spline and kernel, for estimating functions of regression in models of nonparametric regression with both single response and multi-responses.

Estimate regression functions in models of nonparametric regression with single response that uses smoothing techniques, either spline estimator or kernel estimator, depends on smoothing parameters [2, 10, 20, and 39]. In single response spline nonparametric regression, if smoothing parameter value is very small then it will give a very rought estimator of nonparametric regression function [20]. In contrary, if smoothing parameter value is very large then it will give a very smooth estimator of nonparametric regression function [20]. Therefore, we need to select optimal smoothing parameter in order to obtain estimator that is suitable with data. For this need, some researchers have proposed some selection methods for selecting the optimal smoothing parameters, for instance [2] proposed cross validation (CV) method, [10] proposed unbiased risk (UBR) method, and [20 and 39-40] proposed generalized cross validation (GCV) methods.

All researchers who have been mentioned previously, discussed the estimating methods of regression functions in either uniresponse or multiresponse nonparametric regression models only [1-38]. These researchers have not discussed the selecting of smoothing parameters methods. Further, although [2, 10, 20, and 39-40] have discussed the selecting of smoothing parameters methods, but these researchers have not discussed the selecting of smoothing parameters methods for the multiresponse nonparametric regression model by using kernel estimator. Also, previous researchers have not estimated the covariance matrix. Therefore, in this research we discuss method to select optimal smoothing parameters for the considered model by using both smoothing spline and kernel techniques.

2. Results of research and discussion

Here, we provide a mathematical statistics method for selecting the smoothing parameters for the model that we consider through both smoothing spline and kernel techniques.

2.1. *Smoothing parameters selection method by using smoothing spline estimator approach*

Firstly, suppose (y_{ij}, x_{ij}) is data set which accords with the following multiresponse nonparametric regression model:

$$y_{ij} = g_i(x_{ij}) + \varepsilon_{ij} \ , \quad a_i \le x_i \le b_i \, , \quad i = 1,2,...,p \ , \quad j = 1,2,...,n_i \tag{1}$$

where y_{ij} represents i^{th} response of jth observation, unknown smooth regression functions $g_1, g_2, ..., g_p$ are included in Sobolev space $W_2^m[a_i, b_i]$, and independent random errors ε_{ij} have zero-mean and variance σ_{ij}^2 [28 and 37].

Secondly, based on (1) we have $\underset{\sim}{y} = (\underset{\sim}{y}_1, \underset{\sim}{y}_2, ..., \underset{\sim}{y}_p)'$, $\underset{\sim}{f} = (\underset{\sim}{f}_1, \underset{\sim}{f}_2, .., \underset{\sim}{f}_p)'$, $\underset{\sim}{\varepsilon} = (\underset{\sim}{\varepsilon}_1, \underset{\sim}{\varepsilon}_2, ..., \underset{\sim}{\varepsilon}_p)'$, and $\underset{\sim}{x} = (\underset{\sim}{x}_1, \underset{\sim}{x}_2, ..., \underset{\sim}{x}_p)'$ where $\underset{\sim}{y}_i = (y_{i1}, y_{i2}, ..., y_{in_i})'$, $\underset{\sim}{g}_i = (g_i(x_{i1}), g_i(x_{i2}), ..., g_i(x_{in_i}))'$, $\underset{\sim}{\varepsilon}_i = (\varepsilon_{i1}, \varepsilon_{i2}, ..., \varepsilon_{in_i})'$, $\underset{\sim}{x}_i = (\underset{\sim}{x}_{i1}, \underset{\sim}{x}_{i2}, ..., \underset{\sim}{x}_{in_i})'$. So, for $j = 1,2,...,n_i$ and $i = 1,2,...,p$, the equation (1) can be expressed in the following vector notation:

$$\underset{\sim}{y} = \underset{\sim}{g} + \underset{\sim}{\varepsilon} \tag{2}$$

where $\underset{\sim}{\varepsilon}$ have zero-mean and covariance of $[W(\underset{\sim}{\sigma}^2)]^{-1} = diag(W_1(\underset{\sim}{\sigma}_1^2), W_2(\underset{\sim}{\sigma}_2^2), ..., W_p(\underset{\sim}{\sigma}_p^2))$ [28]. Estimation of regression functions $\underset{\sim}{g}$ in (2) through smoothing spline method occurs as a solution to minimization problem of penalized weighted least-square (PWLS). Thus, we must find $\hat{\underset{\sim}{g}}$ that can make the following PWLS minimum:

$$\underset{g_1,\ldots,g_p\in W_2^m}{Min}\{(\sum_{i=1}^{p}n_i)^{-1}(\underset{\sim}{y}_1-\underset{\sim}{g}_1)'W_1(\sigma_1^2)(\underset{\sim}{y}_1-\underset{\sim}{g}_1)+\ldots+(\underset{\sim}{y}_p-\underset{\sim}{g}_p)'W_p(\sigma_p^2)(\underset{\sim}{y}_p-\underset{\sim}{g}_p)+\lambda_1\int_{a_1}^{b_1}(g_1^{(2)}(\underset{\sim}{x}_1))^2d\underset{\sim}{x}_1+\ldots+\lambda_p\int_{a_p}^{b_p}(g_p^{(2)}(\underset{\sim}{x}_p))^2d\underset{\sim}{x}_p\} \quad (3)$$

for pre-specified value of $\underset{\sim}{\lambda}=(\lambda_1,\lambda_2,\ldots,\lambda_p)'$. Equation (3) shows that the first expression represents sum squares of errors (MSE) which penalizes lack of fit, and the second expression with its weight $\underset{\sim}{\lambda}$ represents penalty of roughness. Thus, the penalty of roughness penalizes the curvature of $\underset{\sim}{g}$. Further, we call λ_i $(i=1,2,\ldots,p)$ as parameter of smoothing. If $0<\lambda_i<+\infty$ then the solution of (3) will be different from interpolation to a linear model. Next, equation (3) will be dominated by the roughness penalty, and estimating of smoothing spline will lead to be a constant, if $\lambda_i\rightarrow+\infty$. On the other hand, the penalty of roughness will not occur in (3) if $\lambda_i\rightarrow 0$, and the data will be interpolated by estimating of smoothing spline. It means that smoothing parameter λ_i controls the trade-off between the goodness of fit:

$$(\sum_{i=1}^{p}n_i)^{-1}(\underset{\sim}{y}_1-\underset{\sim}{g}_1)'W_1(\sigma_1^2)(\underset{\sim}{y}_1-\underset{\sim}{g}_1)+\ldots+(\underset{\sim}{y}_p-\underset{\sim}{g}_p)'W_p(\sigma_p^2)(\underset{\sim}{y}_p-\underset{\sim}{g}_p)$$

and smoothness of the estimate:

$$\lambda_1\int_{a_1}^{b_1}(g_1^{(2)}(\underset{\sim}{x}_1))^2d\underset{\sim}{x}_1+\ldots+\lambda_p\int_{a_p}^{b_p}(g_p^{(2)}(\underset{\sim}{x}_p))^2d\underset{\sim}{x}_p\,.$$

The solution to optimization in (3) is a smoothing spline estimator. Function basis of this estimator is a spline of natural cubic where its knots are $x_1,x_2,\ldots,x_{n_i}$ $(i=1,2,\ldots,p)$. It means that interpolation of special structured spline which depends on the smoothing parameter λ_i value selection gets turn into a representative approach of the regression functions g_i $(i=1,2,\ldots,p)$ in model (1). Next, by applying reproducing kernel Hilbert space (RKHS) method, we determine regression function estimate, and its result is:

$$\hat{\underset{\sim}{g}}_\lambda(\underset{\sim}{x})=\left(\hat{g}_{1,\lambda_1}(\underset{\sim}{x}_1),\hat{g}_{2,\lambda_2}(\underset{\sim}{x}_2),\ldots,\hat{g}_{p,\lambda_p}(\underset{\sim}{x}_p)\right)'=K\hat{\underset{\sim}{d}}+\Sigma\hat{\underset{\sim}{c}}=A(\underset{\sim}{\lambda})\underset{\sim}{y} \quad (4)$$

where

$$A(\underset{\sim}{\lambda})=K[K'M^{-1}W(\underset{\sim}{\sigma}^2)K]^{-1}K'M^{-1}W(\underset{\sim}{\sigma}^2)+\Sigma M^{-1}W(\underset{\sim}{\sigma}^2)[I-K(K'M^{-1}W(\underset{\sim}{\sigma}^2)K)^{-1}K'M^{-1}W(\underset{\sim}{\sigma}^2)]$$

is called as "hat matrix" and a non-singular matrix, so it is a invertible matrix [28 and 37], and $\hat{\underset{\sim}{g}}_\lambda(\underset{\sim}{x})$ is estimated regression function based on smoothing spline estimator where its function basis is natural cubic spline with knots $x_1,x_2,\ldots,x_{n_i}$ $(i=1,2,\ldots,p)$, for a fixed parameter of smoothing $\underset{\sim}{\lambda}>\underset{\sim}{0}$. So, $A(\underset{\sim}{\lambda})$ is a smoother matrix that is positive-definite (symmetrical), and it does not depend on $\underset{\sim}{y}$, but it depends on parameter of smoothing $\underset{\sim}{\lambda}$ and knots $x_1,x_2,\ldots,x_{n_i}$ $(i=1,2,\ldots,p)$ [28, 37 and 40].

Furthermore, we may express the estimated regression function of model in (4) as follows:

$$\hat{\underset{\sim}{g}}_\lambda(\underset{\sim}{x})=A(\lambda_1,\ldots,\lambda_p;\underset{\sim}{\sigma}^2)\underset{\sim}{y} \quad (5)$$

where $\underset{\sim}{\sigma}^2=(\sigma_1^2,\ldots,\sigma_p^2)'$. Mean square error (MSE) of (5) can be determined as follows:

$$MSE(\lambda_1,\ldots,\lambda_p;\underset{\sim}{\sigma}^2)=\frac{(\underset{\sim}{y}-\hat{\underset{\sim}{g}}_\lambda(\underset{\sim}{x}))'W(\underset{\sim}{\sigma}^2)(\underset{\sim}{y}-\hat{\underset{\sim}{g}}_\lambda(\underset{\sim}{x}))}{\sum_{i=1}^{p}n_i}$$

$$=\frac{\left(\underset{\sim}{y}-A(\lambda_1,...,\lambda_p;\underset{\sim}{\sigma}^2)\underset{\sim}{y}\right)' W(\underset{\sim}{\sigma}^2)\left(\underset{\sim}{y}-A(\lambda_1,...,\lambda_p;\underset{\sim}{\sigma}^2)\underset{\sim}{y}\right)}{N}$$

$$=\frac{\left[\left(W(\underset{\sim}{\sigma}^2)\right)^{\frac{1}{2}}\left(\left(I_N-A(\lambda_1,...,\lambda_p;\underset{\sim}{\sigma}^2)\right)\underset{\sim}{y}\right)\right]'\left[\left(W(\underset{\sim}{\sigma}^2)\right)^{\frac{1}{2}}\left(\left(I_N-A(\lambda_1,...,\lambda_p;\underset{\sim}{\sigma}^2)\right)\underset{\sim}{y}\right)\right]}{N}$$

$$=\frac{\left\|\left(W(\underset{\sim}{\sigma}^2)\right)^{\frac{1}{2}}\left(I_N-A(\lambda_1,...,\lambda_p;\underset{\sim}{\sigma}^2)\right)\underset{\sim}{y}\right\|^2}{N}$$

where norm $\|\underset{\sim}{a}\|$ is defined as $\|\underset{\sim}{a}\|=\sqrt{a_1^2+a_2^2+...+a_p^2}$ for $\underset{\sim}{a}=(a_1,a_2,...,a_p)'$, $N=\sum_{i=1}^{p}n_i$, I_N is an identity matrix, and $W(\underset{\sim}{\sigma}^2)$ is a diagonal matrix that has been given by [28, 37 and 40], so that $\left(W(\underset{\sim}{\sigma}^2)\right)^{\frac{1}{2}}$ is also a diagonal matrix and it has the square root .

Next, we define a quantity (further it is called as GCV function) as follows:

$$G(\lambda_1,...,\lambda_p;\underset{\sim}{\sigma}^2)=\frac{N^{-1}\left\|\left(W(\underset{\sim}{\sigma}^2)\right)^{\frac{1}{2}}\left(I_N-A(\lambda_1,...,\lambda_p;\underset{\sim}{\sigma}^2)\right)\underset{\sim}{y}\right\|^2}{\left[N^{-1}trace\left(I_N-A(\lambda_1,...,\lambda_p;\underset{\sim}{\sigma}^2)\right)\right]^2}$$

where $(I_N-A(\lambda_1,...,\lambda_p;\underset{\sim}{\sigma}^2)$ is a non-zero matrix, because $\lambda_i>0$, $i=1,2,...,p$.
Finally, we obtain optimal smoothing parameter, namely $\underset{\sim}{\lambda}_{optimal}=(\lambda_{1(optimal)},...,\lambda_{p(optimal)})'$ by taking the solution to the following optimization:

$$G_{optimal}(\lambda_{1(optimal)},...,\lambda_{p(optimal)};\underset{\sim}{\sigma}^2)=\underset{\lambda_1\in R^+,...,\lambda_p\in R^+}{Min}\left\{G(\lambda_1,...,\lambda_p;\underset{\sim}{\sigma}^2)\right\}$$

$$=\underset{\lambda_1\in R^+,...,\lambda_p\in R^+}{Min}\left\{\frac{N^{-1}\left\|\left(W(\underset{\sim}{\sigma}^2)\right)^{\frac{1}{2}}\left(I_N-A(\lambda_1,...,\lambda_p;\underset{\sim}{\sigma}^2)\right)\underset{\sim}{y}\right\|^2}{\left[N^{-1}trace\left(I_N-A(\lambda_1,...,\lambda_p;\underset{\sim}{\sigma}^2)\right)\right]^2}\right\}. \quad (6)$$

2.2. *Smoothing parameters selection method by using Kernel estimator approach.*

In analysis of nonparametric regression, to estimate the function g in model (1) through kernel smoothing technique, basically we use a decreasing distance function in x-space that is a weighted raw data average.

For model given in (1), according to [37] for predicting at x_{ij}, the weighted raw data average scheme by relating it with i^{th}-response observations, namely y_{ik}, is given by:

$$v_{(i)jk}=\frac{K_i(\frac{t_{ij}-t_{ik}}{h_i})}{\sum_{k=1}^{n_i}K_i(\frac{t_{ij}-t_{ik}}{h_i})}=\frac{K_i(u)}{\sum_{k=1}^{n_i}K_i(u)}\ ,\quad i=1,2,...,p \quad (7)$$

where $K(u)$ is called as kernel function which is a decreasing function of u, and $h_i > 0$ is a parameter of smoothing called as bandwidth. The kernel function $K(u)$ should be a symmetric function. It usually give a probability density function for example Gaussian function [34-35].

By considering equation (7), we get the estimated regression function of model given in (1) to estimate function of regression at point of fit x_{ij} which is expressed as:

$$\hat{g}_k(x_{ij}) = \hat{y}_{ij} = \sum_{k=1}^{n_i} v_{(i)jk} y_{ik} = \underset{\sim}{v}'_{ik} \underset{\sim}{y}_i \ , \ i = 1,2,\dots,p \ , \ j = 1,2,\dots,n_i \,. \tag{8}$$

where each point of n_i points is given by a unequal weight $v_{(i)jk}$, $k = 1,2,\dots,n_i$ for any point of fit x_{ij}. Thus, we can express the estimated regression function in (8) as:

$$\underset{\sim}{\hat{g}} = V\underset{\sim}{y} \tag{9}$$

where $V = diag\left(V_{i1}, V_{i2}, \dots, V_{in_i}\right)$, and

$$V_{ij} = \begin{pmatrix} v_{(i)11} & v_{(i)12} & \cdots & v_{(i)1n_i} \\ v_{(i)21} & v_{(i)22} & \cdots & v_{(i)2n_i} \\ \vdots & \vdots & \ddots & \vdots \\ v_{(i)n_i 1} & v_{(i)n_i 2} & \cdots & v_{(i)n_i n_i} \end{pmatrix}.$$

Therefore, we can determine estimator of regression function of model given in (1) based on kernel estimator, and its result is [37]:

$$\underset{\sim}{\hat{g}}(\underset{\sim}{x}) = \left(\hat{g}_1(\underset{\sim}{x}_1), \hat{g}_2(\underset{\sim}{x}_2), \dots, \hat{g}_p(\underset{\sim}{x}_p)\right)' = diag\left(V_{i1}, V_{i2}, \dots, V_{in_i}\right)\left(\underset{\sim}{y}_1, \underset{\sim}{y}_2, \dots, \underset{\sim}{y}_p\right)'. \tag{10}$$

Since the estimated regression function given in (9) depends on smoothing parameter namely bandwidth $h_i > 0$, $i = 1,2,\dots,p$ then we also may write (9) in the following equation:

$$\underset{\sim}{\hat{g}}_h(\underset{\sim}{x}) = B(h_1,\dots,h_p;\underset{\sim}{\sigma}^2)\underset{\sim}{y} \tag{11}$$

Furthermore, based on equations (9)-(11) we can determine the mean square error (MSE) of (11) as follows:

$$\begin{aligned} MSE(h_1,\dots,h_p;\underset{\sim}{\sigma}^2) &= \frac{(\underset{\sim}{y} - \underset{\sim}{\hat{g}}_h(\underset{\sim}{x}))'V(\underset{\sim}{\sigma}^2)(\underset{\sim}{y} - \underset{\sim}{\hat{g}}_h(\underset{\sim}{x}))}{\sum_{i=1}^{p} n_i} \\ &= \frac{\left(\underset{\sim}{y} - B(h_1,\dots,h_p;\underset{\sim}{\sigma}^2)\underset{\sim}{y}\right)' V(\underset{\sim}{\sigma}^2)\left(\underset{\sim}{y} - B(h_1,\dots,h_p;\underset{\sim}{\sigma}^2)\underset{\sim}{y}\right)}{N} \\ &= \frac{\left[\left(V(\underset{\sim}{\sigma}^2)\right)^{\frac{1}{2}}\left(\left(I_N - B(h_1,\dots,h_p;\underset{\sim}{\sigma}^2)\right)\underset{\sim}{y}\right)\right]'\left[\left(V(\underset{\sim}{\sigma}^2)\right)^{\frac{1}{2}}\left(\left(I_N - B(h_1,\dots,h_p;\underset{\sim}{\sigma}^2)\right)\underset{\sim}{y}\right)\right]}{N} \\ &= N^{-1}\left\|\left(V(\underset{\sim}{\sigma}^2)\right)^{\frac{1}{2}}\left(I_N - B(h_1,\dots,h_p;\underset{\sim}{\sigma}^2)\right)\underset{\sim}{y}\right\|^2 \end{aligned}$$

where norm $\|\underset{\sim}{a}\|$ is defined as $\|\underset{\sim}{a}\| = \sqrt{a_1^2 + a_2^2 + \ldots + a_p^2}$ for $\underset{\sim}{a} = (a_1, a_2, \ldots, a_p)'$, $N = \sum_{i=1}^{p} n_i$, I_N is an identity matrix , and $V(\underset{\sim}{\sigma}^2)$ is a diagonal matrix that has been given by [28 and 37], so that $\left(V(\underset{\sim}{\sigma}^2)\right)^{\frac{1}{2}}$ is also a diagonal matrix and it has the square root.

Next, we define a quantity as follows:

$$G(h_1,\ldots,h_p;\underset{\sim}{\sigma}^2) = \frac{N^{-1}\left\|\left(V(\underset{\sim}{\sigma}^2)\right)^{\frac{1}{2}}\left(I_N - B(h_1,\ldots,h_p;\underset{\sim}{\sigma}^2)\right)\underset{\sim}{y}\right\|^2}{\left[N^{-1}trace\left(I_N - B(h_1,\ldots,h_p;\underset{\sim}{\sigma}^2)\right)\right]^2}$$

where $(I_N - B(h_1,\ldots,h_p;\underset{\sim}{\sigma}^2)$ is a non-zero matrix because $h_i > 0$, $i = 1, 2, \ldots, p$.

Finally, we obtain the optimal smoothing parameter called as optimal bandwidth $\underset{\sim}{h}_{optimal} = (h_{1(optimal)}, \ldots, h_{p(optimal)})'$ by taking the solution to the following optimization:

$$G_{optimal}(h_{1(optimal)},\ldots,h_{p(optimal)};\underset{\sim}{\sigma}^2) = \underset{h_1\in R^+,\ldots,h_p\in R^+}{Min}\left\{G(h_1,\ldots,h_p;\underset{\sim}{\sigma}^2)\right\}$$

$$= \underset{h_1\in R^+,\ldots,h_p\in R^+}{Min}\left\{\frac{N^{-1}\left\|\left(V(\underset{\sim}{\sigma}^2)\right)^{\frac{1}{2}}\left(I_N - H(h_1,\ldots,h_p;\underset{\sim}{\sigma}^2)\right)\underset{\sim}{y}\right\|^2}{\left[N^{-1}trace\left(I_N - H(h_1,\ldots,h_p;\underset{\sim}{\sigma}^2)\right)\right]^2}\right\}. \qquad (12)$$

3. Conclusion

The estimated regression functions in model (1) by using both smoothing spline estimator and kernel estimator approaches depends on smoothing parameters $(\underset{\sim}{\lambda})$ and bandwidths $(\underset{\sim}{h})$, respectively. In addition, by minimazing GCV functions given in (6) and in (12) we obtained optimal smoothing parameter for smoothing spline estimator approach and optimal smoothing parameter called as optimal bandwidth for kernel estimator approach, respectively.

Acknowledgement

Many thanks to the Faculty of Science and Technology, Airlangga University, Surabaya, Indonesia which has funded authors to present this paper in the 6th ICRIEMS 2019.

References

[1] Kimeldorf G and Wahba G 1971 Some Result on Tchebycheffian Spline Functions *J. of Math. Anal. and Applications* **33**(1) 82-95.
[2] Craven P and Wahba G 1979 Smoothing Noisy Data With Spline Function: Estimating the Correct Degree of Smoothing by the Method of Generalized Cross Validation Numer. Math. 31 377-403.
[3] Wahba G 1985 A Comparison of GCV and GML for Choosing the Smoothing Parameter in the Generalized Spline Smoothing Problem The Annals of Statistics 13(4) 1378-1402.
[4] Cox D D 1983 Asymptotic for M-type Smoothing Spline The Annals. of Statistics 11(2) 530-551.
[5] Cox D D and O'Sullivan F 1996 Penalized Likelihood Type Estimators for Generalized Nonparametric Regression J. Mult. Anal. 56 185-206.

[6] Wahba G 1983 Bayesian Confidence Intervals for the Cross-Validated Smoothing Spline Journal of the Royal Statistical Society Series B 45 133-150.
[7] Oehlert G W 1992 Relaxed Boundary Smoothing Spline The Annals. of Statistics 20 146-160.
[8] Koenker R, Pin N G and Portnoy S 1994 Quantile Smoothing Splines Biometrics 81(4) 673-680.
[9] Budiantara I N, Subanar and Soejoeti Z 1997 Weighted Spline Estimator Proc. 51st Session of the International Statistical Institute Istanbul 51 333-334.
[10] Wang Y 1998 Smoothing Spline Models With Correlated Random Errors J. Amer. Statist. Assoc. 93(93) 341-348.
[11] Wahba G 2000 An Introduction to Model Building with Reproducing Kernel Hilbert Spaces Technical Report No. 1020, 18 April 2000, Department of Statistics, University of Wisconsin, Madison.
[12] Cardot, Crambes H C, Kneip A and Sarda P 2007 Smoothing Splines Estimators in Functional Linear Regression with Errors-in-Variables Computational Statistics & Data Analysis 51 4832-4848.
[13] Liu A, Tong T and Wang Y 2007 Smoothing Spline Estimation of Variance Functions Journal of Computational and Graphical Statistics 16(2) 312-329.
[14] Wegman E J 1981 Vector spline and estimation of filter function Technometrics 23 83-89.
[15] Miller J J and Wegman E J 1987 Vector function estimation using splines J. Royal Stat. Society, Series B 17 173-180.
[16] Flessler J A 1991 Nonparametric fixed-interval smoothing with vector splines IEEE Trans. Signal Proceeding 39 852–859.
[17] Wahba G 1992 Multivariate functional and operator estimation, based on smoothing spline and reproducing kernels.
[18] Gooijer J G D, Gannoun A and Larramendy I 1999 Nonparametric regression with serially correlated errors 1st Edition Tinbergen Institute Rotterdam pp:19.
[19] Wang Y 2011 Smoothing Splines: Methods and Applications CRC Press New York.
[20] Wahba G 1990 Spline Models for Observational Data SIAM Philadelphia Pennsylvania.
[21] Eubank R L 1999 Nonparametric Regression and SmoothingSpline Marcel Deker New York.
[22] Wang Y, Guo W and Brown W B 2000 Spline smoothing for bivariate data with applications to association between hormones Statistica Sinica 10 377-397.
[23] Chamidah N and Lestari B 2016 Spline Estimator in Homoscedastic Multi-Response Nonparametric Regression Model in Case of Unbalanced Number of Observations Far East Journal of Mathematical Sciences (FJMS) 100 1433-1453.
[24] Lestari B, Anggraeni D, and Saifudin T 2017 Konstruksi dan Estimasi Matriks Kovariansi dalam Model Regresi Nonparametrik Multirespon Berdasarkan Estimator Smoothing Spline untuk Beberapa Kasus Ukuran Sampel Proceeding of National Seminar on Mathematics and Its Applications (SNMA) 2017 Airlangga University Surabaya pp. 238-242.
[25] Lestari B, Fatmawati and Budiantara I N 2017 Estimasi Fungsi Regresi Nonparametrik Multirespon Menggunakan Reproducing Kernel Hilbert Space Berdasarkan Estimator Smoothing Spline Proceeding of National Seminar on Mathematics and Its Applications (SNMA) 2017 Airlangga University Surabaya pp. 243-250.
[26] Lestari B, Anggraeni D and Saifudin T 2018 Estimation of Covariance Matrix based on Spline estimator in Homoscedastic Multi-Responses Nonparametric Regression Model in Case of Unbalance Number of Observations Far East J.Math.Sciences (FJMS) 108(2) 341-355.
[27] Islamiyati A, Fatmawati and Chamidah N 2018 Estimation of Covariance Matrix on Bi-Response Longitudinal Data Analysis with Penalized Spline Regression Journal of Physics: Conf. Series Vol. 979 pp. 012093.
[28] Lestari B, Fatmawati and Budiantara I N 2019 Spline Estimator and Its Asymptotic Properties in Multiresponse Nonparametric Regression Model Songklanakarin Journal of Science and Technology In Press.

[29] Green P J and Silverman B W 1994 Nonparametric Regression and Generalized Linear Models Chapman Hall New York.
[30] Hardle W 1991 Applied Nonparametric Regression Cambridge University Press Cambridge.
[31] Schimek M G 2000 Smoothing and Regression John Wiley & Sons New York.
[32] Watson G S 1964 Smooth Regression Analysis Sankhya Series A 26 359-372.
[33] Nadaraya E A and Watson 1964 On Estimating Regression Theory Prob. Applications 10 186-190.
[34] Yatchew A 2003 Semiparametric Regression for the Applied Econometrician Cambridge University Press Cambridge.
[35] Wand M P and Jones M C 1995 Kernel Smoothing Chapman Hall New York pp. 114-141.
[36] Aydin D 2007 A comparison of the nonparametric regression models using smoothing spline and kernel regression World Acad. Sci. Eng. Technol. 26 730-734.
[37] Lestari B, Fatmawati, Budiantara I N and Chamidah N 2018 Estimation of regression function in multi-response nonparametric regression model using smoothing spline and kernel estimators Journal of Physics: Conference Series 1097 012091.
[38] Lestari B, Chamidah N and Saifudin T 2019 Estimasi fungsi regresi dalam model regresi nonparametrik birespon menggunakan estimator smoothing spline dan estimator kernel Jurnal Matematika, Statistika and Komputasi (JMSK) 15(2) 20-24.
[39] Fernandez F M and Opsomer J D 2005 Smoothing parameter selection methods for nonparametric regression with spatially correlated errors Canadian J. Statistics 33 279-295.
[40] Lestari B, Fatmawati and Budiantara I N 2019 Smoothing spline estimator in multiresponse nonparametric regression for predicting blood pressures and pulse: a theoretically discussion Proceeding of Global Conference on Engineering and Applied Science (GCEAS) July 16-18, 2019 in Sapporo, Hokkaido, Japan 71618 pp. 81-93.

The Infinitely Divisible Characteristic Function of Compound Poisson Distribution as the Sum of Variational Cauchy Distribution

D Devianto*, Sarah, H Yozza, F Yanuar and Maiyastri
Department of Mathematics, Faculty of Mathematics and Natural Sciences, Andalas University, Limau Manis Campus, Padang 25163, Indonesia

*E-mail: ddevianto@sci.unand.ac.id

Abstract. The new particular compound Poisson distribution is introduced as the sum of independent and identically random variables of variational Cauchy distribution with the number of random variables has Poisson distribution. This compound Poisson distribution is characterized by using characteristic function that is obtained by using Fourier-Stieltjes transform. The infinite divisibility of this characteristic function is constructed by introducing the specific function that satisfied the criteria of characteristic function. This characteristic function is employing the properties of continuity and quadratic form in term of real and non-negative function such that its convolution has the characteristic function of compound Poisson distribution as the sum of variational Cauchy distribution.

1. Introduction

A compound Poisson distribution is a sum of independent and identically random variables for any distributions where the number of these random variables has Poisson distribution. The random variable of compound Poisson distribution is defined in the mathematical term as $S=X_1+X_2+...+X_N$ where the random variable N has Poisson distribution and independent and identically random variables $X_{i=1,2,...,N}$ have certain distribution. In the case random variable X_i has variational Cauchy distribution then it is said the random variable S has compound Poisson distribution as the sum of variational Cauchy distribution.

The Poisson distribution and Cauchy distribution are well defined probability distribution where Sato [1] has explained their infinite divisibility. The infinite divisibility can be identified from its distribution, characteristic function and random variable. Steutel and Harn [2] have stated the infinite divisibility of random variable S with distribution function F can be divided into m random variables such that $S=X_1+X_2+...+X_m$ where the random variable $X_{i=1,2,...,m}$ as independent and identically random variables with distribution function F_m, this is well known as m-fold convolution of distribution function of F_m that is $F=F_m*F_m*...*F_m$ for m times. Besides the terminology of a convolution for infinitely divisible distribution, Lukacs [3] has expressed the infinite divisibility by using characteristic function as Fourier-Stieltjes transform, where the characteristic function is defined as $\phi(t)=E(\exp(itX))$ for a random variable X where i as imaginary unit and $t\in(-\infty,\infty)$. The definition of infinitely divisible characteristic function can be also referred from Artikis [4] or Mainardi and Rogosin [5], the characteristic function $\phi(t)$ is said to be infinitely divisible if for every positive integer number m, there exists a characteristic function

$\phi_m(t)$ such that $\phi(t)=(\phi_m(t))^m$. The existence of characteristic function $\phi_m(t)$ could be examined by using necessary and sufficient condition for a function to be characteristic function by Bochner's theorem in Lukacs [3], that is the criteria of $\phi_m(0)=1$ and the function $\phi_m(t)$ non-negative definite, where the condition of non-negative definite is satisfied by continuity and quadratic form of the function $\phi_m(t)$.

The concept of infinitely divisible distribution can be applied to develop the theory of compound distribution by using its property of characteristic function, that is also known as the sum of random variables as compounding distribution constructed as the conditional convolution of m independent and identically random variables. Some theories of a convolution for certain distribution have introduced by Devianto et. al [6,7] as convolution of the generated random variable from exponential and its variation with stabilizer constant. Furthermore, the infinite divisibility of half-Cauchy distribution and its properties explained in Diedhiou [8] and the infinite divisibility of normed product of Cauchy distribution by Takano [9]. Furthermore, Serfozo [10] has discussed the theory of compound Poisson approximation for sums of random variables, while Cekanavicius and Vellaisamy [11] and Barbour et. al [12] have another setting on convergence approximation of compound Poisson distribution via Stein's method.

The property of discrete compound Poisson process on its infinite divisibility and convergence has introduced by Zhang and Li [13]. Although the convergence in distribution of compound Poisson process has introduced as the theoretical approach, but the form of density of compound Poisson distribution also has interested to deliver, where Duval [14] has estimated the density of compound Poisson processes from discrete data, while Lindo et. al [15] has used nonparametric estimation for compound Poisson process via variational analysis on measures.

The property of compound Poisson distribution has introduced by the previous researcher influencing many rules in application particularly on insurance risk models distribution such as Zhang et.al [16] and Hu et. al [17] have introduced discrete compound Poisson model with applications to risk theory. Furthermore, Willmout [18] have reviewed Lundberg approximations for compound distributions with insurance applications. While the application of convolution theory to characteristic function also has explained by Devianto [19] in term of uniform continuity properties of characteristic function from the convolution of exponential distribution with stabilizer constant, and its class of infinitely divisible exponential distribution in Devianto [20]. The theoretical concept and application of compound distribution have the strong position to confirm that characteristic function has important rules on deriving the property of compound distribution especially on its infinite divisibility. However, the summary of recent results does not cover the characterization of specific compound Poisson distribution as the sum of variational Cauchy distribution on its property of characteristic function, then this paper will give exploration on development of characteristic function and its property of infinitely divisible, quadratic form, continuity and no real zeros attributed as the main findings.

2. The Characteristic Function of Compound Poisson Distribution as the Sum of Variational Cauchy Distribution

The property of compound Poisson distribution as the sum of variational Cauchy distribution is constructed by using characterization of its characteristic function. First, it is introduced the Poisson distribution and variational Cauchy distribution in the sense of probability distribution and its characterizations on moment generating function and characteristic function. The Poisson distribution is a discrete probability distribution for counts N of events that occur randomly in a given interval of time with λ is the mean number of events per interval. The probability distribution is given by equation

$$P(N=n)=\frac{e^{-\lambda}\lambda^n}{n!} \tag{1}$$

where $n = 1, 2, 3, \ldots$ and $\lambda > 0$. This distribution is noted as Poisson distribution with a parameter (λ).

The moment generating function of Poisson distribution is obtained as

$$M_N(t)=E(e^{tN})=\exp(\lambda(e^t-1)) \tag{2}$$

for $t \in (-\infty,\infty)$. The most property of Poisson distribution is in the form of characteristic function as follows

$$\phi_N(t) = E(e^{itN}) = \exp(\lambda(e^{it} - 1)) \quad (3)$$

for $t \in (-\infty,\infty)$.

The variational Cauchy distribution is constructed by setting parameter γ multiplied with its random variable in standard Cauchy distribution, so that it can be defined in the following term of probability density function

$$f(x) = \frac{\sqrt{\gamma}}{\pi(1+\gamma x^2)} \quad (4)$$

for $\gamma > 0$ and $-\infty < x < \infty$. This distribution is noted as variational Cauchy distribution with a parameter (γ). The special characterizations of variational Cauchy distribution are undefined for expected value and variance, meanwhile there is also no existance of moment generating. Since the characteristic function is, always exist for every distribution, then the characteristic function of variational Cauchy distribution is obtained as follows

$$\phi_X(t) = E(e^{itX}) = \exp\left(\frac{-|t|}{\sqrt{\gamma}}\right) \quad (5)$$

for $t \in (-\infty,\infty)$.

The characteristic function of Poisson distribution and Cauchy distribution have expressed into canonical representation of Levy-Khinthice in Lukacs [3], then both distributions are infinitely divisible. Furthermore, the infinite divisibility of compound Poisson distribution as the sum of variational Cauchy distribution has to express some important properties on its characteristic function as the main concern on these current results. The characteristic function of compound Poisson distribution as the sum of variational Cauchy distribution is given in the following proposition. The following theoretical concept of compounding distribution is stated by using the random variable $S=X_1+X_2+\ldots+X_N$ as a compound distribution where the random variable N has Poisson distribution and random variables $X_{i=1,2,\ldots,N}$ have variational Cauchy distribution with same parameters. This distribution is noted as compound Poisson distribution as the sum of variational Cauchy distribution with parameters (λ,γ).

Proposition 2.1. Let $S=X_1+X_2+\ldots+X_N$ be a random variable of compound Poisson distribution as the sum of variational Cauchy distribution, then characteristic function of random variable S is in the following form

$$\phi_S(t) = \exp\left(\lambda\left(\exp\left(\frac{-|t|}{\sqrt{\gamma}}\right) - 1\right)\right) \quad (6)$$

for $t \in (-\infty,\infty)$, $\lambda > 0$ and $\gamma > 0$.

Proof. It is used definition of characteristic function and the linearity property of expectation, and then it is obtained the characteristic function of compound Poisson distribution as the sum of variational Cauchy distribution as follows

$$\phi_S(t) = E(\exp(itS)) = E\left(E\left(\exp\left(it\sum_{i=1}^{N} X_i\right)\middle| N\right)\right) = M_N(\ln(\phi_X(t))). \quad (7)$$

It is substituted the characteristic function of variational Cauchy distribution in Equation (5) and the moment generating function of Poisson distribution on Equation (2) to Equation (7), then it is obtained

$$\phi_S(t) = M_N\left(\ln\left(\exp\left(-\frac{|t|}{\sqrt{\gamma}}\right)\right)\right) = \exp\left(\lambda\left(\exp\left(-\frac{|t|}{\sqrt{\gamma}}\right) - 1\right)\right). \quad (8)$$

The Proposition 2.1 has shown that compound Poisson distribution as the sum of variational Cauchy distribution has a unique characteristic function, that is very varying from the characteristic function of the sum of variational Cauchy distribution such as in Devianto [21] or the property of characteristic function as normed product of Cauchy density in Takano [9], and this new type of characteristic function has different property of characteristic function of half-Cauchy distribution in Diedhiou [8] and it is not concluded in generalized Cauchy family of distributions in Alzaatreh et. al [22]. In addition, it is unlikely a compound Poisson process such as in Franciszek et. al [23] or Seri and Choirat [24], but this characteristic function is governed by using Poisson compounding with certain distribution. This is to confirm the uniqueness of characteristic function for difference formula of compound Poisson distribution.

3. The Characterization of Infinitely Divisible Characteristic Function of Compound Poisson Distribution as the Sum of Variational Cauchy Distribution

The main results of this paper are stated in the following proposition and theorem by explaining the characterization of characteristic function and its property of infinite divisibility. The properties of the sum of random variables as compound distribution can also be identified by using its characteristic function in the term of infinite divisibility. Furthermore, the infinite divisibility of compound Poisson distribution as the sum of variational Cauchy distribution is constructed by using property of characteristic function. First, let us define a function $\phi_{Sm}(t)$ as a corresponding function to the random variable S_m with the cumulative distribution function F_{Sm}. Now, it is defined the function $\phi_{Sm}(t)$ as follows

$$\phi_{S_m}(t)=\left(\exp\left(\lambda\left(\exp\left(\frac{-|t|}{\sqrt{\gamma}}\right)-1\right)\right)\right)^{\frac{1}{m}} \tag{9}$$

for any integer number m where λ as parameter of Poisson distribution and γ as parameter of variational Cauchy distribution. The next propositions are properties of function $\phi_{Sm}(t)$ related to necessary and sufficient condition in Bochner's theorem for a function $\phi_{Sm}(t)$ to be a characteristic function.

Proposition 3.1. The function $\phi_{Sm}(t)$ is continuous.

Proof. The continuity of function $\phi_{Sm}(t)$ is explained by using the condition of uniform continuity, that is for every $\varepsilon > 0$ there exists $\delta > 0$ such that $|\phi_{Sm}(t_1) - \phi_{Sm}(t_2)| < \varepsilon$ for $|t_1 - t_2| < \delta$ where δ depends only on ε. Hence, it is used definition of function $\phi_{Sm}(t)$ as ij Equation (9) to have the following equation

$$\left|\phi_{S_m}(t_1)-\phi_{S_m}(t_2)\right|=\left|\left(\exp\left(\lambda\left(\exp\left(\frac{-|t_1|}{\sqrt{\gamma}}\right)-1\right)\right)\right)^{\frac{1}{m}}-\left(\exp\left(\lambda\left(\exp\left(\frac{-|t_2|}{\sqrt{\gamma}}\right)-1\right)\right)\right)^{\frac{1}{m}}\right|. \tag{10}$$

Then let us define $h = t_1 - t_2$ and then it is applied to the Equation (10), so that for $h \to 0$, it is obtained the following limit

$$\lim_{h\to 0}\left|\left(\exp\left(\lambda\left(\exp\left(\frac{-|h+t_2|}{\sqrt{\gamma}}\right)-1\right)\right)\right)^{\frac{1}{m}}-\left(\exp\left(\lambda\left(\exp\left(\frac{-|t_2|}{\sqrt{\gamma}}\right)-1\right)\right)\right)^{\frac{1}{m}}\right|=0. \tag{11}$$

This hold for $\delta < \varepsilon$ where $|\phi_{Sm}(t_1) - \phi_{Sm}(t_2)| < \varepsilon$ for $|t_1 - t_2| < \delta$ and δ depends only on ε. Then $\phi_{Sm}(t)$ satisfied the condition of uniformly continuous, then it is concluded that the function $\phi_{Sm}(t)$ is continuous.

Propositions 3.2. The function $\phi_{Sm}(t)$ satisfies the condition of quadratic form.

Proof. It is used the Equation (9) for variable $t_j - t_l$ where j and l as positive integer numbers such that we have the following doubled summation

$$\sum_{1\le j\le n}\sum_{1\le l\le n} c_j \overline{c_l}\,\phi_{S_m}(t_j - t_l) = \sum_{1\le j\le n}\sum_{1\le l\le n} c_j \overline{c_l}\left(\exp\left(\lambda\left(\exp\left(\frac{-|t_j - t_l|}{\sqrt{\gamma}}\right)-1\right)\right)\right)^{\frac{1}{m}}$$
$$= \sum_{1\le j\le n}\sum_{1\le l\le n} c_j \overline{c_l}\left(\exp\left(\frac{\lambda}{m}\left(\exp\left(\frac{-|t_j - t_l|}{\sqrt{\gamma}}\right)-1\right)\right)\right) \quad (12)$$

for any complex number c_j and c_l with parameters $(\lambda/m,\gamma)$. Next, based on the nature of absolute value of $|t_j - t_l|$ then it can be written the set of the following condition, the first $|t_j - t_l| = t_j - t_l$ for $t_j - t_l > 0$ and the second $|t_j - t_l| = -t_j + t_l$ for $t_j - t_l < 0$. Next, the Equations (12) is divided into two parts as the following forms

(i) The function $\phi^{+}(t_j,t_l) = \sum_{1\le j\le n}\sum_{1\le l\le n} c_j \overline{c_l}\exp\left(\frac{\lambda}{m}\left(\exp\left(\frac{-t_j + t_l}{\sqrt{\gamma}}\right)-1\right)\right)$ for $t_j - t_l > 0$.

(ii) The function $\phi^{-}(t_j,t_l) = \sum_{1\le j\le n}\sum_{1\le l\le n} c_j \overline{c_l}\exp\left(\frac{\lambda}{m}\left(\exp\left(\frac{t_j - t_l}{\sqrt{\gamma}}\right)-1\right)\right)$ for $t_j - t_l < 0$.

Next, it will be explained the real and non-negative property of Equation (12) for both Part (i) and Part (ii). First, it is defined a random variable Y as compound Poisson distribution as the sum of variational Cauchy distribution with parameters $(\lambda/m,\gamma)$, and it is used the property of complex conjugate, then it is obtained the Part (i) from Equation (12) satisfied the following equation

$$\phi^{+}(t_j,t_l) = \sum_{1\le j\le n}\sum_{1\le l\le n} c_j \overline{c_l}\exp\left(\frac{\lambda}{m}\left(\exp\left(\frac{-t_j + t_l}{\sqrt{\gamma}}\right)-1\right)\right)$$
$$= \left(\sum_{1\le j\le n} c_j\, E(\exp(it_j(-Y)))\right)\left(\overline{\sum_{1\le l\le n} c_l\, E(\exp(it_l(-Y)))}\right) \quad (13)$$
$$= \left|\sum_{1\le j\le n} c_j\, E(\exp(it_j(-Y)))\right|^2$$

The property of the real and non-negative in Part (ii) from Equation (12) can be seen in the following equation

$$\phi^{-}(t_j,t_l) = \sum_{1\le j\le n}\sum_{1\le l\le n} c_j \overline{c_l}\exp\left(\frac{\lambda}{m}\left(\exp\left(\frac{t_j - t_l}{\sqrt{\gamma}}\right)-1\right)\right)$$
$$= \left(\sum_{1\le j\le n} c_j\, E(\exp(it_jY))\right)\left(\overline{\sum_{1\le l\le n} c_l\, E(\exp(it_lY))}\right) \quad (14)$$
$$= \left|\sum_{1\le j\le n} c_j\, E(\exp(it_jY))\right|^2$$

From Equation (13) and Equation (14), it is concluded that

$$\sum_{1\le j\le n}\sum_{1\le l\le n} c_j \overline{c_l}\,\phi_{S_m}(t_j - t_l) \ge 0, \quad (15)$$

Then it is proved that the function $\phi_{Sm}(t)$ as real and positively defined function where the quadratic form has non-negative values.

Proposition 3.3. The function $\phi_{Sm}(t)$ is characteristic function.

Proof. Based on Proposition 3.1 and Proposition 3.2, the function $\phi_{Sm}(t)$ satisfied the condition continuity and non-negative definite function, in addition together with condition of $\phi_{Sm}(0)=1$ then the function $\phi_{Sm}(t)$ satisfied the criteria of characteristic function on Bochner's theorem. It is clear that the function $\phi_{Sm}(t)$ is a characteristic function.

The Equation (9) has property of continuity and satisfied the quadratic form such as in Proposition 3.1 and Proposition 3.2 respectively, and then Proposition 3.3 established the function in Equation (9) is characteristic function. The existence of this new particular of compound Poisson distribution as the sum of variational Cauchy distribution has established the property on its characteristic function. This new characteristic function has a specific form that differs from compound Poisson process in Crescenzo et. al [25] or Zhang and Li [13]. The certain distribution of variational Cauchy distribution in Poisson compounding has governed the uniqueness of characteristic function with the property of continuity and quadratic form. These results will be used to establish the property of infinitely divisible of the certain compound Poisson distribution as in Theorem 3.4, and the property of no real zeroes of characteristic function as in the Theorem 3.5.

Theorem 3.4. The compound Poisson distribution as the sum of variational Cauchy distribution is infinitely divisible distribution.

Proof. The property of infinitely divisible characteristic function is shown by using the property of characteristic function $\phi_{Sm}(t)$ such that satisfies $\phi_S(t) = (\phi_{Sm}(t))^m$ for $\phi_S(t)$ as the characteristic function of compound Poisson distribution. Based on Proposition 3.3, there exists a characteristic function $\phi_{Sm}(t)$ such that it is obtained

$$(\phi_{S_m}(t))^m = \left(\left(\exp\left(\lambda \left(\exp\left(\frac{-|t|}{\sqrt{\gamma}} \right) - 1 \right) \right) \right)^{\frac{1}{m}} \right)^m = \exp\left(\lambda \left(\exp\left(\frac{-|t|}{\sqrt{\gamma}} \right) - 1 \right) \right) = \phi_S(t), \quad (16)$$

for any positive integer number m, the function $\phi_S(t)$ is a characteristic function from compound Poisson distribution as the sum of variational Cauchy distribution with parameters (λ,γ), so that it is an infinitely divisible characteristic function.

The random variable S_m has corresponding characteristic function $\phi_{Sm}(t)$ of compound Poisson distribution as the sum of variational Cauchy distribution with the parameter $(\lambda/m,\gamma)$, where λ/m as the mean number of events for every interval and γ as the parameter in variational Cauchy distribution. This result confirmed the compound Poisson distribution as the sum of variational Cauchy distribution is also infinitely divisible, where the property of infinite divisibility also satisfies for Poisson distribution and variational Cauchy distribution.

Theorem 3.5. The characteristic function $\phi_S(t)$ has no real zeroes.

Proof. Based on Theorem 3.4, the characteristic function $\phi_S(t)$ is infinitely divisible, then the function $|\phi_{Sm}(t)|^2$ is a characteristic function satisfied the condition $|\phi_{Sm}(t)|^2 = |\phi_S(t)|^{2/m}$ for any positive integer m. Let us consider the defined function $g(t)$ in the following form

$$g(t) = \lim_{m\to\infty} \left|\phi_{S_m}(t)\right|^2 = \lim_{m\to\infty} \left|\phi_S(t)\right|^{\frac{2}{m}} = \lim_{m\to\infty} \left| \exp\left(\lambda \left(\exp\left(\frac{-|t|}{\sqrt{\gamma}} \right) - 1 \right) \right) \right|^{\frac{2}{m}}, \quad (17)$$

It is assumed that the function $g(t)$ has only two values, that are 0 or 1. Then it is obtained $g(t)= 1$ whenever $\phi_S(t) \neq 0$ and $g(t) = 0$ for all t for which $\phi_S(t) = 0$. Because the characteristic function $\phi_{Sm}(t)$ is continuous based on Proposition 3.1, so that it can easily also be proved that the characteristic function $\phi_S(t)$ is continuous, and as well as it holds for Proposition 3.2, so that $\phi_{Sm}(0) = 1$, then $\phi_S(0) = 1$ therefore $\phi_S(t) \neq 0$.

The property of no real zeroes on characteristic function has spread the characterization of compound Poisson distribution as the sum of variational Cauchy distribution into the specific property. The previous result on Poisson compounding in Zhang and Li [13] and Crescenzo et. al [25] have introduced characterizations of discrete compound Poisson distribution with a Poisson subordinator, but these results does not cover the property of no real zeros of its characteristic function. The result on Theorem 3.5 has completed the new property of this compound Poisson distribution, this characterization is also expressed graphically in Figure 1. The Figure 1 Part (a) describes the characteristic function of compound Poisson distribution as the sum of variational Cauchy distribution with various parameter γ, the curve more platykurtic for γ tends to infinity. The Figure 2 Part (b) describes the curve more leptokurtic for λ tends to infinity, where the both curves have the vertical axis with value 1 for t equals zero. The shape of characteristic function of compound Poisson distribution as the sum of variational Cauchy distribution has exhibited smooth lines as its property of continuity, symmetrical and never has real zeros as infinitely divisible characteristic function.

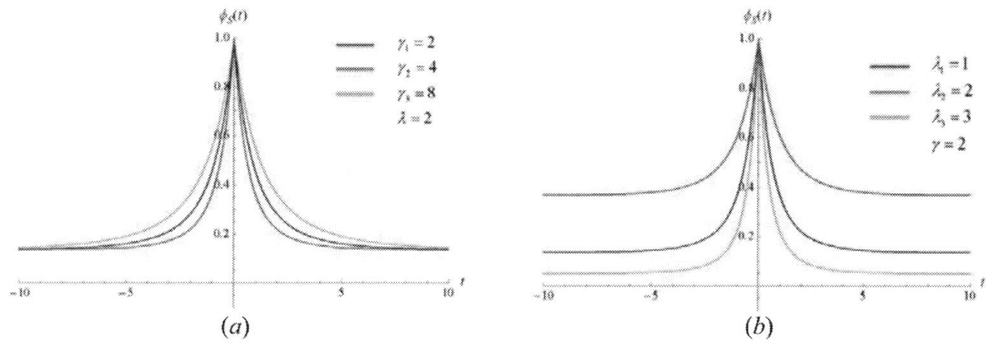

Figure 1. The curves of characteristic $\phi_S(t)$, Part (*a*) the curve with λ=2 and various parameters γ, Part (*b*) the curve with γ=2 and various parameters λ.

4. Conclusion

In this paper it is constructed the new particular compound Poisson distribution as the sum of variational Cauchy distribution with random variable $S=X_1+X_2+...+X_N$ where X_1, X_2, ..., X_N are independent and identically random variables of variational Cauchy distribution and the random variable N has Poisson distribution. The characteristic function of compound Poisson distribution as the sum of variational Cauchy distribution is obtained by using Fourier-Stieltjes transform as $\phi_S(t)$ with parameter (λ,γ) for λ as parameter of Poisson distribution and γ as parameter of variational Cauchy distribution. It is provided a specific function corresponding to a Poisson compounding process for a distribution such that satisfied the criteria of characteristic function employing the properties of uniform continuity and quadratic form in the term of real and non-negative function. This characteristic function $\phi_S(t)$ satisfies the necessary and sufficient conditions to be infinitely divisible of compound Poisson distribution as the sum of variational Cauchy distribution. In addition, the characteristic function of compound Poisson distribution as the sum of variational Cauchy distribution has no imaginary part and no real zeros, it is also more platykurtic for γ tends to infinity and more leptokurtic for λ tends to infinity.

5. References

[1] Sato K 2013 *Levy processes and infinitely divisible distribution* (New York: Cambridge Studies in Advances Mathematics).

[2] Steutel FW and Harn KV 2004 *Infinite divisibility of probability distributions on the real line* (London: Marcel Dekker).

[3] Lukacs E 1992 *Characteristic Function* (London: Hafner Publishing Company).

[4] Artikis T 1983 Constructing infinitely divisible characteristic functions *Arch. Math.* **19**: 57-61.
[5] Mainardi S and Rogosin S 2006 The origin of infinitely divisible distribution: from de Finettis problem to Levy-Khintchine formula *Math. Meth. Econ. Fina.* **1** 37-55.
[6] Devianto D, Oktasari L and Maiyastri 2015 Convolution of generated random variable from exponential distribution with stabilizer constant *Appl. Math. Scie.* **9** (96) 4781-89.
[7] Devianto D, Maiyastri, Oktasari L and Anas M 2015 Some properties of hypoexponential distribution with stabilizer constant *Appl. Math. Scie.* **9** (142) 7063-70.
[8] Diedhiou A 1998 On the self-decomposability of the half-Cauchy distribution *J. Math. Anal. Appl.* **220** 42-64.
[9] Takano K 2003 On infinite divisibility of normed product of Cauchy densities *J. Comp. Appl. Math.* **150** 253-263.
[10] Serfozo RF 1986 Compound Poisson approximation for sums of random variables *Annals of Prob.* **14** (4) 1391-1398.
[11] Cekanavicius V and Vellaisamy P 2018 Approximation by convolution of the normal and compound Poisson laws via Stein's method *Lith. Math. J.* **58** (2) 127-140.
[12] Barbour AD, Louis HY and Liem LW 1992 Compound Poisson approximation for sums of nonnegative random variables via Stein's Method. *Annals of Prob.* **20** (4) 1843-1866.
[13] Zhang H and Li B 2016 Characterizations of discrete compound Poisson distributions *Comm. in Stat: Theo. and Meth.* **45** (22) 6789-6802.
[14] Duval C 2013 Density estimation for compound Poisson processes from discrete data *Stoc. Proc. Appl* **123** (11) 3963-3986.
[15] Lindo A, Zuyev S and Sagitov S 2018 Nonparametric estimation for compound Poisson process via variational analysis on measures *Stat. and Comp* **28** (3) 563-577.
[16] Zhang HY, Liu Y and Li B 2014 Notes on discrete compound Poisson model with applications to risk theory. *Insu: Math. and Econ* **56** 325-336.
[17] Hu X, Zhang L and Sun W 2017 Risk model based on the first-order integer-valued moving average process with compound Poisson distributed innovations *Scan. Actu J.* **2017** 1-14.
[18] Willmot GE and Lin XS 2001 Lunberg approximations for compound distributions with insurance applications (New York: Springer Lecture Note in Statistics).
[19] Devianto D 2016 The uniform continuity of characteristic function from convoluted exponential distribution with stabilizer constant *AIP Conf. Proc.* **1707** 1-5.
[20] Devianto D 2018 On the class of infinitely divisible exponential distribution. *Journal of Physics Conf. Ser.* **1097** 1-6.
[21] Devianto D 2017 The characteristic function property of convoluted random variable from a variaional Cauchy distribution *Proc. 2nd ISI Reg. Stat. Conf. CPS09* 1-6.
[22] Alzaatreh A, Lee C, Famoye F and Ghosh I 2016 The generalized Cauchy family of distributions with applications *J. Stat. Dist. Appl. 3* (12) 1-16.
[23] Franciszek G 2018. Nonhomogeneous compound Poisson process application to modeling of random process related to accidents in the Baltic Sea waters and ports *J.Polish Safety and Reliability Assoc.* **9** (3) 21-29.
[24] Seri R and Choirat C 2015. Comparison of approximation for compound Poisson processes *ASTIN Bull.* **45** (03) 601-637.
[25] Crescenzo AD, Martinucci B and Zacks S 2015 Compound Poisson process with a Poisson subordinator *J. Appl. Prob.* **52** (2) 360-374.

Journal of Physics: Conference Series **1397** (2019) 012066 doi:10.1088/1742-6596/1397/1/012066

Locally stability analysis of the Phytoplankton-Nitrogen-Phosphate-Sediment dynamical system: A study case at Karimunjawa aquaculture system, Central Java

E Triyana[1*], Widowati [2], and S P Putro [3]

[1,2]Department of Mathematics, Faculty of Science and Mathematics, Diponegoro University, Indonesia
[3]Department of Biology, Faculty of Science and Mathematics, Diponegoro University,Indonesia

[*]Corresponding author e-mail: triyanaeka03@gmail.com

Abstract. Disturbance of water environment due to organic enrichment by fish activities may result in the reduction of water quality and sediments. The relative importance of N and P limitation and released from organic sediment is still an open question. The aim of this paper is to analyze the locally asymptotic stable from a dynamical system model in the equilibrium of water ecosystem. Mathematical models are resultant of interaction among four main variables .i.e.nitrogen and phosphate concentration, phytoplankton abundance, and sediment in the water ecosystem at Menjangan Besar, Karimunjawa islands. The four variables are non-linear system differential equation that will form the dynamical system mathematical model. Local stability was determined by using Taylor series and Jacobian matrix. The system will be locally asymptotically stable if eigenvalues are negative. Numerical simulation was used to analyze the dynamic behavior of the system. From numerical simulation results base, it is concluded that equilibrium points. Because all eigenvalues of the Jacobian matrix ware negative, the dynamic system model was locally asymptotically stable.

1. Introduction

Aquaculture can play an important role in meeting human demands in the form of protein-based foods. Indonesia must prepare itself to face the global challenges in the aquaculture industry or also known as fish farming [1]. Aquaculture is carried out in the marine environment and in underwater habitats. Improved fish farming produces a number of major problems for waters. wastewater contains compounds such as suspended solids, total nitrogen, and total phosphorus which can cause algal blooms, fish mortality, and increased sediment [2].

Nitrogen and phosphate are limiting nutrients that limit the growth of phytoplankton. Nitrogen is the basis for making proteins used by aquatic plants in the form of ammonia or nitrate, and plays a role in the formation of phytoplankton biomass as a producer of water, so that it becomes one of the parameters of fertility[3]. Phytoplankton has a direct response to nutrients through rapid changes in biomass, making it good to assess the ecological state of the waters[4]. In the phytoplankton cells, nitrogen and phosphate reflect the relative proportions needed by phytoplankton to grow [16]. A model of phytoplankton, zooplankton, and two complementary nutrients, nitrogen and phosphorus was system nutrients in governing system behavior and in turn in governing water quality control [15]

A complex sediment modeling has been developed to understand the processes that regulate pore water profiles and sediment-water exchange [12]. However, research on the limits of N and P released from sediments has not been done much, because of the complex interactions between the N and P cycles and factors external driving factors lead to dynamics in N or P restriction patterns and internal release that changes with time [13].

Analysis of the model using a linear analysis technique and found that the sinking of phytoplankton could affect the system. Local asymptotic stability around interior equilibrium is studied, which reveals that interior equilibrium loses its stability at some critical value of predation rate [5][6]

This paper aims is to modified the mathematical model from [7][8], to analyze the stability of the mathematical model. The researcher modified the model from the model [9][10].

2. Mathematical model

The The dynamic model for aquaculture has been discussed which explains the influence of growth of phytoplankton and sedimentation with changes concentration of nitrogen and phosphate. The modified model changes phytoplankton, nitrogen, phosphate and sediment can be represented

$$\frac{dP}{dt} = \beta P - (\mu + \gamma) P, \tag{1}$$

$$\frac{dN}{dt} = q\Lambda + \omega\theta S - \gamma N - \alpha\beta P \frac{N}{N+F}, \tag{2}$$

$$\frac{dF}{dt} = (1-q)\Lambda + (1-\omega)\theta S - \gamma F - \alpha\beta P \frac{F}{N+F}, \tag{3}$$

$$\frac{dS}{dt} = \mu P - \theta S, \tag{4}$$

where $P(0) = P_0 \geq 0, N(0) = N_0 \geq 0, F(0) = F_0 \geq 0, S(0) = S_0 \geq 0,$ with parameters β = phytoplankton growth rate (day-1); μ = sedimentation rate of phytoplantonk (day-1); γ =water exchange rate (day-1); q =proportion of N waste entering the water; Λ = total waste input (mg g-1 day-1); ω =proportion of remineralization rate of N in the sludge (with remainder remineralization Phosphate) (day-1); θ =total remineralization sediment. Here P is phytoplankton concentration, N is nitrogen concentration, F is phosphate concentration, S is mass of N in sludge in pond water.

3. Locally stability analysis

Stability analysis is needed to analyze equilibrium points. Let (P^*, N^*, F^*, S^*) be the equilibrium point of the dynamical model nitrogen and phosphate concentration, phytoplankton, and sediment. The equilibrium point is obtained if satisfy $\frac{dP}{dt} = 0, \frac{dN}{dt} = 0, \frac{dF}{dt} = 0, \frac{dS}{dt} = 0$. We obtained the equilibrium point $(P^*, N^*, F^*, S^*) = \left(0, \frac{q\Lambda}{\gamma + v}, \frac{(1-q)\Lambda}{\gamma}, 0\right)$.

Then the stability of non linear system at the equilibrium point (P^*, N^*, F^*, S^*). First, we linearize the nonlinear system at the equilibrium point (P^*, N^*, F^*, S^*) using Taylor series as follows (1), (2), (3), (4)

$$\frac{d\bar{P}}{dt} = (\beta - \mu - \gamma)\, \bar{P}, \tag{5}$$

$$\frac{d\bar{N}}{dt} = \left(-\alpha\beta \frac{N^*}{N^* + F^*}\right)\bar{P} + \left(-\gamma - v - \frac{\alpha\beta P^*}{N^* + F^*} + \left(\alpha\beta P^* \frac{N^*}{(N^* + F^*)^2}\right)\right)\bar{N} + \left(\frac{\alpha\beta P^* N^*}{(N^* + F^*)^2}\right)\bar{F} + \theta\,\bar{S}, \tag{6}$$

$$\frac{d\bar{F}}{dt} = \left(-\alpha\beta \frac{F^*}{N^* + F^*}\right)\bar{P} + \left(\alpha\beta P^* \frac{F^*}{(N^* + F^*)^2}\right)\bar{N} + \left(-\gamma - \frac{\alpha\beta P^*}{N^* + F^*} + \left(\alpha\beta P^* \frac{F^*}{(N^* + F^*)^2}\right)\right)\bar{F} + (1-\omega)\theta\bar{S}, \tag{7}$$

$$\frac{d\bar{S}}{dt}=\mu\bar{P}-\theta\bar{S} \tag{8}$$

Furthermore, the linear differential equation system can be written in the form of a matrix equation as follow

$$\frac{d}{dt}\begin{bmatrix}\bar{P}\\ \bar{N}\\ \bar{F}\\ \bar{S}\end{bmatrix}=\begin{bmatrix}\frac{\partial G_1}{\partial P} & \frac{\partial G_1}{\partial N} & \frac{\partial G_1}{\partial F} & \frac{\partial G_1}{\partial S}\\ \frac{\partial G_2}{\partial P} & \frac{\partial G_2}{\partial N} & \frac{\partial G_2}{\partial F} & \frac{\partial G_2}{\partial S}\\ \frac{\partial G_3}{\partial P} & \frac{\partial G_3}{\partial N} & \frac{\partial G_3}{\partial F} & \frac{\partial G_3}{\partial S}\\ \frac{\partial G_4}{\partial P} & \frac{\partial G_4}{\partial N} & \frac{\partial G_4}{\partial F} & \frac{\partial G_4}{\partial S}\end{bmatrix}\left(P^*,N^*,F^*,S^*\right)\begin{bmatrix}\bar{P}\\ \bar{N}\\ \bar{F}\\ \bar{S}\end{bmatrix},$$

where $\frac{d}{dt}\begin{bmatrix}\bar{P}\\ \bar{N}\\ \bar{F}\\ \bar{S}\end{bmatrix}=J\left(P^*,N^*,F^*,S^*\right)\begin{bmatrix}\bar{P}\\ \bar{N}\\ \bar{F}\\ \bar{S}\end{bmatrix}$, with J is the Jacobian matrix at the equilibrium point $\left(P^*,N^*,F^*,S^*\right)$

$$J=\begin{bmatrix}\beta-\mu-\gamma & 0 & 0 & 0\\ -\alpha\beta\frac{N^*}{N^*+F^*} & -\gamma-v-\frac{\alpha\beta P^*}{N^*+F^*}+\left(\alpha\beta P^*\frac{N^*}{\left(N^*+F^*\right)^2}\right) & \frac{\alpha\beta P^*N^*}{\left(N^*+F^*\right)^2} & \omega\theta\\ -\alpha\beta\frac{F^*}{N^*+F^*} & \alpha\beta P^*\frac{N^*}{\left(N^*+F^*\right)^2} & -\gamma-\frac{\alpha\beta P^*}{N^*+F^*}+\left(\alpha\beta P^*\frac{F^*}{\left(N^*+F^*\right)^2}\right) & (1-\omega)\theta\\ \mu & 0 & 0 & -\theta\end{bmatrix} \tag{9}$$

Furthermore, substitute the equilibrium point $\left(P^*,N^*,F^*,S^*\right)=\left(0,\frac{q\Lambda}{\gamma+v},\frac{(1-q)\Lambda}{\gamma},0\right)$ at Jacobian matrix (9) so that it is obtained.

$$J=\begin{bmatrix}-(-\beta+\mu+\gamma) & 0 & 0 & 0\\ -\alpha\beta\left(\frac{q\Lambda}{(\gamma+v)\left(\frac{q\Lambda}{(\gamma+v)}+\frac{(1-q)\Lambda}{\gamma}\right)}\right) & -\gamma-v & 0 & \omega\theta\\ -\alpha\beta\left(\frac{(1-q)\Lambda}{\frac{q\Lambda}{(\gamma+v)}+\frac{(1-q)\Lambda}{\gamma}}\right) & 0 & -\gamma & (1-\omega)\theta\\ \mu & 0 & 0 & -\theta\end{bmatrix}$$

Then, we obtaind $\det(\lambda I-J)=0$

$$\det(\lambda I-J)=\left|\begin{bmatrix}\lambda & 0 & 0 & 0\\ 0 & \lambda & 0 & 0\\ 0 & 0 & \lambda & 0\\ 0 & 0 & 0 & \lambda\end{bmatrix}-\begin{bmatrix}-(-\beta+\mu+\gamma) & 0 & 0 & 0\\ -\alpha\beta\left(\frac{q\Lambda}{(\gamma+v)\left(\frac{q\Lambda}{(\gamma+v)}+\frac{(1-q)\Lambda}{\gamma}\right)}\right) & -\gamma-v & 0 & \omega\theta\\ -\alpha\beta\left(\frac{(1-q)\Lambda}{\frac{q\Lambda}{(\gamma+v)}+\frac{(1-q)\Lambda}{\gamma}}\right) & 0 & -\gamma & (1-\omega)\theta\\ \mu & 0 & 0 & -\theta\end{bmatrix}\right|=0$$

$$\det(\lambda I - J) = \begin{bmatrix} \lambda + (-\beta + \mu + \gamma) & 0 & 0 & 0 \\ \alpha\beta\left(\dfrac{q\Lambda}{(\gamma+v)\left(\dfrac{q\Lambda}{(\gamma+v)} + \dfrac{(1-q)\Lambda}{\gamma}\right)}\right) & \lambda + (\gamma - v) & 0 & -\omega\theta \\ \alpha\beta\left(\dfrac{(1-q)\Lambda}{\dfrac{q\Lambda}{(\gamma+v)} + \dfrac{(1-q)\Lambda}{\gamma}}\right) & 0 & \lambda + \gamma & -(1-\omega)\theta \\ -\mu & 0 & 0 & \lambda + \theta \end{bmatrix} = 0, \tag{10}$$

From the above characteristic equation, eigenvalues is obtained as follows

$\lambda_1 = -(-\beta + \gamma + \mu)$

$\lambda_2 = -\gamma - v$

$\lambda_3 = -\gamma$

$\lambda_4 = -\theta$

Therefore, the stability of the system of equation (1), (2), (3), (4) depends on the value of the characteristic roots. Based on the stability theory if all eigenvalues of the Jacobian matrix have a negative value, then the equilibrium point $(P^*, N^*, F^*, S^*) = \left(0, \frac{q\Lambda}{\gamma+v}, \frac{(1-q)\Lambda}{\gamma}, 0\right)$ on this system is locally asymptotically stable

4. Results and discussions

The simulation will be used to analize the dynamic system model, to check the optimal concentration of nitrogen and phosphate. Some numerical simulation by using data of nitrogen, phosphate, phytoplankton and sediment by [7] [8]. Research shows that there is a decrease in the concentration of nitrogen and phosphate, sediment, and phytoplankton in aquaculture. This is because the rest of the feed and excretion in the form of nitrogen and phosphate are utilized by marine biota. In order to validate the model equation (1) (2) (3) (4) some data parameter i.e $\Lambda = 3.98, v = 0.05,$ $q = 0.9, \alpha = 0.54, \beta = 0.55, \mu = 0.85, \gamma = 0.068, \theta = 0.06$ Further equation (1), (2), (3), (4) is transformed to equation

$$\frac{dP}{dt} = -0.368P, \tag{11}$$

$$\frac{dN}{dt} = 3.582 + 0.0271S - 0.12N - 0.297P\frac{N}{N+F}, \tag{12}$$

$$\frac{dF}{dt} = 0.398 + 0.0008S - 0.068F - 0.297P\frac{F}{N+F}, \tag{13}$$

$$\frac{dS}{dt} = 0.85P - 0.00187S, \tag{14}$$

We obtained the equilibrium point is $(P^*, N^*, F^*, S^*) = (0, 30.3559, 5.8529, 0)$. The mathematical model, data variable $P(0) = 0.952, N(0) = 0.68, F(0) = 0.56, S(0) = 0.67$ [10][11]. Substitution equilibrium points to Jacobian matrix obtained.

$$J = \begin{bmatrix} -0.368 & 0 & 0 & 0 \\ -0.1628 & -0.2229 & 0.125 & 0.00271 \\ -0.1341 & 0.1029 & -0.193 & 0.1029 \\ 0.85 & 0 & 0 & -0.0018 \end{bmatrix}$$

$\det(\lambda I - J) = 0$

$$\det(\lambda I - J) = \left| \begin{bmatrix} \lambda & 0 & 0 & 0 \\ 0 & \lambda & 0 & 0 \\ 0 & 0 & \lambda & 0 \\ 0 & 0 & 0 & \lambda \end{bmatrix} - \begin{bmatrix} -0.368 & 0 & 0 & 0 \\ -0.1628 & -0.2229 & 0.125 & 0.00271 \\ -0.1341 & 0.1029 & -0.193 & 0.1029 \\ 0.85 & 0 & 0 & -0.0018 \end{bmatrix} \right| = 0$$

$$\det(\lambda I - J) = \begin{vmatrix} \lambda+0.368 & 0 & 0 & 0 \\ 0.1628 & \lambda+0.2229 & -0.125 & -0.00271 \\ 0.1341 & -0.1029 & \lambda+0.193 & -0.1029 \\ 0.85 & 0 & 0 & \lambda+0.0018 \end{vmatrix}$$

$$= (\lambda+0.368) \begin{vmatrix} \lambda+0.2229 & -0.125 & -0.00271 \\ -0.1029 & \lambda+0.193 & -0.1029 \\ 0 & 0 & \lambda+0.0018 \end{vmatrix} - (0) \begin{vmatrix} 0.1628 & -0.125 & -0.00271 \\ 0.1341 & \lambda+0.193 & -0.1029 \\ 0.85 & 0 & \lambda+0.0018 \end{vmatrix}$$

$$+ (0) \begin{vmatrix} 0.1628 & \lambda+0.2229 & -0.00271 \\ 0.1341 & -0.1029 & -0.1029 \\ 0.85 & 0 & \lambda+0.0018 \end{vmatrix} - (0) \begin{vmatrix} 0.1628 & \lambda+0.2229 & -0.125 \\ 0.1341 & -0.1029 & \lambda+0.193 \\ 0.85 & 0 & 0 \end{vmatrix}$$

$$= (\lambda+0.368)(0.00005428 + 0.0309\lambda + 0.4177\lambda^2 + \lambda^3)$$

$$= (\lambda+0.368)(\lambda+0.2229)(\lambda+0.193)(\lambda+0.0018)$$

From the above characteristic equation eigenvalues is obtained as follows $\lambda_1 = -0.368$ $\lambda_2 = -0.2229$ $\lambda_3 = -0.193$ $\lambda_4 = -0.0018$. Based on the theory of stability if the eigenvalues obtained are negative on that the equilibrium point $(P^*, N^*, F^*, S^*) = (0, 30.3559, 5.8529, 0)$ then this system is stable asymptotic. Therefore changes in the concentration of nitrogen and phosphate on sediments and phytoplankton in aquaculture form a stable system. Furthermore, to analyze the stability of the equilibrium point based on parameter values, a stability trajectory is drawn around the equilibrium point.

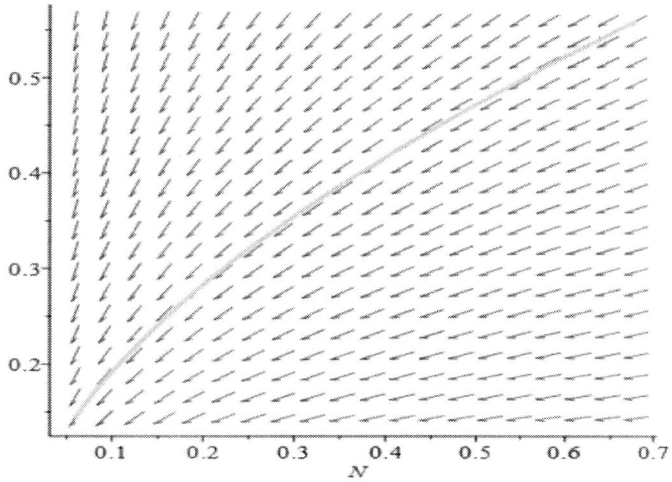

Figure 1. Trajectories stability around equilibrium points $(0, 30.3559, 5.8529, 0)$ in (N, F) plane.

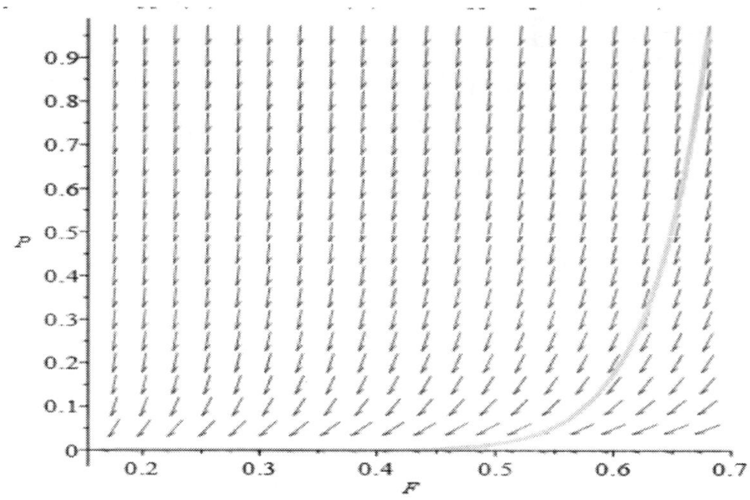

Figure 2. Trajectories stability around equilibrium point $(0,30.3559,5.8529,0)$ in (F,P) plane.

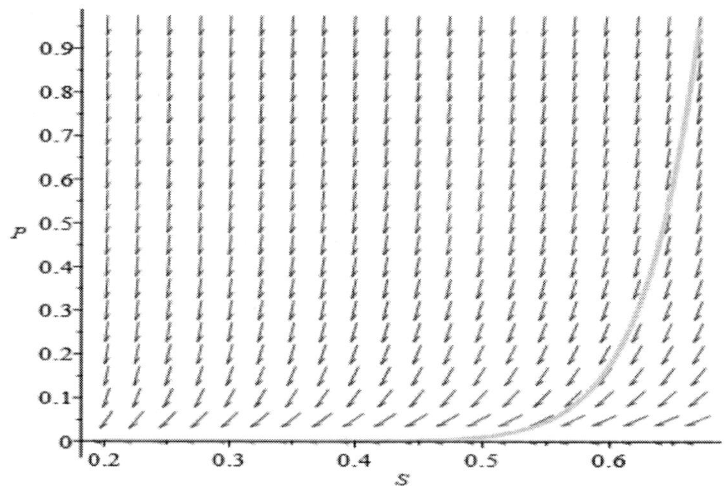

Figure 3. Trajectories stability around equilibrium points $(0,30.3559,5.8529,0)$ in the field (S,P) plane.

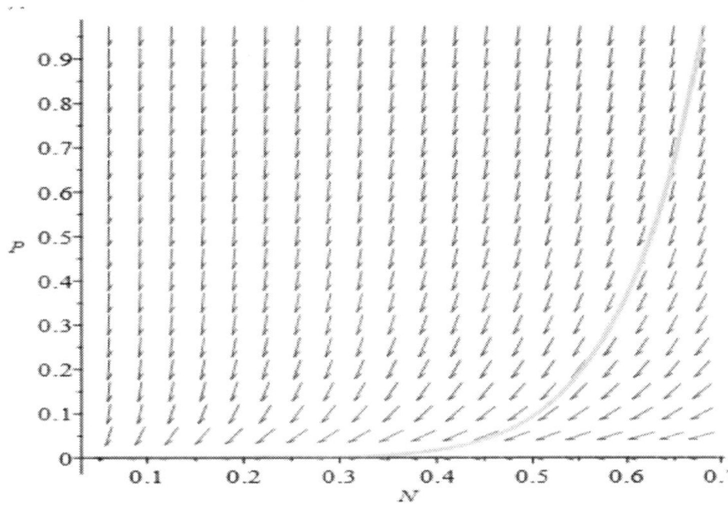

Figure 4 Trajectories stability around equilibrium points $(0,30.3559,5.8529,0)$ in the field (N,P) plane.

Based on Figure 1, Figure 2, Figure 3 and Figure 4 can see the direction of the arrow moving toward the equilibrium point $(P^*,N^*,F^*,S^*)=(0,30.3559,5.8529,0)$. This means that the equilibrium point for the given parameter is stable. The results of the dynamic system model showed that the concentrations of nitrogen and phosphate, phytoplankton and sediment decreased

5. Conclusion

It is concluded that model of nitrogen and phosphate concentration, phytoplankton abundance, and sediment is a non-linear dynamical system model. The numerical simulation concentration of nitrogen and phosphate are an important factor in the system dynamics. Eigenvalues are used to result analysis of the stability from the equilibrium point. Furthermore, analysis of locally stability of the dynamical system was based on the theory of stability. If the eigenvalues are negative, then the equilibrium point is locally asymptotically stable. From numerical simulation results base, it is concluded that equilibrium points and the eigenvalues are negative. Because all eigenvalues of the Jacobian matrix ware negative, the dynamic system model was locally asymptotically stable. Therefore, the dynamical system phytoplankton population experience decline as a result of decreasing concentration N and P from sediment. This model is expected to be implemented to minimize the concentration of nitrogen and phosphorus properly, so that it can be utilized by marine biota.

Acknowledgments

The authors would like to thank Directorate General of Higher Education (Simlitabmas-DIKTI) for financially supporting this research project under the scheme of Excellent Research For National Strategic Grant year 2018 project. The study was supported by Laboratory of Computer Modelling, Mathemathics Departement, Center of Marine Ecology and Biomonitoring for Sustainable Aquaculture (Ce-MEBSA), and Laboratory of Ecology and Biosystematics, Biology Departement, Faculty of Science and Mathemathics, Diponegoro University.

References

[1] Zhang Y, Bleeker A and Liu J 2015 Environ. Res. Lett **10** 045002
[2] Turcios A E and Papenbrock J 2015 Sustainability **6** 836-856
[3] Reynolds C S, Huszar V, Kruk C, Flores L N and Melo S 2002 *Journal of Plankton Research* **24(5)** 417-428.

[4] Rast W, Holland M, Ryding S O 1989 *Eutrophication Management Framework For The Policy-Maker* Paris (FR): United Nations.
[5] Wang Y, Zhao M, Dai C and Pan X 2014 *Hindawi Publishing Corporation Abstract and Applied Analysis* **451757**
[6] Liu C and Liu P 2014 *Mathematical Problems in Engineering* **521917**
[7] Burford M A and Lorenzen K 2004 *Aquaculture* **229** 129–145.
[8] Deng Y, Zhao M, Yu H and Wang Y 2015 *Hindawi Publishing Corporation Discrete Dynamics in Nature and Society* **823026**
[9] Widowati, Putro S P, and Silfiana 2018 *Journal of Physics*: **1025** 012088
[10] Silfiana, Widowati, S P Putro and T Udjiani 2018 *Journal of Physics*: **983** 012122
[11] Testa J M, Brady D C, Toro D M D and Boynton W R 2013 *Estuarine, Coastal and Shelf Science* **131** 245-263
[12] Li Y, Liu Y, Zhao L, Hastings A and Gou H 2015 *Ecological Modelling* **313** 137–148
[13] Widowati, Putro S P, Maan N and Sulpiani R 2019 *J. Phys.: Conf. Ser.* **1217** 012059
[14] Granada L, Lopes S, . Novais S C, Lemos M F L 2018 *Aquaculture* **495** 90–97
[15] Antinios M N 1987 *Mathematical Biosciences* **84** 231-248
[16] Olsen L M, Holmer M, and Olsen Y 2008 *Literature review with evaluated state of knowledge FHF project* 542014

Modeling of hypertension risk factors using local linear of additive nonparametric logistic regression

E Ana[1], N Chamidah[1*], P Andriani[2], and B Lestari[3]

[1]Department of Mathematics, Faculty of Science and Technology, Airlangga University. Jalan Mulyorejo Kampus C UNAIR, Surabaya 60115, Indonesia.
[2]Undergraduate Student of Statistical Study Program, Department of Mathematics, Faculty of Science and Technology, Airlangga University. Jalan Mulyorejo Kampus C UNAIR, Surabaya 60115, Indonesia.
[3]Faculty of Mathematics and Natural Sciences, The University of Jember. 37 Kalimantan Street, Campus of Tegal Boto, Jember 68121, Indonesia.

*Corresponding author's e-mail: nur-c@fst.unair.ac.id

Abstract. Hypertension has become a serious health problem in Indonesia because of its prevalence, however, the causative factors could not be ascertained for about ninety percent of the patients. Various studies have found several risk factors causing hypertension to be obesity, family history, stress levels, heart rate, and an unhealthy lifestyle. In this case, the variables are considered influential on hypertension through a regression curve without a specific pattern. Also, we need to describe the functional relationships between several predictor variables with binary or dichotomous response variables and need to describe locally effect of predictor variables to the response variable. Therefore, in this study, to model the case of hypertension by age, body mass index, heart rate, stress levels we use the additive nonparametric logistic regression approach based on local linear estimators. The results of the study showed that hypertension was most prevalent among respondents over 65 years of age with BMI between 25-30 kg/m2 (obesity) and normal heart rate (60-100) bpm and most of them were found to be experiencing mild stress conditions. The model obtained a classification accuracy of 95 percent (in-sample) and 89.47 percent (out-sample) with a cut off probability value of 0.4.

1. Introduction

Hypertension is often referred to as the silent killer because it takes the life of affected individuals without showing symptoms. However, the factors causing the disease (around 90%) are still unknown. The number of people living with hypertension is predicted to become 1.56 billion worldwide by the year 2025 [1]. The sickness is associated with cardiovascular diseases (CVD) risk factors, incidence, and mortality [2]. It is also found to be prevalent among people of 35 years of age and above, currently smoking, and obese [3]. The Seventh Report on the Joint National Committee on Prevention, Detection, Evaluation and Treatment of High Blood Pressure created a category called "pre-hypertension" which was defined as a systolic blood pressure (SBP) of 120-139 millimeters of mercury (mmHg) and a diastolic blood pressure (DBP) of 80-89 mmHg [1]. Pre-hypertension, even in the low range (SBP: 120-130 mmHg or DBP: 80-85 mmHg), has been confirmed to have a higher risk of developing into hypertension [4].

Hypertension has been associated with increased risk of coronary artery and cardiovascular and cerebrovascular diseases [5-6]. A meta-analysis also reported that lower blood pressure could also lead to cardiovascular and chronic kidney diseases [7-8]. This situation is critical in the Southeast Asian region with studies reporting HTN as an important risk factor for the attributable burden [9-10].
Several studies have found different risk factors for hypertension such as obesity, family history, stress levels, heart rate, and an unhealthy lifestyle. Furthermore, previous research showed that the classification accuracy using binary logistic regression was to be 72.5352 percent greater than the C4.5 algorithm which was 64.0845 percent. Therefore, it could be said that binary logistic regression is better than the C4.5 algorithm [11].

However, the variables used were considered influential on hypertension through a regression curve without a pattern, therefore, a nonparametric approach was implemented. Moreover, logistic regression is a mathematical model used in describing the functional relationships between several predictor variables with binary or dichotomous response variables [12]. Binary response (Y) variables consist of two categories which are success ($y = 1$) and failure ($y = 0$), with respect to the Bernoulli distribution and the predictor variable is assumed to be linear to the parameters [13]. Generalized additive model (GAM) is an approach used in the exploration of nonparametric data [14].

Estimates for the approaches could be conducted by using smoothing techniques [15-22] and a local linear estimator is mostly used because of its ability to estimate functions locally at each point, thereby making the model obtained closer to the actual data pattern [23]. It is also useful because it can model the hypertension by age, body mass index, heart rate, and stress levels effectively due to its mean square error (MSE) value, which is close to zero, making the estimation results close to the observation value, and its determination coefficient which approaches 100 percent. This means that the regression model could describe the variability of risk factors for hypertension close to 100 percent [24]. In addition, the use of local linear in various cases of study for instance in medical and education cases study, have been introduced by researchers who have used local linear in [25-29].

Based on the facts from previous research, the hypertension risk factors of Surabaya Hajj Hospital based on additive nonparametric logistic regression approaches were modeled through the use of local linear estimator. The results of this research could be used to predict the chance of an individual suffering from hypertension showing the risk factors in order to reduce the prevalence of the disease in Indonesia.

2. Literature review

In this section we review some statistical concepts that can be used to support in modeling of hypertension risk factors.

2.1. Nonparametric Regression model

Nonparametric regression model is a model used in determining the relationship pattern between the response variable (y) and the predictor variable (t) where the function of the regression curve or the pattern of the two variables is unknown. The model generally follows the equation below:

$$y_i = m(t_i) + \varepsilon_i \ , \ i = 1, 2, \ldots, n \tag{1}$$

where ε_i is random error assumed as independent with zero mean and variance σ^2, and $m(t_i)$ is the regression function which is to be estimated [30].

2.2. Logistic Regression model

Logistic regression is a mathematical model used in describing the functional relationships between several predictor variables with binary or dichotomous response variables [12]. These response (Y) variables consist of two categories which are success (Y = 1) and fail (Y = 0), with respect to the Bernoulli distribution and the predictor variable is assumed to be linear to the parameters [13]. The model is known as the logit model because it uses logit transformation as a link function from the following equation:

$$g(\mu) = logit\,\mu = log(\mu/(1-\mu)) \tag{2}$$

2.3. Additives nonparametric logistic Regression

The additive model is an illustration of the response variable which depends on the sum of several functions of several predictor variables. It could be expressed as follows:

$$y_i = \sum_{j=1}^{p} m_j(t_{ji}) + \varepsilon_i \ , \quad i = 1,2,...,n \tag{3}$$

Next, by using the generalized additive model (GAM) approach, the equation (3) becomes:

$$log(\mu_i/(1-\mu_i)) = \sum_{j=1}^{p} m_j(t_{ji}) \tag{4}$$

where

$$\mu_i = \frac{e^{\sum_{j=1}^{p} m_j(t_{ji})}}{\left(1+e^{\sum_{j=1}^{p} m_j(t_{ji})}\right)} \tag{5}$$

The general algorithm for obtaining the estimated value of regression functions in the generalized additive model (GAM) is the local scoring algorithm.

2.4. Local linear estimator

The regression function could be estimated by a local linear estimator, through the equation below:

$$\hat{m}(t) = \mathbf{x}(t_o)\hat{\beta}(t_o) \tag{6}$$

where $\beta(t_0)$ is parameter of the model at target point t_0 that is estimated by using weighted least square (WLS) method, i.e., by minimizing the following function:

$$Q(t_0) = (\mathbf{y} - X(t_0)\beta(t_0))^T K_h(t_0)(\mathbf{y} - X(t_0)\beta(t_0))$$

such that we obtain:

$$\hat{\beta}(t_0) = (\mathbf{X}^T(\mathbf{t}_0)\mathbf{K}_\mathbf{h}(\mathbf{t}_0)\mathbf{X}(\mathbf{t}_0))^{-1}\mathbf{X}^T(\mathbf{t}_0)\mathbf{K}_\mathbf{h}(\mathbf{t}_0)\mathbf{y} \tag{7}$$

Based on equations (6) and (7), the linear local estimator $\hat{m}(t)$ could be written as:

$$\hat{m}(t) = t(t_0)(\mathbf{X}^T(\mathbf{t}_0)\mathbf{K}_\mathbf{h}(\mathbf{t}_0)\mathbf{X}(\mathbf{t}_0))^{-1}\mathbf{X}^T(\mathbf{t}_0)\mathbf{K}_\mathbf{h}(\mathbf{t}_0)\mathbf{y} \tag{8}$$

2.5. Selection of optimal bandwidth

Bandwidth is the control of the balance between the smoothness of the function and its goodness of fit to the data [30]. The best estimation of regression function depends on the optimal bandwidth (h_{opt}). The optimal bandwidth can be obtained by getting the solution of the following cross validation (CV) optimization [31]:

$$CV(h) = \min \frac{1}{n}\sum_{i=1}^{n}\left(y_i - \hat{m}_{n,i}^{(h)}(t_i)\right)^2 \tag{9}$$

2.6. Deviance

Deviance is a statistical value used to find the suitability of the model (goodness of fit) by comparing the actual to the expected model. Suppose that we have data set $\{y_i\}$, $i = 1,2,...,n$ which is binomially distributed $BIN(1,\mu_i)$. The deviance $D(y_i,\mu_i)$ could be calculated as follows [32]:

$$D(y_i,\mu_i) = -2\sum_{i=1}^{n}\{y_i \log\mu_i + (1-y_i)\log(1-\mu_i)\} \tag{10}$$

2.7. Accuracy of classification

The accuracy of the classification model is calculated by using the apparent error rate (APPER). The value obtained is used in classifying an object error. The APPER is given by:

$$APPER = \frac{n_{12} + n_{21}}{n_{11} + n_{12} + n_{21} + n_{22}} \times 100\% \qquad (11)$$

where n_{rs} ($r = s = 1, 2$) is the number of observation in cell rs that is intersection between row r and coulomb s. Therefore, the accuracy of classification is given by:

$$\text{Accuracy of Classification} = 100\% - \text{APPER} \qquad (12)$$

3. Methods

The primary data used was obtained through questionnaires and taking of measurements on outpatients of the Heart Poly Surabaya Hajj Hospital. The research made use of 59 respondents consisting of 28 suffering from hypertension and 31 from other ailments with 40 in-sample and 19 out-sample data. The variables in the model used in this study are given in Table 1.

Table 1. Variables in the model

Variable	Description	Scale
Y	Hypertension incidents 0 = not suffering hypertension 1 = suffering hypertension	Nominal
t_1	Age (Year)	Ratio
t_2	Body Mass Index (kg/m^2)	Ratio
t_3	Heart rate (bpm)	Ratio
t_4	Stress level	Ratio

The following steps were taken in analyzing the data:

- Descriptive statistical analysis was conducted on the predictor variables associated with the incidence of hypertension.
- Hypertension risk factors were modeled in Cardiac Poly outpatients in Surabaya Hajj Hospital with an additive nonparametric logistic regression approach based on the local linear local estimator with the following steps:
  - Determining of the kernel functions, i.e., Gaussian kernel functions;
  - Determining of the optimal bandwidth value based on minimum CV on each predictor;
  - Determining of the initial estimation value by using the optimal bandwidth value obtained from step (b);
  - Determining of the estimated value by using local scoring;
  - Testing of suitability of the model;
  - Analyzing and interpreting of the model of hypertension events.

4. Results

In this section we give the results of this study, i.e., description of variables, and modeling of hypertension risk factors.

4.1. Description of variables associated with hypertensive events

The characteristics of hypertension incidence are as shown in Table 2.

Table 2. Percentage of respondents' characteristics of hypertension events in outpatients in heart poly Surabaya Hajj hospital.

Variable		Hypertension Incidents (%)	
		Hypertension	Non-Hypertension
Age (Year)	< 45	7.14	61.3
	45-54	10.71	22.58
	55-64	35.71	12.9
	> 65	46.43	0
BMI (kg/m²)	< 25	35.71	70.97
	25-30	46.43	12.9
	> 30	17.86	16.13
Heart Rate (bpm)	< 60	7.14	3.23
	60-100	85.71	77.42
	>100	7.14	19.35
Stress Level (total score)	0-10	60.71	64.52
	11-16	25	22.58
	≥17	14.29	12.90

4.2. Modelling of hypertension risk factors

The selection of optimal bandwidth based on the minimum CV value was the first step conducted to get the appropriate estimation of hypertension incidence models. The bandwidth and CV for each predictor variable are given in Table 3.

Table 3. Bandwidth and CV for each predictor variable

Predictor Variable	Optimal Bandwidth	Minimum CV
Age	28.3	0.1448348
BMI	8.6	0.2486765
Heart rate	22.1	0.2448074
Stress level	3.2	0.2790422

The results in Table 3 were used to determine the value of the initial estimation of the regression function for each predictor. Each of them was observed with a multi-predictor local linear estimator, and a result of 77 years old, with BMI 26.84 kg/m2, heart rate 84 bpm, and total stress level score 20 was found.

$$\begin{aligned} \eta_1 &= 1.7657713 + 0.1357574(0) + 0.3868306 + 0.1160375(0) + \\ &\quad 0.789679 - 0.077819(0) + 1.5592301 + 0.1534729(0) \\ &= 4.50151156 \end{aligned}$$

Therefore, the $\hat{\mu}_1$ value is:

$$\hat{\mu}_1 = \frac{\exp(4.50151156)}{1+\exp(4.50151156)} = 0.9890295$$

Next, we test the suitability of the model used by calculating the deviance value which is given in (11). Hypothesis used for testing the suitability the model used is as follows:

H_0 : Model is appropriated versus H_1 : Model is not appropriated.

Then, by applying Eq. (11) we get the deviance value $D(y_i, \mu_i) = 7.617064$ which is smaller than the value of the Chi-Square distribution, i.e., $\chi^2_{36;0.05} = 50.9985$. It means that the model used is in accordance with the nonparametric logistic regression model.

For the purpose of classification, a cut off probability was needed such that if the cut off the probability of $\hat{\mu}_i$ is greater, then the prediction result is 1 and if it is less, the prediction result is 0. Based on the output, the best cut off probability value obtained was 0.46. Therefore, the value of classification accuracy in the in-sample data is presented in Table 4.

Table 4. Accuracy of in-sample data classification

Observation	Prediction		Total
	Without Hypertension	With Hypertension	
Not suffering Hypertension	19	1	20
Suffering Hypertension	1	19	20
Total	20	20	40

Based on Table 4, the APPER value could be estimated by applying equation (12) as follows:

$$APPER = \frac{1+1}{40} \times 100\% = 5\% .$$

So, we get the Classification Accuracy $= 100\% - 5\% = 95\%$.
This shows that the estimation of the nonparametric logistic regression model based on the local linear estimator has the ability to explain 95 percent of hypertension incidence in the data in-sample.

Furthermore, 19 data out-sample were also estimated by using the same model used for the in-sample with a cut off probability value of 0.46 and a result of 67 years, with BMI 48.61 kg/m2, heart rate 79 bpm, and stress level scores of 10 were obtained. However,

$$\hat{\mu}_1^* = \frac{\exp(4.458179)}{1+\exp(4.458179)} = 0.988549202$$

The values of the accuracy of classification for the out-sample data is given in Table 5.

Table 5. Classification accuracy of out-sample data

Observation	Prediction		Total
	Without Hypertension	With Hypertension	
Not suffering Hypertension	10	1	11
Hypertension	1	7	8
Total	11	8	19

The APPER value was obtained as follows:

$$APPER = \frac{1+1}{19} \times 100\% = 10.53\%$$

$$\text{Classification Accuracy} = 100\% - 10.53\% = 89.47\%$$

This shows that the estimation of the nonparametric logistic regression model based on local linear estimators has the ability to explain 89.47% of hypertension incidence in the out-sample data.

5. Discussion

Hypertension has become a prevailing health problem in Indonesia because of its prevalence. However, it can be prevented by adequate control of a healthy lifestyle, routine blood pressure measurements, and other effective treatments.

Note that age, BMI, heart rate, and stress level are some of the factors observed to be triggering hypertension. As an example we select a 77 years old of person who represents the pre-very old person. If a person is 77 years old, with a BMI of 26.84 kg/m2, heart rate 84 bpm, and a stress level score of 20, such person has 98.9 percent risk of developing hypertension. This shows that an increase in those factors has a probability of causing the disease. The finding is in accordance with the results of previous studies that increasing age, BMI, smoking, diabetes, and additional salt intake are the common risk factors of hypertension [33-34].

6. Conclusion

Hypertension was found to be generally occurring in persons over the age of 65 years with a BMI between 25-30 kg/m2 and a normal heart rate between 60-100 bpm with most of them experiencing mild stress. Since, the modelling conducted on in-sample data gave a classification accuracy of 95 percent and 89.47 percent for out-sample data with a cut off probability value of 0.46, then it could be concluded that the estimation model by using local linear estimator of additive nonparametric logistic regression is the best estimator for determining the hypertension incidence.

References

[1] Kearney P M, Whelton M, Reynolds K, Muntner P, Whelton P K, and He J 2005 Global Burden of Hypertension: Analysis of Worldwide Data *Lancet* **365** 217–223.

[2] Elliot W J, and Black H R 2007 Prehypertension Nat Clin Pract Cardiovasc Med. 4(10) 538-548.

[3] Sathish T, Kannan S. Sarma P S, Razum O, Amanda G T, and Thankappan K R 2016 Risk Score to Predict Hypertension in Primary Care Settings in Rural India Asia-Pac J Public Health 28 26S-31S.

[4] Israeli E, Korzets Z, Tekes–Manova D, Tirosh A, Schochat T, Bernheim J, and Golan E 2007 Blood-pressure categories in adolescence predict development of hypertension in accordance with the European guidelines American journal of hypertension 20(6) 705-709.

[5] Qureshi A I, Suri M F, Kirmani J F, Divani A A, and Mohammad Y 2005 Is prehypertension a risk factor for cardiovascular diseases ? Stroke 36(9) 1859-63.

[6] Wu S, Huang Z, Yang X, Li S, Zhao H, Ruan C, Wu Y, Xin A, Li K, Jin C, and Cai J 2013 Cardiovascular events in a pre-hypertensive Chinese population: four-year follow-up study International Journal of Cardiology 167(5) 2196-2199.

[7] Huang Y, Wang S, Cai X, Mai W, Hu Y, Tang H, and Xu D 2013 Pre-hypertension and incidence of cardiovascular disease: a meta-analysis BMC Medicine 11 177.

[8] Kim M J, Lim N K, and Park H Y 2012 Relationship between pre-hypertension and chronic kidney disease in middle-aged people in Korea: the Korean genome and epidemiology study BMC Public Health 12 960.

[9] Lim S S, Vos T, and Flaxman A D 2012, A comparative risk assessment of burden of disease and injury attributable to 67 risk factors and risk factor clusters in 21 regions 1990–2010: a systematic analysis for the Global Burden of Disease Study 2010 The Lancet 380(9859) 2224-2260.

[10] Lopez A D, Mathers C D, Ezzati M, Jamison D T, and Murray C J 2006 Global and Regional burden of disease and risk factors 2001: systematic analysis of population health data The Lancet 367(9524) 1747-1757.

[11] Rumaenda W, Wilandari Y, and Safitri D 2016 Perbandingan Klasifikasi Penyakit Hipertensi Menggunakan Regresi Logistik Biner dan Algoritma C4.5 Jurnal Gaussian 3(2) 299-309.

[12] Kleinbaum D G 1994 Logistic Regression Springer New York.

[13] Hosmer D, Lemeshow 2000 Applied Logistic Regression John Wiley and Sons Inc USA.

[14] Hastie T J and Tibshirani R J 1990 Generalized Additive Models Chapman & Hall London.

[15] Chamidah N and Lestari B 2016 Spline Estimator in Homoscedastic Multi-Response Nonparametric Regression Model in Case of Unbalanced Number of Observations Far East Journal of Mathematical Sciences (FJMS) 100 1433-1453.

[16] Lestari B, Anggraeni D, and Saifudin T 2017 Konstruksi dan Estimasi Matriks Kovariansi dalam Model Regresi Nonparametrik Multirespon Berdasarkan Estimator Smoothing Spline untuk Beberapa Kasus Ukuran Sampel Proceeding of National Seminar on Mathematics and Its Applications (SNMA) 2017 Airlangga University Surabaya pp. 238-242.
[17] Lestari B, Fatmawati and Budiantara I N 2017 Estimasi Fungsi Regresi Nonparametrik Multirespon Menggunakan Reproducing Kernel Hilbert Space Berdasarkan Estimator Smoothing Spline Proceeding of National Seminar on Mathematics and Its Applications (SNMA) 2017 Airlangga University Surabaya pp. 243-250.
[18] Lestari B, Anggraeni D and Saifudin T 2018 Estimation of Covariance Matrix based on Spline estimator in Homoscedastic Multi-Responses Nonparametric Regression Model in Case of Unbalance Number of Observations Far East J.Math.Sciences (FJMS) 108(2) 341-355.
[19] Islamiyati A, Fatmawati and Chamidah N 2018 Estimation of Covariance Matrix on Bi-Response Longitudinal Data Analysis with Penalized Spline Regression Journal of Physics: Conf. Series Vol. 979 pp. 012093.
[20] Lestari B, Fatmawati and Budiantara I N 2019 Spline Estimator and Its Asymptotic Properties in Multiresponse Nonparametric Regression Model Songklanakarin Journal of Science and Technology In Press.
[21] Lestari B, Fatmawati, Budiantara I N and Chamidah N 2018 Estimation of regression function in multi-response nonparametric regression model using smoothing spline and kernel estimators Journal of Physics: Conference Series 1097 012091.
[22] Lestari B, Chamidah N and Saifudin T 2019 Estimasi fungsi regresi dalam model regresi nonparametrik birespon menggunakan estimator smoothing spline dan estimator kernel Jurnal Matematika, Statistika and Komputasi (JMSK) 15(2) 20-24.
[23] Hardle W 1990 Applied Nonparametric Regression Cambridge University Press New York.
[24] Adiwati, T and Chamidah, N. 2019 Modelling of Hypertension Risk Factors Using Penalized Spline to Prevent Hypertension in Indonesia IOP Conf. Series: Materials Science and Engineering 546 052003 doi:10.1088/1757-899X/546/5/052003.
[25] Chamidah N, Tjahjono E, Fadilah A R., and Lestari B 2018 Standart Growth Charts for Weight of Children in East Java Using Local Linier Estimator Journal of Physics: Conference Series 1097 012092.
[26] Nidhomuddin, Chamidah N, Kurniawan A 2019 Admission Test Modelling of State Islamic College in Indonesia Using Local Linear for Bivariate Longitudinal Data IOP Conf. Series: Materials Science and Engineering 546 052047 doi:10.1088/1757-899X/546/5/052047.
[27] Darnah, Utoyo M.I and Chamidah N 2019 Modeling of Maternal Mortality and Infant Mortality Cases in East Kalimantan using Poisson Regression Approach Based on Local Linear Estimator IOP Conf. Series: Earth and Environmental Science 243 012023 doi:10.1088/1755-1315/243/1/012023.
[28] Puspitawati A., and Chamidah N 2019 Choroidal Neovascularisation Classification on Fundus Retinal Images Using Local Linear Estimator IOP Conf. Series: Materials Science and Engineering 546 052056 doi:10.1088/1757-899X/546/5/052056.
[29] Chamidah N, Gusti K H, Tjahjono E and Lestari B 2019 Improving of classification accuracy of cyst and tumor using local polynomial estimator TELKOMNIKA, Vol.17, No.3, pp.1492-1500.
[30] Eubank R L 1998 Nonparametric Regression and Spline Smoothing 2nd Edition Marcel Dekker New York.
[31] Gyorfi L, Kohler M, Krzyzak A, and Walk H 2002 A Distribution-Free Theory of Nonparametric Regression Springer.
[32] Collett D 2002 Modelling Binary Data 2nd Edition Chapman & Hall/CRC.

[33] Devi P, Rao M, Sigamani A, Faruqui A, Jose M, Gupta R, Kerkar P, Jain R K, Joshi R, Chidambaram N, Rao D S, Thanikachalam S, Iyengar S S, Verghese K, Mohan V, Pais P, and Xavier D 2012 Prevalence, risk factors and awareness of hypertension in India: a systematic review Journal of Human Hypertension 27(5) 281-287.
[34] Anchala R, Kannuri N K, Pant H, Khan H, Franco O H, Angelantonio E D, and Prabhakaran D 2014 Hypertension in India: a systematic review and meta-analysis of prevalence, awareness, and control of hypertension Journal of Hypertension 32(6) 1170.

Clustering with spatial constraints: The case of diarrhea in Bandung city, Indonesia

I G N M Jaya[1*], B N Ruchjana[2], Y Andriyana[1], and R Agata[1]

[1] Department of Statistics, Universitas Padjadjaran, Indonesia
[2] Department of Mathematics, Universitas Padjadjaran, Indonesia

*Corresponding author e-mail: mindra@unpad.ac.id

Abstract. Clustering with spatial constraints was introduced in disease surveillance to identify high/low risk areas. These improve the geographical pattern of disease clustering. There are two types of spatial constraints, geographical distance and contiguity based. A Ward-like hierarchical clustering algorithm, including the spatial constraints, was employed on diarrheal data in Bandung city. Diarrhea is an infectious disease that causes death. The geographical distance was the best geographical dissimilarity for the diarrheal disease data. A five-cluster solution was determined to be optimal. Cluster two, which consists of three districts (Cibiru, Cinambo, Mandalajati), was considered as a high-risk cluster. It scored high on standardized incidence ratio which is caused by low on healthy house index and water quality index.

1. Introduction

Spatial disease mapping study helps us to understand the geographical distribution and clustering of the disease ([1], [2], [3]). It involves a visual representation of the clustering of high/low-risk areas, which is vital to the development of an etiological hypothesis of risk factors affecting the spatial variation of the disease [2]. Thus, it can be used to develop a strategy for reducing, preventing, and controlling the spread of infectious disease [4]. Spatial disease mapping helps the government in policy making and utility of resources [1].

Diarrhea is one of severe infectious disease cause of death in low-and-middle-income countries like Indonesia. It is the second popular cause of death in children under five years of age [5]. According to UNICEF Indonesia report, 10% of all children under five years of age died of diarrhea in 2016 [6]. Diarrhea, as a health problem, may occurs not only in children under five years, but also in older people of all ages. In Indonesia, clean water supply and sanitation are the significant factors affecting the high number of diarrhea cases in several places such as in the city of Bandung [7].

To identify high/low-risk areas, standardized incidence ratio (SIR) and its statistical significance are commonly presented in a map. However, it fails in informing clustering of high/low-risk areas. Clustering SIR with spatial constraints is needed to identify high/low-risk regions. Spatial clustering method has been developed in disease mapping study. Anderson et al. [8] and Adin et al. [9] developed two stage approach to identify spatiotemporal clusters. They combined Bayesian disease mapping and hierarchical agglomerative clusters (HAC). Choe et al. [10] introduced Bayesian latent approach. Jaya et al. [11] applied local Moran'I and spatial scan statistics. In this paper, we apply hierarchical clustering with spatial constraints to identify diarrhea clustering in the city of Bandung, Indonesia.

The structure of the remainder of this paper is as follows: Section 2 presents the material and methods. Section 3 presents application for diarrheal disease data in Bandung City and Section 4 presents the conclusion.

2. Material and method

2.1. Material

Data were taken from Bandung health surveillance conducted by the Bandung health office in 2017. The data set is an extracted of $p = 4$ quantitative variables for a subsample of n= 30 Bandung districts located on the West Java province, Indonesia. Those quantitative variables are population (number), diarrheal cases (number), healthy house index (%), and water quality index (%).

2.2. Relative risk

Relative risk estimates are typically used to identify the high/low-risk areas. It is developed based on the assumption that number of cases (y_i) follows Poisson distribution with mean and variance are equal to $E_i\theta_i$ ([1] [2] [12] [13]).

$$y_i|E_i\theta_i \sim Poisson(E_i\theta_i) \tag{1}$$

where E_i is expected number of cases and θ_i denotes the relative risk. Expected number of cases E_i is defined as [1]:

$$E_i = N_i \frac{\sum_{i=1}^{n} y_i}{\sum_{i=1}^{n} N_i}; i = 1, \dots, n \tag{2}$$

with N_i is number of population at risk in area i and n denotes number of areas. Parameter of the relative risk θ_i is estimated using maximum likelihood approach given by:

$$\hat{\theta}_i = \frac{y_i}{E_i} \tag{3}$$

which is also called as Standardized Incidence Ratio (SIR). A high-risk area is identified as an area with SIR greater than one.

2.3. Spatial clustering

Mapping SIR measure in disease mapping does not inform the high//low risk disease cluster. We need to identify the high/low risk areas using clustering method. Since diarrhea is an infectious disease, the spatial dependence must be taken into consideration in the clustering method. For this purpose we consider two dissimilarity matrices: (i) $D0$ is the Euclidean distance matrix between 30 districts performed with 3 indicators (i.e., SIR, Healthy House Index, Water Quality Index), and (ii) D1 is a second dissimilarity matrix used to take the geographical proximity between 30 districts into account [14].

2.3.1. The ward-linkage method

Pseudo-inertia. Assume that the regions $\boldsymbol{A} = \{A_1, \dots, A_n\}$ can be partitioned into K clusters, $P_K = (C_1, \dots, C_K)$. The pseudo inertia of a C_k cluster can be generalized into the case of dissimilarity data as [14]:

$$I(C_k) = \sum_{i \in C_k} \sum_{j \in C_k} \frac{w_i w_j}{2\mu_k} d_{ij}^2 \tag{4}$$

where $\mu_k = \sum_{i \in C_k} w_i$ is the weight of C_k. A small value of the pseudo-inertia $I(C_k)$ indicates the observations belong to cluster-k more homogenous [14].
The pseudo within-cluster inertia of the partition P_K is given as follows.

$$W(P_K) = \sum_{k=1}^{K} I(C_k) \tag{5}$$

The total pseudo inertia $I(C_k)$ for K clusters $W(P_K)$ describes the variation between clusters. A small value of $W(P_K)$ indicates variation between the clusters are more homogenous.

2.3.2. Spirit of the ward hierarchical clustering.
The idea to obtain a new partition P_K in K clusters from a given partition P_{K+1} in $K+1$ cluster, is to aggregate the two clusters A and B of P_{K+1} such that the new partition has minimum within-cluster inertia that is [14]:

$$\arg \min_{A,B \epsilon P_{K+1}} W(P_K) \quad (6)$$

where $W(P_K) = W(P_{K+1}) - I(A) - I(B) + I(A \cup B)$
Here we use two different dissimilarity matrices, first is based on diarrheal variables and the second one based on the spatial weight matrix (i.e., distance and contiguity). The partition process needs to consider magnitude of these two different matrices [14]. Here we use a mixed pseudo inertia of the cluster:

$$I_\alpha(C_k^\alpha) = (1-\alpha)\sum_{i\epsilon C_k^\alpha}\sum_{j\epsilon C_k^\alpha}\frac{w_i w_j}{2\mu_k^\alpha}d_{0,ij}^2 + \alpha\sum_{i\epsilon C_k^\alpha}\sum_{j\epsilon C_k^\alpha}\frac{w_i w_j}{2\mu_k^\alpha}d_{1,ij}^2 \quad (7)$$

where $\mu_k^\alpha = \sum_{i\epsilon C_k^\alpha} w_i$ is the weight of C_k^α and $d_{0,ij}^2$ and $d_{1,ij}^2$ is the normalized dissimilarity between observation i and j in D_0 and D_1 respectively.
The spatial clustering in this paper is done using ClustGeo package in R.

2.4. Data processing steps
Clustering data with spatial constrain can be processed using some steps below:
- preparing cluster variables $\mathbf{x} = [\mathbf{x}_1, \ldots, \mathbf{x}_n]'$,
- defining spatial weight matrix based on distance or contiguity,
- estimating the mixed parameter α using e.q (7),
- defining the optimal number of cluster by means dendrogram, and
- constructing the ward agglomerative clustering by minimizing $W(P_K)$ in e.q (5)

3. Result and discussion
Diarrhea data used in this study were consisted 4 variables. The descriptive summary of each variable is given in Table 1.

Table 1. The descriptive statistics summary of diarrheal variables

Variable	Mean	Minimum	Maximum	Standard Deviation
Diarrheal Cases (DC)	1918.000	825.000	3513.000	749.935
Population (P)	83265.000	24872.000	136037.000	31339.075
Healthy House Index (HHI)	79.099	63.239	98.530	8.812
Water Quality Index (WQI)	75.286	0.000	100.000	29.644

Table 1 shows the descriptive statistics of diarrheal variables. We found some districts (i.e. Mandalajati, Cibiru and Cinambo) have water quality index equal to zero due to shortage of clean water supply. To evaluate the risk of diarrheal disease in Bandung City, we calculate the standardized incidence rate (SIR). A cluster with high relative risk and low healthy house and water quality indices are categorized as a high-risk cluster.

3.1. Spatial clustering of diarrheal in Bandung without spatial constraint (use D0 matrix)
First, we consider diarrheal disease spatial clustering without spatial constraints. We use the dendrogram to choose the appropriate number of clusters K. The dendrogram in Fig 1 suggests to retain K=5 clusters. The choropleth map in Fig. 1 is used to visualize the corresponding partition of five clusters.

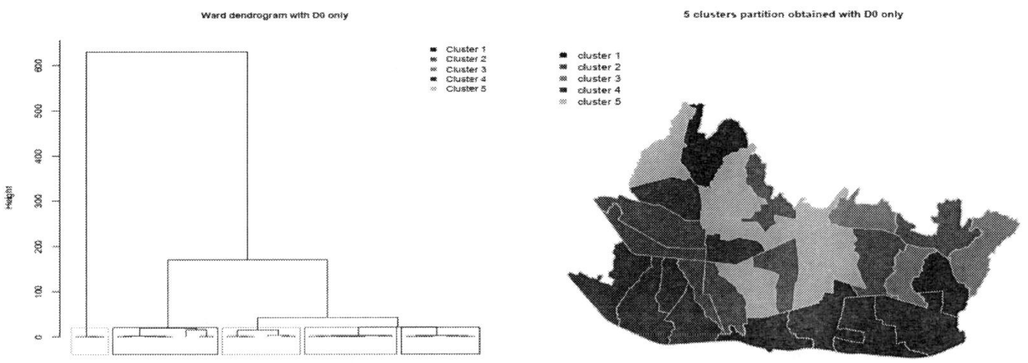

(a) Dendrogram (b) Spatial Clustering

Figure 1. Dendrogram without Spatial Constraint

Figure 1(a) shows the spatial clustering of diarrheal risk in Bandung. However, the map does not indicate spatial clustering. It might be the dissimilarity matrix ignoring the proximity between 30 districts. Cluster 1 is located in the north and southeast region of Bandung. Cluster 4 is also spread. It is separated by cluster 2. It leads to difficulty in defining the high-risk cluster and the districts that belong to this cluster.

3.2. *Spatial clustering of diarrheal in Bandung with spatial constraint (use D1 matrix)*

To solve the problem in previous, we include the spatial constraint in the clustering algorithm. We keep the suitable number of cluster $K = 5$ which is obtained based on dendrogram in Fig 2.

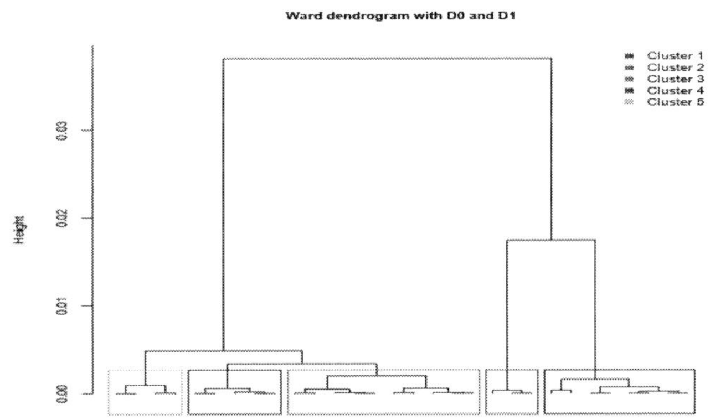

Figure 2. Dendrogram with spatial constraint

Spatial cluster in Fig (1b) can be improved to obtain more geographically compact clusters, by introducing the matrix $D1$ of geographical distances. Figure 2 shows the dendrogram with spatial constraint. We observed there are five clusters constructed. It is easily implemented using `hclustgeo` in package `ClustGeo`. A mixing parameter α is required to improve the geographical cohesion of the 5 clusters without adversely affecting diarrheal cohesion. The mixing parameter $\alpha \in [0, 1]$ defines the importance of $D0$ and $D1$ in the clustering process. If we define $\alpha = 0$, the geographical dissimilarities are not taken into account and when $\alpha = 1$ it is the diarrheal variables distance which are not taken into account and the clusters are obtained with the geographical distances only. We present the trade-off between geographical and diarrheal variables distances by presenting the quality criterion $Q0$ and $Q1$ plot of the partitions $P_K{}^{\alpha}$ in Fig 3.

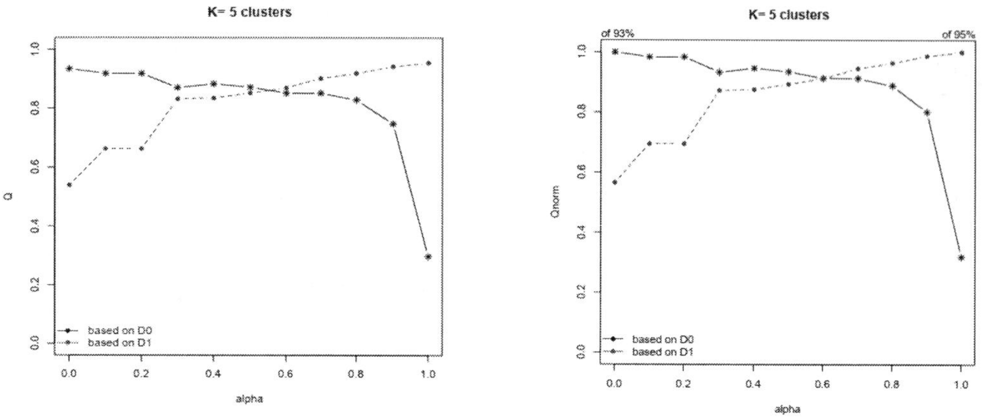

(a) Unstandardized Q (b) Standardized Q

Figure 3. Choice of α for a partition in $K = 5$ clusters when $D1$ is the neighborhood dissimilarity matrix between districts. Left: proportion of explained pseudo-inertias $Q_0(P_K^{\alpha})$ versus α (in black solid line) and $Q_1(P_K^{\alpha})$ versus α (in dashed line). Right: normalized proportion of explained pseudo-inertias $Q^*_0(P_K^{\alpha})$ versus α (in black solid line) and $Q^*_1(P_K^{\alpha})$ versus α (in dashed line)

Table 2. Proportion of explained pseudo-inertia

Alpha	Q		Qnorm	
	Q0	Q1	Q0	Q1
alpha=0.6	0.851	0.869	0.912	0.912
alpha=0.7	0.851	0.901	0.912	0.945
alpha=0.8	0.828	0.918	0.887	0.963

According to the pseudo inertia (Q) and normalized pseudo-inertia (Qnorm), we observed the best choice for alpha is 0.70 with the highest value of Q_0 and high value of Q_1. It indicates, the geographical and diarrheal variables have a high contribution in constructing the clusters. In other words. the geographical and diarrheal variables distances have high quality level with value greater than 0.900. Figure 4 presents the spatial clustering maps for given distance spatial weight matrix.

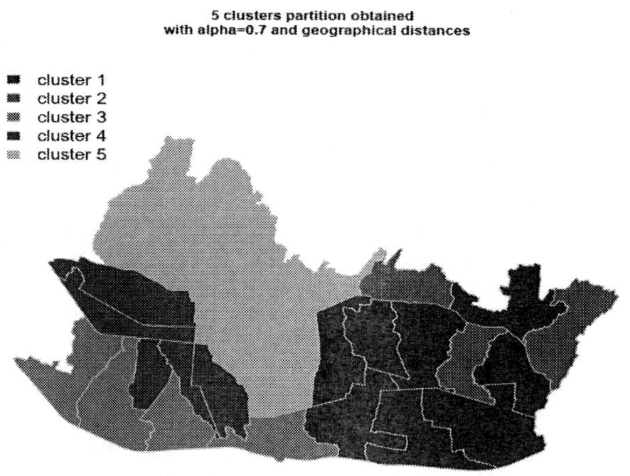

Figure 4. Map of the partition in $K = 5$ clusters based on the diarrheal variables distances D_0 and the "neighborhood" distances of the districts D_1 with $\alpha = 0.7$

Figure 5 shows that clusters with geographical distance are more geographically compact because the geographical distance is involved in clustering process. We also consider geographical dissimilarity based on spatial contiguity. The trade-off geographical dissimilarity based on queen contiguity and diarrheal variables are presented in Fig 5.

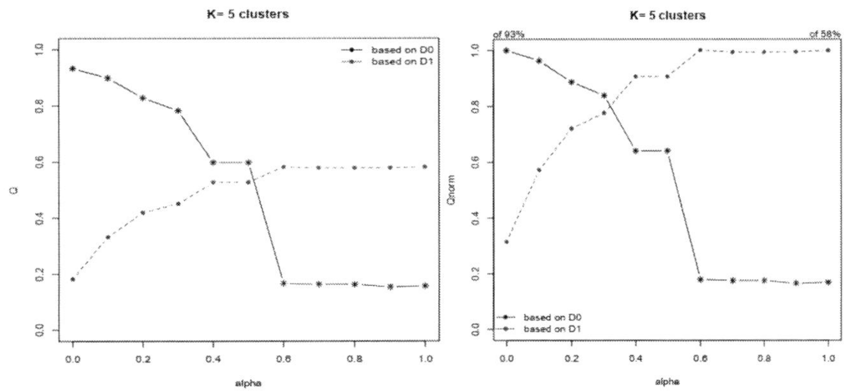

Figure 5. Choice of α for a partition in $K = 5$ clusters when $D1$ is the neighborhood dissimilarity matrix between districts. Left: proportion of explained pseudo-inertias $Q_0(P_K{}^{\alpha})$ versus α (in black solid line) and $Q_1(P_K{}^{\alpha})$ versus α (in dashed line). Right: normalized proportion of explained pseudo-inertias $Q^*{}_0(P_K{}^{\alpha})$ versus α (in black solid line) and $Q^*{}_1(P_K{}^{\alpha})$ versus α (in dashed line)

Table 3. Proportion of explained pseudo-inertia Geographical contiguity

Alpha	Q		Qnorm	
	Q0	Q1	Q0	Q1
alpha=0.2	0.828	0.419	0.887	0.721
alpha=0.3	0.783	0.451	0.839	0.776
alpha=0.4	0.598	0.527	0.641	0.907

The best choice for alpha with considering the trade-off between geographical dissimilarity and diarrheal variables is 0.30. The quality level of Q0 and Q1 based on normalized are 0.839 and 0.776 respectively. The quality of the partition spatial clusters of diarrheal risk with "neighborhood" contiguity is slightly worse than that of partition spatial clusters of diarrheal risk with "neighborhood" contiguity according to criterion Q 0 and Q 1 (83.9% versus 91.2% and 77.7% versus 95.5%).

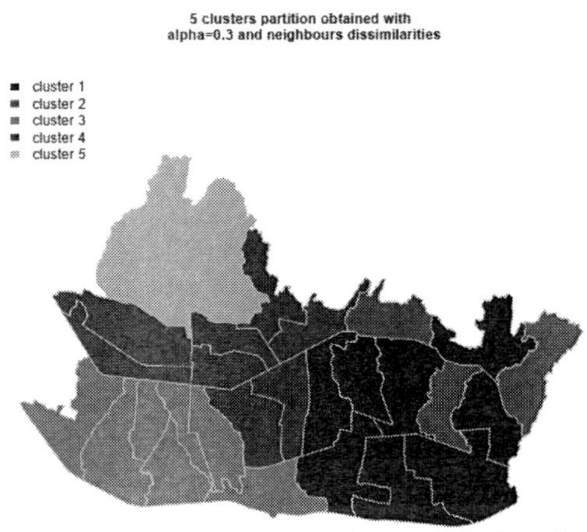

Figure 6. Map of the partition in $K = 5$ clusters based on the diarrheal variables D_0 and the "neighborhood" contiguity D_1 with $\alpha = 0.3$

Figure 6 shows spatial clusters of diarrheal risk with geographical contiguity. It looks different from a spatial cluster based on geographical distance. However, the geographical contiguity is slightly worse than geographical distance. It indicates that the geographical distance in case of diarrheal is a more-appropriate. This might be due to the spread of diarrhea not limited to adjacent locations but also possible in locations that are far apart.

Table 4. Descriptive statistics of diarrheal variables given spatial clusters.

Cluster/District	Mean SIR	Min SIR	Max SIR	Mean HHI	Min HHI	Max HHI	Mean WQI	Min WQI	Max WQI
1 (Low Risk 1)	0.93	0.51	1.61	84.87	72.24	98.53	88.32	66.67	100.00
Antapani	0.83	0.83	0.83	83.34	83.34	83.34	85.71	85.71	85.71
Arcamanik	0.60	0.60	0.60	98.53	98.53	98.53	66.67	66.67	66.67
Buahbatu	0.51	0.51	0.51	88.54	88.54	88.54	100.00	100.00	100.00
Gedebage	1.06	1.06	1.06	83.12	83.12	83.12	100.00	100.00	100.00
Kiaracondong	0.89	0.89	0.89	81.56	81.56	81.56	87.50	87.50	87.50
Panyileukan	1.61	1.61	1.61	83.48	83.48	83.48	100.00	100.00	100.00
Rancasari	0.87	0.87	0.87	88.16	88.16	88.16	100.00	100.00	100.00
Ujungberung	1.09	1.09	1.09	72.24	72.24	72.24	66.67	66.67	66.67
2 (High Risk)	1.56	1.34	1.69	79.40	77.31	81.16	0.00	0.00	0.00
Cibiru	1.34	1.34	1.34	79.73	79.73	79.73	0.00	0.00	0.00
Cinambo	1.66	1.66	1.66	81.16	81.16	81.16	0.00	0.00	0.00
Mandalajati	1.69	1.69	1.69	77.31	77.31	77.31	0.00	0.00	0.00
3 (Low Risk 2)	0.98	0.54	1.61	68.15	63.24	78.72	87.02	62.50	100.00
Astanaanyar	1.36	1.36	1.36	68.89	68.89	68.89	62.50	62.50	62.50
Babakan Ciparay	1.02	1.02	1.02	63.24	63.24	63.24	100.00	100.00	100.00
Bandung Kidul	1.61	1.61	1.61	78.72	78.72	78.72	100.00	100.00	100.00
Bandung Kulon	0.83	0.83	0.83	63.37	63.37	63.37	100.00	100.00	100.00
Bojongloa Kaler	0.88	0.88	0.88	67.65	67.65	67.65	80.00	80.00	80.00
Bojongloa Kidul	0.64	0.64	0.64	66.83	66.83	66.83	100.00	100.00	100.00
Regol	0.54	0.54	0.54	68.38	68.38	68.38	66.67	66.67	66.67
4 (Medium Risk 1)	1.09	0.73	1.92	80.09	71.78	92.10	72.02	50.00	85.71
Andir	1.44	1.44	1.44	80.63	80.63	80.63	50.00	50.00	50.00
Bandung Wetan	1.92	1.92	1.92	84.27	84.27	84.27	83.33	83.33	83.33
Batununggal	0.97	0.97	0.97	75.29	75.29	75.29	71.43	71.43	71.43
Cibeunying Kaler	0.73	0.73	0.73	73.68	73.68	73.68	66.67	66.67	66.67
Cibeunying Kidul	0.73	0.73	0.73	86.82	86.82	86.82	85.71	85.71	85.71
Cicendo	0.99	0.99	0.99	76.17	76.17	76.17	50.00	50.00	50.00
Lengkong	1.01	1.01	1.01	92.10	92.10	92.10	85.71	85.71	85.71
Sumurbandung	0.97	0.97	0.97	71.78	71.78	71.78	83.33	83.33	83.33
5 (Medium Risk 1)	1.08	0.56	1.51	84.50	75.59	88.81	91.67	83.33	100.00
Cidadap	1.15	1.15	1.15	88.81	88.81	88.81	100.00	100.00	100.00
Coblong	1.11	1.11	1.11	87.16	87.16	87.16	91.67	91.67	91.67
Sukajadi	0.56	0.56	0.56	75.59	75.59	75.59	91.67	91.67	91.67
Sukasari	1.51	1.51	1.51	86.43	86.43	86.43	83.33	83.33	83.33

Table 4 presents the descriptive statistics of diarrheal variables given the spatial cluster based on geographical distance and diarrheal variables. According to the results, this high-risk cluster was caused only by low water quality index and a relative low of healthy house index. We observed cluster two was identified as a high-risk cluster which includes three districts: Cibiru, Cinambo, and Mandalajati.

The clustering result can be used to develop an effective strategy to reduce the incidence of diarrhea in Bandung city. Three districts were identified to have a serious problem with healthy house and water quality. The Bandung government should focus on those three districts in order to improve the house quality index and water quality. Cibiru, Cinambo, and Mandalajati are located in the east of Bandung city. In this location, pipe water is very rare. A lot of households used groundwater and they were not aware of the quality of the water.

4. Conclusion

Diarrhea is an infectious disease that can cause death. The transmission of diarrhea is related to water supply and has a spatial phenomenon. Identifying high/low-risk spatial cluster of diarrhea is vital for disease control and prevention. Ward-like hierarchical clustering algorithm, including spatial constraints, were applied. Distance-based spatial weight is more appropriate to use, compared to contiguity-based spatial weight to be a spatial constraint. Spatial constraints must be considered in disease clustering when the disease is categorized as infectious disease because the spread of the disease might depend on the spatial structure. We observed that the diarrheal variables produce five clusters. Cluster two is observed as a high-risk cluster. Three districts belong to high-risk clusters: Cibiru, Cinambo, and Mandalajati. The cluster has high SIR, low healthy house index, and water quality index.

References

[1] Jaya I G N M, Folmer H, Ruchjana B N, Kristiani F and Yudhie A 2017 *Modeling of Infectious Diseases: A Core Research Topic for the Next Hundred Years* ed R Jackson and P Schaeffer (USA: Springer International Publishing) pp 239-254
[2] Lawson A 2013 Bayesian Disease Mapping, Hierarchical Modelling in Spatial Epidemiology 2nd (Boka Raton: CRC Press Taylor and Francis Group)
[3] Kumar S V, Devika S, George S and Jeyaseelan L 2017 Clinical epidemiology and global health 5 87
[4] E Christaki 2015 Virulence 6 558
[5] WHO 2017 Diarrhoeal disease (WHO: https://www.who.int/news-room/fact-sheets/detail/diarrhoeal-disease)
[6] UNICEF 2018 Diarrhoeal disease (UNICEF:https://data.unicef.org/topic/child-health/diarrhoeal-disease/)
[7] Komarulzaman A, Smits J and Jong E D 2016 Global Public Health 12 1141
[8] Anderson C, Lee D and Nema D 2014 Biostatistics 15 457–469
[9] Adin A, Lee D, Goicoa T and Ugarte M D 2018 Statistical Methods in Medical Research 0 1
[10] Choi J, Lawson A B, Cai B, Hossain M M, Kirby R S and Liu J 2012 Statistical Methods in Medical Research 21 445
[11] Jaya I G N M, Tantular B and Zulhanif 2018 International Journal of Engineering Sciences & Management 8 42
[12] Clayton D and Kaldor J 1987 Biometrics 43 671
[13] Wakefield J 2007 Biostatistics 8 158
[14] Chavent M, Kuentz-Simonet V, Labenne A and Saracco J 2018 Comput Stat 33 1799

The existence of clean elements in a matrix ring over integral domain and its connections with $g(x)$-cleanness and strongly $g(x)$-cleanness

IF Ambarsari[1], S Irawati[1,*], I M Sulandra[1], H Susanto[1], A C Y Mui[2] and H Marubayashi[3]

[1] Department of Mathematics, State University of Malang, Semarang Street No. 5, Malang, East of Java, Indonesia
[2] Institute of Mathematical Science, University of Malaya, Malaysia
[3] Department of Mathematics, Naruto University of Education, Naruto, Tokushima, Japan

*santi.irawati.fmipa@um.ac.id

Abstract. An element a in a ring R with unity is called clean, if there exist an idempotent element $e \in R$ and a unit element $u \in R$ such that $a = e + u$. This article aims to show all of clean elements in a certain subring $X_3(R)$ of a matrix ring 3×3 over integral domain R and their connections with $g(x)$-cleanness and strongly $g(x)$-cleanness for some fixed polynomial $g(x)$. To achieve it, we found out unit and idempotent elements in $X_3(R)$ for constructing clean elements and selected some fixed $g(x)$ in the center of R for investigating their relations with $g(x)$-cleanness and strongly $g(x)$-cleanness. In this article, we obtained eight general forms of the clean elements in $X_3(R)$, $g(x)$-clean elements with $g(x) = x^n - x$, which five forms of them were strongly $g(x)$-clean but the other three forms were not. The latter result was shown by providing counter examples.

1. Introduction

In this article, for the 3×3 matrix ring over a ring R is denoted by $M_3(R) = (a_{ij}), a_{ij} \in R$. The set of all units and idempotent elements in R are denoted by $U(R)$ and $id(R)$, respectively, that are, $U(R) = \{a \in R | ab = 1 = ba, \text{for some } b \in R\}$ and $id(R) = \{a \in R | a^2 = a\}$. Following [1] and [2], an element a in a ring R is said to be clean if can be written as the sum of a unit and an idempotent. Many authors ([3], [4] and [5]) investigated the properties of clean ring.

Let $C(R)$ be the center of a ring R and $g(x)$ be a polynomial in $C(R)[x]$. Then following [6], [7] and [8], an element r in a ring R is called $g(x)$-clean if can be written as $r = u + s$ with $u \in U(R)$ and $g(s) = 0, s \in R$. It is clear that $(x^2 - x)$-clean rings are precisely clean rings. Moreover, [9] and [10] proved the relation between clean ring and $(x^n - x)$-clean ring and gived some examples which showed

Journal of Physics: Conference Series **1397** (2019) 012069 doi:10.1088/1742-6596/1397/1/012069

that the relation between clean rings and $g(x)$-clean rings is not two directions. Other authors ([11] and [12]) discussed about $g(x) = x^n - x$ in another classes of clean rings.

Moreover, following this definition, [13] defined that an element $r \in R$ is called strongly $g(x)$-clean if $r \in R$ is $g(x)$-clean and commute. A ring R is called $g(x)$-clean (strongly $g(x)$-clean) if every element in R is $g(x)$-clean (strongly $g(x)$-clean). [13] also gave a property related to strongly $(x^n - x)$-clean rings and an example for the statement strongly clean rings are exactly strongly $(x^2 - x)$-clean rings but there are strongly $g(x)$-clean rings which are not strongly clean and vice versa. Meanwhile, in [10] say the result of [13] generally.

In this article, we consider a subring $X_3(R) = \left\{ \begin{bmatrix} a_{11} & 0 & a_{13} \\ 0 & a_{22} & 0 \\ a_{31} & 0 & a_{33} \end{bmatrix} \middle| a_{ij} \in R \right\}$ of the matrix ring $M_3(R)$. We will show that $X_3(R)$ is not clean. Therefore, we investigate the existence of clean elements in $X_3(R)$ and then analyze $g(x)$-cleanness and strongly $g(x)$-cleanness of those clean elements. Moreover, Inspired by [14], let $C(R)$ be center of a ring R and we will choose a fixed polynomial $g(x) = x^n - x\ in\ C(R)[x]$ which is a slight different polynomial from [14].

In Section 2, we discuss the existence of clean elements in $X_3(R)$ by constructing idempotent and unit elements in $X_3(R)$ for being eight general forms of clean elements and then analyze why $X_3(R)$ needs not be a clean ring.

In Section 3, we focus on $g(x)$-cleanness and strongly $g(x)$-cleanness of the eight general forms of clean elements in $X_3(R)$.

Throughout the paper, we focus on integral domain R with $U(R) = \{-1,1\}$,$Cl(X_3(R))$ denotes the set of all clean elements in $X_3(R)$ and $Sl(X_3(R))$ denotes the set of all strongly clean elements in $X_3(R)$.

2. Clean elements

In order to construct idempotent elements in $X_3(R)$, consider the following subset H of $X_3(R)$

$$H = \left\{ \begin{bmatrix} z & 0 & 0 \\ 0 & c & 0 \\ 0 & 0 & z \end{bmatrix}, \begin{bmatrix} a & 0 & b \\ 0 & c & 0 \\ d & 0 & 1-a \end{bmatrix} \middle| b,d \in R, z,c \in id(R), bd = a - a^2 \right\}$$

Theorem 2.1 H is the set of all idempotent elements in $X_3(R)$, i.e. $H = id(X_3(R))$.

Proof:

Let$A$ be any element in H. We will show that A is idempotent element in $X_3(R)$, that is, $A^2 = A$.

For $A = \begin{bmatrix} z & 0 & 0 \\ 0 & c & 0 \\ 0 & 0 & z \end{bmatrix}$, where $z, c \in id(R)$, we get

$$A^2 = \begin{bmatrix} z & 0 & 0 \\ 0 & c & 0 \\ 0 & 0 & z \end{bmatrix}\begin{bmatrix} z & 0 & 0 \\ 0 & c & 0 \\ 0 & 0 & z \end{bmatrix} = \begin{bmatrix} z^2 & 0 & 0 \\ 0 & c^2 & 0 \\ 0 & 0 & z^2 \end{bmatrix} = \begin{bmatrix} z & 0 & 0 \\ 0 & c & 0 \\ 0 & 0 & z \end{bmatrix} = A$$

For $A = \begin{bmatrix} a & 0 & b \\ 0 & c & 0 \\ d & 0 & 1-a \end{bmatrix}$, where $b, d \in R, c \in id(R), bd = a - a^2$, we get

$$A^2 = \begin{bmatrix} a^2 + bd & 0 & b(a+1-a) \\ 0 & c^2 & 0 \\ d(a+1-a) & 0 & (1-a)^2 + bd \end{bmatrix} = \begin{bmatrix} a & 0 & b \\ 0 & c & 0 \\ d & 0 & 1-a \end{bmatrix} = A$$

So, for any $A \in H$, the matrix A is idempotent elements in $X_3(R)$. It means $H \subseteq id(X_3(R))$.

Conversely, let $A = \begin{bmatrix} a & 0 & b \\ 0 & c & 0 \\ d & 0 & e \end{bmatrix}$ be any idempotent element in $X_3(R)$. We will show that $A \in H$.

Since the matrix A is idempotent element, so we have

$$\begin{bmatrix} a & 0 & b \\ 0 & c & 0 \\ d & 0 & e \end{bmatrix} = A = A^2 = \begin{bmatrix} a & 0 & b \\ 0 & c & 0 \\ d & 0 & e \end{bmatrix}\begin{bmatrix} a & 0 & b \\ 0 & c & 0 \\ d & 0 & e \end{bmatrix} = \begin{bmatrix} a^2+bd & 0 & b(a+e) \\ 0 & c^2 & 0 \\ d(a+e) & 0 & e^2+bd \end{bmatrix}$$

Based on equality of two matrices, we get:

$$c^2 = c \quad (1)$$
$$a^2 + bd = a \quad (2)$$
$$b(a+e) = b \quad (3)$$
$$d(a+e) = d \quad (4)$$
$$e^2 + bd = e \quad (5)$$

From (1), we get $c \in id(R)$ and from (3) we get $b = 0$ or $e = 1 - a$.

Case 1: If $b = 0$, then from (2) and (5) we get $a^2 = a$ and $e^2 = e$, that means $a, e \in id(R)$. Meanwhile, from (4), we get $d = 0$ or $e = 1 - a$.

Case 1.1: If $e \neq 1 - a$, then $d = 0$. Since $a, e \in id(R) = \{0,1\}$, so we get whether $e = 0$ and $a = 0$ or $e = 1$ and $a = 1$. Therefore, we get

$$A = \begin{bmatrix} 0 & 0 & 0 \\ 0 & c & 0 \\ 0 & 0 & 0 \end{bmatrix} \in H \text{ or } A = \begin{bmatrix} 1 & 0 & 0 \\ 0 & c & 0 \\ 0 & 0 & 1 \end{bmatrix} \in H$$

Case 1.2: If $d \neq 0$, then $e = 1 - a$ such that we get whether $e = 0$ and $a = 1$ or $e = 1$ and $a = 0$. Therefore, we get

$$A = \begin{bmatrix} 1 & 0 & 0 \\ 0 & c & 0 \\ d & 0 & 0 \end{bmatrix} \in H \text{ and } A = \begin{bmatrix} 0 & 0 & 0 \\ 0 & c & 0 \\ d & 0 & 1 \end{bmatrix} \in H$$

Case 2: If $e = 1 - a$ or $e + a = 1$, then from (2) and (5) we get $bd = a - a^2$. So we get matrix

$$A = \begin{bmatrix} a & 0 & b \\ 0 & c & 0 \\ d & 0 & 1-a \end{bmatrix}$$

Where $b, d \in R, c \in id(R), bd = a - a^2$, it means $A \in H$.

So, if the matrix A is idempotent elements in $X_3(R)$, then $A \in H$. It means $id(X_3(R)) \subseteq H$.

The following Lemma 2.2 is already proved in [15] and will be used to construct unit elements in $X_3(R)$.

Lemma 2.2 The matrix A is invertible if and only if $\det(A) \in U(R)$.
Proof: See in [15].

In order to construct unit elements in $X_3(R)$, consider the following set $B = B_1 \cup B_2 \subseteq X_3(R)$, where

$$B_1 = \left\{ \begin{bmatrix} y & 0 & 0 \\ 0 & r & 0 \\ 0 & 0 & y \end{bmatrix}, \begin{bmatrix} p & 0 & q \\ 0 & r & 0 \\ s & 0 & -p \end{bmatrix} \middle| y, r \in U(R), qs = 1 - p^2 \right\} \text{ and}$$

$$B_2 = \left\{ \begin{bmatrix} t & 0 & -q \\ 0 & r & 0 \\ -s & 0 & p \end{bmatrix}, \begin{bmatrix} -t_1 & 0 & q_1 \\ 0 & r & 0 \\ s_1 & 0 & -p_1 \end{bmatrix} \middle| r \in U(R), pt - qs = 1, p_1t_1 - q_1s_1 = -1 \right\}$$

Theorem 2.3 B is the set of all unit elements in $X_3(R)$, i.e. $B = U(X_3(R))$.
Proof:
Let A be any element in B. We will show that the matrix A is a unit element in $X_3(R)$, that is there exists $A' \in B$, such that $A.A' = A'.A = I_3$.

For $A = \begin{bmatrix} y & 0 & 0 \\ 0 & r & 0 \\ 0 & 0 & y \end{bmatrix}$ where $y, r \in U(R)$, we can choose $A' = \begin{bmatrix} y & 0 & 0 \\ 0 & r & 0 \\ 0 & 0 & y \end{bmatrix} \in B$, such that

$$A.A' = \begin{bmatrix} y & 0 & 0 \\ 0 & r & 0 \\ 0 & 0 & y \end{bmatrix}\begin{bmatrix} y & 0 & 0 \\ 0 & r & 0 \\ 0 & 0 & y \end{bmatrix} = \begin{bmatrix} y^2 & 0 & 0 \\ 0 & r^2 & 0 \\ 0 & 0 & y^2 \end{bmatrix} = \begin{bmatrix} 1 & 0 & 0 \\ 0 & 1 & 0 \\ 0 & 0 & 1 \end{bmatrix} = I_3$$

For $A = \begin{bmatrix} p & 0 & q \\ 0 & r & 0 \\ s & 0 & -p \end{bmatrix}$ where $r \in U(R)$ and $qs = 1 - p^2$, we can choose $A' = \begin{bmatrix} p & 0 & q \\ 0 & r & 0 \\ s & 0 & -p \end{bmatrix} \in B$, such that

$$A.A' = \begin{bmatrix} p & 0 & q \\ 0 & r & 0 \\ s & 0 & -p \end{bmatrix}\begin{bmatrix} p & 0 & q \\ 0 & r & 0 \\ s & 0 & -p \end{bmatrix} = \begin{bmatrix} p^2+qs & 0 & 0 \\ 0 & r^2 & 0 \\ 0 & 0 & p^2+qs \end{bmatrix} = \begin{bmatrix} 1 & 0 & 0 \\ 0 & 1 & 0 \\ 0 & 0 & 1 \end{bmatrix} = I_3$$

For $A = \begin{bmatrix} t & 0 & -q \\ 0 & r & 0 \\ -s & 0 & p \end{bmatrix}$ where $r \in U(R)$ and $pt - qs = 1$, we can choose $A' = \begin{bmatrix} p & 0 & q \\ 0 & r & 0 \\ s & 0 & t \end{bmatrix} \in B$, such that

$$A.A' = \begin{bmatrix} t & 0 & -q \\ 0 & r & 0 \\ -s & 0 & p \end{bmatrix}\begin{bmatrix} p & 0 & q \\ 0 & r & 0 \\ s & 0 & t \end{bmatrix} = \begin{bmatrix} pt-qs & 0 & 0 \\ 0 & r^2 & 0 \\ 0 & 0 & pt-qs \end{bmatrix} = \begin{bmatrix} 1 & 0 & 0 \\ 0 & 1 & 0 \\ 0 & 0 & 1 \end{bmatrix} = I_3$$

For $A = \begin{bmatrix} -t_1 & 0 & q_1 \\ 0 & r & 0 \\ s_1 & 0 & -p_1 \end{bmatrix}$ where $r \in U(R)$ and $p_1t_1 - q_1s_1 = -1$, we can choose $A' = \begin{bmatrix} p_1 & 0 & q_1 \\ 0 & r & 0 \\ s_1 & 0 & t_1 \end{bmatrix} \in B$, such that

$$AA' = \begin{bmatrix} -t_1 & 0 & q_1 \\ 0 & r & 0 \\ s_1 & 0 & -p_1 \end{bmatrix}\begin{bmatrix} p_1 & 0 & q_1 \\ 0 & r & 0 \\ s_1 & 0 & t_1 \end{bmatrix} = \begin{bmatrix} -(p_1t_1-q_1s_1) & 0 & 0 \\ 0 & r^2 & 0 \\ 0 & 0 & -(p_1t_1-q_1s_1) \end{bmatrix} = \begin{bmatrix} 1 & 0 & 0 \\ 0 & 1 & 0 \\ 0 & 0 & 1 \end{bmatrix} = I_3$$

Similarly, we also get $A'A = I_3$.
So, if the matrix $A \in B$, then A is a unit element in $X_3(R)$. It means $B \subseteq U(X_3(R))$.

Conversely, Let $A' = \begin{bmatrix} p & 0 & q \\ 0 & r & 0 \\ s & 0 & t \end{bmatrix}$ be any unit element in $X_3(R)$. We will show that $A \in B$.

If the matrix A' is unit element in $X_3(R)$, then there exist the following matrix $A \in X_3(R)$, that is the inverse of A'.

$$A = \frac{1}{r(pt-qs)}\begin{bmatrix} tr & 0 & -qr \\ 0 & pt-qs & 0 \\ -sr & 0 & pr \end{bmatrix} = \begin{bmatrix} \frac{tr}{r(pt-qs)} & 0 & \frac{-qr}{r(pt-qs)} \\ 0 & \frac{(pt-qs)}{r(pt-qs)} & 0 \\ \frac{-sr}{r(pt-qs)} & 0 & \frac{pr}{r(pt-qs)} \end{bmatrix}$$

Based on Lemma 2.2, we obtain $\det(A) = r(pt-qs) \in U(R) = \{-1,1\}$,

Case 1.1: If $r(pt-qs) = 1$, then we get

Case 1.1.1: If $r = 1$ and $pt - qs = 1$, then we get

$$A = \begin{bmatrix} t & 0 & -q \\ 0 & 1 & 0 \\ -s & 0 & p \end{bmatrix} \text{ and } A' = \begin{bmatrix} p & 0 & q \\ 0 & 1 & 0 \\ s & 0 & t \end{bmatrix}$$

where $pt - qs = 1$. So $A \in B$.

Case 1.1.2: If $r = -1$ and $pt - qs = -1$, then we get

$$A = \begin{bmatrix} -t & 0 & q \\ 0 & -1 & 0 \\ s & 0 & -p \end{bmatrix} \text{ and } A' = \begin{bmatrix} p & 0 & q \\ 0 & -1 & 0 \\ s & 0 & t \end{bmatrix}$$

where $pt - qs = -1$. So $A \in B$.

Case 1.2: If $r(pt-qs)=-1$, then we get

Case 1.2.1: If $r=1$ and $pt-qs=-1$, then we get

$$A=\begin{bmatrix}-t & 0 & q\\ 0 & 1 & 0\\ s & 0 & -p\end{bmatrix} \text{ and } A'=\begin{bmatrix}p & 0 & q\\ 0 & 1 & 0\\ s & 0 & t\end{bmatrix}$$

where $pt-qs=-1$. So $A \in B$.

Case 1.2.2: If $r=-1$ and $pt-qs=1$, then we get

$$A=\begin{bmatrix}t & 0 & -q\\ 0 & -1 & 0\\ -s & 0 & p\end{bmatrix} \text{ and } A'=\begin{bmatrix}p & 0 & q\\ 0 & -1 & 0\\ s & 0 & t\end{bmatrix}$$

where $pt-qs=1$. So $A \in B$.

If the matrix A' is unit element in $X_3(R)$, then there exist $A \in X_3(R)$ that is the inverse of A', such that $A'.A=I_3$. If$A'=A$, then $A.A=I_3$.

$$A^2=\begin{bmatrix}p & 0 & q\\ 0 & r & 0\\ s & 0 & t\end{bmatrix}\begin{bmatrix}p & 0 & q\\ 0 & r & 0\\ s & 0 & t\end{bmatrix}=\begin{bmatrix}p^2+qs & 0 & q(p+t)\\ 0 & r^2 & 0\\ s(p+t) & 0 & t^2+qs\end{bmatrix}=\begin{bmatrix}1 & 0 & 0\\ 0 & 1 & 0\\ 0 & 0 & 1\end{bmatrix}=I_3$$

Based on equality of two matrices, we get

$$r^2=1 \quad (1)$$
$$q(p+t)=0 \quad (2)$$
$$s(p+t)=0 \quad (3)$$
$$p^2+qs=1 \quad (4)$$
$$t^2+qs=1 \quad (5)$$

From (1), we get $r \in U(R)$ and from (2) we get $q=0$ or $p=-t$.

Case 1: If $q=0$, then from (4) and (5) we get $p^2=1$ and $t^2=1$, that means $p,t \in U(R)$.

Meanwhile, from (3), we get $s=0$ or $p=-t$.

Case 1.1: If $p \neq -t$, then $s=0$. Since $p,t \in U(R)=\{-1,1\}$, so we get whether $p=-1$ and $t=-1$ or $p=1$ and $t=1$. Therefore we get

$$A=\begin{bmatrix}-1 & 0 & 0\\ 0 & r & 0\\ 0 & 0 & -1\end{bmatrix} \in B \text{ and } A=\begin{bmatrix}1 & 0 & 0\\ 0 & r & 0\\ 0 & 0 & 1\end{bmatrix} \in B.$$

Case 1.2: If $s \neq 0$, then $p=-t$. Since $p,t \in U(R)=\{-1,1\}$, so we get whether $p=1$ and $t=-1$ or $p=-1$ and $t=1$. Therefore we get

$$A=\begin{bmatrix}1 & 0 & 0\\ 0 & r & 0\\ 0 & 0 & -1\end{bmatrix} \in B \text{ and } A=\begin{bmatrix}-1 & 0 & 0\\ 0 & r & 0\\ 0 & 0 & 1\end{bmatrix} \in B.$$

Case 2: If $p=-t$, then from (4) and (5) we get $qs=1-p^2$. So, we get matrices

$$A=\begin{bmatrix}p & 0 & q\\ 0 & r & 0\\ s & 0 & -p\end{bmatrix}$$

where $r \in U(R), qs=1-p^2$, it means $A \in B$.

So, if the matrix A is a unit element in $X_3(R)$, then $A \in B$. It means $U(X_3(R)) \subseteq B$.

Finally, based on the constructing idempotent elements in $X_3(R)$ (Theorem 2.1) and unit elements in $X_3(R)$, we obtain clean elements in $X_3(R)$(Theorem 2.3) in the following Table 1.

Table1. Constructing clean elements in $X_3(R)$

$id(X_3(R))$ / $U(X_3(R))$	$\begin{bmatrix} z & 0 & 0 \\ 0 & c & 0 \\ 0 & 0 & z \end{bmatrix}, z, c \in id(R)$	$\begin{bmatrix} a & 0 & b \\ 0 & c & 0 \\ d & 0 & 1-a \end{bmatrix}$ $b, d \in R, c \in id(R), bd = a - a^2$
$\begin{bmatrix} y & 0 & 0 \\ 0 & r & 0 \\ 0 & 0 & y \end{bmatrix}, y, r \in U(R)$	$\begin{bmatrix} z+y & 0 & 0 \\ 0 & c+r & 0 \\ 0 & 0 & z+y \end{bmatrix}$	$\begin{bmatrix} a+y & 0 & b \\ 0 & c+r & 0 \\ d & 0 & (1-a)+y \end{bmatrix}$
$\begin{bmatrix} p & 0 & q \\ 0 & r & 0 \\ s & 0 & -p \end{bmatrix}$ $r \in U(R), qs = 1 - p^2$	$\begin{bmatrix} z+p & 0 & q \\ 0 & c+r & 0 \\ s & 0 & z-p \end{bmatrix}$	$\begin{bmatrix} a+p & 0 & b+q \\ 0 & c+r & 0 \\ d+s & 0 & (1-a)-p \end{bmatrix}$
$\begin{bmatrix} t & 0 & -q \\ 0 & r & 0 \\ -s & 0 & p \end{bmatrix}$ $r \in U(R), pt - qs = 1$	$\begin{bmatrix} z+t & 0 & -q \\ 0 & c+r & 0 \\ -s & 0 & z+p \end{bmatrix}$	$\begin{bmatrix} a+t & 0 & b-q \\ 0 & c+r & 0 \\ d-s & 0 & (1-a)+p \end{bmatrix}$
$\begin{bmatrix} -t_1 & 0 & q_1 \\ 0 & r & 0 \\ s_1 & 0 & -p_1 \end{bmatrix}$ $r \in U(R), p_1t_1 - q_1s_1 = -1$	$\begin{bmatrix} z-t_1 & 0 & q_1 \\ 0 & c+r & 0 \\ s_1 & 0 & z-p_1 \end{bmatrix}$	$\begin{bmatrix} a-t_1 & 0 & b+q_1 \\ 0 & c+r & 0 \\ d+s_1 & 0 & (1-a)-p_1 \end{bmatrix}$

From Table 1 we obtain eight general forms of clean elements in $X_3(R)$. Consider all clean elements in $X_3(R)$ on Table 1. The entry a_{22} in $X_3(R)$ can always be written as $a_{22} = c + r$, with $c \in id(R)$ and $r \in U(R)$. We know that $id(R) = \{0,1\}$ and $U(R) = \{-1,1\}$, so we get $a_{22} \in \{-1,0,1,2\}$. Based on this fact, we conclude the matrix ring $X_3(R)$ is not clean.

3. $g(x)$-cleanness and strongly $g(x)$-cleanness

Inspired by [14], let $C(R)$ be center of a ring R and a fixed polynomial $g(x) = x^n - x$ in $C(R)[x]$, for some positive integer n. In this article, we choose a slight different polynomial from [14]. From previous section, we obtain eight general forms of clean elements in $X_3(R)$. In this section, we will investigate $g(x)$-cleanness and strongly $g(x)$-cleanness of these eight forms of clean elements.

In order to investigate $g(x)$-cleanness, consider the following set $Cl(X_3(R)) \subseteq X_3(R)$, where

$$Cl(X_3(R)) = \left\{ \begin{array}{c|c} \begin{bmatrix} z+y & 0 & 0 \\ 0 & c+r & 0 \\ 0 & 0 & z+y \end{bmatrix}, \begin{bmatrix} a+y & 0 & b \\ 0 & c+r & 0 \\ d & 0 & (1-a)+y \end{bmatrix}, & b, d \in R, z, c \in id(R), \\ \begin{bmatrix} z+p & 0 & q \\ 0 & c+r & 0 \\ s & 0 & z-p \end{bmatrix}, \begin{bmatrix} a+p & 0 & b+q \\ 0 & c+r & 0 \\ d+s & 0 & (1-a)-p \end{bmatrix}, & y, r \in U(R), \\ & bd = a - a^2, \\ \begin{bmatrix} z+t & 0 & -q \\ 0 & c+r & 0 \\ -s & 0 & z+p \end{bmatrix}, \begin{bmatrix} a+t & 0 & b-q \\ 0 & c+r & 0 \\ d-s & 0 & (1-a)+p \end{bmatrix}, & qs = 1 - p^2, \\ & pt - qs = 1, \\ \begin{bmatrix} z-t_1 & 0 & q_1 \\ 0 & c+r & 0 \\ s_1 & 0 & z-p_1 \end{bmatrix}, \begin{bmatrix} a-t_1 & 0 & b+q_1 \\ 0 & c+r & 0 \\ d+s_1 & 0 & (1-a)-p_1 \end{bmatrix} & p_1t_1 - q_1s_1 = -1 \end{array} \right\}$$

Theorem 3.1 Let n be a positive integer and $g(x) = x^n - x$. The matrix $A \in Cl(X_3(R))$ is $g(x)$-clean.
Proof :

Since for every $A \in Cl(X_3(R))$ is a clean element in $X_3(R)$, so A can be written by $A = V + E$, with $V \in U(X_3(R))$ and $E \in id(X_3(R))$. Consider $E \in id(X_3(R))$, since $E^2 = E$ that implies $E^n = E$ such that $g(E) = 0$. We can conclude that the matrix $A \in Cl(X_3(R))$ is $g(x)$-clean.

The multiplication of two matrix A and B is commutative if and only if A or B is a diagonal matrix [15]. This properties will be used to prove the next theorem. In order to investigate strongly $g(x)$-cleanness, consider the following subset $Sl(X_3(R)) \subseteq Cl(X_3(R)) \subseteq X_3(R)$, where

$$Sl(X_3(R)) = \left\{ \begin{array}{c} \begin{bmatrix} z+y & 0 & 0 \\ 0 & c+r & 0 \\ 0 & 0 & z+y \end{bmatrix}, \begin{bmatrix} a+y & 0 & b \\ 0 & c+r & 0 \\ d & 0 & (1-a)+y \end{bmatrix}, \\ \begin{bmatrix} z+p & 0 & q \\ 0 & c+r & 0 \\ s & 0 & z-p \end{bmatrix}, \begin{bmatrix} z+t & 0 & -q \\ 0 & c+r & 0 \\ -s & 0 & z+p \end{bmatrix}, \\ \begin{bmatrix} z-t_1 & 0 & q_1 \\ 0 & c+r & 0 \\ s_1 & 0 & z-p_1 \end{bmatrix} \end{array} \middle| \begin{array}{c} b,d \in R, z,c \in id(R), \\ y,r \in U(R), \\ bd = a-a^2, \\ qs = 1-p^2, \\ pt-qs = 1, \\ p_1t_1 - q_1s_1 = -1 \end{array} \right\}$$

Theorem 3.2 Let n be a positive integer and $g(x) = x^n - x$. The matrix $A \in Sl(X_3(R))$ is strongly $g(x)$-clean.

Proof :

We will show that any $A \in Sl(X_3(R))$ is strongly $g(x)$-clean. Since $Sl(X_3(R)) \subseteq Cl(X_3(R))$ and based on Theorem 3.1, we know that for every $A \in Sl(X_3(R))$ is $g(x)$-clean. Since A can be written by $A = V + S$ with $V \in U(X_3(R))$ and $g(S) = 0$. Furthermore, we will show that for every $A \in Sl(X_3(R)), VS = SV$. The decomposition of element V and S is always V is a diagonal matrix or S is a diagonal matrix. Based on the properties of diagonal matrix, we can conclude that the matrix A is a strongly $g(x)$-clean.

Meanwhile, consider the set of all strongly $g(x)$-clean in $X_3(R)$. The element in $X_3(R)$ is the five general forms of the clean elements in $X_3(R)$. Next, we will show that another three general form of clean elements need not to be strongly $g(x)$-clean as showed the following counter example.

Example 3.3

For $E = \begin{bmatrix} a & 0 & b \\ 0 & c & 0 \\ d & 0 & 1-a \end{bmatrix} \in id(X_3(R))$ and $V = \begin{bmatrix} p & 0 & q \\ 0 & r & 0 \\ s & 0 & -p \end{bmatrix} \in U(X_3(R))$ where $c \in id(R), r \in U(R)$, $bd = a - a^2$ and $qs = 1 - p^2$, such that

$$EV = \begin{bmatrix} ap+bs & 0 & aq-bp \\ 0 & cr & 0 \\ pd+s(1-a) & 0 & qd-p(1-a) \end{bmatrix} = \begin{bmatrix} ap+qd & 0 & bp+q(1-a) \\ 0 & cr & 0 \\ as-pd & 0 & bs-p(1-a) \end{bmatrix} = VE.$$

So we obtain $bs = qd, 2pd = s(2a-1)$ and $2bp = q(2a-1)$. For example, let $a = 2, b = d = 1, p = 3$ and $q = s = 2$, such that $EU = UE$, but $E \notin id(X_3(R))$ and $V \notin U(X_3(R))$. There is a contradiction.

For $E = \begin{bmatrix} a & 0 & b \\ 0 & c & 0 \\ d & 0 & 1-a \end{bmatrix} \in id(X_3(R))$ and $V = \begin{bmatrix} t & 0 & -q \\ 0 & r & 0 \\ -s & 0 & p \end{bmatrix} \in U(X_3(R))$ where $c \in id(R)$, $r \in U(R), bd = a - a^2$ and $pt - qs = 1$, such that

$$EV = \begin{bmatrix} at-bs & 0 & -aq+bp \\ 0 & cr & 0 \\ dt-s(1-a) & 0 & -qd+p(1-a) \end{bmatrix} = \begin{bmatrix} at-qd & 0 & bt-q(1-a) \\ 0 & cr & 0 \\ -as+pd & 0 & -bs+p(1-a) \end{bmatrix} = VE.$$

So we obtain $bs = qd, b(t-p) = q(1-2a)$ and $d(t-p) = s(1-2a)$. For example, let $a = t = 2, b = d = 3$ and $p = q = s = 1$, such that $EU = UE$, but $E \notin id(X_3(R))$ and $V \notin U(X_3(R))$. There is a contradiction.

For $E = \begin{bmatrix} a & 0 & b \\ 0 & c & 0 \\ d & 0 & 1-a \end{bmatrix} \in id\big(X_3(R)\big)$ and $V = \begin{bmatrix} -t_1 & 0 & q_1 \\ 0 & r & 0 \\ s_1 & 0 & -p_1 \end{bmatrix} \in U\big(X_3(R)\big)$ where $c \in id(R)$, $r \in U(R), bd = a - a^2$ and $p_1t_1 - q_1s_1 = -1$, such that

$$EV = \begin{bmatrix} -at_1 + bs_1 & 0 & aq_1 - bp_1 \\ 0 & cr & 0 \\ -dt_1 + s_1(1-a) & 0 & q_1d - p_1(1-a) \end{bmatrix} = \begin{bmatrix} -at_1 + q_1d & 0 & -bt_1 + q_1(1-a) \\ 0 & cr & 0 \\ as_1 - p_1d & 0 & bs_1 - p_1(1-a) \end{bmatrix} = VE.$$

So we obtain $bs_1 = q_1d, b(t_1 - p_1) = q_1(1-2a)$ and $d(t_1 - p_1) = s_1(1-2a)$. For example, let $a = t_1 = 2, b = d = 3$ and $p_1 = q_1 = s_1 = 1$, such that $EU = UE$, but $E \notin id(X_3(R))$ and $V \notin U(X_3(R))$. There is a contradiction.

Acknowledgments

The authors would like to thank Universitas Negeri Malang for the PNBP financial support.

References

[1] Immormino N A and Mcgovern W Wm 2014 Examples of clean commutative group rings *J Algebra* **405** 168–178.

[2] Jamshidvand S, Haj S J H and Vahedian J N 2014 Generalized f -clean rings *J. Linear and Topological Algebra* **3** 55–60.

[3] Han J and Nicholson W K 2001 Extensions of clean rings *Communications in Algebra* **29** 2589-2595.

[4] Diesl A J and Dorsey T J 2014 Strongly clean matrices over arbitrary rings *J Algebra* **399** 854–869.

[5] Călugăreanu G 2017 Clean integral 2 × 2 matrices **2** 1–11.

[6] Sahebi S and Rahmani V 2016 On g (x) -f-clean ring *Palestina Journal of Mathematics* **5** 117–121.

[7] Khashan H A and Handam A H 2017 On weakly g(x)-nil clean rings *International Journal of Pure and Applied Mathematics* **114** 191-202.

[8] Nicholson W K and Zhou Y 2006 Endomorphisms that are the sum of a unit and a root of a fixed polynomial endomorphisms that are the sum of a unit and a root of a fixed polynomial *Canadian Mathematical Bulletin* **49** 265.

[9] Fan L and Yang X 2008 On rings whose elements are the sum of a unit and a root of a fixed polynomial *Communications in Algebra* **36** 269–278.

[10] Haj S J H, Jamshidvand S and Maleki M 2013 On strongly J-clean rings associated with polynomial identity g(x) = 0*J. Linear and Topological Algebra* **2** 71–76.

[11] Ashrafi N and Ahmadi Z 2012 Weakly g (x) -clean rings *Iranian Journal of Mathematical Sciences and Informatics* **7** 83–91.

[12] Khashan H A and Handam A H 2015 g(x)-nil clean rings 1-10.

[13] Fan L and Yang X 2009 On strongly g (x) -clean rings 1–8.

[14] Camillo V 2002 The Nicholson-Varadarajan Theorem on Clean Linear Transformations *Glasgow Mathematical Journal* **44** 365-369..

[15] Brown W C 1993 *Matrices over Commutative Rings* (New York: Marcel Dekker, Inc).

Power law fluid model on wave mitigation, 2D simulation using smoothed particle hydrodynamics

Iryanto[1*], M S Bunga[1], M Mustamiin[1], I Magdalena[2], and N Erwina[2]

[1]Informatics Department, Politeknik Negeri Indramayu, Indonesia.
[2]Industrial and Financial Mathematics Research Group, Faculty of Mathematics and Natural Sciences, Institut Teknologi Bandung, Indonesia.

[*]Corresponding author e-mail: iryanto.math@yahoo.com

Abstract. This article is focused on numerical modelling to describe influence of coastal vegetation in wave mitigation. The model based on Navier-Stokes equations with stress tensor written in power law model. Numerical approach used to solve the problem is SPH (Smoothed Particle Hydrodynamics). Three numerical simulation are conducted; plane Couette-Poiseuille flow, wave mitigation on flat bottom, and wave mitigation on incline bottom. The first simulation shows that our numerical results are in good agreement with analytic solution provided in [13]. The second simulation shows that existence of the coastal vegetation reduce the wave amplitude. Whereas the last simulation shows that the vegetation reduce the run up height. Further, the power law constant influences how high the run up.

1. Introduction

Mangrove has important role in protecting coastal area due to its ability in reducing amplitude and velocity of wave. Moreover, the coastal vegetation can support ecosystem at the coastal area. These advantages often lead mangrove to be as an option to prevent coastal area from devastation.

Resistance of mangrove can be modelled as porous medium as described in [1]. Further, porous medium also can be modelled by adding a certain friction term such as the linearized Dupuit-Forcheimer's formula, see [2] for more detail. Another way to model the resistance is by inserting a friction term directly to the governing equations. Elaboration of the model can be seen in detail in [3]. In the reference, three kind of resistances are used such as Darcy, Manning, and Laminar viscosity resistance.

Formulations of wave interaction with porous medium can use Navier-Stokes type equations for instances see [4], [5], [6] and [7]. The different between those three references is in considering the force acting on the fluid flow. In [4] and [6], the stress tensor is considered whereas in [5] and [7], physical properties of the porous medium such as porosity and intrinsic permeability are considered in friction term.

In this article, the wave mitigation due to the coastal vegetation is modelled using Navier-Stokes type equations. The stress tensor is written in power law model as elaborated in detail in [6]. Whereas the friction term used is Manning resistance given in [3]. In Lagrangian form, the governing equations can be written as follows:

$$\frac{D\rho}{Dt} = -\rho\nabla\cdot\boldsymbol{u},$$
$$\frac{D\boldsymbol{u}}{Dt} = -\frac{1}{\rho}\nabla P + \boldsymbol{g} - \frac{c_f\boldsymbol{u}|\boldsymbol{u}|}{H^{1/3}} + \frac{1}{\rho}\nabla\cdot\tau, \tag{1}$$
$$\frac{D\boldsymbol{x}}{Dt} = \boldsymbol{u},$$

where ρ denotes fluid density, $\boldsymbol{u}$ is flow velocity, P is pressure, $\boldsymbol{g}$ is gravitational force $(0,-9.81)^T$, c_f is the Manning friction coefficient, H is water depth, $\boldsymbol{x}$ is position of particles and τ is the stress tensor defined by

$$\tau_{ij} = 2\eta\zeta_{ij}, \quad \zeta_{ij} = \frac{1}{2}\left(\partial_i\boldsymbol{u}_j + \partial_j\boldsymbol{u}_i\right) \tag{2}$$
$$\eta = \mu\dot{\zeta}^{n-1}, \qquad \dot{\zeta} = \sqrt{2\zeta_{ij}\zeta_{ij}},$$

where n denotes the power law constant (model parameter) and $\mu = 10^{-3}\ Pa.s$ is the molecular dynamics viscosity.

This article is devoted to develop numerical model to describe influence of coastal vegetation in wave mitigation. The model is based on the Navier-Stokes equations with stress tensor written in power law model and with Manning friction term. The governing equations will be solved numerically using Smoothed Particle Hydrodynamics (SPH) method that will be elaborated in Section 2. Three kinds of numerical experiments will be conducted to see the SPH code performance and propagation of waves on flat and incline bottom. Detail and results of the simulation will be presented in Section 3. Finally, some conclusions will be given in Section 4 to close this article.

2. Numerical method

In general numerical method can be grid based or particle based. In this article, the governing equations are solved using particle method namely Smoothed Particle Hydrodynamics (SPH). Briefly explanation about the SPH method will be given in this section, detail description of the method can be found, for instances in [8], [9], [10], and [11]. Here, the explanation of the method is taken from those references.

2.1. SPH fundamentals

Monaghan in [10] described that the SPH method is meshless method based on integral interpolation. In this method, any kind of functions is interpolated by a certain kernel function that has several properties. There are several kernel function that can be used but the common one is cubic spline kernel. Further, here the fluid in the computational domain is approximated by a finite number of particles that have several physical properties such as mass m, velocity **u**, position $\boldsymbol{x}$, density ρ, and other related properties.

The discrete form of the integral interpolation of function $\phi(\boldsymbol{x})$ of point i with smoothing length of the kernel function h can be written as

$$\phi(\boldsymbol{x}_i) = \sum_j m_j\frac{\phi_j}{\rho_j}W(\boldsymbol{x}_i - \boldsymbol{x}_j, h), \tag{3}$$

where the summation is carried out for all particle (j) within compact support area with length $2h$ centered at position of point i and W denotes the cubic spline kernel

$$W(r,h) = \beta\begin{cases} 1 - \frac{3}{2}q^2 + \frac{3}{4}q^3 & 0 \le q < 1 \\ \frac{1}{4}(2-q)^3 & 1 \le q < 2 \\ 0 & q \ge 2 \end{cases} \tag{4}$$

where $\beta = \frac{10}{7\pi h^2}$ denotes the normalization parameter, $q = \frac{r}{h}$ represents the non-dimensional distance between particle i and j and r is the distance between particles.

In the method, the fluid is treated as weakly compressible by adding an equation of state. In this article, the following equation is used to update pressure of the fluid

$$P(\rho) = \frac{c_0 \rho_0}{\gamma}\left[\left(\frac{\rho}{\rho_0}\right)^{\gamma} - 1\right] \tag{5}$$

where $\rho_0 = 1000\ kg/m^3$ denotes the reference density, c_0 is the speed sound of particle at ρ_0, and $\gamma = 7$.

2.2. SPH formulations of the governing equations

Implying the SPH method to the governing equations, we get the following equations.

$$\frac{D\rho_i}{Dt} = \sum_{j=1}^{N} m_j(\boldsymbol{u}_i - \boldsymbol{u}_j) \cdot \nabla W_{ij}, \tag{6}$$

$$\frac{D\boldsymbol{u}_i}{Dt} = -\sum_{j=1}^{N} m_j\left(\frac{P_i + P_j}{\rho_i \rho_j} + \Pi_{ij}\right)\nabla_i W_{ij} + \boldsymbol{g} - \frac{c_f \boldsymbol{u}_i|\boldsymbol{u}_i|}{H_i^{\frac{1}{3}}} + \frac{1}{\rho}(\nabla \cdot \tau)_i. \tag{7}$$

The term Π_{ij} represents the artificial viscosity added to get numerical stability and defined by

$$\Pi_{ij} = \begin{cases} \frac{-\alpha_\Pi \bar{c}_{ij} \psi_{ij}}{\bar{\rho}_{ij}} & \boldsymbol{u}_{ij} \cdot \boldsymbol{x}_{ij} \\ 0 & otherwise \end{cases}, \tag{8}$$

where α_Π is a constant that should be calibrated depend on the problem [12] and

$$\psi_{ij} = \frac{h_{ij} \boldsymbol{u}_{ij} \cdot \boldsymbol{x}_{ij}}{{r_{ij}}^2 + (0.1 h_{ij})^2}, \quad \boldsymbol{u}_{ij} = \boldsymbol{u}_i - \boldsymbol{u}_j, \quad \boldsymbol{x}_{ij} = \boldsymbol{x}_i - \boldsymbol{x}_j$$
$$\bar{c}_{ij} = (c_i + c_j)/2, \quad \bar{\rho}_{ij} = (\rho_i + \rho_j)/2, \quad h_{ij} = (h_i + h_j)/2. \tag{9}$$

Formulations of the stress in SPH terms can be seen in detail in [6], [13], [14], and [15]. Let a and b denote 2D coordinate (x and y direction) and δ^{ba} is delta Dirac function with value 1 if $b = a$ and 0 if $b \neq a$. Following Shaimy in [13], SPH formulations of the stress tensor are given as:

$$\frac{1}{\rho}(\nabla \cdot \tau)_i{}^a = \sum_{j=1}^{N} m_j\left(\frac{\mu_i \dot{\zeta}^{n-1} {\zeta_i}^{ba} + \mu_j \dot{\zeta}^{n-1} {\zeta_j}^{ba}}{\rho_i \rho_j}\right)\frac{\partial W_{ij}}{\partial {\boldsymbol{x}_i}^a},$$
$${\zeta_i}^{ba} = \sum_{j=1}^{N} \frac{m_j}{\rho_j} {\boldsymbol{u}_{ji}}^b \cdot \frac{\partial W_{ij}}{\partial {\boldsymbol{x}_i}^a} + \sum_{j=1}^{N} \frac{m_j}{\rho_j} {\boldsymbol{u}_{ji}}^a \cdot \frac{\partial W_{ij}}{\partial {\boldsymbol{x}_i}^b} - \left(\frac{2}{3}\sum_{j=1}^{N} \frac{m_j}{\rho_j} {\boldsymbol{u}_{ji}}^b \cdot \nabla_i W_{ij}\right)\delta^{ba}. \tag{10}$$

3. Numerical simulation

In this section, three numerical simulations are conducted. The first simulation are carried out to validate our numerical code performance. The second simulation is to see the influence of the coastal vegetation on flat seabed. Whereas the last simulation is conducted to describe propagation of wave on incline seabed and see its run up height. In this article, the Manning coefficient is $c_f = 0.25$.

3.1. Plane Couette-Poiseuille flow

To validate our SPH code of the power-law fluid model, here we solve the Couette-Poiseuille flow problem. In this case, the fluid flows through two parallel plates. The flow is generated by the pressure difference in the gap of the plates and by the movement of one of the plates. In this article, the pressure difference is applied by giving a body force F and the movement of one of plates is generated by giving initial velocity U.

Following Shaimy in [13], the solution of the problem is given as the following equation

$$u(y,t) = \frac{U}{l}y + \frac{-F}{2\mu}y(y-l) + \sum_{i=1}^{\infty}\frac{2U}{i\pi}(-1)^i \sin\left(\frac{i\pi y}{l}\right)e^{-v\left(\frac{i\pi}{l}\right)^2 t} + \sum_{i=0}^{\infty}\frac{4Fl^2}{\mu(2i+1)^3\pi^3}\sin\left(\frac{(2i+1)\pi y}{l}\right)e^{-v\left(\frac{(2i+1)\pi}{l}\right)^2 t}. \tag{11}$$

In [13], detail of derivation of the solution is provided. Here we use the same numerical set up and boundary condition as the reference such as distance between the plates $l = 10^{-3}\, m, F = 4 \times 10^{-4}\, kgm/s^2, U = 5 \times 10^{-5} m/s$. Result of the test case for power-law constant $n = 1$ is given in Figure 1.

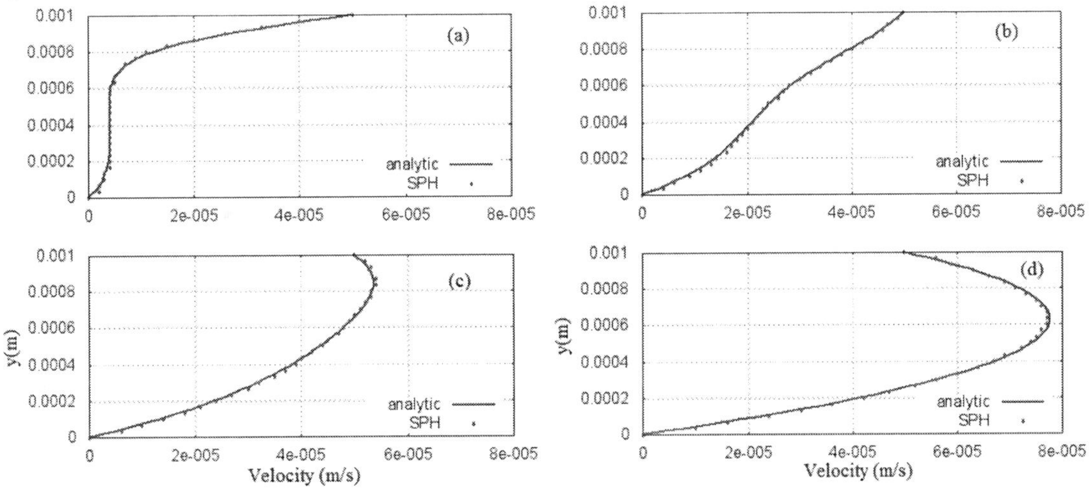

Figure 1. Velocity profile of Couette-Poiseuille flow at time *t = 0.01s* (a), *0.05s* (b), *0.1s* (c), *1s* (d).

Comparison of analytic solution provided by Shaimy in [13] and our numerical solution is presented in Figure 1. Here the velocity profile of numerical solution are plotted together with the analytic solution for several times with power-law constant $n = 1$. The pressure different and movement of the plate will push the fluid particles to move. Velocity profile of the particle in certain area is observed. The results are presented in the figure. As it is shown in the figure, the velocity profile will change until it reach the steady state solution. Moreover, it is hard to distinguish between analytic and numerical solution. The numerical solution is closest enough to the analytic solution. Therefore, it can be said that our numerical solutions are in good agreement with the analytic solution. In other words, our power-law SPH code is rational and can be used.

3.2. *Wave mitigation on flat bottom*

In this subsection, numerical experiment to see influence of the coastal vegetation is carried out. The coastal vegetation are placed at location $3.5\, m \leq x \leq 5\, m$. Piston type wavemaker as used in [3] is adopted to generate periodic waves. At the beginning all particles are set at rest with distance 0.0278 *m*, smoothing length $h = 0.0256\, m$, and placed in front of the wavemaker.

As the time goes higher, the wavemaker will push the water (particles) and create waves. The waves will propagate to the right and enter the vegetation area until they hit the right boundary and bounce back. Influence of the coastal vegetation will spread after the waves enter the area. The simulation is run in computational domain $\Omega = [0,5]$ with time step $\Delta t = 10^{-4}\, s$ and initial water depth $0.5\, m$. Results of the simulation are presented in Figure 2.

It is clearly seen in the Figure 2 that the wavemaker push the water and generate waves that propagate to right side. Figure 2(b) shows that the waves enter the vegetation area. The waves continue to propagate and hit the wall boundary shown in Figure 2(e). Further, the waves bounce back and propagate to the opposite direction due to the hard wall as shown in Figure 2(f). Figure 2 describes the wave generation and propagation over flat bottom, but it is hard to tell the influence of the coastal vegetation. Therefore we conduct the next simulation to see the influence.

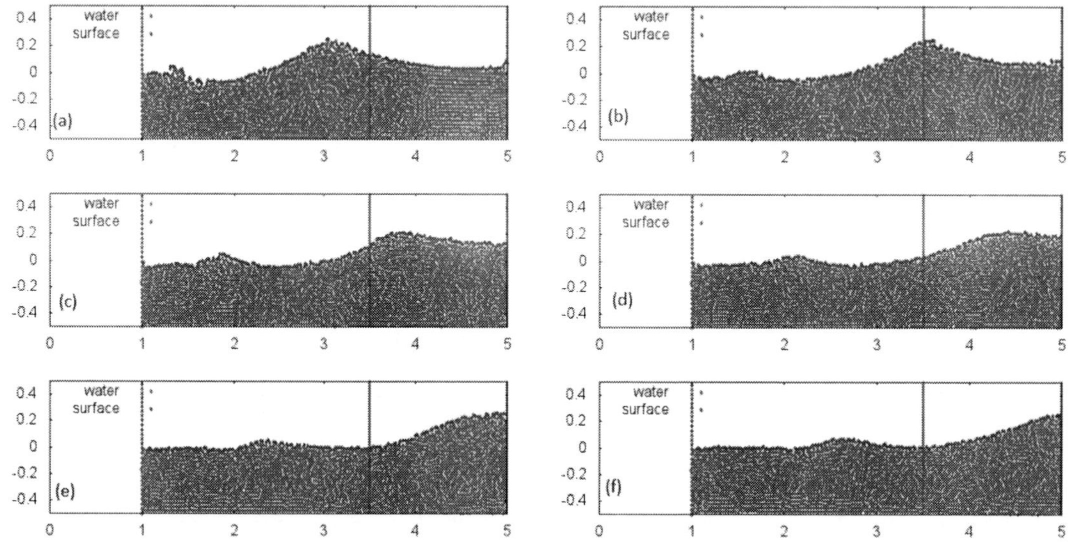

Figure 2. Wave propagation over flat bottom of SPH power-law fluid model with power-law constant $n = 1$ at time t = *0.3s* (a), *0.35s* (b), *0.4s* (c), *0.45s* (d), *0.5s* (e), and *0.55s* (f).

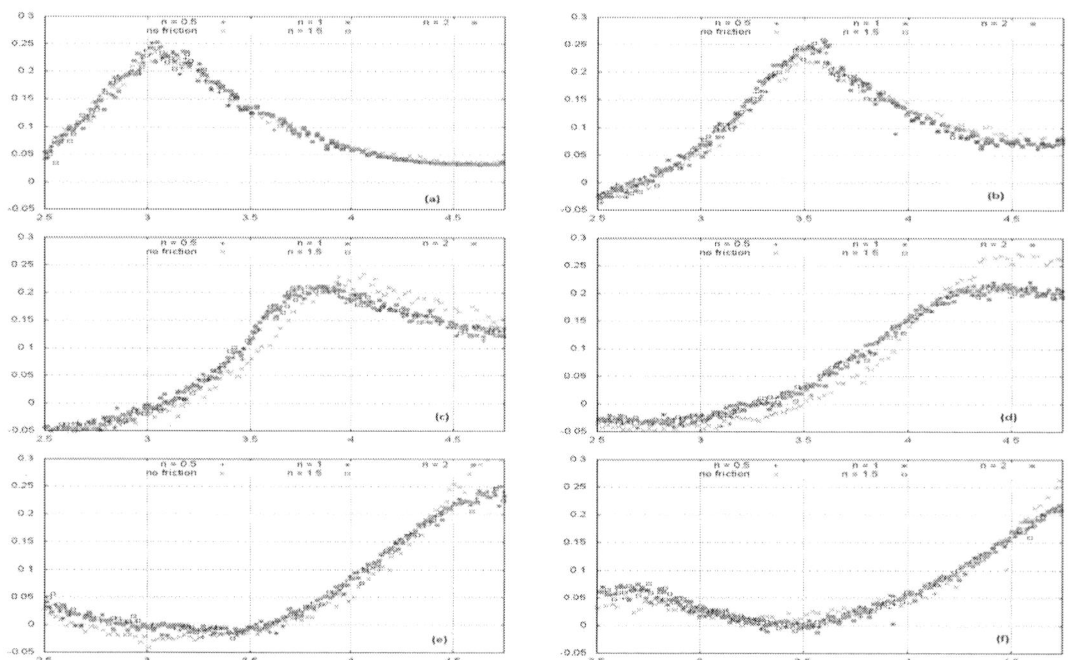

Figure 3. Wave surface of SPH power-law fluid model for several power-law constant n at time t = *0.3s* (a), *0.35s* (b), *0.4s* (c), *0.45s* (d), *0.5s* (e), and *0.55s* (f) propagating over flat bottom. x-axis denotes position in meter and y-axis denotes water surface in meter.

To see effect of power law constant in wave mitigation due to the vegetation, next we simulate the same scenario using several power law constant. In order to see the mitigation, water surface of each power law constant are plotted together with water surface in case there is no vegetation. Snapshots of those scenarios are given in Figure 3. Here we focus on area near the vegetation area. Note that all water surface in current article is detected by using free surface algorithm as described in detail in [16].

Figure 3(a) and 3(b) show the wave enter the coastal vegetation area. In these figures, the reduction of wave amplitude does not happen yet. All scenarios have the same surface with no friction case, in this article the no friction case means there is no coastal vegetation. The reduction of the wave amplitude occurs in Figure 3(c) – 3(f). In these figures, water surface of all scenarios is below the no friction case. Further, in Figure 3(c) we can see that wave profile of no friction case is in front of the other wave profiles. It means that there is reduction of wave velocity due to the coastal vegetation.

3.3. Wave mitigation on incline bottom

In this subsection, numerical simulation is used to see wave mitigation on incline seabed. Initial set up of SPH parameters is same as initial set up in the previous subsection. Results of the simulation are plotted in Figure 4. We capture surface of water for several power-law constant near the vegetation area to see run up height and its result is plotted together with water surface in case there is no vegetation in Figure 5.

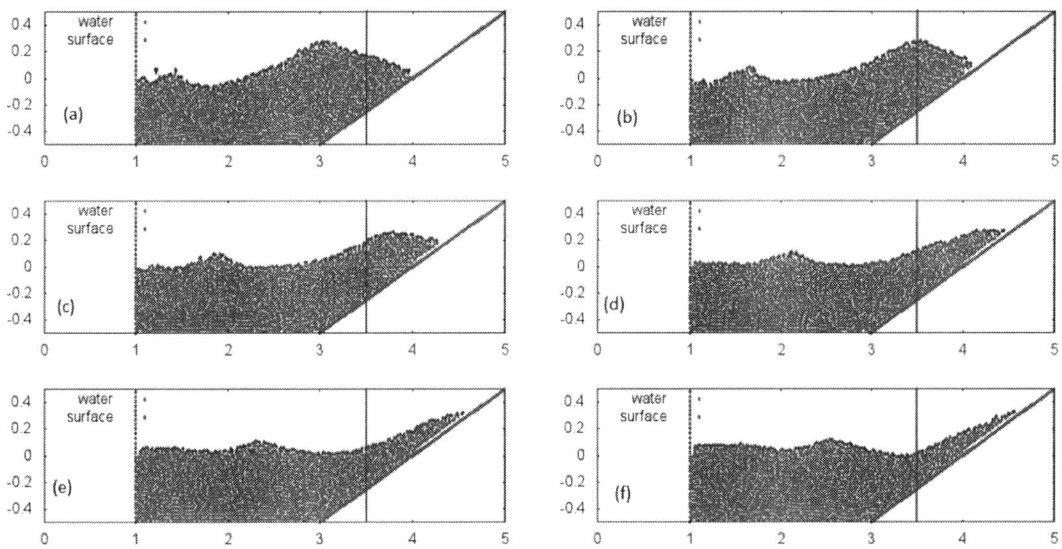

Figure 4. Wave propagation over inclined bottom of SPH power-law fluid model with power-law constant $n = 1$ at time $t = 0.3s$ (a), $0.35s$ (b), $0.4s$ (c), $0.45s$ (d), $0.5s$ (e), and $0.55s$ (f).

Figure 4 shows profile of wave propagations over an inclined bottom. The wavemaker generates wave. The wave propagates to the right, enters the vegetation area, and propagates over the inclined bottom. The recorded run up height is $0.518416\ m$ at position $x = 4.9359\ m$. To see influence of the coastal vegetation in the reduction of wave run up, we compare the results with case of no coastal vegetation. Further, to see effect of the power-law constant, we conduct several scenarios. In this simulation, we choose several values of the constant $n = 0.5, 1.0, 1.5, 2.0$. Results of the scenarios are plotted together in the Figure 5.

Figure 5 is presented comparison of water surface for power-law constant $n = 0.5, 1.0, 1.5, 2.0$ and $n = 1$ in case there is no vegetation. As shown in Figure 5(a) – 5(d) , influence of the vegetation area is seen clearly from the shape of the surface. The reduction of the wave amplitude occurs is shown in the figures. In terms of run up height, Figure 5(d) shows the reduction of the run up height. Here we record the run up height for each scenario and its result is given in Table 1.

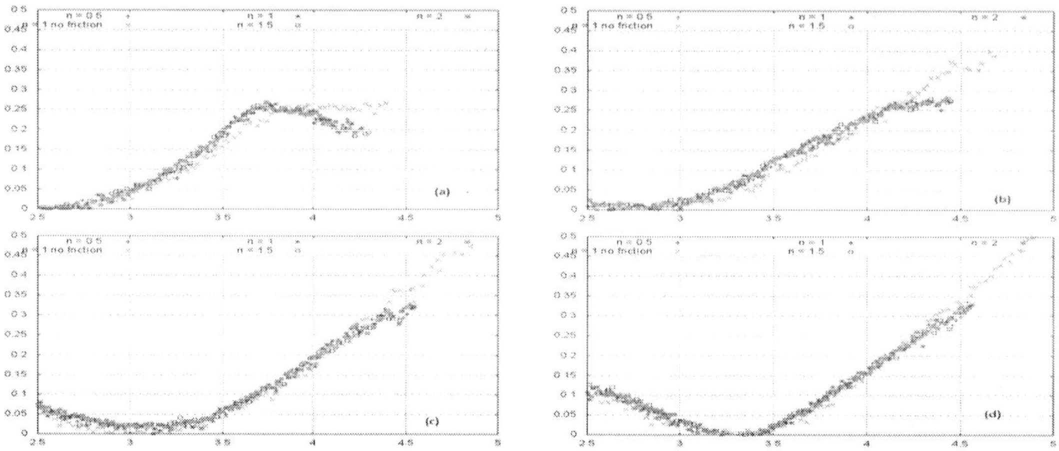

Figure 5. Wave surface of SPH power-law fluid model for several power-law constant n at time $t = 0.4s$ (a), $0.45s$ (b), $0.5s$ (c), and $0.55s$ (d) propagating over incline bottom. x-axis denotes position in meter and y-axis denotes water surface in meter.

Table 1. Run up height for several power-law constant.

n	x (m)	Run up (m)
0.5	4.534083	0.336439
1.0	4.560988	0.329222
1.5	4.527846	0.316588
2.0	4.518687	0.308246

According to Table 1, the highest run up height is obtained when the power law constant is $n = 0.5$. Further the higher the constant is, the smaller run up height is. Based on position in x-axis, the waves get the furthest position when the power law constant is $n = 1.0$. When the constant is less than $n = 1.0$ or greater than $n = 1.0$, the waves reach closer position. In case there is no vegetation and the power law constant is $n = 1.0$, the run up height is $0.518416\ m$ at position $x = 4.9359\ m$. Therefore existence of the vegetation reduce the run up height and the power law constant influences how high the run up could be. It is shown that all run up height of the scenarios in Table 1 are below the run up height of the case of no vegetation.

4. Conclusions

Elaboration of SPH power-law fluid model in wave mitigation due to coastal vegetation has been carried out. In this case, the power-law model is used to model the shear stress tensor whereas the coastal vegetation is modelled directly by adding Manning friction term. Numerical approach used to solve the problem is SPH (Smoothed Particle Hydrodynamics). Three numerical simulation are conducted; plane Couette-Poiseuille flow, wave mitigation on flat bottom, and wave mitigation on incline bottom. The first simulation is used to validate our numerical code by solving the Couette-Poiseuille flow problem. Its results show that our numerical code get results in good agreement with analytic solution. Influence of the coastal vegetation can be seen from the wave profile as shown in the second simulation. From the last simulation, existence of the mangrove reduce the run up height and the power law constant influences how high the run up could be. The highest run up height is obtained when the power law constant is $n = 0.5$ and when value of the constant increase, the height of run-up decrease. Moreover, the waves get the furthest position when the power law constant is $n = 1.0$. When the constant is less than $n = 1.0$ or greater than $n = 1.0$, the waves reach closer position.

Acknowledgement
The authors acknowledge Riset Internal Politeknik Negeri Indramayu for funding this research.

References

[1] Pudjaprasetya S R 2013 *Nonlinear Processes in Geophysics* **20** 1023-1030
[2] Magdalena I, Pudjaprasetya S and Wiryanto L 2014 *Advances in Applied Mathematics and Mechanics* **6** 680-692
[3] Iryanto and P H Gunawan 2016 *J. Phys.: Conf. Ser.* **693** 012013
[4] Fan X J, Tanner R and Zheng R 2010 *Journal of Non-Newtonian Fluid Mechanics* **165** 219-226
[5] Shao S 2010 *Coastal Engineering* **57** 304-316
[6] Vakilha M and Manzari M T 2008 *Transport in Porous Media* **74** 331-346
[7] Khayyer A, Gotoh H, Shimizu Y, Gotoh K, Falahaty H and Shao S 2018 *Coastal Engineering* **140** 1-22
[8] Liu G R and Liu M B 2003 *Smoothed particle hydrodynamics: a meshfree particle method* (World scientific)
[9] Liu M and Liu G 2010 *Archives of computational methods in engineering* **17** 25-76
[10] Monaghan J J 1992 *Annual review of astronomy and astrophysics* **30** 543-574
[11] Monaghan J J 1994 *Journal of computational physics* **110** 399-406
[12] De Padova D, Dalrymple R A and Mossa M 2014 *Journal of Hydraulic Research* **52** 836-848
[13] Shaimy R A F 2013 *Simulation of SPH method for Coutte-Poiseuille flow of power law fluid within a gap filled with obstacles* Master's thesis (Institut Teknologi Bandung, Bandung)
[14] Bayareh M, Nourbakhsh A, Rouzbahani F and Jouzaei V 2019 SN Applied Sciences **1** 1040
[15] Xenakis A, Lind S, Stansby P and Rogers B 2015 *Journal of Non-Newtonian Fluid Mechanics* **218** 1-15
[16] Barecasco A, Terissa H and Naa C F 2013 *arXiv preprint arXiv:1309.4290*

Modeling Poverty Percentages in the Papua Islands using Fourier Series in Nonparametric Regression Multivariable

N P A M Mariati[1,2], I N Budiantara[1,a)], and V Ratnasari[1]

[1]Department of Statistics, Institut Teknologi Sepuluh Nopember, Surabaya, 60111, Indonesia
[2]Universitas Mahasaraswati, Denpasar Bali, 80233, Indonesia

[a)]Corresponding author: i_nyoman_b@statistika.its.ac.id

Abstract. Nonparametric regression has high flexibility in estimating the regression curve. Estimation techniques that are quite popular in nonparametric regression are Fourier Series estimators. Fourier series are trigonometric polynomials that have flexibility, so they can adapt effectively to the local nature of data. This Fourier series estimator is generally used if the data investigated by the pattern is unknown and there is a tendency for repetitive patterns to follow the trend line. The purpose of this study is to examine Fourier Series estimates and apply to cases of poverty in the Papua Islands. The Papua Islands which consist of 2 Provinces namely Papua Province and West Papua Province. Papua Province ranked first and West Papua Province ranked second in poverty in Indonesia. Based on nonparametric regression multivariable modeling using Fourier series, this model is good. This can be seen from the value obtained by Generalized Cross Validation (GCV) = 80,76, Mean Square Error (MSE) is 18,31 and R^2=78,16%.

1. Introduction

Regression analysis has three approaches, namely parametric regression, nonparametric regression and semiparametric regression. Parametric regression approach there is an assumption that is very strong, rigid and regression curve shapes are known. The main objective in regression analysis is to find the estimated form of the regression curve. In contrast to the parametric regression approach, in nonparametric regression the form of the regression curve is assumed to be unknown. Nonparametric regression curves are only assumed to be smooth in the sense contained in a particular function space. Data are expected to seek their own form of the estimate, without being influenced by the subjective factor of the designer of the study. Nonparametric regression approach has high flexibility (Eubank, 1988) [1]. Some nonparametric regression model that is widely used among Kernel (Hardle, 1990) [2], Spline (Wahba, 1990 [3]; Budiantara 2007 [4]), K-Nearest Neighbor (Hardle, 1990) [2], Fourier Series Estimator (Eubank, 1999) [5], Histograms (Green and Silverman, 1994) [6], multivariate adaptive regression splines (MARS), Orthogonal Series, Wavelets, Neural Networks. Nonparametric regression approach that is quite popular is the Fourier series estimator. Fourier series became popular in 1992 popularized by Bilodeau. Fourier series are trigonometric polynomials that have flexibility, so they can adapt effectively to the local nature of data (Asrini, 2014 [7]; Pane et.al., 2014 [8]). Fourier series estimator is generally used when data were investigated patterns are not known and there is a tendency of repeating patterns (a repetition of the response value to the value of the different predictor variables) (Tripena 2007 [9]; Bilodeau, 1992 [10]). Galtchouk (2009) [11] perform adaptive procedure for estimating an unknown nonparametric regression and show that the procedure is efficient asimptotic for quadratic risk. Furthermore, the study was conducted by Pane, et.al (2014) [8] Fourier Series Estimator using Penalized Least Square. Prahutama (2018) [12] conduct research on regression modeling of Fourier series in the case of inflation.

In obtaining Fourier series regression estimates using Least Square method. The advantages of the Least Square method are simple and free of distribution. Nonparametric regression with multivariable if more than one predictor variable. Much of the research conducted a study on mutivariable one of which deals with social issues. A common problem that a lot of population, most of whom have low levels of education will lead to poverty. One of the countries that have problems with poverty percentage is Indonesia. Indonesia is a developing country with a population that is very large, with the fourth largest population in the world. Sustainable development goals are the first goal of poverty alleviation. Economic growth has a negative and significant effect on poverty, meaning that increasing economic growth will reduce poverty. Likewise about education has a negative and significant effect on poverty, meaning that the higher the level of education will reduce poverty. The unemployment rate significantly influence the level of poverty, but it is also related to unemployment and poverty, the government needs to improve the quality of human resources through the community health status and increased access to education to remote areas (Yacoub, 2012) [13]. Hoe, et.al (2017) [14] conducted a research on the factors that influence people's participation in rural poverty reduction program in Malaysia. Research on poverty in Myanmar (Teerawichitchainan, 2018 [15]). Rusdarti (2013) [16] reported that Papua province was ranked first on the poor population percentage in Indonesia based on the depth index. The study was based on the depth index of poverty because the depth index is a measure of the average expenditure gap of each poor population, in this study there is a Poverty Depth Index in Indonesia, Papua Province is first rank depth index. Papua has depth index above the average index of depth in Indonesia [17]. Thus, no one has studied poverty in the islands of Papua (Papua and West Papua Province). This paper uses Fourier series regression for modelling poverty percentage in the Papua island, Indonesia.

2. Model in Nonparametric Regression Multivariable Fourier Series

This section discusses Fourier series estimators in the nonparametric regression multivariable. Fourier series is a function with a cosine or sine database that is owned, so it can adjust effectively to local properties data. Fourier series can be used to describe curves using periodic sine or cosine waves. Given multivariable nonparametric regression model:

$$y_i = \sum_{p=1}^{q} f_p\left(x_{pi}\right) + \varepsilon_i, \quad i = 1,2,\dots,n$$

Furthermore, regression curve f_j approached with Fourier Series functions:

$$f_p\left(x_{pi}\right) = b_p x_{pi} + \frac{1}{2}\alpha_{0p} + \sum_{k=1}^{K} \alpha_{kp} \cos kx_{pi}, \mathrm{p} = 1,2,\dots,q.$$

The estimator of the Fourier Series regression curve is obtained from optimization:

$$\underset{\underset{\sim}{\beta} \in R^{q(K+2)}}{Min} \left\{ \sum_{i=1}^{n} \left(y_i - \sum_{\mathrm{p}=1}^{q} \left(b_p x_{pi} + \frac{1}{2}\alpha_{0p} + \sum_{k=1}^{K} \alpha_{kp} \cos kx_{pi} \right) \right)^2 \right\} = \underset{\underset{\sim}{\beta} \in R^{q(K+2)}}{Min} \left\{ \theta\left(\underset{\sim}{\beta}\right) \right\}$$

In solving the above optimization steps are as follows:

$$\theta\left(\underset{\sim}{\beta}\right) = \sum_{i=1}^{n} \left(y_i - \mu\left(x_{1i}, x_{2i}, \dots, x_{qi}\right) \right)^2$$

$$= \sum_{i=1}^{n} \left(y_i - \sum_{\mathrm{p}=1}^{q} \left(b_p x_{pi} + \frac{1}{2}\alpha_{0p} + \alpha_{1p} \cos x_{pi} + \dots + \alpha_{Kp} \cos Kx_{pi} \right) \right)^2$$

$$= \sum_{i=1}^{n} \left(y_i - \left(b_1 x_{1i} + \frac{1}{2}\alpha_{01} + \alpha_{11} \cos x_{1i} + \ldots + \alpha_{K1} \cos K x_{1i} \right) - \ldots + \right.$$

$$\left. - \left(b_q x_{qi} + \frac{1}{2}\alpha_{0q} + \alpha_{1q} \cos x_{qi} + \ldots + \alpha_{Kq} \cos K x_{qi} \right) \right)^2 \quad (1)$$

Equation (1) above can be written as:

$$\theta(\underset{\sim}{\beta}) = \left(\underset{\sim}{y} - \mathbf{X}(K)\underset{\sim}{\beta}\right)' \left(\underset{\sim}{y} - \mathbf{X}(K)\underset{\sim}{\beta}\right) \quad (2)$$

$$= \underset{\sim}{y}'\underset{\sim}{y} - \underset{\sim}{y}'\mathbf{X}(K)\underset{\sim}{\beta} - \underset{\sim}{\beta}'\mathbf{X}'(K)\underset{\sim}{y} + \underset{\sim}{\beta}'\mathbf{X}'(K)\mathbf{X}(K)\underset{\sim}{\beta}$$

$$= \underset{\sim}{y}'\underset{\sim}{y} - 2\underset{\sim}{\beta}'\mathbf{X}'(K)\underset{\sim}{y} + \underset{\sim}{\beta}'\mathbf{X}'(K)\mathbf{X}(K)\underset{\sim}{\beta}$$

with:

$$\underset{\sim}{y} = [y_1, y_2, \ldots, y_n]', \quad \underset{\sim}{\beta} = \begin{bmatrix} b_1 & \frac{1}{2}\alpha_{01} & \alpha_{11} & \cdots & \alpha_{K1} & | & \cdots & | & b_q & \frac{1}{2}\alpha_{0q} & \alpha_{1q} & \cdots & \alpha_{Kq} \end{bmatrix}' \text{ and,}$$

$$\mathbf{X}(K) = \begin{bmatrix} x_{11} & 1 & \cos x_{11} & \cdots & \cos Kx_{11} & | & \cdots & | & x_{q1} & 1 & \cos x_{q1} & \cdots & \cos Kx_{q1} \\ x_{12} & 1 & \cos x_{12} & \cdots & \cos Kx_{12} & | & \cdots & | & x_{q2} & 1 & \cos x_{q2} & \cdots & \cos Kx_{q2} \\ \vdots & \vdots & \vdots & \ddots & \vdots & | & \ddots & | & \vdots & \vdots & \vdots & \ddots & \vdots \\ x_{1n} & 1 & \cos x_{1n} & \cdots & \cos Kx_{1n} & | & \cdots & | & x_{qn} & 1 & \cos x_{qn} & \cdots & \cos Kx_{qn} \end{bmatrix}$$

Estimation $\hat{\underset{\sim}{\beta}}$ obtained from minimizing $\theta(\underset{\sim}{\beta})$. Partial derivative $\theta(\underset{\sim}{\beta})$ to $\underset{\sim}{\beta}$ obtained:

$$\frac{\partial Q(\underset{\sim}{\beta})}{\partial \underset{\sim}{\beta}} = -2\mathbf{X}'(K)\underset{\sim}{y} + 2\mathbf{X}'(K)\mathbf{X}(K)\underset{\sim}{\beta} \quad (3)$$

If equation (3) is equal to zero, the equation is obtained:

$$\hat{\underset{\sim}{\beta}}(K) = \left(\mathbf{X}'(K)\mathbf{X}(K)\right)^{-1} \mathbf{X}'(K)\underset{\sim}{y}$$

$$= \begin{bmatrix} \hat{b}_1 & \frac{1}{2}\hat{\alpha}_{01} & \hat{\alpha}_{11} & \cdots & \hat{\alpha}_{K1} & | & \cdots & | & \hat{b}_q & \frac{1}{2}\hat{\alpha}_{0q} & \hat{\alpha}_{1q} & \cdots & \hat{\alpha}_{Kq} \end{bmatrix}'$$

Estimator for the regression curve f_p is given by:

$$\hat{f}_p(x_{pi}) = \hat{b}_p x_{pi} + \frac{1}{2}\hat{\alpha}_{0p} + \sum_{k=1}^{K} \hat{\alpha}_{kp} \cos kx_{pi}.$$

After getting the estimator, it will then be applied to cases of poverty in the Papua Islands.

3. Modeling Poverty Percentages in the Papua Islands

Below is given the data in 2017 with the following variables: the response variable is the percentage of Poverty (y) while successive predictor variables Human Development Index (x_1), Open Unemployment Rate (x_2), Old School Expectation Numbers (x_3), School Participation Rates (x_4) and Gross Regional Domestic Product (x_5).

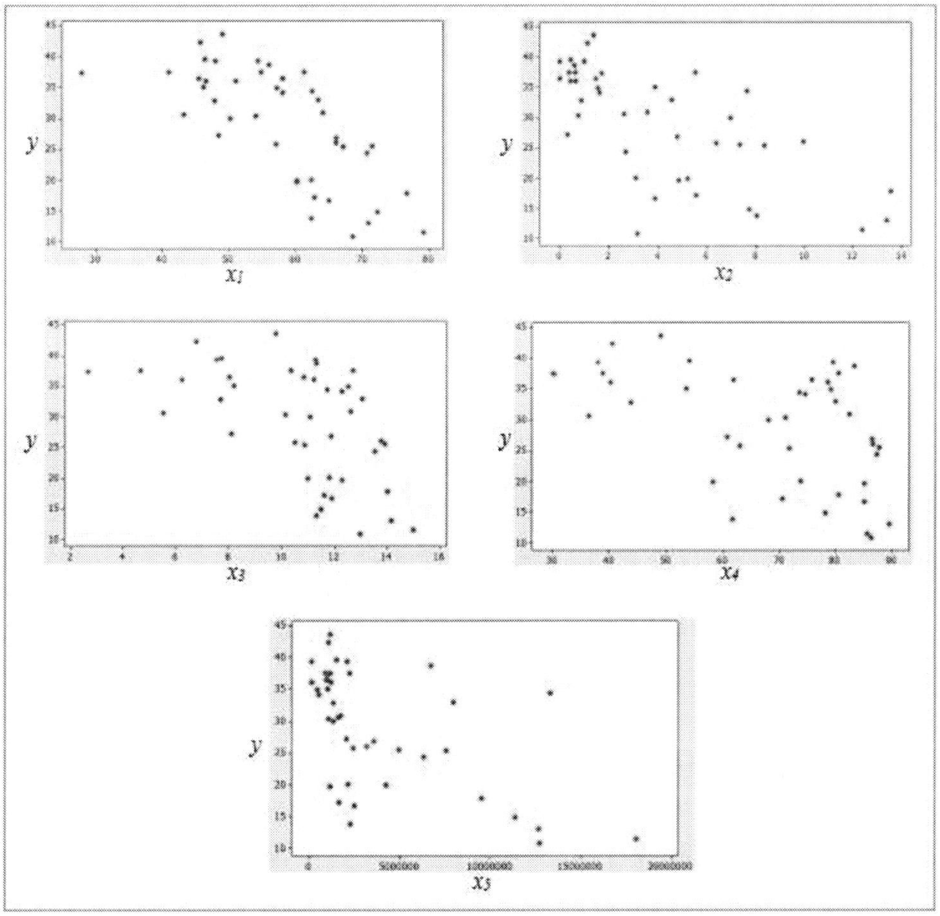

Figure 1. Scatterplot Respon and Predictor

Figure 1, it can be seen that the data are not patterned and tendency to have a repeating pattern following a trend line.

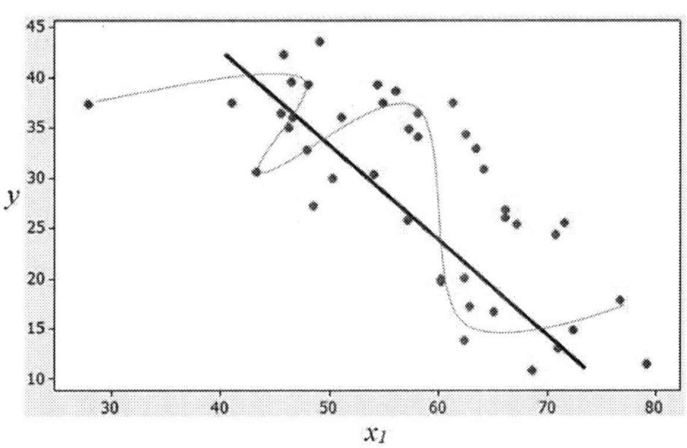

Figure 2. Fourier Series with Trend Line

Based on Figure 2 shows that the relationship with the x_1 variable y is not patterned and the tendency to have recurring patterns follow a trend line. So is the case with other variables. Therefore, for data on poverty in Papua islands used Fourier series approach. Fourier series is a nonparametric regression approach is used when data is investigated the patterns are not known and there is a recurring trend. Nonparametric regression model given Fourier series as follows:

$$\begin{aligned} y_i &= \mu(x_{1i}, x_{2i}, x_{3i}, x_{4i}, x_{5i}) + \varepsilon_i \\ &= \mu(x_{1i}) + \mu(x_{2i}) + \mu(x_{3i}) + \mu(x_{4i}) + \mu(x_{5i}) + \varepsilon_i, i = 1,2,\dots,n \end{aligned}$$

The model used to model the percentage of poverty in the Papua Islands is nonparametric Fourier Series. Estimation models with Fourier series approach is obtained as follows:

$$\hat{\mu}(x_{1i}, x_{2i}, \dots, x_{qi}) = \sum_{p=1}^{q} \left(\hat{b}_p x_{pi} + \frac{1}{2}\hat{\alpha}_{0p} + \sum_{k=1}^{K} \hat{\alpha}_{kp} \cos kx_{pi} \right).$$

Nonparametric regression Fourier series, very dependent on oscillation parameters (K). The oscillation parameter (K) is the sum of the cosine wave oscillations in the model. This research is limited to the oscillation parameter value (K) which is as much as 3. The optimal oscillation parameter selection is done using the Generalized Cross Validation (GCV) method. The advantages of the GCV method are simple, efficient, optimal asymptotic, invariant for transformation and does not require information about σ^2. Fourier series estimator contains K oscillation parameters. Getting the best estimator needs to be selected for optimal oscillation parameters using GCV. The optimal K value chosen is the K value which produces a minimum GCV value. The results of the analysis for the K values are given in Table 1 below:

Table 1. Determination of Optimal Oscillation Parameter (K) Value Based on GCV

Oscillation Parameter (K)	GCV
1	478.10
2	167.15
3	**80.76**
4	90.05

Table 1 shows that the best model is a regression model Fourier series with oscillation parameters (K=3). Thus, the estimated model percentage of poverty in the Papua Islands with Fourier Series nonparametric regression modeling is as follows:

$$\begin{aligned} \hat{y}_i = {} & 71.76 - 1.13x_{1i} - 1.62\cos x_{1i} - 3.98\cos 2x_{1i} - 3.20\cos 3x_{1i} + \\ & -0.61x_{2i} + 1.77\cos x_{2i} - 1.98\cos 2x_{2i} + 0.86\cos 3x_{2i} + \\ & 2.90x_{3i} + 2.78\cos x_{3i} - 0.62\cos 2x_{3i} - 1.53\cos 3x_{3i} + \\ & -0.12x_{4i} + 0.87\cos x_{4i} + 0.96\cos 2x_{4i} + 1.59\cos 3x_{4i} + \\ & 0.0000003x_{5i} + 1.20\cos x_{5i} + 0.82\cos 2x_{5i} - 0.21\cos 3x_{5i}. \end{aligned}$$

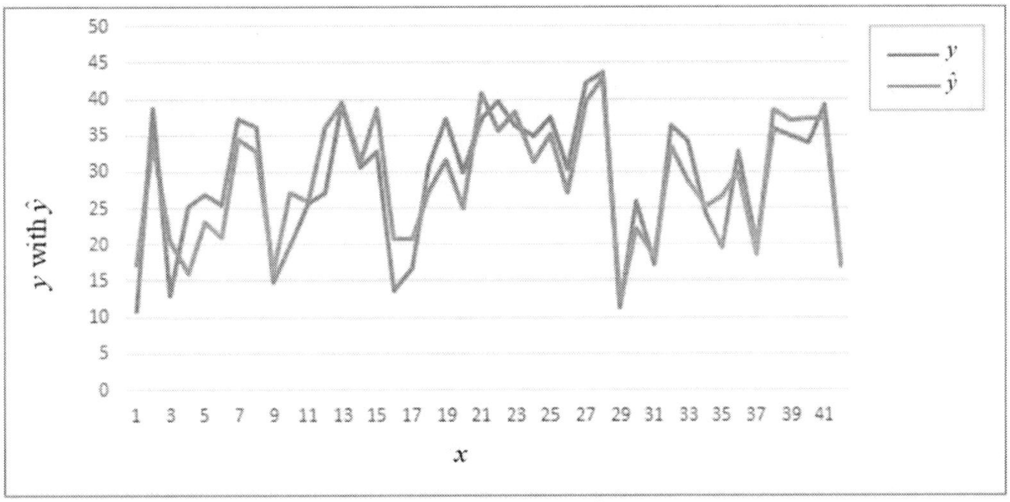

Figure 3. y with $\hat{y}$

It is shown in Figure 3 that the model uses Fourier Series in nonparametric regression both for poverty modeling because it has GCV = 80.76; MSE = 18.31; R^2 = 78.16% and estimates are relatively small.

4. Consclusion

Fourier Series estimator in nonparametric regression multivariable curve $\hat{y}=\sum_{p=1}^{q}\left(\hat{b}_p x_{pi}+\frac{1}{2}\hat{\alpha}_{0p}+\sum_{k=1}^{K}\hat{\alpha}_{kp}\cos kx_{pi}\right)$. This model has been very good for modeling poverty because it has GCV=80,76, R^2=78,16% ; MSE = 18,31 and estimation y with $\hat{y}$ relatively small. The Estimation of the Poverty Model in Papua Islands uses Fourier Series in nonparametric regression

$$\begin{aligned}\hat{y}_i = {} & 71,76-1,13x_{1i}-1,62\cos x_{1i}-3,98\cos 2x_{1i}-3,20\cos 3x_{1i}+\\ & -0,61x_{2i}+1,77\cos x_{2i}-1,98\cos 2x_{2i}+0,86\cos 3x_{2i}+\\ & 2,90x_{3i}+2,78\cos x_{3i}-0,62\cos 2x_{3i}-1,53\cos 3x_{3i}+\\ & -0,12x_{4i}+0,87\cos x_{4i}+0,96\cos 2x_{4i}+1,59\cos 3x_{4i}+\\ & 0,000003x_{5i}+1,20\cos x_{5i}+0,82\cos 2x_{5i}-0,21\cos 3x_{5i}.\end{aligned}$$

References

[1] Eubank R L 1988 *Spline Smoothing and Nonparametric Regression* (New York:Marcel Dekker)
[2] Hardle W 1990 *Applied Nonparametric Regression* (New York:Cambridge University Press)
[3] Wahba G 1990 *Spline Models For Observasion Data* (Pensylvania:SIAM)
[4] Budiantara I N 2007 Inferensi Statistik untuk Model Spline *Jurnal Mat-Stat* **7** 1-14.
[5] Eubank R L 1999 *Nonparametric Regression and Spline Smoothing 2nd* (New York: Marcel Dekker).
[6] Green P J and Silverman B W 1994 *Nonparametric Regression and Generalized Linear Model* (London:Chapman & Hall).
[7] Asrini L J 2014 Fourier Series Semiparametric Regression Models (Case Study: The Production of Lowland Rice Irrigation in Central Java). *ARPN Journal of Engineering and Applied Sciences* **9 (9)** 1501-06.
[8] Pane R, Budiantara I N, Zain I, and Otok B W 2014 Parametric and NonparametricEstimators in Fourier Series Semiparametric Regression and Their Characteristics *Apllied Mathematical*

Sciences **8(102)** 5053-64.

[9] Tripena A and Budiantara I N 2007 Fourier Estimator in Nonparametric Regression *Int. Conf. on Natural Sciences and Applied Natural Scienes* (Yogyakarta: Ahmad Dahlan University).

[10] Bilodeau M 1992 Fourier Smoother and Additive Models *The Canadian Journal of Statistics* **3** 257-69.

[11] Galtchouk L and Pergamenshchikov S 2009 Adaptive Asymptotically Efficient Estimation in Heteroscedastic Nonparametric Regression *Journal of the Korean Statistical Society* **38** 305-22.

[12] Prahutama A, Suparti and Utami T W 2018 Modeling Fourier Regression and Time Series Data a Case Study Modelling Inflation in Foods Sector in Indonesia *Int. Conf. in Mathematics Pure Applied and Computation Journal of Physics* **974** 1-9.

[13] Yacoub Y 2012 Pengaruh Tingkat Pengangguran terhadap Tingkat Kemiskinan Kabupaten/ Kota di Provinsi Kalimantan Barat *Jurnal EKSOS* **8(3)** 176–85.

[14] Hoe K C, Wahab H, Bakar S H A and Islam M R 2017 Community participation for rural poverty alleviation: A case of the Iban community in Malaysia *International Social Work,* **61(4)** 518–36.

[15] Teerawichitchainan B and Knodel J 2018 Long-Term Care Needs in the Context of Poverty and Population Aging: the Case of Older Persons in Myanmar *Journal of Cross-Cultural Gerontolog* **33(2)** 143–62.

[16] Rusdarti and Karolina 2013 Faktor-faktor yang Mempengaruhi Tingkat Kemiskinan di Provinsi Jawa Tengah *Jurnal Economia* **9(1).**

[17] Eddy S S and Agung 2010 Faktor-faktor yang Mempengaruhi Kemiskinan secara Makro di Lima Belas Provinsi Tahun 2007 *Jurnal dan Organisasi Manajemen* **6(2)** 89-100.

Journal of Physics: Conference Series **1397** (2019) 012072 doi:10.1088/1742-6596/1397/1/012072

Estimation of covariance matrix using multi-response local polynomial estimator for designing children growth charts: A theoretically discussion

N Chamidah[1]* and B Lestari[2]

[1]Department of Mathematics, Faculty of Science and Technology, Airlangga University. Jalan Mulyorejo Kampus C UNAIR, Surabaya 60115, Indonesia.
[2]Department of Mathematics, Faculty of Mathematics and Natural Science, The University of Jember. Jalan Kalimantan 37 Kampus Tegal Boto, Jember 68111, Indonesia.

*Corresponding author's e-mail: nur-c@fst.unair.ac.id

Abstract. In statistical analysis for instance regression analysis we always be faced a estimation problem of regression function which draws relationship between variables in the regression model. In the real cases we frequently meet the relationship between one or more response variables and one or more predictor variables where there are correlations between responses that is called a multi-response regression model. There are two approaches to estimate the multi-response regression model, i.e., parametric and nonparametric. One of estimators in nonparametric regression model is local polynomial estimator for estimating the regression function. Since there are correlations between responses then in the estimating of regression function we need a weight matrix that is to be inverse of covariance matrix of error. Therefore, the main objective of this research is to estimate of covariance matrix of error by using multi-response local polinomial estimator. The result of this research is a covariance matrix estimator that is in the future can be used to design children growth charts.

1. Introduction

In statistical modelling there are two approaches for estimating the regression function, i.e., parametric and nonparametric. The nonparametric approach gives more flexibility for the form of the regression function. Many researchers have studied nonparametric regression, i.e., spline estimator by [1-4], local linear estimator by [5-8], and local polynomial estimators with one response and one predictor variables for polynomial order one called as local linear have been discussed by [9] who applied it to spatial regression. Next, [10] have researched local linear estimator and stated that it is better than kernel estimator because it has mean square prediction least than that kernel estimator. Furthermore, [11] applied local polynomial estimator to time series data, and [12] used local polynomial estimator to estimate esophageal pressure on gastroesophageal reflux disease.

In the widely field we frequently be faced cases involving regression models with more than one response variables and there are correlation between responses. So, these regression function problems must be solved by multi-response nonparametric regression or multi-response semiparametric regression. For example [13] used smoothing spline estimator for biresponse that is applied to hormone data; [14-17] have discussed the multi-response nonparametric regression by using spline estimator and [18-20] have studied biresponse local linear regression. Meanwhile, multi-response semiparametric regression in some cases by using spline estimator and local linear estimator have discussed by [21-22] and [23], respectively.

One of real cases that involve regression model with more than one response variables and there are correlations between responses is children growth model [19], [20], [23]. According to [24] children up to one year old grows up quickly and then goes down slowly as long as going up of children's age, so that locally model approach is more suitable for this case. Next, [25] proposed quantile polynomial parametric regression approach to design children growth chart. But, the growth chart given by [25] was too smooth, so that it did not represent the data which has locally pattern and its pattern change is very sharp. In addition, all researchers who have mentioned previously, discussed estimation the regression functions of the nonparametric and semiparametric regression models by assuming the covariance matrix is known, so that the covariance matrix is not need to be estimated. If the covariance matrix is unknown then it must be estimated from the data. Therefore, in this paper we give estimating method of covariance matrix by using multi-response local polynomial estimator.

2. Method

To estimate children growth chart based on anthropometric measures, i.e., weight and height, we use multi-response model by weighted least squares method because there is significant correlation between weight and height of children. Therefore, for estimating the children growth charts we estimate the charts simultaneously and need the weight matrix, i.e., inverse of covariance matrix of error. In next section, we give results and discussion about estimations of the regression function and the covariance matrix methods by using multi-response local polynomial estimator.

3. Results and discussion

3.1. Estimation of Regression function by using multi-response local polynomial estimator

Suppose that the paired observations $(t_i, y_{1i}, y_{2i}, ..., y_{pi})$, $i = 1, 2, ..., n$ follows the following regression model:

$$y_{ri} = f_r(t_i) + \varepsilon_{ri}, \quad r = 1, 2, ..., p \tag{1}$$

where random error ε_{ri} follows the following assumptions:

$$E(\varepsilon_{ri}) = 0, \text{ and } E(\varepsilon_{ri}^2) = \sigma_{ri}^2, \; E(\varepsilon_{ri}\varepsilon_{si'}) = \begin{cases} \sigma_{(rs)i}, & i = i' \\ 0, & i \neq i' \end{cases} \tag{2}$$

for $r \neq s$, $r = s = 1, 2, ..., p$.

Next, based on i-th observation and for p responses, we can elaborate equation (1) as follows:

$$\left.\begin{aligned} y_{1i} &= f_1(t_i) + \varepsilon_{1i} \\ y_{2i} &= f_2(t_i) + \varepsilon_{2i} \\ &\vdots \\ y_{pi} &= f_p(t_i) + \varepsilon_{pi} \end{aligned}\right\} \tag{3}$$

Therefore, we can express equation (3) in the following matrix notation:

$$\underset{\sim}{y}_i = \underset{\sim}{f}(t_i) + \underset{\sim}{\varepsilon}_i \tag{4}$$

where $\underset{\sim}{y}_i = (y_{1i}, y_{2i}, ..., y_{pi})'$ and $\underset{\sim}{f}(t_i) = (f_1(t_i), f_2(t_i), ..., f_p(t_i))'$ is unknown smooth function.

Further, if we elaborate every response in equation (4) for every observation, then equation (4) can be written as follows:

$$\underset{\sim}{y} = \underset{\sim}{f} + \underset{\sim}{\varepsilon}. \tag{5}$$

where $E(\underset{\sim}{\varepsilon})=\underset{\sim}{0}$, $Var(\underset{\sim}{\varepsilon})=\mathbf{V}$, $\underset{\sim}{y}=(\underset{\sim}{y}_1,\underset{\sim}{y}_2,...,\underset{\sim}{y}_p)'$, $\underset{\sim}{f}=(\underset{\sim}{f}_1,\underset{\sim}{f}_2,...,\underset{\sim}{f}_p)'$, $\underset{\sim}{\varepsilon}=(\underset{\sim}{\varepsilon}_1,\underset{\sim}{\varepsilon}_2,...,\underset{\sim}{\varepsilon}_p)'$, $\underset{\sim}{\varepsilon}_r=(\varepsilon_{r1},\varepsilon_{r2},...,\varepsilon_{rn})'$, $\underset{\sim}{y}_r=(y_{r1},y_{r2},...,y_{rn})'$, and $\underset{\sim}{f}_r=(f_r(t_1),f_r(t_2),...,f_r(t_n))'$.

In the model given in (5), there are correlations between responses, so that to get the estimated model by using multi-response local polynomial estimator we must use weights either kernel function or covariance matrix. Note that, statistically, the weight matrix is to be inverse of covariance matrix of error vector $\underset{\sim}{\varepsilon}$. If we notate the covariance matrix as $\mathbf{V}$, then we can determine $\mathbf{V}$ as follows:

$$\mathbf{V}=Var(\underset{\sim}{\varepsilon})=E(\underset{\sim}{\varepsilon}-E(\underset{\sim}{\varepsilon}))(\underset{\sim}{\varepsilon}-E(\underset{\sim}{\varepsilon}))'=E(\underset{\sim}{\varepsilon}\underset{\sim}{\varepsilon}')$$

$$=E\left[\left(\varepsilon_{11},...,\varepsilon_{1n},\varepsilon_{21},...,\varepsilon_{2n},...,\varepsilon_{p1},...,\varepsilon_{pn}\right)'\left(\varepsilon_{11},...,\varepsilon_{1n},\varepsilon_{21},...,\varepsilon_{2n},...,\varepsilon_{p1},...,\varepsilon_{pn}\right)\right].$$

Next, by considering the assumptions in (2), we obtain the covariance matrix $\mathbf{V}$ as follows:

$$\mathbf{V}=E\begin{bmatrix}
\varepsilon_{11}^2 & \varepsilon_{11}\varepsilon_{12} & \varepsilon_{11}\varepsilon_{13} & \cdots & \varepsilon_{11}\varepsilon_{1n} & \varepsilon_{11}\varepsilon_{21} & \varepsilon_{11}\varepsilon_{22} & \cdots & \varepsilon_{11}\varepsilon_{2n} & \cdots & \varepsilon_{11}\varepsilon_{p1} & \cdots & \varepsilon_{11}\varepsilon_{pn}\\
\varepsilon_{12}\varepsilon_{11} & \varepsilon_{12}^2 & \varepsilon_{12}\varepsilon_{13} & \cdots & \varepsilon_{12}\varepsilon_{1n} & \varepsilon_{12}\varepsilon_{21} & \varepsilon_{12}\varepsilon_{22} & \cdots & \varepsilon_{12}\varepsilon_{2n} & \cdots & \varepsilon_{12}\varepsilon_{p1} & \cdots & \varepsilon_{12}\varepsilon_{pn}\\
\vdots & \vdots & \vdots & \cdots & \vdots & \vdots & \vdots & \cdots & \vdots & \cdots & \vdots & \cdots & \vdots\\
\varepsilon_{1n}\varepsilon_{11} & \varepsilon_{1n}\varepsilon_{12} & \varepsilon_{1n}\varepsilon_{13} & \cdots & \varepsilon_{1n}^2 & \varepsilon_{1n}\varepsilon_{21} & \varepsilon_{1n}\varepsilon_{22} & \cdots & \varepsilon_{1n}\varepsilon_{2n} & \cdots & \varepsilon_{1n}\varepsilon_{p1} & \cdots & \varepsilon_{1n}\varepsilon_{pn}\\
\varepsilon_{21}\varepsilon_{11} & \varepsilon_{21}\varepsilon_{12} & \varepsilon_{21}\varepsilon_{13} & \cdots & \varepsilon_{21}\varepsilon_{1n} & \varepsilon_{21}^2 & \varepsilon_{21}\varepsilon_{22} & \cdots & \varepsilon_{21}\varepsilon_{2n} & \cdots & \varepsilon_{21}\varepsilon_{p1} & \cdots & \varepsilon_{21}\varepsilon_{pn}\\
\varepsilon_{22}\varepsilon_{11} & \varepsilon_{22}\varepsilon_{12} & \varepsilon_{22}\varepsilon_{13} & \cdots & \varepsilon_{22}\varepsilon_{1n} & \varepsilon_{22}\varepsilon_{21} & \varepsilon_{22}^2 & \cdots & \varepsilon_{22}\varepsilon_{2n} & \cdots & \varepsilon_{22}\varepsilon_{p1} & \cdots & \varepsilon_{22}\varepsilon_{pn}\\
\vdots & \vdots & \vdots & \cdots & \vdots & \vdots & \vdots & \cdots & \vdots & \cdots & \vdots & \cdots & \vdots\\
\varepsilon_{2n}\varepsilon_{11} & \varepsilon_{2n}\varepsilon_{12} & \varepsilon_{2n}\varepsilon_{13} & \cdots & \varepsilon_{2n}\varepsilon_{1n} & \varepsilon_{2n}\varepsilon_{21} & \varepsilon_{2n}\varepsilon_{22} & \cdots & \varepsilon_{2n}^2 & \cdots & \varepsilon_{2n}\varepsilon_{p1} & \cdots & \varepsilon_{2n}\varepsilon_{pn}\\
\vdots & \vdots & \vdots & \cdots & \vdots & \vdots & \vdots & \cdots & \vdots & \cdots & \vdots & \cdots & \vdots\\
\varepsilon_{p1}\varepsilon_{11} & \varepsilon_{p1}\varepsilon_{12} & \varepsilon_{p1}\varepsilon_{13} & \cdots & \varepsilon_{p1}\varepsilon_{1n} & \varepsilon_{p1}\varepsilon_{21} & \varepsilon_{p1}\varepsilon_{22} & \cdots & \varepsilon_{p1}\varepsilon_{2n} & \cdots & \varepsilon_{p1}^2 & \cdots & \varepsilon_{p1}\varepsilon_{pn}\\
\vdots & \vdots & \vdots & \cdots & \vdots & \vdots & \vdots & \cdots & \vdots & \cdots & \vdots & \cdots & \vdots\\
\varepsilon_{pn}\varepsilon_{11} & \varepsilon_{pn}\varepsilon_{12} & \varepsilon_{pn}\varepsilon_{13} & \cdots & \varepsilon_{pn}\varepsilon_{1n} & \varepsilon_{pn}\varepsilon_{21} & \varepsilon_{pn}\varepsilon_{22} & \cdots & \varepsilon_{pn}\varepsilon_{2n} & \cdots & \varepsilon_{pn}\varepsilon_{p1} & \cdots & \varepsilon_{pn}^2
\end{bmatrix}$$

$$=\begin{bmatrix}\Sigma_1 & \Sigma_{12} & \cdots & \Sigma_{1p}\\ \Sigma_{21} & \Sigma_2 & \cdots & \Sigma_{2p}\\ \vdots & \vdots & \ddots & \vdots\\ \Sigma_{p1} & \Sigma_{p2} & \cdots & \Sigma_p\end{bmatrix} \quad (6)$$

where $\Sigma_r=diag\left(\sigma_{r1}^2,\sigma_{r2}^2,...,\sigma_{rn}^2\right)$, $\Sigma_{rs}=\Sigma_{sr}=diag\left(\sigma_{(rs)1},\sigma_{(rs)2},...,\sigma_{(rs)n}\right)$

Furthermore, by applying Taylor expansion we get $f_r(t)$, $r=1,2,...,p$ as follows:

$$f_r(t)=\sum_{j=0}^{d_r}\frac{f_r^{(j)}(t_0)}{j!}(t-t_0)^j=\sum_{j=0}^{d_r}\beta_{rj}(t_0)(t-t_0)^j,\quad r=1,2,...,p \quad (7)$$

where d_r is order of polynomial, $\beta_{rj}(t_0)=\dfrac{\left(f_r^{(j)}(t_0)\right)}{j!}$ and $t\in(t_0-h,t_0+h)$.

We can express equation (7) in the following matrix notation:

$$\underset{\sim}{f}(t)=X_{t_0}^*\underset{\sim}{\beta}(t_0) \quad (8)$$

where $X_{t_0}^*=diag\left(\underset{\sim}{x}_{1t_0},\underset{\sim}{x}_{2t_0},...,\underset{\sim}{x}_{pt_0}\right)$; $\underset{\sim}{x}_{rt_0}=\left(1\quad (t-t_0)\quad\cdots\quad (t-t_0)^{d_r}\right)$, $r=1,2,...,p$;

$\underset{\sim}{\beta}(t_0)=\left(\underset{\sim}{\beta}_1(t_0)\ \ \underset{\sim}{\beta}_2(t_0)\ \cdots\ \underset{\sim}{\beta}_p(t_0)\right)'$; and $\underset{\sim}{\beta}_r(t_0)=\left(\beta_{r0}(t_0)\ \ \beta_{r1}(t_0)\ \cdots\ \beta_{rd_r}(t_0)\right)'$.

Based on equation (7) and by using n samples of paired observation (t_i , $y_{1i}, y_{2i},..., y_{pi}$), we can write equation (8) as follows:

$$\underset{\sim}{f}(\underset{\sim}{t})=\boldsymbol{X}_{t_0}\underset{\sim}{\beta}(t_0) \tag{9}$$

where $\underset{\sim}{f}(\underset{\sim}{t})=\left(f_1(\underset{\sim}{t}), f_2(\underset{\sim}{t}),..., f_p(\underset{\sim}{t})\right)'$; $f_r(\underset{\sim}{t})=\left(f_r(t_1)\ \ f_r(t_2)\ \cdots\ f_r(t_n)\right)'$;

$$\boldsymbol{X}_{t_0}=diag\left(\boldsymbol{X}_{1t_0},\boldsymbol{X}_{2t_0},...,\boldsymbol{X}_{pt_0}\right) \text{ and } \boldsymbol{X}_{rt_0}=\begin{bmatrix}1 & (t_1-t_0) & \cdots & (t_1-t_0)^{d_r}\\ 1 & (t_2-t_0) & \cdots & (t_2-t_0)^{d_r}\\ \vdots & \vdots & \ddots & \vdots\\ 1 & (t_n-t_0) & \cdots & (t_n-t_0)^{d_r}\end{bmatrix}.$$

Based on (9) we can write equation (1) as follows:

$$\underset{\sim}{Y}=\boldsymbol{X}_{t_0}\underset{\sim}{\beta}(t_0)+\underset{\sim}{\varepsilon} \tag{10}$$

where $\underset{\sim}{Y}=\left(\underset{\sim}{y}_1,\underset{\sim}{y}_2,...,\underset{\sim}{y}_p\right)'$ and $\underset{\sim}{y}_r=\left(y_{r1},y_{r2},...,y_{rn}\right)'$.

In estimating the regression function in model (1) based on multireponse local polynomial, we use weighted least squared optimization as follows:

$$\underset{\underset{\sim}{\beta}(t_0)}{Min}\, Q\left(\underset{\sim}{\beta}(t_0)\right)=\underset{\underset{\sim}{\beta}(t_0)}{Min}\left(\underset{\sim}{Y}-X_{t_0}\underset{\sim}{\beta}(t_0)\right)'\boldsymbol{V}^{-1}\boldsymbol{K}_h(t_0)\left(\underset{\sim}{Y}-X_{t_0}\underset{\sim}{\beta}(t_0)\right) \tag{11}$$

where $\boldsymbol{K}_h(t_0)=diag\left(\underset{\sim}{K}_{h1}(t_0),\underset{\sim}{K}_{h2}(t_0),...,\underset{\sim}{K}_{hp}(t_0)\right)$,

$\underset{\sim}{K}_{hr}(t_0)=\left(K_{hr}(t_1-t_0),K_{hr}(t_2-t_0),...,K_{hr}(t_n-t_0)\right)'$ and matrix $\boldsymbol{V}$ is block diagonal matrix of error given in (6) so it is invertible matrix. So, we get solution of (11) as follows:

$$\hat{\underset{\sim}{\beta}}(t_0)=(X'_{t_0}\boldsymbol{V}^{-1}\boldsymbol{K}_h(t_0)X_{t_0})^{-1}X'_{t_0}\boldsymbol{V}^{-1}\boldsymbol{K}_h(t_0)\underset{\sim}{Y}. \tag{12}$$

Based on (8) and (12) we obtain the estimated of regression function $\underset{\sim}{f}(t)$ as follows:

$$\hat{\underset{\sim}{f}}(t)=X^*_{t_0}\hat{\underset{\sim}{\beta}}(t_0)=X^*_{t_0}\left(X'_{t_0}\boldsymbol{V}^{-1}\boldsymbol{K}_h(t_0)X_{t_0}\right)^{-1}X'_{t_0}\boldsymbol{V}^{-1}\boldsymbol{K}_h(t_0)\underset{\sim}{Y},\ t\in(t_0-h,t_0+h)$$

Therefore, for $t=t_0$ we have $\hat{\underset{\sim}{f}}(t_0)$ as follows:

$$\hat{\underset{\sim}{f}}(t_0)=\boldsymbol{e}\left(X'_{t_0}\boldsymbol{V}^{-1}\boldsymbol{K}_{h_1}(t_0)X_{t_0}\right)^{-1}X'_{t_0}\boldsymbol{V}^{-1}\boldsymbol{K}_h(t_0)\underset{\sim}{Y} \tag{13}$$

where $\boldsymbol{e}=\left(\underset{\sim}{e}_1,\underset{\sim}{e}_2,...,\underset{\sim}{e}_p\right)'$ and $\underset{\sim}{e}_r\ (r=1,2,...,p)$ is row matrix with dimension $\left(p+\sum_{r=1}^{p}d_r\right)$, and d_r is polynomial order of r-th response. For $p=1$, $\underset{\sim}{e}_r$ equals to 1 at the first element, and equals to zero elsewhere. Next for $p\geq 2$, $\underset{\sim}{e}_r$ equals to 1 at $\left[\left(\sum_{r=2}^{p}d_{r-1}\right)+p\right]$-th element, and equals to zero elsewhere. Finally, based on multi-response local polynomial estimator we obtain estimation of regression function $\underset{\sim}{f}(\underset{\sim}{t})=\left(\underset{\sim}{f}(t_1),\underset{\sim}{f}(t_2),...,\underset{\sim}{f}(t_n)\right)'$ as follows:

$$\hat{\underset{\sim}{f}}(\underset{\sim}{t})=A(h)\underset{\sim}{Y} \tag{14}$$

where $\boldsymbol{A}(h)=\left(\boldsymbol{A}_h(t_1),\boldsymbol{A}_h(t_2),\ldots,\boldsymbol{A}_h(t_n)\right)'$ and $A_h(t_i)=\boldsymbol{e}\left(X_{t_i}'\boldsymbol{V}^{-1}\boldsymbol{K}_h(t_i)X_{t_i}\right)^{-1}X_{t_i}'\boldsymbol{V}^{-1}\boldsymbol{K}_h(t_i)$, $i=1,2,\ldots,n$.

3.2. Estimation of covariance matrix by using multi-response local polynomial estimator

In equation (13), $\hat{\underset{\sim}{f}}(t_0)$ contains unknown covariance matrix $\boldsymbol{V}$ which must be estimated. In this section we discuss estimation of covariance matrix. To obtain estimated covariance matrix, firstly we consider the following optimization:

$$\underset{\beta_r^*(t_0)}{Min}\,Q^*(t_0)=\underset{\beta_r^*(t_0)}{Min}\left(\underset{\sim}{Y}_r-X_{rt_0}\underset{\sim}{\beta}_r^*(t_0)\right)'\boldsymbol{K}_{rh}(t_0)\left(\underset{\sim}{Y}_r-X_{rt_0}\underset{\sim}{\beta}_r^*(t_0)\right) \quad (15)$$

So, we get solution of (15) as follows:

$$\hat{\underset{\sim}{\beta}}_r^*(t_0)=(X_{rt_0}'\boldsymbol{K}_{rh}(t_0)X_{rt_0})^{-1}X_{rt_0}'\boldsymbol{K}_{rh}(t_0)\underset{\sim}{Y}_r \quad (16)$$

and

$$E(\underset{\sim}{Y}_r)=\boldsymbol{X}_{rt_0}\underset{\sim}{\beta}_r^*(t_0)=\boldsymbol{X}_{rt_0}\left(\boldsymbol{X}_{rt_0}'\boldsymbol{K}_{rh}(t_0)\boldsymbol{X}_{rt_0}\right)^{-1}\boldsymbol{X}_{rt_0}\boldsymbol{K}_{rh}(t_0)\underset{\sim}{Y}=\boldsymbol{A}_{rt_0}\underset{\sim}{Y}_r$$

where $\boldsymbol{A}_{rt_0}=\boldsymbol{X}_{rt_0}\left(\boldsymbol{X}_{rt_0}'\boldsymbol{K}_{rh}(t_0)\boldsymbol{X}_{rt_0}\right)^{-1}\boldsymbol{X}_{rt_0}\boldsymbol{K}_{rh}(t_0)$.

By assuming $E\left(\varepsilon_{rt_0}^*\right)=0$, $Var\left(\varepsilon_{rt_0}^*\right)=\sigma_{rt_0}^2$, we estimate $\sigma_{rt_0}^2$ that is obtained from $E\left[\underset{\sim}{\varepsilon}_{rt_0}^{*}{}'\underset{\sim}{\varepsilon}_{rt_0}^*\right]$ as follows:

$$\begin{aligned}E\left(\hat{\sigma}_{rt_0}^2\right)&=E\left[\underset{\sim}{\varepsilon}_{rt_0}^{*}{}'\underset{\sim}{\varepsilon}_{rt_0}^*\right]=E\left[\left(\underset{\sim}{Y}_r-E(\underset{\sim}{Y}_r)\right)'\left(\underset{\sim}{Y}_r-E(\underset{\sim}{Y}_r)\right)\right]\\&=E\left[\left(\underset{\sim}{Y}_r-\boldsymbol{A}_{rt_0}\underset{\sim}{Y}_r\right)'\left(\underset{\sim}{Y}_r-\boldsymbol{A}_{rt_0}\underset{\sim}{Y}_r\right)\right]=E\left[\underset{\sim}{Y}_r'\left(\boldsymbol{I}-\boldsymbol{A}_{rt_0}\right)'\left(\boldsymbol{I}-\boldsymbol{A}_{rt_0}\right)\underset{\sim}{Y}_r\right]\\&=E\left[\underset{\sim}{Y}_r'\boldsymbol{Q}_{rr(t_0)}\underset{\sim}{Y}_r\right]\end{aligned} \quad (17)$$

where $\boldsymbol{Q}_{rr(t_0)}=\left(\boldsymbol{I}-\boldsymbol{A}_{rt_0}\right)'\left(\boldsymbol{I}-\boldsymbol{A}_{rt_0}\right)$.

Further, equation can be written as follows:

$$\begin{aligned}E\left(\hat{\sigma}_{rt_0}^2\right)&=tr\left(\boldsymbol{Q}_{(rr)t_0}\sigma_{rt_0}^2\boldsymbol{I}_n\right)+\left(\boldsymbol{A}_{rt_0}\underset{\sim}{Y}_r\right)'\boldsymbol{Q}_{(rr)t_0}\left(\boldsymbol{A}_{rt_0}\underset{\sim}{Y}_r\right)=\sigma_{rt_0}^2tr\left(\boldsymbol{Q}_{(rr)t_0}\right)+\left(\boldsymbol{A}_{rt_0}\underset{\sim}{Y}_r\right)'\boldsymbol{Q}_{(rr)t_0}\left(\boldsymbol{A}_{rt_0}\underset{\sim}{Y}_r\right)\\&=\sigma_{rt_0}^2tr\left(\boldsymbol{Q}_{(rr)t_0}\right)+\underset{\sim}{Y}_r'\boldsymbol{A}_{rt_0}'\boldsymbol{Q}_{(rr)t_0}\boldsymbol{A}_{rt_0}\underset{\sim}{Y}_r=\sigma_{rt_0}^2tr\left(\boldsymbol{Q}_{(rr)t_0}\right)+\underset{\sim}{Y}_r'\boldsymbol{A}_{rt_0}'\left(\boldsymbol{I}-\boldsymbol{A}_{rt_0}\right)'\left(\boldsymbol{I}-\boldsymbol{A}_{rt_0}\right)\boldsymbol{A}_{rt_0}\underset{\sim}{Y}_r\\&=\sigma_{rt_0}^2tr\left(\boldsymbol{Q}_{(rr)t_0}\right)+\underset{\sim}{Y}_r'\boldsymbol{A}_{rt_0}'\left(\boldsymbol{I}-\boldsymbol{A}_{rt_0}\right)'\left(\boldsymbol{I}-\boldsymbol{A}_{rt_0}\right)\boldsymbol{A}_{rt_0}\underset{\sim}{Y}_r\\&=\sigma_{rt_0}^2tr\left(\boldsymbol{Q}_{(rr)t_0}\right)+\underset{\sim}{Y}_r'\boldsymbol{A}_{rt_0}'\left(\boldsymbol{I}-\boldsymbol{A}_{rt_0}\right)'\left(\boldsymbol{A}_{rt_0}-\boldsymbol{A}_{rt_0}\boldsymbol{A}_{rt_0}\right)\underset{\sim}{Y}_r\end{aligned} \quad (18)$$

Since matrix $\boldsymbol{A}_{rt_0}$ is idempotent, i.e.,

$$\begin{aligned}\boldsymbol{A}_{rt_0}\boldsymbol{A}_{rt_0}&=\boldsymbol{X}_{rt_0}\left(\boldsymbol{X}_{rt_0}'\boldsymbol{K}_h(t_0)\boldsymbol{X}_{rt_0}\right)^{-1}\boldsymbol{X}_{rt_0}\boldsymbol{K}_h(t_0)\boldsymbol{X}_{rt_0}\left(\boldsymbol{X}_{rt_0}'\boldsymbol{K}_h(t_0)\boldsymbol{X}_{rt_0}\right)^{-1}\boldsymbol{X}_{rt_0}\boldsymbol{K}_h(t_0)\\&=\boldsymbol{X}_{rt_0}\left(\boldsymbol{X}_{rt_0}'\boldsymbol{K}_h(t_0)\boldsymbol{X}_{rt_0}\right)^{-1}\boldsymbol{X}_{rt_0}\boldsymbol{K}_h(t_0)=\boldsymbol{A}_{rt_0}\end{aligned}$$

then by considering equation (17), we can write equation (18) as follows:

$$E\left(\hat{\sigma}_{rt_0}^2\right)=\sigma_{rt_0}^2 tr\left(\boldsymbol{Q}_{(rr)t_0}\right)+\underset{\sim}{Y}_r' \boldsymbol{A}_{rt_0}'\left(\boldsymbol{I}-\boldsymbol{A}_{rt_0}\right)'\left(\boldsymbol{A}_{rt_0}-\boldsymbol{A}_{rt_0}\right)\underset{\sim}{Y}_r$$
$$=\sigma_{rt_0}^2 tr\left(\boldsymbol{Q}_{rr(t_0)}\right)$$

So, we obtain $\hat{\sigma}_{rt_0}^2=\dfrac{\underset{\sim}{Y}_r'\boldsymbol{Q}_{rr(t_0)}\underset{\sim}{Y}_r}{tr\left(\boldsymbol{Q}_{(rr)t_0}\right)}=\dfrac{\underset{\sim}{Y}_r'\left(\boldsymbol{I}-\boldsymbol{A}_{rt_0}\right)'\left(\boldsymbol{I}-\boldsymbol{A}_{rt_0}\right)\underset{\sim}{Y}_r}{tr\left(\left(\boldsymbol{I}-\boldsymbol{A}_{rt_0}\right)'\left(\boldsymbol{I}-\boldsymbol{A}_{rt_0}\right)\right)}$

On the other hand, we have:

$$E\left(\hat{\sigma}_{(rs)t_0}^2\right)=E\left[\left(\underset{\sim}{\varepsilon}_{rt_0}^*\right)'\left(\underset{\sim}{\varepsilon}_{st_0}^*\right)\right]=E\left[\left(\underset{\sim}{Y}_r-E(\underset{\sim}{Y}_r)\right)'\left(\underset{\sim}{Y}_s-E(\underset{\sim}{Y}_s)\right)\right]=E\left[\underset{\sim}{Y}_r'\left(\boldsymbol{I}-\boldsymbol{A}_{rt_0}\right)'\left(\boldsymbol{I}-\boldsymbol{A}_{st_0}\right)\underset{\sim}{Y}_s\right]$$
$$=E\left[\underset{\sim}{Y}_r'\boldsymbol{Q}_{(rs)t_0}\underset{\sim}{Y}_r\right]$$

where $\boldsymbol{Q}_{(rs)t_0}=\left(\boldsymbol{I}-\boldsymbol{A}_{rt_0}\right)'\left(\boldsymbol{I}-\boldsymbol{A}_{st_0}\right)$.

Therefore, we get:

$$E\left(\hat{\sigma}_{(rs)t_0}\right)=tr\left(\boldsymbol{Q}_{(rs)t_0}\sigma_{(rs)t_0}\boldsymbol{I}_n\right)+\left(\boldsymbol{A}_{rt_0}\underset{\sim}{Y}_r\right)'\boldsymbol{Q}_{(rs)t_0}\left(\boldsymbol{A}_{st_0}\underset{\sim}{Y}_s\right)$$
$$=\sigma_{(rs)t_0}tr\left(\boldsymbol{Q}_{(rs)t_0}\right)+\underset{\sim}{Y}_r'\boldsymbol{A}_{rt_0}'\left(\boldsymbol{I}-\boldsymbol{A}_{rt_0}\right)'\left(\boldsymbol{A}_{st_0}-\boldsymbol{A}_{st_0}\boldsymbol{A}_{st_0}\right)\underset{\sim}{Y}_s$$
$$=\sigma_{(rs)t_0}tr\left(\boldsymbol{Q}_{(rs)t_0}\right)$$

Thus, we obtain:

$$\hat{\sigma}_{(rs)t_0}=\frac{\underset{\sim}{Y}_r'\boldsymbol{Q}_{(rs)t_0}\underset{\sim}{Y}_s}{tr\left(\boldsymbol{Q}_{(rs)t_0}\right)}=\frac{\underset{\sim}{Y}_r'\left(\boldsymbol{I}-\boldsymbol{A}_{rt_0}\right)'\left(\boldsymbol{I}-\boldsymbol{A}_{st_0}\right)\underset{\sim}{Y}_s}{tr\left(\left(\boldsymbol{I}-\boldsymbol{A}_{rt_0}\right)'\left(\boldsymbol{I}-\boldsymbol{A}_{st_0}\right)\right)}$$

Next, for $t_0=t_i$, $i=1,2,\ldots,n$, we have:

$$\hat{\sigma}_{ri}^2=\frac{\underset{\sim}{Y}_r'\boldsymbol{Q}_{rr(t_i)}\underset{\sim}{Y}_r}{tr\left(\boldsymbol{Q}_{rr(t_i)}\right)}=\frac{\underset{\sim}{Y}_r'\left(\boldsymbol{I}-\boldsymbol{A}_{rt_i}\right)'\left(\boldsymbol{I}-\boldsymbol{A}_{rt_i}\right)\underset{\sim}{Y}_r}{tr\left(\left(\boldsymbol{I}-\boldsymbol{A}_{rt_i}\right)'\left(\boldsymbol{I}-\boldsymbol{A}_{rt_i}\right)\right)},$$

and for $r\neq s$, we have:

$$\hat{\sigma}_{(rs)i}=\frac{\underset{\sim}{Y}_r'\boldsymbol{Q}_{(rs)t_i}\underset{\sim}{Y}_s}{tr\left(\boldsymbol{Q}_{(rs)t_i}\right)}=\frac{\underset{\sim}{Y}_r'\left(\boldsymbol{I}-\boldsymbol{A}_{rt_i}\right)'\left(\boldsymbol{I}-\boldsymbol{A}_{st_i}\right)\underset{\sim}{Y}_s}{tr\left(\left(\boldsymbol{I}-\boldsymbol{A}_{rt_i}\right)'\left(\boldsymbol{I}-\boldsymbol{A}_{st_i}\right)\right)}$$

where $r=s=1,2,3,\ldots,p$, $\boldsymbol{A}_{rt_i}=\boldsymbol{X}_{rt_i}\left(\boldsymbol{X}_{rt_i}'\boldsymbol{K}_{rh}(t_i)\boldsymbol{X}_{rt_i}\right)^{-1}\boldsymbol{X}_{rt_0}\boldsymbol{K}_{rh}(t_i)$.

Therefore, we obtain estimation of covariance matrix $\boldsymbol{V}$ as follows:

$$\hat{\mathbf{V}} = \begin{bmatrix} \hat{\Sigma}_1 & \hat{\Sigma}_{12} & \cdots & \hat{\Sigma}_{1p} \\ \hat{\Sigma}_{21} & \hat{\Sigma}_2 & \cdots & \hat{\Sigma}_{2p} \\ \vdots & \vdots & \ddots & \vdots \\ \hat{\Sigma}_{p1} & \hat{\Sigma}_{p2} & \cdots & \hat{\Sigma}_p \end{bmatrix} \quad (19)$$

where $\hat{\Sigma}_r = diag\left(\hat{\sigma}_{r1}^2, \hat{\sigma}_{r2}^2, \ldots, \hat{\sigma}_{rn}^2\right) = diag\left(\frac{\underset{\sim}{Y}_r' \boldsymbol{Q}_{(rr)t_1} \underset{\sim}{Y}_r}{tr\left(\boldsymbol{Q}_{(rr)t_1}\right)}, \frac{\underset{\sim}{Y}_r' \boldsymbol{Q}_{(rr)t_2} \underset{\sim}{Y}_r}{tr\left(\boldsymbol{Q}_{(rr)t_2}\right)}, \ldots, \frac{\underset{\sim}{Y}_r' \boldsymbol{Q}_{(rr)t_n} \underset{\sim}{Y}_r}{tr\left(\boldsymbol{Q}_{(rr)t_n}\right)}\right)$;

$$\hat{\Sigma}_{rs} = \hat{\Sigma}_{sr} = diag\left(\hat{\sigma}_{(rs)1}, \hat{\sigma}_{(rs)2}, \ldots, \hat{\sigma}_{(rs)n}\right) = diag\left(\frac{\underset{\sim}{Y}_r' \boldsymbol{Q}_{(rs)t_1} \underset{\sim}{Y}_s}{tr\left(\boldsymbol{Q}_{(rs)t_1}\right)}, \frac{\underset{\sim}{Y}_r' \boldsymbol{Q}_{(rs)t_2} \underset{\sim}{Y}_s}{tr\left(\boldsymbol{Q}_{(rs)t_2}\right)}, \ldots, \frac{\underset{\sim}{Y}_r' \boldsymbol{Q}_{(rs)t_n} \underset{\sim}{Y}_s}{tr\left(\boldsymbol{Q}_{(rs)t_n}\right)}\right);$$

and $r = s = 1, 2, 3, \ldots, p$.

3. Conclusion

For designing the children growth charts, we estimate them simultaneously and need the weight matrix, i.e., inverse of covariance matrix of error. By using multi-reponse local polynomial, we obtain the estimated regression function of the model as given in equation (14), and we obtain the estimated covariance matrix as given in equation (19). Furthermore, in the next our research, the children growth charts can be used for assessing nutritional status of children.

Acknowledgments

Many thanks to Director of the Directorate of Research and Public Service, the Directorate General of Reinforcing of Research and Development, the Ministry of Research, Technology, and Higher Education of the Republic of Indonesia for funding this research via the University Excellence Fundamental Research Grant (*Hibah Penelitian Dasar Unggulan Perguruan Tinggi – Hibah PDUPT*) in the fiscal year 2019.

References

[1] Adiwati, T and Chamidah, N. 2019 Modelling of Hypertension Risk Factors Using Penalized Spline to Prevent Hypertension in Indonesia *IOP Conf. Series: Materials Science and Engineering* 546 052003

[2] Massaid A Hanif M, Febrianti D and Chamidah N 2019 Modelling of Poverty Percentage of Non-Food Per Capita Expenditures in Indonesia Using Least Square Spline Estimator IOP Conf. Series: Materials Science and Engineering 546 052044

[3] Ramadan W, Chamidah N, Zaman B, Muniroh L, and Lestari B 2019 Standard Growth Chart of Weight for Height to Determine Wasting Nutritional Status in East Java Based on Semiparametric Least Square Spline Estimator IOP Conf. Series: Materials Science and Engineering 546 052063

[4] Oktavitri N.I, Kuncoro E P, Purnobasuki H and Chamidah N 2019 Prediction of suspended and attached process behavior in anaerobic batch reactor using nonparametric regression model approach based on spline estimator Eco. Env. & Cons. 25 (April Suppl. Issue) pp. S96-S1

[5] Chamidah N, Tjahjono E, Fadilah A.R, Lestari B 2018 Standard Growth Charts for Weight of Children in East Java Using Local Linear Estimator. J. Phys: Conf. Ser 1097 012092

[6] Puspitawati A and Chamidah N 2019 Choroidal Neovascularisation Classification on Fundus Retinal Images Using Local Linear Estimator IOP Conf. Series: Materials Science and Engineering 546 052056

[7] Chamidah N, Gusti K H, Tjahjono E, and Lestari B 2019 Improving of Classification Accuracy of Cyst and Tumor Using Local Polynomial Estimator TELKOMNIKA 17(3) pp 1492-1500

[8] Darnah, Utoyo M.I and Chamidah N 2019 Modeling of Maternal Mortality and Infant Mortality Cases in East Kalimantan using Poisson Regression Approach Based on Local Linear Estimator IOP Conf. Series: Earth and Environmental Science 243 012023
[9] Hallin M, Lu Z, and Tran L T 2004 Local Linear Spatial Regression The Annals of Statistics 32 pp 2469-2500.
[10] Baillo A, and Grane A 2009 Local Linear Regression for Functional Predictor and Scala Response Journal of Multivariate Analysis 100 pp 102-111.
[11] Regonda S K, Rajagopalan B, Lall U, Clark M, and Moon Y-I 2005 Local Polynomial Method for Ensemble Forecast of Time Series Nonlinear Processes in Geophysics 12 pp 397-406.
[12] Liang H, and Chen D Z 2005 Assessment of Esophageal Pressure in Gastroesophageal Reflux Disease by Local Regression Annals of Biomedical Engineering 33 pp 847-853.
[13] Wang Y, Guo W and Brown W B 2000 Spline smoothing for bivariate data with applications to association between hormones Statistica Sinica 10 pp 377-397.
[14] Chamidah N and Lestari B 2016 Spline Estimator in Homoscedastic Multi-Response Nonparametric Regression Model in Case of Unbalanced Number of Observations Far East Journal of Mathematical Sciences (FJMS) 100 pp 1433-1453.
[15] Lestari B, Anggraeni D and Saifudin T 2018 Estimation of Covariance Matrix based on Spline Estimator in Homoscedastic Multi-Responses Nonparametric Regression Model in Case of Unbalance Number of Observations Far East J. Math. Sciences (FJMS) 108(2) pp 341-355.
[16] Islamiyati A, Fatmawati and Chamidah N 2018 Estimation of Covariance Matrix on Bi-Response Longitudinal Data Analysis with Penalized Spline Regression Journal of Physics: Conf. Series 979 pp. 012093.
[17] Lestari B, Fatmawati and Budiantara I N 2019 Spline Estimator and Its Asymptotic Properties in Multi-response Nonparametric Regression Model Songklanakarin Journal of Science and Technology In Press.
[18] Welsh A H, and Yee T W 2006 Local Regression for Vector Responses Journal of Statistical Planning and Inference 136 pp 3007-3031.
[19] Chamidah N and Rifada M 2016 Estimation of median growth curves for children up two years old based on biresponse local linear estimator AIP Conference Proceedings 1718 p110001.
[20] Nidhomuddin, Chamidah N, Kurniawan N 2019 Admission Test Modelling of State Islamic College in Indonesia Using Local Linear for Bivariate Longitudinal Data IOP Conf. Series: Materials Science and Engineering 546 052047
[21] Chamidah N, Kurniawan A, Zaman B, and Muniroh L 2018 Least Square-Spline Estimator in Multi-response Semiparametric Regression Model for Estimating Median Growth Charts of Children in East Java, Indonesia Far East J. Math. Sciences (FJMS)107(2) pp 295-307.
[22] Hidayati L, Chamidah N and Budiantara I N 2019 Spline Truncated Estimator in Multiresponse Semiparametric Regression Model for Computer based National Exam in West Nusa Tenggara IOP Conf. Series: Materials Science and Engineering 546 052029.
[23] Chamidah N and Rifada M 2016 Local linear estimator in bi-response semiparametric regression model for estimating median growth charts of children Far East Journal of Mathematical Sciences 99 (8) pp 1233-1244
[24] Narendra M B, Sularyo T S, Soetjiningsih, Suyitno H, and Ranuh I G N G 2002 Tumbuh Kembang Anak dan Remaja CV. Sagung Seto Jakarta.
[25] Chen M-Y, and Peng L 2006 Simple and efficient improvement of multivariate local linear regression Journal of Multivariate Analysis 97 pp 1501-1524.

A comparison of extreme gradient boosting, SARIMA, exponential smoothing, and neural network models for forecasting rainfall data

R Agata* **and I G N M Jaya**

Department of Statistics, Universitas Padjadjaran, Indonesia

*Corresponding author e-mail: ruthagataprb@gmail.com

Abstract. Extreme gradient boosting, is a combination of gradient descent and boosting that can be used to build an optimal model for time series data. This method was used to forecast the rainfall data in city of Bandung for period 2018-2019 and compared to Seasonal Autoregressive Integrated Moving Average (SARIMA) exponential smoothing, and artificial neural network which were used as benchmarks. Data used in this study were monthly rainfall from 2000 through 2017. The extreme gradient boosting had the lowest mean absolute deviance, root mean squared error deviance, and mean absolute percentage error. This indicates the extreme gradient boosting model performed better than the SARIMA, exponential smoothing, and neural network. Based on the extreme gradient boosting model, it is concluded that the highest rainfall will occur between September 2018 and May as a rainy season in Bandung, and the lowest rainfall will occur between June and August as a dry season.

1. Introduction

Rainfall is the most important climate element affecting human daily life with very complex phenomenon and varies by time and space. However, rainfall is predictable [1], [2]. Developing the accurate forecasting methods of rainfall becomes an interesting topic [3] and becomes the biggest difficulty from different areas such as weather and data mining [4], environmental [5], operational hydrology [6], and statistical forecasting [7]. The objective is to obtain the accurate future prediction when the series data have a complex pattern. Small error with high power is important part in evaluating forecasting models. Several methods have been developed for univariate time series data such as Box-Jenkin ARIMA and exponential smoothing models. ARIMA estimates the model based on the autocorrelation function while exponential smoothing models are based on decompose trend and seasonality in the data ([8], [9], [10]). Both methods are developed by using linear models where the prediction is a weighted linear sum of recent past observation or lag [8]. If the deterministic process can be identified easily, the accurate predicted can be obtained.

However, the linear models will be successfully applied to data if data generated from deterministic processes are not a random process [11]. Rainfall data are random process and have non-linear pattern. A non-linear model may be better than ARIMA and exponential smoothing models for forecasting purpose.

A hybrid multi-model method such as artificial neural network (ANN) and gradient boosting were developed to overcome the non-linearity problem and random process in time data [3]. The main advantage of neural network lies in its ability to capture any arbitrary function such as nonlinear function ([12], [13]). However, the "black box" nature in neural network becomes the problem because there is no information provided how and why the neural network came up with a specific output. In some applications the estimation process becomes the most important step. The bad or good result can be identified from the estimation process. Extreme gradient boosting (XGBoost) was developed to captures the interaction and nonlinearity in data structured where the estimation process can be known [14]. Using XGBoost, the most important structure that explains the generating series data can be defined.

The aim of this study is to examine whether extreme gradient boosting can outperform the classical ARIMA and exponential smoothing and the other machine learning artificial neural network for rainfall forecasting. This research provides more evidences as the usefulness of extreme gradient boosting for rainfall forecasting by performing training and testing data method. The comparison method is not done by simulation procedure. We use the real data and compare the model quality based on several criterion including mean absolute deviation (MAD), root mean squared error deviance (RMSED), and mean absolute percentage error (MAPE).

Here we focus on several alternative methods where the software was available in R-library. We believe this research is needed by practitioners due to the limitations in understanding statistics and being better able to operate computer programs.

The paper is structured as follows: in Section 2, we define data and methodology, move on with result and discussion in Section 3. Finally, Section 4 concludes the paper with final remarks.

2. Material and Method

2.1. Material

We obtained historical data of average monthly rainfall for Bandung city over the period of 2000-2015 for the purpose of model identification and those of 2016-2017 for forecast validation of the chosen model from the National Metrological Center, Bandung-Indonesia. The data were presented in Figure 1.

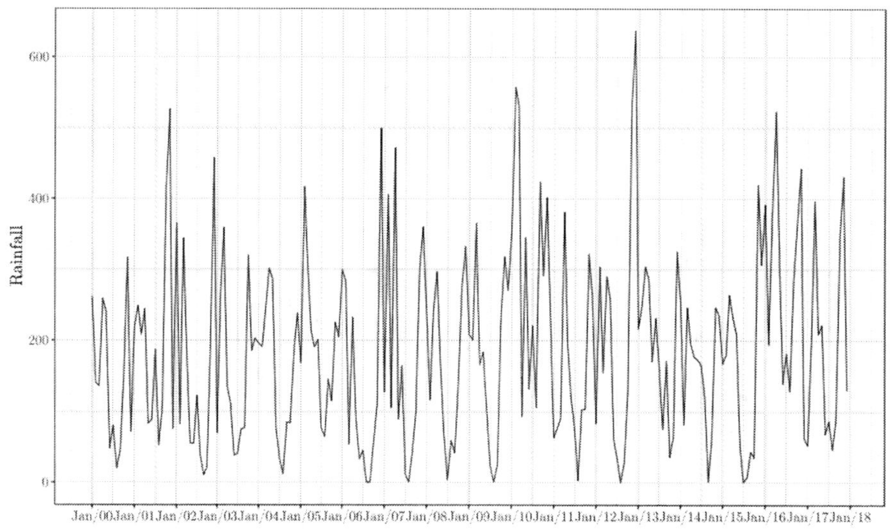

Figure 1. Rainfall data in Bandung city, 2000-2017

Figure 1 presents the temporal pattern of monthly rainfall data, 2000-2017. We observed that there is no temporal trend however, it clearly presents seasonal pattern.

2.2. Method

2.2.1. ARIMA

Box-Jenkins ARIMA is the most popular ordinary time series forecasting technique for stationary and non-stationarity data. The aim is to estimate an autoregressive integrated moving average (ARIMA) model which is used to explains the temporal pattern of rainfall data. The basic Box-Jenkins expressed by the following form [11]:

$$y_t = \alpha_0 + \sum_{i=1}^{p} \alpha_i y_{t-i} - \sum_{j=1}^{q} \beta_j e_{t-j}, i = 1, \dots, p \text{ and j} = 0,1, \dots, q \quad (1)$$

where y_t, denotes a rainfall data, α_0 is a intercept term, and e_t is error term. The error term usually assume to be withe noise. Then, the i and j terms are autoregressive and moving average parts of the model. Eq.1 classified as an $ARMA(p,q)$ model, where p denotes the number of lag, and q is the number of lagged forecast errors. Integrated with number differencing d may be applied to satisfy the stationarity assumption. Eq.1 can be extended for seasonal model becomes $ARIMA(p,d,q)(P,D,Q)[s]$, where P,D,Q are similar to p,d,q but here as a part of seasonal components with s denotes the seasonal order. The t-test is used to evaluate the significance SARIMA parameters.

2.2.2. *Exponential Smoothing (ETS)*

Exponential smoothing (ETS) is the simplest deterministic method is usually used in forecasting with no clear trend or seasonal pattern were identified. The forecast is calculated using a linear weighted average that reduces exponentially as observations are associated with the earliest observation in the past-smallest weights.

$$y_{t+1|t} = \delta y_t + \delta(1-\delta)y_{t-1} + \delta(1-\delta)^2 y_{t-2} \dots \quad (2)$$

where $0 \le \delta \le 1$ denote the smoothing parameter. The forecast for time $t+1$ is a weighted average of all series, $y_1, \dots, y_T$. Parameter δ controls the rate at which the weights decrease [8].

2.2.3. *Artificial Neural Network (ANN)*

The most popular machine learning method for forecasting is artificial neural network (ANN). The ANN uses the technique of error back propagation to train the network configuration. The algorithm of the ANN consists hidden layers, output layer and number hidden layers, and a number of neurons in the input layer. The ANN forecasting model is defined as [3]:

$$y^F_{t+T+(m-1)\tau} = f(y_t, \omega, \theta, m, h) = \theta_0 + \sum_{j=1}^{h} \omega_j^{out} \phi\left(\sum_{i-1}^{m} \omega_{ji} y_{t+(i-1)\tau} + \theta_j\right) \quad (3)$$

where ϕ is the activation function, ω_{ji} denotes the weights defining the link between the i-th input layer and j-th the hidden layer; θ_j are the bias for the j-th node of the hidden layer; ω_j^{out} for the j-th node of the hidden layer and the node of the output layer; and θ_0 is the bias at the output node.

2.2.4. *Extreme Gradient Boosting (XGB)*

Extreme gradient boosting is one of the most loved gradient bosting algorithm. It can be used for regression, classification, and is constructed based on gradient boosting framework. Suppose we have a data set $D = \{x_t : y_t\}$ containing T observation, where x denotes a training variable and y is a target variable. The XGB approach assumes that there are G additive function of boosting as follows [14]:

$$\hat{y}_t = \sum_{g=1}^{G} f_g(x_t) \quad (5)$$

where $\hat{y}_t$ denotes the prediction for time t at g-th boost, where f_g is a tree structured L with leaf j having a weight score ω_j. The estimation process of the XGB approach is based on minimizing the loss function L_g as:

$$L_g = \sum_{t=1}^{T} l(y_t, \hat{y}_t) + \gamma T + \frac{1}{2}\lambda \|\omega\| \quad (6)$$

where l is a differentiable convex loss function that measures the different between prediction $\hat{y}_t$ and target value y_t. T denotes the number of leaves with coefficient γ and λ denotes the smoothing parameters. The following steps are involved in extreme gradient boosting:

(a) Initialize the boosting algorithm

$$F_0(y) = argmin_{(\gamma,\lambda)}\left(\sum_{t=1}^{T} l(y_t, \hat{y}_t) + \gamma T + \frac{1}{2}\lambda\|\omega\|\right) \quad (7)$$

(b) Compute the gradient of the loss function iteratively

$$r_{im} = -\alpha\left[\frac{\partial\left(L(y_t, F(y_t))\right)}{\partial F(y_t)}\right]_{F(y)=F_{m-1}(y)'} \quad (8)$$

where α denotes the learning rate
Fit each $h_m(y) = y - F_m(y)$ on the extreme gradient obtained

(c) Derive the multiplicative factor $(\gamma, \lambda)m$ for each terminal node
Define the boosted model $F_m(y)$ as follows

$$F_m(y) = F_{m-1}(y) + (\gamma, \lambda)_m h_m(y) \quad (9)$$

2.3. Forecast evaluation methods

In order to compare the performance of the extreme gradient boosting, artificial neural network, ARIMA, and exponential smoothing time series model in forecasting rainfall data, several criteria will be used. The first is mean absolute deviance (MAD), as given below [9]:

$$MAD = \frac{1}{H}\sum_{h=1}^{H}|y_{t+h} - \hat{y}_{t+h}| \quad (10)$$

where y_{t+h} denotes data testing at $t + h$ period and $\hat{y}_{t+h}$ is the forecast value for $t + h$ period with H is length of forecast. The second and third criteria are root means squared deviance (RMSED) and mean absolute percentage deviance (MAPE) given below:

$$RMSED = \sqrt{\frac{1}{H-1}\sum_{h=1}^{H}(y_{t+h} - \hat{y}_{t+h})^2} \quad (11)$$

$$MAPE = \frac{1}{H}\left(\sum_{h=1}^{H}\left|\frac{y_{t+h} - \hat{y}_{t+h}}{y_{t+h}}\right|\right)100\% \quad (12)$$

The smallest values of MAD, RMSED, and MAPE indicates the best model.
All computation process will be done by R-software using `forecast` and `xgboost` packages.

3. Results and discussion

Four different models were compared to determine the best for forecasting the rainfall data in city of Bandung, 2018-2019.

3.1. Data exploration

Diagnostic checking is the first step in order to define the temporal pattern in the data and to test the stationarity. We present the ACF plot for evaluating the temporal pattern and the Dicky Fuller test for testing stationarity assumption.

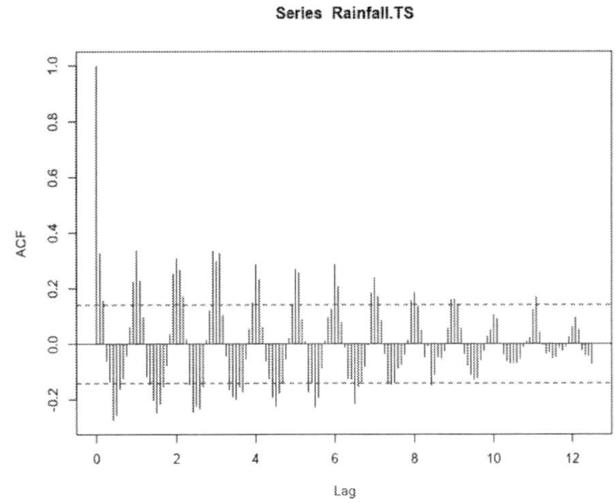

Figure 2. Autocorrelation function plot for training data (2000-2015)

Figure 2 shows the autocorrelation function plot with a clear seasonal pattern at time lag 12. It indicates that the rainfall data in the city of Bandung have seasonal pattern recuring every year (s=12 months).

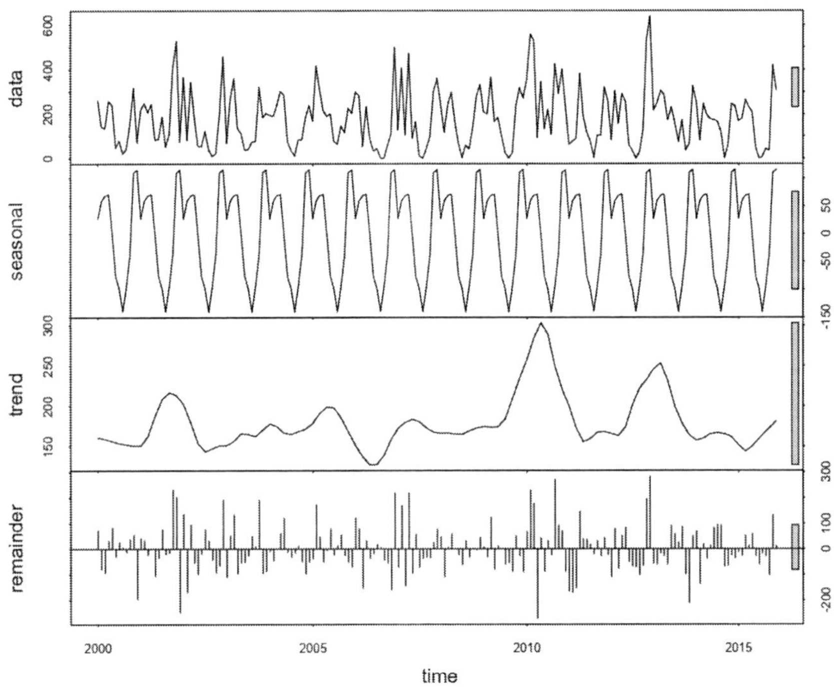

Figure 3. Time series decomposition (2000-2015)

Time series decomposition (Figure 3) supports the idea that the rainfall data have strongly seasonal pattern. There is no clear pattern of linear trend. The Dicky Fuller stationarity test rejects the null hypothesis (H0: data series is not stationary) with a p-value of 0.01. It concludes that the rainfall data were stationary. Based on this diagnostic, the seasonal component have to be included in the models. Using forecast package, we obtained the best model for rainfall data is ARIMA(1,0,0)(2,0,0)[12]. Parameter estimates are presented in Table 1.

Table 1. SARIMA estimates

Parameter	Estimate	Standard error	t-value	p-value
AR(1)	0.1998	0.0755	2.6464	0.0090
SAR(1)	0.2414	0.0709	3.4048	0.0008
SAR(2)	0.1951	0.0753	2.5910	0.0105

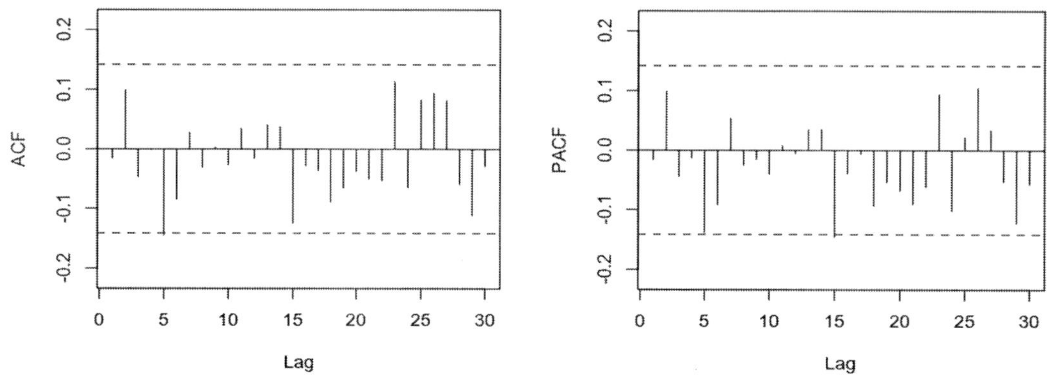

Figure 4. Residual analysis

Figure 4 presents the residual analysis where the ACF and PACF plots show that the residuals satisfy the white noise assumption. This means that the ARIMA (1,0,0)(2,0,0)[12] is the best model for the rainfall training data. The best model of ETS was obtained based on an additive error, without temporal trend and with additive seasonal with smoothing parameter of 1e-04, while the forecast with ANN was obtained based on 20 networks, each of which is a 13-7-1 network with 106 weights and seasonal lag is s=12. The model of extreme boosting gradient is constructed using 28 features. The most important feature with the highest gain is lag 12 and the second one is lag 11. It indicates that there is a seasonal component for order 12 and 11.

We present the model comparison based on the mean absolute deviance (MAD), root mean squared error deviance (RMSED) and mean absolute percentage error (MAPE) of testing data in Table 2.

Table 2. Model comparison

Model	MAD	RMSED	MAPE
SARIMA	116.9017	162.6560	65.7756
ETS	117.1898	158.5143	64.8473
ANN	122.8627	180.1054	69.6140
XGB	112.5225	152.8657	63.7136

Based on the Table 2, the best model for rainfall data is the extreme boosting gradients as it has the smallest MAD, RMSED, and MAPE values.

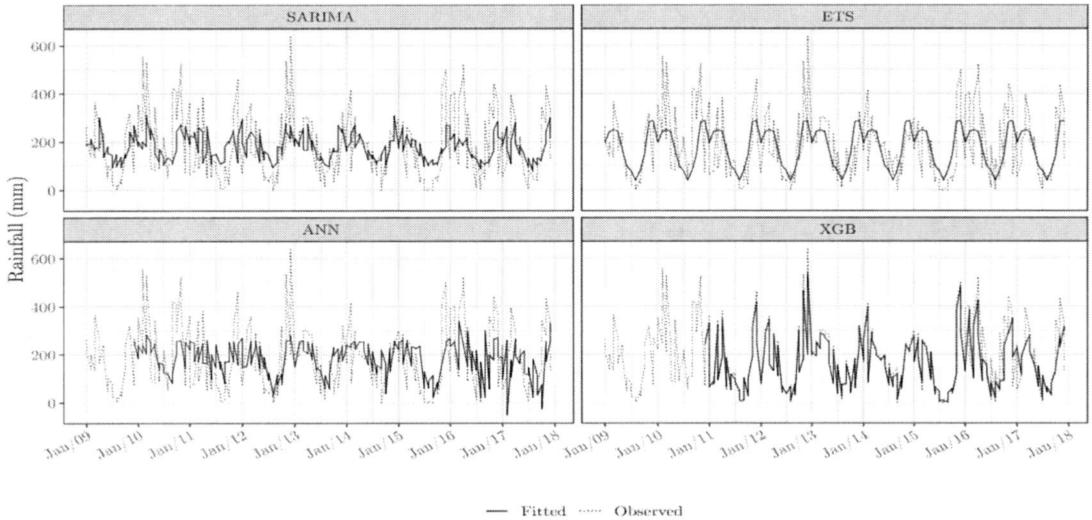

Figure 5. Observed versus fitted rainfall data

Figure 5 shows the observed versus fitted rainfall data by means of SARIMA, EST, ANN and XGB gradient model. Based on this result, the best fitting model is BOOSTING Gradient model where the observed and fitted values are relatively similar.

Forecast by means Extreme Boosting Gradient

Table 3. Forecasted value of rainfall (2018-2019)

Year	Jan	Feb	Mar	Apr	May	Jun	Jul	Aug	Sep	Oct	Nov	Dec
2018	144.73	190.59	279.26	299.08	201.52	83.25	100.52	74.72	147.23	198.06	317.76	178.32
2019	155.02	268.64	278.22	292.42	214.30	67.88	55.63	45.70	28.27	259.71	361.25	246.36

Table 3 shows the forecast values for 2018-2019. The low values of forecast were found around June – September which can be seen clearly in Figure 6.

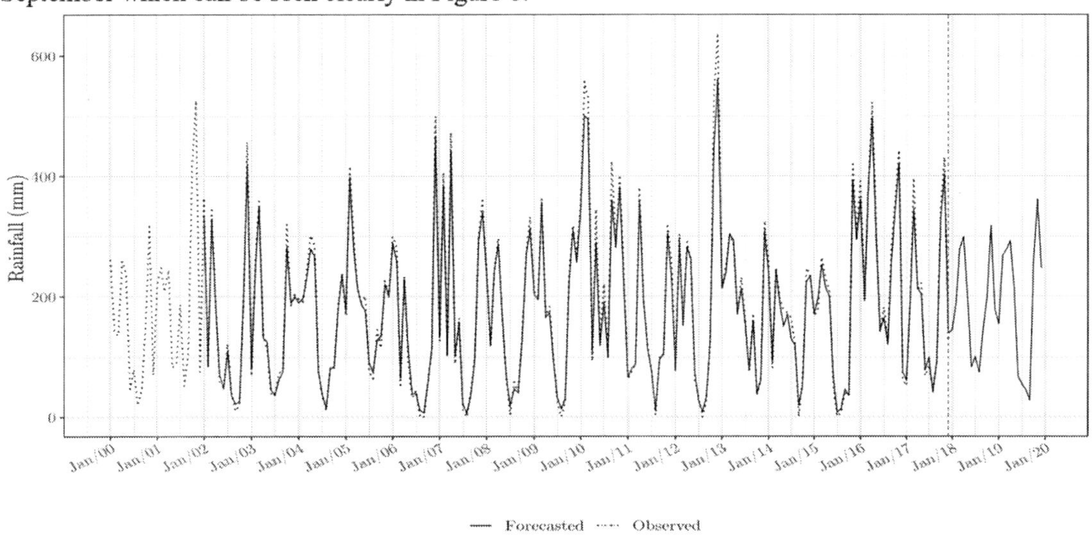

Figure 6. Forecasted rainfall data

Figure 6 shows the predicted (2000-2017) and forecasted value (2018-2019) of rainfall data in city of Bandung.

The forecast values of rainfall seem have similar pattern with rainfall in the previous years. The high rainfall will occur between September and May as rainy season in Bandung and the lowest rainfall will occur between June and August as dry season. The temporal effect has a significant effect on the rainfall. According to Jaya et al. [15], a temporal pattern must be considered in the rainfall modeling.

5. Conclusion

This study compares SARIMA, exponential smoothing (ETS), neural network (NN) and extreme gradient boosting models to forecast monthly rainfall data in Bandung. The result shows that the extreme gradient boosting forecast was considerably more accurate with smallest MAD, RMSED, and MAPE than those of the traditional SARIMA, exponential smoothing, and neural network, which were used as benchmarks. The mean absolute deviance, root mean squared error deviance and mean absolute percentage error were all lower on average for the extreme gradient boosting than for the SARIMA, exponential smoothing, and neural network. We took monthly data from 2000 to 2015 as training data and 2016-2017 as testing data set. The extreme gradient boosting model performed better than the SARIMA, exponential smoothing, and neural network probably because the extreme gradient boosting considers more parameters including lag time and seasonal parameters. This result must be interpreted carefully because the large number of parameters may lead to overfitting problem. The SARIMA forecast is worse than expected. It is might be due to the ignoring model validation. The neural network performs worst due to small number observations.

Reference

[1] Hastenrath S, Greischar L and J Heerden 1995 *American Meteorological Society* **8** 1511
[2] Luk K C, Ball J E, and Sharma A 2001 *Mathematical and Computer Modelling* **33** 883
[3] Sumi M S, Zaman F M and Hirose H 2012 *Int. J. Appl. Math. Comput. Sci.* **22** 841
[4] Yang Y, Lin H, Gou Z and Jiang J 2007 *Computers Geosciences* **33** 20
[5] Hong W 2008 *Applied Mathematics and Computation* **200** 41
[6] Li P and Lai E 2004 *Development* **288** 189
[7] Pucheta J, Patino D and Kuchen B 2009 *Computer and Computing Technologies in Agriculture II vol 2 IFIP Advances in Information and Communication Technology* (Boston: Springer) pp 787–798
[8] Hyndman R J and Athanasopoulos G 2018 *Forecasting: Principles and Practice* (Melbourne: Texts)
[9] Box G E P, Jenkins G M, Reinsel G C and Ljung G M 2016 *Time Series Analysis Forecasting and Control* ed 5 (Hoboken: John Wiley & Sons, Inc.)
[10] Hyndman. R J, Koehler A B, Ord J K and Snyder R D 2008 *Forecasting with Exponential Smoothing The State Space Approach* (Berlin: Springer)
[11] Kohzadi N, Boyd M S, Kermanshahi B and Kaastra I 1996 *Neurocomputing* **10** 169
[12] Cybenko C 1989 *Math. Control Signals Systems* **2** 303
[13] Hornik K 1989 *Neural Network* **2** 359
[14] Chen Z Y, Zhang T H, Zhang R, Zhu Z M, Yang J, Chen P Y, Ou C Q and Guo Y 2019 *Atmospheric Environment* **202** 80
[15] Jaya I G N M, Ruchjana B N, Abdulah A S and Andriyana A 2018 *International Journal of Geoinformatics* **14** 27

Journal of Physics: Conference Series **1397** (2019) 012074 doi:10.1088/1742-6596/1397/1/012074

Comparing the Performance of Seasonal ARIMAX Model and Nonparametric Regression Model in Predicting Claim Reserve of Education Insurance

S M Ulyah*, M F F Mardianto, Sediono
Department of Mathematics, Airlangga University, Surabaya, 60115, Indonesia

*E-mail: maghfirotul.ulyah@fst.unair.ac.id

Abstract. One of the biggest problems in the continuity of one's education is the education fee which is often unaffordable. Therefore, the existence of education insurance is a solution to this problem. Along with increasing public interest in education insurance, insurance companies need to adjust the claims reserves with the number of claims paid to maintain the company's capital. Claim reserves are funds that must be provided by insurance companies to fulfil obligations to policy holders in the future. Losses and inaccuracies in the payment of insurance claims will result in the policy holder and the insurance company itself. Therefore, it is necessary to do a prediction of insurance company's monthly reserve claims. In education insurance, the claim reserve data has seasonal characteristics and the number of educational insurance claims tends to increase at the turn of the school year. These fluctuating patterns are supposed to fit the application of the SARIMA model and the nonparametric regression model with the Fourier series estimator in forecasting. Fourier series is a function that has flexibility in approaching fluctuating, seasonal, and recurring data patterns. The results showed that the prediction accuracy of the SARIMAX model was higher than the nonparametric regression model with MAPE of 15% and 4% respectively.

1. Introduction

Education is one of the components forming the human development index (HDI). The higher the level of education per capita in a country, the higher the HDI of that country. A high level of education is believed to be able to help a person achieve his goals and improve his standard of living. However, the cost of education for children is increasingly high because of inflation. This is one of the serious problems in the continuity of most people's education. Therefore, for the sake of planned education, the existence of education insurance is a solution to these problems.

Education insurance in Indonesia is included in the life insurance category. Education insurance is defined as life insurance whose services focus on education. This type of insurance provides a return value in cash as a consequence of premium payments at the time agreed upon by the insured and the insurer [1]. The development of insurance in Indonesia is very rapid. This can be seen from the number of companies that provide education insurance services. Of the total 50 life insurance companies in Indonesia, 17 companies serve education insurance and 3 of them are under the auspices of State-Owned Enterprises (BUMN).

Along with increasing public interest in education insurance, insurance companies need to adjust the reserve claims associated to the number of claims paid. This is very important to maintain the continuity of the insurance company. Claim reserves are funds that must be provided by insurance companies to fulfil obligations to policy holders in the future. These claim reserves consist of reported but not settled reserves and claims that have already occurred but have not been reported (incurred but not reported claims) [2].

Losses and inaccuracies in the payment of insurance claims will result in the policy holder and the insurance company itself. Therefore, it is necessary to do a prediction or forecasting large reserves of monthly insurance company claims. Previous research on estimating claims reserves deterministically was done by Reference [3] with the Chain Ladder method. Then research on calculating stochastic claims is carried out by Reference [4] for the Bayesian Generalized Linear Model, Reference [5] for the Lognormal Mixed Model method, and Reference [6] for the Hierarchical Generalized Linear Model (HGLM). These studies are based on the Triangle Run-off scheme which takes into account aspects such as the number of claims that occur. In this study, claim reserve data is considered as a single data that has autocorrelation without considering its forming factors. Therefore, the focus in this study is to produce accurate prediction models based on historical data.

In education insurance, data on the amount of claim reserves has seasonal characteristics. In other words, claim reserves will be high in certain periods and tend to be low in a certain period (periodic). If examined further, the number of educational insurance claims tends to increase at the turn of the school year. In addition, there is a tendency for the pattern to decrease and increase significantly when the new school year starts. Based on the fluctuating pattern, two approaches will be used in this study. This approach is a parametric model in the form of Seasonal Autoregressive Integrated Moving Average (SARIMA) and non-parametric models with Fourier series estimators for time series data. At the end, the performance of these two models in predicting claim reserve will be compared.

This paper is organized as follows. The first section presents the background of study. Then, the following section provides some literature reviews including the seasonal ARIMAX and nonparametric models. The material and method of the models are presented in the third section. The fourth section is the section in which the results is elaborated with some discussions. Finally, the fifth section concludes this study.

2. Literature Review

2.1 The Definition of Claim and Claim Reserve

Claims are collateral for the risk or damage that held by the insurance company to insurance customer in accordance with the policy agreement [7]. Incurred claims consist of reported claims and unreported claims. Claim reserves are funds that must be provided by insurance companies to cover obligations to policyholders in the future. As with claims, claim reserves consist of claims reserves in the settlement process (reported claims) and reserve claims that have occurred but have not been reported (unreported claims) [8].

2.2 Autoregressive Integrated Moving Average (ARIMA)

The ARIMA model is a combination of Autoregressive (AR) and Moving Average (MA) models and differencing processes (*d* order for non-seasonal data and *D* order for seasonal data) to time series data [9][10]. Based on seasonality, ARIMA model can be divided into seasonal and non-seasonal ARIMA. There will be an additional term for seasonal ARIMA model, which is the order for seasonality.

In general, non-seasonal ARIMA model can be written as ARIMA (p,d,q) mathematically as follow [9],

$$\phi_p(B)(1-B)^d Y_t = \theta_0 + \theta_q(B)a_t \tag{1}$$

where,

(p,d,q) = AR order (p), differencing order (d), and MA order (q)

$\phi_p(B)$ = the coefficients of non-seasonal AR model with order p

$\phi_p(B) = (1-\phi_1 B - \phi_2 B^2 - \dots - \phi_p B^p)$

$\theta_q(B)$ = the coefficients of non-seasonal AR model with order q

$\theta_q(B) = (1-\theta_1 B - \theta_2 B^2 - \dots - \theta_q B^q)$

a_t = residual at t.

If $d = 0$, then θ_0 is a mean of the process whereas $d \geq 1$ means that θ_0 is the coefficient of deterministic trend.

For multiplicative Seasonal ARIMA Model, it can be written as ARIMA $(P,D,Q)^s$. The *Box-Jenkins multiplicative model for seasonal ARIMA* is presented below [9],

$$\Phi_P(B^S)\phi_p(B)(1-B)^d(1-B^S)^D Y_t = \theta_q(B)\Theta_Q(B^S)a_t \tag{2}$$

where,

(p,d,q)	= AR order (p), differencing order (d), dan MA order (q)
$(P,D,Q)^S$	= AR order (P), differencing order (D), dan MA order (Q), seasonal order (S) for seasonal data
$\phi_p(B)$	= coefficient of non-seasonal AR with order p, where $\phi_p(B) = (1-\phi_1 B - \phi_2 B^2 - ... - \phi_p B^p)$
$\theta_q(B)$	= coefficient of non-seasonal MA with order q, where $\theta_q(B) = (1-\theta_1 B - \theta_2 B^2 - ... - \theta_q B^q)$
$\Phi_P(B^S)$	= coefficient of seasonal AR (S) with order P, where $\Phi_P(B^S) = (1-\Phi_1 B^S - \Phi_2 B^{2S} - ... - \Phi_P B^{PS})$
$\Theta_Q(B^S)$	= coefficient of seasonal MA (S) with order Q, where $\Theta_Q(B^S) = (1-\Theta_1 B^S - \Theta_2 B^{2S} - ... - \Theta_Q B^{QS})$
a_t	= residual at t.

2.3 Autoregressive Integrated Moving Average with Exogenous Variable (ARIMAX)

ARIMAX is an ARIMA model with the addition of particular variables [11]. In this study, the variable of interest is the dummy variables for Indonesia presidential election in 2018. The model in the seasonal ARIMA equation can be rewritten as follows [12],

$$Y_t = \frac{\theta_q(B)\Theta_Q(B^S)a_t}{\Phi_P(B^S)\varphi_p(B)(1-B)^d(1-B^S)^D} \tag{3}$$

The first model is known as ARIMAX model with stochastic trend. In this model, there will be non-seasonal or seasonal differencing order. The model is written below,

$$Y_t = \beta_1 S_{1,t} + ... + \beta_p S_{p,t} + \delta_0 V_t + ... + \delta_j V_{t-j} + \frac{\theta_q(B)\Theta_Q(B^S)}{\phi_p(B)\Phi_P(B^S)(1-B)^d(1-B^S)^D} a_t \tag{4}$$

Another model is ARIMAX model with deterministic trend in which it does not involve differencing order. It can be presented as

$$Y_t = \gamma t + \beta_1 S_{1,t} + ... + \beta_p S_{p,t} + \delta_0 V_t + ... + \delta_j V_{t-j} + \frac{\theta_q(B)\Theta_Q(B^S)}{\phi_p(B)\Phi_P(B^S)^D} a_t \tag{5}$$

where $S_{1,t} - S_{p,t}$ is the seasonal effects, $V_t - V_{t-j}$ is the dummy variable for calendar variation, and γ is the coefficient for trend.

2.4 Nonparametric Regression with Fourier Series Estimator

Nonparametric regression approach is used to determine data patterns that cannot be estimated by parametric curve models. This is because this model will produce large errors and variances. Fluctuating patterns of educational insurance claims are felt to be in accordance with Fourier series applications in forecasting. Fourier series is a function that must be resolved in the form of fluctuating data, taken, and repeated [13]. The process of repeating data patterns occurs at the value of the dependent variable for different independent variables [14].

Given paired data, parametric regression for observations that correspond to the paired data are:

$$y_i = g(t_i) + \varepsilon_i; \quad \varepsilon_i \sim IIDN(0, \sigma^2) \tag{6}$$

where y_i is a response variable, t_i is a predictor variable for nonparametric regression. The function $g(t_i)$ is unknown and estimated by functions in nonparametric regression and ε_i is a random error which is assumed to be identical, independent, and normally distributed with mean 0 and variance σ^2 [15].

The advantage of nonparametric regression is that it has high flexibility. The flexibility in question is that the pattern of data scattered in scatter plots can determine the shape of the regression curve based on estimators in nonparametric regression. One of the estimators used in nonparametric regression is Fourier series [16][17].

This study provides an alternative approach to predict claim reserve data in monthly time series which has a combination of trend and seasonal patterns using nonparametric regression with Fourier series estimators. The approach is appropriate because in nonparametric regression, Fourier series is added with linear functions. It can mathematically accommodate trend and seasonal data patterns in accordance with education insurance claims reserve data.

3. Material and Method

3.1. Data Sources and Research Variables

This study is a quantitative research that focuses on data analysis. The data used is time series data for 12 years (144 months) starting from January 2006 to December 2017. The data are secondary data from one of the state-owned education insurance that has a main branch office in Surabaya. The dependent variable in this study is the amount of claim reserves in billion rupiah, while the independent variable used is the monthly period.

In this study, data is divided into training data and testing data. Training data is used in the estimation process of the model while testing data is used to determine the accuracy of the prediction of the number of claims reserves. Data from January 2006 to December 2015 is considered as training data and data from January 2016 to December 2017 as testing data.

3.2. Steps of Analysis

The steps of analysis in this study are as follows. Firstly, the descriptive statistics analysis is conducted in order to see the characteristic of the returns. This step is followed by the univariate and multivariate modelling using ARIMAX and VARX. ARIMAX modelling involves the following steps: (1) Identifying data patterns through the results of a time series plot, (2) Determining the type of trend, namely deterministic trend, (3) Eliminating the effects of calendar variations of response with deterministic trend fitting equation so that we will get an error, (4) ARIMA modelling of the error if the error does not meet the assumption of white noise that involves identifying the model order by looking at ACF and PACF plots to obtain a significant lag, parameter estimation of ARIMA model with Conditional Least Square estimation, examining the assumption of residuals. If not met, then go back to identification step. If the assumptions are met then we merge ARIMA model with the model equations in step (3). Then, we calculate the goodness criteria of the model obtained by using RMSE. Final model will be used to forecast the data.

After that, we construct nonparametric regression modelling based on Fourier series estimators based on References [15] and [16]. The different thing with these previous studies in this study is that the data used is time series data. The model formation starts by determining the value of Generalized Cross Validation (GCV) based on the input value of the oscillation parameter (k) based on training data with the following formula:

$$GCV\ (K) = \frac{n^{-1}\mathbf{y}^T(\mathbf{I}-\mathbf{A}(K))^T\,(\mathbf{I}-\mathbf{A}(K))\mathbf{y}}{(n^{-1}trace(\mathbf{I}-\mathbf{A}(K)))^2} \tag{7}$$

Then, the optimal oscillation parameter value is determined based on minimum GCV with considering the concept of model simplicity. After that, determining the values of regression parameter estimators in vector form as follows

$$\hat{\boldsymbol{\beta}} = (\mathbf{T}[K]^T\mathbf{T}[K])^{-1}\mathbf{T}[K]^T\mathbf{y} \tag{8}$$

After having the estimator in Equation (4), the equation that contains the Fourier series estimator for nonparametric regression in the following forms:

$$\hat{y}_i = \frac{\hat{a}_0}{2} + \hat{\gamma}t_i + \sum_{k=1}^{K} \hat{a}_k \cos kt_i \tag{9}$$

Then, we determine the goodness of fits such as Mean Square Error (MSE) and coefficient of determination (R^2). Finally, the prediction of claim reserve is obtained.

To compare the prediction accuracy of the two model, Mean Absolute Percentage Error (MAPE) is used. It is defined as

$$MAPE = \frac{100\%}{n}\sum_{i=1}^{n}\left|\frac{y_i - \hat{y}_i}{y_i}\right| \tag{10}$$

where y_i is the actual value, $\hat{y}_i$ is the predicted value, and n is the sample size.

4. Result and Discussion

In education insurance, the claim reserve follows seasonal pattern with increasing trend. Time series plots of 10-year-monthly claim reserve is given in Figure 1. The number is denoted in billion rupiah. The red dashed line shows the position of the seventh month (August).

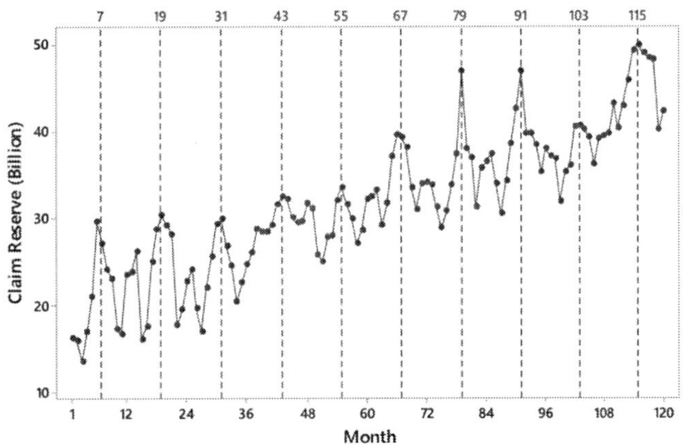

Figure 1. The amount of monthly claim reserve in 10 years

Based on Figure 1, the plot reveals that claim reserve tends to increase in August or a month before it. This month is normally be the start of new educational year in Indonesia, especially for the high school and below. Therefore, it is not surprising for us to get this kind of seasonal pattern as this month is the time for paying the tuition and other fundamental fees (for new student). Moreover, the claim reserve has an increasing trend as the school fee in Indonesia is continuously rocketed. Besides, the raising trend in claim reserve is also caused by the raise in the number of customer.

To model seasonal pattern data, a seasonal ARIMA model is normally used. This model is one of parametric approach group which consider some assumptions that should be satisfied. In some conditions, the assumption is hard to be achieved. Therefore, we will have another model that do not need many assumptions (nonparametric approach). The model used is nonparametric regression model with Fourier series estimator. The fluctuation pattern in Figure 1 supports the use of Fourier series since the series has periodical pattern.

4.1. Modelling with Seasonal ARIMA

The fundamental requirement in ARIMA modelling is to have stationary series. The rounded value of Box-Cox transformation should be unity in order to have stationary-in-variance series. Furthermore, the Augmented Dickey-Fuller (ADF) test was conducted to examine the stationarity in mean of the claim reserve data. The results are given in Table 1.

Table 1. The results of ADF test and Box-Cox Transformation

	ADF		Box-Cox
Dickey-Fuller	-9.6779	Lower CL	0.71
Lag order	8	Upper CL	1.85
P-value	<0.01	Rounded Value (λ)	1.00

Table 1 depicts the p-value of ADF test that is very small and the rounded value is unity. Therefore, the 10-year-claim reserve data are stationary both in mean and variance.

4.1 The ARIMAX Modelling

Although the series is stationary, there is an increasing trend in the data as shown in Figure 1. Thus, this trend will be eliminated by placing trend variable "t", where t = 1, 2, 3, …, and regress it with the claim reserve series (i.e. ARIMAX modelling with *t* as the exogenous variable. Table 2 gives information of the summary of regression of *t* on claim reserve series.

Table 2. The summary of regression coefficients

Series	Variable	Coefficient	P-value
Claim reserve	Constant	19.066	0.000
	T	0.2083	0.000

Based on Table 2, the trend variable significantly affect claim reserve series. Then, diagnostic checking is conducted on the residual. The ACF and PACF plots of residuals are given in Figure 2.

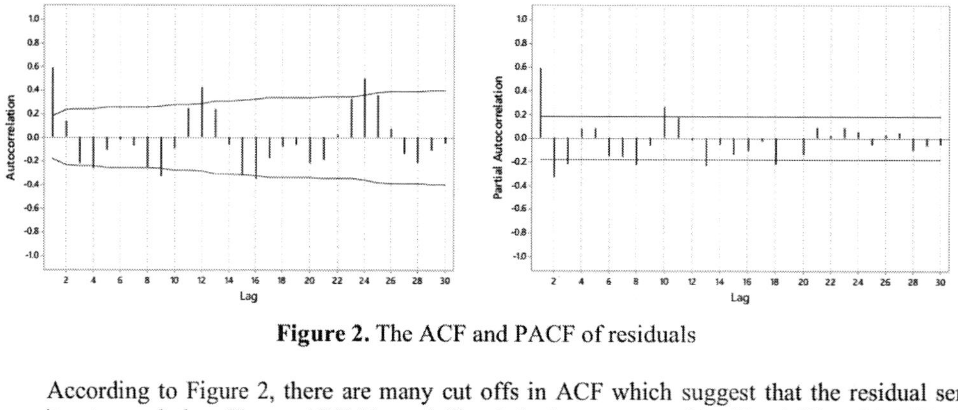

Figure 2. The ACF and PACF of residuals

According to Figure 2, there are many cut offs in ACF which suggest that the residual series commit autocorrelation. Hence, ARIMA modelling is built to capture this. The ACF and PACF plots shows that seasonal model should be used here since the pattern in both plots are periodical. The possible orders for SARIMA are SARIMA$(1,0,0)(1,0,0)^{12}$ and SARIMA$(1,0,0)(1,0,1)^{12}$. After trying to run those models, both order have significant coefficients but only SARIMA$(1,0,0)(1,0,1)^{12}$ with zero mean satisfies the white noise assumption of SARIMA residual (see Table 3 for the model coefficient and Figures 3 and 4 for the residual diagnostic checking).

Table 3. The SARIMA$(1,0,0)(1,0,1)^{12}$ coefficients

Variable	Coefficient	Std. Error	t-Statistic	P-value
AR(1)	0.5163	0.0797	6.48	0.000
SAR (12)	0.9963	0.0152	65.38	0.000
SMA (12)	0.8552	0.0787	10.87	0.000

Table 3 reveals that all of the coefficients of SARIMA$(1,0,0)(1,0,1)^{12}$ model are significant. Moreover, there is no cut off in ACF plot in Figure 3 that means the residual is white noise. In addition, the QQ plot in Figure 4 shows that the residual follows normal distribution as the p-value of Kolmogorov-Smirnov test is quite large.

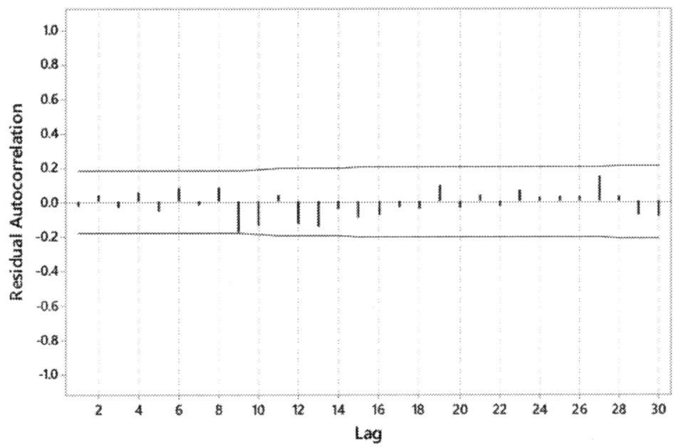

Figure 3. The ACF of residuals of SARIMA$(1,0,0)(1,0,1)^{12}$ model

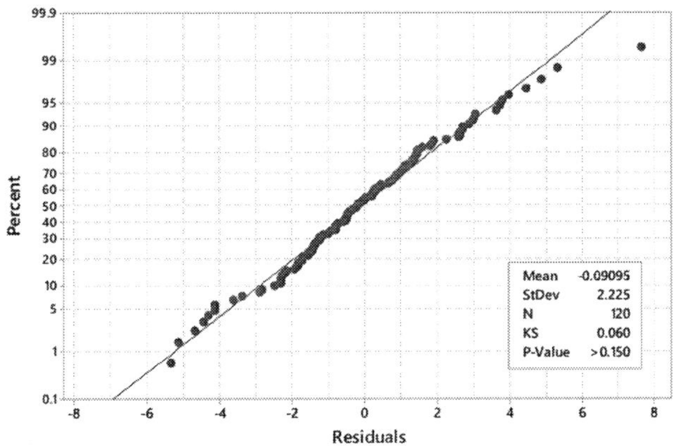

Figure 4. The QQ plot of residual of SARIMA$(1,0,0)(1,0,1)^{12}$

Based on Tables 2 and 3, the final ARIMAX model of claim reserve can be written mathematically as

$$\hat{y}_t = 19.066 + 0.2083\,t + 0.5163\,y_{t-1} + 0.9963\,y_{t-12} - 0.5144\,y_{t-13} - 0.8552\,a_{t-12} + a_t, \text{ where } \varepsilon_t \sim N(\mu, \sigma^2). \quad (11)$$

This model is known as SARIMAX(1,0,0)(1,0,1)^{12} with trend variable as the exogenous variable. The final step is to forecast the claim reserve series using the final model. The forecast results is given in Figure 5.

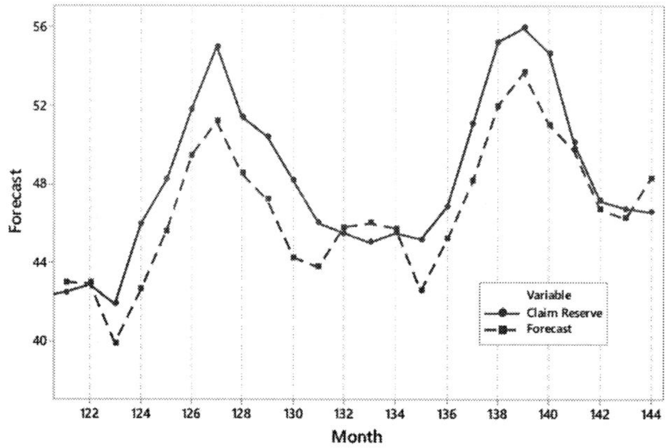

Figure 5. The forecast results of SARIMAX(1,0,0)(1,0,1)^{12} model

Figure 5 provides information on forecasting results of claim reserve based on SARIMAX(1,0,0)(1,0,1)^{12} model. We plotted the forecast with the actual value for the next two year. The figures show that the model can follow the actual pattern closely with MAPE 4.025%.

4.2. Modelling with Nonparametric Regression

In this study, the number of claim reserves is predicted using nonparametric regression with the Fourier series approach. Paired data is given which contains one independent variable, and one dependent variable (t_i, y_i), $i = 1,2, ..., n$ shows the number of observations in the training data. The general form of nonparametric regression equations with Fourier series estimators which refers to References [16] and [17] for paired data are as follows:

$$y_i = \frac{a_0}{2} + \gamma t_i + \sum_{k=1}^{K} a_k \cos kt_i + \varepsilon_i; \varepsilon_i \sim IIDN(0, \sigma^2) \quad (12)$$

where γ, a_0, and a_k are regression parameters whose values are estimated based on the least squares method. The oscillation parameter is denoted by k = 1,2, ..., K, ε_i is an identical random error, independent and normally distributed with zero mean and variance σ^2. In the form of vector equations, Equation (12) can be constructed into

$$\mathbf{y} = \mathbf{T}[K]\,\boldsymbol{\beta} + \boldsymbol{\varepsilon}\,;\ \boldsymbol{\varepsilon} \sim \text{IIDN}\,(\mathbf{0}, \sigma^2\mathbf{I}) \quad (13)$$

where $\mathbf{y} = (y_1, y_2, ..., y_n)^T$, $\boldsymbol{\varepsilon} = (\varepsilon_1, \varepsilon_2, ..., \varepsilon_n)^T$, $\boldsymbol{\beta} = \left(\frac{a_0}{2}, \gamma, a_1, ..., a_K\right)^T$,

$$\text{and } \mathbf{T}[K] = \begin{pmatrix} 1 & t_1 & \cos t_1 & \cdots & \cos Kt_1 \\ 1 & t_2 & \cos t_2 & \cdots & \cos Kt_2 \\ \vdots & \vdots & \vdots & \vdots & \vdots \\ 1 & t_n & \cos t_n & \cdots & \cos Kt_n \end{pmatrix}.$$

Journal of Physics: Conference Series **1397** (2019) 012074 doi:10.1088/1742-6596/1397/1/012074

and I is an identity matrix. Based on the least squares method, the regression parameter estimator vector is obtained according to (6).

By using training data, Equation (5) is used to determine the GCV value for each k. In Equation (5) the numerator is MSE, and $\mathbf{A}(K) = \mathbf{T}[K]\ (\mathbf{T}[K]^T\mathbf{T}[K])^{-1}\mathbf{T}[K]^T$ is the hat matrix. Figure 6 shows GCV values from $k = 1$ to $k = 120$ with $k = 118$ being the optimal k which has minimum GCV value.

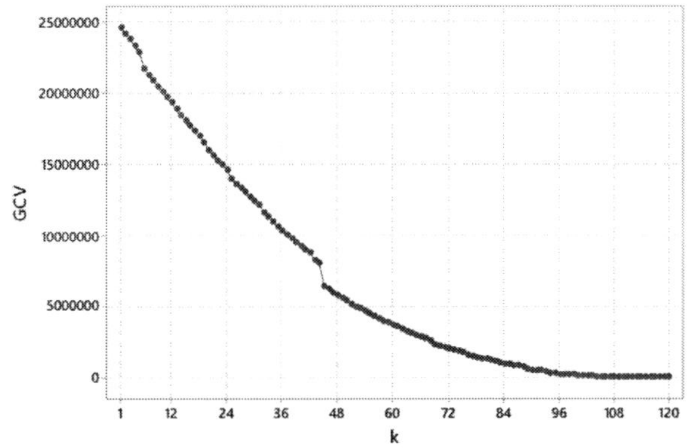

Figure 6. The change of GCV value for k =1 to 120

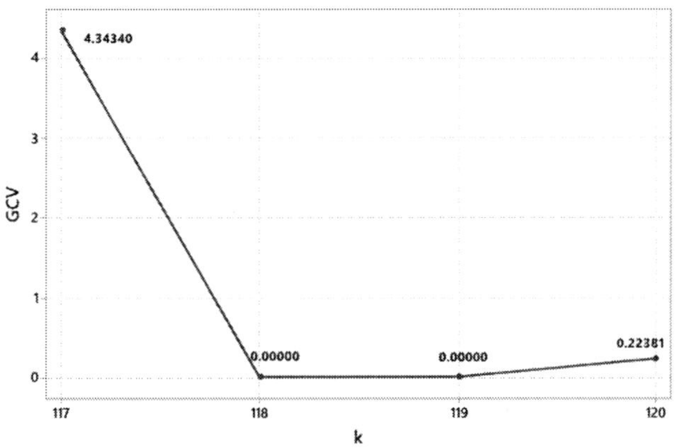

Figure 7. The change of GCV value around the optimal k

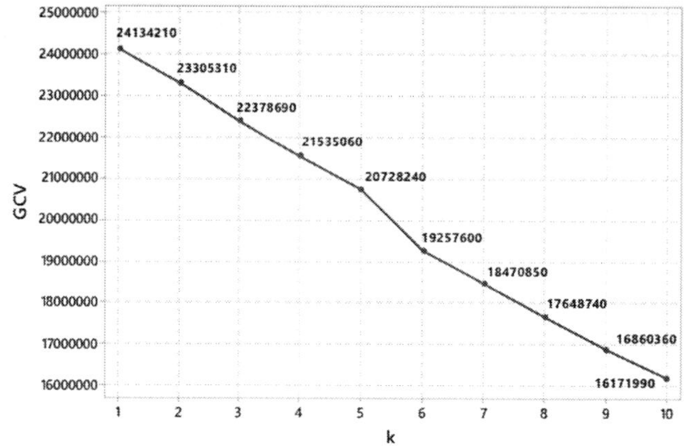

Figure 8. The change of GCV value for k =1 to 10

Figure 6 shows changes in GCV values around the optimal k, while Figure 7 shows changes in GCV values for k = 1 to k = 10. Without considering the simplicity of the model, the Fourier series estimator with k = 118 is selected. However, the selection of k = 118 has a deficiency that is a long form of estimator and has a high computational complexity. Thus, considering the simplicity of the model is needed in selecting oscillation parameters beside the minimum GCV concept. Based on Figure 8, it can be seen that there is a large decrease in GCV value for k = 6. The GCV value for k = 6 is the minimum value compared to k = 1 to 5. Hence, the Fourier series estimator in nonparametric regression with k = 6 was chosen to predict the reserve amount of educational insurance claims. For k = 6, the estimator is obtained as follows:

$$\hat{y}_i = 19{,}0967 + 0{,}2077\, t_i - 0{,}1188 \cos t_i - 0{,}024 \cos 2t_i \\ - \cos 3t_i + 0{,}2221 \cos 4t_i + 0{,}2145 \cos 5t_i + 1{,}016 \cos 6t_i. \quad (14)$$

The Fourier series estimator in nonparametric regression in Equation (14) has MSE of 4.4246 and R^2 of 78.4886%. This R^2 value is quite good, considering that we chose a model based on the principle of simplicity. The value of R^2 is determined by the following formula:

$$R^2 = \frac{(\hat{y}-\bar{y})^T(\hat{y}-\bar{y})}{(y-\bar{y})^T(y-\bar{y})} \quad (15)$$

Then, forecasting is done based on (14) using testing data. Figure 9 gives information about the results of the comparison between testing data and forecast. Although the estimator with the smallest GCV (k = 118) is not selected, by selecting the Fourier series estimator for k = 6 in this model, the predicted value follows the increasing pattern of the testing data but cannot fit it well. The amount of claim reserves equals a tendency to rise over a certain period. In addition, seasonal fluctuations are also predicted to continue, and will reach the highest value in a certain period of one year. The results of the prediction of the reserve amount of educational insurance claims based on the Fourier series estimator in nonparametric regression have a small MAPE value of 15.42%.

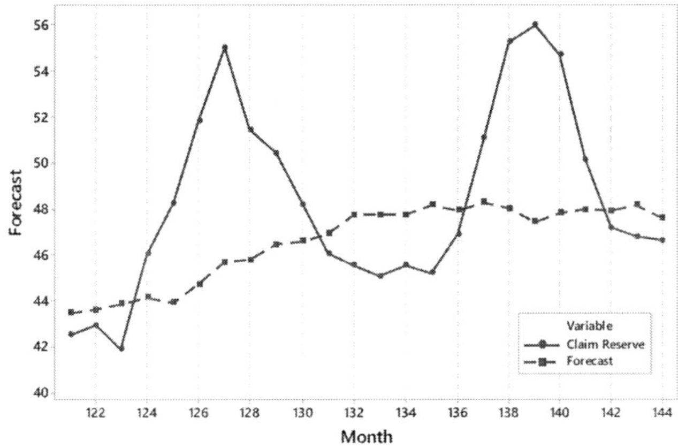

Figure 9. The comparison between testing data (solid line) and forecast (dashed line)

Based on the results on nonparametric regression modeling, it can be concluded that the prediction of the reserve amount of educational insurance claims using the Fourier series estimator in nonparametric regression has good results. The indicator of goodness is shown by the small MSE value with large R^2. In addition, in predicting the reserve amount of educational insurance claims, the Fourier series estimator in nonparametric regression has a small MAPE value. The other indicator is that the difference in the results of the prediction of the data testing is also not too far away, and mostly according to the condition that the number of educational insurance claims has an increasing trend, and will reach the highest value in a certain period of one year. This condition is expected to be anticipated by education insurance companies in Indonesia.

4.3. The Model Comparison

The parametric model outperforms the nonparametric model in prediction accuracy based on MAPE. The MAPE of SARIMAX model is almost one-third of that of nonparametric regression model. Moreover, the prediction result is plotted in Figure 10.

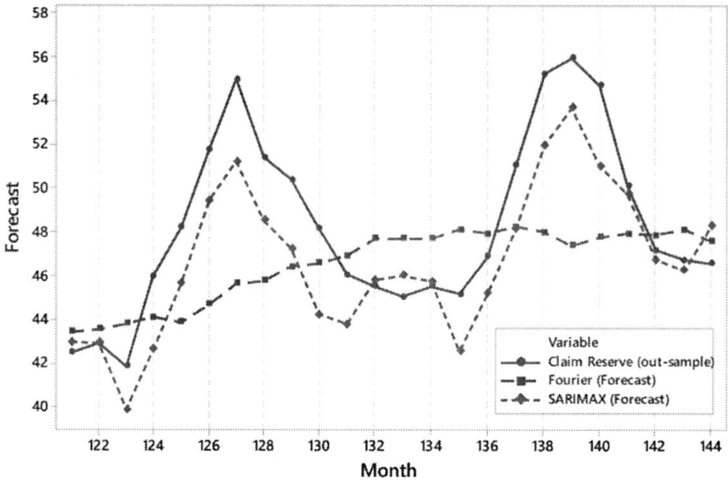

Figure 10. The forecast comparison among two methods

Based on the prediction results in Figure 10, the SARIMAX prediction can follow the pattern of actual data smoothly but the prediction of nonparametric model only follows the increasing trend without capturing the periodical pattern. A raising trend is expected as more people have awareness in their future education. Both models give the same information about this raising trend. However, the nonparametric model in which the selected maximum GCV value is 6, cannot explain seasonal feature in claim reserve. This does not agree with the real pattern of education insurance claim reserve in Figure 1, claim reserve of education insurance has seasonal pattern where the peak is in July or August (the beginning of new educational year in Indonesia).

The reason for this can be explained by the selection of optimal GCV value in nonparametric regression. The optimum GCV is at $k = 118$ but it is bothersome to have that much parameters. Another reason is because our estimator is based on Fourier series which has continuous dimension. When we apply that to time series data, which is discrete, the performance will be not as good as in the continuous case.

5. Conclusion

To conclude, both models (parametric and nonparametric) find the increasing trend and seasonal pattern in the predicted values of claim reserve. However, the parametric model follows the pattern smoother than that of nonparametric model. Furthermore, based on the predicting accuracy using MAPE, the parametric model (SARIMAX) had better performance than nonparametric model.

References

[1] Outreville J F 2015 The relationship between relative risk aversion and the level of education: a survey and implications for the demand for life insurance *Journal of economic surveys* **29**(1) 97-111

[2] Baser F and Apaydin A 2010 Calculating insurance claim reserves with hybrid fuzzy least squares regression analysis Gazi university journal of science **23**(2) 163-170

[3] Mack T and Venter G 2000 A comparison of the stochastic models that reproduce chain ladder reserve estimates Insurance: mathematics and economics **26**(1) 101-107

[4] Verrall R J 2000 An investigation into stochastic claims reserving models. *Insurance: mathematics and economics* **26**(1) 91-99

[5] Antonio K, Beirlant J, Hoedemakers T and Verlaak R 2006 Lognormal mixed models for reported claims reserves *North American Actuarial Journal* **10**(1) 30-48
[6] Gigante P, Picech L and Sigalotti L 2013 Claims reserving in the hierarchical generalized linear model framework *Insurance: Mathematics and Economics* **52**(2) 381-390
[7] Dickson D C, Hardy M, Hardy M R and Waters HR 2013 *Actuarial Mathematics for life contingent risk* (Cambridge: Cambridge University Press)
[8] Lopes H, Barcellos J, Kubrusly J and Fernandes C 2012 A non-parametric method for incurred but not reported claim reserve estimation *International journal for uncertainty quantification* **2**(1)
[9] Wei W W 2006 *Time Series Analysis: Univariate and Multivariate Methods* (United States of America: Pearson Education, Inc.)
[10] Ulyah S M 2019 Forecasting index and stock returns by considering the effect of Indonesia pre-presidential election 2019 using ARIMAX and VARX approaches *Journal of Physics: Conf. Series* **1277** 012053 doi:10.1088/1742-6596/1277/1/012053.
[11] Cryer J D and Chan K S 2008 *Time Series Analysis with Application in R* 2nd Edition (New York: Springer)
[12] Suhartono, Lee M H and Prastyo D D 2015 Two levels ARIMAX and regression models for forecasting time series data with calendar variation effects *AIP conference proceeding* **1691**(1) 050026
[13] Mardianto M F F 2018 Prediksi Cadangan Klaim Asuransi Pendidikan dengan Pendekatan Regresi Nonparametrik Deret Fourier *Proceeding of KNPMP III Universitas Muhammadiyah Surakarta* (ISSN: 2502-6526)
[14] Pane R, Budiantara I N, Zain I and Otok B W 2013 Parametric and Nonparametric Estimators in Fourier Series Semiparametric Regression and Their Characteristics *Applied Mathematical Sciences* **8**(102) 5053-5064 doi: 10.12988/ams.2014.46472
[15] Eubank R L 1999 Nonparametric regression and spline smoothing (CRC press).
[16] Bilodeau M 1992 Fourier Smoother and Additive Models *The Canadian Journal of Statistics* **3**(1) 257-269 doi: 10.2307/3315313.
[17] Prahutama A 2013 Model Regresi Nonparametrik dengan Pendekatan Deret Fourier pada Kasus Tingkat Pengangguran Terbuka di Jawa Timur *Jurnal Nasional Statistika Universitas Diponegoro* **10**(1) 69–76 Retrieved from http://eprints.undip.ac.id/40285/1/A02AlanPrahutama.pdf
[18] Makridakis S and Hibon M 2000 The M3-Competition: result, conclusion and implication. *International Journal of Forecasting* **16** (1) 451-476
[19] Hoch C J 2017 Beyond Fourier *Journal of Magnetic Resonance* **283** 117-123.

Probabilistic multi-objective optimization approach to solve production planning and raw material supplier selection problem under probabilistic demand value

Sutrisno[1]*, P. A. Wicaksono[2], and Solikhin[1]

[1]Department of Mathematics, Diponegoro University, Jalan Prof. Soedarto, SH, Tembalang, Semarang, Indonesia
[2]Department of Industrial Engineering, Diponegoro University, Jalan Prof. Soedarto, SH, Tembalang, Semarang, Indonesia

Corresponding author's e-mail: tresno.math@live.undip.ac.id

Abstract. This article is addressed to study the development of a probabilistic multi-objective optimization model that can be used to optimize the production planning and raw material procurement in a manufacturing industry where the demand value is unknown. First, the unknown demand value is assumed to be a random variable with some known probability distribution. Then, we formulate the multi-objective optimization model with two objective functions which are the total procurement cost that is minimized and the total production number that is maximized. Some related constraints that should be satisfied are also be formulated. We solve this multi-objective optimization problem by finding the Pareto solution. The calculation is performed in LINGO 18.0. To simulate and observe how the optimal decision is made, a computational simulation using generated data was performed. From the results, the optimal decision is obtained (the number of the raw material that should be purchased from each supplier and the number of the product that should be produced).

1. Introduction

Raw material procurement in manufacturing industries plays as an important component in determining of the company's performance. Another component which is also important is production planning. There are many approaches in order to maintain these two components. The most used approaches are mathematical optimization since it can produce a solution which optimizes some objective functions [1], [2]. Then, many research articles have been published about problem solving in raw material procurement and production planning problem by using optimization approach as in [3–7]. For more detail example, an integer linear optimization approach was used in [6] whereas a throughput accounting approach was used in [7]. Besides the development of the mathematical model, some related researches were conducted by case study such as product mixing problem on paper mill [8].

One of the powerful mathematical tools to solve an optimization cases in many fields is mathematical programming. There are several classes on mathematical programming from the simplest linear programming to the most complex multi-objective non-linear programming. The term "multi-objective" means the objective function that will be optimized contains two or more functions. Multi-objective programming approaches have been used in many problems e.g. energy optimization [9]–[14], mechanical vibration optimization [15], optimization on water distribution [16], biopharmaceutical industry [17], MA CFB reactor optimization [18], high-speed train cabin ventilation designing [19], robust scheduling of electric vehicles aggregator [20], etc. A multi-objective optimization problem can be solved by calculating the Pareto solution. To calculate the Pareto solution, the weighting method can be used as a simplest way [21]. Many research articles were published to show how the Pareto solution can be used as a decision. For example, Pareto optimal solution was used in battery cell designing [22], re-insurance optimization [23]–[25] and radar designing [26].

Table 1. Notations used in the model

Category	Notation	Interpretation
Index	r	index of raw material type: 1, 2, ..., R
	s	index of supplier name: 1, 2,..., S
	p	index of product type: 1, 2, …, P
Decision variable	Y_p	(initial decision variable which decided before the demand value is revealed) the number of the product p decided to be produced
	YR_p	(recourse decision variable which decided after the demand value is revealed) the number of the product p decided to be procured
	X_{sr}	Amount of raw material r purchased from supplier s
	Z_s	A binary variable for supplier s (0 if there is no purchased raw material or 1 if there is some purchased raw material)
	S_s	Truck delivery number from supplier s to the manufacturer or warehouse
Parameter	UP_{sr}	Price for one unit of raw material r that purchased from supplier s
	YRP_p	Unit price for product p that should be procured after the demand value is revealed to meet the demand volume
	O_s	Cost for ordering some raw material from supplier s
	TC_s	Transport cost for one truck from supplier s to the manufacturer
	P_{sr}^d	penalty cost for one unit defected raw material r from supplier s
	d_{sr}	Percentage of defected product purchased from supplier s
	P_{sr}^l	Penalty cost for one unit of late delivered raw material r from supplier s
	l_{sr}	Percentage of the late delivered (shortage) raw material r from supplier s
	r_{rp}	The amount of raw material r which required to be used to produce one unit of product p
	$\hat{D}_p$	Demand value of product p which is uncertain parameter approached as a random variable
	MH_{pm}	Duration or hour of machine m which is required to be used to produce one unit product p
	MC_m	Maximum duration capacity of machine m that can be operated for production activities
	C	Truck's capacity to transport raw material from supplier to the manufacturer
	SC_{sr}	Maximum amount of raw material r that can be purchased from supplier s

In this article, we develop a probabilistic multi-objective optimization model which we used to solve an optimization on the production planning and raw material procurement problem that can be applied in many manufacturing industries. The discussed problem is containing an unknown parameter which is demand value for a simple problem, commonly the demand value is known, but for the problem discussed in this paper, the demand value is unknown. But, we assumed that the decision maker has data of this demand value, so the probability distribution can be made. Then, we the solution of the developed model can applied by the decision maker. The solution is the number of the raw material the number of the product that should be produced.

2. Mathematical model

Consider a product mixing and raw material procuring optimization problem with P product types which planned to be produced using R raw material types procured from S supplier alternatives.

2.1. Mathematical notation

Suppose that the decision maker will maximize the amount of the product and minimize the total occurred cost. Let the used notations in the formulated model are shown in table 1.

2.2. Mathematical model

The objectives are maximizing the amount of the planned production volume for each product type and minimizing the total occurred cost from raw material procurement to production cost. Then, we have the following multi-objective optimization problem:

$$\max Z_1 = \sum_{p=1}^{P} Y_p \tag{1}$$

$$\begin{aligned}\min Z_2 = &\sum_{s=1}^{S}\sum_{r=1}^{R}[X_{sr} * UP_{sr}] + \sum_{s=1}^{S}[O_s * Z_s] + \sum_{s=1}^{S}[TC_s * S_s] + \sum_{s=1}^{S}\sum_{r=1}^{R}\left[P_{sr}^{d} * d_{sr} * X_{sr}\right] \\ &+ \sum_{s=1}^{S}\sum_{r=1}^{R}\left[P_{sr}^{l} * l_{sr} * X_{sr}\right] + \sum_{p=1}^{P}\left[YRP_p * YR_p\right]\end{aligned} \tag{2}$$

The first objective function, (1) describes the amount of all products which will be planned to be produced. The second objective function, (2) describes the total occurred cost where the first term is the total buying cost for all raw materials, the second term is the ordering cost for all suppliers, the third term is the raw material's transport cost from all suppliers, the fourth term is the total penalty cost for defected products, the fifth term is the total penalty cost for late delivered products, and the last term is the occurred cost for procuring the shortage product that will be used to meet the demand (if there is some demanded product which is not met by the product produced by the manufacturer). The constraints that should be satisfied are designed as follows:

- The received raw material volume (purchased volume minus defect volume and late delivered volume) has to satisfy (greater or equal to) the raw material volume which is needed by the manufacturer for the production process i.e. to be used to produce the planned product:

  $$\sum_{s=1}^{S} X_{sr} - \sum_{s=1}^{S}[(d_{sr} + l_{sr})X_{sr}] \geq \sum_{p=1}^{P}\left[r_{rp} Y_p\right], \forall r;$$

- The product volume which will be manufactured plus the product volume procured after the demand is revealed has to satisfy the (random) demand value:

  $$Y_p + YR_p \geq \hat{D}_p, \forall p, \forall p;$$

- The machine's working duration has to satisfy it's maximum capacity:

  $$\sum_{p=1}^{P} MH_{pm} Y_p \leq MC_m, \forall m;$$

- The delivery load has to satisfy the truck's capacity:

$$\left\lceil \frac{\sum_{r=1}^{R} X_{sr}}{C} \right\rceil \le S_s, \forall s;$$

- The amount of the raw material purchased from each supplier has to satisfy the supplier's maximum capacity:

$$X_{sr} \le SC_{sr}, \forall s, \forall r;$$

- Auxiliary binary variable indicated a supplier is selected or not is formulated as follows:

$$\sum_{r=1}^{R} X_{sr} \le M * Z_s, \forall s;$$

- The non-negativity and integer constraint are formulated as follows:

$$X_{spr}, Y_p \ge 0 \text{ and integer, } W_s \in \{0,1\}.$$

Consider a two-objectives optimization problem (1)-(2) subject to constraints no. (1)-(7). Since the feasible solution set, if non empty, is closed and bounded then (1) can be replaced by $\min Z_1 \leftarrow (-Z_1) = -\sum_{p=1}^{P} Y_p$. A solution vector x^o is called as Pareto solution if there does not exist another point x so that $Z_i(x) \le Z_i(x^o)$, $\forall i = 1,2$ and $Z_j(x) < Z_j(x^o)$ for $i = 1$ or 2 and the set of all Pareto solution is called Pareto set which can be generated by using weighting method [21]. The Pareto set of this bi-objective optimization problem can be calculated by

$$\min Z = w_1(-Z_1) + w_2 Z_2, \tag{3}$$

subject to: $w_1 + w_2 = 1, 0 \le w_1, w_2 \le 1$.

3. Numerical experiment

Consider the raw material procurement and production planning problem with three raw materials R1, R2 and R3, four suppliers S1, S2, S3 and S4, and three products P1, P2 and P3 which modelled as (1)-(2). The value of parameters are shown in the appendix. Let the probability distribution function of $\hat{D}_p$ is

$$f_{\hat{D}_1}(D_1) = \begin{cases} 0.05, & D_1 = 100; \\ 0.06, & D_1 = 120; \\ 0.10, & D_1 = 140; \\ 0.12, & D_1 = 160; \\ 0.15, & D_1 = 180; \\ 0.20, & D_1 = 200; \\ 0.15, & D_1 = 220; \\ 0.10, & D_1 = 240; \\ 0.05, & D_1 = 260; \\ 0.05, & D_1 = 280; \\ 0.00, & \text{others.} \end{cases} \quad f_{\hat{D}_2}(D_2) = \begin{cases} 0.05, & D_2 = 80 \\ 0.06, & D_2 = 90 \\ 0.10, & D_2 = 100 \\ 0.12, & D_2 = 110 \\ 0.15, & D_2 = 120 \\ 0.20, & D_2 = 130 \\ 0.15, & D_2 = 140 \\ 0.10, & D_2 = 150 \\ 0.05, & D_2 = 160 \\ 0.05, & D_2 = 170 \\ 0.00, & \text{others.} \end{cases} \quad f_{\hat{D}_3}(D_3) = \begin{cases} 0.05, & D_3 = 120 \\ 0.06, & D_3 = 130 \\ 0.10, & D_3 = 140 \\ 0.12, & D_3 = 150 \\ 0.15, & D_3 = 160 \\ 0.20, & D_3 = 170 \\ 0.15, & D_3 = 180 \\ 0.10, & D_3 = 190 \\ 0.05, & D_3 = 200 \\ 0.05, & D_3 = 210 \\ 0.00, & \text{others.} \end{cases}$$

Journal of Physics: Conference Series **1397** (2019) 012075 doi:10.1088/1742-6596/1397/1/012075

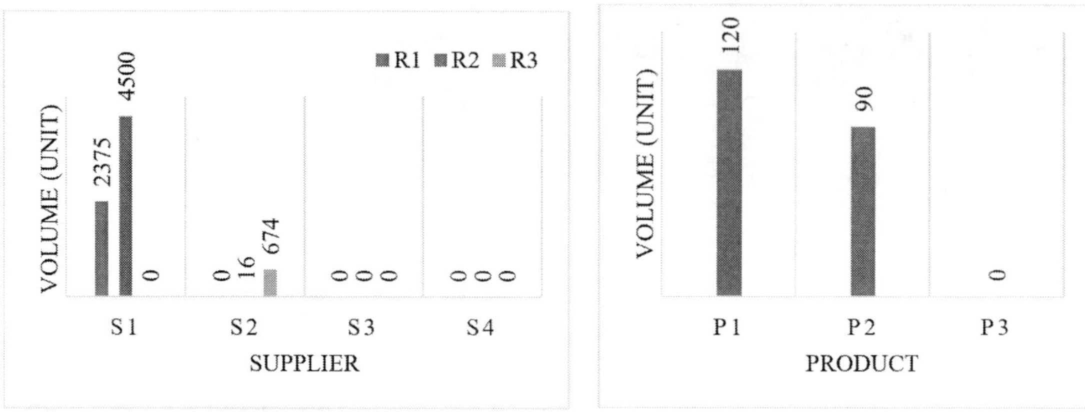

(a) (b)

Figure 1. (a) The optimal decision for raw material procurement for each R1, R2, and R3 (b) The optimal decision for the amount of product P1, P2, and P3 planned to be produced

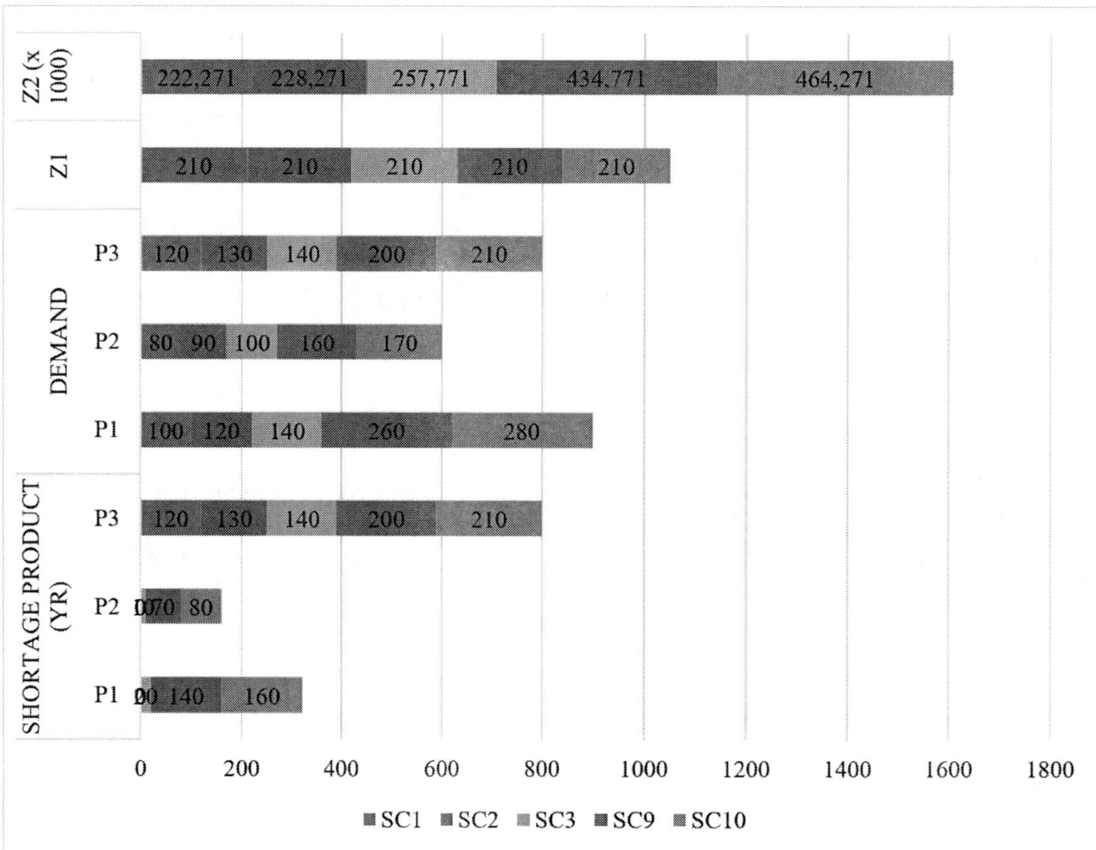

Figure 2. Demand value and product shortage volume for scenarios 1, 2, 3, 9, and 10

We have solved (3) using $w_1 = w_2 = 0.5$ which means that a fifty-fifty impact is applied to the two objective functions (1)-(2). The optimization process was calculated in LINGO 18.0 software interfaced with Ms Excel 2016 as the input and output form. We have solved the problem in a common PC with 2.3 GHz of processor and 6 GB of memory. The computation time is limited to 10 minutes. The value of E[Z] is 164768.5 and the other results are shown in figure 1 and figure 2. It can be seen from figure 1.a that the manufacturer should not purchasing raw material from supplier S3 and S4. The other optimal decision is illustrated in figure 1.b which shows how many units that should be produced for product type P1 and P2 where no unit of product P3 should be produced.

After the demand's value is revealed, the number of the demanded product that is not satisfied by the produced product is revealed and it can be satisfied by purchasing from other manufacturer or it can be released which means that the decision maker is decided to be not satisfying the demanded product. For example, let us see the scenario 1 (SC1) in figure 2. After the demand value is revealed, let the demand value for product P1 is 100 units then it is satisfied by the produced product which is 120 units. The demand value for product P2 is 80 units and it is satisfied by the produced product P2 which is 90 units. The demand value for product P3 is 120 units and it is not satisfied by the produced P2 since there is no P2 which is produced. Then, the options for the decision maker is not satisfying this demand value or satisfying it by purchasing 120 units P2 with unit price 600. The total number of scenarios is ten scenarios where five of them are shown in figure 2.

4. Concluding remarks

In this article, a bi-objective mathematical optimization model was developed that can be used to find the optimal decision for raw material procurement and product mixing planning problem. The optimal decision was calculated by using the Pareto solution scheme based on the weighting method. A numerical problem example was solved and it was resulted the optimal decision for the given problem. For further research, a model for multi-period case will be developed so that it will be a dynamic problem and an option of storing the raw material and product in the warehouse will be discussed. Furthermore, besides the demand value, the other parameters like unit price of raw material or transport cost can also be unknown. It will be a great challenge to solve with many parameters which their values are unknown.

5. Acknowledgment

Authors thank to Diponegoro University for financial support by funding this research via RPP (*Riset Pengembangan dan Penerapan*) 2019 research grant.

References

[1] N. R. Ware, S. P. Singh and D. K. Banwet 2014 A mixed-integer non-linear program to model dynamic supplier selection problem *Expert Syst. Appl.* **41** 671–678
[2] S. S. Kara 2011 Supplier selection with an integrated methodology in unknown environment *Expert Syst. Appl.* **38** 2133–2139
[3] D. Choudhary and R. Shankar 2014 A goal programming model for joint decision making of inventory lot-size, supplier selection and carrier selection *Comput. Ind. Eng.* **71** 1–9
[4] D. Choudhary and R. Shankar 2013 Joint decision of procurement lot-size , supplier selection , and carrier selection *J. Purch. Supply Manag.* **19** 16–26
[5] H. Shavandi, H. Mahlooji and N. Ekram 2012 A constrained multi-product pricing and inventory control problem," *Appl. Soft Comput. J.* **12** 2454–2461
[6] A. Trivedi, A. Chauhan, S. P. Singh and H. Kaur 2017 A multi-objective integer linear program to integrate supplier selection and order allocation with market demand in a supply chain *Int. J. Procure. Manag.* **10** 335–359
[7] O.-P. Hilmola 2005 Product mix decisions and production lot sizes *Int. J. Manuf. Technol. Manag.* **7** 41–51
[8] J. Mattila, V. Hotti and M. Juhola 2011 Product mix and its optimisation in a paper mill according to the profitability computation of process measurements *Int. J. Comput. Aided Eng. Technol.* **3** 155–174

[9] H. M. G. C. Branco, M. Oleskovicz, D. V. Coury and A. C. B. Delbem 2018 Multiobjective optimization for power quality monitoring allocation considering voltage sags in distribution systems *Int. J. Electr. Power Energy Syst.* **97** 1–10
[10] E. B. Schlünz, P. M. Bokov and J. H. van Vuuren 2018 Multiobjective in-core nuclear fuel management optimisation by means of a hyperheuristic *Swarm Evol. Comput.* (in press) 1–19
[11] H. Rashidi and J. Khorshidi 2018 Exergy analysis and multiobjective optimization of a biomass gasification based multigeneration system *Int. J. Hydrogen Energy* **43** 2631–2644
[12] K. Yu *et al.* 2018 Multiobjective optimization of ethylene cracking furnace system using self-adaptive multiobjective teaching-learning-based optimization *Energy* **148** 469–481
[13] F. Ruiming 2019 Multi-objective optimized operation of integrated energy system with hydrogen storage *Int. J. Hydrogen Energy* (in press)
[14] A. Zendehboudi, A. Mota-Babiloni, P. Makhnatch, R. Saidur and S. M. Sait 2019 Modeling and multi-objective optimization of an R450A vapor compression refrigeration system *Int. J. Refrig.* **100** 141–155
[15] J. F. A. Madeira, A. L. Araújo, C. M. Mota Soares and C. A. Mota Soares 2017 Multiobjective optimization for vibration reduction in composite plate structures using constrained layer damping *Comput. Struct.* (in press)
[16] K. Zhang, H. Yan, H. Zeng, K. Xin and T. Tao 2019 A practical multi-objective optimization sectorization method for water distribution network *Sci. Total Environ.* **656** 1401–1412
[17] S. Liu and L. G. Papageorgiou 2018 Multi-objective optimisation for biopharmaceutical manufacturing under uncertainty *Comput. Chem. Eng.* **119** 383–393
[18] P. Chaudhari and S. Garg 2019 Multi-objective optimization of maleic anhydride circulating fluidized bed (CFB) reactors *Chem. Eng. Res. Des.* **141** 115–132
[19] N. Li, L. Yang, X. Li, X. Li, J. Tu and S. C. P. Cheung 2019 Multi-objective optimization for designing of high-speed train cabin ventilation system using particle swarm optimization and multi-fidelity Kriging *Build. Environ.* (in press)
[20] H. Ahmadi-Nezamabad, M. Zand, A. Alizadeh, M. Vosoogh and S. Nojavan 2019 Multi-objective optimization based robust scheduling of electric vehicles aggregator *Sustain. Cities Soc.* **47** 101494
[21] J. Branke, K. Deb, K. Miettinen and R. Slowinski 2008 *Multiobjective Optimization: Interactive and Evolutionary Approaches* (Berlin: Springer-Verlag Berlin Heidelberg)
[22] Y. Hong and C. W. Lee 2018 Pareto fronts for multiobjective optimal design of the lithium-ion battery cell *J. Energy Storage* **17** 507–514
[23] X. Zeng and S. Luo 2017 Stochastic Pareto-optimal reinsurance policies *Insur. Math. Econ.* **53** 671–677
[24] J. Cai, H. Liu and R. Wang 2017 Pareto-optimal reinsurance arrangements under general model settings *Insur. Math. Econ.* **77** 24–37
[25] A. V. Asimit, V. Bignozzi, K. C. Cheung, J. Hu and E. S. Kim 2017 Robust and Pareto optimality of insurance contracts *Eur. J. Oper. Res.* **262** 720–732
[26] C. Niu, Y. Zhang and J. Guo 2018 Pareto optimal layout of multistatic radar *Signal Processing* **142** 152–156

Appendix

Table A. Unit price

Suppliers	Raw Material		
	R1	R2	R3
S1	10	20	40
S2	12	22	40
S3	11	22	41
S4	12	20	45

Table B. Order cost and transport cost

Supplier	Order Cost	Transport Cost
S1	120	100
S2	100	100
S3	100	110
S4	120	110

Table C. Defect product penalty cost

Supplier	Raw Material		
	R1	R2	R3
S1	5	5	4
S2	10	5	8
S3	8	10	12
S4	12	10	12

Table D. Defect rate

Supplier	Raw Material		
	R1	R2	R3
S1	0.02	0.01	0.09
S2	0.03	0.01	0.02
S3	0	0.02	0.03
S4	0.06	0.04	0

Table E. Late delivery rate

Supplier	Raw Material		
	R1	R2	R3
S1	0.02	0.02	0.05
S2	0.01	0.02	0
S3	0.00	0.05	0.05
S4	0.02	0.05	0.03

Table E. Late delivery penalty cost

Supplier	Raw Material		
	R1	R2	R3
S1	5	4	2
S2	5	4	3
S3	4	2	4
S4	3	2	2

Table F. Raw material required to produce the product

Raw Material	Product		
	P1	P2	P3
R1	10	12	10
R2	20	22	20
R3	4	2	2

Table F. Shortage cost

Product	Shortage cost	Raw Material	Short age cost
P1	4	R1	2
P2	3	R2	3
P3	4	R3	2

Table G. Required machine working hour to produce product unit

Product	Machine		
	M1	M2	M3
P1	4	2	2
P2	5	1	2
P3	4	1	2

Table H. Machine working hour max. capacity

M1	M2	M3
8500	7500	7000

Table I. Supplier maximum capacity to supply raw material

Supplier	Raw Material		
	R1	R2	R3
S1	5000	4500	6000
S2	4000	4500	5000
S3	8000	10000	5000
S4	9500	8000	7500

Journal of Physics: Conference Series **1397** (2019) 012076 doi:10.1088/1742-6596/1397/1/012076

Evenly brightening using kurtosis Gaussian pattern to simplify image binarization

T. M. S. Mulyana*, and Herlina

Informatics Engineering Department, Universitas Bunda Mulia, Lodan Raya Street No. 2, North Jakarta, Indonesia

*Corresponding author e-mail: tmulyana@bundamulia.ac.id

Abstract. A problem of separating object from background using thresholding in image segmentation is uneven image brightness. Some binary image objects vanish because the pixel intensity value on the edge side is greater than the pixel's background intensity. The others at the edge of image are covered by background pixel. The background pixel at the edge is bigger than the pixel value of the object in the inside of image. A light source type affects the spread of brightness adding thresholding difficulty. The object's light distribution from the exposed light source follows the Gaussian distribution. A kurtosis value depends on the light source type. Following the kurtosis of standard deviation, the method for creating brightness evenly is designed. The Gaussian pattern is formed from the distribution of Gaussian values in the cells of the Gaussian matrix. The pixel values in the corresponding images will be added to the matrix cells. The pixel value will increase according to the Gaussian matrix pattern. This process results in the highest pixel value not getting added value, while the lowest pixel intensity value will get the greatest value addition.

1. Introduction

A grayscale image using thresholding method is a general process to binarize an image. This process will compares the grayscale value from the image grayscale to the threshold value [1][2][3]. Thresholding process at the image binarization needed evenly bright of image. Unfortunately, the image capture process is due not always at evenly lightening. Uneven image brightness due to poor lighting causes uneven pixel intensity [4][5][6][7]. The part of the image that gets lighter will have a greater pixel intensity value so that it is brighter than other parts that are less light.

When the pixel intensity range is separated by the thresholding operation, the pixel value which is lower than the threshold value is set to 0, while the pixel intensity value that is greater than the threshold value is set to 1. While the background pixel's values are set to 1, some pixels of the object with greater than or equal of the threshold are set to 1 to. In other side, while the object pixels are set to 0, some pixels of background that has values are lower than threshold value is set to 0 too. Setting down threshold value to lower value or setting up threshold value cannot help the segmentation.

Journal of Physics: Conference Series **1397** (2019) 012076 doi:10.1088/1742-6596/1397/1/012076

Figure 1. Picture with lighting at center side

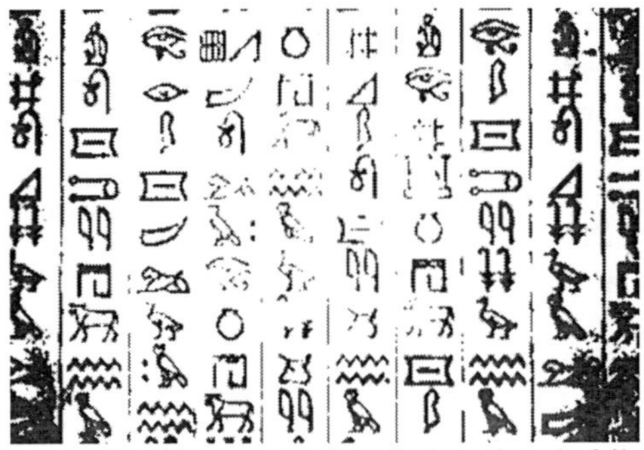

Figure 2. Binary result of the thresholding, some objects disappear, and the other blends with background

For example, the image that shown in Figure 1 has uneven pixel intensity, the center side of the image has higher values than others. After thresholding process, the binary image result can be seen Figure 2. Figure 2 shows that the object pixels on the image are broken as some pixels on the center side of the objects are set to white as background pixels. Other objects at the around side are broken because the background pixels near the objects are set to black as like as the pixels objects.

Because of unevenly of image brightness, the thresholding process to make a binary image cannot work properly. There are some objects disappear and the other are blended with background pixels. The different source of light kind causes the contrast of image part higher. The aim of this research is to simplify the thresholding of image binarization process. The purpose of this study is to simplify the process of image binarization using thresholding. Whereas, the spread conditions image brightness according to Gaussian kurtosis that caused by a single light source.

2. Method design

Generally, the light source has the distribution similar with Gaussian pattern [8][9]. The kurtosis of Gaussian is presenting depend on light source type. The brightening process using Gaussian pattern is the process to evenly the distribution of pixel intensity. Since the distribution of intensity is evenly in every pixel, the thresholding process to make a binary image will easier. Brightening process using Gaussian pattern is depend on the standard deviation value. Gaussian with two variables need the standard deviation to fill the value in each cell of Gaussian matrix. Standard deviation will determine the value of Gaussian in every two variables of x and y as the address of cell Gaussian matrix.

2.1. Standard deviation

The brightening process using the Gaussian pattern is based on two variables as seen in the Equation 1. A G(x,y) function is the distribution function of Gaussian at every cell where the address in row x and column y. The function is filling every cell at the address of x column x and row y with powering the exponential (e) constant value with minus of power x row that subtract with power of y column. Standard of deviation is distribution measure that represent deviation of means. Variant is average measurement deviation power of means [10][11][12].

$$G(x,y) = e^{-(x^2-y^2)/(2\sigma^2)} \tag{1}$$

Where:

G(x,y) : function of Gaussian distribution the cell (x,y),
x : column position
y : row position
e : exponential
σ^2 : a common variance

There are three kind of kurtosis that show in Figure 3 [13]. Zero kurtosis or Gaussian distribution called as Mesokurtic, caused by the standard deviation value is equal with the farthest cell distance from the center of matrix [14]. Positive kurtosis called as Leptokurtic, is caused while the standard deviation value is less than farthest cell distance from the center of matrix. Negative kurtosis called as Platykurtic is caused of the standard deviation value is greater than the farthest cell distance from the center of matrix [15]. The bigger standard deviation value makes the curve more focus, and the smaller of standard deviation value makes the curve flatter [16][17][18]. In other research, Gaussian filter of brightness bi-histogram equalization technique can be used for contrast stretching and brightness preservation to performance evaluation [19] and usually use to remove noise or to image smoothing process. [20][21].

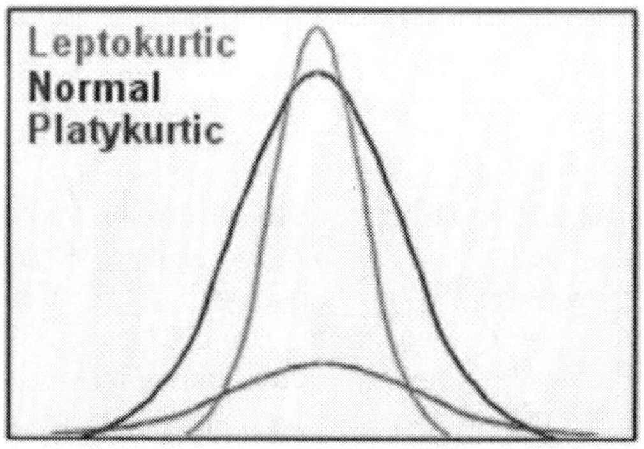

Figure 3. Standard of deviation affect to the kurtosis

2.2. Image evenly brightening and image binarization

Image binarization be processed with thresholding of grayscale intensity value from grayscale image. In this process, the thresholding following the brightness process using gaussian pattern. The brightness using gaussian pattern is due to evenly the distribution of intensity base on area division. This process to evenly the brightness is modeled by gaussian pattern. The gaussian matrix is build and be implemented to controlling distribution of pixels intensity of images. The intensity of light source that different one another at the captured of image process can adapted by standard of deviation that set to follow the gaussian distribution of pixels intensity distribution from the image. After grayscale intensity of every area division is evenly, the binarization process using thresholding process can be implemented.

$$matrixWidth = 2 \times imageWidth \quad (2)$$
$$matrixHeigh = 2 \times imageHeight \quad (3)$$

Flowchart are shown at the figure 4 show the sequence of overall evenly the brightness and binarization process to make a binary image. The brightening image is following the pattern of gaussian matrix. The height and width of image are needed to make the gaussian matrix size. Matrix width is set twice of image width as show on equation (1). Matrix height is set twice of image height as show on equation (2). The matrix size must be fourth of image size to accommodate the light focus of image that can be anywhere at the image. The size of the gaussian matrix which is twice the height and the width of the image to anticipate if the gaussian center is squeezed on one of the corners of the image, then the pixels at the other corner opposite the pixel remain on the matrix gaussians.

The standard of deviation at the gaussian zero mean is the achievement of the center of the matrix to the edge of the matrix. To determine the variant, the Standard of deviation is needed. The gradation values σ^2 in each cell of the gaussian matrix can be determined by variant, to make sure the farthest cell from the center in the gaussian matrix will be zero [22].

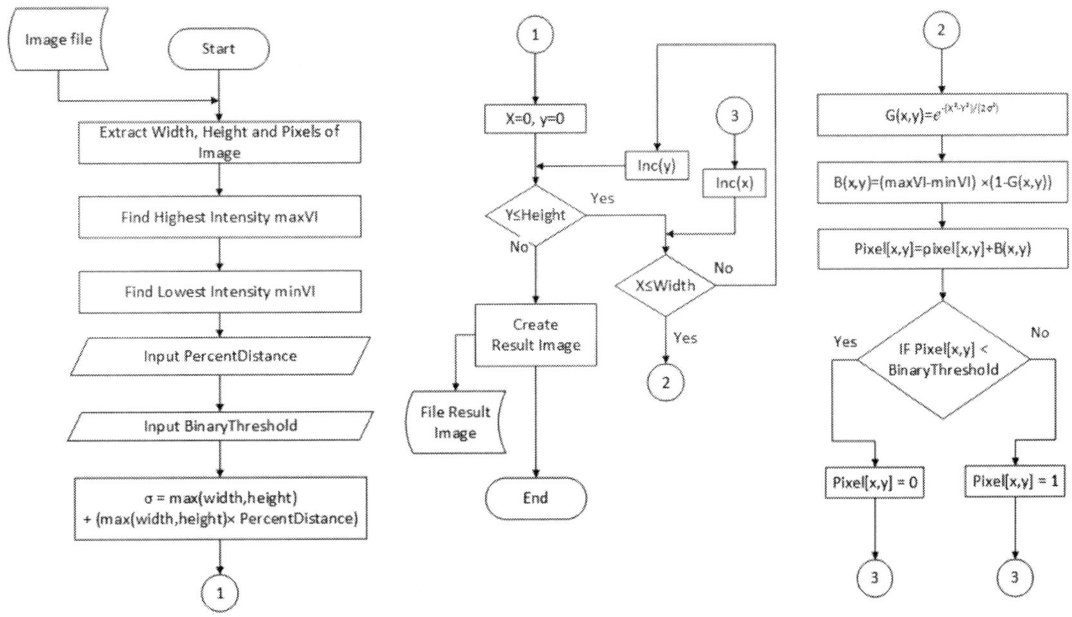

Figure 4. Flowchart process of brightening image using Gaussian pattern and binarization

2.3. *Kurtosis implementation*

The input of this process contains of input of 'percent distance' value to determine the standard of deviation based of the equation (4). The input range of 'percent distance' is from minus 90% to positive unlimited as the needed% to represent the empirical rule kurtosis. The range value standard deviation represents the type of distributed light intensity depends on source of light are used. Based on the above process, the zero mean gaussian with two variables is implemented to the gaussian matrix from the cell (0,0) until cell (2x,2y). Leptokurtic kurtosis is represented while the input 'percent distance' is negative, it means the gaussian matrix is smaller than image size. Mesokurtic kurtosis is represented while the input 'percent distance' is zero. It means that the gaussian matrix size is twice of image width and image height. Platykurtic kurtosis is represented while input 'percent distance' is positive, it means the cell contains of zero value is very far from the center of gaussian matrix.

$$\sigma = max(imageWidth, imageHeight) + max(imageWidth, imageHeight) \times inputPercent \quad (4)$$

$$BrigthnessMatrix(x, y) = (maxIntensity - minIntensity) \times (1 - G(x, y)) \quad (5)$$

Brightness matrix is build based on the Gaussian matrix is illustrated by Figure 5. This matrix is needed to control brightness at every pixel. To fill the brightness matrix cell is needed the distance between most high value and the lowest value of intensity from whole pixels at the image. The distance will reduce depend on the values at every cell of gaussian matrix as the percent to fill every cell at the brightness matrix. Equation (5) shows the process to fill every cell at matrix to control brightness.

Brightness matrix will superimpose to the bitmap, with the center of matrix is placed to the center of brightest pixels and every pixel is add with the value that cell of brightness matrix that superimpose. This process will reduce the pixels value as the pattern of Gaussian distribution.

0.05	0.12	0.19	0.22	0.19	0.12	0.05
0.12	0.26	0.44	0.51	0.44	0.26	0.12
0.19	0.44	0.72	0.85	0.72	0.44	0.19
0.22	0.51	0.85	1	0.85	0.51	0.22
0.19	0.44	0.72	0.85	0.72	0.44	0.19
0.12	0.26	0.44	0.51	0.44	0.26	0.12
0.05	0.12	0.19	0.22	0.19	0.12	0.05

Gaussian Matrix

0.95	0.89	0.81	0.78	0.81	0.89	0.95
0.89	0.74	0.57	0.49	0.57	0.74	0.89
0.81	0.57	0.28	0.15	0.28	0.57	0.81
0.78	0.49	0.15	0	0.15	0.49	0.78
0.81	0.57	0.28	0.15	0.28	0.57	0.81
0.89	0.74	0.57	0.49	0.57	0.74	0.89
0.95	0.89	0.81	0.78	0.81	0.89	0.95

Reversed of Gaussian Matrix

d = 24

22.8	21.2	19.5	18.6	19.5	21.2	22.8
21.2	17.7	13.6	11.7	13.6	17.7	21.2
19.5	13.6	6.79	3.7	6.79	13.6	19.5
18.6	11.7	3.7	0	3.7	11.7	18.6
19.5	13.6	6.79	3.7	6.79	13.6	19.5
21.2	17.7	13.6	11.7	13.6	17.7	21.2
22.8	21.2	19.5	18.6	19.5	21.2	22.8

Gaussian Brightness

Figure 5. Example of Gaussian brightness that build from reversed of Gaussian matrix with distance between minimum and maximum pixel intensity is 24

The input of 'binary threshold' is needed to make the thresholding process after pixels intensity is evenly distributed. If the result pixel is smaller than threshold value, the pixel will be set to 0. If the result pixel is greater than threshold value, the pixel will be set to 1.

3. Result and Discussion

The purpose of enlightenment with a Gaussian pattern is to simplify the process of image binarization. This method is tested for binarized images that contain very bright pixel portions. The results of this process are images that have uniform brigthness in each part of the image, so that the binarization process can be easier because the brightness distribution is almost the same. The sample images in Figures 6 through 9 show how the gaussian pattern brightness process is useful for comparing images. as show as the grayscale image in Figure 1. The central area is very bright, and the surrounding is dark gray.

Figure 6. Raw gray scale image

Figure 7. Gray scale image after brightening with gaussian pattern brightness

Figure 8. Binary image with thresholding until there are no black flood background, many objects at the center are hide

Figure 9. Binary image from result image of Gaussian pattern brightness

The image is showed in Figure 6 is the grayscale image before process by Gaussian pattern brightness. The piksel intesity is unevenly distributed. The image thresholding results of the image are shown at Figure 2 and Figure 8. There are many objects are broken. The center part of the image has the broken objects because some pixels are set to white as like as background pixels. If the threshold value be set to lower value, the binary image will be look like Figure 2. The center part still not so good enough and the corners side are flood by black pixels, so the objects are broken too.

The Image that has been process by gaussian pattern brightness is shown at Figure 7. Because all over part are looks like the same brightness, there is no too light part or too dark part except the objects. The image result of thresholding process is showen at Figure 9. The objects contain of this image separate better and clearly. There are no objects are hide or broke.

Every part of matrix divides by n × n pixels. If the most right and the most bottom area is less then n will be replicated from the most right and most bottom pixel that contains of intensity value of the image. Average of the pixel intensity every part is measured to find out the smallest and biggest average to meassure the success. Distance between the smallest and biggest average from the divided areas will compare to measure the effectiveness of brightness.

The input percent of standard deviation will be trace from +200% to -90% of matrix perimeter to find out the best input. The best input is first time of distance between highest and lowest average value to the lowest value before it raise again. The lowest value after the it raised up again is not useful because it means all pixel became very light until white. If nothing the evenly brightness, the process is noted as fail.

The reason of not use biggest and smallest intensity at the measurement is: because the object pixel must very dark and the background must very bright, so the measurement pixel to pixel cannot represent the evenly of intensity. The using the average intensity of area of n × n pixels are presented the condition of evenly the distribution of pixel intensity. Every part of image will contain of some pixel object and some pixel background.

The experiments will be due depend on kurtosis of leptokurtic, mesokurtic and platykurtic of standard deviation as the input percent to generate the gaussian matrix. To compare how is the kurtosis will affect to the result and to proving the advantage that this process is make binarization simpler, the result of brightening by gaussian will be binarized to the binary images. The measurement of effectively this process is how clear the object binary image that separate from background by the binarization process.

Table 1. The success of brightening using Gaussian pattern brightness

Gaussian Pattern Brightness Kurtosis	Brightening Success	Binarization Success		
		Before Brightening	After Brightening	Increase
Leptokurtic	95%	32%	77%	45%
Platykurtic	92%	50%	100%	50%
Mesokurtic	100%	100%	100%	0%
Average	96%	61%	92%	32%

The success of brightening using Gaussian Pattern Brightness is showed at Table 1. The Brightening process in Leptokurtic pattern has 95% of brightening proses and can increase the success of binarized 45% from 32% binarized success without brightening to 77% binarized success with brightening. The brightening process in Platykurtic pattern has 92% of brightening process and can increase the success of binarized 50% from 50% success binarized without brightening process to 100% success binarized with brightening process. The brightening process in Mesokurtic pattern has 100% of brightening process and can increase the success of binarized 0% because both success binarized without brightening and success brightening process are 100% success. The overall success of brightening is 96%. Overall increasing success of binarized with gaussian pattern brightening is 32% from 61% success dari binarized without brightening of gaussian pattern. The success is caused by the brightness of every part of image are evenly, so that the thresholding value of image binarization at every part are similar.

4. Conclusion
Brightening process at Gaussian pattern is success to evenly the image that intensity distribution is not evenly. The leptokurtic model is the worst model to evenly brightness. Leptokurtic type enlightenment process reaches 95%, but it has effective on binarization is as good as other patterns. Brightening process using Gaussian pattern is effective to increase 32% the success value of binarized from 61% to 92%.

References
[1] R D Atmaja, M A Murti, J Halomoan, F Y Suratman 2016 Journal of Electrical Engineering and Computer Science Vol. 3 No. 2, August 2016, pp. 377 ~ 382, DOI: 10.11591/ijeecs.v3.i2.pp377-382
[2] T R Singh, S Roy, O I Singh, T Sinam, K M Singh 2011 IJCSI International Journal of Computer Science Issues, Vol. 8, Issue 6, No 2, November 2011, ISSN (Online): 1694-0814, p271-277
[3] T M S Mulyana 2016 International Conference on Information & Communication Technology and Systems (ICTS) Proceedings 12-12 Oct. 2016, IEEE, DOI: 10.1109/ICTS.2016.7910287
[4] Moreno, Ivan & Tsai, Chang-Yu & Bermudez, David & Sun, Ching-Cherng 2007 Proceedings of SPIE - The International Society for Optical Engineering. art. no. 66700H p 6670. doi: 10.1117/12.735820.
[5] T R M Filho, K M Weig, M M Werneck, C A C Neto, M F da Costac 2015 Journal of Biomedical Optics May 2015 • Vol. 20(5) p.055005-1:055005-5
[6] Feng Lu, Y Matsushita, I Sato, T Okabe, and Y Sato 2015 IEEE Transactions on Pattern Analysis and Machine Intelligence Volume: 37 , Issue: 10 , Oct. 1 2015 (DOI: 10.1109/TPAMI.2015.2389841) Page(s): 1999 - 2012
[7] I Sato, T Okabe, Y Sato, K Ikeuchi 2005 Proceedings of the IEEE International Conference on Computer Vision - ICCV 2005 October 2005 pp. 325-332,
[8] Y Xu, Y Li and Y Qiu 2009 Journal of Modern Optics - J MOD OPTIC. 56. 2329-2334. 10.1080/09500340903398440.
[9] S Schoenecker and T Luginbuhl 2016 IEEE Signal Processing Letters Volume: 23 , Issue: 5 , May 2016) Page(s): 644 - 647 (DOI: 10.1109/LSP.2016.2537981)
[10] E. L. A. Picasso and J. M. P. Fortes 2008 in Electronics Letters, vol. 44, no. 3, pp. 194-196, 31 January 2008) (doi: 10.1049/el:20082550)
[11] E. A. Aronson 1974 IEEE Transactions on Aerospace and Electronic Systems, vol. AES-10, no. 5, pp. 609-614, Sept. 1974 (doi: 10.1109/TAES.1974.307853)
[12] E. Wong 1966 SIAM Journal on Applied Mathematics, 1966, Vol. 14, No. 6 : pp. 1246-1254 https://doi.org/10.1137/0114099
[13] Petar Cisar and Sanja Maravic Cisar 2010 Acta Polytechnica Hungarica Vol. 7, No. 2 p 95
[14] L T. DeCarlo 1997 Psychological Methods, Vol. 2, No. 3,292-307 1082-989X/97
[15] A. Mansour and C. Jutten 1999 IEEE Signal Processing Letters, vol. 6, no. 12, pp. 321-322, Dec. 1999 doi: 10.1109/97.803435
[16] A H Monahan 2013 J. Climate, 26, p:5563–5577, https://doi.org/10.1175/JCLI-D-12-00424.1
[17] G Marsaglia 2006 Journal of Statistical Software May 2006, Volume 16, Issue 4. http://www.jstatsoft.org/ p1-10
[18] S Mitra, D Mazumdar, K G Corresp., K Bhaumik 2018 PeerJ Preprints 6 Apr 2018 https://doi.org/10.7287/peerj.preprints.26831v1
[19] S.A.T.I, Vidisha, M Pradesh, Y K Jain, G Silakari 2013 International Journal of Emerging Technology and Advanced Engineering Volume 3, Issue 2, February 2013 p:321-327
[20] K R Babu, K.V.N.Sunitha, W G Dist, A Pradesh, G.Narayanamma, Shaikpet 2011 IJCSIT - International Journal of Computer Science and Information Technologies, Vol. 2 No. 1 ISSN:0975-9646 p. 504-511

[21] P A Yirenkyi, J K Appati, I K Dontwi 2016 Open Journal of Applied Sciences, p478-488, http://dx.doi.org/10.4236/ojapps.2016.67048
[22] G Lee, Y Ji Hyun, S Y Ohm, and G C Min 2012 ICHIT, LNCS 7425, pp. 365–372, 2012. © Springer-Verlag Berlin Heidelberg 2012 p:139-774

Evaluation of geographically weighted multivariate negative Binomial method using multivariate spatial infant mortality data

Y S Dewi[1,2], Purhadi[1*], Sutikno[1], and S W Purnami[1]

[1]Department of Statistics, Faculty of Mathematics, Computing and Data Sciences, Institut Teknologi Sepuluh Nopember, Indonesia
[2]Department of Mathematics, Faculty of Mathematics and Natural Sciences, University of Jember, Indonesia

[*]Corresponding author e-mail: purhadi@statistika.its.ac.id

Abstract. Global regression assumes that the relationships being measured are stationary over space or the model is applied equally over the whole region. If there is spatial heterogeneity on the data, then the global model is not suitable to the reality. To overcome multivariate spatial over dispersed negative binomial data, we evaluate geographically weighted multivariate negative binomial (local method) and compare it to the global method (multivariate negative binomial). The results show that the geographically weighted negative binomial performs better than the global method. The log likelihood of the local method is higher than the global method. The deviance and mean square prediction error of the local method are smaller than the global method. Moreover, the prediction of dependent variables of the local method are closer to the observed data than the global method. The estimated coefficients of the local method vary, depending on where the data are observed.

1. Introduction

Spatial data can be found in many fields, such as in transportation [1], [2], mortality [3], economic [4] and demographic [5]. The characteristic of spatial data is the existence of spatial effects on the data, the dependence between observation and location. The spatial effects can be categorized into two types [6], spatial dependence and spatial heterogeneity. In the spatial dependence, there is statistical dependence in a collection of random variables, each of which is associated with a different location. The spatial dependence leads to development of spatial area researches, Conditional Autoregressive, Spatial Autoregressive and Spatial Autoregressive Moving Average Model, among others [4], [6]. The spatial heterogeneity occurs as effects of differentiation between locations. It has the implication on the model vary between locations. It leads to development of Geographically Weighted Regression [7].

The Geographically Weighted Regression is a local regression model used to model spatially varying relationships. The spatial variation in coefficients can reveal interesting pattern contained in the spatial data. Some spatial count data researches were conducted using different techniques. For analyzing univariate spatial count data, Geographically Weighted Poisson Regression (GWPR) is described for analyzing non stationary count data and Semiparametric GWPR (SGWPR) model for mixed model is proposed, where there are global and local parameters in the model [3]. The estimation of parameters was conducted by using maximum likelihood estimation (MLE) method. The application of Generalized Linear Model (GLM) and GWPR on transportation case are described in [2]. The results show that the performance of GWPR model is better than GLM based on corrected Akaike Information Criterion (AICc). However, the GWPR assumes the equidispersion of the data. Unfortunately this assumption is often violated in the real data because data are often overdispersed. One of the methods for overcoming overdispersion is by using a geographically weighted negative binomial model [1].

On the other hand, many researchers have been developing the global regression models for Poisson data with overdispersion. For univariate dependent variable, see [8], [9], [10] and [11]. For more than one dependent variables, the multivariate negative binomial model using copula and MLE method is proposed [12]. The other researches are a robust likelihood approach for the overdispersed correlated count data analysis based on a multivariate negative binomial model using MLE via iterative Newton Raphson algorithm [13], a multivariate generalized Poisson regression model using MLE method [14] and the comparison of two bivariate negative binomial regression models that come from the different distributions derivation [15]. In a similar way to [16], a bivariate negative binomial distribution as a product of negative binomial marginals with a multiplicative factor is defined in [17]. The parameters are estimated using MLE via Newton Raphson algorithm.

There is a relationship between observations from different locations. The near things are more related than the distant things. Accordingly, when we use global method for analyzing spatial data, there is violation to the condition, because the global method assumes that there are no relationships between observations or observations are independent in different locations. Moreover the global regression assumes that the relationship is stationary modeled throughout space, although it might be non-stationary. These conditions are not always suitable for spatial data because there are spatial effects on it. For that reason, if there is non-stationariness of the relationships between locations, the global model is not suitable with reality. Therefore it is necessary to use appropriate method to model such data. Unfortunately, the researches related to multivariate spatial data are less developed. On the other side, many data in real applications are multivariate spatial data. Therefore, the purpose of this research is to evaluate the geographically weighted multivariate negative binomial method on multivariate spatial overdispersed negative binomial data using infant mortality data with spatial effect and overdispersion. We compare it to the multivariate negative binomial (global method). We investigate five aspects in this research, the likelihood, deviance, mean square prediction error, coefficients estimate and closeness of the prediction (means of dependent variables) to the observed data.

2. Geographically weighted multivariate negative Binomial method

The model in this research is built based on multivariate negative binomial distribution , where the observation y_{ij} follow Poisson distribution with assumption $Y_{ij} \mid \varphi_i \sim$ Poisson $(\varphi_i \mu_{ij})$, $i = 1,2,..., n, j = 1,2,...,m$, μ_{ij} is the mean of dependent variable-j on location-i and φ_i is unobservable effect of location-i, φ_i follow gamma distribution or $\varphi_i \sim$ Gamma (τ^{-1}, τ^{-1}), where $E(\varphi)=1$ and $Var(\varphi)=\tau$.

2.1. The model

Let y_{ij} is the observation on location-i, dependent variable-j, where $i = 1,2,\ldots,n$ and $j = 1,2,\ldots,m$. By using the assumption above and letting $\delta = 1/\tau$, the joint probability density function of $(\mathbf{y}_i, \varphi_i)$ is given by

$$f\left(\mathbf{y}_i,\varphi_i;\boldsymbol{\mu},\delta\right)=\frac{\left(\prod_{j=1}^{m}\mu_{ij}^{y_{ij}}\right)\delta^{\delta}}{\left(\prod_{j=1}^{m}y_{ij}!\right)\Gamma(\delta)}\exp\left(-\varphi_i\left(\sum_{j=1}^{m}\mu_{ij}+\delta\right)\right)\varphi_i^{\sum_{j=1}^{m}y_{ij}+\delta-1} \tag{1}$$

By letting $a_i=\varphi_i\left(\sum_{j=1}^{m}\mu_{ij}+\delta\right)$ and integrating the function of a_i with respect to φ_i, the marginal probability mass function of $\mathbf{y}_i$ can be written as follows

$$f\left(\mathbf{y}_i;\boldsymbol{\mu},\delta\right)=\frac{\left(\prod_{j=1}^{m}\mu_{ij}^{y_{ij}}\right)\delta^{\delta}\Gamma\left(\delta+y_{i+}\right)}{\left(\prod_{j=1}^{m}y_{ij}!\right)\Gamma(\delta)\left(\delta+\mu_{i+}\right)^{\delta+y_{i+}}} \tag{2}$$

for i = 1,2,...,n and j = 1,2,...,m, $y_{i+}=\sum_{j=1}^{m}y_{ij}$ and $\mu_{i+}=\sum_{j=1}^{m}\mu_{ij}$. The probability mass function (2) is known as multivariate negative binomial, where

$E\left(Y_{ij}\right)=\mu_{ij}$, $\text{Var}\left(Y_{ij}\right)=\frac{\mu_{ij}^2}{\delta}+\mu_{ij}$ and $\text{Corr}\left(Y_{ij},Y_{il}\right)=\frac{\sqrt{\mu_{ij}\mu_{il}}}{\sqrt{\delta+\mu_{ij}}\sqrt{\delta+\mu_{il}}}$.

The likelihood function of (2) is given by

$$L(\boldsymbol{\theta})=\prod_{i=1}^{n}\frac{\left(\prod_{j=1}^{m}\mu_{ij}^{y_{ij}}\right)\delta^{\delta}\Gamma(y_{i+}+\delta)}{\prod_{j=1}^{m}y_{ij}!\,\Gamma(\delta)(\delta+\mu_{i+})^{(\delta+y_{i+})}} \tag{3}$$

Subsequently, by defining $\mu_{ij}\left(u_i,v_i\right)=t_{ij}e^{\mathbf{x}_i^T\boldsymbol{\beta}_j(u_i,v_i)}$, we build the geographically weighted multivariate negative binomial, where $\mu_{ij}\left(u_i,v_i\right)$ is the mean of dependent variable-j, at location-i, while $\left(u_i,v_i\right)$ is the coordinate of the location-i, t_{ij} is exposure variable of location-i and dependent variable-j, x_{ij} is independent variable at location-i and dependent variable-j, and $\boldsymbol{\beta}_j\left(u_i,v_i\right)$ is the vector coefficients of the regression at location-i and dependent variable-j. Based on [1] and equation (3), the log likelihood function of the local model can be written as

$$\ell\left(\boldsymbol{\theta}^*\right)=\sum_{i=1}^{n}\left\{\left(\sum_{j=1}^{m}y_{ij}\ln\left(t_{ij}e^{\mathbf{x}_i^T\boldsymbol{\beta}_j(u_l,v_l)}\right)\right)-\left(\delta\left(u_l,v_l\right)+y_{i+}\right)\ln\left(\delta\left(u_l,v_l\right)+\sum_{j=1}^{m}\left(t_{ij}e^{\mathbf{x}_i^T\boldsymbol{\beta}_j(u_l,v_l)}\right)\right)+A_i\right\}w_{il} \tag{4}$$

where i,l= 1,2,3,…,n, $A_i=\delta\left(u_l,v_l\right)\ln\delta\left(u_l,v_l\right)-\sum_{j=1}^{m}\ln y_{ij}!+\ln\left\{\frac{\Gamma(y_{i+}+\delta\left(u_l,v_l\right))}{\Gamma(\delta\left(u_l,v_l\right))}\right\}$,

$\boldsymbol{\theta}^*=\left[\boldsymbol{\beta}_1^T\left(u_l,v_l\right)\ \boldsymbol{\beta}_2^T\left(u_l,v_l\right)...\boldsymbol{\beta}_m^T\left(u_l,v_l\right)\ \delta\left(u_l,v_l\right)\right]^T$ and w_{il} is the geographical weight.

2.2. The geographical weight and cross validation

The geographical weight is a value of weights relative to the position of (u_i, v_i) in the study area. The weights themselves are computed from a weighting scheme which is also known as kernel. In this research, we use the Fixed Bisquare Kernel weight, w_{il} as a continuous function of d_{il}. It can be defined as shown in [7],

$$w_{il} = \begin{cases} \left(1-\left(\frac{d_{il}}{b}\right)^2\right)^2 ; & \text{untuk } d_{il} \le b \\ 0 ; & \text{untuk } d_{il} > b \end{cases} \quad (5)$$

where b is referred to as the bandwidth, and d_{il} is the Euclidean distance between point-*i* and *l*.

The optimum bandwidth is obtained by cross-validation using the formula

$$CV(b) = \sum_{i=1}^{n}\sum_{j=1}^{m}\left(y_{ij} - y_{\neq ij}(b)\right)^2 \quad (6)$$

where $\hat{y}_{\neq ij}(b)$ is the fitted value of y_{ij} with the observation for location-*i* omitted from the calibration process, *n* and *m* is the number of location and dependent variables respectively. The optimum bandwidth is the bandwidth with minimum *CV*.

3. Method and algorithm

We use infant mortality data 2014 from East, Middle and West Java, Indonesia [18], [19], [20]. There are three dependent variables (Y_1, Y_2 and Y_3) which have positive correlation between them and six independent variables ($X_1 - X_6$), where they are direct and underlying causes of infant mortality. The description of the variables used in this research is presented in Table 1.

Table 1. Research variables of infant mortality data.

Variable	Description
Y_1	The number of birth deaths
Y_2	The number of neonatal deaths (after birth to 1 month)
Y_3	The number of infancy deaths (1 month to 1 year)
X_1	The percentage of handling obstetric complications
X_2	The percentage of households that have healthy behavior
X_3	The percentage of integrated health posts giving service actively
X_4	The percentage of active family planning participant
X_5	The percentage of prenatal visits to the health worker minimum four times
X_6	The percentage of prenatal getting Fe3 (90 pills)

We use the coordinates of latitude and longitude of the regencies/towns in East, Middle and West Java Indonesia as geographical factor of locations. There are 38, 35 and 27 regencies/towns in East, Middle and West Java, Indonesia respectively. Thus, there are 100 locations in total.

Geographically Weighted Negative Binomial uses MLE method via Newton Raphson algorithm for the estimation of parameters [1]. However, the algorithm has the weakness in dealing with false convergence due to improper initial value. Moreover, when analyzing more than one dependent variables, things are more complicated. In this research, we use MLE method via Nelder Mead algorithm for estimating the mean of dependent variables. Nelder Mead is an alternative algorithm for Newton Raphson. This is free derivative and robust algorithm related to the initial value for geographically weighted multivariate method [21]. The method is nonlinear optimization technique for maximizing or minimizing function on multi dimension. The Nelder Mead algorithm is a popular algorithm, because it is reliable enough even for researches in high dimension.

The algorithm proceeds through operations of the simplex to find local optimum of log likelihood function. Let $\boldsymbol{\Theta}$ is p dimension parameter space and $\ell(\boldsymbol{\theta}^*)$ is a function that we want to maximize (log likelihood function) or $-\ell(\boldsymbol{\theta}^*)$ minimized, where $\boldsymbol{\theta}^* \in \boldsymbol{\Theta}$ and $\boldsymbol{\theta}^* = [\beta_{1,0}(u_l,v_l)\ \beta_{1,1}(u_l,v_l) \ldots \beta_{1,p-1}(u_l,v_l)$ $\beta_{2,0}(u_i,v_i)\ \beta_{2,1}(u_l,v_l) \ldots \beta_{2,p-1}(u_l,v_l)\ \beta_{3,0}(u_l,v_l) \ldots \beta_{m,p-1}(u_l,v_l)\ \delta(u_l,v_l)]$. Each iteration is started from simplex, that is the structure formed by $p+1$ points, not in the same plane, in an p-dimension space, $p = mp + 1$ where m is the number of dependent variables and p is the number of parameters for each dependent variable. Based on the evaluation of $-\ell(\boldsymbol{\theta}^*)$ on the vertex, the vertex having the worst function value is replaced by a new point with a better function value. By assuming that optimum is minimum, the vertex with highest value of the function is replaced by the new point with a lower value of the function through one of reflection, expansion or contraction operations. If all of these operations fail to find new point to replace the worst point, then the simplex will shrink to the vertex with lowest function value. See [22] and [23] for Nelder Mead algorithm in details. The cross validation is conducted using the golden section algorithm to find an optimum bandwidth. The initial estimate of geographically weighted multivariate negative binomial is taken from the coefficients and index of dispersion estimate of multivariate negative binomial for estimating the mean of dependent variables. The data analysis is conducted using R.3.4.3 software with MASS, REdas, spgwr and GW model packages.

4. Result and Discussion

The model used in this research consists three dependent variables. All correlation coefficients between the dependent variables are positive. They are 0.552, 0.449 and 0.706 for the correlation between Y_1 and Y_2, Y_1 and Y_3, and Y_2 and Y_3 respectively. Moreover the variance of dependent variables are much larger than their mean, it indicates the existence of overdispersion in the data. The variance of Y_1, Y_2 and Y_3 are 3507.36, 4444.65 and 547.00, while their means are 94, 117 and 31 respectively. Therefore, we consider multivariate negative binomial distribution for this data.

There is a high correlation between two independent variables X_5 and X_6 (Correlation = 0.909), We omit X_6 from the data analysis for overcoming multicollinearity. The selection of independent variables uses a forward stepwise method. The model with four independent variables (X_1, X_2, X_3 and X_4) has the best performance. Therefore we use those variables for further analysis.

By using four variables (X_1, X_2, X_3 and X_4) and Nelder-Mead algorithms in section 3, we find the optimum bandwidth b = 4.298 and CV = 885910.30. The geographically weighted multivariate negative binomial (local method) performs better than the multivariate negative binomial (global method) as shown in Table 2. Three criteria of the likelihood, deviance and mean square prediction error (MSPE) are used to compare the two methods. The local method has higher log likelihood and smaller deviance than the global method. Moreover, the behavior of the residual of the local method is better than the global method, indicated by a small MSPE value. The MSPE of local method is smaller than the global method.

Table 2. Goodness of fit of the local and global methods.

Method	Log likelihood	Deviance	MSPE
Local	-2049.7	1925.3	2500.8
Global	-2122.2	2073.4	2542.2

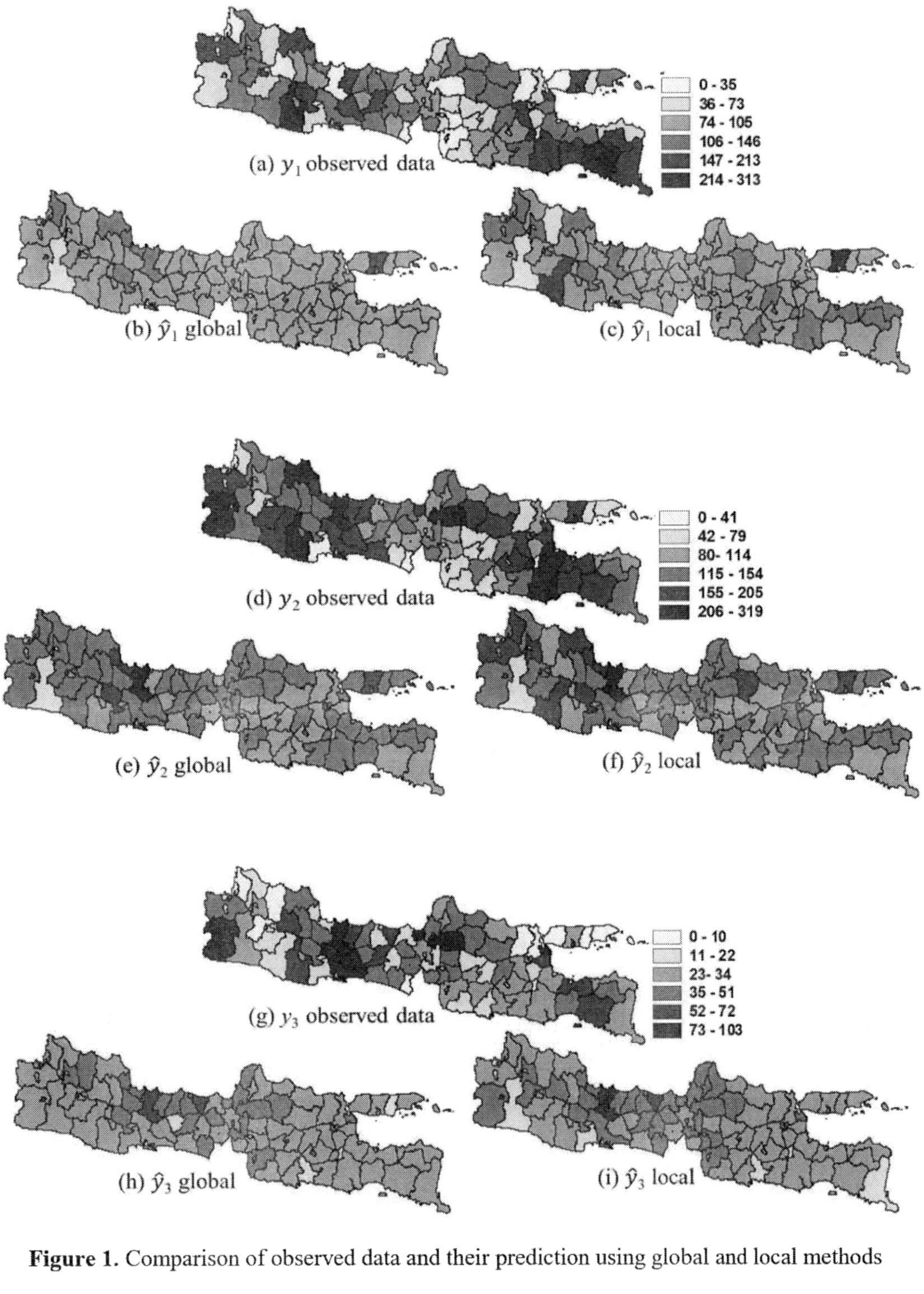

Figure 1. Comparison of observed data and their prediction using global and local methods

The comparison of observed data ($\mathbf{y}_1$, $\mathbf{y}_2$ and $\mathbf{y}_3$) and their prediction are presented in Figure 1. The darker color represents the higher of deaths than lighter color. That figure shows that the prediction of local method tends to close to the observed data than the global method. This indicates that the local method predicts better the dependent variables than the global method. This is in accordance with

MSPE result, that the local method is better than the global method. Although the local method has better prediction than global method, the time for local method is longer than it for global method, mainly for cross validation process. Building one model in local method similar to building n model in global method, where n is the number of locations.

The coefficients estimates of global and local methods and their standard error (in parentheses) are presented in Table 3. They show that there are differences between the coefficients estimate resulted by both methods. This is because the coefficients estimates of spatial data model have the local character, those depend on the location where the data are observed.

Table 3. Estimated coefficients for the three locations using global and local methods.

Coefficient	Global	Local		
		Madiun	Cilacap	Karawang
β_{10}	4.505 (0.459)	4.678 (0.933)	4.440 (0.664)	4.451 (1.877)
β_{11}	0.001 (0.003)	0.004 (0.008)	0.002 (0.004)	-0.002 (0.005)
β_{12}	-0.002 (0.003)	-0.003 (0.005)	-0.002 (0.004)	0.015 (0.007)
β_{13}	-0.003 (0.003)	-0.011 (0.006)	-0.003 (0.004)	-0.008 (0.005)
β_{14}	0.003 (0.003)	0.004 (0.003)	0.003 (0.005)	-0.003 (0.023)
β_{20}	4.657 (0.458)	4.622 (0.931)	4.700 (0.663)	5.622 (1.870)
β_{21}	0.005 (0.003)	0.009 (0.008)	0.005 (0.004)	0.003 (0.005)
β_{22}	0.000 (0.003)	0.002 (0.005)	0.000 (0.004)	0.010 (0.007)
β_{23}	-0.007 (0.003)	-0.015 (0.006)	-0.009 (0.004)	-0.012 (0.005)
β_{24}	0.002 (0.003)	0.002 (0.003)	0.001 (0.005)	-0.015 (0.023)
β_{30}	2.387 (0.471)	2.757 (0.956)	2.466 (0.680)	2.438 (1.930)
β_{31}	0.008 (0.003)	0.009 (0.008)	0.009 (0.004)	0.007 (0.005)
β_{32}	0.007 (0.003)	0.011 (0.005)	0.006 (0.005)	0.008 (0.007)
β_{33}	-0.004 (0.003)	-0.015 (0.006)	-0.005 (0.004)	-0.004 (0.005)
β_{34}	0.001 (0.003)	0.001 (0.003)	0.001 (0.005)	-0.001 (0.024)
τ	0.339 (0.046)	0.396 (0.087)	0.322 (0.062)	0.307 (0.092)
Loglikelihood	-2122.244		-2049.741	
Deviance	2073.387		1925.304	
MSPE	2542.177		2500.843	

5. Conclusion

Geographically weighted multivariate negative binomial is a development method of geographically weighted negative binomial. This method is used when there are multivariate spatial count data having positive correlation between them. The dependent variables are predicted by independent variables, where each location has the local character coefficients based on where the data are observed.

The geographically weighted multivariate negative binomial performs better than the global method (multivariate negative binomial) in predicting the means of multivariate spatial overdispersed data. This is indicated by higher log likelihood and smaller deviance and MSPE. Moreover, the local method predicts the dependent variables better than the global method. The prediction of the local method tends to be closer to the observed data than the global method. However the computational time for local method is longer than the global method.

Acknowledgement

The authors thank East, Middle and West Java Provincial Health Department for providing access to the infant mortality data.

References

[1] da Silva A R and Rodrigues T C V 2014 Geographically Weighted Negative Binomial Regression-incorporating overdispersion *Stat. Comput.* **24** 769–83
[2] Hadayeghi A, Shalaby A S and Persaud B N 2010 Development of planning level transportation safety tools using Geographically Weighted Poisson Regression *Accid. Anal. Prev. J.* **42** 676–88
[3] Nakaya T, Fotheringham A S, Brunsdon C and Charlton M 2005 Geographically weighted Poisson regression for disease association mapping *Stat. Med.* **24** 2695–717
[4] LeSage J and Kelley R 2009 *Introduction to Spatial Econometrics* (Chapman and Hall/CRC)
[5] Mullen W F, Jackson S P, Croitoru A, Crooks A, Stefanidis A and Agouris P 2015 Assessing the impact of demographic characteristics on spatial error in volunteered geographic information features *GeoJournal* **80** 587–605
[6] Anselin L 2001 Spatial econometrics *A companion to Theor. Econom.* 310–30
[7] Fotheringham A S, Brunsdon C and Charlton M 2002 Geographically Weighted Regression-The Analysis of Spatially Varying Relationships
[8] Dean C and Lawless J F 1989 Tests for detecting overdispersion in poisson regression models *J. Am. Stat. Assoc.* **84** 467–72
[9] Consul P C and Famoye F 1992 Generalized poisson regression model *Commun. Stat. - Theory Methods* **21** 89–109
[10] Ridout M, Demetrio C G . and Hinde J 1998 Models for count data with many zeros *Int. Biometric Conf.* 1–13
[11] Famoye F and Singh K P 2006 Zero-Inflated Generalized Poisson Regression Model with an Application to Accident Data *J. Data Sci.* **4** 117–30
[12] Shi P and Valdez E A 2014 Multivariate negative binomial models for insurance claim counts *Insur. Math. Econ.* **55** 18–29
[13] Solis-Trapala I L and Farewell V T 2005 Regression analysis of overdispersed correlated count data with subject specific covariates *Stat. Med.* **24** 2557–75
[14] Famoye F 2015 A Multivariate Generalized Poisson Regression Model *Commun. Stat. - Theory Methods* **44** 497–511
[15] Dewi Y S, Purhadi, Sutikno and Purnami S W 2017 Comparison of Bivariate Negative Binomial Regression Models for Handling Over dispersion *Int. J. Appl. Math. Stat.* **56** 53–62
[16] Lakshminarayana J, Pandit S N N and Srinivasa Rao K 1999 On a bivariate poisson distribution *Commun. Stat. - Theory Methods* **28** 267–76
[17] Famoye F 2010 On the bivariate negative binomial regression model *J. Appl. Stat.* **37** 969–81
[18] Dinas Kesehatan Provinsi Jawa Timur 2015 *Profil Kesehatan Provinsi Jawa Timur 2014*
[19] Dinkes Jawa Tengah 2014 *Profil Kesehatan Provinsi Jateng Tahun 2014*
[20] Dinas Kesehatan Provinsi Jawa Barat 2015 *Profil Kesehatan Provinsi Jawa Barat Tahun 2014*
[21] Dewi Y S, Purhadi, Sutikno and Purnami S W 2019 Comparison of Nelder Mead and BFGS Algorithms on Geographically Weighted Multivariate Negative Binomial *Int. J. Adv. Sci. Enginering Inf. Technol.* **9** 979–87
[22] Nash J C 2014 On best practice optimization methods in R *J. Stat. Softw.* **60** 1–14
[23] Small C G and Wang J 2003 *Numerical methods for nonlinear estimating equations* (Oxford University Press on Demand)

Journal of Physics: Conference Series **1397** (2019) 012078 doi:10.1088/1742-6596/1397/1/012078

Valuation risk adjusted deposit insurance on heston model

N Hariati , M Yunus and E R M Putri

Department of Mathematics, Faculty of Mathematics, Computation and Data Sciences, Institut Teknologi Sepuluh Nopember, Sukolilo- Surabaya 60111, Indonesia

Abstract. Deposit insurance is a tool to stabilize the banking system. In managing deposit insurance in Indonesia, deposit insurance agency applies the flat rate premiun system which is the same premium imposition system for each bank without taking into account the different risk levels of each bank. The implementation of the flat rate, it can still cause moral hazard which can trigger the monetary crisis. To anticipate this problem, it is necessary to design a risk adjusted premium. One of method that can be used in determining deposit insurance premium is Fourier transform. In this study, we analyzed analytical solutions to determine risk adjusted deposit insurance premiums on Heston Model by using Fourier transforms. Based on the simulation results, it was found that the amount of volatility caused the value of deposit insurance premiums getting bigger, the smaller value of interest rates so the greater value of deposit insurance premiums. The high value of deposit insurance premiums is also influenced by the value of debt obligations and dividends value.

1. Introduction

The monetary crisis occurred in the last two decades, namely the monetary crisis that occurred in Southeast Asia and the global monetary crisis. One of the causes of the monetary crisis in various countries is the presence of moral hazard actions from the banking owners. The condition of the banking system is an important factor in the economy of a country. The bank functions as an institution that absorbs public funds in the form of savings, deposits, and so on and distributes these funds in the form of credit, working capital, investment credit, and consumption [1]. At the time of the Southeast Asian monetary crisis, Indonesia bank closed 16 small banks during the 1997 monetary crisis without guarantee of customer deposits. This triggered the occurrence of bank runs in several private banks. To anticipate this, the government established a blanket guarantee policy. The blanket guarantee policy can increase public confidence in bank storage, but it imposes a state budget due to unlimited guarantees. The Government prepares the implementation of the Deposit Insurance System (DIS) on the date through the Deposit Insurance Corporation. Deposit insurance agency applies a flat rate premium system, namely the same premium imposition system for each bank without taking into account the different levels of risk of each bank. The application of flat rate premiums in deposit insurance still poses a burden on the state that must be borne by the government. This is because the amount of premium received by the government is not proportional to the number of guarantee claims that must be paid.

Research on determining deposit insurance premiums was initiated by Merton using option theory but, in his research determining flat premiums [2]. Research on deposit insurance premiums was also carried out by Ronn and Verma using an isomorphic relationship between insurance and put options on the Black-Scholes model [3]. In the Black-Scholes model the underlying asset volatility of option prices is constant during the option period. According to Heston [4] the underlying asset volatility of option prices is not constant but follows a stochastic process with different forms. Therefore, a study was conducted on the determination of risk adjusted deposit insurance premiums in the Heston model.

2. Heston Model

The Heston model is a volatility stochastic model that was first introduced by (Steven L Heston, 1993) [4]. The stochastic process in the Heston model is given as follows:

$$dS(t) = \mu S(t)\, dt + \sqrt{v(t)}\, S(t)\, dW(t)$$

The volatility following the CIR process (Cox-Ingersoll-Ross) as follows:

$$dv(t) = k\big(\theta - v(t)\big)\, dt + \sigma \sqrt{v(t)}\, dZ(t)$$

$W(t)$ and $Z(t)$ are standard Wiener process,

$$dW(t)\, dZ(t) = \rho\, dt$$

The partial differential equations of Heston model in general are as follows:

$$\frac{1}{2} v S^2 \frac{\partial^2 U}{\partial S^2} + \rho\, \sigma v S \frac{\partial^2 U}{\partial S\, \partial v} + \frac{1}{2}\sigma^2 v \frac{\partial^2 U}{\partial v^2} + rS\frac{\partial U}{\partial S} + \big(k(\theta - v) - \lambda(S,v,t)\big)\frac{\partial U}{\partial t} - rU + \frac{\partial U}{\partial t} = 0$$

where model parameters:

U	= option price at the time $t \in [0, T]$	σ	= volatility of volatility
S	= stock price	$\lambda(S, v, t)$	= the price of volatility risk.
θ	= long-term variance	r	= risk free rate
v	= volatility	T	= maturity date
k	= rate of reversion to the long-term price variance	ρ	= correlation coefficient of Wienner prosses is $[-1, 1]$

3. Risk Adjusted Deposit Insurance

Deposit insurance is one of tools used to stabilize the banking system [5]. In mechanism of deposit insurance, the insured risk is the risk of the bank's failure to fulfill its obligations to the customer at maturity date. The bank's failure can be identified from the value of a bank asset that is smaller than a debt obligation.

When a bank asset are less than debt obligations, the bank sells all of their assets. In this case, the guarantor can fulfill the payment promised by the bank to the customer and the bank must hand over the asset to the guarantor. For this opportunity, the guarantor charges a fee for the bank to be insured or premium. According to deposit insurance agency there are two types of premiums that must be paid by the bank, namely risk adjusted premium and premium flat rate. There are 32% of countries that have used a risk adjusted premium system while 68% of countries still use premium flat rate systems, including Indonesia [6]. Economists support the existence of deposit insurance that includes risk because it is considered more equitable and efficient than flat premiums (Ronn and Verma, 1986) [3].

Mechanism for determining the value of deposit insurance is isomorphic correspondence with a European put option [2]. One of model that can be used in calculating option values is the Heston model. Research conducted by Paolo Vanini also aims to determine the price of an option using the Heston model [7]. The stochastic process in the Heston model is given as follows:

$$dS(t) = \mu S(t)\, dt + \sqrt{v(t)}\, S(t)\, dW(t)$$

$$dv(t) = k\big(\theta - v(t)\big)\, dt + \sigma \sqrt{v(t)}\, dZ(t)$$

Next, a portfolio consist of a deposit insurance G, Δ unit of V which is a bank asset with risk factors $(1-\delta)^n$, δ is devidens value, ϕ unit value of the other derivative product used to protect volatility, and U is the value of derivative products whose movements are equal to G. The risk factor for assets is the distribution of dividends to shareholders. The company uses net income and retained earnings to distribute dividends to shareholders, resulting in reduced asset value. Asset value decreases by $V\delta$ for n periods so the bank asset value is equal to $V(1-\delta)^n$, so the portfolio is

$$\Pi = G - \Delta V(1-\delta)^n + \phi U$$

The change in portfolio value is

$$d\Pi = dG - d\Delta V(1-\delta)^n + d\phi U$$

$$d\Pi = \left[\frac{\partial G}{\partial t} + \frac{1}{2}\frac{\partial^2 G}{\partial V^2} v\,(V(1-\delta)^n)^2 + \frac{\partial^2 G}{\partial V \partial v}\rho\,\sigma\,v\,V(1-\delta)^n + \frac{1}{2}\frac{\partial^2 G}{\partial v^2}\,\sigma^2\,v\right] +$$
$$\phi\left[\frac{\partial U}{\partial t} + \frac{1}{2}\frac{\partial^2 U}{\partial V^2} v\,(V(1-\delta)^n)^2 + \frac{\partial^2 U}{\partial V \partial v}\rho\,\sigma\,v\,V(1-\delta)^n + \frac{1}{2}\frac{\partial^2 U}{\partial v^2}\,\sigma^2\,v\right] dt$$

$$d\Pi = [A + \phi B]dt$$

By assuming the portfolio in any bank must be the same with rate of return of a short- term-risk-free bank's assets with interest rate r, and substituting $d\Pi$ it follows that,

$$d\Pi = r\,\Pi\,dt$$
$$[A + \phi B]dt = r(G - \Delta V(1-\delta)^n + \phi U)dt$$

and implies,

$$\frac{\partial G}{\partial t} + \frac{1}{2}\frac{\partial^2 G}{\partial V^2} v\,(V(1-\delta)^n)^2 + \frac{\partial^2 G}{\partial V \partial v}\rho\,\sigma\,v\,V(1-\delta)^n + \frac{1}{2}\sigma^2\,v\frac{\partial^2 G}{\partial v^2} - rG +$$
$$rV(1-\delta)^n\frac{\partial G}{\partial V} + k(\theta - v)\frac{\partial G}{\partial V} = 0 \qquad (1)$$

Ronn and Verma [3] assume that in deposit insurance that includes risk, customers will be entitled to the future value of their deposits, or to a prorated fraction of the value, should the value be less than total debt. In other words, they will receive In other words, they will receive $\min\left\{F_v(B_1), \frac{V_T B_1}{B_1+B_2}\right\}$ after maturity. With $F_v(B_1)$ is the operator of future values and $\frac{V_T B_1}{B_1+B_2}$ is the terminal value of the bank assets recorded. The value of deposit insurance at maturity is $\max\left\{0, F_v(B_1) - \frac{V_T B_1}{B_1+B_2}\right\}$.

4. Solution by Using Fourier Transform

In this section discuss the solution of Heston partial differential equations to obtain a risk adjusted deposit insurance analytic solution using Fourier transforms. To simplify the work, transformation will be carried out using non-dimensional variables as follows:

$$e^{-x} = \left(\frac{\frac{V_T B_1}{B_1 + B_2}}{F_V(B_1)}\right) = \left(\frac{\frac{V(1-\delta)^n\,B_1}{B_1 + B_2}}{B_1 e^{r(T-t)}}\right)$$

$$-x = ln\left(\frac{V(1-\delta)^n}{Be^{r(T-t)}}\right)$$

where:

V	= value of bank assets	r	= interest rate
V_T	= the terminal value of the bank's assets	T	= time until the next bank asset audit
$Fv(B1)$	= future value dari B1	δ	= dividen
B	= $B_1 + B_2$	n	= dividend payment period
B_1	= the face value of the insured deposits	B_2	= the face value of all debt liabilities other than the insured deposits

After transformation, use the non-dimensional variable in equation (1) so that it is obtained,

$$-\frac{\partial G}{\partial t} = \frac{1}{2}v\frac{\partial^2 G}{\partial x} + \left(\frac{1}{2}v - r\right)\frac{\partial G}{\partial x} - \rho\,\sigma\,v\frac{\partial^2 G}{\partial v \partial x} + \frac{1}{2}\sigma^2\,v\frac{\partial^2 G}{\partial v^2} - rG + k(\theta - v)\frac{\partial G}{\partial V} = 0 \qquad (2)$$

Before applying the Fourier transform, the boundary conditions in Equation (2) must be fulfilled as follows,

$$max\{0, B - V\} \leq G_{(x,V,t)} \leq V$$

By using dimensionless no transformation, inequality can be written as follows,

$$max\left\{0, \frac{B_1 e^{-x} B}{V(1-\delta)^n} - B_1 e^{-x}\right\} \leq G_{(x,V,t)} \leq B_1 e^{-x} \quad (3)$$

Based on the assumption that the total debt obligation is greater than the bank asset value $B > V$, then the x value to be proven is only at $x > 0$. The inspection of the boundary conditions is as follows:

$$\lim_{x\to\infty} G_{(x,V,t)} = 0$$
$$\lim_{x\to\infty} G'_{(x,V,t)} = 0$$

(a). $\text{Lim}_{x\to\infty} G_{(x,V,t)} = 0$

Based on Equation (3) and t It follows from Squeeze Theorem in Real Analysis, obtained:

$$\lim_{x\to\infty} max\left\{0, \frac{B_1 e^{-x} B}{V(1-\delta)^n} - B_1 e^{-x}\right\} \leq \lim_{x\to\infty} G_{(x,V,t)} \leq \lim_{x\to\infty} B_1 e^{-x}$$

Because

$\lim_{x\to\infty} max\left\{0, \frac{B_1 e^{-x} B}{V(1-\delta)^n} - B_1 e^{-x}\right\} = 0$, and $\lim_{x\to\infty} B_1 e^{-x} = 0$, so $\lim_{x\to\infty} G_{(x,V,t)} = 0$.

b). $\lim_{x\to\infty} G'_{(x,V,t)} = 0$

Based on Equation (3) and t It follows from Squeeze Theorem in Real Analysis, obtained:

$$\lim_{x\to\infty} max\left\{0, -\frac{B_1 e^{-2x} B}{V(1-\delta)^n} + B_1 e^{-2x}\right\} \leq \lim_{x\to\infty} G_{(x,V,t)} \leq \lim_{x\to\infty} -B_1 e^{-2x}$$

Because

$\lim_{x\to\infty} max\left\{0, -\frac{B_1 e^{-2x} B}{V(1-\delta)^n} + B_1 e^{-2x}\right\} = 0$, and $\lim_{x\to\infty} -B_1 e^{-2x} = 0$, so $\lim_{x\to\infty} G'_{(x,V,t)} = 0$.

Following are the general equations of Fourier transforms,

$$\hat{f}(\omega, v, \tau) = \int_{-\infty}^{\infty} e^{i\omega x} f(x, v, \tau) dx$$

where $\hat{f}(\omega, v, \tau)$ is the result of Fourier transformation. In solving the problem using Fourier transformation, the application of inverses plays an important role. Based on (Rouah F.D., 2013) [8] the inverse Fourier transformation with the fundamental transform is obtained from the multiplication of the pay off transformation with the fundamental solution namely,

$$f(x, v, \tau) = \int_{i\varsigma_i-\infty}^{i\varsigma_i+\infty} e^{-i\omega x} \hat{f}(\omega, v, 0) \hat{f}(\omega, v, \tau) d\omega$$

$\hat{f}(\omega, v, 0)$ is pay off transform and $\hat{f}(\omega, v, \tau)$ is fundamental transform. Following are the general equations of Fourier transforms,

$$\hat{f}_{(\varsigma,v,\tau)} = \int_{\infty}^{-\infty} e^{i\varsigma x} F_{(x,v,t)} dx$$

Next, a Fourier transformation will be applied to in Equation (2) using the properties of Fourier transformation, then obtained,

$\mathcal{F}\left\{-\frac{\partial G}{\partial t}\right\} = -\int_{-\infty}^{\infty} e^{i\omega x} \frac{\partial G}{\partial t} dx = -\frac{\partial \hat{G}}{\partial t}$, $\qquad \mathcal{F}\{G\} = \int_{-\infty}^{\infty} e^{i\omega x} G(x, v, t) dx = \hat{G}$,

$\mathcal{F}\left\{\frac{\partial G}{\partial x}\right\} = \int_{-\infty}^{\infty} e^{i\omega x} \frac{\partial G}{\partial x} dx = -i\varsigma\hat{G}$, $\qquad \mathcal{F}\left\{\frac{\partial^2 G}{\partial x^2}\right\} = \int_{-\infty}^{\infty} e^{i\omega x} \frac{\partial^2 G}{\partial x^2} dx = -\omega^2 \hat{G}$,

$\mathcal{F}\left\{\frac{\partial^2 G}{\partial x \partial v}\right\} = -i\omega \frac{\partial \hat{G}}{\partial v}$, $\qquad \left\{\frac{\partial^2 G}{\partial v^2}\right\} = \frac{\partial^2 \hat{G}}{\partial v^2}$, $\qquad \mathcal{F}\left\{\frac{\partial G}{\partial v}\right\} = \frac{\partial \hat{G}}{\partial v}$.

Next, substitute the Fourier transformation results in equation (2) so obtained,

$$-\frac{\partial \hat{G}}{\partial t} = (-r + ri\varsigma)\hat{G} - \frac{1}{2}v(\varsigma^2 + i\varsigma)\hat{G} + (k(\theta - v) + \rho\sigma v i\varsigma)\frac{\partial \hat{G}}{\partial v} + \frac{1}{2}\sigma^2 v \frac{\partial^2 \hat{G}}{\partial v^2} \quad (4)$$

Based on Lewis [9] the results of Fourier transforms on $G\,(x,v,t)$ are fundamental solutions that can be written in the following form,

$$\hat{G}(x,v,t) = e^{(-r+ri\omega)}\hat{H}(\omega,v,t)$$

where $\tau = T - t$ is the time to maturity date. Function $\hat{H}$ is obtained as the solution of partial differential equations (3). Next, a derivation in the equation $G(x,v,t) = e^{(-r+riç)}$ is expressed as the form of $\hat{H}(\omega,v,t)$.

$$-\frac{\partial\hat{G}}{\partial t} = \frac{\partial\hat{G}}{\partial\tau} = (-r+ri\omega)e^{(-r+ri\omega)}\hat{H}(\omega,v,t) + e^{(-r+ri\omega)}\frac{\partial\hat{H}}{\partial\tau}$$

$$\frac{\partial\hat{G}}{\partial v} = e^{(-r+ri\omega)}\frac{\partial\hat{H}}{\partial v}, \quad \text{and} \quad \frac{\partial^2\hat{G}}{\partial v^2} = e^{(-r+ri\omega)}\frac{\partial^2\hat{H}}{\partial v^2},$$

Next, substitution the results of $\hat{G}$ which is constructed by $\hat{H}$ in equation (4) as follows,

$$\frac{\partial\hat{H}}{\partial\tau} = \frac{1}{2}\sigma^2 v\frac{\partial^2\hat{H}}{\partial v^2} + (k(\theta - v) + \rho\sigma v i\omega)\frac{\partial\hat{H}}{\partial v} - vc\hat{H} \tag{5}$$

where $c = \frac{1}{2}(\omega^2 + i\omega)$.

Next, a solution will be obtained from equation (4). Equation (4) must be changed into a general form of the Riccati equation, defined $t = \frac{1}{2}\sigma^2\tau$,

$$\frac{\partial\hat{H}}{\partial\tau} = \frac{\partial\hat{H}}{\partial t}\frac{\sigma^2}{2} \tag{6}$$

Equation (6) is substituted in equation (5) so that it is obtained

$$\frac{\partial\hat{H}}{\partial t} = v\frac{\partial^2\hat{H}}{\partial v^2} + \tilde{k}(\tilde{\theta} - v)\frac{\partial\hat{H}}{\partial v} - \tilde{c}v\hat{H} \tag{7}$$

where $\tilde{k} = \frac{2(k - i\omega\rho\sigma)}{\sigma^2}$, $\tilde{c} = \frac{2c}{\sigma^2} = \frac{\omega^2 + i\omega}{\sigma^2}$, $\tilde{\theta} = \frac{k\theta}{k - i\omega\rho\sigma}$.

Equation (7) has a solution with the following form $\hat{H}(\omega,v,t) = e^{\,C(t)+D(t)v}$ [8]. Next, find the derivatives of $\hat{H}(\omega,v,t)$and then substitute derivatives of $\hat{H}(\omega,v,t)$ in equation (6), and divide the two segments obtained by $\hat{H}(\omega,v,t)$ so produce,

$$\left(\frac{\partial C}{\partial t} + \frac{\partial D}{\partial t}v\right) = \tilde{k}\tilde{\theta}D + (D^2 - \tilde{k}D - \tilde{c})v \tag{8}$$

From equation (8) is obtained

$$D \quad \frac{\partial D}{\partial t} = D^2 - \tilde{k}D - \tilde{c} \ \text{dan} \ \frac{\partial C}{\partial t} = \tilde{k}\tilde{\theta} \tag{9}$$

The first equation at (8) is the Riccati equation with $P(t) = -\tilde{c}$, $(Qt) = -\tilde{k}$, and $R(t) = 1$. This equation can be solved by reducing to second order linear ordinary differential equations for $w\,(t)$ as follows,

$$w''(t) + \tilde{k}w'(t) - \tilde{c}w(t) = 0 \tag{10}$$

From equation (10) is obtained $\alpha = \frac{-\tilde{k}+\mathrm{j}}{2}$ and $\beta = \frac{-\tilde{k}-\mathrm{j}}{2}$,

where $j = \sqrt{\tilde{k}^2 + 4\tilde{c}}$, the solution to the Riccati equation at (8) is as follows,

$$D = -\left(\frac{K\alpha e^{\alpha t} + \beta e^{\beta t}}{Ke^{\alpha t} + e^{\beta t}}\right)$$

The initial condition $D(0) = 0$ implies that $t = 0$ in the numerator so that it produces $K = -\frac{\beta}{\alpha} = \frac{(\tilde{k}+\mathrm{j})}{-\tilde{k}+\mathrm{j}}$ so that solution D is,

$$D = \frac{\tilde{k}+\mathrm{j}}{2}\left(\frac{1 - e^{jt}}{1 - ge^{jt}}\right)$$

After obtaining the value D, then the value of C. will then be sought for the value $C\,(t)$ can be obtained by integrating $D\,(t)$.

$$\int_0^y D(t)dt = \frac{\tilde{k}+\mathrm{j}}{2}t - \ln\left(\frac{1-ge^{jt}}{1-g}\right)$$

So that $C(t)$ is obtained, that is,

$$C(t) = \tilde{k}\tilde{\theta}\left(\frac{\tilde{k}+\mathrm{j}}{2}t - \ln\frac{1-ge^{jt}}{1-g}\right) \tag{11}$$

After obtaining the values $D(t)$ and $C(t)$, the Fourier transform from the pay off function will be searched as follows,

$$\hat{G}(\omega, v, 0) = \int_{-\infty}^{\infty} e^{i\omega x}\,(B_1 - B_1 e^{-x})dx = \frac{-B_1}{\omega^2 + i\omega} \tag{12}$$

The final step in calculating the value of deposit insurance premiums is to inverse the Fourier transform which can be done by multiplying the transformation results of the pay off function with the basic solution as follows,

$$G(x, v, \tau) = -\frac{1}{2\pi}\int_{\infty}^{\infty} e^{i\varsigma x}\frac{B_1}{\varsigma^2 + i\varsigma}e^{(-r+ri\varsigma)}\hat{H}(\omega, v, t)d\omega$$

$$= -\frac{B_1 e^{-r\tau}}{\pi}\int_0^{\infty} Re\left[\frac{1}{\omega^2 + i\omega}\,\hat{H}(\omega, v, t)\right]d\omega \tag{13}$$

where $P = \ln\frac{V(1-\delta)^n}{B}$.

5. Simulation

In this section numerical simulations will be carried out to find solutions to equation (13) that can be solved by using trapezoidal rules and then simulated using matlab software to represent a general picture of the value of deposit insurance premiums that must be paid by a bank.

5.1. Effects of Volatility on Deposit Insurance Premium Value

Suppose a bank's assets of $V = 100$, interest rate $r = 0.05$, $B_1 = 60$ and dividends $\delta = 0.001\%$ obtained the value of risk adjusted deposit insurance premiums with volatility $v_1 = 0.01, v_2 = 0.03, v_3 = 0.05, v_4 = 0.07, v_5 = 0.09$ as presented in Table 1.

Tabel 1. The Value of Deposit Insurance Premiums for B1= 60

B_1	B_2	$v_1 = 0.01$	$v_2 = 0.03$	$v_3 = 0.05$	$v_4 = 0.07$	$v_5 = 0.09$
60	25	4.2958	4.5518	4.8030	5.0495	5.2912
60	40	7.3755	7.6517	7.9189	8.1777	8.4288
60	55	10.7448	10.9978	11.2416	11.4773	11.7054
60	70	14.0792	14.2872	14.4887	14.6843	14.8747

In this case, there are three conditions; $B < V$ the face value of total debt liabilities is greater than the assets of bank. $B = V$ the face value of total debt liabilities equals to the assets of bank. $B > V$ the face value of total debt liabilities is smaller than assets bank. From Table 1. it can be seen that the value of deposit insurance premiums increases with the amount of the total debt obligation B and the amount value of volatility v. Figure 1 displays is a visualization image of the effect of volatility on the value of deposit insurance premiums with B_1 of 30 and B_2 of 50.

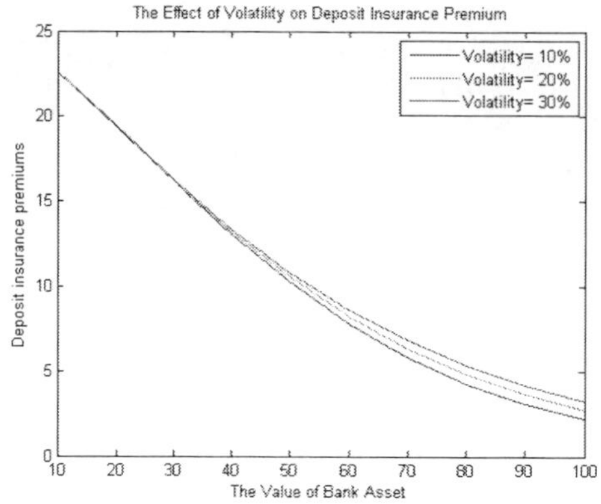

Figure 1. The Effect of Volatility on Deposit Insurance Premium

In Figure 1. it can be seen that the greater the assets of a bank, the value of deposit insurance premiums decrease. While the greater value of volatility (v), the value of deposit insurance premiums increase and it also shows that the value of deposit insurance premiums increases proportional with the amount of insured debt B_1.

5.2. Effect of Interest Rate on Deposit Insurance Premium Value

Interest rate is one of the factors that can affect the value of insurance premiums of deposits from a bank. The following will be presented the effect of vinterest rate on the value of deposit insurance premiums. Suppose a bank's assets are $V = 100$, dividends $\delta = 0.001\%$, $v = 0.05$, and interest rate values are $r_1 = 5\%$, $r_2 = 10\%$, and $r_3 = 15\%$ as Figure 2:

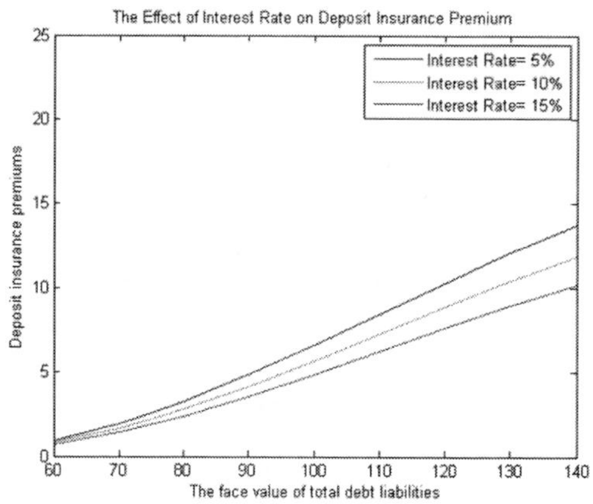

Figure 2. The Effect of Interest Rate on Deposit Insurance Premium

In Figure 2. the greater interest rate causes the value of deposit insurance premiums decrease. Conversely, if the smaller interest rate is causes the value of deposit insurance premiums increase. In 15% interest rate the increase in the value of deposit insurance premium is not too significant when compared to the interest rate 10%. So, it can be said that the value of deposit insurance premiums is inversely proportional to the amount of interest rate.

5.3. Effect of Dividends on Deposit Insurance Premium Value

Dividend is a factor that causes a bank asset to be at risk, so that dividends from banks to shareholders can affect the value of deposit insurance premiums. The following will be shown a description of the effect of dividends on the value of deposit insurance premiums on dividends with the taking of fixed asset V values and different total debt values. Suppose a bank's assets amounting to $V = 100$, interest rate $r = 0.05$, volatility $v = 0.05$ obtained by the value of deposit insurance premiums that include risks with dividends $\delta_1 = 1\%$, $\delta_2 = 5\%$, and $\delta_3 = 10\%$, as Figure 3:

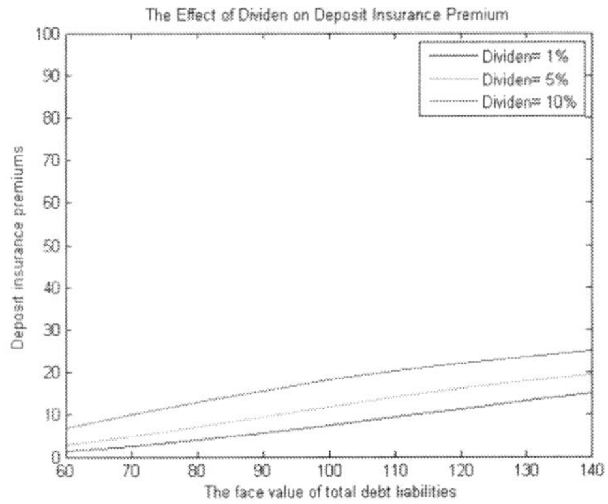

Figure 3. The Effect of Dividends on Deposit Insurance Premium

In Figure 3. in the distribution of dividends of 10%, the value of deposit insurance premiums is greater than the distribution of dividends of 5% and 1%. The greater dividend value causes the value of deposit insurance premiums increase. The greater dividend distributed can caused he greater the reduction in assets so that the value of deposit insurance premiums increases. This is because dividend distribution affects the assets of a bank.

6. Conclusion

Based on the results and discussion can be concluded that the risk adjusted deposit insurance premium value solution is an analytical solution. Based on the simulation it can be concluded that the amount of volatility (v) causes the value of deposit insurance premiums getting bigger. The more debt obligations also causes the value of deposit insurance premium increase. In determining the value of deposit insurance premium, the nominal value of the insured deposit (B_1) is more influential than the nominal of the uninsured debt obligation (B_2). The value of deposit insurance premiums is inversely proportional to the amount of interest rate. The value of deposit insurance premiums increases with the amount of dividends distributed by the bank.

References

[1] King G R dan Levine R 1993 Finance and Growth *The Quarterly Journal of Economics* Vol. 108, No. 3 pp. 717-737.
[2] Merton R C 1977 An Analytic Derivation of the Cost of Deposit Insurance and Loan Guarantees *Journal of Banking and Finance*. Volume 1, Issue 1 (Cambridge:Elsevier) pp 3-11.
[3] Ronn and Verma K 1986 Pricing Risk Adjusted Deposit Insurance: An Option Based Model *Journal of Finance* Vol. 41 No. 4 (America: American Finance Association).
[4] Heston 1993 A Closed-Form Solution for Options with Stochastic Volatility with Applications to Bond and Currency Options *The Review of Financial Studies* Volume 6 No. 2 pp. 327-343.
[5] Asnawi K S 2006 Desain Premi dan Manfaat Asuransi Deposito *Jounal of Business JSB* Vol 11 No 1 (Jakarta: Institute Bisnis dan Informatika Indonesia).
[6] LPS 2015 *Analisis Stabilitas dan Sistem Perbankan* (Jakarta:LPS).
[7] Vanini P and Padovani M 2010 Stochastic Volatility I: Heston Model: Analytics *Journal of Finance* 10.2139/ssrn 1579593
[8] Rouah F D 2013 *Heston Model and Its Extensions in Matlab and C\#* (America: United States of America)

[9] Lewis A L 2000 *Option Valuation Under Stochastic Volatility: With Mathematica Code.* (California: Finance Press) chapter 2.

Journal of Physics: Conference Series **1397** (2019) 012079 doi:10.1088/1742-6596/1397/1/012079

What Difficulties that Students Working in Mathematical Reasoning Questions?

A H Zaini[1] dan H Retnawati[2]

[1]Graduate Program of Education Research and Evaluation, Yogyakarta State University, Yogyakarta, Indonesia
[2]Department of Mathematics Education, Faculty of Mathematics and Natural Sciences, Yogyakarta State University, Indonesia.

arinahusna.2017@student.uny.ac.id

Abstract. Reasoning skill plays an important role in developing all other mathematical skills. However, the results of subhan's research (2017) inform that students are still weak and have difficulty in solving mathematical reasoning questions. Therefore, a study to analyze students' difficulties in solving reasoning questions should be conducted to overcome this problem. Participants in the study were 15 of the 25 public high school students in Bantul district of Yogyakarta who had completed mathematical reasoning questions. They were deliberately chosen because when they completed mathematical reasoning questions from 25 students only 15 students completed mathematical reasoning questions in their entirety. The instruments of mathematical reasoning are as many as five questions in the class XI line material that has been tested for quality.The analysis was performed using types of difficulties adapted from procedure Newman which includes (1) reading, (2) comprehension, (3) transformation, (4) process skill, and (5) encoding. The results of the study indicate to encoding difficulties of 37.93%, comprehension difficulties of 24.14%, transformation difficulties of 19.54%, and process skill difficulties of 18.39%.

1. Introduction

Reasoning is an important ability to train students when learning mathematics. The importance of reasoning ability is evidenced by the inclusion of reasoning in mathematical learning objectives. More broadly, the importance of mathematical reasoning ability was also proven that the National Council of Teachers of Mathematics included reasoning abilities as one of the five standard processes in mathematics learning. Habsah [1] explains that reasoning skills are very important in mathematics learning. This importance is so that Thompson [2] likens reasoning to the pulse of mathematics. Sumarmo in more detail [3] reveals that mathematical reasoning abilities are very important in mathematical understanding, exploring ideas, estimating solutions, and applying mathematical expressions in relevant mathematical contexts, and understanding that mathematics is meaningful. The importance of reasoning in mathematical understanding and exploring ideas shows that reasoning emphasizes mindset. Santrock [4] which explains that reasoning is a logical mindset using induction or deduction to reach conclusions. Lithner [5] explains that reasoning is a mind pattern to produce a statement and conclusion in solving problems. Based on the narrative of the experts above it can be concluded that reasoning is a very important ability in mathematics and emphasizes the logical mind pattern to obtain conclusions.

Logical conclusions in reasoning is one that has very complex implications for students because students must process the mind to produce the right conclusions. This complexity often causes reasoning including abilities that are not easily achieved by students [6]. The scores of the ability of overall high school students or grouped based on the cognitive stages of students in solving mathematical reasoning questions were still low [7]. The study by Priata [8] regarding mathematical reasoning also suggested that the quality of students' mathematical reasoning skill was low, which was 49% of the ideal score. Other studies also showed that students had low mathematical reasoning skill [9].

Other facts show that the importance of mathematical reasoning is not acknowledged with real efforts to improve mathematical reasoning skill of the students. Widdiarto [10] argued that in the learning process, teachers rarely emphasize the process that enables students have a practice on reasoning skill and still use less varied methods to deliver the materials. This results in students' tendency to memorize and think mechanistically which signify low level abilities. In fact, high-order thinking skill should be taught to students at the secondary school level. In reality, the high-order thinking skills of the students are still low. The study by Delima [11] indicated that 6% of students had low reasoning skill. Although students who lack reasoning skill are low in numbers, it should become a concern for researchers to provide solutions so that students are able to increase mathematical reasoning skill.

The low level of mathematical reasoning abilities of students conveyed by various sources above indicate the difficulty of students in completing mathematical reasoning questions. Difficulties in solving mathematical reasoning problems if left unchecked will have a negative impact on students because reasoning is the foundation in learning mathematics [12]. Therefore, early analysis is needed regarding any difficulties experienced by students in solving mathematical problems. By doing so, it is expected that the results of studies can be useful for teachers or other related parties so that they are able to provide proper solutions to improve students' mathematical reasoning skill.

2. Methods

The study aims to identify, describe, and clarify students' difficulties in solving mathematical reasoning questions. Regarding triangulation, researchers used interviews with students as clarification and reinforcement of the findings. In the current study, the researchers identified and clarified what difficulties students faced in solving mathematical reasoning questions. Furthermore, types of difficulties were classified using Newman's Error Procedure (NEP) that consists of reading error, comprehension error, transformation error, process skill error, and encoding error. The participants studied were 15 high school students in grade XI of science program in Bantul Sub-District. Students came from heterogeneous academic skills. Data collected using a test consisting of five items of mathematical reasoning in the form of a description of the sequence material. The five items of reasoning were adopted from the development of mathematical reasoning questions. The items used in this study have been tested for quality through the process of content validation and construct validation.

Data were analyzed qualitatively using Miles & Huberman's model and quantitatively to calculate the percentages of students' difficulties. Miles & Huberman's model consists of three phases which are data reduction, data display, and conclusion drawing. In the data reduction phase, the researchers examined every process of solving mathematical reasoning questions and classified it into correct and incorrect process. Afterwards, in the data display phase, the researchers classified the errors students made based on Newman's Error Procedure. This procedure suggests 5 (five) specific stages that serve as an important guidance to discover where the errors occurred in students' works when solving essay questions, which are: (1) reading, (2) comprehension, (3) transformation, (4) process skill, and (5) encoding. Students' difficulties discovered in this study are described in details in the following table.

Table 1. Description of Difficulties Adapted from Newman

No	Types of Difficulties	Indicators of Difficulties
1	Reading	• Unable to recognize words written in the question • Unable to interpret difficult words in the question
2	Comprehension	• Unable to comprehend certain sentences in the question • Unable to correctly write what is required in the question
3	Transformation	• Transform the information in the question into mathematical words incorrectly • Unable to choose formula or theory correctly
4	Process Skill	• Unable to compute easily • Unable to explain the process of computation correctly in the answer sheet • Unable to continue the solving procedure
5	Encoding	• Unable to interpret the answer correctly (Unable to transform to the beginning form correctly)

Students' errors were classified in accordance with Newman's Error Procedure. Quantitative analysis was performed to calculate the percentage of each type of errors. In each question, a student made more than one error. For example, for the question number 1, a student faced difficulties in comprehension and transformation. To find out the percentage of each type of errors in NEP, the number of students who made each error was divided by the total number of participants. In the conclusion drawing phase, the researchers made conclusions regarding the analysis on students' difficulties in each type of errors.

3. Results and Discussion

3.1. Results

The analysis in the reduction phase informs that the researcher checks the answers of 15 students based on the completion steps in each item. The researcher classified the process of solving students in each item as a wrong and correct process. The results of the analysis inform that there are 87 wrong resolution processes where things indicate student difficulties. Furthermore, in the data presentation phase, 87 students 'difficulties in solving students' mathematical reasoning problems are grouped according to Newman's difficulty procedure. The following is a recapitulation of the percentage of students' difficulties in solving mathematical reasoning questions in each item.

Table 2. Percentages of Students' Difficulties

Question No	Percentages of Difficulties					
	Reading	Comprehension	Transformation	Process Skill	Encoding	Total
1	0(0%)	4 (4.60%)	2 (2.30%)	4 (4.60%)	7 (8.05%)	13(19.54%)
2	0(0%)	6 (6.90%)	5 (5.75%)	2 (2.30%)	6 (6.90%)	19(21.84%)
3	0(0%)	4 (4.60%)	3 (3.45%)	0 (0.00%)	2 (2.30%)	9(10.34%)
4	0(0%)	5 (5.75%)	3 (3.45%)	6 (6.90%)	8 (9.20%)	25(26.29%)
5	0(0%)	1 (2.30%)	4 (4.60%)	4 (4.60%)	9 (11.49%)	20(22.99%)
Total	0(0%)	21(24.14%)	17(19.54%)	16(18.39%)	33(37.93%)	87(100%)

Table 2 illustrates that in general most students experienced difficulties when doing the question number 4 with a percentage of 26.29%. On the other hand, students encountered the least difficulties when doing the question number 3. Regarding each type of students' difficulties, it can be seen that in solving mathematical reasoning questions, students were more likely to experience difficulties in encoding with a percentage of 37.93%, followed by difficulties in comprehension with a percentage of 24.14%, in transformation with a percentage of 19.54%, and process skill with a percentage 18.39%. In the last phase of conclusions, the results of the analysis show that the difficulties of students in completing the biggest mathematical reasoning questions in a row, namely in encording, comprehension, transformation, process skills. Furthermore, analysis of student difficulties in each type of difficulty performed in working on mathematical reasoning questions will be presented in the following section.

3.1.1 Difficulties in Comprehension .Difficulties students faced in comprehending the questions are indicated by the process of solving mathematical reasoning questions which includes if students wrote information or what they knew from the question correctly and understood certain sentences in the question. Based on the analysis results presented in Table 1, it can be seen that the percentage of difficulties in this type of error is 24.14%. The following is an example of students' difficulties in comprehending the question number 2.

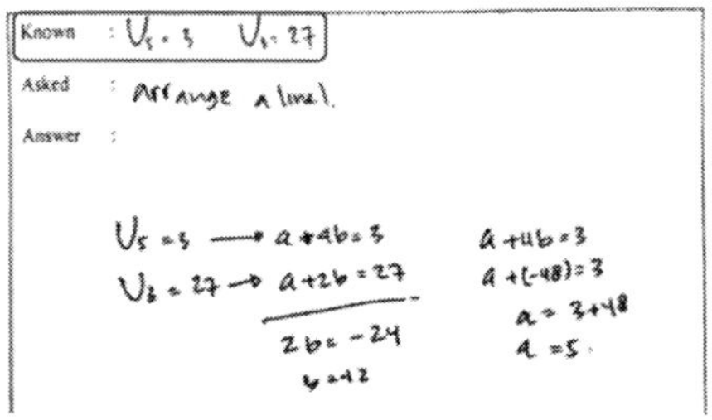

Figure 1. Students' difficulties in comprehending the question number 2

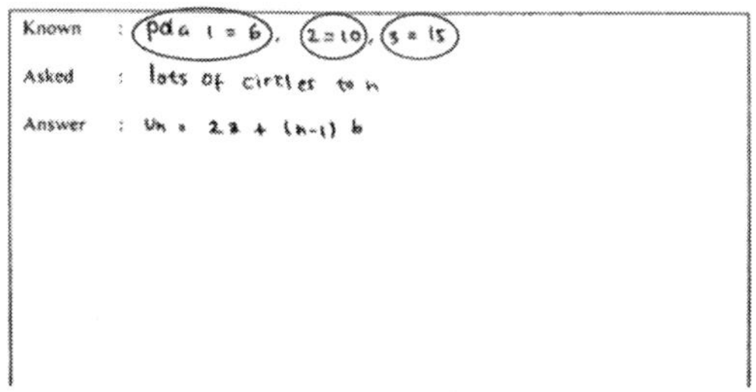

Figure 2. Students' difficulties in comprehending the question number 1

The answer students gave to solve the question number 2 indicates that students found it difficult to comprehend the question. Students did not write the information required in the question completely and transform information correctly. In the question number 2, the two terms have a constant ratio. Ratio means that the sequence is geometric sequence so that students were supposed to find out the ratio and the first term to continue the sequence. In reality, however, students solved the question number 2 by misunderstanding it as an arithmetic sequence. Based on the results of interview, students said that the word "ratio" written in the question was understood as a characteristic of an arithmetic sequence; thus, they used a formula of arithmetic sequence. This finding indicates that students failed to understand the question. This failure is more likely to be caused by the fact that students lacked the ability to comprehend the question well. Therefore, students encountered difficulties to understand the problem given, were not able to write the important information in the question, and failed to understand the requirement of the question.

In Figure 2, it can be seen that students wrote the information contained in the question number 1 by using symbols they made themselves without clear explanation. For example, a student wrote "the pattern 1=6, 2=10, and 3=15". Besides, the information students wrote was not in accordance with the context of the question. This problem is denoted when students wrote the number of pattern 1, pattern 2, and pattern 3 based on the black and white circles. However, students were supposed to write the information regarding the number of pattern 1, pattern 2, and pattern 3 based on the black circles only. The error students made in writing the information in the question number 1 shows that students experienced difficulties in comprehending questions so that they failed to investigate the patterns. The results of interview indicate that students made this error because they thought that the symbols they wrote as the answer already gave information in accordance with the question. Although the symbols the students used were different, they thought that people were still able to understand the answer they wrote. Furthermore, the error in investigating the pattern was caused by the fact that students did not focus their attention on the process of understanding the question. Students were more likely to focus by looking at the pattern presented in the picture without reading the question accurately.

3.1.2 Difficulties in Transformation. Difficulties in transformation occurred when students correctly comprehended what the question required, but were not able to identify the correct mathematical operation or sequence of operation to successfully solve the question. Based on the results of analysis in Table 1, in general the total percentage of this type of difficulties is 19.54%. An example of transformation difficulties is presented in the following figure.

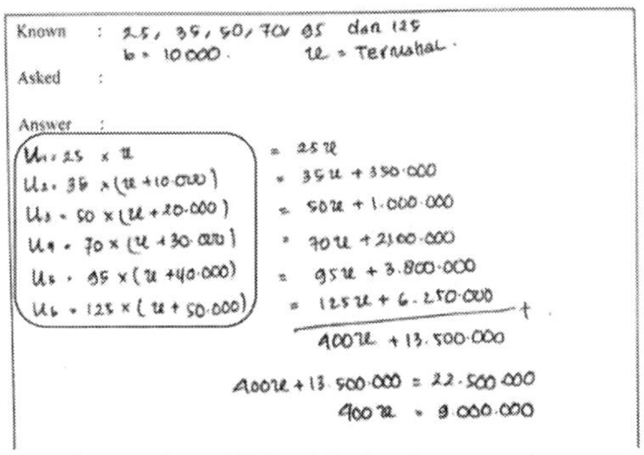

Figure 3. Transformation difficultie in the question number 4

Based on Figure 3, students transformed information in the question number 4 into a correct mathematical sentence. However, one part was not translated correctly. It is shown in the question that the price of the seats in the first row was the most expensive and denoted with variable x. Students wrote the price of the seats in the first row was x and in the next row was x+10,000, and etc. The notation x+10,000 shows that the seats in the second row were more expensive than in the first row. However, this is not in line with the information in the question that x was the most expensive price of seats. Therefore, students' error in this question indicates difficulties in transformation. The results of interview suggest that students understood that the most expensive ticket was x, but due to the incorrect concept, they added 10,000 to the next row.

3.1.2 Difficulties in Process Skill. Difficulties in process skill are shown by students' error in doing mathematical operations, inability to do sequence of operations, or inability to explain the process of operations. Based on the results of analysis in displayed in Table 1, the difficulties in process skill are 18.39%. The following is an example of students' error in process skill for the question number 1.

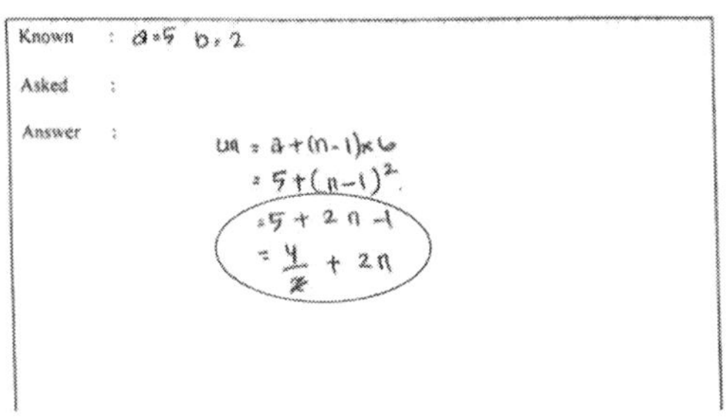

Figure 4. Difficulties in process skill in the question number 1

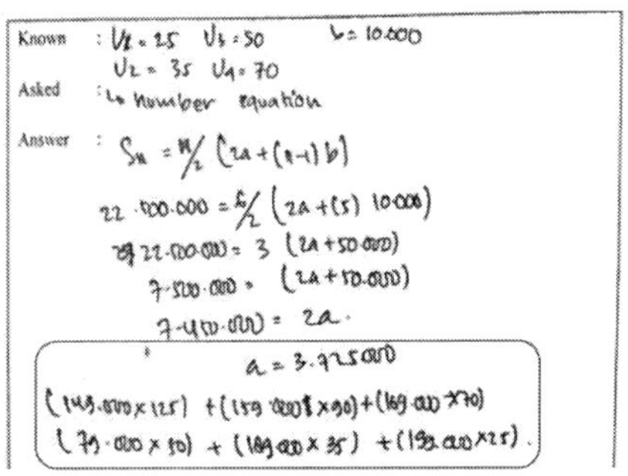

Figure 5. Difficulties in process skill in the question number 4

Figure 4 shows that students make mistakes in the calculation process looking for the number of black spheres in the nth term so that students cannot continue the calculation process performed. In addition, other findings indicate that students do not understand the problem so they do not explain the calculation process. The error was shown when making the operation of "(n-1).2". Students wrote the results of operation of (n-1).2 as 2n-1, although the correct operation was supposed to be 2n-2.

Students' error in this part suggests that students experienced difficulties in process skill of calculating integers. The results of interview indicate that students were confused when doing the calculation in the form of (n-1)2. This form is one of the properties of operations with integers called commutative. When students were given questions related to properties of operations with integers, they were not able to solve them accurately. Therefore, students faced difficulties in process skill due to low comprehension and process skill in calculation.

Figure 5 shows that students did not explain their calculation process when solving the question number 4. After looking for the value of 'a', students wrote one row of numbers and did not give information. No information suggests that students faced difficulties in process skill. The results of interview demonstrate that students made an error in this part because it was difficult for them to make a correct procedure. They had difficulties finding what parts to look for to obtain the right answer.

3.1.3 Difficulties in Encoding. Difficulties in encoding occurred when, despite being able to solve the question, students failed to interpret the answer in accordance with the context, wrote incorrect answers, or even did not write the answers at all. Based on the results of analysis in presented in Table 1, the percentage of difficulties in encoding is 37.93%. The following is an example of students' answers when solving the question number 5 and number 1.

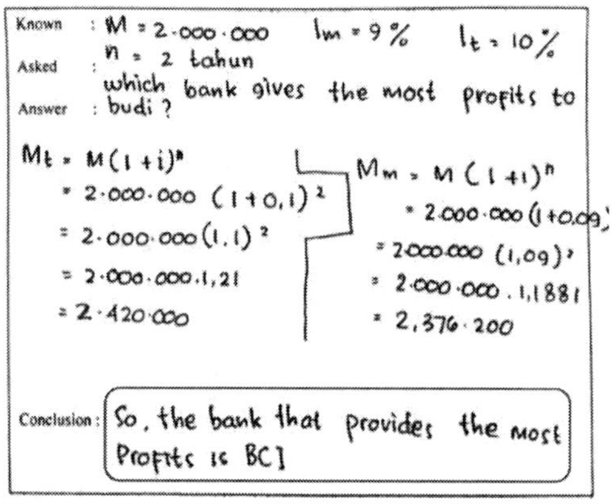

Figure 6. Difficulties in encoding for the question number 5

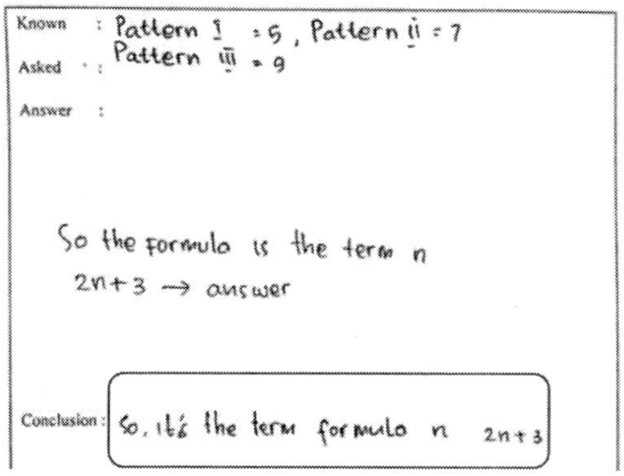

Figure 7. Difficulties in encoding for the question number 1

Students could calculate correctly to solve the question number 5. However, Figure 6 shows that the conclusion students drew was not in accordance with the question. The question number 5 asks which bank Budi has to choose in order to obtain profit complete with the reason. Students, however, wrote that the bank that gave the most profit was BCI. Students were supposed to write "Budi has to choose BMI because it gives the most profit as much as 2,400,000 rupiahs." Based on the results of interview with students, students wrote that way because they thought that their answer was already able to solve the question. Thus, in conclusion drawing, students failed to write the proper and acceptable form of answer. In other words, students were not able to interpret the answer to be in accordance with the question's requirement.

The question number 1 asks how many black circles are in the nth term. Based on Figure 7, it can be seen that students successfully found the number of black circles in the nth term. However, when drawing a conclusion of the answer to the question number 1, the majority of students wrote "the formula for the n term is 2n+3." This is not a proper conclusion because, despite 2n+3 being the correct formula, students were not able to provide the form of answer in accordance with the context. The conclusion students were supposed to write is "the number of black circles in the nth term is 2n+3". Based on the results of interview, it is discovered that students successfully understood the procedure to solve the question number 1, but when giving the final answer, they thought their answer already solved the question without relating it to the context of the question. Students often use this concept; they successfully execute the mathematical processes in order to solve the question, but they often fail to interpret the results of their calculation used as the answer.

3.2. Discussion

Results of the study show that students experienced difficulties in solving mathematical reasoning questions. The errors they made signify the difficulties. The analysis of students' difficulties indicates that when attempting to solve the mathematical reasoning questions, students faced difficulties in comprehension, transformation, process skill, and encoding. This is in line with the results of studies by Prakitipong, et al, [13] and Layn [14], showing that the difficulties students encountered in solving PISA questions included difficulties in comprehension, transformation, process skill, and encoding. Studies by Trapsilo [15] and Dewi, et al, [16] also yield the same result that no students had no difficulties in reading mathematical reasoning questions. This finding signifies that the research participants who were already high school students psychologically had a good reading skill. Besides, the questions tested on students had good quality in terms of content validity so that students were able to understand the information contained in the questions easily.

In solving reasoning questions, the most difficulties faced by students were encoding and comprehension. The difficulties in encoding in this research occurred when students wrote the conclusion or the final result of the calculation. This happened because students did not interpret the answer in accordance with the context of the questions and were more likely to write the answers only in the form of numbers. This is in parallel with the finding of the study by Cahyaningrum [17], showing that students often failed to write the conclusions according to the context in solving mathematical reasoning questions. The difficulties in this process are often faced by students in solving mathematical questions or any problems in daily life. The biggest factors are students are more likely to do mathematical questions in a rush and to think that it should suffice to write the numbers as the answer to the question without adding words to relate it with the context of the question. Rahman and Fauziana [18] argued that such cases can be frequently found in the field. However, these frequent occurrences influence students' learning outcomes and achievements.

This study found that comprehending the mathematical reasoning questions became one of the difficulties most students faced. It was difficult for students to understand the questions or to construct procedures to solve the questions. The study conducted by Phonapichat, et al, [19] found similar result which students did not have patience and like reading the questions. Karimah and Fuad [20] suggested that in solving PISA questions to test reasoning skill, students experienced difficulties in the attempt to comprehend the questions. Students did not write the complete information required in the questions and failed to interpret certain words or sentences. The ability to comprehend questions is a vital aspect of literacy. The results of study by PISA in 2015 [21] signified that mathematical literacy that includes the ability to comprehend mathematical tasks is low. Based on this finding, 20.3% of Indonesian students were below level 1 (the score was below 334.94). The study also reported that Indonesian students were able to comprehend only 30% of the reading material because solving questions that needed comprehension and reasoning skills was difficult for them [22].

Difficulties in transformation were also experienced by students in their attempt to solve mathematical reasoning questions. They failed to use the correct mathematical formulas, to transform information in the questions into mathematical words, and to perform procedures consistent with the questions' requirements. These findings suggest a similarity with the results of the study by Samsul, et al, [23], indicating that in doing transformation, students made errors in determining the formula to solve the mathematical tasks and this has become evidence that not all students have a good understanding of concepts. Furthermore, White [24] argued that regarding difficulties in transformation, students already comprehend the information and the requirement of the questions, but they were not successfully able to identify the operation or sequence of operations necessary to solve the questions. Wijaya, et al, [28], in his study, the lowest difficulty is in process skills. Difficulties in process skills tend to be context-based and therefore do not reach the stage of bringing mathematical procedures out. This can be interpreted that even though students experience difficulties in process skills such as miscalculations or not completing the results of calculations, the resolution process does not deviate from mathematical procedures.

4. Conclusion

Difficulties faced by students in solving mathematical reasoning questions comprise difficulties in comprehension, transformation, process skill, and encoding. Based on quantitative analysis, encoding was the most difficult stage of difficulties with 37.93%, followed by comprehension with 24.14%, transformation with 19.54%, and process skill with 18.39%. Students' difficulties in attempting mathematical reasoning questions were caused by a number of factors. First, students found comprehending the questions difficult. Second, students used words or symbols that are not commonly used so that people did not understand what students wrote and could not focus their attention in comprehending the questions. Third, students' ability to perform mathematical operations and to interpret the questions in accordance with the context was still low. Students' difficulties found in this study suggest that efforts need to be made so that students' abilities to solve mathematical reasoning questions can be improved. Therefore, this study can serve as a start for the next research.

References

[1] Habsah F 2017 Developing teaching material based on realistic mathematics andoriented to the mathematical reasoning and mathematical communication Jurnal Riset Pendidikan Matematika 4(1) 43-55 DOI:10.21831/jrpm.v4i1.10199
[2] Thompson D R. Senk, S. l. & Johnson, G.J (2012) Opportunities to learn reasoning and proof in high school mathematics textbooks Journal for Research in Mathematics Education 43(3) 253-295 Retrieved from https://www.jstor.org/stable/10.5951/jresematheduc.43.3.0253
[3] Sumarmo U 2015 Berpikir dan Disposisi Matematik serta Pembelajarannya Bandung UPI
[4] Santrock John W 2011 Perkembangan Anak Edisi 7 Jilid 2 (Terjemahan: Sarah Genis B) Jakarta Erlangga
[5] Lithner J 2008 *A research framework for creative and imitative reasoning* New York Springer
[6] Sukirwan Darhim and Herman T 2018 Analysis of students' mathematical reasoning Journal of Physics: Conference Series 984 012036
[7] Sulistiawati 2014 Analisis Kesulitan Belajar Kemampuan Penalaran Matematis Siswa SMA Pada Materi Luas Permukaan dan Volume Limas Proceeding National Seminar on Mathematics Education and TIK STKIP. ISBN 978-602-14432-2-4
[8] Sumarmo U 1987 Kemampuan Pemahaman dan Penalaran Matematika dengan Kemampuan Penalaran Logik Siswa dan Beberapa Unsur Proses Belajar Mengajar Doctoral dissertation at PPS UPI Bandung was not published.
[9] Priatna N 2003 Kemampuan Penalaran Matematika Siswa Kelas 3 Sekolah Lanjutan Tingkat Pertama Negeri Di Kota Bandung Doctoral dissertation at PPS UPI Bandung was not published
[10] Somatanaya A A G 2005 Meningkatkan Kemampuan Penalaran Matematis Siswa SLTP melalui Pembelajaran dengan Metode Inkuiri Doctoral dissertation at PPS UPI Bandung was not published.
[11] Widdiarto R 2004 Models of Learning Mathematics SMP [On Line]. Available: http: //p4matematika.org/downloads [March 25, 2019].
[12] Delima M L, Retno M, and Tri Candra W 2017 Analisis kemampuan penalaran matematis peserta didik dalam menyelesaikan soal cerita di sman 6 malang. Pi: Mathematics Education Journal. Vol.1, No.1, Oktober 2017, 27- 33
[13] Prakitipong N and Nakamura S 2006 Analysis of Mathematics Performance of Grade Five Students in Thailand Using Newman ProcedureCICE Hiroshima University Journal of

International Cooperation in Education Vol 9 No 1

[14] Layn M R & Kahar M S 2017 Analisis Kesalahan Siswa DalamMenyelesaikan Soal Cerita Matematika. Math Educator Nusantara Journal (JMEN) A vehicle for the publication of scientific papers in the field of mathematics education

[15] Tarsisius Eko Bagus Trapsilo 2016 Analisis Kesalahan Siswa Menurut Teori Newman Dalam Menyelesaikan Soal-Soal Cerita Materi Persamaan Linier Dua Variabel Pada Siswa Kelas Ix Smp N 1 Banyubiru UKSW Journal

[16] Dewi A I C Zulkardi and Yusuf M 2018 Kesulitan Siswa Dalam Menyelesaikan Soal-Soal Pisa Tahun 2012 Level 4, 5, Dan 6 Di Smp N 1 Indralaya Mathematics education journal vol 11 number 2 July 2017

[17] Cahyaningrum D 2015 Identifikasi Kesulitan Dalam Menyelesaikan Soal Cerita Pokok Bahasan Prisma Dan Limas Siswa Kelas Viii Semester Ii Smp Negeri 4 Delanggu Tahun Ajaran 2014/2015 Surakarta State University Journal

[18] Rahman and Fauziana. 2018 Analysis of Student's Answer Error in Learning Mathematics Using Newman Analysis. *IOSR Journal of Research & Method in Education (IOSR-JRME).* e- ISSN: 2320–7388,p-ISSN: 2320–737X Volume 8, Issue 6 Ver. III. (Nov. – Dec. 2018) PP 77-82

[19] Phonapichat P Wongwanich S and Sujiva S 2014 An analysis of elementary school students' difficulties in mathematical problem solving Procedia Social and Behavioral Sciences 116 3169 3174 doi:10.1016/j.sbspro.2014.01.728.

[20] Aminatul Karimah and Yusuf Fuad 2017 Analisis Kesalahan Siswa Dalam Menyelesaikan Soal Pisa. Jurnal Ilmiah Pendidikan Matematika . *Volume 1 No.6 Tahun 2017 ISSN : 2301-9085*

[21] OECD.2015.Programme for International Student Assessment (PISA) Diakses dari http://www.oecd.org/pisa/PISA-2015-Indonesia.pdf pada tanggal 20 Mei 2019

[22] Basuki I A 2011 Kemampuan Membaca Pemahaman Siswa Kelas Iv Sd Berdasarkan Tes Internasional Dan Tes Lokal Journal Bahasa Dan Seni, Tahun 39, Nomor 2, Agustus 2011

[23] Hadi S Retnawati H Munadi S Apino E Wulandari N F 2018 The Difficulties Of High School Students In Solving Higher-Order Thinking Skills Problems Problems Of Education In The 21st century Vol. 76, No. 4, 2018

[24] Allan L White 2005 Active Mathematics In Classrooms Finding Out Why Children Make Mistakes – And Then Doing Something To Help Them. Journal Square One Vol 15 No 4 December 2005

Why are mathematics teachers advised to use blended learning in the learning process?

Bulan Nuri

Graduate School, Yogyakarta State University

Corresponding author: nuribulan1@gmail.com

Abstract. The success of mathematics learning is strongly influenced by the initiative and strong ability of educators to implement the learning that supports the achievement of the goals of mathematics learning. An effort to help provide an overview to the teacher to find a solution in choosing a learning model that can be applied in mathematics learning, this research was conducted. Focusing on the results of implementing blended learning in all possible learning programs with varying levels of education, 11 relevant articles were obtained to answer the formulation of this problem. The results obtained that the use of blended learning is very possible in mathematics learning. This is based on the results of research relating to the implementation of blended learning which has the potential to improve learning outcomes, activity, interest, motivation, understanding and even provides space for communication skills, teamwork, independence, and student achievement.

1. Introduction

Learning purpose is not to train students' ability to memorize facts and problem solving procedures. Moreover, the purpose of learning is to guide students to understand the concept in order to be able apply concepts in life.To achieve learning outcomes that match expectations and overcome the decline in student learning achievement,the changes in policies and standards in mathematics are some possible solutions to overcome the decline in student achievement in mathematics [1]. These policies and standards can be in the form of developing curriculum and core mathematics standards. The National Mathematics Teachers Council (NCTM) reformed efforts to improve student achievement in mathematics through revising curriculum and core mathematics standards. The revision process is primarily focused on thorough and deeper pedagogical content knowledge from conceptual understanding [2]. Not only working with curriculum and core mathematics standards, but also the effectiveness of teaching approaches and metacognitive strategies, which aims to improve and design mastery of abstract concepts in mathematics is also another problem in achieving success in the mathematics learning [3].

Providing opportunities to explore and practice more deeply about a concept can support student abilities to understand concepts better because conceptual understanding cannot be transferred by the teacher to students, but must be built and formed by students under the guidance of the teacher, and on the basis of prior knowledge and experience. It is same as the learning perspective, "learning is the process of understanding the world and negotiating with others to resolve uncertainties and to achieve an understanding of one's opinion [4]. In addition, scaffolding from the teacher and intensive discussions between students that occur in the learning process can help students understanding. According to Magiera and Zawojewski, the lack of meaningful mathematical discussion is often observed in mathematics classrooms. Students tend to work independently in mathematical tasks and activities, without allowing to communicate and interact with their peers [5].This can be caused by limited time and adequate conditions in the process of learning mathematics even though the socialization aspect of the mathematical communication is fundamental aspects for critical thinking, problem-solving and developing as a mathematician. Therefore, the teacher as an instructor in mathematics learning must be able to choose the appropriate method to overcome these problems. Blended learning as an approach that mixes face-to-face and online learning activities which is considered by researchers as a quite capable approach to overcome these problems.

Blended learning has taken several models, such as combining web-based technology models, pedagogical approaches, teaching technology and the actual work [6].Blended learning is an Internet-based and face-to-face learning environment [7][8] that is a traditional environment in complementary models that uses all possible ways for both of them in the education process [9]. That is to say, traditional and computer-based learning methods run together [10]. In addition, according to Khaled, N R A., Blended E-learning is one of the modern education approaches that rely on the use of information technology in designing new educational contexts that enhance active learning strategies and are strategy-centered [11]. Thus, we can conclude that there is general agreement that the main component of blended learning is face-to-face teaching and online learning. The application of mixed learning can: (1) Expand the space and opportunities available for learning; (2) Supporting course management activities (e.g, Communication, assessment of submission, marking, and feedback); (3) Support the provision of information and resources for students; (4) Involve and motivate students through interactivity and collaboration [12].

Nevertheless the results of a review of the literature on blended learning, Oliver and Trigwell have broadened the notion of what is mixed in identifying seven different mixtures [13]. Which were mixing: (1) E-learning with traditional learning; (2) Online learning with face-to-face; (3) Different media; (4) Different contexts, for example, work and study; (5) Different theories of learning; (6) Learning objectives, for example, those concerning skills as opposed to knowledge; (7) Pedagogic approaches, for example, distance and campus-based learning. As with Oliver and Trigwell, Driscoll and Carliner [14]describe four patterns, each of which they call the form of Blended Learning: (1) a mixture of Web-based technologies; (2) a mixture of various pedagogical approaches (eg, Constructivism, behaviorism, cognitivism); (3) a combination of all forms of technology teaching with face-to-instructor-led conditions; or (4) a combination of instructional technology and actual work tasks to form a mixture of effective and working learning. Blended E-learning is so flexible that it allows students to receive the same message from various sources in different forms in the long run. For example, traditional lessons can be provided and the same material can be presented differently in online [11]. Blended E-learning contributes to mixing learning classrooms with virtual learning to achieve cost balancing in preparing learning programs that help reduce the costs of these programs [15]. Despite the various possibilities of mixing methods in learning, some researchers recommend that the balance in designing educational materials by 50% represent face-to-face activities in the classroom context and 50% including virtual activities on the Internet [16] when designing Blended E-learning environments.

Based on description above, this literature review seeks to examine the benefits of blended learning based on the results of research that has been carried out previously. So based on this review we would like to analyze the correlation between blended learning and mathematics learning. So that we can get an idea of whether blended learning can really be implemented in mathematics learning. The results of this study can be a reference for education practitioners in order to consider the use of blended learning in learning. So that in the future it can produce diverse mathematical competencies for students.

2. Method

Kinds of the procedures and technical use in this literature review method.

2.1 Search procedure
Searching for research articles, the literature analysis was limited to several types of research. The main research analyzed was empirical research to find out more about the effectiveness of blended learning. A literature review was chosen as a relevant study but was not included in the analysis. The formulation of the problem that had been compiled was then searched with several methods that had been chosen. The search procedure was done by searching for as many previous studies as possible. The research sought was about the blended learning model and the effectiveness of blended learning in learning process. In addition, the study of mathematics and e-learning in mathematics learning was also sought through searches in several journals and proceedings.

Looking for several literature studies, the search procedure used was to search databases from Google Scholar, Springer, Z-Library and track citations from a study for follow-up lessons. Some reference search keywords used include blended learning, blended learning in mathematics learning, the influence of the application of blended learning, components, characteristics, advantages and benefits of blended learning, etc. Some journals from search results were also a reference in this literature review. Part of the journal related to blended learning was a reference, including *journals Eurasia, Journal of research in math education*, etc.

2.2 Selection criteria
The results of searches in several journals and articles with relevant keywords specified in the search procedure article, some articles about blended learning were obtained. Then, from the articles selected in accordance with the provisions, namely the effectiveness of blended learning or the successful implementation of blended learning based on previous research conducted by previous researchers. The results of the application of blended learning are not limited only on mathematics learning, because there is still little application of blended learning in mathematics learning especially in junior and senior high school.

Furthermore, there are several criteria in selecting articles that have been searched according to the established procedure. The first criterion is focuses on articles with empirical research that includes information about research subjects. This focus is limited to empirical research because what will be studied is the benefit of blended learning in learning process and the research results of the literature review are used to support the results of the analysis.The second criterion is the limitation of the research subject, namely at the level of primary education to higher education. This is done by researcher for that research can be more comprehensive and come from various levels of education. The college level was chosen because the application of blended learning has so far been more widely used in universities. Whereas students in secondary education are chosen because they can provide a strong argument that the application of blended learning is also possible at a lower level.

The third criterion is selected search results articles published in that period 2000-2018. This is intended to make the study reviewed more expected and more legitimate. However, theoretical studies are not limited by the year of publication of an article. Supporting articles are also used even in longer publishing. Based on the four criteria mentioned above, relevant articles obtained 11 empirical articles of research with various kinds of dependent variables. Differences in the dependent variable were chosen to show the benefits obtained from the application of blended learning in learning.

2.3 Formatting author affiliations
The main characteristics of some articles will be defined, this is based on questions in the formulation of the problem. Articles about the effectiveness or success of blended learning are then selected and grouped in table form. The information entered in the table regarding the selected article contains the following content. (1) Author and year of publication; (2) Number of authors; (3) Dependent variables; (4) Research design; (5) Research subject; (6)Research program. This coding strategy is then used as a method for analyzing articles that have been tracked to facilitate categorization.

Journal of Physics: Conference Series **1397** (2019) 012080 doi:10.1088/1742-6596/1397/1/012080

3. Result

In-depth analysis will be carried out in 11 journal articles that have been previously selected, which are in accordance with the criteria, categories, and procedures that have been set. The information related to the articles can be seen in the table 1 below.

Table 1. Overview of studies on blended learning

Author	Year	Dependent variables	Research design	Subject	Programs
Wahyudi., Waluya, B., Hardi, S &Isnarto	2018	Creative thinking	Pre-experiment design (one group pretest-posttest design)	27 students (one class)	Study program of primary school teacher education
Ya-wen, L., Chin-Lung, T & Po-Jui, C	2016	Blended Learning	Quasi-experiment	Junior high school student	Mathematics
Anwar, M., Indiani, M & Syamsuri	2016	Content Knowledge	Quasi experiment	52 students (30 experiment, 25 control)	Algebra matric
Mulqueeny, K., Victor, K., Ryaan, S & Jaclyn O	2015	Student engagement	Pilot study	3 group students at middle school	Mathematics
Tee, W J., Sharon, L P V., Sarojini, D N & Wong, S Y	2015	Learning management, social learning and video	Quantitative approach (Survey)	35 university students	RMIIT degree programs
Marc, T K	2014	Student outcomes	Quasi experiment (non-equivalent control group)	38 students (blended learning) and 28 students (traditional approach)	Public health course
Hamidreza, K., Zaleha, I and Yudariah, M Y	2012	Communication skill and team work	Survey	62 first year students	Mathematics department
Sulihin, B S	2012	Motivation and learning outcome	Quasi experiment	62 students	Vocational high School
M. Victoria, L-P., M. Carmen, P-L & Lazaro, R-A	2011	Perceptions and outcomes	Quantitative (Final Mark)	985 valid samples	Economy and business
Aynur, G & Funda, D	2012	Student experience	Survey	67 students	Departments of mathematics and primary school education
Vanessa, W., Andreas, H., Klaus S & Cord, S	2009	Blended learning and problem based learning	a mixed-method study.	185 third-year students and 14 tutors took part in the study	Aachen faculty Model Curriculum Medicine

Based on the results of the selection of journals that supporting this research. In-depth discussions related to the results of the research from the article will be conducted, so that it will be seen related to mathematics learning. First, research conducted by Wahyudi, Waluya, Hardi and Isnarto, obtained results that blended learning effectively improves students' creative thinking abilities in solving problems[17]. The reasons why this can occur because learning gives students the opportunity to think systematically by starting and criticizing interesting contextual problems and ending with meaningful reflections with adequate learning resources, both when face to face and online. This research implies that it is true that student creativity can be improved by giving problems to students to find solutions, including in mathematics learning. Student's creative ability is one problem that requires a solution to be solved.

Second, the research of Ya-wen, Chin-Lung and Po-Jui, showed that blended learning not only gave positive results to student learning outcomes but also had an impact on attitudes towards mathematics. In addition, specifically, the results show that male and high-ability students are more motivated in learning with the blended learning model [18]. This provides reinforcement for the author's initial argument that blended learning can be carried out at a lower level of education, as well as in mathematics learning. The application of blended learning can make efforts to improve mathematics learning outcomes and eliminate negative perceptions of mathematics. Third, research conducted by Anwar, M., Indiani, M &Syamsuri; it is about the strategy to improve content knowledge. Overview of the results from the data showed that the learning outcomes of the blended learning class were better than ordinary classes [19]. In addition, students seem more active in doing assignments than ordinary classes.

Fourth, Mulqueeny, Victor, Ryaan and Jaclyn O's research shows that when using the reasoning mind, blended learning program students show more than 65% of the concentration involved and more than 85% of the time working on the task. This high level of involvement continues for one hour of math lessons [20]. Continuous involvement creates greater opportunities for student learning outcomes. Research on these lower grade of educations shows that there are great opportunities for learning mathematics with blended learning by mathematics teachers in junior high schools. Hopefully it can improve student learning achievement.

Fifth, research conducted by Tee, Sharon, Sarojini and Wong shows that blended learning can increase students' interest in learning [21]. The increased interest is one positive factor that will result in increased understanding. A high interest in learning will increase the intensity of learning, thus leading to increased understanding. Understanding is one of the cognitive components of mathematics learning goals [22], therefore it is seen that there is a positive correlation in the use of blended learning in mathematics learning.

Sixth, based on the results of a study conducted by Marc that for students in health science courses can be concluded that the blended learning approach can be an effective way to optimize student learning and improve student performance [23]. This shows that it is possible to apply blended learning in the hope of obtaining results in accordance with this study. This is because in previous studies related to mathematics learning, blended learning also gave positive results.

Seventh, research examining communication skills and teamwork, conducted by Hamidreza, Zaleha and Yudariah. The results obtained that blended learning in multivariable calculus programs has a positive impact on students' communication skills and teamwork [24]. Paul's research also shows that most students (95%) report that they have positive results with blended learning approaches. They prefer active learning methods or peer learning rather than traditional lecture-based instruction, and technology-mediated instruction is very effective in helping them build knowledge [25]. Both of these studies show that through active interaction of students in learning, for example through peer interaction, students' communication and cooperation abilities increase. Then refer to this study, blended learning can be used as a tools to improve the ability of student communication and teamwork on mathematics learning. Communication skill is one of the components of ability that students must have in learning mathematics because one of the objectives of mathematics learning is to grow mathematical communication skills. And this is also in line with one of the demands of 21st-century capabilities, namely communication.

Eighth, the research conducted by Sulihin, found that the motivation and learning outcomes of students had differences between the control class and the experimental class. Further data are obtained as follows. 1) There are different kinds of learning motivation between students taught blended learning compared to students taught conventional learning with values 0.012 with an average of 4.74 and there are differences in learning outcomes with sig values 0,000 with an average of 13.39. 2) There is an increase in student learning motivation due to the application of blended learning with sig 0,000 average increase of 13.55 and there is an increase in student learning outcomes with sig values 0,000 average increase of 38.23 [26]. The difference in motivation and improvement in student learning outcomes can be a reference or can be an illustration that there are teachers who have applied the blended learning method in mathematics learning and it works. So there is no reason for mathematics teachers to lack teaching method to apply, to overcome the low motivation and learning outcomes of mathematics students.

To promote students' motivation, various strategies must be used to encourage them to be actively involved in the learning process [27].Learning motivation can be classified into two types: intrinsic and extrinsic. Intrinsic motivation refers to the performance of an activity because of satisfaction attached to it from external reasons such as rewards, while extrinsic motivation can be defined as carrying out activities to achieve results [28]. The two types of motivation greatly affect students' learning outcomes. Donnelly revealed that increasing the level of interaction in learning has become a major factor for increasing students' motivation and for creating positive attitudes, which consequently improves their performance [29]. In addition to increasing motivation, blended learning that is designed must also aim to overcome students' learning difficulties [30].

Ninth, according to what was done by Victoria, Carmen and Lazaro, utility level, motivation, and high satisfaction were felt from blended learning, which could lead students to have a positive attitude towards learning [31]. In addition, it was shown that online activities including those useful for students, who can have a beneficial influence on the work they do independently. A positive attitude towards learning helps to form good interest and motivation in learning so it is strongly recommended that blended learning is applied in learning process. In addition, blended learning trains students to do their jobs independently.

Furthermore, the research conducted by Aynurand Funda shows that applying for courses especially with electronic activities has a positive effect on students from a learning and evaluation perspective [32]. Students state that a mixed learning environment supports their active participation in course activities and shows that following the course content, homework and online projects is interesting and useful. Because of that, blended learning is one of the supporting facilities to increase students' active participation in learning. Hopefully students will have their own initiative in seeking new knowledge and information related to learning. This is supported by the results of research conducted by Paul [25]. The results of the study show that the blended learning approach leads to more conceptual change, acquisition of more skills, and higher performance. He also added that this discovery could be a reference for the application of blended learning in subsequent learning.

The latest is research by Vanessa, Andreas, Klausand Cord showing that motivation, subjective learning gains, and satisfaction achieved increase significantly with blended problem-based learning compared to learning problem-based learning in a rational manner [33]. In addition, he adds that according to log-file analysis, web-based learning models are often used and can enhance collaboration during self-learning. Based on the results of these studies, it can be analyzed that learning on the basis of giving problems using blended learning is more effective. Because learning mathematics that often uses problems to find or construct a concept in students can combine problem-based learning and blended learning in order to increase the effectiveness in achieving its implementation goals.Based on the results of the analysis that has been done, it can be seen that it is possible to apply blended learning in mathematics learning.

The analysis of 11 articles above shows the positive impact of applying blended learning in various study programs including mathematics. The analysis is in accordance with the results of research conducted by Wee L Ng, who stated that there is encouraging positive impact of the blended learning on students' learning of mathematics [34]. The positive impact of blended learning can result in reducing dropout rates and increasing test scores. In addition, students' perceptions of blended learning are interrelated, with their final grades dependent on mixed learning activities, as well as on students' age, background, and attendance in the classroom [31]. All these aspects will affect students' attitudes toward learning. The use of technology at lower education can be through simple and easy technology that is and can be understood by students. But at a higher level of education an increase in learning outcomes can be obtained from the use of new information and communication technology (ICT); this offers a new way of producing, distributing and receiving university education [35]. Learning with the blended learning model also encourages students to be more active in working on assignments than the ordinary class [19]. At a higher level, the difficulties experienced in mathematics learning are certainly different from other lower levels. According to Christina, there are two requirements to counteract the difficulties of mathematics in higher education: the need for personalization (related to the needs of each individual) and the need for collaboration (according to the theory of social construction of competence) [36].

The mathematics that is lot of abstract concepts will be greatly helped by the use of technology in its integration. Moreover, e-learning offers opportunities to use and coordinate multiple representations (graphical, visual, symbolic, verbal, etc.) and to plan activities that are compatible with a pragmatic approach [37]. For students from aspects of the blended learning environment, it provides a great opportunity for students to be free in learning. The online component of blended learning increases students' agency (allows them to control both instructional and sequence steps), reduce typical disturbances in classrooms or lecture halls, increase assignment time, and improve students' performance [38]. Blended learning implementation that turns out to be not only beneficial for students, but blended learning is also beneficial for teachers or staff. For example in the case of weekly online practice quizzes to support lecture material and textbooks using automatic tagging functionality that generates direct and automatic feedback to each student about understanding their concepts and practicality avoiding a continuing workload for a gift in staff. In addition, the use of blended learning by teachers is an attempt to expand and add to classroom learning with online exploration and discussion. [39].

The results of the analysis that have been carried out provide a very clear explanation, how the positive impacts that have been obtained from the implementation of blended learning in various types of study programs including mathematics. Therefore, the application of blended learning in mathematics learning for teachers is no longer an impossible thing to do while considering the related aspects of learning fluency, in order to overcome difficulties in understanding concepts, time constraints and disruption of the learning environment that is often experienced by students. In addition, the results of the above studies are also in line with the mathematical components that need to be taught to students, namely critical thinking, creative thinking, conceptual understanding, problem-solving skills and mathematical communication. Thus, it can be stated that blended learning is in accordance with mathematical abilities.

4. Conclusion

Based on a study of 11 articles above, it is concluded that in several programs and levels of education, blended learning has a positive impact. The successful application of blended learning does not only have an impact on students' achievement but also has an impact on students' attitudes towards learning. This success is also demonstrated in the application of blended learning to mathematics learning. In addition the results show that blended learning can be used at various levels of education. Therefore, the application of blended learning in mathematics learning needs to be considered.

References

[1] Woodward J 2004 *J of Learn Disabilities* **37** 16
[2] National Council of Teacher of Mathematics (NCTM) 2000 *Principles and standards for school Mathematics* (Reston, VA: Author)
[3] National Mathematics Advisory Panel 2008 *Foundations for success: The final report of the National Mathematics Advisory Panel* (Washington DC: U.S. Department of Education)
[4] Duffy, T M and Cunningham, D J 1996 Constructivism: Implications for the design and delivery of instruction. In D. Jonassen (Ed.), *Handbook of research for educational communication and technology* (New York: Macmillan)pp170–198
[5] Magiera M T and Zawojewski J 2011 *J for Research in Math Educ* **42** 486
[6] Driscoll M 2002 *eLearning* **1** 1

[7] Graham C R 2006 *Blended learning systems: Definition, current trends and future directions*. In C. J. Bonk & C. R. Graham (Eds.), *The handbook of blended learning*: Global perspectives, local designs (San Francisco: Pfeiffer) pp3–21
[8] Garrison D R and Kanuka H 2004 *JInternet and Higher Education* **7**95
[9] Littlejohn A andPegler C 2007 *Preparing for Blended eLearning* (London: Routledge)
[10] Mitchell P and Forer P 2010 *J of Geography in Higher Educ* **34** 77
[11] Khaled N R A 2013 *Education Journal* **2** 176
[12] Debra B and John B 2010 Getting started with belnded learning (Autralia :Griffith University)
[13] Oliver M and Trigwell K 2005 *E-Learning* **2** 17
[14] Driscoll M and Carliner S 2005 *Advanced web-based training strategies. Blended learning as a curriculum design strategy* (New York: ASTD Press) pp87–116
[15] Twigg C 2003 *Educause Review*pp 28-38
[16] Korkmaz Ö and Karakusm U 2009 *TOJET* **8**51
[17] Wahyudi., Waluya, B., Hardi, S andIsnarto 2018 *Advances in Social Science, Education and Humanities Research (ASSEHR)* 247
[18] Ya-wen L, Chih-Lung T and Po-Jui C 2017 *EURASIA J of Math Science and Technol Educ* **13** 741
[19] Anwar M, Indiani M andSyamsuri 2016 *Cakrawala pendidikan* **XXXV**no 1
[20] Mulqueeny et al 2015*Inter J of STEM Educ* **2** 1
[21] Tee W J, Sharon L P, SarojiniD N and Wong S Y *Teach and Learn* Conference 2014 Proceedings
[22] Kilpatrick, Swafford, danFindell 2001 *Adding It Up: Helping Children Learn Mathematics* (National Academy Press)
[23] Marc T K 2014 *BMC Med Educ* **14** 1
[24] Hamidreza, K., Zaleha, I and Yudariah, M Y 2012 *Proc. Soc & Behav Science* **56** 341
[25] Paul B and Tenzin D 2018 *J Computer Educ* 1
[26] SulihinB S 2012 *Jurnal Pendidikan Vokasi* **2** 368
[27] Carroll L and Leander S 2001 *Improving student motivation through the use of active learning Strategie* (Saint Xavier University: Master of Arts Action Research Project)
[28] Ryan R and Deci E 2000 Contemporary Educational Psychology **25** 54
[29] Donnelly R 2010 *Computers & Education* **54**350
[30] Saleh A 2016 Active Blended Learning to Improve Students' Motivation in Computer Programming Courses: A Case Study. In: M. Abdulwahed et al. (eds)*Advances in Engineering Education in the Middle Eastand North Africa* (Switzerland: Springer) pp187-204
[31] M. Victoria L-P, M. Carmen, P-L and Lazaro R-A 2011 *Computer&Educ***56** 818
[32] Aynur G and Funda D 2013 Education science: theory & practice**12** 438
[33] Vanessa W, Andreas, H, Klaus S and Cord S 2009 *Adv in Health SciEduc***14** 725
[34] Wee L Ng, Beng C T, Joseph B W, Weng K H and Kok M T 2019 Use of Technology in Mathematics Education. In: Toh T., Kaur B., Tay E. (eds) Mathematics Education in Singapore. Mathematics Education – An Asian Perspective (Singapore:Springer) pp314-348
[35] Orton-Johnson K 2009 *British J of EducTechnol* **40** 837
[36] Cristina B and Pietro D M 2012 *ZDM Mathematics Education* **44** 787
[37] Bardelle C and Ferrari P L 2010 *IMSCI-EISTA* **1**
[38] Allen I E and Seaman J 2010 *Class Differences: Online Education in the United States* (Sloan Consortium: Newburport)
[39] Marcelo C B, Petek A, Johann E, George G, Salvador L and Mario S A 2016 *ZDM Mathematics Education* **48** 589

The Profile of Junior High School Students' Mathematical Creative Thinking Skills in Solving Problem through Contextual Teaching

Cholis Sa'dijah[1,*], Ucik Fitri Handayani[1], Sisworo[1], Sudirman[1], Susiswo[1], Ety Tejo Dwi Cahyowati[1], Mukhtamilatus Sa'diyah[1]

[1] Mathematics Education, Faculty of Mathematics and Science, Universitas Negeri Malang, Jalan Semarang No. 5, Malang 65145, Indonesia

* cholis.sadijah.fmipa@um.ac.id

Abstract. The purpose of this study was to describe the profile of students' mathematical creative thinking skills in solving problems through contextual learning based on high, medium, or low initial abilities. The method used in this study is descriptive qualitative. There were 6 research subjects of junior high school students in the border area of Malang and Blitar Regencies. Data collection was done by observation, tests, and interviews. The results of this study are (1) High-ability students can fulfill three indicators; those are fluency, flexibility, and novelty according to the material, so that they are in the very creative category. There are high ability students who fulfill two indicators; those are fluency and flexibility well based on the material, so that they are in the creative category. (2) Middle-ability students can fulfill an indicator namely flexibility according to the material, so that they are in the category of quite creative. (3) Low ability students can fulfill two indicators, namely fluency and flexibility well based on the material, so that they are in the creative category. (4) The profile of students' mathematical creative thinking skill in solving problems through contextual learning based on early mathematical abilities was different. The profile of students' creative thinking skills is expected to be a teacher's reference to determine appropriate assignments and assessments to be given to students.

1. Introduction

The challenges of the 21st century in the field of education require students to be able to have various abilities including creativity, critical, communication, and collaboration. As explained that education in the 21st century carries a learning paradigm that requires graduates to have high-level thinking skills those are logical, critical, creative and innovative thinking [1]. As a result, there is a change in the education curriculum in order to realize 21st century competence for each subject in school, including mathematics. Mathematics is one of the various main subjects that must be taught by teachers to students [2]. Therefore, it is proper that in achieving the learning objectives, the learning of Mathematics in schools needs to be carried out full of totality.

In the 2013 Curriculum, it is clearly written about the purpose of mathematics learning at school. The purpose of learning, especially junior high school level based on the 2013 Curriculum for mathematics subject, one of which requires students to have the ability to think creatively. Mathematical creative thinking focuses on the ability of students to produce a new variety of correct answers to solve open mathematical problems. Lince [3] explains that if students can solve mathematical questions that are classified as non-routine questions using a different strategy than the teacher's explanation and example in the learning process is called creative. Creative thinking is also closely related to solving problems in mathematics [4]. Therefore, the ability to think creatively in solving problems is one of the abilities that must be mastered by students at every level of education.

Creative thinking shown by students in solving problems need to be developed most. Sriraman et al. [5] stated that problem solving is part of creative thinking. According to Sa'dijah [6] mathematics learning in schools can encourage creative thinking skills of students to solve mathematical problems and in deciding a choice. The role of the mathematics teacher is needed in developing students' creativity through learning [7].

The ability of Indonesians' students is still not optimal if it is associated with the mathematical creative thinking skills. The results of the 2015 Program for International Student Assessment (PISA) competition with a sample of junior high school / MTs students aged around 15 years old stated that Indonesia was ranked 63rd out of 72 countries[8]. Gurria [9] added that one of the characteristics of the questions in the PISA competition is requiring basic skills in finding a variety of solutions / answers and resolution / solving strategies, and the results are surprising, that was the percentage of students who cannot solve the easiest math problems is 32%. This is in accordance with what is stated in the 2013 Curriculum, students' achievement in the PISA competition survey report are not encouraging, because the material and questions tested in the PISA competition are not contained in the Indonesian curriculum [10].

Observations conducted at one class of junior high school in the border area of Malang and Blitar Regencies. It showed that many passive students in the teaching learning process, students' creative thinking ability was relatively low, indicated by the results of the initial test answers when given contextual problems. Among 23 students who took the initial test, all students only wrote one possible correct answer and one strategy in solving the problem.

Various ways can be done in order to familiarize the ability to think creatively. In the mathematics learning process, it is necessary to develop material and problems through contextual learning. Samo, et al. [11] states that by modifying and innovating activities in mathematics learning in the classroom can develop the ability to think creatively. Creative learning strategies encourage students to produce new ideas more deeply [12]. Sarassanti, et al. [13] also shows that the improvement of students' mathematical creative thinking skills using contextual learning is better than conventional learning in terms of initial mathematical abilities (high, medium, and low). Sugandi [14] in his research also states that contextual learning had a large effect compared to conventional learning, school level, and students' initial mathematical abilities towards achieving high level mathematical thinking skills.

The mathematical creative thinking skill of each student is not the same because they have different knowledge and experiences in solving problems in the contextual learning process based on students' initial mathematical abilities which are high, medium, and low abilities. The existence of differences in students' Mathematical Creative Thinking skills in solving problems through contextual learning made researchers interested in conducting a study about junior high school students' Mathematical Creative Thinking skills in solving problems through contextual learning. This research aims to describe the profile of students' mathematical creative thinking skills in solving problems through contextual learning based on high, middle, or low initial abilities. It is based on research by Sarassanti, et al. [13] which aimed to improve mathematical creative thinking skills of junior high school students and research by Kadir & Masi [15] which discussed about creative thinking skills of the junior high school in Kendari city. Furthermore, those researches are connected to research of Sugandi [14] which aimed to determine the effects of the contextual approach to the ability of high-level mathematical thinking.

2. Method

This research is a descriptive qualitative research. The subjects of this study were students of junior high school in the border area of Malang and Blitar Regencies. Class selected was chosen because

Journal of Physics: Conference Series **1397** (2019) 012081 doi:10.1088/1742-6596/1397/1/012081

students have a diverse ability and recommended by mathematics teacher in junior high school. Before determining the subject of the study, students are grouped based on their initial mathematical abilities on the teacher's list of scores. Then 6 research subjects were selected, those were 2 students of high skill group, 2 students of middle skill group, and 2 students of low skill group. Data collections were done through observation, tests, and interviews. The main instruments in this study were the researchers themselves and assisted with other instruments including Lesson Plan (LP), Student Worksheets (SW), Group Worksheets (GW), creative thinking ability tests, teacher's and students' activity observation sheets, and interview guidelines. The tests consists of a problem with two questions about number pattern based on 3 indicators creative thinking skills those are fluency, flexibility, and novelty. It is used to describe student's mathematics creative thinking skills in solving contextual problems.

In this study, there are contextual mathematical problems that students must work on. The analysis of the students' mathematical creative thinking skill tests was done based on indicators of mathematical creative thinking skills; those are fluency, flexibility, and novelty then the categorization of students' Mathematical Creative Thinking Levels is in Table 1. Fluency indicators cn be seen from how students can write various solutions / answers correctly and smoothly without difficulty. Flexibility indicators are seen from how students write one problem solving strategy or various problem-solving strategies in sequence and can explain the process of solving it. Novelty indicators cn be seen from how students can write solutions / answers / mathematical problem-solving strategies that look different from those used regularly in classroom learning.

Table 1. Guidelines and Descriptions of the Mathematical Creative Thinking (MCTL) Level

MCTL	Fluency	Flexibility	Novelty
Level 4 (Very Creative)	√	√	√
Level 3 (Creative)	√	-	√
	√	√	-
Level 2 (Creative Enough)	-	√	-
	-	-	√
Level 1 (Less Creative)	√	-	-
Level 0 (Not Creative)	-	-	-

Data analysis from the test results of students' mathematical creative thinking skills is used to categorize the level of creative thinking skill based on indicators that can be achieved. Data analysis techniques include data reduction, data presentation, and conclusions. Checking the validity of the data is using triangulation techniques.

3. Result

There are 24 students who are grouped based on their initial mathematical abilities in the teacher's list of score. The results of initial abilities grouping test is showed in Table 2.

Table 2. The Result of Initial Mathematical Abilities Grouping

First Group	Percentage (%)
High	8,3%
Middle	25 %
Low	66,7%

Table 2 shows that 8.3% of students have high ability, 25% of students have middle ability and 66.7% of students have low ability. After grouping, students are given an initial test that is used to provide initial information about students' mathematical creative thinking skills. Then contextual learning was carried out for 8 meetings by mathematics teachers. Then students were given the final test of students' mathematical creative thinking skills. Based on the results of the tests, it was selected 6 research subjects; those were group of 2 students with high skill, group of 2 students with middle skill, and group of 2 students with low skill. Then conducted the interviews with the six students based on the test results. The results of students' work were analyzed based on indicators of students' mathematical creative thinking skills. Analysis of the data from the test results of students' mathematical creative thinking skills was used to categorize the level of creative thinking skill based on the indicators achieved. The following are the results of the learning process exposure and the mathematical skills of each student.

3.1 Data Exposure of Contextual Learning Process

Contextual learning in this study is carried out based on the lesson plan that has been made. The lesson plan consists of three main activities, those are preliminary activities, core contextual learning activities, and closing activities. There are 7 basic principles that are carried out in contextual learning, including:

3.1.1 Constructivism Principal. The teacher shows the photos to students through a slide that contains the patterns in the surrounding environment, and students observe it each meeting. Then the teacher provides the opportunity for students to be able to build their own new knowledge based on the photos displayed, and even connect them with the students' experiences and understanding in real life and previous experiences.

3.1.2 Inquiry Principle. The teacher gives student's worksheets to students. Then students work on the worksheet about finding patterns of numbers independently. In this case the teacher plays important role in planning and designing activities that focus on inquiry activities in the material of numbers patterns.

3.1.3 Questioning Principle. The teacher and students do question and answer about the material of the number pattern. Students are given the opportunity to ask questions that have not been understood. Then the teacher must answer them clearly. In this case asking questions in contextual learning can be used by the teacher to encourage, guide and assess students' abilities, so students will be able to find a variety of new information that is not yet known.

3.1.4 Learning Community Principle. The learning community in this study consists of a study group that conducts discussion activities to find solutions in solving contextual problems in group worksheets relating to number patterns. During the discussion, the teacher helps students work together in completing or solving contextual problems on the group worksheet if there are difficulties.

3.1.5 Modelling Principle. The teacher models the number patterns that exist in everyday life by giving and explaining examples of surrounding problems related to number patterns. In this case, students more understand the relation of the material with the daily real life.

3.1.6 Reflection Principle. The teacher asks students to make a reflection of today's learning about the difficulties experienced and the material that has been understood. Then let students interpret the experience independently, so that they can summarize their experiences independently.

3.1.7 Authentic Assesment Principle. Teacher conducts authentic assessment by guiding students while sharing or discussing in their group and when the students present the result of their discussion in front of the class.

3.2 Profile of High Ability Students' Mathematical Creative Thinking Skills

Based on the results of the Subject 3's (S-3's) work on fluency indicators, he/she can write more than one possibility answer, those are four possible sizes accompanied by the number of ceramics in each pattern correctly and well. The number of ceramics written by the S-3 on each pattern of possibilities 3 and 4 are correct, but in the possibility of 1 and 2 there is an error calculation in the number of ceramics in pattern 4 and 5. S-3 can also provide explanations and reasons stating the number of ceramics in each pattern that forms a number pattern. Furthermore, based on S-23 that can write more than one answer, those are two possible sizes completed by the number of ceramics on each pattern correctly and smoothly. Furthermore, based on observations, test results, and interviews of S-3 and S-23, the subjects fulfilled fluency indicators.

S-3, on the flexibility indicator, can write two problem solving strategies correctly, those are by writing down the whole information then using the method of dividing and manipulating objects or examples using a comparison method to find the price of each required ceramic in order and clearly accompanied by an explanation. S-23 can write one problem-solving strategy that is writing down the whole information by using the distribution method in sequence and correctly. Furthermore, based on observations, test results and interviews, S-3 and S-23 meets the indicators of flexibility.

S-3, on the novelty indicator, writes one problem solving strategy that is slightly different, that is by manipulating objects or making an example using a comparison method. Furthermore, based on observations, test results and interviews, the S-3 meets the novelty indicator. S-23 solved the problem in the final test but did not write a problem-solving strategy that was different from the others. This is because other students do not write a comparison method in solving the problem given. Furthermore, based on observations, test results and interviews, the S-23 did not meet the novelty indicator.

Based on the results of research and data analysis, the results obtained through contextual learning, high ability group students can meet the indicators of fluency, flexibility, and novelty. So, based on the Mathematical Creative Thinking Level (MCTL), he/she is in MCTL 4 which means that S-3 is including as very creative student. Furthermore, there are subjects who through contextual learning can meet the indicators of fluency, flexibility, and not yet fulfilled the novelty indicator, so the S-23 is in MCTL 3 which means that S -23 is including as creative students.

3.3 Profile of Middle Ability Students' Mathematical Creative Thinking Skills

Subject 2 (S-2) and Subject 11 (S-11) can write only one possible size answer, accompanied by the number of ceramics in each pattern correctly. However, they can provide an explanation and reason that states the number of ceramics in each pattern that forms a pattern of numbers orally. Furthermore, based on observations, test results, and interviews, the S-2 and S-11 do not meet fluency indicators.

S-2 and S-11 can write one problem solving strategy correctly, in order, and clearly. The subjects write a problem-solving strategy that is writing the entire information using a division method to find the price of each required ceramic. Furthermore, based on observations, test results, and interviews, S-2 and S-11 meet the indicators of flexibility.

On the test of students' mathematical creative thinking skill through contextual learning S-2 and S-11 do not write problem solving strategies that were different from the others, so based on observations, the results of tests and interviews showed that the S-2 and S-11 were not fulfilling novelty indicator.

Based on the results of the research and analysis of test data, it obtained that, through contextual learning, middle ability group students did not meet fluency indicators, fulfilled flexibility indicators, and did not meet the novelty indicator, so the S-2 and S-11 are in MCTL 2 which means that the S-2 and S-11 are including as quite creative students.

3.4 Profile of Low Ability Students' Mathematical Creative Thinking Skills

The Subject 5 (S-5) can write two possible answers of sizes completed with the number of ceramics on each pattern correctly and smoothly. The S-5 can also provide explanations and reasons that stating the number of ceramics in each pattern to form a pattern of number orally. Furthermore, based on observations, test results, and interviews, S-5 meets fluency indicator.

Based on the work results, it shows that the Subject 14 (S-14) can write two possible answers of sizes accompanied by the number of ceramics on each pattern correctly and smoothly. S-14 can also provide explanations and reasons that stating the number of ceramics in each pattern that forms a pattern of numbers. Furthermore, based on observations, test results, and interviews, S-14 meet the fluency indicator.

S-5 and S-14 can write one strategy correctly. The S-5 uses a strategy by writing down the entire information using a distribution method to find the price of each required ceramic. Furthermore, based on observations, test results, and interviews of S-5 and S-14, they meet flexibility indicators. On the test of students' mathematical creative thinking ability through contextual learning, S-5 and S-14 did not write different problem-solving strategies with others, so based on observations, the results of tests and interviews showed that the S-5 and S-14 do not fulfill novelty indicator.

Based on the results of the research and data analysis, the results of the low ability students group meet the fluency indicator, fulfill the flexibility indicator, and de not meet the novelty indicator, so the S-5 and S-14 are in MCTL 3 which means that S-5 and S-14 are including as creative students.

4. Discussion

4.1 The result Analysis of Contextual Learning Process

The instructional strategy used in this study is contextual learning that links subject matter with real things that exist in the surrounding environment / around the students, so that students can understand better the importance of number patterns matter and the benefits of mathematics in everyday life. This arrangement aims to make effective learning because the basis of effective learning is internal and external representation [16]. Through this learning, teacher helps students to develop their analogical reasoning because they understand the matter by relating it to real life [17]. The teacher can also guide students to develop various representation forms in problem solving as suggested by Sa'dijah **Error! Reference source not found.**. According to Lailiyah et al. [19], analogical reasoning can help students to improve their creativity. Evidently when contextual learning takes place students look more active, even actively in asking questions, arguing, discussing, presenting the results of discussions, and so forth. In line with Danis et al. [20] which explains that one of the learning models that actively involves students in learning is contextual learning.

Contextual learning can help students recognize non-routine problems that cannot be solved procedurally. The students' unfamiliarity in solving problems makes it difficult for students to develop students' creative thinking skills [21]. This goal is in line with Suryawati, et al. [22] which explains that contextual learning is one of student-centered learning that emphasizes the development of high-level thinking. In contextual learning students are given Student Worksheets (SW) by the teacher to be done independently. In addition, students are also given a Group Worksheet (GW) by the teacher to work in groups. By having discussion, students are required to be able to solve contextual problems that have been given. Then, students make other contextual problems about the pattern of numbers being studied. Therefore, through the SW and GW that have been given, the teacher can train students' creative mathematical thinking skills. In an effort to train / familiarize mathematical creative thinking skills, in this learning students are required to be able to write various kinds of solutions / answers / ways of solving, especially various kinds of number patterns that are different and correct.

4.2 The Analysis Results of Students' Mathematical Creative Thinking Ability Through Contextual Learning

In this study, through contextual learning students are given contextual problems on the open tests of mathematical creative thinking skills with various answers / solutions. The test results show that high ability group students, through contextual learning, fulfilled three indicators of mathematical creative thinking skills, those are indicators of fluency, flexibility, and novelty, so they are in MCTL 4 (very creative). This result is in line with the results of research by Mustakim & Budiarto which shows that students with high mathematical abilities have achieved optimal indicators of fluency, flexibility, and novelty [23]. The results of this study agree with Damayanti & Sumardi [24] in their research which explains that indicators of fluency and flexibility can be achieved by students with high and middle abilities, but indicators of novelty can only be achieved by high ability students.

High-ability students are in the MCTL category 4 (very creative) and MCTL 3 (creative). Although the subjects of the high ability group are in the same group but have a different category of mathematical creative thinking abilities. Both of these subjects have the characteristics of different answers on flexibility indicators. This result is in line with the research conducted by Motaharah, et al. [25] where students who have the same level of mathematical skill have different creative thinking skill. Utami & Kuneni [26] stated that the higher the initial ability of students, the higher the level of creative thinking skill. Leikin & Lev [27] also added that there is a relationship between students who have high mathematics and mathematical creative thinking skills. This can be seen in the difference of MCTL in high ability students with middle ability students. Middle ability group students, through contextual learning, based on the test result are in MCTL 2 (quite creative) by only meeting the flexibility indicators and not meeting the indicators of fluency and novelty.

Low ability group students, through contextual learning, are in MCTL 3 (creative) by fulfilling fluency and flexibility indicators. These results are in line with the results of Sarassanti, et al. [17], which states that the increase in students' creative thinking skills using contextual learning is better than students who use conventional learning in terms of their initial mathematical abilities (high, medium, low. In addition, the results of this study are also in line with Winarti [28] which explains that effective contextual learning is used to improve mathematical creative thinking skills. The difference in the level of creative thinking skill of each student is in line with the results of Mursidik's research, et al. [29] which shows that the creative thinking skill of each student in solving open contextual mathematical problems will vary depending on the knowledge and abilities possessed by each student.

5. Conclusions and Suggestions

Based on theoretical studies, data exposure from research and discussion results, the researcher can draw conclusions as follows: (1) High-ability students can fulfill three indicators namely fluency, flexibility, and novelty well according to the material, so that it is in the very creative category. There are high ability students who fulfill two indicators, namely fluency and flexibility well according to the material, so that it is in the creative category. (2) Middle ability students can fulfill an indicator namely flexibility according to the material, so that it is in the category of quite creative. (3) Low ability students can fulfill two indicators, namely fluency and flexibility well according to the material, so that it is in the creative category. (4) The profile of students' mathematical creative thinking skill in solving problems through contextual learning based on initial mathematical abilities is different.

The profile of students' creative thinking skills is expected to be a teacher's reference to determine appropriate assignments and assessments to be given to students. In addition, it can be seen that even though students have the same mathematical abilities, they have different levels of creative thinking. It is hoped that there will be further research to find out the things that affect these differences.

6. Acknowlegements

The authors are very grateful to PNBP Universitas Negeri Malang on research funding.

7. References

[1] Nasution T K, Surya E, Asmin, and Sinaga B 2017 An Analysis of Student's Mathematical Creative Thinking Ability Senior High School on Geometry *International Journal Of Advance Research And Innovative Ideas In Education* **3(2)** 3860-6

[2] Kemendikbud 2016 *Permendikbud Nomor 21 Tahun 2016 tentang Standar Isi Sekolah Dasar dan Menengah*

[3] Lince R 2016 *Journal of Education and Practice* **7(6)** 1-13

[4] Nadjafikhah M, Yaftian N, and Bakhshalizadeh S 2012 *Procedia - Social and Behavioral Sciences* **31(2012)** 285–91

[5] Sriraman B, Haavold P, and Lee K 2013 *ZDM Mathematics Education* **45** 215–25

[6] Sa'dijah C 2013 *Jurnal Pendidikan dan Pembelajaran* **20(2)** 222–227

[7] Suastika K 2017 *IEJME — Mathematics Education* **12(6)** 569–77

[8] OECD 2016 *PISA 2015 Assessment and Analytical Framework: Science, Reading, Mathematic and Financial* (Paris: OECD Publishing, Ed)

[9] Fitri 2013 *Skor PISA: Posisi Indonesia Nyaris Jadi Juru Kunci* Retrieved from https://bit.ly/1Vec0Yk

[10] Kemendikbud 2014 *Permendikbud Nomor 58 tentang Kurikulum 2013 Sekolah Menengah Pertama/Madrasah Tsanawiyah*

[11] Samo D D, Darhim, and Kartasasmita B 2017 *International Education Studies* **10(2)** 17-29
[12] Daud A M, Omar J, Turiman P, Osman K 2012 *Procedia - Social and Behavioral Sciences* **59 (2012)** 467-74
[13] Sarassanti Y, Prabawanto S, and Cahya MA E 2017 Penerapan Pembelajaran Kontekstual Untuk Meningkatkan Kemampuan Berpikir Kreatif Matematis Siswa SMP *Seminar Nasional Pendidikan MIPA dan Teknologi IKIP PGRI* (Pontianak:) pp 323-40
[14] Sugandi A I 2014 Pengaruh Penerapan Pendekatan Kontekstual Terhadap Kemampuan Berpikir Matematis Tingkat Tinggi Siswa *Prosiding Seminar Nasional Pendidikan Matematika STKIP Siliwangi* (Bandung: Prodi Pendidikan Matematika STKIP Siliwangi) pp. 23-31
[15] Kadir and Masi L 2014 Mathematical Creative Thinking Skills Of Students Junior High School In Kendari City *Int. Conf. on Innovation in Mathematics and Mathematics Education 1st ISIM-MED 2014* (Yogyakarta State University: Departement of Mathematics Education) pp. 295-306
[16] Utami A. D, Sa'dijah C, Subanji, Irawati 2019 Students' Pre-Initial Mental Model: The Case of Indonesian First-Year of College Student *International Journal of Instruction* **12(1)** 1173-88
[17] Davtyan R 2014 Contextual Learning *ASEE 2014 Zone I Conference*
[18] Sa'dijah C, Nurrahmawati, Sudirman, Muksar M and Anwar L 2018 *the 2nd International Conference on Education and Multimedia Technology (Okinawa, Japan)* (New York: The Association for Computing Machinery (ACM))
[19] Lailiyah S, Nusantara T, Sa'dijah C, Irawan E B, Kusaeri, and Asyhar A H 2018 Structuring students' analogical reasoning in solving algebra problem *The Cons. Of Asia-Pacific Education Universities* (IOP Publishing)
[20] Danis A, Perangin R B, and Milfayetty S 2017 *IOSR Journal of Research & Method in Education (IOSR)* **7(6)** 24-33
[21] Hidajat F A, Sa'dijah C, Sudirman, Susiswo *International Journal of Instruction* **12(2)** 573-86
[22] Suryawati Evi, Osman Kamisah 2018 *EURASIA Journal of Mathematics, Science, and Technology Education* **14(1)** 61-76
[23] Mustakim and Budiarto M T 2017 *Jurnal Ilmiah Pendidikan Matematika* **2(6)** 266-8
[24] Damayanti H T and Sumardi 2018 *Journal of Research and Advances in Mathematics Education* **3(1)** 36-45
[25] Muthaharah Y A, Kriswandani, and Prihatnani E 2018 *Jurnal Mitra Pendidikan (JMP Online)* **2(1)** 63-75
[26] Utami A K and Kuneni E 2016 Analisis Tingkat Kemampuan Berpikir Kreatif Pada Materi Geometri ditinjau Dari Kemampuan Awal (Pada Siswa Kelas VII SMP Negeri 2 Kedu Kabupaten Temanggung Tahun Pelajaran 2014/2015) *Prosiding Seminar Matematika dan Pendidikan Matematika* (FKIP UNS:) pp 351-61
[27] Leikin R and Lev M 2013 *ZDM - The International Journal on Mathematics Education* **45(2)** 183–97
[28] Winarti 2015 *JPFK* **1(1)** 1-8
[29] Mursidik E M, Samsiyah N, and Rudyanto H E 2014 *JURNAL LPPM* **2(1)** 7-13

Journal of Physics: Conference Series **1397** (2019) 012082 doi:10.1088/1742-6596/1397/1/012082

The Effect of Problem Based Learning on Mathematical Critical Thinking Skills of Junior High School Students

Fitriana Yolanda

Program Studi Pendidikan Matematika Universitas Islam Riau. Jalan Kaharuddin Nasution No.113, Marpoyan, Pekanbaru, Riau, Indonesia.

fitrianayolanda@edu.uir.ac.id

Abstract. This research aims to know the effect of problem based learning on mathematical critical thinking skills of junior high school students. This research is a quasi-experimental study with a non-equivalent control group design. This research was carried out in class VIII in one of the junior high school in Bandung, West Java Province, academic year 2014/2015 on relations and functions. Determination of the sample was conducted with purposive sampling and two classes were selected as the research sample, namely class VIII.C as the experimental class that received problem based learning and class VIII.F as the control class that received learning with the scientific approach. The results showed that the average post-test for problem based learning was 42.16 and the average posttest of the learning class with a scientific approach was 35.27. Then after being analyzed using the t-test obtained a significance value of $0.024 < 0.05$. It can be concluded that there is an effect of problem based learning on mathematical critical thinking skills of junior high school students.

1. Background

Critical thinking is one of the high-level thinking skills that needs to be developed in mathematics learning from elementary to high school. Critical thinking is a directed and clear process used in mental activities such as solving problems, making decisions, persuading, analyzing assumptions, and conducting scientific research [1]. A person who has critical thinking skills is able to think of ways to solve problems with certain steps. People who have critical thinking skills will always be active in understanding and analyzing the information obtained so that critical thinking skills include intellectual processes related to the process of understanding, analyzing, synthesizing and evaluating and applying information based on experience and observation. Therefore a systematic process related to critical thinking encourages students to formulate and evaluate students' self-beliefs and opinions.

Critical thinking is the basis of students' ability to focus and clarify, analyze, understand and estimate a solution to problems with self-regulatory and regular assumptions [2]. In addition, based on the results of Asiedu [3] research states that students with low mathematical skills are a result of the inability of students to think critically and analyze mathematical concepts systematically. In line with this opinion, according to Juano and Pardjono [4] states that critical thinking guides students in determining which information is acceptable and which information is not acceptable, so students who think critically are able to distinguish which ones are trustworthy. According to Ruseffendi [5], it cannot be denied that the presumption that is currently developing in most students is mathematics is a difficult and unpopular field of study, only a few are able to explore and understand mathematics as a science that can improve critical thinking skills. Karim [6] in his research revealed that the average critical thinking ability of junior high school students is in poor qualification. From this fact it can be generalized that critical thinking skills in students both at the elementary and secondary levels are still has some problem.

If you look further at mathematics learning that occurs in classes, generally it has not yet presented learning that fosters the ability to think critically. Learning in the classroom is still focused on the teacher as the only source of knowledge with the lecture method is still the main choice of learning methods. The learning process that occurs in one direction which only emphasizes the cognitive aspects of students only while the affective and psychomotor aspects of students are not being prioritized. Students only "know" and do not "experience" what they are learning. Such learning does not develop students' high-level thinking skills. The impact of students' attitudes toward lessons, especially mathematics, tends to be negative, which ultimately can lead to low student learning outcomes.

Based on preliminary studies conducted on SMP Negeri 15 Bandung towards 37 students, students' ability to generalize from function problems is still on a sufficient scale. The following questions are given: Consider the table below:

X	...	2	3	4	...
Y	...	10	14	18	...

Based on the table above, create a rule that states the relationship between x and y, if it is known that the function is a linear function! Determine the value of y for x = 15!

Based on the research, it was found that students were not able to generalize the function problems correctly. As a result the solution given to answer the problem is not right. In addition according to observation in classes also showed that some of the students who faced with the problem of thinking critically not eager to solve it and tend to give up without trying first. Based on the results of these studies it can be concluded that students' critical thinking skills are still low.

Based on the above studies, it can be said that students' critical thinking skills in Indonesia still have not achieved satisfactory results. Therefore we need a learning that can develop students' mathematical critical thinking skills, namely problem based learning. According to Rusman [7] Problem Based Learning is learning that uses real (authentic) unstructured problems and is open as a context for students to develop problem solving and critical thinking skills while building new knowledge.

Problem based learning starts with presenting the problem whose solution requires student collaboration [8]. Problem Based Learning or PBL is used to support high-level thinking (HOT or higher order thinking) in problem oriented situations, including learning "how to learn" [9]. In line with the above opinion, Fathurrohman [10] states that Problem Based Learning is a learning model that involves students to solve a problem through the stages of the scientific method so that students can learn knowledge related to the problem and at the same time have the skills to solve problems. The advantages of the problem-based learning model according to Hamdayama [11] are that students are involved in learning activities so that their knowledge is properly absorbed, students are trained to cooperate with other students, and students can obtain knowledge from various sources.

Sianturi el al [12] in his research revealed that several critical thinking indicators used were: (1) Elementary clarification, (2) Basic support, (3) Inference and (4) Strategies and tactics. Furthermore, the formulation of the problem used is the mathematical critical thinking ability of students who take learning Problem Based Learning (PBL) models higher than students who take conventional learning. Furthermore, research conducted by Fakhriyah [13] revealed that some weaknesses in problem-based learning are the independent learning activities of each student is highly demanded at the implementation stage of the learning process, some stages of learning cannot be done in a relatively short time and in the learning process some students are still some rely on friends in their respective groups.

The difference in research conducted by the author with other research that is in this study the authors use five indicators of mathematical critical thinking ability consisting of (1) looking for similarities and differences, (2) making generalizations, (3) making and considering decisions and applying principles, (4) ability to give reasons, and (5) identify problems. While the formulation of the problem in this study is whether there is an effect of problem-based learning on the mathematical critical thinking ability of junior high school students. During the learning process, student activities are quite conducive and effective. Students also look active and enthusiastic in solving problems, discussing, asking questions, asking friends for an explanation when discussing, giving different opinions, or refuting a friend's answer by giving reasons. This is what adds to the more conducive and enthusiastic learning atmosphere, although there are some students who are often silent, ashamed to give ideas, just write without commenting or giving opinions.

According to Ibrahim and Nur [14], problem based learning is not designed to help provide as much information as possible to students and aims: (a) Helping students development: (1) ability or thinking skills; (2) problem solving skills; (3) intellectual skills; (b) make students learn various roles of adults (learning to be) with their involvement in real experiences or simulations; (c) make students as autonomous and independent learners. Based on the description above, the authors hope that problem-based learning influences students' critical thinking skills mathematically. So the tittle of this research is "The Effect of Problem Based Learning on Mathematical Critical Thinking Skills of Junior High School Students"

2. Research Methods

This research was conducted in class VIII in one of Junior High Schools (SMP) in Bandung, West Java Province in the 2014/2015 academic year. Determination of the sample is done by purposive sampling, namely sampling techniques based on certain considerations [15], namely classes that have equivalent academic characteristics and abilities, so that two classes are selected as research samples namely class VIII.C as the experimental class and class VIII .F as a control class. In the experimental class, problem based learning was carried out. In the control class learning was carried out with a scientific approach. This research is a Quasi Experimental study consisting of two research groups. The research design used in this study is the design of non-equivalent control groups [5]. The instruments in this study were obtained through tests of students' mathematical critical thinking skills (pretest and posttest) arranged in the form of descriptions. In this study, before the instrument was used, the instrument was first tested on students who had obtained material related to what they were going to study. The test results of students' mathematical critical thinking ability tests were then analyzed to determine the validity of the items, the reliability of the test, the distinguishing power of the test items, and the level of difficulty of the test items. This trial was conducted to obtain a measuring instrument that was as expected. Then the results of tests of students' mathematical critical thinking skills (pretest and posttest) were analyzed using the normality test, the variance homogeneity test and the t-test. This research uses SPSS software program version 21 for windows for processing data.

3. Result and Discussion

Result

3.1. Mathematical Critical Thinking Ability

Table 1. Pretest-Postest Descriptive Statistics Mathematical Critical Thinking Ability

Value	**Problem Based Learning**					**Learning with Scientific Approach**				
	N	X_{min}	X_{maks}	$\overline{x}$	Sd	N	X_{min}	X_{maks}	$\overline{x}$	Sd
Pretest	37	0	25	7,43	6,934	37	0	25	6,76	5,678
Posttest	37	15	70	42,16	13,720	37	10	60	35,27	11,899

Ideal maximum score= *100*

The table above illustrates that the average pretest of students' critical thinking skills in the problem-based learning class and learning class with the scientific approach is not much different. Where the average pretest of mathematical critical thinking skills of students in the problem-based learning class is 7.43 while the average pretest of mathematical critical thinking skills of students in the learning class with the scientific approach is 6.76. So that the average pretest of mathematical critical thinking skills of the students of the two classes is relatively same. However, the posttest average in the problem-based learning class is greater than the posttest average in the class with scientific approach.

3.2. Analysis of Students' Mathematical Critical Thinking Ability Pretest Score

Analysis of the pretest score was done to determine the students' critical thinking skills before being given treatment both to students in the class who received problem-based learning as well as to students who received learning with the scientific approach. To determine the right statistical test, the normality test and homogeneity test are first carried out. If the data meets the requirements for normality and homogeneity, the average similarity test uses the t-test, whereas for data that does not meet the normality requirements, it uses a non-parametric test.

Table 2 Mann-Whitney Test Results Pretest Score Mathematical Critical Thinking Ability

	Overall pretest
Mann-Whitney U	667,500
Wilcoxon W	1370,500
Z	-,190
Asymp. Sig. (2-tailed)	,849

Based on the table above it can be seen that the value of Sig. (2-tailed) Mathematical Mann-Whitney test of pretest critical thinking ability data is equal to 0.849 which is greater than 0.05, then H0 is accepted. This means that there are no differences in the students' mathematical critical thinking skills pretest between both classes

3.3. Analysis of Students' Mathematical Critical Thinking Ability Posttest Scores

The average posttest difference test was conducted to test whether the sample in the problem-based learning class and the class with the scientific approach had different final abilities. The posttest difference test uses 2 Independent Sample test at a significant level $\alpha = 0.05$. The criteria used are if the sig.2-tailed value > α then $\boldsymbol{H_0}$ is accepted and the opposite statement $\boldsymbol{H_0}$ is rejected. The following are the test results for the average post-test score difference:

Table 3. Posttest Score Test Results Problem Based Learning Class and Learning Class with Scientific Approach

		Levene's Test for Equality of Variances		t-test for Equality of Means				
		F	Sig.	T	Df	Sig. (2-tailed)	Mean Difference	Std. Error Difference
Overall Posttest	Equal variances assumed	,657	,420	2,308	72	,024	6,892	2,986
	Equal variances not assumed			2,308	70,587	,024	6,892	2,986

Based on the table above, the value of Sig. (2-tailed) t-test posttest mathematical thinking ability is 0.024 $<\alpha$, so it can be concluded that there is the effect of problem based learning on mathematical critical thinking skills of junior high school students

Discussion

This research was conducted in one of the junior high school in Bandung in the 2014/2015 academic year, precisely in class VIII (eight). The learning process is conducted in the experimental class using problem-based learning and the control class with the scientific approach. Arend [16] says that there are five stages of problem-based learning, including: (1) Student orientation to the problem; (2) Organizing students to learn; (3) Guiding individual and group investigations; (4) Develop and present the work; (5) Analyze and evaluate the problem solving process.

The learning process in both classes runs in accordance with the lesson plans that have been prepared before learning is conducted. Different treatment gives a significant impact on learning outcomes in both classes. In problem-based learning students are grouped in group learning consisting of 4-5 students, aiming that student can play an active role in the thought process and in teaching and learning activities. According to Trianto [8] states that learning by involving groups will make it easier to find and understand concepts if they discuss with each other. At this stage, the teacher is not just grouping students and let students work together, but the teacher must continue to encourage each student to fully participate in group activities and take responsibility for their work. The teacher ensures that each student is responsible for their respective group work and encourages students to be able to express ideas and opinions and think critically in solving the given problem. Research related to group-based learning was conducted by Somakim [17] who suggested that students' critical thinking skills improved significantly compared to the usual approach.

Ibrahim [14] added that problem-based learning was not designed to help provide as much information to students as possible and aimed at: (1) Helping students' development: (a) Ability or thinking skills; (b) Problem solving skills; (c) Intellectual skills; (2) Making students learn various adult roles (learning to be) by engaging in real or simulated experiences; (3) Making students autonomous and independent learners. From the beginning students who obtain problem-based learning have been directed to be able to think critically and be able to obtain and use learning resources appropriately so they can find concepts, procedures or mathematical principles both individually and in groups. So it is very possible that students who get problem-based learning have a better mathematical critical thinking skills campare to students who get learning with a scientific approach. This can be seen based on the results of the students' mathematical posttest data, where the posttest average of students who receive problem-based learning is higher than the posttest average of students who receive learning with the scientific approach.

Unlike the experimental class that received problem-based learning, students in the control class applied learning with a scientific approach. The lesson plans that are designed and adopted based on the book are then outlined in the form of worksheets. Learning with a scientific approach used group and individual settings. Learning with scientific approach describes the learning steps into five, namely: observing, asking, trying, associating and communicating [18]. In problem based learning students have the freedom to explore their abilities. Here students are given the freedom to determine ways and solutions to problem solving so that what is obtained and what is used is based on their abilities, whereas in the scientific approach there is a tendency in the process of systematic discovery so that students are not given the freedom to express their ideas.

The mathematical critical thinking abilities measured by the indicators include: (1) Finding similarities and differences; (2) Make generalizations; (3) Making and considering decisions and applying principles; (4) Ability to give reasons; (5) Identifying problems. Based on preliminary data the results of the acquisition of pretest scores, students in both classes had the same mathematical critical thinking skills. After being given problem-based learning in the experimental class and learning with a scientific approach in the control class, there was a critical thinking skills improvement in the problem-based learning class compared to the class with a scientific. This shows that problem-based learning can influence students' critical thinking skills mathematically. From the results of the pretest of the two classes it was found that the average problem-based learning class pretest was 7.43 with the highest score of 25 and the lowest score of 0 with a standard deviation of 6.934. While the average acquisition pretestscore of the class with a scientific approach of 6.76 with a standard deviation of 5.678 and a maximum value of 25 and a minimum value of 0.

Although descriptively there are differences, but based on statistical tests it can be concluded that there is no difference between the average initial ability of students' mathematical critical thinking in both classes, because the values obtained significance of 0.849 greater than $\alpha = 0.05$. This means that the initial mathematical thinking ability of the two classes is not significantly different. Based on descriptive statistics, it was found that the average posttest of problem-based learning class of 42.16 was greater than the average posttest of the learning class with a scientific approach of 35.27. The results of the above data are also needed by testing the difference in the average of the two classes. The average difference test results showed that the average of mathematical critical thinking skills posttest score of students who received problem-based learning was higher than the posttest average of students who received learning with a scientific approach with a significance value of 0.024 smaller than $\alpha = 0.05$. Thus it can be concluded that there is the effect of problem based learning on mathematical critical thinking skills of junior high school students. This achievement is strengthened by research conducted by [4] which states problem posing learning affect students' critical thinking skills.

4. Conclusion

Based on the results of the analysis, findings and discussion described in the previous chapter, it was found that the average post-test for problem-based learning class was 42.16 greater than the average posttest of the class with a scientific approach with 35.27. Furthermore, after using posttest data analysis using the t-test, a significance value of 0.024 was smaller than $\alpha = 0.05$. The conclusions from this research show that there is the effect of problem based learning on mathematical critical thinking skills of junior high school students. Based on the conclusions obtained, it is advisable for teachers to pay attention to time management in order to be as efficient as possible, because problem-based learning requires considerable time. Furthermore, students should think more critically to improve their ability to solve mathematical problems during the learning process in class. The subject that was developed was only at the junior high school level, namely relations and functions, so there is still an opportunity to conduct further research at the level and other material.

References

[1] Johnson, E. B. 2006. *Contextual Teaching and Learning*. California: Corwin Pres, Inc

[2] Masek, A., and Yamin, S. 2011. "The Effect of Problem Based Learning on Critical Thinking Ability: A Theoretical and Empirical Review". *International Review of Social Sciences and Humanities* **2** 1 215-221

[3] Asiedu, O. 2007. Councelling Implications on The Role of The New Mathematics Teacher in The Teaching and Learning of Mathematics. *Internasional Journal of Educational Research* **3**113-20

[4] Juano, A and Pardjono. 2016. Pengaruh Pembelajaran Problem Posing Terhadap Kemampuan Berpikir Kritis dan Komunikasi Matematis Siswa Kelas V SD. *Jurnal Prima Edukasia.* **4** 1 46-53

[5] Ruseffendi, H. E. T. 2006. *Pengantar kepada Membantu Guru Mengembangkan Kompetensinya dalam Pengajaran Matematika untuk Meningkatkan CBSA* (Bandung: Tarsito)

[6] Karim, A. 2010. *Meningkatkan Kemampuan Penalaran dan Berpikir Kritis Matematis Siswa SMP Melalui Pembelajaran Model Reciprocal Teaching*. Thesis at (SPs UPI Bandung: Not published).

[7] Rusman. 2011. *Model-Model Pembelajaran Mengembangkan Profesionalisme Guru*. Jakarta: PT Raja Grafindo Persada)

[8] Trianto. 2011. *Pembelajaran Inovatif Berorientasi Konstruktivistik.* (*Konsep, Landasan Teori, Praktis dan Implementasinya.* Jakarta: Prestasi Pustaka Publisher

[9] Wisudawati, W.A and Sulistyowati, E. 2014. *Metodologi Pembelajaran IPA*. Jakarta: Bumi Aksara

[10] Fathurrohman, M. 2015. *Paradigma Pembelajaran Kurikulum 2013 (Strategi Alternatif Pembelajaran Di Era Global)*. Yogyakarta: KALIMEDIA

[11] Hamdayama, J. 2016. *Metodologi Pengajaran*. Jakarta: Bumi Aksara

[12] Sianturi, A el al. 2018. Pengaruh Model Problem Based Learning (PBL) Terhadap Kemampuan Berpikir Kritis Matematis Siswa SMPN 5 Sumbul. *Jurnal Pendidikan Matematika*. **6** 1 29-42

[13] Fakhriyah, F. 2014. Penerapan Problem Based Learning Dalam Upaya Mengembangkan Kemampuan Berpikir Kritis Mahasiswa. *Jurnal Pendidikan IPA Indonesia*. **3** 1 95-101

[14] Ibrahim, M., and Nur, M. 2000. *Pembelajaran Berdasarkan Masalah*. Surabaya: UNESA University Press.

[15] Sugiyono. 2013. *Memahami Penelitian Kualitatif.* Bandung: Alfabeta.

[16] Arend, R.I. 2007. *Learning to Teach.* New York: McGraw Hill.

[17] Somakim. 2010. *Peningkatan Kemampuan Berpikir Kritis dan Self-Efficacy Matematik Siswa Sekolah Menengah Pertama dengan Penggunaan Pendekatan Matematika Realistik.* Dissertation at (SPs UPI Bandung: Not published).

[18] Kementerian Pendidikan dan Kebudayaan direktorat Jenderal pendidikan Menengah. 2013. *Pembelajaran Berbasis Kompetensi Mata Pembelajaran Matematika (perminatan).* Direktorat PSMA.

The Types and Factors of Error of Elementary School Students in Solving Mathematical Word Problems: An Analysis Using the Fong's Method

Hikmah Ramdhani Putri 1)*, Djamilah Bondan Widjajanti 1)**

1) Yogyakarta State University, Yogyakarta, Indonesia

Corresponding author: *hikmah.ramdhani2015@student.uny.ac.id
**djamilah_bw@uny.ac.id

Abstract. This research aims to identify types of error and factors that cause 6th-grade elementary school students in rural and urban areas to make errors in solving mathematical word problems based on Fong's Schematic Model for Error Analysis. This research is a survey with a quantitative approach. The research population is all of the 6th-grade elementary school students in Kebumen, Jawa Tengah, Indonesia, totaling 1.928 students. The sample was 522 students consisting of 316 students from urban areas and 206 students from rural areas. The instrument of data collection is an essay test in mathematical word problems form. The instrument is declared valid by the content and construct validity. The instrument reliability coefficient is 0.83. The results of the study show that 6th-grade elementary school students in the urban area make errors incomplete schema with errors (E5) category while rural area students make errors in the category of irrelevant procedures (E2) and complete schemes with errors (E5). Furthermore, the cause of 6th-grade elementary school students in urban areas to make mistakes is a language factor, while in rural areas students make language and operational factors simultaneously.

1. Introduction

There are five standard mathematical abilities to be possessed by students, namely problem-solving, communication, connection, reasoning, and representation [1]. The problem-solving ability of Indonesian students is still low. It can be seen from the results of tests by two international studies, namely the Program for International Student Assessment (PISA) and Trends in the International Mathematics and Science Study (TIMSS). According to the PISA report in 2015, the math score of Indonesian students was 386 points and it was in position 63 of 70 participating countries [2]. In the 2015 TIMSS report, it is stated that Indonesian students' score was 397 and it was in position 44 out of 49 participating countries [3].

Problems in mathematics can be presented in verbal, images, symbols or combinations of the three [4]. Mathematical problems presented in verbal cannot be separated from the application of mathematics in everyday life. Mathematical problems in the verbal form are often known as word problems or story problems.

The most common problem solving is presented in word problems [5]. Word problems here is a problem where mathematical concepts and principles are expressed in daily language [6,7]. The solution to the word problems is not only concerned about the final answer, but the process of the solution must also be considered. In solving word problems students must focus on what mathematical operations are needed to solve the word problem. Students must also be able to change the story questions in the form of mathematical sentences, then do calculations [8]. Students are expected to solve word problems step by step through the process so that the flow of thinking shown. Steps in solving word problems consist of: determining what is known and what is asked; making mathematical models; solving mathematical models according to mathematical rules so that they get answers to what is asked and returning the answer in the context of the question being asked.

Mathematical word problem is a difficult problem for most students [6, 7, 9, 10]. Error in solving word problems is still mostly done by most elementary school students. Errors in solving word problems need to be analyzed so that where the error happens, the type of the error and what cause the errors in completing the story problems can be clearly known. The types and causes of errors made by students in solving word problems can be different. This can be caused by the location of the school or the student environment.

Fong's schematic model for error analysis is a method to analyze the error in solving word problems. Fong's error analysis is based on a schematic approach to know student's strategy in solving word problems [11]. Fong classifies the error into two levels. The first level is categorized, based on the schematic approach, into five categories (Table 1).

Table 1. Fong's Type of Error

Type of error	Description
(E1) No Solution	Students do not answer the questions or just write what is known and what is asked without a solution.
(E2) Irrelevant Procedure	Students cannot process the information provided so that the problem-solving steps do not lead to the correct answer.
(E3) Incomplete Schema with No Errors	Students do not write the complete steps of solution and do not make any mistakes in the solving process.
(E4) Incomplete Schema with Errors	Students do not write the complete steps of solution and make any mistakes in the solving process.
(E5) Complete schema with errors	Students write the complete solution but make any mistakes in the solving process.

Category E2, E4, and E5 of errors are analyzed at the second level. A second level analysis was conducted to determine the cause of the errors/mistakes made by students. Fong classifies three factors causes of errors namely language, operational and conceptual. Every cause of the error has certain indicators (Table 2).

Table 2. The Indicators of the Cause of Error

Language	1. students cannot read the words, units, or symbols correctly 2. students do not write what is known 3. students write down what is known but not right 4. students do not write what is asked 5. students write down what is asked but not right 6. students do not write the conclusion 7. students write down the conclusion but not right
Operational	1. students are wrong in choosing the operation used in solving the problem 2. students incorrectly use the correct mathematical rules 3. students do not further process solutions to problem-solving 4. students make errors in calculations
Conceptual	students do not operate according to the level in the sequence of work (e.g. do the addition (+) or substraction (-) before multiplication (×) or division (÷)

The purpose of this paper is to identify the types and causes of errors made by rural and urban grade VI elementary school students in solving word problems based on the Fong method.

2. Method

The research population was all of the 6th-grade elementary school students in Kebumen, Jawa Tengah, Indonesia, totaling 1.928 students. The sample was 522 students consisting of 316 urban students and 206 rural students.

The instrument for data collection was an essay test in mathematical word problems form. The word problems consist of word problems with two unequal operations and word problems with three or more operations (Table 3).

Table 3. Test Instrument

Basic Competency	Word Problems	No item
Word problems with two unequal operations	The temperature in the refrigerator is -8^0C. After being turned off, the temperature inside rises 2^0C every minute until it reaches room temperature. After reaching room temperature, the temperature in the refrigerator is fixed. What is the temperature in the refrigerator after being turned off for 9 minutes?	1
	Diki brings money to the stationery shop. With that money, Diki bought 4 notebooks, 2 pencils and got a change of IDR 2000. If the price of a book is IDR 2500 and the price of a pencil is IDR 1500 then how much money does Diki bring?	3
	Father filled up as much as 30 liters of gasoline. Every day the car is used to go to the office. After three weeks, the remaining gasoline is 3 liters. If the father works six days a week, how many liters of gasoline are used to go to the office every day?	4
Word problems with three or more operations	Exam questions consist of 25 items. The assessment of each item is as follows: correct answer gets a score of 2, the wrong answer gets a score of -1, does not answer the question gets a score of 0. Student A answer 12 questions correctly, 11 wrong, and the rest are not answered. Student B answers 9 questions correctly, 4 wrong, and the rest are not answered. Who is the student who has a higher score?	2
	Lily and Rose are playing games. In the game, each player has the first 0 points and the player with the most points will be the winner. Lily gets 5 points twice, loses 11 points, and gets 3 points. Rose lost 3 points twice, lost 1 point, get 4 points twice. Who is the winner in the game?	5
	There are 27 students in 6th grade of Taruna elementary school and Mrs. Nisa is the teacher. She has 35 pieces of books. She bought another 10 packs of books. Each pack contains 10 books. All books will be distributed equally to students. How many books does each student get?	6

The steps taken in the data analysis process in this study were as follows: (1) identifying errors made by students in solving word problems based on student answers; (2) classifying types of error based on Fong's Schematic Model for Error Analysis. Errors made by students were classified by referring to the error indicator in each type of error; (3) calculating the types of error percentage and then sort from the largest percentage to the smallest; (4) identifying causes of mistakes made by students. To identify the causes of errors in solving word problems, the types E2, E4, and E5 of errors made by students need to be analyzed; and (5) determining the estimated proportion for each type of error. The estimation of this proportion was carried out to determine the percentage of each type of error made by the sixth-grade elementary school students in Kebumen Sub-District. If $\hat{p}$ is the proportion type of error in n random sample and $\hat{q} = 1 - \hat{p}$, then the confidence interval is approximately$(1 - \alpha)100\%$for binom p parameters given by

$$\hat{p} - z_{\frac{\alpha}{2}}\left(\frac{\hat{p}\hat{q}}{n}\right)^{\frac{1}{2}} < p < \hat{p} + z_{\frac{\alpha}{2}}\left(\frac{\hat{p}\hat{q}}{n}\right)^{\frac{1}{2}}$$

$z_{\frac{\alpha}{2}}$ is the z value in a standard normal table with an area of the right side equal to $\frac{\alpha}{2}$ [12].

3. Result and Discussion

Analysis of solving word problems in grade 6th elementary school students in urban and rural areas is intended to see the percentage of each type of error made by elementary students in urban and rural areas. In general, the results of the analysis of the solving of the word problems in elementary school students in urban areas can be seen in Table 4 and rural areas in Table 5.

Table 4. Result of analysis solving word problems 6th-grade student in the urban area ($n = 316$)

No item	Correct Answer		Types of Error									
			E1		E2		E3		E4		E5	
	∑	%	∑	%	∑	%	∑	%	∑	%	∑	%
1	130	41,14	2	0,63	72	22,78	4	1,27	17	5,38	91	28,80
2	93	29,43	18	5,70	45	14,24	2	0,63	13	4,11	145	45,89
3	144	45,57	1	0,32	13	4,11	11	3,48	32	10,13	115	36,39
4	69	21,84	6	1,90	101	31,96	38	12,03	31	9,81	71	22,47
5	92	29,11	15	4,75	20	6,33	1	0,32	11	3,48	177	56,01
6	117	37,03	16	5,06	51	16,14	5	1,58	12	3,80	115	36,39

Table 5. Result of analysis solving word problems 6th-grade student in rural area $n = 206$

No item	Correct Answer		Types of Error									
			E1		E2		E3		E4		E5	
	∑	%	∑	%	∑	%	∑	%	∑	%	∑	%
1	50	24,27	2	0,97	67	32,52	2	0,97	6	2,91	79	38,35
2	30	14,56	10	4,85	32	15,53	3	1,46	12	5,83	144	69,90
3	69	33,50	3	1,46	17	8,25	14	6,80	30	14,56	73	35,44
4	21	10,19	5	2,43	126	61,17	11	5,34	18	8,74	25	12,14
5	36	17,48	16	7,77	30	14,56	7	3,40	4	1,94	113	54,85
6	51	24,76	9	4,37	38	18,45	13	6,31	7	3,40	88	42,72

Based on Table 4 and Table 5, the results of the analysis generally show that the percentage of correct answers made by urban students is greater than those made by students in rural areas. It means that rural area students make more error than urban students. This is comparable with the learning achievement of rural area students that is lower than urban area students in Kebumen Sub-District, Jawa Tengah, Indonesia. School location has a significant effect on student achievement, where rural student's achievement is not as good urban student achievement [13]. The following is a discussion of each type of error made by students in urban and rural areas.

No Solution (E1). Based on Analysis result, not many students make the E1 error category. But if we compare, rural area students made the E1 error category more than urban area students. On average, there are 3.06% of urban students and 3.64% of rural students make E1 error category. The most percentage of errors with category E1 is in item number 5 for urban area students and item number 2 for rural area students. The cause of students making this error cannot be determined specifically because there is no written answer.

Irrelevant procedure (E2). Irrelevant procedures become a category of errors that are mostly carried out by both urban and rural area students. But if we compare, rural area students made the E2 error category more than urban area students. On average, there are 15.93% of urban students and 25.08%

of rural students make E2 error category. The most percentage of errors with category E2 is in item number4 for both urban and rural area students. The cause of the error is analyzed with the second level analysis according to the students' answers.

Incomplete Schema with No Errors (E3). Not many students make the E3 error category. But, rural area students made the E3 error category more than urban area students. On average, there are 3.21% of urban students and 4.04% of rural students make an error in this category. The most percentage of errors with category E3 is in item number 4 for urban area students and item number 3 for rural area students. The cause of students making this error cannot be determined because there is no error in the second level.

Incomplete Schema with Errors (E4). This error is done a lot enough by urban and rural students. On average, there are 6.11% of urban students and 6.23% of rural students make an error in this category. The most percentage of errors with category E4 is in item number 3 for both urban and rural area students. The cause of the error is analyzed with the second level analysis according to the students' answers.

Complete schema with errors (E5). The E5 category is the most error category carried out by urban and rural students. In almost every item, the error category E5 is the most percentage of the error category. On average, there are 37.66% of urban students and 42.23% of rural students make an error in this category. The most percentage of errors with category E5 is in item number 5 for urban area students and item number 2 for rural area students. The cause of the error is analyzed with the second level analysis according to the students' answers.

Based on Table3 we can also see that the dominant type of error that makes by urban students in solving word problems with two unequal operations (Item 1, 3 and 4) and word problems with three or more operations (item 2, 5, and 6) is complete schema with no errors (E5). In addition, based on Table 4, we can find out what kind of mistakes rural students make when solving word problems with two unequal operations (Items 1, 3 and 4) which are irrelevant procedures (E2) and word problems with three or more operations (items 2, 5 and 6), namely a complete scheme without error (E5).

The second level analysis was used to find out the cause of students making errors in solving word problems. The analysis on the second level is only in categories E2, E4, and E5. There are 3 factor that causes the errors, namely language, operational and conceptual. Based on the analysis, both urban and rural students made an error caused by language and operational. The 6th-grade elementary school students in Kebumen Sub-district, Central Java, Indonesia have understood the concept of mixed counting operations. Based on the second level analysis of urban and rural students, there are seven dominant indicators that cause errors, they are (1) students cannot read the words, units, or symbols correctly, (2) students write down what is known but not right, (3) students do not write the conclusions, (4) students are wrong in choosing the operation used in solving the problem, (5) students incorrectly use the correct mathematical rules, (6) students do not further process solutions to problem-solving, and (7) students make errors in calculations.

The main cause of students making E2 category errors is that students choose the wrong operation to solve the problem. Every student who makes an E2 category error, made an error in choosing the operation used in solving the problem. Figure 1 is an example of student error E2 category.The student is wrong in choosing the operation so that the solving of the model does not lead to correct answers. Students only use the numbers in the question then arrange them into mathematical models that are not relevant to the question.

The main cause of students making E4 category errors is that students do not write conclusions or students cannot continue the problem-solving process. This is because writing conclusion is part of the scheme. Therefore, if the student does not write the conclusion then his scheme incomplete. Similarly, if students do not finish in processing the solution, the student scheme is also incomplete. Besides that students also make other mistakes, such as writing down what is known but not right, using incorrect rules or mathematical rules and choosing the wrong operation, or miscalculating. Figure 2 is an example of student error E4 category.

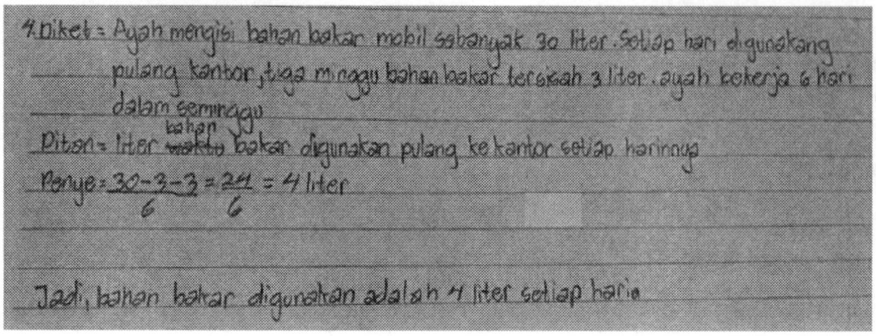
4.Diket = Ayah mengisi bahan bakar mobil sebanyak 30 liter. Setiap hari digunakang pulang kantor, tiga minggu bahan bakar tersisah 3 liter. ayah bekerja 6 hari dalam seminggu
Ditan = liter bahan bakar digunakan pulang ke kantor setiap harinnya
Penye = 30-3-3 / 6 = 24 / 6 = 4 liter
Jadi, bahan bakar digunakan adalah 4 liter setiap hari.

Figure 1. An Example of a student's answer in E2 Category

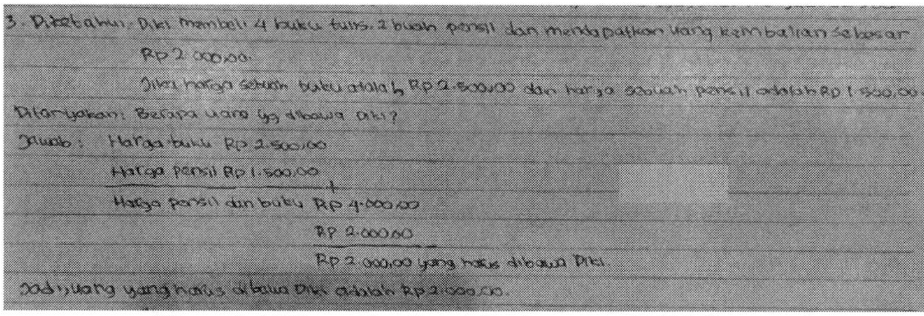
3. Diketahui: Diki membeli 4 buku tulis, 2 buah pensil dan mendapatkan uang kembalian sebesar Rp 2.000,00.
Jika harga sebuah buku adalah Rp 2.500,00 dan harga sebuah pensil adalah Rp 1.500,00.
Ditanyakan: Berapa uang yg dibawa Diki?
Jawab: Harga buku Rp 2.500,00
Harga pensil Rp 1.500,00 +
Harga pensil dan buku Rp 4.000,00
Rp 2.000,00
Rp 2.000,00 yang harus dibawa Diki.
Jadi, uang yang harus dibawa Diki adalah Rp 2.000,00.

Figure2. An Example of a student's answer in E4 Category

Schema to solve item no 3 is a-b-c-d-e where: a) Write down what is known and asked, b) Determine the price of 4 books and 2 pencils, c) Determine the amount of money to be paid, d) Determine how much money is used to pay, e) Returns the answer to the original question. Based on Figure 2, student's scheme is (a)-c-(d)-e. The student did not write one of the steps of solution namely, step b andmake a second level error in steps a and d. In step a, students do not complete the writing of what is known and in step d, students are wrong in choosing the operation used in solving the problem.

There are several dominant factors cause students to make E5 category error, i.e. students write down what is known but not right, students cannot read the words, units, or symbols correctly, students incorrectly use the correct mathematical rules and students make errors in calculations. Figure 3 is an example of student error E5 category.

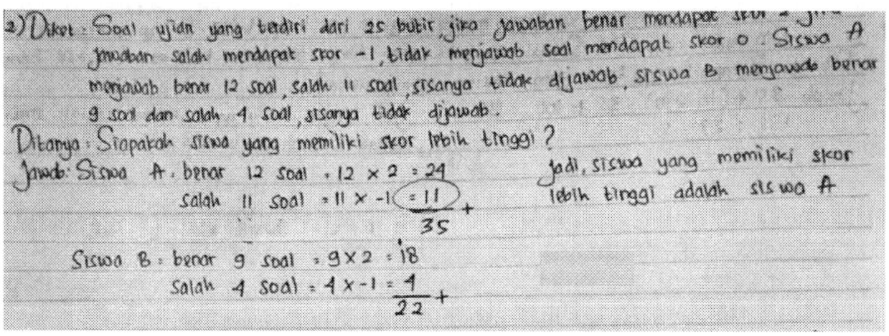
2) Diket: Soal ujian yang terdiri dari 25 butir jika jawaban benar mendapat skor 2 jika jawaban salah mendapat skor -1, tidak menjawab soal mendapat skor 0. Siswa A menjawab benar 12 soal, salah 11 soal, sisanya tidak dijawab. Siswa B menjawab benar 9 soal dan salah 4 soal, sisanya tidak dijawab.
Ditanya: Siapakah siswa yang memiliki skor lebih tinggi?
Jawab: Siswa A. benar 12 soal = 12 x 2 = 24
salah 11 soal = 11 x -1 = 11 +
35
Siswa B = benar 9 soal = 9 x 2 = 18
Salah 4 soal = 4 x -1 = 4 +
22
Jadi, siswa yang memiliki skor lebih tinggi adalah siswa A

Figure3. An Example of a student's answer in E5 Category

Schema to solve item no 2 is a-b-c, where: a) Write down what is known and asked, b) determine each score of student A and B, c) Return the answer to the original question. Based on Figure 3, student's scheme is a-(b)-c. Student's scheme is complete, but he makes an error in step b, that makes an error in calculation.

The most dominant factor causing errors in rural areas are language and operations that are carried out simultaneously. Meanwhile, the cause of errors in urban students is language or operations.

Overall, urban area students are better able to understand questions than rural area students. Understanding the problem is the initial key for students to solve word problems. It is impossible for a problem solver to solve a problem without beginning with a representation of the problem. [14]. In this case, students cannot solve the word problem if the student is unable to understand the problem. In several studies, problem-solving skills can be improved only by providing training for understanding strategies [15,16].

Operational factor that cause the error that rural students do are error in choosing an operation to solve word problems and error in calculation. In the problem-solving phase revealed by Polya, error in choosing operation occurred at the stage of making a mathematical model. Scaffolding can help reduce student errors when making mathematical models [16]. Therefore, to reduce rural students' errors in solving word problems in Kebumen Sub-District, the teacher must provide many exercises in word problem so that students are accustomed to understanding various types of problems in various stories. Second, during the process of the exercises, the teacher must guide students until they can really solve the word problems without the help of the teacher.

Urban student's error in solving word problem caused by incorrect use of the correct mathematical rules. To minimize these errors the teacher needs to teach mathematical rules that students must and cannot do when solving mathematical problems. It is very important to give logical and easy accepted explanations so that students really understand and they do not forget easily.

Determining the percentage of errors made by 6th-grade elementary school students in Kebumen Sub-Districtis done through an estimate for the proportion of the population. Estimates made on proportions use estimates of confidence intervals. Table 6 shows the number of students who make mistakes for each category for the whole sample (522 students) in each item.

Table 6. Number of students who make mistakes on each item and category

No item	E1	E2	E3	E4	E5
1	4	139	6	23	170
2	28	77	5	25	259
3	4	30	25	62	188
4	11	227	49	49	96
5	31	50	8	15	290
6	25	89	18	19	203
average	17	102	19	32	201

Based on Table 6 we can find out the average number of students who make mistakes in each category of errors. From these data, we specify a 95% confidence interval for the proportion of types of errors students makes in solving story problems in each category. Table 7 shows the 95% convention interval for each category of errors.

Table 7. 95% Confidence interval for the proportion of error types (p) in each error category

JK	Amount	n	$\hat{p}$	$\hat{q}$	$z_{\frac{\alpha}{2}}\left(\frac{\hat{p}\hat{q}}{n}\right)^{\frac{1}{2}}$	Confidence interval for the proportion
E1	17	522	0,0326	0,9674	0,0152	$0{,}0173 < p_1 < 0{,}0478$
E2	102	522	0,1954	0,8046	0,0340	$0{,}1614 < p_2 < 0{,}2294$
E3	19	522	0,0364	0,9636	0,0160	$0{,}0203 < p_3 < 0{,}0525$
E4	32	522	0,0613	0,9387	0,0205	$0{,}0407 < p_4 < 0{,}0819$
E5	201	522	0,3851	0,6149	0,0417	$0{,}3433 < p_5 < 0{,}4268$

4. Conclusion

Based on the results and discussion above, several conclusions were made. First, an error made by grade 6 elementary school students in Kebumen Sub-district in urban areas in solving word problems with two unequal operations and the word problem with three or more operations is a complete scheme with errors (E5). Second, errors made by 6th-grade elementary school students in Kebumen Sub-district in rural areas in solving word problems with two unequal operations is irrelevant procedures (E2). Then, error in solving word problems with three or more operations are complete schemes with errors (E5). Third, the factor that caused errors made by 6th-grade elementary school students in Kebumen Sub-district in urban areas is a language, namely writing what is known but incomplete or operational, namely students incorrectly use mathematical rules and students make an error in calculations. Fourth, the factors that caused errors made by 6th-grade elementary school students in Kebumen Sub-district in rural areas are language and operational that are simultaneously do. Language factors that rural students do are students cannot read the words, units, or symbols correctly, students write down what is known but not right, andstudents do not write the conclusions. Operational factors that rural students do are students are wrong in choosing the operation used in solving the problem, students incorrectly use the correct mathematical rules, students do not further process solutions to problem-solving, and students make errors in calculations. Fifth, the percentage of each type of error E1, E2, E3, E4 and E5 conducted by 6th-grade elementary school students in Kebumen Sub-district respectively is $1{,}73\% < p_1 < 4{,}78\%$, $16{,}14\% < p_2 < 22{,}94\%$, $2{,}03\% < p_3 < 5{,}25\%$, $4{,}07\% < p_4 < 8{,}19\%$, and $34{,}33\% < p_5 < 42{,}86\%$

References

[1] NCTM 2000 Principles and standards for school mathematics (Reston: NCTM)
[2] OECD 2016 PISA 2015 Results (Vol I): Excellence and Equity in Education (Paris: OECD Publishing)
[3] Mullis I VS, Martin MO, Foy P, & Hooper M 2016 TIMSS 2015 International Results in Mathematics (Retrieved from Boston College, TIMSS & PIRLS International Study)
[4] Chapman O 2006 *Educational Studies in Mathematics*, **2** 211-230.
[5] Alter P 2012 *Preventing School Failure*, **1** 55-64
[6] Awofala AO 2016 *Turkish Journal of Computer and Mathematics Education, 3*, 486-509.
[7] Leh J 2011 Mathematics word problem solving: An investigation into schema-based instruction in a computer-mediated setting and a teacher-mediated setting with mathematically low-performing students (Available from Education Database, Lehigh University)
[8] Burns M 1992 About teaching mathematics a k-8 resource (Sausalito, CA: Math Solution Publication)
[9] Vula E, Avdyli R, Berisha V, Saqipi B, & Elezi S 2017 *International Electronic Journal of Elementary Education***1** 49-60
[10] Newton, N 2017 Math Problem Solving in Action: Getting Students to Love Word Problems, Grades 3-5 (London: Routledge)
[11] Fong H 1993 Schematic model for categorizing children's errors in mathematics. International Seminar on Misconceptions and Educational Strategies in Science and Mathematics, Misconceptions (Ithaca, NY: Misconception Trust)
[12] Walpole Robert, E 1995 Pengantar statistika edisi ke-3 (Jakarta: Gramedia Pustaka Utama)
[13] Young DJ 1998 *School Effectiveness and School Improvement*, **4** 386-418
[14] Montague M 2005 Math problem solving for primary elementary students with disabilities (American Institue for Research)
[15] Hite S 2009 Improving Problem Solving by Improving Reading Skills (Math in the Middle Institute Partnership Summative Projects for MA Degree. University of Nebraska, Lincoln)
[16] Ulu, M 2017*International Electronic Journal of Elementary Education*, **3**553-580

The mathematics education department students' ability in mathematical literacy for the change and relationship problems on the PISA adaptation test

Antonius Yudhi Anggoro, Hongki Julie, Febi Sanjaya, M Andy Rudhito, and Dewa Putu Wiadnyana
Sanata Dharma University, Jl. Affandi Tromol Pos 29, Sleman, Yogyakarta 55002 Indonesia

E-mail: hongkijulie@yahoo.co.id

Abstract. One of goals of this research was to describe the mathematics education department students' ability in mathematics literacy for change and relationship problem on Programme for International Students Assessment (PISA) test. The procedures of this research were (1) adapt the PISA test, (2) validate the PISA adaptation test, (3) ask seven students from mathematics education department to solve PISA adaptation test, and (4) describe bachelor students' solution profile. There were (1) three change and relationship problems, (2) four space and shape problems, (3) two uncertainty problems, and (4) four quantity problems. The type of this research is a design research. Subjects of this research were seven bachelor students of mathematics education department. The research results were as follows: (1) level four achieved by one student (14.29%) in problem number 2b.4; (2) level three achieved by (a) six students at problem number 2a, (b) five students at problem number 2b.2; and (c)three students at number 2b.3 and 3; and (3) level two achieved by three students at number 3.

1. Introduction

In the 21st century, human needs 21st century skills for survive. Those skills include critical thinking and problem solving, creativity and innovation, communication and collaboration, flexibility and adaptability, initiative and self-direction, social and cross-cultural, productivity and accountability, leadership and responsibility, and information literacy [1, 2, 3, 4]. One of components that needed to build 21st century skills is mathematical literacy [5]. Mathematical literacy is an individual's ability to identify and understand mathematics role in the world, to make an accurate assessment, use and involves mathematics in various ways to fulfill the individual needs as a reflective, constructive and filial citizen [2, 6].

Unfortunately, mathematical literacy of Indonesian students was not good as expected. It could be seen from Indonesian ranking in PISA test. In 2015, Indonesia achieved ranking 63 from 70 countries and the average score for mathematics is 386. In 2012, Indonesia achieved ranking 65 from 65 countries, and the average score for mathematics is 375 [7, 8].

PISA test consist of four contents namely (1) the quantity, (2) space and shape, (3) change and relationship, and (4) uncertainty and data [5, 6, 9]. In the PISA test, there are six levels related to mathematical literacy of students [5, 6, 9].

According to Campbell et al., mathematical abilities of primary teachers who teach in the elementary and junior high school are related directly and positively with the students' achievement [10]. It means that a teacher who has good mathematical literacy will impact positively on improving student's mathematical literacy. This result gave idea to us how to improve our students' mathematics ability. If the schools want to improve students' mathematical literacy, they need to improve their teachers' mathematical literacy. It means universities who produced the teacher candidates has an obligation to improve the quality of prospective teachers it produced. How is our teacher's and teacher candidates' mathematical literacy? Do their mathematical literacy is good or not? Unfortunately, research in this area is still limited. So that's why this research is important to do. Because our university as one of universities producing candidate of teachers, we had obligation to know about the students' mathematics literacy and improve it. The one of research aims was to describe mathematics education department students' ability in mathematics literacy for change and relationship problem on PISA adaptation test.

2. Research Methodology

The type of this research was design research. Subjects of this research were seven mathematics education department students. They were chosen randomly from mathematics education department students in one of private university. The goal of this study was achieved by using Akker's design research procedure [11]. There were three steps in the design research, i.e. (1) construct the design, in this research, the researchers constructed the PISA adaptation test, (2) try out the design, in this case, the researchers asked seven mathematics education students to do the test, and (3) do a restrospective analyze, in this case, the researchers analyze the test result base on the three qualitative data analyze steps developed by Miles and Huberman [12]. This procedures was described in the following diagram.

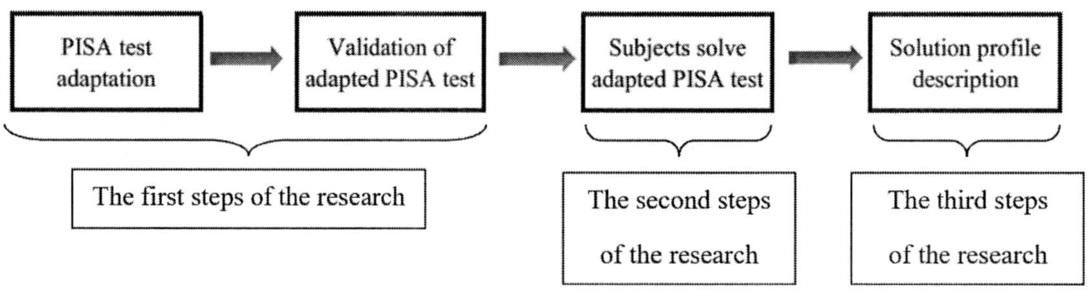

Figure 1. The research procedures

There were four types of the problems in the PISA adaptation test which resulted from this research, i.e. (1) three change and relationship problems, (2) four space and shape problems, (3) two uncertainty problems, and (4) four quantity problems. The languge that used in the test is Indonesian, but for the benefit of scientific publications, the test was translated into English.

Journal of Physics: Conference Series **1397** (2019) 012085 doi:10.1088/1742-6596/1397/1/012085

3. Result and Discussion

Problem 1

This graph shows how the speed of a racing car varies along a flat 3 kilometer track during its second lap.

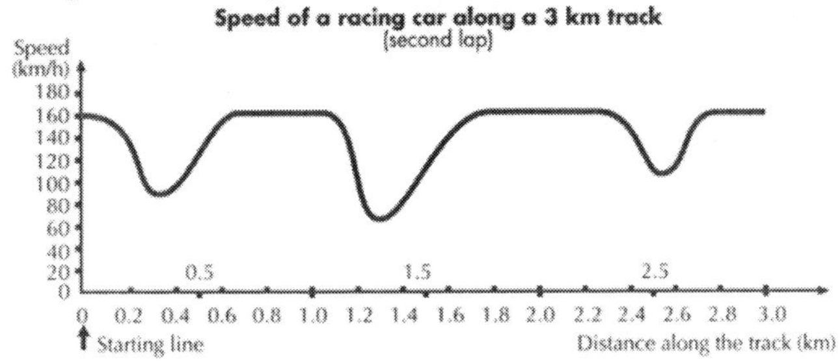

a. Where was the lowest speed recorded during the second lap?
   A. at the starting line. C. at about 1.3 km.
   B. at about 0.8 km. D. halfway around the track.
b. What can you say about the speed of the car between the 2.6 km and 2.8 km marks?
   A. The speed of the car remains constant. C. The speed of the car is decreasing.
   B. The speed of the car is increasing. D. The speed of the car cannot be determined from the graph.

Figure 2. The problem 1 in the PISA adaptation test

In the graph, the vertical axes represent car's speed and the horizontal axes represent distance along the track. The deepest valley in the graph indicated the lowest car's speed. In this case, it happened at about 1.3 km. So, the answer of problem 1a is C. At interval (2.6, 2.8), the graph was increasing monotone. It indicated that car's speed increase at that interval. Thus, the answer of problem 1.b is B. Note that all relevant information was given in the problem, and the questions were defined clearly. All subjects could answer both problems correctly. It means, the subjects could answer questions involving familiar contexts where all relevant information is present and the questions are clearly defined. Thus, base on the PISA classification [5, 6, 9], the students' mathematics literacy were classified in the level 1.

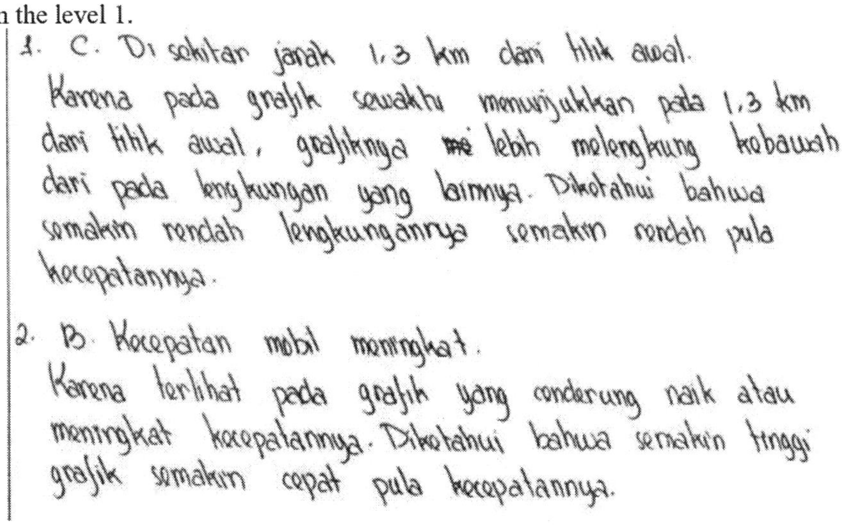

Figure 3. The one example of subjects answer for problem 1

Problem 2

People living in an apartment building decide to buy the building. They will put their money together in such a way that each will pay an amount that is proportional to the size of their apartment. For example, a man living in an apartment that occupies one fifth of the floor area of all apartments will pay one fifth of the total price of the building.

a. There were three apartments in the building. The largest, apartment 1, has a total area of 95 m^2. Apartments 2 and 3 have areas of 85 m^2 and 70 m^2 respectively. The selling price for the building is 30 billion rupiah. How much should the owner of apartment 2 pay? Show your work.
b. Circle Correct or Incorrect for each of the following statements:

Statement	Correct / Incorrect
A person living in the largest apartment will pay more money for each square meter of his apartment than the person living in the smallest apartment.	Correct / Incorrect
If we know the areas of two apartments and the price of one of them we can calculate the price of the second.	Correct / Incorrect
If we know the price of the building and how much each owner will pay, then the total area of all apartments can be calculated.	Correct / Incorrect
If the total price of the building were reduced by 10%, each of the owners would pay 10% less.	Correct / Incorrect

Figure 4. The problem 2 in the PISA adaptation test

Generally, all subjects used the same technique using ratio concept to solve this problem. First technique, they counted the total area of the apartment, that is 95 m^2 + 85 m^2 + 70 m^2 = 250 m^2. After that, they used ratio to count the price of apartment 2. They used ratio because the price of the apartment is proportional to the size of the apartment according to the problem. There were five subjects who use this technique.

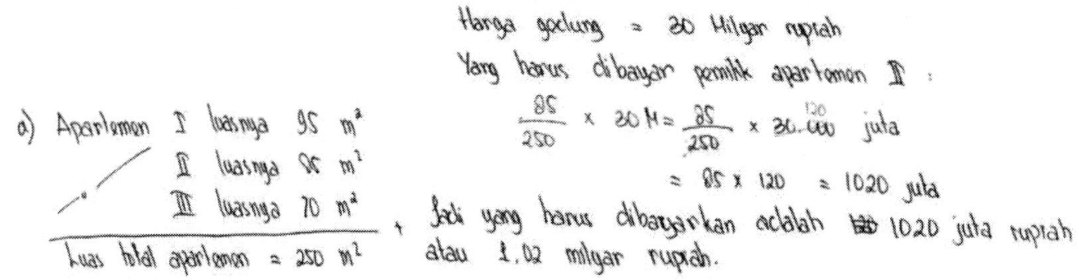

Figure 5. One of subject's answer for problem 2a using the first technique

Let us pay attention to the answer in figure 5. Subjects wrote $\frac{85}{250}\times 30M = \frac{85}{250}\times 30000\ juta$ .The letter M represents billions in Indonesian. To simplify calculation, subject converted "30 billion" to "30000 juta" and then divided 30000 with 250. "30.000 juta" was not common way to represent 30 billion, but mathematically it is true.

The second technique to solve problem 2a is the subject simplified 95:85:70 by divided it with five. The subject got 19:17:14. After that, subject counted the sum of 19+17+14 = 50 and counted the price of the apartment by using ratio. From the subject's answer, it can say that the subject can interpret agreement between apartment buyer and seller and represent it by using ratio. Thus, six subjects using the first and the second technique were classified by PISA classification [5, 6, 9] in level 3. One another subjects could not solve it.

Subjects answer for problem no 2b.1: All subjects claimed that the statement is correct. This claim is false because everyone would pay the same amount of money for each meter square. This was a consequence of the rule that payment was proportional to the size of apartment. Thus, all students were classified by PISA classification [5, 6, 9] in level 1, becuse they used relevant information but not in the true way.

Subjects answer for problem no 2b.2: Five subjects claimed that the statement is correct. Thus 71.43% subjects were classified by PISA classification [5, 6, 9] in level three because they could communicate their interpretation of given information in the problem, the result of their thinking, and the reason of their answer. Two subjects did not answer the question.

Subjects answer for problem no 2b.3: Three subjects answered "incorrect" for this question. It showed that subjects could communicate their interpretation of given information in the problem, the result of their thinking, and the reason of their answer. Thus, they were classified by PISA classification [5, 6, 9] in level three. Three subjects answered "correct" for this problem. Thus, these students were classified by PISA classification [5, 6, 9] in level 1, becuse they used relevant information but not in the true way. One subject didn't answer the question.

Subjects answer for problem 2b.4: Only one student answered "correct" for this problem. This answer is true. Thus, she was classified by PISA classification [5, 6, 9] in level 4, because she could construct and communicate the reasons why they answer "correct" based on their interpretation of proportional understanding. Four subjects answered "incorrect"and they were classified by PISA classification [5, 6, 9] in level 1, becuse they used relevant information but not in the true way. Two subjects did not answer the question.

Problem 3

In 1998 the average height of both young males and young females in the Netherlands is represented in this graph. According to this graph, on average, during which period in their life are females taller than males of the same age?

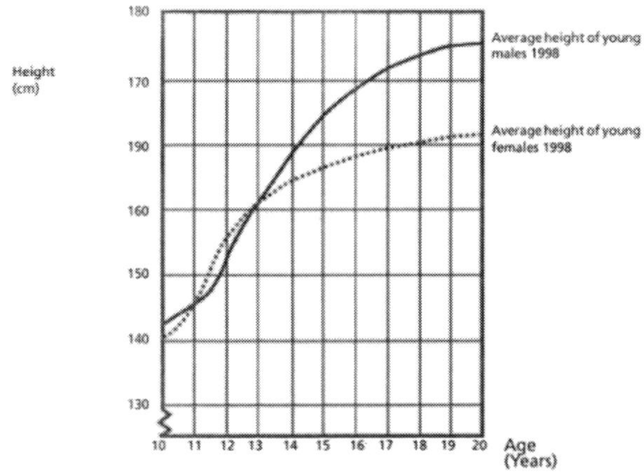

Figure 6. The problem 3 in the PISA adaptation test

In above graph, the vertical axis represent high, horizontal axis represent age, the dash line represent average high of young females and straight line represent average high of young males. Three students answered that young females were taller than young males during age period 11 – 13 year. These students were able to execute clearly described procedures, including those that require sequential decisions. They also were able to interpret and use representations based on different information sources and reason directly from them. Thus, they were classified by PISA classification [5, 6, 9] in level three. One student answered that young females were taller than young males during age period 11 – 12 year. One student answered that young females were taller than young males during age period 12 – 13 year. One student answered that young females were taller than young males when they are 12 year old. These three students were able to extract relevant information from a single source and make use of a single representational mode. Thus, they were classified by PISA classification [5, 6, 9] in level two. One student answered that young females were taller than young males during age period 11 – 14 year. This student was not able to extract relevant information from a single source and make use of a single representational mode, so this student was classified by PISA classification [5, 6, 9] in level 1.

The following table is the summary of subject's level

Table 1. the summary of subject's level

Problem	Subject's Achievement Level	Reason	The number of subject	Percentage
1a	Level 1	Subjects could answer questions involving familiar contexts where all relevant information is present and the questions are clearly defined.	7	100 %
1b	Level 1	Subjects could answer questions involving familiar contexts where all relevant information is present and the questions are clearly defined.	7	100 %
2a	Level 3	Subjects could interpret agreement between apartment buyer and seller and represent it by using worth comparison.	6	85,71 %
	Didn't answer the question	-	1	14.29 %
2b.1	Level 1	-	7	100 %
2b.2	Level 3	they can communicate their interpretation of given information in the problem, the result of their thinking, and the reason of their answer	5	71.43 %
	Didn't answer the question	-	2	28.57 %
2b.3	Level 3	Subjects can communicate their interpretation of given information in the problem, the result of their thinking, and the reason of their answer.	3	42.86 %
	Level 1	-	3	42.86 %
	Didn't answer the question	-	1	14.29 %
2b.4	Level 4	Subjects could construct and communicate the reasons why they answer "correct" based on their interpretation of proportional understanding.	1	14.29 %
	Level 1	-	4	57.14 %
	Didn't answer the question		2	28.57 %
3	Level 3	Subjects were able to execute clearly described procedures, including those that require sequential decisions. They also were able to interpret and use representations based on different information sources and reason directly from them.	3	42.86 %
	Level 2	Subjects able to extract relevant information from a single source and make use of a single representational mode	3	42.86 %
	Level 1	-	1	14.28 %

4. Conclusion

According to above results and discussion, we can take some conclusion. Only one or 14.29% subject achieved level four in problem number 2b.4. Six or 85.71 % subjects achieved level three at problem number 2a. Five students or 71.43 % achieved level three at problem number 2b.2. Three subjects achieved level three in the problem 2b.3 and 3. Three subjects achieved level two in the problem three.

From the results of this study, researchers suggest that there needs to be concrete steps from the mathematics education department, to improve mathematical literacy skills of the students. One of steps that can be taken is to change the learning method and evaluation system in the mathematics education department. Learning methods that should be used are no lecturer-centered, but student-centered. The evaluation system with the non-test method in the lecture needs to be increased in frequency of use. If the evaluation system uses a test method, the evaluation questions in the lecture must begin to be improved in quality to be equivalent to the questions in level four, five, and six in PISA calcification.

Acknowledgments

We thank the Ministry of the Research, Technology and Higher Education for funding this research via "Penelitian Strategis Nasional Institusi Tahun 2018" grant distributed through the Sanata Dharma University.

References

[1] Stacey K 2011 The PISA view of mathematical literacy in Indonesia *Journal Mathematics Education* **2** 95
[2] Ariyadi W 2016 Students' information literacy: A perspective from mathematical literacy *IndoMS Journal Mathematics Education* **7** 73
[3] Ojose B 2011 *Mathematics* literacy: Are we able. To put the *mathematics*, we learn into Everyday use? *Journal of Mathematics Education* **4** 89
[4] Christiansen I B 2006 Mathematical literacy as a school subject: Failing the progressive vision? *Pythagoras* **64** 6
[5] Julie H and Marpaung Y 2012 PMRI dan PISA: Suatu usaha peningkatan mutu pendidikan matematika di Indonesia *Widya Dharma* **23**
[6] Julie H 2017 The elementary school teachers' ability in adding and subtracting fraction, and interpreting and computing *International Journal of Science and Applied Science: Conference Series* **1** 55
[7] OECD 2013 *PISA 2012 Results: What students know and can do. Student Performance in mathematics, reading, and science* (Paris: OECD)
[8] OECD 2012 *Assessment Framework. Key Competencies in Reading, Mathematics, and Science* (Paris: OECD)
[9] Hongki et all. 2017 The students' ability in mathematical literacy for the quantity, and the change and relationship problems on the PISA adaptation test *IOP Conf. Series: Journal of Physics: Conf. Series* **890** 012089
[10] Campbell, Patricia F, et. all 2014 The relationship between teachers' mathematical content and pedagogical knowledge, teachers' perceptions, and student achievement *Journal for Research in Mathematics Education* **45** 419
[11] Akker J V D, Gravemeijer K McKenney S and Nieveen N 2006 *Educational Design Research* (New York: Taylor and Francis Group).
[12] Miles M B and Huberman A M 1994. Qualitative Data Analysis. London: Sage Publications.

Journal of Physics: Conference Series **1397** (2019) 012087 doi:10.1088/1742-6596/1397/1/012087

The Lost Process of Mathematical Literacy on Excellent Students at MAN 2 Kudus

Lulu Choirun Nisa, Ainal Inayah

UIN Walisongo Semarang

lulu.choirunnisa@walisongo.ac.id

Abstract. Data showed that Indonesian mathematical literacy from 2006 to 2015 was always below the PISA average score. Previous research showed that student on Regular Science Program in MAN 2 Kudus has a better level of mathematical literacy than students on Excellent Science Program. This is a case study on Excellent Science Program MAN 2 Kudus as the subject of the research, 27 students. Data was taken with adjusted PISA tests and interviews then analyzed by Miles and Huberman method. The results showed that 36% of students at the Regular Science Program and 41% of students at the Excellent Science Program had literacy level below level 1. An in-depth study of the mathematical literacy process in the Excellent Science Program shows that 11% of students did not understand the problem. Of the 89% of students who understand the problem, 42% of them cannot translate the context into mathematical form, 29% of students only reach the Formulate process, 4% of students reach the Employ and Interpret process, and 21% of students reach the Evaluate process. Thus, to improve mathematical literacy, understanding contextual problems and improving skills in formulating problems into mathematical languages or models must be strengthened.

1. Introduction

Civilization has developed increasingly rapidly, starting with the era of the industrial revolution which brought about disrupted technology. Learning no longer can only be seen as a transfer of knowledge, but about how students make and develop their knowledge. Learning outcomes are no longer seen as scores or credit, sent more to mastery of competencies. In mathematics learning, one of the competencies as a learning outcome is mathematical literacy skills. Mathematical literacy that connects mathematics learned in the classroom with various kinds of real-world lessons [1]. Mathematical literacy shows what students see and what they can do they do with mathematics. High or low level of mathematical literacy shows an individual's ability to reason mathematics and use mathematics in various contexts, using abilities that are needed in life. - well prepared to meet future challenges.

Data on the measurement of mathematical literacy by PISA for Indonesian students themselves still show unsatisfactory results, where the average math literacy score is below the average.

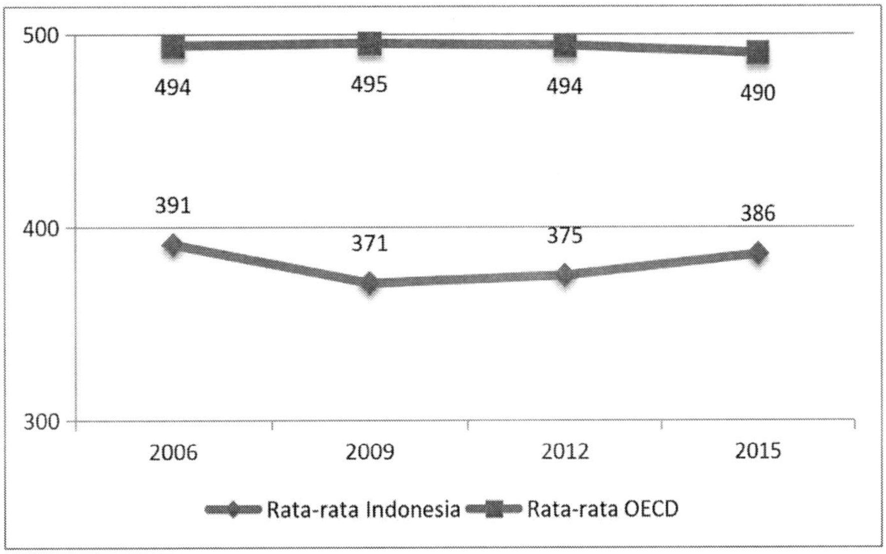

Source: [2], [3], [4], [5]

Figure1. The average Score of Literacy Mathematics PISA for Indonesia, 2006 – 2015

Apart from the score, PISA also categorizes literacy skills into level 1 to level 6. In terms of levels, the PISA results show that mathematical literacy is the lowest compared to science literacy and reading literacy. This can be seen in the following graph.

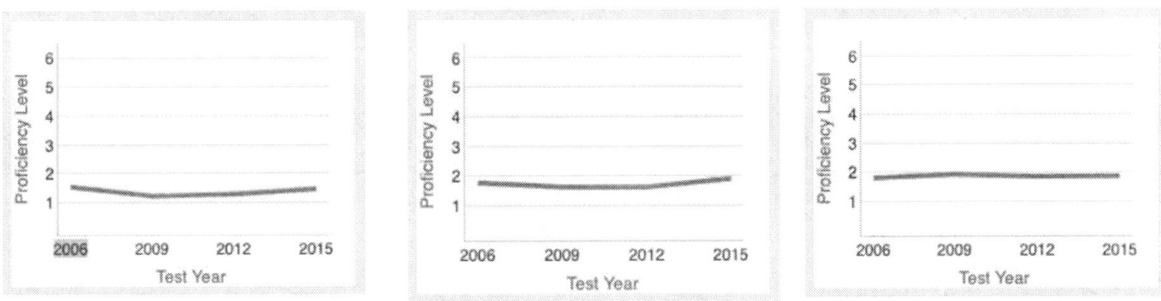

a b c

Source: [6]

Figure 2. The Proficiency Level of (a) Mathematical Literacy, (b) Science Literacy, and (c) Reading Literacy for Indonesian Students since 2006 - 2015

The data above shows that an in-depth study of mathematical literacy is needed so that students can improve their mathematical literacy. The study of efforts to improve mathematical literacy has been carried out for various levels, for example with the Problem Based Learning learning model [7], Creative Problem Solving [8], or the Indonesian Mathematics Realistic approach combined with Edmodo implementation [9]. And although PISA is intended for junior high school students, the study of mathematical literacy is also conducted at the Senior High School level. Shava, for example, did a research to find didactical practice to improve students' mathematical literacy in high school [10]; andSari, et.al who studied about mathematical literacy abilities of high school students in Yogyakarta which is show that the abilities are in a very low category [11]. In her research, Sera, et.al analyzed quantitatively the achievements and literacy processes carried out by 15 high school students. Different with that, this research would describe and explore qualitatively the mathematical literacy process experienced by students from the best Senior High School in Kudus, MAN 2 Kudus. Our previous study in MAN 2 Kudus showed that students in the Regular Science Program have a better level than students in the Excellent Science program, as shown in Table 1 below.

Table 1. Level of Mathematical Literacy of MAN 2 Kudus Students

Program	Below Level 1	Level 1	Level 2	Level 3	Level 4	Level 5	Level 6
Regular Science	36%	8%	17%	17%	17%	6%	0%
Excellent Science	41%	15%	22%	4%	11%	0%	7%
Social Science	88%	8%	5%	0%	0%	0%	0%
Total	57%	10%	14%	7%	9%	2%	2%

The table above shows that the majority of students are at levels below 1, and in the Excellent Science Program, more than 50% of students cannot reach level 2, and only 2 students or as many as 3% of students reach level 6.Excellent class students have failed to acvhieve the performance of mathematical literacy better thanRegular Science or the Social Science Program. The comparison of level distribution for Regular, Excellent and Social Science programs is described in Figure 3.

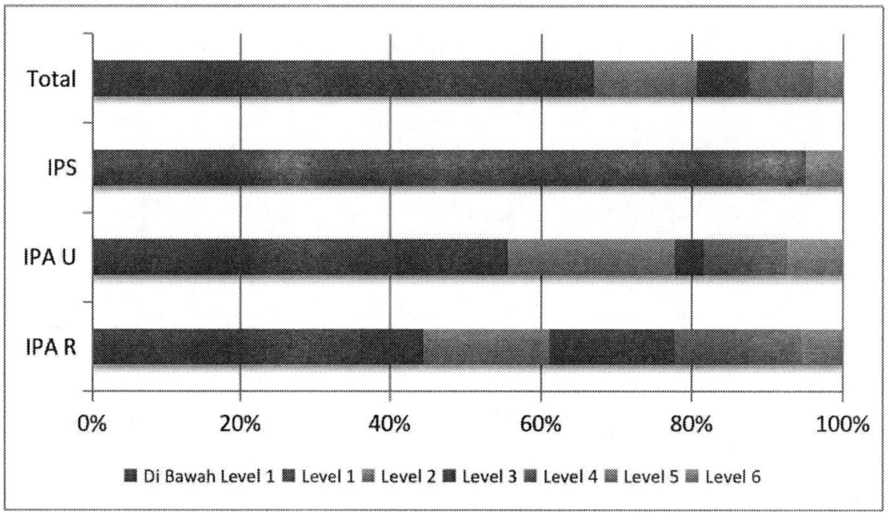

Figure 3. The level distribution of mathematical literacy for Regular Science Program (IPA R), Excellent Science Program (IPA U), and Social Science Program (IPS).

The graph above shows that the achievement of Regular Science is better than the achievement of the level of the Excellent Science Program, where about 40% of students in the Regular Science Program are at levels 4.5, and 6, while for the Excellent Science Program is less.So further studies then are conducted to reveal on what process students in the Superior class cannot reach the highest level of mathematical literacy.

2. PISA Mathematical Literacy

OECD defines mathematical literacy as

> Mathematical literacy is an individual's capacity to formulate, employ and interpret mathematics in a variety of contexts. It includes reasoning mathematically and using mathematical concepts, procedures, facts and tools to describe, explain and predict phenomena. It assists individuals to recognize the role that mathematics plays in the world and to make the well-founded judgments and decisions needed by constructive, engaged and reflective citizens [12]

In every PISA mathematical literacy problem, there are three domains, i.e content, context and process. The domain of mathematical content consisting of Quantity, Uncertainty, and Data, Change and relationship, Space and Shape. The context domain relates to the situation where the assessment is carried out, consisting of Personal, Societal, Occupational and Scientific. The process domain explains how students connect the context of problems with mathematics and then solve them. To solve the PISA problem, students need to formulate mathematical problems into mathematical forms, using them to solve problems, interpret the results according to the context, and evaluate the results [12]. From the three domains above, this study will be focused on the process domainbecause this is the domain related to the construction of student knowledge and skill of mathematics.

The process of mathematical literacy can be described in the following flow.

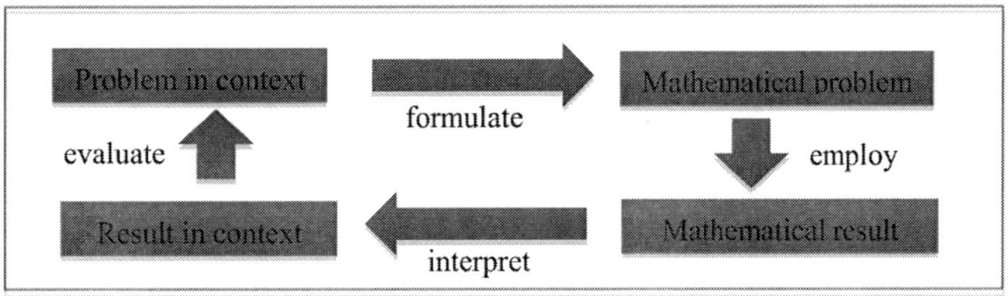

Figure 4. Mathematical Literacy Process

3. Data and Discussion

The study was conducted at MAN 2 Kudus, Central Java, Indonesia. This school has the strength of a research-based school and it is one of the best schools in Central Java. There are a lot of achievements both nationally and internationally, such as the special International Award Exhibition of Young Inventor (2014), Finalists of the ASEAN Student Science Project Competition (2016), The Outstanding Teacher in ASEAN, Melayu Day, Yala (2017), The 1st Winner of Developer Augmented Reality game in Southeast Asia, and the Winner of Applicative Game Developer SEAMEO (2017).

This is a case study to explore why the students in the Excellent Science Program have a lower level of mathematical literacy than students in the Regular Science Program as mention in Table 1 and Figure 3. The subjects of the study were the students of grade X from Excellent Science Program. Data was taken using anadjusted PISA test and interviewed. Adjustments to the instrument of thePISA test are carried out by changing the context to suit the child's life and presenting it in Indonesian.

Journal of Physics: Conference Series **1397** (2019) 012087 doi:10.1088/1742-6596/1397/1/012087

There are six questions in the test given to all of the students. Table 1 shows thatthe highest gap of the result for the Regular and the Excellent Science Program is for problem number 3, which represents the mathematical literacy level 3. There are 17% of students in the Regular Science program can solve this problem, and only 3% of students in the Excellent Science program can solve it. Problem number 3 is:

Seorang petani menanam Apel dalam pola persegi. Untuk melindungi pohon apel dari angin, la menanam pohon Conifer di sekitar kebuh buah. Di bawah ini adalah diagram pola pohon Apel dan pohon Conifer untuk sejumlah (*n*) deretan pohon Apel :

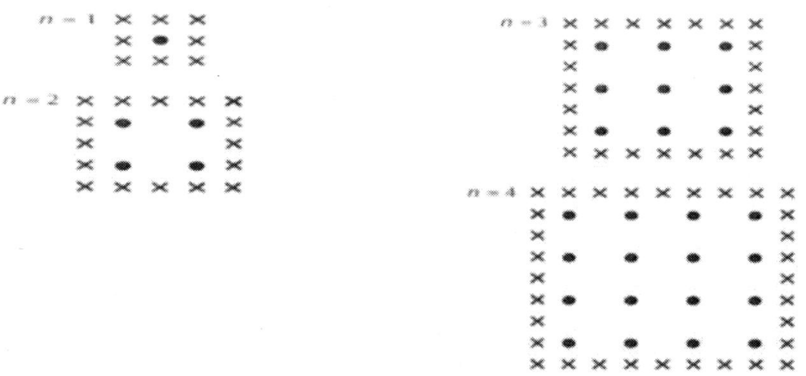

Keterangan : • = Pohon Apel x = Pohon Conifer

Bila *n* adalah jumlah deretan pohon Apel, pada nilai *n* keberapakah jumlah pohon Apel sama dengan jumlah pohon Conifer? Jelaskan dan berikan alasannya.

Figure 5. Question number 3.

Based on the results of theanalysis of the answer sheet and interviews with the students, it was known that only 5 out of 27 students in the Excellent Science Program or as many as 19% were able to go through all the literation processes to evaluate. There are 3 students did not understand the problem or identify what is asked. 10 students couldunderstand the problem but couldn't formulate the problem into a mathematical form. There are7 students could formulate the problem into mathematical forms or models, but failed to connect or use these formulations to solve the problem. One student could reach the employ stage, which means he could use formulas to solve problems, but couldn't interpret them into context. Another student could interpret it into context but couldn't evaluate it. The detail of the result is described in Table 2.

Table 2. The Highest Mathematical Literacy process of Excellent Science Students Program

Proses	*f*	%
Didn't understand the problem	3	11%
Understand the problem	24	89%
Can't formulate	10	42%
Formulate	7	29%
Employ	1	4%

Interpret	1	4%
Evaluate	5	21%

Table 2 shows that 42% of students understand the problem, but they could not formulate the problem into mathematical language. There were 58% of students passed the process of formulating, but only 21% of students could achieve the process of evaluating. It means that the most process that lost from the students of the Excellent program is formulated the problems in context into a mathematical problem.

From Figure 3 above, the process of formulating was the first process in mathematical literacy. So it means that to improve the students' mathematical literacy, the teacher needsto give some scaffoldingsin the learning process to formulate the problems in context into mathematical problems, so they can continue the process of mathematical literacy.

4. Conclusion

Based on the result above, it can be concluded that the difficulty of students in solving problems is to understand the contextual problems and how to make mathematical models of the problems presented. This needs the attention of mathematical educators that students need to get reinforcement in formulating problems into mathematical problems.

References

[1] Masjaya, Wardono. The Importance of Mathematical Literacy Ability to Increase Mathematical Ability in Improving the Quality of Human Resources. *Prism Pros Semin Nas Mat* 2018; 1: 568–574.
[2] OECD. *PISA 2006 Science Competencies for Tomorrow's World.* Paris, http://www.oecd-ilibrary.org/education/pisa-2006_9789264040014-en (2007).
[3] Organisation for Economic Co-operation and Development-OECD. *PISA 2009 Results: Executive Summary*, http://www.oecd.org/pisa/pisaproducts/46619703.pdf (2010).
[4] Organisation for Economic Co-operation and Development (OECD). *PISA 2012 Assessment and Analytical Framework: Mathematics, Reading, Science, Problem Solving and Financial Literacy*. 2013.
[5] OECD. *PISA 2015 Results*. 2016.
[6] The Centre for Internet Research. Open PISA, http://openpisa.org (2019).
[7] Firdaus FM, Herman T. Improving primary students ' mathematical literacy through problem based learning and direct instruction. 2017; 12: 212–219.
[8] Ovan, Waluya SB, Nugroho SE. Analysis mathematical literacy skills in terms of the students' metacognition on PISA-CPS model. *J Phys Conf Ser*; 983.
[9] Wardono, Waluya SB, Mariani S, et al. Mathematics Literacy on Problem Based Learning with Indonesian Realistic Mathematics Education Approach Assisted E-Learning Edmodo. *J Phys Conf Ser*; 693.
[10] Mavugara-shava FM. *Teaching for Mathematical Literacy in Secondary and High School in Lesotho*. the University of The Free State, 2005.
[11] Sari RHN, Wijaya A. Mathematical Literacy of Senior High School Students in Yogyakarta. 2017; 4: 100–107.
[12] OECD. *PISA 2015 Assessment and Analytical Framework.* 2017.

Exploration of teachers' "knowledge of students" in study-based teaching on polyhedron material

Ma'rufi[1], M Ilyas[2], R F Pasandaran[3], Salwah[4]
Universitas Cokroaminoto Palopo[1,2,3,4]

marufi.ilyas@gmail.com[1], muhamadilyas949@yahoo.com[2], riolovemath@gmail.com[3], salwa_gama@yahoo.com[4]

Abstract. Teachers' knowledge in "knowing" student problems is an important factor to support the improvement of the Mathematics learning process in the classroom. Therefore, this research aims to explore the teacher's strategies to overcome students' errors/misconceptions in the material of polyhedron. This is an explorative study and carried out with a qualitative approach so that it contains a set of credible facts that were revealed in depth. Some instrument that used are; field notes, observation sheets for teacher activities, student activities, and learning transcripts A phase lesson study scheme was also applied to create coherent learning scenarios. This combination resulted in findings that had an impact on improving teacher skills in dealing with the errors made by students in terms of concept, procedure, and initial knowledge systematically. The findings of this study were; the teacher can describe the types of knowledge about students' errors/miskonceptions, and then developed strategies to deal with errors/misconceptions effectively.

1. Introduction

Pedagogical Content Knowledge (PCK) in the current educational paradigm plays an important role to advance the quality of learning. Some dimensions of PCK including knowledge of the subject matter, knowledge of curriculum, knowledge of pedagogic and knowledge of students are the teacher's strategic needs in fulfilling their duties and functions. A teacher must understand his quality and capacity as the organizer of learning and this must be done continuously to increase his professionalism.

Lesson study is a vehicle for building teacher professionalism to implement the PCK concept. Both have the same final goal, but different forms and roles. PCK is the concept of identity of each teacher in advancing his learning, while lesson study is a forum for actualizing the concept of the PCK. If combined, we can hypothesize it as "quality learning". To support the quality of mathematics learning, teachers need to know the personal and intellectual characteristics of students, as well as their conceptions and misconceptions about certain topics to be taught. Teacher flexibility must include the ability to adapt learning to meet student needs, ways to overcome students' difficulties in understanding teaching material and eliminate their misconceptions effectively. The teacher also needs to arrange a topic based on the curriculum to achieve the learning goals. Therefore, it is important to have not only knowledge of teaching material and pedagogic necessary, but also knowledge about students and curriculum of PCK [1].

The existence of PCK has shifted the classical and traditional ways of learning. PCK allows each teacher to design innovative learning scenarios which have distinctive characteristics, and of course based on student learning needs. One important aspect of PCK is the knowledge of students (KS). Knowledge about students includes ways that teachers use to know their students' knowledge, how to diagnose potential problems that may arise, as well as how to solve them. In other words, this process can be categorized as teacher metacognition of students. The ability of teachers to understand students' thinking processes allows the dynamics of learning, considering that each student has different abilities and has the potential to raise different problems. This challenge is very interesting to explore more deeply to project all the actions that teachers can take in dealing with cases of classroom learning. To formulate the problem in this article, several focus issues can be made, including;

a. How is the description of teachers' *knowledge of students* (KS) in *Lesson Study*- based learning in polyhedron?
b. What are teachers' strategies to overcome the errors/misconception made by students in polyhedron?

The purpose of this study are:

a. To describe the teachers' *knowledge of students* (KS) in polyhedron.
b. To formulate the teachers' strategies to overcome the errors/misconception made by students in polyhedron.

2. Literature Review

2.1 Pedagogical Content Knowledge (PCK)

Studied how many elements exist in the teacher's mathematical knowledge. They categorize it into some different views. Shulman proposed a framework for analyzing teacher knowledge that distinguishes different knowledge categories: content knowledge, general pedagogic knowledge, curriculum knowledge, pedagogical content knowledge (PCK), knowledge of students, knowledge of the educational context and final knowledge, goals, and values of education. PCK is a key aspect to overcome the problem of teaching-learning [2].

High-quality teachers and experts not only understand the subject matter but also know how to teach certain knowledge called PCK [3]. [4] agreed that being an effective math teacher requires knowledge of mathematical content and an understanding of the teaching process which is needed to efficiently transfer this knowledge to students. He chose seven expert teachers, each of whom had taught for 12-25 years, for a case study. To determine which teacher was an expert teacher, Leinhardt traced the increase in students to standardized mathematics tests over a period of 5 years. The principal and supervisor were also asked to review each mathematics teacher and to advise outstanding mathematics teachers. In the case study, expert teachers were observed and recorded to find out what factors lead them to be high-quality teachers in teaching mathematics. The results of the study indicated that PCK teachers were very important for their success. The aspects involved in this study to measure PCK teachers were knowledge of teacher teaching materials and teacher's pedagogic knowledge (for example, patterns used in developing lessons and academic involvement using various teaching materials). [5] supported that a teacher with a mathematical PCK tends to display high-quality mathematics teaching. Their study proved that teachers who have an understanding of how to support children's problem-solving abilities in mathematics tend to support children's problem-solving strategies more often than teachers who don't.

2.2 Knowledge of Students (KS)

Knowledge of students is generally defined as an understanding of the characteristics of a particular group of students, how to build an appropriate classroom environment and learning planning skills to meet student needs [6]. Student learning needs can be met if the teacher can develop the planning in an effective way related to teaching certain topics, assessing students' understanding, and overcoming their difficulties.

PCK is a specific knowledge which integrates knowledge of mathematics concept, knowledge of pedagogic, and knowledge of students. The PCK is an important knowledge in teaching because it can help teacher to anticipate students' learning misconception and mistake and be ready to give alternative model or explanation to overcome those misconception and mistake [7]. Teachers should understand their teaching material comprehensively and be aware of the learning process to understand what students understand and what is difficult for them to understand [2]. Teachers need to create meaningful learning for students by connecting students 'previous knowledge with the new one(s) through various representations, examples, and manipulations and focusing on students' conceptual understanding of procedures or rules. To avoid the occurrence of conceptual misconceptions of students, the teacher must identify opportunities for the emergence of misconceptions correctly and eliminate these misconceptions with questions that are investigating or leading students' thoughts in building an idea.

This can affect the preparation of the teaching plan (RPP). The teacher is expected not only to help students overcome misconceptions but also to have special skills to develop RPPs that contain the possible sources of students' difficulties and errors and some steps to anticipate them effectively.

2.3 *Lesson Study*

Lesson Study is a model of professional educator development through a collaborative and sustainable study of learning based on collegiality principles that help each other in learning to build learning communities. Based on the definition of Lesson Study, it has 7 keywords, namely professional development, learning study, collaborative, sustainable, collegiality, mutual learning, and learning communities. Lesson Study aims to carry out professional development of educators in a sustainable manner so that there is an increase in educator professionalism continuously. If there is no continuous coaching, professionalism can decrease as time goes by. It could be built by studying continuous and collaborating learning. Learning how to teach must be conducted regularly. The learning assessment cycle is carried out in three stages. Conventional training is a top-down process, meaning that the training material has been prepared and provided by the instructor, whereas training through Lesson Study is bottom-up because the problem-based training material found by teachers in schools is then studied collaboratively and sustainably.

Lesson Study is sustainably carried out in three stages, namely planning, implementing, and reflecting. In other words, Lesson Study is a way to improve the quality of education that never ends (continuous improvement). Lesson Study has a considerable role in making changes systemically. The five paths that Lesson Study can take are 1) presenting the goals of educational standards to the real world in the classroom, 2) promoting improvements based on a database, 3) targetting various achievements based on student qualities that influence learning activities, 4) creating the demands for learning and 5) upholding the teachers' value [8]. There are 7 stages or steps included in Lesson Study, namely: forming a Lesson Study team, focusing on Lesson Study, arranging a lesson plan, preparing for observation, carrying out teaching and observations, carrying out question-answer/discussion and reflecting and planning the next stage.

3. Research Method

This research is a qualitative case study conducted on the material of polyhedron taught at SMP 8 Palopo in February-April 2019. The research data are in the form of statements and arguments derived from observations and field notes. The main instrument of research is the researchers themselves, as collectors, processors, and interpreters of data. In addition, supporting instrument are also used to collect data such as; field notes, observation sheets for teacher activities, student activities, and learning transcripts. This research was conducted in 3 cycles of lesson study in each class. Each cycle consisted of planning, implementation and reflection stages. Researcher observed the learning three times. The observation result are recorded in observation sheet, field notes, and video recording. The notes are then reduced to learning transcript. Researcher analyzed learning transcript to find out the types of teacher's knowledge that used to overcome the student's errors/misconception. Furthermore, the results of this analysis are discussed and used as research findings.

Data was obtained through learning observation, by recording all learning activities from the beginning to the end. After the data was collected, it was validated through triangulation by looking at the compatibility of data from each research instrument. The data collected will be reviewed for consistency. Data which was consistent/relatively the same is regarded to be valid and can be analyzed. Data analysis was performed using Miles and Huberman techniques (Data Collection, Data Reduction, Data Display, Data Verification) interactively and continuously to obtain patterned data forms, with the following steps.

a. Data collection through observation sheet, field notes, learning transcript, and video recording
b. Data reduction is the process of selecting, abstracting, and transforming raw data. Data reduction in this study is done by making a summary that consist of; essence, process, and statements that are consistent with the research objectives. Validation technique used in this study is an observation extension.
c. Data presentation which includes data classifications and identifications. The set of Organized and categorized data are written to produce valid data.
d. Data presentation and interpretation contains discussion of valid data. The aim is to produce established research findings.

4. Findings and Discussion

Exploration was carried out by observing learning in the lesson study material based on lesson study prism. Observations were focused on the forms of conceptual errors/student misconceptions that arose during the learning process. The following presents the process of research findings based on the type of student's errors and the solution strategies.

Table 1. Findings Research

Kinds of Misconception	Cognitive Levels	Misconception Domain	Teachers' Knowledge to Overcome Misconception
Students could not differentiate between a plane figure and a solid figure	C2	**Understanding of concept**	a. Providing the student with models of the plane figure and a solid figure. b. Asking the student to identify their criteria. c. Writing their differences.
Students could not mention the formula of areas of plane figures forming the prisms	C1	**Prior Knowledge**	a. Asking students to read a case made by a teacher related to the area of a plane figure. b. Asking students to rewrite the area formulas that have been identified.
Students could not mention kinds of plane figures forming prisms	C1	**Prior knowledge**	a. The teacher gave an example of some plane and non-plane figures. b. Students were asked to write the similarities and differences of some plane figures.
Students could not differentiate between lateral face and base	C2	**Understanding the concept**	a. The teacher asks students to define the polyhedron and base of a solid figure.

Kinds of Misconception	Cognitive Levels	Misconception Domain	Teachers' Knowledge to Overcome Misconception
of prims			b. Students discussed those definitions.
Students could not sum all areas of plane figures forming prisms	C3	**Applying the procedure**	a. The teacher had to prepare an additional meeting to explain the algebra operation and its properties.
Students could not simplify the algebra of area of prisms surface	C3	**Applying the procedure**	
Students could not conclude the formula of area of prisms surface.	C1, C2, C3	**Prior knowledge, understanding concept, and applying procedure**	a. Developing students worksheet (LKS) which specifically presents the three domains through lesson study-based teaching b. Making a prop which specifically presents the three domains through lesson study-based teaching

5. Discussion

The learning material was the area of prisms surface. It was designed with cooperative learning models through guided discovery techniques. To achieve this, the lecturer and teacher designed the learning to allow students to be actively involved through student worksheets (LKS). LKS contains a number of instructions and entries that must be followed and completed by students. The learning instructions were expected to guide students independently without direct guidance from the teacher, while the questionnaire was provided with the aim of building students' reasoning on an ongoing basis, by making simple information connections into complex information. In addition, manipulative props were also provided in the form of beam and cubes models from cardboard.

The use of props can minimize the level of abstractness of the prism concept. In line with the results of research which explains that the application of learning methods using teaching aids especially in the field of mathematics is based on the fact that in that subject there are many subjects that need tools to describe them, including on prism material [9]. Through teaching aids, both teachers and students interact when asking and providing assistance. Teacher's technique to help students is one form of knowledge of students.

The combination of worksheet and teaching aids was expected to stimulate student participation in learning because the main trigger of low mathematics learning outcomes in these schools is the students' relatively low activity. In addition, the lecturers and teachers also designed several types of evaluation tools in the form of problem-solving tests integrated into the student worksheets and individual tests that are given at the end of learning. Both types of tests had different objectives. Tests in LKS aimed to train students to work in groups to apply the concepts they found in problem-solving, while individual tests aimed to test the understanding of their concepts independently. Overall, the planning stage was oriented towards learning that was expected to help students build and develop their ideas on an ongoing basis. The assessment was not based on the final results but rather focused on changing student behavior in expressing their ideas in depth.

When starting learning, the teacher first conveyed the learning objectives, namely finding the surface area of the prism and using it in problem-solving. When performing apperception, the teacher did not explain directly about the facts related to the prism but used examples and questions that guided students to construct definitions of several terms in prisms, such as edge and plane. but students could not distinguish them. The teacher overcame this by taking a triangular prism and pointing to the edge of the prism and calling its term. The sides we were shown by rubbing the planes that make up the prism. The teacher also asked a number of questions as follows.

G : "how many plane figures am I touching?"
S : " there are five, Bu!" (students answered it simultaneously)
G : "How about the number of the edges?"
S : (silent).

This silence shows that they could not determine the number of edges precisely. After that, the teacher showed a video showing prisms and ask students to pay attention to below prisms picture.

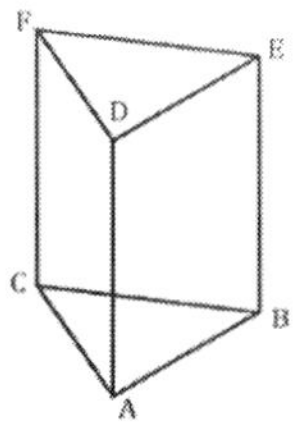

Picture 1. Triangular Prisms

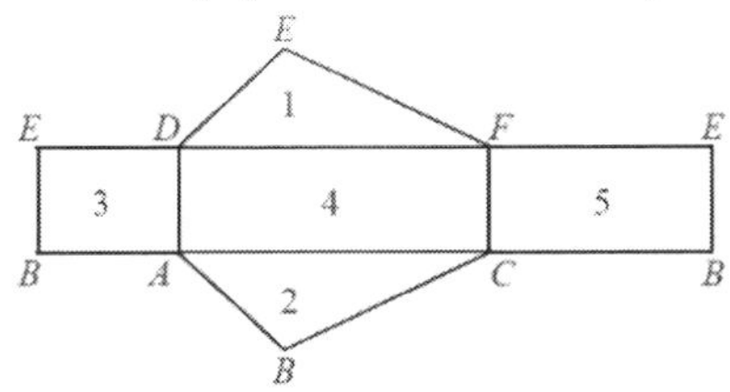

Picture 2. Triangular Prisms

The teacher shows an edge of the prisms (AD). This example enabled students to mention other edges i.e. AB, BC. CA, CF, BE, ED, DE, and FE. Then teacher re-asked students;

G : "how many edges am I touching?"
S : "There are 9, Bu !"

Then the teacher asked students to unwrap the prisms model in their group so it shows the net. The teacher took one net and showed it to the students. She gave some instructions as follows:

G : "this is called as prisms net. Is this net a solid figure of a solid figure?"
S : "Solid figure, Bu!" (students answered simultaneously) , some students whispered softly "that's a solid figure, Bu!"

The teacher was shocked and again gave some assistance through questions such as;

G : "What figures can you observe through this object?"
S : "triangular and square.....some students mentioned triangular and rectangle!"
G : "Nah,,,triangular, square, and rectangle, are they solid figure or solid figure?"
S : "planes, Bu!"
G : "Correct! So, what is the difference between prisms and its net?"
S : "prism has space and its net is a solid figure, Bu?"
G : Good ! Correct answer!

Based on the above script, the teacher was guiding students to find a concept by provoking inquiry questions. Students were involved in the observation to determine the correct conclusions from the results of his observations. Subsequent investigations were carried out on the formula for the area of the planes forming prisms. The teacher asked the formula for the area of a triangle and the area of a rectangle through the following conversation.

G : "How many triangles can you see in a prisms net?"
S : "There are 2, Bu!"
G " How about rectangle, how many?"
S : "There are 3, Bu!"
G : "Is there any square?"
S : "There is no, Bu!"
G : "Alright, so, the net of a prisms consists of ....?"
S : "two triangles and three rectangles, Bu!"
G : "Anyone remembers the formula of the areas of triangle and rectangle??"
S : *" base is multiplied by height, and the length is multiplied by the width bu"!*
G : *"How about the formula of the circumference of both figures?*
S : *silent..............*

Students recorded the material in slides in their books, then the teacher asked students in the group to work on the worksheets according to the instructions listed on them. But before that, the teacher again showed students about the process of unfolding a prism into nets and how to calculate the total area of the net. At the same time, the teacher game affirmations such as;

G : "remember, our goal is to identify the area of each figure forming the prisms and sum them."!
S : "Alright, Bu!"

This affirmation serves as a reminder for students so that they can describe the process of finding the prism surface area correctly. In the learning process, visual media was used at the starting step, in this case, the stage of delivering information. Concrete media was used in the main learning activities. It was distributed to each group with the aim of encouraging student activities in learning. The activities included holding the prism model, cutting, and measuring it. During her involvement in learning activities, the teacher did not explain the material/instructions on the board but always visited each group even though the work instructions had been written on the worksheet.

The teacher instructed each student to investigate the prism net model, but not all students did so. Therefore the teacher provided limited assistance by scaffolding in parts that were difficult for students to do. Teacher activities were still dominant so the guided discovery that was planned did not work as it should because students were not used to working independently and just waited for instructions from the teacher. This process required a lot of time so that the time allocation was not in accordance with the plan. As a result, individual and group assessments could be done because time was up. The following were the results of student entries in the LKS.

The results of the apperception about the area and circumference formula yielded results. Students understood when and how the formula works. Difficulties actually occurred when students added up the area of each plane forming a prism to be a formula for prism surface area. The problem lied in the student's ability of algebraic manipulation which was, especially in using distributive properties. Students could only calculate the size of the area of each field but could not yet write it into an algebraic form.

The difficulties experienced by students were an indication that they were unable to remember the concept, the inability to deduce useful information from a concept and the lack of understanding skills (schematic knowledge) which was indicated by the difficulty experienced in writing formulas [10]. The teacher handled it by giving an example of a rectangle, then students were asked to follow the existing examples.

The area of each plan could be determined by students based on the area of plane I which had been written in the LKS. Another problem occurred in the determination of the total plane area. Students could not simplify the following algebra formula.

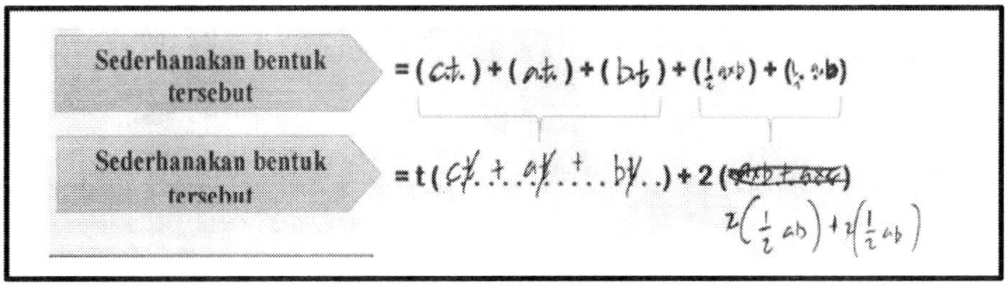

Picture 3. Students work in worksheet

The teacher overcomes this by asking students to write each formula that represents each plane, then to determine the factors of the formula of those five figures. Those factors were the same variable, which was always attached to each algebra term. The teacher gave the following instructions.

G : "what letters do always exist in each formula you are writing?"
S : "T and AB, Bu!"
G : "Extract those letter from the bracket!"

After being simplified, students made another mistake by crossing the variable 't' and distribute the number '2' in each formula of the triangle area. Based on this mistake, the teacher re-instructed students to pay attention to the formula simplification process which can be seen in the following transcript.

G :"Why do you cross the letter 't'?"
S : "because 't' has been removed from the bracket"
G : "does 't' disappear or ......?"
S : "No, Bu. "t' was only removed from the bracket, therefore I crossed it"
G : "well, remember that the 't' does not disappear, so there it shouldn't be crossed because there is no reduction operation"!
S : "Alright, Bu!"
G : "How about the number '2' which is inserted into the triangle formula?"
S : silent..........................

Therefore, the teacher then wrote $\frac{1}{2}ab + \frac{1}{2}ab = ...$ on the board and then asked students the following questions.

G : " how is the result?"
S : silent.........................
G : "how many fraction ½ , letter a, and letter b do you see?
S : "they are two, Bu"

Then the teacher completed the note on the board to be $\frac{1}{2}ab + \frac{1}{2}ab = 2\left(\frac{1}{2}ab\right)$, then pointed one student to continue it. He/she completed it to be $\frac{1}{2}ab + \frac{1}{2}ab = 2\left(\frac{1}{2}ab\right) = 1(ab) = ab$. The teacher then gave appreciation to that student. The formula of the prism surface area can be proven by using concrete props (prism nets), visual props (powerpoint and prism video) and LKS. The teacher's role also strongly supports this activity. Even though she dominated the learning, the teacher seemed astute in overcoming each student's difficulties in various ways. She utilized some questions that aimed to lead students to understand and reason, through the expression of words or terms that can reduce the level of abstractness of a symbol / algebraic form. Everything was performed systematically even though it was not yet fully constructivist. From this study, there was at least an important note that teacher flexibility in using approaches in guiding students was a characteristic of knowledge of students in PCK. In line with this, [11] explained that changes in teaching strategies are natural, especially when addressing the demands of a dynamic student situation. Unexpected student reactions or responses, such as discussions that cannot work well, none of the questions prepared can be solved by students, or no students willing to explain their answers in front of the class need to be anticipated quickly by the teacher. The sensitivity of the teacher in reading the classroom situation is an indication that the teacher has the right knowledge about the condition of his students.

The teaching strategy employed by the teacher must also be tailored to the needs and situation of students. On the other hand, the concept reconstruction process must also go according to the Guided reinvention principle which strongly recommends that the mathematical process be carried out inductively. Learning Mathematics with the teacher must be carried away just as mathematics is found. The process of discovering the concept of prism surface area above is a representation of a concept built through the activities of investigation and discovery.

In the future, teachers must be more courageous in providing more opportunities for students to take part in learning. Aspects of student activity must be built through a series of investigative activities. Systematic investigations will give rise to a variety of solutions and can build students' divergent thinking skills.

On the other hand, lesson study helps the teacher to do reflection through continuous student interaction. The details of the observation allow the teacher to map out the forms of learning interactions so that they can activate each student and also accommodate their learning needs. The teacher PCK described through Lesson Study illustrates it influences their mathematical teaching. [12] explained that the form of influence can be seen not only from the teaching object, the structure of teaching ideas but also in terms of education, teaching emotions, teaching design, teaching language, students' mathematical thinking, student learning attitudes and so on.

6. Conclusion

Teacher's knowledge about student in polyhedral include:

a. The teacher identifies some students who cannot distinguish between plane figure and solid figure Guru.
b. The teacher finds some students who cannot mention formulas of surface area of plane figure that build a prism.
c. The teacher reveals the mistake of students who cannot add the surface area of plane figure that build the prism.
d. The teacher reveals the mistake of student in simplifying the algebraic form in surface area of prism.
e. The teacher reveals the mistake of student that cannot write the formula of surface area of prism.

Strategies that can be done to overcome the student's errors/misconception are;

a. Shows the concrete models of plane figures and solid figures, give student to opportunity to observe the solid figures and ask them to write down their characteristics.
b. Asks to student to reread the surface area material and then write down the formula of the surface area in student's worksheet.
c. Asks student to show nets of prism and then guide student to add every surface area of plane figure.
d. Provide examples of algebraic additional operation on the board or on the student's worksheet.
e. In student's worksheet, teacher lists the simple algebraic characteristics that students can use in formulating surface area of prism.

References

[1] Kilic, H. (2011). Preservice Secondary Mathematics Teacer's knowledge of Students. *Turkish Online of Qualitative Inquiry*. **Vol. 02**. No. 02, Pages 2.

[2] Shulman. (1987). *Knowledge and Teaching. Harvard Educational Review*. Foundations of The New Reform. Harvard Educational Review. United States of America.

[3] Kennedy, M. (1998). Education Reform and Subject Matter Knowledge. *Journal of Research in Science*. **Vol. 35**. No. 03. Pages 253.

[4] Leinhardt, G & Smith, D. A. (1986). *Expertise in Mathematics Instruction*. Subject Matter Knowledge. Journal of Educational Psychology. **Vol. 77**. No. 3. Pages 248.

[5] Carpenter. (1989). Using Knowledge of Children's Mathematics Thinking in Classroom Teaching : An Experimental Study. *American Educational Research Journal*. **Vol. 26.** No. 04. Pages 520.

[6] Fennemma, E., & Franke, M. (1992). *Teacher's Knowledge and Its Impact.* New York: Handbook of Research on Mathematics Teaching and Learning. United States of America.

[7] Ma'rufi. (2018). Pedagogical Content Knowledge: Teacher's Knowledge of students in Learning Mathematics on Limit of Function Subject. *Journal of Physics IOP Conference Series*. **954** pages 2.

[8] Lewis, C. (2002). *Lesson Study: A handbook of teacher Instructional Change.* Philadelphia: Research for Better Schools.

[9] Yensy. B, N. A. (2012). Penerapan Model Pembelajaran Kooperatif Tipe *Examples Non Examples* dengan Menggunakan Alat Peraga untuk Meningkatkan Hasil Belajar Siswa di Kelas CIII SMPN 1 Argamakmur. *Jurnal Exacta,* **Vol. X**, No. 1. Pages 26.

[10] Tanjungsari, R. D., Soedjoko, E., & Mashuri. (2012). Diagnosis Kesulitan Belajar Matematika SMP Pada Materi Persamaan Garis Lurus. *Unnes Journal of Mathematics Education. UJME* **1 (1)** (2012). Pages 53.

[11] Mahmudi, A. (2009). Mengembangkan Kompetensi Guru Melalui *Lesson Study. Forum Kependidikan,* **Vol.28, No. 02.** Pages 2-6

[12] Li, M., & Yu, P. (2009). Study on Effect of Mathematics Teachers Pedagogical Content Knowledge on Mathematics Teaching. *Journal Of Mathematics Education.* **Vol.02, No. 1.** Pages 66-67.

Students metacognitive skill in learning mathematics through cooperative based emotional intelligence

M Ilyas[*], Ma'rufi, and F Basir

Cokroaminoto University of Palopo, Indonesia

[*]Corresponding author e-mail: muhammadilyas949@yahoo.com

Abstract. This type of research is quasi-experimental research which aims to determine whether mathematics learning through cooperative based emotional intelligence can improve the metacognitive skills of students Class X MIA SMAN 4 Palopo. The experimental design was one group pretest-posttest design. The population in this study were all class X SMAN 4 Palopo whose samples were determined using purposive sampling technique. The instrument of this study used a test of metacognitive skills and emotional intelligence questionnaire which were then analyzed with descriptive statistics and inferential statistics. The results of this study indicate that the average of students' metacognitive skills is 85.67 and the results of hypothesis testing indicate that there is an increase of students' metacognitive skills in learning mathematics through cooperative based emotional intelligence.

1. Introduction

Mathematical learning is often seen as one of the difficult subjects for students because it is abstract and has many formulas. Mathematics is a frightening spectre for most students. Students are afraid, because it is difficult to solve problems. Most of the students' difficulties come in the form of understanding the purpose of the given problem, understanding what is being asked about or students' lack of understanding of the material related to the issue. In addition, sometimes students make errors in their calculations [1]. One of the causes that is currently considered the most influential and can explain this fact is the low thinking ability including students' metacognitive abilities.

Metacognitive is interpreted as awareness and control upon the cognitive process, or thinking about thinking, as well as process to know and to monitor thinking process or cognitive process and employing strategies to enhance and problem solve solutions when there is understanding failure [2]. The results of the study found that students whose on mathematical learning processes contained metacognitive skills realized their cognitive processes during solving the problem; they look more likely to be able to capture their thinking processes regularly and the process seems to contribute to students' success in solving problems [3]. The components of metacognition are planning, monitoring and reflection. One important component is the planning of problem solving steps with suitable mathematical tools. Second, the use of the tools has to be controlled, an analysis of the latest state of what has been achieved is necessary, a comparison with the goals set has to be made. The administration of this controlling and comparison is called monitoring. A third component is reflection, for example, on the given problem, on the understanding of mathematical concepts and tools, or on the discrepancy between representations and conceptions [4]. Metacognitive skills play a role in guiding students to realize and control the process of interaction in thinking. Internally students will build knowledge by interacting ideas in their minds based on their prior knowledge and externally students build knowledge through interaction with their environment including their friends to achieve a more perfect understanding. Thus the learning process will be more effective in achieving goals.

Meanwhile, to create a mathematical learning process by optimizing the students' potential, then emotional intelligence that owned by students needs to be a concern. Considering emotional in learning mathematics in a special way may be a little more helpful in accepting mathematics, in the midst of the assumption that is still believed by most students, that mathematics is a difficult subject. Thus, the presence of emotional intelligence can be seen as an aspect that needs to be considered, it can even be used as a basis for following a problem-based learning process well so that the achievement of mathematics learning results as a whole.

Emotional intelligence has the meaning to get and manage emotions in oneself and others [5]. Emotional intelligence as a potential to understand what influences thoughts, feelings and responses and how one can manipulate information for oneself and establish harmonious relationships with others [6].

The highest intellectual intelligence contributes approximately 20% to the factors that determine individual success in life while 80% are filled by other forces including emotional intelligence [7]. One characteristic of children who do not have good emotional intelligence in learning mathematics is that children are less passionate or lackluster, are not critical and only think about and focus on the final results or answers [8]. Emotional intelligence is one of the intelligence possessed by everyone. Given the importance of the application of emotional intelligence in this era of development, then in the world of education a teacher must also be able to manage the emotional intelligence possessed by his students. There are times when knowledge of emotional intelligence is applied in teaching and learning activities. Although mathematics is a subject that is related to intellectual intelligence, it will be very interesting if the teacher can also know the relation between mathematics and emotional intelligence.

One way to maximize metacognitive skills and emotional intelligence is to apply innovative learning models and can help students understand the mathematical concepts that was cooperative learning models. Learning patterns that are able to develop emotional intelligence and children's thinking are learning patterns that are socially nuanced, namely learning patterns that involve interactive learning communities [9]. Cooperative learning models provide opportunities for students to think in answering questions or completing individual or group tasks. Then the students are given the opportunity to reflect on answers or ideas that are delivered both in class discussions and presentations. These activities provide opportunities for students to identify concepts or subject matter that has been known or not yet known. Thus, Mathematics learning through cooperative based emotional intelligence has the potential to empower students' metacognitive skills.

Mathematics learning through cooperative based emotional intelligence is the learning method which combine cooperative learning and emotional intelligence in learning mathematics intelligence or integrating emotional intelligence into the mathematics learning through cooperative to increase students' metacognitive skill.

These are the steps of the learning method,

- Teacher asks students to consider him while explain the topic of learning, to consider other students who have a question or an argument, and asks students to not mock each other but to support each other.
- Teacher asks students to work cooperatively and to be responsible for their work
- Teacher motivates students to be patient and concentrate during the learning process,
- Teacher asks students how to control feelings of anger, how to discuss their problems and how to overcome the problems they face,
- Motivate students that the topic of learning is important to be learned, Asking students to present their work, and give them reward.

Based on the description above the purpose of the study is to increase students' metacognitive skill in learning mathematics through cooperative based emotional intelligence.

Journal of Physics: Conference Series **1397** (2019) 012089 doi:10.1088/1742-6596/1397/1/012089

2. Method

This is a quasi-experimental research using a one group pretest-posttest design. The subjects in this study will get treatment, namely Cooperative Based Emotional Intelligence. Then students are given a test before treatment and after treatment.

Table 1. Design of cooperative learning based on emotional intelligence

Pretest	Treatment	Posttest
O_1	X	O_2

Information :
X : Mathematics Learning Through Cooperative Based Emotional Intelligence
O_1 : Pretest of Metacognitive Skills Test
O_2 : Posttest of Metacognitive Skills Test

This research was conducted in Class X MIA SMAN 4 Palopo Even Semester Academic Year 2018/2019. The instrument used was a test of metacognitive skills and Emotional Intelligence Questionnaire which was then analyzed using descriptive statistical analysis and inferential statistical analysis. Inferential analysis uses the normality test and hypothesis testing with one sample t test.

3. Result and discussion

3.1. Result

The results of the descriptive statistical analysis show the characteristics of the score distribution of each variable and at the same time the answers to the descriptive problems formulated in this study.

The results of the emotional intelligence questionnaire is to identify which student have high emotional intelligence and then put each one of them into the each group to help teacher develops emotional intelligence in the group. The result of the emotional intelligence questionnaire is shown below.

Table 2. Students' emottional intelligence

Category	Criteria	Frequence	Percentage (%)
Low	< 84.29	8	28.57
Medium	$84.29 \leq x \leq 99.71$	14	50
High	> 99.71	6	21.43
Total		28	100

Based of the table above there are six students that have high category of emotional intelligence. The sixth students then put on each group to help teacher develops emotional intelligence of the students.

The results of the tests of students' metacognitive skills in Mathematics Learning Through Cooperative Based Emotional Intelligence which can be seen in the following table.

Table 3. Frequency

Statistic	Pretest	Posttest
Sample	28	28
Maximum Score	54	98
Minimum Score	30	74
Mean	42.60	88.67
Median	40	89
Mode	39	90
Range	24	24
Varians	43.35	34.81
Standard Deviation	6.58	5.90

The above results show the average score of students' metacognitive skills on pretest 42.60 and posttest 88.67. It means that there was an increase of the students' metacognitive skill.

Score of the increase of students' metacognitive skill that learned by mathematics learning through cooperative based emotional intelligence can be seen by using gain normalized classification that shown below.

Table 4. Score of gain normalized of students' metacognitive skill

Coefficient of gain normalized	Classification	Frequence	Percentage (%)
$g < 0,3$	Low	0	0
$0,3 \leq g < 0,7$	Medium	20	71,43
$g \geq 0,7$	High	8	28,57
Total		28	

Based on the table above 71.43 % students at medium category and 28,57% at high category. The average of the gain normalized is 0.67 or at medium category. It means that the increase of the students' metacognitive skill at the medium category.

Based on indicators of metacognitive skills scores obtained for each indicator can be seen in the following table.

Table 5. Score of students metacognitive skill

Metacognitive Skill	Score
Planning	82.25
Monitoring	77.71
Evaluation	72.05

In Table 5 shows that the scores of students' metacognitive skills for aspects of planning, monitoring and evaluation are 82.25, 77.71 and 72.05.

The results of inferential statistical analysis based on Kolmogorov-smirnov probability value of the student's gain normalized score is 0.200. This means that $p > 0.05$, Based on the results of the normality test it can be concluded that the distribution on students' metacognitive skills in mathematics learning through cooperative based emotional intelligence is normally distributed. While the results of the t test for one sample obtained a probability value (2-tailed) 0,000. Because the probability value (2-tailed) <0.05 (0,000 <(0.05)) then H_0 is rejected and H_1 is accepted, such that there is increase of students' metacognitive skill in learning mathematics through cooperative based emotional intelligence.

3.2. Discussion

Based on the results of the study, first it was found that there is an increase of students' metacognitive skill after taught by mathematics learning through cooperative based emotional intelligence. Based on the results of descriptive statistics it can be seen that the average of students' metacognitive skill that taught by mathematics learning through cooperative based emotional intelligence is 42.60 on pretest and increase to be 88.67 on posttest. It can be said that mathematics learning through cooperative learning based emotional intelligence increases students' metacognitive skills. Cooperative learning is created to influence the pattern of student interaction in learning by giving students the opportunity to think individually, in groups, and share throughout the class while mathematics learning based emotional intelligence builds a conscious effort for students in developing strategies to solve mathematical problems.

The combination will have an effect together in developing students' thinking skills and learning skills. Students that taught by cooperative learning based emotional intelligence can practice thinking about their own thinking processes, and practice applying certain learning strategies to complete their learning tasks. Mathematics learning through cooperative based emotional intelligence is based on constructivistic strategies that give students the opportunity to construct their own knowledge through their active involvement in learning. The process of knowledge construction can be achieved by students through a process of discussion, presentation and collaboration between students in the completion of their learning assignments that have been designed in the worksheet. Thus mathematics learning through cooperative based emotional intelligence provides an opportunity for students to develop their thinking skills and learning skills.

This research supported by other research said that Cooperative learning can be successfully used to promote student' performance in mathematics. It was found that students' performance in mathematics and attitudes towards mathematics were affected by exposure to the cooperative learning. Students seem to prefer learning mathematics by sharing knowledge. They feel contented when they can function effectively in the group work [10]. For the successful completion of a complex problem a variety of metacognitive processes is needed. A successful problem solver realizes that they can guide their efforts by finding and recognizing steps that were previously done by combining and connecting information between past knowledge and problem situations. The less experience a problem solver has, he cannot monitor the resolution process effectively, even though they can continue the problem solving process but maybe the strategy used is wrong [11].

In general, metacognitive theory focuses on: (1) the role of awareness and the management of thought processes; (2) differences in self-assessment and management of cognitive development; (3) main knowledge and abilities that develop through experience; and (4) strategic and constructive thinking [12]. Metacognitive strategies can help students to solve problems through effective design, which involves the process of knowing problems, determining the characteristics of problems that need to be solved and understanding effective strategies to solve problems [13]. Based of those then it can be said that the mathematics learning through cooperative based emotional intelligence increase students' metacognitive skill.

4. Conclusion

The results of this study indicate that mathematics learning through cooperative based emotional intelligence increase students' metacognitive skill.

References

[1] Mulbar, U., Rahman, A., & Ahmar, A. (2017). Analysis of the ability in mathematical problem-solving based on SOLO taxonomy and cognitive style.

[2] Corebima, A. D. (2009). Metacognitive skill measurement integrated in achievement test. *State University of Malang.*

[3] Teong, S.K. (2000). The Effects of Metacognitive Training in Mathematical Word Problem Solving in a Computer Environment. Proceedings International Group for The Phsycology of Mathematics Education. Jepang: Hiroshima University.

[4] Cohors-Fresenborg, E., Kramer, S., Pundsack, F., Sjuts, J., & Sommer, N. (2010). The role of metacognitive monitoring in explaining differences in mathematics achievement. *ZDM*, *42*(2), 231-244.

[5] Mortiboys, A. (2005). Teaching with emotional intelligence. New York: Taylor & Francis e-Library.

[6] Bowkett, S. & Percival, S. (2011). Coaching emotional intelligence in the classroom. New York: Taylor & Francis e-Library

[7] Goleman, Daniel. 1998. Emotional Intelligence Kecerdasan Emosional: Mengapa EI Lebih Penting Daripada IQ. Jakarta: PT. Gramedia Pustaka Utama.

[8] Skovsmose, O. (1994). Towards a critical mathematics education. *Educational studies in mathematics*, *27*(1), 35-57.

[9] Covey, Stephen R. (2008). The 8th HABIT. Melampaui Efektivitas, menggapai Keagungan. Jakarta: Gramedia Utama.

[10] Hossain, A., & Tarmizi, R. A. (2013). Effects of cooperative learning on students' achievement and attitudes in secondary mathematics. *Procedia-Social and Behavioral Sciences*, *93*, 473-477.

[11] Panaoura, A. & Philippou, G. 2004 The Measurement of Young Pupils' Metacognitive Ability in Mathematics: The Case of Self-Representation and Self-Evaluation. Jurnal CERME 4. Diambil dari http://Cerme4.crm.es/papers%20 definities /2/ panaoura.philippou.pdf

[12] Paris, S.G., & Winograd, P. 1990. How metacognition can promote academic learning and instruction. In B. F. Jones & L. Idol (Eds.), Dimensions of thinking and cognitive instruction (pp. 15-51). Hillsdale, NJ: Erlbaum

[13] Davidson, J.E., Deuser, R., & Sternberg, R.J. 1994.The role of metacognition in problem solving. In J. Metcalf & A. P. Shimamura (Eds.), Metacognition. Boston, MA: The MIT Press.

How Students Build Their Mathematical Dispositions towards Solving Contextual and Abstract Mathematics Problems

Mukhtamilatus Sa'diyah[1], Cholis Sa'dijah[1,*], Sisworo[1], UcikFitri Handayani[1]

[1] Mathematics Education, Faculty of Mathematics and Science, Universitas Negeri Malang, Jalan Semarang No. 5, Malang 65145, Indonesia

* cholis.sadijah.fmipa@um.ac.id

Abstract. Mathematical dispositions affect students' performance when studying mathematics. Teachers should have high awareness of students' mathematical dispositions when solving mathematical problems. This research identifies students' viewpoints about problems in mathematics. Understanding students' viewpoints would assist teachers to elevate students' mathematical dispositions by developing more efficient test items. This is a case study research using qualitative method, of which students who have high motivation in learning mathematics were asked to answer questions related to their mathematical dispositions. Three main focuses observed on this research were students' perceptions, attitudes, and efforts. The data show that the students build their mathematical dispositions through some process. Although they have mathematical dispositions, their preference in mathematics problems is different. Most of the students prefer solving abstract problem while others prefer solving contextual problems. This difference affected their performance in solving mathematics problems. This means that teachers should pay more attention to the form and level of problems given to the students, so that students would have less effort to understand the problems and emphasize their works by thinking more deeply when solving problems.

1. Introduction

The most critical barrier when students learning mathematics is the difficulties in resolving the given mathematical problems [1]. As a concept, the word "problem" has multiple academic interpretations. A number of researchers in the mathematic education field outline "problem" as all practice test items given for train students' ability related to a newly learned materials[2][3]. Others define "problem" as non-routine and complex test item which require application and association of mathematical concept[4][5]. In this research, the word "problem" refers to all mathematics test items aimed at training students' competence. There are two type of problems, which are abstract and contextual mathematics problems.

Abstract mathematics problems are defined as mathematics problems that use mathematical symbols and operations. Athough it is easy to solve the problems using mathematics procedure, students sometimes have difficulty in understanding the context of usage of the problems. Whereas contextual mathematics problems are defined as mathematics problems that are closely related to students' everyday life situation and most likely to be experienced by the students in person. Students would have better understanding about the context of the given problems[1]. However, some students are more likely to be confused when distinguishing important information of the problems.

Not all of mathematics problems could be solved immediately by applying the given formulas. A considerable number of problems require further analysis and implementation of multiple steps of solving methods to obtain correct answers. The skill to resolve the problems is called the problem-solving ability. Problem solving is essential for mathematics learning, as it help students to develop their cognitive process and openness when solving more complex problems[6].

Researchers had formulated strategies to reduce students' difficulties when solving mathematics problems by developing learning methods[7][8], learning media[9][10][11], and also evaluation test items[12][13]. However, it is also constructive to shape students' attitude towards mathematics. A number of researches had supported the importance of students' attitudes and desires to learn mathematics[10][14][15][16]. These attitude and desire are known as mathematical dispositions.

Students' mathematical dispositions refer to students' tendency to perceive mathematics as useful, reasonable, and important material, aside from their confidences of their skills and efforts[17].A research conducted by Beyers[18] has introduced mathematical disposition division into three assessment domains, of cognitive, affective, and conative. The cognitive domain refers to students' awareness when learning new knowledge or concepts. The affective domain refers to students' emotional responses. The conative domain focuses on students' efforts to develop and utilize their knowledge when solving mathematical problems. These three domains, then became assessment references to learn about students' mathematical dispositions.

A considerable number of enquiries have been conducted to discover the significance of students' mathematical dispositions[17][18][19][20]. Some examined the influence of students' mathematical dispositions towards performance[21], students' achievements[22], and students' problem-solving ability[16]. Students with mathematical dispositions tend to have more confidence about their skills, appear to be more challenged to solve mathematical problems, and realize how mathematics contributes to their daily life problem solving. When students' mathematical dispositions are improved, students' performance and learning interest would be improved, then students could easily and happily take part in mathematics learning at school[21][22]. However, there has been inadequate number of researches conducted to learn the construction process of mathematical dispositions and its relations to mathematics problems. In fact, the construction process of students' mathematical dispositions had a significant influence towards students' perceptions about mathematics in the future[20]. Students would assume that mathematics is important, if they feel attracted and challenged to solve given problems[16].

Furthermore, improving mathematical dispositions should be supported by selecting proper problems. Identifying preferred and suitable problems could also help students increasing their performance in learning mathematics. Investigating students' mathematical dispositions when solving problems would help teachers depicting students' affective, cognitive, and conative viewpoints[18]. Researches related to mathematical dispositions mostly use systems of multiple-choice questionnaire[17][23]. The use of questionnaire is limited to statements written or chosen, and unable to disclose in-depth meaning. It also holds back students' flexibility to state their standpoints. By selecting open-ended questions, the answers given by students would be more diverse and receptive[14].

This research discloses variations of test items that appear more attractive to the students. In addition, teachers would have further references to elevate their students' mathematical dispositions by observing the process of solving mathematical problems and the type of test items preferred by the students. The results would also be implemented by teachers to develop teaching methods aiming at improving students' mathematical dispositions.

2. Method

This is a qualitative descriptive research[24]. Research subjects were selected from 8th graders in Malang who took extra mathematics lesson independently. The researcher gave questionnaire inquiring their mathematics learning motivation. Through the questionnaire, the researcher collected five studentswho had high motivation in learning mathematics as subjects. This research identifies how they built their mathematical dispositions towards solving mathematics problems and showing their preferred problem.Instrument utilized was essay questions including rationale behind students' motivation to learn mathematics, students' knowledge about activities related to mathematics, forms of mathematical problems, and their difficulties when solving problems. There were five students, namely S1, S2, S3, S4, S5 who had been asked to answer the questionnaires. The problem given was being connected with general activities which is more likely to be experienced by students. Figure 1. and Figure 2. exhibit test items given to the students.As the possibilities for short and lack of information answers provided in the questionnaires, the researchers had prepared interview sessions to obtain in-depth responses and further confirmations.

Test Item A

Enter data on the graph based on the given table

a	b	c	d	e	f	g
5	$2a$	$3b$	$a+10$	$b-5$	b	$c-10$

Of which $a-g$ is value is the most? What is the value?
Which one has the smallest value? What is the value?
Explain how you got the answer.

Figure 1. A mathematics problem using an approach of abstract representation

Test Item B

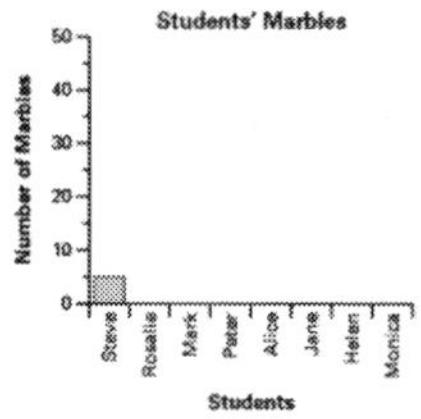

From the diagram, it appears that Steve has 5 marbles. If it it known that

- Rosalie has twice as many as what Steve has
- Mark has 4 times marbles more than Rosalie
- Peter has 4 times marbles more than Steve
- Alice has as many as what Steve has
- Jane has 5 times marbles less than Rosalie
- Helen lost all marbles she had
- Monica has as many marbles as Rosalie.

Finish the bar chart above based on the given information!	Who has the most marbles? How many marbles does she/he have? Who has the less marbles? How many marbles does she/he have? Explain how you got the answer.

Figure 2. A mathematics problem using an approach of contextual representation

At the end of data collection phase, students' worksheets and interview transcripts were identified based on Beyer's indicators[18] of students' mathematical dispositions, of: perception, interest, and effort when solving contextual mathematical problems. The data were analysed thematically, based on students' dominant answers and other unique findings. Afterwards, conclusion was drawn by considering variations of students' answers.

3. Result

3.1. Students' self-perception towards mathematics

The first question in the essay was related to students' early opinion about mathematics. Of all 5 students, there were two students who claimed that they have pleasant experiences about mathematics, which were S3 and S4. S3 stated that the mathematics teacher had been teaching with clarity making them more interested in studying mathematics. Meanwhile, S4 wrote that the high mark on mathematics at the previous education level was the main factor for loving mathematics. This made the S4 keeping at studying mathematics. The answers imply the important role of the teachers to build students' confidence towards mathematics, either from test items formulation, or the teaching methods. These two side would further assist students to build their motivations to learn mathematics.

On the other side, three other students asserted that bad experiences when learning had caused them dislike mathematics. For S1, the difficulties were in understanding the given problems and memorizing related formulas. S2 had a similar experience by stating that hardships when memorizing necessary formulas had withheld the subject from learning mathematics, but the subject express the willing to learn mathematics further due to the consciousness about the importance of mathematics. Whereas, S5 experienced mood swings when learning mathematics, depending on the learned topics.

All the answers imply the complexity of mathematics as the main set back from becoming interested in learning mathematics.

In the next question, students were asked about activities and games related to mathematics. All the five students wrote activities related to numbers and basic calculations, whereas the mathematics materials that they had learned in the junior high school have not been exemplified in the daily life. On talking about games, three students, which were S1, S2, and S4 mentioned number-related games, while the two others admitted that they did not have references about games that are closely related with mathematics. This signifies that the students could not implement the use and advantages of mathematics in their daily life situations. In other words, mathematics is perceived as a learning subject in formal education settings of all levels. The games that the subjects familiar with were popular online games such as Hago, Line Gets Rich, PUBG, and CS:60. Skills learned from mathematics should be applicable for these games. Further, teachers also should be aware that these games could be utilized as a means of attracting students' interest in learning mathematics. Students' lack of comments on the relations between their popular games and mathematics exhibits the fact that mathematics problem solving given had failed depicting students' everyday life situations. As a result, students were less aware about the importance of mathematics in their life.

3.2. Students' interest in solving mathematical problems

In order to understand students' interest in solving mathematics problems, two test items were given to the research subjects with different representations. One test item was an abstract problem with numbers of mathematical operations and symbols. While the other test item was more contextual.

When students were given a choice of questions, four out of five students chose the abstract mathematical problems compared to the contextual one. But interestingly, the reasons the had given were anonymous, which was: it is easy. Students who preferred abstract problems stated than they were easier as they were short and not confusing. Students claimed that they had only needed to apply learned formulas or resolution procedures. These students stated that, to solve problems, they need to memorize and study the formulas they had been taught before. This had also become the reason why they dislike contextual problems. According to their statements, contextual problems were more complicated as it has too many words and given information.Similar justifications were given by students who preferred solving contextual problems. It was easier to solve as the students had a better frame of reference and it was relatable. This helped the students to map out problem solving procedures based on information obtained from their daily life experiences.

Their answers to the interviews were also noteworthy. The word "easy" refers also to the procedure of problem solving. When solving abstract problems, they just had to utilize general and standard procedure to solve the problem. Contextual problems appeared to be more difficult to be solved as narrative test items contain too many words, which confused them to choose the proper information. On the contrary, students who preferred solving contextual problems admitted that they could be more flexible when solving the problems as they did not need to be fixated on any problem-solving procedures. When students happened to forget the formula needed, their experience would become practical for seeking resolution and problem-solving strategies for given test items. Futhermore, the problems had become more familiar and relevant for the students.

3.3. Students' effort to solve mathematical problems

To discover students' effort, firstly, the students were asked about their difficulties when solving mathematics problems. Figure 1. and Figure 2. have relatively simple problems, but it appeared that students had been still encountering obstacles to finish the test item. When solving the abstract problem, they were frequently asking about required formulas and resolving procedures. Conversely, when solving the contextual problems, they were frequently asking about relevant information for answering the given questions. The type of difficulties faced by the students was a difficulty in basic work. As a matter of fact, the problems given on this research instrument was a simple one, which did not require multiple formulas to finish. There are more complex and require more advanced level of thinking to be solved.

Regarding their efforts, students were asked about their attempts to solve the problems. Their answers were divided into two. S1 and S2 had been relying to other people's assistance when solving problems. They confessed that they consulted the teacher or friends about the necessary procedures to solve the problems, aside from peer-comparing their answers. On the contrary, S3, S4, and S5 had preferred to re-read the problems and tried to better understand the given information. The different efforts demonstrate contrast in the learning methods. A number of students tend to be depended to other people to help them, while the other half prefer to be independent in figuring out the answers for the given problems.

Lastly, the students were given questions about their motivations to learn mathematics. S1, S2, and S3 had an inclination for good marks to be easier to embark upon higher level of education at a reputable school. These students were grade-oriented, so the type of mathematics problems that could motivate them to a greater extent is the solvable and easy problems. S4 dan S5 had given different answers. S4 explained that mathematics is needed in all fields, so that the subject tried to learn mathematics in discrete ways. Although S4 showed preference for solving abstract problems, the subject was aware about the importance of studying how to solve contextual problems, as it requires higher level of comprehension. S5 also answered in accordance to S4. The subject was willing to learn further the problem that are challenging, by mastering the materials. This indicates that the student has a high motivation for learning and becoming an expert about mathematical concepts.

3.4. Students' insight about contextual problems

It emerges that most of the students tend to avoid solving contextual problems. It has become a common assumption that contextual problems should be easier for students, because the problem is adapted from students' daily lives, as stated by S2. However, the other four students, although admitting the closeness between the test item and everyday life situations, had argued that the narration was too long and complicated, thus become hard to be understood. S1 asserted that the main trouble was interpreting abundant words and given information. The subject felt lazy to read the problem. In addition, S2 had been avoiding contextual problems for its complexity, as S4 and S5 also disclosed in the interview. Meanwhile, S3 articulated that contextual problems have not made him more interested in studying mathematics as the problems given by the teacher were trivial and unattractive. This made students become reluctant in solving contextual problems.

4. Discussion

4.1. Students' perception

The students' answers had revealed that they are fond of mathematics they had built a favourable history with mathematic since they were in the elementary school. Especially when they got good grades on tests or examinations. Their passions towards mathematics then has been continued until their junior high schools, and make them like activities related to mathematics, one of which was their choice of games.It also has become apparent that students' good grades at the elementary school has affected their positive preferences for mathematics. The preceding impression of mathematics affects students' attitudes at the next level of education[20]. The high standard achievements foster students' strong confidence when learning mathematics. On the contrary, students with bad grade would have a propensity to feel reluctant to study mathematics at the next level of education.

Students' achievements in Indonesia are measured at the final examination, especially at the end of a level of education. Students' statements confirmed that the final examination namely national exam had been an indication of students' confidence level. If their national exam grades are good, they would be highly motivated to learn mathematics further at the next level of education. This should be a significant consideration for teachers to further build students' confidence both in the learning and in the evaluation processes. Teacher should be able to adjust test items given to their students accordingly to their level of competence to avoid scepticism due to hardships when solving mathematics problems. The level of difficulties should be tailored fitting to students' level of competences. The same treatment should be given on the evaluation phase. A good mathematical problem is characterized with a complexity, but still solvable if students spare more efforts. The accommodation of ideal problems would help students improving their learning motivations.

It is obligatory for teacher to build their students' level of confidence by providing comprehensive test items that also able to stimulate students' deep thinking. Not only that students would be more interested in solving the problems given, but also would have opportunities to practice their abilities gradually. This idea is supported by Lee[25], who states that an ideal problem should have an easy entry method, but requires higher mathematical abilities to find the solutions. Besides, in the problem learning process, teachers should provide varied forms of problem-solving representations, as proposed by Sa'dijah[26], that introducing multiple forms of representations would support students to adjust their interpretations with their learning styles.

4.2. Students' Motivation

Students' priorities towards mathematical problems are discrete, depending on their perceptions about the problems. Either students who preferred solving abstract problems or the ones who preferred to solve contextual problems, had provide the same reason with different angles of "easy." This signifies that there are no differences in the level of difficulties between abstract and contextual problems[4][27]. But it is recognized that the purpose of the two distinctive type of problems are unalike. Abstract problems are given to introduce students towards mathematical concepts and mathematical forms that simplified their efforts to learn mathematical formulas and creating a network of mathematical concepts. Whereas, contextual problems aim at introducing the implementation of mathematics in students' everyday life situations. This differences also refer to the concepts that are being taught and the approaches that are being applied. In addition, it is more likely that students' age and their prior achievements in mathematics might affect the role of this abstract and contextual problems[27].

The students who showed preferences towards abstract problems expressed that formulas and procedures are significant for them. Listening procedures for solving problems have become habitual for these students. It might be caused by the tendency for their teachers to provide descriptive explanations. This is supported by a research conducted by Murtafiah et al.[28] towards pre-service teachers.By memorizing problem solving procedures, students assumed that they would be able to find the answer for the test items. However, there are setbacks that occurred once they were given more complex problems. In addition to learn about procedures, students should practice their abilities to connect their learned concepts and procedures to systematically solve problems. This could be trained to the students when they solve contextual mathematical problems.

Contextual mathematical test items introduce closely related problems of students' life. This situational approach assists students to identify which mathematical concepts relevant for solving given problems. Further, students would be able to use their knowledge of everyday life situations to determine applicable problem-solving procedures[29]. But when solving contextual problems, it is suggested that the students are equipped with capabilities to analyze relevant information to solve the problems. These skills were less mastered by the students, so that they felt only a little interested in solving contextual problems.

4.3. Students' Efforts

It is emerged from this research that the first thing came up to these students' mind when solving mathematical problems is formulas. This has also become the main reason why students prefer abstract problems to contextual ones, as abstract problems are more likely to be directly solved procedurally by following formulas that were previously introduced to the students. Whereas, it is assured that to solve contextual problems, students should understand the framework before finding resolutions for the problems[2]. In fact, there are mathematical problems that require simple solutions without applying any formulas.

The contrasting standpoints stated by the students were strongly influenced by the way they manage information. Students who showed preferences towards solving abstract problems tend to experience difficulties in sorting and managing information obtained. They were confused in determining which information that is relevant for solving the problems. The amount of information given in contextual problems made them avoiding the problems. On the contrary students who had preference in solving contextual problems take the advantage of their everyday situations to solve problems.

4.4. Contextual mathematical problems

Contextual mathematical problems define as problems that require less procedural problem solving and have tendency to be solved by informal strategies[30]. Students who preferred solving contextual mathematical problems expressed little interest towards standard formulas and problem-solving procedures. They claimed that they only need basic concepts to solve problems. Majority of these students rarely used formulas given by the teachers in class. They tend to trust their cognitive comprehension about the problems. Mostly, procedures that were implemented by the students are different with ones that are taught by the teacher.

Students' habit when experiencing difficulties in solving problems also need to be highlighted. The strategy chosen by the students in finding resolutions would affect their performance when solving problems[31]. A number of students have been independently seeking resolutions by learning the situations or looking for examples of problem-solving strategies from books or their teaching handouts. Others would prefer to ask completion strategies to their teacher or their fellow students.The variance of information management and problem-solving strategies amongst students are influenced by their cognitive styles[32]. The difference of cognitive styles affects students' ability to solve problems[31]. There are students who prefer to work individually, and others who prefer to work in groups. There are students who are capable to directly understand the given problems, while others are dependently in need of assistance. These differences also cause varied time of completion amongst the students, and also the gap of thinking. It is strongly suggested to the teacher to be conscious about these gaps and provide adjustable treatments without exhibiting their preferences. It is obliged for teacher to accommodate students' cognitive style differences to further elevate students' mathematical dispositions.

5. Conclusions and Suggestions

The construction of these students' mathematical dispositions was started in their perceptions of problems given to them. They choose problems which they think are easier to solve. According to the research, some students were fixated towards mathematical formulas and representations, so that they were displaying a positive preference towards solving abstract problems. Other students who prefer contextual problems enjoy solving problems related to their daily life. Thus, in promoting the establishment of students' mathematical dispositions, it is critical for teacher to introduce varied forms of problems with gradual level of difficulties. Once students' perceptions are formed, students' motivation could be further improved by introducing various mathematical problems that are closely related to their lives. By cultivating students' interest in solving mathematical problems, students would display more efforts in finding solutions for the problems.

The tendency towards these students' preference of either abstract of contextual problems are caused by students' comprehension abilities and their information management towards given problems. This is also related to the students' cognitive styles. Further research on the connection between this research and students' cognitive styles is recommended to find out students' attitude when solving problems. This would further help teacher to determine appropriate problems that are also interesting for the students.

6. Acknowlegements

The authors are very grateful to Director of Directorate of Research and Community Service (DRPM), Ministry of Research and Technology and Higher Education of the Republic of Indonesiaon research funding.

7. References

[1] Novriani M R and Surya E2017 *IJSBAR* **33 (3)** 63–75
[2] Haghverdi M and Wiest L R2016 *The Math Educator* **25 (1)** 56–73
[3] YewW T, Lian L H and Meng C C2017 *J. of Edu. and Practice* **8 (15)** 136–40
[4] Tural G2012 *Procedia - Social and Behavioral Sci* **46 (2012)** 3609–13
[5] Xenofontos C and Andrews P2014 *Mathematics Edu. Research J.* **26 (2014)** 279–99
[6] Khotimah R P andMasduki2016 *J. of Research and Advances in Math. Edu.***1 (1)** 1–13
[7] CancoyO andDarbas S2010 *J. of Edu.* **38 (1)** 11–24
[8] Retnowati E, Ayres P andSweller J2017 *J. Edu. Psychology* **109** 666–79
[9] Cope L2015 *Delta J. of Edu.* **5(1)** 10–9
[10] Kontas H2016 *J. of Edu. and Learning* **5(3)** 10–20
[11] Laski E V, Jor'dan J R andDaoust C 2015 *SAGE Open*, 1–8
[12] Brehmer D, Ryve A and Van SteenbruggeH 2015 *Problem solving in Swedish mathematics textbooks for upper secondary school*(Sweden: Mälardalen University Press)
[13] Tolar T D, Fuchs L, Cirino P T, Fuchs D, Hamlett C L and Fletcher J M2012 *J. Educational Psychology* **104 (4)** 1083–93
[14] Martino P D and Zan R2010 *J. Math. Teacher Edu.* **13** 27–48
[15] Mensah J K, Okyere M andKuranchie A2013 *J. of Edu. and Practice* **4 (3)** 132–9
[16] Mutai J K 2010 *Attitudes Towards Learning and Performance in Mathematics among Students in Selected Secondary Schools in Bureti District, Kenya* (Kenya)
[17] Su W L and Wen C T 2016 *Universal J. of Educational Research* **4 (8)** 1903–11
[18] Beyers J 2011*Int. J. of Business and Social Sci.* **2(16)** 20–32
[19] Cai J, Moyer J, Wang N andNie B2012 *Annual Meeting of the American Educational Research Association(Canada)* (United States: AERA Online Paper Repository)
[20] Feldhaus C A2014 *American Int. J. of Contemporary Research* **4(6)** 91–7
[21] Venkatesh kumar G and Karimi A2010 *J. of The Indian Academy of Applied Psychology* **36(1)** 147–50
[22] Zakaria E, Zain N M, Ahmad N A andErlina A2012 *American J. of Applied Sciences* **9(11)** 1828–32
[23] Code W, Merchant S, Maciejewski W, Thomas M and Lo J2016 *IJMEST* **47(6)** 917–37
[24] Creswell J W2012 *Educational Research: Planning, Conducting, and Evaluating Quantitative and Qualitative Research* 4th ed.(Boston: Pearson)
[25] Lee P Y (2012). *The Int. Seminar and Workshop: The Use of Contextualized Tasks to Foster Mathematics Learning(Singapore)*
[26] Sa'dijah C, Nurrahmawati, Sudirman, Muksar M and Anwar L 2018 *the 2nd International Conference on Education and Multimedia Technology(Okinawa, Japan)* (New York: The Association for Computing Machinery (ACM))
[27] Jones I, Inglis M, Gilmore C and Bisson MJ2016 *Measuring Conceptual Understanding: The Case of Teaching with Abstract and Contextualised Representations* (Executive Summary) (Loughborough: Mathematics Learning Support Centre, Loughborough University) Retrieved from www.nuffieldfoundation.orgpp. 1–18
[28] Murtafiah W, Sa'dijah C, Candra T D, SusiswoandAs`ari A R2018 *J. on Math. Edu.* **9(2)** 259–70
[29] Hanah R, Muhsetyo G andSisworo2016 *Jurnal Pendidikan Teori, Penelitian, dan Pengembangan* **1 (5)** 927–39
[30] Widjaja W 2013 *IndoMS-JME* **4 (2)** 151–9

[31] Amanah S, Sa`dijah C and Sudirman2016 *Jurnal Pendidikan Teori, Penelitian, dan Pengembangan* **1(2)** 237–45

[32] BurnettW C 2010 *Cognitive Style: A Meta-Analysis of The Instructional Implications for Various Integrated Computer Enhanced Learning Environments* Dissertation from Indiana University of Pennsylvania Retrieved from http://knowledge.library.iup.edu/etd/814

Journal of Physics: Conference Series **1397** (2019) 012091 doi:10.1088/1742-6596/1397/1/012091

Development of student worksheets based on Realistic Mathematics Education (RME) oriented to mathematical reasoning

N Arsoetar[1], and S Sugiman[2]

[1]Postgraduated of Mathematics Education, Universitas Negeri Yogyakarta, Indonesia
[2] Department of Mathematics Education, Faculty of Mathematics Education and Natural Science, Universitas Negeri Yogyakarta, Indonesia

*Corresponding author e-mail: niraarsoetar@gmail.com

Abstract. Student worksheets in schools have not been fully developed based on contextual problems that are close with student environment. Therefore, students thinking skills have not facilitated. Based on these problems, the purpose of this study is develop student worksheets based on RME that can facilitate students to improve mathematical reasoning. This research is development research that uses Plomp development model. The subject trial of this study are 31 students class VIII of Junior high school in Wonosobo. Validity assessment instrument is a validation sheet. Practical assessment instruments are teacher and student assessment sheets, also learning implementation observation sheets. The effectiveness assessment instrument is mathematical reasoning test. Data analysis was carried out by converting quantitative data into qualitative data. The results showed that student worksheet validity score was 95. Practicality based on the teacher's assessment with a score of 37, the results based on students assessment have an average score of 62.35. Average of learning achievement is 95.37% and the results of mathematical reasoning test which shows that 96.55% of students complete the minimal mastery criteria. Therefore, student worksheet who developed is suitable for mathematics learning activities and can be used to improve students mathematical reasoning.

1. Introduction

Learning in junior high schools contains activities with various characteristics. In mathematics learning based on Minister of Education and Culture Regulation Number 22 of 2016, states that learning with interactive and inspirational characteristics makes learning fun and can motivate students to actively participate in learning. It aims to develop learning that makes students feel that mathematics learning is meaningful and enjoyable learning.

Student participation in learning makes students easier to understand the concept of knowledge. Active involvement of students can help them to discover their understanding[1]. Students who have understood themselves can evaluate their understanding and will be able to put themselves in learning related to the knowledge they get. Students with mathematical abilities also have other critical abilities such as reasoning, abstraction, patterns and recognition of relationships and conceptual thinking[2]. In mathematics learning, one of the thinking skills related to understanding is reasoning ability. Mathematical reasoning is a element key of mathematics and becomes a center for learning mathematics[3].

Mathematical reasoning as a basis for learning practice is useful to make a conclusion. In mathematics learning, arguments are also have an important part in making a conclusion. Reasoning ability allows students to give reasons and actions to make conclusions[4]. Student can practice making conclusions through problem solving. Problem solving can improve mathematical reasoning that used to make a conclusion[5]. The relationship between mathematical ideas is very influential in mathematical reasoning.

Mathematical reasoning is very important in forming patterns from mathematical event. Pattern identification is a mathematical reasoning process by looking for similarities and differences, concluding the narrative about recursive relations between objects[6]. Mathematical reasoning will be very effective if it is well understood by students and can be applied in mathematics learning. However, in reality, mathematical reasoning in Indonesia is not good, it is seen based on the results of TIMSS 2011 on the cognitive domain for 8th grade students that the lowest ability of Indonesian students in the domain of cognitive processes was founded in reasoning aspect (17%), and at the international level just 30% [7].

When ongoing observation, it have been seen that students are only given mathematical problems that focus on problem solving without being associated with contextual problems. In the learning process, an appropriate approach is needed so that it can make students easy to understand mathematical concepts that are appropriate to their surroundings. Approach through contextual problems is needed to built students mathematical knowledge. In relation to contextual problems, RME is one approach that based on the context of daily life. Theory of RME that proposed by Hans Freudenthal in 1970s, shows that RME involves students in learning mathematics by starting with meaningful context. Context is seen as a starting point and as a source for learning mathematics[8].

Based on interviews, the use of student worksheets has not maximal. This happens because the content of the student worksheets emphasizes more on the collection of questions and instructions that are less directing students, so that mathematical reasoning skills can't develop. While, the student worksheet preferably contains a clear procedure, about what students should do, interesting as well as the age of the student and provides continuity of directed training rather than continuation of learning[1]. Teacher also said that she has never compiled a student worksheet before.

In connection with previous research, there are several studies that have been conducted in developing student worksheets based on RME to improve mathematical reasoning. Student worksheets who developed based on RME can improve mathematical reasoning in the marine context[9]. The difference between this research and that research is that research based on the marine context while this research based on the mountain context. In addition, the development of student worksheets on Algebra material with RME approach can improve mathematical reasoning[10]. The difference is that research on algebra material while this research on probability material.

Based on the description above, it is necessary to make improvements in mathematics learning, considering mathematical reasoning is element key in mathematics learning. Developing student worksheets that are practical, valid, and effective using a realistic mathematics education approach are needed to improve their reasoning and application of mathematical concepts about probability can be mastered well.

The purpose of this study is to produce and describe the characteristics of probability students worksheets with RME approach. Students worksheets was developed expected can be improve students mathematical reasoning.

2. Methods

This research is research development. The development model used in this study was the Plomp development model which consisted of three stages, namely preliminary research phase, development or proto typing, and assessment phase[11].

Subjects trial in this study were students of class VIII Junior high school in Wonosobo consisted of 31 female students with an average age of 13-14 years old.

Preliminary research phase was preparation phase to develop the product. In this phase, identification and study about students condition in mathematics learning that took place in schools. It was viewed from the mathematical reasoning and the probability material.

Second phase was the development or proto typing. In this phase, the activity of designing and compiling student worksheet based on RME approach which was containing problems that could improve mathematical reasoning. In compiling student worksheet, the researcher made map of the needs of the student worksheet, made framework of student worksheet, and made student worksheet.

The assessment phase toward student worksheet was developed, would be rate by two expert judgments of the UNY Postgraduate Mathematics Education program lecturers, to find out the Validation. Furthermore, the practicality of student worksheet was developed was assessed based on teacher assessment sheet, students assessment, and observation sheet of the implementation learning process. Meanwhile, to measure the effectiveness of student worksheet which was developed could be seen from the results of the students mathematical reasoning test.

Data collection techniques in this study consisted of (1) giving of validation sheet consisting of 26 statements to measure the validity of the student worksheet which was developed, (2) teacher assessment sheet consisted of 14 statements, (3) student assessment sheet consisted of 16 statements, (4) Observation sheet implementation of learning used to measure the practicality of the student worksheet developed, and the test of mathematical reasoning consisted of 5 questions in essay form.

The data analysis technique was used in this study was to make the conversion table into the standard value of five criteria and converted the assessment score data into the standard value of five criteria. The basis that was used as reference was adapted from Widoyoko[12].

Validity of student worksheet was developed belong in valid criteria if the worksheets assessment results were in the minimum category "high". The student worksheets was developed said to be practical if the teacher and students assessment results in a "practical" minimum category and average percentage of learning implementation was more than 80%. Validity and practical categories were shown in Table 1 below.

Tabel 1. Validity and practical category of student worksheet

Validity Category		Practical Category		
		Interval Score		Category
Score	Category	Teacher assessment	Student assessment	
$X > 109,2$	Very High	$X > 33,6$	$X > 67,2$	Very practical
$88,4 < X \leq 109,2$	High	$27,2 < X \leq 33,6$	$54,4 < X \leq 67,2$	Practical
$67,6 < X \leq 88,4$	Moderate	$20,8 < X \leq 27,2$	$41,6 < X \leq 54,4$	Pretty practical
$46,8 < X \leq 67,6$	Low	$14,4 < X \leq 20,8$	$28,8 < X \leq 41,6$	Less practical
$X \leq 46,8$	Very Low	$X \leq 14,4$	$X \leq 28,8$	Very less practical

Effectiveness analysis was carried out to determine the extent to which student worksheets that have been made meet the effective criteria determined based on the results of a mathematical reasoning test. The criteria for the effectiveness of student worksheets were developed based on mathematical reasoning tests if the percentage of student learning completeness more than equal to 75%.

3. Result and discussion

The results of the initial research phase shown that students still depend on teacher, so that students not accustomed to built their own understanding, students feel difficult to generate mathematical ideas, students have not been very good in abstract thinking in solving problem, student worksheet has not facilitated students to construct their knowledge. Teacher have never compiled worksheets in mathematics learning and students participation in the process of mathematics learning still low.

In the development phase, development of student worksheet refers to the basic competition and indicators. It was suitable for probability topic and based on mathematical reasoning indicator. Student worksheets were developed with the RME approach which adopted the characteristics of the RME approach with a context that was close to the student environment, namely those related to mountain area because Wonosobo is the mountain area. Distribution of probability material on student worksheets in each meeting can be seen in Table 2.

Table 2. Distribution of probability material on student worksheets in each meeting

	Probability material
Student worksheet 1	Empirical probability
Student worksheet 2	Sample space and sample point
Student worksheet 3	Theoretical probability
Student worksheet 4	Differences between empirical probability and theoretical probability

In the field of mathematics, when students do mathematical activities, many aspects of reasoning are involved in it. A lot of reasoning is involved in doing mathematics even though it is hidden from view[13]. Some aspect that relate to mathematical reasoning such as make a conclusion. Reasoning ability is defined as the ability to justify choices and conclusions[14]. Other than that, argument used to make conclusion. Mathematical reasoning is defined as an argument produced to convince oneself or another about the truth of a statement[15]. So that, we must have an idea or can generate a connection between them. Mathematical reasoning is the ability to see connections/relations between mathematical ideas and apply these understandings as solutions to new problems[16]. If students can use their mathematical reasoning exactly, they can solve every problem. So that we must facilitate student to develop their mathematical reasoning in order to make they understood about mathematics concept. Problem solving is important for training students reasoning which can be used to measure mathematical concepts they have understood[17]. After student understood about mathematics concept, and they accustomed to solving problems, so that they know the pattern of a problem they solve. Mathematical pattern used in a problem supports students arguments in convincing each other that the solution of the pattern used is valid[18]. The indicators of mathematical reasoning are shown in Table 3.

Table 3. Indicators of mathematical reasoning

Aspects	Indicator
Pattern	Determines the pattern of mathematical events
Conjecture	Make conjecture
Conclusion	Make conclusion based on logical arguments

In order to developed student worksheet that can facilitate mathematical reasoning, need an approach who suitable. RME aimed to create mathematics learning more interesting and meaningful for students who are taught through contextual problems where the problem is close to the experience and knowledge of students[19]. RME uses contextual problems that can be imagined as a way into mathematics to support students as they develop mathematical abilities[20]. Providing contextual problems facilitates students to use their own models in solving mathematical problems[21]. Context are not added at the end of learning, but serve as realistic starting point to support student engagement and elicit student thinking[22]. The main purpose of the RME is to describe mathematics education to students by increasing the rediscovery of their mathematics[23].

Student worksheets were developed with the RME approach which adopted the characteristics of the RME approach with a context that was close to the student environment, namely those related to mountainous regions because Wonosobo is the mountain area. Design of student worksheets based on characteristics of RME can be shown in Table 4.

Table 4. Design of student worksheets based on characteristics of RME

Characteristics of RME	Features of student worksheets	Design
Use of Context	Use of contextual problems	Understand the problem contains context in accordance with the material to be studied in "Come to Understand"
Use of a model for progressive mathematizing	Using a mathematical model	Activity contains the process of developing a mathematical model. Students are directed to develop informal models in their own way or understand models that have been developed
Utilization of students construction results	Learning activities facilitate students to write formal results from the discussions conducted	In the conclusions and exercises section, students are given the opportunity to write down the formal results that have been done during the discussion.
Interactivity	Learning is designed in groups/ discussions	In the student activities, students work in groups to determine the problem solving of a given problem so that it is expected that there will be an exchange of ideas in each child in each group or even between groups.
Linkages	Learning activities facilitate students to use interrelations between concepts or knowledge	The relationship between learning activities facilitates students to use the interrelationship between concepts or knowledge In the part of student activities, student activities use the connection of concepts / knowledge possessed to solve problems

However, student worksheet is a tool for organizing learning activities[24]. So that suggestion or comment to better worksheet are needed. The expert assessment aims to explore comments and suggestions, both in writing and verbally as material for improvement of the student worksheets developed. Expert judgment is carried out by the validator. An assessment of the student worksheets developed was also carried out by the teacher and students to see the practicality of the product being developed. Effectiveness of the student worksheet developed by looking at the results of the test of mathematical reasoning ability. The results of the trial of student worksheets on validity, practicality, and effectiveness of student worksheets developed after analysis are described in Table 5.

Based on Table 5, the average score for validator assessment is 95 $(88.4 < X \leq 109.2)$ which is in the valid category. This shown that the student worksheet was developed has reached a valid category.

Practicality assessment of student worksheets that was developed in terms of teacher assessment sheets, student assessment sheets, and observation sheets for learning implementation. The results of the teacher's assessment of the student worksheet that was developed received a total score of 37 $(x > 33.6)$. While, the results of the students assessment for student worksheet that was developed have an average score 62,35 that belong in practical category. This shown that the student worksheets was developed meet the practical criteria according to the teacher's assessment. The learning implementation observation sheet was filled by observers to find out the percentage of learning implementation have an average 95,37% of compliance.

Table 5. The results of the validator's assessment of student worksheets

Assessment Indicator	Validator assessment score	
Content	50	56
Construction	28	24
Language	16	16
Sums	94	96
Average	95	
Category	Valid	

Furthermore, analysis of the implementation of learning based on RME steps will be presented in Table 6 below.

Table 6. Observations of learning outcomes based on RME steps

Steps in RME	Percentage of implementation (%)
Understanding contextual problems	100
Developing informal models	100
Using linkages between concepts	100
Interactivity and formalization	100
Applying formal models to other problems	100

Based on these results it can be concluded that the student worksheet was developed meets the practical criteria, it has seen from the results of the analysis of teacher assessment, student assessment, and observation of learning achievement.

Criteria for effectiveness, if the result of mathematical reasoning test more than equal with 75%. The result of mathematical reasoning test analysis are presented in Table 7.

Table 7. The results of mathematical reasoning test in the probability material

Information	Class VIII F
Top Rated	100
Lowest Score	29.17
Average	89.66
The number of student complete minimal mastery criteria	28
The number of student not compete minimal mastery criteria	1
Passage percentage	96.55%

Based on Table 7 it is known that as many as 28 students were completed with a percentage of 96,55% completeness. This shows that the student worksheet that was developed fulfills the criteria effectively, it seen from the results of the mathematical reasoning ability test because it has fulfilled the completeness criteria that more than equal with 75%.

The student worksheets was developed studied based on validity, practicality and effectiveness of student worksheets that had been developed. Quality development products meet the criteria for validity, practicality and effectiveness of Nieveen[25].

Product validity based on an assessment conducted by experts, it shown that the average score on the assessment of student worksheets was 95 from the score range 26-130. Validation results have been shown to be valid. The student worksheets that has been developed has fulfilled the valid category. In addition, the student worksheets were developed are valid because the assessment results according to the aspects of assessment which include aspects of content, construction, and language have also been included in the valid category. Thus, it can be concluded that student worksheet based on RME oriented towards mathematical reasoning ability has fulfilled the validity aspect and is suitable for use.

The practicality of student worksheets were developed in this study is based on the teacher's assessment in the very practical category with the acquisition of scores 37 as much as the score range 8-40. Furthermore, in the field trial also obtained student assessment results after participating in learning using student worksheets that have been developed into the practical category with an average score 62,35 on the range 16-80. Next, based on the results of observations of the implementation of learning that was carried out as many as six meetings using the developed student worksheets, the average learning achievement exceeded the specified target is 95,37% which indicated that the student worksheets could be implemented well and easy to use. Based on the results of the teacher's assessment, student assessment and observations of learning feasibility shown that the student worksheets based on RME oriented towards mathematical reasoning abilities had fulfilled the practical aspects and were worthy of use.

The effectiveness of the product were developed after the implementation of the field trial is based on the results of a mathematical reasoning ability test. Based on the results of field trials, obtained the results of mathematical reasoning ability tests which show that as many students complete the minimum completeness criteria, so that it has fulfilled the aspect of effectiveness and is suitable for use. Furthermore, effectiveness based on each indicators of mathematical reasoning ability is presented in Table 8 below.

Tabel 8. Test results of mathematical reasoning seen from each indicators

	Mathematical reasoning		
	Pattern	Conjecture	Conclusion
Top Rated	100	100	100
Lowest Score	66,67	50	18,18
Average	96,55	86,21	89,59
The number of student complete minimal mastery criteria	26	25	28
The number of student not compete minimal mastery criteria	3	4	1
Passage percentage	89,66	86,21	96,55

Based on Table 8, the lowest aspect of student mathematical reasoning is on find a mathematics conjecture, so that from the results the teacher can pay more attention to this aspect when wanting to develop a student worksheet based on RME oriented to mathematical reasoning. The percentage of students completeness can be seen from each aspect that has exceeded the effectiveness criteria that are more than 75%.

Student worksheets were developed, contributing to the teaching and learning of mathematics in relation to mathematical reasoning because at the beginning of each material the contextual problem is given as a starting point in learning which is characteristic of the RME approach. Furthermore, students are directed to develop an informal model facilitated to solve problems that exist in contextual problems using their own methods. So students can imagine the contextual problems given according to what they see in everyday life to be modeled into mathematics, this can grind the way students reason

In the use of student worksheets, students work in groups to determine problem solving of the given problems so that it is expected that there will be an exchange of ideas for each other in each group or even between groups. This includes the characteristics of RME, interactivity. So that in the process students are guided to discuss well.

In group discussions, when the exchange of ideas has been carried out students will find agreement on ideas that they think are the most correct. In the process of exchanging ideas, to solve problems, this activity also aims to improve students mathematical reasoning. The context presented encourages students to find a mathematics conjecture, find patterns and represent their mathematical ideas. It also contains the characteristics of RME, namely the relationship, after students are guided to discuss properly students will use the knowledge or concepts that have been possessed to be linked to problem solving from the context presented in order to reach an agreement of ideas to be made.
Furthermore, in writing the conclusions, students are given the opportunity to write down the formal results that have been done during the joint discussion. Then from the conclusions made by students, they will practice doing the questions in the exercises to explore their understanding and deepen the knowledge they already have from the results of the construction on student activities. So that it is expected that when students do the exercises the construction of students understanding that has been possessed can be stated in this section. This section which contains the characteristics of the RME (student construction utilization).

Student worksheets were developed based on the contextual issues that came from the context of mountain area, because Wonosobo was a mountain area. So that contained of context problems is close to the students environment, such as agricultural products and vegetables. Those things will make easier for students to develop models when contextual problems are given, therefore students can improve their mathematical reasoning.

4. Conclusion
Based on the study description above, it can be concluded that student worksheets based on RME oriented to mathematical reasoning have fulfilled the aspects of validity, practicality, and effectiveness. Therefore, student worksheet who developed is suitable for mathematics learning activities and can be used to improve students mathematical reasoning.

References

[1] Arends R I and Killcher A 2010 *Teaching for Student Learning: Becoming An Accomplished Teacher* (New York: Routledge)
[2] Wilis J 2010 *Learning to Love Math: Teaching Strategies That Change Students Attitudes and Get Results* (USA: ASCD Publications)
[3] Brodie K 2010 *Teaching Mathematical Reasoning in Secondary School Classroom* (New York: Springer)
[4] Fox S and Surtees L 2010 *Mathematics Across The Curriculum: Problem Solving, Reasoning and Numeracy in Primary Schools* (New York: Continuum)
[5] Vale C, Widjaja E, Herbert S, Bragg L A, and Loong E Y K 2017 Int J of Sci and Math Edu, **15(5)** 873894
[6] Jeannotte D, and Kieran C 2017 Edu Studies in Math **96(1)** 10
[7] Mullis I V S et. al. 2012 *TIMSS 2011 International Mathematics Report: Findings from IEA'S Trends in International Mathematics and Science Study at The Fourth and Eight Grades* (Chestnut Hill, MA: TIMSS and PIRLS International Study Center, Boston College)
[8] Treffers A 1989 *PsycCRITIQUES* **34** 656
[9] Ralmugiz U 2017 *Perangkat Pembelajaran Matematika SMK Kelautan Kelas X Semester Genap dengan Pendekatan Realistic Mathematics Education Berorientasi pada Penalaran, Koneksi, dan Rasa Ingin Tahu Siswa.* (Universitas Negeri Yogyakarta, Yogyakarta: eprints UNY)
[10] Kusumaningrum B 2017 *Pengembangan Perangkat Pembelajaran dengan Pendekatan RME dalam Pembelajaran Aljabar di SMP yang Berorientasi pada Kemampuan Penalaran Matematis, Kemampuan Penyelesaian Masalah, dan Kemandirian Belajar Matematika.* (Universitas Negeri Yogyakarta, Yogyakarta: eprints UNY)

[11] Plomp T *2013Educational Design Research: An Introduction*. T Plomp and N Nieveen (Eds), Educational Design Research Part A: An Introduction pp 10-51 (Enschede, OV: SLO)
[12] Widoyoko E P 2016 *Evaluasi Program Pembelajaran: Panduan Praktis Bagi Pendidik* dan *Calon Pendidik* (Yogyakarta: Pustaka Pelajar)
[13] Nickerson R S 2011 *Mathematical reasoning: Patterns, problems, conjectures, and proofs* (New York, N. Y: Taylor and Francis e-Library)
[14] Jonsson B, Norqvist M, Liljekvist Y, and Lithner J 2014 J. Math. Behav. **36** 2032
[15] Øystein, H. P 2011 *What Characterises High Achieving Students Mathematical Reasoning? in Sriraman B. and Lee K. H. (Eds)* (Netherlands: Sense Publishers) p 193–215
[16] English L D 2004 *Mathematical and Analogical Reasoning in Early Childhood* (Mahwah :Lawrence Erlbaum Assosiatec)
[17] Hunt J, Westenskow A, and Moyer-Packenham P 2017 *Edu Sci* **7(1)** 37
[18] Maher C A, Palius M F, Maher J A, Hmelo-Silver C E, and Sigley R 2014 Issues in Teacher Edu **23(1)** 3147
[19] Arsaythamby V and Zubainur C M 2014 *Procedia-Social and Behav Sci* **159** 309313
[20] Stemn B S 2017 Childhood Education **93(5)** 388393
[21] Fauzan A, Plomp T, and Gravemeijer K 2013 *The Development of an Rme-Based Geometry Course for Indonesian Primary Schools*. Plomp T and Nieveen N (Eds.) Educational design research-Part B:Illustrative cases pp 159-178 (Enschede, the Netherlands: SLO)
[22] Webb D C, Van der Kooij H, and Geist M R 2011 J of Math Edu at Teachers College **2** 4752
[23] Özdemir B G 2017 European J of Edu Studies **3(5)** 405431
[24] McArdle 2010 *Instructional Design for Action Learning (*New York: CerMcArdle)
[25] Nieveen N 1999 *Prototyping to reach product* quality (London:Kluwer Academic Publisher)

A Construct Validity of Spatial Literacy Instrument

R Anindyarini[1] and R Rosnawati[2]

[1]Undergraduate Program of Evaluation and Eduacational Research, Yogyakarta State University
[2]Department of Mathematics Education, Faculty of Mathematics and Natural Sciences, Yogyakarta State University

rosyitaanindyarini@gmail.com

Abstract. To measure spatial literacy skills, a good instrument is needed. Therefore, this study aims to 1) compile the construct of a spatial literacy ability test instrument for prospective mathematics teacher students, 2) to do validation of the construct to see the index of validity and reliability. Content validation using the V Aiken formula and the Cronbach Alpha coefficient were used to test the validity and reliability. The results showed that: 1) the construct of spatial literacy in the form of operational definitions, six indicators, 17 sub indicators and 34 questions in the form of multiple choices has shown a form of spatial concept integration with literacy, 2) spatial literacy test instruments are feasible because they meet index validity requirements V Aiken between 0.67 - 0.89. The test kit also states fit for 34 items, because it is in the acceptance limit range of >0.77 to <1.30. In addition, the level of difficulty on the test using a logit scale that produces 13 categorized items is very easy, 6 items are easy, and 15 items are difficult. In the study, the value of reliability was 0.704 (> 0.5). While the Standard Error of Measurement (SEM) which is a value that measures how precise the value obtained is equal to 1.287. Henceforth, this test kit can be used by the academic community at universities or other agencies concerned to measure the level of spatial literacy of prospective mathematics teachers.

1. Introduction

In the early development, literacy was defined as the ability to use language and images in various forms to read, write, listen, speak, see, present and think critically about ideas [1]. Furthermore, mathematical literacy is considered as important as reading literacy and writing literacy is considered dominant. If literacy is the basis for all learning, mathematics literacy is also important if someone wants to understand the information that exists in society [2].

Based on De Lange's theory, mathematical literacy abilities are not only limited to aspects of numeracy. He revealed that mathematical literacy includes spatial literacy, numerical and quantitative literacy [3]. Of the three aspects, research on spatial literacy considered as not yet done. This is one of them because the instrument for measuring spatial literacy is not yet available. Some mathematical literacy research more often leads to quantitative literacy. One of them was done by Setiani and Rafianti, they conducted a study to determine the effect of the level of visual-spatial intelligence on the quantitative literacy of prospective mathematics teacher students [4]. They use the notion that quantitative literacy is similar to mathematical literacy, where the equation between quantitative literacy and mathematical literacy is not only focused on minimal knowledge in mathematics but also includes "doing mathematics" that is using mathematical concepts in other fields and in aspects of daily life [5]. The instrument of quantitative literacy questions he developed was in the form of a test in the form of a description so that researchers could easily see the process of answering the test questions.

In addition, Firnanda et al. has also conducted research to identify the quantitative literacy abilities of students studied from the aspect of content change and relationship in algebra in junior high school [6]. In his research, he only made a written test amounting to 2 questions, which then the results of student work were analyzed specifically examined from the four aspects of quantitative literacy, that is interpretation, representation, calculation, and communication. Similar to the study, Oktaviani et al. also conducted student quantitative literacy analysis on statistical material but in aspects of content uncertainty and data [7]. Hence, the instruments he developed were two questions that were adopted and adapted from the PISA questions in 2009 and 2012 related to the content uncertainty and data aspects.

Related to spatial, there is a test that has been developed and is often used by several studies. However, this measuring instrument known as the TOSA (Test of Spatial Assembly) was developed to measure spatial abilities for children aged three years [8]. Thus, the usefulness of this test is only valid for spatial growth and development in three-year-olds. The research related to spatial literacy has been done by Maharani and Maryani. They conducted experiments using media maps to improve the spatial literacy of fourth grade students in elementary school. The instruments used were in the form of spatial literacy tests consisting of pre-test (initial test) and post-test (final test) [9].

If the context is observed, the study is more focused on the geographic subjects of elementary school students. Maps are considered as media that are able to help students understand spatial knowledge. Meanwhile, in mathematical literacy, building space and space become a whole aspect that can support understanding of geometry. This correct understanding of geometry can be built by mastering spatial literacy. Thus, spatial literacy is considered important to be held by prospective mathematics teacher students. Bernarz also mentioned that the United States National Academy of Sciences found that spatial thinking is a key component in spatial literacy, where spatial literacy has been considered to be at the heart of several major discoveries in the world of science and has become a fundamental component of several professions and career [10]. However, so far, the spatial literacy test is less available because the development of spatial literacy constructs is considered still small.

De Lange's definition of spatial literacy has actually been revealed, where spatial literacy is the ability to use spatial thinking skills to visualize ideas, situations, and problems in everyday life and the world around us [11]. This definition is used as a conceptual definition in this study. Meanwhile, in this case, the researcher seeks to establish operational definitions and constructs of spatial literacy from the results of observations on some conceptual definitions of literacy and spatial ability. There are three definitions of literacy and twelve definitions of spatial ability from twelve experts. Some of these definitions have mentioned the dimensions or indicators contained therein, but there are also definitions that have not mentioned the indicators contained therein. Thus, that researchers estimate the indicator points of the definition of spatial ability.

Mizzi also observed a number of definitions found in spatial red threads that focus on spatial ability like navigating and manipulating space. Spatial tasks also require students to develop ways to solve spatial problems [12]. There are no definitions and dimensions/indicators that have been stated by spatial experts previously that are integrated with the concept of literacy where several items of tests become contextual and the problems presented occur in everyday life. Thus, based on the observations of researchers, an operational definition of spatial literacy is formed where spatial literacy is the ability of a person to use spatial thinking to build, visualize, understand, create, change the space well in the mind, and communicate it to solve problems in life, which is shown in the results of spatial literacy tests. After estimating the indicator points from the twelve definitions of spatial ability, the researcher takes the slices of these indicators. The result is that there are 6 (six) main indicators that are considered capable of forming spatial literacy skills in a person. The six indicators include: 1) Spatial Perception, 2) Spatial Visualization, 3) Mental Rotation, 4) Spatial Relations, 5) Spatial Orientation, and 6) Spatial Communication. Furthermore, these six indicators produce 34 items in the form of multiple choices. Multiple choice which has five A, B, C, D, and E answer options, it is necessary to conduct a feasibility test before it is used to measure.

In several studies, it was said that a good test instrument is an instrument whose points are well designed and empirically evaluated to ensure accuracy and user information. Miller, Linn, &Gronlund stated that good tests must fulfill three characteristics, which are validity, reliability, and reusability [13]. Validity is if the instrument actually measures what should be measured. The empirical validity test consists of three groups of validation procedures; content validation, construct validation and validation based on criteria [14].

Besides that, the reliability of the test also needs to know. Reliability is one of the criteria in the development of both test and non-test instruments. If the reliability of the instrument is high, then an instrument is very good for use in measurement. Instruments are considered consistent if the instrument is carried out from time to time but has the same value. In this internal consistency approach, the reliability estimation procedure is applied through an analysis of the distribution of grain scores or the score distribution of the item groups, not an analysis of the test scores. The several ways of estimating reliability with internal consistency include: 1) test division, 2) brown spearman formula, 3) alpha formula, 4) rulon formula, 5) several formulas for Richardson brothers, 6) halves with different lengths, 7) formula Kristof for split three, fund tau 8) hoyt formula [14]. Thus, the validity and reliability test needs to be done to determine the feasibility level of the spatial literacy test instrument for prospective mathematics teacher students.

2. Methods

This study involved 150 students majoring in mathematics education for empirical trials, which were devoted to those who were studying in the first year. The researcher considered that mathematics education majors who were in their first year had taken geometry subjects, especially building space, during junior high and high school, and had not gained much geometry during the lecture. Those who study in mathematics education are also considered to have an interest in mathematics and a good initial spatial ability capital than other majors.

The feasibility test of the construct of the spatial literacy test instrument used in this study is the test of content validity and reliability. Content validity is the extent to which the elements of a measuring instrument are truly relevant and represent representations of constructs that are in accordance with the objectives of measurement [15]. One of the statistics that shows the validity of item content is by using V Aiken's analysis [16] [17]. Validation involving 3 (three) experts in the field of mathematical geometry and measurement with 4 (four) scale validation instructions, which are very appropriate, appropriate, quite appropriate, and not appropriate. The data from the evaluation of the three validators were analyzed by calculating the content validity coefficient using the following V Aiken formula:

$$V = \frac{\sum S}{[n(c-1)]}$$

Where V is the rater agreement index regarding item validity, S is the score set for each rater minus the lowest score in the category used (s = r - lo, r = score of rater pickup category, lo lowest score in the scoring category), n is the number of rater and c is the number of categories the validator can choose. An item or device can be categorized as validity based on its index. If the index is less than or equal to 0.4 it is said that the validity is lacking; 0.4 - 0.8 said validity is moderate; and if greater than 0.8 is said to be valid [18].

The reliability test used in this study is to use the Alpha formula which will produce the Cronbach's Alpha coefficient. The procedure for estimating reliability with Cronbach's Alpha computational coefficients can be directly processed with SPSS from item score distribution data without dividing or dividing the items into groups. In addition, in this study also tested the suitability of items with the Rasch model using Quest software to determine the level of difficulty of the item. In this case, the researcher determined the difficulty level category according to Sumintono and Widhiarso with the provisions presented in table 1 below [19].

Table 1. Categories of Item Difficulty Index Based on Logit Scale

Index of Difficulty	Category
$x > 1$	Very Difficult
$0 \leq x \leq 1$	Difficult
$-1 \leq x \leq 0$	Easy
$x < -1$	Very Easy

3. Result and Discussion

3.1. Construct of Spatial Literacy

Based on the operational definition of spatial literacy that is formed, that is the ability of a person to use spatial thinking to build, understand, create, change, and visualize the building of space well in the mind, and communicate it to solve problems in daily life; developed constructs of spatial literacy instruments consisting of 6 (six) indicators, 17 sub-indicators, and 34 questions. Thus, the grid of spatial literacy test instruments is presented in the following Table 2.

Table 2. Spatial Literacy Test Instrument Framework

Theory	Indicator	Sub Indicator	Item Number
Spatial Literacy of Candidates for Mathematics Teachers	A.Spatial Perception	A.1 Determine the position or quantity of depth of a building space (in the context of personality, education, work, public, or scientific)	9
		A.2 Determine the number of compilers to build space that forms a building space	15
		A.3 Determine the position and content of a building space against vertical / horizontal references (in the context of personality, education, work, public, or scientific)	1, 17, 22
	B.Spatial Visualization	B.1 Determine an irregular shape building if the nets are known	25
		B.2 Determine an irregular shape space if the nets are known	2, 24
		B.3 Determine the shape of the net if a regular spatial structure is known	23
		B.4 Determine the shape of the net if an irregularly shaped facet is known	5, 14
		B.5 Mention some of the building spaces that make up another building	11
	C. Mental Rotation	C.1Determine the shape of the building if the wake is rotated	7, 12, 30, 33
	D. Spatial Relation	D.1 Connecting the visual parts on the side of the building in a regular aspect space if it is transformed into two dimensions	3, 29
		D.2 Connecting the visual parts to the side of the building in an irregular aspect space if it is transformed into two dimensions	10, 18, 32

E. Spatial Orientation	E.1 Determine the surface area of the building	6, 28
	E.2 Determine the shape of the space formed by the intersection/ intersection of two fields in a building space	34
	E.3 Determine a visual object that is contextual when viewed from various perspectives	19, 27
	E.4 Determine irregular visual objects in building space when viewed from various points of view	21, 26
F. Spatial Communication	F.1 Determine the position of objects against other objects in open or closed spaces (in the context of personality, education, work, public, or scientific)	4, 13, 20
	F.2Describe the route of an object on a map / floor plan with the help of wind direction and clockwise (in the context of personality, education, work, public, or scientific)	8, 16, 31

3.2. Content Validity

Data from the assessment of three predefined validators were analyzed by calculating the content validity coefficient using the V Aiken formula. The following is a graph of the distribution of V Aiken values for 34 items.

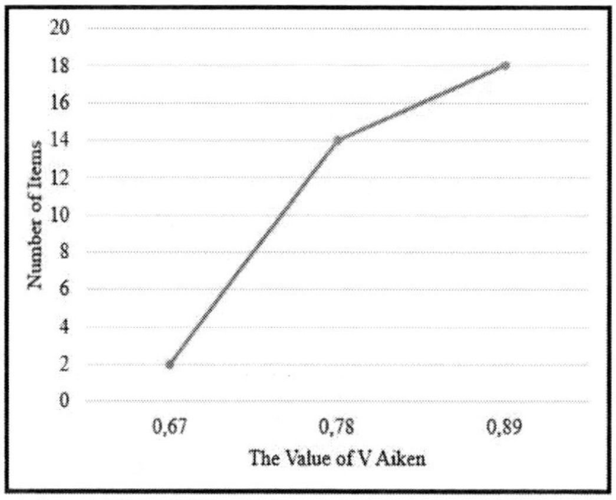

Figure 1. Graph of V Aiken' Items

Based on Figure 1 above, it can be seen that the V Aiken value for 34 items in the spatial literacy test is in the range 0.67 - 0.89. There are 16 items that have a validity index of 0.67 and 0.78 included in the medium valid category. While the remaining 18 items have a validity index of 0.89, which means that it belongs to the high validity category. When referring to the Aiken's V table, the minimum value of Aiken's V resulting from the evaluation of three validators on a scale of four is between 0 and 1.00 [16]. On the other hand, Azwar said that the number 0.67 obtained from the calculation of the V Aiken formula can be interpreted as an adequate coefficient for the item, and is considered to have good content validity and supports the value of the validity of the overall test contest [14].

3.3. Item Fit Analysis

The next step is to test the compatibility of the items with the model (fit test). The item is fit if MNSQ's INFIT value is in the range ≥0.77 to ≤1.30 [20] [21]. The following are the fit test results for 34 items on the spatial literacy ability test.

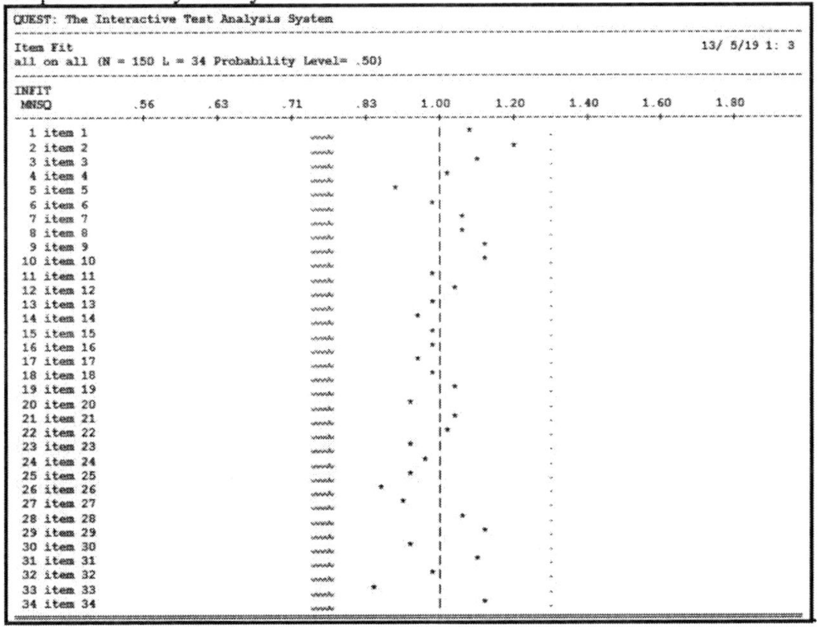

Figure 2. Result of Item Fit Analysis on Spatial Literacy Test Instrument

Based on the picture above, it appears that 34 multiple choice questions on the spatial literacy ability test instrument for mathematics teacher candidate students are declared fit or suitable with Rasch models with acceptance limits of≥0.77 to ≤1.30. Hence, all items have been declared valid.

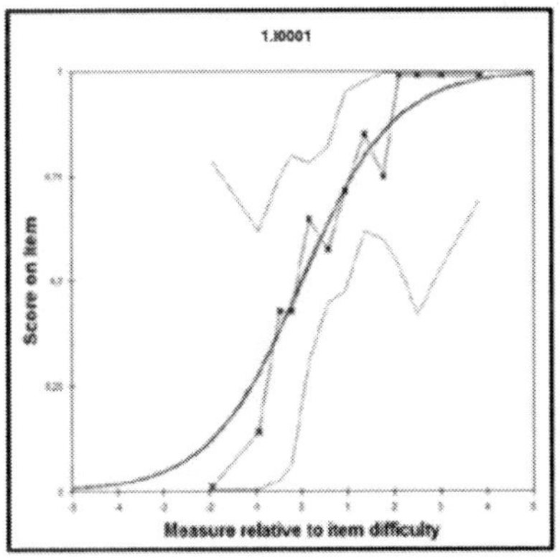

Figure 3. Graph of ICC for Item Number 1

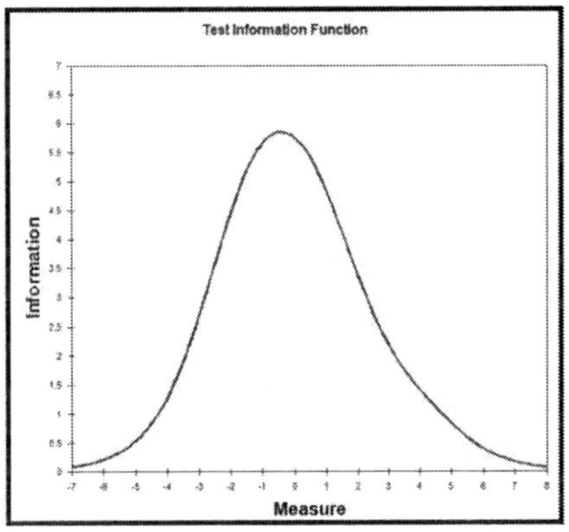

Figure 4. Graph of *Test Information Function* (TIF) for Spatial Literacy Test Instrument

Figure 3 shows an analysis of grain compatibility using Winstep also provides illustrations in curve graph form. The graph illustrates the answer patterns of prospective mathematics teacher students. For example, the graph is the answer pattern of prospective mathematics teacher students in answering item number 1 on the spatial literal ability test. The answer pattern is in the infit reception and outfit space. The graph of item suitability analysis with the Rasch model above shows that the items developed provide a good answer pattern. While, the graph in figure 4 shows that the curve is increasingly pointing towards the center. Sumintono and Widhiarso explained that the more curves pointing towards the center, indicating that the instrument was considered to be good for measuring the level of spatial literacy abilities of the average student candidate for mathematics [19]. Based on this explanation, test instruments provide optimal information when given to prospective mathematics teacher students with an average level of spatial literacy ability.

3.4. Level of Difficulty Item

The level of difficulty in the spatial literacy ability test instrument consists of 2 (two) discussions, that is the level of difficulty of each item and the level of difficulty of the test based on indicators of spatial literacy ability. Because the items in the spatial literacy ability test match the Rasch model, the grain characteristics that appear are the difficulty level of each item. The level of item difficulty in the Rasch model is the same as the level of difficulty of classical test theory, that is the comparison between the correct answer and the number of questions tested (odd-ratio). The difference is that this opportunity value is scaled by entering a logarithmic function. The logit estimation result of the odd-ratio is called logit or W Score or Measure value. Thus, information covering the level of difficulty of the items on the spatial literacy test instrument is summarized in Figure 5 below.

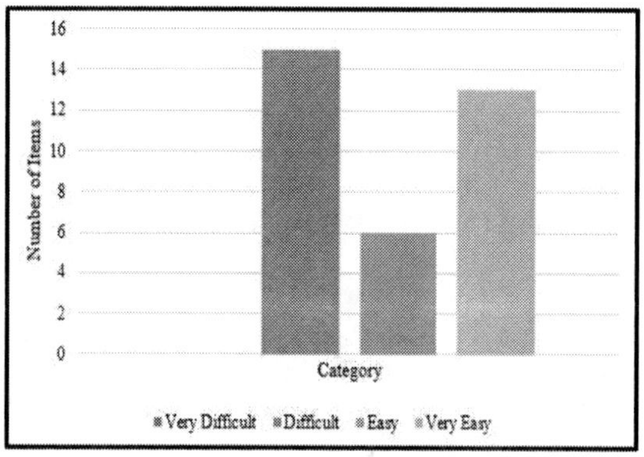

Figure 5. Distribution of the Difficulty Level of Question Items on Spatial Literacy Test Instruments

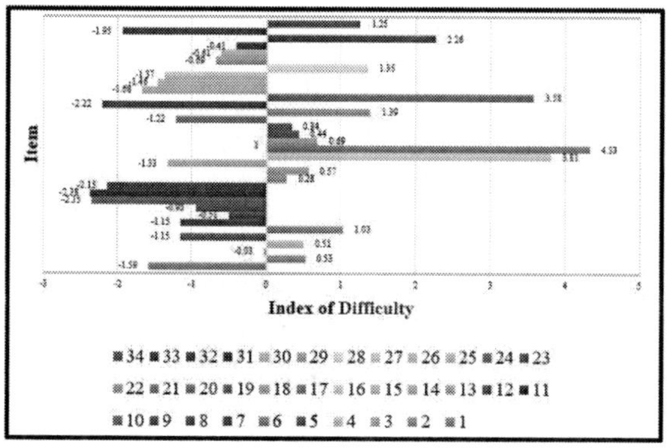

Figure 6. Index of Difficulty for All Item

If observed in Figure 5 above, then there are 13 categorized items very easily, 6 items are easy, 15 items are categorized as difficult and no items categorized are very difficult in this test. For more detailed information, the index of the difficulty of the test items is presented in Figure 6 below.While on Figure 6, it can be seen that the items in the test instrument for spatial literacy ability have difficulty levels which are in the range -1.465 up to 0.886. It means that the level of difficulty of the items in the test of spatial literacy abilities is different. In addition to each item, the average score for the difficulty level of spatial literacy items if viewed from the indicators can be seen in Figure 7 below.

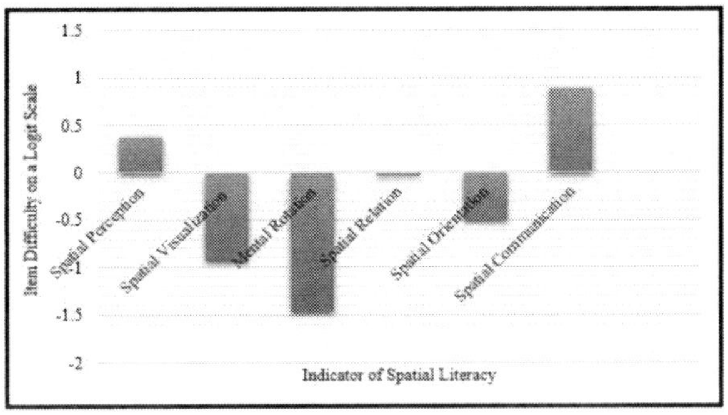

Figure 7. Index of Difficulty for All Indicator of Spatial Literacy

Figure 7 above shows a comparison of the level of difficulty of the items from each indicator of spatial literacy. From the picture, it can be observed that the items on the mental rotation indicator are considered the easiest among the items on the other indicators. According to the results of interviews with two test participants, in general mental rotation indicators are indeed easy because they only imagine the rotation of the objects listed in the problem. Even though the objects contained in the question are irregular spaces, to choose the answer to the option is not difficult.

3.5. Reliability

The technique used to calculate the estimated reliability coefficient in this study is to use the Cronbach's Alpha coefficient. The following results of the reliability analysis of the instrument test spatial literacy abilities.

Table 3. Estimated Reliability of Instruments for Spatial Literacy Test

Reliability Statistics		
Cronbach's Alpha	Cronbach's Alpha Based on Standardized Items	N of Items
,704	,710	34

Based on the above calculation results, it is known that the spatial literacy test instrument has a reliability index of 0.704. According to Allen and Yen, using Cronbach's Alpha, the reliability can be said to be high if more than 0.7 [22]. Thus, the spatial literacy test instrument in this study has a high degree of permanence and can be used to collect data. If the reliability index has been obtained, the next step is to find the value of the Standard Error of Measurement (SEM). From the SEM value, it can be concluded whether the test that measures the spatial literacy of students can be reused or not. SEM calculations use the following formula:

$$SEM = SD\sqrt{1-r_{11}}$$

The SD value is the standard deviation test. In this case, the standard deviation value has been obtained from SPSS output which is the result of analysis of 34 items that have been fit, where the standard deviation of the spatial literacy test instrument is 4.34937. The following is a statistical scale output, which contains the standard deviation value.

Table 4. Statistic Scale

Mean	**Variance**	**Std. Deviation**	**N of Items**
23,6800	18.917	4,34937	34

Thus, the calculation of the Standard Error of Measurement (SEM) value for test questions that measure the spatial literacy of mathematics education students is as follows.

$$SEM = SD\sqrt{1-r_{11}}$$
$$= (4.34937)\sqrt{1-0.704}$$
$$= 1.28741352 \approx 1.287$$

Based on the SEM values above, it can be said that if the test questions that measure spatial literacy are used again, then the score that will be obtained by the respondent that is the student lies in the interval up to $X_r - 1.287$ and up to $X_r + 1.287$.

4. Conclusion

This study produced a spatial literacy construct in the form of operational definitions, six indicators, 17 sub-indicators and 34 multiple choice questions. This spatial literacy test device has been tested theoretically and empirically, so it is feasible to use it to measure spatial literacy abilities. Spatial literacy includes six indicators, namely spatial perception, spaisal visualization, mental rotation, spatial relations, spatial orientation, and spatial communication. The construction of a spatial literacy construct is the result of a study of some of the theories of previous spatial experts. McGee emphasized that spatial ability has two main factors, that is spatial visualization and spatial orientation [23]. While Guay divides spatial abilities only into three dimensions, which are development, view, and rotation [24]. The instruments that he has developed are 36 items in the form of multiple choices and all of them have no context. Maier argues that spatial literacy includes four major aspects that intersect, mentally dynamic, mentally static, mental position and mental processes. Of the four aspects, it includes five indicators as compiled by researchers, except for spatial communication indicators [25]. Unfortunately, there are no dimensions/indicators that have been suggested by previous spatial experts that are integrated with the concept of literacy where some of the measuring test items are contextual. So the preparation of this structure has shown a form of integrating spatial concepts with literacy.

Based on the content validity test with V Aiken from 3 (three) validators, the validation value was 0.67 to 0.89. The test kit also states fit for 34 items, because it is in the acceptance limit range of >0.77 to <1.30. In addition, the level of difficulty on the test using a logit scale that produces 13 categorized items is very easy, 6 items are easy, and 15 items are difficult. In the study, the value of reliability was 0.704 (> 0.5). Meanwhile, the Standard Error of Measurement (SEM) which is a value that measures how precise the value obtained is equal to 1,287. Based on the description, it can be said that the spatial literacy construct is feasible because it meets the requirements of validity and reliability. For further research, it is expected to be able to test the feasibility of spatial literacy test instruments in a larger population, so that the value of validity and consistency of the test is getting better. Then, this test tool can be used by the academic community at universities or other agencies with an interest to know how the level of spatial literacy skills of a person, especially mathematics education students, who in fact are prospective mathematics teachers.

References

[1] Abidin Y, Mulyati T., and Yunansah H 2017 Pembelajaran Literasi Stratesi Meningkatkan Kemampuan Literasi Matematika, Sains, Membaca, dan Menulis(Bandung: Bumi Aksara)

[2] Abdussakir 2018LiterasiMatematisdanUpayaPengembangannyadalamPembelajaran di Kelas(Prosidingdalam Seminar PendidikanMatematika "MenanamkanPendidikanKarakter (AkhlaqulKarimah) danKesadaranLiterasiMatematikaSiswaMelaluiPembelajaranMatematika)(Sumenep : Prodi PendidikanMatematika STKIP PGRI)

[3] De Lange J 2006 Mathematical Literacy for Living from OECD-PISA Perspective *Tsukuba. Journal.Of. Educational. Study.In. Mathematics*.**25**13–35

[4] Setiani, Y. and Rafianti I 2018Pengaruh Tingkat Kecerdasan Visual – Spasial terhadap Literasi Kuantitatif Mahasiswa Mahasiswa calon Guru Matematika. *Kreano – Jurnal. Matematika Kreatif.Inovatif.* **9** 38-46
[5] Kusumah Y S 2011 Literasi Matematis (Prosiding Seminar Nasional Pendidikan MIPA: Pengembangan Pembelajaran MIPA Berorientasi Soft Skill) Bandar Lampung
[6] Firnanda P, Sugiatno, and Nursangaji A 2015 Literasi Kuantitatif Siswa dikaji dari Aspek Content Change and Relationship dalam Aljabar di SMP. *Jurnal.Pendidikan.Dan.Pembelajaran.* **4** 1-11
[7] Oktaviani AC, Jamiah Y, and Suratman D 2017Analisis Literasi Kuantitatif Siswa Aspek Konten *Uncertainty and Data* pada Materi Statistika. *Jurnal.Pendidikan.Dan.Pembelajaran.* **6**1-11.
[8] Verdine B N, Do link off R M, Pasek K H, and Newcombe N S 2017 Spatial Skill, Their Development, and Their Link To Mathematics. *Monographs.Of. The. Society.For. Research.In. Child. Development.* **8** 7-30
[9] Maharani W and Maryani E 2015 Peningkatan *Spatial Literacy* Peserta Didik Melalui Pemanfaatan Media Peta. *Jurnal Pendidikan Geografi.***15** 46-54
[10] Berdnarz S W and Kemp Karen 2011 Understanding and Nurturing Spatial Literacy. *Procedia. Social.And. Behavioral. Sciences.***21** 18-23
[11] De Lange J 2006 Mathematical Literacy for Living from OECD-PISA Perspective. *Tsukuba. Journal.Of. Educational. Study.In. Mathematics****25*** 13–35
[12] Mizzi A 2017 *The Relationship between Languange and Spatial Ability: An Analysis of Spatial Languange for Reconstructing the Solving of Spatial Task* (Essen: Springer Nature)
[13] Miller M D, Linn R L, and Gronlund N E 2009*Measurement and Assessment in Teaching* (Upper Saddle River: Pearson Educational)
[14] Azwar S2017) *Reliabilitas dan Validitas* (Yogyakarta: Pustaka Pelajar)
[15] Haynes S N, Richard D C S, and Kubany E S1995 Content Validity in Psychological Assessment: A Functional Approach to Concepts and Methods.*Psychological. Assessment,* **7** 238-247
[16] Aiken L R 1985 Three Coefficients for Analyzing the Reliability and Validity of Ratings. *Educational.And. Psychological. Measurement,* **45** 131-134
[17] Yadiannur, Mitraand Supahar 2017 Mobile Learning based Worked Example in Electric Circuit (WEIEC) Application to Improve the High School Students' Electric Circuits Interpretation Ability. *International. Journal.Of. Environmental.And. Science. Education* **12** 539-558
[18] Retnawati H 2016 *Validitas Reliabilitas dan Karakteristik Butir* (Yogyakarta: NuhaMedika)
[19] Sumintono B and Widhiarso W 2015 *Aplikasi Pemodelan RASCH pada Assessment Pendidikan* (Cimahi: Tim Komunikata Publishing House)
[20] Subali B and Suyata P 2012 *Pengembangan Item Tes Konvergen Dan Divergen: Penyelidikan Validitasnya Secara Empiris* (Yogyakarta: Diandra Pustaka Indonesia)
[21] Supahar 2014 The Estimation Of Inquiry Performance Test Items Of High School Physics Subject With Quest Program (*Proceedings In International Conference On Research, Implementation And Education Of Mathematics And Sciences*) (Yogyakarta: Yogyakarta State University)
[22] Allen M J and Yen WM 1979 *Introduction to Measurement Theory* (Monterey: Brooks/ Cole)

Why is Mathematical Representation Difficult for Students?

S Saifiyah[1] and H Retnawati[2]

[1]Postgraduate of Research and Education Evaluation, Universitas Negeri Yogyakarta, Indonesia
[2]Department of Mathematics, Faculty of Mathematics and Natural Science, Universitas Negeri Yogyakarta, Indonesia

sofisaifiyah20@gmail.com

Abstract. Mathematical representation is one of the mathematical skills that can help students present mathematical problems in various forms. However, students often face difficulties in doing so. This study aims to investigate the causes of difficulties faced by students in solving mathematical representation questions on trigonometry of the 10th grade. This is a qualitative case study. Data were collected both by documentation of research results in the form of 6 essay questions on mathematical representation and by interviews. The research participants were 10 high school students in the 10th grade of science program in Cirebon Regency, West Java, Indonesia. Through this case study, found the causes of difficulties experienced by students in solving mathematical representationquestions. Most difficulties (10 students) occurbecause students cannot represent mathematical expressions/symbol to solve questions in the form of images. On the other hand, the least difficulties (6 students) occurbecausestudents cannot using image representation to solve written text questions.

1. Introduction

Mathematical skills that students should learn include: (1) problem solving, (2) reasoning and proof, (3) communication, (4) connections, and (5) representation [1]. In line with that, Callison revealed that students need to master mathematical skills, such as solving problems, making connections, understanding various representations of mathematical ideas, communicating ideas, and explaining reasoning [2]. Mathematical representation is one of the important mathematical skills that students need to master because it serves as the foundation for students in order to comprehend and utilize mathematical ideas. Therefore, it is necessary to investigate if students find mathematical representation challenging [3]. NCTM argued that mathematical representation is the most important and the most needed skill to be developed as mathematical representation functions as the basic skill in learning mathematics [4].

Representation refers to the results of students' interpretation of a problem used to help find solutions [5]. Aspects of mathematical representation include pictures, graphic, tables, equations, mathematical symbols, words, and writings. Due to a lot of aspects of mathematical representation, students need to repeatedly learn how to solve problems containing those aspects with the purpose of developing their ability to organize ideas when solving the problems. Students can use representation to find and create a way of thinking in communicating their mathematical ideas from an abstract concept into a concrete one that enables them to understand the problems more easily [6].

According to NCTM, one of the students' weaknesses in learning mathematics is that they are not able to connect mathematical concepts with daily experiences. Learning mathematics is often viewed to be too formal and not to really associate with meaning, understanding, and application of mathematical concepts [7]. These perceptions signify that problems occur when students attempt to solve applicative essay questions, especially questions about mathematical representation. To find out problems and difficulties that students encounter, students are given several questions that tested students' mathematical representation skill. The test is a set of instruments used to take measurements [8]. The measurements in this regard refer to the measurements of mathematical representation skill to identify difficulties that students face in solving mathematical representation questions. If no actions are taken to tackle these difficulties, it is possible that students will experience more difficulties in other areas of mathematics and mathematical skills. Some studies mention that students face mathematical difficulties because they often miscalculate [9], are unable to comprehend mathematical symbols [10], lack counting skills [11], and lack skills to understand mathematical language [12].

The appropriate way to investigate students' difficulties is to examine students' answers to the questions in a test. After the students' difficulties are successfully examined, the results can be used by teachers to review the points of difficulties. The study by Fauzan that aims to discover the points of students' difficulties in high school mathematics subject found that students were detected to have difficulties in geometry as they considered it the most difficult area of mathematics [13]. The difficulties were specifically related to the concepts and procedures in solving mathematical questions.

The results of interview with 10 students in the 10th grade indicated that besides geometry, another difficult area of mathematics for them was trigonometry. Trigonometry is a lesson discussed in compulsory mathematics for the 10th grade students. The topics include trigonometric ratios in various quadrants, trigonometric ratios in related angles, laws of sines and cosines, and trigonometric functions. Based on the interview results, the most difficult topic was trigonometric functions, although some students argued that all topics of trigonometry were difficult to be learned.Based on the explanation above, this study aims to examine why students experience difficulties in solving essay questions about mathematical representation. Furthermore, this study also investigates further the types of difficulties that students face in solving essay questions about mathematical representation.

2. Methods

A case study uses a qualitative approach in order to obtain information about a case thoroughly and deeply [14]. The research participants were 10 high school students in 10th grade in Cirebon Regency, West Java. The participants were selected based on the criteria according to the study [15], then taking the subject using a purposive sampling technique. The criteria were the 10th grade students who had already learned trigonometry and experienced difficulties in solving mathematical representation questions, specifically about trigonometry. Data were collected by observation of the results of

Table 1. Indicators of Mathematical Representation Skill.

No	Indicators of Representation
1	Use pictures in solving expressions/mathematical symbols (Mathematical Expressions/ Symbols → Pictures)
2	Use pictures in solving questions in the form of written texts (Written Texts of Mathematics → Pictures)
3	Use expressions/mathematical symbols in solving questions in the form of pictures (Pictures→ Mathematical Expressions/ Symbols)
4	Explain expressions/mathematical symbols in solving questions in the form of expressions/mathematical symbols (Mathematical Expressions/ Symbols → Mathematical Expressions/ Symbols)
5	Use expressions/mathematical symbols in solving questions in the form of mathematical texts (Written Texts of Mathematics → Mathematical Expressions/ Symbols)

students' tests on mathematical representation and interviews. Data were later analyzed using Miles & Huberman theory that consists of three stages which are data reduction, data display, and conclusion drawing [15]. The following are indicators of mathematical representation ability used to identify difficulties experienced by students in working on mathematical representationproblems.

3. Results

Students' difficulties were observed and measured by using the indicators of mathematical representation skill on each question given to 10 students of science program in 10th grade. The questions were developed by the researchers. Table 2 below illustrates the numbers of difficulties experienced by 10 students in solving mathematical representation questions.

Table 2. Numbers of 10 Students' Difficulties in Mathematical Representation.

Number	Indicators of Competence	Indicators of Representation	Difficulties
1A	Illustrate the values of trigonometric ratios in quadrant III	Mathematical Expressions/ Symbols → Pictures	7 students
1B	Determine the values of cosecant and cotangent ratios if the value of secant ratios is known in quadrant III	Mathematical Expressions/ Symbols → Mathematical Expressions/ Symbols	7 students
2	Determine law of cosines if two sides and an angle of a triangle are known	Written Texts of Mathematics → Mathematical Expressions/ Symbols	8 students
3	Determine the values of trigonometric functions using a unit circle with a >360° angle	Mathematical Expressions/ Symbols → Mathematical Expressions/ Symbols	7 students
4	Illustrate contextual problems of trigonometric ratios	Written Texts of Mathematics → Pictures	6 students
5	Analyze the changing graphs of trigonometric functions as a result of changes in constants in the function of y=a sin nx	Pictures →Mathematical Expressions/ Symbols	10 students

Table 2 shows that the question number 5with the indicator of mathematical representation "using mathematical expressions/symbols in solving problems in the form of pictures" is the most difficult question for students because all students experience difficulties. On the other hand, the least difficult question about mathematical representation for students is the question number 4. In question number 4, a student faced a difficulty in using mathematical expressions/symbols. The following is the explanation of students' difficulties in solving mathematical representation questions based on the indicators of mathematical representation employed in this study.

3.1. Using pictures to solve mathematical expressions/symbols

The difficulties in the indicator of the first representation are presented in Figure 1(a) and 1(b) for the question number 1A. The indicator of competence is to determine the values of cosecant and cotangent ratios if the value of secant ratio in quadrant III is known. The instruction of the question number 1A is the value of sec $\alpha = -25/7$ and $\pi < \alpha < 3/2\pi$. Based on the mathematical expressions/symbols in the question, students were asked to illustrate the comparative value of secant α in the Cartesian plane. However, students were not able to write mathematical expressions/symbols as instructed by the question. This difficulty happened to 7 out of 10 students. Therefore, when students are instructed to illustrate the value of trigonometry comparisons of the information contained in the question, 7 students experienced difficulties in drawing it as seen in Figure 1(b).

Figure 1(a) shows students' works indicating that they faced difficulties in writing mathematical expressions/symbols. In this regard, students made mistakes or were careless in writing mathematical expressions/symbols. Another difficulty, students find it difficult to understand the problem of mathematical expressions/ symbols is that students did not write any mathematical symbols for the instruction "known". This occurred because students are not accustomed to do essay questions in a structured and organized manner in which students are required to write the complete answer based on what is known in and asked by the question. According to previous studies, students faced difficulties because they lacked comprehension of the basic concepts of trigonometry and considered the materials difficult [16].

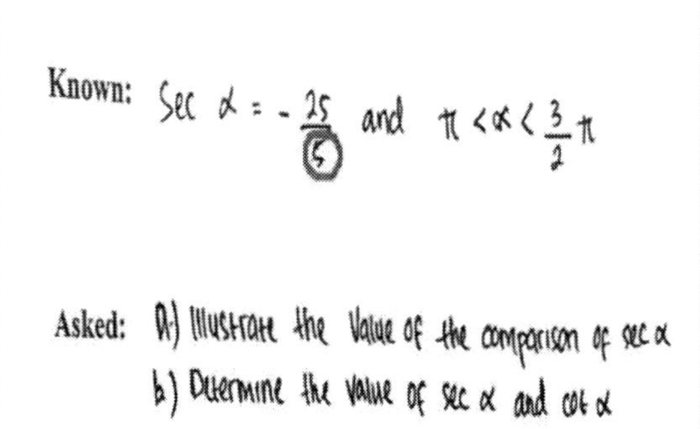

Figure 1(a). Difficulties in using mathematical expressions/symbols.

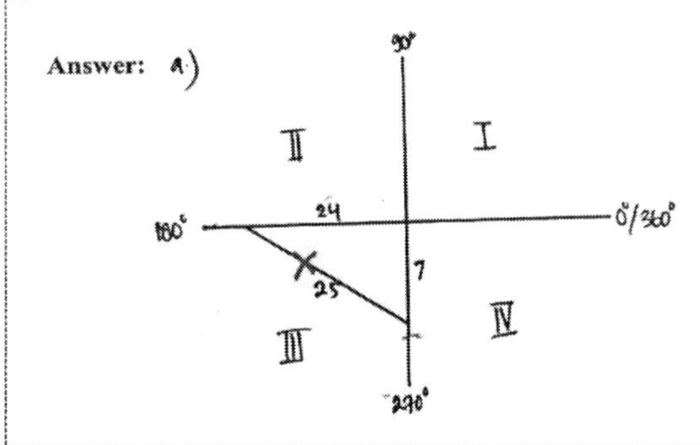

Figure 1(b). Difficulties in drawing mathematical expressions.

Figure 1(b) shows students' works in drawing mathematical expressions/symbols based on what is known in the question. In that figure, students faced a difficulty in drawing the comparative value of secant α. The figure made by students is not seen in accordance with the concept of trigonometry. As a result of this difficulty, students cannot use representation rules for the next questions related to this question [17]. Based on the results of the study by Usman, students should be given adequate opportunity to learn how to do questions in a structured manner with the purpose of helping them improve their mathematical representation skill [18].

3.2 *Describing mathematical expressions/symbols in solving questions in the form of mathematical expressions / symbols*

Based on the second indicator, the point of difficulty is in the question number 1B. The indicator of trigonometric competence in the question number 1B is to determine the values of cosecant and cotangent ratios if the value of secant ratio in quadrant III is known. The instruction of the question number 1B is "determine the values of cosecant α and cotangent α if sec α = $-25/7$ and $\pi < \alpha < 3/2\pi$". Before determining the values of cosecantα and cotangentα, students wrote the information of expressions/symbols in the "known" part. However, 7students did not write complete information as seen in Figure 2(a).

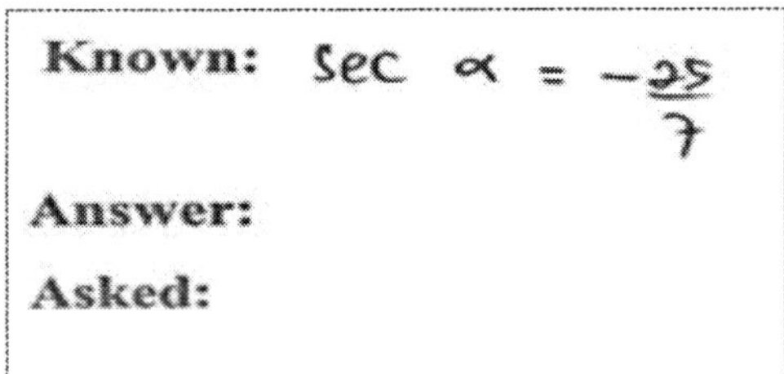

Figure 2(a). Difficulty in describing mathematical expressions/symbols.

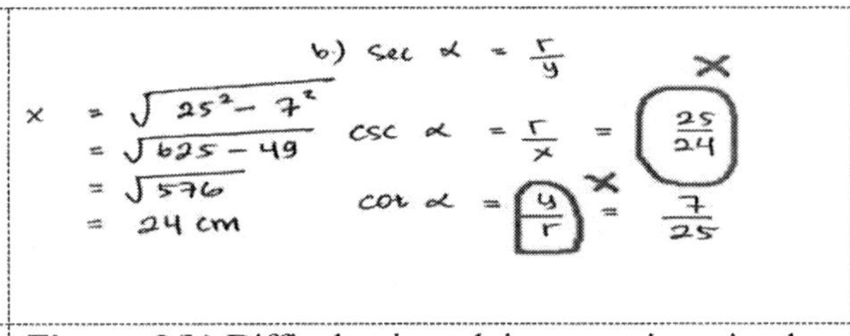

Figure 2(b). Difficulty in solving questions in the form of mathematical expressions/symbols.

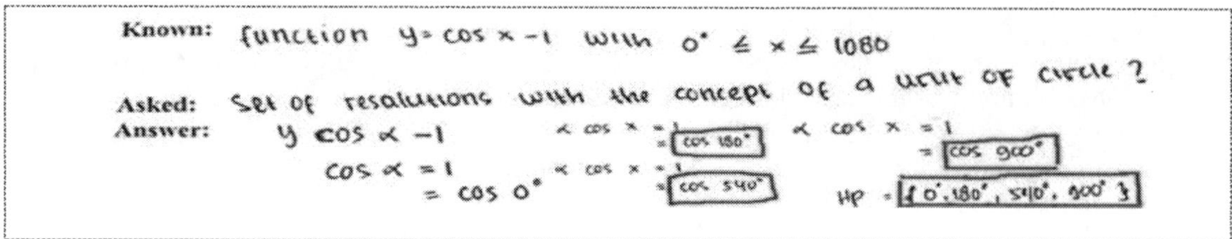

Figure 2(c). Difficulty in comprehending the topic of trigonometric function with more than one unit circle.

Figure 2(a)shows that students faced difficulties in writing mathematical expressions/symbols. Thus, they wrote only some parts of the expressions/symbols. Besides, to calculate the values of cosecant α and cotangent α, students first determine the length of an unknown side of the triangle based on the information given in the question. The unknown side is the front leg because it is known that the value of sec α is $-25/7$in quadrant III. The comparative secant value is the hypotenuse and the side leg; thus, the length of the front leg needs to be found. After the length of the front leg is found, the values of all three sides are used to determine the values of cosecant α and cotangent α.Students' difficulty in solving the question number 1B is presented in Figure 2(b). It can be seen that students have low comprehension of the basic concepts of trigonometry and the signs of +/- in each quadrant so that they obtained incorrect comparative value of cotangent α.

The question number 3 contains the same indicator with the question number 1B. The difficulties that occur in questions number 3 and 4 arethe same problem can give rise to errors from different sources, and the same error can arise from different problem-solving processes [19]. There were 7 students who experienced difficulties in this question. The question number 4 gives information of the function y= cos x – 1 with 0°≤x≤1080°. Furthermore, students are asked to determine the set of functions using the unit circle concept. The difficulty encountered by students is illustrated in Figure 2(c). It can be seen that students faced difficulties in comprehending the topic of trigonometric function with more than one unit circle. When looking for the first set of functions, students gave the right answer that cos x=1; for x=0°. However, students' further answers were incorrect; the value of cos 180° is 1, but it is supposed to be –1. The same case happened to other answers. Therefore, the set of resolutions of the question number 3 is {0°, 360°, 720°, 1080°}.

3.3 Using mathematical symbols in solving problems in the form of pictures

The question number 5 contains the indicator of the third representation. The indicator of competence is to analyze the changes of the graph of trigonometric function as a result of the change of constant in the function y= a sin nx. The question number 5 shows the graph changes so that the two graphs of functions can be determined. The following Figure 3(a) is the graph used in the question number 5.

The difficulties most students faced occurred to the question number 5 in which 10 students did not write anything on the answer sheet. Students understood the graph, but they were not able to represent mathematical expressions/symbols given in the graph. One of the difficulties is presented in Figure 3(b) below.

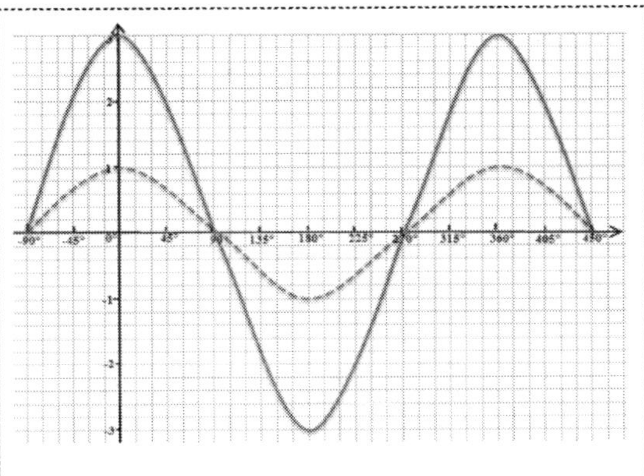

Figure 3(a). Graph of the Function y = a sin nx.

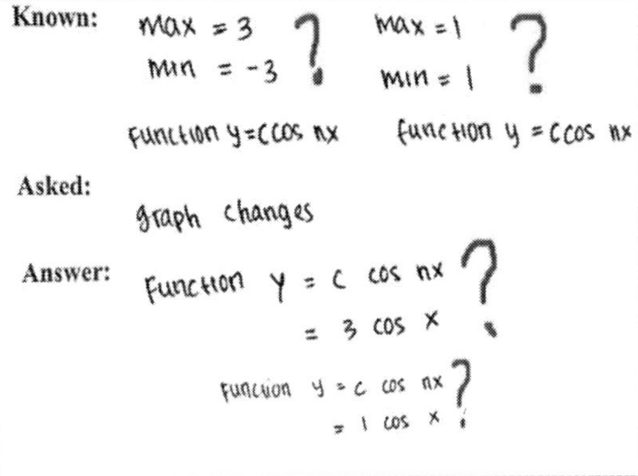

Figure 3(b). Difficulty in representing mathematical symbols from the problem in the form of pictures.

Out of 10 students, all haddifficultyin solving the question number 5 shown in Figure 3(b). Only a small number of students did as in Figure 3(b) and the rest did not answer. In this regard,students were not able to mention the reason of graph changes, even though they were able to read what the graph means and to understand the changes of both functions. However, they were not able to explain the reason or give evidence that the functions are y=cos x and y=3 cos x. One of the reasons to prove that the function is y=cos x is if x=0°, then f(x)= cos 0°=1 and if x=90°, then f(x)= cos 90°=0. On the other hand, one of the reasons to prove that the function y = 3 cos x is if x = 0°, then f(x) = 3 cos 0° = 3 × 1= 3 and if x = 90°, then f(x) = 3 cos 90° = 3 × 0 = 0. These two reasons are already in accordance with the graph that has dotted lines.

3.4 Using pictures as representation in solving questions in the form of written texts

The indicator of the fourth representation is contained in the question number 4 with the indicator of competence "to illustrate the contextual problem of trigonometric ratios of a right triangle". There were 6 students who faced difficulties in solving this question. The question reads "Father leaned a ladder right at the top of the wall. The length of the ladder is 6 m, while the angle between the floor and the bottom end of the stairs is 60 °. Illustrate this information with a picture." Out of 6 students who faced difficulties, two of them illustrated the information as presented in Figure 4(a) and 4(b) below.

Figure 4(a). Difficulty in representing pictures from a written text question.

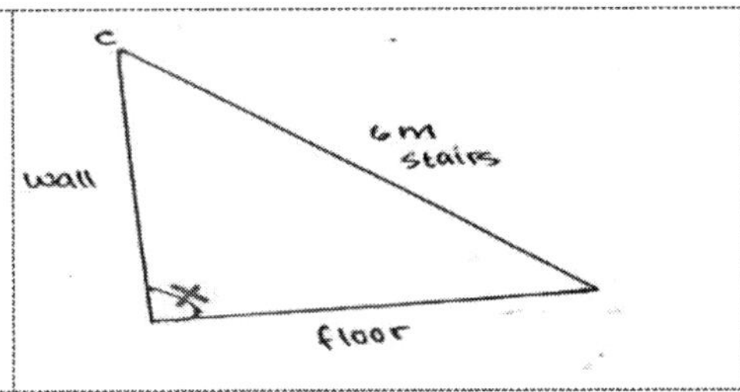

Figure 4(b). Difficulty in representing pictures from a written text question.

In Figure 4(a) and 4(b), students faced the same difficulty that is to represent pictures in order to solve a question in the form of written texts. The concept in this question relates to the concept of a triangle. students 'difficulties in this concept are in accordance with the students' responses from Akhter research that, introduction to coordinate geometry and Parallelograms and triangles is difficult for students [20]. Figure 4(a) shows that it was difficult for students to locate the size of the known angle. The question tells that the angle of 60° is located between the bottom end of the ladder and the

Journal of Physics: Conference Series **1397** (2019) 012093 doi:10.1088/1742-6596/1397/1/012093

floor, but students located the angle of 60° between the top of the ladder and the wall. On the other hand, based on Figure 4(b), students represented only the length of the ladder. They found it difficult to represent the size of the known angle. Both cases occurred because students are unfamiliar with questions in the form of written texts or with essay questions represented in pictures.

3.5 *Using mathematical expressions/symbols in solving questions in the form of mathematical texts*

The question number 2 contains the indicator of representation of using mathematical expressions/symbols in order to solve questions in the form of mathematical texts. The indicator of its competence is to determine law of cosines if two sides and an angle are known. There were 8 students faced difficulties in regard to this indicator. The question number 3 reads "An ABC triangle has sides a, b, and c. The triangle applies the formula: $a^2 = b^2 + c^2 - (\sqrt{2})(bc)$ and $c^2 = a^2 + b^2 - (\sqrt{3})(ab)$, then determine the size of angle B of the triangle." A student solved the question as presented in Figure 5, but the other 7 students did not write anything or faced difficulties in solving this question.

In one of the students' answers presented in Figure 5, the student used the right concept, but after the value of cos $A=\sqrt{2}/2$ was obtained, he was not able to finish it until the end that is to find the size of angle A. in an attempt to find the size of angle B, students only did the question until they obtained the value of cos $C=\sqrt{3}/2$, but they could not find the size of angle C. After the sizes of angle A and C were obtained, students were supposed to be able to find the size of angle B by subtracting the total of three angles of a triangle from the total of angles A and C. When doing the question number 2, students were not able to use mathematical expressions/symbols in order to solve questions in the form of mathematical text/essay.

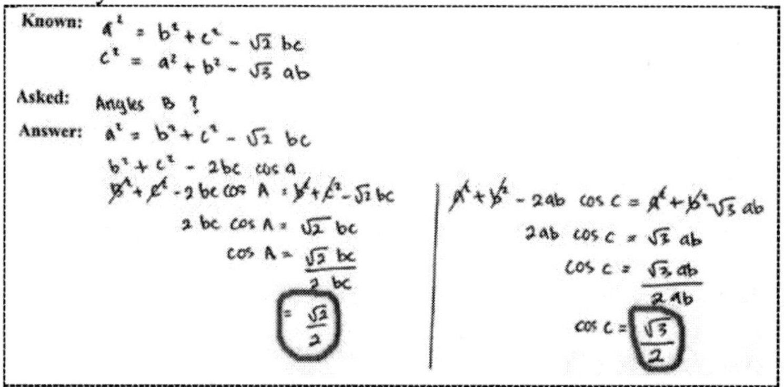

Figure 5. Difficulty in representing mathematical expressions in solving a question in the form of written texts.

4. Discussion

Based on the results, in general, students still feel that mathematical representation skill is difficult. All five indicators of representation are considered difficult by students. Given the indicator of using pictures to represent mathematical symbols, students did not write complete and proper mathematical symbols. This means students faced difficulties in using pictures to represent symbols. As a result, their pictures were not in accordance with the concept of trigonometry. Another factor is that students experienced misunderstanding of the theory and did not comprehend mathematical symbols [10]. Furthermore, students also encountered difficulties in writing symbols from the comparison concept of trigonometry. This resulted in errors in calculation [12] and in using representation of the symbols (+) and (-) in regards to the comparison concept of trigonometry in each quadrant. The third indicator is using mathematical symbols to represent pictures. In this indicator, it was difficult for students to read and comprehend the pictures of graph changes of trigonometric function which caused students to unable to use mathematical symbols to represent pictures [12].The difficulty in using pictures to represent written text questions happened to students when doing essay questions or questions related to daily life. The last indicator is to use mathematical symbols to represent written text questions. In this indicator, students felt difficult to comprehend written text questions so that they were not able to use mathematical symbols to represent written text questions. Other students faced difficulties in certain stages; thus, they could not solve the problem until the final process [10]. This occurred because students lacked skills in calculation and comprehension of mathematical symbols.

The types of difficulties found in this study are (1) difficulty in using pictures to represent mathematical symbols, (2) difficulty in using mathematical symbols to represent mathematical symbols, (3) difficulty in using mathematical symbols to represent pictures, (4) difficulty in using pictures to represent written text questions, and (5) difficulty in using mathematical symbols to represent written text questions. Besides, students consider trigonometry as one of the materials in mathematics that is difficult and challenging to be learned. Given the difficulties found in this study, the information in this regard is expected to be useful for teachers to discover the proper ways to overcome difficulties experienced by students [10], because the ability of representation has a very important role in the learning of mathematics that need to be owned by each student.

5. Conclusion

Based on the results and the discussion above, it can be concluded that students experienced difficulties in solving mathematical representation questions. Students' difficulties were apparent from students' works in solving the questions given by the researchers. In addition, the results of interviews also suggest that it was difficult for them to do questions about mathematical representation. The questions designed and given to students were considered more difficult to be solved than the questions students usually received in the learning process. This happens because students lack practice in doing mathematical skill-based questions.

All indicators of representation skill contained in each question were considered difficult by students. Mathematical representation is difficult for students because of five reasons, including because students have difficulty in representing mathematical expressions/symbols in the form of images (7 students), students have difficulty in representing questions in the form of mathematical expressions/symbols in the form of expressions/symbols mathematics (7 students), students have difficulty in representing written text questions in the form of mathematical expressions/symbols (8 students), students have difficulty representing mathematical written text in the form of images (6 students), and students have difficulty in representing questions in the form of images in the form of mathematical expressions/symbols (10 students).After finding the cause of the difficulties felt by students in solving mathematical representation questions, it can be using as a teacher as a remidial or stabilization relating to trigonometric mathematical representation.

References

[1] NCTM 2000 *Principles and Standards for School Mathematics.*United States of America: The National Council of Teachers of Mathematics, Inc.

[2] Callison D 2013 *Common Core for Mathematics in School Library Monthly* Vol 29 (5): 21-24 Santa Barbara: Libraries Unlimited, Inc.

[3] Dahlan J A 2011 *Materi Pokok Analisis Kurikulum Matematika* Jakarta: Universitas Terbuka

[4] NCTM 2003 *Standards for Secondary Mathematics Teacher.* United States of America: The National Council of Teachers of Mathematics, Inc

[5] Sabirin M 2014 Representasi dalam Pembelajaran Matematika *JPM IAIN Antasari, 1(2): 33-44*

[6] Effendi A L 2012 Pembelajaran Matematika dengan Metode Penemuan Terbimbing untuk Meningkatkan Kemampuan Representasi dan Pemecahan Masalah Matematis Siswa SMP *Jurnal Pendidikan Indonesia, 13(2):2*

[7] NCTM 2014 *Principles to Actions: Ensuring Mathematical Success for All.* Reston:nctm.org

[8] Mardapi D 2017 *Pengukuran, Penilaian dan Evaluasi Pendidikan* Parama Publising Yogyakarta

[9] Lewis C H and Walker P 1994 The Prevalence of Specific Arithmetic Difficulties and Specific reading in 9-to 10-year old Boys and Girls. *Journal of Child Psychology and Psychiatry, 35,*

pp 283-292

[10] Wood D 2007 *Kiat Mengatasi Gangguan Belajar* Yogyakarta: Katahati

[11] Bzufka M W, Hein J & Neumarker K J 2000 Neuropsychological Differentiation of Subnormal Arithmetic Ablities in Children *Europaean Child and Adolescent Psychiatry, 9, pp 65-76*

[12] Jane G & Zakaria E 2012 Mathematical Difficulties Among Primar School Students *Journal Faculty of Education, Universiti Kebangsaan Malaysia, 7, pp. 1086-1092*

[13] Fauzan 2010 Pengembangan Tes Diagnostik Kesulitan Belajara Matematika di SMA *Disertasi tidak dipublikasi Universitas Negeri Yogyakarta, Yogyakarta*

[14] Creswell J W 2007 *Qualitative Inquiry & Research Design: Choosing Among Five Approaches* (Thousand Oaks, California: Sage Publication, Inc)

[15] Miles M B and Huberman A M 1994 *Qualitative Data Analysis: An Expanded Sourcebook* (Thousand Oaks, California: Sage Publications, Inc)

[16] Miksalmina 2013 Penguasaan Siswa Pada Materi Trigonometri Di MAN Darussalam Aceh Besar *Vol 4 No 2 101-110*

[17] Magara K and Shindo T 2015 Why Are Rule Representations Difficult to Form? *Japanese Journal of Educational Psychology 63(3): 267-278*

[18] Usman M H and Hussaini M M 2017 Analysis of Student's Error in Learning of Trigonometry Among Senior Secondary School Students in Zaria Metropolis, Nigeria *IOSR Journal of Mathematics (IOSR-JM) Volume 13 No 2 1-4*

[19] Hendrik R 1979 Error Analysis in Mathematics Education *Journal for Research in Mathematics Education: Vol. 10, No. 3 163-172*

[20] Akhter N 2018 Learning in Mathematics: Difficulties and Perceptions of Students *Journal of Educational Research, Dept. of Education, IUB, Pakistan Vol. 21(3)*

Improving Students' Mathematical Literacy Skills Through Multiple Intelligences Approach in Problem Based Learning

Dini Arrum Putri[1] and Djamilah Bondan Widjajanti[2]

[1]Graduate School Program, Universitas Negeri Yogyakarta, Indonesia
[2]Department of Mathematics Education, Faculty of Mathematics and Natural Sciences Universitas Negeri Yogyakarta, Indonesia

diniarrum.2018@student.uny.ac.id

Abstract. This literature study was focused on Multiple Intelligences approach in Problem Based Learning to improve Students' Mathematical Literacy Skills. There are some relevant researches to describe Multiple Intelligence approach in Problem Based Learning and improve mathematical literacy skills. Now, the implementation of mathematics education should emphasize how the concepts learned by students can be applied in the real life context. This means that students are expected to not only able to solve problems at school, but also the problems outside of school. This ability is called mathematical literacy skills. The implementation of Problem-Based Learning methods can be applied by teachers to be able to improve students' mathematical literacy skills. Problem-Based Learning is a student-centered learning method in which students actively participate in discussion groups where they discuss problem solving strategies. However, not every student is logically intelligent. Therefore, the implementation of Problem Based Learning must be combined with multiple intelligence approaches that can enhance each type of students' intelligence.

1. Introduction

Mathematics is one of the branches of science that becomes one part of the education curriculum and has a big role in transforming human civilization, thus, it needs to be mastered by students from elementary school level [1]. Mathematics is not only taught in schools but also implemented in daily activities. Therefore, the method of engaging mathematics in schools with everyday life has become a priority in designing school mathematics curriculum [2].

Mathematical literacy has become an important topic in mathematics education. Various studies on mathematical literacy have been conducted. Based on the resolution proposed by PISA, mathematical literacy is the ability of individuals to implement the knowledge and concept of mathematics to solve the problems in daily life. It includes the ability in reasoning mathematically by using mathematical concepts, procedures, facts, and mathematical tools [3].

In Indonesia, Mathematics teacher face some problems as a result of there are still many students who are anxious in facing mathematics problems [4]. To be able to facilitate the students to be literate in mathematics, teachers cannot only use traditional methods that are often implemented in schools. Traditional method will only give the students a little understanding of what they are doing or why they are doing it. Students only learn to memorize procedures and to match the same problems in exams. Therefore, they never learn to think mathematically and they assume that mathematics is a boring subject, not applicable, useful in real life, and increasingly unable to be understood [5]. Hence, the efforts to create the optimal learning process by using the Problem Based Learning method are definitely needed. Problem Based Learning is the method that emphasizes in giving mathematical problem at the beginning. According to Denise [6], by implementing this method in the classroom, teachers can help students to find and analyze information critically to solve the problems.

Levy [7] stated that students have different abilities, learning styles, and personalities when they entering the classroom. It should also be noted that not every child is intelligently good at logical mathematics so that Problem-Based Learning approach needs to be combined with multiple intelligence approaches in order to enhance and improve each type of students' intelligence. Furthermore, Eissa and Mostafa [8] stated that students come to school with various abilities: low, medium, and high. Teachers must realize and focus what was important for students to know in studying mathematics and find ways of accomplishing teaching. Gardner [9] defines intelligence as the ability to solve a problem, create products that have a culture value, and divide intelligence into existential, lingusitic, spatial, logical, kinesthetic, musical, naturalist, interpersonal and intrapersonal intelligences. Regarding those nine intelligences, the teacher needs to consider the characteristics of students and use them as a benchmark so that learning can take place according to the interests and needs of students.

2. Method

This article is a theoretical study that can provide information or descriptions of learning approaches that have potential to improve mathematical literacy skills theoretically. Data were collected and analyzed in the form of literature obtained from the review 19 of journal articles and 9 relevant research results from related scientific journals that related to Problem-Based Learning with multiple intelligences approach. First, the author collected a theoretical study of problem based learning methods, multiple intelligences approaches and mathematical literacy abilities. Second, the author sought and collected some relevant research results. The results of various literature reviews will be used to see the correlation between Multiple Intelligence in Problem-Based Learning and Mathematical Literacy and to identify potential research of Multiple Intelligences approach in Problem Based Learning that can improve mathematical literacy skills. Last, the author drew conclusion that the approach of multiple intelligences has the potential to improve the ability of mathematical literacy.

3. Result and Discussion

3.1. Problem Based Learning

Problem-Based Learning is one of the learning methods that has been widely known in the education world. That method is a method where practically in the learning process, students are encouraged to study independently to build their own knowledge while the teacher's role is as a facilitator. Barrows and Kelson [10] define Problem-Based Learning as a curriculum and learning process that is designed with problems that require students to acquire important knowledge, in order to make them proficient in problem solving. In addition, Savery [11] stated that problem-based learning emphasizes a process in empowering students to conduct research, apply knowledge and skills in order to get the great solutions to solve problems. Further, Roh [12] suggested that problem based learning defines as a learning process where learning begins with a problem to be solved by students. So, students need to gain new knowledge before they can solve the problem.

According to Barrows and Kelson's theory [10, 13], there are four characteristics of Problem-Based Learning, namely: (1) learning is student-centered, the learning process with problem based learning emphasizes students as learners. This learning is supported by the theory of constructivism where students are encouraged to be able to develop their knowledge individually; (2) authentic problems form the organizing focus for learning, the topics of the problem that are presented to the students are authentic issues so that students are able to easily understand the problem and implement it in daily life; (3) new information is acquired through self-directed learning, in the process of problem solving, students are considered not yet aware of and understand all concepts that lead them to find information through various sources independently; (4) learning occurs in small groups, in Problem-Based Learning, students discuss in a heterogeneous small group where there is an interaction in solving a problem; and (5) teachers act as facilitators, teachers only act as mentors who monitor the improvement of student activities and encourage students to achieve goals they set.

While Barret [14] explained the steps of problem based learning method into: (1) teacher give the problems to students; (2) students have discussion in small groups such as clarifying cases of problems given, defining problems, exchanging ideas based on their knowledge, and deciding things needed to solve the problem; (3) students do independent study materials about the problems that must be solved, and they can do it by looking for references or making observations; (4) students return to their previous group to exchange information and cooperate in solving problems; (5) students tell the solutions they find; and (6) students have evaluation related to all learning activities with the teacher. This includes the extent of the knowledge obtained by students and the roles of each student in the group.

Some of the findings of Problem-Based Learning method in mathematics learning studies have been conducted in the world of education. Research conducted by Astriani, Surya and Syahputra [15] concluded that Problem Based Learning has a good effect for improving students' mathematical problem solving abilities. Also, Tandililing's findings [16] concluded that Problem Based Learning method is effective in terms of students' mathematical representation abilities. Not only able to improve problem solving abilities and mathematical representations, it is also proven to be able to improve students' critical thinking skills in mathematics [17]. In addition, Masitoh and Fitriyani [18] stated in their findings that Problem-Based Learning method can improve students' self-efficacy in learning mathematics, while Hendriana, Johanto, and Sumarmo [19] stated that this can help students improve their self-confidence in mathematics. Further, Santria and Jailani's findings [20] concluded that Problem-Based Learning helped students to improve interpersonal skills.

From several studies conducted previously, Problem-Based Learning has a positive effect if implemented in a process of learning mathematics. By implementing this method in the classroom, students have the opportunity to develop and improve their own mathematical concepts in solving mathematic problems. Thus, Problem Based Learning is potentially able to help students develop and improve their skills and abilities in learning mathematics.

3.2. Multiple Intelligences

Howard Gardner (1983) introduced the existence of multiple intelligence theories. In his book entitled Frames of Mind: The Theory of Multiple Intelligences, Gardner [9] defined intelligence as the ability to solve a problem and create products that have cultural value. Gardner divided the intelligence category into nine categories that presented in the following table [9, 21]:

Table 1. Howard Gardner's Multiple Intelligences

Multiple Intelligences	Description
Verbal-linguistic intelligence	The ability of individuals to use language and words in both spoken and written expressions can be reasoned abstractly, and have conceptual patterns.
Logical-mathematical intelligence	The ability of individuals to process numbers accurately and correctly, well-reasoned. This includes the sensitivity to patterns and logical relationships that can convey causal relationship
Musical intelligence	The ability of individuals to create, produce and express is related to tone, music, rhythm, rhythm.
Spatial intelligence	The ability of individuals to understand, recognize and create all forms of images, space, and direction.
Body-kinesthetic intelligence	The ability of individuals to optimize the body work. Someone in this intelligence prefers to do physical activity directly.
Intrapersonal intelligence	The ability of individuals to reflect and analyze him/herself. Someone in this intelligence is more likely to understand themselves than others.

Multiple Intelligences	Description
Interpersonal intelligence	The ability of individuals to work with others in small groups, as well as the ability to communicate verbally and nonverbal.
Naturalistic intelligence	The ability of individuals to recognize and classify certain objects related to their patterns in nature. Someone with this intelligence tends to have activities outdoors/deal directly with nature
Existensial intelligence	The ability of individuals to answer problems about human existence.

In Gardner's perspective, each of the nine intelligences has an important role in learning and multiple intelligence theory can motivate teachers to use and implement various methods to help their students in mathematics [9, 22]. This theory does not only discuss intelligence, but also explains how children learn and can help their teachers in teaching practices and how to manage classes well [22]. Furthermore, Armstrong [23] mentioned the four main components of multiple intelligences theory namely: (1) everyone has each of nine types of intelligence; (2) intelligence in each individual can be developed; (3) intelligence usually works in a complex scope; (4) there are many ways to be smart in each category.

3.3. Mathematical Literacy
Mathematical literacy can be defined as the ability of students to implement mathematical concepts and knowledge that they have learned in class into everyday life. These abilities are in the forms of the ability to employ, formulate and interpret concepts in real life contexts. Ojose [24] the ability of mathematical literacy is defined as knowledge to know and implement the basis of mathematics in daily life. There are seven indicators of literacy skills that students need to have [25]. The indicators are: (1) **communicating**, students are able to identify and understand a problem such as reading, translating, and interpreting statements where questions and solutions can be presented; (2) **mathematizing**, students are able to compile, conceptualize, make predictions and formulate mathematical models and solve the problems by using mathematical solutions; (3) **representation**, students are able to represent objects and mathematical situations involving interpretation, translation, use of various representations in the form of tables, graphs, images and diagrams to interact with various problems; (4) **reasoning and arguments**, students are able to involve logical processes to reason, analyze, and argue in exploring problems to draw conclusions and examine the truth of a statement; (5) **devising strategies for solving problems**, students are able to find and create the solution for solving mathematical problems, (6) **using symbolic**, by involving understanding, students are able to interpreting, manipulating and utilizing symbolic expressions by understanding definitions, rules and algorithms; and (7) **using mathematical tools**, students are able to use mathematical tools as measuring instruments.

In addition, Magen [26] stated that mathematical literacy includes two aspects of individual abilities, those are: (1) aspects of understanding, the ability in knowing facts, concepts and interpreting mathematics into various contexts; (2) aspects of application, this aspects include where students can use the ability in using concepts, facts, and procedures for solving mathematical problems; (3) aspects of reasoning, the ability to reason and analyze by providing explanations and justifications; and (4) aspects of communication, the ability to communicate or explain the explanations and problem solving.

3.4. The Correlation between Multiple Intelligence in Problem-Based Learning and Mathematical Literacy
Arends [27] explains that Problem-Based Learning presents real life problems and is useful for students. Consequently, this can be students' milestones to observe and explore their knowledge. Problem-Based Learning is an implementation of learning that has been widely used as a research material or implemented in the learning process in schools. However, teacher must realize that every student has a various intelligences in class so Problem-Based Learning need to be combined with

multiple intelligence approaches in the learning process. Teachers need to understand and to be aware of the students' characteristics and intelligence. Supartono and Waluyo [28] also explains that basically, one of the most important thing in the learning process was how the teacher can deliver any information well called teaching styles. While the students should be able to receive the information from the teacher well called learning styles.

Mathematical literacy has seven aspects namely communication, mathematizing, representation, reasoning-argument, devising strategies for solving problems, using symbolic, and using mathematical tools [25]. The stages of multiple intelligences approach in Problem-Based Learning can potentially improve students' mathematical literacy skills. Students can discuss in groups and present solutions with various ways of thinking. When concluding the results of the discussion, students can present the results in a verbal form, diagram or other media. Also, at this stage, students are expected to be able to use symbols or mathematical tools so that, unconsciously, they can practice communicating and presenting the results of the discussions in accordance with the intelligence or knowledge they posses. As stated by Saeidi [29], multiple intelligence theories can be used as a tool to create various tasks and activities that are meaningful and interesting for students as it is in line with their interests.

At the stage of determining the topic, identifying, analyzing and collecting information from various types of problems, students perform the mathematizing process. Students discuss in groups to design or set a completion strategy in solving problems. In this process, students can reason and argue with various intelligences and ways of thinking through group discussions to solve problems. Gouws [20] stated that paying attention to students' intelligence is as same as creating opportunities for students and encouraging them to think about making the learning process better and assessing it in different ways in order to meet the students' needs of various intelligences. In addition, by applying the multiple intelligences approach in the classroom, will make students feel comfortable and students will appreciate their intelligence more [30].

Research conducted by Kartikasari and Widjajanti [31] concluded that the multiple intelligence approach in Problem-Based Learning was effective in terms of connecting ability and students' self-esteem. Not only in terms of connecting ability and students' self-esteem, but also multiple intelligence approach in Problem-Based Learning has a good effect to improve the learning achievement. Furthermore, the results of the research conducted by Rafianti [32] showed that there was an increase in conceptual comprehension and students' self confidence in mathematics learning bases Multiple Intelligences. Research conducted by Septiana and Ikhsan [33] also showed the same results that Problem-Based Learning with the multiple intelligences approach can improve students' ability to understand concepts and think creatively. All the research concluded that multiple intelligences in Problem Based Learning was better to improve students' ability than the traditional method in learning mathematics.

Regarding some of the results of the research described previously, implementing Problem- Based Learning with the multiple intelligences approach can improve the components of mathematical literacy itself such as formulating, interpreting, reasoning, interpreting, communicating, and explaining problem solving in everyday life effectively and efficient. Based on the stages of Problem-Based Learning methods that facilitate the abilities and skills of students like problem solving skills, mathematical reasoning, mathematical connections, mathematical communication and mathematical representation. Where from these abilities is a component of mathematical literacy. In addition, the multiple intelligences approach can be used as a foundation of the learning process in the classroom. The activities of students in the classroom can be tailored to their interests and needs so that the optimal learning process can be created. Therefore, it can be concluded that Multiple intelligences approach in Problem-Based Learning can potentially improve students' mathematical literacy skills. In short, the relationship between Problem-Based Learning and multiple intelligences approaches with mathematical literacy skills is explained by the illustration below.

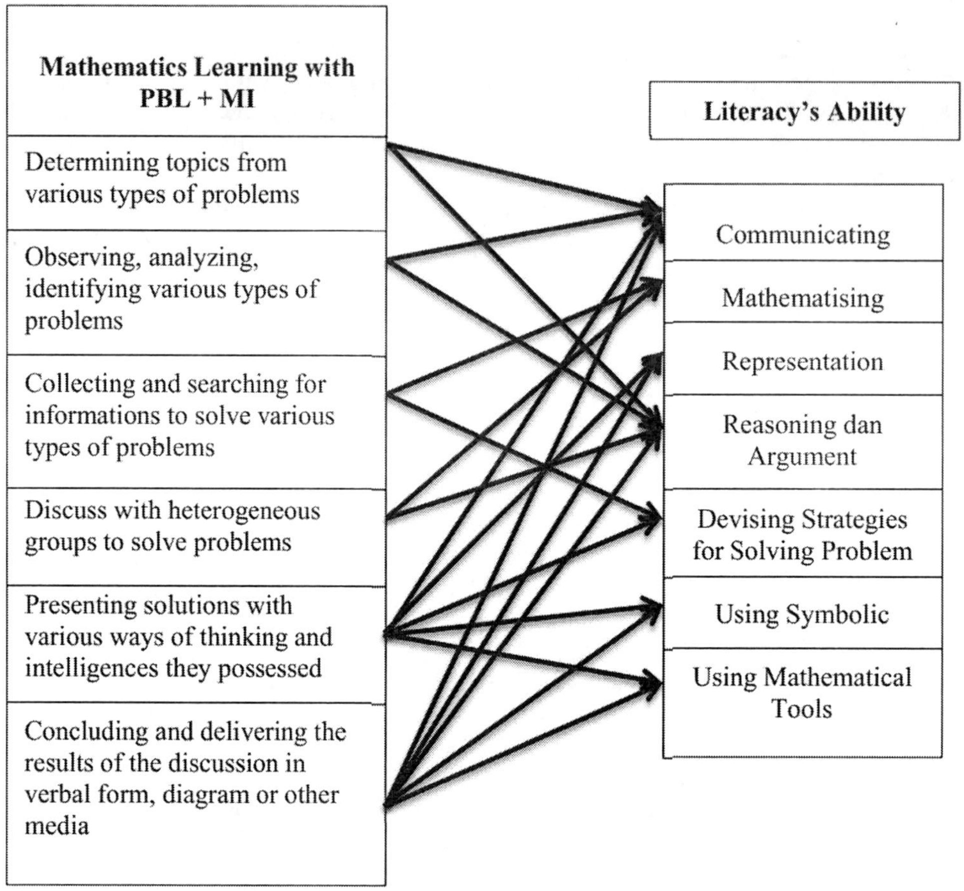

Figure 1. The Correlation between Multiple Intelligence in Problem-Based Learning and Mathematical Literacy

4. Conclusions

According to the studies described previously, theoretically, multiple intelligences approach in Problem-Based Learning has the potential to improve students' mathematical literacy skills. By implementing Problem-Based Learning methods, students can actively develop their skills and abilities in solving mathematical problems through the stages of Problem-Based Learning. In addition, students are able to develop their own mathematical concepts through group discussions. Teaching by paying attention to each student's ntelligence and characteristics can also help teachers manage the class well. Learning can be tailored to the students' interests and needs as multiple intelligences approach can enhance each type of students' intelligence.

References

[1] Kusmaryono I 2014 *Int. Conf. on Math. Sci. Educ.* (Semarang State University) p 35
[2] Maulina Y and Retnawati H 2017 *ICRIEMS Proceedings of fifth annual conference* (Yogyakarta State University) p 467
[3] OECD 2010 *PISA 2015 Draft Mathematics Framework.*
[4] Djamilan Bondan Widjajanti 2013 *Journal Technology* **2** 39
[5] Haglund R 2004 *Humanistic Mathematics Network Journal.* **1** 2
[6] Denise M 2014 *Master of Education Program Theses.* Paper 53 (Dordt University) p 5
[7] Levy R 2008 *Meeting the needs of all students through differentiated instruction: Helping every child reach and exceed standards* (*Electronic Version* vol 4) *Clearing House* p 161

[8] Eissa Mourad Ali and Mostafa Amaal A 2013 *International Journal of Psycho Educational Sciences.* **2** 34
[9] Gardner H 2011 *Frames of Minds: The Theory of Multiple Intelligences* (New York, NY: Basic Books)
[10] Barrows H and Kelson A C 1995 *Problem-Based Learning in Secondary Education and the Problem-Based Learning Institute (Monograph 1)* (Springfield: IL)
[11] Savery J R 2006 *The Interdisciplinary Journal of Problem-based Learning* **1** **9**
[12] Roh K H 2003 *Clearinghouse for Science, Mathematics, and Environmental Education* (Arizona State University) p 1
[13] Sari M P and Khiyarunnisa *Proceedings National Conference Mathematics and Mathematics Education* (Yogyakarta State University) p 444
[14] Barret T, Mac Labhrainn, Fallon H 2005 *Handbook of Enquiry and Problem Based Learning* vol 2, (Galway: CELT) p 15
[15] Nurullita Astriani, Surya E and Syahputra E 2017 *International Journal Of Advance Research And Innovative Ideas In Education* vol 3 (Medan State University) pp. 3441-3446
[16] Tandililing E 2015 *Proceeding of International Conference On Research, Implementation And Education Of Mathematics And Sciences* (Yogyakarta State University) pp. 147-152
[17] Arviana R Irwan and Dewi M P 2018 *Journal of Computational and Theoretical Nanoscience* vol 24 (American Scientific Publishers) pp. 1-3
[18] Fitriana L and Fitriyani H 2018 *J. Math. Learn.* (Malikusalleh University) pp. 26-30
[19] Hendriana H, Johanto T, and Sumarmo U 2018 *J. Math. Educ.* (Sriwijaya University) pp. 291-300
[20] Santria U and Jailani J 2018 *ICRIEMS Proceedings of fifth annual conference* (Yogyakarta State University) pp. 37-44
[21] Gouws F E 2008 *Africa Educ. Rev.* **2** 63
[22] Niroo M, Haji G Nejhad H, and Haghani M 2012 *Soc. Behav. Sci.* **47** 2170
[23] Armstrong T 2003 *Multiple intelligences in the classroom. 3rd* ed Alexandria (VA: Association for Supervision and Curriculum Development) p 76
[24] Ojose B 2011 *J. Math. Educ.* **4** 90
[25] Turner R Dossey, J Blum W, and M Niss 2013 *Using mathematical competencies to predict item difficulty in PISA* (New York: Springer).
[26] Noga Magen Nagar 2016 *Int. J. of Research in Educ. and Sci.* **2** 2
[27] Arends R I 2008 *Learning to Teach (*New York: The McGraw-Hill Campanies, Inc)
[28] Rico Prasetya K, Budi W, and Supartono 2015 *Int. Conf on Math. Sci. Educ.* (Semarang University) p 24
[29] Saeidi M 2016 *Journal of Teaching English as a Foreign Language and Literature* **1** 111
[30] Widjajanti D B 2012 *Proceedings National Conference on Education Research and MIPA* Yogyakarta: FMIPA UNY)
[31] Kartikasari and Widjajanti D B 2017 *J. Phys.:Conf. Ser* **812** 012097
[32] Rafianti I 2013 *Journal of Mathematics Research and Learning* (Sultan Agung Tirtayasa University)
[33] Septiana K G and Ikhsan J *J. Educ. Math. Sci* (Madiun State University) pp. 44-52

Journal of Physics: Conference Series **1397** (2019) 012095 doi:10.1088/1742-6596/1397/1/012095

The characteristics of final mathematics test items based on classical test theory

A T Panjaitan[1*] and H Retnawati[2]
[1]Department of Mathematics Education, Graduate School Program, Yogyakarta State University, Indonesia
[2]Department of Mathematics Education, Yogyakarta State University, Indonesia

*Corresponding author: agnesteresa.2018@student.uny.ac.id

Abstract. Final Examination is the part of an evaluation that exists in school to assess student's knowledge. This research aimed to show the characteristic of the final mathematics test in Junior High school, grade 8. This research was an explorative descriptive using a quantitative approach. Data collection was gathered using the documentation of mathematics' test in Binjai, North Sumatera including student's answers. The characteristics of test items were analyzed by expert judgment using the classical test theory approach. The average of content analysis was 0.81, there were three items without key answers so it can not be analyzed. The reliability average was 0.84 which has very good quality and difficulty index indicated the difficulty level of questions was on three categories (easy, moderate, and difficult). The difficulty is in the cartesian position, the relation of two sets, determining gradient, determining the solution of linear equation systems of two variables.

1. Introduction

Assessment in education has been an interesting topic at many levels. There are many kinds of assessments that the teacher used to evaluate a student's mathematical learning. Not only for evaluating student's mathematical learning but also for providing student grades, national accountability, tracking on the right system, allocating resources within a district, regulating involvement, upgrading the class' activities, or giving suggestion and response to students and their guardians [1]. Assessing performance is the influential activity for any subjects, it will facilitate formative report to the pupil and educator to guide future study and occasional summative feedback [2]. Therefore, to present the real achievement of student's learning, it is essential to give the best qualification of the test [3].

NCTM [4] noted that an assessment instrument containing many computational items and relatively few problem-solving questions is poorly aligned with a curriculum that stresses problem solving and reasoning. It also highlights the integration of mathematical knowledge must contain tasks, mathematical power, a task with a non-unique solution. Setting the standard of the tests is not as simple as the expectation, there are some factors to make the good qualification of tests such as the technique chosen by test makers.

Moreover, the criterion shown by the score from the other examination results is similar. It can be concluded that the test can be replaced by other tests, such as school performance tables also impose the same requirement. Besides, the main purpose of the public examination is to provide information for future educational system and vocational selection decisions, and it is a feature of this selection that information from different examinations is combined and compared. As a result, it is critical to arrange the good qualification of questions, with regard to the judicial impartiality which required examination as the selection, to show the same common scale at the similar point on other kind of tests in the same level [5]. Therefore, it is needed to put on the valid score for any examination the students face.

Thorndike [6] stated that the implementation of measurement error diagnosis by discrepancy measurement. the reliabilities of the tests, their correlation, and the size of the score difference involved as the aspects of the tests. Furthermore, Thorndike mentioned that it is clearly impossible to develop the accurate diagnostic interpretation of score differences by the existing instrument. If the diagnostic interpretations used correctly by the educators, it would affect the learning process. The result of the score reports the learning process and the way of student's thinking. Thus, the test must have been considered as valid and in the good categories. The importance of the valid test is not only to monitor the students learning but also to enhance the learning method, the proposition of items, and the standard of any selection.

One of the best ways to know the quality of the tests is by using the analysis of item characteristics [7]. In general, the analysis of item characteristic facilitates the teacher to know how well-qualified the test is [8]. Lamentably, the researchers have found that one of the junior high schools in Binjai designed the test without analyzing the characteristic of the items for Mathematics Final Examination. The tests were designed by multiple choice tests. The quality of the good test such as the discrimination index. Index of difficulty, distractor effectiveness has not known. However, the characteristic of a good test should be developed by the teacher.

The test item discrimination index shows the ability of an item in recognizing the participants having the high grades or the low grades, the test item discrimination index distinguishes the students who do the test well or not. Hence the discrimination index is divided into three categories: positive, negative, and zero. The difficulty index describes the percentage of students answered correctly [8]. The quality of the test can be found after the analysis has been done. Based on the literature review, the analysis of item characteristics will show the qualification of the test. Accordingly, this research aims to show the quality of The Junior High School Mathematics school examination test in Binjai. Furthermore, this research is expected to develop a well-qualified test instrument of mathematics in the future.

2. Method

The research was a document analysis of the descriptive quantitative approach. Due to the existence of the data before the research, the data functioned as the secondary data. The data used for this study was the mathematics final examination with the answer sheets. The multiple choice test had been answered by 100 students in the 8th grade. There are four choices, one of them will be the key answer while the others became the distractors. The procedures might be described as the following:

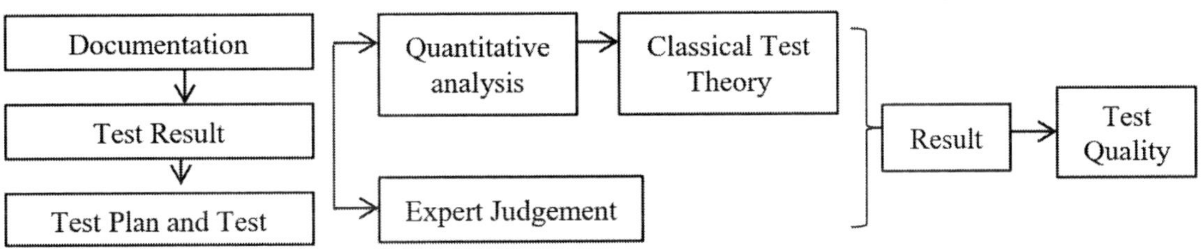

Figure 1. Research procedures

After collecting the documentation, The first analysis regarding to the test material will show the validity of the content. Allen [9] stated content validity is denoted through a rational analysis of the content of a test and based on individual, subjective judgement. According to Aiken [10] to determine the validity index as follows:

$$V = \frac{\sum s}{n(c-1)}$$

The index of validity of the test item is termed by V. the experts' assigned score minus the lowest score in that category is symbolized as S (where: s= r-r0, r is the assigned score by experts and r0¬ is the lowest score in the category). n is the number of experts and c is the number of categories that might be selected by an expert. The quantitative analysis approach to classical test theory consisted of reliability, discrimination index, and distractor effectiveness. According to Miller, Linn & Gronlund [11], The reliability index category should be based on the correlation coefficients in Table 1.

Table 1. Reliability criteria

Reliability Index	Criteria
0.81 - 1.00	Very Good
0.61 - 0.80	Good
0.41 - 0.60	Quite
0.21 - 0.40	Poor
0.00 – 0.20	Very Poor

The discrimination index that the researcher implemented was the biserial correlation point. Mardapi notes the discrimination index can be presented in the following table [12].

Table 2. Discrimination index criteria

Discrimination index	Criteria
> 0.30	Good and Acceptable
0.20 – 0.30	Quite Good and need repairing
< 0.20	Not Good and not acceptable

And, the measurement of difficulty index test items can be seen in the table below.

Table 3. Difficulty index criteria

Difficult index	Criteria
> 0.70	Easy and Not Good
0.30 – 0.70	Medium and Good
< 0.30	Hard and not Good

According to Attali & Bar-Hillel [8] the distractor is known as qualified, were it selected at least by 5 % of all the participants' test and had a negative in biserial correlation point selection. Thus, if the distractors are said as inaffective, it should be changed by the others, then the mastered students would be attracted to select the options.

3. Result and discussion

3.1. The qualitative analysis based on the quality of the test

The results of the test plan qualitative analysis for every question of Junior high School Mathematics Examination in Binjai had grouped as Quite Good and Good. The results provided by the table below:

Table 4. Qualitative analysis result

Criteria	Number of items
Good	1, 2, 3, 4, 5, 6, 7, 10, 11, 12, 13, 14 ,15, 16, 17, 19, 20, 22, 23, 24, 25, 27, 28, 29, 30, 31, 32, 33, 34
Quite Good	8, 9, 18, 21, 26, 35

The result above showed that there had been 82.85 % items of the 35 numbers have good quality and 17.15 % items test that had quite good quality with minor revisions in terms of indicator suitability).

3.2. The quantitative analysis based on the quality of the test
The validity test was conducted by correlating the test to the other standardized test using the QUEST program. Concerning to the criteria validity test, the Aiken index proved the mean of reliability is equal to 0.84 which means The Very High category, bigger than 0.81 (>0.81). Based on the output generated by the QUEST program, the discrimination index for every item of Junior High School Mathematics Examination conducted in Binjai can be read from the correlation point biserial. The discrimination index would be presented in the table below:

Table 5. The discrimination index

Criteria	Number of Items
Good	7, 9 , 10, 11, 13, 15, 16, 17, 18, 19, 20, 21, 22, 23, 25, 26, 27, 28, 29 , 32, 34
Quite Good	3
Not Good	1, 2, 4, 5, 6, 8, 12, 14, 24, 30, 31, 33, 35

The discrimination index average score of 35 items belongs to the "Good category" (0,3). The difficulty index of every Junior High School Mathematics Examination might be seen from the proportion of correct answer in the following table:

Table 6. The difficulty index

Criteria	Number of Items
Easy	1, 2, 5, 12
Medium	3, 4, 6, 7, 9, 10, 11, 13, 14, 15, 16, 18, 19, 20, 21, 22, 23, 27, 29, 32, 35
Hard	8, 17, 24, 25, 26, 34

Based on The difficulty index of every Junior High School mathematics Examination item, four items belong to the easy category, it is around 11. 42% among all. The indicator for the question is determining the number pattern, fibanocci sequence, and locating the cartesian coordinate. The medium category belonged to 21 items among 35, it was about 60% item is Medium. It was represented by five topics examined such as number pattern, cartesian diagram, relation and function, equation of the straight line, linear equation in two variables. The five materials used in the test were in the medium category at most. The hard category for the items belonged to 6 numbers. The questions determined about relation and function, equation of the straight line, linear equation in two variables. It means 17.14% of the items are hard for the students.

Instead of all the categories, there are 3 questions without any key answer. The numbers are 30, 31, 33. The questions can be seen below:

30. The price of one dress and three clothes is IDR.185.000,- and the price for three dresses and three clothes for the same label is IDR.275.000. Then, the price of 2 dresses and one cloth is

a. IDR. 107.000,- c. IDR. 170.000,-
b. IDR. 115. 000,- d. IDR. 195.000,-

The indicator for this question is solving the problems about the Linear Equation in Two variables in real life. The price of a dress is IDR.46.666 and the cloth is IDR. 46.000. Then, the price of 2 dresses and one cloth is IDR. 136.666.

The other question is in the same topic by another kind of the linear equation in two variable application.

31. If the sum of the two numbers is 72 and the difference is 6. The multiplication of those numbers is

a. 429 b. 249 c. 78 d. 43

The reason why the solution cannot be found is the answer must be 1287, due to the value of x is 39 and the value of y is 33. The last number is still in the same topic, yet the question is about finding the value of the linear equation in two variables, after finding the value of x and y from the question. The reason why the solution cannot be found is the answer must be 1287, due to the value of x is 39 and the value of y is 33.

The last number is still in the same topic, yet the question is about finding the value of the linear equation in two variables, after finding the value of x and y from the question. The other question is in the same topic by another kind of the linear equation in two variable applications.

31. If 3x-y = 15 dan x + y = 3. Thus the result of 2x – y is equal to
a. 10 b. 12 c. -10 d. 15

The solution of the question above is -19,5 where the x = 4.5 and y = 28,5. To find the result of the last question participant should substitute the value of each variable. Then, after finding the result there is no answer to be found.

3.3. Discussion

Based on the analysis of classic theory, The Junior High School Mathematics Test in one of school in Binjai has 60% medium category among 35 questions. This result is one of the purposes of the analysis item characteristic. It can present the quality of the questions that can enlarge the quality of the test by revising the "not good" category to be a good category. Then the questions can be in the good category for all items [8]. Three of the questions cannot be solved because the choice did not provide the answer. This result cannot be said as an adequate qualification of the test. Moreover, the examination is viewed as the report of students mathematical thinking.

The Difficulty index for the test has the representative for five subjects that were used. The distractor effectiveness from the table above shows for the easy category, each distractor was not chosen by most of the participants. For the medium category, The distractor effectiveness shows each distractor has more probability chosen by the participants. For the hard category, the distractor has the most number chosen by the participants compared to the key answer.

Kriswantoro [13] states that to make meaningful questions, the teacher should consider the distractors for all items. The items with "not good" distractors have the biggest chance to answer by guessing due to the four choices the participants have. Conversely, the distractor that is close enough to the answer will make the question in the hard category. Furthermore, the existence of the unlogic distractor can make the students decide the correct answer.

According to The discrimination index of the test can conclude as a good category. Due to 19 questions belonged to the good category, one is good enough and 13 is the "not good" category. The three questions without answer key cannot be measured. Some factors that put the "not good" category are the difficulty index is too easy and too difficult for students [14].

The analysis of classic theory can increase the quality of the test by checking the indicator of the learning with the items so the question is able to evaluate the student's learning process. The questions without the answer will increase the probability for the students for guessing the answer. Therefore, the teacher should check the distractor and the questions for the test. Moreover, the distractor must be applicable considering Mathematics can be implemented in daily activities[15]. Nevertheless, there are the other factors that affect the examinee's measurement such as the internal(pyshcology) and external (environment) factors [16].

4. Conclusions

The result shows the content analysis(aiken index) is 0.81 and the reliability of the test is 0,84. There are three questions without the key answer. Thus, the questions that can be analyzed in this research are 33 questions. Six of the question considered a hard category that represents four topics of five topics that are used in the test. The distractor work mostly for the hard category. Based on the quantitive analysis by employing classical test theory approach, all of the test items in the Junior High School Mathematics Examination in Binjai had a good category. Nonetheless, the three questions without the answer key should be revised by the teachers to make the test is adequate to be answered by the participants.

According to the studies described previously, theoretically, multiple intelligences approach in Problem-Based Learning has the potential to improve students' mathematical literacy skills. By implementing Problem-Based Learning methods, students can actively develop their skills and abilities in solving mathematical problems through the stages of Problem-Based Learning. In addition, students are able to develop their own mathematical concepts through group discussions. Teaching by paying attention to each student's intelligence and characteristics can also help teachers manage the class well. Learning can be tailored to the students' interests and needs as multiple intelligences approach can enhance each type of students' intelligence.

References

[1] Newton P 2007 *Assessment in Education Journal.* (Informa UK limited) pp 149-170
[2] Swan M and Burkhardt H 2012 *Journal of the International Society of Design and Development in Education* (*Educational Designer volume 2*) pp 1-41
[3] Kartowigran B, Munadi S, Retnawati H, and Apino E 2017 *SHS Web of Conferences (GC TALE vol 42)* p 6
[4] NCTM 1989 *Curriculum and Evaluation Standards for School Mathematics.*
[5] Creswell J W 2015 *Penelitian kualitatif dan Desain Riset.* (Yogyakarta: Pustaka Pelajar).
[6] Bejar I 1984 *Journal of Educational Measurement.* **21** pp 175-189
[7] Santoso A 2018 *Pendidikan Matematika dan Sains Journal* (Yogyakarta State University) **6** p 159
[8] Sutrisno 2016 *Riset Pendidikan Matematika Journal* (Yogyakarta State University) pp 162-177
[9] Allen M J and Yen W M 1979 *Introduction Of Measurement theory* (California USA: Brooks/Cole Publishing Company).
[10] Aiken L R 1980 *SAGE Journal* **40** pp 955-959
[11] Miller M D, Linn R L and Gronlund N D 2009 *Measurement and Assessment in Teaching* (New Jersey USA: Pearson).
[12] Mardapi D 2012 *Pengukuran penilaian & evaluasi pendidikan* (Yogyakarta: Nuha Litera).
[13] Amelia R N and Kiswantoro K 2017 JKPK journal **2** pp 1-12
[14] Thorndike 2005 *Measurement and Evaluation Education* 2005 (New Jersey USA: Pearson Education).
[15] Maulina Y and Retnawati H 2017 *ICRIEMS Proceedings of fifth annual conference* (Yogyakarta State University) p 467
[16] Bichi, Ado A 2016 *IJSS journal* **2** p 32

Acknowledgements

Special thanks should be given to Jesus Christ, without His blessing, the process of this research would not have been completed and special thanks should and always be given from the author to the sponsor of the author's education funding, LPDP. This research would also never have been possible without the support and guidance of various people at the Yogyakarta State University. Mrs. Dr. Heri Retnawati as a lecturer who gave ideas, advice and moral support in writing this article. Members of my class, thanks for being one of my support systems and giving extra information about Quest Program. Further, the author would like to express my appreciation to Ms. Naila Muroda, S.Pd who became my proofreader for checking the English grammar in this writing. Lastly, a special note of thanks should also be given to my parents and my sisters, Sonia Natasya Panjaitan and Sintya Monica Panjaitan. Your unconditional love, devotion, and optimism was greatly appreciated.

AUTHOR INDEX

AUTHOR INDEX

AUTHOR INDEX